KIRK-OTHMER

ENCYCLOPEDIA OF CHEMICAL TECHNOLOGY

THIRD EDITION

VOLUME 8

DIURETICS
TO
EMULSIONS

A WILEY-INTERSCIENCE PUBLICATION

John Wiley & Sons

NEW YORK • CHICHESTER • BRISBANE • TORONTO

Copyright © 1979 by John Wiley & Sons, Inc.

All rights reserved. Published simultaneously in Canada.

Reproduction or translation of any part of this work
beyond that permitted by Sections 107 or 108 of the
1976 United States Copyright Act without the permission
of the copyright owner is unlawful. Requests for
permission or further information should be addressed to
the Permissions Department, John Wiley & Sons, Inc.

Library of Congress Cataloging in Publication Data:

Main entry under title:
  Encyclopedia of chemical technology.

  At head of title: Kirk-Othmer.
  "A Wiley-Interscience publication."
  Includes bibliographies.
  1. Chemistry, Technical—Dictionaries. I. Kirk,
Raymond Eller, 1890–1957. II. Othmer, Donald Frederick,
1904–   III. Grayson, Martin. IV. Eckroth, David.
V. Title: Kirk-Othmer encyclopedia of chemical technology.

TP9.E685   1978   660'.03   77-15820
ISBN 0-471-02044-3

Printed in the United States of America

CONTENTS

Diuretics, 1
Driers and metallic soaps, 34
Drycleaning and laundering, 50
Drying, 75
Drying agents, 114
Drying oils, 130
Dye carriers, 151
Dyes and dye intermediates, 159
Dyes, anthraquinone, 212
Dyes, application and evaluation, 280
Dyes, natural, 351
Dyes, reactive, 374
Dyes, sensitizing, 393
Economic evaluation, 409

Eggs, 429
Elastomers, synthetic, 446
Electrical connectors, 641
Electrochemical processing, 662
Electrodecantation, 721
Electrodialysis, 726
Electroless plating, 738
Electrolytic machining methods, 751
Electromigration, 763
Electrophotography, 794
Electroplating, 826
Electrostatic sealing, 870
Embedding, 877
Emulsions, 900

EDITORIAL STAFF FOR VOLUME 8

Executive Editor: **Martin Grayson**
Associate Editor: **David Eckroth**
Production Supervisor: **Michalina Bickford**
Editors: **Galen J. Bushey**   **Caroline L Eastman**   **Anna Klingsberg**
        Leonard Spiro

CONTRIBUTORS TO VOLUME 8

C. B. Anderson, *Mobay Chemical Corp., Union, New Jersey,* Dyes, application and evaluation

Morton Antler, *Bell Telephone Laboratories, Columbus, Ohio,* Electrical connectors

Joseph P. Ausikaitis, *Union Carbide Technical Center, Tarrytown, New York,* Drying agents

Oliver Axtell, *Celanese Chemical Company, Dallas, Texas,* Economic evaluation

F. P. Baldwin, *Exxon Chemical Co., Linden, New Jersey,* Butyl rubber under Elastomers, synthetic

D. W. Bannister, *Toms River Chemical Corporation, Toms River, New Jersey,* Dyes and dye intermediates

P. Bass, *Mobay Chemical Corp., Union, New Jersey,* Dyes, application and evaluation

R. G. Bauer, *The Goodyear Tire & Rubber Company, Akron, Ohio,* Styrene-butadiene rubber under Elastomers, synthetic

Dwight H. Berquist, *Henningsen Foods, Inc., Omaha, Nebraska,* Eggs

E. L. Borg, *Uniroyal Chemical Co., Naugatuck, Connecticut,* Ethylene-propylene rubber under Elastomers, synthetic

Frederick J. Buono, *Tenneco Inc., Piscataway, New Jersey,* Driers and metallic soaps
D. A. Chung, *The Goodyear Tire & Rubber Company, Akron, Ohio,* Thermoplastic elastomers under Elastomers, synthetic
Rack Hun Chung, *General Electric Co., Waterford, New York,* Dyes, anthraquinone
Charles L. Cormany, *PPG Industries, Barberton, Ohio,* Drycleaning under Drycleaning and laundering
John C. Cowan, *Bradley University, Peoria, Illinois,* Drying oils
Donald Danly, *Monsanto Chemical Intermediates Co., Pensacola, Florida,* Electrochemical processing, organic
François d'Heurle, *IBM, Yorktown Heights, New York,* Electromigration
Anthony DeMaria, *Tanatex Chemical Co., Wellford, South Carolina,* Dye carriers
John Elliott, *Toms River Chemical Corporation, Toms River, New Jersey,* Dyes, reactive
Russell E. Farris, *Sandoz Colors & Chemicals, Martin, South Carolina,* Dyes, anthraquinone and Dyes, natural
Peter W. Feit, *Leo Pharmaceutical Products, Ballerup, Denmark,* Diuretics
Martin L. Feldman, *Tenneco Inc., Piscataway, New Jersey,* Driers and metallic soaps
A. F. Finelli, *The Goodyear Tire & Rubber Company, Akron, Ohio,* Thermoplastic elastomers under Elastomers, synthetic
William C. Griffin, *ICI United States, Inc., Wilmington, Delaware,* Emulsions
Charles A. Harper, *Westinghouse Electric Corporation, Baltimore, Maryland,* Embedding
W. Heise, *Mobay Chemical Corp., Union, New Jersey,* Dyes, application and evaluation
James P. Hoare, *General Motors Research Laboratories, Warren, Michigan,* Electrolytic machining methods
Allan G. Holcomb, *3M Company, St. Paul, Minnesota,* Fluorinated elastomers under Elastomers, synthetic
S. E. Horne, Jr., *The BFGoodrich Company, Brecksville, Ohio,* Polyisoprene under Elastomers, synthetic
Paul R. Johnson, *E. I. du Pont de Nemours & Co., Inc., Wilmington, Delaware,* Chlorosulfonated polyethylene; Neoprene both under Elastomers, synthetic
W. J. Kelly, *The Goodyear Tire & Rubber Company, Akron, Ohio,* Polybutadiene under Elastomers, synthetic
J. C. King, *Mobay Chemical Corp., Union, New Jersey,* Dyes, application and evaluation
Gerald A. Krulik, *Borg-Warner Corporation, Des Plaines, Illinois,* Electroless plating
R. Kuehni, *Mobay Chemical Corp., Union, New Jersey,* Dyes, application and evaluation
L. J. Kuzma, *The Goodyear Tire & Rubber Company, Akron, Ohio,* Polybutadiene under Elastomers, synthetic
Mitchell A. LaBoda, *General Motors Research Laboratories, Warren, Michigan,* Electrolytic machining methods
Frederick A. Lowenheim, *Consultant, Plainfield, New Jersey,* Electroplating
R. A. Marshall, *The Goodyear Tire & Rubber Company, Akron, Ohio,* Thermoplastic elastomers under Elastomers, synthetic
Paul Y. McCormick, *E. I. du Pont de Nemours & Co., Inc., Wilmington, Delaware,* Drying
James E. McGrath, *Virginia Polytechnic Institute and State University, Blacksburg, Virginia,* Survey under Elastomers, synthetic
Wayne A. McRae, *Research Ionics, Inc., Watertown, Massachusetts,* Electrodialysis
Eilert A. Ofstead, *The Goodyear Tire & Rubber Company, Akron, Ohio,* Polypentenamers under Elastomers, synthetic
A. D. Olin, *Toms River Chemical Corporation, Toms River, New Jersey,* Dyes and dye intermediates

David T. Peterson, *Iowa State University, Ames, Iowa,* Electromigration
R. E. Phillips, *Mobay Chemical Corp., Union, New Jersey,* Dyes, application and evaluation
Donald A. Reich, *PPG Industries, Barberton, Ohio,* Drycleaning under Drycleaning and laundering
J. M. Robertson, *Celanese Chemical Company, Corpus Christi, Texas,* Economic evaluation
Peter M. Robertson, *Eidgenössische Technische Hochschule, Zurich, Switzerland.* Electrochemical processing, introduction and inorganic
H. W. Robinson, *Uniroyal, Inc., Naugatuck, Connecticut,* Nitrile rubber under Elastomers, synthetic
R. H. Schatz, *Exxon Chemical Co., Linden, New Jersey,* Butyl rubber under Elastomers, synthetic
Joseph C. Sherrill, *Sherrill Associates, Homewood, Illinois,* Laundering under Drycleaning and laundering
Paul Stamberger, *Crusader Chemical Co., Baltimore, Maryland,* Electrodecantation
H. A. Stingl, *Toms River Chemical Corporation, Toms River, New Jersey,* Dyes and dye intermediates
David Sturmer, *Eastman Kodak Co., Rochester, New York,* Dyes, sensitizing
H. Teich, *Mobay Chemical Corp., Union, New Jersey,* Dyes, application and evaluation
Harold Tucker, *The BFGoodrich Company, Brecksville, Ohio,* Polyisoprene under Elastomers, synthetic
F. Tweedle, *Mobay Chemical Corp., Union, New Jersey,* Dyes, application and evaluation
E. J. Vandenberg, *Hercules Incorporated, Wilmington, Delaware,* Polyethers under Elastomers, synthetic
T. M. Vial, *American Cyanamid Company, Bound Brook, New Jersey,* Acrylic elastomers under Elastomers, synthetic
George Wallis, *P. R. Mallory & Co., Inc., Burlington, Massachusetts,* Electrostatic sealing
Robert Wannemacher, *Tantatex Chemical Co., Wellford, South Carolina,* Dye carriers
John W. Weigl, *Xerox Corporation, Rochester, New York,* Electrophotography
Arthur C. West, *3M Company, St. Paul, Minnesota,* Fluorinated elastomers under Elastomers, synthetic
Nikolaus E. Wolff, *Consultant, Hanover, New Hampshire,* Electrophotography
Patrick P. Yeung, *Toms River Chemical Corporation, Toms River, New Jersey,* Dyes, reactive

NOTE ON CHEMICAL ABSTRACTS SERVICE REGISTRY NUMBERS AND NOMENCLATURE

Chemical Abstracts Service (CAS) Registry Numbers are unique numerical identifiers assigned to substances recorded in the CAS Registry System. They appear in brackets in the *Chemical Abstracts* (CA) substance and formula indexes following the names of compounds. A single compound may have many synonyms in the chemical literature. A simple compound like phenethylamine can be named β-phenylethylamine or, as in *Chemical Abstracts,* benzeneethanamine. The usefulness of the Encyclopedia depends on accessibility through the most common correct name of a substance. Because of this diversity in nomenclature careful attention has been given the problem in order to assist the reader as much as possible, especially in locating the systematic CA index name by means of the Registry Number. For this purpose, the reader may refer to the CAS Registry Handbook-Number Section which lists in numerical order the Registry Number with the Chemical Abstracts index name and the molecular formula; eg, **458-88-8,** Piperidine, 2-propyl-, (S)-, $C_8H_{17}N$; in the Encyclopedia this compound would be found under its common name, coniine [*458-88-8*]. The Registry Number is a valuable link for the reader in retrieving additional published information on substances and also as a point of access for such on-line data bases as Chemline, Medline, and Toxline.

In all cases, the CAS Registry Numbers have been given for title compounds in articles and for all compounds in the index. All specific substances indexed in *Chemical Abstracts* since 1965 are included in the CAS Registry System as are a large number of substances derived from a variety of reference works. The CAS Registry System identifies a substance on the basis of an unambiguous computer-language description of its molecular structure including stereochemical detail. The Registry Number is a machine-checkable number (like a Social Security number) assigned in sequential order to each substance as it enters the registry system. The value of the number lies in the fact that it is a concise and unique means of substance identification, which is independent of, and therefore bridges, many systems of chemical nomenclature. For polymers, one Registry Number is used for the entire family; eg, polyoxyethylene (20)sorbitan monolaurate has the same number as all of its polyoxyethylene homologues.

Registry numbers for each substance will be provided in the third edition index (eg, Alkaloids shows the Registry Number of all alkaloids (title compounds) in a table in the article as well, but the intermediates will have their Registry Numbers shown only in the index). Articles such as Absorption, Adsorptive separation, Air conditioning, Air pollution, Air pollution control methods have no Registry Numbers in the text.

Cross-references have been inserted in the index for many common names and for some systematic names. Trademark names appear in the index. Names that are incorrect, misleading or ambiguous are avoided. Formulas are given very frequently in the text to help in identifying compounds. The spelling and form used, even for industrial names, follow American chemical usage, but not always the usage of *Chemical Abstracts* (eg, *coniine* is used instead of *(S)-2-propylpiperidine*, *aniline* instead of *benzenamine*, and *acrylic acid* instead of *2-propenoic acid*).

There are variations in representation of rings in different disciplines. The dye industry does not designate aromaticity or double bonds in rings. All double bonds and aromaticity will be shown in the *Encyclopedia* as a matter of course. For example, tetralin has an aromatic ring and a saturated ring and its structure will appear in the

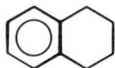

Encyclopedia with its common name, Registry Number enclosed in brackets, and parenthetical CA index name, ie, tetralin [*119-64-2*] (1,2,3,4-tetrahydronaphthalene). With names and structural formulas, and especially with CAS Registry Numbers the aim is to help the reader have a concise means of substance identification.

CONVERSION FACTORS, ABBREVIATIONS, AND UNIT SYMBOLS

SI Units (Adopted 1960)

A new system of measurement, the International System of Units (abbreviated SI), is being implemented throughout the world. This system is a modernized version of the MKSA (meter, kilogram, second, ampere) system, and its details are published and controlled by an international treaty organization (The International Bureau of Weights and Measures) (1).

SI units are divided into three classes:

BASE UNITS

length	meter[†] (m)
mass[‡]	kilogram (kg)
time	second (s)
electric current	ampere (A)
thermodynamic temperature[§]	kelvin (K)
amount of substance	mole (mol)
luminous intensity	candela (cd)

[†] The spellings "metre" and "litre" are preferred by ASTM; however "-er" will be used in the Encyclopedia.

[‡] "Weight" is the commonly used term for "mass"

[§] Wide use is made of "Celsius temperature" (t) defined by

$$t = T - T_0$$

where T is the thermodynamic temperature, expressed in kelvins, and $T_0 = 273.15$ K by definition. A temperature interval may be expressed in degrees Celsius as well as in kelvins.

xiv FACTORS, ABBREVIATIONS, AND SYMBOLS

SUPPLEMENTARY UNITS

plane angle   radian (rad)
solid angle   steradian (sr)

DERIVED UNITS AND OTHER ACCEPTABLE UNITS

These units are formed by combining base units, supplementary units, and other derived units (2–4). Those derived units having special names and symbols are marked with an asterisk in the list below:

Quantity	Unit	Symbol	Acceptable equivalent
*absorbed dose	gray	Gy	J/kg
acceleration	meter per second squared	m/s^2	
*activity (of ionizing radiation source)	becquerel	Bq	1/s
area	square kilometer	km^2	
	square hectometer	hm^2	ha (hectare)
	square meter	m^2	
*capacitance	farad	F	C/V
concentration (of amount of substance)	mole per cubic meter	mol/m^3	
*conductance	siemens	S	A/V
current density	ampere per square meter	A/m^2	
density, mass density	kilogram per cubic meter	kg/m^3	g/L; mg/cm^3
dipole moment (quantity)	coulomb meter	C·m	
*electric charge, quantity of electricity	coulomb	C	A·s
electric charge density	coulomb per cubic meter	C/m^3	
electric field strength	volt per meter	V/m	
electric flux density	coulomb per square meter	C/m^2	
*electric potential, potential difference, electromotive force	volt	V	W/A
*electric resistance	ohm	Ω	V/A
*energy, work, quantity of heat	megajoule	MJ	
	kilojoule	kJ	
	joule	J	N·m
	electron volt[†]	eV[†]	
	kilowatt-hour[†]	kW·h[†]	

[†] This non-SI unit is recognized by the CIPM as having to be retained because of practical importance or use in specialized fields (1).

Quantity	Unit	Symbol	Acceptable equivalent
energy density	joule per cubic meter	J/m³	
*force	kilonewton	kN	
	newton	N	kg·m/s²
*frequency	megahertz	MHz	
	hertz	Hz	1/s
heat capacity, entropy	joule per kelvin	J/K	
heat capacity (specific), specific entropy	joule per kilogram kelvin	J/(kg·K)	
heat transfer coefficient	watt per square meter kelvin	W/(m²·K)	
*illuminance	lux	lx	lm/m²
*inductance	henry	H	Wb/A
linear density	kilogram per meter	kg/m	
luminance	candela per square meter	cd/m²	
*luminous flux	lumen	lm	cd·sr
magnetic field strength	ampere per meter	A/m	
*magnetic flux	weber	Wb	V·s
*magnetic flux density	tesla	T	Wb/m²
molar energy	joule per mole	J/mol	
molar entropy, molar heat capacity	joule per mole kelvin	J/(mol·K)	
moment of force, torque	newton meter	N·m	
momentum	kilogram meter per second	kg·m/s	
permeability	henry per meter	H/m	
permittivity	farad per meter	F/m	
*power, heat flow rate, radiant flux	kilowatt	kW	
	watt	W	J/s
power density, heat flux density, irradiance	watt per square meter	W/m²	
*pressure, stress	megapascal	MPa	
	kilopascal	kPa	
	pascal	Pa	N/m²
sound level	decibel	dB	
specific energy	joule per kilogram	J/kg	
specific volume	cubic meter per kilogram	m³/kg	
surface tension	newton per meter	N/m	
thermal conductivity	watt per meter kelvin	W/(m·K)	
velocity	meter per second	m/s	
	kilometer per hour	km/h	
viscosity, dynamic	pascal second	Pa·s	
	millipascal second	mPa·s	
viscosity, kinematic	square meter per second	m²/s	

xvi   FACTORS, ABBREVIATIONS, AND SYMBOLS

Quantity	Unit	Symbol	Acceptable equivalent
	square millimeter per second	mm^2/s	
volume	cubic meter	m^3	
	cubic decimeter	dm^3	L(liter) (5)
	cubic centimeter	cm^3	mL
wave number	1 per meter	m^{-1}	
	1 per centimeter	cm^{-1}	

In addition, there are 16 prefixes used to indicate order of magnitude, as follows:

Multiplication factor	Prefix	Symbol	Note
10^{18}	exa	E	
10^{15}	peta	P	
10^{12}	tera	T	
10^{9}	giga	G	
10^{6}	mega	M	
10^{3}	kilo	k	
10^{2}	hecto	h[a]	
10	deka	da[a]	
10^{-1}	deci	d[a]	
10^{-2}	centi	c[a]	
10^{-3}	milli	m	
10^{-6}	micro	μ	
10^{-9}	nano	n	
10^{-12}	pico	p	
10^{-15}	femto	f	
10^{-18}	atto	a	

[a] Although hecto, deka, deci, and centi are SI prefixes, their use should be avoided except for SI unit-multiples for area and volume and nontechnical use of centimeter, as for body and clothing measurement.

For a complete description of SI and its use the reader is referred to ASTM E 380 (4) and the article Units and Conversion Factors which will appear in a later volume of the *Encyclopedia*.

A representative list of conversion factors from non-SI to SI units is presented herewith. Factors are given to four significant figures. Exact relationships are followed by a dagger. A more complete list is given in ASTM E 380-76(4) and ANSI Z210.1-1976 (6).

Conversion Factors to SI Units

To convert from	To	Multiply by
acre	square meter (m^2)	4.047×10^{3}
angstrom	meter (m)	1.0×10^{-10}†
are	square meter (m^2)	1.0×10^{2}†
astronomical unit	meter (m)	1.496×10^{11}
atmosphere	pascal (Pa)	1.013×10^{5}
bar	pascal (Pa)	1.0×10^{5}†

† Exact.

To convert from	To	Multiply by
barrel (42 U.S. liquid gallons)	cubic meter (m^3)	0.1590
Bohr magneton μ_β	J/T	9.274×10^{-24}
Btu (International Table)	joule (J)	1.055×10^3
Btu (mean)	joule (J)	1.056×10^3
Btu (thermochemical)	joule (J)	1.054×10^3
bushel	cubic meter (m^3)	3.524×10^{-2}
calorie (International Table)	joule (J)	4.187
calorie (mean)	joule (J)	4.190
calorie (thermochemical)	joule (J)	4.184†
centipoise	pascal second (Pa·s)	1.0×10^{-3}†
centistoke	square millimeter per second (mm^2/s)	1.0†
cfm (cubic foot per minute)	cubic meter per second (m^3/s)	4.72×10^{-4}
cubic inch	cubic meter (m^3)	1.639×10^{-5}
cubic foot	cubic meter (m^3)	2.832×10^{-2}
cubic yard	cubic meter (m^3)	0.7646
curie	becquerel (Bq)	3.70×10^{10}†
debye	coulomb·meter (C·m)	3.336×10^{-30}
degree (angle)	radian (rad)	1.745×10^{-2}
denier (international)	kilogram per meter (kg/m)	1.111×10^{-7}
	tex‡	0.1111
dram (apothecaries')	kilogram (kg)	3.888×10^{-3}
dram (avoirdupois)	kilogram (kg)	1.772×10^{-3}
dram (U.S. fluid)	cubic meter (m^3)	3.697×10^{-6}
dyne	newton (N)	1.0×10^{-5}†
dyne/cm	newton per meter (N/m)	1.00×10^{-3}†
electron volt	joule (J)	1.602×10^{-19}
erg	joule (J)	1.0×10^{-7}†
fathom	meter (m)	1.829
fluid ounce (U.S.)	cubic meter (m^3)	2.957×10^{-5}
foot	meter (m)	0.3048†
footcandle	lux (lx)	10.76
furlong	meter (m)	2.012×10^{-2}
gal	meter per second squared (m/s^2)	1.0×10^{-2}†
gallon (U.S. dry)	cubic meter (m^3)	4.405×10^{-3}
gallon (U.S. liquid)	cubic meter (m^3)	3.785×10^{-3}
gallon per minute (gpm)	cubic meter per second (m^3/s)	6.308×10^{-5}
	cubic meter per hour (m^3/h)	0.2271
gauss	tesla (T)	1.0×10^{-4}
gilbert	ampere (A)	0.7958
gill (U.S.)	cubic meter (m^3)	1.183×10^{-4}
grad	radian	1.571×10^{-2}
grain	kilogram (kg)	6.480×10^{-5}
gram force per denier	newton per tex (N/tex)	8.826×10^{-2}
hectare	square meter (m^2)	1.0×10^4†
horsepower (550 ft·lbf/s)	watt (W)	7.457×10^2

† Exact.
‡ See footnote on p. xii.

FACTORS, ABBREVIATIONS, AND SYMBOLS

To convert from	To	Multiply by
horsepower (boiler)	watt (W)	9.810×10^3
horsepower (electric)	watt (W)	$7.46 \times 10^{2\dagger}$
hundredweight (long)	kilogram (kg)	50.80
hundredweight (short)	kilogram (kg)	45.36
inch	meter (m)	$2.54 \times 10^{-2\dagger}$
inch of mercury (32°F)	pascal (Pa)	3.386×10^3
inch of water (39.2°F)	pascal (Pa)	2.491×10^2
kilogram force	newton (N)	9.807
kilowatt hour	megajoule (MJ)	$3.6^\dagger$
kip	newton (N)	4.48×10^3
knot (international)	meter per second (m/s)	0.5144
lambert	candela per square meter (cd/m^2)	3.183×10^3
league (British nautical)	meter (m)	5.559×10^3
league (statute)	meter (m)	4.828×10^3
light year	meter (m)	9.461×10^{15}
liter (for fluids only)	cubic meter (m^3)	$1.0 \times 10^{-3\dagger}$
maxwell	weber (Wb)	$1.0 \times 10^{-8\dagger}$
micron	meter (m)	$1.0 \times 10^{-6\dagger}$
mil	meter (m)	$2.54 \times 10^{-5\dagger}$
mile (U.S. nautical)	meter (m)	$1.852 \times 10^{3\dagger}$
mile (statute)	meter (m)	1.609×10^3
mile per hour	meter per second (m/s)	0.4470
millibar	pascal (Pa)	1.0×10^2
millimeter of mercury (0°C)	pascal (Pa)	$1.333 \times 10^{2\dagger}$
minute (angular)	radian	2.909×10^{-4}
myriagram	kilogram (kg)	10
myriameter	kilometer (km)	10
oersted	ampere per meter (A/m)	79.58
ounce (avoirdupois)	kilogram (kg)	2.835×10^{-2}
ounce (troy)	kilogram (kg)	3.110×10^{-2}
ounce (U.S. fluid)	cubic meter (m^3)	2.957×10^{-5}
ounce-force	newton (N)	0.2780
peck (U.S.)	cubic meter (m^3)	8.810×10^{-3}
pennyweight	kilogram (kg)	1.555×10^{-3}
pint (U.S. dry)	cubic meter (m^3)	5.506×10^{-4}
pint (U.S. liquid)	cubic meter (m^3)	4.732×10^{-4}
poise (absolute viscosity)	pascal second (Pa·s)	$0.10^\dagger$
pound (avoirdupois)	kilogram (kg)	0.4536
pound (troy)	kilogram (kg)	0.3732
poundal	newton (N)	0.1383
pound-force	newton (N)	4.448
pound per square inch (psi)	pascal (Pa)	6.895×10^3
quart (U.S. dry)	cubic meter (m^3)	1.101×10^{-3}
quart (U.S. liquid)	cubic meter (m^3)	9.464×10^{-4}
quintal	kilogram (kg)	$1.0 \times 10^{2\dagger}$

† Exact.

To convert from	To	Multiply by
rad	gray (Gy)	1.0×10^{-2}†
rod	meter (m)	5.029
roentgen	coulomb per kilogram (C/kg)	2.58×10^{-4}
second (angle)	radian (rad)	4.848×10^{-6}
section	square meter (m²)	2.590×10^{6}
slug	kilogram (kg)	14.59
spherical candle power	lumen (lm)	12.57
square inch	square meter (m²)	6.452×10^{-4}
square foot	square meter (m²)	9.290×10^{-2}
square mile	square meter (m²)	2.590×10^{6}
square yard	square meter (m²)	0.8361
stere	cubic meter (m³)	1.0†
stokes (kinematic viscosity)	square meter per second (m²/s)	1.0×10^{-4}†
tex	kilogram per meter (kg/m)	1.0×10^{-6}†
ton (long, 2240 pounds)	kilogram (kg)	1.016×10^{3}
ton (metric)	kilogram (kg)	1.0×10^{3}†
ton (short, 2000 pounds)	kilogram (kg)	9.072×10^{2}
torr	pascal (Pa)	1.333×10^{2}
unit pole	weber (Wb)	1.257×10^{-7}
yard	meter (m)	0.9144†

† Exact.

Abbreviations and Unit Symbols

Following is a list of commonly used abbreviations and unit symbols appropriate for use in the *Encyclopedia*. In general they agree with those listed in *American National Standard Abbreviations for Use on Drawings and in Text (ANSI Y1.1)* (6) and *American National Standard Letter Symbols for Units in Science and Technology (ANSI Y10)* (6). Also included is a list of acronyms for a number of private and government organizations as well as common industrial solvents, polymers, and other chemicals.

Rules for Writing Unit Symbols (4):

1. Unit symbols should be printed in upright letters (roman) regardless of the type style used in the surrounding text.

2. Unit symbols are unaltered in the plural.

3. Unit symbols are not followed by a period except when used as the end of a sentence.

4. Letter unit symbols are generally written in lower-case (eg, cd for candela) unless the unit name has been derived from a proper name, in which case the first letter of the symbol is capitalized (W,Pa). Prefix and unit symbols retain their prescribed form regardless of the surrounding typography.

5. In the complete expression for a quantity, a space should be left between the numerical value and the unit symbol. For example, write 2.37 lm, *not* 2.37lm, and 35 mm, *not* 35mm. When the quantity is used in an adjectival sense, a hyphen is often used, for example, 35-mm film. *Exception:* No space is left between the numerical value and the symbols for degree, minute, and second of plane angle, and degree Celsius.

6. No space is used between the prefix and unit symbols (eg, kg).

7. Symbols, not abbreviations, should be used for units. For example, use "A," not "amp," for ampere.

8. When multiplying unit symbols, use a raised dot:

$$N \cdot m \text{ for newton meter}$$

In the case of W·h, the dot may be omitted, thus:

$$Wh$$

An exception to this practice is made for computer printouts, automatic typewriter work, etc, where the raised dot is not possible, and a dot on the line may be used.

9. When dividing unit symbols use one of the following forms:

$$m/s \; or \; m \cdot s^{-1} \; or \; \frac{m}{s}$$

In no case should more than one slash be used in the same expression unless parentheses are inserted to avoid ambiguity. For example, write:

$$J/(mol \cdot K) \; or \; J \cdot mol^{-1} \cdot K^{-1} \; or \; (J/mol)/K$$

but *not*

$$J/mol/K$$

10. Do not mix symbols and unit names in the same expression. Write:

$$\text{joules per kilogram} \; or \; J/kg \; or \; J \cdot kg^{-1}$$

but *not*

$$\text{joules/kilogram} \; nor \; \text{joules/kg} \; nor \; \text{joules} \cdot kg^{-1}$$

ABBREVIATIONS AND UNITS

A	ampere	AIChE	American Institute of Chemical Engineers
A	anion (eg, HA)	AIP	American Institute of Physics
a	atto (prefix for 10^{-18})	alc	alcohol(ic)
AATCC	American Association of Textile Chemists and Colorists	Alk	alkyl
		alk	alkaline (not alkali)
ABS	acrylonitrile–butadiene–styrene	amt	amount
		amu	atomic mass unit
abs	absolute	ANSI	American National Standards Institute
ac	alternating current, *n.*		
a-c	alternating current, *adj.*	AO	atomic orbital
ac-	alicyclic	APHA	American Public Health Association
ACGIH	American Conference of Governmental Industrial Hygienists	API	American Petroleum Institute
ACS	American Chemical Society	aq	aqueous
AGA	American Gas Association	Ar	aryl
Ah	ampere hour	*ar*-	aromatic

as-	asymmetric(al)	d	differential operator
ASH-RAE	American Society of Heating, Refrigerating, and Air Conditioning Engineers	d-	dextro-, dextrorotatory
		da	deka (prefix for 10^1)
		dB	decibel
		dc	direct current, n.
ASM	American Society for Metals	d-c	direct current, adj.
ASME	American Society of Mechanical Engineers	dec	decompose
		detd	determined
ASTM	American Society for Testing and Materials	detn	determination
		dia	diameter
at no.	atomic number	dil	dilute
at wt	atomic weight	dl-; DL-	racemic
av(g)	average	DMF	dimethylformamide
bbl	barrel	DMG	dimethyl glyoxime
bcc	body-centered cubic	DOE	Department of Energy
Bé	Baumé	DOT	Department of Transportation
bid	twice daily	dp	dew point; degree of polymerization
BOD	biochemical (biological) oxygen demand		
		dstl(d)	distill(ed)
bp	boiling point	dta	differential thermal analysis
Bq	becquerel	(E)-	entgegen; opposed
C	coulomb	ϵ	dielectric constant (unitless number)
°C	degree Celsius		
C-	denoting attachment to carbon	e	electron
		ECU	electrochemical unit
c	centi (prefix for 10^{-2})	ed.	edited, edition, editor
ca	circa (approximately)	ED	effective dose
cd	candela; current density; circular dichroism	EDTA	ethylenediamine tetraacetic acid
CFR	Code of Federal Regulations	emf	electromotive force
cgs	centimeter-gram-second	emu	electromagnetic unit
CI	Color Index	eng	engineering
cis-	isomer in which substituted groups are on same side of double bond between C atoms	EPA	Environmental Protection Agency
		epr	electron paramagnetic resonance
cl	carload	eq.	equation
cm	centimeter	esp	especially
cmil	circular mil	esr	electron-spin resonance
cmpd	compound	est(d)	estimate(d)
COA	coenzyme A	estn	estimation
COD	chemical oxygen demand	esu	electrostatic unit
coml	commercial(ly)	exp	experiment, experimental
cp	chemically pure	ext(d)	extract(ed)
cph	close-packed hexagonal	F	farad (capacitance)
CPSC	Consumer Product Safety Commission	f	femto (prefix for 10^{-15})
		FAO	Food and Agriculture Organization (United Nations)
D-	denoting configurational relationship		

FACTORS, ABBREVIATIONS, AND SYMBOLS

fcc	face-centered cubic	IUPAC	International Union of Pure and Applied Chemistry
FDA	Food and Drug Administration		
FEA	Federal Energy Administration	IV	iodine value
		J	joule
fob	free on board	K	kelvin
fp	freezing point	k	kilo (prefix for 10^3)
FPC	Federal Power Commission	kg	kilogram
frz	freezing	L	denoting configurational relationship
G	giga (prefix for 10^9)		
g	gram	L	liter (for fluids only) (5)
(g)	gas, only as in $H_2O(g)$	l-	*levo-*, levorotatory
g	gravitational acceleration	(l)	liquid, only as in $NH_3(l)$
gem-	geminal	LC_{50}	conc lethal to 50% of the animals tested
glc	gas-liquid chromatography		
g-mol wt; gmw	gram-molecular weight	LCAO	linear combination of atomic orbitals
		lcl	less than carload lots
grd	ground	LD_{50}	dose lethal to 50% of the animals tested
Gy	gray		
H	henry	liq	liquid
h	hour; hecto (prefix for 10^2)	lm	lumen
		ln	logarithm (natural)
ha	hectare	LNG	liquefied natural gas
HB	Brinell hardness number	log	logarithm (common)
Hb	hemoglobin	LPG	liquefied petroleum gas
HK	Knoop hardness number	ltl	less than truckload lots
HRC	Rockwell hardness (C scale)	lx	lux
HV	Vickers hardness number	M	mega (prefix for 10^6); metal (as in MA)
hyd	hydrated, hydrous		
hyg	hygroscopic	M	molar
Hz	hertz	m	meter; milli (prefix for 10^{-3})
^{i}(eg, Pr^i)	iso (eg, isopropyl)		
i-	inactive (eg, i-methionine)	m	molal
IACS	International Annealed Copper Standard	m-	meta
		max	maximum
ibp	initial boiling point	MCA	Manufacturing Chemists' Association
ICC	Interstate Commerce Commission		
		MEK	methyl ethyl ketone
ICT	International Critical Table	meq	milliequivalent
ID	inside diameter; infective dose	mfd	manufactured
IPS	iron pipe size	mfg	manufacturing
IPT	Institute of Petroleum Technologists	mfr	manufacturer
		MIBC	methylisobutyl carbinol
ir	infrared	MIBK	methyl isobutyl ketone
IRLG	Interagency Regulatory Liaison Group	MIC	minimum inhibiting concentration
ISO	International Organization for Standardization	min	minute; minimum
		mL	milliliter

MLD	minimum lethal dose	NSF	National Science Foundation
MO	molecular orbital	NTA	nitrilotriacetic acid
mo	month	NTSB	National Transportation Safety Board
mol	mole		
mol wt	molecular weight	O-	denoting attachment to oxygen
mom	momentum		
mp	melting point	o-	ortho
MR	molar refraction	OD	outside diameter
ms	mass spectrum	OPEC	Organization of Petroleum Exporting Countries
mxt	mixture		
μ	micro (prefix for 10^{-6})	OSHA	Occupational Safety and Health Administration
N	newton (force)		
N	normal (concentration)	owf	on weight of fiber
N-	denoting attachment to nitrogen	Ω	ohm
		P	peta (prefix for 10^{15})
n (as n_D^{20})	index of refraction (for 20°C and sodium light)	p	pico (prefix for 10^{-12})
		p-	para
		p.	page
n (as Bu^n),		Pa	pascal (pressure)
		pd	potential difference
n-	normal (straight-chain structure)	pH	negative logarithm of the effective hydrogen ion concentration
n	nano (prefix for 10^{-9})		
na	not available	pmr	proton magnetic resonance
NAS	National Academy of Sciences	POP	polyoxypropylene
		pos	positive
NASA	National Aeronautics and Space Administration	pp.	pages
		ppb	parts per billion
nat	natural	ppm	parts per million
NBS	National Bureau of Standards	PPO	poly(phenyl oxide)
		ppt(d)	precipitate(d)
neg	negative	pptn	precipitation
NF	*National Formulary*	Pr (no.)	foreign prototype (number)
NIH	National Institutes of Health	pt	point; part
NIOSH	National Institute of Occupational Safety and Health	PVC	poly(vinyl chloride)
		pwd	powder
		qv	quod vide (which see)
nmr	nuclear magnetic resonance	R	univalent hydrocarbon radical
NND	New and Nonofficial Drugs (AMA)		
		(R)-	rectus (clockwise configuration)
no.	number		
NOI- (BN)	not otherwise indexed (by name)	rad	radian; radius
		rds	rate determining step
NOS	not otherwise specified	ref	reference
nqr	nuclear quadrople resonance	rf	radio frequency, *n*.
NRC	Nuclear Regulatory Commission; National Research Council	r-f	radio frequency, *adj*.
		rh	relative humidity
		RI	Ring Index
NRI	New Ring Index	RT	room temperature

xxiv   FACTORS, ABBREVIATIONS, AND SYMBOLS

s (eg, Bu^s);		T	tera (prefix for 10^{12}); tesla (magnetic flux density)
sec-	secondary (eg, secondary butyl)	t	metric ton (tonne) temperature
S	siemens	TAPPI	Technical Association of the Pulp and Paper Industry
(S)-	sinister (counterclockwise configuration)	tex	tex (linear density)
S-	denoting attachment to sulfur	TGA	thermogravimetric analysis
		THF	tetrahydrofuran
s-	symmetric(al)	tlc	thin layer chromatography
s	second	TLV	threshold limit value
(s)	solid, only as in H$_2$O(s)	trans-	isomer in which substituted groups are on opposite sides of double bond between C atoms
SAE	Society of Automotive Engineers		
SAN	styrene–acrylonitrile		
sat(d)	saturate(d)	TSCA	Toxic Substance Control Act
satn	saturation		
SCF	self-consistent field	Twad	Twaddell
Sch	Schultz number	UL	Underwriters' Laboratory
SFs	Saybolt Furol seconds	USDA	United States Department of Agriculture
SI	Le Système International d'Unités (International System of Units)		
		USP	*United States Pharmacopeia*
sl sol	slightly soluble	uv	ultraviolet
sol	soluble	V	volt (emf)
soln	solution	var	variable
soly	solubility	*vic-*	vicinal
sp	specific; species	vol	volume (not volatile)
sp gr	specific gravity	vs	versus
sr	steradian	v sol	very soluble
std	standard	W	watt
STP	standard temperature and pressure (0°C and 101.3 kPa)	Wb	Weber
		Wh	watt hour
SUs	Saybolt Universal seconds	WHO	World Health Organization (United Nations)
syn	synthetic		
t (eg, Bu^t), *t-,*		wk	week
		yr	year
tert-	tertiary (eg, tertiary butyl)	(Z)-	zusammen; together

Non-SI (Unacceptable and Obsolete) Units   *Use*

Å	angstrom	nm
at	atmosphere, technical	Pa
atm	atmosphere, standard	Pa
b	barn	cm^2
bar†	bar	Pa

† Do not use bar (10^5Pa) or millibar (10^2Pa) because they are not SI units, and are accepted internationally only for a limited time in special fields because of existing usage.

Non-SI. (Unacceptable and Obsolete) Units		Use
bhp	brake horsepower	W
Btu	British thermal unit	J
bu	bushel	m³; L
cal	calorie	J
cfm	cubic foot per minute	m³/s
Ci	curie	Bq
cSt	centistokes	mm²/s
c/s	cycle per second	Hz
cu	cubic	exponential form
D	debye	C·m
den	denier	tex
dr	dram	kg
dyn	dyne	N
erg	erg	J
eu	entropy unit	J/K
°F	degree Fahrenheit	°C; K
fc	footcandle	lx
fl	footlambert	lx
fl oz	fluid ounce	m³; L
ft	foot	m
ft·lbf	foot pound-force	J
gf den	gram-force per denier	N/tex
G	gauss	T
Gal	gal	m/s²
gal	gallon	m³; L
Gb	gilbert	A
gpm	gallon per minute	(m³/s); (m³/h)
gr	grain	kg
hp	horsepower	W
ihp	indicated horsepower	W
in.	inch	m
in. Hg	inch of mercury	Pa
in. H₂O	inch of water	Pa
in.·lbf	inch pound-force	J
kcal	kilogram-calorie	J
kgf	kilogram-force	N
kilo	for kilogram	kg
L	lambert	lx
lb	pound	kg
lbf	pound-force	N
mho	mho	S
mi	mile	m
MM	million	M
mm Hg	millimeter of mercury	Pa
mμ	millimicron	nm
mph	miles per hour	km/h
μ	micron	μm
Oe	oersted	A/m
oz	ounce	kg
ozf	ounce-force	N
η	poise	Pa·s
P	poise	Pa·s
ph	phot	lx
psi	pounds-force per square inch	Pa
psia	pounds-force per square inch absolute	Pa
psig	pounds-force per square inch gage	Pa

qt	quart	m^3; L
°R	degree Rankine	K
rd	rad	Gy
sb	stilb	lx
SCF	standard cubic foot	m^3
sq	square	exponential form
thm	therm	J
yd	yard	m

BIBLIOGRAPHY

1. The International Bureau of Weights and Measures, BIPM, (Parc de Saint-Cloud, France) is described on page 22 of Ref. 4. This bureau operates under the exclusive supervision of the International Committee of Weights and Measures (CIPM).
2. *Metric Editorial Guide (ANMC-75-1)*, American National Metric Council, 1625 Massachusetts Ave. N.W., Washington, D.C. 20036, 1975.
3. *SI Units and Recommendations for the Use of Their Multiples and of Certain Other Units (ISO 1000-1973)*, American National Standards Institute, 1430 Broadway, New York, N. Y. 10018, 1973.
4. Based on *ASTM E 380-76 (Standard for Metric Practice)*, American Society for Testing and Materials, 1916 Race Street, Philadelphia, Pa. 19103, 1976.
5. *Fed. Regist.*, Dec. 10, 1976 (41 FR 36414).
6. For ANSI address, see Ref. 3.

R. P. LUKENS
American Society for Testing and Materials

D *continued*

DIURETICS

The word diuretic originally referred solely to the ability of agents to increase the rate of urine formation. In a modern therapeutic sense it is generally used to describe all drugs that act on the kidney to promote the urinary excretion of water and electrolytes, particularly sodium ion. Additionally, the term saluretic is used for those drugs that exert their diuretic effect by primarily increasing the excretion of sodium chloride. It should be emphasized that these valuable clinical tools are heterogenous, not only in respect to chemical structure but, even more important, in their mode and site of action in the kidney. This results both in various diuretic profiles and in entirely different pharmacodynamic behavior. The differences are of importance to the physician in allowing selection of the most appropriate agent, depending on the clinical situation and the patient's need. The main and traditional use of diuretics is in the treatment of edematous states. Diuretic treatment permits, in most cases, satisfactory mobilization and subsequent prevention of edema, ascites, and pleural effusion. Such symptomatic treatment appears obvious, as the pathogenesis of local or general accumulation of extracellular fluid comprises either a secondary or primary increase in renal water and sodium chloride retention (1). The underlying disease might be of cardiac, hepatic or renal origin, or the retention might be associated with other clinical situations such as inflammation, hypersensitivity, and premenstrual tension.

During the past decade, diuretics have been increasingly employed in nonedematous conditions, usually unaccompanied by sodium and water retention. Thus their efficacy in the treatment of hypertension, a well established use of certain diuretics, is not closely related to their natriuretic effect and the choice of the agent depends on the nature of the hypertension. Powerful diuretics may be used in the treatment of drug overdose to produce an enhanced clearance of salicylates and some barbiturates

2 DIURETICS

by forced alkaline diuresis. Even the altered renal handling of calcium ion, which might be considered to be a side effect of diuretic treatment, may be utilized. Certain diuretics exert an inhibitory effect upon renal calcium excretion and thereby decrease the likelihood of renal stone formation, and others show an increased rate of calcium excretion, causing a fall in serum calcium levels valuable in the management of hypercalcemia. Various clinical applications are described in more detail in recent reviews (2-3). In addition to the desired effects in the human organism, any diuretic may cause a variety of unwanted or adverse effects. These adverse effects are discussed only incidentally in this survey. Information on this subject can be obtained from reviews (4-5).

It is remarkable that even now, with few exceptions, significant progress in diuretic research is taking place without a knowledge of the precise biochemical mode of action of the diuretic drugs. Important developments have been achieved despite only a limited understanding of the physiology of the target organ. Conversely, extensive studies on diuretics have had a valuable effect on the investigation of the mechanisms of renal water and solute excretion. The observed diuretic profile does, of course, reflect both the primary action and its physiological consequences. Certain classes of diuretics might exert their action by a common underlying biochemical mechanism, and the variation in their effects could depend on differences in intrarenal distribution, leading to different sites of action in the kidney. A list of diuretics mentioned in the text with their CAS Registry Numbers is given at the end of the article.

Renal Function

The principal function of the kidney is to maintain homeostasis of the body's extracellular fluid composition. It does this by simultaneous regulation of water and multiple solute excretion, and this is of particular interest in regard to diuretic drugs.

The functional units of the kidney are approximately 1 million almost identical nephrons (6). The individual nephron consists of a glomerulus, proximal convoluted tubule, loop of Henle, distal convoluted tubule, and a multibranched collecting duct, which is common for several nephrons. The glomeruli and both the proximal and distal convoluted tubules are confined to the isotonic cortex, the distal tubules terminating in the collecting ducts which extend down through the medulla. The loop of Henle is subdivided into a thick descending limb, a thin descending and ascending segment, and a thick ascending limb which passes on to the distal convoluted tubule. Depending on their place in the cortex, an anatomical division is made into cortical and juxtamedullary nephrons (Fig. 1). They differ further in the length of the loop of Henle. The juxtamedullary nephrons have longer loops, arising primarily from the juxtamedullary region of the cortex and descending deep into the inner medulla along its osmolaric gradient. Although the blood supply and distribution of these two types of nephrons may change as part of a regulatory mechanism, in principle they operate equally.

Under normal conditions, ca 25% of the total renal blood flow is filtered in the glomeruli, resulting in ca 120 mL of glomerular filtrate (ultrafiltrate) per minute. Owing to the nature of the glomerular membrane, the filtrate contains all the plasma constituents except lipids, proteins, and protein-bound substances. The driving force is the blood pressure within the glomerular capillaries. On its way through the nephron,

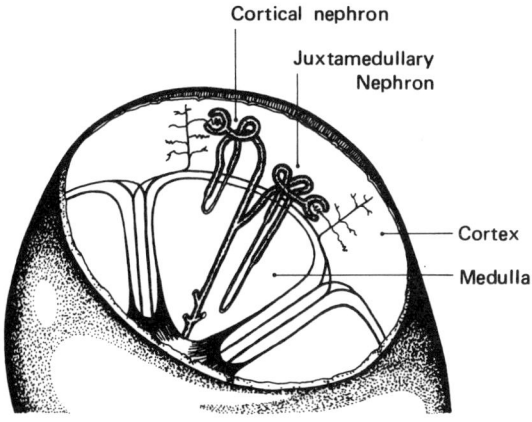

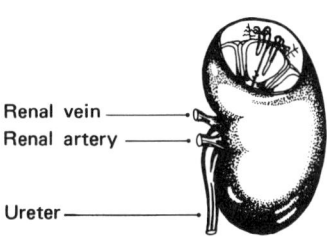

Figure 1. Position of the two types of nephrons (not drawn to scale) in the kidney.

where it ends up as urine, the ultrafiltrate rapidly loses its identity by both passive and active transport processes based on physical forces or involving consumption of energy from cellular metabolism, respectively. The final urine volume is approximately 1% of the ultrafiltrate. It is virtually cleared of filtered glucose and amino acids, and contains, with respect to the main plasma electrolytes, only ca 1% of the filtered sodium and chloride ions, together with traces of bicarbonate ion. Substantial fractions of the remaining solutes are reabsorbed and exogenic organic acids and bases are added to the tubular fluid by secretion.

The net effect of various diuretics on the urinary excretion of H_2O, Na^+, K^+, Cl^-, and HCO_3^- is a consequence of their interference with transport processes and is highly dependent on their site of action in the nephron. The transport processes for relevant electrolytes along the nephron are schematically depicted in Figure 2. Under normal circumstances, ca 60% of the ultrafiltrate is isotonically reabsorbed in the proximal tubule and the descending thick segment of the loop of Henle, where both active and passive transport of Na^+ and passive reabsorption of Cl^- occur. Almost all the bicarbonate is reabsorbed, depending on cellular carbonic anhydrase. Substantial parts of the filtered load of K^+ are likewise reabsorbed in the proximal tubule by processes that may be both active and passive (7). Owing to a high water and low NaCl permeability of the thin descending limb, the tubular fluid, when it enters the ascending limb, has become hypertonic by osmotic equilibration along the osmolaric gradient in the medulla. This effect is not simply reversed when the fluid passes the ascending limb, as the latter is impermeable to water. On the contrary, as the tubular fluid moves up

4   DIURETICS

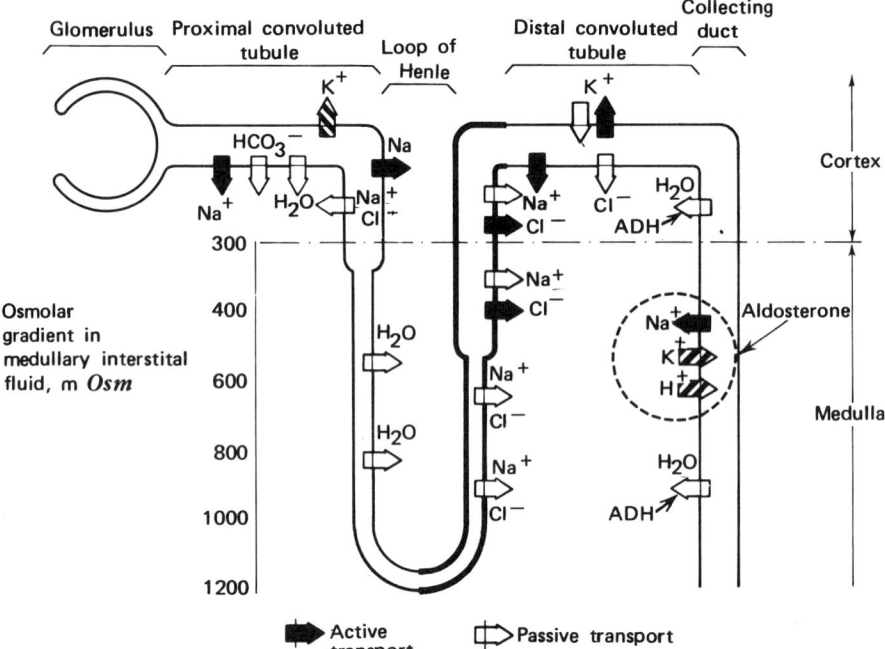

Figure 2. Schematic representation of relevant electrolyte transport through the renal tubule. m Osm = milliosmolar-milliosmoles per liter. An osmole equals a mole of solute divided by the number of ions formed per molecule of the solute. Thus one mole of sodium chloride is equivalent to two osmoles, ie, 1 M NaCl = 2 Osm NaCl. ADH = antidiuretic hormone.

the ascending limb it decreases in osmolarity owing to Na^+ and Cl^- reabsorption. Active Cl^- reabsorption, with passive Na^+ reabsorption in the thick ascending loop, provides a hypotonic fluid that moves into the distal convoluted tubule (8). The reabsorbed NaCl is part of the solute supply to the interstitium and is necessary for the osmolaric gradient in the medulla. The principle of fluid concentration and dilution in the loop of Henle is known as countercurrent multiplication. In the distal convoluted tubule, additional Na^+ and Cl^- are abstracted by active Na^+ reabsorption, whereas K^+ might be both secreted and reabsorbed, depending on the K^+ balance of the organism. Finally, in the collecting duct, Na^+ is reabsorbed under the influence of aldosterone. At this part of the nephron, the rate of Na^+ reabsorption is probably intimately connected with both potassium and carbonic anhydrase-dependent hydrogen ion secretion. The hypothesis that an exchange of sodium for potassium and hydrogen ions occurs in a 1:1 ratio has recently been modified (7). The collecting ducts are distinguished from other segments of the nephron by changing their water permeability as a result of the action of antidiuretic hormone (ADH) (9). ADH is released from the pituitary gland into the circulation in hydropenic states. Maximal action makes these segments freely permeable to water, which allows back-diffusion of water into the hypertonic interstitial tissue, resulting in a urine of high osmolarity. On the other hand, the impermeability to water during hydration (ADH absence), leads to water diuresis.

Detailed discussions of the effects of diuretics on the normal transport phenomena and renal physiology are presented in recent reviews (2,10–11). The results from multiple experimental techniques employed to elucidate these topics have been interpreted in various ways, especially concerning the distal part of the nephron.

Diuretic Compounds

The diuretic compounds discussed below are divided into the following groups: osmotic diuretics, sulfamoyl diuretics, newer loop diuretics, potassium-sparing diuretics, xanthines, pyrimidines, organomercurials, and aryloxylacetic acids. Consideration is mainly given to diuretic compounds available as drugs or of potential clinical interest. It is likely that new diuretic drugs will result from the large number of new structures with diuretic activity reported in recent years. Some of these structures have been briefly reviewed (12–13). With the exception of carbonic anhydrase inhibitors, the current knowledge of the biochemical action of diuretics does not allow us to draw any final conclusions and, for this reason this subject will not be considered. The still incomplete and fragmentary views on the molecular mode of action are presented in ref. 14.

Classification and Terminology. Logically, diuretics should be classified from a therapeutic point of view, according to their diuretic profile and the maximal natriuretic effect that can be achieved. This reflects their sites of action in the nephron. In describing the following diuretic compounds, however, related chemical structures or structural features are discussed together whenever possible. For the characterization of diuretic behavior in a general sense, however, reference is made to the following classes.

Osmotic diuretics act at multiple sites, primarily in response to the osmotic load in the tubule. They provide a mild saluretic action and an enhanced urine formation, probably by decreasing the medullary osmolar gradient.

Carbonic anhydrase inhibitors act in the proximal convoluted tubule by decreasing carbonic anhydrase enzyme activity responsible for sodium and bicarbonate reabsorption in this segment of the nephron. Additionally, hydrogen ion secretion in the distal tubule is impaired, leading to increased potassium excretion. The maximal obtainable natriuretic effect—their natriuretic efficacy—is highly dependent on the acid–base status of the organism. It is generally low, as in the more distal part of the tubule, the higher load of sodium results in increased compensatory sodium reabsorption.

Thiazide-type diuretics act mainly at the earlier cortical part of the distal tubule. Their action leads to excretion of about 8% of the filtered load of sodium chloride. Similarly to the loop diuretics, potassium excretion is enhanced by the distal tubule. This type of diuretic action does not impair the concentrating ability of the kidney and, consequently, provides a urine of relatively high osmolarity. The group name is derived from chlorothiazide which was the first diuretic of this type.

Loop diuretics act mainly at the medullary and cortical thick ascending loop of Henle by decreasing active chloride reabsorption. In providing high maximal saluresis and possessing high-ceiling diuretic activity, they are the most efficient diuretic compounds available to date. The particular site of action leads to excretion of ca 20% of the filtered load of sodium chloride. A diminished medullary osmolar gradient owing to a wash-out of the medullary solute results, furthermore, in a dilute urine even in the presence of antidiuretic hormone (ADH). The increased sodium delivery to the distal tubule is responsible for kaliuretic effects.

6 DIURETICS

Potassium-sparing diuretics decrease aldosterone-dependent sodium–potassium exchange at the distal site of the nephron, the collecting duct. They act either directly on the tubular cells or indirectly by interfering with aldosterone synthesis or as competitive antagonist of this mineralocorticoid. Natriuretic efficacy is relatively low, reaching about 3% of the filtered load while, on the other hand, potassium excretion is unaffected or decreased. In the concurrent administration with more efficient saluretics this important feature is utilized for counteracting potassium loss of the more proximally acting drugs.

Miscellaneous diuretics show a diuretic profile in which the main site of action is not common to any of the above classes. The xanthines are representative of this particular group.

Independent of this classification is, of course, both the duration of action, which is dictated by the pharmacokinetic behaviour of the compound in question, and the potency. The words potent and potency are used exclusively in the context of their relationship to activity per unit weight.

Osmotic Diuretics. In principle, every nontoxic compound showing a low apparent volume of distribution and being freely filtered in the glomerulus and poorly reabsorbed by the renal tubule, could act as an osmotic diuretic due to the osmotic load of the nephron. A rise in the osmotic pressure in the tubules diminishes reabsorption of water by the tubular cells, resulting in increased urine formation. Although the diuretic effect has classically been ascribed solely to the physical forces of the increased osmolarity of the tubular fluid, osmotic diuretics appear to have multiple sites and mechanisms of action. Their major component of action is probably a decrease of the medullary osmolaric gradient (11). However, these effects might be consequences of the osmotic load and evoked by involvement of renal hormones. Although the introduction of modern saluretics has superseded the osmotic diuretics, there has been a revival of interest in their use (15). In some cases they are advantageously administered in conjunction with other diuretics.

The use of neutral salts and various nonelectrolytes, such as urea and sugars, as osmotic diuretics, is almost only of historical interest. The clinical application is now concentrated on D-mannitol (1) and isosorbide (2) (1,4-3,6-dianhydro-D-glucitol) from the group of sugar alcohols and their anhydrides, respectively (see Alcohols, polyhydric). Owing to the nature of their primary action, high dosage is required. Mannitol has to be given intravenously, preferably by infusion, because of unpredictable absorption from the gastrointestinal tract and partial metabolism after oral administration (16). Oral isosorbide is well absorbed in quantities required for osmotic diuresis (17). Essentially, it is not metabolized and shows a pharmacokinetic behavior ideal for this category of agents (drugs) (18). In a comparative study of oral isosorbide and intravenous mannitol, both drugs were found to be equally effective (19).

Sulfamoyl Diuretics. *Carbonic Anhydrase Inhibitors.* In 1937 Southworth (20) reported that treatment of bacterial infections with high doses of sulfanilamide (**3**), a *p*-aminobenzoic acid antagonist and as such the active component of the antibacterial sulfonamides, produced a metabolic acidosis (a decrease of plasma pH below the normal range). Subsequently, this side effect was found to be owing to an inhibition of the enzyme carbonic anhydrase, which catalyzes the reversible hydration of carbon dioxide ($CO_2 + H_2O \rightleftharpoons H_2CO_3 \rightleftharpoons H^+ + HCO_3^-$), and that the unsubstituted sulfamoyl group appeared to be directly concerned with this inhibitory effect (21). After it was shown that this enzyme was present in renal tissue (22), the mechanism of hydrogen ion excretion of the excess of metabolically formed acid was elucidated (23). Carbonic acid was the probable source of hydrogen ions and the secretion of these ions by the renal tubular cells into the tubule involved exchange with a monovalent cation, primarily Na^+. The carbonic anhydrase inhibitor, sulfanilamide (**3**), decreases acid excretion in acidotic dogs. The major credit for the introduction of carbonic anhydrase inhibitors for clinical use must go to Schwartz (24). He took advantage of these findings and treated patients with congestive heart failure with sulfanilamide. The resulting increased excretion of sodium, potassium, and water, and furthermore, the rise in urinary pH reflected the effects of carbonic anhydrase inhibition. This inhibition leads to diminished sodium bicarbonate reabsorption in the renal proximal tubule, impaired hydrogen ion excretion and, consequently, increased distal sodium–potassium exchange. However, sulfanilamide was too toxic for prolonged clinical use in the doses required (see Antibacterial agents, synthetic–sulfonamides).

A variety of other aromatic sulfonamides were found to possess *in vitro* carbonic anhydrase inhibitory effects (25). The most potent compounds were a series of heterocyclic sulfonamides (26). Among these, acetazolamide (**4**) was intensively studied (27) and soon became the most widely used diuretic of this type. It exerted a safe but rather mild diuretic effect following oral application, lasting up to 16 h. Owing to its nature of action, no chloruresis occurs and undesired potassium excretion is rather high. Prolonged treatment leads to acidosis, which drastically decreases the diuretic efficacy, thus demonstrating the pronounced dependence of renal carbonic anhydrase inhibitory effect on the acid–base status. Introduction of the thiazide-type saluretics gradually superseded the use of acetazolamide as a diuretic.

Thiazides and Thiazide-Type Diuretics. The parent compound of all thiazide diuretics, chlorothiazide (**5**), was the result of efforts to improve the diuretic properties of sulfanilamide. At Merck Sharp & Dohme, Beyer (28) investigated in dogs the renal electrolyte response of *in vitro* carbonic anhydrase inhibitors, amongst these being *p*-sulfamoylbenzoic acid (**6**), a close congener of sulfanilamide. He paid attention to the fact that this compound produces a small but significant increase in chloride excretion besides the expected diuretic profile of carbonic anhydrase inhibition. This particular effect could be seen independently in humans (29–30), although (6) was not suitable for clinical use. Encouraged by this exceptional observation, Sprague and co-workers, together with Beyer and co-workers, subsequently examined various aromatic sulfonamides in a cooperative chemical and pharmacological study. They found that introduction of chloro groups and a second sulfamoyl group, as shown in clofenamide (7) and dichlofenamide (8) caused a substantial increase in renal chloride excretion, despite a considerable carbonic anhydrase inhibitory effect *in vitro*. Further research on these lines led to chlorodisulfamoylaniline (9) (31). The diuretic picture of this compound, in respect to sodium chloride excretion in experimental animals and man (32), revealed a clear reduction of *in vivo* carbonic anhydrase inhibition.

The final step leading to 6-chloro-7-sulfamoyl-1,2-4-benzothiadiazine 1,1-dioxide (chlorothiazide) (5) was taken when (9) was cyclized with formic acid (33). With (5), a valuable new type of diuretic action was discovered (34). Although the compound displayed *in vitro* carbonic anhydrase inhibitory activity, *in vivo* it caused an excretion of nearly equivalent amounts of sodium and chloride. In this respect, it resembles the mercurial diuretics, but is distinguished from the latter in being effective under conditions of both acidosis and alkalosis. The main site of action in the nephron is in the cortical part of the distal tubule, with an additional effect in the proximal tubule. The concentration ability of the kidney is not involved, so that the saluretic effect is accompanied by excretion of an osmotic equivalent of water. Owing to the increased amount of sodium reaching the distal tubular site of sodium–potassium exchange, potassium excretion is enhanced, although when compared with carbonic anhydrase inhibitory action, a much more favorable sodium–potassium ratio is achieved. A large amount of clinical work has shown chlorothiazide (5) to be an orally well-absorbed and safe drug in the treatment of various edematous states and hypertension.

The valuable properties of this new drug, which became an immediate success in the clinic, initiated a renewed interest in sulfamoyl compounds. Shortly after the introduction of chlorothiazide, the corresponding dihydro compound, hydrochlorothiazide (10), was shown to have a similar diuretic profile, and was about ten times more potent (35). The corresponding 6-trifluoromethyl derivative, hydroflumethiazide (11), was shown to possess diuretic properties and potency comparable to that of (10) (36).

Since then, many hundreds of 7-sulfamoylbenzothiadiazines (thiazides) and 7-sulfamoyldihydrobenzothiadiazines derivatives have been evaluated. In addition, structural alterations of the thiadiazine ring and its replacement by various cyclic systems has been performed. Furthermore, the structures of the monocyclic aromatic sulfamoyl compounds have been extensively modified. With respect to biological evaluation, the interest in carbonic anhydrase inhibition was obviously replaced by interest in the *in vivo* diuretic activity. The tremendous amount of chemical and pharmacological research that ensued is described in detail in several reviews (37–41) and resulted in the introduction of a large number of thiazide-type diuretics into clinical practice. They are all closely related chemically to either hydrochlorothiazide or hydroflumethiazide on the one hand, or clofenamide on the other. The designation thiazide-type diuretic, however, refers to the similarity of diuretic profile with chlorothiazide, probably resulting from the same site of action in the nephron. The formulae (12–18) depict the most important drugs related to the bicyclic hydrothiazide structure, the hydrothiazides. Apart from the introduction of different side chains (R and R'), the thiadiazine ring has been modified in (19) and (20) by replacement of sulfonyl (SO_2) by carbonyl (CO), the latter being 1,2,3,4-tetrahydro-4-quinazolinone derivatives.

Some diuretics related to clofenamide (7) are shown in the formulas (21–24). Various side chains lead to diuretic activity and, as before, SO_2 may be replaced by CO. The 1-isoindolinone derivative, chlorexolone (25), might be regarded as occupying an intermediate position. Indapamide (S 1520) (23), is structurally closely related to clopamide (22) and is placed among the thiazide-type diuretics, although investigations in dogs seem to indicate that its site of action is predominantly in the proximal part

(12)	methyclothiazide,	R = CH_3,	R' = CH_2Cl,	X = Cl
(13)	trichlomethiazide,	R = H,	R' = $CHCl_2$,	X = Cl
(14)	polythiazide,	R = H,	R' = $CH_2SCH_2CF_3$,	X = Cl
(15)	althiazide	R = H,	R' = $CH_2SCH_2CH=CH_2$	X = Cl
(16)	cyclothiazide,	R = CH_3,	R' = —CH_2–(cyclohexenyl),	X = Cl
(17)	cyclopenthiazide,	R = H,	R' = CH_2–(cyclopentyl),	X = Cl
(18)	bendroflumethiazide,	R = H,	R' = $CH_2C_6H_5$,	X = CF_3

(19) quinethazone, R = H, R' = C_2H_5

(20) metolazone, R = —(o-tolyl), R' = CH_3

10 DIURETICS

of the renal tubule (42). By having a length of action of up to 36 h in man (43), this drug appears to be the longest acting diuretic so far available. All these thiazide-type diuretics only differ clinically in two major respects; their potencies vary over a range of 1–2000, and perhaps more important with respect to clinical use, their duration of action varies from ca 6 to >24 h.

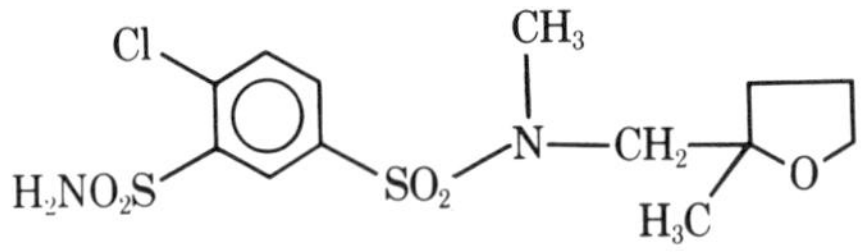

(21) mefruside

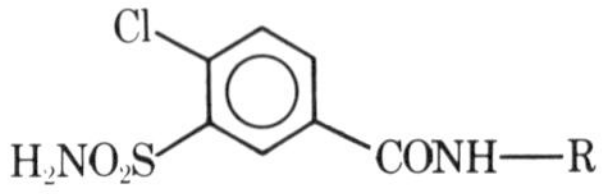

(22) R =

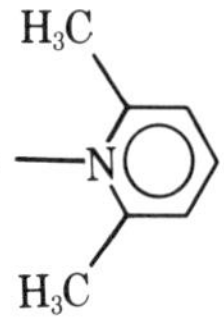

(23) R =

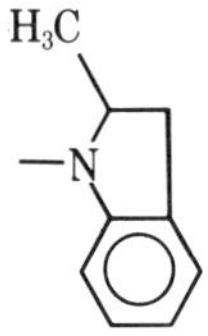

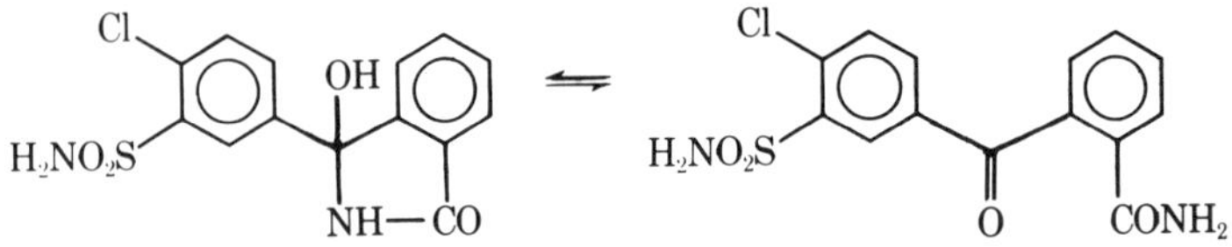

(24) chlorthalidone

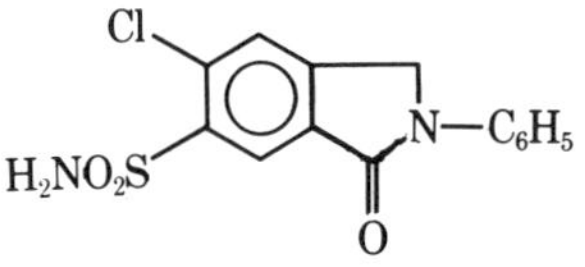

(25)

With respect to the structural features of the aromatic disulfonamides and the thiazides and related structures leading to a high order of diuretic activity, Sprague (44) has made some broad generalizations that still hold for the sulfamoyl thiazide-type diuretics. Common to all these compounds are: the SO_2NH_2 group and an activating group X ortho to this sulfamoyl (26). Outstanding X groups are halogens and a pseudo halogen, eg, trifluoromethyl. In a position meta to the sulfamoyl there must be an electronegative group Z, eg, SO_2NH_2, SO_2HNR, COR, CONHR, or CONRR'. This electronegative group may be incorporated into a cyclic structure (27).

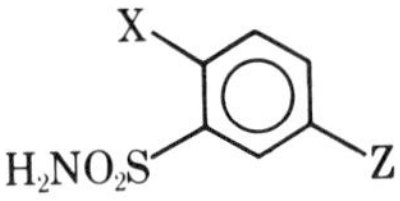

(26)

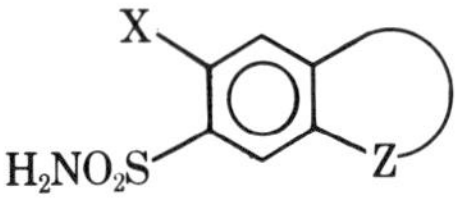

(27)

Sulfamoyl Loop Diuretics. Coincidental with the development of the loop diuretic, ethacrynic acid (**29**), a similar type of diuretic action was independently discovered at Farbwerke Hoechst A.G., FRG. Among various 5-sulfamoylanthranilic acids synthesized by Sturm and co-workers (45), the 4-chloro-N-(2-furylmethyl) derivative (furosemide, frusemide) (**28**) was found to be the most potent member of this series. A pharmacological investigation by Muschaweck and Hajdú (46) and a clinical study by Kleinfelder (47), revealed that this compound displayed the now well-known characteristics of those nonmercurial loop diuretics that are carboxylic acids. In contrast to the thiazide-type diuretics, it showed a shorter duration of action, a steeper dose-response curve, and a greater maximal saluretic effect (48). These properties led to extended use in clinical practice. The drug is active after both oral and intravenous administration, similar to ethacrynic acid (**29**). The principal site of action has been found by several investigators to be the ascending limb of the loop of Henle, a finding which has been confirmed recently by *in vitro* microperfusion techniques (49).

Since the publication of a clinical study with 4-chloro-3-sulfamoyl-2'6'-dimethylsalicylanilide (xipamide) (**30**) (50), more information has been made available. The compound, made by Beiersdorf A.G., FRG, was selected from a series of salicylic acid derivatives (51). Pharmacological investigations (52) proved that the compound differed from the thiazides in its site of action in the renal tubule. Reference has been made to an action in the ascending limb of the loop of Henle (53). A smooth onset of action and, when compared with furosemide, a prolonged action, has been observed in dog (52) and man (50).

Being a loop diuretic, furosemide deviates substantially from the thiazide-type diuretics but in respect to chemical structure it still fits the empirical rules for the diuretic activity of sulfamoyl compounds given by Sprague and mentioned earlier. Owing to this peculiarity extensive structural alterations in the furosemide molecule have been made by Leo Pharmaceutical products in order to try and find new structures with loop diuretic activity. This line of research resulted in numerous 5-sulfamoylbenzoic acids possessing considerable diuretic activity (54–59). The most potent members of this new series are depicted in the general formula (**31**), which includes compounds shown to be approximately three hundred times more potent than furosemide (59).

(31) X = NH, O, S, SO, CH₂, CO
R in 2 position = NHR'
R in 3 position = NHR', OR', SR', CH₂R'
R' = n-butyl, benzyl,
furylmethyl, thienylmethyl, etc

The most striking feature is that various 4-substituents contribute to high potency and that the departure from the 4-chloro substituent, even in furosemide (28), could enhance the diuretic potency dramatically (55). Consequently, the question arose as to whether the new 4-substituents should be regarded as specific for loop diuretic activity or new activating groups in the sense of Sprague's prediction. The answer is that these substituents are rather specific. Substitution of phenoxy or phenylthio groups in place of the chloro group in several thiazide-type diuretics leads to a loss of diuretic activity (60). Bumetanide (32) was selected for pharmacological studies (61) and ultimately for clinical use (62). The drug was shown to be a potent loop diuretic in dogs and a valuable alternative to furosemide in clinical practice. When compared with furosemide, it shows a significantly higher kidney–plasma concentration ratio (63) which, however, does not solely account for the higher potency. At Farbwerke Hoechst A.G., FRG, a congener of bumetanide has been evaluated. Piretanide (33), derived from bumetanide by replacing the 4-n-butylamine side chain with the pyrrolidino substituent, was compared with bumetanide and found to be more potent in rats with a different regression of the dose response (64). However, the fact that bumetanide has a rather low potency in this animal owing to rapid and dose dependent metabolism, mainly by attack at the n-butylamino side chain (65), should explain the great difference in potency. More recently it was discovered that neither the sulfamoyl group nor the carboxylic function, is a prerequisite for the diuretic activity of bumetanide (32) and some of its congeners, (66–68).

(32) R = NH(CH₂)₃CH₃
(33) R = N⟨⟩

Newer Loop Diuretics. *2-Methylenthiazolidone Derivatives.* In 1977 Satzinger reported on a new approach in the search for diuretics (69). The fact that various diuretics increase bile production (choleretic activity) without affecting the electrolyte concentration in the bile after intravenous administration in high doses, suggested a mechanism of action which could have similarities with renal cellular liquid transport.

The question arose as to whether the structure of choleretics capable of increasing bile secretion without affecting its osmolarity, could serve as model for diuretic active compounds. The results obtained with 2-carboxymethylene-4-thiazolidones answered this question. The replacement of an ethyl group by a methyl group in the choleretic piprozolin (34) (70) led to the diuretic etozolin (35).

(34) R = C$_2$H$_5$
(35) R = CH$_3$

Both drugs have been introduced by Gödecke A.G., FRG, as selected compounds of this series where the choleretic and diuretic components, respectively, could be optimalized. Etozolin, synthesized in 1963 (71), has since been pharmacologically evaluated (72–75). The results indicate that this drug is a loop diuretic and its duration of action appears to be more than 10 h following oral administration. Compared to other loop diuretics in use, it is striking that etozolin loses its effect in patients with impaired renal function and a glomerular filtration rate of less than 20 mL/min (76). However, at normal renal function this rather low potency drug appears to possess the high efficacy seen with other loop diuretics.

3-Aminopyrazolinones. In 1976 muzolimine (36) emerged from chemical and pharmacological work at Bayer A.G., FRG, in the field of 1-substituted 3-aminopyrazolinones. It is a new loop diuretic (77–79) different in chemical structure from other diuretics. It was shown in a clinical trial (80) to be an orally effective diuretic in short-time studies which included patients suffering from severe renal insufficiency with substantial reduced glomerular filtration rates. Although providing almost the same diuretic profile as seen with other loop diuretics available for clinical use, the effect of muzolimine is longer lasting. At normal renal function the effect lasts 6–8 h and is prolonged to more than 24 h at low glomerular filtration rates (81). Muzolimine has not yet been introduced into clinical practice.

(36)

Salicylamines. At Merck Sharp & Dohme a large series of salicylamines was screened from which 2-aminomethyl-4-*tert*-butyl-6-iodophenol (MK 447) (37) and its hydrochloride (38) were selected as promising new diuretic compounds (82). It has been reported as being superior to existing loop diuretics in its ceiling effect. Results

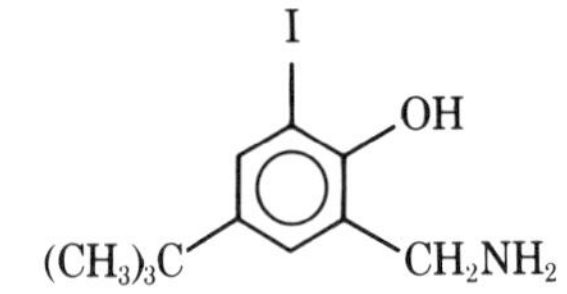

14 DIURETICS

(37)
(38) . HCl

in volunteers show a profuse diuresis and natriuresis without significant change in potassium excretion (83). On theoretical grounds, it was suggested that MK 447 acted by stimulating prostaglandin E_2 biosynthesis (see Prostaglandins). Further studies are needed to establish its sites of action; to show whether the results from the volunteer study can carry over to patients; and to show whether the interference with prostaglandin synthesis is relevant to its diuretic action.

Potassium-Sparing Diuretics. The mineralocorticoid aldosterone (39), synthesized by the adrenal cortex, is released by this gland in response to a variety of stimuli. It plays an important role in homeostatic mechanisms in situations where conservation of sodium by the body is required. Elevated aldosterone levels increase the sodium–potassium exchange in the distal tubule (Fig. 2), thus leading to sodium retention and enhancement of potassium excretion. Aldosterone levels are raised in several chronic edematous states and, owing to a high sodium load in the distal renal tubule, in prolonged treatment with powerful saluretics (secondary aldosteronism), which complicates successful treatment and produces an undesirable potassium loss. Consequently efforts have been directed towards counteracting this hormonal effect: by inhibiting aldosterone synthesis; by antagonizing endogenous mineralocorticoid activity competitively at its renal receptor site; and by decreasing sodium/potassium exchange with direct action at the distal portion of the renal tubule.

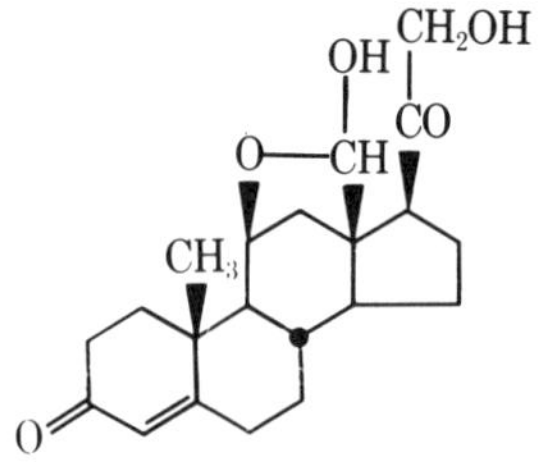

(39) (hemiacetal form)

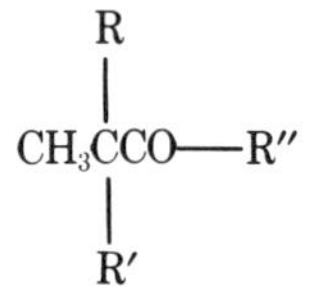

(40) R = R' = —⟨◯⟩—NH_2, R'' = CH_3
(41) R = CH_3, R' = R'' = pyridyl

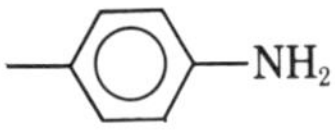

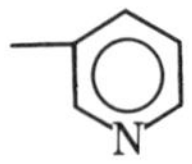

Inhibition of Aldosterone Biosynthesis. Amphenone B (40), synthesized in 1950 (84), was the first compound to interfere with the biosynthesis of adrenal corticosteroids (see Hormones; Steroids). The correct chemical structure was recognized several years later to be 3,3-bis(*p*-aminophenyl)-2-butanone (85). It was shown to inhibit aldosterone excretion in man (86–87) but owing to serious side effects and toxicity, clinical use was abandoned. From the subsequent investigation of a series of α-dimethyldesoxybenzoins and 3,3-diaryl-2-butanones (88), special attention was paid to 1,2-bis(3-pyridyl)-2-methyl-1-propanone, metyrapone (41), from Ciba Pharmaceutical Company (89). By inhibiting hydroxylation of the steroid molecule at the 11-α position, this drug not only decreases aldosterone synthesis, but also diminishes the production of hydrocortisone. Diminution of hydrocortisone levels, on the other hand, stimulates the pituitary gland to to release adrenocorticotropin. This results in an outpouring of 11-deoxyhydrocortisone and 11-deoxycorticosterone by the adrenal gland, both of which are themselves potent salt retaining mineralocorticoids. This latter effect can be prevented by combined administration of metyrapone with prednisolone or dexamethasone, leading to a diuresis in edematous patients with hyperaldosteronism (90). Treatment with aldosterone biosynthetic inhibitors is no longer of importance and has been superseded by potassium-sparing diuretics of other types.

Aldosterone Antagonists. The initial steps towards the synthesis of aldosterone antagonists were taken in 1957, when Cella and Cagawa from G. D. Searle & Co. synthesized the androstenone spirolactone (**42**) (91). This compound, as well as its 19-nor analogue (**43**) suppresses the effect of aldosterone when administered subcutaneously to adrenalectomized rats (92).

(**43**) was shown to antagonize the effect of intravenously infused aldosterone in humans (93). Subsequently, a large series of spirolactones was investigated. The proposed structural requirements for optimal activity have been summarized (94). A valuable structural modification, leading to spironolactone (**44**), was obtained by the introduction of the 7α-acetylthio group (95), which contributes to an improvement in absorption after oral administration. In a search for more potent analogues, it was later found that replacement of the acetylthio group by a carbalkoxy group (—CO_2CH_3 (**45**), —$CO_2C_2H_5$ (**46**), —CO_2Pr^i (**47**)) resulted in good oral and subcutaneous activity (96). The same year the structure of spironolactone was altered to potassium 3-(17β-hydroxy-6β,7β-methylene-3-oxo-4-androsten-17-yl)propionate (**47**) (prorenoate), which has the characteristics of an aldosterone antagonist and is several times more potent than spironolactone (97). However, spironolactone remains the only drug of this particular class of diuretics in clinical use to date. Obviously, owing to its nature of action, spironolactone reduces potassium excretion and sodium reabsorption depending on aldosterone levels. As the aldosterone-induced regulatory sodium reabsorption in the renal tubule is only responsible for a few percent of the total amount reabsorbed, the natriuretic effect at the normal range of aldosterone is relatively small.

In the 1970s syntheses were reported of 2,3-disubstituted 1,6- and 1,8-naph-

(**42**) R = CH_3, R′ = H
(**43**) R = R′ = H
(**44**) R = CH_3, R′ = $SCOCH_3$
(**45**) R = CH_3, R′ = CO_2CH_3
(**46**) R = CH_3, R′ = $CO_2C_2H_5$
(**47**) R = CH_3, R′ = CO_2Pr^i

thyridines as potential diuretic agents, structurally related to diuretically active pteridines (98–100). Several of these compounds showed diuretic natriuretic activity, but no kaliuretic effect in the saline-loaded rat. The most potent compound of this series was 2-amino-1,8-naphthyridine-3-carboxamide (49), the hydrochloride of which (50) (39A69) has been investigated in more detail (101).

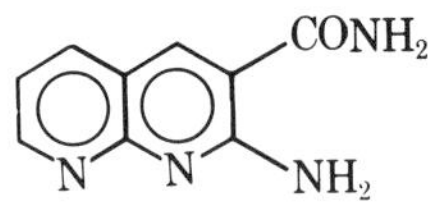

(49)
(50) . HCl

This compound possesses an extended action up to 30 h following intraperitoneal administration to rats, but is inactive when administered to adrenalectomized rats. Thus the compound is clearly distinguished from pteridine diuretics. It might interfere with aldosterone synthesis or possess aldosterone antagonizing properties. A direct effect at the renal distal tubule site can, however, be excluded.

Direct Action on the Distal Renal Tubule. *Symmetric Triazines.* The observation in the early 1940s that the grouping =N—C—N= is a constituent part of the diuretically active urea, biuret, and xanthine compounds, encouraged Lipschitz and Hadidian (102) at that time to investigate various compounds possessing this particular structural feature. A diuretic effect was noted in the rat with triamino-s-triazine (melamine) (51) and 2,4-diamino-s-triazine (formoguanamine) (52). The interest at that time in nonmercurial diuretics led to more extensive investigations of the formoguanamine derivatives by several research groups, and revealed that N-substitution had a favorable effect on potency in the rat. The p-chlorophenyl derivative, chlorazanil (53), has since been widely studied in several animal species and in man (103). The pharmacological picture of this drug appears complex (104), and the diuretic potency found in animal studies does not carry over, in full, in humans. Clinical use of this drug has, however, been established. Chlorazanil has never found a real clinical application but is of interest in being the first compound investigated to interfere with the mineralocorticoid-regulated potassium secretion (105), at least in part by a possible direct action at the renal distal tubule.

Pteridines. The diverse potential for influencing biological events with various ptéridines, led Smith Kline and French to tests of a series of pteridines for diuretic effect in rats (106). Diuretic properties were discovered for many aminopteridines. The structure–activity relationships of these have been summarized and discussed

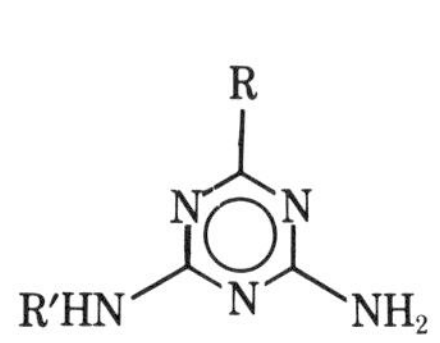

(51) R = NH$_2$, R' = H
(52) R = R' = H
(53) R = H, R' = —⟨phenyl⟩—Cl

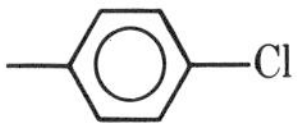

by the investigators retrospectively, and comparative data is given for three principal classes, deviating in structural features and electrolyte excretion patterns (107). These are the 6,7-dialkyl-2,4-diaminopteridines, the 2,7-diamino-2-arylpteridino-6-carboxamides, and the 6-aryl-2,4,7-triaminopteridines. The last group includes triamterene (54), which was extensively investigated and ultimately introduced for clinical use.

(54)

This compound, originally synthesized in 1954 (108) in a series of compounds with possible antifolic acid activity, was found to be outstanding with respect to diuretic properties. The moderate natriuretic action and potassium preserving properties after oral and parenteral administration appear qualitatively similar to those of spironolactone (44). However, pharmacological investigations in animals and humans clearly demonstrate that triamterene is not an aldosterone antagonist but exerts its effect by a direct action at the distal tubule independently of the presence of mineralocorticoids (106,109–110). Although many patients respond satisfactorily to triamterene, its main applications lie in the combination with natriuretics having a more proximal site of action, such as thiazide-type diuretics. Concurrent administration augments the sodium excretion and prevents potassium loss (111–115). However, even when appropriately combined, the risk of hyperkalemia does exist.

Pyrazines. An important contribution to the field of potassium-sparing diuretics was made when Cragoe and co-workers at Merck Sharp & Dohme discovered that N-amidino-3-amino-6-halogenopyrazinecarboxamides possessed saluretic activity while leaving unaffected, or even suppressing, potassium excretion (115). Comparative studies in rats performed with a representative of this chemically new class of diuretics, the 6-chloro compound (55), demonstrated a 12.5-fold increase in potency over spironolactone (44) (116). However, as the compound exerts its natriuretic and antikaliuretic effect in the absence of mineralocorticoid and acts directly at the distal site of the renal tubule, it should be placed in the same group as triamterene (54). In subsequent investigations on several hundred congeners the structure–activity relationships of the piperazinecarboxamides were elucidated (117–122). This inhibitory effect on the renal action of exogenous desoxycorticosteroneacetate (as mineralocorticoid) in the adrenalectomized rat was used as a measurement for diuretic activity. This assay, however, does not strictly parallel the effects on the intact rat. In summary, 6-chloro substitution, especially when combined with 5-amino, methylamino, or dimethylamino groups, contributes to high potency. Other alkylations of the 5-amino group reduce potency. Various alkylations at the terminal nitrogen of the amidino moiety generally appear not to influence potency. With the exception of hydroxylation, replacement of the unsubstituted 3-amino group by alkylamino, alkoxy, mercapto or alkylthio groups, leads to compounds with diminished potency. N-Amidino-6-chloro-3,5-diaminopyridazinecarboxamide (amiloride) (56) is one of the most potent compounds of this series. Its hydrochloride (57) has been chosen for pharmacological studies and is now available for clinical use in various countries. Remarkably, the

18 DIURETICS

$$\text{Cl}-\underset{6}{\overset{1}{N}}\underset{2}{=}\text{CONH}-\underset{\parallel}{\overset{NH}{C}}-\text{NH}_2$$
(with R at position 5, NH_2 at position 3)

(55) R = H
(56) R = NH$_2$
(57) R = NH$_2$·HCl

(58) R = N(CH$_3$)$_2$·HCl
(59) R = NH(CH$_3$)

corresponding 5-dimethylamino analogue (58) was found to be more potent and showed a slow onset and a prolonged action. A stepwise metabolic dealkylation to the monomethyl (59) compound and to amiloride (56) in rat, dog, and man (123), probably explains this behavior and confines the proper 5-substituent for high potency to the unsubstituted amino group.

Glitzer and Steelman (124) reported that amiloride possessed natriuretic and antikaliuretic properties and was about fifteen times more potent than its 5-desamino analogue (55) and one hundred and eighty times as potent as spironolactone (44). From data from adrenalectomized rats untreated with mineralocorticoids, it was concluded that this drug was not acting by way of competitive aldosterone inhibition. Further investigations demonstrated an increased concentration of sodium, a decreased concentration of potassium, and a fall in pH in the distal renal tubule (125). The direct action at this segment of the tubule markedly enhanced sodium excretion and depressed potassium excretion in hydrochlorothiazide (10)-treated rats. Thus, with respect to its diuretic profile and main site of action, amiloride resembles triamterene. It appears, however, to exert an additional effect in the distal convoluted tubule as well as the proximal site (11). This might explain the clinical findings that the natriuretic and antikaliuretic efficacy exceed those of triamterene (126–127).

Imidazolidinones. Interesting diuretic properties of 2-imino-3-methyl-1-phenyl-4-imidazolidinone (60) (azolimine, CL-90748) have been reported recently (128).

(60)

This compound was selected from a series of imidazolidinones synthesized by Hanifin and Shepherd at Lederle Laboratories. Pharmacological action has not yet been fully explored and species differences are seen. Azolimine shares in some ways the diuretic profile of other potassium-sparing diuretics. It deviates from spironolactone (44) in not being a pure competitive antagonist of mineralocorticoids, as in large doses the compound is shown to exert a moderate natriuretic effect and to elevate the urinary sodium–potassium ratio in the absence of mineralocorticoid (128). In showing a greater efficacy in the presence of mineralocorticoid, it is also distinguished from drugs like triamterene (54) and amiloride (55) that have a direct action at the distal renal tubule. Experiments in rats, under conditions of high mineralocorticoid levels, revealed that azolimine, when combined with thiazide-type or loop diuretics, could enhance sodium excretion and improve the sodium–potassium ratio. There was no indication of frank potassium retention and hyperkalemia.

Xanthines. The diuretic properties of coffee and tea were well known in antiquity but only in the second half of the nineteenth century has caffeine (**61**) (1,3,6-trimethylxanthine) been identified as the active component of coffee and its pharmacological behavior investigated (129–130) (see Alkaloids). From the subsequently discovered natural dimethylxanthines, theobromine (**62**) and theophylline (**63**), theophylline was shown to have the most favorable ratio between its diuretic effect and other well-known pharmacological properties such as central nervous system stimulation.

(**61**) R = R′ = CH$_3$
(**62**) R = H, R′ = CH$_3$
(**63**) R = CH$_3$, R′ = H

It has been formulated in various pharmaceutical preparations of its water soluble salts and double salts (131). The formulation with ethylenediamine (theophylline ethylenediamine, aminophylline) has found the widest clinical use. Its importance as a diuretic has been, however, gradually reduced by the introduction of more efficient compounds. Nevertheless, it is remarkable that theophylline has remained the only diuretic exerting its mild diuretic effect as a result of a rise in the filtration rate, although an additional direct effect on the renal tubule probably exists. This might explain the renewed interest in its use, especially in concurrent administration with other diuretics, when an additional effect is required owing to changes in renal pharmacodynamics after long term diuretic treatment. Recently, a supra-additive or synergistic effect of aminophylline with the loop diuretic bumetanide (**32**), has been demonstrated (132). This effect, only observed in patients pretreated with the loop diuretic for a long period of time, is tentatively ascribed to an inhibition of the enhanced sodium reabsorption at the distal sites of the nephron, the result of a compensatory homeostatic mechanism. The search for new xanthine derivatives with improved diuretic properties remains unsuccessful to date, but has led to the discovery of other heterocyclic diuretics.

Pyrimidines. The diuretic effect of 5-methyluracil (**64**) (thymine) was noted as long ago as 1907 (133). The weak diuretic effects of uracil (**65**) and 5-aminouracil (**66**) compared with theophylline (**63**) were reported in 1944 (102). More potent and efficient compounds were found when various 1,3-dialkyl-6-aminouracils were synthesized as intermediates in the preparation of alkylated xanthines (134). Among these, passing clinical interest was paid to aminometradine (**67**) (135) and amisometradine (**68**) (136). Although these drugs replaced other diuretic therapy in the late 1950s or at least allowed a reduced administration of organomercurials in cardiac patients (137), their efficacy and safety were by no means comparable to the more modern saluretics discovered at that time.

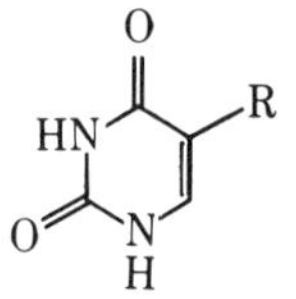

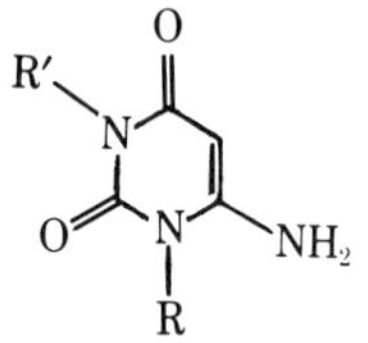

20 DIURETICS

(64) R = CH₃
(65) R = H
(66) R = NH₂

(67) R = CH₂CH=CH₂, R' = C₂H₅
(68) R = CH₂—C(CH₃)(CH₂)CH₃, R' = CH₃

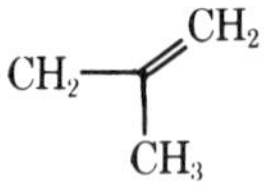

Organomercurials. By noticing an unexpectedly enhanced urine volume after parenteral administration of merbaphen (**69**) to a nonedematous girl with congenital neurosyphilis, Vogel initiated results into the area of organomercurial diuretics (138). A subsequent study on edematous patients showed, in respect to its diuretic action, that this mercury-containing drug was superior to existing diuretics at that time (139).

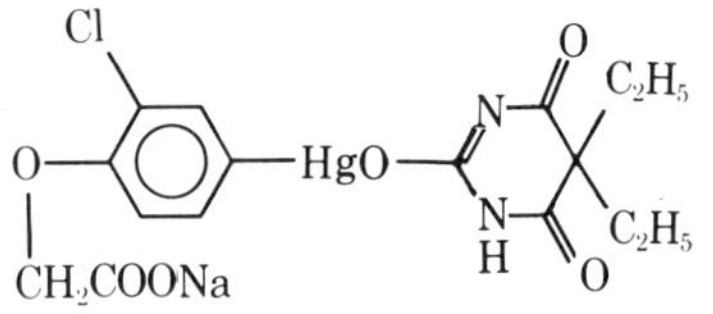

(69)

Merbaphen, whose structure was only recently proved unequivocally (140), was, however, abandoned because of its toxicity and replaced by more suitable agents. One of these was mersalyl (**70**) which was introduced for parenteral administration in 1924

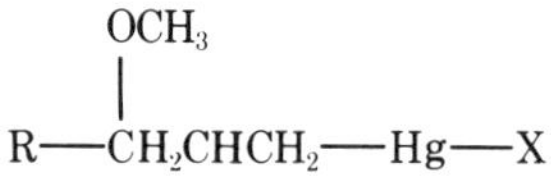

R—CH₂CH(OCH₃)CH₂—Hg—X

(70) R = —HNC(O)—C₆H₄—O—CH₂CO₂H , X = OH

(71) R = —HNC(O)NC(O)(CH₂)₂CO₂H , X = OH
           H

(72) R = —HNC(O)—[2,2,3-trimethylcyclopentane-CO₂H] , X = SCH₂CO₂H

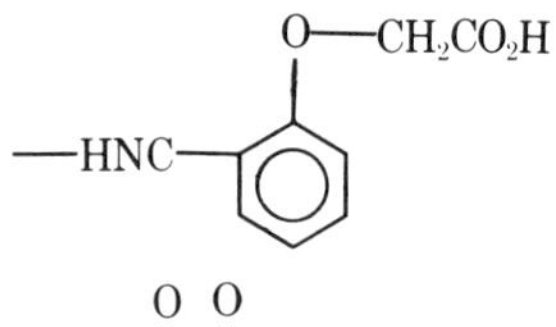

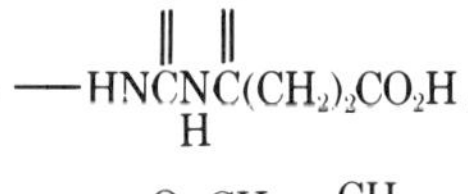

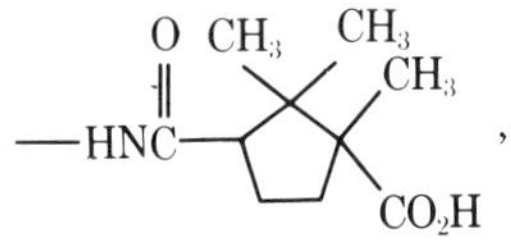

(73) R = —HNC(O)NH₂ , X = Cl.

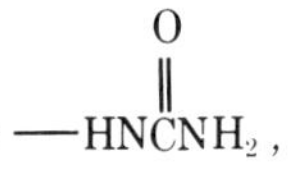

(141). The general formula for (70–73) covers almost all the mercurial diuretics that have a clinical application; meralluride (71) and mercaptomerin (72) are still available for parenteral use in several countries (71 in the United States).

The chemistry and structure–activity relationships of the various mercurial diuretics are described extensively in several reviews (142–144). The syntheses of many new organic mercurials in the 1920s to the late 1950s was mainly directed at decreasing their toxicity, rather than improving their diuretic activity. By parenteral administration, unpredictable and poor oral absorption could be overcome and systemic toxicity decreased to an acceptable level. Variation of the substituent R in (70–73) not only had an impact on toxicity, but provided a few agents with sufficient absorption after oral administration, as eg, is seen with chlormerodrin (73) (145). X is easily interchangable, thus allowing the compounds to exist in their monoionic form $RCH_2CH(OCH_3)CH_2$—Hg^+. When X is a theophylline moiety, an additional diuretic activity is probably achieved (146).

Although experimental proof of an intrarenal action of the mercury compounds became evident earlier (147–148), conflicting results have been obtained with respect to the sites of action in the kidney (11). More recent investigations, however, support the indication of action in the ascending limb of the loop of Henle (149). In 1973, by means of the microperfusion of an isolated nephron, mersalyl (70) was shown to inhibit chloride transport in the thick ascending limb (150). The organic mercurials appear to have the same principal site of action as the modern loop diuretics, and this is a logical explanation for the high efficacy of both, which has not yet been reached by any other type of diuretic. Obviously, owing to a greatly diminished toxicity and a good oral absorption, independence of the organism's acid–base status, and an immediate onset of action when given intravenously, the nonmercuric loop diuretics have now almost completely overshadowed the organic mercurials.

The clarification of the mechanism of action of the mercurial diuretics has been a challenge to many research workers, but the mechanism is not yet understood. It is generally accepted that preferential binding proceeds with an essential —SH group at the diuretic receptor. The mechanism of binding, however, is still controversial. Two hypotheses prevail (see 151–152): on the basis of structure–activity relationships, a specific lock-and-key type interaction of the intact ionized mercurial molecule with the receptor was proposed (153). On the other hand, it has been suggested that the organomercurials were transported as such to the target organ, where a rupture of the C—Hg bond occurred, releasing the inorganic mercuric ion to be combined with the receptor (154–155). This mercuric ion hypothesis has apparently been supported by the fact that diuretically active organic mercurials are, in vitro, acid labile at the C—Hg bond, thus explaining the potentiation of mercurial diuresis in acidotic states. Using ^{203}Hg- and ^{14}C-labeled mersalyl (70), a C—Hg bond rupture was detected in acidotic, but not in alkalotic, dogs (156). However, it was found more recently that inhibition of active chloride transport in the thick ascending limb of the loop of Henle is not influenced by pH change from 6–7.4 (150). In addition, the citrate anion, whose renal concentration is dependent on the acid–base balance and high only under alkaline conditions, probably moderates the potency of mercurial diuretics by competing with the receptor (157). In summary, the mechanism of action of this class of diuretics appears to be of a more complex nature than was first supposed and still remains unsolved.

Aryloxyacetic Acids. The unique properties of mercurial diuretics, especially the combination of high efficacy with minimal potassium loss, is a continuous challenge for medicinal chemists to search for nonmercury-containing structures that possess a similar diuretic profile. On the basis of the concept that activated double bonds, capable of reacting with mercapto groups, could mimic mercurial diuretics in reacting with essential —SH at the receptor site, and that a carboxylic function would facilitate the transport to the renal target, Sprague and co-workers at Merck Sharp & Dohme examined various α,β-unsaturated ketones of this type (158–159). These investigations resulted in the important group of α,β-unsaturated (acylphenoxy)acetic acids of the general formula (**74**), which includes ethacrynic acid (**29**). This was one of the first nonmercuric loop diuretics active in humans (160–161) and is now in use in clinical practice. The structure–activity relations of this series and closely related compounds have been reviewed (162). Maximum potency requires a terminal methylene group and that the ortho position of the unsaturated keto function is occupied by halogen or the methyl group.

$$\text{RCCO}-\underset{\underset{R'}{\overset{\|}{\text{CH}}}}{\overset{\overset{X\quad Y}{|\quad|}}{\bigcirc}}-\text{OCH}_2\text{CO}_2\text{H}$$

(**74**) R = CH$_3$, C$_2$H$_5$; R' = H, CH$_3$; X = Cl, CH$_3$; Y = H, Cl, CH$_3$

In vivo experiments demonstrate that diuretically active members of this series react with the —SH of mercaptoacetic acid and cysteine. Considering that such a model does not account for distribution and metabolic fate *in vivo,* it appears that the reactivity correlates sufficiently with the diuretic activity in dogs.

With ethacrynic acid (**29**), other similarities to the mercurial diuretics were shown. It reduces protein-bound SH in the dog kidney during maximal diuresis, whereas no reduction was shown in rats in which both categories of drugs are inactive (163–165). Competition for the same receptor site has been shown recently (166). The renal effect of ethacrynic acid has been intensively investigated (167). The drug shows a rapid onset of action, and being well absorbed orally, it can be administered both orally and parenterally. It is excreted as both the intact drug and as metabolites, including the cysteine adduct. Unlike the mercurials, it is effective both in conditions of alkalosis and acidosis. Under conditions of natriuresis, it causes a slight potassium loss, probably by leaving unaffected the exchange of sodium for potassium in the distal part of the nephron. Its diuretic properties are further distinguished from those of the mercurials by a different action in the proximal tubule which results in uric acid retention (168). As suggested by human studies (169), the major site of action of ethacrynic acid is found in the ascending limb of the loop of Henle, which is reflected in the high ceiling in terms of NaCl and water output. By means of microperfusion techniques, the action of ethacrynic acid in this segment of the nephron has been confirmed, and inhibition of active chloride transport has been observed (170). A remarkable observation was that the cysteine adduct of ethacrynic acid was more active in the lumen than the drug itself. The thiol adducts of ethacrynic acid have diuretic activity (171). These were shown, under appropriate conditions *in vitro,* to liberate ethacrynic acid by release of the corresponding thiol. There could be conjugation *in vivo* with an endogenous thiol such as cysteine (171).

A series of investigations at Merck Sharp & Dohme has contributed to the question of whether thiol addition reactivity is strictly correlated with the diuretic effect of unsaturated phenoxyacetic acid derivatives. A series of ethacrynic acid congeners, namely various (vinylaryloxy)acetic acids were synthesized and evaluated. In these congeners, the double bond was activated towards nucleophilic attack by different electron withdrawing groups (172–175). For many of these compounds, remarkable diuretic activity and both a diuretic profile and a ceiling effect comparable to those of ethacrynic acid was found. Some of the most potent members of these series are depicted in formulae (**75–78**), and they are about five times as potent as ethacrynic acid. Diuretic activity was also demonstrated with thiol adducts. However, *in vivo* diuretic activity of the different types of unsaturated aryloxyacetic acids is now recognized to correlate poorly with the *in vitro* reaction with thiols (176–177). It has been concluded recently that the role of thiol binding is of secondary importance in the mechanism of action. On the other hand, by saturation of the double bond, diuretic potency is markedly reduced, but diuretic activity still remains (176). In the case of dihydroethacrynic acid, a shift from uric acid retention to a uricosuric effect is achieved. From all these observations it might be suggested that in the unsaturated series, two types of active structure are superimposed.

Following the principle of the activated double bond and using the capability of the previously mentioned α,β-unsaturated (acylphenoxy)acetic acids to undergo cyclization to the corresponding alkylated (1-oxo-5-indanyloxy)acetic acids by treatment with sulfuric acid, Cragoe and co-workers (176–177) directed their initial efforts towards the introduction of a double bond in the 2,3- or 2,2α-position of these compounds (**79**). Most surprisingly, substantial diuretic and uricosuric effect was found within the saturated indanone derivatives (**80**). In contrast to ethacrynic acid, activity was observed even when they were administered to rats. Subsequent to this observation, a large series of these structures was evaluated, the structure–activities of which have been, at least in part, reported (177). For many of the dichloro-substituted

(**75**) R = R′ = COCH$_3$
(**76**) R = COCH$_3$, R′ = CH$_3$
(**77**) R = NO$_2$, R′ = CH$_3$
(**78**) R = R′ = CN

(**79**)

(**80**) R, R′, and R″ = H or alkyl group
(**81**) R = —⟨phenyl⟩ , R′ = H, R″ = CH$_3$

24 DIURETICS

Table 1. United States Retail Market as Measured by Total Prescriptions 1976–1981

		\multicolumn{4}{c}{Prescriptions in millions}
		\multicolumn{2}{c}{1976}	\multicolumn{2}{c}{1981[a]}
Diuretics	Structures	Amount	% of total	Amount	% of total
Thiazides and thiazide-type diuretics		41.4	49.3	37.5	38.3
Hydro Diuril (Merck)	(10)	14.8	17.6	14.0	14.3
Diuril (Merck)	(5)	7.4	8.8	6.5	6.6
Hygroton (Revlon)	(24)	6.8	8.1	7.5	7.7
Esidrex (Ciba-Geigy)	(10)	3.6	4.3	3.0	3.1
Enduron (Abbott)	(12)	2.0	2.4	2.0	2.0
Zaroxolyn (Pennwalt)	(20)	0.8	1.0	1.5	1.5
Ticrynafen (Smith Kline)	(82)			1.5	1.5
others		6.0	7.1	3.0	3.1
Loop diuretics		19.9	23.7	26.5	27.0
Lasix (Hoechst)	(28)	19.6	23.3	22.5	23.0
Edecrin (Merck)	(29)	0.3	0.4		
MK-196 (Merck)	(81)			1.0	1.0
Muzolimine (Delbay)				1.0	1.0
Bumetanide (Roche)	(32)			0.5	0.5
Potassium-sparers[b]		22.3	26.5	31.5	32.1
Dyazide (Smith Kline)	(10, 54)	13.3	15.8	17.0	17.3
Dyrenium (Smith Kline)	(54)	0.5	0.6	1.0	1.0
Aldactazide (Searle)	(10, 44)	6.2	7.4	7.0	7.1
Aldactone (Searle)	(44)	2.2	2.6	3.0	3.1
Moduretic (Merck)	(10, 57)			3.0	3.1
other		0.1	0.1	0.5	0.5
Others		0.5	0.6	2.5	2.5
Total		84.0		98.0	

[a] Kidder, Peabody & Co. Incorporated estimates (186).
[b] Including potassium-sparing diuretics sold in combination with hydrochlorothiazide (10).

(indanyloxy)acetic acids high potency and efficacy could be accounted for and, using selected compounds, confirmed in man. It was possible to design compounds by structural variation in which either diuretic or uricosuric effect predominated or both effects were optimal (177). Finally, the racemic form of [6,7-dichloro-2-methyl-1-oxo-2-phenyl-5-indanyl]oxyacetic acid (MK-196) (**81**) was chosen for further evaluation. Studies in animals (178–180) and clinical investigations confirmed this compound to be a loop diuretic with uricosuric properties. Results in volunteers after single doses of 10 or 20 mg suggested a sustained diuretic activity over several hours and a slow return to base line (181). The prolonged action could make this loop diuretic suitable for use in the treatment of hypertension. This is an area in which MK-196 has been shown to be effective and, in comparison with hydrochlorothiazide (**10**), to produce less alteration in serum uric acid (182).

Independently of the above developments, the Centre de Recherches et de

(**82**)

Table 2. Estimated World Markets in 1976 by Type of Diuretic[a], in Millions of U.S. Dollars

	Thiazides[b]	K-sparing	Loop	Other	Total
United States	115	125	80	5	325
Western Europe	50	125	60	10	245
Japan	25	15	15		55
rest of world	15	30	45		90
Total	*205*	*295*	*200*	*15*	*715*
	\multicolumn{5}{c}{Percent of total}
United States	56.1	42.4	40.0	33.3	45.5
Western Europe	24.4	42.4	30.0	66.7	34.3
Japan	12.2	5.1	7.5		7.7
rest of world	7.3	10.2	22.5		12.6

[a] Kidder, Peabody & Co. Incorporated estimates (186).
[b] Including thiazides sold in combination form with potassium.

Table 3. Estimated World Sales of Diuretics by Manufacturer, 1976–1981[a], in Millions of U.S. Dollars

	United States	Japan	Western Europe	Rest of world	Total	Share of market, %
1976						
Hoechst	80	10	55	40	185	25.9
G. D. Searle	66	15	46	10	137	19.2
Merck	55	10	35	15	115	16.1
Smith Kline	58		9	5	72	10.1
Ciba-Geigy	8	5	20	10	43	6.0
Boehringer Mannheim			40		40	5.6
others	58	15	40	10	123	17.2
Total	*325*	*55*	*245*	*90*	*715*	*100.0*
1981						
Hoechst	100	20	80	70	270	25.0
G. D. Searle	79	30	55	15	179	16.6
Merck	68	20	90	25	203	18.8
Smith Kline	101	5	25	15	146	13.5
Ciba-Geigy	7	10	25	20	62	5.7
Boehringer Mannheim			35		35	3.2
others	70	35	50	30	185	17.1
Total	*425*	*120*	*360*	*175*	*1,080*	*100.0*
compound growth rate 1976–1981	5.5%	16.9%	8.0%	14.2%	8.6%	

[a] Kidder, Peabody & Co. Incorporated estimates (186).

Pharmacologie Albert Roland (CREPHAR), in France, altered the structure of ethacrynic acid by syntheses of several 4-heterocyclic acylated (2,3-dichlorophenoxy)acetic acids (183). From this series, in which reaction with sulfhydryl groups is likewise precluded, tienilic acid (ticrynafen) (**82**) emerged (184–185) and was introduced for clinical use in France in 1976.

This drug was evaluated in cooperation with Smith Kline and French. It is of moderate potency but combines uricosuric with saluretic properties and is distinguished from ethacrynic acid and others of its congeners by not being a loop diuretic.

26 DIURETICS

With respect to diuretic effect, the main action appears to be inhibition of sodium reabsorption in the cortical segment of the distal nephron. Thus it has to be classified as a thiazide-type diuretic. However, as the first diuretic of this type possessing substantial uricosuric properties, thienilic acid might well find its place as a diuretic in clinical practice, especially when hyperuricemia is involved.

Economic Aspects

The market for diuretic drugs has been growing in the last decade in most countries. This might be a consequence of the improvement of the health service, certain progress in the development of new drugs, and a better knowledge of their proper use. Table 1 shows the major diuretics in use in the United States in 1976 and estimates their use in 1981.

Table 2 is a list of estimated world markets in 1976 by type of diuretic and Table 3 shows estimated world sales in 1976 and 1981.

Drug Regulatory Aspects

Legislation governing the evaluation of the safety and efficacy of drugs now exists in most countries (see Regulatory agencies). In general, no compound or any of its pharmaceutical preparations can be given to humans unless registration approval has been given by the health authorities of the government in question, eg, the FDA. To reach the point of official approval (ie, a diuretic drug) is a time-consuming process, taking many years. Scientific material on the following matters must be provided in order to allow a judgment on the utility and the balance between benefit and risk: synthesis, purification, and identification of the new chemical; biological and pharmacological investigations with respect to desired and potential undesired effects; establishment of quality specifications and stability of both the active ingredient and its pharmaceutical preparation; extensive preclinical studies including acute, short-term, and long-term toxicological studies in several animal species; pharmacokinetic investigations (absorption, distribution in the organism, biotransformation, and excretion); and human studies, step by step, from the first pharmacological investigation of single dose to controlled therapeutical trials in order to elucidate a potential beneficial effect and any possible adverse or side effects.

A numerical list of diuretics referred to in the text.

Structure no.	*Diuretic*	*CAS Registry No.*
(1)	mannitol	[69-65-8]
(2)	isosorbide	[652-67-5]
(3)	sulfonilamide	[63-74-1]
(4)	acetazolamide	[59-66-5]
(5)	chlorothiazide	[58-94-6]
(6)	p-sulfamoylbenzoic acid	[138-41-0]
(7)	clofenamide	[671-95-4]
(8)	dichlofenamide	[120-97-8]
(9)	chlorodisulfamoylaniline	[121-30-2]

(10)	hydrochlorothiazide	[58-93-5]
(11)	hydroflumethiazide	[135-09-1]
(12)	methyclothiazide	[135-07-9]
(13)	triclomethiazide	[133-67-5]
(14)	polythiazide	[346-18-9]
(15)	althiazide	[5588-16-9]
(16)	cyclothiazide	[2259-96-3]
(17)	cyclopenthiazide	[742-20-1]
(18)	bendroflumethiazide	[73-48-3]
(19)	quinethazone	[73-49-4]
(20)	metolazone	[17560-51-9]
(21)	mefruside	[7195-27-9]
(22)	clopamide	[636-54-4]
(23)	indapamide	[26807-65-8]
(24)	chlorthalidone	[77-36-1]
(25)	clorexolone	[2127-01-7]
(28)	furosemide	[54-31-9]
(29)	ethacrynic acid	[58-54-8]
	dihydroethacrynic acid	[5378-94-9]
(30)	xipamide	[14293-44-8]
(32)	bumetanide	[28395-03-1]
(33)	piretanide	[55837-27-9]
(34)	piprozolin (choleretic)	[17243-64-0]
(35)	etozolin	[73-09-6]
(36)	muzolimine	[55294-15-0]
(37)	MK 447	[58456-91-0]
(38)	MK 447 hydrochloride	[38946-46-2]
(39)	aldosterone (endogenous hormone)	[52-39-1]
(40)	Amphenone B	[2686-47-7]
(41)	metyrapone	[54-36-4]
(42)	(17α)-pregn-4-ene-21-carboxylic acid, 17-hydroxy-3-oxo-γ-lactone	[976-70-5]
(43)	(17α)-19-norpregn-4-ene-21-carboxylic acid, 17-hydroxy-3-oxo-γ-lactone	[1722-54-9]
(44)	spironolactone	[52-01-7]
(45)	($7\alpha,17\alpha$)-pregn-4-ene-7,21-dicarboxylic acid, 17-hydroxy-3-oxy-γ-lactone methyl ester	[41020-65-9]
(46)	($7\alpha,17\alpha$)-pregn-4-ene-7,21-dicarboxylic acid, 17-hydroxy-3-oxo-γ-lactone, ethyl ester	[41020-77-3]
(47)	($7\alpha,17\alpha$)-pregn-4-ene-7,21-dicarboxylic acid, 17-hydroxy-3-oxo-γ-lactone, isopropyl ester	[41020-79-5]
(48)	prorenoate	[49847-98-5]
(49)	2-amino-1,8-naphthyridine-3-carboxamide	[15935-96-3]
(50)	39A69	[42373-29-5]
(51)	melamine	[108-78-1]
(52)	formoguanamine	[504-08-5]
(53)	chlorazanil	[500-42-5]
(54)	triamterene	[396-01-0]

28 DIURETICS

(55)	N-amidino-3-amino-6-chloropyrazinecarboxamide	[1203-87-8]
(56)	amiloride	[2609-46-3]
(57)	amiloride hydrochloride	[2016-88-8]
(58)	N-amidino-3-amino-6-chloro-5-(dimethylamino)-pyrazinecarboxamide hydrochloride	[2235-97-4]
(59)	N-amidino-3-amino-6-chloro-5-(methylamino)-pyrazinecarboxamide	[1140-83-6]
(60)	azolimine	[40828-45-3]
(61)	caffeine	[58-08-2]
(62)	theobromine	[83-67-0]
(63)	theophylline	[58-55-9]
(64)	thymine	[65-71-4]
(65)	uracil	[66-22-8]
(66)	5-aminouracil	[932-52-5]
(67)	aminometradine	[642-44-4]
(68)	amisometradine	[550-28-7]
(69)	merbaphen	[36568-91-9]
(70)	mersalyl	[486-67-9]
	mersalyl, sodium salt	[492-18-2]
(71)	meralluride	[140-20-5]
	meralluride, sodium salt-theophyllinate	[8069-64-5]
	meralluride, sodium salt	[7620-30-6]
(72)	mercaptomerin	[20223-84-1]
	mercaptomerin, disodium salt	[21259-76-7]
(73)	chlormerodrin	[62-37-3]
(75)	[4-(2-acetyl-3-oxo-1-butenyl)-2,3-dichlorophenoxy]acetic acid	[17796-20-2]
(76)	[2,3-dichloro-4-(2-methyl-3-oxo-1-butenyl)-phenoxy]acetic acid	[25195-03-3]
(77)	[2,3-dichloro-4-(2-nitropropenyl)phenoxy]acetic acid	[16861-25-9]
(78)	[2,3-dichloro-4-(2,2-dicyanovinyl)phenoxy]acetic acid	[29360-56-3]
(81)	MK-196	[56049-88-8]
(82)	tienilic acid	[33504-51-7, 40180-04-9]

Definition of Terms

Acidosis: (adj. acidotic)	pH of blood or plasma below normal.
Alkalosis: (adj. alkalotic)	pH of blood or plasma above normal.
Ascites:	Accumulation of fluid in the peritoneal cavity.
Clearance experiments: (in connection with the	Calculation of the free water clearance and free water reabsorption from the urinary volume and osmolarity of plasma and urine. Determination

determination of a diuretic's site of action in the nephron)	of volume and osmolarity are performed over a period of time without and under the influence of the diuretic, and both in the absence of antidiuretic hormone and under its maximal influence. Free water clearance and free water reabsorption are defined as the amount of water that would have to be taken from or added to the urine respectively in order to make the urine isotonic with plasma.
Clearance (renal):	A calculated volume (mL) of blood (plasma, serum) which is cleared of a compound per minute by renal elimination.
Collecting duct:	Part of the nephron, see Fig. 2. (In the original definition of the nephron the collecting duct is excluded).
Cortex: (adj. cortical)	Outer part of the kidney, see Figure 1.
Diuretic profile:	The pattern of urinary volume and electrolyte excretion.
Endogenous:	Originating within the organism.
Exogenous:	Originating outside the organism.
Glomerulus: (adj. glomerular)	Part of the nephron, see Figure 2.
Homeostasis: (adj. homeostatic)	Tendency to uniformity or stability in the normal body states of the organism.
Hydropenia: (adj. hydropenic)	Water deficiency.
Hypercalcemia:	Serum calcium levels above normal.
Hyperkalemia:	Serum potassium levels above normal.
Hypernatremia:	Serum sodium levels above normal.
Hypertonic:	Tonicity (osmolarity) more than isotonic (qv).
Hyperuricemia:	Serum uric acid levels above normal.
Hypocalcemia:	Serum calcium levels below normal.
Hypokalemia:	Serum potassium levels below normal.
Hyponatremia:	Serum sodium levels below normal.
Hypotonic:	Tonicity (osmolarity) less than isotonic (qv).
Interstitium:	The interspace of a tissue, eg of the kidney.
Isotonic:	Having the same tonicity (osmolarity). Usually in comparison with plasma or other body fluids.
Juxtamedullary:	Close to the medulla.
Loop of Henle:	Part of the nephron, see Figure 2.
Medulla: (adj. medullary)	Inner part of the kidney, see Figure 1.
Microperfusion technique:	A technique where various measurements are made on a perfused isolated part of the nephron.
Nephron:	The functioning unit of the kidney.
Parenteral:	Introduced by any other route than by way of the digestive tract.
Pharmacodynamics: (adj. pharmacodynamic)	The physiological effects of medicines (drugs).
Pleural effusion:	The presence of fluid in the pleural space.
Reabsorption:	Absorption of a substance previously secreted, diffused or filtered.
Secretion:	Physiological release of compounds from cells or glands.
Systemic:	Relating to the body as a whole.
Tubule, distal convoluted:	Part of the nephron, see Figure 2.
Tubule, proximal convoluted:	Part of the nephron, see Figure 2.
Uricusoric:	Increasing the excretion of uric acid.
Volume of distribution, apparent:	The volume of fluid in which any compound appears to be distributed in the organisms; calculated from the amount the compound administered and its plasma concentration extrapolated to zero time (time of administration).

30 DIURETICS

BIBLIOGRAPHY

"Diuretics and Antidiuretics" in *ECT* 1st ed., Vol. 5, pp. 188–194, by Erwin Di Cyan, Di Cyan & Brown; "Diuretics" in *ECT* 2nd ed., Vol. 7, pp. 248–271, by George deStevens, Ciba Pharmaceutical Co.

1. R. E. Weston, "Pathogenesis and Treatment of Edema with Special Reference to Use of Diuretics" in M. H. Maxwell and C. R. Kleeman, eds., *Clinical Disorders of Fluid and Electrolyte Metabolism*, 2nd ed., McGraw-Hill, Inc., New York, 1972, Chapt. 6, p. 163.
2. D. L. Davies and G. M. Wilson, *Drugs* **9,** 178 (1975).
3. A. F. Lant, *Practitioner* **219,** 519 (1977).
4. A. F. Lant and G. M. Wilson, "Diuretics" in D. Black, ed., *Renal Diseases*, 3rd ed., Blackwell Scientific Publications, Oxford, Eng., 1972, p. 684.
5. Ref. 2, p. 209.
6. R. F. Pitts, *Physiology of the Kidney and Body Fluids*, Year Book Medical Publishers, Inc., Chicago, Ill., 1963, Chapt. 1, p. 11.
7. R. G. Schultze, *Arch. Int. Med.* **131,** 885 (1973).
8. A. S. Rocha and J. P. Kokko, *J. Clin. Invest.* **52,** 612 (1973).
9. P. Brazeau, "Agents Affecting the Renal Conservation of Water" in L. S. Goodman and A. Gilman, eds., *The Pharmacological Basis of Therapeutics*, 5th ed., Macmillan Publishing Co., Inc., New York, 1975, Chapt. 40, p. 848.
10. G. H. Mudge, "Drugs Affecting Renal Function and Electrolyte Metabolism," introduction in ref. 9, p. 809.
11. H. R. Jacobson and J. P. Kokko, *Ann. Rev. Pharmacol.* **16,** 201 (1976).
12. E. M. Schultz, R. L. Smith, and O. W. Woltersdorf, Jr., *Ann. Rep. Med. Chem.* **10,** 71 (1975).
13. R. L. Smith, O. W. Woltersdorf, Jr., and E. J. Cragoe, Jr., *Ann. Rep. Med. Chem.* **11,** 71 (1976).
14. B. R. Nechay, *J. Clin. Pharmacol.* **17,** 626 (1977).
15. Ref. 4, p. 675.
16. S. M. Nasrallah and F. L. Iber, *J. Am. Med. Sci.* **258,** 80 (1969).
17. J. F. Treon, L. E. Gongwer, and W. H. C. Rueggeberg, *Proc. Soc. Exp. Biol. Med.* **119,** 39 (1965).
18. J. H. Nodine and co-workers, *Clin. Pharmacol. Ther.* **14,** 196 (1973).
19. O. Gagnon, P. M. Gertman, and F. L. Iber, *Am. J. Med. Sci.* **254,** 284 (1967).
20. H. Southworth, *Proc. Soc. Exp. Biol. Med.* **36,** 58 (1937).
21. T. Mann and D. Keilin, *Nature* **146,** 164 (1940).
22. H. W. Davenport and A. E. Wilhelmi, *Proc. Soc. Exp. Biol. Med.* **48,** 53 (1941).
23. R. F. Pitts and R. S. Alexander, *Am. J. Physiol.* **144,** 239 (1945).
24. W. B. Schwartz, *New Engl. J. Med.* **240,** 173 (1949).
25. H. A. Krebs, *Biochem. J.* **43,** 525 (1948).
26. R. O. Roblin, Jr., and J. W. Clapp, *J. Am. Chem. Soc.* **72,** 4890 (1950).
27. T. H. Maren, E. Mayer, and B. Wadsworth, *Bull. Hopkins Hosp.* **95,** 199 (1954).
28. K. H. Beyer, *Arch. Int. Pharmacodyn.* **98,** 97 (1954).
29. J. P. Merrill, *Am. J. Med.* **14,** 519 (1953).
30. A. E. Lindsay and H. Brown, *J. Lab. Clin. Med.* **43,** 839 (1954).
31. J. M. Sprague, *Ann. N.Y. Acad. Sci.* **71**(4), 328 (1958).
32. A. Lund and K. Størling, *Acta Pharmacol. Toxicol.* **15,** 300 (1959).
33. F. C. Novello and G. M. Sprague, *J. Am. Chem. Soc.* **79,** 2028 (1957).
34. K. H. Beyer, *Ann. N.Y. Acad. Sci.* **71**(4), 363 (1958).
35. G. deStevens and co-workers, *Experientia* **14,** 463 (1958).
36. W. Kobinger and F. Lund, *Acta Pharmacol. Toxicol.* **15,** 265 (1959).
37. G. deStevens, *Diuretics, Chemistry and Pharmacology*, Academic Press, Inc., New York, 1963, Chapt. 4, pp. 58–117.
38. J. M. Sprague, "Diuretics" in J. L. Rabinowitz and R. M. Myerson, eds., *Topics in Medicinal Chemistry*, Vol. 2, John Wiley & Sons, Inc., New York, 1968, pp. 27–41.
39. J. G. Topliss, "Diuretics" in A. Burger, ed., *Medicinal Chemistry*, John Wiley & Sons, Inc., New York, 1970, pp. 987–1000.
40. G. Peters and F. Roch-Ramel, "Thiazide Diuretics and Related Drugs" in H. Herken, ed., *Handbook of Experimental Pharmacology*, Springer Verlag, Berlin, 1969, p. 257.
41. K. H. Beyer and J. E. Baer, *Pharmacol. Rev.* **13,** 517 (1961).
42. Y. Suzuki, Y. Hamaguchi, and I. Yamagami, *Jpn. J. Pharmacol.* **26**(Suppl.), 161 P (1976).
43. B. Goldberg and K. I. Furman, *S. Afr. Med. J.* **48,** 113 (1974).
44. Ref. 38, p. 41.

45. K. Sturm and co-workers, *Chem. Ber.* **99,** 328 (1966).
46. R. Muschaweck and P. Hajdú, *Arzneim.-Forsch.* **14,** 44 (1964).
47. H. Kleinfleder, *Dtsch. Med. Wschr.* **88,** 1695 (1963).
48. R. J. Timmerman, F. R. Springman, and R. K. Thoms, *Curr. Ther. Res.* **6,** 88 (1964).
49. M. Burg and co-workers, *Am. J. Physiol.* **225,** 119 (1973).
50. R. Fischer and A. Lenhartz, *Med. Welt,* 270 (1970).
51. W. Liebenow and F. Leuschner, *Arzneim.-Forsch.* **25,** 240 (1975).
52. F. Leuschner, W. Neumann, and H. Bahrmann, *Arzneim.-Forsch.* **25,** 245 (1975).
53. K. H. G. Piyasena, C. W. H. Havard, and J. C. P. Weber, *Curr. Med. Res. Opin.* **3,** 121 (1975).
54. P. W. Feit, *J. Med. Chem.* **14,** 432 (1971).
55. P. W. Feit and O. B. T. Nielsen, *J. Med. Chem.* **15,** 79 (1972).
56. P. W. Feit, O. B. T. Nielsen, and N. Rastrup Andersen, *J. Med. Chem.* **16,** 127 (1973).
57. O. B. T. Nielsen, Ch. K. Nielsen, and P. W. Feit, *J. Med. Chem.* **16,** 1170 (1973).
58. P. W. Feit, O. B. T. Nielsen, and H. Bruun, *J. Med. Chem.* **17,** 572 (1974).
59. O. B. T. Nielsen and co-workers, *J. Med. Chem.* **18,** 41 (1975).
60. P. W. Feit, O. B. T. Nielsen, and H. Bruun, *J. Med. Chem.* **15,** 437 (1972).
61. E. H. Østergaard and co-workers, *Arzneim.-Forsch.* **22,** 66 (1972).
62. *Postgr. Med. J.* **51**(Suppl. 6), (1975).
63. M. R. Cohen and co-workers, *J. Pharmacol. Exp. Ther.* **197,** 697 (1976).
64. W. Merkel and co-workers, *Eur. J. Med. Chem.* **11,** 399 (1976).
65. M. P. Magnussen and E. Eilertsen, *Arch. Pharmacol.* **297**(Suppl. 2), R39 (1977).
66. P. W. Feit and O. B. T. Nielsen, *J. Med. Chem.* **19,** 402 (1976).
67. P. W. Feit and O. B. T. Neilsen, *J. Med. Chem.* **20,** 1687 (1977).
68. P. W. Feit and O. B. T. Nielsen, *Abstr. Pap. Am. Chem. Soc.* **174** Meet., MEDI 2 (1977).
69. G. Satzinger, *Arzneim.-Forsch.* **27,** 1742 (1977).
70. *Ibid.,* p. 463.
71. G. Satzinger, *Liebigs Ann. Chem.* **665,** 150 (1963).
72. O. Heidenreich and co-workers, *Arzneim.-Forsch.* **14,** 1242 (1964).
73. O. Heidenreich and L. Baumeister, *Klin. Wschr.* **42,** 1236 (1964).
74. M. Herrmann and co-workers, *Arzneim.-Forsch.* **27,** 1745 (1977).
75. J. Greven and O. Heidenreich, *Arzneim.-Forsch.* **27,** 1755 (1977).
76. E. Scheitza, *Arzneim.-Forsch.* **27,** 1807 (1977).
77. D. Loew and K. Meng, *Pharmatherapeutica* **1,** 333 (1977).
78. E. Möller, H. Horstmann, and K. Meng, *Pharmatherapeutica* **1,** 540 (1977).
79. K. Meng and co-workers, *Curr. Med. Res. Opin.* **4,** 555 (1977).
80. K. J. Berg, S. Jørstad, and A. Tromsdal, *Pharmatherapeutica* **1,** 319 (1976).
81. G. Hoppe-Seyler and co-workers, *Pharmatherapeutica* **1,** 422 (1977).
82. R. L. Smith, G. E. Stokker, and E. J. Cragoe, Jr., *Abstr. Pap. Am. Chem. Soc.* **174** Meet., MEDI 7 (1977).
83. M. B. Affrime and co-workers, *Clin. Pharmacol. Ther.* **21,** 97 (1977).
84. M. J. Allen and A. H. Corwin, *J. Am. Chem. Soc.* **72,** 114, 117 (1950).
85. W. L. Bencze and M. J. Allen, *J. Med. Pharm. Chem.* **1,** 395 (1959).
86. G. W. Thorn and co-workers, *New Engl. J. Med.* **254,** 547 (1956).
87. A. E. Renold and co-workers, *J. Clin. Invest.* **35,** 731 (1956).
88. Ref. 37, p. 136.
89. W. L. Bencze and M. J. Allen, *J. Am. Chem. Soc.* **81,** 4015 (1959).
90. S. Shaldon and J. R. McLaren, *Lancet* **2,** 1330 (1960).
91. J. A. Cella and C. M. Kagawa, *J. Am. Chem. Soc.* **79,** 4808 (1957).
92. C. M. Kagawa, J. A. Cella, and C. G. van Arman, *Science* **126,** 1015 (1957).
93. E. J. Ross and J. E. Bethune, *Lancet* **1,** 127 (1959).
94. J. A. Cella in J. H. Moyer and M. Fuchs, eds., *Edema-Mechanism and Management,* Saunders, Philadelphia, Pa., 1960, p. 303.
95. J. A. Cella and R. C. Tweit, *J. Org. Chem.* **24,** 1109 (1959).
96. R. M. Weyer and L. M. Hogmann, *J. Med. Chem.* **18,** 817 (1975).
97. L. M. Hofmann and co-workers, *J. Pharmacol. Exp. Ther.* **194,** 450 (1975).
98. E. M. Hawes, D. K. J. Gorecki, and D. D. Johnson, *J. Med. Chem.* **16,** 849 (1973).
99. D. K. J. Gorecki and E. M. Hawes, *J. Med. Chem.* **20,** 124 (1977).
100. E. M. Hawes, D. K. J. Gorecki, and R. G. Gedir, *J. Med. Chem.* **20,** 838 (1977).
101. J. Trethewey and co-workers, *Can. J. Pharmac. Sci.* **12,** 12 (1977).

102. W. L. Lipschitz and Z. Hadidian, *J. Pharmacol. Exp. Ther.* **81,** 84 (1944).
103. O. Clauder and G. Bulcsu, *Magy Kem. Foly* **57,** 68 (1951); *Chem. Abstr.* **46,** 4023 (1952).
104. K. N. Modi, C. V. Deliwala, and U. K. Sheth, *Arch. Int. Pharmacodyn.* **151,** 13 (1964).
105. H. E. Williamson, F. E. Shideman, and D. A. LeSher, *J. Pharmacol. Exp. Ther.* **126,** 82 (1959).
106. V. D. Wiebelhaus and co-workers, *J. Pharmacol. Exp. Ther.* **149,** 397 (1965).
107. J. Weinstock and co-workers, *J. Med. Chem.* **11,** 573 (1968).
108. R. G. W. Spickett and G. M. Timmis, *J. Chem. Soc. (London),* 2887 (1954).
109. G. W. Liddle, *Metabolism* **10,** 1021 (1961).
110. W. I. Baba, G. R. Tudhope, and G. M. Wilson, *Br. Med. J.* **2,** 756 (1962).
111. *Ibid.,* p. 760.
112. S. Shaldon and J. A. Ryder, *Br. Med. J.* **2,** 664 (1962).
113. K. B. Hansen and A. D. Bender, *Clin. Pharmacol. Ther.* **8,** 392 (1967).
114. R. M. Stote and co-workers, *J. Int. Med. Res.* **2,** 379 (1974).
115. J. B. Bicking and co-workers, *J. Med. Chem.* **8,** 638 (1965).
116. M. S. Glitzer and S. L. Steelman, *Proc. Soc. Exp. Biol. Med.* **120,** 364 (1965).
117. E. J. Cragoe, Jr. and co-workers, *J. Med. Chem.* **10,** 66 (1967).
118. J. B. Bicking and co-workers, *J. Med. Chem.* **10,** 598 (1967).
119. J. H. Jones, J. B. Bicking, and E. J. Cragoe, Jr., *J. Med. Chem.* **10,** 899 (1967).
120. J. H. Jones and E. J. Cragoe, Jr., *J. Med. Chem.* **11,** 322 (1968).
121. K. L. Shepard and co-workers, *J. Med. Chem.* **12,** 280 (1969).
122. J. H. Jones, W. J. Holtz, and E. J. Cragoe, Jr., *J. Med. Chem.* **12,** 285 (1969).
123. C. B. Jones, H. F. Russo, and A. Zacchei, *Fed. Proc.* **25,** 197 (1966).
124. M. S. Glitzer and S. L. Steelman, *Nature* **212,** 191 (1966).
125. J. E. Baer and co-workers, *J. Pharmacol. Exp. Ther.* **157,** 472 (1967).
126. W. I. Baba and co-workers, *Clin. Pharmacol. Ther.* **9,** 318 (1968).
127. A. F. Lant, A. J. Smith, and G. M. Wilson, *Clin. Pharmacol. Ther.* **10,** 50 (1969).
128. R. Z. Gussin and co-workers, *J. Pharmacol. Exp. Ther.* **195,** 8 (1975).
129. Koschlakoff, *Virchows Arch. Pathol. Anat.* **31,** 438 (1864); A. Vogl, *Diuretic Therapy,* Williams and Wilkins, Baltimore, Md., 1953.
130. Von Schröder, *Arch. Exp. Pathol. Pharmacol.* **22,** 39 (1887).
131. J. Murdoc-Ritchie, "The Xanthines" in ref. 9, Ch. 19, p. 374.
132. B. Sigurd and K. H. Olesen, *Am. Heart J.* **94,** 168 (1977).
133. P. A. Levene, *Biochem. Z.* **4,** 316 (1907).
134. V. Papesch and E. F. Schroeder, *J. Org. Chem.* **16,** 1879 (1951).
135. A. Kattus, T. M. Arrington, and E. V. Newman, *Am. J. Med.* **12,** 319 (1952).
136. E. Settel, *Postgrad. Med. J.* **21,** 186 (1957).
137. M. S. Belle, *Am. Heart J.* **55,** 114 (1958).
138. A. Vogl, *Am. Heart J.* **39,** 881 (1950).
139. P. Saxl and R. Heilig, *Wien. Klin. Wochenschr.* **33,** 943 (1920).
140. E. M. Schultz and co-workers, *J. Med. Chem.* **14,** 998 (1971).
141. W. Wobbe, *Arch. Pharm.* **262,** 70 (1924).
142. H. L. Friedman, *Ann. N.Y. Acad. Sci.* **65**(5), 461 (1957).
143. G. deStevens, *Diuretics, Chemistry and Pharmacology,* Academic Press, Inc., New York, 1963, Ch. 4, p. 36.
144. J. M. Sprague, "Diuretics" in J. L. Rabinowitz and R. M. Myerson, eds., *Topics in Medicinal Chemistry,* Vol. 2, John Wiley & Sons, Inc., New York, 1968, p. 7.
145. W. A. Leff and H. E. Nussbaum, *Br. Med. J.* **1,** 883 (1959).
146. K. H. Beyer and J. E. Baer, "Newer Diuretics" in E. Jucker, ed., *Progress in Drug Research,* Vol. 2, Birkhäuser Verlag, Basel, Switz., 1960, p. 30.
147. P. Govaerts, *C. R. Soc. Biol.* **99,** 647 (1928).
148. E. A. Bartram, *J. Clin. Invest.* **11,** 1197 (1932).
149. D. W. Seldin and co-workers, *Ann. N.Y. Acad. Sci.* **139,** 328 (1966).
150. M. Burg and N. Green, *Kidney Int.* **4,** 245 (1973).
151. Ref. 4, p. 658.
152. T. W. Clarkson and J. J. Vostal in A. F. Lant and G. M. Wilson, eds., *Modern Diuretic Therapy in the Treatment of Cardiovascular and Renal Disease,* Excerpta Medica, Amsterdam, The Netherlands, 1973, p. 229.
153. R. H. Kessler, R. Lozano, and R. F. Pitts, *J. Clin. Invest.* **36,** 656 (1957).

154. G. H. Mudge and I. M. Weiner, *Ann. N.Y. Acad. Sci.* **71,** 344 (1958).
155. I. M. Weiner, R. I. Levy, and G. H. Mudge, *J. Pharmacol. Exp. Ther.* **138,** 96 (1962).
156. E. J. Cafruny and co-workers in ref. 152, p. 124.
157. V. Nigrovic, *J. Clin. Pharmacol.* **17,** 642 (1977).
158. E. M. Schultz and co-workers, *J. Med. Pharm. Chem.* **5,** 660 (1962).
159. J. M. Sprague, "Excursions in Medicinal Chemistry—Renal Agents" in C. K. Cain, ed., *Ann. Rep. Med. Chem.,* XI (1969).
160. P. J. Cannon, R. P. Ames, and J. H. Laragh, *J. Am. Med. Ass.* **185,** 854 (1963).
161. K. E. W. Melvin, R. O. Farrelly, and J. D. K. North, *Br. Med. J.* **1,** 1521 (1963).
162. Ref. 144, p. 15.
163. R. M. Komorm and E. J. Cafruny, *Science* **143,** 133 (1964).
164. R. M. Komorm and E. J. Cafruny, *J. Pharmacol. Exp. Ther.* **148,** 367 (1965).
165. R. J. Gussin and E. J. Cafruny, *J. Pharmacol. Exp. Ther.* **149,** 1 (1965).
166. V. Nigrovic, D. A. Koechel, and E. J. Cafruny, *J. Pharmacol. Exp. Ther.* **186,** 331 (1973).
167. K. H. Beyer and co-workers, *J. Pharmacol. Exp. Ther.* **147,** 1 (1965).
168. R. J. Sperber, A. C. DeGraff, and A. F. Lyon, *Am. Heart J.* **69,** 281 (1965).
169. M. Goldberg and co-workers, *J. Clin. Invest.* **43,** 201 (1964).
170. M. Burg and N. Green, *Kidney Int.* **4,** 301 (1973).
171. D. A. Koechel and E. J. Cafruny, *J. Med. Chem.* **16,** 1147 (1973).
172. J. B. Bicking and co-workers, *J. Med. Chem.* **19,** 530 (1976).
173. *Ibid.,* p. 544.
174. E. M. Schultz and co-workers, *J. Med. Chem.* **19,** 783 (1976).
175. O. W. Woltersdorf Jr. and co-workers, *J. Med. Chem.* **19,** 972 (1976).
176. E. J. Cragoe Jr. and co-workers, *J. Med. Chem.* **18,** 225 (1975).
177. O. W. Woltersdorf, Jr. and co-workers, *J. Med. Chem.* **20,** 1400 (1977).
178. R. McKenzie, T. Knight, and E. J. Weihman, *Proc. Soc. Exp. Biol. Med.* **153,** 202 (1976).
179. G. M. Fanelli, Jr., D. L. Bohn, and A. G. Zacchei, *J. Pharmacol. Exp. Ther.* **200,** 413 (1977).
180. L. S. Watson and co-workers, "New Antihypertensive Saluretic-Uricosuric Indanone" in A. Scriabine and Ch. S. Sweets, eds., *New Antihypertensive Drugs,* Spectrum Publications, Inc., New York, 1976, p. 307.
181. K. F. Tempero and co-workers, *Clin. Pharmacol. Ther.* **19,** 116 (1976).
182. K. F. Tempero and co-workers, *Clin. Pharmacol. Ther.* **21,** 119 (1977).
183. G. Thuillier and co-workers, *Eur. J. Med. Chem.* **9,** 625 (1974).
184. *J. Pharmacol. Clin.* (Fr.) (Special No.), 1 (Jan. 1976).
185. M. Nemati, M. C. Kyle, and E. D. Freis, *J. Am. Med. Assoc.* **237,** 652 (1977).
186. *The Drug Industry: Diuretics,* Kidder, Peabody & Co., Inc., Sept. 26, 1977.

PETER W. FEIT
Leo Pharmaceutical Products, Denmark

DRIERS AND METALLIC SOAPS

Metallic soaps are a group of water-insoluble compounds containing alkaline earth or heavy metals combined with monobasic carboxylic acidsof 7 to 22 carbon atoms. They can be represented by the general formula $(RCOO)_x M$ (neutral soap), where R is an aliphatic or alicyclic radical and M is a metal with valence x. They differ from ordinary soap by composition and their insolubility in water. Their solubility or solvation in a variety of organic solvents accounts for their many and varied uses (see Table 1). Driers are the most important group of metallic soaps promoting or accelerating the drying, curing, or hardening of oxidizable coating vehicles such as paints (see Paint).

Metal soaps were used by the ancient Egyptians. Deposits taken from the axle of a chariot dating to ca 1400 BC contain quartz, iron, and a sufficient quantity of fat and lime to indicate the use of a lime soap, and there is evidence of the use of lead in combination with oils in mummification. Galen, about 200 AD, reported that linseed oil could be coagulated by the use of lead oxide.

Vegetable oil coatings were first used in conjunction with pigments (1). Later small quantities of those oils were added, not for coloring purposes but to improve drying properties. In the 18th century Macquer reported a type of metallic soap that was a combination of fatty oil and calyx of lead; and Liardet patented a combination of red lead with drying oil to be used as a grease. In the 19th century a new development was the substitution of zinc oxide for white lead as a prime white pigment (1). Later on fused linoleates and resinates based on lead and manganese dissolved in linseed oil, naphtha, and turpentine were employed as paint driers. Cobalt soaps were introduced about 1900, followed by naphthenic acid soaps.

Composition and Properties

The acid or anion portion of a metal soap can be varied. Typical anions currently used are rosin and tall oil fatty acids, saturated and unsaturated naturally-occurring long-chain monocarboxylic fatty acids with 7 to 22 carbon atoms, and naphthenic, 2-ethylhexanoic, and the newer synthetic tertiary acids. Water-soluble products (ie, salts of acids with fewer than seven carbon atoms) and salts of dicarboxylic acids are not included in this discussion.

Acid soaps contain free acid (positive acid number), whereas neutral (normal) soaps contain no free acid (zero acid number). That is, the ratio of acid equivalents to metal equivalents is greater than one in the acid soap and equal to one in the neutral soap. The basic soap is characterized by a higher metal-to-acid equivalent ratio than the normal metal soap. Particular properties are obtained by adjusting the basicity.

Table 1. Applications of Metal Soaps

Application	Metal
stabilizers for plastics	Ba, Cd, Sn, Sr
fungicides	Cu, Hg, Zn
catalysts	Co, Cu, Mo, Mn, Cr, Ni
driers	Ca, Co, Fe, Pb, Mn, Zn, Zr
fuel additives	Ba, Fe, Mg, Mn, Pb

Properties are furthermore determined by the nature of the organic acid, the type of metal and its concentration, the presence of solvents and additives, and the method of manufacture. Generally, metals of low atomic weight form soaps with higher melting points (eg, lithium). Higher melting points are also characteristic of soaps made of high molecular weight, straight-chain, saturated fatty acids. Branched-chain, unsaturated fatty acids form soaps with lower melting points. Light powders are characteristic of precipitated soaps, whereas soaps made by fusion are usually hard solids. Liquids and pastes are mixtures or solutions of soaps in hydrocarbon and/or nonvolatile solvents. Liquid soaps are manufactured as fluid as possible compatible with maximum metal content. Newer techniques (eg, overbasing with CO_2, see below) are available in which liquid soaps of very high metal content (18–24%) are obtained which still possess excellent fluidity. Both liquid and paste soaps are manufactured to strict viscosity and flow specifications.

The odor of a soap is determined by its organic constituents. Stearates, laurates, and palmitates are almost odor free. Naphthenates, linoleates and tall oil soaps have characteristic odors that are not objectionable since they are used at low levels. These soaps are approved by the FDA for use in paints and inks, and food-related usage.

Color, another important property of metal soaps, is determined by the type and amount of metal used, the color and quality of the organic raw materials, the method of manufacture, and the care in processing. Soaps that are water-white in liquid form and light colored in powdered form are usually obtained with alkaline earths, lithium, and lead. Copper soaps are green, iron soaps dark reddish-brown, cobalt soaps are blue, violet, or red, and manganese soaps reddish-brown. Liquid soaps darken with increasing metal content. Fused soaps are usually darker than soaps made by precipitation. In general, good quality control of the raw materials gives soaps for certain applications. For example, iron soaps used as driers are limited to black and brown coatings. However, the highly colored but very active cobalt and manganese soaps can, with care, be used in white and pastel coatings in small amounts (0.02–0.05% as metal). Zinc soaps (eg, Zn naphthenate) are used as fungicides where green copper soaps would be objectionable, although the zinc soaps are much less active.

Tables 2 and 3 give the properties of commercial liquid and solid soaps, respectively.

Manufacture

Metallic soaps are manufactured by three methods: the precipitation or double-decomposition process; the fusion process; and the direct metal reaction (DMR) process. The choice of process and variation depends upon the metal, the desired form of the product, the desired purity, raw material availability and cost, metal content, etc. The desired final composition determines the stoichiometry of the starting materials, ie, whether an acid, neutral, or basic soap is obtained.

In the case of metallic soap solutions, the metal is assumed to be the active part and the acid portion serves as carrier for the metal, conferring oil solubility, water insolubility, and compatibility with the other components of the system in which it is to be used. Therefore, it is economically advantageous to incorporate as much metal per unit of acid as possible, providing the resulting soap is oil soluble and fulfills the requirements of the end-use.

Table 2. Typical Properties of Commercial Liquid Metal Soaps (Driers)[a,b]

Compound	CAS Registry No.	Sp gr at 25°C	Weight, kg/L	Color[c]	Content, % Metal	Content, % Solids	Viscosity[d] At 25°C	Viscosity[d] At −7°C
2-ethylhexanoates								
lead	[16996-40-0]	1.120	1.08–1.11	3	24.0	62	A-1	V
cobalt	[136-52-7]	0.895	0.86–0.90	violet–blue	6.0	38	A-3	H
cobalt		1.060	1.01–1.06	violet–blue	12.0	75	E	
manganese	[15956-58-8]	0.910	0.87–0.91	17	6.0	46	A-2	A
zinc	[136-53-8]	0.900	0.86–0.90	1	8.0	40	A-3	A
calcium	[136-51-6]	0.920	0.89–0.92	4	5.0	61	B	Y
iron	[19583-54-1]	0.905	0.88–0.90	dark brown	6.0	48	A-3	
zirconium	[22464-99-9]	0.880	0.84–0.88	2	6.0	28	A-5	A-2
naphthenates								
lead	[61790-14-5]	1.115	1.05–1.11	7	24.0	64	A-2	K
cobalt	[61789-51-3]	0.960	0.91–0.96	violet–blue	6.0	67	A	K
manganese	[61788-57-6]	0.975	0.94–0.97	17	6.0	65	A	K
zinc	[12001-85-3]	0.950	0.86–0.95	7	8.0	59	A-2	C
zinc		1.000	0.96–1.00	8	10.0	75	A	Z-1
calcium	[61789-36-4]	0.925	0.89–0.92	7	4.0	63	D	X
calcium		0.950	0.91–0.95	8	5.0	68	D	T
iron	[1338-14-3]	0.980	0.93–0.98	dark	6.0	67	M	
cerium		0.935	0.90–0.93	17	6.0	57	A-1	D
tallates								
lead	[61788-54-3]	1.140	1.10–1.14	10	24.0	64	A-1	I
cobalt	[61789-52-4]	0.965	0.93–0.96	red–violet	6.0	72	C	Z-2
manganese	[61788-58-7]	0.980	0.94–0.98	17	6.0	76	I	Y
calcium	[61789-35-3]	0.925	0.89–0.92	9	4.0	66	B	Z-5
iron	[61788-81-6]	0.980	1.00–1.04	dark brown	6.0	75	J	

neodecanoates							
calcium	[27253-33-4]	0.896		1	5	71	F	Y
cobalt	[27253-31-2]	1.030		blue	12	68	A	A
lead	[27253-28-7]	1.100		2	24	61	A-Z	Y
manganese	[27253-32-3]	0.876		10	6	44	D	H
synthetic acids[e]								
lead		1.365	1.32–1.37	4	36.0	80	D	L
lead		0.095	1.06–1.10	3	24.0	55	A-2	A
cobalt		1.060	1.01–1.06	blue–violet	12.0	75	E	C
cobalt		0.905	0.86–0.90	blue–violet	6.0	38	A-3	A
manganese		1.060	1.03–1.06	dark brown	12.0	100	W	Z-2
manganese		0.985	0.94–0.98	17	9.0	70	C	Y
manganese		0.910	0.87–0.91	17	6.0	46	A-2	M
zinc		1.050	1.01–1.05	2	16.0	72	A-1	
zinc		0.900	0.86–0.90	1	8.0	40	A-3	
calcium		0.960	0.92–0.96	5	6.0	75	N	
calcium		0.920	0.89–0.92	4	5.0	61	B	
calcium		0.900	0.86–0.90	3	4.0	46	A-1	
zirconium		1.255	1.23–1.25	5	24.0	84	J	B
zirconium		1.120	1.08–2.19	4	18.0	70	A-1	A
zirconium		0.990	0.95–0.99	4	12.0	50	A-2	A-2
zirconium		0.880	0.84–0.88	2	6.0	28	A-5	A-2
iron		0.945	0.94–0.98	dark brown	9.0	65	A	

[a] Tenneco Chemicals, product specifications.
[b] All soaps listed have a flash point (Pensky-Marten closed cup) at 40°C.
[c] Gardner.
[d] Gardner-Holdt.
[e] Synthetic acids vary according to availability, whereas the metal content is always constant as stated here. Synthetic acids include neodecanoic, isononanoic, and heptanoic, and their mixtures.

Table 3. Typical Properties of Solid Metal Soaps[a]

Compound	CAS Registry No.	Total ash	Free fatty acid	Sp gr	Mp, °C	Color	Fineness, μm (U.S. standard sieve)
stearates							
Al	[637-12-7]	5.5–16.0	3.0–3.5	1.01	110–150	white	95–98% <74 (−200)
Ba	[6865-35-6]	19–28	0.5–1.0	1.23	dec	white	<44 (−325)
Ca	[1592-23-0]	8.8–10.6	0.5	1.12	145–160	white	<44 (−325)
Cd	[2223-93-0]	19.0	0.5	1.21	104	white	99% <74 (−200)
Co	[13586-84-0]	8.2	2.0	1.13	140	violet	99% <74 (−200)
Cu	[660-60-6]	14	1.0	1.10	112	blue–green	99% <44 (−325)
Fe	[2980-59-8, 555-36-2]	13	4.0	1.12	100	red–brown	99% <74 (−200)
Pb	[7428-48-0]	30.2–57.0	0.1–0.6	1.34–2.0	103–dec	white	<74 (−200)
Li	[4485-12-5]	2.5	nil	1.01	212	white	99% <149 (−100)
Mg	[557-04-0]	8.0	0.5	1.03	145	white	<44 (−325)
Mn	[10476-84-3]	12.5	1.0	1.22	110	light brown	99% <74 (−200)
Ni	[2223-95-2]	9.4	5.2	1.13	180	green	<44 (−325)
Sr	[10196-69-7]	17.5	0.5	1.03	155	white	<44 (−325)
Zn	[557-05-1]	13.5–17.7	0.5–0.9	1.09–1.11	120	white	<44 (−325)
octanoate							
Al	[6028-57-5]	15.7	3.0	1.03	dec	white	98% <74 (−200)
Li	[16577-52-0]	4.7	nil	1.01	dec	white	<149 (−100)
oleate							
Al	[688-37-9]	10.0	8.5	1.01	134	cream	85% <74 (−200)
palmitate							
Al	[555-35-1]	8.3	12.5	1.01	120	white	97% <74 (−200)
Zn	[4991-47-3]	14.7	1.4	1.12	123	white	<44 (−325)

[a] Ref. 2.

Precipitation Process. In the precipitation or double-decomposition process, the sodium salt of the fatty acid and an appropriate metal salt react in a hot aqueous solution. The metal soap precipitate is filtered, washed, dried, and milled. This process is probably the most widely used for commercial metallic soaps, particularly stearates. Proper control of temperature, agitation, rate of addition, and concentration of reactants gives very fine crystals, best for milling. Preparation, washing, and drying operations are expensive and time-consuming, increasing the production costs. The process is used to make the high-melting, powdered soaps of aluminum, calcium, magnesium, zinc, cadmium, strontium, and barium, as well as the lower-melting solid soaps of cobalt and lead.

The reaction and isolation are generally carried out in batches of up to 3.6 metric tons. Recent efforts have led to continuous reaction and isolation processes for the production of zinc, lead, calcium, barium, cadmium, and aluminum stearates (3).

Fusion Process. In the fusion process, metal oxides, hydroxides, or carbonates and other salts react directly at 150–200°C with the fatty acids, which may be melted. The reaction is carried to completion by removal of the water formed during reaction. The melted soap is treated with filter aids and filtered while still hot. It is packaged directly or dissolved in suitable solvents. The products obtained are generally hard, glassy, and difficult to grind. The process is particularly useful when a fine particle size is not required.

In a modification, a liquid fatty acid reacts with an aqueous slurry of the oxide, hydroxide, or carbonate. Organic solvents such as mineral spirits or naphtha may be used. The water of reaction is azeotropically removed, followed by clarification. The soap solution can then be concentrated to give a liquid soap or completely stripped to give a solid.

In a combination of the precipitation and fusion processes recently developed, the metal oxide, hydroxide, or carbonate reacts with an aqueous solution of ammonium or alkylamine carboxylates (4). The reactants are heated to a temperature at which the amine complex is dissociated and soap formed. The volatile amine can be recovered and recycled.

A solid fusion reaction has also been reported (5). Here, metal soaps of higher fatty acids are prepared in the absence of water at temperatures below the melting point of the metallic soap. The process has been used with calcium hydroxide and lead oxide to prepare stearates, with ammonium carbonate as catalyst. A free-flowing dry powder is obtained.

Direct Metal Reaction. The DMR process is carried out over a catalyst with fatty acids in a melted state or dissolved in hydrocarbons. The acid reacts directly with the metal, supplied in a finely divided state, producing the metal soap and, in some cases, hydrogen. Catalysts include water, aliphatic alcohols, and low molecular weight, organic acids.

The DMR process has no aqueous effluent, gives high-purity products, and is less expensive. However, if hydrogen is produced, it has to be removed carefully and should not reach explosive limits. Not all metals are sufficiently reactive to be suitable for the DMR process.

Overbasing. Basic metallic soaps are prepared by overbasing. That is, a normal metallic soap is treated in the presence of excess metal, metal oxide, hydroxide, or various salts with reactive species, such as CO_2 or SO_2, capable of forming covalent bonds with the metal. The resulting moiety generally contains metal–oxygen bonds free of carboxy groups.

Low viscosity, overbased metallic soap solutions of aluminum, barium, copper, iron, magnesium, zirconium, zinc, manganese, calcium, lead, and cobalt have been prepared by treatment with carbon dioxide (6). Overbasing is also accomplished by the reaction of glycol ethers or polyols with the corresponding metal derivatives (7), giving, eg, a low viscosity, 48% lead composition. Borate esters have also been used (8).

Overbasing can be carried out simultaneously with the general manufacturing techniques although the fusion and DMR processes are preferred; it can also be used as a post treatment.

Shipping and Storage

Liquid, paste, and solid metal soaps such as driers are packaged and shipped in unlined or lined (usually Heresite), sealed, nonreturnable, 45–180 kg, steel drums. Liquid soaps are also shipped in tank wagons. Powdered metal soaps such as stearates are packaged and shipped in sealed, 18–23 kg, paper bags or sealed, corrugated, 23–45 kg, cardboard drums.

Metal soaps are subject to deterioration by solvent loss, oxidation, and hydrolysis unless properly stored in a cool, dry area.

Economic Aspects

Tables 4 and 5 give prices and production figures for the commercially important metal soaps.

Analysis and Specifications

For liquid and paste metal soaps, the metal content, color, specific gravity, and viscosity are usually reported (see Table 2). For the solid metal soaps, total ash, moisture, specific gravity, and fineness are reported (see Table 3). The metal is determined as the oxide by standard procedures or by weighing the ash directly. Direct methods include EDTA titration and, more rarely, atomic absorption spectrophotometry (10). Free acids are extracted and titrated; moisture is determined by drying (11).

Viscosity and color specifications are qualitative. The Gardner-Holdt bubble–tube method is used to measure viscosities, which are indicated on an increasing scale from A to Z (12–13). The Gardner scale is used for color comparison of liquid driers (14), ranging from no. 1 (water-white) to no. 18 (reddish-brown).

Standard specifications for liquid paint driers are available in ASTM D600-73 and Federal Specifications TT-D-643d and 651e.

Health and Safety Factors

The hazards encountered in the manufacture, processing, handling, and use of metal soaps are associated for the most part with the inherent toxicity of the metals and solvents present. In general, the acid portion of the metal soap is low in toxicity. However, naphthenic acid is highly irritating to the skin on prolonged contact. High concentrations have a narcotic effect.

Ingested metal soaps are usually enzymatically hydrolyzed to the metal ion and free acid, most of which is excreted. The toxicity of minor absorbed amounts is associated with the toxicity of the metal (15).

Few studies of the toxic effects of metal soaps are available. General studies (17) on calcium, cadmium, barium, lead, and zinc stearates have shown acute oral toxicity decreasing in the following order: Cd > Co > Ba > Pb > Zn > Ca. The pathological response of stearates is detected primarily in the lungs. Acute oral toxicity of several metal naphthenates, Co, Ca, Cu, Pb, Mn, and Zn, reveal a similar order of toxicity with cobalt and lead being significantly more toxic than calcium, copper, manganese, and zinc (17). Intraperitoneal administration of the metal soaps increases the toxicity approximately tenfold.

The oral toxicity of cadmium soaps is high, LD_{50} of less than 500 mg/kg of body weight (18). However, when these materials are ingested, the irritation and emetic action is so violent that little of the cadmium is absorbed and fatal poisoning does not as a rule ensue. Cadmium compounds are carcinogens of the connective tissue, lungs, and liver; barium and copper soaps are less toxic (18). Pathogenic responses to barium and copper poisoning include nausea, severe abdominal pains, anemia, vomiting, dizziness, and rapid pulse. Oral toxicity of cobalt compounds is low, but exposure can cause severe liver and kidney damage (18). Cobalt soaps can be dermatitic and cobalt compounds are suspected carcinogens of the connective tissue and lungs. Manganese

Table 4. Typical Metal Soap Prices[a] and Metal Concentrations[b]

Metal	Percent	Price, $/kg
synthetic acids[c]		
zirconium	24	4.21
zirconium	18	3.68
zirconium	12	2.71
zirconium	6	1.63
lead	36	1.52
lead	24	1.51
cobalt	12	3.95
cobalt	6	2.34
manganese	12	2.07
manganese	9	1.68
manganese	6	1.28
zinc	17	1.81
zinc	8	1.17
calcium	8	1.87
calcium	6	1.54
calcium	5	1.34
calcium	4	1.19
iron	9	1.79
2-ethylhexanoates		
lead	24	1.28
cobalt	6	2.49
cobalt	12	4.30
manganese	6	1.41
zinc	8	1.28
calcium	5	1.54
iron	6	1.52
zirconium	6	1.63
naphthenates		
lead	24	1.19
cobalt	6	2.36
manganese	6	1.32
zinc	8	1.21
zinc	10	1.48
calcium	4	1.23
calcium	5	1.54
iron	6	1.39
cerium	6	3.17
rare earth	4	2.05
copper	8	1.57
tall oils		
lead	24	1.43
cobalt	6	2.56
manganese	6	1.54
calcium	4	1.46
stearates		
aluminum		1.90
calcium		1.59
magnesium		1.70
zinc		1.76

[a] Average prices as of Jan. 1978, Tenneco Chemicals.
[b] Tenneco Chemicals, 1978 specifications.
[c] Synthetic acids vary according to availability, whereas the metal content is always constant as stated here. Synthetic acids include neodecanoic, isononanoic, heptanoic acids, and their mixtures.

Table 5. 1977 Production of Driers and Metal Soaps[a]

Compound	Production, t
2-ethylhexanoates	
calcium	903
cobalt	2,000
lead	860
manganese	635
zinc	710
zirconium	1,207
all others	1,070
Total	*7,285*
stearates	
aluminum (distearate)	1,363
aluminum (tristearate)	135
barium	361
calcium	22,923
cobalt	164
lead	656
magnesium	2,264
zinc	12,102
all others	1,716
Total	*41,684*
naphthenates	
calcium	508
cobalt	1,694
copper	579
lead	2,223
manganese	698
zinc	590
all other	1,287
Total	*7,579*
tallates[b]	
calcium	56
cobalt	166
lead	144
manganese	79
Total	*445*

[a] Ref. 9.
[b] 1976 figures.

soaps attack the central nervous system (18). Although nickel soaps do not cause systemic poisoning, they are suspected carcinogens of the nasal cavity, para-nasal sinuses, and lungs.

The use of solvents in the liquid metal soaps poses problems of skin and inhalation toxicity as well as fire hazards. Inhalation of concentrated vapors causes nausea and headache. Neoprene gloves, goggles, and respirators should be routinely employed for protection against hazardous solvents.

Solid metal soaps, when finely divided, are capable of spontaneous combustion.

Uses

Driers. Driers (19) are added to coatings (paints, varnishes, inks) to bring about a physical change from the liquid to solid state, in a reasonable time (see Coatings). This change is effected by an oxidative cross-linking mechanism accelerated by the metallic cation of the added metal soap.

There are two categories of driers: primary or active driers, including cobalt, manganese, iron, and cerium; and auxiliary driers, including lead, calcium, zinc, and zirconium, which are used in conjunction with the active drier to improve the final properties of the coatings. Lead, which has been the most important auxiliary drier, especially in applications requiring drying at low temperatures ($\leq 10°C$), is no longer used in the United States. Government legislation now limits the lead content in residential paints to 0.06%. It had been used at levels of 0.1–0.8%, based upon vehicle solids. In general, lead has been replaced by combinations of zirconium and calcium. For interior and moderate exterior temperatures (21°C), these auxiliary driers work well with common active driers. However, below 10°C reformulation may be required. Bismuth soaps are comparable to lead in drying times and film properties at ambient and low temperatures (20).

Cobalt 2+ is a strong oxidant and the most active drier metal. It effects rapid surface drying and is generally used in conjunction with auxiliary driers. It is usually added at 0.005–0.4%, based on vehicle solids. Manganese, another active drier metal and strong oxidant, promotes polymerization to a greater degree than cobalt. It is often used alone in baking finishes and in combination with lead and cobalt in air-dry applications (0.01–0.1% manganese, based on vehicle solids). Iron, whose dark color limits its utility, promotes rapid drying by polymerization. It is widely used in baking finishes and for eliminating tack in air-dry coatings containing oils high in nondrying constituents such as fish oils. Cerium, generally supplied with lanthanum and traces of other rare earth metals, is recommended for baking finishes, particularly whites and overprint varnishes where color retention is vital. It is particularly useful with alkyd and epoxy resins, or in combinations of these with amine resins.

Lead, where permitted, is used with active or top driers. It promotes hard, thorough drying throughout the entire film and flexibility. Calcium is often used with or instead of lead as, eg, in fume-resistant paint or toy finishes. Calcium also improves the solubility of lead in certain vehicles having poor lead tolerance. It frequently performs better than other auxiliaries in baking finishes. Zinc, like lead, produces a hard, thoroughly dried film, but does not promote drying at low temperatures. Its naphthenate soap also serves as a good wetting agent and often contributes to improved gloss. Zirconium appears to be the best lead replacement. It shows good gloss, gloss retention, color, and color retention compared to lead but does not perform as well under adverse drying conditions of low temperature and high humidity. In combinations with calcium, zirconium achieves adequate drying times with less sensitive finishes.

Improved drying properties are claimed for alkoxyaluminum compounds (21–22) that do not dry by the conventional oxidative mechanism but rather by a cross-linking reaction.

The drier metals, in order to be effective, have to be in the correct oxidation state and in suitable form. The preferred form in the paint industry is the homogeneous, low viscosity liquid, whereas liquids and pastes are used in the ink industry. Solid forms of driers are also available; however, they are almost always used for applications other than drying. The organic portion of the drier metal soap is chosen on the basis of availability, cost, color, and metal soap properties required (see Table 1). Mixtures of acids are also used. The effect of the different acids on the drying properties of the metal soaps has been discussed for some time. A recent study (23), using a controlled medium-oil alkyd enamel, demonstrated significantly better drying times with cobalt neodecanoate and tallate than with the cobalt soaps of six other common drier acids. In another study (24), the effect of acid type was found to be minimal, following vehicle type, temperature, and shelf aging in importance.

Legislative developments caused the change from nonaqueous inks and coatings in order to eliminate volatile organic solvents. Thus water-reducible or soluble resin systems are receiving increased attention, requiring driers that can be readily incorporated and retain their activity. Careful formulation of acids and dispersants gives driers that are readily dispersed in aqueous coatings with maximum hydrolytic stability. The latter property is important because many of the standard driers precipitate the metallic hydroxide on prolonged contact with water.

Many coatings suffer from a phenomenon called loss of dry, caused by systems containing pigments of high surface area. These systems lose their drying characteristic with time because of adsorption of the metal soap onto the pigment surface. Loss of dry is controlled by adding an inhibitor (25), sometimes referred to as a feeder drier, which shows a higher affinity for the pigments than a standard drier, and is thought to be preferentially adsorbed. The standard driers thereby remain serviceable over long periods of time.

Waterproofing Agents. Because of their hydrophobic nature, metallic soaps are good waterproofing agents. The longer chain fatty acids, such as stearic, oleic, and palmitic, are especially useful—their metal soaps being water insoluble and nonwettable. They can be applied in a variety of ways depending upon the nature of the substrate or product to be waterproofed, such as fabric, paper, masonry, concrete, mineral wool, and metals. Concrete can be sealed to moisture with a coating containing zinc oleate (26) or by incorporation of calcium, zinc, and aluminum stearates during manufacture (27) (see Cement). These stearates can be added to dry portland cement or applied directly to masonry as dispersions in water-thinnable paints (Al plus Zn) or to masonry and fabrics as solutions (nongelling Al stearate) in hydrocarbon solvents (27). Lithium and complex aluminum soaps have also been used as hydrophobic chemical thickeners for masonry sealing compositions (28); stearates of magnesium (29–31) and barium (32) have been used to coat inorganic salts to prevent caking. Aluminum, calcium, magnesium, and zinc stearates have been added to floor and wall covering adhesives as waterproofing agents (33). Strong, humidity-resistant, heat-cured urethane adhesives are prepared with manganese, zinc, iron, or lead naphthenates, tallates, or octanoates (34) (see Waterproofing).

Fuel Additives. With the increasing emphasis on energy conservation and environmental considerations, additives for fuels that can correct combustion-related problems have aroused considerable interest (see Air pollution control methods). Many commercial fuel additives are combinations of organometallics (qv), dispersants (qv), emulsifiers, and carrier solvents. The organometallic, often a metal soap, acts as a combustion catalyst, increasing efficiency with reduction of smoke, deposits, and corrosion. Although the organometallic need not be a metal soap, they are often used. Soaps of sodium, potassium, titanium, manganese, iron, nickel, chromium, cobalt, and molybdenum increase the thermal yield and decrease exhaust gases of heavy petroleum fuels (35). Exhaust gas pollution (NO$_x$, SO$_x$, and/or CO) can be reduced by addition of (1) calcium soaps (36); (2) naphthenates of sodium [61790-13-4], potassium, manganese, lead, and aluminum [61789-64-8] (37); and (3) naphthenates of magnesium, zinc, manganese, calcium, or barium [61789-67-1], or octanoates of magnesium [3386-57-0], zinc [557-09-5], manganese [6535-19-9], calcium [6107-56-8], or barium [4696-54-2]. Barium and manganese naphthenates reduce exhaust smoke from aviation diesel fuel combustion (39). Copper naphthenate [1338-02-9] prevents deposits in furnaces and burners fired with coal or heavy oil (40), whereas molybdenum naphthenate added to jet fuel minimizes manganese deposits on jet engine surfaces (41). Calcium tallate reduces smoke in diesel fuels (42).

Rubber. The rubber industry utilizes metallic soaps as both processing aids and for the improvement of properties (see Rubber chemicals). The soaps can either be prepared *in situ* by the reaction of the metal oxide or hydroxide with the organic acids used in rubber processing (eg, stearic) or by incorporating metallic soap solutions or solid soaps in the rubber formulation. Zinc soaps act as softeners, dispersants for pigments, reinforcing agents, and vulcanization catalysts. They are also used externally as dusting powders on coatings (zinc stearate in particular) for mold–mill release. Both calcium and magnesium soaps are used to prevent scorching. Lead soaps, like zinc soaps, catalyze vulcanization and thus decrease curing time.

Metal soaps promote adhesion of rubber to steel as required in steel-belted radial tires and similar applications. However, the amount has to be carefully controlled because many metallic soaps accelerate oxidative degradation. Cobalt or nickel naphthenate [61788-71-4] used with various phenols improve the adhesion of sulfur-vulcanized rubbers to plain, brass-plated, or galvanized steel (43). The adhesion of vulcanized rubber to zinc-plated steel after aging is improved by cobalt stearate and magnesium oxide (44). Cobalt naphthenate, in conjunction with conventional adhesion promoters or alone, significantly improves rubber-to-metal bonds (45–47). Cobalt soaps also reduce damage to tires during sudden deflation (48).

Greases, Lubricants, and Chemical Thickeners. Metal soaps effect the gelation of oils and polar organic solvents when heated with those liquids or formed *in situ*. Upon cooling, the resulting solutions set to a gel. Many of these gels are used in greases, lubricants, and other products (see Lubrication and lubricants).

An adhesive grease is prepared from naphthenic and paraffinic oils by the *in situ* preparation of a complex calcium soap salt from calcium hydroxide, 12-hydroxystearic acid, and acetic acid (49). Lubricating oils give greases by thickening with calcium, lithium, or sodium soaps of dibasic acids prepared from monohydroxy fatty acids (50). Aluminum complex soap greases are formed *in situ* by the reaction of stearic and benzoic acids with aluminum isopropoxide, removing the isopropanol as it is formed (51). Lubricating greases having extreme pressure properties consist of oil containing aluminum, barium, calcium-complex, lithium, or sodium soaps with about 1% fish oil-based lead soap and about 0.5% sulfur-modified oil (52). Soaps of cadmium, cobalt, magnesium, nickel, mercury, and strontium have also been used.

Soaps of calcium, aluminum, magnesium, and zinc are used alone or in combination with other materials as metal-working lubricants. A formulation especially useful for drawing aluminum tubing is prepared by mixing sulfur-free wool wax, aluminum stearate, castor oil wax, and 1,1,1-trichloroethane (53). In the cold shaping of steel and nonferrous metals, mixtures of alkali soaps, C_{15-18} fatty acid glycerides, and inorganic salts have been used as 10–15% aqueous solutions (54).

Metal carboxylates (lower acids, aromatic acids, fatty acids) are used in unsaturated polyesters as lubricants and chemical thickeners (55–56).

Gasolines gelled by aluminum soaps have been used as wartime incendiaries (Napalm). These gels are prepared by mixing hydroxyaluminum-bis(2-ethylhexanoate) [30745-55-2] with gasoline and either water or a nonionic surfactant (57–58) (see Chemicals in war).

Plastics. Metal soaps are used by the plastics industry as lubricants and mold release agents and by the rubber industry (see Abherents). They are also used as thermal and uv stabilizers (qv) in vinyl polymers and polyolefins. The color stability of PVC is enhanced by the admixture of the 2-ethylhexanoates of barium [2457-01-4], cadmium, and zinc (59). The efficiency of vinyl stabilizers often depends upon the ratio of metal soaps used and the nature of the anion (60). When cadmium neodecanoate [61951-96-0] and zinc neodecanoate [27523-29-8] stabilizers were compared to the corresponding 2-ethylhexanoates, the neodecanoates showed better plastisol viscosities, air-release time, clarity, plate-out resistance, and uv stability in PVC compositions (61). Calcium naphthenate imparts thermal stability to polypropylene (62) (see Heat stabilizers).

Polyethylene and PVC were rendered mold resistant by a synergistic combination of lead naphthenate and a halogenated pesticide, with no adverse effect on mechanical or electrical properties (63).

Most of the formulations used in the plastics industry are proprietary.

Catalysts. The drying of a paint film aided by metallic soap driers is an example of their catalytic action. Metallic soaps are also used as catalysts in various specific reactions.

Molybdenum naphthenate catalyzes the liquid-phase reaction of propylene with an organic peroxide (64) to give propylene oxide (Oxirane process). Metallic soaps catalyze the air oxidation of organic compounds in many commercial processes. Thus toluene and p-xylene are converted to benzoic and terephthalic acids, respectively, by cobalt soaps. Naphthenates of cobalt and manganese catalyze the liquid-phase oxidation of isopropylbenzene and tetrahydronaphthalene (65), whereas those of copper, cobalt, iron, and manganese catalyze the oxidation of cumene and tetralin (66). Similarly, the laurates of copper [19807-12-6], cobalt [24212-14-4], nickel [22993-28-8], manganese [24670-78-8], and iron [21248-71-5] are used in the oxidation of 1,3,5-trimethylcyclohexane (67).

Methyl methacrylate is polymerized by radical initiation with cobalt, manganese, cerium, calcium, and zinc naphthenates (68); zirconium 2-ethyl-hexanoate polymerizes formaldehyde (69) (see Initiators).

Cobalt soaps are widely used as accelerators for peroxide-initiated curing of unsaturated polyesters. A combination of vanadium naphthenate and tin(II) octanoate [1912-83-0] accelerates polyester cure (70); tin(II) octanoate accelerates the curing of poly(dimethyl siloxane) prepolymers (71). A catalyst modifier containing zinc 2-ethylhexanoate was used to prepare epoxy resins with good optical properties (72).

Fungicides. The metallic soaps of copper, zinc, and mercury are fungicidal and extend the life of wood and cellulosic fibers (see Fungicides). They are usually applied in petroleum solvents or as the ammonium complex of copper naphthenate in aqueous solution. The insoluble soaps remain on the fibers following solvent evaporation.

During shortages of creosote, wood is pressure-impregnated with copper naphthenate. Solutions of both copper and zinc naphthenates are used in brush and dip treatments of lumber that is used in contact with the soil. These soaps are also used to treat ropes, fishing nets, and canvas for tarpaulins.

Cosmetics and Pharmaceuticals. In the cosmetic, toilet, and pharmaceutical industries, alkali and heavy metal soaps are employed as ingredients of face powders, talcum powders, medicated dusting powders, tablet formulations, and creams. Lithium, calcium, magnesium, zinc, and aluminum stearates increase the smoothness, adhesion, and water repellancy of cosmetics and are used as gellants. Strict control in the manufacture of the soaps is maintained, and the soaps have to be finely divided and white. Zinc undecylenate [557-08-4] is used in the treatment of athlete's foot (see Cosmetics).

Miscellaneous. When added to waxes in candle and crayon manufacture, metallic soaps increase the melting point and prevent softening and sagging at higher temperatures. They add lubricity and smoothness to pencil leads. Melting and glazing compounds are formulated with metallic soaps for increased binding and water repellancy. Copper oleate [10402-16-1] is useful in the chemical analysis of organic sulfur compounds in oils and gasoline (73). Improved abrasive and polishing pastes are obtained by incorporation of magnesium, aluminum, or zinc palmitates or stearates (74). Barium soaps and magnesium and zinc naphthenates are added to lubricants and greases for anticorrosive properties (75–76). Metal soaps are dispersed in stearyl stearate along with insecticide to give sustained release without absorption of water vapors (77).

BIBLIOGRAPHY

"Driers and Metallic Soaps" in *ECT* 1st ed., Vol. 5, pp. 195–206, by S. B. Elliott, Ferro Chemical Corporation; "Driers and Metallic Soaps" in *ECT* 2nd ed., Vol. 7, pp. 272–287, by G. C. Whitaker, Harshaw Chemical Company.

1. W. J. Stewart in W. H. Madson, ed., *Paint Driers and Additives, Federation Series of Coatings Technology,* Federation of Societies for Paint Technology, Philadelphia, Pa., 1969, Unit 11, pp. 1–26.
2. Stearate product specifications, Tenneco Chemicals, Inc., Piscataway, N.J.; *Witco Metallic Stearates, Bulletin 55-4R-5-63,* Witco Chemical Corp., New York, May 1963.
3. Fr. Pat. 1,581,786 (Sept. 19, 1969), P. P. Tamkovich.
4. Brit. Pat. 1,353,320 (May 15, 1974), J. H. W. Turner (to Hardman and Holden Ltd.).
5. U.S. Pat, 3,476,786 (Nov. 4, 1969), R. E. Lally and J. Cunder (to Diamond Shamrock Corp.).
6. Brit. Pat. 1,236,085 (June 16, 1971), D. K. V. Steel and D. Smith (to ICI Ltd.).
7. Brit. Pat. 1,430,347 (Mar. 31, 1976) A. V. Collins and R. E. Pearl (to Mooney Chemicals, Inc.).
8. Brit. Pat. 972,804 (Oct. 14, 1964), J. H. W. Turner, E. W. Downs, and S. E. Harson (to Hardman & Holden, Ltd.).
9. *Synthetic Organic Chemicals, U.S. Production and Sales, 1977, USITC Publication 920,* U.S. Tariff Commission, U.S. Government Printing Office, Washington, D.C., 1978.
10. *Analytical Procedures, E24.02, 40.07, 55.06, 56.00, 59.05, 65.09, 91.03,* Tenneco Chemicals, Inc., Piscataway, N.J., Jan. 1967.
11. *Witco Metallic Stearates, Bulletin #170,* Witco Chemical Corp., New York, Sept. 1970.
12. *Standard Specification for Liquid Driers, D600-73,* American Society for Testing and Materials, Philadelphia, Pa.
13. *Standard Methods of Testing Liquid Driers, D564-47,* American Society for Testing and Materials, Philadelphia, Pa.
14. *Color of Transparent Liquids, D1544-68,* American Society for Testing and Materials, Philadelphia, Pa.
15. R. A. Goyer and J. J. Chisolm in D. H. K. Lec, ed., *Metallic Contaminants and Human Health,* Academic Press, Inc., New York, Proceedings No. 9, pp. 57–86, 1972.

16. N. Y. Tarasenka, L. P. Shabalina, and V. S. Spiridonova, *Int. Arch. Occup. Environ. Health* **37**, 179 (1976).
17. W. T. Rockhole, *AMA Arch. Ind. Health* **12**, 477 (1955).
18. N. I. Sax, *Dangerous Properties of Industrial Materials,* 4th ed., Van Nostrand Reinhold Co., New York, 1975, pp. 432, 504, 565, 574, 594, 855, 956.
19. W. J. Stewart, *Off. Dig. Fed. Paint Varn. Prod. Clubs* **26**, 413, 1954.
20. U.S. Pat. applied for M. L. Feldman and M. Landau (to Tenneco Chemicals, Inc.).
21. Fr. Demande 2,213,973 (Aug. 9, 1974), (to Hardman & Holden Ltd.).
22. Ger. Offen. 2,402,039 (July 18, 1974), J. H. W. Turner (to Hardman & Holden Ltd.).
23. J. R. Kiefer, Jr., *J. Paint Technol.* **45**, 48 (1973).
24. J. R. Garland, *J. Paint Technol.* **41**, 623 (1969).
25. Ger. Offen. 2,502,282 (July 24, 1975), R. T. Gottesman, G. Kagan, and J. Fath (to Tenneco Chem. Inc.).
26. Ger. Offen. 2,503,871 (Aug. 5, 1976), H. Lehmann and S. Riethmayer (to Industrie Automation G.m.b.H. & Co.).
27. O. Knott, *Am. Paint J.* **54**(36), 54 (1970).
28. U.S. Pat. 3,546,007 (Dec. 8, 1970), G. G. Douglas (to Sun Oil Co.).
29. U.S. Pat. 2,631,977 (Mar. 17, 1953), S. E. Allen and D. Levin (to C-O Two Fire Equipment Co.).
30. U.S. Pat. 2,539,012 (Jan. 23, 1951), H. W. Diamond and F. G. Miller (to General Foods Corp.).
31. U.S. Pat. 2,683,664 (July 13, 1954), A. E. Green.
32. U.S. Pat. 2,768,952 (Oct. 30, 1956), C. Anthony, Jr., and R. Thomman (to Specialties Development Corp.).
33. Ger. Offen. 2,056,566 (May 27, 1971), B. G. Manson and co-workers (to ICI, Ltd.).
34. U.S. Pat. 3,510,439 (May 5, 1970), T. F. Kaltenbach and P. E. Wright (to General Motors Corp.).
35. Fr. Pat. 1,540,340 (Sept. 27, 1968), A. J. F. Guillermic and J. Rouet (to Gamlen-Maintre et Cie).
36. Jpn. Kokai 75 117,805 (Sept. 16, 1975), H. Sano (to Agency of Industrial Sciences and Technology).
37. K. I. Ivanov and co-workers, *Teploenergetika,* **10**, 85 (1973).
38. H. A. Friedrichs and co-workers, *Arch. Eisenhuettenuv.* **44**(2), 125 (1973).
39. T. A. Rogimova, *Azerb. Neft. Khoz.* **52**(3), 37 (1972).
40. Jpn. Kokai 7,122,740 (June 29, 1971), H. Ogawa.
41. U.S. Pat. 3,718,444 (Feb. 27, 1973), V. F. Hrizda (to Ethyl Corp.).
42. Fr. Pat. 1,594,227 (July 10, 1970), H. J. Andress, Jr. (to Mobil Oil Corp.).
43. Ger. Offen. 2,303,674 (Apr. 4, 1974), H. Hirakawa (to Yokahama Rubber Co.).
44. Ger. Offen. 2,363,655 (July 18, 1974), T. Fujimura and co-workers (to Grigestone Tire Co., Ltd.).
45. Jpn. Kokai 7,420,074 (May 22, 1974), K. Kishida and K. Makagawa (to Toyo Rubber Industry Co.).
46. U.S. Pat. 3,897,583 (July 29, 1975), C. Bellamy (to Uniroyal S.A.).
47. Jpn. Kokai 7,633,131 (Mar. 22, 1976), E. Kondo (to Honny Chemicals Co., Ltd.).
48. Ger. Offen. 2,544,168 (Apr. 22, 1976), J. W. Messerly (to B.F. Goodrich Co.).
49. U.S. Pat. 3,642,627 (Feb. 15, 1972), R. L. Waring (to Cato Oil and Grease Co.).
50. U.S. Pat. 3,828,086 (Aug. 6, 1974), H. E. Kenney and E. T. Donahue (to U.S.D.A.).
51. U.S. Pat. 3,591,505 (July 6, 1971), A. T. Polishuk (to Sun Oil Co.).
52. Brit. Pat. 1,236,140 (June 23, 1971), K. S. Bergeron.
53. U.S. Pat. 3,692,678 (Sept. 19, 1972), T. F. Stiffler and A. M. Murphy.
54. East Ger. Pat. 73,596 (June 5, 1970), E. Will and W. Miereke.
55. U.S. Pat. 3,557,042 (Jan. 19, 1971), A. J. Dalhuisen (to Merck and Co.).
56. C. W. Fletcher and J. W. Waldeck, *Proc. Ann. Conf. Reinf. Plast. Comp. Inst. SPI,* **30**, 8A (1975).
57. U.S. Pat. 3,539,310 (Nov. 10, 1970), L. Cohen and W. T. Gregory (to U.S. Dept. of the Army).
58. U.S. Pat. 3,539,311 (Nov. 10, 1970), L. Cohen and W. T. Gregory (to U.S. Dept. of the Army).
59. J. Szekelyhidi and co-workers, *Nuanyog Gurni,* **8**(9), 281 (1971).
60. *Ibid.,* **8**(5), 153 (1971).
61. M. Fefer and G. Rubin, *Mod. Plast.* **47**(4), (1970).
62. D. A. Akhmedzade and co-workers, *Issled. Obl. Sint. Polim. Monomernykh Prod.* 103 (1974).
63. Jpn. Kokai 7,030,570 (Oct. 3, 1970), T. Shiina (to Hitachi Cable, Ltd.).
64. R. Ference, *Oil Gas J.,* 16 (Feb. 13, 1978).
65. I. M. Reibel and co-workers, *Dokl. Nanch. Konf. Molodykh Uch. Kishinev. Sel'skokhoy. Inst. Plodovinogradnyi Fak.* **1967**, (Publ. 1968).
66. I. M. Reibel and co-workers, *Tr. Kishinev. Sel-Khoz. Inst.* (67), (1971).
67. B. Jadrnicek and K. Vesely, *Coll. Czech. Chem. Commun.* **35**(1), (1970).

68. S. Aoki and co-workers, *Nippon Secchaku Kyokai Shi* **6**(1), (1970).
69. Jpn. Kokai 7,036,316 (Nov. 18, 1970), E. Ajisaka and T. Matsumoto (to Japan Catalytic Chemical Industry Co., Ltd.).
70. U.S. Pat. 3,914,336 (Oct. 21, 1975), J. G. Baker (to PPG Ind. Inc.).
71. U.S. Pat. 3,578,616 (May 11, 1971), L. D. Harry and G. A. Sweeney (to the Dow Chemical Co.).
72. U.S. Pat. 3,565,825 (Feb. 23, 1971), M. S. Antelman (to Wells-Benrus, Inc.).
73. G. R. Bond, *Ind. Eng. Chem. Anal. Ed.* **5**, 257 (1933).
74. Ger. Pat. 1,519,089 (Apr. 8, 1971), H. Mueller (to Dursol-Fabrik, Otto Durst).
75. U.S. Pat. 3,539,514 (Nov. 10, 1970), A. F. Strouse.
76. Fr. Demande 2,011,476 (Feb. 27, 1970), (to Esso Research & Engineering Co.).
77. Ger. Offen. 2,439,811 (Mar. 4, 1976), U. Mache.

General References

T. A. Girard, M. Beispiel, and C. E. Bricker, "The Mechanism of Cobalt Drier Action," *J. Am. Oil Chem. Soc.* **42**, 828 (Oct. 1965).
D. J. Love, "The Effects of Oxidative and Coordination Drier Systems on Film Properties," *J. Oil Colour Chem. Assoc.* **60**, 214 (1977).
A. Fischer, *J. Am. Oil Chem. Soc.* **43**, 469 (1966).
A. Fischer and S. E. Harran, "Driers Based on Tall Oil—A Review," *J. Am. Oil Chem. Soc.* **42**(2), 109 (1965).
H. Rackoff, W. F. Kwolek, and L. E. Gast, "Drier Composition and Yellowing of Linseed Oil Films," *J. Coat. Technol.* **50**(637), 51 (1978).
K. C. Salooja, "Burner Fuel Additives," *J. Inst. Fuel,* 37 (Jan. 1972).
W. Büttner, "Application of Fuel Oil and Flu Gas Additives in Large Boilers," *Combustion* **49**(6), 22 (Dec. 1977).
H. Pilpel, "Metal Stearates in Pharmaceuticals and Cosmetics," *Manuf. Chem. Aerosol News,* 37 (Oct. 1971).
J. Skalsky, "Preparation and Application of Drying Agents in Paints," *Prog. Org. Coat.* **4**, 137 (1976).
T. Wexler, "Metallic Soaps in Industry," *Chim. Ind. Genie Chim.* **103**, 1657 (1970).

FREDERICK J. BUONO
MARTIN L. FELDMAN
Tenneco Inc.

DRILLING FLUIDS. See Petroleum.

DRILLING MUDS. See Petroleum.

DRYCLEANING AND LAUNDERING

Drycleaning, 50
Laundering, 68

DRYCLEANING

Drycleaning can be defined as the cleaning of fabrics in a substantially nonaqueous, liquid medium. This process has evolved during the last 60–70 years into a highly effective, low-cost, safe method of removing soils from all types of textiles.

There are no clear records indicating the origin of the drycleaning process. Knowledge of the cleaning properties of turpentine is quite old. Improvements in the quality of petroleum distillates during the latter half of the nineteenth century resulted in their wide use as drycleaning solvents. Jolly Belin, a member of a dyeing and cleaning firm in Paris, is credited with the accidental discovery that spilled lamp oil removed the soil from a tablecloth (1). Pullar's, a cleaning firm in Perth, Scotland, introduced solvent cleaning in the British Isles in 1866 (2). Drycleaning began to be used extensively in the United States in the 1910–1920 period. This new industry developed as the result of the large-scale production of gasoline needed for the expanding automobile market.

Gasoline was the major drycleaning solvent until 1928 when a less flammable, odor-free solvent was developed, largely through the efforts of W. J. Stoddard, president of the National Institute of Drycleaning (now the International Fabricare Institute, IFI). This petroleum fraction was designated Stoddard Solvent in recognition of his work (see Solvents, industrial). Standardized specifications were adopted by the United States Department of Commerce in 1928 (3). The industry enjoyed steady growth until the introduction of wash-and-wear garments, and after a period of adjustment, again started to expand and diversify. Recent growth has been aided by the development of new equipment, cleaning techniques, and applications.

The original "dry solvent" technique removed oil-soluble soils; current methods remove oils, many water-soluble, and some insoluble materials with the help of detergents and various agents. Although drycleaning solvents do not appear to wet the fabric like water, textile fibers are wetted even more readily by the substantially nonaqueous systems than by water alone (see Textiles).

The important distinction between dry and wet solvents is that the latter (principally water, glycols, and other hydroxylic compounds) swell the hydrophilic textile fibers, but the drycleaning solvents do not. Dimensional changes that fibers undergo as they swell in water are transmitted throughout the textile structure and can cause serious fabric damage. These changes can be local distortions (wrinkles) or more extensive, causing shrinkage. Drycleaning solvents do not swell the textile fibers and thus do not cause wrinkles or shrinkage. This is one of the major advantages of drycleaning processes over laundering. Another advantage is that drycleaning solvents remove oily soils at low temperatures, whereas a high temperature, colloidal suspension mechanism is needed in wet laundering processes.

These advantages have established drycleaning as a safe process for cleaning delicate textiles. Wool (qv) is not normally thought of as a delicate fiber, but in respect to shrinkage during normal laundering processes, it is very delicate unless the wool has been treated for dimensional stability. The ease of removing oily soils can be used to advantage to clean less delicate garments and new fabrics not sensitive to wet solvents.

The main difference between drycleaning and laundering processes is the solvent used. Both techniques use one or more of the processes listed above. The nature of the textile item to be cleaned, its response to water, or the processing costs of the drycleaning solvents determine which type of cleaning shall be used. Since both methods remove soil by the same mechanisms, similar operations are used. Garments are tumbled in the cleaning liquid in horizontal washers to separate soil and fabric as thoroughly as possible without damage. Detergents are used in both cases to emulsify or solubilize the soils insoluble in the cleaning liquid. Special detergent ingredients peptize and suspend insoluble soils.

In laundering, bleaching agents are added, and strong alkalies can be used to saponify fats. Saponification is not used in drycleaning, and until recently bleaches were applied only in spotting procedures. Techniques have now been developed that permit the use of some bleaches in drycleaning (4).

Laundering has little, if any, cost advantage over drycleaning. In the past, the types of garments drycleaned required special handling and finishing procedures, increasing the costs of this process over laundering. However, reduction of solvent costs, changes in handling procedures, and new, larger equipment have increased the use of drycleaning solvents in the industrial-cleaning area, a field that was normally serviced by laundering.

Soils and Soil Removal

An understanding of the nature and properties of textiles and soils is necessary to understand cleaning technology.

Detergent chemists classify garment soils into three specific categories: (1) solvent-soluble soils, which are insoluble in water; (2) the water-soluble soils, which are insoluble in drycleaning solvents; and (3) soils that are insoluble in drycleaning solvents and water.

Cleaning techniques have been developed to remove most types of soils. Solvent-soluble soils (5–9) consist of a variety of oils, fats, waxes, hydrocarbons, and fatty acids. The major source of solvent-soluble garment soils is human sebum (7); therefore, these soils contain the ingredients of sebum or their degradation products.

Water-soluble soils originate chiefly from perspiration and food and beverage contamination. Salts, carbohydrates, and proteins are the main components of this category.

Insoluble soils are principally air-borne dust, such as lint, clay, silt, ash, smoke (carbon), iron oxide (2,7), and paint residues.

Soils, defined as any unwanted material on a garment, cover a spectrum as broad as chemistry itself. Nevertheless, these three categories provide adequate understanding of the drycleaning process.

Basically different cleaning mechanisms are available for soil removal, including (1) mechanical, such as vacuum and ultrasonic cleaning (see Ultrasonics); (2) colloidal suspension used for insoluble soils; (3) emulsification; (4) solubilization for soils insoluble in the solvent being used, but soluble in solvents of opposite character; (5) solution for soluble soils; (6) chemical reaction, notably bleaching; and (7) enzymes (qv). Chemicals and enzymes are used to convert insoluble to soluble soils.

The Drycleaning Operation

The following sequence of operations is used in the normal drycleaning process:

(1) Soiled garments are marked and sorted.

(2) Prespotting is performed when required.

(3) Garments are rotated in a tumble-type washer containing the drycleaning solvent.

(4) Solvent is drained from the tumbler and most of the residual solvent removed by centrifugal extraction.

(5) The small amount of remaining solvent is removed in heated dryers.

(6) Dirty solvent from the wash cycles is continuously passed through diatomaceous earth and activated carbon or disposable cartridge filters to remove as much of the fugitive dyes and insoluble soils as possible.

(7) Soils not removed by filters, diatomaceous earth, or activated carbon can be removed from the solvent by distillation.

(8) Dry, solvent-free garments are inspected and spot cleaned a second time, if necessary.

(9) A garment may be "wet cleaned" at this point, but this is usually omitted when operating a charged solvent system (see below).

(10) Clean garments are finished on suitable presses, puff irons, steam tunnels, or on adjustaforms.

(11) After a final inspection, the garments are assembled according to customer's order.

(12) Garments are bagged and placed on racks ready to be picked up. Mechanical conveyors aid in fast recovery of a customer's garments.

Many drycleaning operations provide special services, such as garment repair, dyeing, and moth- and waterproofing treatments. Many operators add optical brighteners, antistats, and sizing to improve the brightness and feel (hand) of a garment (see Brighteners, fluorescent; Antistatic agents).

Operating Steps

Marking. Each article to be cleaned is identified in the marking department by a tag or stamping a number on an unexposed portion of the garment. The marked garments are placed in movable hampers holding about 45 kg of garments. Each load is weighed before being drycleaned.

Various types of garments are sorted in different hampers so that each hamper contains only one class of garment. The six major classifications are: (1) white and light-colored dresses; (2) dark-colored dresses; (3) white and light-colored wools; (4) dark wools; (5) draperies and furniture covers; and (6) raincoats.

In the marking department, articles from one customer are tied together or placed in bags. Each garment is handled individually while being marked. During this handling, the amount of regain moisture in the garment approaches that which is in equilibrium with the relative humidity and temperature of the marking room. The moisture of garments in a filled hamper responds very slowly to changes in relative humidity. Therefore, garments being drycleaned contain approximately the same amount of regain moisture as that which was in equilibrium with the air in the marking room.

Prespotting. This consists of the application to heavily soiled areas of various solvents or solutions that can be removed during the drycleaning process. Prespotting methods are discussed in refs. 10–12.

Volatile dry solvents, such as amyl acetate, are used to remove oil-base stains and soluble plastics. Nonvolatile oily paint removers are used to lubricate stains, such as paint and varnish. Emulsifying agents mixed with solvent and water to form a water-in-oil emulsion remove water-soluble soils, such as perspiration and foodstuffs (see Emulsions). They are often applied by a compressed air spray, and are commonly called spray spotters.

Other types of prespotters are known as digestive agents. They are mixtures of enzymes capable of digesting or solubilizing some food and albuminous stains (13).

Cleaning Cycle. In general, a shorter cleaning cycle is used with halogenated hydrocarbon solvents than with petroleum solvents.

The start of a day's operation depends upon the type of filtration equipment employed. With a replaceable cartridge filter system, filter pressures and solvent dye content must be monitored. With other types of pressure filters, a charge of filter aid is placed in the washer, the washer is started, and the filter solvent circulation is started to carry the filter powder onto the filter surface. This initial charge, known as the precoat, forms the filtering medium for the day's operation. The quantity of filter aid used varies; producers of diatomite-type powder usually recommend a precoat of 2.4–3.4 kg of filter aid for each 10 m^2 filter area. For the removal of solubilized dye or for reduction of solvent color, a solvent slurry of activated carbon at concentrations recommended by the carbon producers is added to the button trap during formation of the precoat (see Diatomite; Carbon).

After the filter has been precoated, the washer is stopped and the solvent flow automatically by-passes the washer allowing the work load to be placed in the cleaning cylinder. The washer and filter circulation through the washer are started, and the soil flushed from the work load deposited on or in the filter. The washing time and, in some instances, rinsing time, vary according to the solvent flow rate and the size and type of work load. It can vary from 5 to 45 min, depending on fabric composition, construction, and degree of soil. At the beginning of each cleaning cycle, for units operating with filter powder, a small additional charge is introduced to minimize the build-up of filter pressure. At the end of the day, the precoat is broken from the filter by a solvent back-wash, by air bubbling, or by some form of mechanical action. The soil and filter aid are allowed to settle to the bottom of the filter. Some filters drain soil, filter aid, and remaining solvent to a subcontainer for later solvent recovery.

In many cases, the filter residues are filtered through a bag and the reclaimed solid "filter muck" discarded; perchloroethylene can be reclaimed in a muck cooker.

Extraction. After the garments have been drycleaned, excess solvent is removed by centrifugation, referred to in the industry as extraction (see Centrifugal separation).

Today's drycleaning machines perform the washing and extraction cycles in the same cylinder. After the cleaning operation, the solvent is drained from the wheel, which is then rotated at high speed to remove excess solvent. Handling of wet garments and exposure of the operators to high solvent vapor concentrations is thus eliminated. Some older type petroleum washers do not extract. Either an extended draining period or a separate extractor is used.

54 DRYCLEANING AND LAUNDERING

Drying. After centrifugation of excess solvent, the garments are dried in a revolving perforated cylinder contained in a tumbler. As air enters the tumblers, it first passes through steam-heated coils, then through the garments in the revolving cylinder, and finally through an exhaust fan. Tumblers are usually equipped with manually operated dampers for temperature control. If the dampers are closed so that all air passes through the heating coils, the temperature is usually about 71°C.

Garments that might be damaged by tumblers may be dried by a fan in a drying cabinet (see Drying).

In the United States, little effort is made to recover petroleum solvent from exhaust vapors, but drycleaning machines using chlorinated hydrocarbons permit solvent vapor condensation and recovery. In many such machines, washing, extraction, and drying are all carried out in the same cylinder. After extraction, the cylinder slows down to approximately the speed used in washing and the air is circulated by a fan through the heating coils, the garments, and a condenser, and then again through the fan. Some drycleaning units for use with chlorinated hydrocarbons contain separate drying equipment. This resembles the tumbler used with petroleum solvents, except that the vapor-laden air is not exhausted, but is passed through a condenser for recovering the solvent and then through a fan and a heater, and back again through the garments (see Fig. 1).

Studies have been made on removal of Stoddard solvent from textiles in conventional tumblers. It has been found that during the normal drying operation, the concentration of solvent vapors in the exhaust stack increases rapidly at first, but does not quite reach a concentration equivalent to one half of the lower limit of the explosive range, ie, the concentration of vapors at the flash point of the solvent (14). After 4–5 min, the vapor concentration decreases; after 20–25 min, solvent has been completely removed from the fabric. The air temperature in the tumbler is considerably above the flash point of the solvent, but the concentration of vapors is kept low by constant dilution with fresh air. If the garments contain a large amount of solvent when placed in the tumbler, the vapor concentration during the first few minutes of tumbling is considerably higher. If the garments are not extracted but are taken directly from the

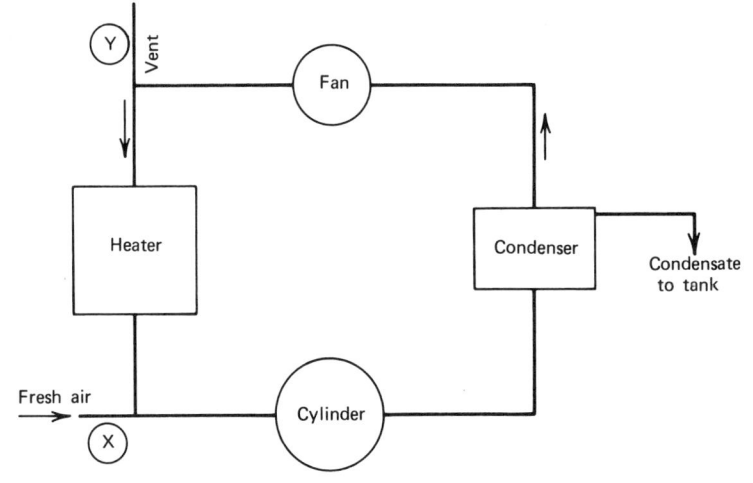

Figure 1. Flow of air in recovery-type drycleaning unit.

drainboard of the washer to the dryer, the vapor concentration soon reaches the explosive limit (see Fig. 2).

The dilution of the hydrocarbon vapors with fresh air is the main safety factor in tumbling, and if the air speed drops there is danger of fire from sparking in the tumbler. Diminished air speed is usually caused by an accumulation of lint in the exhaust stack, or by a clogged air filter. Many tumblers for use with 140-F solvent (see below) are designed to prevent tumbler fires by maintaining the air temperature at 57°C (below the flash point of the solvent) for the first 10 min of tumbling. Thus the most volatile portion of the solvent is vaporized before higher-temperature air is used. There are also tumblers designed to shut down automatically if air speed falls too low.

During drying, garment temperatures are kept below the air temperature by evaporation (15), finally reaching the air temperature when drying is completed. Excessively high temperatures can cause shrinkage in some wool fabrics, fabrics made of heat-sensitive fibers, and fabrics not treated for dimensional stability in textile production.

Moisture content drops rapidly during tumbling. A 20% moisture content in wool falls below 6% within the first 5 min of drying (15). Even 5 min of warm air (37.7°C) at the start of tumbling reduces the moisture content considerably. After tumbling, wool fabrics contain approximately 2% moisture on a dry-weight basis.

Spotting. At one time, spotting was a skilled operation performed in separate departments in most plants. However, this has been greatly reduced by the advent of the charged system.

Special spotting tables are equipped with steam guns, compressed air, vacuum,

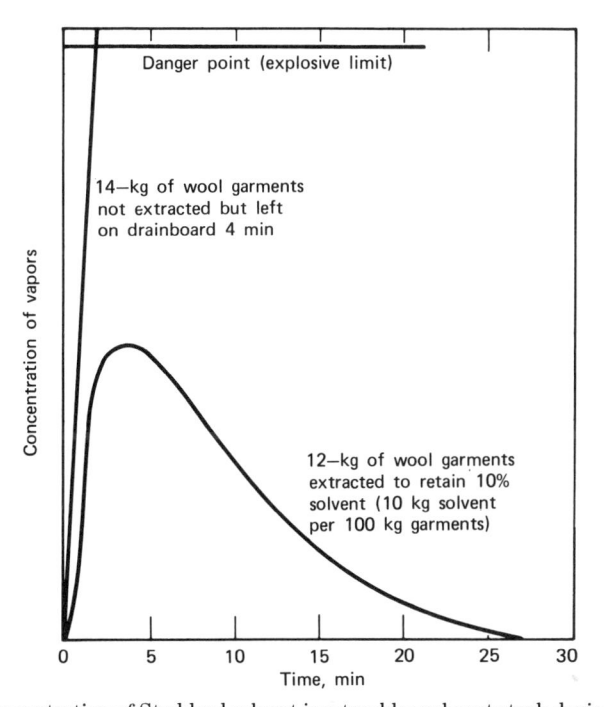

Figure 2. Concentration of Stoddard solvent in a tumbler exhaust stack during drying (18).

and spot-removal chemicals. Numerous books describe these procedures in detail (12,16–19).

Finishing. Finishing is the process of restoring a garment as nearly as possible to its original size, shape, feel, and appearance. The degree to which this restoration is possible is governed by the history of the garment, the quality of the fabric, the dyes, and the quality of the tailoring. There are numerous styles of finishing steam-air forms, but only two basic types of finishing presses. The grid-head press is used on woolens, the hot-head press on silks and silklike fabrics.

The grid-head press has a perforated head, the hot-head press a smooth, nonporous, stainless-steel head. It is heated internally by steam to surface temperatures of 149°C. Both types of presses have perforated bucks, as the bottom platen is called. The garment is placed on the buck and pressure exerted by lowering the head. Steam or vacuum suction can be applied to heat, moisten, or dry the fabric. The proper use of these presses is a highly skilled art, and finishing is the most expensive single operation in drycleaning. A "quality cleaner" is judged more by the finishing than by his cleaning standards. Textile fabrics can be repeatedly ruined in shape and appearance, and yet be completely restored.

The IFI has issued a number of publications on garment finishing (20–22).

Inspection, Sewing, and Assembly. Garments are inspected at each stage of the cleaning process, and finally sent to a seamstress for minor repairs. Then they are bagged or packaged.

Related Processes. In addition to drycleaning and finishing, drycleaners generally offer other services. These include water-repellent treatment for rainwear (see Waterproofing), moth-repellent treatments for wools (see Textiles), leather cleaning, and resizing. Large-volume cleaners may have separate departments for bedspreads, slipcovers, drapes, blankets, etc.

Certain processes are now handled by specialists rather than drycleaners. These include rug, hat, and fur cleaning, and garment dyeing.

Evaluation. The many variables in the drycleaning process make it difficult to evaluate its performance. In part, this difficulty stems from the complex nature of soils and the problem of devising a standard soiled cloth. Many model soils have been proposed or used but there is no general agreement on a standard, and the industry has been forced to use arbitrary test procedures. The seven tests used (23–24) at the IFI are soil redeposition on cotton, rayon and wool; removal of sodium chloride, glucose, and water-soluble food dye from rayon; and removal of an oily carbon black soil from wool. Standard test swatches are available from IFI that can be placed in a cleaning cycle and returned for rating.

Materials

Detergents. Drycleaning solvents remove oily soils easily by simple solution, and hence detergents are not required. However, addition of detergents greatly enhances the removal of the many other soils that are generally present. The solvents dissolve oily soils, releasing the insoluble or pigmentary soils. These insoluble soils can redeposit on the fibers unless a suspending agent is added. The rather high interfacial energy between most fiber surfaces and the drycleaning solvents can be lowered by orienting adsorption of soils at these fiber interfaces. This is probably the mechanism of soil redeposition (25), although mechanisms have been postulated based on zeta potential and other surface electrical considerations (26). Detergents probably inhibit redeposition by being adsorbed on the soil particles to cause a lowering of the interfacial energy between them and the solvents, at the same time increasing the interfacial energy between particle and fiber surface. The fact that all effective drycleaning detergents are surface-active materials is consistent with this interpretation despite the fact that the lowering of surface tension is of no importance in drycleaning. Drycleaning solvents have very low surface tensions.

Water-soluble soils are also removed by detergent action in drycleaning, including some types of hydrophilic food stains that are not soluble in water alone. Water wets and swells the stain, loosening its bond with the fabric so it can be detached by the mechanical action of the washer. A first attempt to remove water-soluble soils by drycleaning was the addition of water-in-oil emulsions to the solvents. Much of the emulsified water was picked up by the garments, and accelerated the removal of water-soluble soil by weakening the soil–fiber bond. Some soil may be retained in the emulsion droplets remaining in the solvent.

The older emulsion systems had serious limitation because they were difficult to filter. They have been superseded by a modern procedure called the *charged system,* eliminating filtering problems. Precharged solvents are available from some producers (see Surfactants and detersive systems).

The Charged System. In this system a detergent is used that forms a stable colloidal solution in the solvent. The term colloidal solution, as used here, means a solution containing dispersed aggregates called micelles (27–32).

Micellar solutions of this kind have unusual properties; the hydrophilic interior of these micelles is capable of dissolving water that can now dissolve water-soluble soils (33), a process called solubilization (27). The key to the success of the charged system is the reduced effective vapor pressure of water dissolved in detergent micelles. This prevents the solubilized water from being completely removed from solution by the garment. Under optimum conditions, the water is distributed between textile and micelles to reach equilibrium. In fact, at equilibrium the vapor pressure of water in the solution is equal to that of the water adsorbed on the garments and in the atmosphere (34–36).

At equilibrium, if an atmosphere of 50% rh exists over a drycleaning solution of the micellar-type containing fabric, this fabric contains the same amount of water as if hanging in the air over the solution (34–35). This solution is said to have an rh of 50%. The term solvent relative humidity (solvent rh) has been used to describe the activity of water in such solutions. By maintaining the solvent rh below 75%, most textiles can be processed without serious wrinkling or shrinkage. Water-soluble or hydrophilic soil is best removed at the highest safe solvent rh, but significant amounts can be removed at lower rh. The IFI set a practical solvent rh of 70% based on their research (35,37–38).

Types of Detergents. Before 1950 only two types of drycleaning detergents were in general use, true soaps and filter soaps. True soaps were colloidal sols or gels composed of soap and fatty acid mixtures. Filter soaps were petroleum and other sulfonates, called filter soaps because they are soluble in the solvent, and consequently pass through the unit's filter (see also Soaps).

Sulfonates of mixed petroleum hydrocarbons, the original charged-system detergents, were used increasingly after the expiration of the Reddish patent in 1950 (39). They have now been replaced by purer synthetic detergents, particularly the amine and sodium salts of alkylarenesulfonic acids. Drycleaning detergents are not formulated products similar to household laundry detergents but principally surfactants or a mixture of surfactants, with concentrations of active ingredients ranging from 40 to 90% (23). Major manufacturers and IFI provide test performance data and special testing information. However, the final determination of the usefulness of a detergent depends on its performance in the drycleaner's own operation. The following detergent classes (1) soap-fatty acid mixtures; (2) "mahogany" or petroleum sulfonates; (3) sulfosuccinic acid salts; (4) alkylarenesulfonic acid salts; (5) amine alkylarenesulfonates; (6) fatty acid esters of sorbitol; (7) ethoxylated alkanolamides; (8) ethoxylated phenols; (9) ethoxylated phosphate esters have been identified in commercial drycleaning detergents by IFI (40–41).

Water Control. Sensitive and reliable control systems are required to maintain the 70% rh in a drycleaning unit. These controls are designed to measure the moisture, inject water into the system, and close the injection valve at optimum rh. Some water also enters the unit with the garments. In temperate climates, garments being cleaned do not contain enough adsorbed water to prevent its transfer from the solvent to the fabric. The controls actuate the water addition system when this condition exists. There is no provision for automatic removal of excessive moisture if the garments begin losing water to the dryer solvent. In some cases the moisture content of the load can be reduced to a safe level if the solvent contains a strong charge of detergent that has a relatively high water-holding capacity. However, it is not possible to remove excessive water from the system automatically.

At excessive water levels (>75% rh), serious wrinkling and shrinkage can occur, especially with rayon and wool. Loosely woven wool fabrics even shrink at 70% rh.

Two types of water control systems are widely used. The direct type uses an electric hygrometer to measure relative humidity in the atmosphere above the solution. Careful engineering is needed to maintain the hygrometer at the temperature of the solvent and protect it against contact with liquid droplets.

The Dunmore hygrometer has two branches of an electrical circuit separated by lithium chloride crystals in a plastic matrix (42). The electrical conductivity of the encapsulated lithium chloride film between the wire is proportional to the rh of the surrounding air. Obviously, this system depends on equilibrium conditions, and there is a lag between atmospheric and solvent rh under normal operating conditions. To overcome this lag and other disadvantages of having the control device remote from the critical point in the process, many attempts have been made to devise sensing elements that can be immersed in the solvent to obtain a direct rh measurement. Units based on dimensional changes of various hydrophilic organic films when exposed to different rh were not very successful, because of hysteresis in the rh–dimensional change curve. Frequent recalibration was necessary because of creep of the elastic films.

An indirect method based on the electrical conductivity of the solution has proved the most successful (43–44). This is an indirect method because the electrical conductivity of hydrocarbon and chlorinated hydrocarbon solutions of detergents varies directly with water concentration rather than the partial pressure or activity of water. The relative humidity of such solutions, at constant temperature, depends on the ratio of water-to-detergent concentration. Detergent concentration must be maintained fairly constant since the observed water concentration is dependent on this water-to-detergent ratio. The control of the detergent concentration is complicated because garments may contain some residual detergent from previous cleanings and losses occur during the cleaning cycle.

Despite the success of the charged system with rh control, in recent years there has been a revival of interest in the batch system of adding water to the detergent solution in a manner that produces a water-in-oil emulsion rather than a colloidal solution. The amount of water added is controlled manually rather than automatically. In the charged system, a compromise must be made between good water-soluble soil removal and wrinkle-free garments. In the modified batch system, water added in emulsion form to the washer is quickly absorbed by the load where it presumably attacks and removes water-soluble soil. Soon, however, the water becomes diluted and evenly dispersed throughout the system. By the time the load is extracted, the actual rh may be down to 50 or 60%, so that wrinkling is avoided (45).

So far, no scientific study of this emulsion system has been published, and the above mechanism represents only the unsupported claims of its proponents. The system is definitely better than the batch system used before 1950 because the emulsion is apparently formed at the point of water injection in the line and can be absorbed by the load as it enters the washer. For this reason there is apparently no problem with filtration. In the older system, a short batch run was necessary to enable the load to pick up the water. Otherwise most of the water and detergent would have been picked up at the filter, and filter pressure problems would have developed quickly.

Many small synthetic solvent drycleaning operations are currently using a low detergent charge with no water addition or rinse cycle. The one-batch systems are less efficient but acceptable.

Solvents. An acceptable drycleaning solvent must be an effective solvent for fats and oils, sufficiently volatile to permit easy drying, easily purified, and of low toxicity. Solvents should not weaken, dissolve, or shrink the ordinary textile fibers or cause bleeding of dyes. They must be noncorrosive to metals commonly used in drycleaning machinery and have a flash point above 37.7°C or be nonflammable. These severe requirements have restricted the commercially successful solvents to a few petroleum fractions (see Solvents, industrial), some chlorinated hydrocarbons, and chlorofluorocarbons. The more important chemicals that have been or are being used are discussed below.

Petroleum Solvents *Stoddard solvent.* It is estimated that about 29% of the volume of drycleaning in the United States is carried out in Stoddard solvent, a petroleum fraction (see Petroleum). The solvent must have the following specifications: It must be clear, water-white, free of rancid and objectionable odors, and give a negative "doctor" test for mercaptans; a copper strip corrosion test for 3 h at 100°C must be negative; maximum absorption by sulfuric acid must be 5% or less; and the minimum flash point has to be 37.7°C. In a distillation test, not less than 50% must be recovered by 177°C, not less than 90% by 190°C, and the end point not exceed 210°C. Residue from the distillation test may not be over a certain amount and should not show an acid reaction to methyl orange (3). Most of the Stoddard solvent now used is of the fast-drying type, similar to regular Stoddard solvent, but with a much narrower boiling range. In a distillation test, the end point of a fast drying solvent is as low as 168°C. Articles drycleaned in a solvent having a low end point dry in about half the time as those cleaned in regular Stoddard solvent (46).

140-F Solvent. A small amount of the drycleaning in the United States is carried out in 140-F solvent, so named because its flash point is 60°C (140°F). This is a petroleum distillate with specifications similar to those of Stoddard solvent, except for the distillation range, which corresponds to the upper half of Stoddard solvent (176–210°C). When used with special equipment, 140-F solvent may be employed in many locations where fire restrictions would prohibit the use of Stoddard solvent. It is more expensive and slower drying than Stoddard solvent.

Halogenated Hydrocarbons. *Chlorinated hydrocarbons.* The amount of drycleaning carried out with chlorinated hydrocarbon solvents in the United States is estimated to be 70% of the business (47). Their main advantage over the petroleum solvents is their nonflammability, permitting their use in many areas where zoning laws prohibit the use of a flammable solvent. This is also an advantage in locations where the use of a petroleum solvent would increase insurance rates (see Chlorocarbons and chlorohydrocarbons).

It is difficult to compare the efficiency of chlorinated hydrocarbon solvents with petroleum solvents, but both types can give excellent results.

Carbon tetrachloride was the first chlorinated hydrocarbon solvent used for drycleaning. It was introduced in Europe because of the high cost of petroleum solvents and later became popular in the United States, being widely used as late as the 1950s. Because of its toxicity and corrosiveness it has been replaced by perchloroethylene.

Trichloroethylene is less toxic and more stable than carbon tetrachloride. Its major disadvantage is that it causes bleeding of many acetate dyes. Therefore, its use in the United States has been limited. Some trichloroethylene is still used in Europe and in a few industrial cleaning plants in the United States.

Perchloroethylene (tetrachloroethylene) is the most widely used drycleaning solvent. It has a relatively high boiling point (121°C) and specific gravity (1.623 at 15.6°C). The vapors are denser than air, facilitating vapor removal by floor-level collection systems. Perchloroethylene is not easily hydrolyzed, and therefore, under normal conditions, contact with moisture does not yield corrosive amounts of hydrochloric acid. Thus, it may be used with moisture-containing detergents. Since it does not evaporate rapidly, hazardous vapor concentrations do not occur with properly maintained equipment. Perchloroethylene has less tendency to bleed dyes than trichloroethylene. For these reasons, it is now used in almost all synthetic solvent plants in the United States.

1,1,1-Trichloroethane is being evaluated in textile and drycleaning applications. It is more easily hydrolyzed than perchloroethylene, but has a lower boiling point (74°C) and greater solvent power. This can be an advantage in some cleaning operations, such as spotting (48). It is being investigated in Europe and the United States. Recently, very encouraging results have been obtained from an epidemiologic study of 151 textile workers, which showed "no related adverse effects on health" from exposure to this solvent (49).

Fluorinated chlorohydrocarbons. 1,1,2-Trichloro-1,2,2-trifluoroethane is one of the newer drycleaning solvents (50) (see Fluorine compounds, organic). It has several advantages, such as greater compatibility with some specialty fabrics, dyes, and garment accessories; shorter drying time because of its very low boiling point (47°C); high stability; and lower toxicity than perchloroethylene. However, it must be used in carefully constructed and maintained equipment to minimize solvent losses. Several companies have made professional and coin-operated units for this solvent. Stain removal is less effective and special spotting agents and techniques are needed for optimum cleaning. Normally, the control of water in cleaning baths is more difficult and the removal of water-soluble stains troublesome (51).

Solvent Purification, Distillation. The principal method of purifying drycleaning solvents in the United States is distillation, which has displaced the caustic clarification process. Distillation under reduced pressure (13.5 Pa or 0.01 mm Hg, with a rotary displacement pump) lowers the distillation temperature to the designed limits of a shell boiler. Petroleum stills operating under atmospheric pressure often yield distillates with undesirable odors. The distillate, therefore, usually passes through a small filter containing damp cotton rags to remove any traces of rancid materials. Vacuum stills are usually designed for automatic solvent feed, with the level regulated by a float valve. Before entering the still pot, the solvent passes through a preheater.

The stills are heated by steam, with heating elements consisting of horizontal tubes through which the steam passes. Some stills contain a heating element of vertical tubes with the steam outside the tubes. A commercial still has been designed which contains a heating element somewhat similar to a long-tube evaporator in which the solvent rises vertically through a long tube. About 3.6 kg of steam are usually required to distill 18.9 L of petroleum solvent (approximately 0.45 kg of steam for 1.8 kg of solvent) (52).

Chlorinated hydrocarbon solvents can be readily distilled at atmospheric pressure; no undesirable materials are formed. The feed for chlorinated hydrocarbon solvent stills is regulated by a float valve or batch-charge operation. The designs of heating elements and condensers used in such stills vary with the different manufacturers of equipment.

In most atmospheric stills, the heat is supplied by steam coils. The vapors rising from the boiling liquid are condensed by the water-cooled tubes of the condenser. The distillate passes through a water separator and then usually through a small rag filter. The perchloroethylene and water separate easily.

Recommended practice in distilling drycleaning solvent is based on the amounts cleaned and generally varies from 41.7 to 125.2 L/100 kg of clothes. The IFI recommends 50 L/100 kg, based upon a compromise of quality and economy. Obviously, quality would be best if all the solvent were distilled after each load, but this would mean the loss of all the detergent in the system, at prohibitive cost.

There has been an unfortunate tendency in recent years for cleaners to omit distillation and depend entirely upon activated charcoal and fatty acid-adsorbing sweetener powders for solvent purification. Lower quality cleaning results from this cost-cutting practice.

Carbon Recovery. Devices, called sniffers, should be an essential part of any drycleaning operation to protect workers from exposure to drycleaning solvents and increase operating life. The exhaust ducts from suction fans on perchloroethylene drycleaning units are connected to the carbon recovery unit. Pick-up ducts at floor level are also connected to this recovery unit. A fan in the carbon recovery unit forces the air and any perchloroethylene vapors through a bed of activated carbon which absorbs these solvent vapors.

Solvent absorbed on the carbon is recovered by passing low-pressure steam through the bed. Steam and solvent vapors are passed through a water-cooled condenser and the condensate collected in a phase separator. The carbon is then dried and ready for reuse. Recovered perchloroethylene is returned to the drycleaning system.

Equipment

Figure 3 depicts a drycleaning unit using a two-bath system with two filters (10). Auxiliary equipment, stills, and carbon recovery units are not shown.

Filters. Pressure filters have an outer shell in which filter tubes, cartridges, or screens are mounted. Tubes or screens of fine mesh, braid, or coil configurations are usually made of Monel metal wire. The screens or tubes hold the filter cake, usually a drycleaning grade of diatomaceous earth (see Diatomite). A precoated filter is somewhat more effective than paper or fiber glass cartridges, but requires precoating, extra maintenance, and subsequent processing of the used filter powder.

Several methods are used to remove the deposit of filter powder and trapped soils. Mechanical scrapers are used on some types, others use a back wash technique. Coil springs and braided tubes may be flexed by mechanical means to shake and flush out the filter aid and soil automatically after each load. Most modern units use self-priming centrifugal pumps to circulate the solvent through the filters as shown for the two-batch system with two filters in Figure 3 (10) (see Filtration).

A muck cooker, which operates like a still but with steam injection, is used to recover perchloroethylene from the used filter powder. It may also be used to recover solvent from the residue from the distillation unit.

Recently disposable filter cartridges of fiber glass or paper have been introduced. They eliminate the need for filter powders yet retain the option of using an activated carbon core for control of solubilized dyes and color (see Carbon).

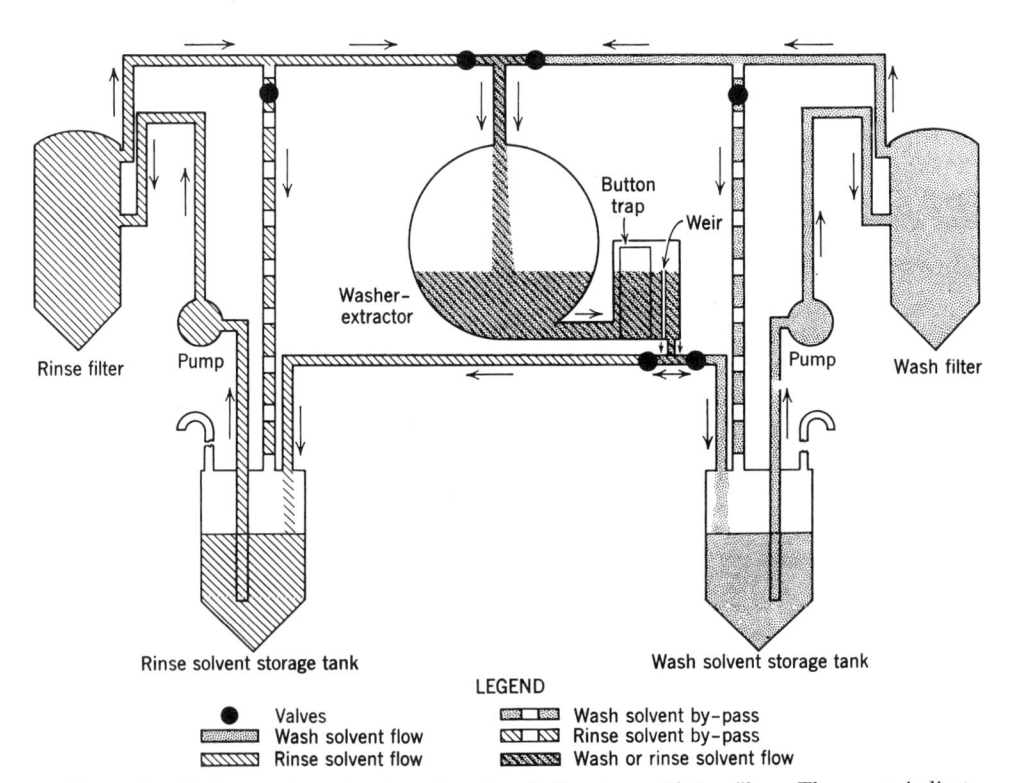

Figure 3. Modern washer-extractor using a two-bath system with two filters. The arrows indicate the direction of solvent flow.

Machines. Drycleaning washers consist of a metal shell and a rotatable perforated inner cylinder or wheel. The shell is a container for solvent, whereas the wheel holds the garment load. The dimensions of the cylinders vary widely, typically from 75 × 75 cm to 132 × 175 cm. As the quality of cleaning can be significantly affected by cylinder diameter, volume, and loading factor, the IFI recommends a load factor of no more than 48 kg work load per m^3 cylinder volume. However, most machine manufacturers use loading factors of 64 kg/m^3 in computing the capacity of their machines.

The washer cylinder or wheel rotates on a horizontal axis during the cleaning cycle to supply the necessary mechanical action. Wheels may be open or divided into two or three compartments. In open types, inside ribs lift and drop the work loads as the wheel rotates. Compartmented wheels separate work loads and isolate garments to minimize tangling and mechanical damage. Most modern washers have automatic controls.

Economic Aspects

Between 1963 and 1978, progress in the manufacture and marketing of synthetic textile fibers and the development of wash-and-wear fabrics was reflected in the drycleaning industry by a decrease in the number of professional retail plants throughout the United States. From a high of 36,500, the number dropped to 25,000. During the same period, the number of coin-op stores peaked, and then leveled off at 15,000, and industrial drycleaning grew to 540 major plants (47).

Industry wide, employment is estimated at 140,000, of which 123,000 are in synthetic solvent operations. Of the latter, approximately 50% are working in plants having five or fewer employees (47).

Annual solvent consumption by professional drycleaning plants is approximately 98 × 10^6 L (24 million gal) of perchloroethylene and 87 × 10^6 L (23 million gal) of petroleum solvents; 7.5 × 10^6 L (2 million gal) of chlorofluorocarbons (53) are used for processing about 589,670 metric tons of fabrics. This represents a 1.773 billion dollar volume of which 70% or 433,180 metric tons are processed in perchloroethylene; 156,352 metric tons in petroleum solvents, and 114 metric tons in chlorofluorocarbons (47).

Solvent loss rates are about 3.4 kg petroleum solvent, 20.7 kg chlorofluorohydrocarbon, and 12 kg perchloroethylene per liter of drycleaning solution with additional recovery of 40% in plants operating activated carbon vapor-recovery units.

Total processing costs have followed inflationary trends, with even greater increases in petroleum and perchloroethylene prices. Under current market conditions, the cost advantage has swung toward the use of synthetic solvents by a 2:1 margin (54).

Coin-op drycleaning (see below) is provided at approximately 15,000 locations equipped with 30,000 perchloroethylene and 930 chlorofluorocarbon units. Nearly 109,770 metric tons or 97% of the total is cleaned in 13.3 × 10^6 L (3.5 million gal) of perchloroethylene, annual receipts are 105.7 million dollars. About 3300 metric tons of chlorofluorocarbon is used, annual receipts are 3.2 million dollars (47). No petroleum solvent is used in coin-op drycleaning (53).

The gross annual sales for industrial drycleaning are not available, but the best estimate by IFI is that approximately 217,700 metric tons are cleaned in about 540 plants, divided evenly between petroleum solvents and perchloroethylene (47).

Health and Safety Factors

Environmental effects of all chemicals, including drycleaning solvents, are receiving more attention since the enactment of the Occupational Safety and Health Act of 1970. Many studies of the long-term (chronic) toxicity of solvents have been performed. Only perchloroethylene, the petroleum solvents, and the chlorofluorocarbon 1,1,2-trichloro-1,2,2-trifluoroethane are discussed since all U.S. drycleaning processes now use one of these solvents. The IFI has been working with various government agencies since 1973 to develop realistic safety standards for the drycleaning industry (47).

Stoddard Solvent. The current TLV for Stoddard solvent is 200 ppm. OSHA lists a time-weighted average (TWA) limit of 500 ppm. A NIOSH criteria document, *Refined Petroleum Solvent*, recommends lowering the exposure level to 85 ppm (55).

Data from human and animal studies indicate exposure to Stoddard solvent vapors can cause eye and throat irritation, dizziness, and slight anesthesia (56). Persons exhibiting these symptoms should be removed from the contaminated area and given medical attention if the symptoms persist.

Stoddard and F-140 solvents are flammable, requiring special handling. The National Fire Protection Association has established standards for drycleaning plants (57).

Solvent-resistant gloves, aprons, and face protection should be worn and good ventilation should be provided.

Perchloroethylene. The current (OSHA) recommended TWA for perchloroethylene is 100 ppm for an 8-h day and a 15-min acceptable ceiling concentration of 200 ppm (58). A NIOSH *Criteria for a Recommended Standard Occupational Exposure to Tetrachloroethylene* (Perchloroethylene) recommends reducing the TWA for a 10-h day to 50 ppm (59); it provides a good summary of biological and environmental studies. NIOSH intelligence bulletin #20 and a recent publication by the Dow Chemical Company provide additional information (60–61).

The major hazard from perchloroethylene is vapor exposure which can cause lightheadedness, dizziness, eye and throat irritation, and mental dullness. Anyone showing these symptoms should be moved from the contaminated area immediately and given medical attention. Anesthetic effects may be observed. Contact with the liquid can cause skin burns, blistering, and erythema (62). Perchloroethylene can be absorbed through the skin (63).

Special safety garments and protective equipment should be provided. Labels containing safety information can be obtained from perchloroethylene manufacturers.

Regulations applying to perchloroethylene are currently being reviewed by the Government's Interagency Regulatory Liaison Group (IRLG) (64). Regulations for drycleaning plants may be changed in the near future.

Special drycleaning equipment that eliminates or significantly reduces operator exposure to solvents was exhibited at the 6th Internationale Austellung von Wäscherei-Chemische Reinigung in Frankfurt, Federal Republic of Germany (65).

1,1,2-Trichloro-1,2,2-trifluoroethane. This chlorofluorocarbon has low toxicity (1000 ppm) and is classified as nonflammable and nonexplosive by the Underwriters' Laboratory. Vapors exposed to an open flame produce toxic decomposition products with a warning irritating odor. Eye and skin irritation can result from contact with the liquid.

Although low in toxicity, its concentration in air cannot exceed 25% by volume without reducing the oxygen concentration to a hazardous level. This solvent has a higher vapor pressure than other drycleaning solvents, making ventilation more important.

Neoprene gloves, splash-proof goggles, and a protective apron should be worn.

The Government's IRLG is currently reviewing information on the nonaerosol use of chlorofluorocarbons for possible regulatory action (64) (see also Aerosols). More precise measurements are being made of the concentration of chlorocarbons and chlorofluorocarbons in the troposphere and lower stratosphere (66–67). The best equipment that is economically tolerable should be used to reduce release of solvents to the atmosphere.

Self-Service Drycleaning

Self-service or coin-op drycleaning was introduced in 1960 by the Whirlpool Corporation in Benton Harbor, Michigan. The public reaction was so enthusiastic that dozens of manufacturers entered the market within a year. This mode of drycleaning grew rapidly in the next three years, reaching an estimated 7300 stores by January, 1963. Despite slower subsequent growth, there were 15,000 stores in 1978 (47). Many coin-op stores employ full-time operators; this provides improved service and protection against equipment abuse and vandalism.

Detailed descriptions of coin-op machines are found in the trade literature. They resemble commercial machines but are much smaller, accommodating loads of 3.6 to 9 kg, instead of over 100 kg.

They may also differ in the filtration system. Some early manufacturers used a regular commercial filter serving all the machines in the store. Others had an individual automatic filter on each unit. Some machines were designed to use both diatomaceous earth filter powder and activated carbon precoats. Most current units are equipped with paper or fiber glass cartridge filters which eliminates the need for filter powders, yet retains the option of using activated carbon core cartridges for control of fugitive dyes and solvent color. Another difference in coin-op machines is that the load is cleaned and dried in the same cylinder, whereas many commercial machines have separate dryers in order to speed production. The coin-op machines generally have refrigeration units to cool the solvent, a necessary item in any machine incorporating a drying cycle. Perhaps the most significant difference is that coin-op systems do not use water and cannot remove water-soluble soils and stains. Some stores provide a water-spray device for prespotting stained areas.

Most coin-op systems use perchloroethylene, although some machines use the more volatile, 1,1,2-trichloro-1,2,2-trifluoroethane. This solvent provides cycle times of 15–20 min compared with 45–50 min for the early perchloroethylene models. Newer perchloroethylene machines have reduced cycle time to approximately 25 min. For this reason interest in the fluorinated solvent has subsided, although it has the additional advantages of being less toxic than perchloroethylene and less harmful to leather and vinyl fabrics. The high cost of the fluorinated solvents and machines that use it are serious handicaps.

Coin-op drycleaners use the same detergents as commercial cleaners, but always as a one-bath (no rinse) operation.

In general, the coin-op machines compare favorably with commercial drycleaning despite the above limitations (68). In general, coin-op machines have excellent filtration and rapid solvent turnover.

Industrial Drycleaning

In the last decade, a new service has become significant in the drycleaning industry, known as industrial drycleaning. With improvement in fabric appearance, color retention, hand, odor, and moisture penetration, customer resistance to fabrics of synthetic fibers has disappeared. Rental and maintenance service of career apparel, industrial garments, and linen supply items became economically feasible. In the most successful processes, high concentrations of water were added to the cleaning solvent or a 100% water cycle incorporated.

In most cases, the machines used are scaled-up modifications of the standard retail unit with faster solvent distillation, incorporation of water cycles, and facilities for separation and recovery of the solvent-water emulsions or suspensions. Some units retain filtration, others rely solely on distillation for solvent maintenance and recovery. Both hot units (drying in washer cylinder) and transfer units (separate tumblers) are used in perchloroethylene operations.

The work loads of these high-volume machines are usually finished in steam-air tunnel units. Garments are hung on hangers, conveyed through saturated steam, and dried by hot compressed air. Durable-press garments thus treated are wrinkle-free and uniform in appearance.

A typical solvent industrial drycleaning cycle, requiring from 30 to 50 min processing time, depending upon type and degree of soil, might be as follows:

(1) Solvent break-run, ie, initial short contact with solvent or water to remove bulk of the soils (used solvent transferred to still).

(2) Solvent drycleaning (used solvent transferred to still).

(3) Water-bath cycle (water to solvent-phase separator).

(4) Solvent rinse (solvent retained for next break-run).

(5) Dry.

(6) Deodorize.

(7) Cool work load.

BIBLIOGRAPHY

"Dry Cleaning" in *ECT* 1st ed., Vol. 4, pp. 214–231, by G. P. Fulton, National Institute of Cleaning and Dyeing; "Dry Cleaning" in *ECT* 2nd ed., Vol. 7, pp. 307–326 by A. R. Martin, and co-workers, National Institute of Drycleaning.

1. E. M. Michelsen, *Remember The Years: 1907–1957, National Institute of Drycleaning,* (NID), National Institute of Drycleaning now International Fabricare Institute (IFI), Silver Spring, Md., 1957.
2. W. Brown, *Modern Dyeing and Cleaning Practice,* Heywood & Co., London, 1937.
3. *Commercial Standard CS3-41,* National Bureau of Standards, U.S. Department of Commerce, Washington, D.C., 1928.
4. P. Garner, *Power Laundry and Cleaning News* (International Issue), 12 (Oct. 13, 1978).
5. C. B. Brown, *Research* **1,** 46 (1947).
6. H. L. Sanders and J. M. Lambert, *J. Am. Oil Chem. Soc.* **27,** 153 (1950).
7. W. C. Powe, *Text. Res. J.* **29,** 879 (1959).
8. W. C. Powe and W. L. Marple, *J. Am. Oil Chem. Soc.* **37,** 136 (1960).
9. K. H. Bey, *Fette Seifen Anstrichm.* **65,** 611 (1963); *Am. Perfum.* **79** (8), 35 40 (1964).
10. E. R. Phillips, *Drycleaning,* National Institute of Drycleaning, 1961.

11. C. B. Randall, *The Drycleaning Department,* 2nd ed., National Institute of Drycleaning, 1940.
12. C. H. Covington, *The Spotting Department,* National Institute of Drycleaning, 1940.
13. *Bulletin F-7,* National Institute of Drycleaning, 1947.
14. G. P. Fulton, *Technical Bulletin T-161,* National Institute of Drycleaning, 1946.
15. G. P. Fulton, *Technical Bulletin T-194,* National Institute of Drycleaning, 1948.
16. A. O. Fligor and P. C. Trimble, *The Spotting Manual of the Drycleaning Industry,* 3rd ed., The National Cleaner and Dyer, R. H. Donnelley Corp., New York, 1948.
17. I. Mellan and E. Mellan, *Removing Spots and Stains,* Chemical Publishing Co., Inc., New York, 1959.
18. J. C. Randlett and W. J. Nicklaw, *Spotting,* National Institute of Drycleaning, 1956.
19. A. J. E. Moss, *Stain Removal: The Technique of Spotting,* Trader Publishing Co., London, 1950.
20. G. P. Fulton, *Technical Bulletin T-167* 1946, *T-185* 1947, National Institute of Drycleaning.
21. *Bulletin F-17,* National Institute of Drycleaning, 1957.
22. J. H. Maher, *Silk Finishing,* National Institute of Drycleaning, 1943.
23. A. R. Martin, A. C. Lloyd, and I. Burkartsmaier, *Technical Bulletin T-407,* National Institute of Drycleaning, 1964.
24. J. R. Wiebush and E. R. Clark, *Technical Bulletin T-376,* National Institute of Drycleaning, 1958.
25. J. Berch, H. Peper, and G. L. Drake, *Text. Res. J.* **34,** 29 (1964).
26. A. R. Martin and G. P. Fulton, *Drycleaning: Technology and Theory,* Textile Book Publishers, Inc., New York, 1958.
27. J. W. McBain, *Adv. Colloid Sci.* **1,** 99 (1942).
28. E. Gonick, *J. Coll. Sci.* **1,** 393 (1946).
29. R. W. Mattoon and M. B. Matthews, *J. Chem. Phys.* **17,** 496 (1949).
30. M. van der Waardn, *J. Colloid Sci.* **5,** 448 (1950).
31. C. R. Singleterry, *J. Am. Oil Chem. Soc.* **32,** 446 (1955).
32. K. Shinoda and co-workers, *Colloidal Surfactants,* Academic Press, Inc., New York, 1963.
33. C. M. Aebi and J. R. Wiebush, *J. Colloid Sci.* **14,** 161 (1959).
34. G. P. Fulton and A. R. Martin, *Technical Bulletin T-268,* National Institute of Drycleaning, 1951.
35. G. P. Fulton and co-workers, *Technical Bulletin T-290,* National Institute of Drycleaning, 1952.
36. G. P. Fulton and co-workers, *American Society Testing Materials Bulletin No. 192,* 1953, pp. 63–68.
37. G. P. Fulton and co-workers, *Technical Bulletin T-309,* National Institute of Drycleaning, 1953.
38. G. P. Fulton and co-workers, *Technical Bulletin T-325,* National Institute of Drycleaning.
39. U.S. Pat. 1,911,289 (May 30, 1933), W. T. Reddish (to Emery Industries, Inc., Cincinnati, Ohio).
40. A. R. Martin, *Technical Bulletin T-403,* National Institute of Drycleaning, 1964.
41. *Technical Bulletin IC-56,* Geigy Chemical Corp., Ardsley, N. Y., 1957.
42. U.S. Pat. 2,285,421 (June 9, 1942), F. W. Dunmore (to the Government of the United States as represented by the Secretary of State).
43. M. B. Matthews and E. Hirschborn, *J. Colloid Sci.* **8,** 86 (1953).
44. *Bulletin F-14,* National Institute of Drycleaning, 1954.
45. *DCRO Monograph No. 6,* Dyers & Cleaners Research Organization, 1962.
46. *Bulletin F-20,* National Institute of Drycleaning, 1960.
47. *Fabricare News,* International Fabricare Institute, (IFI), Vol. 7. No. 11, Nov. 1978.
48. Jpn. Pat. 73 43711 (1973) (to Central Glass Ltd.).
49. *Chem. Eng. News,* 20, (Dec. 11, 1978); *Arch. of Environ. Health,* in press.
50. U.S. Pat, 3,042,479 (July 3, 1962), A. H. Lawrence, Jr. and J. H. Dowling (to E. I. du Pont de Nemours & Co., Inc.).
51. *Drycleaning World,* 52 (Feb. 1972).
52. *Technical Bulletin T-94,* National Institute of Drycleaning, 1940.
53. J. Woolsey, International Fabricare Institute, private communication, Dec. 1978.
54. *Fabricare News,* International Fabricare Institute, June 1978, p. 3.
55. National Institute for Occupational Safety and Health *Criteria Document (NIOSH 77-192),* 1977.
56. C. P. Carpenter and co-workers, *Toxicol. Appl. Pharmacol.* **32,** 282 (1975).
57. *Fire J,* FIJOAO, formerly National Fire Protection Association, NFPA, No. 32, 470 Atlantic Ave., Boston, Mass. 02210, 1974.
58. "Occupational Safety and Health Standards Subpart Z Toxic and Hazardous Substances," *Occupational Safety and Health Reporter,* Vol. 99, Nov. 23, 1978.

59. National Institute for Occupational Safety and Health, *Criteria for a Recommended Standard Occupational Exposure to Tetrachloroethylene (Perchloroethylene) PB 266 583* July, 1976.
60. J. C. Parker and co-workers, *Tetrachloroethylene (Perchloroethylene), Current NIOSH Intelligence Bulletin #20,* March, 1978.
61. V. L. Stevens and R. H. Lalk, *Modern Paint and Coatings* **68**(9), 59 (1978).
62. S. Ling and W. A. Lindsay, *Br. Med. J.* (Corres.) **3,** 115 (1971).
63. R. D. Stewart and H. C. Dodd, *Am. Ind. Hyg. Assoc. J.* **25,** 439 (1964).
64. *Chem. Week,* 24 (Dec. 6, 1978).
65. P. Garner, *Canadian Cleaner and Launderer* **22**(5), 45 (Sept./Oct. 1978).
66. D. R. Cronn and co-workers, *J. Geophys. Res.* **82**(37), 5935 (1977).
67. J. E. Lovelock, *Ecotoxicology and Environmental Safety,* **1,** 399 (1977).
68. *Bulletin F-21,* National Institute of Drycleaning, 1961.

DONALD A. REICH
CHARLES L. CORMANY
PPG Industries

LAUNDERING

Laundering is the process in which soils and stains are removed from textiles in an aqueous medium. Together with drycleaning, which accomplishes a similar result in nonaqueous solvents, commercial laundering constitutes a multi-billion dollar industry. However, the major portion of consumer textiles is laundered in the home.

A very early reference to laundering is found in the book of Malachi (5th century BC). The ancients recognized the efficacy of animal fats mixed with wood ashes in cleansing; this was identified in modern times as the reaction of caustic potash in ashes with triglycerides in fats to form crude soap.

Commercial Laundering

Commercial laundering was started only about 125 years ago. Many changes in techniques have occurred since then. For example, in the 1950s home washing machines had to be adapted to handle nylon, polyester, acrylic fabrics and their blends. Later on the appearance of polyester/cotton blended fabrics changed commercial laundry finishing techniques (extracting, drying, ironing) completely.

Commerical laundry practice differs in many ways from home laundering. For example, in the commercial and industrial laundry plant ion exchange is always used as water softener. Caustic and silicated alkalies are usually employed as detergent builders, and tallow soap is much more widely used than in home laundering.

Classification of articles is generally by soil intensity, according to the following list:

1. light — hotel and hospital linens
2. medium — family work, colored table linens
3. heavy — white table linen, white wearing apparel
4. very heavy — kitchen towels and aprons, grill pads, wet mops, and shop towels

Other criteria also govern sorting in commercial operations: eg, polyester/cotton garments are always washed separately because of the tendency of polyester to pick up loose soil.

Home Laundering

The laundry is first sorted, primarily by color. White fabrics should be washed separately in order to avoid dye pick-up from colors that bleed into the washing solution. This is especially important for nylon and polyester fabrics, which have a strong affinity for dyes and pick up minute amounts of color in mixed loads. Colored articles can be further separated into light and dark colors.

Further sorting by type of fabric and garment construction is advisable. Loosely knit or woven and sheer fabrics require gentle handling, and mechanical action has to be kept to a minimum. Hard-to-remove stains, such as ground-in grease and dirt, body soils at collars and cuffs, rust, ink, or grass have to be removed by special procedures.

Pretreatment is often effective, loosening stains that might be firmly set at the higher temperatures employed in the regular washing cycle (1). Enzyme-containing detergents and presoaks are particularly effective against proteinaceous and albuminous stains (eg, baby formula, chocolate, and blood). They must be used at temperatures below those used in the regular washing cycle for maximum effectiveness. Pretreatment of oily or greasy soils by adding concentrated detergent directly to the dry fabric is advisable.

The regular washing cycle for white and colored fabrics, from heavily soiled work cloths to lightly soiled lingerie, normally employs a heavy-duty detergent. A variety of formulations is available containing one or more surfactants, builders (either phosphate or carbonate) for hardness sequestration or precipitation and reserve alkalinity (2), suds control agents, corrosion inhibitors, anti-redeposition agents, fluorescent whiteners, and perfumes. They may also contain oxygen bleach, enzymes, bluing, and bacteriostatic agents. Heavy-duty liquid detergents, heavy-duty soap products, and light-duty powders and liquids are also available (see Surfactants and detersive systems).

Water softening agents are added in areas where water supplies are unusually hard, in addition to bleaches such as: liquid sodium hypochlorite, dry chlorine bleach, or oxygen bleach (see below). Chlorine bleach recommended for white fabrics normally provides about 200 ppm solution basis (3) (see Bleaching agents).

Fabric softeners impart softness and fluffiness to finished fabrics and reduce static cling, especially in synthetic fabrics. Softeners are now available for use in the washer (usually in the rinse cycle, since most softeners are incompatible with anionic surfactants) or in the dryer, in the form of softener impregnated on nonwoven fibers. Starches impart body to fabrics, especially cottons.

Materials

Textiles. Textiles are either from natural or synthetic fibers (see Fibers, man-made and synthetic; Textiles). More than 90% of the natural fibers encountered in laundering are cotton (qv); the only other natural fiber laundered of any importance is wool. The chemical constitution of these fibers is quite different, and therefore they must be handled differently in laundering.

Hot water strengthens cotton by causing it to swell. Cotton is resistant to alkalies, except that at temperatures close to boiling, cotton loses strength in caustic solutions if air is not excluded. Cotton is resistant to such organic acids as formic acid, but mineral acids attack the fiber even at low temperatures. The action of sodium hypochlorite varies with concentration, temperature, and alkalinity. It is used extensively in cotton bleaching and, properly controlled, is effective in stain removal and sterilization. Excessive hypochlorite, however, rapidly oxidizes the cellulose in cotton resulting in weakening of the fabric and its eventual destruction.

Cotton wrinkles easily unless treated with resin finishes which make it wrinkle resistant. In general resin finishes impart many desirable properties to fabrics, such as crease resistance, wrinkle recovery, soil resistance, flame retardance, and luster (see Flame retardant textiles).

Any tension in cotton is relaxed when the fabric is wet, giving rise to so-called relaxation shrinkage, which is reversible.

Wool generally requires more careful handling. Wool is also subject to reversible relaxation shrinkage, but, through the application of mechanical action and heat, it also may mat or felt, which is irreversible shrinkage. Although wool is generally much more sensitive to alkalies than cotton, developments in wool finish permit laundering of wool without shrinkage.

Nylon is inert to organic acids, but mineral acids attack it. It is very resistant to alkali and bleach. Polyester, the most commonly encountered fiber in wearing apparel, is durable, resilient, hydrophobic, and of high tensile strength. It has outstanding resistance to most chemicals, although it may be weakened by caustic alkalies at high temperatures. Polyester and its blends have good resistance to wrinkling and exhibit excellent crease retention.

Water. Water is the most important material used in laundering, assisted by detergents. Water provides the medium through which mechanical action is transmitted to the fabrics to be cleansed and serves as a wetting agent to penetrate the soil-fiber interface and displace the soil from fibers. It carries chemicals to the soiled area and removes suspended soil.

The quality of the water used for washing is crucial to the quality of the finished product (see Water). Impurities in water that effect the final quality of the product are those which cause hardness and alkalinity. Dissolved iron and chlorine also create problems when present at concentrations exceeding about 0.5 ppm.

Hardness of water is a function of the concentration of dissolved calcium and magnesium salts. Soap cannot be used in hard water because of the formation of insoluble curds of lime and magnesium soap (see Soap):

$$RCOONa + M(HCO_3)_2 \rightarrow (RCOO)_2M \downarrow + 2\,NaHCO_3 \quad (M = Ca \text{ or } Mg)$$

Lime soap (eg, calcium stearate) has no detergent value. It precipitates from solution, occluding soil particles onto fabrics being washed, and causing graying of white fabrics and dulling of colors. In addition, lime soap deposits on machine parts.

Most synthetic detergents do not produce insoluble curds, but their efficiency is reduced by hard water. Commercial laundries as a general practice soften water with a hardness above 35 ppm, using ion-exchange resins. Phosphate-containing detergent formulations soften hard water, with increasing quantities required with increasing water hardness. Water softeners, either in bulk or in individual packages, can also be used directly in the washer with detergents. The softeners may be of the sequestering type, where hardness elements are merely tied up, or of the precipitating type.

In water softening by ion exchange certain minerals or synthetic resins replace the calcium and magnesium ions with sodium ions present in the ion exchanger. The ion-exchange resin can be regenerated, that is, the process is reversed using brine as the sodium ion source. Ion-exchange materials generally are styrene–divinylbenzene copolymers. Their useful softening capacity ranges from 0.04 to 0.06 kg/m^3 resin, depending on the quantity of salt used in regeneration (see Ion exchange; Molecular sieves).

Chemicals

Chemicals used in laundering are classified as follows: alkalies, detergents, bleaches, sours, and finishing specialties.

Alkalies, sometimes referred to as builders, for the most part are alkaline salts formed by the neutralization of a strong alkali with a weak acid (4–5). In solution these compounds produce an abundance of hydroxide ions:

$$Na_2SiO_3 + H_2O \rightarrow 2\,Na^+ + OH^- + HSiO_3^-$$
sodium
metasilicate

Detergency is accomplished most effectively in an alkaline medium, especially when tallow soap is used (6). Soap reverts to the fatty acid from which it was produced with resultant loss in detergent properties unless acting in an alkaline medium.

Detergents. Detergents may be anionic or nonionic and a great variety is available to the commercial launderer. The most widely used in commercial and home laundering are listed in Table 1.

With few exceptions detergents are combined with alkalies for use in the laun-

Table 1. Laundry Detergents

Surfactant	Formulas	R derived from
Anionic		
soap	RCOONa	tallow or tallow fatty acids
alkyl sulfates	ROSO$_3$Na	fatty alcohols (by reduction)
alkylbenzene sulfonates	R–⌬–SO$_3$Na	ethylene, propylene
Nonionic		
ethoxylated alkyl phenols	RO–⌬–(CH$_2$CH$_2$O)$_n$H	ethylene, propylene
ethoxylated alcohols	RO(CH$_2$CH$_2$O)$_n$H	ethylene, propylene

dering process; the combinations are called built detergents, or built soaps. In commercial laundry practice, the builder is not a complex phosphate or carbonate (although these may be present in a heavy-duty product), but usually sodium orthosilicate or metasilicate. Tallow soap, properly built, is used in commercial laundries, that employ ion-exchange water softening.

The synthetic anionics, ethoxylated phenol derivatives, and ethoxylated alcohol derivatives are widely used in both commercial and home laundering as built powders or as unbuilt liquid concentrates.

Alkyl sulfates are not used in laundering, but in specialties such as shampoos (7).

Surface-Active Agents (Surfactants). Surface-active agents or surfactants are substances that concentrate at solution surfaces. Detergents contain surfactants in addition to builders, anti-redeposition agents, corrosion inhibitors, fluorescent whiteners (see Brighteners), and perfume.

In the washing machine, surfactants orient themselves at the solution–air interface, on fiber surfaces, on the interior surfaces of the machine, and around soil particles. In the surfactant molecule a hydrophilic fragment is joined to a hydrophobic fragment. The hydrophilic portion of the molecule seeks and enters the water phase, whereas the hydrophobic portion separates from it.

Detergency can be defined as the removal of foreign substances from a surface. In laundering, the detergent function is one of solubilization, saponification, emulsification, and soil suspension (8–9) (see Surfactants and detersive systems). Particulate soils attached to textile fibers frequently are surrounded by oil or grease that act as a penetration barrier. The surfactant, soluble in both water and oil, lowers the surface and interfacial tension in the washing solution, allowing penetration of the oil barrier; the mechanical action of the process provided by the tumbling or oscillating motion of the washing machine then can free particulates from the fiber. Once freed from the fiber surface, the oily soil particle moves through the detergent solution, governed by gravity and the turbulence of the washing solution. These particles, comprising the oily soil particle associated with surfactant molecules, are exceedingly finely divided and repel each other because of the ionic character of anionic surfactants. Typical ionizing surfactants are shown below:

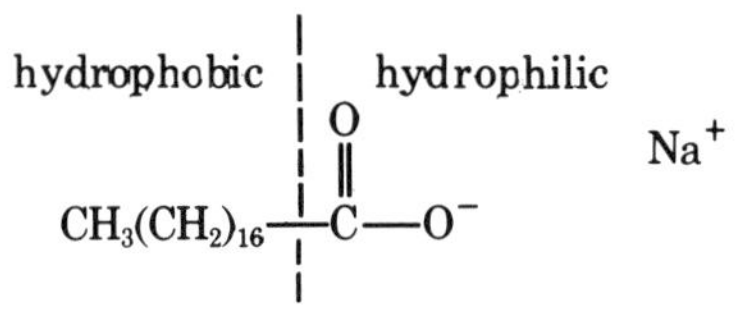

$$\underset{\text{hydrophobic}}{CH_3(CH_2)_{16}} - \underset{\text{hydrophilic}}{\underset{\|}{\overset{O}{C}} - O^-} \quad Na^+$$

sodium stearate

$$\underset{\text{hydrophobic}}{CH_3(CH_2)_{11} - C_6H_4} - \underset{\text{hydrophilic}}{SO_3^-} \quad Na^+$$

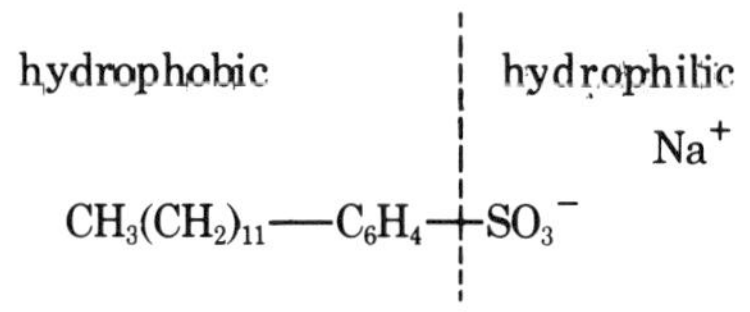

sodium dodecylbenzene sulfonate

Soil particles surrounded by anionic surfactant molecules thus repel each other, thereby holding soil particulates in suspension in the washing solution. Nonionic surfactants do not ionize in solution and hence do not exhibit this action. Generally, they are less effective than anionics in suspending soil.

Bleaches. Bleaches decolorize many stains that resist hot alkaline soap, and function as antibacterial agents. They may also assist in the maintenance of fabric whiteness (see Bleaching agents).

The following bleaching agents are currently used in commercial and home laundering:

(1) Sodium hypochlorite solution (NaOCl) is supplied for commercial use as a 10 to 15% solution and is available to consumers as a 5.25 wt % solution (see Chlorine oxygen acids).

(2) Dry bleach is available for commercial and home laundering as a sodium dichloroisocyanurate formulation, which decomposes in solution to release sodium hypochlorite (see Cyanuric and isocyanuric acids).

(3) Hydrogen peroxide is used to some extent in commercial laundering and by dry cleaners on fine fabrics and colored fabrics requiring bleaching (see Hydrogen peroxide).

(4) Sodium perborate is used for colored and fine fabrics in home laundering. In solution it releases hydrogen peroxide which is the bleaching agent (see Peroxides).

Properly balanced washing formulas are the key to whiteness retention; bleach has no role in particulate or oily soil removal. Laboratory investigations have identified the following critical factors (10):

(1) A bleach bath concentration not exceeding 100 ppm available chlorine.

(2) Temperature between 60 and 70°C.

(3) pH between 10.2 and 10.8.

In home laundering, bleaching normally occurs at temperatures of 50°C or lower and at a pH of 10.2 or lower. The bleaching is quite mild under these circumstances, at least insofar as white cotton-blended fabrics are concerned. In commercial practice antichlors are used in the final rinsing, removing residual chlorine.

Finishing Specialties. *Sours* are mild acids or acid salts used to neutralize residual alkali, which if left in fabric can cause yellowing, dulling, odors, and skin irritation. Sours may be formic or acetic acid, or sodium or ammonium silicofluoride or bifluoride.

Fabric softeners are usually quaternary ammonium compounds, for example:

$$\left[\begin{array}{c} CH_3(CH_2)_{17} \\ CH_3(CH_2)_{17} \end{array} N \begin{array}{c} CH_3 \\ CH_3 \end{array} \right]^+ Cl^-$$

Since cotton is negatively charged when wet, cationic fabric softeners are substantive. Nonionic formulations may also be used. Softeners provide softness and fluffiness in addition to static control (see also Textiles).

Bacteriostatic agents reduce ammonia formation in diapers and control mildew and mold. Compounds used as bacteriostatic agents include heavy metal compounds (organic tin salts), carbanilides and salicylanilides, chlorinated phenolics, iodophors, and quaternary ammonium compounds, among others. Quaternary ammonium bacteriostatic agents are similar to quaternary ammonium softeners, but contain only one fat-derived group attached to the nitrogen atom instead of two as in the fabric softeners.

Table 2. Typical Washing Formula

Operation	Temperature, C°	Time, min	pH	Ingredient	Analysis
break	60–71	6–8	11.0–12.0	alkali soap/detergent	600 to 900 ppm as Na_2O
carryover	60–71	4–6		soap/detergent	
bleach	60–71	6–8	10.2–10.8	bleach	
rinse	49–60	1–2			
rinse	40–46	1–2			
rinse	40–46	1–2			
finish	40–46	4–6	5.0–6.0	sour[a]	50 to 120 ppm as HCO_3^-

[a] May include softener and bacteriostat.

Optical brighteners are colorless dyes that absorb ultraviolet radiation and emit bluish light. They are used in the final rinse to impart greater apparent whiteness to laundered textiles.

Starch and bluing may also be used for special finishing effects. Not all articles are subjected to the same washing procedure. Hotel and motel sheets, eg, require less severe treatment than restaurant linens or garage uniforms and wipers. Basic washing procedures called formulas have been developed for each soil classification. A typical washing formula for medium soil used for family laundry is given in Table 2.

BIBLIOGRAPHY

"Laundering" in *ECT* 1st ed., Vol. 8, pp. 208–216, by Lee G. Johnston, American Institute of Laundering; "Laundering" in *ECT* 2nd ed., Vol. 12, pp. 197–207, by Lee G. Johnston, American Institute of Laundering.

1. *Learning About Laundering*, Proctor and Gamble, 1977, p. 19.
2. *Detergents in Depth*, Soap and Detergent Association, 1974, p. 13.
3. Ibid., 1976, p. 23.
4. P. B. Mack and J. Sherrill, *Laundering Chemistry*, Linen Supply Association of America, Hallandale, Fla., 1952, p. 17.
5. H. Cohen and G. E. Linton, *Laundry Chemistry*, Interscience Publishers, Inc., New York, 1961, p. 12.
6. C. L. Riggs and J. C. Sherrill, *Textile Laundering Technology*, Linen Supply Association of America, Hallandale, Fla., 1979, p. 44.
7. A. M. Schwartz and J. W. Perry, *Surface Active Agents*, Interscience Publishers, Inc., New York, 1949, p. 445.
8. J. W. McBain, *Colloid Science*, D. C. Heath and Company, Boston, Mass., 1950, p. 271.
9. W. W. Niven, Jr., *Industrial Detergency*, Reinhold Publishing Corp., New York, 1950, p. 5.
10. R. B. Smith, *Washroom Methods and Practice in the Power Laundry*, Laundry Age Publishing Company, New York, 1939, p. 82

JOSEPH C. SHERRILL
Sherrill Associates

DRYING

Drying is an operation in which a volatile liquid is separated from a solid or semisolid material by vaporization. In dehydration, vegetable and animal products are dried to less than their natural moisture contents, or water of crystallization is removed from chemical compounds. In freeze drying, the wet material is cooled to freeze the liquid; vaporization then occurs by sublimation of ice. Evaporation differs from drying in that feed and product are both pumpable fluids; different equipment is employed (see Evaporation). Gas drying means the separation of a condensable vapor from noncondensable gases by adsorption on solid surfaces or absorption in a solution (see Adsorptive separation; Absorption; Drying agents).

Reasons for drying include customer convenience or preference, reduction of shipping costs, maintenance of product stability, and removal of toxic or noxious liquids. Waste recycling and disposal are growing applications. Most manufacturing operations that produce solid products include drying steps; eg, pigments, paper, polymers, ceramics, leather, wood, and foods. Drying operations involving flammable and toxic liquids are becoming common. For these applications, gas-tight, inert-gas recirculating equipment is manufactured with integral dust collectors, vapor condensers, and recycle-gas heaters.

When a material dries, heat is transferred to evaporate the liquid, and mass is transferred in the form of liquid and vapor. In most dryers, heat is transferred first to the surface of the material and then into the interior. The usual heat transfer mechanisms are: (1) convection from a hot gas that is brought in contact with the material; (2) conduction from a hot surface that contacts the material; and (3) radiation from a hot gas or hot surface that either contacts or is in close proximity to the material (see Heat transfer technology). In dielectric heating, a fourth but less common method, the energy source is an alternating electric field that generates heat inside the material by molecular friction. High internal vapor pressures may develop, and the temperature inside the material may be greater than at the surface (see Microwave technology). Many dryers employ more than one of these four mechanisms. However, any industrial dryer can usually be characterized by one predominate heat transfer mechanism. Mass transfer (qv) during drying involves the removal of vapor from the material surface and the movement of internal moisture to the surface. Material structure and the mechanisms of internal liquid and vapor flow usually control the drying rate when internal moisture is removed during drying.

Because all drying operations require the handling of solids, the capability of the equipment to handle solids is of greatest importance. The material must reliably enter one end, leave the other, and not linger too long in passage. If possible, drying should be preceded by a mechanical separation operation to minimize the liquid quantity that has to be vaporized (see Crystallization; Filtration; Centrifugal separation; Sedimentation). Liquid separation from a solid without vaporization is always less expensive than drying. Evaporators with low equipment and energy costs are also used to reduce dryer vaporization.

Drying costs are determined by labor costs, equipment, materials of construction, fuel, and plant size. Generally, the product form desired, the chemical and physical properties of the liquid and solid, and process options dictate drying method and equipment, not cost. However, continuous dryers are less expensive than batch dryers.

Drying costs rise rapidly as plant size decreases below about 500 metric tons per year. Atmospheric-pressure batch dryers are about one third as costly as vacuum batch dryers; freeze drying of a liquid is twice as costly as vacuum drying. Once-through air dryers are about one half as costly as recirculating inert-gas dryers. Dielectric dryers are the most expensive in investment and operating costs. The cost difference between direct- and indirect-heat dryers is minimal. The cost of drying particulate solids at 1000–50,000 t/yr rates is about the same for rotary, fluid-bed, or pneumatic-conveyor dryers. However, there are only few applications for which all three are equally suited. An analysis of capital and operating costs of drying equipment used in the food industry is presented in ref. 1.

Terminology

Bound moisture is liquid held by a material that exerts a vapor pressure less than that of the pure liquid at the same temperature. Liquid may be bound by solution in cell or fiber walls, homogeneous solution throughout the material, and by chemical or physical adsorption on solid surfaces. Bound moisture can be removed only under specific conditions of humidity and temperature in the external surroundings.

Capillary flow is the flow of liquid through the interstices and over the surfaces of a solid, caused by liquid–solid molecular attraction.

Constant rate period is the drying period during which the rate of liquid removal per unit of drying surface is constant.

Critical moisture content is that obtained when the constant rate period ends and the falling rate periods begin.

Direct-heat dryer is one of a class of drying equipment in which heat is transferred to the material being dried by direct contact with the heating medium. The latter is usually a hot gas and the heat transfer mechanism is convection.

Dry basis indicates the moisture content of a wet material as the weight of moisture per unit weight of dry material. The advantage of using this basis is that the moisture change per unit weight of dry material is obtained simply by subtracting the moisture content before and after drying.

Dryer efficiency is the fraction of the total energy supplied used to heat and evaporate the liquid.

Equilibrium moisture content is that to which a given material can be dried under specific conditions of gas temperature and humidity.

Evaporative efficiency in a direct-heat dryer compares the amount of evaporation actually obtained with that which would obtain if the drying gas were saturated adiabatically.

Falling rate period is a drying period during which the instantaneous drying rate per unit surface or weight of dry material continually decreases.

Fiber saturation point is the bound moisture content of cellular materials (wood, etc), at which the cell walls are completely saturated, whereas the cavities are liquid-free. It is the equilibrium moisture content occurring when the humidity of the surrounding atmosphere approaches saturation.

Free moisture content is the liquid content that is removable at a given temperature and humidity. Free moisture may include both bound and unbound moisture, and is equal to the total average moisture content minus the equilibrium moisture content for the prevailing conditions of drying.

Funicular state is the condition that occurs while drying a porous body when capillary action causes air to be sucked into the pores.

Humidity denotes the amount of vapor actually present in a gas and is generally expressed as weight of vapor per unit weight of dry gas.

Indirect-heat dryer is one of a class of drying equipment in which heat is transferred primarily by conduction and the heating medium is physically separated from the material by a wall.

Internal diffusion occurs in a material during drying when liquid or vapor flow obeys the fundamental laws of diffusion.

Moisture gradient refers to the moisture profile in a material at a given moment in a drying process. The nature of the moisture gradient depends on the mechanism of moisture flow inside the material.

Pendular state is the state of liquid in a porous body when a continuous film of liquid no longer exists around and between discrete particles, so that flow by capillarity cannot occur. This state succeeds the funicular state.

Percent saturation is the ratio of the partial pressure of a condensable vapor in a gas to the vapor pressure of the liquid at the same temperature, expressed as a percentage. For water vapor in air, this is percent relative humidity.

Unaccomplished moisture change refers to the ratio of the free moisture present at any time to that initially present.

Unbound moisture in a hygroscopic material is the moisture in excess of the equilibrium moisture content corresponding to saturation humidity in the surrounding atmosphere. All water in a nonhygroscopic material is unbound moisture.

Wet basis expresses the moisture in a material as a percentage of the weight of wet material. This basis is less satisfactory for drying calculations than the dry basis for which the percentage change of moisture per unit weight of dry material is constant for all moisture contents. Figure 1 shows the relationship between dry and wet basis, and indicates that when the wet basis is used to express moisture content a 2 or 3% change at high moisture contents (above 70%) actually represents a 15–30% change in evaporative load per unit weight of dry material.

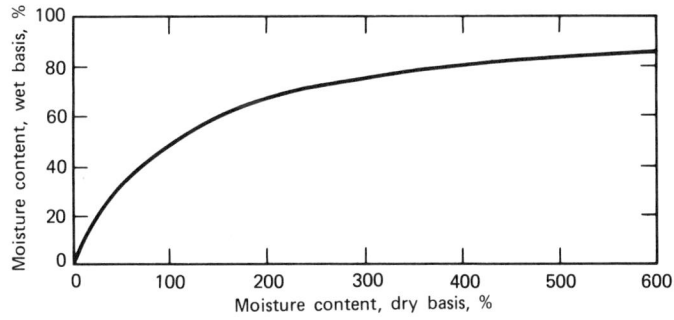

Figure 1. Relationship between dry- and wet-weight basis. Dry basis = kg liquid/kg dry material. Wet basis = kg liquid/kg (dry material + liquid).

78   DRYING

Psychrometry

Before drying can begin, a wet material must be heated to a temperature such that the vapor pressure of the liquid content exceeds the partial pressure of vapor already present in the surrounding atmosphere. The effect of a dryer's atmospheric vapor content on drying rate and material temperature can conveniently be studied by construction of a psychrometric chart for the vaporizing liquid such as shown in Figure 2 (see also Air conditioning).

The wet-bulb or saturation-temperature curve indicates the maximum weight of vapor that can be carried by a unit weight of dry air. For any air temperature on the abscissa, saturation humidity can be found by reading up to the saturation temperature curve, then to the ordinate, ie, kg moisture content/kg dry air. At saturation humidity, the dry-bulb and wet-bulb temperatures are identical and the partial pressure of vapor in air equals the vapor pressure of water at the same temperature:

$$H_w = \frac{p_w}{(P - p_w)} \cdot \frac{18}{28.9} \tag{1}$$

where H_w = saturation humidity, kg moisture/kg air; P = total system pressure, kPa (kPa × 7.5 = mm Hg); p_w = water vapor pressure at the gas temperature, kPa; and 18/28.9 = molecular weight ratio of water, 18, and air, 28.9. The humidity at any condition less than saturation is expressed similarly:

$$H = \frac{p}{(P - p)} \cdot \frac{18}{28.9} \tag{2}$$

where p = partial pressure of the vapor in air, kPa; and H = humidity, kg moisture/kg air.

The percent relative humidity curves indicate percent saturation and are related to vapor pressure:

$$\% \, (\text{rh}) = 100 \, \frac{p}{p_w} \tag{3}$$

where % (rh) = percent relative humidity.

The lines designated as volume in m³/kg dry air indicate humid volume, which is equal to the specific volume of dry air plus the volume of vapor it contains.

Enthalpy at saturation data are precise only at the saturation temperature and humidity. For air–water vapor mixtures, the diagonal wet-bulb temperature lines are approximately the same as constant–enthalpy, adiabatic cooling lines. The latter are based on the relationship:

$$(H_s - H) = \frac{C_s}{L_s} (t - t_s) \tag{4}$$

where H_s and t_s = adiabatic saturation humidity, kg/kg, and temperature, K, respectively, corresponding to the air conditions represented by H and t; C_s = humid heat for humidity H, kJ/(kg·K); L_s = latent heat of vaporization at t_s, kJ/kg (J/4.184 = cal). The slope of a constant–enthalpy, adiabatic cooling line is C_s/L_s, which is the relationship between the temperature and humidity of air passing through a totally adiabatic dryer.

The humid heat of an air–water vapor mixture per unit weight of dry air is determined by the relationship:

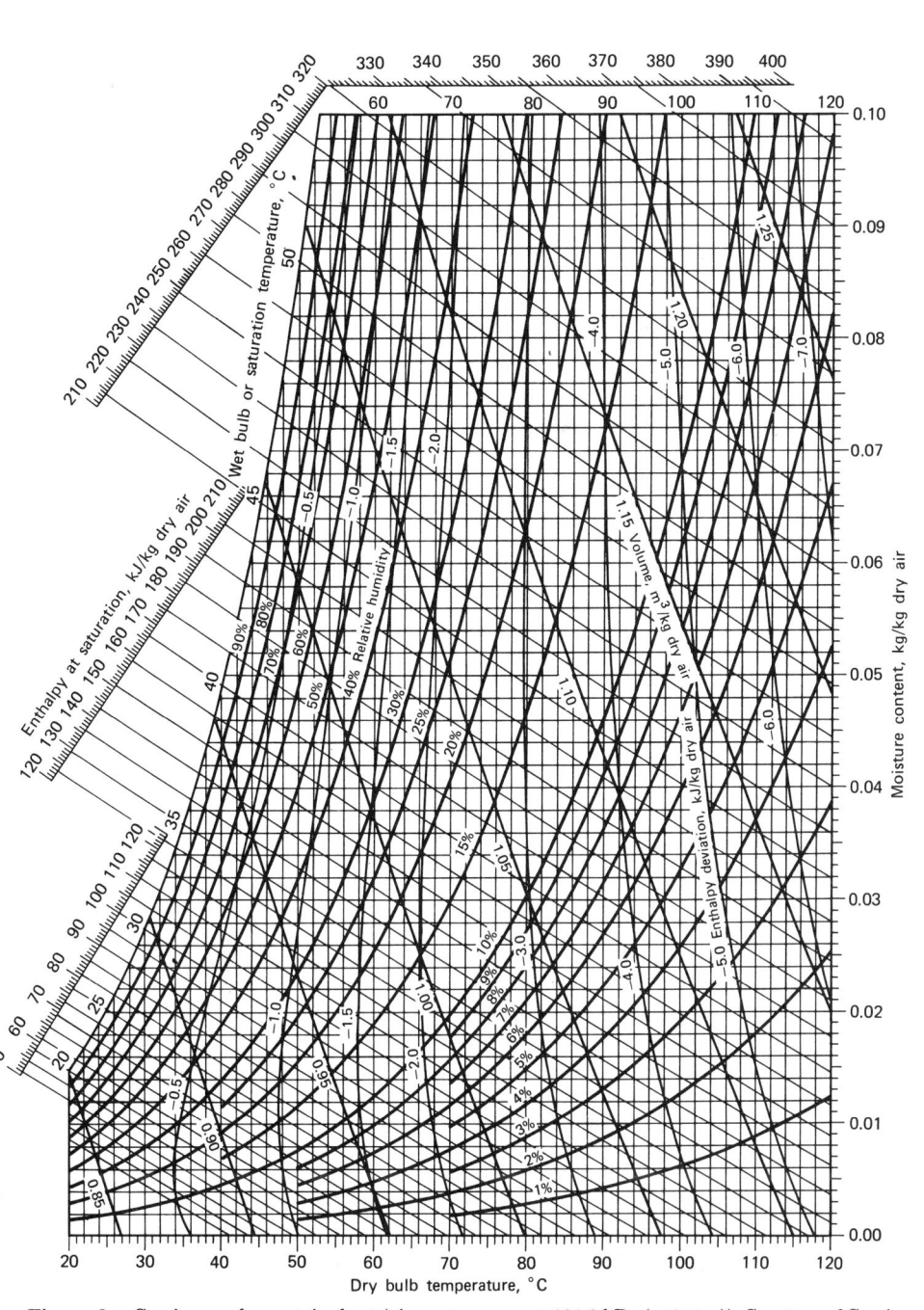

Figure 2. Carrier psychrometric chart (air–water vapor at 101.3 kPa (= 1 atm)). Courtesy of Carrier Corporation (2). To convert J to cal, divide by 4.184.

$$C_s = 1.0 + 1.87 H \tag{5}$$

where 1.0 = specific heat of dry air, kJ/(kg·K); 1.87 = specific heat of water vapor, kJ/(kg·K).

The wet-bulb temperature of a gas is established by a steady-state relationship between heat transfer and mass transfer when a liquid evaporates from a small mass, eg, the wet bulb of a thermometer, into a sufficiently large mass of gas so that the latter undergoes no temperature or humidity change. Steady-state conditions are expressed in simplified form by the relationship:

$$k'L_w(H_w - H) = h_c(t - t_w) \tag{6}$$

where k' = mass transfer coefficient, kg/(s·m²·kg/kg); h_c = heat transfer by convection, kW/(m²·K); t = gas dry-bulb temperature, K; t_w = gas wet-bulb temperature, K; H = humidity at t, kg/kg; and H_w = saturation humidity at t_w, kg/kg. For air–water vapor mixtures, it happens that $h_c/k' = C_s$, approximately; therefore, as the ratio $(H_w - H)/(t - t_w) = h_c/k'L_w$, the slope of the wet-bulb temperature line in equation 6, it is also approximately equal to C_s/L_s, the slope of the constant–enthalpy, adiabatic cooling line in equation 4, and t_w is approximately the same as t_s. *Enthalpy deviation curves* in Figure 2 permit enthalpy corrections for humidities less than saturation and indicate the extent to which the wet-bulb temperature lines do not precisely coincide with constant–enthalpy, adiabatic cooling lines. For thorough treatment of the principles of wet-bulb thermometry, see refs. 3–4.

If a system pressure is different from 101.3 kPa (1 atm), the humidity at measured wet-bulb and dry-bulb temperatures can be determined by the following relationship:

$$H = H_o + p_w\left[\frac{1}{(P - p_w)} - \frac{1}{(101.3 - p_w)}\right] \cdot \frac{18}{28.9} \tag{7}$$

where H = humidity of air at pressure P, kg/kg; H_o = humidity of air as read from a humidity chart based on 101.3 kPa total pressure and at the measured dry- and wet-bulb temperatures, kg/kg; P = total pressure at which the dry- and wet-bulb temperatures were taken, kPa; p_w = vapor pressure of water at the observed wet-bulb temperature, kPa. Similar relationships can be derived to correct specific volume, saturation humidity, and relative humidity.

To illustrate changes that occur in temperatures and humidities as air passes through a direct-heat dryer, several profiles are shown on Figure 3. Line AB is a wet-bulb temperature line. Air that enters the dryer at H_1 and t_1 cools and humidifies along line a. If the air leaves at t_2, its humidity is H_2; the wet-bulb and adiabatic saturation temperatures throughout are approximately identical and equal to t_w, and the maximum humidity change, were the air cooled to saturation, could be $(H_w - H_1)$, kg/kg, at t_w. The percentage ratio of $(H_2 - H_1)/(H_w - H_1)$ is the evaporative efficiency of a direct-heat dryer. If there is a heat loss through the dryer enclosure, the air temperature change is not accounted for completely by a humidity gain; the outlet humidity is less than the adiabatic cooling humidity at t_2, whereas the air profile follows the path of line b or c, depending on the nature of the loss. In dryers that employ internal, steam-coil reheaters, drying air passes several times over the material and through the heaters before it leaves the dryer; line d shows a characteristic temperature–humidity profile.

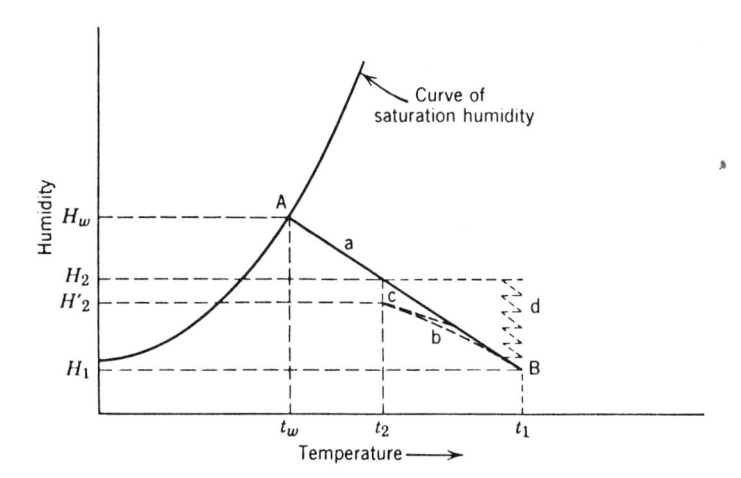

Figure 3. Humidity chart to illustrate changes occurring in air temperature and humidity conditions in direct-heat dryers.

Adiabatic cooling lines for organic solvents differ significantly from wet-bulb temperature lines because the values of $h_c/k'C_s$ do not approximately equal one. The $h_c/k'C_s$ values for carbon tetrachloride, benzene, and toluene are 0.51, 0.54, and 0.47, respectively. Psychrometric charts for these solvents have been published (5–6).

Drying Mechanisms

Drying Periods. Although it is sometimes possible to select a suitable drying method simply by evaluating variables such as humidities and temperatures, the goal of many operations is not only to separate a volatile liquid, but also to produce a dry solid of specific size, shape, porosity, density, texture, color, or flavor; an understanding of liquid and vapor mass transfer mechanisms is essential for quality control. Mass transfer mechanisms are best identified by measuring drying time and drying-rate behavior under controlled conditions. No two materials behave alike and a change in material handling method or any operating variable, such as heating rate, may affect mass transfer. For example, sand in a static layer on a moving belt does not exhibit the same drying profile as sand in a rotary dryer.

Figure 4 shows drying time profiles for three materials dried in layers. The drying rate profiles derived from Figure 4 are shown in Figure 5. These demonstrate that several distinct periods may occur during drying: (1) an initial induction period, not shown, during which the wet material is heated to drying temperature; (2) a period of constant rate drying, shown by horizontal portions of the rate profiles in Figure 5, during which the drying rate per unit area is constant; (3) a period of decreasing rate, shown by the sloping portions of two rate profiles, during which the drying rate appears proportional to moisture content; and (4) a period of decreasing rate, shown by the curved portions of two rate profiles, during which the drying rate is evidently a more complex function of moisture content than simple proportionality.

The moisture content obtained at the end of the constant rate drying period is designated the critical moisture content. Drying periods following the constant rate period are the falling rate periods. The curved portion of profile B (Fig. 5) is a second

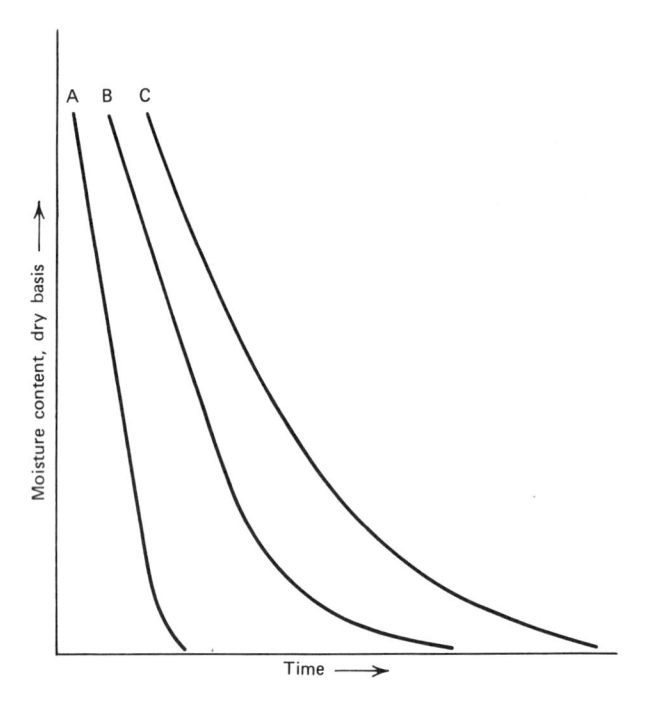

Figure 4. Drying time profiles for various materials. A, granular material during constant rate drying; B, porous material during constant rate drying followed by periods of capillarity and vapor diffusion; C, soluble material during falling rate drying and diffusion.

falling rate period; the moisture content obtained at the second break in profile B is the second critical moisture content. Profile C shows that a drying operation may occur almost entirely in a falling rate period and a slight change in required final moisture content may have a major effect on drying time. These profiles could also represent individual solid particles and either hygroscopic or nonhygroscopic materials.

Constant Rate Drying. During the constant rate period, vaporization occurs as from a free liquid surface of constant composition and vapor pressure; material structure has no influence. In an atmospheric-pressure dryer, drying proceeds by diffusion of vapor from the wet surface through a gas film into the environment. In a vacuum or superheated vapor dryer in which there is only condensable carrier gas, evolved vapor passes into an identical vapor atmosphere. Moisture movement from within the material is fast enough to maintain a completely wet surface. The vaporization rate is controlled by the heat transfer rate to the evaporating surface; the mass transfer rate adjusts to the heat transfer rate, and the liquid surface reaches a steady-state temperature. The drying rate is, therefore, constant as long as the heat transfer rate is held constant by external conditions. If heat is transferred solely by convection from a gas, the steady-state surface temperature is the gas wet-bulb temperature. When conduction or radiation contribute to heat transfer, eg, the material rests on a warm surface or receives radiation from a warm enclosure, a surface temperature somewhere between the wet-bulb temperature and the liquid's boiling point is obtained. In indirect-heat and radiant-heat dryers where there are high rates of conduction and radiation heat transfer, the surface liquid may boil as in a pan of water

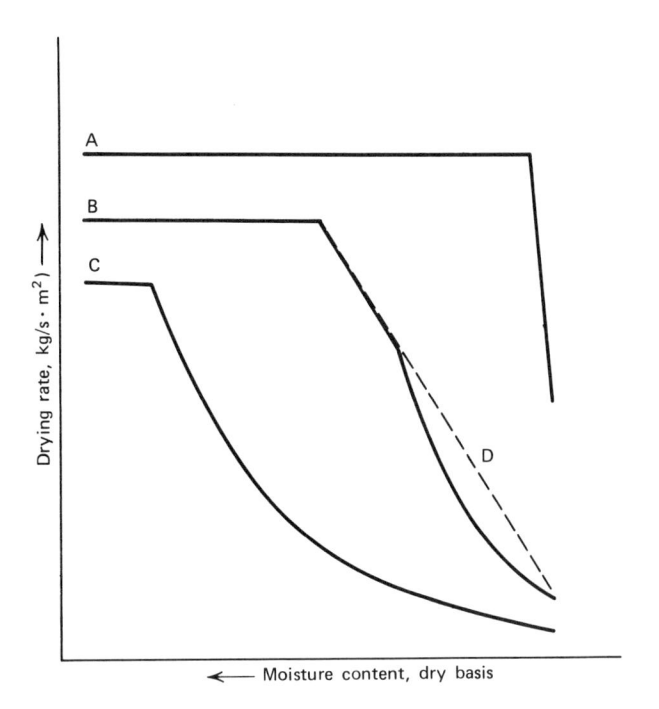

Figure 5. Drying rate profiles for various materials. A, granular material during constant rate drying; B, porous material during constant rate drying followed by periods of capillarity and vapor diffusion; C, soluble material during falling rate drying and diffusion; D, assumed profile for falling rate drying of a porous material.

on a stove regardless of humidity or temperature in the environment. During constant rate drying, therefore, a material temperature can more easily be controlled in a direct-heat dryer than in an indirect- or radiant-heat dryer because in a direct-heat dryer the material temperature does not exceed the gas wet-bulb temperature as long as all surfaces are wet.

All principles relating to simultaneous heat and mass transfer between gases and liquids apply to constant rate drying. The steady-state relationship obtained between heat transfer and mass transfer at the liquid surface can be represented in the following manner:

$$-\frac{dw}{d\theta} = \frac{h_t A}{L'_s}(t - t'_s) = k'_a A(p'_s - p) \tag{8}$$

where $dw/d\theta$ = drying rate, kg/s; h_t = the sum of all convection, conduction, and radiation components of heat transfer, kW/(m²·K); A = surface area for vaporization and heat transfer, m²; L'_s = latent heat of vaporization at t'_s, kJ/kg; k'_a = mass transfer coefficient, kg/(s·m²·kPa); t = average source temperature for all components of heat transfer, K; t'_s = liquid surface temperature, K; p'_s = liquid vapor pressure at t'_s, kPa; p = partial pressure of vapor in the gas environment, kPa. It is sometimes useful to express this relationship in terms of moisture change rather than the liquid quantity evaporated. For vaporization from a layer of wet material:

$$-\frac{dw'}{d\theta} = \frac{h_t}{p_m d_m L'_s}(t - t'_s) \tag{9}$$

where $dw'/d\theta$ = drying rate, kg liquid/(s·kg dry material); p_m = dry material bulk density, kg/m³; d_m = layer thickness, m. A similar equation can be written for through circulation drying in which gas flows through a deep bed of particles:

$$-\frac{dw'}{d\theta} = \frac{h_t a}{p_m L'_s}(t - t'_s) \qquad (10)$$

where a = heat transfer area per unit volume of bed, 1/m. As a practical matter, p_m can be measured, but the heat transfer area in a particle bed is usually difficult to estimate except by actual drying tests from which a value for the quantity $h_t a$ is calculated by inserting temperature and drying rate data in equation 10. For a few uniform shapes, for which bed void fraction can be estimated from particle specific gravity and bulk density, the effective drying surface may be approximated as follows: for spherical particles,

$$a = \frac{6(1-F)}{(D_p)_m} \qquad (11)$$

and for uniform cylinders,

$$a = \frac{4(0.5 D_c + Z)(1-F)}{D_c Z} \qquad (12)$$

where F = void fraction; $(D_p)_m$ = mean diameter of spherical particles, m; D_c = cylinder diameter, m; Z = cylinder height, m. An adaptation of equation 10 may be used to describe any system in which drying gas flows among a mass of particles, such as direct-heat rotary, fluid-bed, pneumatic-conveyor, and spray dryers. In these dispersed-particle systems, where the particle concentration per unit volume of dryer is constantly changing and varies from point to point, no effort is made to determine particle heat transfer area per unit volume. Tests are conducted in the prototype of a commercial dryer and scale-up is based on average values. The designer's concern is that the quality of particle dispersion and surface exposure in the scaled-up equipment duplicates the prototype or that a proper allowance is made for differences and an average, particle drying surface per unit volume of dryer is assumed.

$$U_a = h_t a \qquad (13)$$

where U_a = the average, volumetric heat transfer coefficient, kW/(m³ dryer volume·K). For constant rate drying in all dispersed-particle dryers, a general relationship is:

$$-\frac{dw}{d\theta} = \frac{U_a V \Delta t_m}{L'_s} \qquad (14)$$

where V = dryer volume, m³; Δt_m = log-mean temperature difference between all convection, conduction, and radiation heat sources and the material surface, K; L'_s = latent heat of vaporization at the material surface temperature, kJ/kg.

Critical Moisture Content. The critical moisture content, which is the average material moisture content at the end of the constant rate drying period, is a function of material properties, the constant drying rate, and particle size. The critical moisture content is difficult to determine without a prototype drying test. When drying a wet material during the constant rate period, it is assumed that the drying rate is increased by increasing the heat transfer rate, eg, by increasing the gas velocity across the surface. The moisture gradient below the surface which causes liquid flow to replenish the surface must become more steep. As the surface approaches dryness, the average moisture content throughout the material is greater than it would have been had the drying rate not been increased. Critical moisture content data may be misleading, therefore, unless the drying method and exact conditions are known. Approximate values for materials dried in layers are available in refs. 5–7.

Particle size distribution determines surface-to-mass ratios and the distance internal moisture must travel to reach the surface. Large pieces have higher critical moisture contents than fine particles of the same material. Pneumatic-conveyor dryers work because very fine particles with low critical moisture contents are dried. Many materials that take hours to dry in static layers can be dried in a few seconds.

Case hardening describes a situation in which a large mass of nonporous, soluble, or colloidal material is dried at such a high rate during the constant rate period that the surface overheats and shrinks. Because diffusivity decreases with moisture concentration, the barrier of the overdried surface prevents moisture flow from the interior of the mass. Case hardening of nonporous materials can be minimized by increasing the relative humidity of the gas environment to hold a high, surface equilibrium moisture content until internal moisture has escaped.

Equilibrium Moisture Content. Equilibrium moisture content is a steady-state equilibrium obtained by the gain or loss of moisture when a material is exposed to an environment of specific temperature and humidity. This equilibrium condition is independent of drying rate or method but is a material property. Only hygroscopic materials have equilibrium moisture contents. Clean beach sand is nonhygroscopic and has an equilibrium moisture content of zero. A hygroscopic material retains a fixed percentage of moisture under specific conditions of temperature and humidity. At constant temperature, if the humidity of the atmosphere surrounding the material increases or decreases, an increase or decrease in material moisture content follows. The retained moisture is called equilibrium moisture because it is held in vapor pressure equilibrium with the partial pressure of vapor in the atmosphere. The reason it is retained, even when the atmosphere is quite dry, is that the mechanism of retention reduces the effective liquid vapor pressure. It is bound moisture because it is bound to the material in solution or as an adsorbed surface layer. Bound moisture behaves as if the atmosphere were saturated even though the atmosphere is not saturated relative to the pure liquid exerting its normal vapor pressure. Chemically combined water may or may not behave like bound water depending on the nature of the chemical bond.

Because equilibrium moisture is affected by both the partial pressure of vapor in the atmosphere, and its own effective vapor pressure which varies with temperature, any correlation must take both humidity and temperature into account. For many materials in the 15–50°C temperature range, this can be done by plotting equilibrium moisture content vs percent relative humidity; the equilibrium moisture content then appears independent of temperature. The retained moisture follows Henry's law:

$$p = H'(x) \tag{15}$$

where p = partial pressure of vapor in the atmosphere, kPa; H' = Henry's constant; x = dry basis, moisture content, kg/kg. Henry's constant is a function of the pure liquid's vapor pressure:

$$H' = i(p_w) \tag{16}$$

where i = a constant that is independent of temperature; p_w = the pure liquid's vapor pressure at any temperature, kPa; therefore, $p = i(p_w)(x)$, and since percent relative humidity = $100(p/p_w)$:

$$100(p/p_w) = 100i(x) \qquad (17)$$

At any given percent relative humidity, x is constant. In a typical silica-gel–air–water-vapor system at 5–50°C, $p = 1.79(p_w)(x)$, where p = partial pressure of vapor in air, kPa; x = gel moisture content, kg/kg; and p_w = water vapor pressure at the adsorption temperature, kPa (5). For most materials, as the temperature rises above 50°C, the equilibrium moisture content decreases even at constant relative humidity. For water vapor in air at atmospheric pressure and above 100°C, there is no saturation humidity and relative humidity has no meaning. At high temperatures, hygroscopic materials can often be dried in 100%, superheated vapor atmospheres.

Figure 6 is a profile of equilibrium moisture content vs percent relative humidity. The profile is not necessarily a straight line because, at high humidities, a porous material may collect condensed moisture. At low humidities, the retained moisture may be adsorbed as a single molecular layer on surfaces in the smallest capillaries. A difference between adsorption and desorption profiles can have several causes. When a material initially dries, shrinkage often closes many small capillaries and these do not readsorb moisture when the material is rewetted. An important reason for freeze-drying solid foods is to prevent shrinkage and capillary closing so that the dry residue can easily be reconstituted to its natural moisture content. Figure 6 also identifies the maximum bound moisture a material can hold. Any moisture content greater than that indicated by the intersection of the equilibrium profile with the 100% relative humidity ordinate is unbound moisture that exerts the same vapor pressure as pure liquid. A moisture content less than the 100% saturation content represents bound moisture that exerts reduced vapor pressure.

Falling Rate Drying. The principal internal mass transfer mechanisms that control falling rate drying are: (*1*) liquid diffusion in continuous, homogeneous materials; (*2*) capillarity in porous and fine granular materials; (*3*) gravity flow in granular materials; (*4*) vapor diffusion in porous and granular materials; (*5*) flow caused by shrinkage-

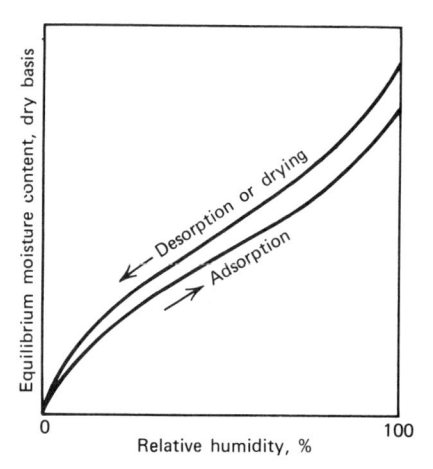

Figure 6. Typical equilibrium moisture content profiles.

induced pressure gradients; and (6) pressure flow of liquid and vapor when a material is heated on one side and vapor escapes from the other.

Liquid flows by diffusion through materials in which the liquid is soluble; eg, single-phase systems like soap, gelatin, and glue. Movement of bound moisture by liquid diffusion may occur, but the mechanism is probably more complicated. Vapor flows by diffusion in the gas phase when liquid is vaporized in isolated pockets in porous and granular materials, eg, in wood and other cellular materials at moisture contents less than fiber saturation, and in the last drying stages of textiles, paper, clay, and hydrophilic solids. Diffusion-controlled mass transfer is assumed when liquid or vapor flow appears to conform to Fick's second law of diffusion (see Diffusion separation methods; Mass transfer).

$$\frac{\partial c}{\partial \theta} = D_{AB} \frac{\partial^2 c}{\partial z^2} \qquad (18)$$

which is the unsteady-state diffusion equation in mass transfer notation and where c = concentration of one component in a two-component phase of A and B; θ = diffusion time; z = distance in the direction of diffusion; D_{AB} = binary diffusivity of the phase A–B. This equation applies to diffusion in solids, stationary liquids, and stagnant gases. An analogous equation applies to heat conduction in solids.

In porous and granular materials, liquid movement may occur by capillarity and gravity provided the pores are continuous. Capillarity applies to liquids that are not held in solution by the material, moisture contents greater than fiber saturation in cellular materials, unbound moisture in other hygroscopic materials, and all moisture in nonhygroscopic materials.

When clays and similar materials are dried, a pressure gradient is developed by the forces of repulsion between particles as shrinkage brings the particles close together (8). This gradient forces liquid toward the surface, and the resulting moisture profiles resemble those characteristic of diffusion.

Only one mass transfer mechanism usually predominates at any given time during drying, although several may occur together. In most solids, the mechanisms of internal liquid and vapor flow during falling rate drying are complex. Simultaneous heat transfer is an important variable, and falling rate drying can rarely be described with mathematical precision. Data for numerous materials have been published (3,5–18), but final decisions are almost always based on performance during drying tests. In the absence of tests, the falling rate drying periods are usually studied on the assumption that any material is either porous or nonporous, and internal mass transfer is controlled by either capillarity or diffusion.

Diffusion. Characteristic drying time and drying rate profiles for diffusion are shown as profile C in Figures 4 and 5. When liquid diffusion occurs because the material and liquid are mutually soluble, there is usually only a brief constant rate drying period during which the surface dries to the material's equilibrium moisture content in the drying atmosphere. For evaluation of falling rate drying, an integration of equation 18 can be employed provided several assumptions are made: (1) diffusivity is independent of liquid concentration, (2) initial liquid distribution is uniform, (3) material size, shape, and density are constant, and (4) the material's equilibrium moisture content is constant. For a material in slab form:

$$\frac{W_\theta - W_e}{W_c - W_e} = \frac{8}{\pi^2} \left\{ \sum_{n=0}^{n=\infty} \frac{1}{(2n+1)^2} \exp\left[-(2n+1)^2 D\theta(\pi/2d)^2\right] \right\} \qquad (19)$$

where W_θ, W_c, W_e are the moisture content at any time, that at the critical moisture content, and that in equilibrium with the environment, respectively, kg/kg; D = liquid diffusivity, m²/s; θ = drying time, s; d = one half the slab thickness for drying from two sides, or total thickness for one-side drying, m. For long drying times, when $D\theta/d^2$ is greater than about 0.1,

$$\frac{W_\theta - W_e}{W_c - W_e} = \frac{8}{\pi^2} \{\exp[-D\theta(\pi/2d)^2]\} \qquad (20)$$

A drying rate expression can be derived from equation 20:

$$-\frac{dw'}{d\theta_f} = \frac{\pi^2 D}{4d^2}(W_\theta - W_e) \qquad (21)$$

where $dw'/d\theta_f$ = drying rate, kg/(s·kg), dry basis. When internal diffusion predominates for long periods, the drying rate is proportional to free moisture content and liquid diffusivity, and inversely proportional to the square of the material thickness. If equation 19 is plotted on semilogarithmic paper, a straight line is formed when values of $(W_\theta - W_e)/(W_c - W_e)$ are less than 0.6; equation 21 represents the straight-line portion of this plot. An approximate relationship for the falling rate drying time for materials that exhibit rate profiles characteristic of diffusion is:

$$\theta_f = \frac{4d^2}{\pi^2 D} \ln\left[\frac{8(W_c - W_e)}{\pi^2(W_\theta - W_e)}\right] \qquad (22)$$

where θ_f = the falling rate drying time, s. Equations 19–22 apply to materials in slabs. Equation 18 has been solved for numerous other shapes (4) and for many analogous heat conduction problems (19–20). These diffusion equations may also be used to study vapor diffusion in porous materials; however, all estimates based on relationships that assume constant diffusivity are approximations. Liquid diffusivity in solids usually decreases with moisture concentration; liquid and vapor diffusivity also change as material shrinks during drying.

Capillarity. The outer surface of a porous solid has pore entrances of various sizes. As surface liquid is evaporated during the constant rate period, a liquid meniscus is formed across each entrance and capillary forces are set up by interfacial tension between the liquid and solid. These forces draw liquid from the interior to the outer surface. At the critical moisture content, some of the menisci begin to retreat into the pores. Wetted outer surface gradually decreases and, although the drying rate per unit area of wetted outer surface remains constant, drying rate based on total outer surface decreases. This stage of unsaturated surface drying of porous materials is the first falling rate drying period and is represented by the straight line, decreasing rate portion of profile B in Figure 5. During this period, gas begins to enter the pores, but the internal liquid phase is continuous; ie, the liquid is in the funicular state. If the pores are small and not of uniform size, the retreat of the liquid phase is not uniform. Small pores produce stronger capillary forces than large pores; thus small pores draw liquid out of large pores. As more liquid is removed, the outer surface dries completely and a point is reached where there is not enough internal liquid left to maintain a continuous phase linking all interior pores. Interfacial tension breaks; residual liquid retreats to isolated pockets in the smallest pores, and gas enters far enough into the material so that gas forms the continuous phase. The liquid then obtains the pendular state. This point is called the second critical moisture content in porous materials because there is a second break in the drying rate profile. All heat for vaporization must now pass by conduction through the material to the liquid pockets; the temperature rises, and residual liquid vaporizes and diffuses in the gas phase to the surface. In many porous materials during the second falling rate drying period, the drying rate profile has the curved, concave-upward shape of a diffusion-controlled process. In granular materials, where pores are large and capillary forces weak, gravity contributes to retreat of the liquid surface and both falling rate profiles may be straight lines (3). For simultaneous heat, mass, and momentum transfer in porous solids, see ref. 21.

Drying of materials by capillarity and vapor diffusion may be approximated by assuming that falling rate drying follows the straight line, profile D in Figure 5.

$$-\frac{dw'}{d\theta_f} = K(W_\theta - W_e) \qquad (23)$$

where K is a function of the constant rate drying rate at the critical moisture content; ie,

$$-\frac{dw'}{d\theta_c} = K(W_c - W_e) \qquad (24)$$

For a layer of wet material, using equation 9,

$$K = \frac{h_t}{p_m d_m L'_s}\left(\frac{t - t'_s}{W_c - W_e}\right) \qquad (25)$$

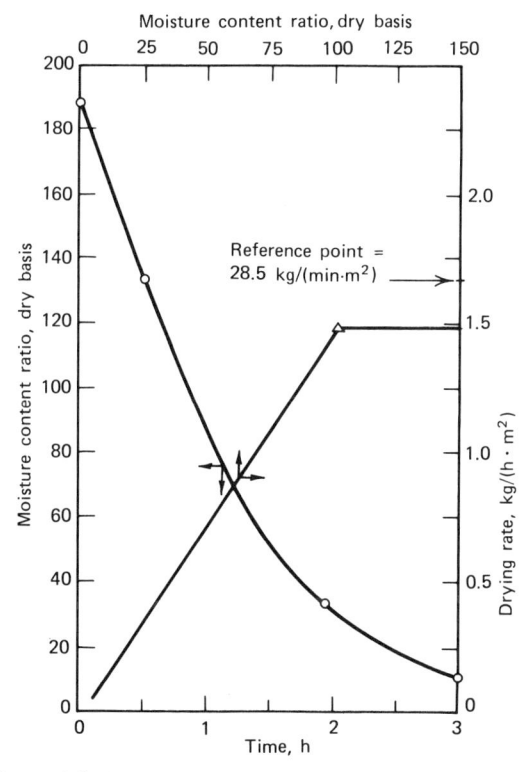

Figure 7. Experimental drying time and rate profiles for leather pasted on glass plates and dried in two temperature stages. Falling rate drying is proportional to residual moisture content. Air velocity = 5.0 m/s (across the leather); $t = 71°C$ at first stage; $t = 57°C$ at second stage. To convert kg/(h·m²) to lb/(h·ft²), divide by 4.88.

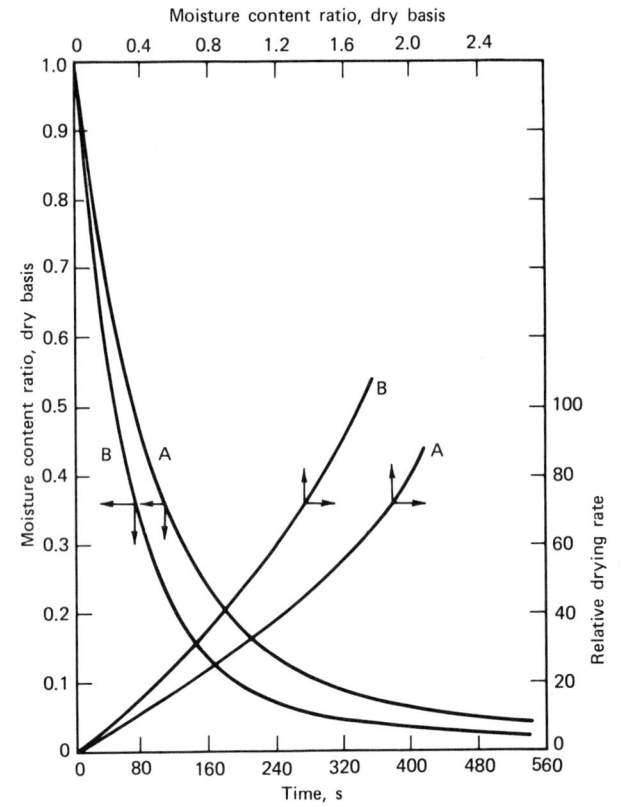

Figure 8. Drying profiles for single drops of whole milk.

Curve	Initial drop diameter, μm	Initial moisture, kg/kg	Air temperature, °C	Relative air velocity, m/s
A	1900	2.6	94	0.6
B	1470	2.4	94	0.6

and

$$-\frac{dw'}{d\theta_f} = \frac{h_t(t - t'_s)}{p_m d_m L'_s}\left(\frac{W_\theta - W_e}{W_c - W_e}\right) \quad (26)$$

For materials that follow equation 26, the drying rate is inversely proportional to the first power of material thickness, and the falling rate drying time θ_f can be estimated as follows:

$$\theta_f = \frac{p_m d_m L'_s (W_c - W_e)}{h_t(t - t'_s)} \ln\left(\frac{W_c - W_e}{W_\theta - W_e}\right) \quad (27)$$

A relationship for through circulation drying, analogous to equation 26, using equation 10, is:

$$-\frac{dw'}{d\theta_f} = \frac{h_t a(t - t'_s)}{p_m L'_s}\left(\frac{W_\theta - W_e}{W_c - W_e}\right) \quad (28)$$

Drying Profiles. Figures 7 to 11 show that both constant rate and falling rate drying may occur regardless of the drying method. Figure 7 is the drying profile of a sheet of leather pasted on a glass plate across both sides of which air was circulated. During the drying cycle, two air–temperature stages were used; heat was transferred by convection, conduction through the plate, and radiation; and both constant and falling rate drying periods were significant.

The data in Figure 8 were obtained by suspending single drops of whole milk in a heated air stream. Because of the rapid formation of a surface film, drying was mostly by diffusion. The drying characteristics of drops of numerous materials have been studied (9). Periods of drying occur during freeze-drying as shown by the data in Figure 9 (13); ice sublimes to vapor, and the mechanism can be visualized as the retreat of a rigid ice phase into the interior of the material as drying proceeds. Heat transfer is by conduction through the solid from a radiant source. The data in Figure 10 were obtained during the spray drying of sodium sulfate in a 12-m tower. The spray was sampled below the nozzle and at two points farther down the tower. These are typical spray drying profiles which, as in Figure 8, show that drying is mostly by diffusion. Figure 11 is a drying profile for corn kernels in a fluid-bed dryer. Moisture flow was by diffusion, and diffusivity in the corn was a function of moisture content. A long drying cycle was needed because of the low air temperature employed.

Dryer Classification

A batch dryer is one into which a charge is placed, then removed after the dryer runs through its cycle. A continuous dryer operates under steady-state conditions, with continuous feed-in and product removal. Table 1 gives a classification of industrial dryers according to their predominate heat transfer mechanisms. An alternative classification is based on material handling characteristics (5–6) and specific problems.

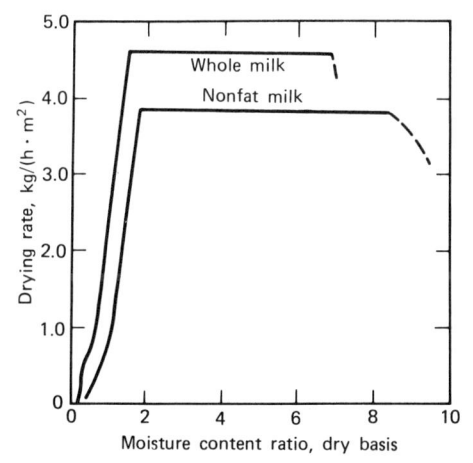

Figure 9. Vacuum freeze-drying data (13). Heat was transmitted by radiation from heated wires to a frozen solid resting in a transparent plastic tray. Total pressure was 33 Pa (250 μm Hg). Total drying time is: whole milk, 9.1 h to 0.025 kg/kg (dry solid); nonfat milk, 7.7 h to 0.07 kg/kg (dry solid). To convert kg/(min·m^2) to lb/(min·ft^2), divide by 4.88.

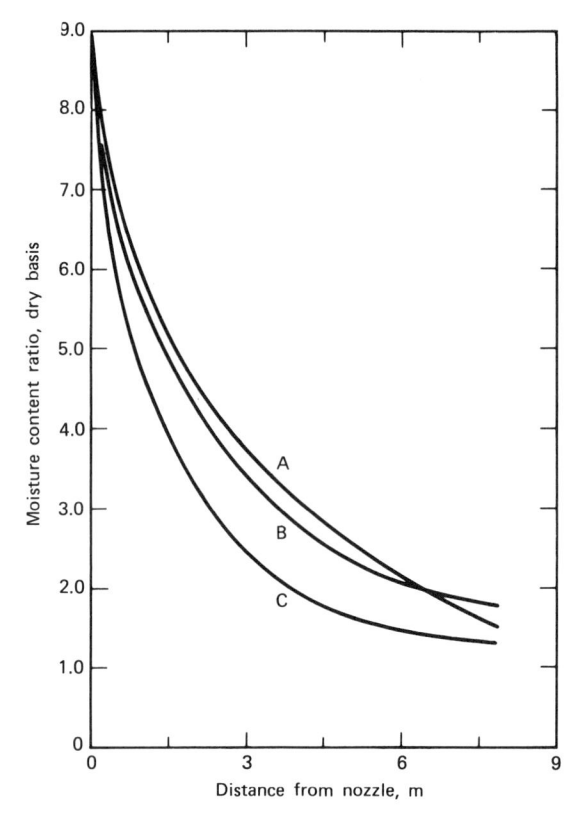

Figure 10. Drying profiles for spray drying sodium sulfate solutions in a 12 m tower. The initial evaporation rates were estimated to be: A, 26.4 kg/(s·kg); B, 31.4 kg/(s·kg); C, 40.3 kg/(s·kg). Nozzle pressure is 8.273 MPa (ca 1200 psi).

Curve	Air rate at 20°C, m³/s	Feed rate, cm³/s	Inlet air, °C
A	2.64	86.1	122
B	1.68	93.2	167
C	1.10	94.2	245

Direct-Heat Dryers

In direct-heat dryers, heat is transferred to the material from a hot gas. Steam coils may be used for gas heating up to about 200°C; electric heaters are added for higher temperatures, and combustion products are used for all temperatures. If a material must be protected from contact with combustion products, drying gas may be heated indirectly by passing it through tubes in a furnace. A clean, high temperature gas is obtained in this manner; however, fuel efficiency is about twice as high when combustion products are used directly. Unless the metal surfaces in a dryer are protected by insulating or refractory brick, the maximum permissible gas temperature is about 1000°C.

For low temperature drying, the drying gas may be dehumidified by refrigeration (qv) or gas drying. It is usually more economical to recirculate dehumidified gas back through the dehumidifier after each dryer pass than to dehumidify fresh gas. Hygroscopic polymers that must be dried to low moisture contents for extrusion often require dehumidified drying gas regardless of drying temperature.

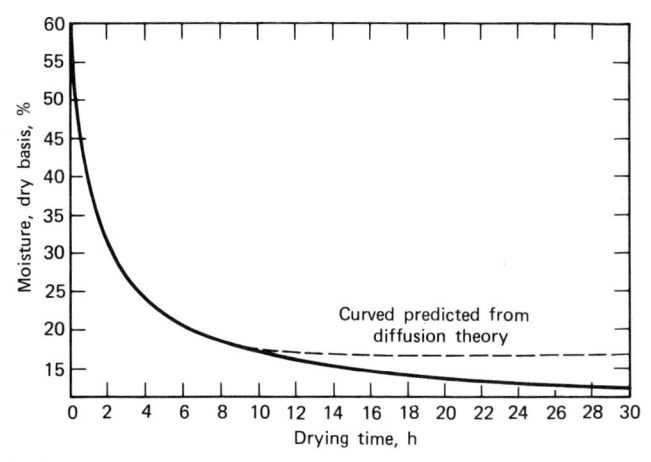

Figure 11. Drying curve for corn kernels. Drying is by diffusion. The dashed curve would be obtained if diffusivity were constant and independent of moisture content. Air temperature = 38°C.

Table 1. Industrial Dryers

Heat-transfer mechanism	Continuous dryers	Batch dryers
direct	tray	parallel-flow tray
	tunnel	through-circulation tray
	turbo-tray	fluid-bed
	through circulation	spouted-bed
	suction-drum	rotary
	foam-mat	
	festoon	
	multipass-loop	
	tenter-frame	
	float	
	rotary	
	fluid-bed	
	spouted-bed	
	vibrating-conveyor	
	pneumatic-conveyor	
	spray	
indirect	steam-tube	vacuum
	rotary	shelf
	fluid-bed	freeze
	screw-conveyor	rotating
	high-speed agitator	rotary
	drum	pan
	can	fluid-bed
	belt	rotary
	tunnel	screw-conveyor
radiant	rf[a]	rf[a]
dielectric	microwave	microwave

[a] rf = radio frequency.

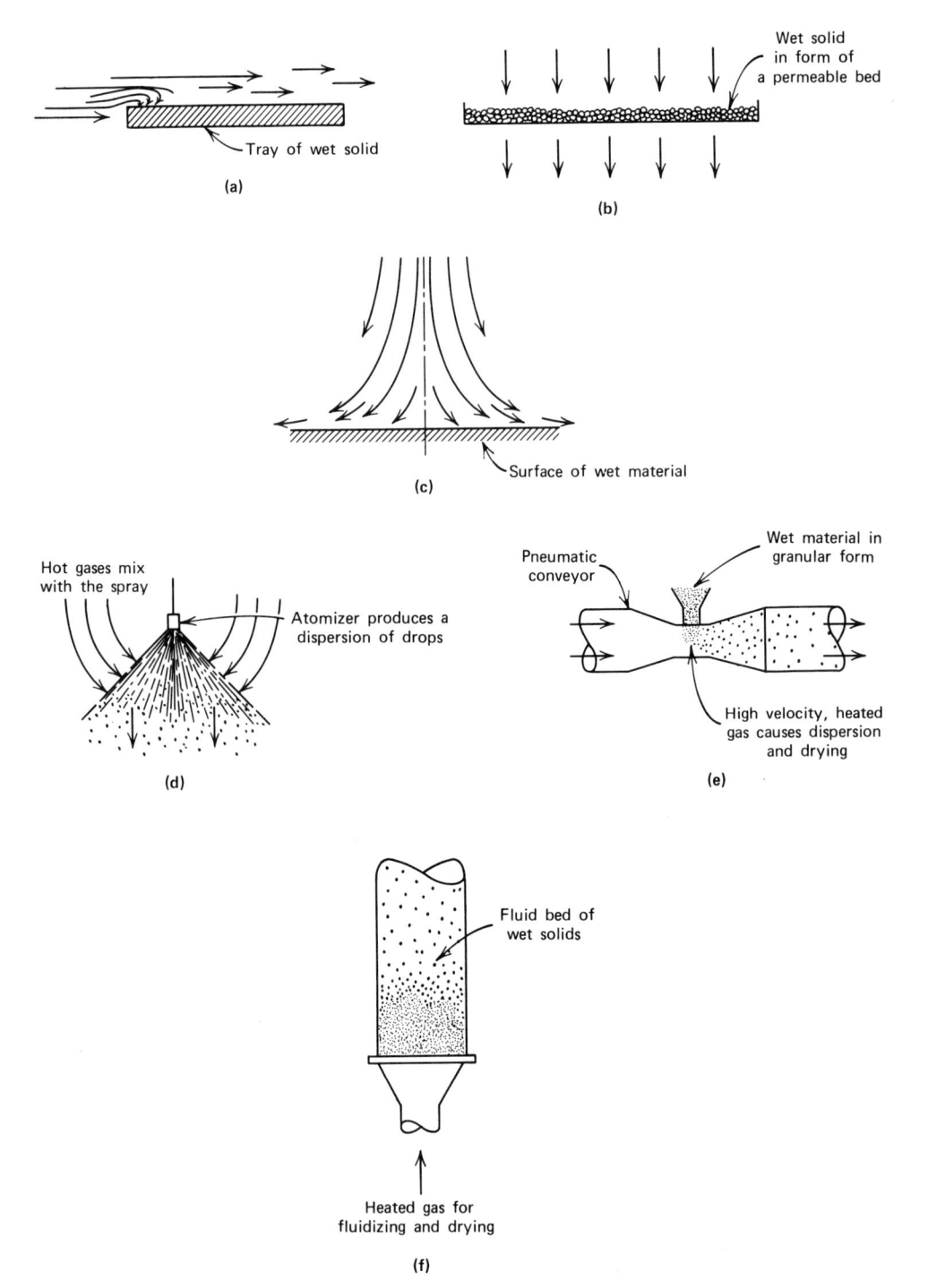

Figure 12. Gas-contacting methods for direct-heat drying. (**a**) Parallel flow; (**b**) through circulation; (**c**) impinging flow; (**d**) spray atomization; (**e**) pneumatic conveying; (**f**) fluidization.

In most direct-heat dryers, more gas is needed to transport heat than to purge vapor. Therefore, larger dust collection systems are needed than for indirect-heat dryers performing the same work. With stricter pollution regulations, the cost advantage once enjoyed by direct-heat dryers is disappearing. Today, the dust collection system needed on a spray dryer may cost as much as the dryer.

Generally, the greater the gas velocity over, through, or impinging upon a material, the greater is the convection heat-transfer coefficient (5–6). Furthermore, the better the solids are dispersed in a gas for surface exposure, the greater is the heat transfer rate. Figure 12 shows several common methods for the contacting of gas and solids in direct-heat dryers. Gas and solids flowing in the same direction in a continuous dryer constitute cocurrent flow; gas flowing opposite to the solids is countercurrent flow. Gas flow perpendicular to the path of solids flow is crossflow, and gas flow directed against a solid surface is impinging flow.

Batch Dryers. *Trays, Compartments.* A tray or compartment dryer is an insulated housing in which particulate solids are placed in tiers of trays, and large objects may be stacked in piles or on shelves. Gas usually flows across the dryer parallel to tray surfaces and around large objects. Figure 13 is a schema of a tray dryer. Unless the material is very dusty, gas is recirculated through an internal heater to improve the dryer's thermal efficiency and only enough purge gas is exchanged to maintain the needed internal humidity. Because manual labor is needed for dryer loading and unloading, batch compartment dryers are usually economical only for a single-product rate of less than 500 metric tons per year, large objects that require drying cycles longer than eight hours, drying several products in one dryer, or keeping batches distinct. Tray loading depth, tray spacing, and the gas velocity across all solids surfaces must be uniform. Tray loading depth is usually 2–10 cm, gas velocity 1–10 m/s. Deep tray loading reduces labor cost, but dryer capacity is reduced because critical moisture content increases with depth. Shallow loading reduces drying time, but care is needed to assure depth uniformity and more tray handling is involved. Metal trays enhance heat transfer through the tray bottoms; screen-bottom trays permit vapor escape through the bottoms. Overall drying rates are 0.2–2.0 kg water/(h·m^2 material surface); thermal efficiency is 20–50%.

Through Circulation. Through-circulation dryers employ perforated- or screen-bottom trays and suitable baffle arrangements so that gas is forced through the material, as in Figure 12**b**. If a material is not inherently permeable, it may be mechani-

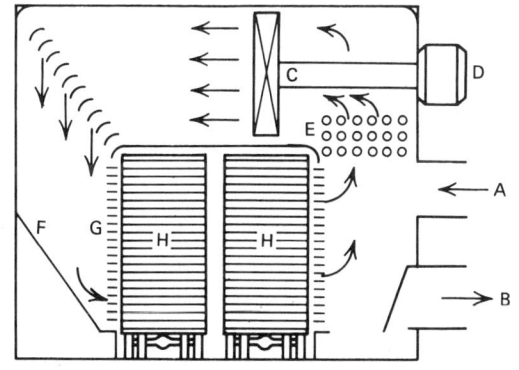

Figure 13. Two-truck tray dryer; A, air inlet duct; B, air exhaust duct with damper; C, axial flow fan; D, fan motor, 2–15 kW; E, steam heaters; F, air distribution plenum; G, air distribution slots; H, trucks and trays.

cally shaped into noodles, pellets, or briquets. These dryers are used to dry explosives, foods, and preformed filter cakes. Overall drying rates are 1–10 kg water/(h·m² tray surface); thermal efficiency is about 50%.

Continuous Dryers. *Conveyors.* This group is characterized by continuous material flow without mixing, and thus, dryer residence time is uniform for all material increments. Tray and tunnel dryers consist of long, insulated housings through which material is moved on trucks or trays fastened together. Gas flow is usually parallel to material surfaces; it may be cocurrent, countercurrent, or it may flow from side to side and be recirculated through heaters and fans mounted on each side of the housing. Conveyor movement may be continuous or semicontinuous. Performance is comparable to that of batch compartment dryers. The drying profiles in Figure 7 show the performance of a tunnel dryer used for leather in which the glass plates are suspended vertically and conveyed by an overhead chain to conserve space.

The turbotray dryer is a continuous-tray dryer consisting of a stack of circular trays that is carried inside a vertical, cylindrical housing on a slowly rotating center shaft. Each tray has uniformly spaced radial slots through which material is discharged by a stationary plough once each revolution. Material falls through a slot, accumulates on the tray below, and is leveled to uniform depth by a rake; after another revolution, the discharging and leveling actions are repeated, and material fed onto the top tray gradually moves down the stack and is discharged at the bottom. Circulating fans are mounted on the center shaft and gas heaters in the housing walls. Gas flows across the material as in a batch tray dryer. However, for materials that are not sticky, not dusty, and free flowing, the turbotray dryer provides more rapid and uniform drying than a static-bed dryer.

Foam-mat drying is a process in which a liquid is converted to a stabilized foam by injection of an inert gas. The porous foam structure is dried in trays in a tunnel or on a belt. A free-flowing powder capable of rapid rehydration by cold water often results; many fruit juices (qv) have been successfully dried by this process (see also Food processing).

Continuous-web dryers are used for polymer films, nonwoven fabrics, printed cloth, and paper. Gas is circulated over or against a moving sheet, called a web, that may be supported by various methods. In a festoon dryer, the web is festooned over sticks that are carried on a chain through a heated enclosure; gas movement about the material must be comparatively moderate to avoid tangling adjacent festoons. In the multipass-loop dryer, the web is conveyed either horizontally or vertically over and under a series of rolls, some of which are driven and some free turning. Web tension is controlled by the location of the driven rolls and differential driving speeds. Because the web does not hang free, a high gas velocity may be used or gas impingement nozzles may be installed on one or both sides of the web. For one-side drying of coated or printed webs, roller conveyors are often installed in housings hung from a ceiling to conserve floor space. Tenter-frame dryers restrain the web in two dimensions during drying. These are employed to control shrinkage or stretch the web as drying occurs; they present an ideal arrangement for two-side, gas impingement drying. They are, however, expensive compared to other conveyor dryers. Float dryers employ specially designed nozzles on both sides of the web to support, convey, and dry with minimum web tension. For materials that need long drying cycles, a festoon or multipass-loop dryer is usually the most economical choice because a great length of web can be contained in a tall enclosure that occupies little floor space.

Through-Circulation Dryers. In these dryers a bed of permeable material is conveyed through an enclosure on one or more perforated plate or screen conveyors. The enclosure is a series of independent compartments, each having its own fans and heaters. Figure 14**a** shows a section of one compartment; Figure 14**b** shows an assembled dryer. Humid gas is removed below the conveyor and fresh gas enters under the heaters. Weak points in the gas-circulation loop are the moving material guards and seals on each side of the conveyor; these tolerate a maximum pressure drop of about 0.50 kPa (4.0 mm Hg) through the material bed before letting a significant quantity of gas bypass. Conveyor widths are 0.5–5.0 m; length may be 50 m. Fibers, elastomers, vegetables, plastics, and centrifuge and filter cakes are dried in 2–20-cm deep layers. Drying rates up to 50 kg water/(h·m² conveyor surface) are obtained at temperatures up to 400°C; thermal efficiency is 50–70%. Centrifuge and filter cakes may be preformed for through-circulation drying by granulation in knife mills or extruded to form short noodles. Thin pastes and slurries are predried on finned-drum dryers to form short sticks. Thixotropic materials are scored and cut into small pieces,

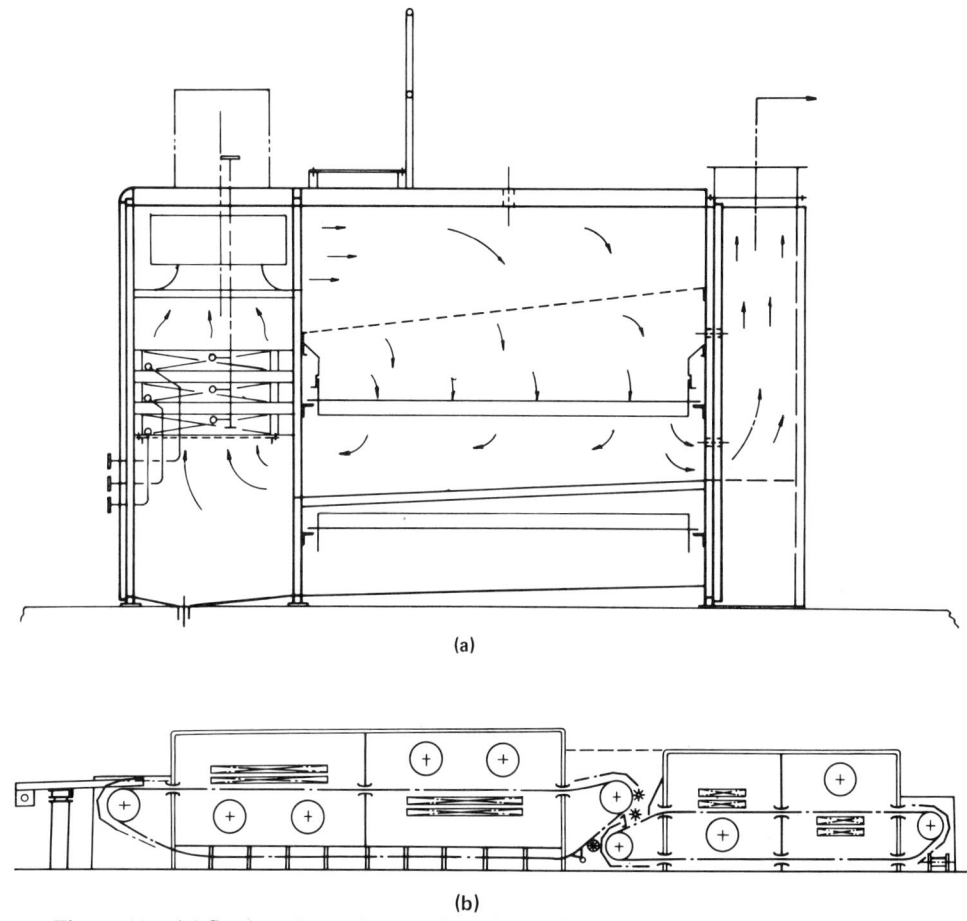

Figure 14. (**a**) Section of a continuous, through-circulation, conveyor dryer; (**b**) assembly of a two-stage, continuous, through-circulation dryer. Courtesy of the National Drying Machinery Co., Philadelphia, Pa.

and powders are briquetted or pelleted in pan pelletizers (see Pelleting and briquetting). Pin elevator feeders open fibrous clumps and lay a uniform bed on the conveyor. Fiber tow is distributed by oscillating chutes. Materials that shrink during drying are redistributed on a second conveyor, as shown in Figure 14b, after partial drying.

Suction-drum dryers are through-circulation dryers especially suitable for fiber staple, tow, and nonwoven fabrics. The material is conveyed on a series of perforated, screen-covered suction drums through compartments that are similar in design to the conveyor-type dryer. Drying is more uniform because the material is turned over as it passes from one drum to the next. Drying rates are greater because material guards and seals are not needed. Therefore, a greater pressure drop and gas velocity through the material can be employed.

Mechanically- and Fluid-Agitated Dryers. A through-circulation dryer obtains a relatively high drying rate because drying gas flows through, rather than in parallel flow above and below the material. However, if bed depth and porosity are not uniform, gas channels through thin spots and larger passages and all particle surfaces are not contacted equally. A better arrangement is one in which the particles are completely separated in such a manner that the gas can flow freely among them. The drying rate for all particles of a given size should then approach a constant. The purpose of various forms of mechanically- and fluid-agitated dryers is to obtain optimum conditions of particle separation and surface exposure for products of various drying and material handling characteristics. The latter usually govern the choice of dryer. A disadvantage of agitated compared to tunnel dryers is the loss of the plug-flow property. Agitated dryers produce a statistical distribution of material residence time. An average particle residence time may be calculated from feed rate, dryer fillage, and material density. However, the residence time for any one particle may deviate from the average in the manner shown in Figure 15. Operating methods can be used to minimize plug-flow deviation which cannot, however, be completely eliminated.

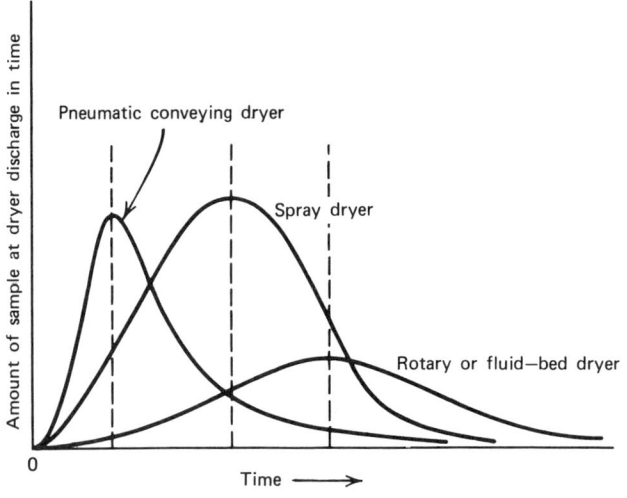

Figure 15. Residence-time distribution profiles for various dryers.

Rotary Dryer. A direct-heat, rotary dryer is a horizontal, cylindrical, rotating shell through which gas is blown to dry material that is tumbled inside. Shell diameters are 1–6 m. Batch dryers are usually one or two diameters long, continuous dryers four to ten diameters long. At each end, a stationary hood is joined to the shell by a rotating seal. These hoods carry the inlet and exit gas nozzles and the feed and product conveyors; one hood also houses the gas heater. For continuous drying, the shell may be slightly inclined to cause material flow in a predictable manner. Gas flow may be countercurrent or cocurrent. An array of lifting flights of various shapes is attached to the inside of the shell to shower material for gas contacting. Figure 16 shows a rotary dryer cross-section. Shell knockers are used to dislodge wet material that may stick to metal surfaces. A ring of flights is usually 0.5–2.0 m long and adjacent rings of flights are usually offset to improve gas contacting. If a material is too fluid or sticky for good dispersion, the dry product may be recycled for feed conditioning. Slurries may be sprayed inside a shell in a manner such that the feed falls on, and is mixed with a moving bed of dried particles. A continuous dryer is filled with material to 10–15% of shell volume. A greater percentage is not dispersed and tends to flush toward the discharge end. Cocurrent dryers are preferred for heat-sensitive materials because gas and product leave at about the same temperature. To prevent material blow-through, a cocurrent shell may be given a negative inclination. Countercurrent dryers are multistage vessels, ie, the exit gas temperature is usually lower than the product temperature. Thermal efficiency can be as high as 80%. Some dryers have enlarged shell sections at the product end to increase material holdup and reduce gas velocity. Indirectly heated tubes are installed in some shells for additional heat transfer. Rotary dryers are usually operated with a negative internal pressure of 0.05–0.10 kPa (0.4–0.8 mm Hg) to prevent dust and vapor escape through the rotating seals. Most particulate solid materials can be dried in direct-heat, rotary dryers. They provide reasonably good gas contacting, positive material conveying without serious back-mixing, good

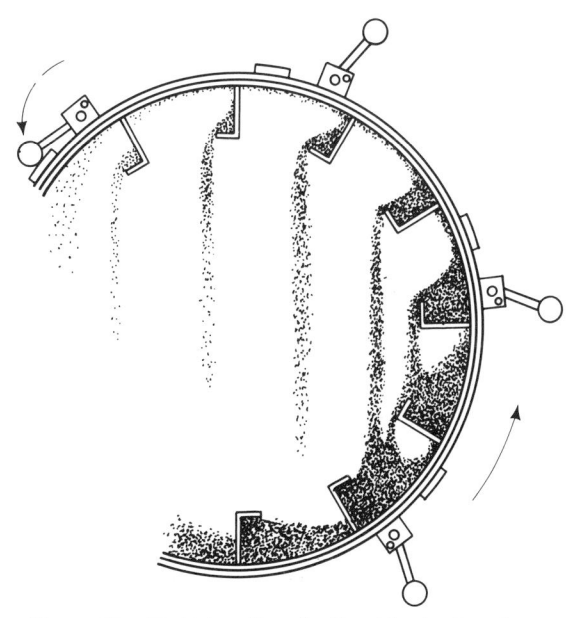

Figure 16. Partial section of a direct-heat rotary dryer.

100 DRYING

thermal efficiency in either cocurrent or countercurrent use, and great flexibility for control of gas velocity and solids residence-time. Drying rates are 10–50 kg water/(h·m^3 shell volume); thermal efficiency is 50–80%.

Fluid- and Spouted-Bed Dryers. A fluid bed of solids is produced by introducing gas through a perforated plate, nozzles, or a ceramic grid under a static bed of solids in such a manner that the solids are lifted and mixed; the gas and particles together then behave like a boiling fluid (see Fluidization). For drying, the average upward gas velocity is usually less than the final settling velocities of the particles and gas does not convey many particles out of the bed. At the same time, gas bubbles rise fast enough to lift particles above them. Particle motion is violent, and a fluidized-bed surface exhibits intensive splashing. A substantial freeboard is needed above the bed in the fluidized tank to prevent blow-through. Nevertheless, particle attrition is usually moderate, probably because of the gas cushion that surrounds each particle as it moves through the mass. Figure 17 shows particle-gas motion near a single fluidizing orifice.

Spouted beds are used for coarse particles that do not fluidize well. A single, high velocity gas jet is introduced under the center of the static bed. This jet conveys a stream of particles up through the bed into the tank freeboard where the jet expands, loses velocity, and allows the solids to become disentrained. The solids fall back into the bed and gradually move downward with the peripheral mass until reentrained. Particle-gas mixing in a spouted bed is less uniform than in a fluid bed.

Fluid and spouted beds offer nearly ideal conditions for gas contacting and drying provided a feed material is consistently suitable for fluidization or spouting. If a drying operation is preceded by mechanical separation, eg, filtration, use of fluid and spouted beds should be considered with caution. Fluid and spouted beds do not tolerate sticky materials and large lumps. Their major drying applications are particulate and pelleted polymers, grain, coal, sand, and mineral ores.

Fluid- and spouted-bed dryers are attractive for inert gas and solvent recovery operations, compared to rotary dryers, because their tanks are stationary. The volumetric drying capacity of a fluid bed may be one hundred times that of a rotary dryer. Gas flowing through a rotary dryer moves between a series of particle curtains in which it must be entrained and mixed to contact particle surfaces, whereas gas in the form of small bubbles enters a fluid bed and immediately penetrates and mixes with a cloud

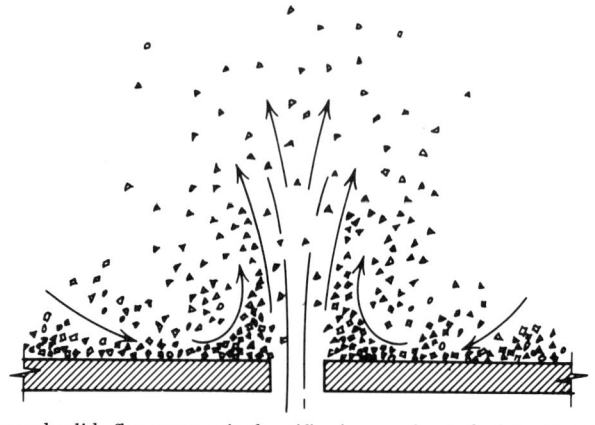

Figure 17. Gas and solids flow near a single orifice in a perforated-plate, fluid-bed distributor.

of particles. Figure 18 shows that the thermal efficiency of a fluid bed is comparable to that of a cocurrent, rotary dryer. Each is a single-stage vessel. To approach the thermal efficiency of a countercurrent rotary dryer, two or more fluids beds with countercurrent gas flow must be operated in series. Figure 19 shows one form of two-stage, fluid-bed dryer.

Vibrating-Conveyor Dryer. A vibrating conveyor dries by conveying the wet material on a screen-covered, perforated deck. Gas is blown up through the material as it is conveyed, and a particle dispersion much like that in a fluid bed may be produced. Both fluid and mechanical energy contribute to mixing, and bed depths rarely exceed 5.0 cm because mechanical energy cannot be effectively transmitted through a deep gas–solids mixture above the deck. A vibrating conveyor is a form of crossflow dryer; there is little end-to-end particle mixing and several temperature stages may be incorporated in a single conveyor. Maximum material residence time on a single conveyor is about 5 min. The feed material must be free flowing and nonsticky for satisfactory conveying.

Pneumatic-Conveyor Dryer. Pneumatic conveyors can be adapted for drying by heating the conveying gas. However, the gas-to-solids ratio needed for drying is usually

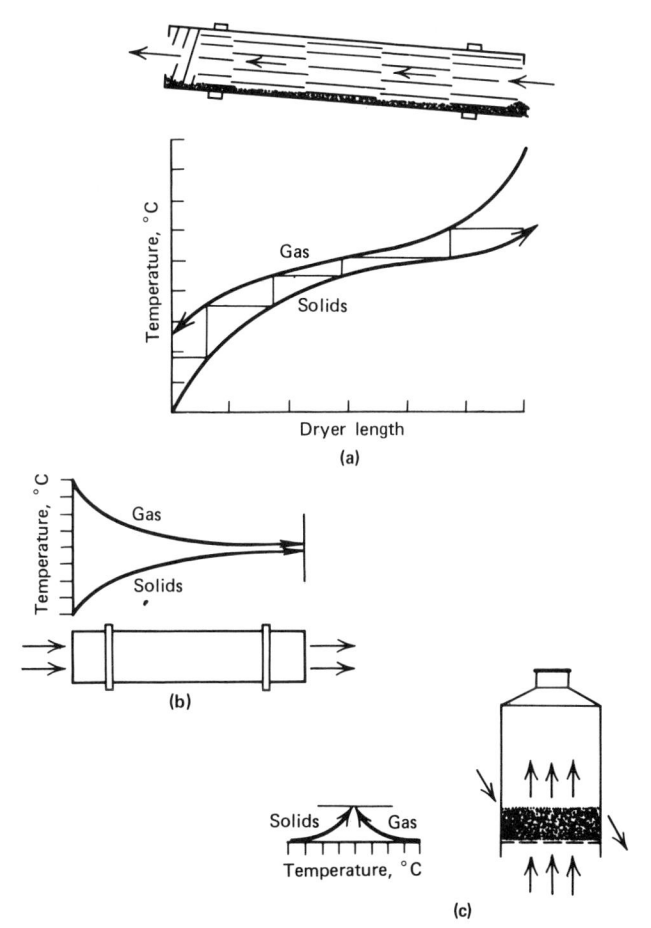

Figure 18. Temperature profiles for gas and solids in direct-heat dryers. (**a**) Countercurrent, rotary dryer; (**b**) cocurrent, rotary dryer; (**c**) single-stage fluid bed.

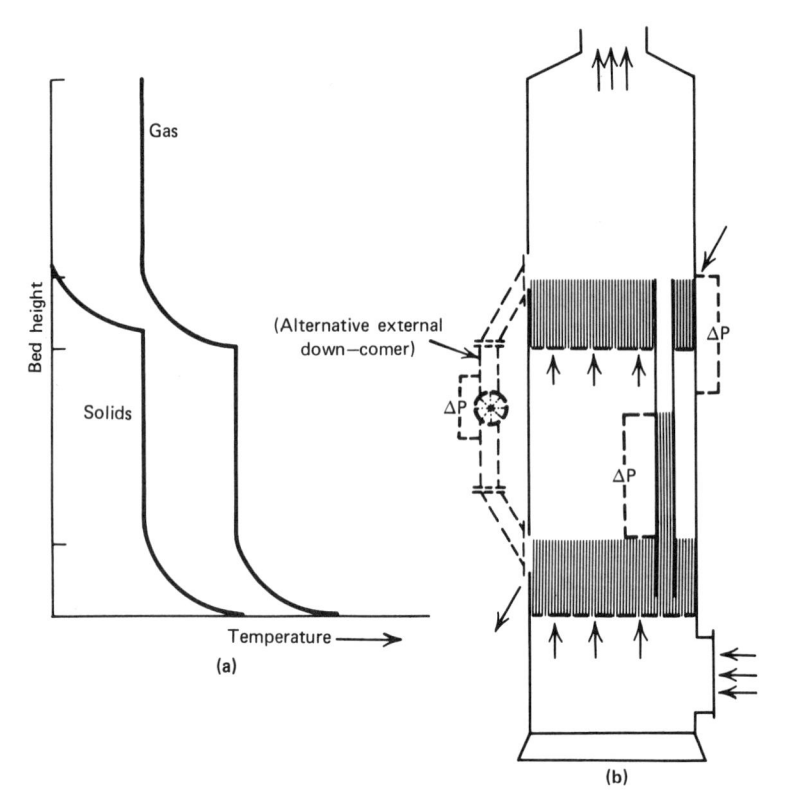

Figure 19. Two-stage fluid-bed dryer: (**a**) gas and solids temperature profiles; (**b**) bed arrangement; ΔP = pressure drop through upper stage distributor and bed.

much greater than that sufficient merely for conveying. Particle residence time is only a few seconds. In fact, most of the drying occurs near the feed point where the velocity difference between gas and solids is greatest. Conveying tubes are rarely more than 50 diameters long. For falling rate drying, two or three pneumatic-conveyor dryers may be operated in series, and the dried product may be recycled to the first stage for feed conditioning. Figure 20 shows a simple dryer consisting of a venturi feeder, tube, and product collector. Knife, hammer, and roller mills are alternative feed dispersion devices, and a paddle conveyor with its paddles inclined to retard material flow may be installed in place of the venturi to increase residence time or dispersion of material during drying. Tube gas velocity is 20–50 m/s. A pneumatic conveyor is a low grade, fluid energy mill, and particle attrition may be severe. The dryer is single-stage and cocurrent, like a cocurrent, rotary dryer. Evaporative capacity is fixed by gas quantity, inlet gas temperature, and exit gas and product temperatures which are usually about the same. Indirect-heat drying may be combined with direct-heat drying in a pneumatic conveyor by enclosing the tube in a jacket or installing heating elements inside the tube. The principal applications of pneumatic-conveyor dryers are for materials that are not sticky and can be thoroughly dispersed in the drying gas so that drying is mostly at a constant rate. If significant falling rate drying is required, a cocurrent, rotary dryer is usually preferred.

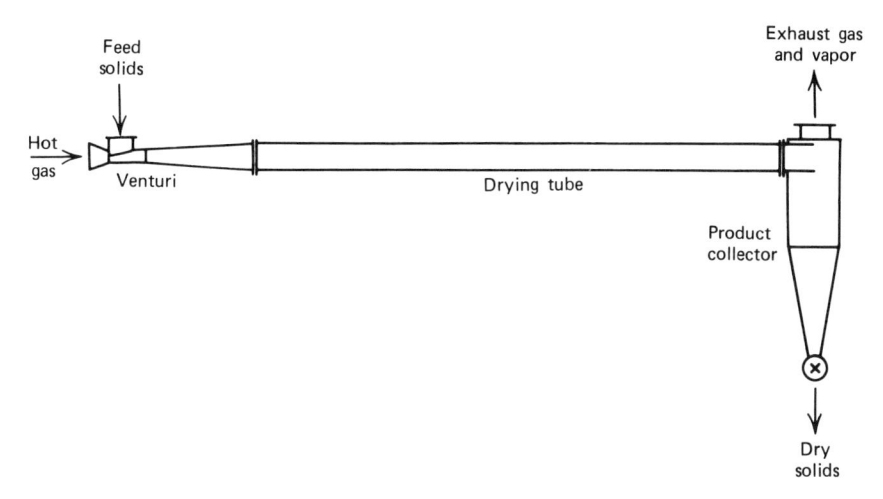

Figure 20. Single-stage, pneumatic-conveyor dryer with a venturi for solids–gas mixing.

Spray Dryer. A spray dryer is a large, usually vertical, chamber through which a hot gas is blown, and into which a liquid, slurry, or paste is sprayed by a suitable atomizer. There are three forms of atomizers commonly employed: (*1*) two-fluid pneumatic nozzles for small drops at rates less than 2 t/h; (*2*) single-fluid pressure nozzles for large drops to make dust-free products; and (*3*) high speed, disk atomizers for various drop sizes and rates up to 50 t/h. The largest spray-dried particle is about 1 mm; the smallest about 10 μm. The characteristics and performance of various fluid atomizers are described in refs. 14–15. All drops must be dried until no longer sticky, before they strike the chamber wall; therefore, the largest drop produced by a given atomizer determines the size of the spray chamber, and chamber shape is fixed by spray pattern. Nozzle chambers are tall towers, whereas disk-atomizer chambers are short, but have large diameters. For a given evaporative load, a chamber volume may be estimated by assuming a probable inlet gas temperature, an exit gas at 100°C, and a gas residence time of 20 s. A spray dryer may be cocurrent, countercurrent, or mixed flow. Countercurrent dryers yield high bulk density products and minimize the production of hollow, dry particles. Cocurrent dryers are used for heat sensitive materials because inlet gas temperatures up to 800°C may be used whereas gas and product leave the chamber at 90–120°C, and usually the material temperature never exceeds the exit gas temperature. Figure 21 shows a cocurrent, disk-atomizer chamber with a pneumatic conveyor employed for cooling. Figure 22 gives three cocurrent chamber-discharge arrangements, two of which accomplish particle size-classification. Spray dryers are often followed by fluid-bed dryers for second-stage, falling rate drying or cooling. Common spray dryer applications include soluble coffee, detergents, milk, agglomerated instant food products, pigments, dyes, and inorganic chemicals.

Indirect-Heat Dryers

Although heat is transferred mostly by conduction, radiation heat transfer may be effective when the conducting surface temperature exceeds 150°C. Steam is the most common surface-heating medium. Hot water is used for low temperature heating,

104 DRYING

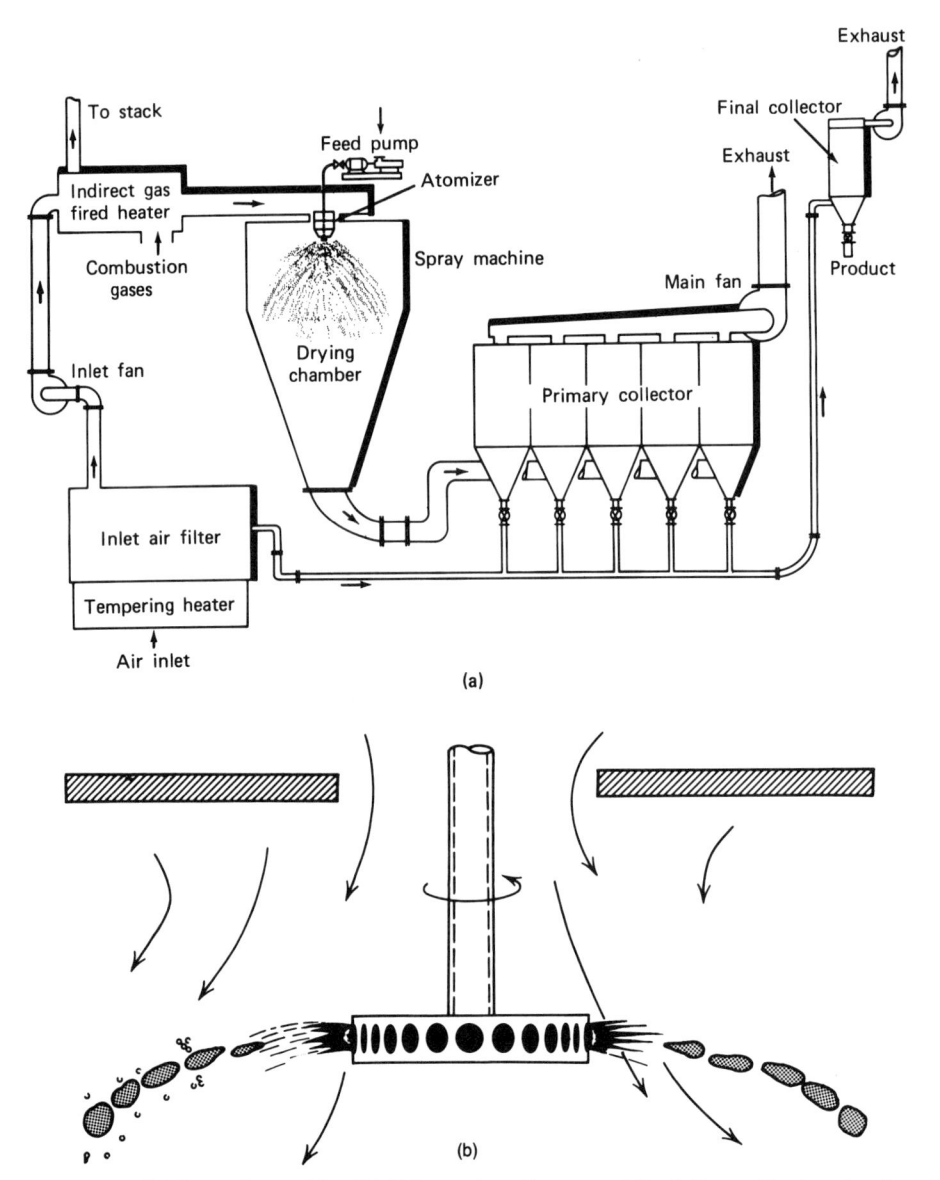

Figure 21. (a) Spray dryer with a (b) disk atomizer. Courtesy of Stork-Bowen Engineering, Inc.

and heat transfer oils are used for high temperatures. An organic vapor is an excellent high temperature heating medium except for rotating vessels, because rotary joints are not leak-proof. The film resistance to heat transfer of a liquid is usually much greater than that of a condensing vapor. Therefore, when a liquid medium is employed, the overall heat transfer coefficient through the conducting surface to the material is greatly reduced. For drying calculations, the film resistance of condensing steam may be assumed as essentially zero. Rotary dryers operated at temperatures exceeding 200°C are usually enclosed in a furnace and the shells are heated externally by electric heaters, radiant burners, or combustion products; temperature control is maintained by monitoring the shell temperature.

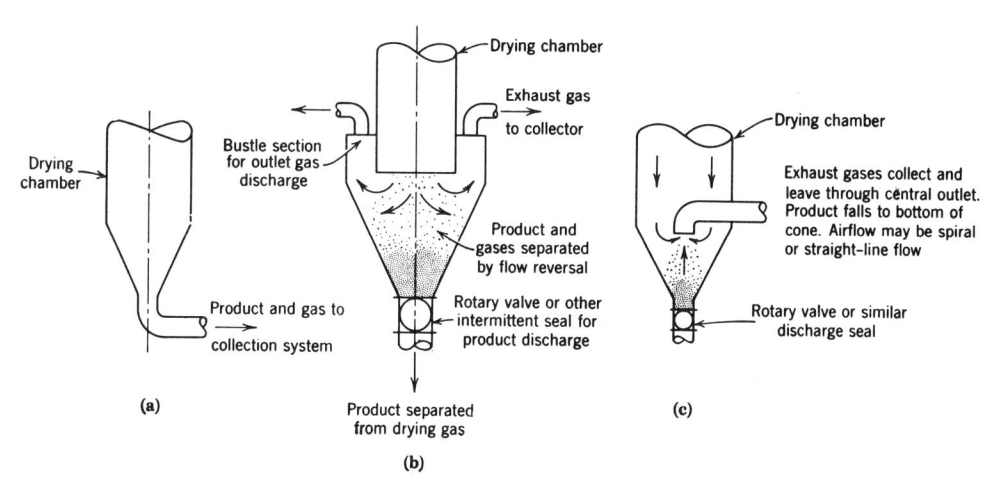

Figure 22. Product removal arrangements in the cones of cocurrent spray dryers. (**a**) Conventional cocurrent; (**b**) agglomeration; (**c**) dust-free product separation.

Continuous dryers are usually countercurrent so that the purge gas can be humidified to 80–90% saturation and its quantity can be minimized. For dusty materials, indirect-heat dryers offer advantages over direct-heat dryers because of lower gas requirements. For low temperature, indirect-heat drying operations, vacuum dryers are usually employed.

The heat transfer coefficient between particulate solids and a conducting surface is affected by particle size, shape, specific gravity, and the intensity of particle mixing near the surface. The heat transfer coefficient usually increases with specific gravity and bulk density, decreases as particle size increases, and, except in vacuum dryers operated at extremely low pressures, increases with the intensity of particle agitation and mixing (22).

Wet scrubbers are commonly employed for dust recovery. Dry dust collectors must be insulated to prevent condensation from high humidity purge gases. If a dry collection system is installed outdoors, external surfaces must be steam-traced or electrically heated.

Atmospheric-Pressure Dryers. *Steam-Tube Dryer.* A steam-tube (rotary) dryer is a horizontal, cylindrical, rotating shell in which are installed one or more circumferential rows of steam-heated tubes. These tubes extend axially the length of the shell and are connected to a combined steam-and-condensate manifold at one end. Figure 23 shows a steam-tube dryer that incorporates a dry product recycle system for feed conditioning. A cross-section of the dryer with two rows of tubes is included in Figure 23. Shell diameters are 0.5–4.0 m. Shell length may be 50 m. A large dryer having 1500 m^2 of tube surface in three rows of tubes may evaporate 8 t water/h. A nominal evaporation rate in a dryer operated at 1.0 MPa (10 atm) steam pressure is 5.0 kg water/(h·m^2 tube surface). For the high temperature dehydration of inorganic salts, dryers are built for steam pressures up to 3.5 MPa (35 atm). Steam is introduced and condensate is removed through a rotary joint attached to the manifold located at the product discharge and purge gas inlet end of the shell. Feed is introduced and purge gas is removed through a stationary throat piece that is fitted to the shell by a sliding seal at the end opposite the manifold. The dryer may be slightly inclined to aid product

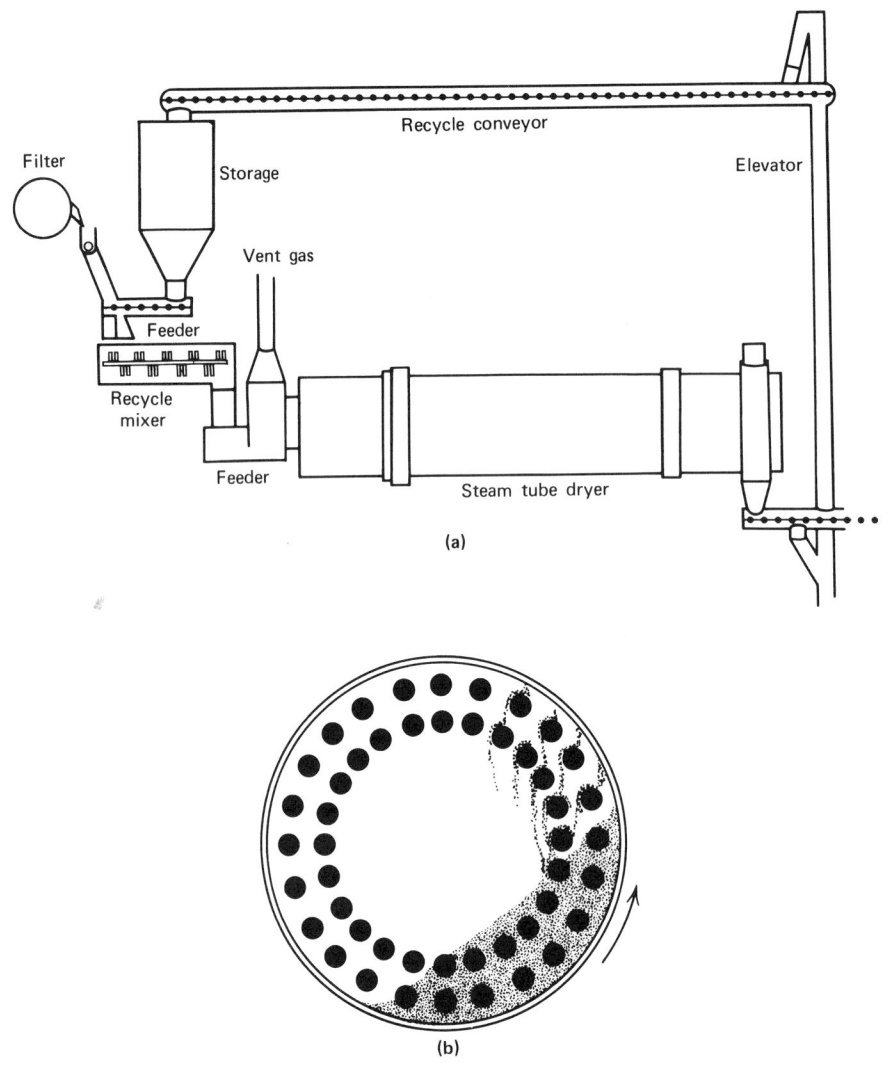

Figure 23. (a) Steam tube dryer with dry product, recycle system; (b) section through the steam tube dryer.

flow and tube drainage, but the amount of material is maintained at 15–20% of shell volume to assure that all tubes are submerged during each shell revolution. To prevent dust and vapor escape, dryers are operated slightly below atmospheric pressure. Steam-tube dryers are suitable for any particulate material that does not stick to metal when dry; they are probably the most commonly used form of indirect-heat dryer. Gas-tight seals are sometimes provided for inert-gas service, but these are intricate and expensive. A stationary, indirect-heat dryer, like a fluid-bed or screw-conveyor dryer, is preferred for inert-gas operations.

Rotary Dryer. This dryer resembles the direct-heat rotary dryer. However, (*1*) short turning bars are used in the place of lifting flights, hence material rolls in contact with the shell; (*2*) the shell is enclosed in a furnace and heat is transferred by convection and radiation to the shell, then by conduction and radiation to the material; (*3*) gas flowing through the shell is only that needed to purge vapor; and (*4*) the shell volume is filled 5–8% to prevent formation of an unheated core in the rolling material bed. By choosing suitable alloys, shells can be fabricated for temperatures up to 1000°C for drying materials that are too dusty for direct-heat dryers at temperatures that are too high for steam-tube dryers. Evaporative rates are 5–15 kg water/(h·m^2 shell surface).

Fluid-Bed Dryer. For indirect-heat drying, pipe or plate coils are installed in the fluid bed. Plate coils set vertically can provide up to 8 m^2 surface/m^3 bed volume. Excellent heat transfer coefficients are obtained in an environment of intense particle agitation and mixing, and drying rates may be 10–25 kg water/(h·m^2 surface). The fluidizing gas is also the vapor purge gas. An indirect-heat fluid-bed dryer approaches the ideal for solvent-recovery and inert-gas operations.

Screw-Conveyor and Disk Dryers. The indirect-heat, screw-conveyor dryer has a double-wall, steam- or oil-heated screw. The conveyor trough is also jacketed, but the trough represents only a small fraction of the total heating surface. Screw-conveyor dryers may contain up to four, parallel, heated screws. A common arrangement, used for both batch and continuous drying, consists of two screws that convey in opposite directions in a double U-shaped trough; this internal recycle system is used for drying slurries and liquid solutions. A bed of dry particles is loaded, circulated, and heated. Feed material is sprayed onto the moving bed on one side of the dryer, mixed, and dried. The product is removed through an adjustable gate on the opposite side. Particle recycle rate is controlled by gate height and screw speed. In batch operation, a heel of dry material is left in the trough after each batch for start-up of the subsequent batch. Purge gas is circulated through a vapor hood that covers the trough. Heated paddle conveyors may be used in the trough for drying lumpy materials, and these dryers may be used for any material that does not stick to metal when dry. A disk dryer is a U-shaped trough in which a series of closely spaced, parallel, heated plates is carried on a slowly rotating, heated shaft. A single, quadruple-screw conveyor dryer may contain 150 m^2 of heated surface, whereas a single disk dryer may carry 400 m^2 of heated surface on one shaft 15 m long with disks of 2 m diameter. Drying capacity of indirect-heat, screw-conveyor and disk dryers operated at 1.0 MPa (10 atm) steam pressure is about 15 kg water/(h·m^2 heated surface).

High Speed, Agitator Dryers. Both theory and empirical data demonstrate that the higher the mixing rate of wet, particulate solids in contact with a heated surface, the higher the rate of heat transfer to the solids. As system pressure is reduced, the benefit of mixing is reduced and at pressures less than 0.13 Pa (1.0 µm Hg) mixing appears to offer no benefit (22). Figure 24 shows drying-time profiles for sawdust in vacuum rotary dryers operated at moderate vacuum and otherwise similar conditions except for agitator peripheral speed. High speed agitator dryers operate at atmospheric pressure and take advantage of the higher heat transfer rates that can be obtained by intense gas-particle mixing near a heated surface. A typical dryer consists of a stationary, horizontal, jacketed shell inside which is installed an array of paddles mounted on a high speed shaft. Shell diameter is 0.5–2.0 m; shell length is 2–15 m, and paddle tip speed is about 15 m/s. The largest dryer has about 100 m^2 of jacket heating surface. Paddle inclination may be adjustable to control residence time. In continuous operation, residence time rarely exceeds 5 min. Particle attrition may be severe. Evaporative rates are 10–25 kg water/(h·m^2 jacket surface). To handle fluid or sticky

108 DRYING

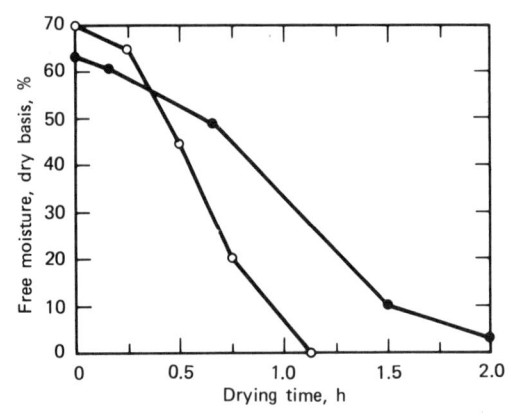

Figure 24. Effect of agitator speed on drying rates of sawdust in rotary vacuum dryers. Jacket temperature, 150°C; internal pressure, 6.7 kPa (50 mm Hg); ratio of heating surface to volume is 6. ●, 38 cm dia, 20 rpm; ○, 50 cm dia, 124 rpm.

feed materials, some agitators are provided with jacket scrapers. In any event, these dryers are suitable only for materials that do not stick to metal when dry.

Drum Dryers. Drum drying is effected by applying a thin film of slurry or solution to the outer surface of a rotating heated drum by means of applicator rolls, spray nozzles, or by dipping the drum into a pan containing the liquid. Drum rotational speed is such that drying occurs in a few seconds. The product may be heated to the drum temperature, but usually little thermal damage is experienced. Acceptable milk powders have been produced on drum dryers. Single-, twin-, and double-drum dryers are used; choice depends on material properties and application method. Any dryer may be operated at atmospheric pressure or, for heat sensitive materials, enclosed in a vacuum housing. A twin-drum dryer consists of two single drums operated in parallel, usually with a common feed system. On a double-drum dryer, feed is retained and partially concentrated in the reservoir formed by the nip between the drums. The clearance between the drums is adjusted to control film thickness. Liquid is conveyed through the clearance, dried, and the product is removed from the back sides of the drums by spring-loaded doctor knives. Single and twin-drum dryers may also be provided with nip-feed systems. However, separate applicator and smoothing rolls are needed to form the reservoirs and control film thickness (see also Coating processes). Drum dryer evaporative capacities are 50–80 kg water/(h·m² drum surface). The finned-drum dryer is a preforming device for materials intended for through-circulation drying. A thick slurry or paste is forced into circumferential grooves machined in the outer drum surface, partially dried, and removed in the form of short sticks by a finger scraper. Complete drying is rarely attempted. Horizontal-belt dryers are also used to dry solutions and slurries. The material is heated by radiant heaters mounted above and below the belt. Residence time can be longer than on drum dryers; thicker films can be dried, and the temperature is more easily controlled than on a drum dryer.

Cylinder Dryers. Also called can dryers, cylinder dryers are used to dry plastic, textile, and paper webs that are not self-supporting and cannot be handled in festoon, multipass-loop or tenter-frame dryers (see above under Continuous Dryers). A cylinder dryer may be one 3–5 m diameter drum, such as the Yankee dryer, or a number of heated drums arranged so the web passes over them in series. To improve conduction heat transfer, the web may be held in drum contact by an endless fabric belt that also absorbs liquid and is separately dried. Alternatively, high velocity, air-impingement nozzles may be mounted above and close to the exposed web surface to provide convection heat transfer. Humid air is removed through hoods placed over the drums. The drying of paper is probably the largest, single application of cylinder dryers (see Paper).

Vacuum Dryers. *Shelf Dryer.* Wet material is spread on trays that rest on heated shelves in a pressure-tight insulated chamber. The shelves are heated by steam, oil, or hot water, and the vacuum source is a pump or a series of steam jets (see Vacuum technology). Heat is conducted through the tray bottom into the material; some heat may also be transferred by radiation to the exposed material surface from the shelf above. However, heat transfer rates are low because there is rarely uniform contact between the tray bottom and its supporting shelf. On shelves heated by 200 kPa (ca 2 atm) steam, in a chamber maintained at 1 kPa (7.5 mm Hg) pressure, drying capacity is 1–4 kg water/(h·m^2 contacting shelf surface); however, because of the low drying-rate, dust loss is negligible and hence vacuum shelf dryers are useful for drying small quantities of valuable materials. They are rarely economical for production rates exceeding 200 t/year.

Freeze Dryer. The original form of the freeze dryer is a vacuum shelf dryer operated at less than 100 Pa (750 μm Hg) pressure, so that water is sublimed from ice. The purpose is to protect heat-sensitive materials from thermal damage and prevent shrinkage of porous materials such as freeze-dried foods (see Food processing). Wet material is frozen before being placed in the dryer, and the drying rate is controlled by the porosity and thermal conductivity of the frozen solid. Therefore, the best drying rate is obtained when all material surfaces are heated uniformly. Figure 25 shows the evolution of tray support methods. When a tray rests on a heated shelf, the bottom of the material is heated more rapidly than the upper surface. The shelf temperature must, therefore, be kept low to prevent melting of the bottom. Furthermore, sublimate must force its way through the frozen solid to escape. Rib trays conduct heat from the

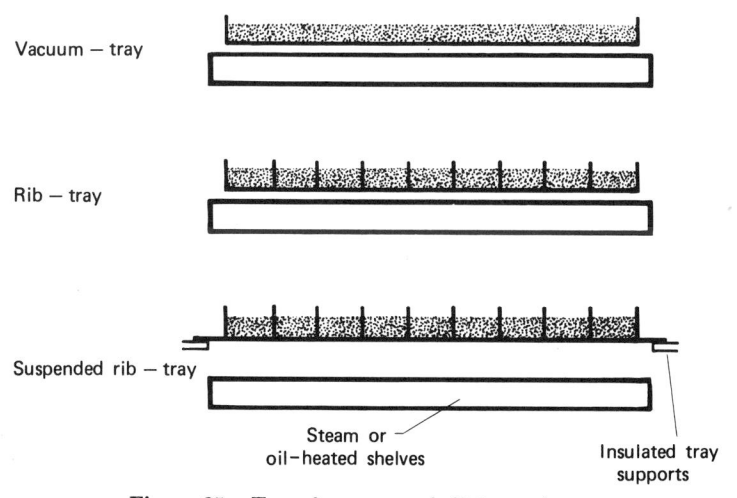

Figure 25. Trays for vacuum-shelf, freeze dryers.

tray bottom into the material and may offer channels for vapor escape. However, the bottom may overheat and, unless tray-to-shelf contact is uniform, hot spots may develop. The suspended rib tray depends entirely on heat transfer by radiation. Higher shelf temperatures can be used without product overheating and both top and bottom surfaces of the material are uniformly heated. Freeze dryers that are used for pharmaceuticals dried in vials include stoppering mechanisms to permit the vials to be sealed under vacuum. Drying capacity of a shelf-type, freeze dryer is 0.1–1.0 kg water/(h·m^2 exposed material surface). Freeze drying is also carried out in agitated, vacuum pan dryers, vibrating conveyors, and fluid beds. Dielectric heating has been successfully used for freeze drying (see Microwave technology).

Rotating Dryer. A rotating vacuum dryer is formed by providing a double-cone mixer with a jacket and a stationary, internal vapor tube passing through a rotary joint mounted on one trunnion. Volume may be 0.1–30 m^3 (25–8000 gal), filled 50–70% with material; internal pressure is 1–10 kPa (7.5–75 mm Hg). Rotating vacuum dryers are suitable for materials that do not stick to metal and do not pelletize during drying. Ratios of jacket surface to dryer volume in large dryers are low and cause drying cycles to be long. Large, rotating vacuum dryers often include internal heating surface in the form of pipe and plate coils to overcome this disadvantage. Drying capacity of a dryer heated with 200 kPa (ca 2 atm) steam, with an internal pressure of 1 kPa (7.5 mm Hg), is 4–5 kg water/(h·m^2 total heating surface).

Rotary and Pan Dryers. The rotary dryer is a stationary, horizontal, jacketed cylinder with an internal ribbon agitator. The agitator shaft and ribbon supports may also be heated if a material sticks to metal at no time during its drying cycle. For wet materials that are sticky, shell scrapers may be mounted on the ribbon. In any event, these dryers are suitable only for materials that do not stick to metal when dry. A typical dryer has a volume of 0.5–20 m^3 (130–5300 gal) to be filled 50–70%; internal pressure is 1–10 kPa (7.5–75 mm Hg). Drying capacity, when heated with 200 kPa (ca 2 atm) steam, with an internal pressure of 1 kPa (7.5 mm Hg), is 5–6 kg water/(h·m^2 total heating surface). Vacuum bag collectors are often mounted on these dryers to recover dust from the vapors before it reaches a wet condenser.

A pan dryer is a stationary, vertical, jacketed cylinder with a jacketed bottom and scraper agitator designed to handle doughlike materials that would overload or break the ribbon agitator in a rotary dryer. The bottom of the pan provides most of the effective heating surface. The slowly rotating agitator stirs the material in the bottom of the pan while the material gradually dries and breaks down to a granular or powderlike product. The largest pan dryer is about 4 m in diameter and holds a 10 m^3 (2600 gal) batch. Drying capacity when heated with 300 kPa (ca 3 atm) steam, with an internal pressure of 1 kPa (7.5 mm Hg), is 2–4 kg water/(h·m^2 total heating surface). These dryers are not suitable for materials that stick to metal when dry.

Radiant-Heat Dryers

Heat transfer by radiation is controlled by the radiant-source temperature and the radiation-adsorption properties of the material. For drying, the source may consist of a number of incandescent lamps, reflector-mounted quartz tubes, electrically heated ceramic surfaces, or ceramic-enclosed gas burners. The source temperature may be 800–2500 K. Radiant energy does not penetrate most solid surfaces and, under some conditions, the temperature of the material surface may rise above the liquid boiling point. Radiant-heat dryers are not suitable for large objects and deep layers of material in which the drying rates are controlled by internal heat and mass transfer mechanisms. When drying heat sensitive materials, low temperature sources should be used. The best applications for radiant-heat dryers are thin coatings and paint films on metal, polymers, wood, and paper during continuous and uniform conveying. Banks of radiant sources are placed above and on both sides of the material in an enclosed tunnel that is designed to minimize reflection losses to the outside. When materials are dried that may degrade or burn if exposed too long, provisions are made to shutter the radiant sources in the event material flow is interrupted. Purge gas must be provided to remove vapors from the tunnel. A common practice is to pass the gas behind the source enclosures to preheat the gas and cool the enclosures. Radiant heaters are used in multipass-loop, tenter-frame, and roller-conveyor dryers. Drying rates are 10–100 kg water/(h·m^2 exposed material surface).

Dielectric Dryers

Dielectric dryers include radio-frequency dryers that operate in the range of 3–150 MHz, and microwave dryers that operate at 915 and 2450 MHz. As depicted in Figure 26, a radio-frequency dryer may consist of two flat, metal plates between which a wet material is placed or conveyed. This arrangement forms a capacitor, the plates of which are connected to a high frequency generator. During one half of an electrical cycle, one plate has a positive charge and the other has a negative charge. A half cycle later, the polarities are reversed, which reverses the molecular stress in the material. It is this continuous, rapid reversal of polarities that generates heat in the material. One explanation may be that heating is caused by a phenomenon analogous to frictional stress; ie, molecular friction. In any event, because the electrical field is uniform throughout the material, heating is also uniform; ie, the material is heated as rapidly at the center as at the surface. For drying objects of irregular shape and to avoid hot spots, electrodes are needed that conform to the material surfaces. Heating thick materials too rapidly may generate sufficient internal steam pressure so as to rupture the product. Dielectric dryers are used for drying rayon cakes, sand molds, lumber, and food products. Generally, because of high capital and operating costs, a dielectric dryer is best suited for removing a small quantity of moisture that is difficult to reach by conventional, surface heating methods. Currently available microwave generators for drying are limited to about 30 kW output; ie, about 30 kg water per hour of evaporation. At present, microwave dryers are only used for removing small quantities of water from materials in situations in which the energy must be accurately applied; eg, the drying of glue lines on envelopes. Response to dielectric energy is a function of the material's dielectric constant and dissipation factor; the product of these two

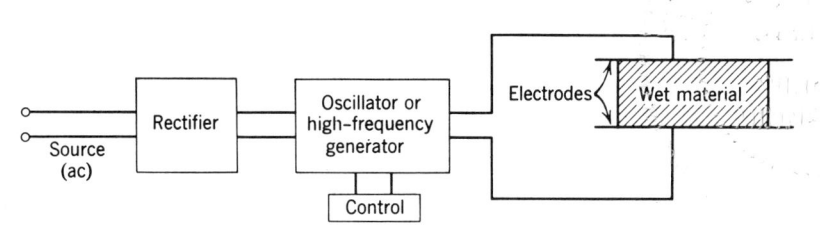

Figure 26. Diagram of a dielectric dryer.

112 DRYING

quantities is the loss factor. Any material having a loss factor greater than 0.05 is a potential candidate for dielectric heating. Water and many alcohols are good energy absorbers. Most hydrocarbon liquids are poor absorbers. A properly designed dielectric dryer for a water-containing material evaporates about 1.0 kg/kWh energy input. The generator consumes about 1.5 kWh/1.0 kWh energy input to the material (see Furnaces, electric; Microwave technology).

Nomenclature

a	= heat transfer area per unit volume of bed, 1/m
A	= surface area for vaporization and heat transfer, m²
c	= concentration of one component in a two-component phase of A and B
C_s	= humid heat, kJ/(kg·K)
d	= one half the slab thickness for drying from two sides, or total thickness for one-side drying, m
d_m	= layer thickness, m
$dw/d\theta$	= drying rate, kg/s
$dw'/d\theta$	= drying rate, kg liquid/(s·kg), dry material
$dw'/d\theta_f$	= falling rate, drying rate, kg/(s·kg), dry basis
D	= liquid diffusivity, m²/s
D_{AB}	= binary diffusivity of the phase A–B
D_c	= cylinder diameter, m
$(D_p)_m$	= mean diameter of spherical particles, m
F	= void fraction
h_c	= heat transfer by convection, kW/(m²·K)
h_t	= the sum of all convection, conduction, and radiation components of heat transfer, kW/(m²·K)
H	= humidity, kg moisture/kg air
H'	= Henry's constant
H_o	= humidity of air as read from a humidity chart based on 101.3 kPa total pressure and at the measured dry- and wet-bulb temperatures, kg/kg
H_s	= adiabatic saturation humidity, kg/kg
H_w	= saturation humidity, kg moisture/kg air
i	= a constant that is independent of temperature
k'	= mass transfer coefficient, kg/(s·m²·kg/kg)
k'_a	= mass transfer coefficient, kg/(s·m²·kPa)
K	= a function of the constant rate drying rate at the critical moisture content
L_s	= latent heat of vaporization at t_s, kJ/kg
L'_s	= latent heat of vaporization at t'_s, kJ/kg
p	= partial pressure of the vapor in air, kPa
p_m	= dry material bulk density, kg/m³
p'_s	= liquid vapor pressure at t'_s, kPa
p_w	= water vapor pressure at the gas temperature, kPa
P	= total system pressure, kPa
% (rh)	= percent relative humidity
t	= gas dry-bulb temperature, K
Δt_m	= log-mean temperature difference between all convection, conduction, and radiation heat sources and the material surface, K
t_s	= adiabatic saturation temperature, K
t'_s	= liquid surface temperature, K
t_w	= gas wet-bulb temperature, K
U_a	= the average, volumetric heat transfer coefficient, kW/(m³ dryer volume·K)
V	= dryer volume, m³
W_c	= the critical moisture content, kg/kg, dry basis
W_e	= the moisture content in equilibrium with the environment, kg/kg, dry basis

W_θ = the moisture content at any time, kg/kg, dry basis
x = dry basis, moisture content, kg/kg
z = distance in the direction of diffusion
Z = cylinder height, m
θ = drying time
θ_f = the falling rate drying time, s

BIBLIOGRAPHY

"Drying" in *ECT* 1st ed., Vol. 5, pp. 232–265, by W. R. Marshall, Jr., University of Wisconsin; "Drying" in *ECT* 2nd ed., Vol. 7, pp. 326–378, by W. R. Marshall, Jr., University of Wisconsin.

1. S. F. Sapakie, D. R. Mihalik, and C. H. Hallstrom, "Drying in the Food Industry," *Chem. Eng. Progr.* **75**(4), 44 (1979).
2. *Carrier Psychrometric Chart, Catalog No. 794-005,* copyrighted by Carrier Corporation, Syracuse, N.Y., 1975.
3. W. L. McCabe and J. C. Smith, *Unit Operations of Chemical Engineering,* 3rd ed., McGraw-Hill, New York, 1976.
4. T. K. Sherwood, R. L. Pigford, and C. R. Wilke, *Mass Transfer,* McGraw-Hill, New York, 1975.
5. J. H. Perry, *Chemical Engineers' Handbook,* 3rd ed., McGraw-Hill, New York, 1950.
6. R. H. Perry and C. H. Chilton, *Chemical Engineers' Handbook,* 5th ed., McGraw-Hill, New York, 1973, Section 20.
7. R. B. Keey, *Drying: Principles and Practice,* Pergamon Press, New York, 1972.
8. H. H. Macey, *Trans. Br. Ceram. Soc.* **41,** 73 (1942).
9. D. H. Charlesworth and W. R. Marshall, *AIChE J.* **6**(1), 9 (1960).
10. J. Crank and G. S. Park, *Diffusion in Polymers,* Academic Press, Inc., New York, 1968.
11. F. Kneule, *Das Trocknen* Sauerlander, Aarau, FRG, 1975.
12. O. Krischer, *Die Wissenschaftlichen Grundlagen der Trocknungstechnik,* Springer Verlag, Berlin, Wilmersdorf, FRG, 1963.
13. J. Lambert and W. R. Marshall, *Conference on Freeze-Drying of Foods,* National Academy of Sciences, Nat. Res. Council, 1962.
14. W. R. Marshall, *Chem. Eng. Prog. Monogr. Ser.* **50,** 2 (1954).
15. K. Masters, *Spray Drying,* 2nd ed., John Wiley & Sons, Inc., New York, 1976.
16. G. Nonhebel and A. A. H. Moss, *Drying of Solids in the Chemical Industry,* CRC Press, Cleveland, Ohio, 1971.
17. W. B. Van Arsdel and M. J. Copley, *Food Dehydration,* Avi, Westport, Conn., 1963.
18. A. Williams-Gardner, *Industrial Drying,* CRC Press, Cleveland, Ohio, 1971.
19. H. S. Carslaw and J. C. Jaeger, *Conduction of Heat in Solids,* 2nd ed., Oxford University Press, London, Eng., 1959.
20. J. Crank, *The Mathematics of Diffusion,* 2nd ed., Oxford University Press, London, Eng., 1964.
21. J. P. Hartnett and T. F. Irvine, Jr., *Advances in Heat Transfer,* Vol. 13, Academic Press, New York, 1977.
22. N. H. Afgan and E. U. Schlünder, *Heat Exchangers,* McGraw-Hill, New York, 1974.

General References

B. Gebhart, *Heat Transfer,* 2nd ed., McGraw-Hill, New York, 1971.
W. M. Kays, *Convective Heat and Mass Transfer,* McGraw-Hill, New York, 1966.
D. B. Spalding, *Convective Mass Transfer,* McGraw-Hill, New York, 1963.
F. A. Zenz and D. F. Othmer, *Fluidization and Fluid-Particle Systems,* Reinhold, New York, 1960.
R. B. Keey, *Introduction to Industrial Drying Operations,* Pergamon Press, Elmsford, N.Y., Oxford, Eng., 1978, 376 pp.

PAUL Y. MCCORMICK
E. I. du Pont de Nemours & Co., Inc.

DRYING AGENTS

There are many substances that take up (sorb) water from their surroundings by one or more of a number of different physical or chemical mechanisms, or both. These substances are widely used for removing water from gases, liquids, and solids. For this reason they are called drying agents or desiccants. Drying agents may be liquids or solids. They may be used repetitively by regenerating the desiccant after use to return it to its active state. In some cases they are used only once, and the spent desiccant is discarded. Drying agents are used either in a static (batch-wise) or dynamic (continuous or semicontinuous) mode.

In the last twenty years the use of drying agents has grown rapidly. For example, dehydration of process streams to extremely low water contents has become much more prevalent partly because of the popularity of cryogenic processing for the separation of gas mixtures (see Cryogenics). Other examples of the industrial uses of drying agents, designated as being either a dynamic or static application, are given in Table 1. The list is not all inclusive and ignores various laboratory uses.

Drying agents have different fundamental characteristics in terms of water capacity and the rate of water sorption. The degree of water removal achieved is usually expressed in terms of the water content remaining in the substance that has been dried. This can be expressed in several ways, such as relative humidity (at atmospheric pressure only), relative saturation (at elevated pressure), dew point (above 0°C), ice point (below 0°C), or parts per million (ppm by weight or volume). The effectiveness of any drying agent can be measured in terms of its water capacity. In static applications this is usually the true equilibrium or stoichiometric capacity. In dynamic systems, the rate of water removal must be taken into account. Usually, to allow for mass transfer zones, additional drying agent is used (see Mass transfer). In these instances, the dynamic capacity (also termed breakthrough capacity) falls short of the true equilibrium capacity (5).

Table 1. Applications of Drying Agents

Industry	Application	Classification
compressed air	prevent freeze-ups and corrosion in air actuated components	dynamic
air separation	prevent ice formation in heat exchangers before cryogenic distillation	dynamic
natural gas	prevent corrosion and hydrate formation in pipelines	dynamic
	remove water before cryogenic hydrocarbon recovery	
	dry LP gas to prevent freeze-ups during vaporization	
petrochemical	remove moisture before low temperature fractionation	dynamic
	remove moisture during the rejuvenation/burnoff of spent catalysts	
	prevent side reactions during catalytic refining	
chemical	remove water which is a diluent or contaminant of some finished product	static or dynamic
heavy chemical	prevention of caking and corrosion	static or dynamic
storage and shipping	prevention of food deterioration and corrosion of equipment relative humidity control	static
moisture vapor control	lower the dew point in hermetically sealed systems where moisture condensation could occur	static
refrigeration	remove moisture from refrigerant loops	dynamic

Mechanism

The dehydration mechanism of drying agents may be classified as follows:

Type 1—Chemical reaction: (a) formation of a new compound; (b) formation of a hydrate.

Type 2—Physical absorption with constant relative humidity (solid + water → saturated solution).

Type 3—Physical absorption with variable relative humidity (a solid or liquid + water → diluted solution).

Type 4—Physical adsorption.

These mechanisms are characterized by the relative magnitudes of the heats of reaction, solution, or adsorption (see Adsorptive separation). Phosphorus pentoxide is a Type 1 drying agent and reacts with water to form a polyphosphoric acid (6).

$$x\ P_2O_5 + (2x+1)\ H_2O(g) \rightarrow HO\underbrace{\left(\begin{array}{c}O\\ \|\\ -P-O-\\ |\\ OH\end{array}\right)}_{2x} H \quad \Delta H = 109.5\ \text{kJ/mol}\ (26.17\ \text{kcal/mol})$$

Type 1 drying agents liberate the largest amounts of heat and have to be used with appropriate care.

Calcium chloride is a Type 1 drying agent and reacts with water to form a hydrate.

$$CaCl_2 + H_2O(g) \rightarrow CaCl_2 \cdot H_2O \quad \Delta H = 70\ \text{kJ/mol}\ (16.7\ \text{kcal/mol})$$

The more highly complexed hydrates of calcium chloride ($CaCl_2 \cdot n\,H_2O$ where $n > 2$) may also exhibit the characteristics of a Type 2 drying agent, since the hydrated species can physically absorb additional water to form a saturated solution. The term absorption (qv) is used to describe the phenomenon that occurs when a gas or vapor penetrates the solid structure, producing a saturated solution.

$$CaCl_2 \cdot 6H_2O + n\ H_2O(g) \rightarrow \text{saturated solution}$$

Because there are a number of hydrates of calcium chloride, the one that is in equilibrium with a saturated solution is a function of the temperature. In this case, the solid is dissolved as it absorbs water to form the saturated solution, and three phases are present: solid, saturated solution, and vapor. Systems having these three phases, or two solids and a vapor phase, have a constant vapor pressure at a given temperature. Therefore, Type 2 drying agents can be used to maintain a constant relative humidity.

Ethylene glycol is an example of a Type 3 drying agent. Because the solution produced is unsaturated, only two phases exist, solution and vapor.

$$\text{ethylene glycol} + n\ H_2O(g) \rightarrow \text{dilute solution}$$

For this system, the vapor pressure is a function of both temperature and the concentration of water in the dilute solution.

Molecular-sieve zeolites are an example of a Type 4 drying agent. Water is removed by physical adsorption but at no time is there a phase change or solution of the adsorbent.

$$\text{molecular-sieve adsorbent} + H_2O(g) \rightarrow H_2O\ (\text{adsorbed})$$

116   DRYING AGENTS

Adsorption is a phenomenon whereby molecules in a fluid phase spontaneously concentrate on a solid surface without any chemical change (see Adsorptive separation). The adsorbed molecules are bound to the surface by weak interactions between the solid and gas, similar to condensation (van der Waals) forces. Since adsorption is a surface phenomena, all practical adsorbents possess large surface areas relative to their mass.

Because moisture can contaminate a drying agent during storage, they should be analyzed before use. If possible, the materials should be reactivated (regenerated) before putting them in service.

Static Drying Agents

Many liquids are dried batchwise rather than continuously. The drying agent is added to the liquid, and sufficient time is allowed to dry the product. The liquid is then separated from the drying agent by filtration, decantation, or distillation. Drying agents employing Type 1 or 2 mechanisms are generally used for these applications. The most commonly used, from a long list of these materials, are discussed below.

Barium Oxide. Barium oxide [1304-28-5] (Type 1, nonregenerative) is used primarily as a laboratory drying agent (7–8). It is the only drying agent that continues to dry even at red heat. Barium oxide is relatively expensive and cannot be regenerated by conventional methods. Therefore it is not used extensively in commercial applications even though it is one of the most efficient drying agents (see Barium compounds).

Calcium Chloride. Calcium chloride [10043-52-4] (Type 1 or 2, regenerative) can either be a solid or liquid drying agent (7,9). The major advantage is that its cost is low enough to permit discarding after use in small units. Commercial, anhydrous calcium chloride is available in a range of compositions from $CaCl_2.0.05H_2O$ to $CaCl_2.0.25H_2O$. When used in a desiccator, its action is rapid but its inherent desiccating ability is such that it does not produce a "bone dry" condition.

Figure 1 depicts the phase diagram for anhydrous calcium chloride. The right hand portion shows an area represented by straight horizontal lines. These represent two solid phases and a gas phase for vertical line intersections. In addition, a solid phase, saturated solution, and a vapor phase occur in the regions between the vertical lines. The lower left-hand corner represents the ice solution line. The region in between with skewed isothermal lines represents unsaturated solutions; the vapor pressure varies as a function of temperature (see Calcium compounds).

Calcium Oxide. Calcium oxide [1305-78-8] (Type 1, nonregenerative), also termed lime (qv) or quicklime (7,10), is relatively inexpensive. It is prepared by roasting calcium carbonate (limestone) and is available in a soft and a hard form according to the way in which it was burned. For desiccant service, soft-burned lime should always be used. Calcium oxide is most commonly used to dehydrate liquids and is most efficient when it can be heated to speed the reaction rate. The reaction product is calcium hydroxide which crumbles as it picks up moisture.

Calcium Sulfate. Calcium sulfate [7778-18-9] (Type 1, regenerative) is commonly known under the trade name Drierite (7–8,11). It occurs in nature in the anhydrous form ($CaSO_4$) and in the hydrated form ($CaSO_4.2H_2O$), commonly known as gypsum. When prepared properly, the pore volume within the granules creates small capillaries

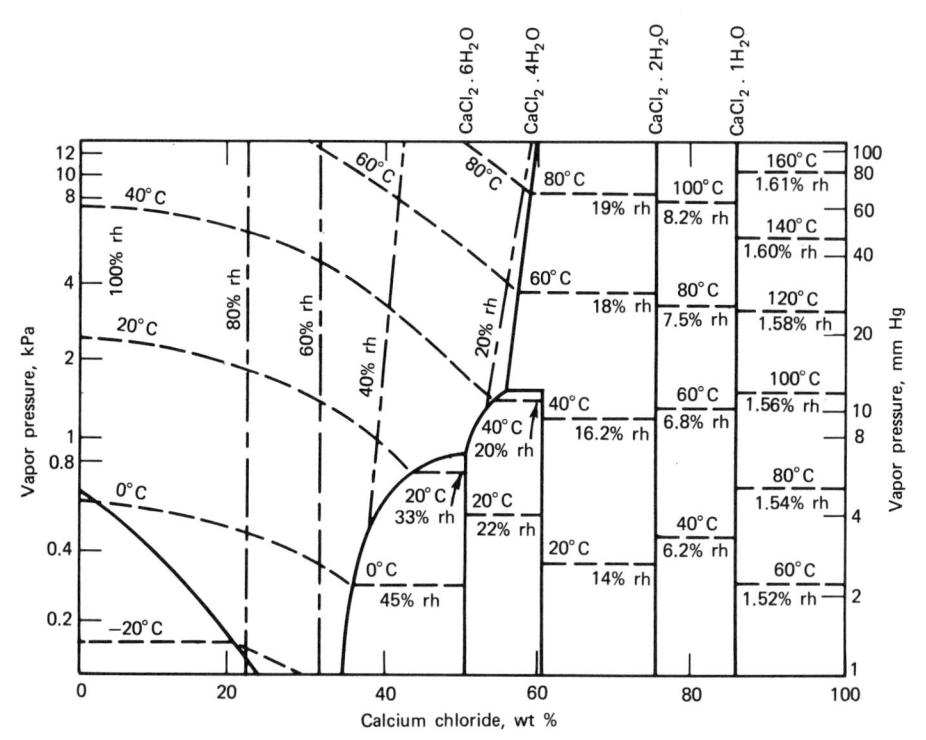

Figure 1. Vapor pressure and relative humidity over CaCl$_2$ solutions and solids.

which increase this material's somewhat low water capacity by sorbing additional moisture. The first stages of water removal occur via adsorption and a chemical reaction to form a hemihydrate [10034-76-1] (CaSO$_4$·½H$_2$O). The material can be regenerated repeatedly by heating to about 200°C. However, above 300°C it loses some of its desiccating power. Calcium sulfate is used extensively because it is chemically inert to most materials, reusable, and inexpensive.

Lithium Chloride. Of the metal halides CaBr$_2$ [7789-41-5], ZnCl$_2$ [7646-85-7], and CaCl$_2$, lithium chloride [7447-41-8] (Type 1, nonregenerative) is the most effective for water removal (7,9). All are available in the form of deliquescent crystals. The hydrates of lithium chloride are LiCl·nH$_2$O where n = 1, 2, or 3. Lithium chloride is very stable in air and does not form a carbonate as readily as the other metal halides (see Lithium compounds).

Perchlorates. The three common perchlorates (Type 1, nonregenerative) used as drying agents are barium perchlorate [13465-95-7], Ba(ClO$_4$)$_2$, lithium perchlorate [7791-03-9], LiClO$_4$, and magnesium perchlorate [10034-81-8], Mg(ClO$_4$)$_2$. The last is the most efficient with drying action above 100°C. Even the higher hydrate form has good drying capacity. Perchlorates are strong oxidizing agents and should never be used in the presence of organic compounds since the mixture is highly explosive. For this reason, perchlorates are usually not regenerated (see Chlorine oxygen acids and salts).

Phosphorus Pentoxide. Phosphorus pentoxide [1314-56-3] (Type 1, nonregenerative) is made by burning phosphorus in dry air. It is considered the reference material by which other drying agents are judged (6–7). It removes water first by adsorption, followed by the formation of several forms of phosphoric acid. Phosphorus pentoxide has a high vapor pressure and should be used only below 100°C. Its main drawback is that as moisture is taken up, the surface of the granules becomes wetted impeding further moisture removal. For this reason, phosphorus pentoxide is sometimes mixed with an inert material (see Phosphoric acids and phosphates).

Sodium and Potassium Hydroxides. Sodium hydroxide [*1310-73-2*] and potassium hydroxide [*1310-58-3*] (Type 1, nonregenerative) are commonly used when moisture and carbon dioxide or hydrogen sulfide must be removed simultaneously (7). Fused sticks or solutions of the alkali hydroxides are frequently used. These materials must be handled with care in order to prevent serious skin burns.

Capacity and Efficiency. Figure 2 shows the drying capacity of a group of drying agents as a function of relative humidity. The higher capacity agents go through the various hydrate levels yielding fairly broad ranges of constant relative humidities as moisture is picked up. However, these compounds do not provide very low relative humidities or dew points. Figure 3 shows the water vapor pressure over a few drying agents as a function of temperature. Addition of more drying agent lowers the vapor pressure. The best performance, or lowest dew point, occurs with excess drying agent. The minimum dew points (or ice points) attainable at room temperature are given (12–13) in Table 2 for atmospheric pressure.

The design of static drying systems is usually straightforward. The total amount of moisture to be removed must first be calculated from the volume of the system and

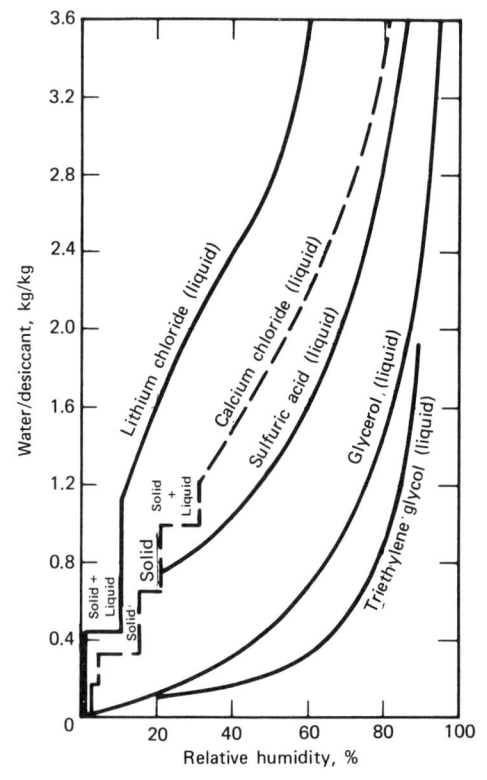

Figure 2. Drying capacity of selected drying agents.

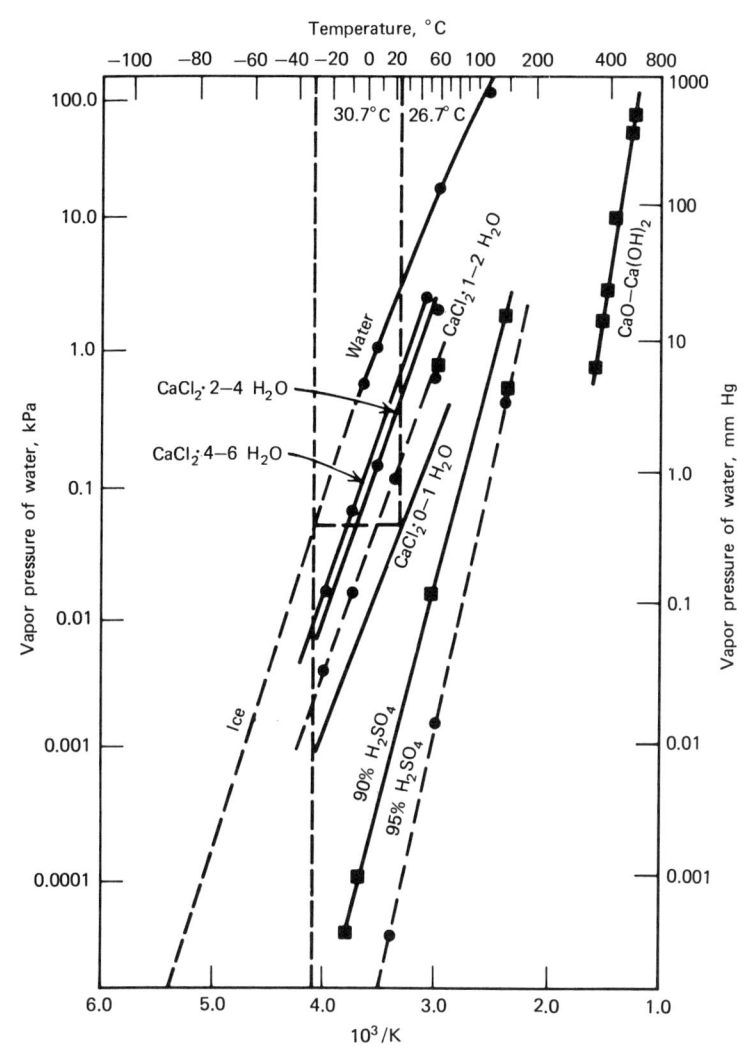

Figure 3. Water vapor pressure over ice and several drying agents.

the initial water concentration. Depending on the final moisture content desired, a drying agent can be selected based on its compatibility with the material to be dried and its ability to produce the final dew point. The equilibrium capacity of the drying agent must be determined at the system temperature and the final water concentration. An amount of drying agent must be used so that the total amount of water to be removed does not exceed the predetermined equilibrium capacity of the agent. The agent and the material are then placed in intimate contact. After sufficient time, the system is in equilibrium with the drying agent.

Dynamic Drying Agents

Continuous drying is employed in many operations where it is not practical to dry a volume of gas or liquid in a batchwise fashion. When a solid dynamic drying agent

120 DRYING AGENTS

Table 2. Performance of Chemical Drying Agents at Room Temperature

Substance	Minimum humidity[a]	Minimum dew point at 101.3 kPa[b], °C
P_2O_5	0.0193	−98
BaO	0.503	−80.5
KOH fused	1.546	−73
CaO	2.32	−70.5
H_2SO_4	2.32	−70.5
$CaSO_4$ anhyd.	3.87	−67
Al_2O_3	3.87	−67
KOH sticks	10.83	−60
NaOH fused	123.7	−40
$CaBr_2$	139.2	−39
$CaCl_2$ fused	262.9	−33
NaOH sticks	618.6	−25
$Ba(ClO_4)_2$	634.0	−24.5
$ZnCl_2$	657.2	−24
$ZnBr_2$	896.9	−21
$CaCl_2$ granular	1159.8	−18

[a] mg H_2O/kg dry air.
[b] To convert kPa to mm Hg, multiply by 7.5.

is used, the flowing stream is passed over a fixed bed of drying agent which must have the physical properties to allow the flowing stream to pass readily through. When liquids are used, the drying is usually achieved by countercurrent contact of the gas (flowing up) against the liquid drying agent (flowing down). These drying agents are usually regenerable.

Liquid Agents. *Sulfuric Acid.* Sulfuric acid [7664-93-9] (Type 3, regenerative) is used extensively throughout the chemical industry to dry acidic and corrosive gases. It has good capacity and drying capability as illustrated by the vapor pressure curves in Figure 4. At 25°C the dew point attainable in gases dried with 95% sulfuric acid is less than −75°C.

Sulfuric acid is used in circulating towers like those depicted in Figure 5, with the gas flowing countercurrent to the acid. The spent acid is removed from the primary contactor at a 50% H_2SO_4 content and may be used for chemical manufacturing or recycled after reconcentration. The make-up or recycled acid is usually introduced at 93% H_2SO_4 (sp gr 1.84, 66° Bé or higher) because it is this moisture content that establishes the final moisture level of the dried product gas. However, sulfuric acid is highly corrosive and protective clothing and eye protection must be provided.

Glycerol, Glycol, and Other Polyhydric Alcohols. These substances (Type 3, regenerative) are widely used to dry gases (14). However, they can only produce dew points in the range of −15 to 0°C. They have somewhat lower capacity than sulfuric acid, as shown in Figure 2, but are very effective when either injected or employed in a multistage contactor to achieve dew point depression.

Ethylene glycol [107-21-1], diethylene glycol [111-46-6], and triethylene glycol [112-27-6] are used extensively in the natural gas industry to inhibit hydrate formation (15). Diethylene glycol (DEG) is the most widely used and the dew point that can be achieved in the treated gas (16) is a function of the percentage of glycol present and the contact temperature, as shown in Figure 6. The major advantages of glycol dehy-

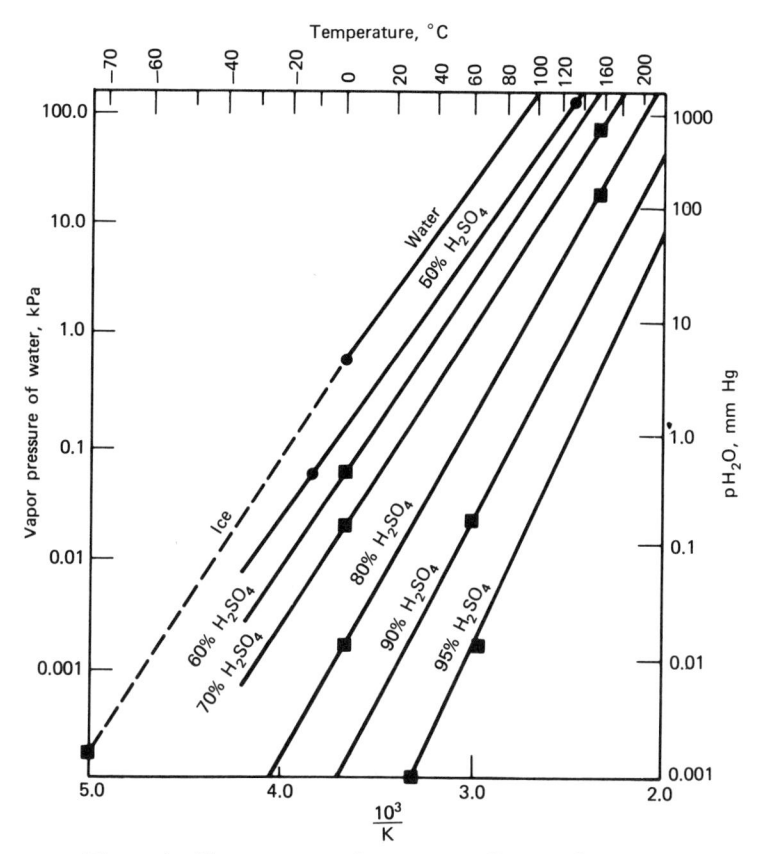

Figure 4. Vapor pressure of water over sulfuric acid solutions.

dration are low cost, ease of regeneration, and minimal losses of the drying agent caused by solubility or vapor pressure during subsequent recovery or reclamation.

A typical flow diagram of a glycol dehydration plant is shown in Figure 7. The absorber, or gas–liquid contactor, operates with glycol flowing downward, countercurrent to the gas stream. The regenerator, or stripping column, operates at low pressures in order to keep the column temperature below the decomposition temperature of the glycol. The regenerated glycol is then recirculated to the absorber (see Alcohols, polyhydric; Glycerol; Glycols).

Solid Agents. Solid drying agents used in dynamic applications are most commonly termed adsorbents. Because they are used in large packed beds through which the gas or liquid to be treated is passed, the adsorbents are formed into solid shapes which allow them to withstand the static (fluid plus solid head) and dynamic (pressure drop) forces imposed on them. The most common shapes are granules, pellets or beads. As the particle size increases for a fixed fluid flow rate, the pressure drop decreases. However, as the particle size increases, the mass transfer resistance increases and the adsorbent takes longer to achieve its final equilibrium water capacity. As a result, larger amounts of desiccant are required. Therefore, the optimal particle size is a compromise between pressure drop (energy) and amount of desiccant.

122   DRYING AGENTS

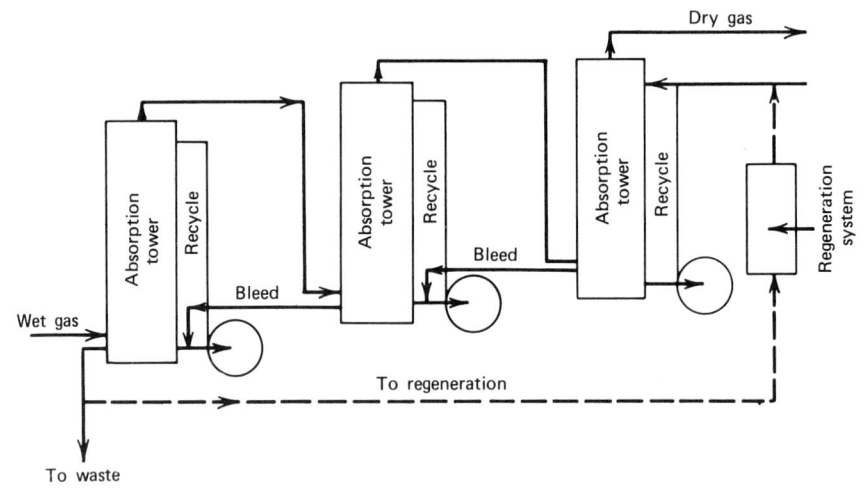

Figure 5.   Continuous sulfuric acid drying system.

Activated Alumina. This material (Type 4, regenerative) (17–18) is made by the calcination of an alumina gel or aluminum oxide trihydrate [*21645-51-2*] (Al$_2$O$_3$.3H$_2$O) into crystallized phases of transition aluminas. Depending on the manufacturing procedure and starting material, the final product has different degrees of specific surface and pore volume. There are many manufacturers and grades; the physical properties of a typical Grade A activated alumina are given in Table 3.

The water removal mechanism is adsorption as it is for all Type 4 drying agents. The capacity of such materials is often shown in the form of adsorption isotherms as depicted in Figure 8**a** and **b**. The initial adsorption mechanism at low concentrations of water is believed to occur by monolayer coverage of water on the adsorption sites. As more water is adsorbed, successive layers are added until condensation or capillary action takes place at water saturation levels greater than about 70% relative humidity. At saturation, all the pores are filled, and the total amount of water adsorbed, expressed as a liquid, represents the pore volume of the adsorbent.

In order to regenerate activated alumina, heating to a temperature of 150–200°C is sufficient to recover nearly all of the initial water capacity. This is usually accomplished by passing a heated gas through the adsorbent bed. Unless the surface is fouled by the gas or liquid being dried, the life of activated alumina is good and usually depends on the number of regeneration cycles (see Aluminum compounds).

Silica Gel. Silica gel [*7631-86-9*] (19–21) (Type 4, regenerative) is made by dehydrating high-purity silica hydrosol. The final product is high in purity (99.7% SiO$_2$) which contributes to its chemical inertness. The typical physical properties are listed in Table 3. The pore size of silica gel has the broadest range of the three major solid desiccants. Therefore, it can adsorb larger molecules in addition to water (critical diameter = 0.265 nm). The average pore diameter is ca 2.5 nm.

The capacity of silica gel is shown on Figures 8**a** and **b,** and the shape of the isotherm is similar to activated alumina. At saturation ($p/p° = 1.0$), silica gel has the highest capacity of the desiccants shown, taking up 40 kg H$_2$O/100 kg of adsorbent. However, some high-capacity silica gels tend to shatter in the presence of liquid water. In applications where liquid water may be present, a lower-capacity water-resistant silica gel must be used.

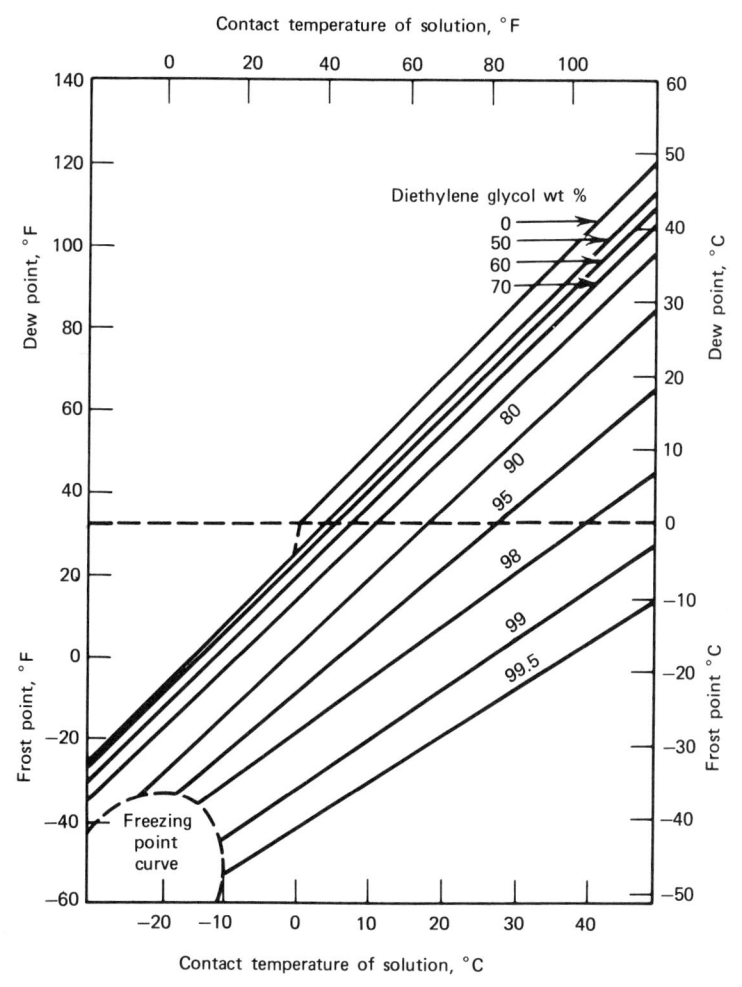

Figure 6. Dew points of aqueous diethylene glycol solutions.

The normal regeneration temperature for silica gel is 175°C. In hydrocarbon service, higher temperatures (225–275°C) are recommended to desorb heavy hydrocarbons which tend to foul the adsorbent during prolonged use (see Silicon compounds).

Molecular Sieves. These materials (Type 4, regenerative) (22) are called zeolites, and some occur in nature. Commercial molecular sieve zeolites are usually synthetic (see Molecular sieves). They are crystalline framework aluminosilicates containing alkali metal cations. The structure extends in three dimensions by a network of AlO_4 and SiO_4 tetrahedra linked to each other by sharing of oxygen atoms. Molecular sieves possess the high porosity that is characteristic of all adsorbents. In addition, the ordered crystalline structure of the molecular sieve provides pores of a constant size. In contrast, activated alumina and silica gel do not have an ordered crystal structure and, consequently, the pores are nonuniform in size, as shown in Table 3.

The pore size of molecular sieves can be enlarged or diminished by appropriate cation exchange. Therefore many commercial types are available with pore openings

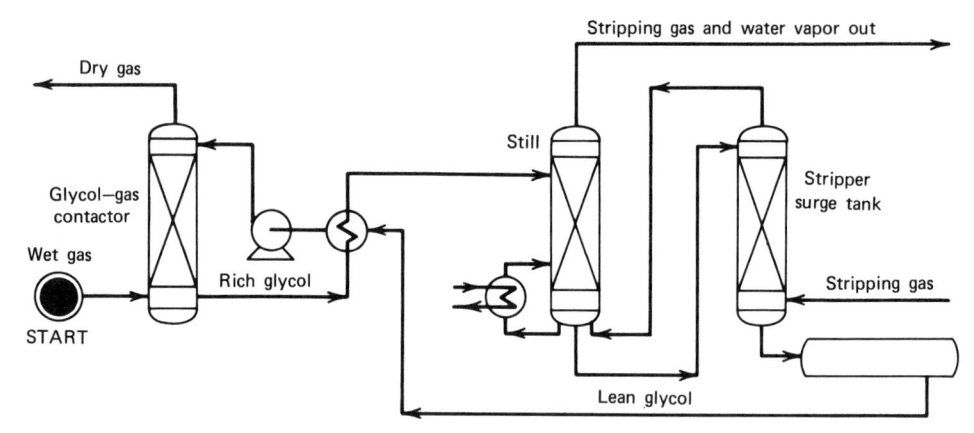

Figure 7. Natural gas diethylene glycol dehydration system.

Table 3. Physical Properties of Solid Adsorbents

Property	Grade A activated alumina	Silica gel	Type 4A molecular sieves
surface area, m^2/g	320	832	750
bulk density, kg/m^3	800	720	670
maximum heat of adsorption, J/g H$_2$O[a]	ca 1395	ca 930	ca 4180
specific heat, J/(kg·K)[a]	1005	921	1046
reactivation temperature, °C	150–315	125–275	200–315
pore volume, % of total	ca 50	ca 55	ca 48
pore size, nm	1–7.5	1–40	0.42
pore volume, cm^3/100 g	40	43	28.9

[a] To convert J to cal, divide by 4.184.

ranging from 0.3 nm to about 1.0 nm. This property gives molecular sieves unique advantages in certain applications because many gases or liquids can be excluded from the microporous structure, thus the term molecular sieves. This can prevent fouling of the adsorptive surface by high molecular weight compounds. Larger-pore molecular sieves can also be used to simultaneously dry and purify streams, eg, by the adsorption of carbon dioxide or sulfur compounds in addition to water.

Because of the ordered structure, molecular sieves have excellent capacity at low water concentrations and do not exhibit a capillary condensation pore-filling mechanism at high water concentrations. The desiccating properties of the material are still good at elevated temperatures as shown in Figure 9. A dew point of −75°C can be obtained in a gas dried at 90°C with a molecular sieve having water adsorbed to the level of 1 wt %. In normal operations at ambient temperature, dew points of <−100°C have been measured.

Molecular sieves are also inert to most fluids and are physically stable when wetted with water. Strong inorganic acids or alkalies should be avoided as well as temperatures above 700°C. Mildly acidic streams can be dried with an acid-resistant type molecular sieve (23).

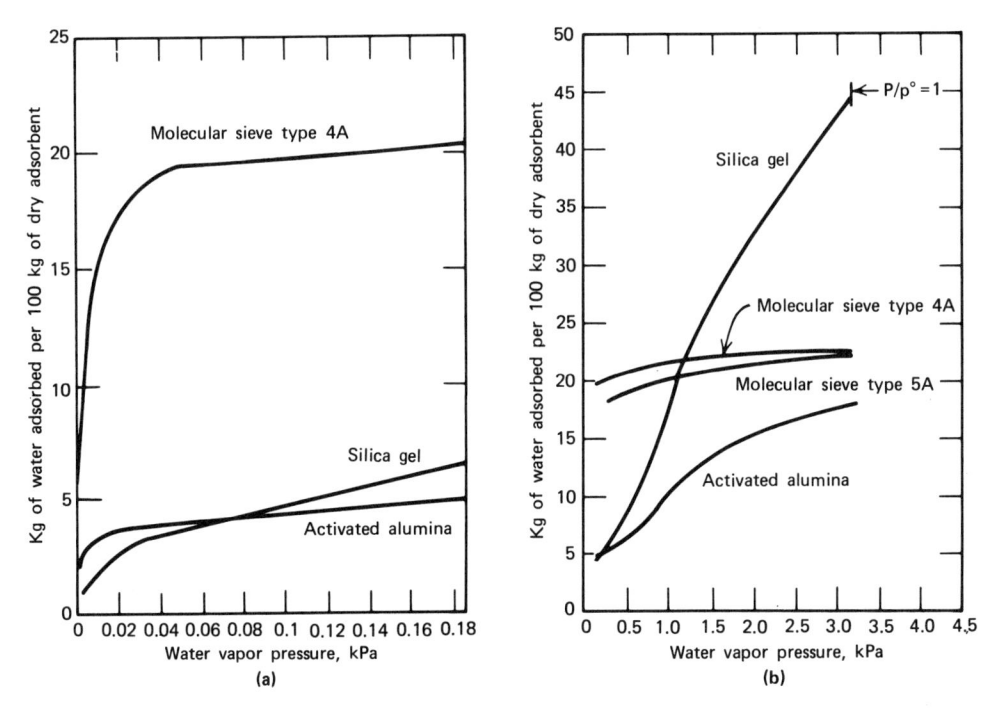

Figure 8. Water adsorption isotherms at 25°C (to convert kPa to mm Hg, multiply by 7.5).

Design of Adsorption Drying Systems

Adsorbent drying systems (13–14) are typically operated in a regenerative mode with an adsorption half-cycle to remove water from the process stream and a desorption half-cycle to remove water from the adsorbent and to prepare it for another adsorption half-cycle. Usually, two beds are employed to allow for continuous processing. In most cases, some residual water remains on the adsorbent after the desorption half-cycle because complete removal is not economically practical. The difference between the amount of water removed during the adsorption and desorption half-cycle is termed the differential loading. This is the working capacity available for dehydration.

The two most common types of drying systems operate on either a pressure swing cycle or a thermal-swing cycle. A pressure-swing cycle uses a high pressure adsorption half-cycle and a low pressure desorption half-cycle and does not require an elevated temperature for regeneration. However, some amount of thermal energy can be added, if desired, during the desorption half-cycle in order to increase the desorption efficiency. This type of system operates with small differential water loadings. Therefore, short adsorption and desorption half-cycle times are used (1 to 60 min). Higher effluent dew points are characteristic of this type of dehydration operation. A thermal-swing cycle requires an elevated temperature during the desorption half-cycle. A depressurization can also be made during this step in order to improve regeneration efficiency. This type of cycle operates with the highest differential loadings and longer cycle times are normally used (4 to 24 h). If very low dew points are required, a thermal-swing cycle should be employed.

126   DRYING AGENTS

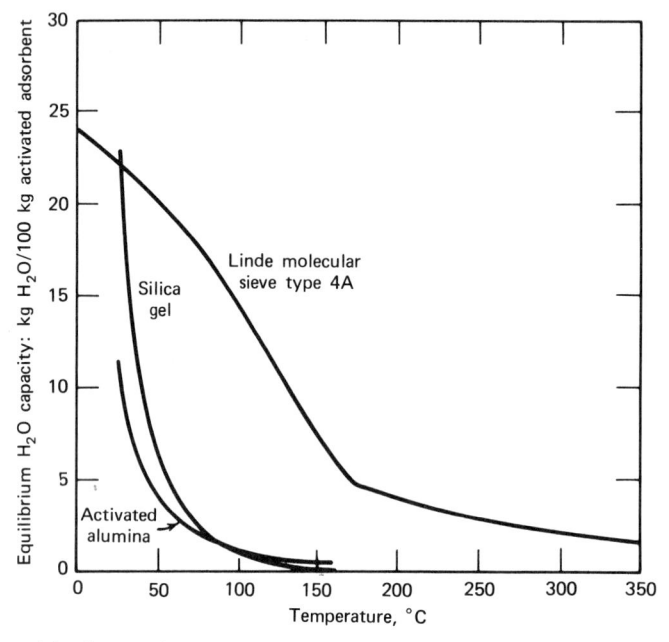

Figure 9. Adsorbent isobars (p_{H_2O} = 1.38 kPa; to convert kPa to atm, divide by 101.3).

In many cases a two-bed dehydrator system can process all the fluid to be dried; with one bed on adsorption and the other on regeneration. Figure 10 is a schematic of a simple two-bed natural gas dehydrator (the same scheme could be used to dry air). Natural gas is first passed through a gas liquid separator to ensure single-phase op-

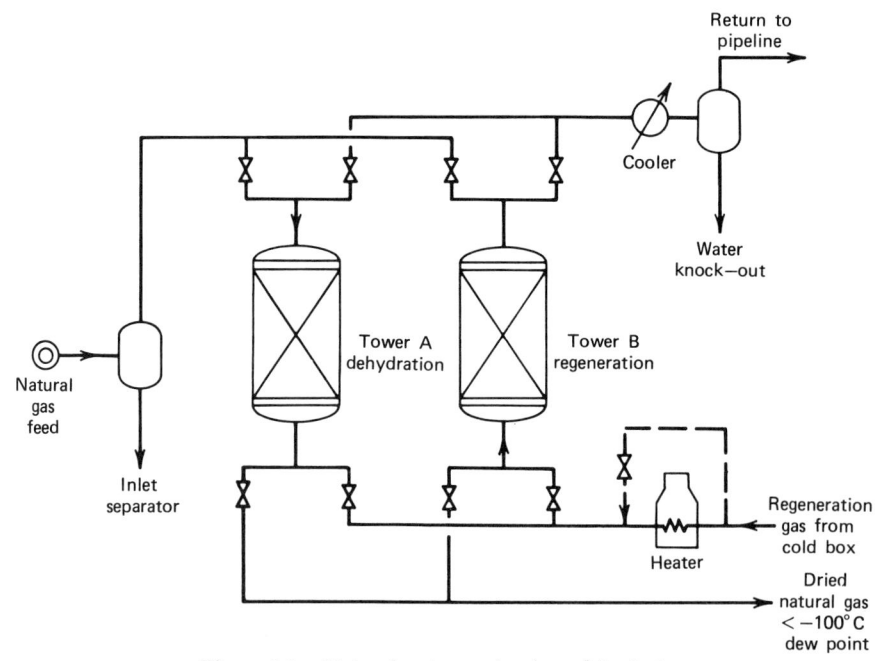

Figure 10. Molecular sieve natural gas dehydrator.

eration. The gas passes downflow through molecular-sieve drying tower A. The dried product natural gas may then be sent for further processing, eg, cryogenic hydrocarbon recovery of liquefied petroleum (LP) condensates. The residue or lean gas from the cold section of such a plant is commonly used for regeneration. The gas is heated and passed through the exhausted adsorbent bed (tower B) with the flow countercurrent to the gas drying step. Tower B is then cooled to feed temperature by flowing gas around the heater directly to the bed. The spent regeneration gas then passes through a cooler and liquid water is removed before it is returned to the pipeline. Depending on the process conditions, the amount of regeneration gas required is about 5 to 10% of the total amount of gas dried (see Liquefied petroleum gas; Gas, natural; Cryogenics).

Adsorption Plots

Among the methods of plotting adsorption data, isotherm plots are the most common, wherein the water vapor pressure in equilibrium with the adsorbent is plotted vs the adsorbed water content of the adsorbent with lines of constant temperature being shown. Another commonly used plot is called an isostere plot. In this case, temperature is plotted vs the water vapor pressure in equilibrium with the adsorbent, with lines of constant adsorbed water content on the adsorbent. Figure 11 depicts

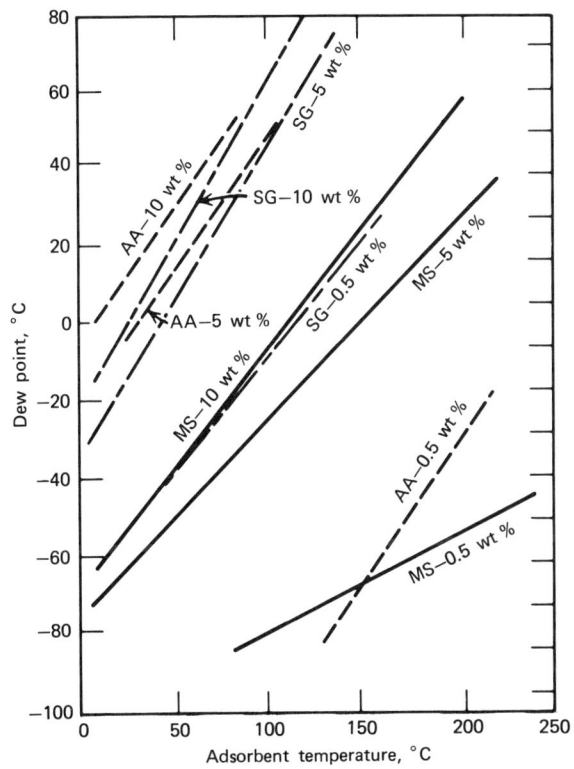

Figure 11. Adsorbent isosteres for activated alumina (AA), silica gel (SG), and molecular sieves (MS).

128 DRYING AGENTS

isosteres for the three major adsorbents described previously. In this case, the dew points for the three adsorbents are plotted at 0.5, 5, and 10 kg H_2O/100 kg of adsorbent. At equilibrium and at a given adsorbed water content on the adsorbent, the dew point that can be obtained in the treated fluid is a function only of the adsorbent temperature. By comparing the slope of the isosteres, it can be seen that molecular sieves are less temperature sensitive.

The term useful capacity, also referred to earlier as breakthrough capacity, differs from the equilibrium capacity shown on Figures 8a and 8b. The useful capacity is a measure of the total moisture picked up by a packed bed of adsorbent to the point where moisture begins to appear in the effluent. Thus, the drying process cycle must be stopped before the adsorbent is fully saturated. The portion of the bed that is not saturated to an equilibrium level is called the mass-transfer zone (see Mass transfer).

The parameters affecting the size and shape of a mass transfer zone are adsorbent type, adsorbent equilibrium capacity, flow rate, packed-bed depth, adsorbent particle size, physical properties of the carrier fluid, temperature, pressure, and the concentration of water in the carrier fluid. If conditions are chosen that are favorable to mass transfer (eg, long contact time), then the mass transfer zone is small when compared to the total amount of packed bed employed in drying service. In this case, bed utilization is more efficient and the breakthrough (or useful) capacity closely approaches the true equilibrium capacity. More often, conditions cannot be optimized based on the adsorbent needs but are fixed by the drying process needs. This may dictate unfavorable mass-transfer conditions when practical packed-bed diameters and depths must be employed. Bed utilization is then less efficient and the breakthrough capacity falls short of the equilibrium capacity.

Figure 12 depicts the location of the water front in packed beds of adsorbents at a short, but typical, contact time for dehydration. Because of the different mass-transfer characteristics of the adsorbents, the percentage of useful capacity as compared to equilibrium capacity is about 50% for activated alumina and silica gel, and

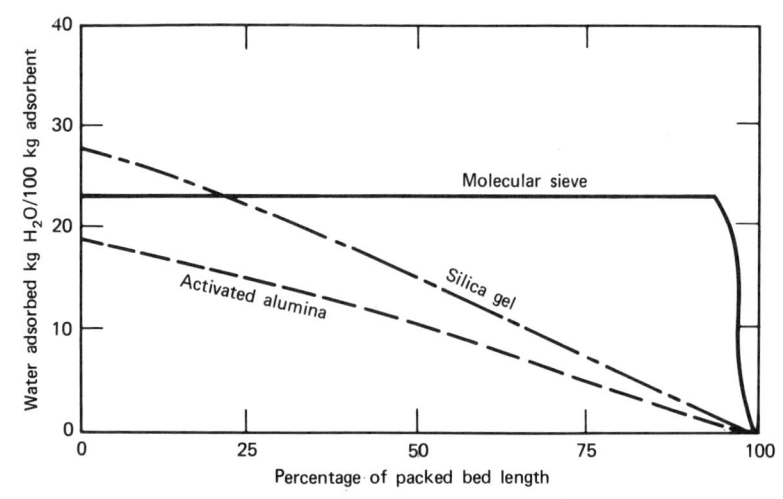

Figure 12. Position of water front in packed bed of absorbent during dynamic dehydration. Conditions: 50% rh; 10.2 cm/s air; particle size = ca 0.167 cm; temperature = 25°C; contact time = 1.7 s.

over 90% for the molecular sieve. This demonstrates the point that the highest equilibrium capacity adsorbent may not always result in the highest useful capacity in a dynamic system. If higher moisture contents can be tolerated in the effluent, then a larger fraction of the mass transfer zone can be allowed to leak through into the effluent stream. This improves bed utilization at the expense of an increase in the effluent dewpoint.

Cost Effectiveness of Adsorbents. The cost effectiveness of a dynamic drying agent can be evaluated in three respects: (a) the water capacity, (b) the water adsorption rate, and (c) the life of the drying agent for the service intended. This assumes that the drying agent selected is capable of achieving the required water content in the fluid being dried.

BIBLIOGRAPHY

"Drying Agents" in *ECT* 1st ed., Vol. 5, pp. 266–276, by J. F. Skelly, The M. W. Kellogg Company; "Drying Agents" in *ECT* 2nd ed., Vol. 7, pp. 378–398, by B. K. Beecher, Wyandotte Chemicals Corporation.

1. R. B. Scott, *Cryogenic Engineering,* Van Nostrand Reinhold, Princeton, N.J., 1959.
2. E. O. Patterson, Jr., *Oil Gas J.* **67,** 108 (March, 1969).
3. G. H. Weyemuller, J. D. Horlan, and D. Roberts, *Chem. Process.* 29, (Nov. 1966).
4. P. L. Young Jr., *Molecular Sieve Adsorption in LNG Plant Purification,* presented to the Distribution Conference American Gas Assoc., Houston, Texas, May 6–9, 1968.
5. J. J. Collins, "The LUB/Equilibrium Section Concept for Fixed-Bed Adsorption," *Chem. Eng. Prog.-Sym. Series,* **63**(74), 31 (1967).
6. A. D. F. Toy, *Phosphorus Chemistry in Everyday Living,* American Chemical Society, Washington, D.C., 1976, pp. 156–161.
7. O. A. Hougen and F. W. Dodge, *The Drying of Gases,* Edwards Brothers, Inc., Ann Arbor, Mich., 1947.
8. S. H. Ibrahim and N. R. Kuloor, *Chem. Age India,* **17,** 876 (1956).
9. U.S. Pat. 3,390,511 (July 2, 1968), O. C. Norton (to Van Products Co.).
10. R. S. Boynton, *Chemistry and Technology of Lime and Limestone,* Wiley Interscience, New York, 1966.
11. Drierite Equipment Catalog, *Bulletin No. 236,* W. A. Hammond Drierite Co., Xenia, Ohio, 1977.
12. M. Shepherd, "Relative Efficiencies of Drying Agents," *International Critical Tables,* Vol. 3, McGraw Hill, Inc., New York, 1928.
13. K. G. Davis and K. D. Manchanda, *Chem. Eng.* **81,** 102 (Sept. 16, 1974).
14. A. L. Weiner, *Chem. Eng.* **81,** 92 (Sept. 16, 1974).
15. L. D. Polderman, *The Glycols as Hydrate Point Depressants in Natural Gas Systems,* presented at the Gas Conditioning Conference held at the University of Oklahoma, March 5–6, 1958.
16. *Treating Chemicals, Technical Bulletin F-41335 B,* Gas Treating Chemicals, Union Carbide Corporation, New York, 1978.
17. *Activated Alumina and its Properties,* Aluminum Corporation of America, Chemical Division, Pittsburgh, Pa.
18. *Rhone-Poulenc Activated Alumina, Technical Documentation Bulletin CF-G-6-86-3,* Rhone-Poulenc Ind, Paris, France, 1976.
19. *Davison Granular Silica Gels, Technical Bulletin 303,* Davison Chemical Co., W. R. Grace, Baltimore, Md.
20. *Davison Silica Gels, Technical Bulletin 2M ADS-IS-675,* Davison Chemical Co., W. R. Grace, Baltimore, Md., 1975.
21. E. A. Hauser, *Silic Science,* Van Nostrand Co., New York, 1955.
22. D. W. Breck, *Zeolite Molecular Sieves,* Wiley-Interscience, New York, 1974.
23. P. N. Kraychy and A. Masuda, *Oil Gas J.* **64,** 66 (Aug. 8, 1966).

General References

R. A. Ford and A. J. Gordon, *The Chemist's Companion: A Handbook of Practical Data, Techniques, and References,* John Wiley & Sons, Inc., New York, 1972, pp. 445–447.
R. B. Scott, *Cryogenic Engineering,* Van Nostrand Reinhold, Princeton, N.J., 1959.
O. A. Hougen and F. W. Dodge, *The Drying of Gases,* Edwards Brothers, Inc., Ann Arbor, Mich., 1947.
M. Shepherd, "Relative Efficiencies of Drying Agents," *International Critical Tables,* Vol. 3, McGraw Hill, Inc., New York, 1928.

JOSEPH P. AUSIKAITIS
Union Carbide Corporation

DRYING OILS

Since prehistoric times, human beings have attempted to leave some record of their accomplishments or of objects in the environment. Many of these efforts made use of paints. Although paints are often associated with drying oils such as linseed, the early paints used in the cave paintings in France or in the pyramids probably contained no drying oil. They were mainly water-based with egg albumin, gum arabic, gelatin, or casein as the binder.

Linseed oil may have been used to some extent during the days of the Roman Empire but even in the 13th century egg albumin was the binder of choice. By the 16th century, oleoresinous varnishes (combinations of resin and oil) had become common. During the 19th and early 20th century, drying oils became the primary vehicle for coatings, but since 1945 oils have gradually been displaced in many of their traditional uses by water-based latex emulsions.

One goal of modern coating chemists is to find both vehicles and pigments that bond more strongly with the substrate and thus last for longer periods. How big a part drying oils will play in these new developments remains to be determined. It is, however, clear that drying oils are declining in use in the United States and elsewhere. For example, domestic use of linseed oil in drying oil products slipped from 267,000 metric tons in 1955 to 57,000 t in the 1975 calendar year. But use rose to 84,000 t in the 1975–1976 crop year, beginning in October, and reached 90,000 t in the 1976–1977 marketing year. New technology in the petroleum and coating industry and competition for land from food crops is expected to continue to decrease use of linseed oil. The use of all fats in drying oil products from 1960–1978 is shown in Table 1. The decline in oil used per capita serves to emphasize the loss of market that has occurred (1) (see Fats and fatty oils; Vegetable oils).

Table 1. Use of Fats and Oils in Drying Oil Products[a]

Calendar year	Drying oil products, thousands of metric tons	Per capita use, kg
1961	384	2.1
1964	403	2.1
1967	383	2.0
1970	271	1.4
1973	308	1.5
1974	253	1.2
1975	193	0.90
1976[b]	201	0.94[c]
1977[b]	193	0.88[c]
1978[b]	187	0.86[c]

[a] Ref. 1.
[b] Ref. 2.
[c] Ref. 3 for population estimates.

Occurrence

Linseed oil is produced from the seeds of the common flax plant, *Linum usitatissimum*. This plant was probably first cultivated in the Mediterranean area and brought to the United States during colonial times. Although flax has been grown in Texas and the Imperial Valley of California, the Dakotas and Minnesota are the main area of production at present. Flaxseed for oil is also grown extensively in Argentina, Canada, India, and the USSR.

The light-blue-flowered flax plant produces globular pods containing about 10 long flat elliptical seeds that vary considerably in size. A typical seed is 3–4 mm long, 2–3 mm wide and about 0.5 mm thick and weighs 3–9 mg. The seeds contain about 33–43% oil with some mucilaginous matter in their outer layer. Like soybeans, they apparently contain little if any starch. The major acid component is linolenic acid (9,12,15-octadecatrienoic acid). The oil of commerce must have an iodine value of at least 170, but it is usually 175–185. Climatic conditions at the time the oil is laid down in the seed affects the iodine value; the lower the temperature, the higher the iodine value or extent of unsaturation (4).

Soybean oil is obtained from the almost spherical seeds of the legume *Glycine max* (L) Merrill. It apparently was first domesticated in the 11th century BC in the eastern half of North China (5). Soybeans are widely cultivated throughout the world but grow best between 30 and 45° latitude. The United States, Brazil, and China are major producers of soybeans. Oil content varies from about 18–22% with an iodine value of 128–135. The oil, which is more saturated than linseed oil, is classified as a semidrying oil. It is seldom the only drying component in coating (see Soybeans; Alkyd resins).

Tung or Chinawood oil comes from the seed kernels of the *Aleurites fordii* tree. Prior to World War II and the Japanese invasion of China in 1937, tung oil was widely used in varnishes and enamels. Because of its scarcity in the late 1930s and early 1940s synthetic products were developed to replace tung oil coatings, and its use has declined.

132 DRYING OILS

Also, the nuts are very hard and difficult to process for oil. Agricultural subsidies helped to create a crop in the United States in the Gulf states, but repeated freezes and hurricanes have limited production there. No domestic tung mills operated between 1973 and 1976. With the expiration of mandatory price supports in 1977, it appears unlikely that tung will return as a United States farm crop. The annual use in the United States has averaged about 14,000 t in the last 7 years. The oil contains a very high percentage of the conjugated fatty acid eleostearic acid (9,11,13-octadecatrienoic acid). This fatty acid reacts rapidly with oxygen or resinifies upon heating in an inert atmosphere.

Certain varieties of fish, such as menhaden, pilchard, and sardine are high in oil content. The meal commands a relatively high price because it contains very nutritious protein and vitamins that are sometimes difficult to obtain elsewhere. For coating use, the oils must be cooled to remove saturated glycerides which are present in relatively large amounts. The oil with the high melting glycerides removed is said to be winterized. The oils also must be relatively free of fish odor to be acceptable. Because the winterized oil produces brittle films and yellow on aging, its use is declining.

Oiticica and isano oil are somewhat similar to tung oil. Oiticica contains a conjugated keto fatty acid (licanic acid, 4-keto-9,11,13-octadecatrienoic acid) and isano oil contains conjugated acetylenic bonds such as found in isanic acid [506-25-2] (9,11,17-octadecadiynoic acid). Both oils are extremely reactive, isano reacting even faster than tung. Whereas tung oil gels in 3 min when heated rapidly to 300°C, isano oil explodes under the same conditions. Oiticica oil is derived from the fruit of *Licania rigidia*, a tree that grows in northeastern Brazil, and isano oil is obtained from the fruits of *Onguekoa gore* and was imported for a short period from Zaire. Oiticica oil continues to be imported on a relatively small scale, but importation of isano oil has apparently ceased.

Castor oil (qv) derived from the plant *Ricinus communis* is not a drying oil but can be converted to dehydrated castor oil which produces superior alkyd resins. The beans contain 48–53% oil of which about 80% is ricinoleic acid [141-22-0] (12-hydroxy-9-octadecenoic acid) which can be dehydrated to octadecadienoic acids.

There are a number of minor oils that have been used commercially for limited periods as drying or semidrying oils. They include safflower, perilla, sunflower, walnut, tobacco seed, lumbang, candlenut, and kamale seed. Safflower oil is obtained from the seed of *Carthamus tinctorius,* which is native to the mountainous areas of southwest Asia. Domestically it is grown mainly in the western half of the United States where dry weather prevails in the late growing season, eg, the Sacramento valley. Although once used by a major paint company as the oil component of exterior paints, safflower oil is now used mainly as an edible oil. Perilla oil is derived from the seeds of *Perilla aicymoides*. The oil is similar to linseed with a higher (65%) content of linolenic acid. The remainder of the oils are used only to a very minor extent in the United States and in local areas in the world. Sunflower oil, like soybean and safflower oils is used primarily in edible products.

An extensive search for potential new oilseeds launched by the USDA in 1958 has led to much new information on oils containing epoxy groups (6), hydroxy conjugated dienes (7), conjugated trienes (8), and potential replacements for sperm oil (9). Some of these may become commercially available, and others now available may be able to continue to compete successfully with petroleum-based products (10).

Preparation

Drying oils are usually separated from their sources by one or more of the following operations: (a) hydraulic pressing, (b) continuous screw pressing, sometimes called expelling, and (c) solvent extraction. Soybeans and sources containing a low percentage of oil are extracted. Hexane, a mixture of petroleum hydrocarbons, is the solvent of choice except in hot weather when a slightly higher boiling hydrocarbon such as heptane may be used. Fish oils are separated by treatment with steam. Flaxseed is usually reduced to 20–30% oil in a screw press and then extracted with hexane (see Extraction).

Cold pressed oils need no refining to remove minor components that may interfere with their use in subsequent processing into paints or varnishes, but they are seldom used in paints because of their higher price. Degumming or another method of removing phosphatides is usually necessary with linseed and soybean oils (11). Water is added to the warm oil, and the oil is separated by centrifugation (12–13) after a thorough mixing. Sometimes the oils are degummed and then alkali-refined, but alkali refining may also be carried out without prior degumming. Acid refining gives linseed oils with relatively good color and oils that are more suitable for grinding with pigments.

Some oils are winterized by cooling to precipitate the more saturated glycerides and waxes followed by filtration in a plate-and-frame filter press. This process may be carried out with (14) or without a solvent. Two other processes can be used to separate the more saturated from the polyunsaturated glycerides. Furfural (15–16) and liquid propane (17) have been used commercially for separations. The furfural extract contains the more unsaturated glyceride components, whereas liquid propane dissolves the more saturated. Furfural was used commercially for a short period on linseed and to a lesser extent on soybean oil, and liquid propane on soybean oil. The propane operation was not used to produce a drying oil product commercially but to obtain sterols, particularly sitosterol (see Steroids). The process is no longer in operation. Removal of saturates from soybean oil is best achieved by directed transesterification and winterization (18).

Functionality

Chemical Composition. Oils are generally divided into three classes, drying, semidrying, and nondrying, according to their iodine values. Different authors have defined the iodine value limits differently. Rheineck and Austin (19) define them as follows: drying oils, iodine value >140; semidrying, 125–140; and nondrying, <125. Actually, any spreadable liquid that reacts with the oxygen of the air to form a comparatively dry film would be classified as a drying oil. Although a solvent might be used to improve the spreadability, the film-forming product should itself be liquid before the film sets. This definition includes a wide range of products, but chemists generally define oil more narrowly as a liquid triacyl glycerol or triglyceride. Thus, most oils are the mixed esters of glycerol with fatty acids. Glycerol (qv) is the alcoholic component and a variety of monobasic fatty acids, usually C_{18}, are the acidic part of the esters. They are low melting oils rather than fats when the amount of unsaturation is high enough. Table 2 shows the important fatty acids present in representative commercial drying oils (19–20). The shortened notation indicates the length of the carbon chain plus the number and location of double bonds.

134 DRYING OILS

Table 2. Fatty Acid Composition of Drying Oils, %

Name	CAS Reg. No.	Notation[a]	Linseed	Soybean	Safflower	Tung	Oiticica	Menhaden
oleic	[112-80-1]	18:1 (9c)	17	24	13	4	6	19
linoleic	[60-33-3]	18:2 (9c, 12c)	14	51	77	8	10	2
linolenic	[463-40-1]	18:3 (9c, 12c, 15c)	60	9				
α-eleostearic[b]	[506-23-0]	18:3 (9c, 11t, 13t)				82		
licanic[b]	[17699-20-6]	18:3 (9c, 11t, 13t) 4C=O					74	
docosahexenoic	[2091-24-9]	22:6 (4c, 7c, 10c, 13c, 16c, 19c)						14
palmitic	[57-10-3]	16:0	6	12	7	4	5	23
stearic	[57-11-4]	18:0	3	4	3	2	5	4

[a] The notation indicates: (a) the number of carbon atoms in the acid (C16 or C18); (b) the position and geometry of the double bonds (0 = none, c = cis, t = trans).
[b] Various materials such as sulfur isomerize 18:3 (9c, 11t, 13t) to 18:3 (9t, 11t, 13t).

For example, the unsaturated fatty acids of the linseed oil, also found in safflower, soybean, and other minor drying oils, all have 18-carbon chains with unsaturation at the following positions: oleic at 9, linoleic at 9 and 12, and linolenic at 9, 12, 15. Linoleic acid is an essential fatty acid and precursor of the prostaglandins (qv). The unsaturation of these fatty acids give the drying oils their reactivity. Linoleic and linolenic acids are particularly reactive because of the 1,4-pentadiene structure, —CH=CHCH$_2$CH=CH—, or the *cis*-methylene-interrupted unsaturation. Other oils contain reactive conjugated unsaturation such as in α-eleostearic acid and licanic acid. The conjugated acids contain both cis and trans isomers but treatment with sulfur can convert them to all trans. Such reaction with tung oil changes it to a solid that requires special heating equipment for handling (see also Carboxylic acids).

Reactivity. Glycerol (qv) has three hydroxy groups, giving it a functionality of three for reactions with organic acids. In nonenzymatic chemical reactions, the differences between the hydroxyls are of minor importance. Unsaturated double bonds in the acyl glycerols (glycerides) have some variable functionality depending on the reaction and conditions therein. Their functionality can be expressed as follows (19): f = (number of esterifiable hydroxyls) × (average number of ethylene groups per acyl chain). Linseed oil with an iodine value of 180 has an apparent functionality of 6. Conversion to the pentaerythritol ester raises the apparent functionality to 8. Such a change increases the rate of film formation and cross-linking when a liquid film of the ester is spread uniformly. In polymerization by heat alone, linoleate has a functionality of about 1.5, since dimers and trimers are formed. Both soybean and linseed oils can be gelled but heat polymerization or bodying of these oils in the absence of air will give highly branched and very viscous polyesters with average molecular weights of 2500 and 3000, respectively. This polymerization may involve oleic acid esters as well as linoleic esters. Certain catalysts such as boron trifluoride give mainly trimeric products with methyl linoleate. At 200°C, oleate reacts with 1 mole of maleic anhydride, linoleate with 2 moles, and linolenate with 2.5 (21). Functionality in drying oils is thus at best only a guide to behavior since it varies with reaction and reaction conditions.

Minor components may have some effect on the film-forming properties of an oil. Some linseed oils of high polyunsaturation may not dry as fast as similar oils of lower polyunsaturation. Varietal differences can account for higher unsaturation and for the presence of higher amounts of antioxidants.

The distribution of the fatty acids on glycerol was originally believed to be even (22) and then thought to be random (23). The work of Matson and Lutton (24), however, showed that in vegetable fats the saturated fatty acids are predominantly on the primary hydroxyls of glycerol. No general pattern prevails among animal fats although lard is known to have saturated fatty acid predominantly on the secondary hydroxyl or the 2-position. Linseed oil has a close to random distribution (25) with considerably more trilinolenyl glycerol (trilinolenin) than would be expected for an even distribution.

Reactions of Drying Oils

For a comprehensive review of the reactions of the fatty acids present in drying oils as glycerol esters or acyl glycerols, see reference 26.

Autoxidation and Polymerization. The reaction with oxygen is probably the most important one that drying oils undergo. Oxidation gives rise to trans isomers, polymers, and cleavage of the carbon–carbon chain with the formation of volatile products. These reactions in a liquid film convert it to a solid containing some soluble liquid material. The autoxidation is catalyzed by certain metallic salts called driers (qv). They are usually the cobalt, manganese, or lead derivatives of organic acids, such as naphthenic, linoleic, or 2-ethylhexanoic acids. These catalysts, particularly cobalt, exert a push-pull action once hydroperoxides are formed as shown in equations 1 and 2 (27–28).

$$Co^{2+} + ROOH \rightarrow Co^{3+} + RO\cdot + OH^- \qquad (1)$$
$$Co^{3+} + ROOH \rightarrow Co^{2+} + RO_2\cdot + H^+ \qquad (2)$$

The free radicals formed in the oil molecule can react to yield larger molecules. Many and varied products are formed. Farmer and co-workers (29–30) made the first breakthrough on the actual mechanism of autoxidation when they applied Criegee's findings on the formation of cyclohexene hydroperoxide (31) to esters from drying oils. Oleic ester gives four different hydroperoxides with the hydroperoxide located at C-8, C-9, C-10, and C-11 with unsaturation at C-10, C-9 and C-8 (32) (eq. 3):

$$RCH_2CH=CHCH_2R' + O_2 \rightarrow$$

oleic ester

$$\begin{array}{cc}
RCH=CHCHCH_2R' & RCHCH=CHCH_2R' \\
\;\;\;\;\;\;\;\;\;\;\;\;\;\;| & | \\
\;\;\;\;\;\;\;\;\;\;\;\;\;\;OOH & OOH
\end{array} \;\; + $$

methyl 9-hydroperoxy-10-octadecenoate   methyl 11-hydroperoxy-9-octadecenoate
[13045-58-1]   [13045-55-1]

$$\begin{array}{cc}
RCH_2CHCH=CHR' & RCH_2CH=CHCHR' \\
| & | \\
OOH & OOH
\end{array} \qquad (3)$$

methyl 10-hydroperoxy-8-octadecenoate   methyl 8-hydroperoxy-9-octadecenoate,
[13045-57-3]   [13045-56-2]

where R is $(CH_2)_6CH_3$ and R', $(CH_2)_6CO_2CH_3$

With linoleic esters, the reaction must proceed somewhat differently because only two monohydroperoxides and a cyclic diperoxide can be isolated (eqs. 4 and 5):

R″CH=CHCH$_2$CH=CHR‴ + O$_2$ →

methyl linoleate [112-63-0] oxidation

$$\underset{\substack{|\\ \text{OOH}\\ \text{methyl 13-hydroperoxy-9,11-octadecadienoate}\\ [14606-80-5]}}{\text{R″CHCH=CHCH=CHR‴}} \quad + \quad \underset{\substack{|\\ \text{OOH}\\ \text{methyl 9-hydroperoxy-10,12-octadecadienoate}\\ [13974-49-7]}}{\text{R″CH=CHCH=CHCHR‴}} \quad (4)$$

$$\downarrow O_2$$

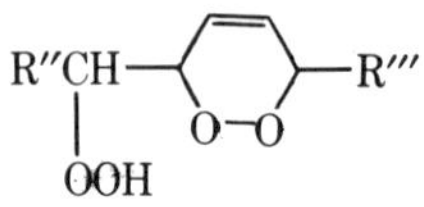

$$\underset{\substack{|\\ \text{OOH}}}{\text{R″CH}}\!\!-\!\!\overset{\frown}{\underset{O-O}{}}\!\!-\!\!\text{R‴} \qquad (5)$$

where R″ = (CH$_2$)$_4$CH$_3$ and R‴ = (CH$_2$)$_7$CO$_2$CH$_3$

The corresponding cyclic peroxide with the hydroperoxide at C-9 is formed. Oxidation to form polymers appears to follow some scheme such as follows:

Initiation   RH + O$_2$ --→ R· + ·OOH   (6)

Propagation   R· + O$_2$ --→ ROO·   (7)

ROO· + RH --→ ROOH + R·   (8)

See also reactions 1 and 2 of Co^{2+} and Co^{3+} (driers)

Termination   ROO· + R· --→ ROOR   (9)

ROO· + ROO· --→ ROOR + O$_2$ (10)

RO· + R· --→ ROR   (11)

R· + R· --→ RR   (12)

Polyperoxides probably result when films are formed and decompose to give polyethers. New carbon–carbon bonds may be formed, but because oxygen is so reactive and its concentration relatively high during film formation, very few new carbon-carbon bonds are likely. Vinyl monomers form relatively high polymers when a small amount of free radical source is introduced. Drying oils that contain the active methylene between two double bonds inhibit such polymerization. Blown oils of varying viscosity are available by reaction of air with oil at slightly elevated temperatures (see Hydrocarbon oxidation).

The steps in film formation with linseed oil can be summarized as follows: (a) an induction period in which little visible change in chemical or physical properties occurs but oxygen is absorbed to destroy antioxidants; (b) a substantial increase in oxygen uptake with the appearance of hydroperoxides and conjugated dienes; (c) hydroperoxides form free radicals; the reaction becomes autocatalytic; and (d) onset of cleavage and polymerization. High-molecular weight cross-linked polymers form as well as low molecular weight cleavage products including carbon dioxide and water. Absorption of oxygen attains a maximum at about the time the film sets and continues but at a much slower rate. Chipault and co-workers (33) indicate that comparative rates of autoxidation for triolein [122-32-7], trilinolein [537-40-6], and trilinolenin [14465-68-0] are 1, 120, and 330. Conjugated trienes oxidized faster than their methylene-interrupted nonconjugated isomers but conjugated dienes oxidize somewhat more slowly than their nonconjugated isomers (34). During film formation, epoxy and hydroxyl groups are presumably formed in part by reaction of hydroperoxides with unsaturation. In addition to water and carbon dioxide, other volatile and nonvolatile carbonyl compounds, acids, esters, alcohols, and hydrocarbons are formed. Conjugation appears in nonconjugated oil films and takes part in the reaction with oxygen.

Heat Polymerization. When heated in the absence of air, both conjugated and nonconjugated oils react to form polymeric products. If heating is carried out long enough, sufficient cross-linking results to form gels. The reaction forms new carbon–carbon bonds. The resultant films have better alkali resistance than films formed by oxidation alone. Gas proof or wrinkle-proof films result when a conjugated oil like tung oil is heat-polymerized properly. Heat polymerization has been studied extensively (35–38), and a large number of chemicals catalyze the reaction. Wheeler outlines the reaction for dienoic esters by a reaction of nonconjugated esters to give conjugated esters (eq. 13) and followed by a Diels-Alder reaction of the conjugated with the nonconjugated ester (eq. 14) (37). The kinetics of the polymerization appear to be first order during the first half of the reaction but become more complicated thereafter and approach second order.

$$CH_3(CH_2)_4CH=CHCH_2CH=CH(CH_2)_7CO_2CH_3 \quad (13)$$

methyl linoleate [112-63-0]

↓

$$CH_3(CH_2)_5CH=CHCH=CH(CH_2)_7CO_2CH_3 \quad (14)$$

methyl 9,11-octadecadienoate [17675-24-0]

↓

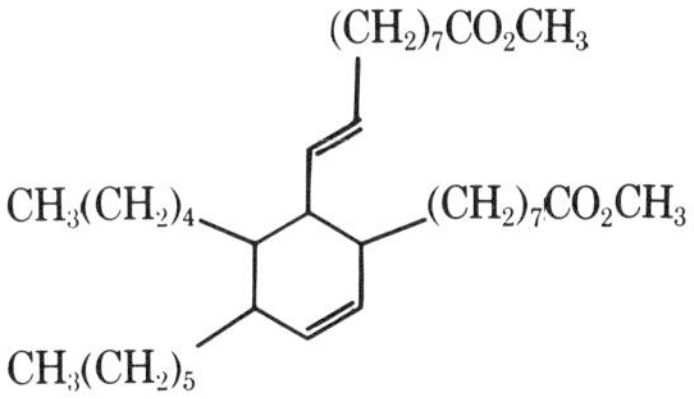

methyl dilinoleate [19680-96-7]

Dimers from linolenic ester may undergo internal cyclization to give bicyclic dimers. Other reactions include formation of cyclic monomer (eq. 17), particularly with linolenic ester.

$$CH_3CH_2CH=CHCH_2CH=CHCH_2CH=CH(CH_2)_7CO_2CH_3 \rightarrow \quad (15)$$
$$CH_3CH_2CH_2CH=CHCH=CHCH=CH(CH_2)_8CO_2CH_3 \rightarrow \quad (16)$$

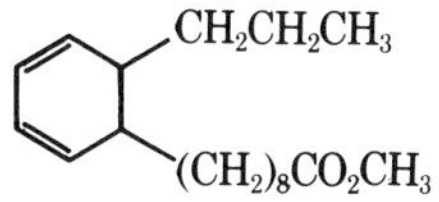

$$(17)$$

Methyl 9-(2-n-propyl-3,5-cyclohexadienyl)nonanoate

After prolonged heating, linseed oil contains comparatively large amounts of cyclic monomeric, dimeric, and trimeric products (39) as shown in Table 3. Among the effective catalysts for polymerization are anthraquinone (40), boron trifluoride (41), sulfur dioxide (42), anthrones, and peroxides such as di-*tert*-butyl peroxide (43). When the mixture of dimer fatty acids is dehydrogenated, an aromatic product is obtained. Subsequent oxidation gives 1,2,3,4-tetracarboxybenzene and confirms that such products are formed (38).

Castor Oil Dehydration. The discovery that proper polymerization of tung oil led to so-called gas proofing of the oil for varnishes and enamels increased the use of this oil throughout the world, particularly after World War I. In the United States, total use in drying oil products rose to a maximum of 67,000 metric tons per year in 1937. The superior properties of tung oil led many chemists in industrial companies, government research laboratories, and universities to seek replacements. Research was accelerated after the Japanese invasion of China in 1937 and by World War II. The most successful of all of these attempts based on oils was the dehydration of castor oil and conversion of the product to an alkyd resin (see Castor oil; Alkyd resins).

Castor oil contains about 87% ricinoleic ester which can be dehydrated to dienoic esters. Organic and inorganic acids were found effective for dehydration. Among the most useful compounds are acetic, phthalic, rosin, and fatty acids and some of their anhydrides, sulfuric, sulfonic, phosphoric, and silico-tungstic acids. Dehydration occurs by esterification and pyrolytic cleavage to give an unsaturated bond. The released acid reesterifies another hydroxyl group and cleavage is repeated. Much of the dehydration forms nonconjugated as well as conjugated diene structures in the dehydrated esters. One major side reaction is esterification to give a polyester. Table 4 gives data for different methods of dehydration (44).

Isomerization-Conjugation. The same events that encouraged research on the dehydration of castor oil led research chemists to seek methods of shifting the nonconjugated methylene-interrupted double bonds in linoleic and linolenic acids to conjugated dienoic and trienoic products. A number of potentially commercial procedures were found: alkali hydroxides in glycols at elevated temperatures (45) and in aqueous solutions under pressure (46), iodine compounds (47–48), nickel-on-carbon catalyst (49), aluminum oxide (50), and sulfur dioxide (42,51) all convert the methylene-interrupted dienes and trienes to conjugated products. Yields of conjugated dienes were reasonably good, but yields of conjugated trienes from nonconjugated triene were comparatively low. Alkali saponifies the oil, and the fatty acid must be esterified again with glycerol or other polyol or converted into some other suitable

Table 3. Composition of Heat-Polymerized Linseed Oils[a]

Fatty esters, %	Methyl esters from M-37[b]	M-55[b]
16:0; 18:0	8	8
18:1	28	27
18:cyc.	25	18
36+:1 to 3	35	43

[a] Ref. 39.
[b] Viscosities in minutes as measured in the Gardner viscosity tubes.

Table 4. Composition and Properties of Dehydrated Castor Oils[a]

Property	Al$_2$O$_3$	Synthenol[b]	Dehydrol[c]	Bodied Synthenol
viscosity, mPa·s (= cP)	110	180	720	4200
iodine value	183	156	116	133
refractive index	1.487	1.482	1.484	1.486
hydroxyl value	5	25	15	12
percent conjugated diene	52	18	18	29
gelation time, min		67	41	24
drying time, min	30	120	108	54

[a] Ref. 44.
[b] Spencer-Kellog.
[c] Sherwin-Williams.

vehicle such as an alkyd resin. Iodine compounds were recently reported as conjugating catalysts in a reaction involving *in situ* condensation of acrylic acid with linoleic acid to form a new dibasic acid (52). Table 5 reports percentage conjugation achieved by different procedures.

Most of these conjugated oils prepared from nonconjugated materials exhibit after-tack. This phenomenon of drying to a film with a minimum of tackiness and of the film later becoming tackier was found in most of the products made from a synthetically conjugated oil. The higher the temperature during isomerization the greater was the after-tack. Nickel-on-carbon catalyst at 200–230°C gave large amounts of conjugated isomers as well as a trans isomer. Trans monoenes do not react as fast as cis monoenes in film formation, and they would undoubtedly oxidize slowly after the film is formed. Rapid film formation followed by later slower oxidation suggests that volatile oxidation products are trapped and plasticize the film causing the after-tack (34,47,53).

Alkali isomerization of linolenic acid at elevated temperatures rapidly converts

Table 5. Isomerization of Drying Oil for Conjugation of Unsaturation[a]

Oil	Derivative	Method	Conjugation % Diene	Triene	Total
linseed	acid	NaOH at 225°C	40	13	53
soybean	acid	NaOH at 225°C	44	2	46
linseed	methyl ester	Ni on C			46
linseed	glyceryl ester	Ni on C			33
linseed	PE ester	Ni on C			10
linseed	polymer (oil)	Ni on C			1
soybean	glyceryl ester	AlI$_3$	28	5	33
soybean	glyceryl ester	Ni on C			36
dehydrated castor	methyl ester	comm.	31		31
dehydrated castor	methyl ester	Ni on C	44		44
dehydrated castor	methyl ester	NaOH at 225°C	56		56
tung	glyceryl ester	natural		80	80

[a] Ref. 47

the conjugated trienoic acid to a cyclic acid (eq. 17) containing a six-membered ring (54). This acid also forms during polymerization with heat. Attempts to develop this product commercially have been unsuccessful.

Maleic Adducts. In recent years, the maleic anhydride adducts of linseed oils have had considerable use in coatings of curing for concrete (55) and in various vehicles for coatings, as in polymers used for electrodeposition of coatings (56). The latter have proved useful in coating hard-to-cover hidden areas in automobile and other frames (see Maleic anhydride).

Methyl oleate undergoes an ene reaction readily at 200°C to form the succinyl derivative. Location of the succinyl group is similar to the location of hydroperoxides at positions 8, 9, 10, and 11 with the double bond in 9, 11, 8, and 9, respectively (57–58). The linoleic and linolenic ester react in linseed oil with 2 and 2.5 moles of maleic anhydride respectively (21). Reaction of 2 moles with linoleate is easy to visualize since formation of the succinyl derivative would give a conjugated intermediate (eq. 18). The latter undergoes a Diels-Alder reaction (eq. 19):

$$RCH=CHCH_2CH=CHR' + \text{(maleic anhydride)} \xrightarrow{200\ °C} RCHCH=CHCH=CH-R' \quad (18)$$

$$+ \text{(maleic anhydride)} \rightarrow \text{(adduct)} \quad (19)$$

$$R=(CH_2)_4CH_3$$
$$R'=(CH_2)_7CO_2CH_3$$

Teeter and co-workers (59) found that monomeric and polymeric adducts were formed by reaction of maleic anhydride with the methyl esters of monomeric distillate from dimerization of linoleate. Both monoadducts and polymeric adducts were formed when less than 2 moles of maleic anhydride was present per double bond. Rheineck and Khoe (60) studied the reaction with all three cis configurations, eg, with methyl oleate, methyl linoleate, and methyl linolenate. These workers confirmed Bickford's work and found no polymeric products when an excess of maleic anhydride was used. Rheineck and Khoe found succinyl adducts formed both with and without the movement of the double bonds. Thus a variety of products were found including those shown in equations 18 and 19. These products include succinyl adducts at carbons 9 and 13 with shifts of the double bonds to conjugated positions as well as succinyl adducts at 14 and 8 with no movement of double bonds. Formation of disuccinyl adducts at carbons 14 and 10 with the movement of the C-9 double bond to C-8 was suggested. Linolenate behaved similarly giving both succinyl adducts without and with double bond movement. Tri-adducts were apparently formed from Diels-Alder reactions between maleic anhydride and disuccinyl adducts containing a conjugated diene as shown in equation 20:

$$\text{maleic anhydride} + C_2H_5CHCH=CHCH=CHCH_2CHCH=CH(CH_2)_6CO_2CH_3$$
$$\underset{R}{|} \qquad \underset{R}{|}$$

$$\longrightarrow C_2H_5CH-\underset{R\ O=\underset{O}{\bigcirc}=O}{\bigcirc}-CH_2CHCH=CH(CH_2)_6CO_2CH_3 \quad (20)$$
$$\underset{R}{|} \qquad \underset{R}{|}$$

$$R = -\underset{O}{\bigcirc}_O$$

These adducts contain six carboxylic acid groups plus one carboxylic ester per 30 carbons. Their ammonium or amine salts are soluble in solvents containing water and migrate to steel when used as the anode in an electric cell (61).

Polyols Other Than Glycerol. Linseed oil has an apparent functionality of six when each double bond has a functionality of one. A highly cross-linked three-dimensional polymer should result on high temperature bodying or oxidation of the film. Linseed oil dries to a characteristic film with oil tackiness. Soybean oil with an apparent functionality of 4.5 dries on oxidation to a soft, tacky film that is inferior in hardness, gloss, alkali resistance, etc. Improvement in drying and film properties can be effected by replacing the glycerol with a polyol of higher functionality, such as pentaerythritol, dipentaerythritol, sorbitol, poly(vinyl alcohol), α-methyl-O-glucoside, and poly(allyl alcohol). By far the most commercially successful synthetic oil has been one made with pentaerythritol containing some dipentaerythritol (see Alcohols, polyhydric). Drying times (free from tack on the Sanderson Tester) of unbodied linseed esters of the glycerol are more than 24 h. Those of the mono-, di-, and tripentaerythritol esters are 6, 5.5 and 3 hours, respectively (62).

Poly(allyl esters) of soybean and linseed oils set to touch in 75 and 45 min, respectively. Esterification of poly(vinyl alcohol) by the acids derived from the drying oil might be considered the ultimate in increasing the functionality of the drying oil. This esterification was effected by Rheineck (63) by use of a phenolic solvent and by Eckey (64) with poly(vinyl acetate) by ester interchange with methyl esters of fatty acids. Films derived from soybean oil-poly(vinyl alcohol) dry tack-free in 20–30 min but are not hard. After 24 h, the films become hard and tough and resemble vinyl plastics. Poly(vinyl oleate) prepared by these methods becomes tack-free in several hours. The polymerization of vinyl linoleate or of vinyl oleate containing small amounts of vinyl linoleate does not proceed readily because the methylene-interrupted diene structure of the linoleate inhibits free radical polymerization. Teeter (65) has reviewed vinyl esters and ethers from fats and oils. Vinyl stearate will polymerize and copolymerize with other vinyl monomers, such as vinyl acetate. Such copolymers are suitable for latex paints (66). Vinyl oleate (67) retards free radical polymerization and as little as 5% of oleate reduces yields in polymerization of other vinyl compounds (see Vinyl polymers).

Copolymers of Drying Oils. Cyclopentadiene (68), styrene (69), and a variety of other reactive unsaturated compounds appear to copolymerize with both conjugated and nonconjugated oils. More reactive monomers, such as styrene, probably do not copolymerize with soybean or linseed oil, but they do copolymerize with conjugated oils to give relatively short polymers (70). The methylene-interrupted dienes in linoleate esters serve as chain transfer or chain termination reagents with either the initiating free radical source or the growing chain of the vinyl monomer. Consequently, styrenated oils prepared either by solvent or by bulk methods are mainly solutions of relatively short-chain styrene polymers in the soybean or linseed oil. Properly prepared mixtures give excellent films. The presence of α-methylstyrene, divinylbenzene or related materials modifies the polymerization. Some fatty ester is probably incorporated into the polymer molecule by the chain-terminating reaction and with such catalysts as boron trifluoride. Data on some styrenated products is given in Table 6 (68).

Styrene (qv) in the presence of conjugated dienoic esters reacts to form a dibasic ester from one molecule of styrene and two of the ester. When it reacts with a polyol this dibasic ester gives polyesters (qv). These convert readily to urethane foams (70) (see Urethane polymers; Styrene plastics).

Dicyclopentadiene reacts with most drying oils when heated under reflux or pressure to retain the diene. Above 170°C, dicyclopentadiene undergoes a reverse Diels-Alder reaction to give cyclopentadiene. This diene and dicyclopentadiene react with the nonconjugated or conjugated oil esters to form bicyclic adducts, with a gradual increase in viscosity of the reaction mixture; reaction continues until cross-linking and gelation occurs. Reaction proceeds faster and at lower temperatures with conjugated oils. Properties of some linseed-cyclopentadiene copolymers are reported in Table 7 (68) (see Cyclopentadiene).

Uses

Commercial Drying Oils. Drying oils are available in a variety of viscosities for use in paints, enamels, and varnishes. The characteristics of a number of commercially available linseed oils are listed in Table 8 and those of some other drying oils are shown in Table 9. Table 10 shows the decline in volume of fats and oils in the drying oil industries of about one-half in the period 1951–1978. Although the use of linseed oil in crop year 1978–1979 was expected to reach ca 84,000 t, this amount still represents

Table 6. Comparison of Styrenated Products[a]

Base oil	Viscosity, cm^2/s (= St)	Peroxide value	Process	Time, h	Temp, °C	Drying time, h	Hardness Sward, 2 wk
linseed	30	70	bulk	4.5	170	0:30	30
soybean	30	90	bulk	2.5	170	0:45	22
dehydrated castor	14		solution	64	145	0:10	26
dehydrated castor	30	40	bulk	14	170	0:30	32
tung-soybean	2		solution	90	143	0:12	26

[a] Ref. 68.

Table 7. Properties of Some Linseed Dicyclopentadiene Copolymers[a]

Reaction conditions Temp, °C	Time, h	Diene % combined	Color[b]	Viscosity, cm²/s (= St)	Drying time, h[c]	Rocker hardness 7 d
288	7	14	9	103	1:20	28
285	4.5	20	12+	2900	1:15	52
282	2	34	6	3600	0:50	66

[a] Ref. 68.
[b] Gardner.
[c] Sanderson.

Table 8. Characteristics of Some Commercial Linseed Oils[a]

Linseed oils	Viscosity, mPa·s (= cP)	Color Gardner, max	Iodine value	Acid value, max	Density, g/mL
nonbreak	50	12	180	7	0.92
alkali-refined	50	6	180	0.5	0.94
acid-refined	50	6	180	6	0.94
polymerized oils	1,760	8	118	7	0.96
polymerized oils	2,700	8	118	9	0.96
polymerized oils	6,340	8	118	9	0.96
polymerized oils	59,000	8	118	9	0.97
boiled	50–65	7	175	5	0.94
blown	3,620–6,340	10	125	11	0.99
treated with					
dicyclopentadiene	4,630–6,340	10	156	4	0.98
styrene	3,620–6,340	10	70	4	0.96

[a] Ref. 71.

Table 9. Some Chemical and Physical Properties of Drying Oils[a]

Property	Soybean	Safflower (nonbreak)	Tung	Dehydrated castor oil	Oiticica (raw)	Menhaden
specific gravity 25/25°C	0.920	0.924	0.935	0.931	0.970	0.926
viscosity, mPa·s (= cP)	50	50	165–250	165–200	solid	50
iodine value	120–140	142–150	158–166	136	148	170–178
acid value	0.5–1.6	0.2–0.5	0.3–4.0	4	4	<8
$n_D^{25°C}$	1.473	1.475	1.5150	1.482	1.510	1.480
unsaponifiable, %	1.5		0.4		1	1.2
color (Gardner)		9–10	6–10	4	8–10	5–7

[a] Ref. 17.

a decline from the 1960s of about 75,000 t. The wide swings in the price of linseed and other drying oils are in part responsible for this decline, but the improvement in petroleum-based products particularly for exterior paints has also been a major factor. Table 11 gives average prices for December 1978 as well as December 1970 and 1975 (1,73). Flax is a crop of second or third choice for many farmers in the Northern growing

144 DRYING OILS

Table 10. Fats and Oils Used in Drying Oil Industries 1951–1976[a], Thousands of Metric Tons

Calendar year	Linseed oil	Soybean oil	Tung oil	Fish oil	Castor oil	Oiticica oil	Other primary oils	Secondary[b] fatty materials	Tall oil	Total[b]
1951	302	88	28	13	17	5	5	26	38	522
1954	226	94.9	22	10	17	4	7	27	53	461
1957	198	86.7	22	15	36	5	6	42	58	468
1960	159	78.1	17	22	39	7	3	14	39	378
1963	176	80.8	14	10	46	2	8	14	47	399
1966	140	97.1	15	59	40	4	0.5	11	52	419
							combined			
1969	123	86.6	14	29	28	5		10	40	335
1972	93	64	14	15	25			15	33	259
1974	88	78	16	12	12			19	30	255
1975	57	65.3	15	7.3	2			24	25	195
1976[c]	81	72	6.8	3.8	3.3				6.4	201
1977[c]	79	75	6.5	4.2	3.5				6.4	193
1978[c]	80	73	5.1	5.7	5.2				5.4	187

[a] Ref. 72.
[b] Fatty acids used in drying oil products not included.
[c] Combines paints and varnish and resins and plastics (2).

Table 11. Prices of Drying Oil Materials, $/kg

Oil	Dec. 1970[a]	Dec. 1974[a]	Dec. 1978[b]
castor oil, Brazilian, tanks, New York	0.392	0.84	0.83
fish oil refined, alkali, tanks, New York	0.29	0.82	0.59
linseed oil raw, tank car, Minneapolis	0.194	1.07	0.53
oiticica oil, drums, New York	0.397	0.73	1.52
soybean, crude, tanks, Decatur	0.273	0.875	0.57
tung oil (imported) tanks, New York	0.328[c]	0.88	2.60

[a] Ref. 1.
[b] Ref. 73.
[c] Annual average.

area but becomes their crop of choice if the season is unduly delayed. This situation makes wide fluctuations in the available supplies of linseed oil.

The choice of a drying oil for a particular use depends on price and availability, reactivity toward resins, or drying characteristics and color retention. Oil-modified resins are now used extensively in latex paints to improve adhesion on chalky surfaces, and water-reducible alkyds are used in paints for chemical applications.

Paints. For many years, the main drying oil product (vehicle) in paints has been linseed oil, either raw or boiled, or as a combination of raw and boiled or, as bodied oil diluted with a hydrocarbon solvent (see Paint). Other oils such as safflower, soybean modified with maleic anhydride, perilla and long-oil alkyd resins from soybean, linseed, and other oils have been used extensively. Linseed oil emulsions (74) and water-soluble linseed oil (eg, Linaqua, Spencer Kellogg Division of Textron) vehicles have competed to some extent with latex emulsions, but they have generally not proven as acceptable to the consumer as the latex emulsions based on acrylate copolymers because of the latter's easy clean-up and faster drying rate (see also Latex technology).

Varnishes. When drying oils are heated with many resins, they form mixtures or solutions of the resin in the oil. Some chemical reactions also occur, such as esterification, diene polymerization, and oxidation. Solvent is often added to the combination of resin and oil to reduce the viscosity and improve the ease of application. Such solutions are called varnishes. Resins used for varnishes may be classified as follows: fossil resins (see Resins, natural; Gums) including congo, kauri, and amber; oleoresins such as damar and ester gum (glycerol esters of gum rosin); and synthetic resins such as phenolics, coumarone-indenes (see Hydrocarbon resins), urethanes, and polyesters. A spar varnish may be obtained by heating tung oil to 302°C and adding one-third as much ester gum by weight. The initial heating keeps the tung oil from wrinkling when it dries, and the addition of ester gum cools the oil so that the polymerization reaction does not proceed to gelation. After cooling, the product can be diluted with the solvent of choice to the proper viscosity. Linseed oil takes much longer to polymerize. Linseed varnishes containing ester gum require a longer heating period, the resin often being heated with the oil. For example, 58 parts of alkali-refined linseed oil kept at 296°C for 75 min with 10 parts of p-phenylphenol–formaldehyde resin gives a varnish with an initial viscosity of 5 Pa·s (50 P). Soybean oil requires over 2 hours to give a product of a similar viscosity. Table 12 lists drying oils in the order of their hardness, and resins in the order of their toughness for various applications (75) (see Coatings, industrial; Coatings, resistant).

Drying oils can be bodied to viscosities of 100 Pa·s (1000 P) for use in printing inks. Short-oil varnishes are often used on high speed presses; the printed sheet is heated for a short time to remove solvent. Table 13 gives ranges for the amounts of oil found in various oil-length varnishes (19).

Table 14 lists some of the ASTM and U.S. specifications for oils used in drying-oil products and some of the products made from them.

Linseed Oil on Concrete. Each year in the United States, millions of dollars are spent to repair concrete that has scaled or spalled. This surface deterioration of concrete results from repeated freezing and thawing. It is particularly noticeable on

Table 12. Drying Oils and Resins for Varnishes[a]

Oils in order of decreasing hardness	Resins in order of decreasing toughness
tung	cellulose esters
oiticica	vinyls (acetate, chloride)
fish (high IV fraction)	styrene
perilla	phenolic
linseed	phthalate esters
dehydrated castor	rosin with phenolic or maleic
corn	coumarone–indene
soya	congo
	rosin ester
	rosin

Driers
cobalt
manganese
lead (use now limited)

[a] Ref. 73.

Table 13. Oil Length in L/kg[a] of Resin[b]

short oil	<6.32
medium	6.32–9.48
long	9.48–15.79
very long	>15.79

[a] To convert L/kg to gal/100 lb, multiply by 11.98.
[b] Ref. 19.

Table 14. ASTM and Federal Specification Citations

Oil	ASTM[a]	Federal[a]	AASHO[b]
boiled linseed	D260-61, type I, (1974)	TT-L-190D, amend-1, (July 1977)	M126-65 (replaced by AASHTO, M233-72)
raw linseed	D234-22	TT-L-215D (April 1977)	
heat polymerized		TT-L-201 (May 1958)	
linseed mixture thinned		Cancelled: TT-L-00371 (July 1967)	
alkali-refined linseed		TT-L-1155 (Oct. 1967)	
no. 10 and 20 grinding oil		TT-P-1046A (May 1975)	
degummed soybean	D124-70	TT-S-600, type II	
refined soybean	D1462-70 (1976)	TT-S-600, type I	
soybean (organic coatings)		TT-S-600 (April 1975)	
raw tung	D12-75		
tung (organic coatings)		TT-T-775, amend-Z (May 1966)	
safflower	D1392-70 (1976)		
dehydrated castor	D961-75	MIL-C-15179B(OS) (Jan. 1971)	
raw castor (C.P. castor)	D960-73	JJJ-C-86A (Oct. 1970)	
oiticica (permanently liquid)	discontinued D60-55		
varnish-spar		TT-V-121G, amend-1, (June 1977)	
varnish-interior		TT-V-71G, amend-1, (June 1977)	

[a] See new specs, 0.5% maximum lead has been changed to reflect 0.06% in many specifications; eg, in TT-L-190D.
[b] American Association of State Highway Officials.

bridges, pavements, and related structures where deicing salts have been used. Boiled linseed oil diluted with an equal volume of a suitable solvent such as mineral spirits was used to treat cured concrete to retard this action. After 1941, the increasing use of air-entraining agents substantially improved the resistance of concrete to freeze-thaw cycles and linseed oil use on concrete was discontinued in most states (76).

With the increased road and bridge usage and repair on the interstate highway system, it became evident that scaling and spalling particularly of bridges, continued to be a major problem. In the early 1960s, the National Flaxseed Processors Association began promoting a mixture of linseed oil and mineral spirits to protect the concrete after it was cured (77). Work at Kansas State University, under contract from the Northern Regional Research Center, USDA, Peoria, Illinois, showed that linseed oil used as an antiscaling compound applied after the concrete was properly cured for about 28 d did improve resistance to freeze-thaw cycles and to the effect of deicing chemicals (78). Repeated application after 50 freeze-thaw cycles further improved resistance, and application of boiled linseed oil to concrete that had started to scale helped to prevent continued deterioration. Emulsions of equal parts of aqueous phase and boiled linseed oil containing 3% tallow alcohols also proved effective antiscaling compounds (79). These emulsions were approved and widely used as curing agents for concrete in several states.

When bodied linseed oil with a viscosity of 59 Pa·s (590 P) constituted 20% of the oil phase, water retention of the concrete was improved over that of emulsion containing only boiled linseed oil (80). Water loss in the ASTM test (81) was substantially below the required maximum, 55 mg/(cm^2·72h) (see Cement).

Studies showed that the boiled oil emulsion and solutions of boiled oil in mineral spirits penetrated the concrete after 35 d of curing with polyethylene, wax, or resin composition (82). Penetrations were 1.5–3 mm when either emulsion or solution was spread at the rate of 87 g/m^2 (0.16 lb/yd^2). Emulsions containing the bodied linseed oil penetrated the concrete surface at least 1 mm during curing. Improved resistance to freeze-thaw cycles and abrasion using the linseed oil products as curing and antispalling agents was found by Best and co-workers (83). Rheineck and Heskins (55) found that properly formulated maleic anhydride adducts gave a better curing agent, and the companies belonging to the National Flaxseed Processors Association developed a different formulation based on maleic adducts (84).

Miscellaneous. Soybean and linseed oil use is covered in many patents and articles. Some applications are in epoxy plasticizers and stabilizers for poly(vinyl chloride) (85,86), for epoxy derivatives converted to coatings (87), in baits to attract ants (88), in coatings for electrodeposition, in variations in the preparation of alkyd resins, polyester-amide resins (89), and urethane oils, in preparation of aldehyde oils (90) and carboxy derivatives of fatty acids (91). An excellent comprehensive source of references is ref. 92.

Research related to the use of linseed oil has declined recently but discontinued studies at the Northern Regional Research Center, SEA, USDA, Peoria, Illinois, should be mentioned. A study on the aging of films from latex emulsion paints and linseed oil paints showed no major differences in durability after 2 (93) and 4 years (94). Improvement of the linseed oil emulsions appears feasible by reversed encapsulation (qv) (95), and use of cationic emulsifiers (96). Studies on the swelling of linseed oil films in buffered solutions suggest that pigmented films swell because the weakly acidic nature of linseed oil is insufficient to overcome excess alkalinity contributed by pigments such as zinc oxide in the presence of water (97).

Linseed oil yellows in the dark but bleaches in sunlight. Rakoff and co-workers (96,98) confirmed that the yellowing correlated with linolenate content and original color. Ozonized monoolein inhibited after yellowing in line with the reports of Privett and co-workers (99), but 0.5% ozonized monoolein achieved a greater effect than acetoacetic ester at any level (100). Drier effects appeared to be additive, and paints with optical brighteners gave significantly lower yellow indexes than paints without brighteners. Paints with brighteners appeared to be less yellow to the human eye (101) (see Brighteners, fluorescent).

BIBLIOGRAPHY

"Drying Oils" in *ECT* 1st ed., Vol. 4, pp. 277–299, by O. Grummitt, Western Reserve University, H. J. Lanson, Mastercraft Paint Manufacturing Company, and A. E. Rheineck, Hercules Powder Company; "Drying Oils" in *ECT* 2nd ed., Vol. 7, pp. 398–428 by O. Grummitt, Western Reserve University, J. Mehaffy, Sherwin-Williams Co., A. E. Rheineck, North Dakota State University, and H. J. Lanson, Lanson Chemicals Corporation.

1. *U.S. Fats & Oils Statistics 1961–1976, Statistical Bulletin No. 574* Economic Research Service, U.S. Department of Agriculture, 1977.
2. *Fats and Oils: Production, Consumption, and Stocks,* Bureau of Census, U.S. Department of Commerce, Monthly reports, Jan. 1978–Dec. 1978, including summary for 1977.
3. *Statistical Abstracts of the United States, 1978,* Bureau of Census, U.S. Department of Commerce, 1978, p. xiii.
4. A. C. Dillman and T. H. Hopper, *Effect of Climate on the Yield and Oil Content of Flaxseed and on Iodine Number of the Oil, U.S. Department of Agriculture Technical Bulletin 844,* 1943, 69 pp.
5. T. Hymowitz, *Econ. Bot.* **24,** 408 (1970).
6. F. R. Earle, *J. Am. Oil Chem. Soc.* **47,** 510 (1970).
7. R. E. Knowles and co-workers, *J. Agric. Food Chem.* **12,** 390 (1964).
8. F. R. Earle, I. A. Wolff, and C. A. Glass, *J. Am. Oil Chem. Soc.* **39,** 381 (1962).
9. H. Gisser, J. Messina, and D. Chasan, *Wear* **34**(1), 53 (1975).
10. L. H. Princen, *J. Coat. Technol.* **49**(635), 88 (1977).
11. J. C. Cowan, *J. Am. Oil Chem. Soc.* **53,** 344 (1976).
12. R. A. Carr, *J. Am. Oil Chem. Soc.* **53,** 347 (1976).
13. B. Brae, *J. Am. Oil Chem. Soc.* **53,** 353 (1976).
14. H. P. Kreulen, *J. Am. Oil Chem. Soc.* **53,** 393 (1976).
15. S. W. Gloyer, *Ind. Eng. Chem.* **40,** 228 (1948).
16. R. L. Kenyon, S. W. Gloyer, and C. C. Georgian, *Ind. Eng. Chem.* **40,** 1162 (1948).
17. A. W. Hixson and J. B. Bockelmann, *Trans. Am. Inst. Chem. Engrs.* **38,** 891 (1942).
18. E. W. Eckey, *Ind. Eng. Chem.* **40,** 1183 (1948).
19. A. E. Rheineck and R. O. Austin in R. R. Myers and J. S. Long, eds., *Film-Forming Compositions,* Marcel Dekker, Inc., New York, Vol. 1, No. 2, 1968, p. 238.
20. F. D. Gunstone, *Chemistry and Biochemistry of Fatty Acids and Their Glycerides,* 2nd ed., Chapman and Hall, Ltd., 1967.
21. W. G. Bickford, P. Krauzunas, and D. H. Wheeler, *Oil Soap* **19,** 23 (1942).
22. T. P. Hilditch, *The Chemical Constitution of Natural Fats,* 3rd ed. (revised), John Wiley & Sons, Inc., New York, 1956, p. 16.
23. H. J. Dutton and C. R. Scholfield in R. T. Holman, W. O. Lundberg, and T. Malkin, eds., *Progress in the Chemistry of Fats and Other Lipids,* Vol. 6, Macmillan Co., New York, 1963, p. 314.
24. F. H. Mattson and E. S. Lutton, *J. Biol. Chem.* **233,** 868 (1958).
25. H. J. Dutton and J. A. Cannon, *J. Am. Oil Chem. Soc.* **33,** 46 (1956).
26. R. J. Harwood, *Chem. Rev.* **62,** 99 (1962).
27. C. E. H. Bawn, *Discuss. Faraday Soc.* **14,** 181 (1953).
28. G. A. Russell, *J. Chem. Ed.* **36**(3), 111 (1959).
29. E. H. Farmer, *Trans. Faraday Soc.* **38,** 356 (1942).
30. E. H. Farmer and D. A. Sutton, *J. Chem. Soc.,* 10 (1946).
31. R. Criegee, *Ann.* **522,** 75 (1936).
32. J. Ross, A. I. Gebbart, and J. F. Gerecht, *J. Am. Chem. Soc.* **71,** 282 (1949).
33. J. R. Chipault, E. E. Nickell, and W. O. Lundberg, *Off. Dig. Fed. Paint Varn. Prod. Clubs* **322,** 740 (1951).
34. A. E. Rheineck and D. D. Zimmerman, *Fette, Seifen, Anstrichm.* **71,** 869 (1969).
35. T. F. Bradley and W. B. Johnson, *Ind. Eng. Chem.* **32,** 802 (1940); *Ind. Eng. Chem.* **33,** 86 (1941).
36. D. H. Wheeler, *Off. Dig. Fed. Paint Varn. Prod. Clubs* **322,** 661 (1951).
37. D. H. Wheeler and J. White, *J. Am. Oil Chem. Soc.* **44,** 298 (1967).

38. J. C. Cowan, *J. Am. Oil Chem. Soc.* **39,** 534 (1962).
39. L. E. Gast and co-workers, *J. Am. Oil Chem. Soc.* **40,** 287 (1963).
40. R. P. A. Sims, *J. Am. Oil Chem. Soc.* **32,** 94 (1955).
41. C. B. Croston and co-workers, *J. Am. Oil Chem. Soc.* **29,** 331 (1952).
42. H. I. Waterman, C. van Vlodrop, and J. Hannevijk, *Verfkroniek* **13,** 130, 180 (1940).
43. R. W. Tess and H. Dannenberg, *Ind. Eng. Chem.* **48,** 339 (1956).
44. A. E. Rheineck in R. T. Holman, W. O. Lundberg, and T. Malkin, eds., *Progress in the Chemistry of Fats and Other Lipids,* Vol. 5, Pergamon Press, New York, 1958, pp. 170, 176.
45. J. P. Kass and G. O. Burr, *J. Am. Chem. Soc.* **61,** 482, 3292 (1939).
46. T. F. Bradley and G. H. Richardson, *Ind. Eng. Chem.* **34,** 237 (1942).
47. J. C. Cowan, *J. Am. Oil Chem. Soc.* **27,** 492 (1950).
48. U.S. Pats. 2,411,111-3 (Nov. 12, 1946), A. W. Ralston and O. Turinsky (to Armour & Co.).
49. S. B. Radlove and co-workers, *Ind. Eng. Chem.* **38,** 997 (1946).
50. A. Turk and P. D. Boone, *Oil Soap (Chicago)* **21,** 231 (1944).
51. D. Cannegeiter, *Paint, Oil Chem. Rev.* **110**(4), 17 (1947).
52. Brit. Pat. 1,373,316 (Nov. 6, 1974), B. F. Ward (to Westvaco Corp.); *Chem. Abstr.* **83,** 58269 (1975).
53. J. C. Cowan, *Ind. Eng. Chem.* **41,** 294 (1949).
54. C. R. Scholfield and J. C. Cowan, *J. Am. Oil Chem. Soc.* **36,** 631 (1959).
55. A. E. Rheineck and R. A. Heskin, *J. Paint Technol.* **42,** 299 (1970).
56. U.S. Pat. 3,366,563 (Jan. 30, 1968), D. P. Hart and R. M. Christenson (to the Pittsburgh Plate Glass Co.).
57. J. Ross, A. I. Gebhart, and J. F. Gerecht, *J. Am. Chem. Soc.* **68,** 1373 (1946).
58. W. G. Bickford and co-workers, *J. Am. Oil Chem. Soc.* **25,** 254 (1948).
59. H. M. Teeter, M. J. Geerts, and J. C. Cowan, *J. Am. Oil Chem. Soc.* **25,** 158 (1948); H. M. Teeter and co-workers, *J. Am. Oil Chem. Soc.* **26,** 660 (1949).
60. A. E. Rheineck and T. H. Khoe, *Fette, Seifen, Anstrichm.* **71,** 644 (1969).
61. A. E. Rheineck and A. M. Usmani, *J. Paint Technol.* **41**(538), 597 (1969).
62. A. E. Rheineck and R. M. Brice, *J. Am. Oil Chem. Soc.* **31,** 306 (1954).
63. A. E. Rheineck, *J. Am. Oil Chem. Soc.* **28,** 456 (1951).
64. U.S. Pat. 2,558,548 (June 26, 1951), E. W. Eckey.
65. H. M. Teeter, *J. Am. Oil Chem. Soc.* **40,** 143 (1963).
66. W. S. Port, F. A. Kinel, and D. Swern, *Off. Dig. Fed. Paint Varn. Prod. Clubs* **26,** 408 (1954).
67. W. S. Port and co-workers, *J. Polym. Sci.* **7,** 207 (1951).
68. L. I. Hansen, J. C. Konen, and M. W. Formo, *Preprint Booklet* Division of Paint, Varnish and Plastics Chemistry, American Chemical Society, Sept. 1949, p. 57.
69. D. H. Hewitt and F. Armitage, *J. Oil Colour Chem. Assoc.* **29,** 109 (1946).
70. F. Weghorst and J. Baltes, *Fette, Seifen, Anstrichm.* **67,** 447 (1965).
71. A. E. Rheineck and R. O. Austin in R. R. Myers and J. S. Long, eds., *Film-Forming Compositions,* Vol. 1, No. 2, Marcel Dekker, Inc., New York, 1968, p. 210.
72. *Agricultural Statistics (1972 and 1976),* U.S. Government Printing Office, Washington, D.C.
73. *Fats and Oils Situation, FOS 296,* Economics, Statistics and Cooperative Service, USDA, May 1979, pp. 16, 22.
74. U.S. Pat. 3,333,976 (Aug. 1, 1967) C. E. Penoyer (to Sherwin Williams).
75. J. C. Weaver in W. Fischer, ed., *Paint and Varnish Technology,* Reinhold Publishing Corp., New York, 1948, p. 314.
76. M. A. Swayze, *Eng. News Record* **126,** 946 (1941).
77. C. E. Morris, *J. Am. Oil Chem. Soc.* **38,** 24 (1961).
78. C. H. Scholer and C. H. Best, *Special Report Number 60, Kansas State University Bulletin,* Manhattan, Kansas, 1965.
79. W. L. Kubie and J. C. Cowan, *J. Am. Oil Chem. Soc.* **44,** 194 (1967).
80. U.S. Pat. 3,873,326 (Mar. 25, 1975), W. L. Kubie (to USDA).
81. "ASTM C 156-65–Water Retention of Liquid Membrane Forming Compounds and Impermeable Sheet Materials for Curing Concrete" *1969 Books of Standards, Part 10, Concrete and Mineral Aggregates,* American Society Testing Materials, Philadelphia, Pa., 1969, pp. 12–15.
82. L. E. Gast, W. L. Kubie, and J. C. Cowan, *J. Am. Oil Chem. Soc.* **48,** 807 (1971).
83. C. H. Best and co-workers, *Transportation Research Record* **504,** 63 (1974).
84. Write to former member companies, Archer Daniels Midland, Cargill, or Spencer Kellogg Division of Textron for information.

85. B. Phillips, F. C. Fostick, Jr., and P. S. Starcher, *J. Am. Chem. Soc.* **79,** 5982 (1957); see also U.S. Pat. 2,804,473 (Aug. 27, 1957) (to Union Carbide Corp.).
86. British Pat. 934,689 (Aug. 21, 1963) (Swift & Co.).
87. J. D. Von Mikusch, *Farbe Lack* **77**(12), 1173 (1971).
88. J. M. Cherrett, *Trop. Agri. (London)* **46**(2), 81 (1969).
89. L. E. Gast, W. J. Schneider, and F. L. Baker, *J. Coat. Technol.* **49**(624), 57 (1977).
90. U.S. Pat. 3,112,329 (Nov. 26, 1963), E. H. Pryde and D. E. Anders (to USDA).
91. E. N. Frankel, F. L. Thomas, and W. F. Kwolek, *J. Am. Oil Chem. Soc.* **51,** 393 (1974).
92. N. S. Baer and N. Indicator, "Linseed Oil and Related Materials: An Annotated Bibliography," *Art and Archaeology Tech. Abstr.* **10**(1), Supp. 155–256, 1973.
93. L. H. Princen, *Paint Varn. Prod.* **64**(11), 24 (1974).
94. L. H. Princen, private communication.
95. L. H. Princen, J. A. Stolp, and R. Zgol, *J. Colloid Interface Sci.* **28,** 466 (1968).
96. H. Rakoff, F. L. Thomas, and L. E. Gast, *J. Coat. Technol.* **48**(619), 55 (1976).
97. R. L. Eissler and L. H. Princen, *J. Paint Technol.* **42**(542), 155 (1970).
98. H. Rakoff, W. F. Kwolek, and L. E. Gast, *J. Coat. Technol.* **50**(637), 51 (1978).
99. O. S. Privett and co-workers, *J. Am. Oil Chem. Soc.* **38,** 22 (1961).
100. H. Rakoff, F. L. Thomas, and L. E. Gast, *J. Coat. Technol.* **49**(628), 48 (1977).
101. H. Rakoff and L. E. Gast, *J. Coat. Technol.* **50**(642), 84 (1978).

General References

Federation Series on Coatings Technology, 25 volumes, Federal Society Paint Technology, Philadelphia, Pa., 1965–1977

C. R. Martens, *Technology of Paints, Varnishes and Lacquers,* Reinhold, N. Y., 1968, reprinted 1974, R. E. Krieger, Huntington, N. Y.

Film-Forming Compositions, R. R. Myers and J. S. Long eds., Marcel Dekker, New York, Vol. 1–5, partially completed, 1967–1975; see reference 17.

J. C. Cowan "Chemistry and Technology of Drying Oils" in J. K. Craver and R. W. Tess, eds., *Applied Polymer Science,* Organic Coatings and Plastics Chem., Division of American Chemical Society, Washington, D.C., 1975, p. 512.

A. G. Roberts, *Organic Coatings,* U.S. Government Printing Office, Washington, D.C., 1968.

J. J. Mattiello, ed., *Protective and Decorative Coatings,* Vol. I–V, John Wiley & Sons, Inc. New York, 1941–1946.

"Lectures of the 1950 Short Course on Drying Oils," *J. Am. Oil Chem. Soc.* **27,** 433 (1950).

"Lectures of the 1959 Short Course on Drying Oils," *J. Am. Oil Chem. Soc.* **36,** 477, 565 (1959).

"Lectures of the 1962 Short Course on Developments in Fat Chemistry," *J. Am. Oil Chem. Soc.* **39,** 448, 480, 521 (1962).

JOHN C. COWAN
Bradley University

DUST, ENGINEERING ASPECTS. See Air pollution control methods.

DUST, HYGIENIC ASPECTS. See Air pollution.

DYE CARRIERS

Dye carriers are used to achieve complete dye penetration of polyester fibers. They loosen the interpolymer bonds and allow the penetration of water insoluble dyes into the fiber.

Carriers are not needed for the dyeing of natural fibers because natural fibers have an open, partially hollow structure and are easily penetrated by dye solutions in water.

It is difficult for dye solutions in water to penetrate synthetic fibers such as polyester, cellulose triacetate, polyamides, and polyacrylics which are somewhat hydrophobic. The rate of water imbibition differs with each fiber as shown in Table 1 as compared to viscose (see Rayon), which imbibes water at the rate of 100% (1). The reason for the low imbibition rate is the tight fiber structure obtained when the polymeric fibers are drawn. During this drawing operation the polymer chains become highly oriented and tightly packed, forming a structure practically free of voids.

Of the various synthetic fibers, the polyamide and polyacrylic contain chemically reactive groups. These can be anionic or cationic in nature. Under certain conditions of temperature and pH, these fibers can be dyed with soluble anionic or cationic dyes.

The polyester polymer does not contain an ionic group and is hydrophobic. Therefore, it cannot be dyed with water-soluble ionic dyes. It can be dyed, however, with certain water-insoluble dyes called disperse dyes (see Fibers, man-made and synthetic; Polyester fibers).

Disperse dyes are nonionic in nature and dye the polyester fiber through a diffusion mechanism. Prolonged boiling of the dyebath loosens the forces binding the polymer chains to each other causing the fiber to swell. This allows a limited penetration of the fiber surface by the dye. The rate of absorption or diffusion of disperse dyes in polyester is much lower than that on nylon (see Polyamides) or cellulose triacetate (qv) fibers. This low dyeing rate is too costly to meet the economic requirements of industrial processing. In addition, deep shades are difficult to achieve and the final dyeing does not meet the minimum fastness required by commercial standards (see Dyes, application and evaluation).

Table 1. Rate of Water Imbibition of Fibers Compared to Viscose [a]

Substrate, fiber	Water imbibition %
viscose	100
acetate	25
triacetate	10
polyamide	11–13
polyacrylic	8–10
polyester	3
polypropylene	0

[a] Ref. 1.

152   DYE CARRIERS

Deep shades and full fastness properties on polyester can be achieved using disperse dyes and carriers, or temperatures over 100°C with or without carriers.

Dye carriers, occasionally called dyeing accelerants, are used on cellulose triacetate fibers but have found their greatest use in the dyeing of polyester. Many theories have been advanced to explain the mechanism of carrier dyeing. One of these is based on the ability of carriers to solubilize disperse dyes. This theory is untenable because many organic solvents do this and yet do not exhibit carrier properties. Another theory suggests that the carrier coats the individual fibers, forming a layer through which the dye transfers to the fiber. Still another suggests that the carrier loosens the binding forces holding the polymer chains together, thus providing suitable spaces for the dye molecules.

An excellent discussion of these theories is given by a research committee of the American Association of Textile Chemists and Colorists (2). Although a much better understanding of the subject has been gained, no universal agreement exists on the mechanism of carrier dyeing.

Many substances show carrier behavior. Of these, some have found more acceptance than others for various reasons among which are: availability, cost, toxicity, ease of handling, odor, etc.

Most carriers are aromatic compounds and have similar solubility parameters to the poly(ethylene terephthalate) fibers and to some disperse dyes (3).

There are many chemicals theoretically suitable as carriers. The boiling point is one of the major criteria in selection. If it is too low, the compound will evaporate from the dyebath at dyeing temperatures and will be lost before it is effective in its role as a carrier. It may also steam distill, condense on the cooler parts of the equipment, and cause drips which will spot the fabric. On the other hand, if the boiling point is too high, the compound cannot be removed from the fabric under normal plant drying conditions and will affect lightfastness of finished goods, leave residual odor, and possibly cause skin irritation to the wearer.

o-Phenylphenol was one of the earliest carrier-active compounds of industrial use. Originally it was used as its water-soluble sodium salt (4). By lowering the pH of the dyebath, the free phenol was precipitated in fine form and made available to the fiber. However, proprietary liquid preparations containing the free phenol are available today that afford a greater ease of handling.

Table 2 lists the four major groups of some of the compounds most commonly used as dye carriers.

In order for these compounds to act effectively as carriers, they must be homogeneously dispersed in the dyebath. Because the carrier-active compounds have little or no solubility in water, emulsifiers are needed to disperse these compounds in the dyebath (see Emulsions).

Proper emulsification is essential to the satisfactory performance of a carrier. A well-formulated carrier will readily disperse when poured into water. It will form a milky emulsion upon agitation or steaming. It should not cause oil separation upon heating or crystallization and sedimentation upon cooling.

Many proprietary carriers are available as solids (flakes or pellets) or in preemulsified form. These present some difficulties in the dyehouse. The former require dispersion in water through steam injection and addition to a preheated dyebath. The latter suffer from short storage life owing to separation of the emulsion. Currently the industry prefers clear products easily emulsified by premixing with water at the time of use.

Table 2. Compounds Most Commonly Used as Dye Carriers

	CAS Registry No.	mol wt	bp, °C
Phenolic compounds			
o-phenylphenol	[90-43-7]	170.2	280–284
p-phenylphenol	[92-69-3]	170.2	305–308
methyl cresotinate	[23287-26-5]	166.0	240
Chlorinated aromatic compounds			
o-dichlorobenzene	[95-50-1]	147.0	172–178
1,3,5-trichlorobenzene	[108-70-3]	181.45	214–219
Aromatic hydrocarbons and ethers			
biphenyl	[92-52-4]	154.2	255.9
methylbiphenyl	[28652-72-4]	168.24	255.3
diphenyl oxide	[101-84-8]	170.0	259.0
1-methylnaphthalene	[90-12-0]	142.2	244.6
2-methylnaphthalene	[91-57-6]	142.2	241
Aromatic esters			
methyl benzoate	[93-58-3]	136.14	198–200
butyl benzoate	[136-60-7]	178.22	250
benzyl benzoate	[120-51-4]	212.24	323–324
Phthalates			
dimethyl phthalate	[131-11-3]	194.18	298
diethyl phthalate	[84-66-2]	212.18	298
diallyl phthalate	[131-17-9]	246.25	290
dimethyl terephthalate	[120-61-6]	194.18	284

Manufacturing of the flake and pellet forms requires melting of the various components. The mass solidifies upon cooling and can be flaked or pelletized according to need.

The preemulsified carriers contain water. These products usually require homogenization through colloidal mills or similar equipment to reduce the particle size and ultimately stabilize the product. The preemulsified as well as the clear self-emulsifying products require the use of a solvent when the carrier-active material is a solid.

Carrier Formulation

The formulation of a carrier depends on three basic ingredients: (1) the carrier-active chemical compound; (2) the emulsifier; and (3) special additives. Additional parameters to be considered in the formulation of a carrier product with satisfactory and repeatable performance are provided by the equipment in which the dyeing operation is to be carried out. The choice of equipment is usually dictated by the form in which the fiber substrate is to be processed (eg, loose fiber, staple, continuous or texturized filament, woven or knit fabric, yarn on packages or in skeins) (see Textiles).

The carrier-active chemical is selected according to its effectiveness at various temperatures. Members of the phenolic group (Table 2), considered to be stronger carriers, are employed for formulations to be used in open equipment at the boil. Weaker carriers such as the members of the aromatic ester group, are utilized generally for high temperature dyeing.

154   DYE CARRIERS

The emulsifier is seldom a single surfactant but a blend selected according to the hydrophilic-lipophilic balance (HLB) required by the chemicals employed. The performance of the final carrier preparation will be limited to the effectiveness of its emulsification system. It provides the initial fine dispersion throughout the dyebath. It prevents the carrier from separating and causing spotting or uneven dyeing throughout the dyeing cycle. It supplies effective reemulsification of the steam-distilled chemicals that drop back into the dyebath and onto the fabric. Finally, it produces the foaming characteristics of the product which determine the suitability of the carrier for use with different types of equipment. For example, a carrier intended for use on jet dyeing equipment is formulated with low foaming nonionic surfactants (see Surfactants).

Special additives are often included in a carrier formulation to provide specific properties such as foam control, stability, and fiber lubrication during dyeing. Most important are the solvents used to solubilize the solid carrier-active chemicals. These often contribute to the general carrier activity of the finished product. For example, chlorinated benzenes and aromatic esters are good solvents for biphenyls and phenylphenols. Flammable compounds (flash point below 38°C) should be avoided.

Basis for Carrier Selection

A carrier is selected by the dyer according to various criteria. The type of equipment and conditions under which it is to be used have been already mentioned. Other considerations are:

Color Yield. The amount of dye successfully removed from the dyebath by the fiber results in a specific depth of shade or intensity of coloration (see Color). This is commonly called color yield. The color-building properties of a carrier are influenced by other factors such as the fiber to liquor ratio (its concentration in the dyebath), the maximum temperature that can be reached on the chosen machine and other dyebath additives. Most carriers give optimum dye utilization in atmospheric dyeing (95–100°C) at the 8–12% level calculated on the weight of the fiber (owf). In low-pressure dyeing equipment (106–110°C), 4–6% owf is usually required. In high pressure equipment (126–132°C), the amount of carrier used can be reduced to 1–3% owf. Above these limits of optimum carrier concentrations, the equilibrium between fiber, dye, and water is switched in favor of the dyebath, and higher amounts of carriers act as shade-reducing agents as shown in Figure 1.

Dye Migration. Dye carriers promote dye migration and transfer, thus producing level and satisfactory dyeings.

The dyestuff is exhausted out of the dyebath and into the fiber with the assistance of carriers and heat. This process can be reversed by the use of an excessive amount of carrier: the depth of shade can be greatly reduced and the dyestuff returned to the dyebath (stripping effect). This phenomenon is utilized to level an unevenly dyed fabric. Some carriers are more effective than others in promoting dye migration and equalizing the dye distribution throughout the fiber substrate. Defectively dyed fabrics showing streaks or unlevelness (shade variations from side to center or from one end of a length of fabric to the other) are usually repaired by treating them with two or three times the amount of carrier normally required in dyeing.

Barré is a fabric defect that is not recognized until the fabric is dyed. It appears as a repetitive characteristic pattern of varying intensity that is easily recognizable.

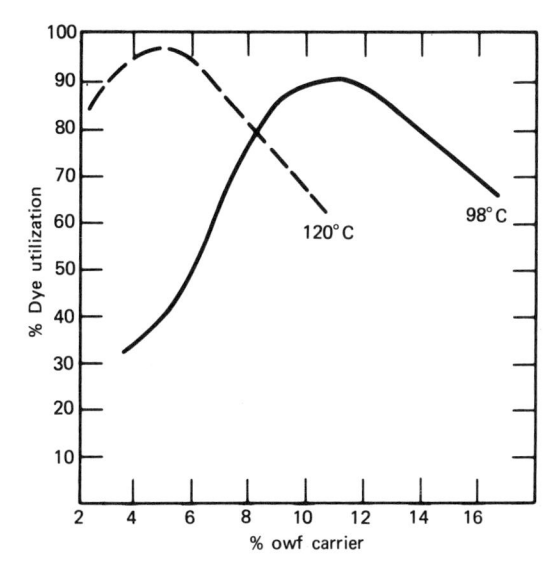

Figure 1. Dyestuff utilization as a function of carrier concentration.

Its causes may be mechanical or chemical: variations in knitting tension cause lack of uniformity of fabric density that in dyeing produce the barré effect, and different degrees of crystallinity of the polymeric fiber owing to uneven drawing or subsequent heat treatments of the filament are also responsible for barré. Dye carriers in combination with selected dyes are instrumental in overcoming this condition.

Product Stability and Emulsion Stability. These properties are not necessarily related but are both highly prized in the selection of a carrier. The first refers to the storage or shelf stability of the product. Many carrier preparations are not properly balanced, or unsuitable emulsifiers have been used. Upon storing, these products separate in layers, particularly when exposed to temperature changes.

Emulsion stability refers to the stability of the emulsion in water. It must withstand various conditions such as:

The dilution factor of the dyebath. The amount of water in a dyebath may vary from 5–25 times the weight of fiber to be dyed, according to the capacity and type of equipment employed. The larger water-to-fiber ratio dilutes the emulsifier, thus reducing its effect.

Elevated temperatures of the dyebath. Emulsions that are stable in cold or warm water lose their stability at higher temperatures. The carrier–emulsifier equilibrium undergoes stress particularly when the time at high temperature is prolonged.

Agitation and shear action of the dyebath. In some equipment fabrics are rotated through the dyebath, thus keeping it in constant agitation. In other equipment the dyebath is circulated through the fabric or yarn. In jet equipment the high speed circulation of the liquid is responsible for the rotation of the fabric. Although agitation contributes to the stability of certain emulsions, the high shear of the pumps used in the equipment with forced circulation is often detrimental.

pH and electrolyte content of the dyebath. A pH of 4–5 provides the best conditions for dyeing polyester with disperse dyes. A carrier emulsion must be stable under this condition. In addition, when cellulose is present in a fiber blend with polyester

156 DYE CARRIERS

and it is to be dyed in the same dyebath with direct or fiber-reactive dyes, the carrier emulsion needs to have considerable stability to large amounts of inorganic salts. Sodium chloride or anhydrous sodium sulfate are employed for this purpose in 10–50% owf.

Other Considerations. Some carrier-active products, especially o-phenylphenol and methylnaphthalenes, have an adverse effect on the lightfastness of the finished dyeing. The reason for this is not clear but the effect is readily established. This problem is overcome by submitting the dyed material to temperatures higher than those normally required in drying. Under the conditions (150–175°C) that are usually required to heat-set dyed fabrics or to cure resins applied in finishing operations, the residual carrier is volatilized.

The cost of a carrier, in addition to its satisfactory performance in dyeing, is often a considerable factor in selection. The rising cost of petroleum-derived chemicals is a factor in the price structure of carrier-active chemicals and most carriers, unfortunately, fall under this category.

Dyeing Procedures

Dyeing procedures vary according to the fiber content of the textile material and the equipment to be used. Examples of basic carrier dyeing procedures are as follows:

For 100% Polyester Fabric Dyed Atmospherically. Prepare the dyebath with water conditioning chemicals as needed to control the hardness of the water being used. Disperse the dyes properly by pasting them with cold water and diluting them with warm water (70°C). Add the dye dispersion to the bath and run equipment for 10 min. Add 5–10 g/L of carrier following the dilution procedures recommended by the manufacturer. Adjust pH to ca 5 with acetic acid. Bring to a boil in 30–45 min. Hold the boil for at least 1 h before checking shade. Cool slowly–rinse completely. Drop the bath and "afterscour" as required to remove residual carrier and unfixed dye.

For 100% Polyester (Textured or Filament) Dyed Under Pressure. Set the dyebath (50°C) with water conditioning chemicals as required, acetic acid to ca 5 pH, properly prepared disperse dyes and 1–3 g/L carrier. Run 10 min. Raise temperature at 2°C/min to 88°C and seal the equipment. Raise temperature at 1°C/min to 130°C. Hold max temperature for $\frac{1}{2}$–1 h according to the fabric and depth of shade required. Cool to 82°C at 1–2°C/min, depressurize machine and sample. Correct shade if needed. Cool slowly to avoid shocking and setting creases into the fabric. Afterscour as needed.

Health and Safety Factors

Most carrier active compounds are based on aromatic chemicals with characteristic odor. An exception is the phthalate esters which are often preferred when ambient odor is objectionable or residual odor on the fabric cannot be tolerated. The toxicity of carrier-active compounds and of their ultimate compositions varies with the chemical or chemicals involved. The environment surrounding the dyeing equipment where carriers are used should always be well-ventilated and operators should wear protective clothing (eg, rubber gloves, aprons, and safety glasses or face shields).

Specific handling information can be obtained from the supplier or manufacturer.

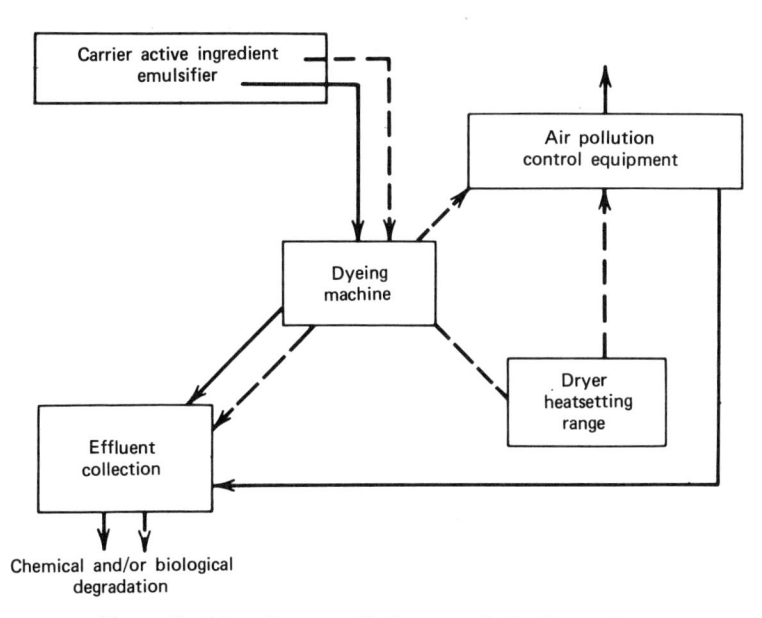

Figure 2. Air and water pollution controls for dye carriers.

Table 3. LD$_{50}$ and Threshold Limit Values for Compounds Most Commonly Used as Dye Carriers [a]

	TLV (in air)[b]	LD$_{50}$ mg/kg[c]
Phenolic compounds		
o-phenylphenol		2700
p-phenylphenol		
methyl cresotinate		
Chlorinated aromatic compounds		
o-dichlorobenzene	50 ppm	500
trichlorobenzene	5 ppm	756
Aromatic hydrocarbons and ether		
biphenyl	0.2 ppm	3280
methylbiphenyl		
diphenyl oxide	1 ppm	3370
methylnaphthalene		4360
Aromatic esters		
methyl benzoate		2170
butyl benzoate		5140
benzyl benzoate		1700
Phthalates		
dimethyl phthalate	5 mg/m^3	6900
diethyl phthalate	5 mg/m^3	5058
diallyl phthalate		770
dimethyl terephthalate		4390

[a] Data extracted from the Registry of Toxic Effect of Chemical Substances. Published by NIOSH, U.S. Department of Health, Education and Welfare.

[b] TLV is a measure of air pollution. It indicates the upper limit to which most workers can be exposed without adverse effect.

[c] LD$_{50}$ is a measure of toxicity. It represents the calculated dose of a substance expected to cause the death of 50% of a defined number of animals tested. All LD$_{50}$ figures are for rat-oral (ingestion) except for diethyl phthalate which is rat-introperitoneal.

OSHA and EPA have established exposure limits that must be carefully considered in relation to the waste disposal method available and the environment in which dye carriers are to be used. Some are only skin irritants, whereas others can contribute to air and water pollution. Atmospheric dyeing equipment should be exhausted to scrubbers that wash the vapors into the general plant effluent for chemical or biological degradation. The same system applies to drying and heat-setting equipment where residual carriers are completely volatilized out of the dyed fibers. A schematic illustration for the control of air and water pollution is shown in Figure 2 (see Air pollution control methods).

Threshold Limit Values and LD_{50}s are given on Table 3 for the substances for which data are available. Additional information is continuously being developed to provide guidelines for the safe handling of dye carriers and carrier active chemicals.

The increasingly stringent government regulations and the introduction of carrierless-dyeable polyester have not substantially affected the use of carriers. The factor with the greatest impact on their use has been provided by the spectacular technological advances in dyeing equipment that have taken place in the last 5 to 10 yr. High temperature dyeing has greatly reduced the time element (dyeing cycle) and the need for carriers.

BIBLIOGRAPHY

1. K. Tandy, Jr., "Characteristics of Polyester Homopolymer Fiber Which Affect Dyeing Properties," paper presented at the *14th AATCC New England Regional Technical Conference,* New Hampshire, May 19–21, 1977.
2. S. Salvin and co-workers, *Am. Dyest. Rep.* (22) (Nov. 2, 1959).
3. C. M. Hansen, *J. Paint Technol.* **39,** 104 (1967).
4. M. C. Keen and R. J. Thomas, "Absorption Properties of Latyl Disperse Dyes on Application to Dacron Polyester Fibers," *Dyes and Chemicals Technical Bulletin,* E. I. du Pont de Nemours & Co., Inc., Organic Chemicals Dept., Wilmington, Del.

ROBERT WANNEMACHER
ANTHONY DEMARIA
Tanatex Chemical Co.

DYES AND DYE INTERMEDIATES

Dyes are intensely colored substances used for the coloration of various substrates, including paper, leather, fur, hair, foods, drugs, cosmetics, waxes, greases, petroleum products, plastics, and textile materials. They are retained in these substrates by physical adsorption, salt or metal-complex formation, solution, mechanical retention, or by the formation of covalent chemical bonds. The methods used for the application of dyes to the substrates differ widely, depending upon the substrate and class of dye. It is by application methods, rather than by chemical constitutions, that dyes are differentiated from pigments. During the application process dyes lose their crystal structures by dissolution or vaporization. The crystal structures may in some cases be regained during a later stage of the dyeing process. Pigments, on the other hand, retain their crystal or particulate form throughout the entire application procedure. They are usually applied in vehicles, such as paint or laquer films, although in some cases the substrate itself may act as the vehicle, as in the mass coloration of polymeric materials.

The optical properties of dyes are determined by electronic transitions between the various molecular orbitals of the dye molecules which absorb some, but not all, of the incident radiation. These properties are defined by the terms color, intensity, and brightness. The color, also frequently referred to as the shade or hue, of a dye is determined by the energy differences between the molecular orbitals. The intensity, strength, or saturation is determined by the probability of the electronic transitions and the amount of dye present. The brightness or purity depends upon the width of the waveband absorbed by the dye molecules (see Color).

The energy, probability, and distribution of the electronic transitions are, to a large extent, governed by the chemical constitution of the molecules. The chemical constitution also determines the other properties of a dye, such as the suitability for dyeing a specific substrate and the fastness properties of dyeings produced by the application of the dye to the substrate. A large number of dyes, with widely differing properties, is necessary because of the great variety of materials to be dyed. There are at present some 1200 different commercial dyes manufactured in the United States, and a further 800 are imported (1–2). On a worldwide basis over 8000 chemically different dyes have achieved commercial significance (3). To assist both the dye users and dye-manufacturers, dyes are classified into groups in two ways. The first method of classification is by chemical constitution, in which the dyes are grouped according to the chromophore or color-giving unit of the molecule. The second method is based on the application class or end-use of the dye. The first, from a chemical standpoint, satisfies the needs of the manufacturer, and the second is used by the dyer.

The art of dyeing textile materials has been practiced for nearly 5000 years. Natural dyes (see Dyes, natural), such as indigo, Tyrian purple, alizarin, and logwood, were used exclusively until the discovery of Mauve or Mauveine by W. H. Perkin in 1856. Although the structure of the compound was not known at the time, the discovery is of particular importance. It marked the beginning of the dye manufacturing industry, from which grew many of the major chemical companies of today, and provided impetus to the study of organic chemistry. The elucidation of the structure of benzene by Kekulé in 1865 laid the foundation for the present concepts of structural organic chemistry. P. Griess discovered the diazotization reaction of aromatic amines in 1856, and the coupling reaction of diazonium compounds in 1864 (see Azo dyes). Thus the

160 DYES AND DYE INTERMEDIATES

first synthetic dyes of known constitution were prepared, in contrast to the empirical products, such as Magenta, Rosaniline Blue, Methyl Violet, Hofmann's Violet (see Triphenylmethane dyes), and Aniline Black, which were available in the decade following Perkin's initial discovery.

Chemists of the period applied their energy not only to the synthesis of new compounds but also to the elucidation of the structure of natural coloring materials. The structure of indigo, which is still the largest volume dye used in the world, was determined in 1880 by A. von Bayer. A commercial synthesis of this dye, by K. Heumann of the Badische Anilin- und Soda-Fabrik in 1890, made possible the large-scale synthetic production of a compound which was previously only available from natural sources (see Dyes, natural).

British preeminence in the manufacture of dyes, resulting from Perkin's initial discovery, was lost to Germany and, to a lesser extent, Switzerland during the latter part of the nineteenth century. By the beginning of World War I in 1914, Germany was manufacturing 87% of the world's dyes. The war effectively cut off supplies of German and Swiss dyes from the rest of the world. This stimulated domestic dye manufacturers and induced new companies to begin manufacture of both dyes and dye intermediates in Europe and the United States. After the war the domestic industries were protected by tariffs and changes in patent laws. From strength in the general organic chemistry of dyes and dye intermediates, the dye companies expanded their activities to other fields, eg, pharmaceuticals. For example, the sulfa drugs, which were the first group effective for the treatment of bacterial infections, are related directly to compounds that were already in use as dye intermediates (see Antibacterial agents, synthetic, sulfonamides).

Although dyes today represent less than 1% of the total sales of organic chemicals on a worldwide basis, this measure does not give a true picture of the importance of dyes in everyday life, in the chemical industry, and in organic chemistry.

Classification and Nomenclature of Dyes

The dual classification system used in the *Colour Index* (CI) is accepted internationally throughout the dye-manufacturing and dye-using industries (3). Adopted in the second edition published in 1956, it was retained in the third edition published in 1971. In this system dyes are grouped according to chemical class with a Colour Index number for each chemical compound and according to usage or application class with a CI name for each dye, whether made by one or several different manufacturers. The CI numbers, the CI names, and the commercial names of the different dye-manufacturers are cross referenced. The CI name for a dye is derived from the application class to which the dye belongs, the shade or hue of the dye, and a sequential number, eg, CI Acid Yellow 3, CI Acid Red 266, CI Basic Blue 41, and CI Vat Black 7. The commercial names of dyes are also usually made up of three parts. The first is a trademark used by the particular manufacturer to designate both the manufacturer and the class of dye, the second is the color, and the third is a series of letters and numbers used as a code by the manufacturer to define more precisely the hue, and also to indicate important properties which the dye possesses. The code letters used by different manufacturers are not standardized. A few of the more widely used designations include R for reddish, B for bluish, and G for greenish shades; L for light, S for sublimation, and W for wash are used for fastness properties. Dyeing properties are indicated by

N for neutral, and E for exhaust. There are instances where one manufacturer may designate a bluish-red dye as Red 4BL and another manufacturer uses Violet 2RL for the same dye. The CI system overcomes this uncertainty by the use of a standard hue-indication chart in the shape of a hexagon with the six primary and secondary hues (yellow, orange, red, violet, blue, and green) each occupying one segment of this hexagon (see Color). The tertiary hues (brown and black) are inside the areas of the primary hues. This system is one of several that have been developed for the mathematical definition of color (4). A five-digit CI number is assigned to a dye when its chemical structure has been made known by the manufacturer. Table 1 gives the CI number-ranges that have been assigned to each chemical class. The characteristic structural unit and an example of a dye with its CAS Registry Number are shown together with application classes and references.

The principal usage or application classes of dyes, accounting for 85% of production in the United States, are as follows:

Acid dyes are water-soluble anionic dyes for application to nylon, wool, silk, and modified acrylics. They are also used to some extent for paper, leather, food, and cosmetics. The original members of this class all had one or more sulfonic or carboxylic acid groups in their molecules. This characteristic probably gave the class its name. Chemically, the acid dyes consist of azo (including preformed metal complexes), anthraquinone, and triarylmethane compounds with a few azine, xanthene, ketone imine, nitro, nitroso, and quinoline compounds.

Basic dyes are water-soluble cationic dyes for application to modified acrylics, modified nylons, modified polyesters and unbleached paper. Their original use was for silk, wool, and tannin-mordanted cotton when brightness of shade was more important than fastness to light and washing. Basic dyes are water soluble and yield colored cations in solution. For this reason they are frequently referred to as cationic dyes. The principal chemical classes are azo (including thiazole-azo), anthraquinone, triarylmethane, methine, thiazine, oxazine, acridine, and quinoline. Some basic dyes show biological activity and are used in medicine as antiseptics (see Disinfectants and antiseptics).

Direct dyes are water soluble anionic dyes which, when dyed from aqueous solution in the presence of electrolytes, are substantive to cellulosic fibers. The principal use is the dyeing of cotton and regenerated cellulose, paper, leather and, to a lesser extent, nylon. Most of the dyes in this class are azo compounds with some stilbenes, thiazoles, phthalocyanines and oxazines. After-treatments, frequently given to the dyed material to improve washfastness properties, include development by a diazotization and coupling procedure on the fiber, chelation with salts of metals (usually copper or chromium), and treatment with formaldehyde or a cationic dye-complexing resin.

Disperse dyes are substantially water insoluble nonionic dyes for application to hydrophobic fibers from aqueous dispersion. They are used on polyester, nylon, cellulose diacetate, cellulose triacetate, and acrylic fibers. Because of their use on synthetic materials they are the fastest growing class of dyes in both number and tonnage. A recently developed application method, transfer printing, in which the dyes are printed on to paper and subsequently transferred to the fiber by a dry-heat process, represents a new and important use for selected members of this class.

Fluorescent brighteners are colorless compounds that absorb incident uv light and re-emit it in the visible (blue) region of the spectrum. Although strictly speaking,

Table 1. Chemical Classification of Dyes

Class	Example with CI names and CI number	CAS Registry Numbers	Characteristic structural units	Dyeing classes	Relevant articles in ECT
1. nitroso CI 10000–10299	CI Acid Green 1; CI Pigment Green 12 CI 10020	[10401-67-9, 19381-50-1]	C=N—OH ortho to C=O	acid, disperse, mordant	Dyes, application and evaluation
2. nitro CI 10300–10999	CI Disperse Yellow 42 CI 10338	[5124-25-4]	aromatic —NO₂ ortho or para to an amino, hydroxyl or ether group	acid, disperse, solvent	Nitrobenzene and nitrotoluenes
3. azo monoazo CI 11000–19999	CI Disperse Red 1; CI 11115	[3180-81-2]	—N=N—	acid, basic, direct, disperse, mordant, pigment reactive, solvent, leather, FDC	Azo dyes
disazo CI 20000–29999	CI Acid Blue 116 CI 26380	[6406-30-0]	—N=N—	acid, basic, direct, disperse, mordant, pigment reactive, solvent, leather, FDC	
trisazo CI 30000–34999	CI Direct Black 78 CI 30015	[8003-79-0]	—N=N—	acid, basic, direct, disperse, mordant, pigment reactive, solvent, leather, FDC	

Class	CAS No.	Chromophore	Application	Related Articles
polyazo CI 35000–36999 CI Acid Brown 120 CI 35020	[6428-26-8]	—N=N—	acid, basic, direct, disperse, mordant, pigment reactive, solvent, leather, FDC	
4. azoic CI 37000–39999 CI Azoic Diazo Component 24 CI 37155	[6268-05-9]	—N=N—	azoic	Azo dyes
CI Azoic Coupling Component 2 CI 37505	[92-77-3]	—N=N—	azoic	
5. stilbene CI 40000–40799 CI Fluorescent Brightening Agent 34 CI 40605	[6416-25-7]	stilbene group with SO$_3$Na	direct, fluorescent brightening agents	Stilbene derivatives
6. carotenoid CI 40800–40999 CI Food Orange 5 CI 40800	[7235-40-7]	(C=C)$_n$	FDC	Dyes, natural; Colorants for food, drugs, and cosmetics
7. diphenylmethane (ketone imine) CI 41000–41999 CI Basic Yellow 37 CI 41001	[6358-36-7]	HN=C	basic	Dyes, application and evaluation; Amines, methylenedianiline

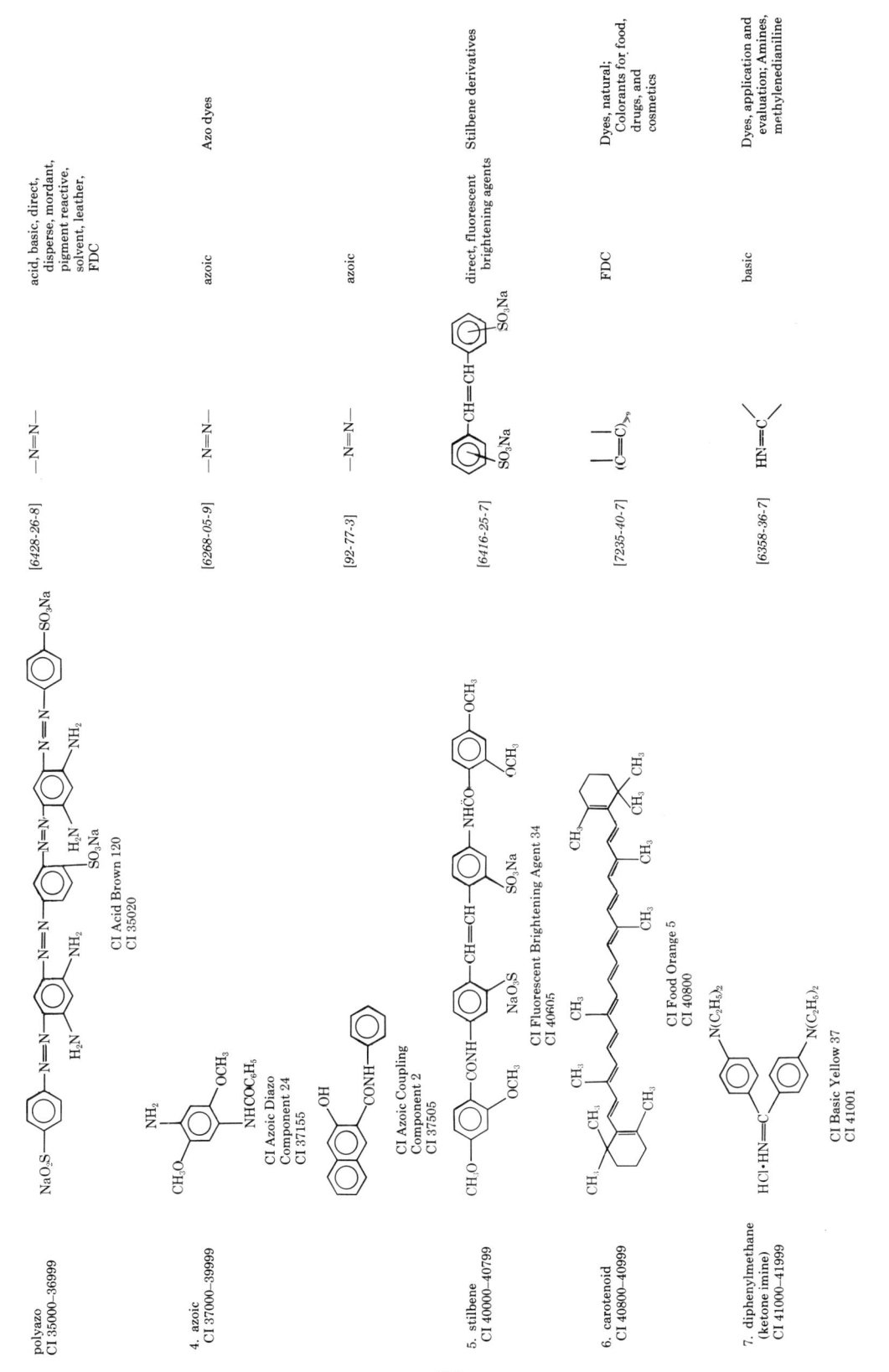

Table 1. (*continued*)

Class	Example with CI names and CI number	CAS Registry Numbers	Characteristic structural units	Dyeing classes	Relevant articles in *ECT*
8. triarylmethane CI 42000–44999	CI Acid Green 3 CI Food Green 1 CI 42085	[4680-78-8]	—X=, X = O,N	acid, basic, solvent, FDC	Dyes, application and evaluation; Triphenylmethane and related dyes
9. xanthene CI 45000–45999	CI Acid Yellow 73 CI 45350	[518-47-8]		acid, basic, solvent	Xanthene dyes
10. acridine CI 46000–46999	CI Basic Orange 14 CI 46005	[65-61-2]		basic	Azine dyes; Pyridine and pyridine derivatives
11. quinoline CI 47000–47999	CI Acid Yellow 5 CI 47035	[1324-04-5]		acid, direct, disperse	Quinoline dyes

12. methine and polymethine CI 48000–48999	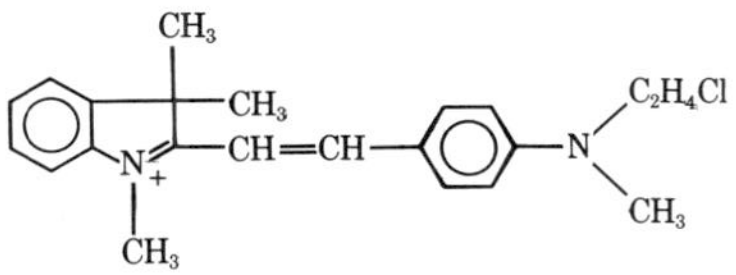 [3648-36-0]	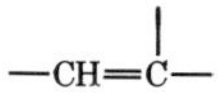 —CH=C—	basic	Cyanine dyes; Polymethine dyes
13. thiazole CI 49000–49399	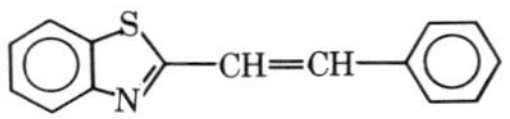 [1483-30-3]	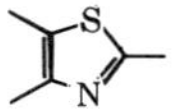 (thiazole ring)	basic, fluorescent brightener	Thiazole dyes, Brighteners, fluorescent
14. indamine and indophenol CI 49400–49999	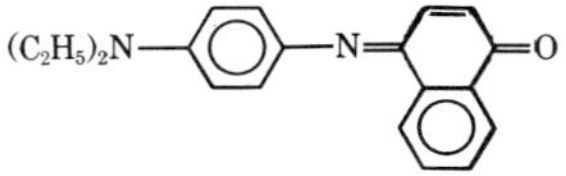 [2363-99-7]	 —N=⟨=⟩=X— ; X = O, N	intermediates	Amines, phenylenediamines
15. azine CI 50000–50999	[6378-87-6]	(pyrazinium ring)	acid, basic, solvent	Azine dyes
16. oxazine CI 51000–51999	[6598-58-9]	(oxazinium ring)	basic, direct, mordant, pigments	Azine dyes

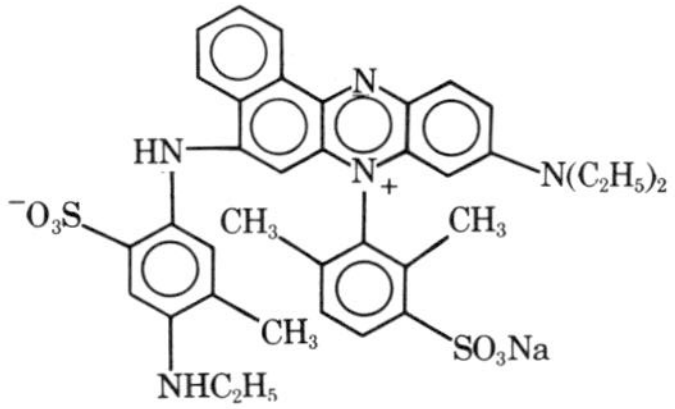

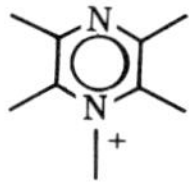

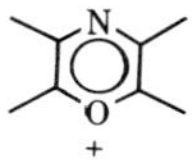

12. methine and polymethine CI 48000–48999

CI Basic Red 13
CI 48015

13. thiazole CI 49000–49399

CI Fluorescent Brightener 41
CI 49015

14. indamine and indophenol CI 49400–49999

CI Solvent Blue 22
CI 49705

15. azine CI 50000–50999

CI Acid Blue 121
CI 50310

16. oxazine CI 51000–51999

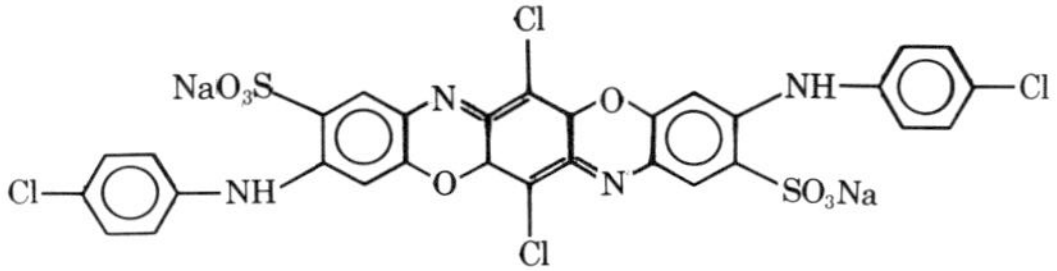

CI Direct Blue 190
CI 51305

165

Table 1. (continued)

Class	Example with CI names and CI number	CAS Registry Numbers	Characteristic structural units	Dyeing classes	Relevant articles in *ECT*
17. thiazine CI 52000–52999	(CH₃)₂N—[structure]—N(CH₃)₂ Cl⁻ CI Basic Blue 9 CI Solvent Blue 8	[61-73-4]	[thiazinium structure]	basic, mordant, direct, solvent	Azine dyes
18. sulfur CI 53000–53999	final forms of dyes are of indeterminate constitution and based on a variety of thionated aromatic substrates, eg NHOOCH₃, NO₂, NO₂ + Na₂S + S →180°C CI Sulfur Yellow 9, CI 53010 [carbazole structure] + [phenol-NH structure] + Na₂Sₓ →107°C CI Vat Blue 43, CI Vat Blue 47	[1326-40-5] [1397-79-3] [86-72-6]	[phenothiazine/phenoxathiine structures] or [thiazole structure]	sulfur and vat	Sulfur dyes
19. aminoketone CI 56000–56999	[anthraquinone-pyridine fused structure] CI Vat Yellow 27 CI 56080	[6252-73-9]	[acyl-aniline structure] [quinone structure]	disperse, fluorescent brighteners	Brighteners, fluorescent

20. hydroxyketone
CI 57000–57999

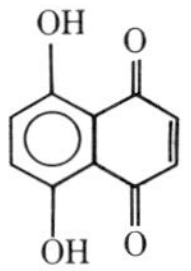

O=C—C=C—C=O and O=C—C=C—C=OH ⟶ mordant

[475-38-7]

Dyes, anthraquinone; Azo dyes

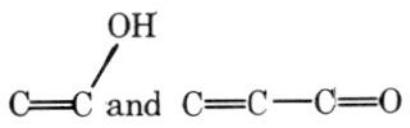

CI Mordant Black 37
CI 57010

21. anthraquinone
CI 58000–72999

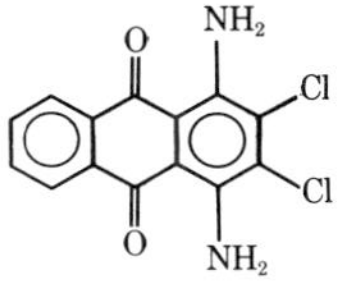

[quinoid structure] or extended quinoid

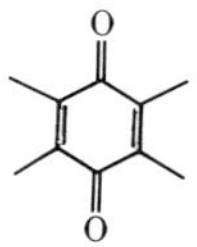

[81-42-5]

acid, mordant, vat, disperse, basic, direct, reactive, pigment

Dyes, anthraquinone

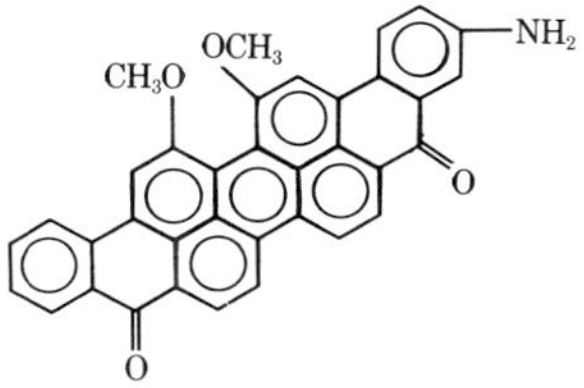

CI Disperse Violet 28
CI 61102

CI Vat Green 1
CI 59825

[128-58-5]

22. indigoid
CI 73000–73999

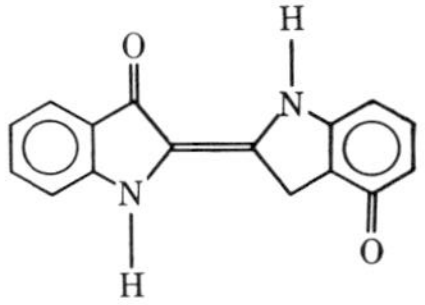

O=C—C=C—C=O

[482-89-3]

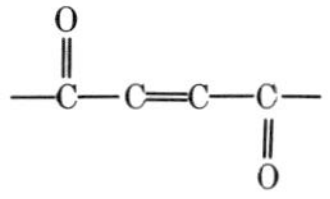

vat, acid, pigment, FDC

Dyes, natural

CI Vat Blue 1
CI 73000

167

Table 1. (*continued*)

Class	Example with CI names and CI number	CAS Registry Numbers	Characteristic structural units	Dyeing classes	Relevant articles in *ECT*
23. phthalocyanine CI 74000–74999	CI Direct Blue 86 CI 74180	[1330-38-7]	tetrabenzoporphyrazine	acid, pigment, reactive, solvent	Dyes, natural; Colorants for foods, drugs, and cosmetics
24. natural organic coloring matters CI 75000–75999	CI Natural Red 6 CI 75330	[72-48-0, 8005-35-4]	various aromatic conjugated systems	direct, mordant, vat	Dyes, natural; Azo dyes
25. oxidation bases CI 76000–76999	CI 76000	[62-53-3] [55330-79-5]	incompletely characterized oxidation products from amines, diamines and aminophenols		Aminophenols; Hair preparations
26. inorganic coloring matters CI 77000–77999	KFe[Fe(CN)₆] CI Pigment Blue 27 Prussian Blue CI 77520	[15418-51-6, 25869-98-1]			Pigments

these compounds are not dyes, because of their widespread application to textile and other materials, a CI classification has been assigned. These compounds produce their optical effect by adding light to the visible part of the radiation reflected from or transmitted by the substrate. In the case of dyes, the optical effect is due to the removal, by absorption, of some wavelengths of the incident light (see Brighteners, fluorescent).

Reactive dyes are dyes that form a covalent bond with the fiber, usually cotton, wool, or nylon. This class of dyes, first introduced commercially in 1956, made it possible to achieve extremely high washfastness properties by relatively simple dyeing methods. A marked advantage over direct dyes is that their chemical structures are much simpler, their absorption spectra show narrower absorption bands, and the dyeings are brighter. The principal chemical classes of reactive dyes are azo, anthraquinone, and phthalocyanine (see Dyes, reactive).

Sulfur dyes are dyes that are applied to cotton from an alkaline reducing bath with sodium sulfide as the reducing agent. Numerically this is a relatively small group. However, the low cost and good washfastness properties of the dyeings make this class important from an economic standpoint (see Sulfur dyes).

Vat dyes are water insoluble dyes that are applied mainly to cellulosic fibers as soluble leuco-salts after reduction in an alkaline bath, usually with sodium hydrosulfite. Following exhaustion on to the fiber, the leuco forms are reoxidized to the insoluble keto forms and aftertreated, usually by soaping, to redevelop the crystal structure (5). The principal chemical classes of vat dyes are anthraquinone and indigoid.

Table 2 is arranged according to the CI application classification. It shows the major substrates, the method of application, and the representative chemical types for each application class. The CI application classification, from which the CI name is derived, is also used by Chemical Abstracts Service. Chemical Abstracts Service Registry Numbers are usually assigned to compounds on the basis of their chemical names. A final complication in nomenclature is the trivial names of some of the older dyes. Names such as Congo Red, Zambezi Black, Malachite Green Tartrazine, and Bindschedler's Green may still be found in some literature, although their use is becoming infrequent. The following example illustrates the methods used to characterize a dye:

Chemical structure $C_{36}H_{20}O_4$

Chemical names
　Chemical Abstracts:　　　　　16,17-dimethoxydinaphtho[1,2,3-*cd*:3′,2′,1′-*lm*]perylene-5,10-dione
　commonly used:　　　　　　　16,17-dimethoxyviolanthrone
　commonly used:　　　　　　　16,17-dimethoxydibenzanthrone

Table 2. Usage Classification of Dyes

Class	Major substrates	Method of application	Chemical types	Relevant articles in *ECT*
acid	nylon, wool, silk, paper, inks, and leather	usually from neutral to acidic dyebaths	azo, including premetallized dyes, anthraquinone, triphenylmethane, azine, xanthene, nitro, and nitroso	Azo dyes
azoic components and compositions	cotton, rayon, cellulose acetate, and polyester	fiber impregnated with coupling component and treated with a solution of stabilized diazonium salt	azo	Azo dyes
basic	acrylic, modified nylon and polyester, paper, and inks	applied from acidic dyebaths	methine, diphenylmethane, triarylmethane, azo, azine, xanthene, thiazole, acridine, oxazine, and anthraquinone	Azine dyes; Azo dyes
direct	cotton, rayon, paper, leather, and nylon	applied from neutral or slightly alkaline baths containing additional electrolyte	azo, phthalocyanine, stilbene, oxazine, and thiazole	Azine dyes; Azo dyes
disperse	polyester, polyamide, cellulose, acetate, acrylic, and plastics	fine, aqueous dispersions often applied by higher temperature–pressure or lower temperature carrier methods; dye may be padded on cloth and baked on or thermofixed	azo, anthraquinone, nitro, and methine	Azo dyes; Dyes, anthraquinone
fluorescent brighteners	soaps and detergents, all fibers, oils, paints, and plastics	from solution, dispersion or suspension in a mass	stilbene, azoles, coumarin, pyrazine, and naphthalimides	Brighteners, fluorescent
food, drug, and cosmetic	foods, drugs, and cosmetics		azo, anthraquinone, carotenoid, and triarylmethane	Colorants for foods, drugs and cosmetics
mordant	wool, leather, and anodized aluminum	applied in conjunction with chelating Cr salts	azo and anthraquinone	Azo dyes; Dyes, applications and evaluation

natural		applied as mordant, vat, solvent, or direct and acid dyes	anthraquinone, polymethine, ketone imine, flavones, indigoids, quinones, chlorophylls, etc	Dyes, natural
oxidation bases	hair, fur, and cotton	aromatic amines and phenols oxidized on the substrate	aniline black and indeterminate structures	Azine dyes
pigments	paints, inks, plastics, and textiles	printing on the fiber with resin binder or dispersion in the mass	azo, basic, phthalocyanine, quinacridone, oxazine, anthraquinone, and indigoid	Pigments
reactive	cotton, rayon, wool, silk, and nylon	reactive site on dye reacts with functional group on fiber to bind dye covalently under influence of heat and proper pH	azo, anthraquinone and phthalocyanine	Dyes, reactive
solvent	gasoline, varnish, lacquer, stains, inks, fats, oils, and waxes	dissolution in the substrate	azo, triphenylmethane, anthraquinone and phthalocyanine	Azo dyes; Dyes, anthraquinone; Phthalocyanine compounds
sulfur	cotton and rayon	aromatic substrate vatted with sodium sulfide and reoxidized to insoluble sulfur containing products on the fiber	indeterminate structures	Sulfur dyes
vat	cotton, rayon, and wool	H_2O insoluble dyes solubilized by reducing with sodium hydrosulfite, then exhausted on fiber and reoxidized	anthraquinone (including polycyclic quinones) and indigoids	Dyes, anthraquinone

CI name:	CI Vat Green 1
CI number:	CI 59825
Application class:	vat
Chemical class:	anthraquinone
Chemical Abstracts Service Registry Number:	[128-58-5]
Commercial names:	
Calcoloid Jade Green M,NC	American Cyanamid Company (U.S.)
Caledon Green XBN,XBNQF	Imperial Chemical Industries Ltd. (U.K.)
Cibanone Brilliant Green BF,2BF,BFD	Ciba-Geigy S.A. (Switz.)
Indanthrene Brilliant Green B,FB	Badische Anilin- und Soda-Fabric A.G. (FRG)
Solanthrene Brilliant Green B,FF	Produits Chimiques Ugine Kuhlman (Fr.)
Trivial name:	
Jade Green	

Dye Intermediates

The precursors of dyes are called dye intermediates. They are obtained from simple raw materials, such as benzene and naphthalene, by a variety of chemical reactions. Usually, the raw materials are cyclic aromatic compounds. They are derived from two principal sources, coal tar and petroleum (qv).

Sources of Raw Materials. Coal tar results from the pyrolysis of coal and is obtained chiefly as a by-product in the manufacture of coke for the steel industry (see Coal, carbonization). Products recovered from the fractional distillation of coal tar have been the traditional organic raw material for the dye industry. Among the most important are benzene, toluene, xylene, naphthalene, anthracene, acenaphthene, pyrene, pyridine, carbazole, phenol, and cresol.

The petroleum industry is now the major supplier of benzene, toluene, the xylenes and naphthalene. Petroleum displaced coal tar as the primary source for these aromatic compounds after World War II because it was relatively cheap and abundantly available. Table 3 illustrates the importance of petroleum products as a source of primary raw materials (6).

Lists of Intermediates; Nomenclature; Auxiliary Agents. The United States International Trade Commission lists ca 1000 compounds under the heading of cyclic intermediates (6). They include cycloaliphatic and heterocyclic species among a majority of carbocyclic aromatic compounds. United States production and sales data are published for little more than 60 of them (Table 4), most of which are used in the dye industry. The very large volume items among them have a principal use outside the dye industry. For example, the five volume leaders, terephthalic acid–dimethyl terephthalate, phenol, isocyanates, phthalic anhydride, and cyclohexanone, are almost exclusively consumed by the fiber and plastics industries, and a large portion of the sixth, aniline, is used by the rubber industry.

Most intermediates in Table 4 are listed by their internationally recognized systematic names as used by *Chemical Abstracts*. These systematic names define the chemical structure and composition of any compound unequivocally but alternatives are also in use. Some are preferred over others in the dyestuff intermediates field for reasons of practicality or custom (eg, anthraquinone was substituted for 9,10-dihy-

Table 3. United States Production of Primary Raw Materials[a], 1000 m³ [b]

Chemical	1967	1975	1976
tar	2954	2443	2409
benzene			
coke-oven operators	343	246	229
petroleum refiners	3326	3629	5166
Total	*3669*	*3875*	*5395*
toluene			
coke-oven operators	73.3	37.2	33.4
petroleum refiners	2364	2631	3748
Total	*2437*	*2668*	*3781*
xylene			
coke-oven operators	20.8	7.1	5.7
petroleum refiners	1701	2412	2727
Total	*1722*	*2419*	*2733*
naphthalene, 1000 t			
crude	236		
petroleum naphthalene (all grades)	171		255
Total	*407*		

[a] Ref. 6; Data reported by tar distillers are not included for benzene, toluene and xylene because publication would disclose the operations of individual companies. Crude naphthalene data for 1975 and 1976 are omitted for the same reason.
[b] To convert m³ to liq gal, multiply by 264.2.

dro-9,10-dioxoanthracene). Systematic names of multiple substitution products and polycyclic compounds can become unwieldy and cumbersome to recognize. Trivial names have, therefore, always found ready acceptance and are still widely used. Lists of trivial names and their corresponding chemical names are available (3,7).

In addition to cyclic intermediates, a great variety of aliphatic reagents and inorganic chemicals are used in the dye industry. These include formaldehyde, acetic acid, sodium acetate, acetic anhydride, ethyl chloride and bromide, dimethyl and diethyl sulfate, sodium methoxide, ethylene oxide, acrylonitrile, alkyl- and alkanolamines, etc; sulfuric acid and oleum for sulfonation, nitric acid for nitration, chlorine and bromine for halogenation, caustic soda and caustic potash for fusion and neutralization, and sodium nitrite for diazotization, as well as ammonia, hydrochloric acid, chlorosulfonic acid, thionyl chloride, phosphorus tri-, oxy-, and pentachlorides, phosphorus pentoxide, calcium carbonate, magnesium oxide, sodium carbonate, bicarbonate and sulfate, sodium and potassium chlorides, sodium bisulfite, hydrosulfite, sulfide and sulfhydrate, sulfur, sodium hypochlorite and chlorate, sodium dichromate, manganese dioxide, iron powder, copper and cuprous salts, aluminum chloride, zinc chloride, and many others.

Equipment and Manufacture. The reactions for the production of intermediates and dyes are carried out in kettles made from cast iron, stainless steel, or steel lined with rubber, glass (enamel), brick, or carbon blocks. These kettles have capacities of 2–40 m³ (ca 500–10,000 gal) and are equipped with mechanical agitators, thermometers or temperature recorders, condensers, pH-probes, etc, depending upon the nature of the operation. Jackets or coils serve for heating and cooling by circulating through them high-boiling fluids (eg, hot oil, or Dowtherm), steam or hot water to raise the

Table 4. United States Production and Sales in 1976 of Some Cyclic Intermediates[a]

Cyclic intermediates	CAS Registry No.	Production, 1,000 kg	Sales Quantity, 1,000 kg	Sales Value, $1,000	Unit value[b], $/kg
acetoacetanilide	[102-01-2]		1754	2,759	1.56
o-acetoacetanisidide	[92-15-9]		489	1,785	3.66
o-acetoacetotoluidide	[93-68-5]		328	738	2.25
4'-aminoacetanilide (acetyl-p-phenylenediamine)	[122-80-5]	166			
4-amino-4'-nitro-2,2'-stilbenedisulfonic acid	[119-72-2]	61			
p-[(p-aminophenyl)azo]benzenesulfonic acid	[104-23-4]	186			
aniline (aniline oil; aminobenzene)	[62-53-3]	246,658	71,077	42,895	0.60
anilinomethanesulfonic acid (aniline omega acid)	[103-06-0]	211			
1-anthraquinonesulfonic acid and salt (gold salt)	[82-49-5]	487			
benzaldehyde, tech.	[100-52-7]	3,758	3,979	5,124	1.28
benzoic acid, tech. (benzene carboxylic acid)	[65-85-0]	36,131	14,949	7,561	0.51
2-benzothiazolethiol, sodium salt	[1321-08-0]		1,366	1,912	1.39
biphenyl	[92-52-4]	25,807	6,549	4,073	0.62
chlorobenzene, mono	[108-90-7]	149,267	30,797	16,786	0.55
4-chloro-3-nitrobenzenesulfonamide	[121-18-6]	316			
4-chloro-3-nitrobenzenesulfonyl chloride	[97-08-5]	238			
cresols, total[c]	[8003-33-6], [8006-62-0]	45,466	93,176	44,876	1.04
o-cresol	[95-48-7]	10,064	9,404	8,785	0.93
all other[d]		35,392	33,773	36,091	1.06
cresylic acid, refined[c]	[1319-77-3]	25,904	14,113	11,814	0.84
cyclohexanone	[108-94-1]	290,664			
cyclohexylamine	[108-91-8]		2,742	4,327	1.59
1,4-diamino-2,3-dihydroanthraquinone (leucamine)	[81-63-0]	241			
o-dichlorobenzene	[95-50-1]	22,042	10,939	6,999	0.64
p-dichlorobenzene	[106-46-7]	16,647	16,992	8,545	0.51
2,4-dichlorophenol	[120-83-2]		1,991	2,444	1.24
dicyclohexylamine	[101-83-7]		313	570	1.83
N,N-diethylaniline	[91-66-7]	903	606	1,350	2.23
1,4-dihydroxyanthraquinone (quinizarin)	[81-64-1]	779	76	344	4.54
1,8-dihydroxy-4,5-dinitroanthraquinone (4,5-dinitrochrysazine)	[81-55-0]	114			
N,N-dimethylaniline	[121-69-7]	6,151	3,685	4,601	1.26
N,N-dimethylbenzylamine	[103-83-3]	84	49	198	4.01
N,N-dimethylcyclohexylamine	[98-94-2]	1,827			
4,4'-dinitro-2,2'-stilbenedisulfonic acid	[128-42-7]	5,030			
2,4-dinitrotoluene	[121-14-2]	148,773			
2,4- (and 2,6-) dinitrotoluene	[606-20-2]	179,788			
dodecylbenzene	[121-01-3]	239,810	189,654	105,523	0.55
Ṅ-ethylaniline, refined	[103-69-5]	476	413	830	2.01
2-(N-ethylanilino)ethanol (hydroxyethylaniline)	[92-50-2]	132			
hydroquinone, tech. (p-dihydroxybenzene)	[123-31-9]		4,666	17,299	3.70
isocyanic acid derivatives, total		430,138	385,034	362,280	0.95

Table 4 (*continued*)

Cyclic intermediates	CAS Registry No.	Production, 1,000 kg	Sales Quantity, 1,000 kg	Sales Value, $1,000	Unit value[b], $/kg
polymethylene polyphenylisocyanate	[9016-87-9]	141,584	117,606	115,037	0.97
toluene 2,4- and 2,6-diisocyanate (80:20 mixture)	[26471-62-5]	255,718	241,579	201,431	0.84
other isocyanic acid derivatives		32,649	25,814	45,812	1.76
4,4'-isopropylidenediphenol (bisphenol A)	[80-05-7]	203,590	51,344	41,470	0.82
melamine	[6147-18-8]	57,265	36,342	26,781	0.73
d,l-p-mentha-1,8-diene	[138-86-3]	5,068	2,757	798	0.29
metanilic acid (m-aminobenzenesulfonic acid)	[121-47-1]	723			
4,4'-methylenedianiline (p,p'-diaminodiphenylmethane)	[101-77-9]		533	1,716	3.22
3-methyl-1-phenyl-2-pyrazolin-5-one (developer Z)(phenylmethylpyrazolone, PMP)	[12235-58-4]		6	29	4.92
α-methylstyrene	[98-83-9]	27,834	21,854	7,363	0.33
3'-nitroacetanilide	[122-28-1]	16			
nitrobenzene	[98-95-3]	185,533	8,865	4,521	0.51
5-nitro-o-toluenesulfonic acid (SO_3H = 1) (2-methyl-5-nitrobenzenesulfonic acid)	[121-03-9]	3,357			
nonylphenol	[1300-16-9]	35,369	16,217	10,729	0.51
1-[(7-oxo-7H-benz[de]anthracen-3-yl)-amino]anthraquinone (Indanthrene Olive Green B Base)	[3271-76-9]	108			
phenol, total[c]	[108-95-2]	962,282	439,664	214,580	0.48
from cumene	[98-82-8]	873,816	404,066	195,560	0.48
other		88,466	35,597	19,020	0.53
2,2-[(phenyl)imino]diethanol (N-phenyldiethanolamine; dihydroxydiethylaniline)	[120-07-0]	225	137	200	.66
phthalic anhydride	[85-44-9]	409,320	239,859	118,136	0.48
2-picoline (α-picoline; α-methylpyridine)	[109-26-8]	189	193	401	1.46
piperidine (hexahydropyridine)	[110-89-4]	231			
salicylaldehyde	[90-02-8]	2,002	1,299	7,173	5.51
salicylic acid, tech. (o-hydroxybenzoic acid)	[69-72-7]	14,182	1,800	3,298	1.83
terephthalic acid and its dimethyl ester (dimethyl terephthalate, DMT)	[100-21-0] [120-61-6]	3,270,734			
toluene-2,4-diamine (4-m-tolylenediamine)	[95-80-7]	105,736			
7,7'-ureylenebis[4-hydroxy-2-naphthalenesulfonic acid] (J acid urea)	[134-47-4]	153			
all other cyclic intermediates		1,817,088	1,847,882	1,289,882	0.70
Total		*8,979,238*	*3,476,192*	*2,386,993*	*0.69*

[a] Ref. 6; listed below are all cyclic intermediates for which any reported data on production and/or sales may be published. Missing data were not reported or may not be published.

[b] Calculated from rounded figures.

[c] Does not include data for coke ovens and gas-retort ovens, reported to the Division of Fuels Data, U.S. Bureau of Mines.

[d] Figures include (o, m, p)-cresol from coal tar and some m-cresol and p-cresol.

temperature, and air, cold water, or chilled brine to lower it. Unjacketed kettles are often used for aqueous reactions, where heating is effected by direct introduction of steam and cooling by addition of ice or by heat exchangers.

Products are transferred from one piece of equipment to another by gravity flow, pumping, or by blowing with air or inert gas. Solid products are separated from liquids in centrifuges, on filter boxes, on continuous belt filters, and perhaps most frequently in various designs of plate-and-frame or recessed plate filter presses. The presses are dressed with cloths of cotton, Dynel, polypropylene, etc. Some provide separate channels for efficient washing, others have membranes for increasing the solids content of the presscake by pneumatic or hydraulic squeezing. The plates and frames are made of wood, cast iron, or now usually hard rubber, polyethylene and polyester.

When possible, the intermediates are taken for the subsequent manufacture of other intermediates or dyes without drying. Where drying is required, air or vacuum ovens (in which the product is spread on trays), rotary dryers, spray dryers, or less frequently drum dryers (flakers) are used. Spray dryers have become increasingly important. They need little labor and accomplish rapid drying by blowing concentrated slurries, eg, reaction masses, through a small orifice into a large volume of hot air. Dyestuffs, especially disperse dyes, which require wet-grinding as the penultimate step, are now often dried this way. In this case their final standardization, ie, addition of desired amounts of auxiliary agents and solid diluents, is achieved in the same operation (see Drying).

The comparatively small tonnages of the numerous intermediates needed exclusively by the dye industry have made their manufacture by continuous process uneconomical or impractical. Batch processes remain the rule but the progress in computer science and technology, electronic circuitry, and instrumentation has led to a growing use of automatic process control in recent years. Automated plants for dyes and dye intermediates are now in operation (11).

Automation has been applied to the weighing of the entire kettle; to the charging of solid reactants and the metering of preanalyzed solutions or reactant slurries; to the control of addition, heating and cooling via feedback signals, ie, by presetting limits and monitoring reaction parameters, such as temperature, pressure, concentration of hydrogen ions (pH), and others (with pH-probes and ion-selective electrodes (qv)), changes of redox potential, conductivity, etc; and to mass transfer (qv), phase separation, filtration (qv) (especially on continuous filters) and discharge.

Direct labor costs probably still represent a significant if diminishing portion of the final dye price in many plants, and the costs for energy, equipment, raw materials (except for temporary oversupply situations), more stringent safety and health measures, much stricter environmental control and expansive reporting requirements to government agencies have increased substantially.

The Chemistry of Dye Intermediates. Organic dyes and their intermediates range in structure from simple substitution products of benzene, naphthalene and anthraquinone to fairly complex ring systems. Their prime building block is the hexagonal skeleton of benzene, itself the prototype of aromatic compounds. Typical for the reactivity of an aromatic compound is the presence of delocalized π-electrons in the ring (9). These permit facile polarization by certain reactants so that electrons become readily available for electrophilic substitution reactions such as sulfonation, nitration, halogenation and acylation, which are indispensable for the manufacture of synthetic organic dyes and intermediates.

Electrophilic Substitution. An electrophilic substitution is initiated by the attack of a positively charged (cationic) species X^+ or by a polarized neutral species $X^{\delta+}$—$Y^{\delta-}$ which combines with the aromatic molecule to form an intermediate. The intermediate then reverts to the stable aromatic state by expulsion of a positively charged hydrogen ion. Once a hydrogen atom has been replaced, the substituent modifies the ease of further substitution depending on its electron-attracting (deactivating) or electron-donating (activating) tendencies by way of inductive and mesomeric effects (10). The substituent's electronic effects will also direct a second substituent to certain preferential positions on the ring, which are determined by the degree of charge stabilization possible for the respective reaction intermediate.

Orientation rules have been derived for benzene that may be summarized as follows: (1) electron-attracting substituents such as CF_3, $(CH_3)_3N^+$, NO_2, SO_3H, SO_2Cl, SO_2CH_3, CN, CHO, $COCH_3$, COC_6H_5, $COOH$, $COOCH_3$, $CONH_2$, H_3N^+, CCl_3, I, Br, Cl and NO make the introduction of a second substituent more difficult and (with the important exception of the ortho–para-directing halogens) direct it to the meta position on the benzene ring; and (2) electron-donating substituents such as O^-, $N(CH_3)_2$, NH_2, OH, $NHCH_3$, OCH_3, $NHCOCH_3$, $OCOCH_3$, SH, SCH_3, SCN, N=N-(azo), CH_3, and C_6H_5 facilitate the introduction of a second substituent and direct it to the ortho and para position on the benzene ring.

The mesomeric or conjugation effect (resonance) dominates in substituents that carry at least one unshared electron pair on the atom attached to the ring. The unshared electron pair contributes to the charge stabilization of the reaction intermediate in resonance structures, which are only possible for ortho and para substitution.

The six hydrogen atoms of benzene are equivalent owing to the symmetry of the molecule. Monosubstitution products are unique and have no isomers. Naphthalene, however, carries two nonequivalent sets of hydrogen atoms. Therefore, two isomers of every monosubstituted naphthalene are known. The same holds true for monosubstituted anthraquinones.

Halogenation and nitration of naphthalene yield the α-substitution products almost exclusively, as does sulfonation with sulfuric acid in the cold, which is kinetically controlled. β-Sulfonation is predominant above 120°C. An α-sulfo group migrates to the β-position above 160°C to form the thermodynamically more stable β-sulfo isomer.

Anthraquinone is manufactured by oxidation of the tricyclic raw material anthracene with retention of carbon skeleton and aromatic character; or from phthalic anhydride (an oxidation product of naphthalene or o-xylene) and benzene by a Friedel-Crafts (qv) acylation and a ring closure. Newer schemes for its synthesis include the addition of butadiene to naphthoquinone (also an oxidation product of naphthalene) in a Diels-Alder reaction, or the catalytic dimerization of styrene. Anthra-

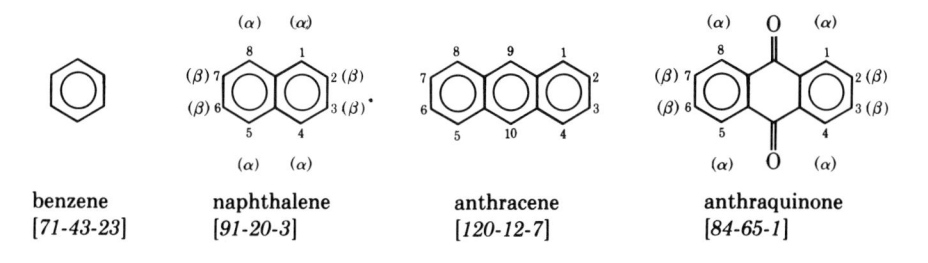

benzene
[71-43-23]

naphthalene
[91-20-3]

anthracene
[120-12-7]

anthraquinone
[84-65-1]

quinone is less reactive towards electrophilic substitution than benzene or naphthalene. Sulfonation with oleum attacks the β-position and produces a mixture of 2-anthraquinonesulfonic acid with 2,6- and 2,7-anthraquinone disulfonic acid. Nitration occurs in the α-position and leads readily to a mixture of 1,5-, 1,8-, and 1,6-dinitroanthraquinones, which can be separated. The 1,5-, 1,8- mixture is often processed as such. α-Sulfonation is achieved in the presence of mercury salts.

In summation, the first step in the manufacture of dye intermediates from aromatic raw materials is usually a direct electrophilic substitution. Additional substituents take preferred positions as suggested by the orientation rules. Other synthetic routes must be sought if different isomers are desired.

Transformations of Primary Substitution Products. A variety of reactions permits conversion of the primary substitution products to dye intermediates. They can be grouped under the following headings: (1) chemical change of a substituent, (2) removal of a substituent, (3) replacement of a substituent with another by nucleophilic substitution, (4) special reactions and rearrangements, (5) condensations and ring closures, usually done to form polycyclic or heterocyclic structures, (6) dimerizations, and (7) oxidations. The common transformations delineated in the first three categories are summarized below. Specific examples of the most important reactions are given subsequently.

Chemical Change of Substituents. (R = alkyl or aryl.) The reaction types of importance are:

reduction	$NO_2 \rightarrow NH_2$; $NO \rightarrow NH_2$; $N_2^+ \rightarrow NH—NH_2$; $N=N$(azo) $\rightarrow 2 NH_2$; $SO_2Cl \rightarrow SO_2R \rightarrow SH$; $COR \rightarrow CH_2R$
alkylation and arylation	$NH_2 \rightarrow NHR \rightarrow NR_2$; $SO_2H \rightarrow SO_2R$; $SH \rightarrow SR$; $OH \rightarrow OR$; $NH_2 \rightarrow N(C_2H_4CN)_2$ with acrylonitrile or $\rightarrow N(C_2H_4OH)_2$ with ethylene oxide or ethylene chlorohydrin
acylation	(see ester and amide formation below)
oxidation	$CH_3 \rightarrow CHO$ or $COOH$; $COCH_3 \rightarrow COOH$; $SH \rightleftharpoons S—S$ and $\rightarrow SO_3H$ or $\rightarrow SO_2Cl$; $NHOH \rightarrow NO \rightarrow NO_2$
halogenation	(1) with Cl_2 + light: $CH_3 \rightarrow CH_2Cl \rightarrow CHCl_2 \rightarrow CCl_3$
	(2) with halogenating agents such as $POCl_3$: $C_2H_4OH \rightarrow C_2H_4Cl$; $COOH \rightarrow COCl$; $SO_3H \rightarrow SO_2Cl$; $OH \rightarrow Cl$ in some heterocycles
hydrolysis	$CH_2Br \rightarrow CH_2OH$; $CHCl_2 \rightarrow CHO$; $CCl_3 \rightarrow COOH$; $SO_2Cl \rightarrow SO_3H$; $NHCOR \rightarrow NH_2$; CN or $CONH_2 \rightarrow COOH$
ester formation	$COOH \rightarrow COCl \rightarrow COOR$; $SO_2Cl \rightarrow SO_2OR$; $OH \rightarrow OSO_2R$; $OH \rightarrow OSO_3H$; $OH \rightarrow OCOR$
amide formation	$COCl \rightarrow CONH_2$; $SO_2Cl \rightarrow SO_2NH_2$, SO_2NHR or SO_2NR_2; $NH_2 \rightarrow NHSO_2R$; $NH_2 \rightarrow NHCOR$
dehydration	$COONH_4 \rightarrow CONH_2 \rightarrow CN$
diazotization and azo coupling	$NH_2 \rightarrow N_2^+ \rightarrow N=N$

Removal of Substituents.

desulfonation $H_2O + SO_3H \rightarrow H + H_2SO_4$
decarboxylation $COOH \rightarrow H + CO_2$
reductive decomposition of diazonium salts $NH_2 \rightarrow N_2^+ \rightarrow H + N_2$
reductive dehalogenation Cl or $Br \rightarrow H$

Replacement of Substituents. Nucleophilic substitution is the most prevalent mechanism. A negatively charged (anionic) species or a neutral species with a lone pair of electrons combines with the aromatic molecule to form a temporary intermediate, which regenerates the aromatic state through loss of an anion (11).

Alkali fusion $SO_3H \rightarrow OH$

$$\underset{}{\bigcirc}-X + Y^- \rightarrow \underset{X}{\overset{Y}{\bigcirc}} \rightarrow \underset{}{\bigcirc}-Y + X^-$$

or $\underset{}{\bigcirc}-X + HY \rightarrow \underset{X}{\overset{\overset{+}{Y}H}{\bigcirc}} \rightarrow \underset{}{\bigcirc}-\overset{+}{Y}H + X^- \rightarrow \underset{}{\bigcirc}-Y + HX$

Replacement of activated Cl or Br } $\quad$ Cl $\rightarrow$ OH, OR, NH$_2$, NHR, NR$_2$,
(via Meisenheimer complexes) $\quad\quad$ S—S, SO$_3$H (with Na$_2$SO$_3$)

Introduction of OH by alkaline hydrolysis of
NH$_2$ $\quad\quad\quad\quad\quad\quad$ under pressure; also acid,
Cl, Br $\quad\quad\quad\quad\quad\quad$ strong alkali, temp. >300°C; via benzyne,
N$_2^+$ $\quad\quad\quad\quad\quad\quad\quad$ aqueous, heat; via aryl cation,
SO$_3$H, NO$_2$, Cl $\quad\quad\quad$ aqueous, heat; if ortho in diazonium compounds

The Sandmeyer reaction, a versatile method with few commercial applications, can serve to introduce a number of substituents via decomposition of diazonium salts in the presence of Cu(I) salts:

$$NH_2 \rightarrow N_2^+ \xrightarrow[-N_2]{Cu^+} Cl, Br, CN, SCN, SH, F \text{ (with BF}_4\text{), NO}_2, \text{etc (probably via phenyl radicals)}$$

Replacement of halogen in the presence of Cu(I) salts or copper:

$$Br \xrightarrow[pyridine]{Cu_2(CN)_2} CN \text{ (especially ortho to azo or nitro group),}$$

$$Cl, Br \xrightarrow{Cu\ catalyst} NH_2, NHR \text{ (eg, } \rightarrow p\text{-aminophenol), (also important in vat dye synthesis)}$$

Other replacements, with principal use in the anthraquinone series (marked with * if also used in benzene or naphthalene derivatives):

SO$_3$H $\rightarrow$ OR, SR $\rightarrow$ 1-OCH$_3$ and 1,5-di-OCH$_3$;
$\quad\quad$ $\rightarrow$ Cl, Br, at the boil, with Br$_2$ or with conc HCl + NaClO$_3$;
$\quad\quad$ $\rightarrow$ CN
$\quad\quad$ these replacements are facile in 2-position if 1- and 4-positions are substituted with OH, NH$_2$, and NO$_2$
$\quad\quad$ $\rightarrow$ NH$_2$*, NHC$_6$H$_5$*, OH* usually near 200°C but $\rightarrow$ α-NHCH$_3$ < 128°C; add H$_3$AsO$_4$ or Revatol (m-nitrobenzenesulfonic acid); high pressure with NH$_3$
$\quad\quad$ $\rightarrow$ NO$_2$* (with HNO$_3$); mainly in benzene and naphthalene derivatives
NO$_2$ $\rightarrow$ Cl* rare, eg, m-dinitro $\rightarrow$ m-dichlorobenzene
$\quad\quad$ $\rightarrow$ NH$_2$, NHC$_6$H$_5$, NR$_2$; OR*, SR*
Cl, Br $\rightarrow$ OH acid; facile in α-position if para to NH$_2$ or OH;
$\quad\quad$ $\rightarrow$ OR alkaline;
OH $\quad$ $\rightarrow$ NH$_2$, NHR important in 1,4-dihydroxy anthraquinones, after reduction to the leuco form (catalyzed by H$_3$BO$_3$ or small amounts of strong acids), eg,

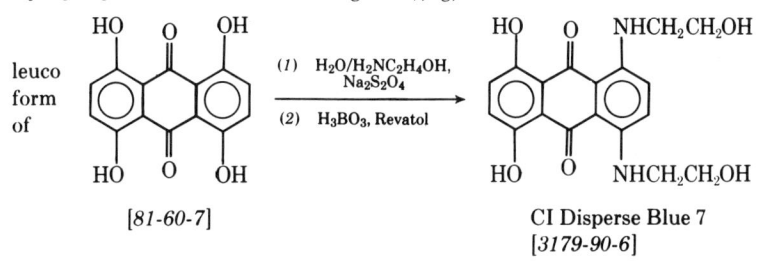

leuco form of

[81-60-7]
$\xrightarrow{\substack{(1)\ H_2O/H_2NC_2H_4OH,\\ Na_2S_2O_4 \\ (2)\ H_3BO_3,\ Revatol}}$
CI Disperse Blue 7
[3179-90-6]

180   DYES AND DYE INTERMEDIATES

Examples of the Most Important Reactions. *Sulfonation.* The main sulfonating agents are: concentrated sulfuric acid (96% H_2SO_4) or monohydrate (100% H_2SO_4); oleum, a solution of SO_3 in H_2SO_4 (commercially available in 25% and 65% strength); sulfur trioxide, SO_3 (marketed under the trademark Sulfan, containing stabilizer); chlorosulfonic acid, $ClSO_3H$ (also used to introduce the SO_2Cl group). $SO_3(^{\delta+}SO_3^{\delta-})$ or SO_3H^+ is believed to be the active species in most sulfonations:

[reaction scheme: benzene + SO_3 → arenium intermediate → C_6H_5-SO_3H (in H_2SO_4)] [98-11-3]

[reaction scheme: C_6H_5-SO_3H + SO_3H^+ → intermediate → H^+ + disulfonated benzene -SO_3H (in oleum)] [98-48-6]

β-naphthol [135-19-3] → Schaeffer's acid [93-01-6] → depending on reaction conditions →

G acid [118-32-1]   and   R acid [148-75-4]   important couplers for azo dyes

Tobias acid [81-16-3] → [55524-84-0] → (acid hydrolysis, desulfonation) → Amino J acid [86-65-7]

Sulfonation is a reversible reaction. Desulfonation occurs when sulfonic acids are heated, eg, with steam or diluted sulfuric acid. Alkaline reductions in the anthraquinone series can also cause desulfonation.

sulfonation

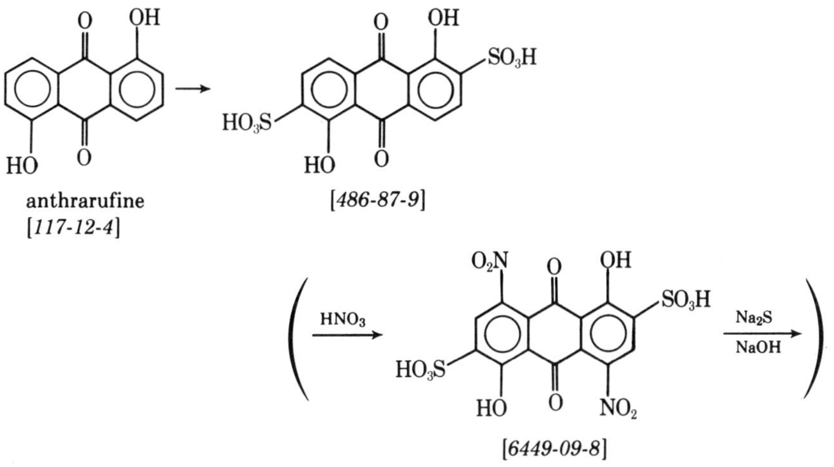

anthrarufine
[117-12-4]

[486-87-9]

$$\left(\xrightarrow{HNO_3} \quad [6449\text{-}09\text{-}8] \quad \xrightarrow[NaOH]{Na_2S} \right)$$

desulfonation

CI Acid Blue 45
[2861-02-1]

$$\xrightarrow[\substack{H_3BO_3 \\ 130°C.}]{92\% \; H_2SO_4}$$

CI Acid Blue 43
[2150-60-9]

$$\xrightarrow[\substack{Na_2S_2O_4 \\ 104°C}]{H_2O, \; NaOH}$$

1,4,5,8-tetrahydroxyanthraquinone (leuco)
(used to make CI Disperse Blue 7)

[81-60-7]

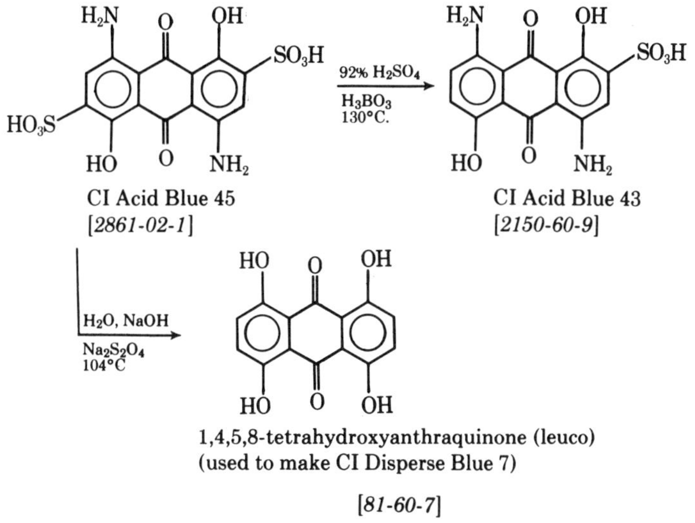

The removal of an SO₃H group is often desired after it has served to block a certain position and has directed other substituents to the intended positions.

Aromatic sulfonic acids are important starting materials for phenols and naphthols. Other methods for the introduction of an SO₃H group are described later.

Nitration. Nitric acid, HNO₃, is the customary nitrating agent, usually mixed with sulfuric acid (mixed acid). The nitronium ion NO_2^+ is considered to be the active nitrating species.

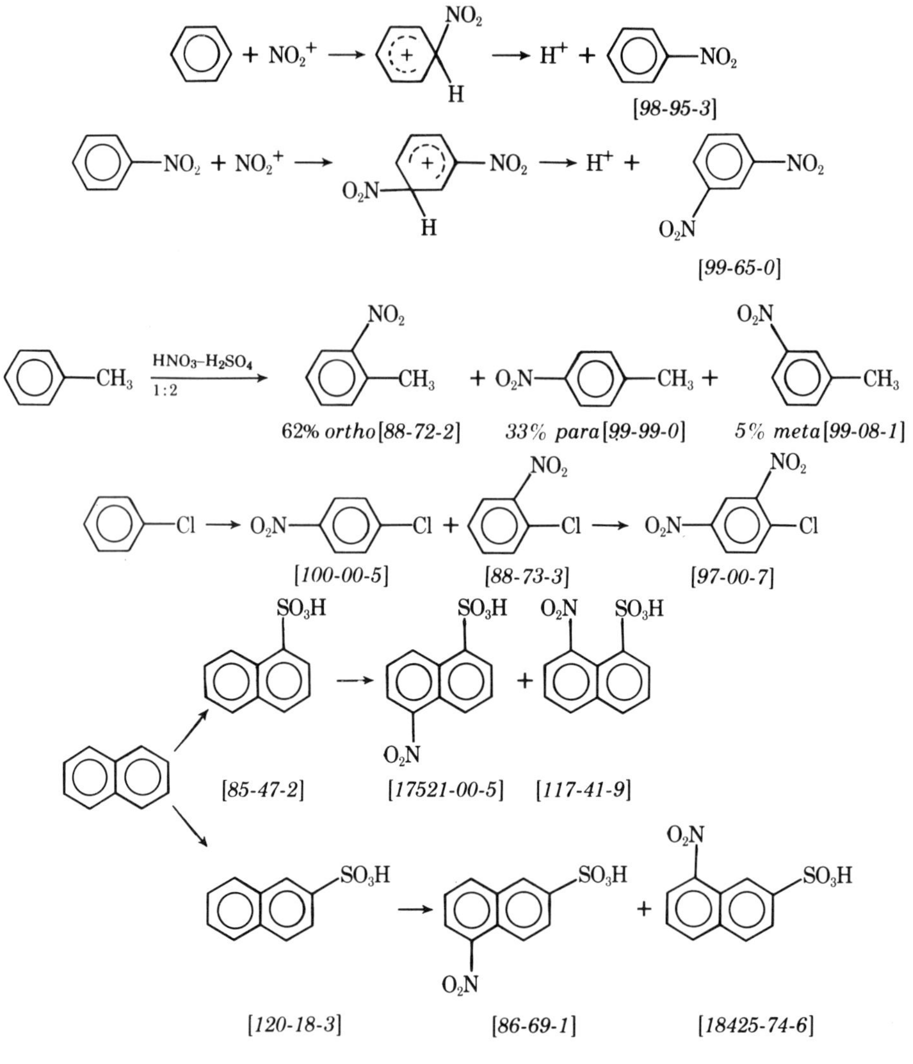

182 DYES AND DYE INTERMEDIATES

[Reaction schemes showing nitration of benzene to nitrobenzene [98-95-3], then to m-dinitrobenzene [99-65-0].]

[Nitration of toluene giving 62% ortho [88-72-2], 33% para [99-99-0], 5% meta [99-08-1].]

[Nitration of chlorobenzene giving p-nitrochlorobenzene [100-00-5], o-nitrochlorobenzene [88-73-3], and 2,4-dinitrochlorobenzene [97-00-7].]

[Sulfonation/nitration of naphthalene: naphthalene-1-sulfonic acid [85-47-2], 5-nitronaphthalene-1-sulfonic acid [17521-00-5], 8-nitronaphthalene-1-sulfonic acid [117-41-9]; naphthalene-2-sulfonic acid [120-18-3], 5-nitronaphthalene-2-sulfonic acid [86-69-1], 8-nitronaphthalene-2-sulfonic acid [18425-74-6].]

Isomers in the benzene series can usually be separated by distillation or crystallization.

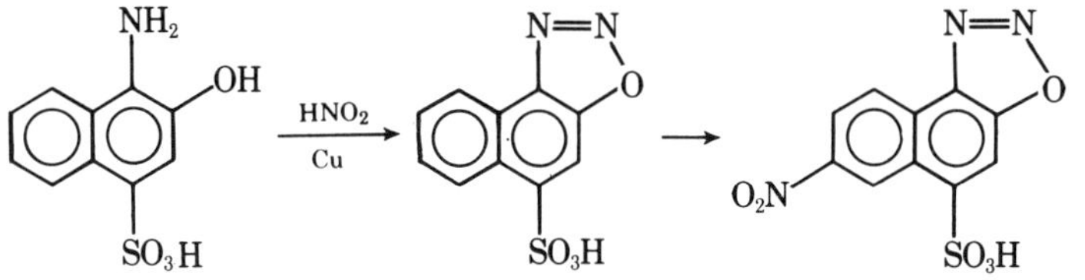

[Reaction scheme: 1,2,4-acid [116-63-2] → (HNO2, Cu) → diazo acid [84-23-1] → nitrodiazo acid [84-91-3, 130-59-6].]

1,2,4-acid
[116-63-2]

diazo acid
[84-23-1]

nitrodiazo acid
[84-91-3, 130-59-6]

Isomers in the naphthalene series are often separated after reduction.

The reduction of nitro compounds is the major source of the important amino compounds in the benzene and the naphthalene series. Nitrated chlorobenzenes are the starting materials for nitroanilines, nitrophenols and aminophenols.

Nitrosation. A carbon–nitrogen bond is easily formed by the action of nitrous acid on compounds with a strong electron-donating group such as $N(CH_3)_2$ or OH, for which nitration would be too severe. Nitroso compounds are reactive and are used for condensations as well as for transformation into amines.

α-nitroso-β-naphthol
1-nitroso-2-naphthol
[131-91-9]

CI Basic Blue 3
[2787-91-9]

Halogenation. Cl_2 and Br_2 are most frequently used to halogenate the aromatic ring, often in the presence of an electron-accepting metal halide as catalyst to polarize the halogen molecule, as in $Br^{\delta+}$—$Br^{\delta-}$----$FeBr_3$.

184 DYES AND DYE INTERMEDIATES

Sometimes the halogen is generated from HCl or HBr and an oxidizing agent; hypochlorites are used in alkaline solutions.

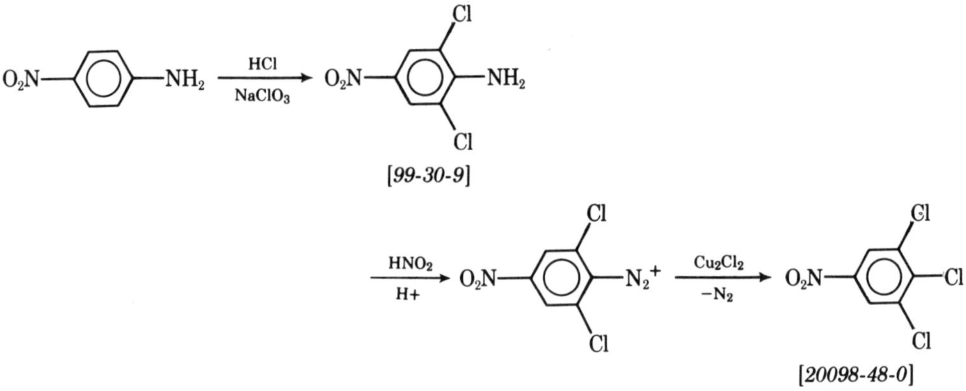

[Reaction scheme: p-nitroaniline + HCl/NaClO$_3$ → 2,6-dichloro-4-nitroaniline [99-30-9]; then HNO$_2$/H+ → diazonium salt; then Cu$_2$Cl$_2$, −N$_2$ → 1,2,3,4-tetrachloro-5-nitrobenzene [20098-48-0]]

The sequence in parentheses is an application of the Sandmeyer reaction as an example of halogen-introduction by replacement.

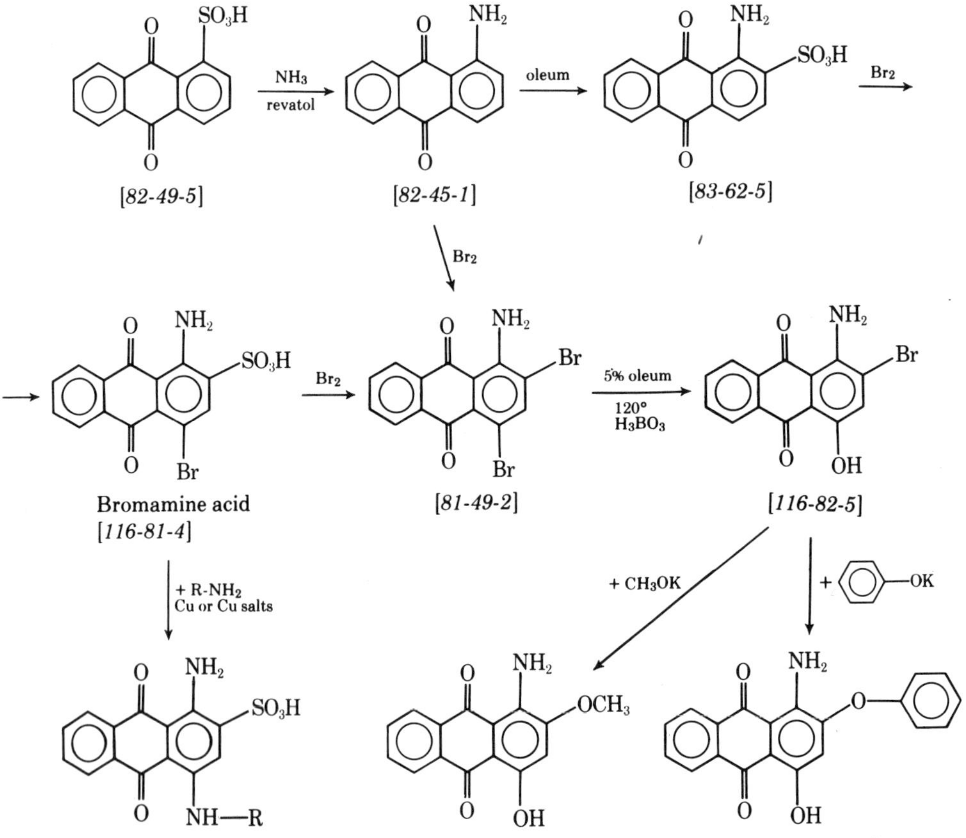

[Reaction scheme showing anthraquinone-1-sulfonic acid [82-49-5] → (NH$_3$, revatol) 1-aminoanthraquinone [82-45-1] → (oleum) 1-amino-2-sulfo anthraquinone [83-62-5] → (Br$_2$) ... Bromamine acid [116-81-4]; [81-49-2]; (5% oleum, 120°, H$_3$BO$_3$) → [116-82-5]]

Bromamine acid
[116-81-4]

+ R-NH$_2$
Cu or Cu salts

+ CH$_3$OK

+ C$_6$H$_5$OK

CI Acid Blue 25 R = C$_6$H$_5$ [6408-78-2]
40 R = p-CH$_3$CONHC$_6$H$_4$ [6424-85-7]
62 R = cyclohexyl [4368-56-3]

CI Disperse Red 4
[2379-90-0]

CI Disperse Red 60
[12223-37-9]

Halogenation in the benzene series often produces mixtures that have to be separated. Naphthalenes are seldom halogenated directly because the reaction is difficult to control.

The sequence of reactions followed to produce 1-amino-4-bromo-2-anthraquinonesulfonic acid (bromamine acid) and 1-amino-2-bromo-4-hydroxyanthraquinone illustrates not only halogenation in the anthraquinone series but also four of the replacement reactions listed on page 179. Bromamine acid, also called bromaminic acid, is an important intermediate for the manufacture of blue acid dyes, and 1-amino-2-bromo-4-hydroxyanthraquinone is the starting material for a series of brilliant pink disperse dyes.

Side chain halogenation, the introduction of halogen into alkyl groups, occurs with halogens under the influence of ultraviolet irradiation (sunlight) by a radical mechanism.

186   DYES AND DYE INTERMEDIATES

Friedel-Crafts Acylation. Acid halides or anhydrides are polarized with anhydrous aluminum chloride, AlCl$_3$, to form the reactive species.

Reduction. Conversion of the nitro to the amino group is a key reaction in dye intermediate synthesis. Aromatic amines are indispensible for most organic dyes. Iron in dilute hydrochloric or in acetic acid (Béchamp reduction) and alkali sulfides (Na$_2$S, Na$_2$S$_x$, NaHS) are the prevalent reducing agents. The sulfides are employed to advantage in the partial reduction of polynitro compounds. Catalytic reduction with hydrogen is very effective and increasingly used.

Ph–NO$_2$ $\xrightarrow{\text{Fe, HCl}}$ Ph–NH$_2$;   (O$_2$N)Ph–NO$_2$ [99-65-0] $\xrightarrow{\text{Fe, HCl}}$ (H$_2$N)Ph–NH$_2$ [108-45-2]

R-substituted (O$_2$N)Ph–NO$_2$ R = OH [51-28-5] $\xrightarrow{\text{NaHS, H}_2\text{O, 95°}}$ R-substituted (O$_2$N)Ph–NH$_2$ R = OH [99-57-0]

(Ph–NO$_2$) $\xrightarrow{\text{oleum}}$ (HO$_3$S)Ph–NO$_2$ Revatol [98-47-5]   →   (HO$_3$S)Ph–NH$_2$ metanilic acid [121-47-1]

Naphthalene derivatives: [17521-00-5] → Laurent's acid [84-89-9] + Peri acid [82-75-7]

Naphthalene derivatives: [86-69-1] + [56721-33-6] → 1,6 Cleve's acid [119-79-9] + 1,7 Cleve's acid [119-28-8]

Separation of the isomers is achieved through solubility differences of the salts. Some mixtures are often used as such, eg, mixed Cleve's acid.

Alkali sulfite serves well for the reduction of diazonium salts to hydrazines and of sulfonyl chlorides to sulfinic acids.

⟨O⟩—N₂⁺ → ⟨O⟩—NH—NH₂   (form pyrazolones on condensing with
                  [100-63-0]   acetoacetic ester, etc, and ring closure)

Cl—⟨O⟩—SO₂Cl → Cl—⟨O⟩—SO₂H   (from sulfones on alkylation)
  [98-60-2]   [100-03-8]

Nitrobenzene can also be reduced to phenylhydroxylamine or hydrazobenzene.

Alkali Fusion. The arylsulfonic acid is heated with NaOH, KOH or a mixture of the two; a widely used reaction, especially in the naphthalene series.

fusion: ⟨O⟩—SO₃H + 3 NaOH → ⟨O⟩—ONa + Na₂SO₃ + H₂O

drowning: ⟨O⟩—ONa + Na₂SO₃ $\xrightarrow[(H_2O)]{+3\,H^+}$ ⟨O⟩—OH + SO₂ + H₂O + 3 Na⁺

m-benzenedisulfonic acid $\xrightarrow{330°}$ resorcinol;   naphthalene-SO₃H → β-naphthol
   [98-48-6]   [108-46-3]   [120-18-3]   [135-19-3]

Amino J acid   J acid
[86-65-7]   [87-02-5]

Amino G acid   Gamma acid
[86-65-7]   [12235-59-5]

important azo dye intermediates

188   DYES AND DYE INTERMEDIATES

Koch acid [117-42-0] → (25–50% NaOH, 178°C, 0.5–0.6 MPa (5–6 atm)) → H acid [90-20-0] → (higher temp) → chromotropic acid [148-25-4]

Nucleophilic Replacement of Activated Cl. A variety of nucleophilic reagents are effective. All carry either a negative charge or a lone pair of electrons. HO^-, CH_3O^-, $C_6H_5O^-$, NH_3, RNH_2, and $RNHR'$ are employed most frequently (symbolized by Y:). The aromatic halogen is activated by electron-attracting substituents in the ortho and para positions, especially NO_2. Such substituents stabilize the intermediate addition product and lower the energy of the transition states through which it forms.

(Contributory resonance structures like (3) or (4) are not possible with an electron attracting substituent in the meta position, and activation is negligible in that case.)

Some intermediate complexes have actually been isolated by Meisenheimer (11).

Special Reactions and Rearrangements. *Baking* is the process of heating an amine-sulfate paste to 100–300°C in order to effect sulfonation through an intramolecular rearrangement. para-Sulfonation is much preferred but ortho-sulfonation occurs in high yield when the para position is occupied.

[606-35-9] → (C_2H_5OK) → a Meisenheimer complex ← (CH_3OK) → [4732-14-3]

Benzidine Rearrangement. The benzidine (qv) rearrangement is the acid-catalyzed intramolecular rearrangement of hydrazobenzene. A few substituted benzidines are also made commercially in the same manner. Hydrazobenzenes are obtained from unsubstituted or substituted nitrobenzene by alkaline reduction with Zn, Fe, HCHO, etc, or by electrolytic reduction.

R = H   benzidine [92-87-5]
R = CH₃   o-tolidine [119-93-7]
R = OCH₃   o-dianisidine [119-90-4]
R = Cl   3,3′-dichlorobenzidine [1331-47-1]

These compounds were extremely important as starting materials for numerous substantive cotton dyes but the manufacture and use of benzidine itself has been largely discontinued because it is a carcinogen.

A notable intermolecular rearrangement occurs with yet another reduction product of nitrobenzene:

phenylhydroxylamine [100-65-2]   p-aminophenol [123-30-8]

Bucherer Reaction. This is the reversible interconversion of the OH and NH₂ groups with the help of sulfite. It has practical applications in the naphthalene series, where tetralonesulfonic acids have been identified as intermediates (12).

naphthionic acid [84-86-6]   [70729-34-9]   [70729-35-0]

Nevile and Winther's acid (N & W acid) [84-87-7]

Similarly, α-naphthylamine is converted to α-naphthol:

[134-32-7]   [90-15-3]

Other examples are the manufacture of Tobias acid and the amination of G acid to Amino G acid under pressure.

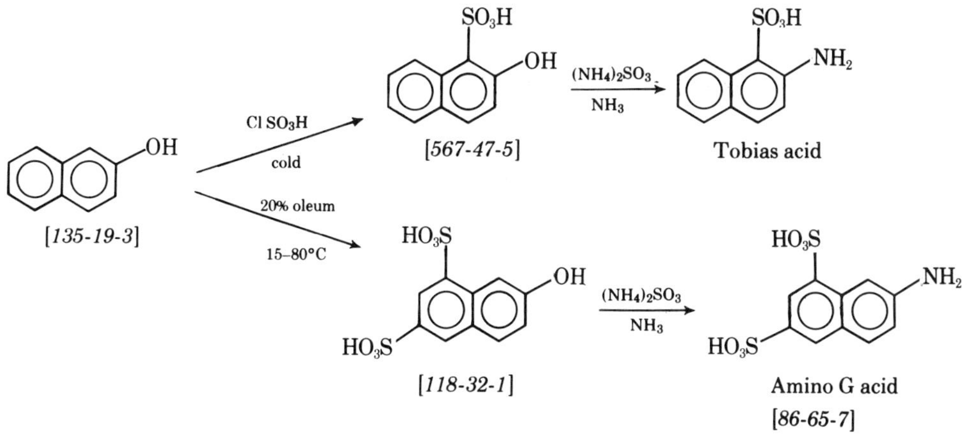

[Scheme: 2-naphthol [135-19-3] → (ClSO₃H, cold) → 1-sulfo-2-naphthol [567-47-5] → (NH₄)₂SO₃/NH₃ → Tobias acid (1-amino-2-naphthalenesulfonic acid); 2-naphthol → (20% oleum, 15–80°C) → 6-hydroxy-1,3-naphthalenedisulfonic acid [118-32-1] → (NH₄)₂SO₃/NH₃ → Amino G acid [86-65-7]]

Tobias acid is used for conversion to J acid via sulfonation and alkali fusion. It has also replaced the extremely carcinogenic β-naphthylamine in azo coupling reactions which in this case proceed displacement of the sulfo group.

The reaction sequence for the manufacture of 1-amino-2-naphthol-4-sulfonic acid bears a similarity to the Bucherer reaction but here the sulfo group remains in the molecule and a nitroso group is reduced by the excess of sulfite.

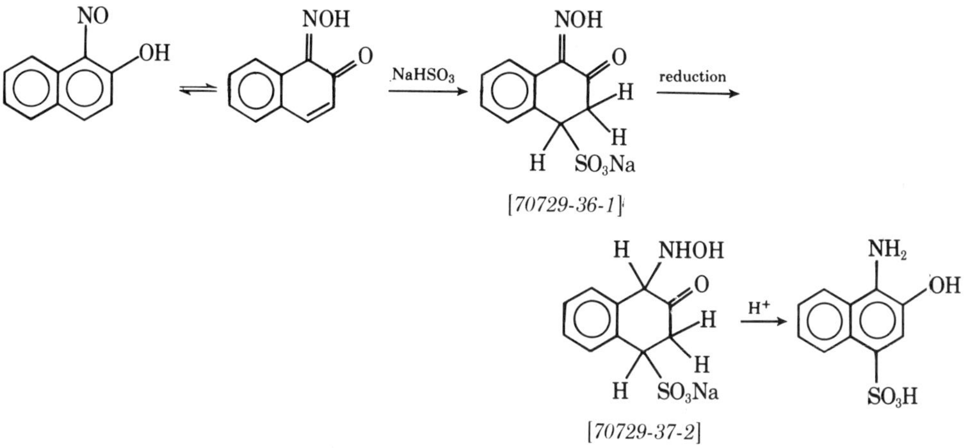

[Scheme: 1-nitroso-2-naphthol ⇌ oxime tautomer → (NaHSO₃) → sulfonated intermediate [70729-36-1] → reduction → NHOH intermediate [70729-37-2] → (H⁺) → 1-amino-2-naphthol-4-sulfonic acid]

Diazotization of the 1,2,4-acid yields a stable diazoxide called diazo acid (naphth[1,2-d] 1,2,3 oxadiazole-5-sulfonic acid), which is used as such or after nitration for the manufacture of important black chrome dyes.

Kolbe-Schmitt Reaction. o-Hydroxycarboxylic acids are obtained by heating sodium phenolate or β-naphtholate with CO₂ under pressure.

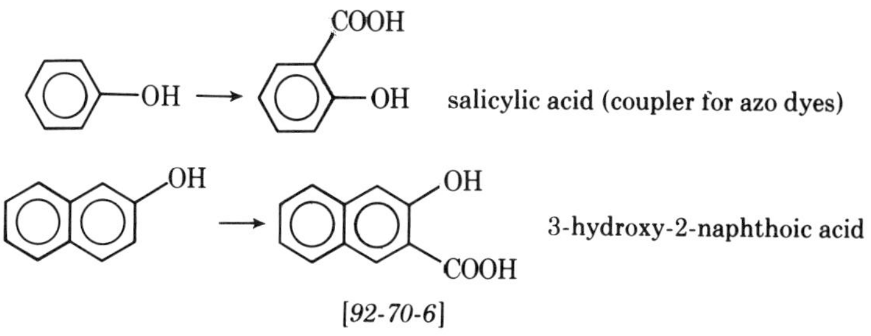

[Scheme: phenol → salicylic acid (coupler for azo dyes); 2-naphthol → 3-hydroxy-2-naphthoic acid [92-70-6]]

192   DYES AND DYE INTERMEDIATES

Arylamides of 3-hydroxy-2-naphthoic acid are used as Naphthol AS couplers in azoic dyes and azo pigments.

Condensation, Dimerization, and Ring Closure. Connections are made between two or more aromatic molecules through bonds or linking groups by various reactions to form larger systems. The products can be intermediates or dyes.

Activated para positions react with formaldehyde or acetone under acid conditions to afford methylene-bis and 2,2-propylene-bis compounds, respectively.

$$2\ R_2N\text{-Ar} \xrightarrow{\text{HCHO, HCl}} R_2N\text{-Ar-}CH_2\text{-Ar-}NR_2 \quad R = CH_3\ [28213\text{-}81\text{-}2],\ R = H\ [26446\text{-}71\text{-}9]$$

$$2\ HO\text{-Ar} \xrightarrow{\text{CH}_3\text{COCH}_3, \text{HCl}} HO\text{-Ar-}C(CH_3)_2\text{-Ar-}OH \quad [80\text{-}05\text{-}7]$$

Difunctional compounds, especially diamines, are used as links with aromatic acids or acid chlorides.

$$2\ Cl\text{-}(O_2N)\text{C}_6H_3\text{-}SO_2Cl + H_2N\text{-}A\text{-}NH_2 \longrightarrow Cl\text{-}(O_2N)\text{C}_6H_3\text{-}SO_2\text{-}NH\text{-}A\text{-}NH\text{-}SO_2\text{-}\text{C}_6H_3(NO_2)\text{-}Cl$$

(A = alkylene or arylene, eg, —CH$_2$—CH$_2$— or Ar—B—Ar where B = CH$_2$, CO, CH=CH, O, SO$_2$, NH, etc)

$$2\ \text{(3-hydroxy-2-naphthoic acid)} + H_2N\text{-Ar(OCH}_3\text{)-Ar(OCH}_3\text{)-}NH_2 \longrightarrow$$

naphthol-CONH-Ar(OCH$_3$)-Ar(OCH$_3$)-NHCO-naphthol

CI Azoic Coupling Component 3
[91-92-9]

Aromatic amines are joined together with reactive dichloro compounds. The most important linking agents are phosgene (COCl$_2$) to make direct dyes, and cyanuric chloride (C$_3$N$_3$Cl$_3$, 2,4,6-trichloro-*s*-triazine) or similar halogenated heterocycles to make reactive dyes (qv) and other products (see Cyanuric acids).

The formation of biaryl compounds is accomplished by heating aryl halides with copper powder (Ullmann reaction), or by the decomposition of diazonium compounds in the presence of copper powder or cuprous salts.

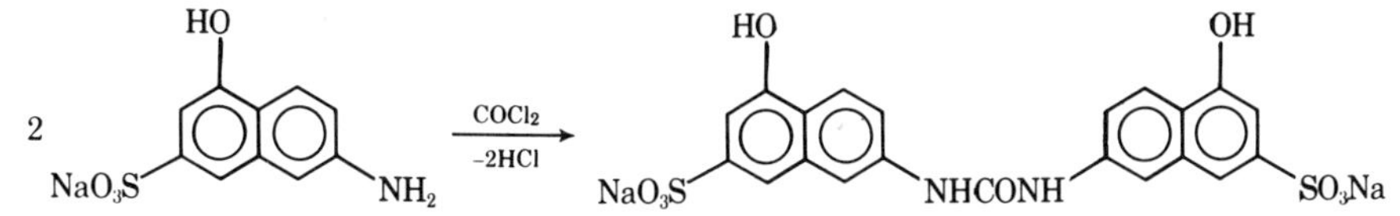

[20074-69-5]

J Acid urea
[134-47-4]

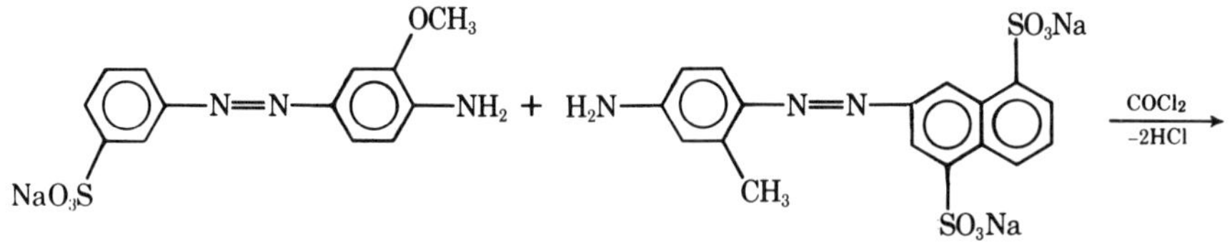

[138-28-3]

[6629-26-1]

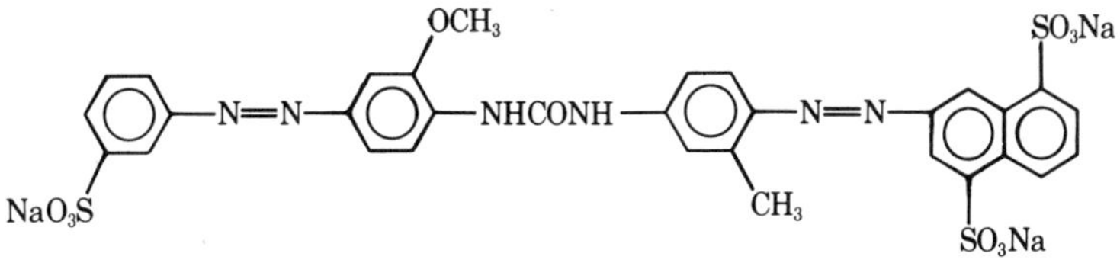

CI Direct Yellow 84
[12222-65-0]

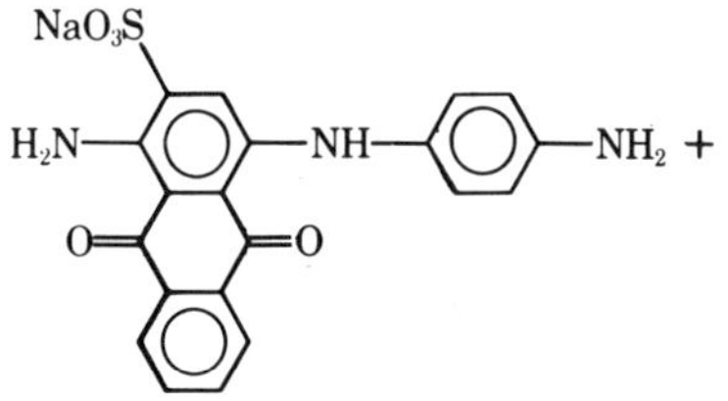

blue
[40951-76-6]

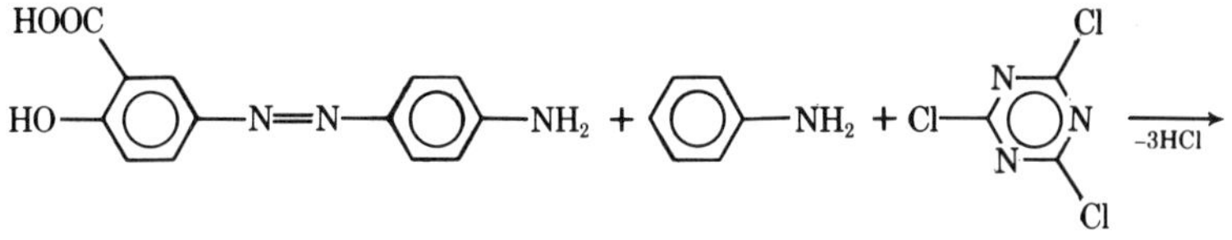

yellow
[101-51-9]

[108-77-0]

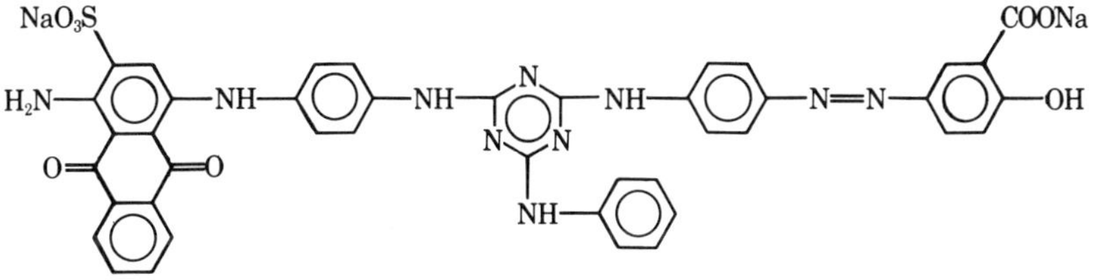

CI Direct Green 28
[6471-09-6]

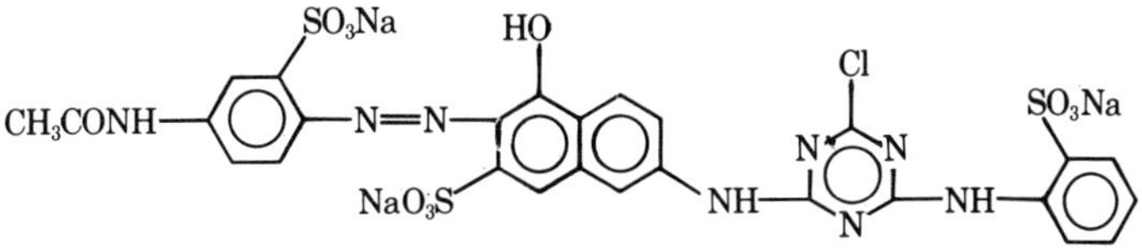

CI Reactive Red 9
[14826-60-9]

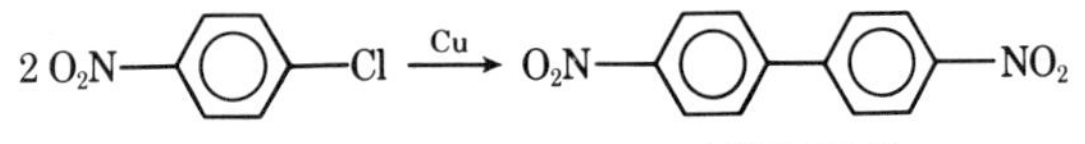

[1528-74-1]

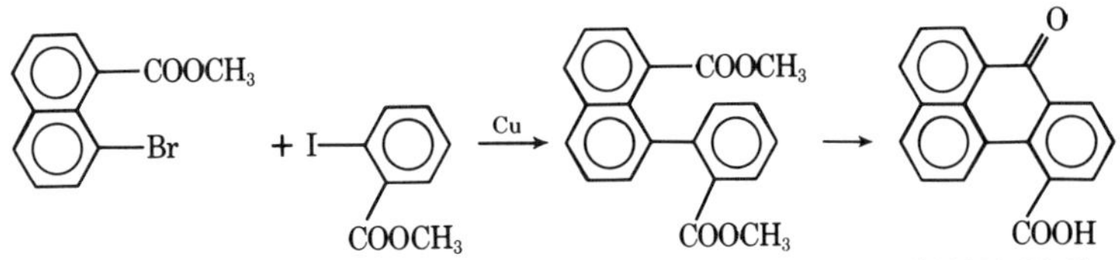

[38058-95-6] [610-97-9] [10383-78-5] [10360-92-6]

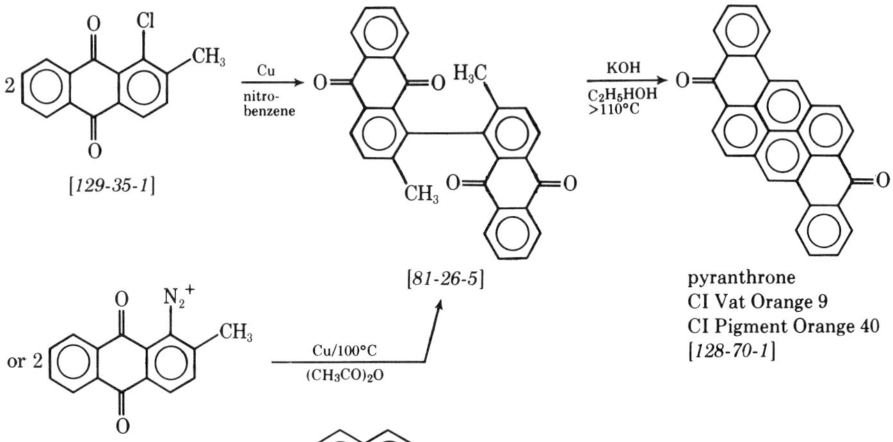

[129-35-1]

[81-26-5]

pyranthrone
CI Vat Orange 9
CI Pigment Orange 40
[128-70-1]

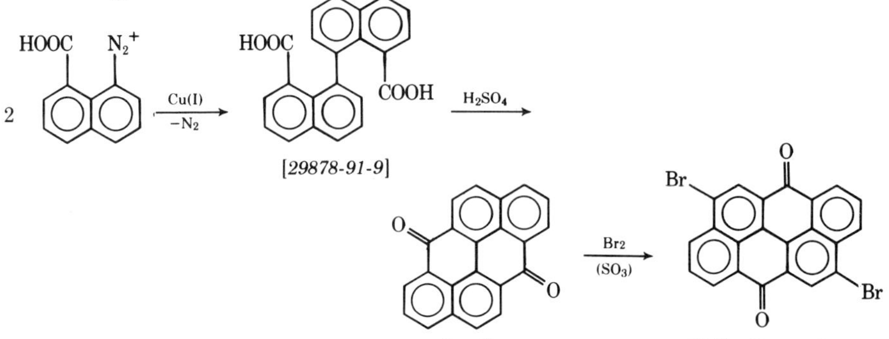

[29878-91-9]

anthanthrone
[641-13-4]

CI Vat Orange 3
CI Pigment Red 168
[4378-61-4]

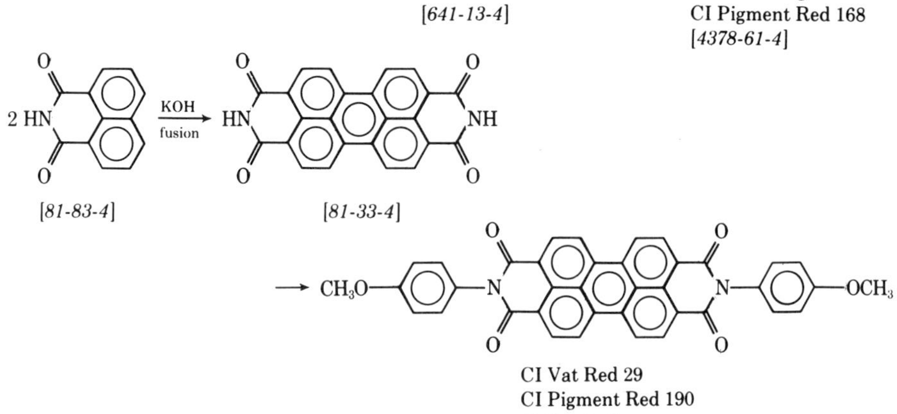

[81-83-4] [81-33-4]

CI Vat Red 29
CI Pigment Red 190
[6424-77-7]

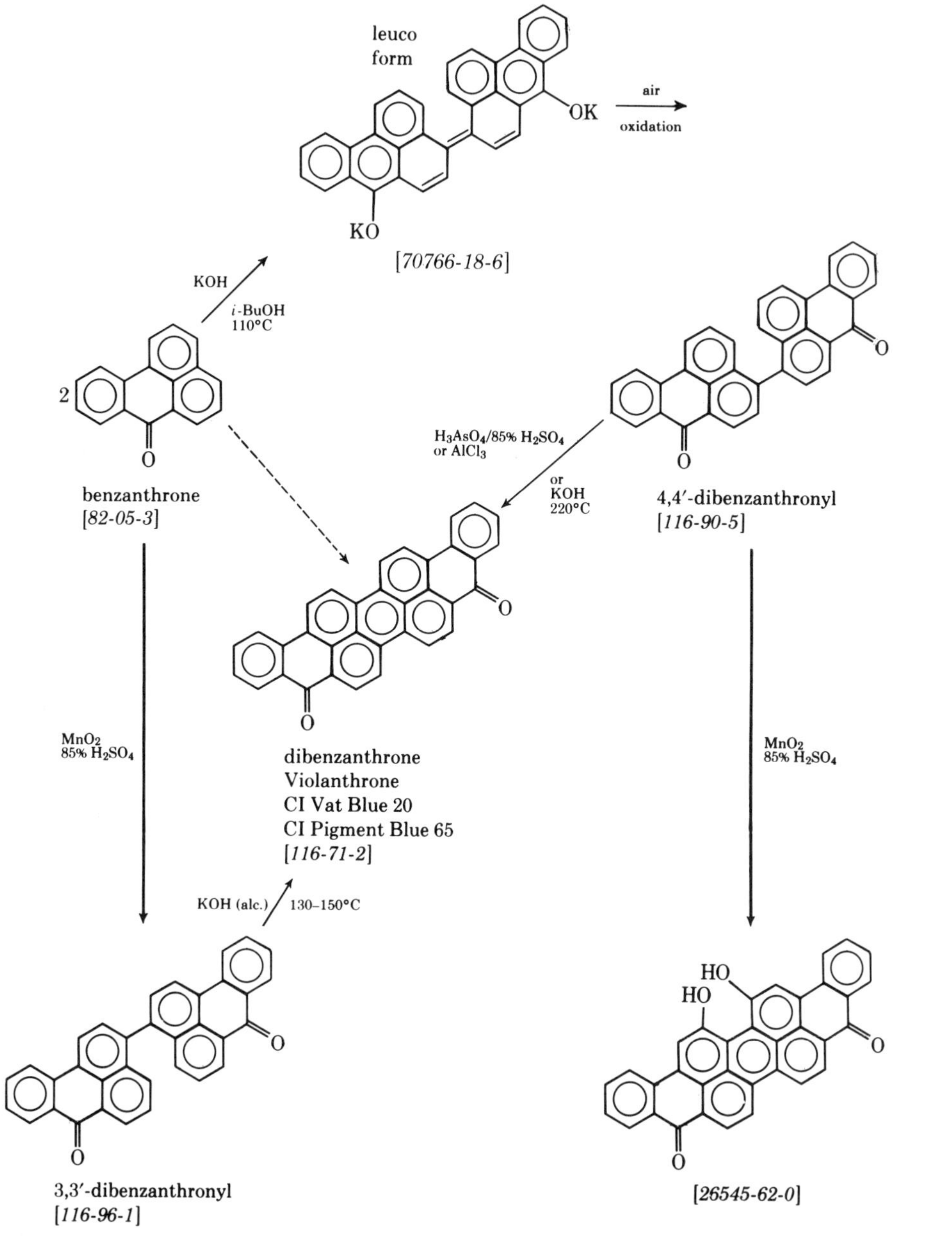
leuco
form
OK
air
oxidation
KO
[70766-18-6]
KOH
i-BuOH
110°C
2
O
benzanthrone
[82-05-3]
H3AsO4/85% H2SO4
or AlCl3
or
KOH
220°C
O
O
4,4′-dibenzanthronyl
[116-90-5]
O
O
dibenzanthrone
Violanthrone
CI Vat Blue 20
CI Pigment Blue 65
[116-71-2]
MnO2
85% H2SO4
MnO2
85% H2SO4
KOH (alc.)
130–150°C
O
O
3,3′-dibenzanthronyl
[116-96-1]
HO
HO
O
O
[26545-62-0]

Some ring systems are heated with oxidizing agents to achieve dimerization and ring closure (oxidative coupling). The same is often accomplished by alkaline fusion. Ring closures are also effected by heating of aromatic biaryl or aromatic benzoyl compounds in the presence of $AlCl_3$ (see Friedel-Crafts reactions). The benzoyl groups furnish the carbonyl functions for the condensed ring system. Properly located carboxylic acid groups do the same on dehydration with concentrated H_2SO_4.

Specialized methods are often required for the synthesis of the different het-

196   DYES AND DYE INTERMEDIATES

erocyclic intermediates, used by the dye industry in large amounts. Only a few are mentioned here. Nitrogen is the most common heteroatom.

Pyrazolones (2-pyrazolin-5-ones, usually 1-phenyl and 3-methyl or 3-carboxy substituted) are coupling compounds for yellow and red azo dyes. They are made from hydrazones of β-ketoesters.

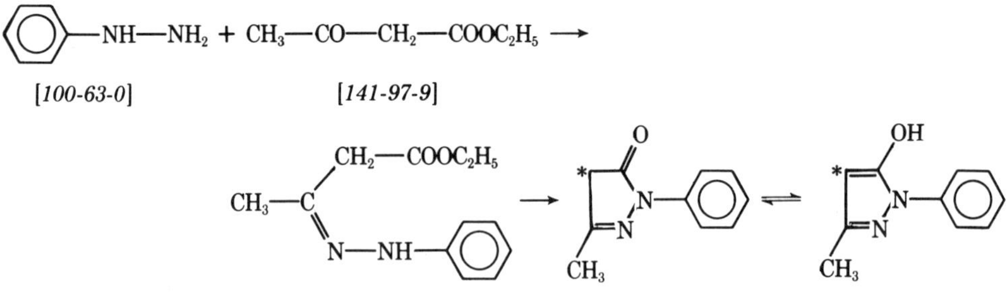

$$C_6H_5-NH-NH_2 + CH_3-CO-CH_2-COOC_2H_5 \rightarrow$$

[100-63-0]   [141-97-9]

$$CH_3-C(CH_2-COOC_2H_5)=N-NH-C_6H_5$$

[6078-46-2]   [89-25-8]

(The asterisk marks the point of attack of the diazonium cation in an azo coupling reaction.)

Phenylhydrazones of aldehydes and ketones lose one nitrogen on treatment with $ZnCl_2$ and form indoles.

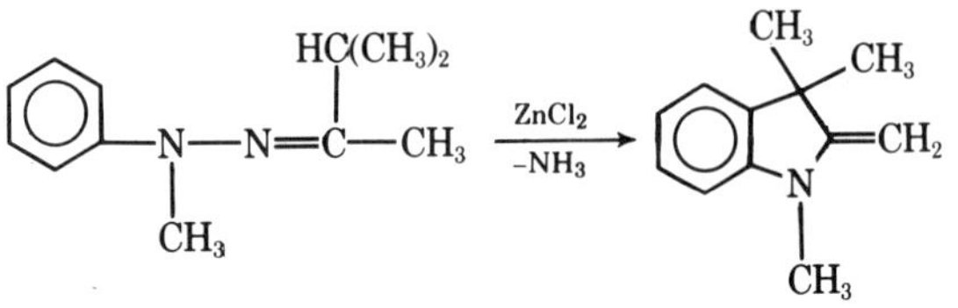

[15754-48-0]
(N-methylation can follow the ring closure)

Fischer base
[118-12-7]

Fischer base and Fischer aldehyde are two of the more important heterocyclic intermediates used to make basic dyes.

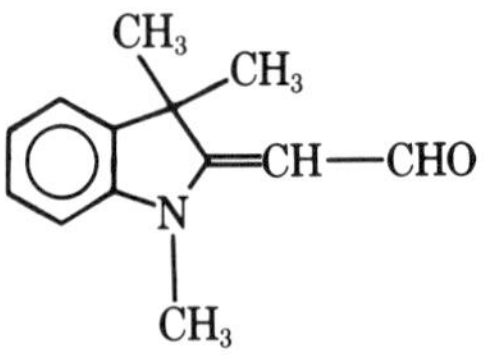

Fischer aldehyde
[84-83-3]

Isatin, another derivative of indole, is obtained in high yield by the oxidation of indigo or synthetized from aniline via isonitroso-acetanilide. It is transformed into a derivative of quinaldine (2-methylquinoline) for use in quinophthalones, which are a small group of valuable yellow dyes, eg, CI Disperse Yellow 54.

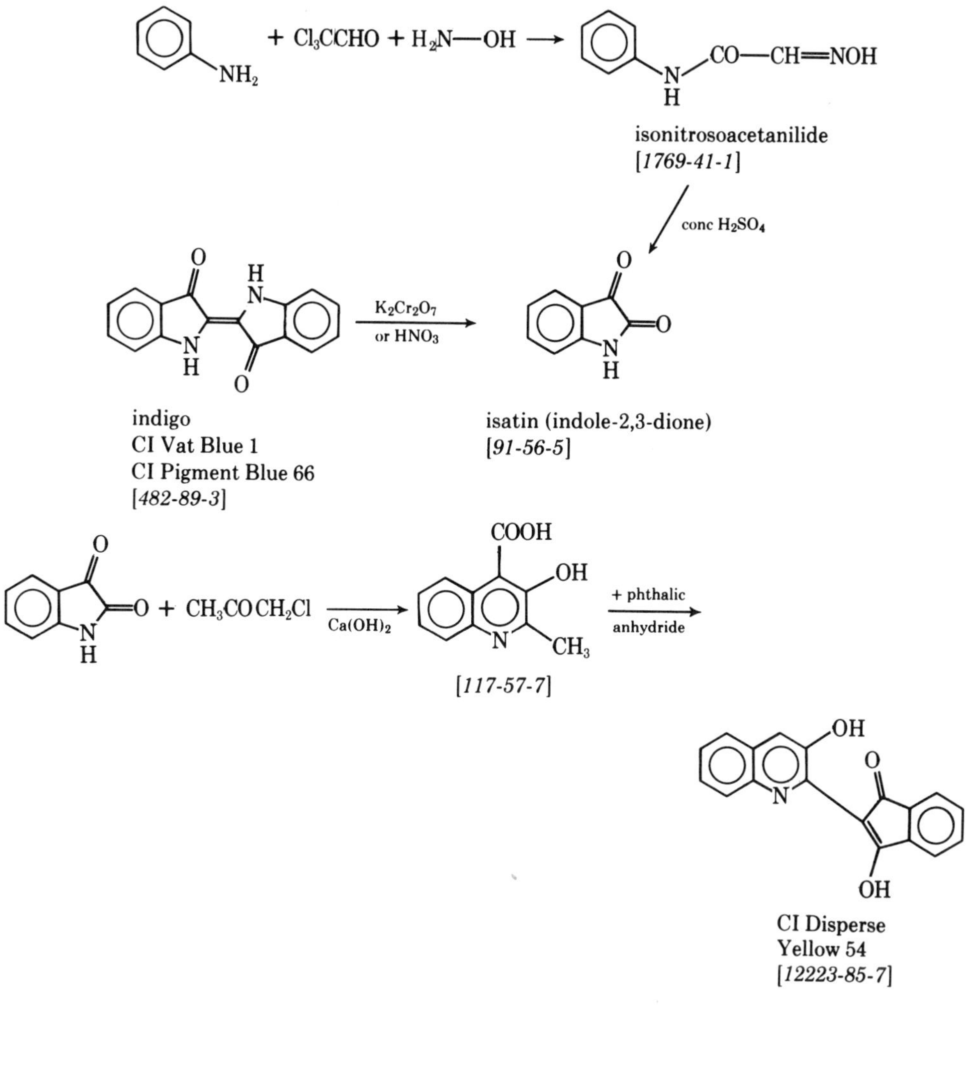

$$\text{C}_6\text{H}_5\text{NH}_2 + \text{Cl}_3\text{CCHO} + \text{H}_2\text{N—OH} \rightarrow \text{C}_6\text{H}_5\text{N(H)—CO—CH=NOH}$$

isonitrosoacetanilide [1769-41-1]

↓ conc H$_2$SO$_4$

indigo
CI Vat Blue 1
CI Pigment Blue 66
[482-89-3]

$\xrightarrow{\text{K}_2\text{Cr}_2\text{O}_7 \text{ or HNO}_3}$

isatin (indole-2,3-dione) [91-56-5]

isatin + CH$_3$COCH$_2$Cl $\xrightarrow{\text{Ca(OH)}_2}$ [117-57-7] $\xrightarrow{\text{+ phthalic anhydride}}$

CI Disperse Yellow 54 [12223-85-7]

The incorporation of sulfur into the ring of an aromatic structure is shown with the synthesis of three aminobenzothiazoles, which find use in azo dyes.

198   DYES AND DYE INTERMEDIATES

[Reaction scheme: 4-ethoxyaniline sulfate [55993-36-7] + NH$_4$SCN → ethoxyphenylthiourea [880-79-5] → (S$_2$Cl$_2$) → 6-ethoxy-2-aminobenzothiazole [94-45-1] → red disperse azo dyes]

[Reaction scheme: p-toluidine [106-49-0] + S(melt) (+Na$_2$CO$_3$) → dehydrothio-p-toluidine [92-36-4] → (65% oleum, 30°C) → thiotol acid [130-17-6] → CI Acid Yellow 186 / CI Direct Yellow 8 [10130-29-7]]

Some other heterocycles are formed by oxidation.

Oxidation. Oxidation is used in intermediate and dye synthesis at many stages and few generalizations are appropriate. A wide range of oxidizing agents are employed but air and a catalyst (V$_2$O$_5$, Al$_2$O$_3$–ZnO, various metal salts) have become the mainstay of the large-scale continuous processes that supply starting materials to the dye and other industries. The oxidations of naphthalene and anthracene are catalytic air oxidations, and so are the side-chain oxidations of o-xylene, p-xylene, and toluene, which yield phthalic anhydride, terephthalic acid, and benzoic acid, respectively. The oxidation of methyl groups with manganese dioxide (MnO$_2$), potassium dichromate (K$_2$Cr$_2$O$_7$), chromic acid (CrO$_3$), selenium dioxide (SeO$_2$), potassium permanganate (KMnO$_4$), and others also lead to carboxylic acids.

Large amounts of phenol and acetone are manufactured from cumene (qv) via air oxidation to a peroxide.

[Reaction scheme: benzene + CH$_3$CH=CH$_2$ / H$_3$PO$_4$ → cumene → O$_2$ → cumene hydroperoxide [80-43-5] → phenol + CH$_3$COCH$_3$]

Aniline is commercially oxidized with MnO$_2$ in acid medium to give 1,4-benzoquinone [106-51-4],

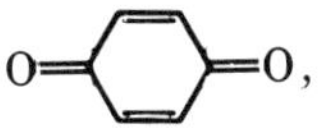

$$O=\!\!\!\bigcirc\!\!\!=\!O,$$

but oxidizing agents such as sodium dichromate (Na$_2$Cr$_2$O$_7$), ferric chloride (FeCl$_3$), lead dioxide (PbO$_2$) or sodium chlorate (NaClO$_3$) also work.

The incomplete oxidation of aniline or *p*-phenylenediamines with chromic acid, chlorates, etc, produces indamine structures

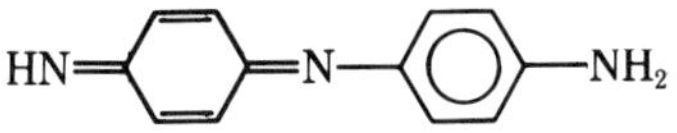

$$HN=\!\!\!\bigcirc\!\!\!=\!N\!-\!\!\!\bigcirc\!\!-NH_2$$
[101-78-0]

as intermediate stages of dyes called Aniline Blacks, which are derivatives of phenazine [92-82-0],

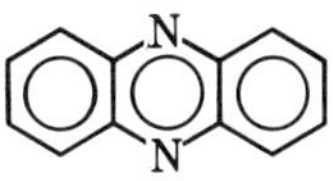

The very first synthetic dye, Perkin's Mauve, was also a phenazine derivative. It was obtained by the oxidation of crude aniline (containing toluidine) with potassium dichromate.

Indophenols

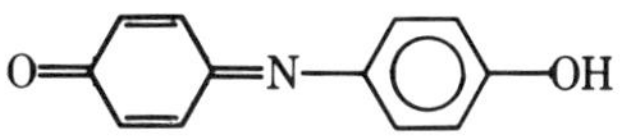

$$O=\!\!\!\bigcirc\!\!\!=\!N\!-\!\!\!\bigcirc\!\!-OH$$
[500-85-6]

and

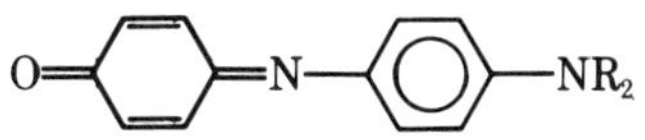

$$O=\!\!\!\bigcirc\!\!\!=\!N\!-\!\!\!\bigcirc\!\!-NR_2$$
R = H [6245-87-0]

are formed by oxidizing mixtures of *p*-aminophenol and a phenol or an aromatic amine with bichromate or hypochlorite. They are used to manufacture sulfur dyes (qv). These generally need an after-oxidation, usually an aeration, before they are isolated. Such sulfur dyes are derivatives of thiazone (3*H*-isophenothiazin-3-one [581-30-6]).

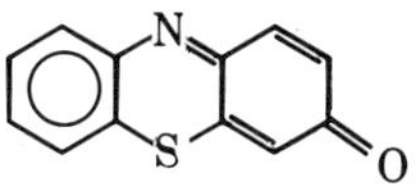

200 DYES AND DYE INTERMEDIATES

The formation of triarylmethane dyes (see Triphenylmethane dyes) also requires an oxidation step and reagents like $Na_2Cr_2O_7$, PbO_2, MnO_2, H_2O_2 (hydrogen peroxide) or $Na_2S_2O_8$ (sodium persulfate) are used for it.

Sodium hypochlorite (NaOCl) is rather inexpensive in the form of an aqueous alkaline solution (bleach) and is therefore often used for oxidations. For example, it promotes the formation of dinitrostilbenedisulfonic acid from 4-nitrotoluene-2-sulfonic acid.

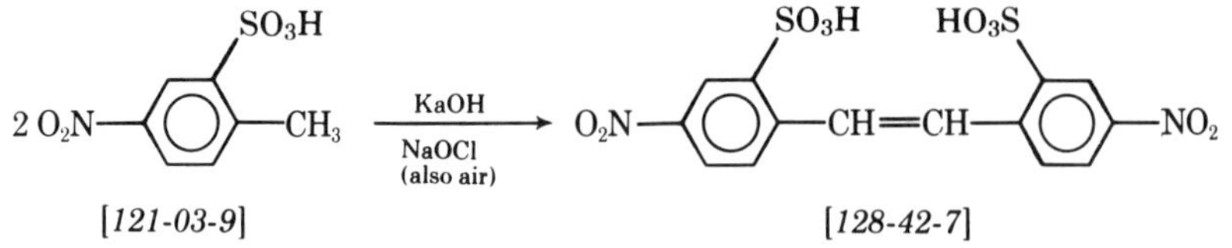

$$2\ O_2N\text{-}C_6H_3(SO_3H)\text{-}CH_3 \xrightarrow[\text{(also air)}]{KaOH,\ NaOCl} O_2N\text{-}C_6H_3(SO_3H)\text{-}CH=CH\text{-}C_6H_3(SO_3H)\text{-}NO_2$$

[121-03-9]   [128-42-7]

(Subsequent treatment of the product with glucose or other mild reducing agents produces yellow azo–azoxy dyes for cotton.)

Another example of hypochlorite oxidation is the manufacture of Chloramine Yellow from dehydrothio-*p*-toluidinesulfonic acid. Many vat dyes receive a hypo-treatment to free them from oxidizable impurities.

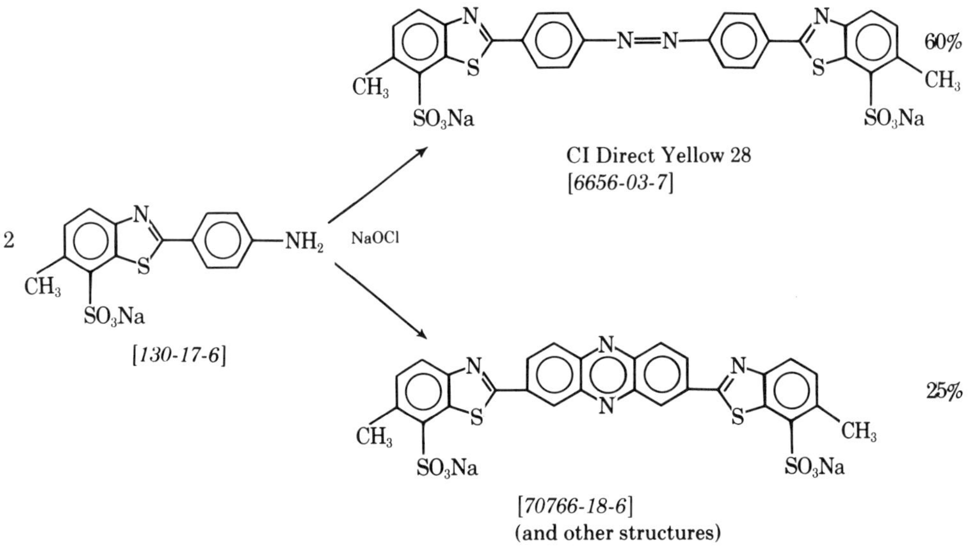

CI Direct Yellow 28 [6656-03-7] — 60%

[130-17-6]

[70766-18-6] (and other structures) — 25%

Bleach will transform *o*-aminoazo compounds into triazoles, which find use as brighteners or in certain direct azo dyes. This reaction is also effected with Cu(II) salts, or by copper-catalyzed air oxidation.

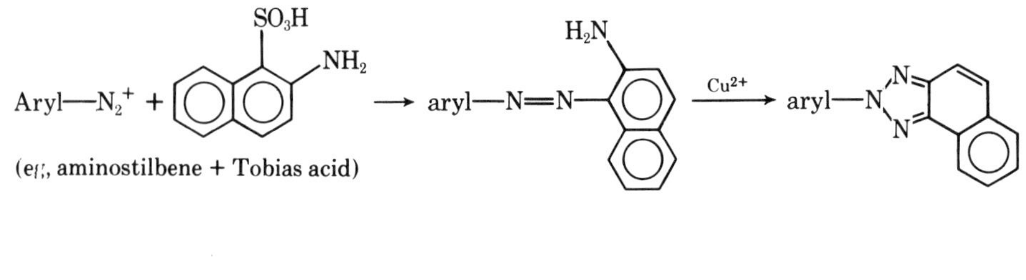

Aryl—N_2^+ + (naphthalene-SO$_3$H, NH$_2$) → aryl—N=N—(naphthalene with H$_2$N, NH$_2$) $\xrightarrow{Cu^{2+}}$ aryl—N (triazole-naphthalene)

(e/;, aminostilbene + Tobias acid)

Sometimes direct hydroxylation of the aromatic skeleton is possible with oxidizing agents such as MnO_2, $KClO_3$, oleum, $NaNO_3$, etc (see Dyes, anthraquinone).

[84-48-0] →(NaOH–NaNO₃, 200°C)→ Alizarin (See CI Pigment Red 83) [72-48-0]

→(MnO₂, H₂SO₄, 20°C)→ Purpurine [81-54-9]

→(80% oleum, 28°C)→ Quinalizarin (see CI Mordant Violet 26) [81-61-8]

[82-05-3] →(KOH, H₂O, KClO₃, 225°C)→ [33705-86-1]

Diazotization and Azo Coupling. A principal use of primary aromatic amines is their conversion to diazonium compounds (diazotization) and the subsequent coupling with reactive aromatic ring systems (couplers) to form aromatic azo compounds. This reaction sequence is the chemical foundation of the azo dye industry.

In the dye industry, diazotizations and couplings are carried out in aqueous medium whenever possible. Sodium nitrite, $NaNO_2$, and a mineral acid, such as muriatic acid (aq HCl) or dilute sulfuric acid, are used for the diazotization in a majority of cases. Nitrosylsulfuric acid, HSO_4NO, prepared by dissolving $NaNO_2$ in concentrated or 100% H_2SO_4, is employed for amines of low basicity, eg, those with more than one electron-withdrawing group, which will not diazotize in water. The temperature is usually kept at 0–5°C to avoid decomposition (loss of N_2). Some diazos can tolerate temperatures up to 20° or even 30°C, and a few must be held below 0°C. Since the diazotization reaction is strongly exothermic, cooling is provided by addition of ice or externally with brine. Continuous diazotization (and coupling) schemes with short contact times at higher temperatures have been reported (13). Stoichiometrically, a diazotization is expressed by the following equation:

$$\text{Ar—NH}_2 + 2\,\text{H}^+ + \text{NaNO}_2 \rightarrow \text{Ar—N}_2^+ + \text{Na}^+ + 2\,\text{H}_2\text{O}$$

It is a quantitative reaction and frequently used for the analytical determination of aromatic amines.

The diazotization mechanism is complex. Several active species may act as nitrosating agents, eg, ON^+, $ON—\overset{+}{O}H_2$, $ON—Cl$, $ON—NO_2$ (N_2O_3), and $ON—OH$ (HNO_2, nitrous acid). They attack the free amine, whose concentration decreases through salt formation at lower pH values where the more active nitrosating species (ON^+, $ON—\overset{+}{O}H_2$) prevail. The following two equations give an accepted representation of diazotization (14).

$$ON—NO_2 + Ar—NH_2 \rightarrow Ar—\overset{+}{N}H_2—NO + NO_2^-$$

$$Ar—\overset{+}{N}H_2—NO \xrightarrow[-H_2O]{\text{fast}} Ar—\overset{+}{N}\equiv N\ (Ar—N_2^+)$$

Equilibria for the active nitrosating species may be formulated as follows:

$$NaHO_2 \xrightarrow{+H^+} HONO \xrightarrow{+H^+} H_2O—NO \longrightarrow H_2O + ON^+;$$

$$2\,HONO \xrightarrow[-H_2O]{} N_2O_3,\ (ON—NO_2 \rightarrow ON^+ + NO_2^-)$$

Most diazonium salts are quite sensitive to heat, light, and alkali. They are unstable when dry and usually earmarked for azo coupling (or reduction) without isolation soon after manufacture. There are, however, certain stable diazonium salts and stabilized diazo preparations on the market for the azoic dye or ice-color trade.

Diazonium ions are electrophilic but only moderately so. They require relatively reactive partners with positions of high electron density for a successful coupling reaction.

The commercially most important groups of couplers are the acetoacetanilides, the phenylmethylpyrazolones and their sulfonic acids, the 6-hydroxy-2-pyridone derivatives, the phenols, the naphthols and their sulfonic acids, the 3-hydroxy-2-naphthoic arylamides, the anilines, the N-alkylated anilines, the naphthylaminesulfonic acids, and the aminonaphtholsulfonic acids. Phenols and enols couple rapidly in a pH range of 7–11 as phenolates and enolates. The rate of coupling increases with rising alkalinity up to about pH 10 where the concentration of diazonium ions decreases sharply due to an equilibrium with diazo hydroxides and diazotate ions, which do not couple. Aromatic amines are coupled at pH 3–7 where the concentration of the free amine is high relative to the inactive amine salt. However, the more reactive N,N-dialkylated anilines and their derivatives with electron donating substituents couple fast enough at lower pH values, especially with the more reactive diazonium compounds which carry nitro and other electron-withdrawing groups. Azo compounds with a primary amino group can be rediazotized and coupled again, giving rise to dis-, tris-, and polyazo compounds. [Although all the aromatic azo compounds manufactured are strongly colored substances which absorb light waves selectively in the range of ca 400–750 nm (visible light) and would color certain materials, only those that are sold for application to textiles and other substrates are called dyestuffs, whereas the others are still considered intermediates. The same is true for many anthraquinone derivatives.]

The stoichiometric equation of a typical coupling reaction is:

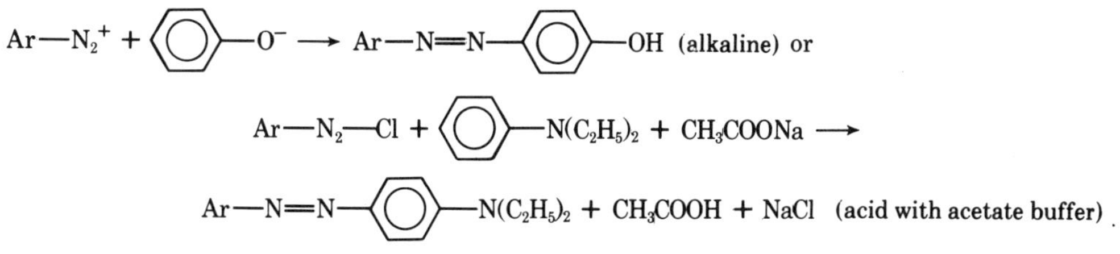

$$Ar-N_2^+ + \langle\bigcirc\rangle-O^- \rightarrow Ar-N=N-\langle\bigcirc\rangle-OH \quad \text{(alkaline)} \quad \text{or}$$

$$Ar-N_2-Cl + \langle\bigcirc\rangle-N(C_2H_5)_2 + CH_3COONa \rightarrow$$

$$Ar-N=N-\langle\bigcirc\rangle-N(C_2H_5)_2 + CH_3COOH + NaCl \quad \text{(acid with acetate buffer)}$$

The coupling mechanism is that of an electrophilic substitution (E = electron donating group; Ar = aryl).

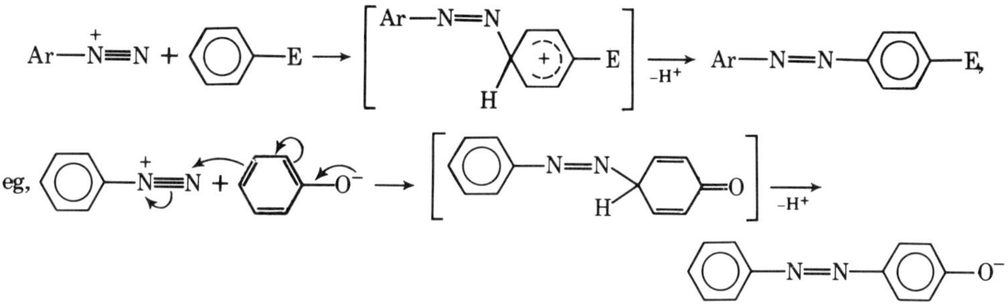

[mechanism schemes]

[18938-15-3]

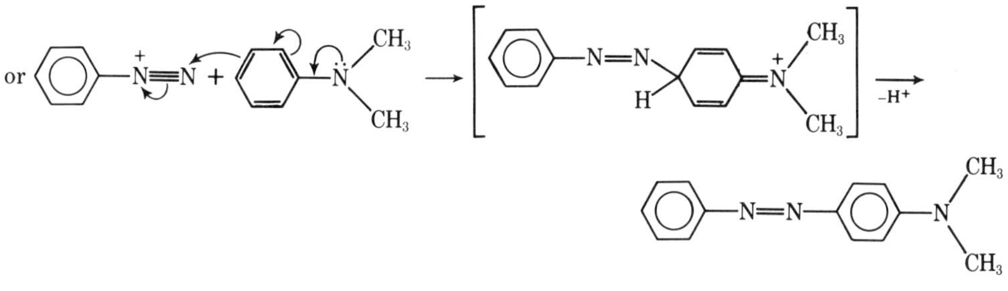

[60-11-7]

Economic Aspects

Table 5 gives statistics for the production of dyes, in the years 1960, 1974, and 1975, for countries in the Organization for Economic Cooperation and Development (OECD). Annual production figures for the United States are shown in Table 6. In both cases the increase in the number of metric tons of dyes produced from 1960 to 1974 represents an average annual rate of growth for the industry of 3–5%. The sharp drop in production in 1975 reflects the general worldwide recession of that year. During the last fifteen years the Federal Republic of Germany has again emerged as the major producer of dyes, a position it has not occupied since World War I.

The increasing importance of dyes for synthetic fibers is shown by a comparison of the statistics for the production and sales of dyes in the United States by class of application. Table 7 gives the figures for 1961 and 1976. From 1961 to 1976 the pro-

204 DYES AND DYE INTERMEDIATES

Table 5. Production of Dyes in OECD Countries[a,b]

Country	1960 Thousands of metric tons	1960 Millions of U.S. $	1974 Thousands of metric tons	1974 Millions of U.S. $	1975 Thousands of metric tons	1975 Millions of U.S. $
Belgium	1.19	2.3	5.64	23.2	1.68	10.2
France	14.1	39.8	31.5	147.5	19.9	122.5
FRG	46.1	172.0	147.9	976.1	86.8	693.8
Italy	17.7	25.5	14.5	61.5	14.0	67.4
Japan	na[c]	na[c]	68.5	282.2	56.9	277.1
Netherlands	4.88	6.5	2.20	8.1	2.20	8.1
Spain	3.87	5.4	15.4	22.2	11.0	39.9
Switzerland	19.9	92.6	30.3	370.2	20.1	296.5
United Kingdom	37.8	103.6	53.6	267.7	43.2	243.9
United States	70.7	192.1	125	556.2	93.5	475.6
Total	216	639.8	495	2,714.9	349	2,235.0

[a] OECD = Organization for Economic Cooperation and Development.
[b] Ref. 15.
[c] na = not available.

Table 6. Annual Production of Dyes in the United States[a]

Year	Production, 10^3 t	Year	Production, 10^3 t
1955	76.2	1972	119
1960	70.7	1973	129
1965	94.0	1974	125
1970	106	1975	93.5
1971	111	1976	116

[a] Ref. 6.

duction of acid dyes, used mainly for dyeing nylon, has increased by 87%. Basic dyes used for acrylic fibers, have increased by 97%, and disperse dyes for polyester and nylon show an increase of 490%. Vat dyes, on the other hand, where the use is confined to dyeing cotton, show a production increase of only 10%. The sales of vat dyes, however, show a 29% increase over the same period. The difference between the production and sales figures for this class of dyes is due to large imports of indigo required to meet the demand of the fashion trend for denim fabrics. This demand reached its peak in 1976. It is unusual for a fashion trend to be so clearly discernible in dye production statistics. As a rule the large number of classes of dyes and shades in each class will almost completely mask such effects.

A summary of trade between the OECD countries is given in Tables 8 and 9. The principal exporting countries are the Federal Republic of Germany, Switzerland, and the United Kingdom. The United States plays a relatively minor role in the export trade but represents a substantial market for foreign supplies. Dyes imported into the United States are classified into two groups by the U.S. International Trade Commission. Competitive dyes that are equivalent to a product manufactured in the United States carry a tariff based on the American selling price. Noncompetitive dyes carry a duty assessed on the United States invoice value. Table 10 gives the United States imports of dyes for 1976 by both class of application and by competitive status.

Table 7. 1961 and 1976 United States Production and Sales of Dyes by Application Class[a]

Class of application	Production, metric tons 1961	Production, metric tons 1976	Sales (1961) Quantity, t	Sales (1961) % of total	Sales (1961) Value, $1000	Sales (1961) Unit value, $/kg[b]	Sales (1976) Quantity, t	Sales (1976) % of total	Sales (1976) Value, $1000	Sales (1976) Unit value, $/kg[b]
acid	6,870	12,810	6,560	9.1	27,268	4.17	12,250	11	87,108	7.12
azoic dyes and components	3,190	863	3,020	4.7	11,038	3.57	824	0.7	2,561	3.62
basic	3,350	6,620	2,890	4.4	14,824	5.13	6,750	5.7	49,770	7.36
direct	10,350	15,210	10,620	13.7	35,144	3.30	14,340	13.0	78,772	5.49
disperse	3,620	17,740	3,260	4.8	17,354	5.34	16,460	15.3	138,019	8.38
fiber-reactive	550	1,590	450	0.7	4,172	9.26	1,810	1.4	21,876	12.10
fluorescent brightening agents	4,400	19,700	4,130	5.8	20,772	5.03	17,210	16.9	55,464	3.22
food, drug, and cosmetic colorants	1,150	2,610	1,030	1.5	9,514	9.24	2,320	2.2	31,754	13.69
mordant	1,750	300	1,860	2.3	5,291	2.84	300	0.3	2,149	7.23
solvent	2,950	5,420	2,520	3.9	9,210	3.66	5,220	4.6	35,341	6.77
sulfur	15,450	c	14,680	20.4	8,673	0.60	c	c	c	c
vat	21,860	24,150	20,780	28.9	49,386	2.38	26,800	20.7	86,876	3.24
all other[d]	60	9,230	37	0	432	11.75	9,070	7.9	30,604	3.37
Total	75,550	116,243	71,830	100	213,078	2.98 (av)	113,300	100	620,294	5.47 (av)

[a] Refs. 6 and 16.
[b] Calculated from rounded figures.
[c] Statistics for sulfur dyes may not be published separately because publication would disclose information received in confidence.
[d] The data include oxidation base, ingrain dyes, sulfur dyes (for 1976), and miscellaneous dyes.

Table 8. 1975 Imports of Dyes[a]

Country	Western Europe Metric tons	Western Europe Millions of U.S. $	Canada and U.S. Metric tons	Canada and U.S. Millions of U.S. $	Japan Metric tons	Japan Millions of U.S. $	From other countries Metric tons	From other countries Millions of U.S. $	Totals Metric tons	Totals Millions of U.S. $
Belgium	5,494	37.7	881	4.0	48	0.3	241	1.1	6,664	43.0
France	10,578	88.9	1,435	3.4	16	0.1	150	0.7	12,179	93.0
FRG	12,433	77.7	548	2.8	826	2.7	441	2.2	14,248	85.4
Italy	14,355	85.9	1,061	4.8	279	1.4	821	3.4	16,516	95.5
Japan	6,464	54.0	494	3.0			226	0.9	7,184	57.9
Netherlands	6,566	37.7	451	1.2	74	0.4	456	2.2	7,547	41.8
Spain	2,922	35.8	148	1.5	90	0.8	49	0.4	3,209	38.5
Switzerland	8,561	68.9	360	2.6	348	2.9	441	1.8	9,298	76.2
United Kingdom	7,538	102.1	1,473	7.1	189	1.1	455	1.9	9,655	112.2
United States	8,335	71.6	411	1.5	651	5.3	246	1.5	9,643	79.9

[a] Ref. 15.

Table 9. 1975 Exports of Dyes[a]

Country	Western Europe Metric tons	Western Europe Millions of U.S. $	To OECD member countries Canada and U.S. Metric tons	Canada and U.S. Millions of U.S. $	Japan Metric tons	Japan Millions of U.S. $	To other countries Metric tons	To other countries Millions of U.S. $	Totals Metric tons	Totals Millions of U.S. $
Belgium	2,927	16.5	6	na	6	na[b]	480	3.1	3,419	19.6
France	8,705	65.1	380	2.4	391	3.1	3,347	32.4	12,823	103.0
FRG	32,194	246.7	4,810	33.8	4,190	33.2	23,770	248.4	64,964	562.1
Italy	3,351	18.5	419	1.9	120	0.6	2,236	10.0	6,126	31.0
Japan	991	4.6	3.8	1.6			7,536	50.9	8,982	57.3
Netherlands	3,746	14.1	0.1		42	0.3	1,874	7.9	5,673	22.3
Spain	1,424	4.5	17	0.1			273	1.2	1,735	5.8
Switzerland	14,176	172.1	2,713	28.3	2,639	30.0	13,782	191.7	33,310	422.6
United Kingdom	17,557	101.4	2,573	12.1	2,447	15.3	12,286	83.0	34,863	211.8
United States	7,564	29.6	5,942	18.9	1,310	7.4	7,511	30.2	22,328	86.1

[a] Ref. 15.
[b] na = not available.

DYES AND DYE INTERMEDIATES

Table 10. 1976 United States Imports of Dyes by Application Class[a]

Class of application	Total imports, t	Percent of total	Competitive status Competitive, t	Noncompetitive t
acid	1410	13.4	373	1037
azoic components	1430	13.6	1280	150
basic	616	5.9	355	261
direct	559	5.3	135	424
disperse	1690	16.1	183	1507
fiber-reactive	646	6.2	17.1	629
fluorescent brightening agents	681	6.5	39.4	642
food, drug and cosmetic colorants	59.7	0.6	58.9	0.85
ingrain	0.46	0.005		0.46
mordant	83.3	0.8	20.5	62.8
solvent	533	5.1	408	125
vat	2302	21.9	1930	372
miscellaneous	488	4.6	12.9	475
Total	10,500	100.0	4810	5690
total (invoice value), $	76,290,130		21,533,612	54,131,646
average unit value, $/kg	7.27		4.48	9.51

[a] Ref. 2.

Three classes of dyes account for over half the total imports. For acid dyes, 73% of the imports are noncompetitive, and for disperse dyes the corresponding figure is almost 90%. In the case of vat dyes, imports of indigo (CI Vat Blue 1) are 1540 t or 61% of the total. Exclusive of indigo, 50% of the imported vat dyes are noncompetitive.

Health and Safety Aspects

Ecology. Biological treatment, either by the individual factory or in conjunction with a regional or municipal system, is the method most widely used for the treatment of effluent from dye and dye intermediate factories. Treatment of such effluent does, however, present unusual problems. Control of pH, by the addition of lime, is a required pretreatment step. The effluents are usually colored and in most cases the bacteria in the biological treatment are not effective in degrading the dyes present. Although most dyes are relatively nontoxic to fish (17), the discharge of colored wastes is esthetically displeasing. If the receiving water cannot offer a sufficiently high dilution factor, a tertiary treatment with activated carbon or a polymeric material may be used to decrease the color value (18–20). The batch nature of the industry, with a large number of products and, in many cases, only short production runs, makes the bacteria particularly liable to shock. Large equalization basins or holding tanks are used to reduce this effect. The presence of inorganic salts in relatively high concentration may also disrupt the treatment process and cause difficulties for other downstream users of the water supply. Finally, there is the problem of the presence of specific pollutants as defined by the regulatory agencies of the various states and countries of the world. The limitations placed upon the discharge of these materials, which include certain heavy metals as well as specified organic compounds, are being met by changes in processing technology or, in cases where this is not possible, by the development of special treatment methods for individual process streams. An example of the latter is the special treatment methods that have been developed for the removal of mercury from the manufacture of anthraquinone dyes (21–22).

Toxicology. The toxic nature of some dyes and intermediates has long been recognized. Acute, or short term, effects are generally well known. They are controlled by keeping the concentration of the chemicals in the workplace atmosphere below prescribed limits and avoiding physical contact with the material. Chronic effects, on the other hand, frequently do not become apparent until after many years of exposure. Statistically higher incidences of benign and malignant tumors, especially in the bladders of workers exposed to certain intermediates and dyes, were recorded in dye-producing countries during the period 1930–1960. The specific compounds involved were: 2-naphthylamine [*91-59-8*], 4-aminobiphenyl [*92-67-1*], benzidine [*92-87-5*] (4,4'-diaminobiphenyl), fuchsine [*632-99-5*] (CI Basic Violet 14), auramine [*2465-27-2*] (Solvent Yellow 2). There is considerable evidence that metabolites of these compounds are the actual carcinogenic agents (23). Strict regulations concerning the handling of known carcinogens have been imposed in most industrial nations. In the United States the regulations (24) caused virtually all the dye companies to discontinue use of the compounds.

Other actual or suspected carcinogens, such as the nitrosamines, or *N*-nitroso-compounds (25) (see *N*-Nitrosamines), polycyclic hydrocarbons, alkylating agents (26), and other individual compounds, such as the dichromates, should be considered in the wider context of industrial chemistry rather than as dye intermediates (see Colorants for foods, drugs, and cosmetics; Industrial hygiene and toxicology).

Literature

The most important reference work is the *Colour Index*, third edition (3). Dyes are cross indexed by commercial names and Colour Index name. For each entry, chemical, physical, and fastness properties are given. Names are cross indexed with CI numbers and chemical constitutions, where these have been made known by the manufacturers. In these cases, an outline of the method of preparation and references to patents or published procedures for manufacture are included. CI names of dyes have been accepted by the Chemical Abstracts Service as generic names for the allocation of Registry Numbers. Both the CI generic name and the commercial name are related to the registry number in ref. 27.

PB Reports. Although these documents consist of information captured 35 years ago from the German chemical industry, they are still very valuable. The I.G. Farbenindustrie dye companies were visited by teams of Allied investigators who reported extensively from company files and interviews with key personnel, on dye and intermediates research, development, and plant technology. It is probably fair to say that there is hardly a dye manufacturer today who is not using processes essentially derived from this information. A list of PB reports for the manufacture of important dyes and intermediates is available (28).

Patents. Much of the literature of dyes and their intermediates is in patents. The most valuable source for the years beginning with 1877 is the large collection of German patents by Friedländer (29). Unfortunately, this work is indexed only with a general subject index, and by German patent numbers and their foreign equivalents. To search for German patents after World War II, and for foreign patents in general, the patent system of the particular country must be consulted.

Until recently each country's classification was somewhat different and it was necessary, eg, to know that U.S. dye patents were classified under section 260 or 8, British under group IV, French under class XIV and Swiss under section G-37. Most of the other Western European nations listed their dye patents under class 22.

Currently, patents are listed under the designation of the world patents index. The major countries are divided into 2 groups: (1) faster issuing—Belgium; France; Japan; Netherlands; South Africa; and the FRG; and (2) slower issuing—Canada; The Democratic Republic of Germany (DDR); Soviet Union; Switzerland; United Kingdom; and United States. The IPC (International Patent Class) codes are shown below.

IPC Code	Category
C0914—industrial organic	
C09 b	organic dyestuffs
C09 cd	paints, pigments, varnishes
D textiles; flexible materials	
D06 b-1	finishing, laundering, bleaching
mn	other textile treatment
pq	dyeing, printing textiles

Although the principal patents are ultimately abstracted by *Chemical Abstracts,* early alerting is important to those actively engaged in the field. The Official Gazette of the U.S. Patent and Trademark Office is published weekly and is a source of current U.S. patents in the dye field. All the other patents are best reviewed by an international patent alerting service such as the British Derwent Publication's Ltd. which provides quick abstracts, indexes, tapes, and a variety of computerized searches.

BIBLIOGRAPHY

"Dyes and Dye Intermediates" in *ECT* 1st ed., Vol. 5, pp. 327–354, by G. E. Goheen and Jesse Werner, General Aniline & Film Corporation, and August Merz, American Cyanamid Company; "Dyes and Dye Intermediates" in *ECT* 2nd ed., Vol. 7, pp. 462–505, D. W. Bannister and A. D. Olin, Toms River Chemical Corporation.

1. *Technical Manual and Year Book,* American Association of Textile Chemists and Colorists, New York, 1977.
2. *Imports of Benzenoid Chemicals and Products, United States International Trade Commission Publication 828,* U.S. Government Printing Office, Washington, D.C., 1976.
3. *Colour Index,* 3rd ed., Society of Dyers and Colourists, Bradford, Eng., and American Association of Textile Chemists and Colorists, Lowell, Mass., 1971.
4. W. D. Wright, *The Measurement of Colour,* 4th ed., Van Nostrand Reinhold Inc., New York, 1969.
5. J. Wegmann, *Am. Dyest. Rep.* **51**(8), 46 (1962).
6. *Synthetic Organic Chemicals—U.S. Production and Sales, 1976, United States International Trade Commission Publication 833,* U.S. Government Printing Office, Washington, D.C., 1977.
7. H. A. Lubs, ed., *"The Chemistry of Synthetic Dyes and Pigments," ACS Monogr.* **127,** 689 (1955).
8. *Int. Dyer Text. Printer* **149,** 240 (1973).
9. J. D. Roberts and M. C. Caserio, *Basic Principles of Organic Chemistry,* 2nd ed., W. A. Benjamin Inc., New York, 1977, pp. 1037–1068.
10. P. Rys and H. Zollinger, *Fundamentals of the Chemistry and Application of Dyes,* John Wiley & Sons, Inc., New York, 1972.
11. Ref. 9, pp. 552–557.
12. H. Seeboth, *Angew. Chem. Int. Ed.* **6,** 307 (1967).

13. H. Kindler and D. Schuler, *Chem. Ing. Tech.* **37,** 402 (1965).
14. H. Zollinger, *Azo and Diazo Chemistry,* Interscience Publishers, Inc., New York, 1961, p. 32.
15. *The Chemical Industry, 1975,* Organization for Economic Cooperation and Development (Annual Publication) Paris, Fr., 1977.
16. *Synthetic Organic Chemicals—U.S. Production and Sales, 1961, United States International Trade Commission Publication 833,* U.S. Government Printing Office, Washington, D.C., 1962.
17. L. W. Little and J. C. Lamb, *Dyes and the Environment,* Vol. 1, American Dye Manufacturers Inc., 1973, Ch. 5.
18. P. B. Dejohn and R. A. Hutchins, *Text. Chem. Color.* **8,** 34 (1976).
19. S. L. Rock and B. W. Stevens, *Text. Chem. Color.* **7,** 34 (1975).
20. R. A. Montanaro and A. H. Noble, *The Textile Industry and the Environment,* American Association of Textile Colorists and Chemists Symposium No. 12, 1973.
21. U.S. Pat. 3,873,581 (Mar. 25, 1975), J. W. Fitzpatrick, C. J. Berninger, and D. O. Lewis (to Toms River Chemical Corporation).
22. Brit. Pat. 1,420,191 (Jan. 7, 1976), (to Imperial Chemical Industries Ltd.).
23. W. C. Hueper, *Occupational and Environmental Cancers of the Urinary System,* Yale University Press, New Haven, Conn., 1969, p. 216.
24. *Fed. Reg.* **38,** 10929, (1973).
25. "Chemical Carcinogens," *ACS Monograph* **173,** 491 (1977).
26. *Ibid.,* p. 83.
27. E. Derz, *Chem. Buy Direct, Chem Product Index Section,* E. de Gruyter, Berlin, New York, 1976.
28. Ref. 7, p. 692.
29. P. Friedlander, ed., *Fortschritte der Teerfarbenfabrikation und verwandter Industriezweige,* Vols. I–XXV, Julius Springer, Berlin, 1888–1942.

General References

R. Allen, *Colour Chemistry,* Appleton-Century-Crofts, New York, 1971.
E. Abrahart, *Dyes and Their Intermediates,* 2nd ed., Edward Arnold (Publishers) Ltd., London, Eng., 1977 (1st American ed., Chemical Publishing Co., Inc., New York, 1977).
W. Beech, *Fibre-Reactive Dyes,* SAF International, New York, 1970.
W. Bradley, *Recent Progress in the Chemistry of Dyes and Pigments,* The Royal Institute of Chemistry, London, Eng., 1958.
E. Clayton, *Identification of Dyes on Textile Fibres,* 2nd ed., Society of Dyers and Colourists, Bradford, Eng., 1963.
N. Donaldson, *The Chemistry and Technology of Naphthalene Compounds,* Edward Arnold, London, Eng., 1958.
H. E. Fierz-David and L. Blangey, *Fundamental Processes of Dye Chemistry,* Interscience Publishers, Inc., New York, 1949 (transl. from the 5th Austrian ed. by P. W. Vittum).
L. F. Fieser and M. Fieser, *Organic Chemistry,* 3rd ed., Reinhold Publishing Corp., New York, 1956, Ch. 36.
T. S. Gore and co-workers, *Recent Progress in the Chemistry of Natural and Synthetic Colouring Matters and Related Fields,* Academic Press, Inc., New York, 1962.
E. Gurr, *Synthetic Dyes in Biology, Medicine, and Chemistry,* Academic Press, London, New York, 1971.
H. A. Lubs, ed., "The Chemistry of Synthetic Dyes and Pigments," *ACS Monogr.* **127,** (1955).
M. Ranney, *Dyeing of Synthetic Fibers; Recent Developments,* Voyer Data Corp., Park Ridge, N.J., 1974.
I. Rattee and M. Breuer, *The Physical Chemistry of Dye Adsorption,* Academic Press, London, New York, 1974.
P. Rys and H. Zollinger, *Fundamentals of the Chemistry and Application of Dyes,* Wiley-Interscience, London, Eng., 1972.
H. R. Schweizer, *Künstliche Organische Farbstoffe und ihre Zwischenprodukte,* Springer-Verlag, Berlin, Göttingen, Heidelberg, 1964.
K. Venkataraman, *The Chemistry of Synthetic Dyes,* Vols. I–VII, Academic Press, New York, 1952–1974.

D. W. BANNISTER
A. D. OLIN
H. A. STINGL
Toms River Chemical Corporation

DYES, ANTHRAQUINONE

Anthraquinones are among the earliest natural products to be used as dyestuffs. The characteristic chromophore of the anthraquinone series consists of one or more carbonyl groups in association with a conjugated system. Amino or hydroxyl groups, as well as their substituted forms, and more complex heterocyclic systems may be present as auxochromes. In a number of cases the parent carbonyl compound of complex fused-ring systems, such as pyranthrone, violanthrone, and dibenzpyrenequinone, is colored even though no auxochrome is present.

Anthraquinone (1), the parent compound for the carbonyl dyes, exhibits a long uv band (λ_{max} = 327 nm, in CH_2Cl_2) which extends into the visible spectrum producing a faint yellow color; however, it is not a dye. The introduction of relatively simple electron donors to the basic chromophore structure, creates compounds that absorb at various regions of the visible spectrum depending upon the strength of the electron donor (OH < NH_2 < NR_2 < NHAr) (Table 1).

Substituents in the α-position (eg, 1-, 1,4-, or 1,5-) cause large bathochromic shifts. These simple substituted compounds are frequently used directly as dyes or serve as

(1)

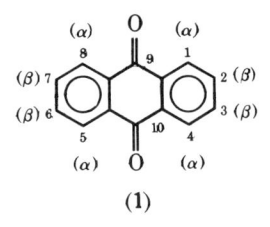

Table 1. Absorption of Substituted Anthraquinones

Substituent	λ_{max}, nm
1-hydroxy	405
1-amino	465
1-methylamino	508
1-hydroxy-4-amino	520
1,4-diamino	550
1,5-diamino	480
1,4,5,8-tetraamino	610
1,4-dianilino	620

intermediates for the manufacture of colored polycondensed carbonyl compounds (eg, indanthrone). The position of the substituents and the consequent formation of hydrogen bonds between substituent and carbonyl group influence not only the absorption maximum but also other properties such as the resistance to sublimation of disperse dyes, affinity for the substrate, lightfastness, and vatting properties.

Anthraquinone derivatives with β-hydroxyl or β-amino groups are capable of forming intermolecular hydrogen bonds and generally exhibit better resistance to sublimation, better solubility, and better affinity for textile substrates than the α-substituted compounds. On the other hand, intramolecular hydrogen bonds reduce the acidity of the carbonyl groups in the peri-position (more positive redox potential) which generally is advantageous with respect to wash- and lightfastness. X-ray data have shown that the relative strengths of inter- and intramolecular hydrogen bonds with different hydroxyl- and aminoanthraquinones are reflected in their melting points. Intramolecular hydrogen bonds can raise the melting point as much as 100°C. Analogous relations between chemical structure and gas chromatography retention times supply information on the formation of intermolecular hydrogen bonds between substrates and simple anthraquinone dyes.

Disperse Dyes

The anthraquinone disperse dyes have their major use in coloring synthetic fabrics. The present day trend toward the use of polyesters (qv) with truly hydrophobic fibers has created new problems in dyeing and has nurtured interest in the disperse dyes. The problem of preparing stable dispersions of insoluble or sparingly soluble dyes, required for such fabrics, has been met with the use of more powerful dispersing agents, with improved methods of dispersion, and with the inclusion of alkanol, carboxyl, amide, and similar functional groups in the dye molecules. Aside from making dyes more readily dispersible such functionality in the molecule has also increased affinity, improved leveling, and rendered minor modifications in shade.

The violet, blue, and green anthraquinone dyes used on cellulose acetate and polyester fibers are commonly synthesized from 1,4-diaminoanthraquinone, employing two different alkyl or aryl groups. These substituted diamino compounds have good affinity and level-dyeing power as well as excellent fastness. In addition, it has been found that in coloring polyester and rayon acetate fibers blue with these dyes, a mixture of two or three such dyes gives brighter and more intense shades than any one of the mixture components alone. It should be noted that these anthraquinones are primarily used for violet, blue, and green shades, and the yellow to red shades are commonly, though not exclusively, obtained by means of azo dyes. With the exception of pure green, however, the anthraquinones for synthetic fibers cover the spectrum from yellow through violet. Table 2 lists the more important, commercially manufactured anthraquinone disperse dyes.

The affinity of a dye for cellulose acetate and polyester fibers is proportional to the simplicity of the molecule and approximately proportional to its basicity. The ideal dye for this type of synthetic fiber would demonstrate good dispersibility, substantivity, light- and sublimation-fastness, as well as resistance to gas-fading and washing. Although these may be the goals for a disperse dye, this, in practicality, is not always the case. For example, in the darker shades of anthraquinone dyes, blues and violets, the dyes have a tendency to undergo an irreversible color change. Many of the an-

Table 2. Major Commercially Produced Anthraquinone Disperse Dyes

CI Name	CAS Registry No.	CI No.	Method of preparation or structure	1976 U.S. sales Metric tons	1976 U.S. sales Thousands of dollars	1976 Import, t
Yellows						
Disperse Yellow 13	[3688-79-7]	58900	treat 3-bromobenzanthrone with sodium methoxide in methanol			0.75
Disperse Yellow 65	[69912-86-3]					13.50
Disperse Yellow 92	[14179-99-8]	60635				
Disperse Yellow 127	[3627-47-2]	65410	acetylate 1-aminoanthraquinone with isophthaloyl chloride			
Oranges						
Disperse Orange 11	[82-28-0]	60700	nitrate 2-methylanthraquinone and reduce with aqueous sodium sulfide			
Reds						
Disperse Red 3	[4465-58-1]	60507	1-(2-hydroxyethylamino)anthraquinone			
Disperse Red 4	[2379-90-0]	60755	methylate 4-aminoanthrapurpurin with dimethyl sulfate in acetone-water			3.26
Disperse Red 9	[82-38-2]	60505	treat 1-chloroanthraquinone or 1-anthraquinone sulfonic acid with methylamine under pressure in the presence of an oxidizing agent			
Disperse Red 11	[2872-48-2]	62015	condense 1-amino-4-bromo-2-anthraquinonesulfonic acid with *p*-toluenesulfonamide, hydrolyze, replace the sulfonic acid group by OH, and methylate			2.12
Disperse Red 15	[116-85-8]	60710	nitrate 1-hydroxyanthraquinone in sulfuric acid in the presence of boric acid; reduce the 1-hydroxy-4-nitroanthraquinone with aqueous sodium sulfide			19.05
Disperse Red 22	[2944-28-7]	60510	1-anilinoanthraquinone			2.76
Disperse Red 55	[12223-36-8]					2.30
Disperse Red 59	[12731-57-6]					
Disperse Red 60	[17418-58-5]	60756	condense 1-amino-2-bromo-4-hydroxyanthraquinone with phenol in an alkaline medium	1,025	7,651	14.99
Disperse Red 86	[12223-43-7]			12.25	189	4.55
Disperse Red 91	[12236-10-1]					28.59
Disperse Red 92	[12236-11-2]					4.54
Disperse Red 93	[12236-12-3]					2.20
Disperse Red 121	[12223-62-0]					1.02
Disperse Red 132	[12223-67-5]					1.05
Disperse Red 133	[12223-68-6]					2.27
Disperse Red 204	[61968-57-8]					

Violets

Disperse Violet 1	[128-95-2]	61100	condense quinizarin with aqueous ammonia and sodium hydrosulfite under pressure; oxidize the leuco-1,4-diaminoanthraquinone in a mixture of o-dichlorobenzene and nitrobenzene	15.88	179	0.96
Disperse Violet 4	[1220-94-6]	61105	methylate leuco-1,4-diaminoanthraquinone with methanol in sulfuric acid in the presence of chlorine			0.40
Disperse Violet 6	[6471-02-9]	61140	deacetylate 1,4-diacetamidoanthraquinone under controlled conditions			5.00
Disperse Violet 8	[82-33-7]	62030	nitrate 1,4-diaminoanthraquinone in sulfuric acid			0.42
Disperse Violet 23	[12217-94-6]					
Disperse Violet 27	[53989-04-1]	60724	1-anilino-4-hydroxyanthraquinone	44.00	247	
Disperse Violet 28	[81-42-5]	61102	chlorinate 1,4-diaminoanthraquinone with sulfuryl chloride in nitrobenzene			
Disperse Violet 31	[12239-36-0]					1.36

Blues

Disperse Blue 1	[2475-45-8]	64500	acylate 1,5-diaminoanthraquinone with oxalic acid, nitrate in sulfuric acid, hydrolyze and reduce (10)			2.62
Disperse Blue 3	[2475-46-9]	61505		397.8	2,556	5.40
Disperse Blue 5	[4486-13-9]	62035	convert 1-amino-4-methylamino-2-anthraquinone sulfonic acid to 1-amino-4-methylanimo-2-anthraquinonecarbonitrile by the action of potassium cyanide; hydrolyze the product with monohydrate			0.05
Disperse Blue 6	[3443-93-4]	62050	treat 1-amino-4-bromo-2-anthraquinonesulfonic acid with cyclohexylamine; treat this with sodium cyanide and ammonium bicarbonate under pressure to form 1-amino-4-cyclohexylamino-2-anthraquinonecarbonitrile and hydrolyze			
Disperse Blue 7	[3179-90-6]	62500	condense leuco-1,4,5,8-tetrahydroxyanthraquinone with ethanolamine in the presence of hydrosulfite; oxidize this in the presence of ammonia and boric acid or caustic soda and piperidine			
Disperse Blue 9	[27341-33-9]	61115	1-methoxyanilino-4-aminoanthraquinone			0.05
Disperse Blue 14	[2475-44-7]	61500				15.53
Disperse Blue 19	[4395-65-7]	61110	condense 1-amino-4-hydroxyanthraquinone with cyclohexylamine			5.40
Disperse Blue 22	[6373-16-6]	60715	treat leuco-quinizarin with methylamine in methylated spirit and oxidize (9)			
Disperse Blue 23	[4471-41-4]	61545	brominate 1-methylaminoanthraquinone and condense with aniline			
Disperse Blue 24	[3179-96-2]	61515	treat 1,5-dihydroxy-4,8-dinitroanthraquinone with methylamine			59.00
Disperse Blue 26	[3860-63-7]	63305				
Disperse Blue 27	[15791-78-3]	60767	condense p-aminophenethyl alcohol with 1,8-dihydroxy-4,5-dinitroanthraquinone			

215

Table 2 (continued)

CI Name	CAS Registry No.	CI No.	Method of preparation or structure	1976 U.S. sales Metric tons	1976 U.S. sales Thousands of dollars	1976 Import, t
Disperse Blue 28	[6408-79-3]	62065	convert 1-amino-4-bromo-2-anthraquinonecarboxylic acid to the ethanolamide via the acid chloride, and condense with aniline			
Disperse Blue 31	[1328-23-0]	64505	acylate 1,5-diaminoanthraquinone with oxalic acid, nitrate in sulfuric acid, hydrolyze and reduce; methylate this 1,4,5,8-tetraaminoanthraquinone with methanol in sulfuric acid			
Disperse Blue 34	[4424-82-2]	61510	(12)			0.12
Disperse Blue 35	[12222-75-2]					
Disperse Blue 40	[3178-78-7]	63302	1-amino-4,8-dihydroxy-5-(m-sulfonamido)anilinoanthraquinone			2.99
Disperse Blue 55	[12217-78-6]					8.40
Disperse Blue 56	[12217-79-7]	63285	monochloro adduct of 4,8-diamino-1,5-dihydroxyxanthraquinone			0.20
Disperse Blue 60	[12217-80-0]			212.3	1,230	
Disperse Blue 64	[12222-77-4]					0.10
Disperse Blue 72	[12217-81-1]					408.80
Disperse Blue 73	[12222-78-5]					12.00
Disperse Blue 83	[12222-81-0]					3.29
Disperse Blue 84	[12222-82-1]					9.62
Disperse Blue 87	[12222-85-4]					4.34
Disperse Blue 93	[12235-95-9]					23.73
Disperse Blue 95	[12235-97-1]					4.29
Disperse Blue 127	[12270-40-5]					2.79
Disperse Blue 134	[14233-37-5]					0.20
Disperse Blue 154	[56509-54-7]					0.09
Disperse Blue 180	[69912-84-1]					
Disperse Blue 185	[61968-36-3]					7.90
Disperse Blue 296	[69912-85-2]					2.10

thraquinone dyes also show poor sublimation-fastness; this has been an especially significant problem in the yellow and red shades because of the rather simple molecular structure of the dye. This problem is less severe for the darker colored anthraquinone dyes, since their structures usually include polar functional groups, such as —CONHCH$_2$CH$_2$OH, —N(R)CH$_2$CF$_2$H, —N(R)CH$_2$CH$_2$OH, —SO$_2$NHCH$_2$CH$_2$OH, and —N(R)CH$_2$CHOHCH$_2$OH, that have been found to improve sublimation fastness.

Aminoanthraquinones. As is the case with many of the different types of anthraquinones, the disperse dyes are frequently prepared from 1,4-diaminoanthraquinones, leucoquinizarin, and quinizarin. Disperse dyes derived from such starting materials cover a wide variety of structures and colors.

The series of 1-hydroxy-4-amino- (alkylated or arylated)anthraquinones has been used effectively on acetate rayon and polyester fiber for pink to violet shades. These dyes are usually prepared by the well-known process of condensing equimolar quantities of the amine (or ammonia), with a quinizarin–leucoquinizarin mixture under mild conditions. Such compounds range from the very simplest, (2), where R = H, through an entire array of alkyl and aryl amines (3–8) (1).

The alkylamino derivatives are principally used on cellulose acetate. All of them, with the exception of the sulfonyl derivative (8), demonstrate inferior lightfastness on polyesters (2).

The arylamino derivatives usually absorb at longer wavelength and usually show poor affinity for cellulose esters. They are used primarily on polyesters even though their lightfastness is at times inferior to that of other anthraquinones. The presence of alkanol, carboxyl, and other such groups on substituted aminoanthraquinones facilitate the dispersion of these dyes in water.

At the next level of complexity in these disperse dyes are the 1,4-diaminoanthraquinones. The customary methods of preparing these are: (1) reaction of amines and ammonia with quinizarin; (2) replacement of labile groups such as halo and nitro with an amine; (3) reduction of nitro groups; and (4) reaction of 1-amino-2-sulfo (or halo)-4-halo-anthraquinone with an amine followed by replacement or elimination of the groups in the 2-position. The 1,4-diamino derivatives are usually bright blue.

For many years, blue shades have been derived from a group of 1,4-bis(alkylamino)anthraquinones. Recently these dyes have become popular on nylon carpeting

(2) R = H [116-85-8]
(3) R = CH$_2$CH$_2$CO$_2$C$_2$H$_5$ [70321-15-2]
(4) R = CH$_2$CH$_2$CH(CN)CH$_3$ [70321-16-2]

R = —⟨phenyl⟩—R′

(5) R′ = H [19286-75-0]
(6) R′ = OCH$_3$ [23552-76-3]
(7) R′ = OH [6409-73-0]

(8) R = —⟨ring with SO$_2$⟩ [70321-21-0]

because of their good barré coverage. The disadvantage of these dyes include less than desirable lightfastness and resistance to atmospheric fading.

Analogous to those discussed above, CI Disperse Blue 23 (9) is obtained by heating quinizarin with a 30% solution of monoethanolamine. The alkanol groups in this symmetrical molecule assist in dispersion of this dye.

The use of this dye in combination with CI Disperse Blue 3 (10), the main constituent, and CI Disperse Blue 14 (11) is an example of obtaining a tinctorial strength that is far stronger than any one of its components alone.

1,4-Bis-(arylamino) anthraquinones have also been prepared by this method. Those in which both aryl groups bear —SO$_2$NH$_2$, —SO$_2$NH alkyl (3), or —CH$_2$O-CH$_2$CH$_2$OH (4) substitutions exhibit bright blue shades of good fastness on polyester.

Preparation of unsymmetrically substituted 1,4-diaminoanthraquinone derivatives such as CI Disperse Blue 34 (12) is accomplished by the two-step replacement of the hydroxyl groups in quinizarin.

Structures (13–14) exemplify a group of green or greenish-blue shades in which one amino group is substituted by a hydroxyalkyl group and the other by a hydroxy- or aminophenyl group. Treatment of primary or secondary amines, such as that of the aminophenyl group above, as well as 1,4-diaminoanthraquinone derivatives, with ethylene oxide or ethylene chlorohydrin results in β-hydroxyethyl group alkylation of the nitrogen atom. For example, reaction of the green dye (15) with epichlorohydrin produces a new dye (16) which has greater affinity for and better fastness on acetate silk than does the parent dye.

	R	R'
(9)	NHCH$_2$CH$_2$OH	NHCH$_2$CH$_2$OH
(10)	CH$_3$	NHCH$_2$CH$_2$OH
(11)	CH$_3$	CH$_3$
(12)	CH$_3$ or CH$_2$CH$_2$OH	CH$_2$CH$_2$OCH$_2$CH$_2$OH

(13) R″ = OH [70321-22-1]
(14) R″ = NH$_2$ [62758-81-0]

(15) R = H [4221-67-4]
(16) R = CH$_2$CH(OH)CH$_2$Cl [70321-23-2]

Similar dyes may be prepared from 1-amino-4-arylaminoanthraquinones with ethylene oxide or chlorohydrin; from 4-halogen substituted 1-arylaminoanthraquinones by means of hydroxyalkylamines; or from leuco-1,4-diaminoanthraquinones by condensation with a hydroxyalkylamine and an arylamine simultaneously or successively.

One of the newer dyes in this series is (17). It has been used to dye and transfer print polyamide fibers fast blue shades (5).

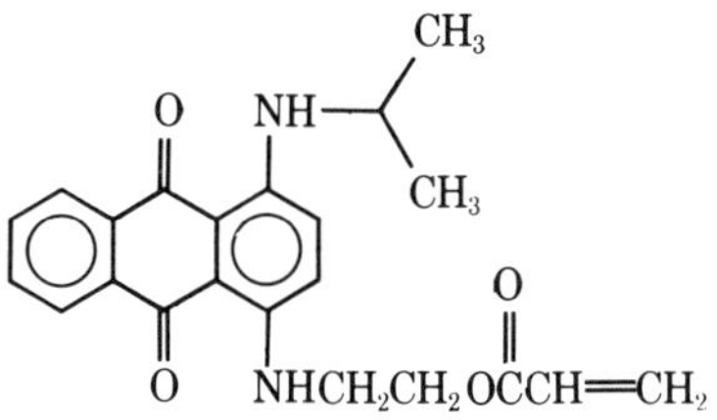

(17) [61734-61-0]

Aside from the amine moiety of the 1,4-diaminoanthraquinones, substitutions may be made to several different positions on the anthraquinone nucleus in order to alter the shade, affinity, or fastness qualities of a dye. This point is illustrated by the family of compounds 1-amino-2-X-4-hydroxyanthraquinones, where X is frequently an alkyloxy, aryloxy, or thio substituent. They are used for their bright red and pink shades on cellulose acetate, polyamides, and polyesters. One of the simpler derivatives is the red dye obtained by methylating 1-amino-4-hydroxy-2-mercaptoanthraquinone; it demonstrates good fastness to gas fumes. Frequently, these types of dyes are prepared as follows:

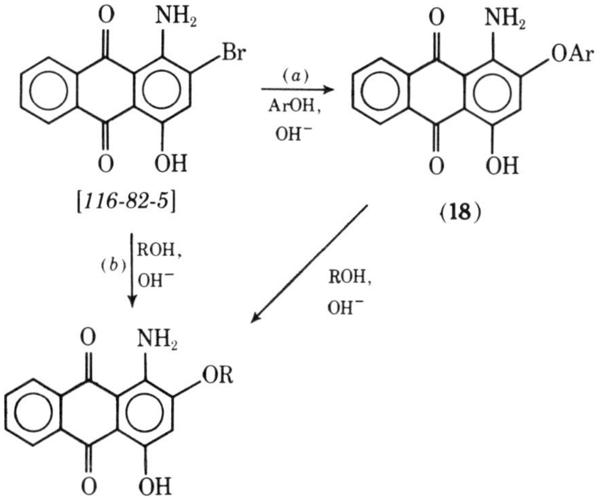

Although route (a), through dye (18), requires an additional step in the preparation of (19), the product is usually of better quality than that obtained via route (b) (6). These dyes demonstrate excellent tinctorial value and lightfastness, but have the disadvantage of inadequate sublimation-fastness. Attempts to remedy this problem have resulted in compounds (20–24) (7), as well as some more exotic compounds like the bicyclic cyano-derivative (25) (8).

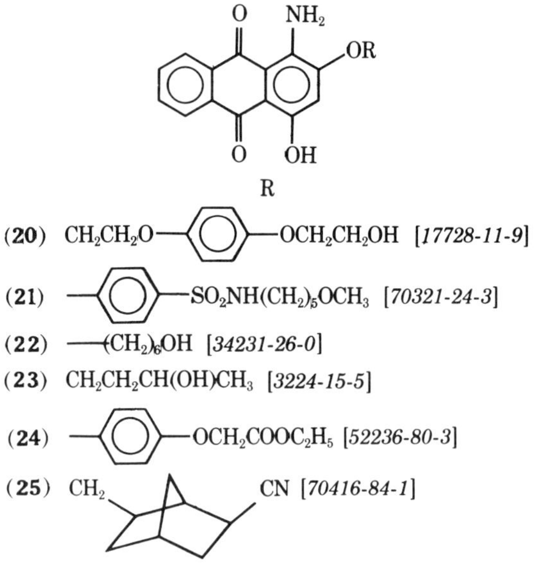

220 DYES, ANTHRAQUINONE

$$\text{[1-amino-2-OR-4-hydroxyanthraquinone structure]}$$

R

(20) CH₂CH₂O—⟨C₆H₄⟩—OCH₂CH₂OH [17728-11-9]

(21) —⟨C₆H₄⟩—SO₂NH(CH₂)₃OCH₃ [70321-24-3]

(22) —(CH₂)₆OH [34231-26-0]

(23) CH₂CH₂CH(OH)CH₃ [3224-15-5]

(24) —⟨C₆H₄⟩—OCH₂COOC₂H₅ [52236-80-3]

(25) CH₂—[norbornyl]—CN [70416-84-1]

Dye (21) is prepared by chlorosulfonating 1-amino-2-phenoxy-4-hydroxyanthraquinone and treating that product with 3-methoxypropylamine (9). The other compounds shown are obtained by displacing a halogen atom in position 2 by means of alcohol and alkali as illustrated above.

Among the methods developed for improving sublimation-fastness are the use of a 4-(hydroxymethyl)cyclohexylmethoxy group in the 2-position, as well as the reaction of 2-aryloxy (or thio) groups with N-methylamides or imides (eg, N-hydroxymethylphthalimide and N-hydroxymethylpyrrolidone). Two newer dyes of these types are (26) (10) and (27) (11).

The brilliant bluish-red dye (26) for polyester fabric, possesses a phthalimide group on the substituent at the 2-position which contributes significantly to the excellent fastness demonstrated by this dye.

Dye (27) also exhibits good fastness, in part owing to the substituted cyclohexyl group at the 2-position. This brilliant pink dye has proved versatile in its use on acetate, polyamide, polyester and polyurethane fibers.

Members of this family have been reported that possess such substitutions at the 2-position of 1-amino-4-hydroxyanthraquinone as —OCH₂CH₂X, —SCH₂CH₂X, —OCH₂CH₂O-alkyl, —O(CH₂CH₂O)n-alkyl, and —S-alkylene—X, where X represents a variety of functional groups (12). Two of the newer thio-substituted dyes are (28) (13) and (29) (14). Both of these demonstrate good fastness properties on polyester fibers. Furthermore, it has been found that such thio-alkyl groups, or sulfite groups, as in (30) (15), induce a bluish tint to the color of the dyes. It is thus not surprising that (28–30) are bluish red in color.

The 1,4-diaminoanthraquinones have also found a place in this family of dyes. Blue dyes with good fastness can be obtained by introducing any one of a variety of substituents into the 2-position of this parent compound; these have included halogen, —OR, —SR, —O₂SR, CN, NO₂ (16), and —CO₂R (17). From the analogous sulfonic acid (31), the thio derivatives, such as (32), have been prepared.

(26) OCH₂CH₂OCCH₂—N[phthalimide] [63224-84-0]

(27) OCH₂—[cyclohexyl]—CN [70321-25-4]

(28) SCH₂CH(OH)—[phenyl] [60615-97-6]

(29) SCH₂CH₂O—[phenyl]—CH₂NHCCH₃ [23478-50-4]

(30) OSO₂—[phenyl]—CH₃ [16517-78-5]

(31) → (32) [HSCH₂CH₂OH]

Typically, these dyes possess an —NHSO₂(alkyl) (or aryl) group in the 4-position (as exemplified in (33–34) (18)), —NH₂, —OH or —NHSO₂R in the 1-position, and —H, —CH₃, —O-alkyl, or —O-aryl in the 2-position.

(33) R = H [2907-79-1]
(34) R = OCH₃ [2382-33-9]

1,4-diaminoanthraquinones have also been used to synthesize 2,3-disubstituted dyes of the type (35–36) and (37–38).

Beginning with those that have identical substituents at the 2- and 3-positions, 1,4-diamino(alkyl or arylamino, etc)-2,3-dicyano derivatives have been prepared from the reaction of CuCN with the corresponding dichloro derivatives (19).

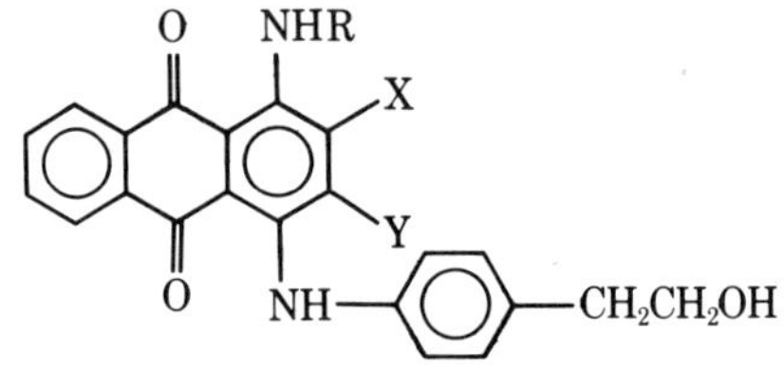

(35) R = H [70321-17-4]
(36) R = CH₃ [70321-18-5]
X = Y = Br
R = H
X = OCH₂CH₂OH
(37) Y = H [70321-26-5]
(38) Y = Cl [70321-27-6]

Anthrarufin and Chrysazin Derivatives. Another large class of disperse dyes derived from anthraquinones are those in which all of the α-positions are occupied, primarily by hydroxy, nitro, and amino groups, and their derivatives. Dinitroanthrarufin (39) and dinitrochrysazin (40) are two such dyes which have been in use for many years. These two, however, have shown poor affinity for synthetic fibers. To improve the dyeing qualities of dinitroanthrarufin, one nitro group is replaced with an arylamino bearing hydroxyalkyl, hydroxyalkoxy, etc, group. This produces blue dyes for acetate and polyester with high affinity and outstanding fastness to light and gas (20). Similar dyes derived from dinitrochrysazins are slightly greener and faster to light than the 1,5-hydroxy isomers (anthrarufin [117-12-4] derivatives). Examples of such arylamino compounds are (41) and (42).

These condensations are easily performed in various organic solvents or in the presence of excess amine at temperatures up to 130°C. The second nitro group is replaced only with considerable difficulty; in fact, when the aromatic amine is deactivated by electronegative groups, replacement of even one nitro group may be difficult. The replacement of both nitro groups in dinitroanthrarufin and dinitrochrysazin by hydroxyalkylanilino, —NHC₆H₄SO₂alkyl, or —NHC₆H₄N(alkyl)SO₂alkyl has been reported (21). Amines that have been used for the mono-condensation include aniline, anisidine, and toluidine, among others. The dyes thus obtained have satisfactory lightfastness for polyester (22) and if the remaining nitro group is reduced, the dye

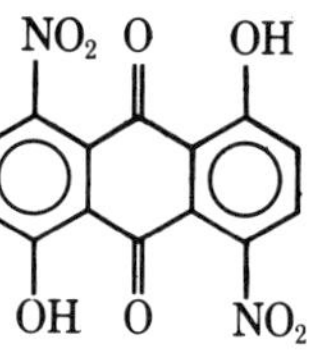

(39) [128-91-6]

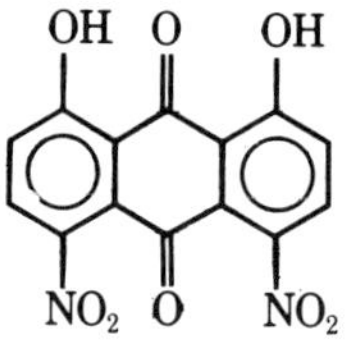

(40) [81-55-0]

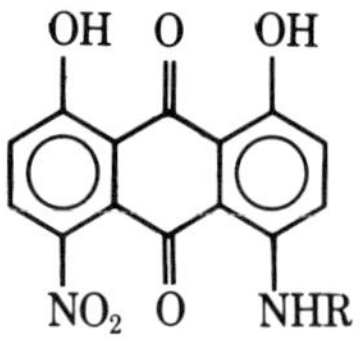

(41) [15791-78-3]

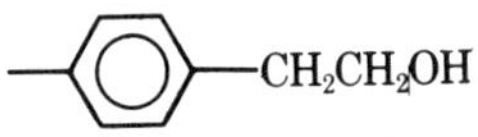

(42) [70321-19-6]

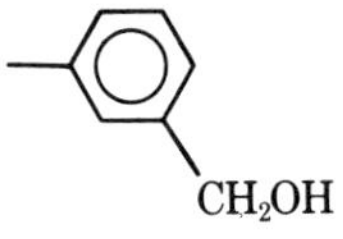

produced has brighter colors and slightly better dyeing properties. On cellulose acetate the dyes exhibit reduced gasfastness, but improved fastness to cold water bleeding. Dyes with excellent fastness properties have been derived by condensing dinitroanthrarufin or dinitrochrysazin with 3-aminosulfolane, 2-aminopyrimidines, and anilines bearing triazole, piperidone, pyrrolidine, etc, groups (23).

Intermediates, such as dinitrochrysazin (40), used to prepare these dyes are customarily synthesized by either of two routes. One method is by direct nitration of the respective dihydroxy compound (24). A second route, which usually produces better yields is given in the sequence below (25).

Other dyes in this series have been obtained from diaminoanthrarufin and diaminochyrsazin. When these compounds are monoarylated and treated with formaldehyde in order to hydroxymethylate one or both of the amino groups, increased tinctorial strength is conferred on the dyes (26).

As with many of the anthraquinone parent compounds, diaminoanthrarufin and diaminochrysazin can be substituted in the 2- and/or 6-position in order to improve the qualities of the dyes. These are frequently derived from the 2,6-disulfonic acid derivatives (43) thru the monosulfonic acid (44) as illustrated below.

Fastness properties of this type of dye on polyester have recently been improved by substituting the phenylalkoxy group in position 2 as seen in (45) (27).

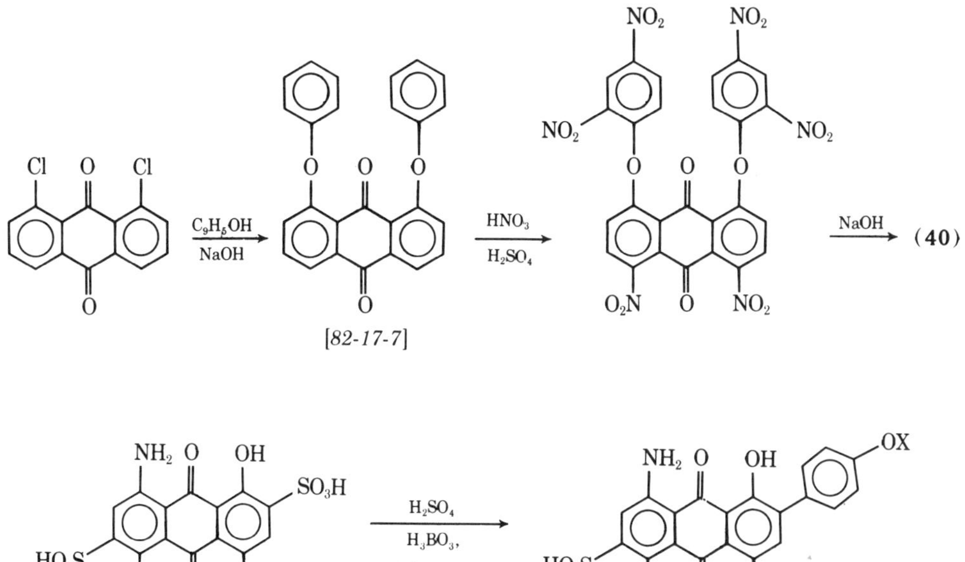

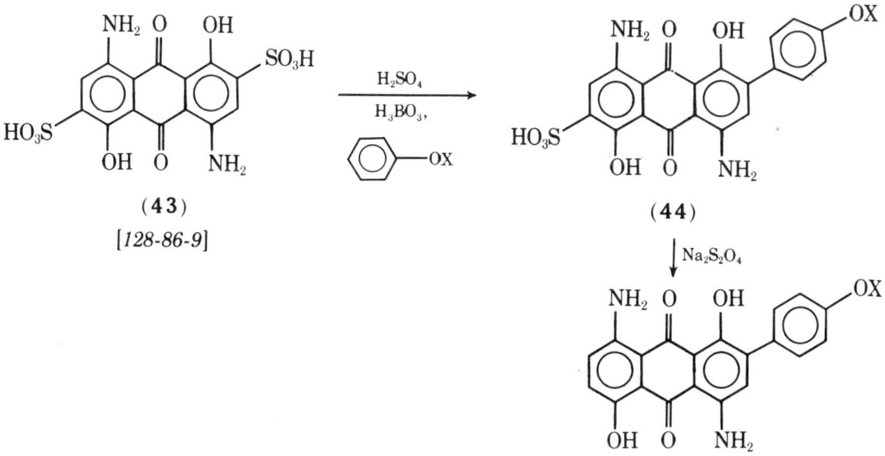

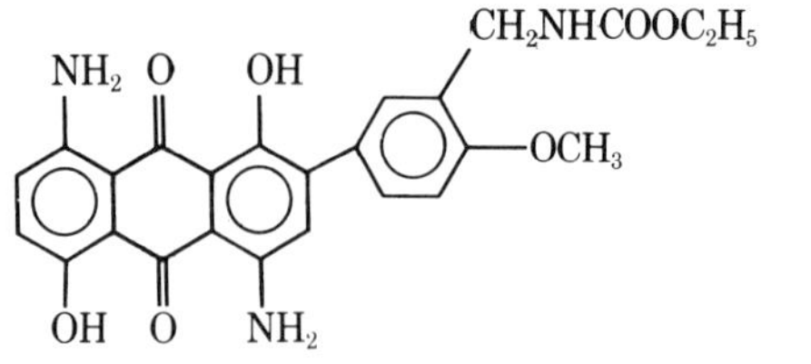

(45) [23613-88-9]

When vigorous hydrolysis is performed on the monosulfonic acid (46), loss of the amino groups occurs, yielding the red–violet tetrahydroxy dye (47), which can then be treated with a primary amine, as shown in the sequence below, to yield the disubstituted amino blue dye (48).

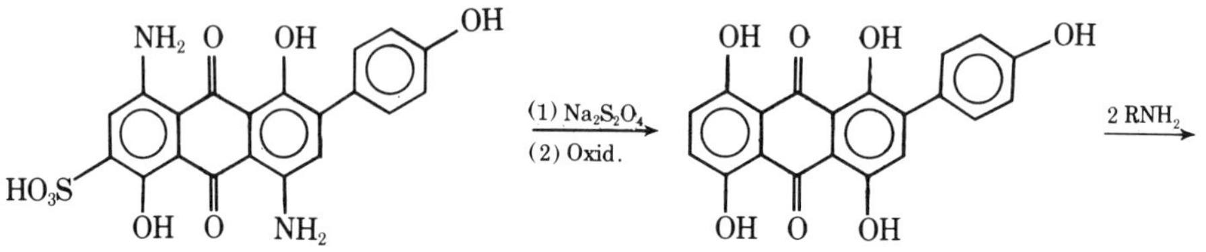

(46) [26869-99-8]   (47) [19249-08-2]

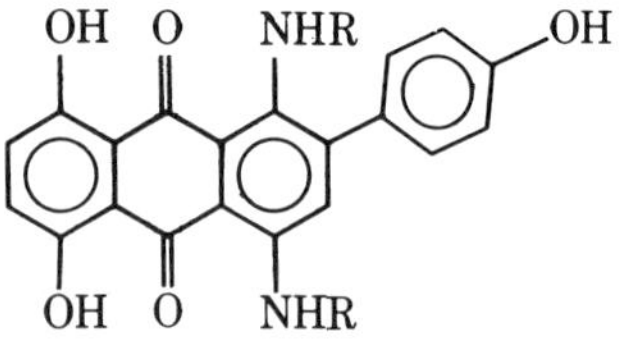

(48)

Diaminoanthrarufin and diaminochrysazin have also been used in other types of reactions, in order to prepare 3-phenol and 3-phenyloxy diaminoanthrarufins (49) (28). The parent compound is treated with phenols or their ethers in the presence of sulfuric and boric acids.

Furthermore, when halogenated with 0.4–1.5 moles of halogen, particularly bromine, bright reddish-blue polyester dyes with good build-up and tinctorial strength are obtained from diaminoanthrarufin and diaminochrysazin (29).

Recently, a wide variety of substituents and substitution patterns on the anthraquinone nucleus have been developed. The manipulation of position and type of substituent has proved a useful tool in improving the qualities of these dyes; eg, structure (50) dyes clear blue on polyester with good sublimation and lightfastness (30).

Especially beautiful are the blue to turquoise heterocyclic dyes derived from

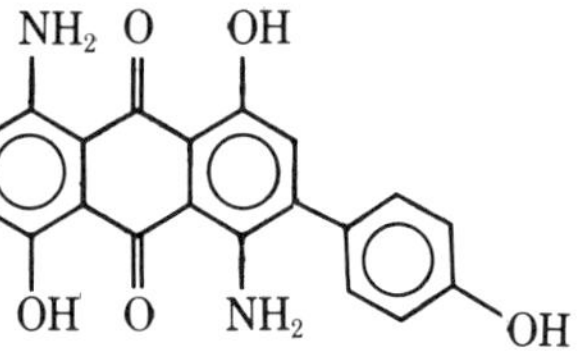

(49) [7098-08-0]

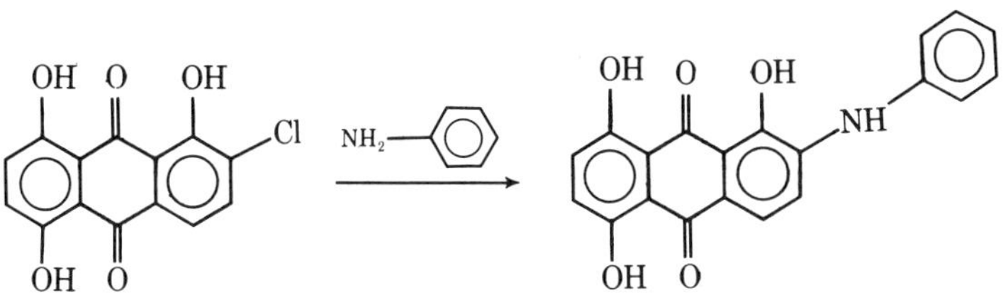

(50) [58376-69-5]

1,4-diaminoanthraquinone-2,3-dicarboximides (1,3-dioxo-4,7-diamino-5,6-phthaloyl isoindolines) (52). Sulfuric acid treatment of the 2,3-dicyano (31) (or dicarbamoyl (32))-1,4-diaminoanthraquinone (51), sometimes in the presence of a secondary alcohol, yields the greenish-blue dyes (52–53) (33). Derivatives of (52), such as (54), may also be obtained by reaction of the former with primary amines.

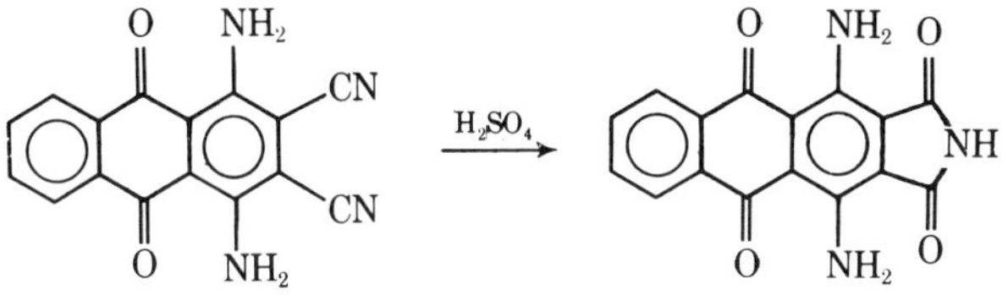

(51) [81-41-4]

(52) [128-81-4]

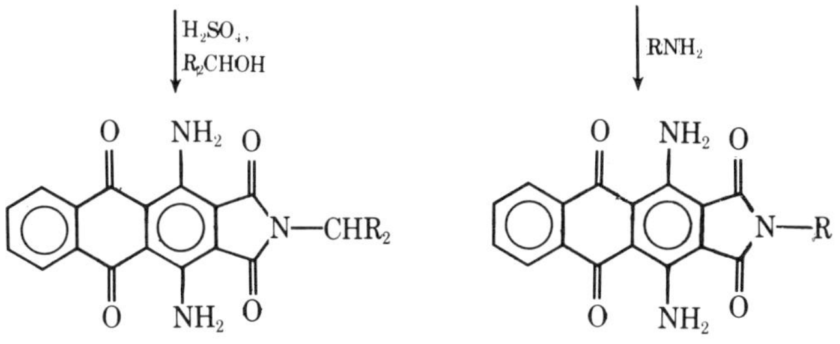

(53)

(54)

The dinitriles and diamides used for the preparation of these green-blue dyes may be synthesized by reaction of metallic cyanides on the corresponding sulfonic acids (34) or by reaction of amides with the dicarboxylic acid chlorides (35).

A closely related family of dyes, such as (56) are derived from 1,4-diaminoanthraquinonecarboxamides (55) in the presence of oxidizing agents and inorganic cyanides. 1-Amino-4-nitrocarboxyl chloride (57) has also been used with a primary amine and inorganic cyanide in the same type of reaction. This route proceeds via 1-amino-4-nitrocarboxamide utilizing the nitro group as an oxidizing agent to (58) (36). Organic cyanides have been found to improve the process in both cases (37).

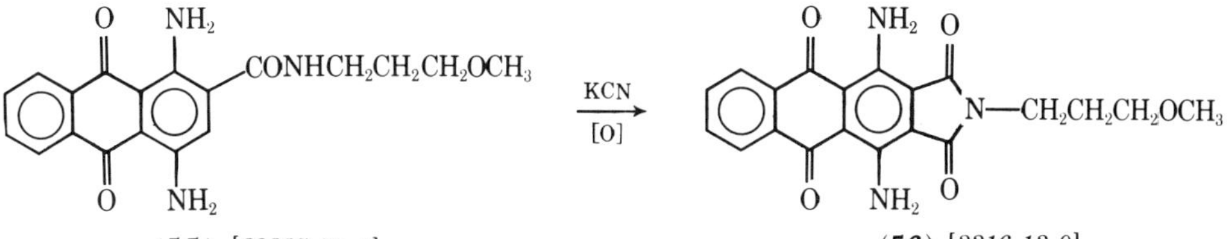

(55) [62335-58-4]

(56) [3316-13-0]

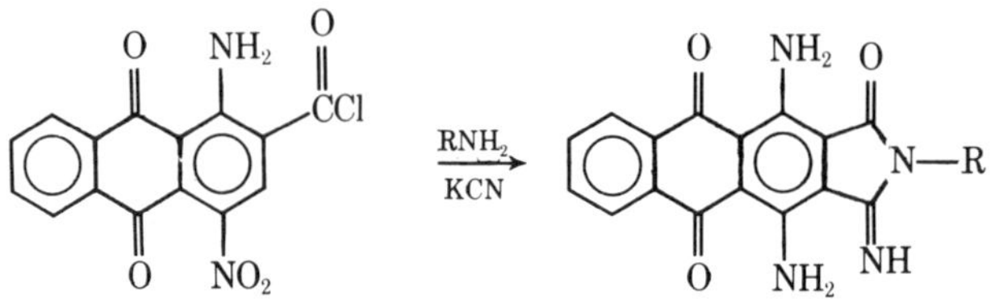

226 DYES, ANTHRAQUINONE

(57) [25665-00-3]   (58)

Blue to bluish-green shades for polyesters have been obtained from at least two different series of anthraquinone heterocycles. The first are members of the anthraquinoid-1,2-acridones (phthaloylbenzacridones) (59). Beginning with 1-amino-2,4-dibromoanthraquinone or 1-amino-4-bromoanthraquinone-2-sulfonic acid, nucleophilic substitution of the 4-bromo atom is accomplished with o-aminobenzoic acid, followed by cyclization with polyphosphoric acid and finally amide formation at the 2-position using CuCN and sulfuric acid.

The second type, (60), is derived from leucoquinizarin or 2-bromoquinizarin by reaction with 2-aminothiophenol as illustrated below (38).

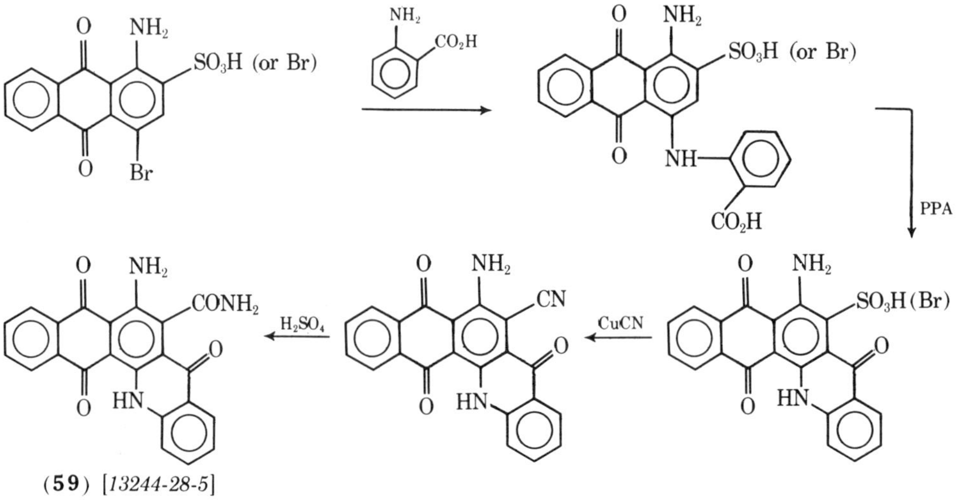

(59) [13244-28-5]

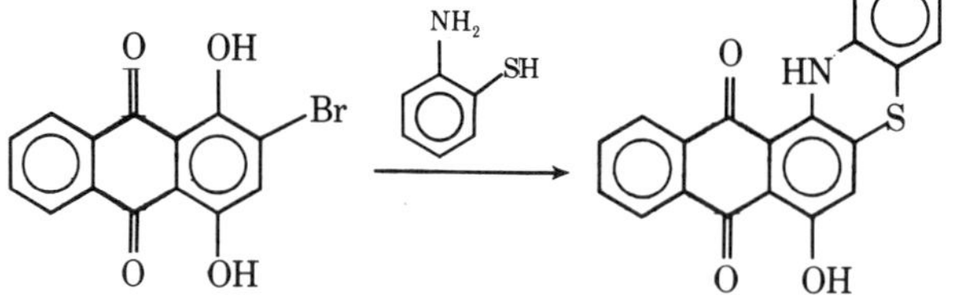

(60) [13289-47-9]

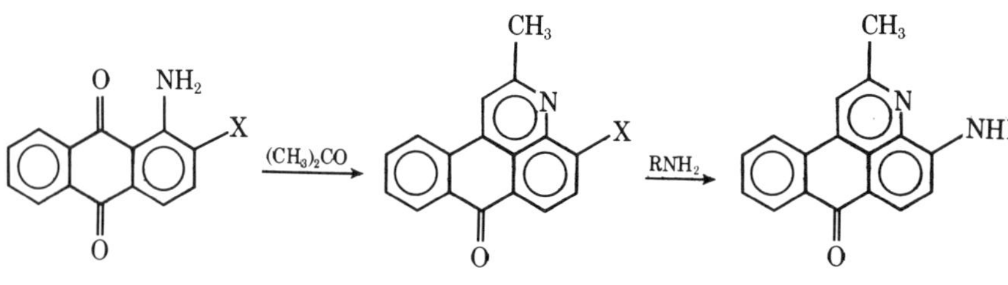

(61)

Shades in the yellow range of the spectrum have also been synthesized as anthraquinone heterocycles. These include derivatives of the greenish yellow dye obtained from 5-(3-chloropropionamido)pyrazoleanthrone (61), and the bright yellow dye of comparable structure (62) (39).

(62) [53057-10-6]

This dye has demonstrated very good fastness to daylight, sublimation, laundering, and crocking.

Vat Dyes

Despite their relatively high cost and difficult methods of application, anthraquinone vat dyes are one of the most important groups of synthetic dyes because of their all-around superior fastness.

One of the earliest of these dyes was blue vat indanthrone (84) (CI Vat Blue 4). It was prepared in 1901 by the cyclization of β-aminoanthraquinone and chloroacetic acid with caustic alkali, in an attempt to obtain an anthraquinone analogue of indigo. Subsequently, a wide range of shades have become available in this series. The anthraquinoids are rich in blues, greens, browns, khakis, and blacks, but a serious defect of many of the yellow and orange dyes is their property of accelerating the degradative action of light and bleaching agents on cellulose. The vats undergo characteristic coloration in sulfuric acid and nitric acid, as well as in concentrated sulfuric acid upon the addition of small amounts of potassium persulfate, or divanadyl trisulfate (40). The dyed fiber, treated successively with acidified permanganate and hydrogen peroxide, undergoes color changes that are useful as supplementary tests, particularly for certain groups of dyes such as the halogenated indanthrone and benzanthrone derivatives.

The shade of many of these dyes also tends to change by hot-pressing and water-spotting, but the change is reversible and the original color can be restored. These changes have been attributed to a variety of reactions including hydration and dehydration, changes in the state of aggregation of the pigment and oxidation (41). Furthermore, the effects of each process can be reduced by vigorous soaping of the dyed material, which removes surface color and assists in crystallization of the dye within the fiber.

An important factor influencing the ease of vatting and affinity for the fabric is the physical form of an anthraquinone vat dye. A simple and effective method of producing vat dyes in a very highly dispersed form is to oxidize a solution of the leuco salt or the leuco sulfuric ester in the presence of a protective colloid. For high tinctorial power, vat dyes may be prepared by adding a surface-active reagent such as an alkylnaphthalenesulfonic acid, diethylene glycol, or various substances (see Surfactants and detersive systems). Such agents improve solubility, assist in the vatting process, increase penetration, and produce higher color value as well as brighter, clearer prints.

The anthraquinone vat dyes are frequently sold in the insoluble, oxidized form, sometimes as dry solids but more often as aqueous pastes. They are applied in the reduced form as vats, the reduction being effected by the use of sodium hydrosulfite under strongly alkaline conditions. As a consequence of this alkalinity, the use of vat dyes is mainly confined to cellulosic fibers, but under suitable conditions they may also serve as pigments.

As mentioned above, these dyes are applied to cotton and cellulose fibers from a hydrosulfite–caustic soda vat (42). The alkali concentration is perhaps the most significant factor in view of its effect on the fiber and its influence on the rate of absorption of the vatted dye, but temperature must also remain a consideration.

During the oxidation of the leuco derivative to the vat dye on the fiber, by means of air or other oxidizing agents, severe degradation of the cellulose can occur as mentioned earlier with respect to the yellow and orange dyes. This is mainly true of the photochemically active vat dyes, but under certain conditions of pH and redox potentials, any anthraquinoid vat dye can exhibit this undesirable property. The oxidation stage of the vat dyeing process must, therefore, be handled with care if an oxidizing agent other than atmospheric oxygen is employed.

The anthraquinone vat dyes are, in general, difficult to obtain in analytically pure form because of their high molecular weights and sparing solubility in organic solvents. However, procedures such as precipitation from sulfuric acid, vatting and reoxidation and oxidation with sodium hypochlorite or with dichromate and sulfuric acid, are useful methods of purification. In addition, many may be crystallized from high boiling solvents such as nitrobenzene, o-dichlorobenzene, trichlorobenzene, phenol, cresols, o-chlorophenol, and quinoline.

The fastness of the dyes to alkali renders them suitable for coloring paper pulp (eg, for soap wrappers and printed wall papers), washable distempers and cement. As a consequence of their excellent lightfastness and great stability, the anthraquinone vat dyes are of interest as pigments, but their use is limited by the cost as well as the relatively low covering power of many of them. They are useful in coloring plastics (including rubber) since they can withstand the temperatures used in molding and the chemicals used in vulcanization. They may be used for producing photographic prints on textiles by padding them successively with a dispersion of a vat dye and a solution of ferric ammonium citrate. Vat dyes manufactured primarily for dyeing and printing textiles, as well as vat dyes made especially for use as pigments have been employed also in the coloration of paints, varnishes, enamels and like materials. The more important, commercially manufactured anthraquinone vat dyes are given in Table 3 (see Colorants for plastics).

Acylaminoanthraquinones. The substance 1-aminoanthraquinone, although it is colored, has found no real use as a dye because its leuco compound has little or no affinity for cotton and the reddish dyeing obtained with it is weak and unsatisfactory. At the next level of complexity among the anthraquinone–vat dyes are the acylaminoanthraquinones. They are noted for their simplicity both of structure and method of preparation. As a group they are easily vatted and, in general, give level shades. Furthermore, they have good fastness to light and chemical agents and their substantivity has been found to increase with increases in molecular size.

Although the aliphatic carboxylic acids are unable to impart the necessary intensity of color and substantivity for technically valuable dyes, almost all of the benzoylated α-aminoanthraquinones can be used practically. The most useful dyes of this type, however, have been derived from diamino- and tetraaminoanthraquinones or from 1-aminoanthraquinone condensed with an aromatic dicarboxylic acid.

The benzamido group has found a major role among the anthraquinone vat dyes. In addition to its use with simple aminoanthraquinones, it has been found sufficient in preparing vat dyes from the more complex anthraquinone derivatives such as the pyrimidoanthrones. Furthermore, the introduction of a benzamido group provides a simple and convenient method for modifying the shades and improving the dyeing properties of other types of anthraquinone vat dyes. The range of shades covered is limited mainly to yellows and reds except for two important violet dyes.

Among some of the simpler examples of the anthraquinone vat dyes, aside from benzamidoanthraquinone itself, are those prepared from substituted acid derivatives. Dye (**63**) has been derived by the condensation of a 2-hydroxy-3-naphthoic acid derivative with 1-(*p*-aminobenzamido)anthraquinone. This bright, lemon-yellow vat dye is of particular interest because it is capable of coupling with diazonium salts.

In the category of substituted diaminoanthraquinone vat dyes, the benzamido derivatives also maintain an important position. 1,4-Dibenzamidoanthraquinone is CI Vat Red 42 (**64**), which has excellent lightfastness and very good fastness to chlorine. Similarly CI Vat Yellow 3 (**65**) is obtained by benzoylating 1,5-diaminoanthraquinone. The customary method of commercial preparation for both these dyes is treatment of the aminoanthraquinone with benzoyl chloride in boiling nitrobenzene or *o*-dichlorobenzene. In certain procedures, acid binding agents such as sodium acetate may be included. Other routes for synthesis include heating the aminoanthraquinone with an ester or amide, or condensing a halogenated anthraquinone with an acid amide in the presence of a copper catalyst and sodium acetate. As an example, pure α-benzamidoanthraquinone can be obtained by heating α-chloroanthraquinone with benzamide in the presence of potassium carbonate and copper(II) bromide (43).

By the addition of another benzamido group to the 2-position of dye (**64**), (**66**) is obtained. The hypsochromic effect of the group results in an orange dye. If a third benzamido group is introduced to the 5-position of a 1,4-bisbenzamidoanthraquinone, the shade created is a bluish red. The resulting dye is CI Vat Red 19 (**67**).

A variety of agents have been employed in order to couple aminoanthraquinones. Many of these agents are dicarboxylic acid derivatives which act by simultaneously acylating two molecules of an aminoanthraquinone. For example, by condensing two moles of α-aminoanthraquinone with succinic acid, the greenish-yellow dye (**68**) (CI 65400) is obtained. Likewise, CI Vat Yellow 26 (**69**) and dye (**70**), which have gained recognition as being the only pure yellow dyes among the acylamidoanthraquinones, are derived from the condensation of 1-aminoanthraquinone with isophthalic acid and terphthalic acid, respectively.

The diaminoanthraquinones have also been used in the preparation of such compounds. The symmetrical CI Vat Yellow 12 (**71**) is obtained by condensing 1-amino-5-benzamidoanthraquinone with oxalyl chloride in nitrobenzene. The compound dyes reddish-yellow from a gray–green vat.

A rather interesting dye of this type is CI Vat Yellow 10 (**72**) which can be synthesized by condensation of two moles of 1-amino-5-benzamidoanthraquinone with one mole of 2-azodiphenyl-4′,4″-dicarboxylic acid. Its lightfastness is better than that of the other yellow acylamidoanthraquinones but the dye has little commercial value because of its high cost of preparation. A notable feature of (**72**) is the stability of its azo group which is unreactive to hydrosulfite and alkali at 60°C.

Aside from the diamide type of linkages used to couple to aminoanthraquinones, the dyes in this series have been prepared by using an aminoanthraquinone-carboxylic

Table 3. Major Commercially Produced Anthraquinone Vat Dyes

				1976 U.S. sales		1976 Import, t
CI Name	CAS Registry No.	CI No.	Method of preparation or structure	Metric tons	Thousands of dollars	
Yellows						
Vat Yellow 1	[475-71-8]	70600	**(86)**			2.00
Vat Yellow 1 Sol.	[6487-09-8]	70601	treat leuco CI Vat Yellow 1 with chlorosulfonic acid in the presence of pyridine; add to alkali, filter the insolubles, remove pyridine and salt out			
Vat Yellow 2	[129-09-9]	67300	**(100)**	179	686	1.60
Vat Yellow 2 Sol.	[13109-68-7]	67301	prepared as CI Vat Yellow 1 with the leuco sulfuric ester of 67300			
Vat Yellow 3	[82-18-8]	61725	**(65)**			
Vat Yellow 3 Sol.	[6711-83-7]	61726	prepared as CI Vat Yellow 1 with the leuco sulfuric ester of 61725			9.30
Vat Yellow 4	[128-66-5]	59100	**(120)**			
Vat Yellow 4 Sol.	[3564-70-3]	59101	prepared as CI Vat Yellow 1 with the leuco sulfuric ester of 59100			
Vat Yellow 7 Sol.	[3956-62-5]	60531	condense 1-aminoanthraquinone with 4-biphenylcarboxylic acid in the presence of thionyl chloride; prepare the solubilized form as in CI Vat Yellow 1 with the leuco sulfuric acid			0.27
Vat Yellow 8 Sol.	[6535-53-1]	60605	couple diazotized 1-aminoanthraquinone with *p*-chloro acetoacetanilide and convert to the disodium salt of the disulfuric ester of the leuco			
Vat Yellow 9	[6370-85-0]	66510	heat 2-methylanthraquinone with benzidine and sulfur at 250°C			
Vat Yellow 10	[2379-76-2]	65430	**(72)**			
Vat Yellow 11	[6451-11-2]	70405	treat 2-amino-1-mercaptoanthraquinone with terephthaloyl chloride			
Vat Yellow 12	[6370-75-8]	65405	**(71)**			
Vat Yellow 13	[6417-50-1]	65425	condense 1-amino-5-benzamidoanthraquinone with terephthaloyl chloride			
Vat Yellow 18	[6370-83-8]	66500	fuse 2-methylanthraquinone with *p*-phenylenediamine and sulfur			
Vat Yellow 20	[4216-01-7]	68420				19.10
Vat Yellow 21	[1328-40-1]	69705	heat 2-(chloromethyl)anthraquinone with sulfur at low temperatures and treat the product with sodium hypochlorite			

Name	CAS	CI	Note	Value1	Value2
Vat Yellow 26	[3627-47-2]	65410	(69)		
Vat Yellow 28	[4229-15-6]	69000	(90)		
Vat Yellow 29	[4216-00-6]	68400	(107)		
Vat Yellow 31	[6871-93-8]	68405	(108)		
Vat Yellow 33	[12227-50-8]				
Vat Yellow 44	[70356-08-0]	59110	chlorinate and then brominate CI Vat Yellow 4 in the presence of iodine		6.90
Vat Yellow 45 Sol.	[12213-66-0]				0.02
Vat Yellow 46	[12237-50-2]				3.60
Vat Yellow 47 Sol.	[12226-71-0]				0.13
Oranges					
Vat Orange 1	[1324-11-4]	59105	(121)		
Vat Orange 1 Sol.	[1324-15-8]	59106	prepared as CI Vat Yellow 1 with the leuco sulfuric ester of 59105		0.02
Vat Orange 2	[1324-35-2]	59705	(89)	336	47.20
Vat Orange 2 Sol.	[1324-37-4]	59706	prepared as CI Vat Yellow 1 with the leuco sulfuric ester of 59705		
Vat Orange 3	[4378-61-4]	59300	(124)		
Vat Orange 3 Sol.	[10290-03-6]	59301	the disodium or aliphatic amine salt of the leuco sulfuric acid of 59300		
Vat Orange 4	[1324-33-0]	59710	brominate pyranthrone in nitrobenzene or chlorosulfonic acid		
Vat Orange 4 Sol.	[1324-36-3]	59711	the leuco sulfuric ester of 59710		
Vat Orange 7	[4424-06-0]	59700	(88)		7.40
Vat Orange 9	[128-70-1]	59701	the leuco sulfuric ester of 59700		
Vat Orange 9 Sol.	[70356-06-8]	70805	(98)		
Vat Orange 11	[2172-33-0]	70806	the leuco sulfuric ester of 70805		
Vat Orange 11 Sol.	[10169-29-6]	67820	(133)	100	
Vat Orange 13	[6417-38-5]	69025	(92)		0.78
Vat Orange 15	[2379-78-4]	69540	condense 1-chloro (or nitro)-2-anthraquinonecarboxylic acid with 2-aminoantrhaquinone and cyclize	947	0.34
Vat Orange 16	[10142-57-1]				
Vat Orange 17	[6370-77-0]	65415	condense isophthaloyl chloride with 1-amino-4-benzamidoanthraquinone and 1-amino-5-benzamidoanthraquinone		
Vat Orange 18	[6370-81-6]	65705	condense cyanuric chloride with 2 moles 1-amino-4-methoxyanthraquinone		
Vat Orange 19	[1324-02-3]	59305	(123)		
Vat Orange 20	[6370-69-0]	65025	(80)		

231

Table 3 (*continued*)

CI Name	CAS Registry No.	CI No.	Method of preparation or structure	1976 U.S. sales Metric tons	1976 U.S. sales Thousands of dollars	1976 Import, t
Vat Orange 21	[1328-39-8]	69700	heat 2-methylanthraquinone with sulfur at high temperature; treat the product with sodium hypochlorite with or without pretreatment with sulfuric acid			
Reds						
Vat Red 10	[2379-79-5]	67000	condense 1-nitro-2-anthraquinonecarbonyl chloride with 2-amino-3-hydroxyanthraquinone and ring close to the oxazole; reduce the nitro group			32.30
Vat Red 10 Sol.	[10126-90-6]	67001	the leuco sulfuric ester of 67000			
Vat Red 13	[4203-77-4]	70320				0.06
Vat Red 14	[8005-56-9]	71110	heat 1,4,5,8-naphthalenetetracarboxylic acid with *o*-nitroaniline, reduce the nitro group and cyclize (**105**)			
Vat Red 15	[4216-02-8]	71100	treat CI Vat Red 14 with alcohol and caustic potash, filter, distill the alcohol and acidify			5.60
Vat Red 18	[6409-68-3]	60705	treat 1-amino-2-anthraquinonecarboxaldehyde or anthraquinone 1,2-isoxazole with hydrazine hydrate (**67**)			
Vat Red 19	[4392-71-6]	64015				
Vat Red 20	[6371-49-9]	67100	condense 2-amino-3-chloroanthraquinone with 1-nitro-2-anthraquinonecarboxylic acid, reduce, convert to the mercapto derivative, and cyclize			
Vat Red 23	[5521-31-3]	71130	fuse *N*-methylnaphthalimide with caustic alkali			
Vat Red 28	[6370-82-7]	65710	condense 2-amino-4,6-dichloro-1,3,5-triazine with two moles of 1-amino-4-methoxyanthraquinone			
Vat Red 29	[6424-77-7]	71140	condense 3,4,9,10-perylenetetracarboxylic acid with *p*-anisidine			
Vat Red 32	[2379-77-3]	71135	condense 3,4,9,10-perylenetetracarboxylic acid with *p*-chloroaniline (**127**)			
Vat Red 35	[3737-76-6]	68000				
Vat Red 37	[70356-07-9]	59310	treat dibromoanthrone (CI Vat Orange 3) with iodine and arsenic pentoxide (**131**)			6.00
Vat Red 38	[6219-99-4]	67810				

232

Vat Red 39	[6373-40-6]	67900	convert 2-methyl-1-nitroanthraquinone to 1-amino-2-anthraquinonecarboxylic acid, condense with aniline, cyclize and chlorinate	
Vat Red 40	[3333-61-7]	68300	condense 1,2-(dihydroxypyrazino)anthraquinone with *m*-toluidine	
Vat Red 42	[2987-68-0]	61650	(64)	
Vat Red 48	[4478-06-2]	65205	(83)	
Violets				
Vat Violet 1	[1324-55-6]	60010	(115)	
Vat Violet 1 Sol.	[1324-57-8]	60011	the leuco sulfuric ester of 60010	
Vat Violet 6 Sol.	[1324-18-1]	59316	ring-close dimethoxy(1,1′-binaphthalene)-8,8′-dicarboxylic acid; make the leuco sulfuric ester of this	
Vat Violet 7 Sol.	[1324-23-8]	59321	ring-close diethoxy(1,1′-binaphthalene)-8,8′-dicarboxylic acid; make the leuco sulfuric ester of this	
Vat Violet 9	[1324-17-0]	60005	(116)	
Vat Violet 10	[128-64-3]	60000	(114)	14.40
Vat Violet 13	[4424-87-7]	68700	(132)	23.20
Vat Violet 14	[6373-31-5]	67895	(129)	
Vat Violet 15	[6370-58-7]	63355	(78)	
Vat Violet 16	[4003-36-5]	65020	condense 1-amino-4-benzamidoanthraquinone with 2-chloroanthraquinone in naphthalene in the presence of copper powder and sodium acetate	
Vat Violet 17	[3076-87-7]	63365	(79)	
Blues				
Vat Blue 4	[81-77-6]	69800	(84)	92.90
Vat Blue 4 Sol.	[2747-19-5]	69801	the leuco sulfuric ester of 69800	
Vat Blue 6	[130-20-1]	69825	(85)	59.50
Vat Blue 6 Sol.	[2519-28-0]	69826	the leuco sulfuric ester of 69825	
Vat Blue 7	[6505-58-4]	70305	condense 3-mercaptobenzanthrone with chloroacetic acid, fuse with caustic potash, and oxidize	
Vat Blue 7 Sol.	[10126-85-9]	70306	the leuco sulfuric ester of 70305	
Vat Blue 10	[8005-28-5]	69830	chlorinate indanthrone with chlorine gas in the presence of sulfuric acid and manganese dioxide	

233

Table 3 (continued)

CI Name	CAS Registry No.	CI No.	Method of preparation or structure	1976 U.S. sales Metric tons	1976 U.S. sales Thousands of dollars	1976 Import, t
Vat Blue 11	[130-19-8]	69815	condense 2-amino-1,3-dibromoanthraquinone in the presence of cupric chloride			
Vat Blue 12	[1324-28-3]	69840	treat indanthrone with sulfuric acid in the presence of boric acid			
Vat Blue 13	[6871-71-2]	69845	condense 1-amino-2-bromo-4-hydroxyanthraquinone in presence of copper or cupric chloride in an inert solvent (eg, naphthalene)			5.50
Vat Blue 14	[1324-27-2]	69810	chlorinate indanthrone with sulfuryl chloride			
Vat Blue 15	[1328-43-4]	70010	treat CI Vat Blue 11 with formaldehyde			
Vat Blue 16	[6424-76-6]	71200	add 16,17-dihydroxyviolanthrone to ethylene dibromide or 2-chloroethyl p-toluenesulfonate in the presence of trichlorobenzene			44.60
Vat Blue 18	[1324-54-5]	59815	chlorinate violanthrone with the introduction of at least three chlorine atoms			17.60
Vat Blue 19	[1328-18-3]	59805	(113)			14.90
Vat Blue 20	[116-71-2]	59800	condense 9-anthrol with glycerol in the presence of sulfuric acid; fuse the benzanthrone formed with potassium hydroxide			8.10
Vat Blue 21	[6219-97-2]	67920	condense 1-amino-4-bromo-2-anthraquinonesulfonic acid with 2-amino-4-trifluoromethylbenzoic acid. Cyclize, simultaneously eliminating the sulfonic acid group, and benzoylate			0.25
Vat Blue 22	[6373-20-2]	59820	(112)			
Vat Blue 25	[6247-39-8]	70500	condense 3-bromobenzanthrone with anthrapyrazole and melt the product with alcoholic caustic potash			0.29
Vat Blue 26	[4430-55-1]	60015	(117)			
Vat Blue 28	[1328-42-3]	70005	condense indanthrone with paraformaldehyde			
Vat Blue 30	[6492-78-0]	67110	(102)			
Vat Blue 31	[6371-50-2]	67105	condense 1-amino-4-nitro-2-anthraquinonecarboxylic acid with 2-amino-3-chloroanthraquinone; form the mercaptan; cyclize; reduce; benzoylate			
Vat Blue 32	[6219-98-3]	67910	condense 1-amino-4-bromo-2-anthraquinonesulfonic acid with 2-amino-4-chlorobenzoic acid and cyclize, simultaneously eliminating the sulfonic acid group			

Name	CAS	CI	Method	n1	n2	n3
Vat Blue 33	[4215-99-0]	67915	(130)			
Vat Blue 51	[1328-20-7]	59856	treat violanthrone with hydroxylamine in conc. sulfuric acid			
Vat Blue 53	[1324-70-5]	71205	dibrominate CI Vat Blue 16 in chlorosulfonic acid			51.60
Vat Blue 66	[57456-24-3]					
Greens						
Vat Green 1	[128-58-5]	59825	(118)	487	2,034	63.50
Vat Green 1 Sol.	[2538-84-3]	59826	the leuco sulfuric ester of 59825			
Vat Green 2	[25704-81-8]	59830	brominate CI Vat Green 1 in oleum in the presence of sodium nitrite			
Vat Green 2 Sol.	[1324-72-7]	59831	the leuco sulfuric ester of 59830			
Vat Green 3	[3271-76-9]	69500	(96)	836	3,477	12.20
Vat Green 3 Sol.	[4471-37-8]	69501	the leuco sulfuric ester of 69500			
Vat Green 4	[6472-75-9]	59835	oxidize benzanthrone with manganese dioxide in sulfuric acid; condense with alcoholic caustic potash and ethylate			
Vat Green 4 Sol.	[10126-86-0]	59836	the leuco sulfuric ester of 59835			
Vat Green 5	[1328-37-6]	69520	condense 3-bromobenzanthrone with 1,5-diamino-anthraquinone in boiling naphthalene with copper oxide and sodium carbonate; cyclize; chlorinate; benzoylate			
Vat Green 6	[6871-70-1]	69855	condense 1-amino-4-anilino-2-bromoanthraquinone in presence of a copper catalyst and an acid binding agent			
Vat Green 7	[1328-12-7]	58825	heat anthracene with sulfur			
Vat Green 8	[14999-97-4]	71050				
Vat Green 9	[6369-65-9]	59850	nitrate a suspension of violantrone in glacial acetic acid or 40% chloroacetic acid			
Vat Green 11	[1328-41-2]	69850	condense 1,4-diamino-2,3-dichloro (or dibromo) anthraquinone in presence of a copper catalyst and an acid binding agent			
Vat Green 12	[6661-46-7]	70700	(134)			
Vat Green 14	[1328-13-8]	58830	treat anthracene with sulfur chloride at elevated temperatures			
Vat Green 17	[6871-80-3]	69010	condense 4,4'-diamino-1,1'-dianthrimide with 2 moles of 2-anthraquinonecarbonyl chloride; cyclize			
Vat Green 21 Sol.	[6054-59-7]	59051	oxidize hexachloropyrene with nitric acid to form 3,5,8,10-tetrachloro-1,6-pyrenedione and condense with aniline; form the leuco sulfuric ester by treating with chlorosulfonic acid in the presence of pyridine			
Vat Green 28	[61931-58-6]					0.17
Vat Green 48	[69912-82-9]					0.85

Table 3 (*continued*)

CI Name	CAS Registry No.	CI No.	Method of preparation or structure	1976 U.S. sales Metric tons	1976 U.S. sales Thousands of dollars	1976 Import, t
Browns						
Vat Brown 1	[2475-33-4]	70800	(97)			1.60
Vat Brown 1 Sol.	[23725-15-7]	70801	the leuco sulfuric ester of 70800			0.02
Vat Brown 3	[131-92-0]	69015	(93)			2.90
Vat Brown 3 Sol.	[10127-24-9]	69016	the leuco sulfuric ester of 69015			
Vat Brown 8	[6487-07-6]	70810	condense 2 moles of 1-benzamido-4-chloroanthraquinone with 1,5-diaminoanthraquinone and cyclize			
Vat Brown 9	[6424-83-5]	71025	treat 1-chloroanthrone with glyoxal sulfate; condense the product with 2 moles of 1-amino-5-benzamidoanthraquinone and cyclize			
Vat Brown 14	[8005-79-6]	71120	condense 1,4,5,8-naphthalenetetracarboxylic acid with 4-ethoxy-*o*-phenylenediamine			
Vat Brown 16	[6449-16-7]	70910	condense 6,7,9,12-tetrachlorophthaloylacridone with 2 moles 1-aminoanthraquinone and cyclize with aluminum chloride			
Vat Brown 22	[8605-30-9]	71115	condense 1,4,5,8-naphthalenetetracarboxylic acid with 4-chloro-*o*-phenylenediamine in glacial acetic acid or pyridine			
Vat Brown 23	[1328-26-5]	65250	treat 2-aminoanthraquinone with sulfuric acid and copper powder			
Vat Brown 25	[6247-46-7]	69020	condense 1-amino-4-benzamidoanthraquinone with 1-benzamido-5-chloro-4-methoxyanthraquinone and cyclize			
Vat Brown 26	[6717-38-0]	70510	condense 1-amino-5-benzamidoanthraquinone with 6,10,12-trichlorophthaloylacridone and cyclize			
Vat Brown 31	[6451-09-8]	70695	condense 2 moles of 1-nitro-2-anthraquinonecarboxylic acid with 1 mole of benzidine and cyclize in sulfuric acid; decarboxylate			
Vat Brown 33	[61902-46-3]	59270	condense the monobromo derivative of CI Vat Yellow 4 with 1-amino-2-anthraquinonecarboxyaldehyde			0.92
Vat Brown 39	[1328-15-0]					
Vat Brown 44	[2475-33-4]	70802	condense 1,4-diaminoanthraquinone with 2 moles 1-chloro (or bromo) anthraquinone and fuse the product with potassium hydroxide			

Vat Brown 45	[6424-51-7]	59500	treat 2-chloroanthrone with glyoxal sulfate; fuse with alcoholic potash; treat with sulfuric acid	
Vat Brown 55	[4465-47-8]	70905	condense 6-bromo-10,12-dichlorophthaloylacridone with 2 moles 1-amino-5-benzamidoanthraquinone and cyclize	
Blacks				
Vat Black 7	[6369-65-9]	59850	nitrate a suspension of violanthrone in glacial acetic acid or 40% chloroacetic acid; the mononitroviolanthrone is reduced in the dye liquor to give the amino compound which may be reoxidized on the fiber to give the black	
Vat Black 8	[2278-50-4]	71000	condense together 3,9-dibromobenzathrone anthrapyrazole, and 1-aminoanthraquinone in presence of caustic potash	
Vat Black 8 Sol.	[34599-04-7]	71001	the leuco sulfuric ester of 71000	
Vat Black 9	[1328-25-2]	65230	condense tetrabromopyranthone in nitrobenzene with aminoviolanthrone and 1-aminoanthraquinone in the presence of copper oxide and sodium acetate	
Vat Black 16	[1328-19-4]	59855	treat violanthrone with hydroxylamine in conc sulfuric acid	
Vat Black 23	[61868-00-6]	65260	fuse 1,5- (or 1,8-) diaminoanthraquinone with potassium hydroxide	
Vat Black 25	[4395-53-3]	69525	condense 2 moles of 1-aminothraquinone with 3,9-dibromobenzanthrone in boiling naphthalene with copper oxide and sodium carbonate; cyclize with caustic potash in boiling isobutyl alcohol	905 3,329
Vat Black 27	[2379-81-9]	69005	(94)	108
Vat Black 28	[128-79-0]	65010	nitrate 1,1'-dianthrimide; reduce with sodium sulfide; benzoylate	731
Vat Black 29	[6049-19-0]	65225	condense 4,10-dibromoanthraquinone with 2 moles of 1-amino-4-benzamidoanthraquinone	

238 DYES, ANTHRAQUINONE

(63) [70321-29-8]

	R	R'	R''
(64)	H	NHCOC₆H₅	H [2987-68-0]
(65)	H	H	NHCOC₆H₅ [82-18-8]
(66)	NHCOC₆H₅	NHCOC₆H₅	H [70321-13-0]
(67)	H	NHCOC₆H₅	NHCOC₆H₅ [4392-71-6]

	R	R'
(68)	—CH₂CH₂—	H [4375-89-7]
(69)	(m-phenylene)	H [3627-47-2]
(70)	(p-phenylene)	H [70321-14-1]
(71)	nothing	NHCOC₆H₅ [6370-75-8]
(72)	—C₆H₄—C₆H₄—N=N—C₆H₄—C₆H₄—	NHCOC₆H₅ [2379-76-2]

acid, with the carboxyl group at positions 2,3- or 6-, or a substituent at these positions to which a carboxyl group is attached, in order to acylate a second-aminoanthraquinone. Examples of such dianthraquinones may possess a simple amide bridge as in dye (73). These 1,4-bisaroylamido-8-aminoanthraquinones form a secondary amide linkage with the carboxyl group of a 1-aminoanthraquinone-2-carboxylic acid molecule, providing brilliant bordeaux shades with fine fastness to light and wet treatment (44). Similarly, red to red–brown dyes have been prepared by condensing 1-amino-4-benzamidoanthraquinone or 1,4-diaminoanthraquinone with 1,4-bisaroylamidoanthraquinone-6-carboxylic acid (45). The anthraquinonyl benzothiazole carboxylic acids, eg, (74) that yield bordeaux shades can also be used in such bridging anthraquinone molecules (46).

Red to violet dyes are obtained by treating a 4,8-diamino-1-aroylamidoanthraquinone with two moles of 1-nitro-, amino-, or chloroanthraquinone-2-carboxylic acid. Likewise, red to orange vat dyes of very good fastness have been synthesized by condensing a heterocylic acid like 1,9,5,10-anthradiisothiazole-3,8-dicarboxylic acid with aminoanthraquinones.

These trimer linkages have also been created through the use of phosgene or cyanuric chloride. In the former, the resulting urea derivatives such as CI Vat Dye (CI 66000) (**75**) are obtained.

In the latter, the cyanuric chloride reacts readily with two moles of an aminoanthraquinone and the third chloride is replaced by amino-, anilino-, or phenoxy groups. Although the third aminoanthraquinone residue can be introduced, it is generally more difficult to add and does not have a proportionate effect on the properties of the dye.

Returning to the simple aroylaminoanthraquinones, the introduction of hydroxyl and methoxy groups to the anthraquinone nucleus produces a marked bathochromic effect. The methoxy compound is the vat dye (CI 60750) (**76**) which dyes scarlet and the 4,5,8-trihydroxy derivative of that nucleus is the vat dye (CI 60770 (**77**).

The introduction of a hydroxyl group to the 8-position, in addition to the one at position 4, gives a violet colored dye as demonstrated by CI Vat Violet 15 (**78**). Furthermore, the shade and fastness properties are improved in such dyes by replacing the benzamido groups with anisic acid amides. This is illustrated by a second important violet dye CI Vat Violet 17 (CI 63365) (**79**), the dianisoyl derivative of 4,8-diaminoanthrarufin.

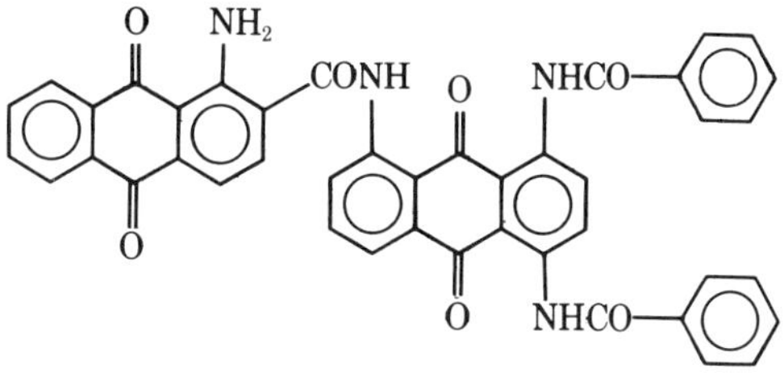

(**73**) [70321-30-1]

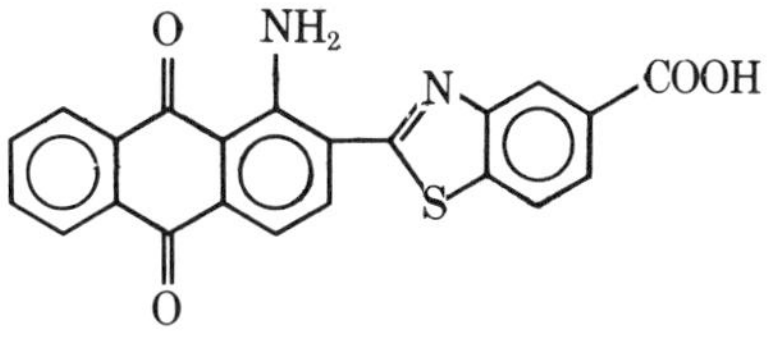

(**74**) [70321-31-2]

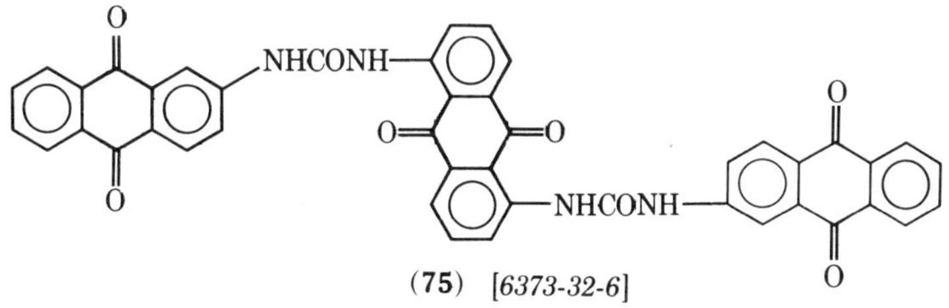

(**75**) [6373-32-6]

The fastness to washing and to chlorine of these dyes is below the standard of other anthraquinone vat dyes but they have excellent fastness to light. The hydroxyl groups are believed to be responsible for the lower fastness to alkali of these dyes.

Anthrimides. In general, α,α-dianthraquinonylamines exhibit poor affinity for fiber and are not fast to light when employed as vat dyes. α,β-Condensates, on the other hand, yield very satisfactory vat dyes. The anthrimides dye somewhat dull shades of orange, red, bordeaux, and gray and display excellent fastness properties.

The linkage between anthraquinone nuclei may be either a direct —NH— type,

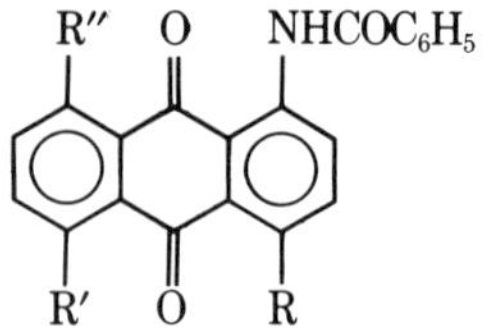

240 DYES, ANTHRAQUINONE

	R	R'	R''
(76)	OCH₃	H	H [6409-75-2]
(77)	OH	OH	OH [6409-79-6]
(78)	OH	NHCOC₆H₅	OH [6370-58-7]

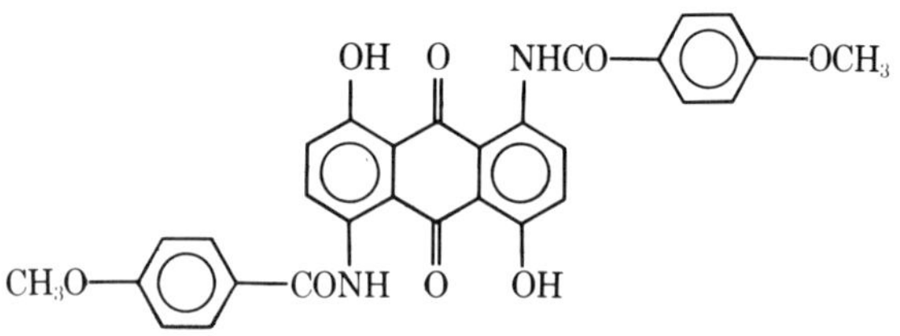

(79) [3076-87-7]

or one which involves two —NH— groups separated by one of several possible residues, frequently aromatic hydrocarbons. The simplest anthrimide used practically is CI Vat Orange 20 (80). It is prepared by the condensation of 1-benzamido-5-chloroanthraquinone with 1-aminoanthraquinone and illustrates the direct —NH— type linkage. The second type of linkage is exemplified by dyes (81) and (82). In neither case does the additional aryl group destroy any of the compound's dyeing character.

More useful dyes have been obtained by introducing benzamido groups to one or both of the anthraquinone nuclei and by increasing the number of anthraquinone residues in order to form trianthrimides. The trianthrimides most widely used in recent years are: di-β-anthraquinonyl-1,5-diaminoanthraquinone [6417-48-7] (Vat Dye CI 65200), di-α-anthraquinonyl-2,6-diaminoanthraquinone (83) (CI Vat Red 48) and di-4'-methoxy-α-anthraquinone-2,6-diaminoanthraquinone [6370-71-4] (Vat Dye CI 65210). All of these are used to dye cotton shades of red from cold, weakly alkaline hydrosulfite vats. They have good all-around fastness but, as with other yellow and red anthraquinone vat dyes, possess the disadvantage of causing the dyed cotton to deteriorate, especially on exposure to sunlight (tendering).

Indanthrones. Indanthrone (CI Vat Blue 4) (84) was discovered by R. Bohn in 1902 and was the first anthraquinone vat dye used for cotton. It is still employed today largely for dyeing cotton but the use of indanthrone blues also includes dyeing and printing on linen and rayon. In addition, they function as pigments in graphic art and in the lacquer industry.

As is customary with the anthraquinone vat dyes, the leuco compound is used for actual dyeing. Although indanthrone is composed of two anthraquinone residues, it requires only two hydrogen atoms (or two reducing equivalents, such as hydrosulfite and alkali) to convert it into a leuco compound. The stable character of the blue dihydro derivative of indanthrone has stimulated special interest. Resonance of the ion through such structures as the three illustrated below have been employed to account for the stability.

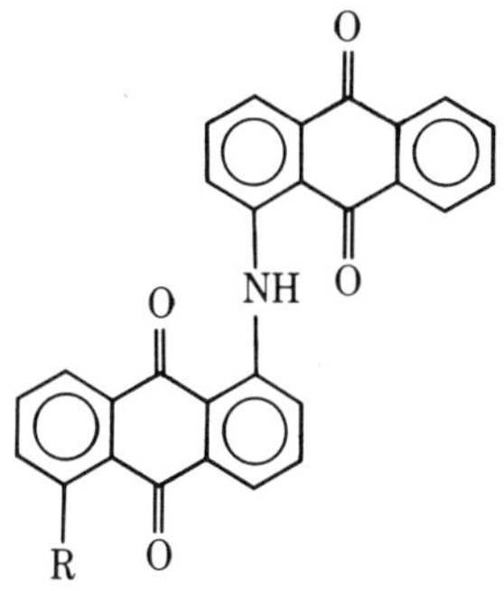

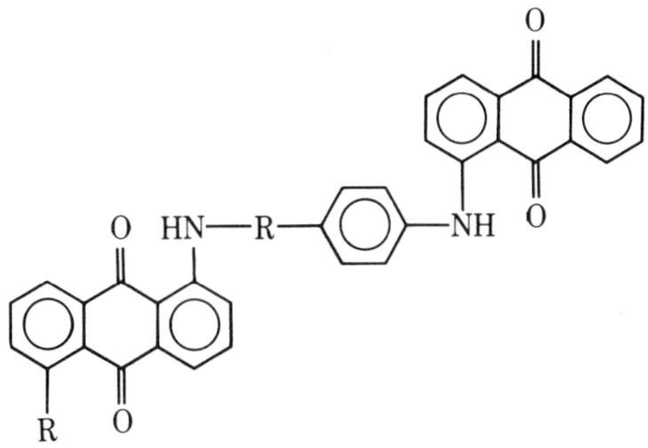

(80) 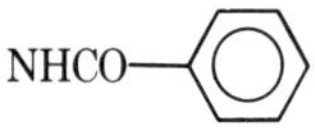 R = NHCO—⟨⟩ [6370-69-0]

(81) R = nothing [70416-85-2]

(82)  R = —⟨⟩— [54939-98-9]

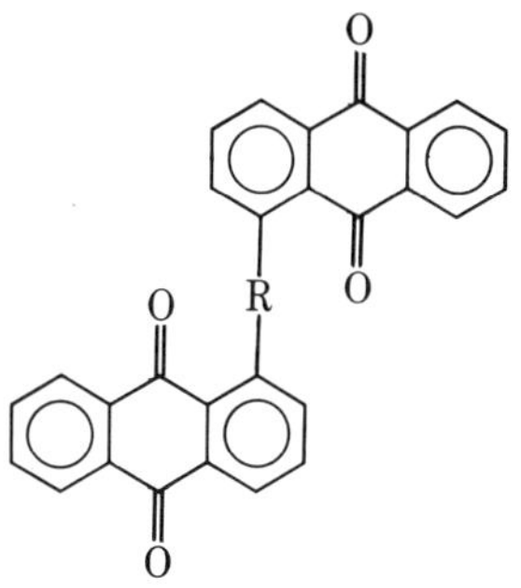

(83) 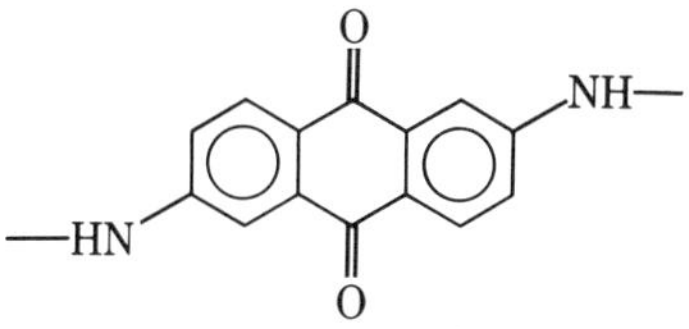 [4478-06-2]

Preparation of the dyeing vat and application of the dye to the fiber necessitates careful control of dye and hydrosulfite–caustic soda concentrations, as well as temperature. The disodium salt of dihydro (or normal leuco) indanthrone is sparingly soluble in vat solution and consequently may separate if its concentration is too high. Temperatures are maintained ca 50–60°C because, if higher, duller shades and lower color values result from over reduction. Excess alkali must also be monitored carefully. If present at the end of the dyeing process after air oxidation has been completed, it must be removed immediately to avoid further oxidation of the dye to the azine with loss of the beautiful reddish-blue shades.

The shades of blue produced by indanthrone are much faster than indigo to light and washing, and in fact, all of the indanthrones possess good lightfastness, as well as fastness to boiling alkali. The imino groups of these dyes are very sensitive to oxidizing agents (eg, hypochlorites and chlorine) since their dihydrazine rings are readily and reversibly oxidized to yellow azine derivatives. Despite the problems associated with indanthrone, however, it is used extensively because of its beautiful shades for which there is no other equivalent available.

In the normal vatting process, only two carbonyl groups are reduced; a more vigorous reduction yields undesired side products. This tendency for further reduction

(84) X = H [81-77-6]
(85) X = Cl [130-20-1]

can be considerably diminished by the introduction of two chloride atoms into the anthraquinone nucleus. Such a chlorination is accomplished by the action of sulfuryl chloride on indanthrone in an inert solvent or by passing chlorine through a solution of indanthrone in concentration sulfuric acid containing a small amount of manganese dioxide. The resulting dye, CI Vat Blue 6 (85) is used in large quantities for the production of bright blue dyes which possess outstanding lightfastness and very good general fastness, even to bleach.

The halogenated indanthrones closely resemble the parent fastness properties. Their preparation is accomplished by halogenation of the parent dye in a variety of media including thionyl chloride, sulfuryl chloride, antimony(II) chloride, bromine, aqua regia, etc, or by construction from suitably substituted anthraquinones as starting materials. Dyeing conditions for these indanthrones requires much more careful control than that used for the parent dyes in order to obtain level shades.

Flavanthrone. The only significant member of this group is flavanthrone itself (CI Vat Yellow 1) (86). It is a by-product of the potash fusion of 2-aminoanthraquinone during the manufacture of indanthrone, but is usually prepared in an Ullmann reaction by heating two equivalents of 2-amino-1-chloroanthraquinone in the presence of copper powder. The resulting dianthraquinonyldiamine is then condensed with two equivalents of benzaldehyde to yield the dianil derivative, or more frequently, with phthalic anhydride to afford the phthalimido derivative (47).

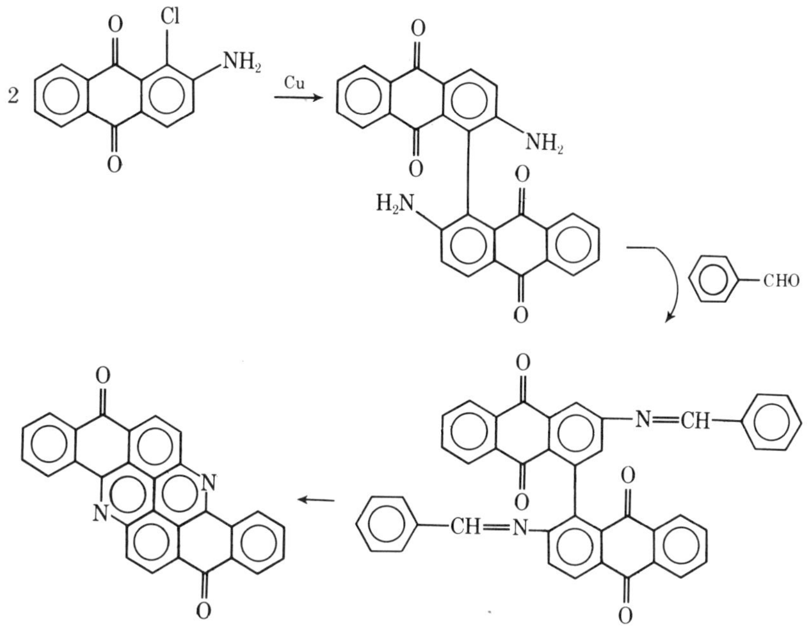

(86) [475-71-8]

Several other methods of preparation have been developed and reported (48).

Flavanthrone dyes yellow–orange shades from a blue alkaline hydrosulfite vat. If more drastic reduction is carried out the result is a brown vat which may undergo facile reoxidation. Flavanthrone serves as a useful yellow component of lightfast greens, olives, etc, and is phototropic, ie, undergoes a change in shade by severe exposure to light. This light change is also readily reversible by contact with air or by a mild soap treatment. In general, flavanthrone is an important dye because it is a nontenderer and has good lightfastness. On the other hand, since it exhibits reducibility by cellulose and boiling alkali, it is not recommended for goods that are bleached after dyeing.

Derivatives of flavanthrone, include a new nitro derivative (49), α-tetrahydro- (87), α-hexahydroflavanthrone, and 3,3′-halo-flavanthrone.

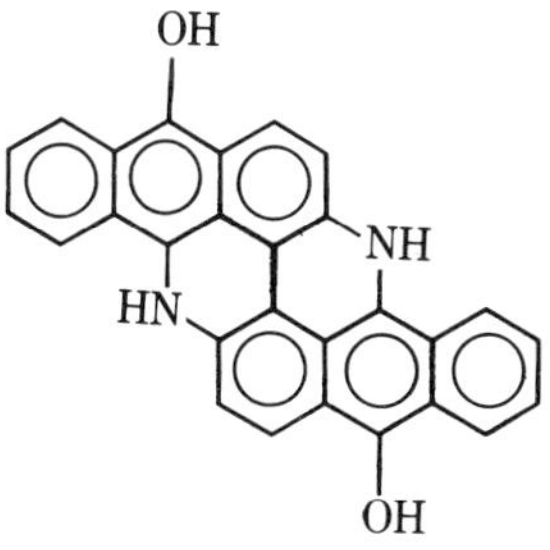

(87) [70321-32-3]

The reduced derivatives, tetra- and hexahydroflavanthrone, are derived from reaction of the parent dye with zinc dust in caustic soda at 70–80°C under an atmosphere of hydrogen. The former is a dark, greenish-blue color; the dihalide derivative has demonstrated improved fastness, but redder and less attractive shades.

Pyranthrone. Pyranthrone is closely related to flavanthrone, with two methine groups in place of flavanthrone's nitrogen atoms. It has gained great commercial importance under the designation CI Vat Orange 9 (**88**).

X
(**88**) H [128-70-1]
(**89**) Br [1324-35-2]

Preparation is accomplished by two possible routes: (*1*) from 2,2′-dimethyl-1,1′-dianthraquinone, by heating with zinc chloride or alcoholic potassium hydroxide (50), or (*2*) condensation of pyrene with benzoyl chloride and aluminum chloride to give dibenzoylpyrene, which upon heating with aluminum chloride provides pyranthrone. This dye produces bright orange shades, which show remarkable fastness to chlorine and alkali washes, from a cherry-red vat.

The action of bromine in chlorosulfonic acid at 60°C with a catalytic amount of iodine on pyranthrone affords 4,12-dibromopyranthrone (CI Vat Orange 2) (**89**). This dye, which produces a reddish-orange shade, exhibits excellent lightfastness and very good all-around fastness.

Anthraquinone Carbazoles. 1,1′-Anthrimide is itself of no value as a vat dye, but its treatment with aluminum chloride or caustic yields the useful carbazole derivative (**90**) (CI Vat Yellow 28). As a dye, this particular heterocyclic compound colors bright orange–yellow and is a tenderer.

As a class these compounds include some of the most important anthraquinone vat dyes. They produce orange, olive, khaki, and brown shades with good overall fastness properties when applied by the hot-dyeing process. These carbazoles have been prepared in an angular form such as (**90**) as well as in a linear arrangement as the *N*-ethyl derivative (CI Vat Dye 69400 (**91**). The latter is obtained by condensing phthalic anhydride with *N*-ethylcarbazole followed by cyclization to the bisanthraquinone. When applied to cotton from a brown vat, it exhibits yellow shades which are extremely fast, especially to light.

Anthrimides containing benzamido groups in the 4- or 5-position, with reference to the imino group, are cyclized much more readily than the parent anthrimides, providing brighter dyes with better tinctorial power. CI Vat Orange 15 (**92**) is one such dye which represents a series of colorants distinguished for their excellent fastness. Other examples are CI Vat Brown 3 (**93**) and CI Vat Black 27 (**94**).

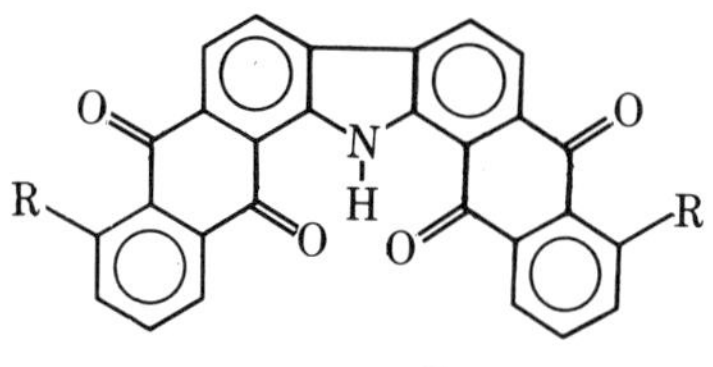

	R	
(90)	H	[4229-15-6]
(92)	NHCO—C₆H₅	[2379-78-4]

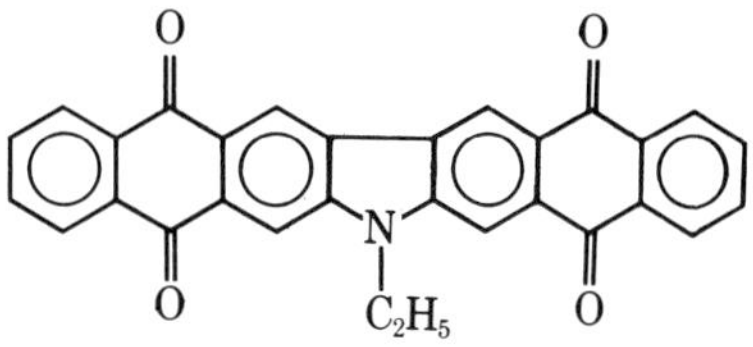

(91) [6871-78-9]

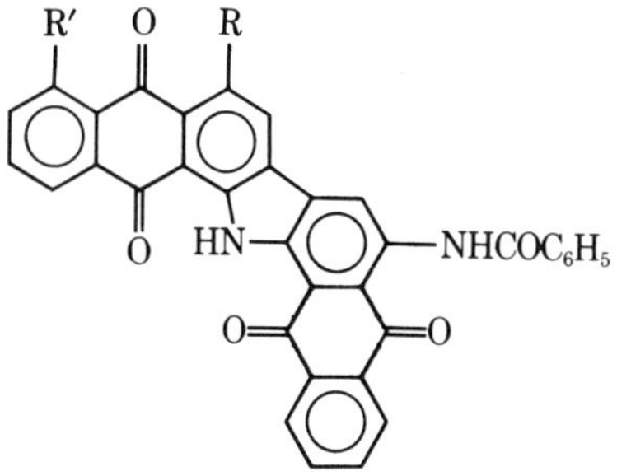

	R	R'
(93)	H	NHCOC₆H₅ [131-92-0]
(94)	NHCOC₆H₅	H [2379-81-9]

Benzamido derivatives which color red–brown to violet, have also been prepared from dyes such as (95). These carbazole vat dyes are derived from two anthraquinone molecules combined with one molecule of a naphthalene derivative such as 2,6-dibromonaphthalene.

The condensation of 3-bromobenzanthrone with 1-aminoanthraquinone in the presence of sodium carbonate in copper oxide in boiling naphthalene followed by ring closure with potassium hydroxide in isobutanol affords CI Vat Green 3 (96). It too has outstanding lightfastness and is used extensively for curtains and furnishings.

Trianthrimides and their benzamido derivatives yield carbazoles which are a useful series of vat dyes. Members of this group include CI Vat Brown 1 (97), CI Vat Brown 44, CI Vat Orange 11 (98) and CI Vat Brown 8. Compound (97) is an important vat dye obtained by the condensation of 1,4-diaminoanthraquinone with two moles

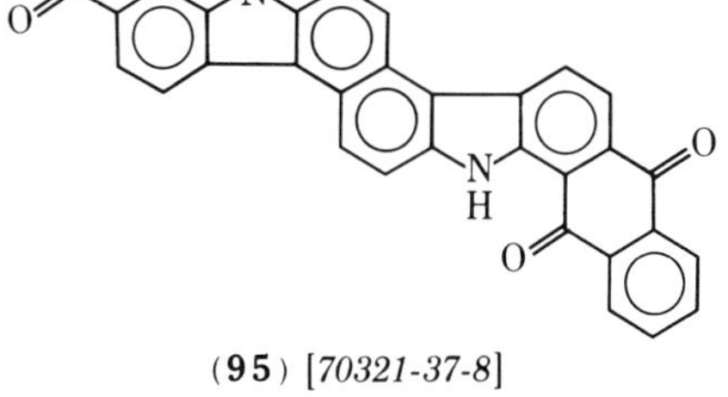

(95) [70321-37-8]

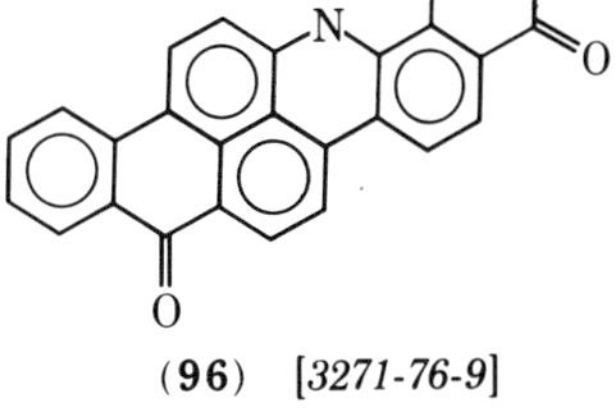

(96) [3271-76-9]

(97) [2475-33-4]   (98) [2172-33-0]

of 1-chloroanthraquinone followed by ring closure in aluminum chloride or a caustic potash melt, and finally oxidation with hypochlorite. CI Vat Brown 8 is the carbazolyation product of a benzamidoanthrimide derived from one mole of 1,5-diaminoanthraquinone condensed with two moles of 1-chloro-4-benzamidoanthraquinone.

An even larger molecule in this class of dyes is CI Vat Green 8 (99) prepared by the carbazolyation of a pentanthrimide (51).

(99) [14999-97-4]

Anthraquinone Thiazoles. Anthraquinones with condensed thiazole rings have found only limited value as vat dyes. The oldest and most important member of this class is CI Vat Yellow 2 (100). It is prepared by reaction of benzotrichloride and sulfur with 2,6-diaminoanthraquinone, and yields bright greenish-yellow shades which are not especially fast to light.

The symmetrically coupled dianthraquinone (CI Vat Dye 70400) (101) is obtained by the action of glyoxal on 2-amino-1-mercaptoanthraquinone in sulfuric acid solution followed by hypochlorite oxidation. It is a tenderer with inferior lightfastness for a vat dye but is used as a shading color because of its facile applicability by all of the vat dyeing methods.

CI Vat Blue 30 (102) is another thiazole dye which demonstrates excellent all-around fastness properties. It possesses a very stable trifluoromethyl group that, as has been found in other dyes, increases the brightness and tinctorial value as well as fastness to light and chlorine.

Anthrapyrazolones. Pyrazolanthrone (103) itself can be converted to a yellow vat dye (CI Vat Dye 70315) (104) by fusion with potassium hydroxide. This dye, possessing two anthraquinone nuclei, dyes, from a blue vat, bright yellow shades which exhibit fastness to chlorine, but has poor fastness to alkali because of its two acidic hydrogen atoms. Replacement of the two acidic hydrogen atoms with ethyl groups affords the more useful dye CI Vat Red 13 (105).

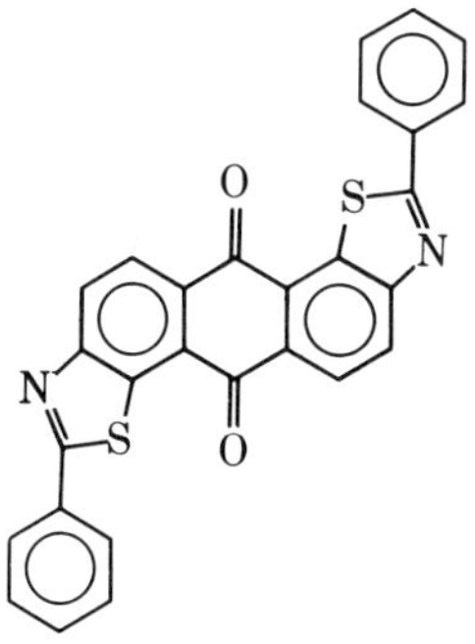

(100) [129-09-9] 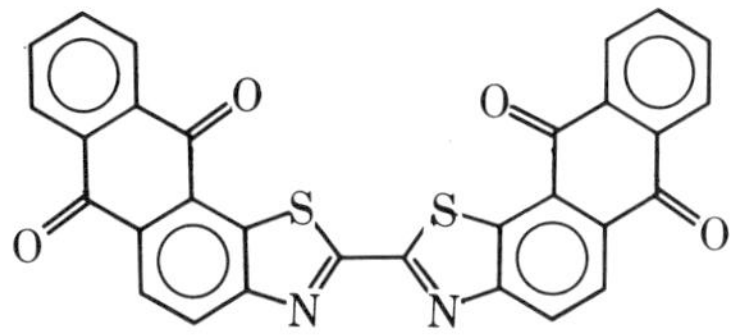
(101) [6451-12-3] 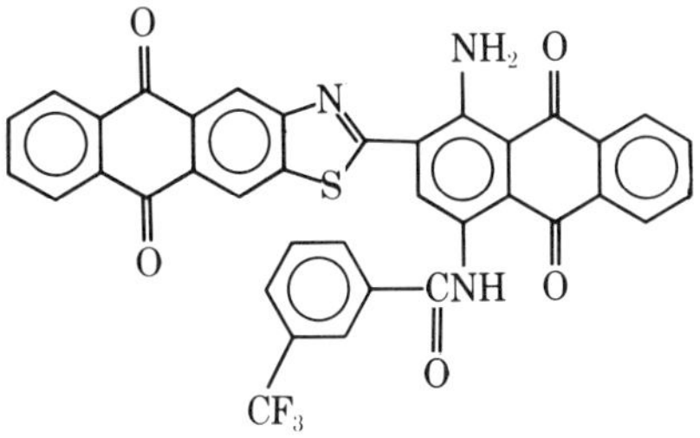
(102) [6492-78-0]

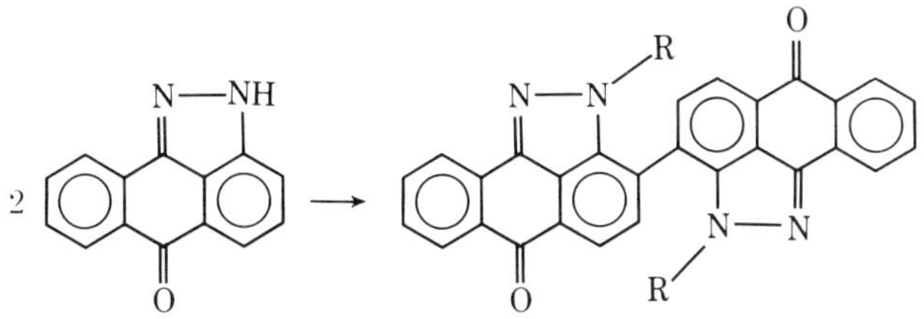

(103) [129-56-6]

(104) H [129-54-4]
(105) CH₂CH₃ [4203-77-4]

If one or both of the alkyl groups are replaced by an alkoxyalkyl group, such as β-methoxyethyl, improved red vat dyes are obtained. The solubility of the corresponding leuco compound tends to be improved as does the fastness to soaping while the colors appear somewhat yellower.

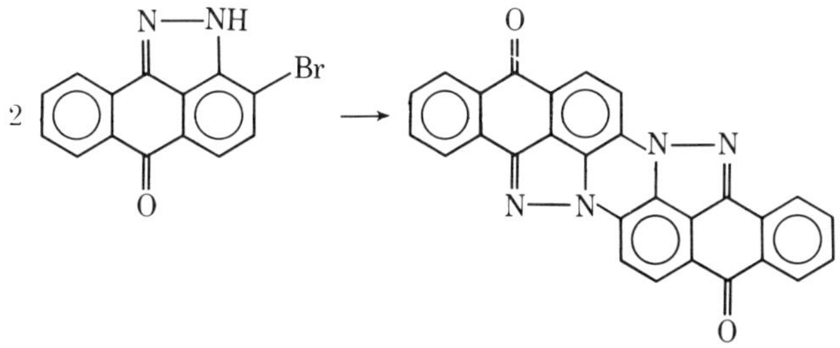

(106) [70321-38-9]

248 DYES, ANTHRAQUINONE

Self-condensation of 2-bromopyrazolanthrone by means of copper and potassium acetate affords the complex polycyclic dye (**106**), which gives fast red shades from a blue vat.

Anthrapyrimidines. Many yellow anthraquinone vat dyes possess tendering properties; dyes of the anthrapyrimidine group are an exception, they are nontendering. The best representatives of this class are CI Vat Yellow 29 (**107**) and CI Vat Yellow 31 (**108**).

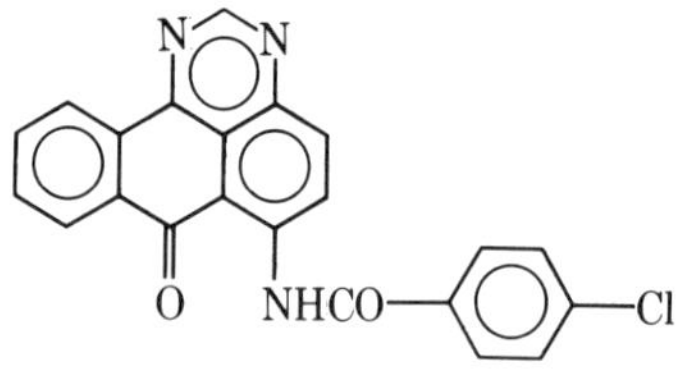

(**107**) [4216-00-6] 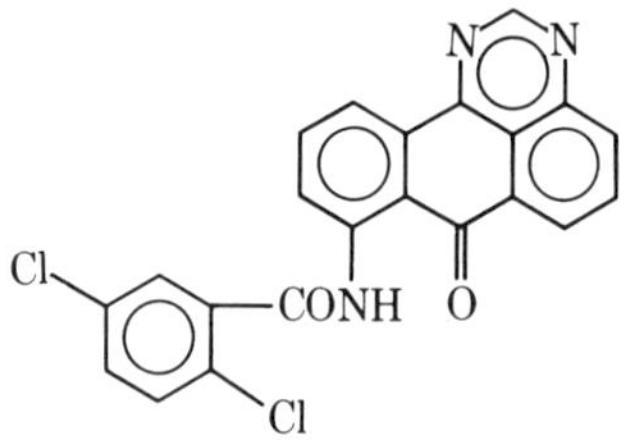   (**108**) [6871-93-8]

As vat dyes, the anthrapyrimidines are unique in their possession of a single carbonyl group. The vattability of the pyrimidanthrone nucleus is owing to the increased acid strength of dihydropyrimidoanthrone by resonance stabilization of the anions:

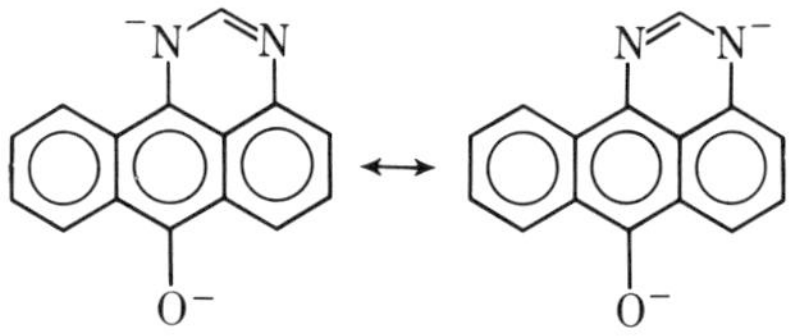

These dyes, therefore, form an alkali-soluble leuco derivative on reduction; when a benzamido group is present, the affinity of the leuco compound for cellulose is adequate for dyeing purposes.

Besides the yellow pyrimidanthrones, a series of violet, blue, gray, and black dyes can be prepared. They are characterized by high tinctorial power, excellent fastness to light and wet treatment, and stability for printing.

Dibenzanthrones. Benzanthrone (**109**) is a yellow compound devoid of dyeing properties. The fusion of two molecules with potassium hydroxide, however, yields a dark blue dyestuff (**110**) of excellent fastness to light, chlorine, and washing. It is used to dye cotton from a strongly alkaline hydrosulfite vat. From a technical and commercial point of view, the series of dibenzanthrones which are derivatives of (**110**), known as violanthrone (CI Vat Blue 20), and its isomer isoviolanthrone (see structure (**114**) are some of the most interesting polycyclic vat dyes available.

Although certain of these dyes are poor levelers, the introduction of halogen atoms can be used to remedy this defect. Examples of such halogenated dibenzanthrones are CI Vat Dye 59810 (**111**), CI Vat Blue 22 (**112**), and CI Vat Blue 19 (**113**); their colors tend to be redder than the parent compound.

If chlorinated benzanthrones are subjected to fusion with potassium hydroxide, the result is a halogen-free vat dye, isomeric with dibenzanthrone which produces decidedly redder colors. The polycyclic dye obtained is isodibenzanthrone.

CI Vat Violet 10 (**114**) or isoviolanthrone, its dichloro derivative CI Vat Violet 1 (**115**), and the monobromo compound, CI Vat Violet 9 (**116**), are all extraordinarily fast dyes.

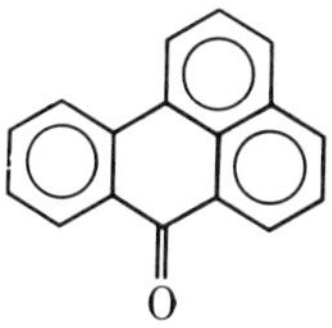

(109) [82-05-3]

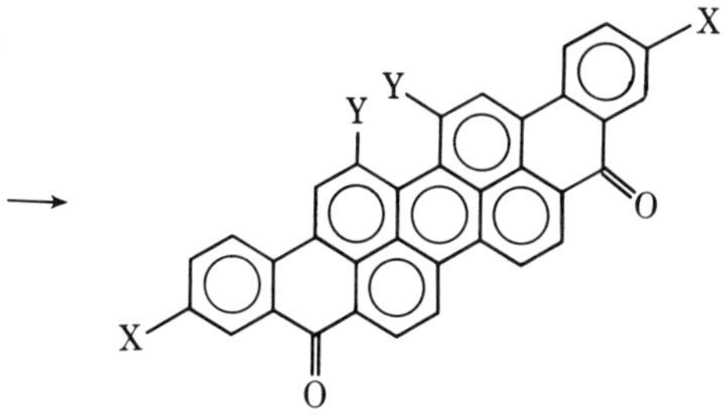

	X	Y
(110)	H	H [116-71-2]
(111)	H	Cl [1324-56-7]
(112)	Cl	Cl [6373-20-2]

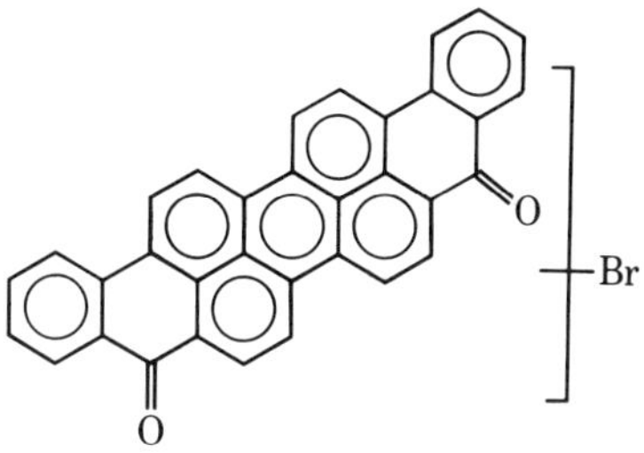

(113) [1328-18-3]

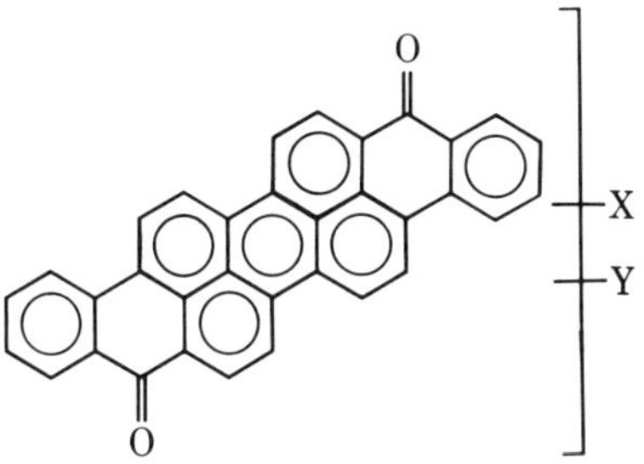

	X	Y
(114)	H	H [128-64-3]
(115)	Cl	Cl [1324-55-6]
(116)	Br	H [1324-17-0]

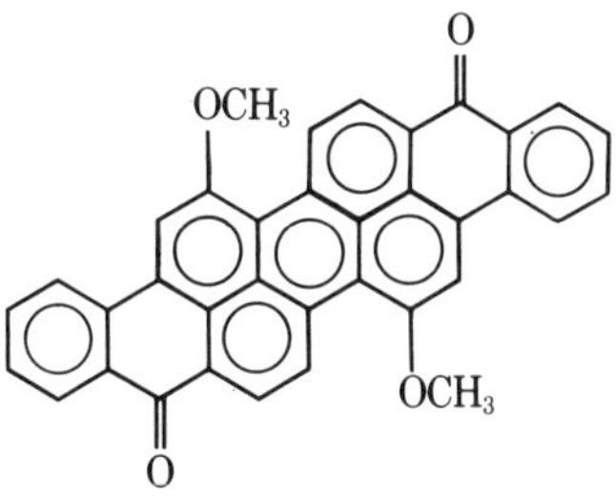

(117) [4430-55-1]

When the isobenzanthrone is oxidized and methylated, 6,15-dimethoxyisoviolanthrone (CI Vat Blue 26) (117) is obtained.

Oxidation with nitric acid or manganese dioxide in concentrated sulfuric acid produces quinones which may be reduced to green dihydroxybenzanthrones. Of special commercial interest is dihydroxybenzanthrone, methylated in a nitrobenzene suspension of methyl sulfate and sodium carbonate. The result is a blueish-green dye (CI Vat Green 1 (118) that dyes cotton from a strongly alkaline blue hydrosulfite vat at 45°C. It is one of the fastest dyes known and is particularly valuable since it fills a gap in the range of vat colors which is weak in satisfactory green shades.

3-Bromobenzanthrones may also be condensed with various aminoanthraquinones

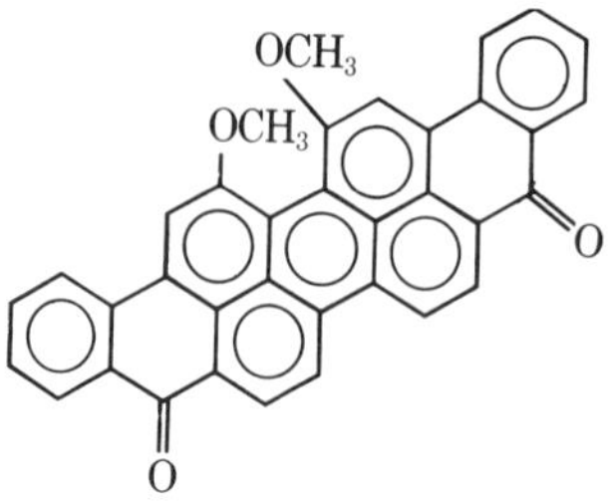

(118)
[128-58-5]

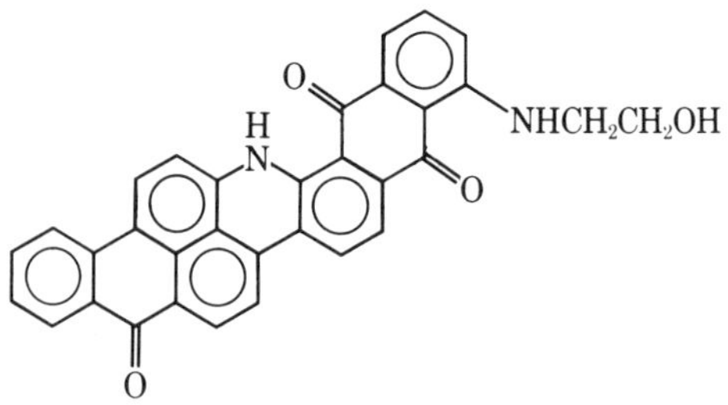

(119)
[62359-42-6]

to yield valuable dyes such as the above mentioned CI Vat Green 3 (96), and (119). Compound (119) is a gray to black vat dye for cotton and is also used as a fast pigment for coatings and plastics (52).

Dibenzopyrenquinones. Dibenzopyrenquinone (CI Vat Yellow 4) (120) is a dye closely related to anthanthrone. It is prepared by application of the Scholl cyclization with aluminum chloride to 3-benzoylbenzanthrone or 1,5-dibenzoylnaphthalene.

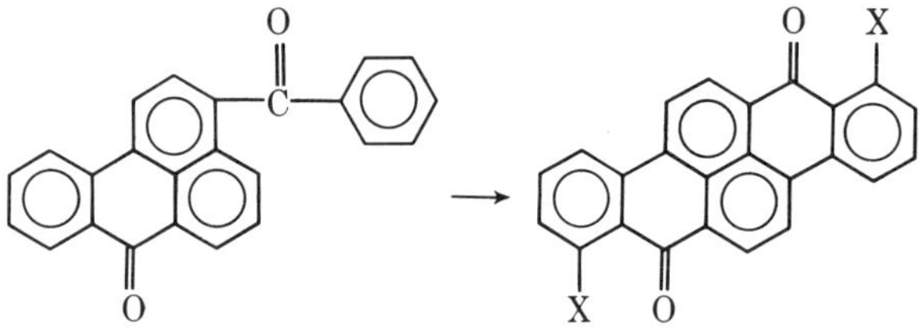

(120) H [128-66-5]
(121) Br [1324-11-4]

Structure (120) exhibits only moderate fastness, but bromination of it results in improved fastness to light, washing, and bleach.

The dibromodibenzopyrenquinone derived from this procedure is CI Vat Orange 1 (121) which provides a redder color than its parent compound.

Anthanthrones. Anthanthrone (122) itself has the properties of a vat dye, but, because of its poor tinctorial power, is of no practical value. The orange compound is synthesized by the dehydrative cyclization of 1,1'-dinaphthyl-8,8'-dicarboxylic acid with sulfuric acid (53).

The dichloro and dibromo derivatives, on the other hand, are valuable level dyes—CI Vat Orange 19 (123) and CI Vat Orange 3 (124). The bright orange shades obtained from a red violet vat have very good lightfastness and good fastness to both chlorine and soda. Their light-tendering activity is low. The dichloro derivative (123) is prepared by cyclizing 1,1'-dinaphthyl-8,8'-dicarboxylic acid with 96% sulfuric acid at 50°C and passing chlorine into the solution in the presence of a small amount of ferrous sulfate.

Anthraquinone Acridones. The anthraquinone acridones are an important group of dyes with relatively simple molecular structure. They cover a wide range of hues including orange, red, violet, blue, brown, and khaki, with good general fastness especially to light. Dyeing is accomplished from a cold, readily soluble vat with good

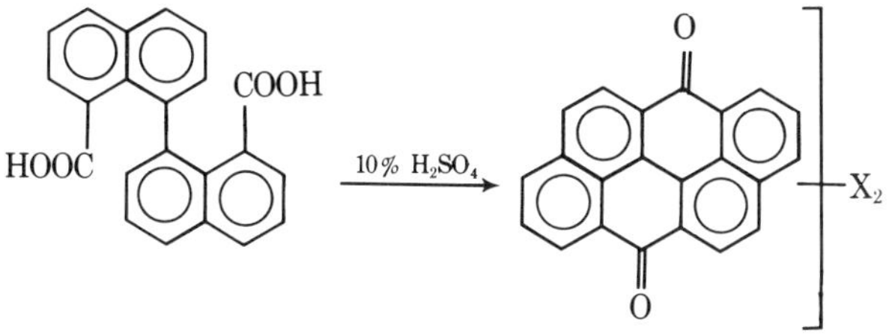

(122) H [641-13-4]
(123) Cl [1324-02-3]
(124) Br [4378-61-4]

substantivity, but the dyes themselves demonstrate only moderate affinity for fabrics.

The main synthetic method for preparation of the 2,1-acridone (125) consists of condensing α-chloroanthraquinone with anthranilic acid in the presence of copper and sodium acetate followed by cyclization with sulfuric acid or aluminum chloride. The isomeric 1,2-acridones (126) have no value as vat dyes.

A simple analogue of acridone with one additional aromatic ring is CI Vat Red 35 (127). Its synthetic route involves condensing 2-amino-1-naphthalenesulfonic acid with 1-chloroanthraquinone-2-carboxylic acid, ring-closing with concomitant elimination of the sulfuric acid group by heating to 200–210°C.

2,1-Anthraquinoneacridone is itself a rather weak violet-red shade but demonstrates very good lightfastness. The monochloro compound (128) formed by chlorinating the parent dye with iodine–chloride and sulfuryl chloride is a member of the halogen derivatives which have proven more valuable to the dyeing industry than the parent dyes.

When 1-anilinoanthraquinone-2-carboxylic acid is chlorinated in sulfuric acid solution, cyclization and halogenation occur simultaneously and the trichloro compound, CI Vat Violet 14 (129) is obtained.

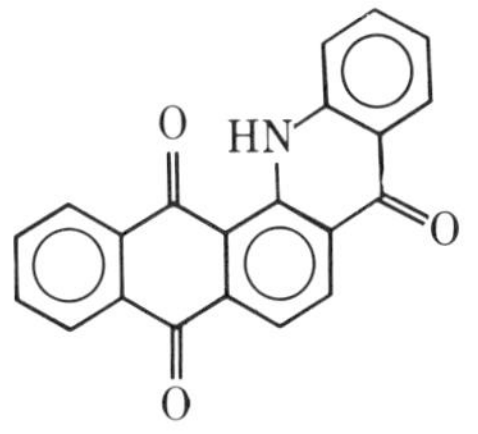

(125) [3669-01-5]

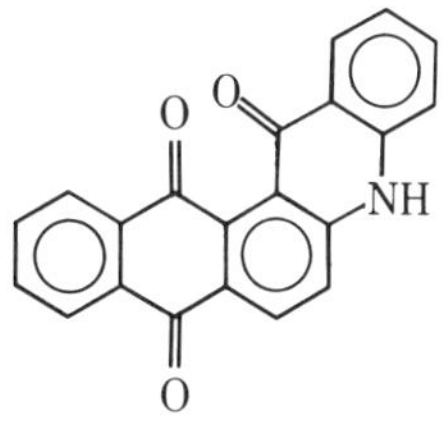

(126)

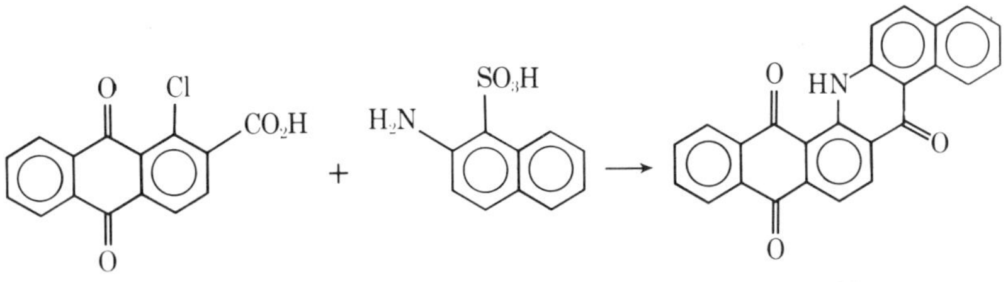

(127)

[3737-76-6]

252　DYES, ANTHRAQUINONE

Another important type of modification to the anthraquinone acridone base is the introduction of an amino group to its 6-position. CI Vat Blue 33 (**130**) is prepared from bromamine acid and 3,5-dichloroanthranilic acid. It provides a greenish-blue color with good fastness to soda boil.

Red to violet dyes are obtained by cyclizing 1-(2'-phenoxy)-anilinoanthraquinone-2-carboxylic acid, which demonstrates good fastness to light and boiling. If two chloro atoms are included in its structure, a pink dye, CI Vat Red 38 (**131**), which has better fastness to light and is less of a tenderer, is obtained. Preparation of (**131**) is given below.

Although many of the acridones discussed thus far have been obtained from 1-chloroanthraquinone, several other anthraquinone nuclei have been used as starting materials for such preparations.

1,5-Dichloroanthraquinone has been condensed with two moles of potassium

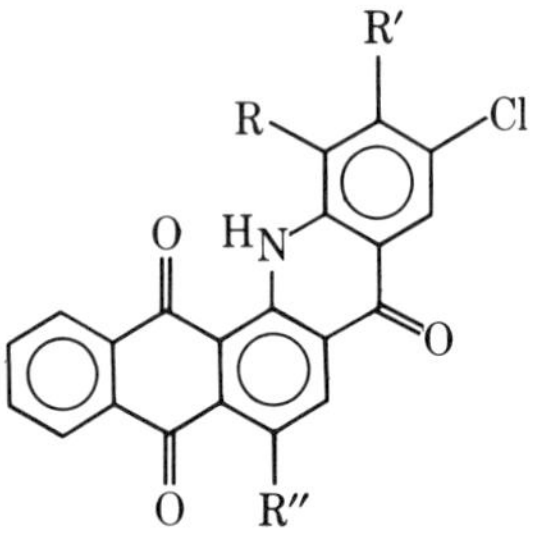

	R	R'	R"
(128)	H	Cl	H [70321-39-0]
(129)	Cl	H	Cl [6373-31-5]
(130)	Cl	H	NH$_2$ [4215-99-0]

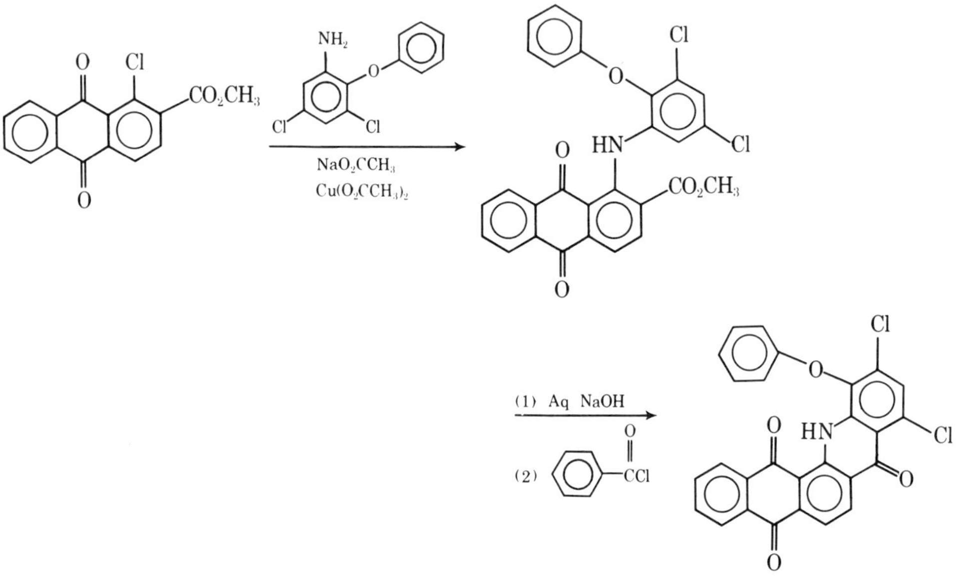

(**131**)

[6219-99-4]

anthranilate by heating with copper powder. Ring closure of the resultant intermediate by heating with sulfuric acid and chlorosulfonic acid affords CI Vat Violet 13 (132).

1-Nitroanthraquinone-2-carboxylic acid has also been a useful starting material for dyes. CI Vat Orange 13 (133), which has excellent all-around fastness and is free from tendering problems, has been obtained by condensing 1-nitroanthraquinone-2-carboxylic acid with p-aminobenzoic acid in aqueous solution in the presence of MgO. The acridone is cyclized, its 10-carbonyl group converted to the chloro derivative by means of thionyl chloride, and finally condensed with 1-amino-5-benzamidoanthraquinone.

The more deeply shaded acridones are produced by the addition of other heterocyclic systems, such as quinazoline and carbazole, to the structure; improved alkali fastness is frequently a result of such additions also. Examples of two such dyes are CI Vat Green 12 (134) and CI Vat Brown 55 (135). Dye (134) has stimulated interest because it is a quinazoline derivative (54). It dyes green from a brown hydrosulfite vat with excellent fastness.

Dye (135) contains an acridone and two carbazole ring systems. This dye has the maximum all-around fastness of all the acridones.

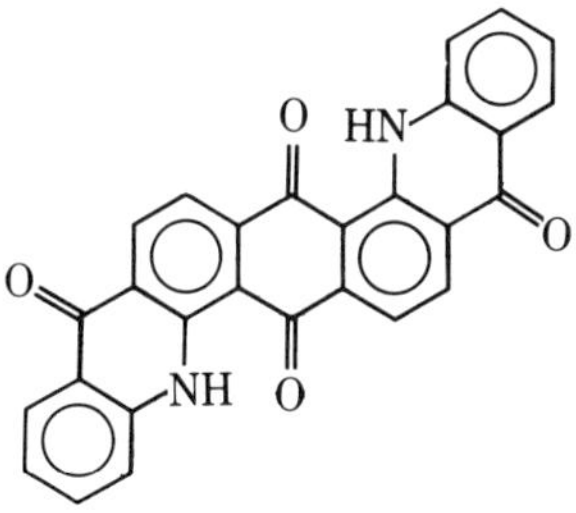

(132)
[4424-87-7]

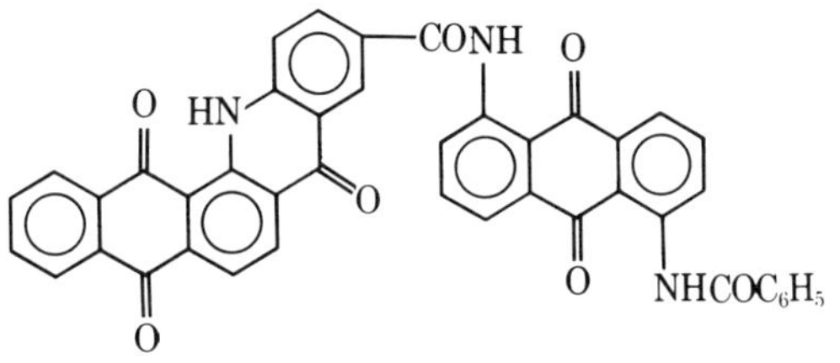

(133)
[6417-38-5]

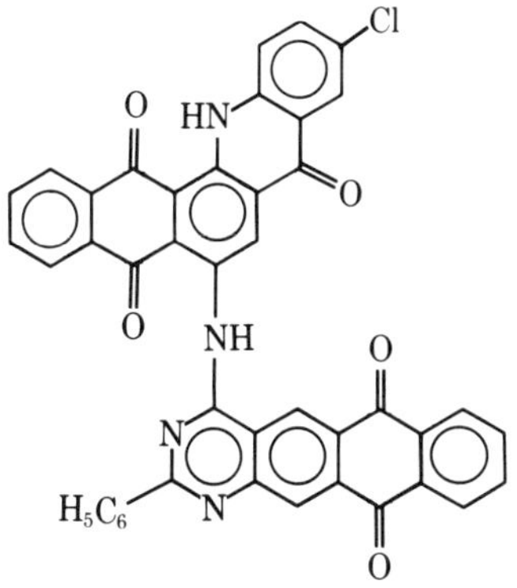

(134)
[6661-46-7]

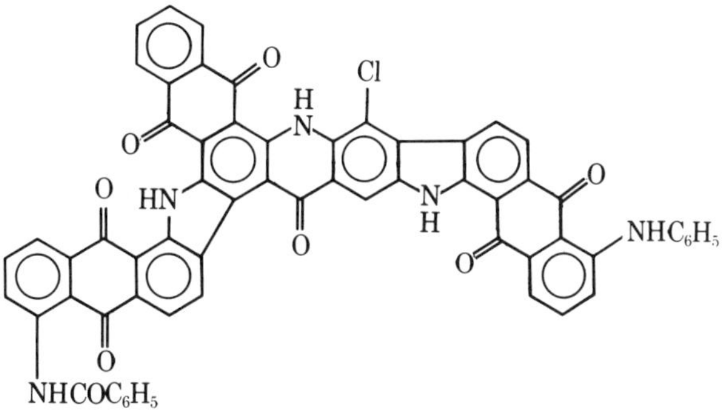

(135)
[4465-47-8]

Solubilized Vat Dyes. Soluble esters of the leuco compounds of the anthraquinone dyes can be made directly from the parent dyes. In order to accomplish this, the dyestuff is treated with an alkyl ester of chlorosulfonic acid or with oleum in the presence of a reducing metal, such as iron, zinc, or copper, and a tertiary base such as pyridine or dimethylaniline. During application, the original vat dyes are regenerated on the fiber by means of an acid oxidizing agent, eg, nitrous acid or ferric chloride.

These soluble vat dyes are generally sold as spray-dried powders and, because of their cost, are used mainly for pale shades and special applications.

Vat Dyestuffs Prepared from Crude Aminoanthraquinone Mixtures. 1-Aminoanthraquinone is a major starting material in the manufacture of many dyes, particularly the vat dyes. The quality of this compound used in such productions must be as pure and free of impurities, ie, 2-aminoanthraquinone and diaminoanthraquinones, as possible, according to many observers. It has been reported that the impurities reduce the quality of dyes derived from the mixture. Several obstacles and disadvantages, however, have been encountered with the methods used to prepare the necessary high-purity aminoanthraquinone.

Recently, in contradiction to the above reports, it has been found that mixtures of 65–80% aminoanthraquinone, and 20–35% 2-aminoanthraquinone and 1,5- and 1,8-diaminoanthraquinones could be used very effectively in preparing vat dyestuffs as a substitute for the purer 1-aminoanthraquinone (55). Furthermore, the vat dyes obtained with this mixture demonstrated greater affinity for certain fibers like polyester–cotton blends than those prepared from the purer 1-aminoanthraquinone and the quantitative yield of dye from both aminoanthraquinone preparations is almost equal. Even more important, the dyestuff from the crude mixture is claimed to possess the same or better dyeing strength at the same dyestuff level.

The mixture of aminoanthraquinones can be prepared by any of the known simple synthetic routes; nitration of anthraquinone can be accomplished by treating anthraquinone with nitric acid in the cold, either in the presence (56) or absence (53) of sulfuric acid. The 1-nitroanthraquinone is then reduced to the amino compound by a reaction such as treating a slurry of the nitroanthraquinone mixture with sodium sulfide at elevated temperature. The product mixture thus prepared can be condensed with an equimolar amount of 1-chloroanthraquinone to form a dianthrimide, nitrating, reducing, benzoylating, and ring-closing to yield (**136**).

(**136**) [*2379-81-9*]

Similarly, the aminoanthraquinone mixture can be used to prepare many of the anthraquinone vat dyes in use today.

Anthraquinonesulfonic Acid Dyes

The introduction of sulfonic groups to anthraquinones produces the anthraquinone acid dyes. The predominant shades in this series of dyes are blues and greens, not available among the azo dyes and unequalled by any other class of dye. They are characterized as a group by their excellent fastness to light, which falls within the range 5–7 (9 being the maximum rating) and a net-fastness which is moderate to good.

Several studies have been performed correlating structure changes with dyeing properties (57). Frequently these dyes are sulfonic acids of 1,4-arylaminoanthraquinones. These produce the blue and green shades. If, however, the amino groups in the 1,4-positions are unsubstituted, or if only one is an arylamino group, a shift occurs toward the violet shades. Replacement of one arylamino group in the 1- or 4-position by an acylic amino group results in dyes of increased brightness. Furthermore, methylation of the amino groups causes a shift from the green range to a definite blue.

The presence of sulfonic acid groups in the molecule can be achieved by one or a combination of three possible routes: (1) by condensation of bromamine acid with aromatic amines which results in a sulfonic acid group on the same ring of the anthraquinone system as contains the auxochromes; (2) by sulfonation of the anthraquinone nucleus itself with oleum; or (3) when sulfonation is the final stage, with strong sulfuric acid or weak oleum which sulfonates the arylamino group.

With the exception of azo dyes, the anthraquinonesulfonic acids constitute the most important group of dyes for wool and silk. Besides their additional use in dyeing nylon, they have also demonstrated the advantage of leaving cellulose fibers unaffected and have, therefore, been found useful for cross-dyeing union materials of the cellulose fibers and wool or silk. These dyes are usually sold and used as the sodium salts or as other alkali metal salts. A summary of the more important, commercially manufactured anthraquinone acid dyes is given in Table 4. When available from U.S. International Trade Commission reports, tonnage and economic data are given.

Bromamine Acid. Bromamine acid (1-amino-4-bromo-2-anthraquinonesulfonic acid) is one of the most important intermediates in the synthesis of acid dyes. Aside from quinizarin and leucoquinizarin, more acid dyes have been produced from bromamine acid by replacement of the bromine atom with arylamino groups than from any other anthraquinone intermediate. The reaction is customarily run in boiling aqueous alkaline solution in the presence of a copper catalyst. The simplest dye derived by this procedure is CI Acid Blue 25 (**137**); it was the first arylaminoanthraquinone dye possessing a sulfonic group in the anthraquinone portion of the molecule.

(**137**)
[6408-78-2]

Table 4. Major Commercially Produced Anthraquinone Acid Dyes

CI Name	CAS Registry No.	CI No.	Method of preparation or structure	1976 U.S. sales Metric tons	1976 U.S. sales Thousands of dollars	1976 Import, t
Reds						
Acid Red 80	[4478-76-6]	68215	(186)			
Acid Red 81	[6846-33-9]	68200	(187)			
Acid Red 82	[2611-80-5]	68205	condense 4-bromo-N-methyl-1(N),9-anthrapyridone with aniline, disulfonate with oleum and convert to the sodium salt			
Acid Red 83	[6871-98-3]	68220	(188)			
Violets						
Acid Violet 34	[6408-63-5]	61710	condense 1,5-dinitro (or dichloro) anthraquinone with p-toluidine, sulfonate and convert to the sodium salt			0.45
Acid Violet 36	[1323-87-1]	62010	chlorinate 1,4-diaminoanthraquinone, sulfonate with oleum in the presence of boric acid and convert to the sodium salt			0.32
Acid Violet 39	[6871-89-2]	68500	(189)			0.11
Acid Violet 41	[6408-71-5]	62020	(180)			
Acid Violet 42	[6408-73-7]	62026	(181)			
Acid Violet 43	[4430-18-6]	60730	(159)			
Acid Violet 47	[12235-16-4]					
Acid Violet 51	[1324-52-3]	62165	(157)			0.20
Acid Violet 63	[6460-06-6]	62160	condense 1-amino-4-bromo-2-methylanthraquinone with N-1-methylmetanilide, sulfonate with weak oleum and convert to the sodium salt			
Acid Violet 103	[12220-59-6]					0.45
Acid Violet 109	[12220-63-2]					0.22
Blues						
Acid Blue 8	[8005-16-1]	58800	heat 1,5- and 1,8-dinitroanthraquinone in 30% oleum with sulfur in the presence of boric acid			
Acid Blue 14	[12234-61-6]					
Acid Blue 23	[33340-33-9]	61125	condense 1-amino-4-bromo-2,5-anthraquinonedisulfonic acid with p-aminoacetanilide			9.50
Acid Blue 25	[6408-78-2]	62055	(137)	286	3,725	
Acid Blue 27	[6408-51-1]	61530	(168)	19.1	226	
Acid Blue 35	[1324-61-4]	61560	(166)			0.42

256

Name	CAS	Index	Method	Ref	Val1	Val2	Val3
Acid Blue 40	[6424-85-7]	62125		(141)			1.50
Acid Blue 41	[2666-17-3]	62130		(139)			
Acid Blue 43	[2150-60-9]	63000		(173)			
Acid Blue 45	[2861-02-1]	63010		(174)			1.25
Acid Blue 47	[4403-89-8]	62085		(154)			0.10
Acid Blue 49	[1324-06-7]	62095	condense 1-amino-4-bromo-2-methylanthraquinone with *p*-toluidine, disulfonate, and convert to the sodium salt				
Acid Blue 51	[6424-88-0]	62145		(140)			
Acid Blue 53	[6424-87-9]	62135	treat 1-amino-4-bromo-2-anthraquinonesulfonic acid with *m*-aminobenzonitrile in the presence of copper and sodium carbonate				
Acid Blue 55	[6370-91-8]	63315	treat 1,5-dihydroxy-4,8-dinitroanthraquinone with methylamine; condense this with formaldehyde in sulfuric acid; split off one sulfonic acid group and convert to the sodium salt				
Acid Blue 56	[6408-69-1]	62005	reduce quinizarin with ammonia and hydrosulfite; follow with oxidative sulfonation		307	3,207	
Acid Blue 62	[4368-56-3]	62045		(138)			4.11
Acid Blue 68	[8005-74-1]	63330	condense ethylamine with 1,5-dibromo-4,8-dihydroxyanthraquinone in the presence of copper and sulfonate				
Acid Blue 78	[6424-75-5]	62105		(155)			11.5
Acid Blue 80	[4474-24-2]	61585		(163)			1.35
Acid Blue 81	[8005-76-3]	64515		(177)			0.50
Acid Blue 96	[1323-98-4]	62110	condense 1-amino-2,4-dibromoanthraquinone with *p*-toluidine; sulfonate with monohydrate and convert to the sodium salt; sulfonate with oleum in the presence of boric acid to form the disulfonic acid and convert to the sodium salt				
Acid Blue 111	[6424-90-4]	62155	condense 1-amino-4-bromo-2-anthraquinonesulfonic acid with α-(4-amino-*m*-tolyl)-*o*-toluic acid ethyl ester in the presence of cuprous chloride and acid binding agents and convert to the sodium salt				
Acid Blue 124	[6370-63-4]	64005	condense 5-acetamido-1-amino-4-bromo-2-anthraquinonesulfonic acid with cyclohexylamine and convert to the sodium salt				
Acid Blue 127	[6471-01-8]	61135	condense 2 moles of 1-amino-4-bromo-2-anthraquinonesulfonic acid with 1 mole of 4,4′-isopropylidenedianiline and convert to the sodium salt				9.17
Acid Blue 127:1	[12237-86-4]						
Acid Blue 129	[6397-02-0]	62058					3.23
Acid Blue 138	[1324-53-4]	62075	condense bromamine acid with 2,4,6-trimethylaniline	(156)			7.40
Acid Blue 143	[61723-95-3]						

Table 4 (continued)

CI Name	CAS Registry No.	CI No.	Method of preparation or structure	1976 U.S. sales Metric tons	1976 U.S. sales Thousands of dollars	1976 Import, t
Acid Blue 145	[6408-80-6]	62070	treat 1-amino-2-bromo-4-(2-sulfo-p-toluidino)anthraquinone with sodium sulfite			
Acid Blue 150	[6408-47-5]	61130	reaction of 2 moles of 1-amino-4-anilino-2-anthraquinone sulfonic acid with manganese dioxide			
Acid Blue 175	[12219-25-9]					2.90
Acid Blue 204	[61724-00-3]					0.70
Acid Blue 205	[12238-92-5]					0.16
Acid Blue 208	[69912-83-0]					1.25
Acid Blue 239	[12219-42-0]					7.50
Acid Blue 245	[12219-46-4]					0.30
Acid Blue 247	[12219-48-6]					1.70
Acid Blue 258	[61847-68-5]					6.96
Acid Blue 261	[61847-66-3]					1.90
Acid Blue 264	[39315-90-7]					2.74
Acid Blue 266	[53168-71-1]					0.80
Acid Blue 277	[61967-93-9]					66.5
Acid Blue 278	[61931-04-2]					1.30
Acid Blue 288	[61967-95-1]					7.24
Acid Blue 290	[39280-53-0]					9.15
Acid Blue 294	[61931-06-4]					1.00

Greens

Acid Green 25	[4403-90-1]	61570	(161)	0.85
Acid Green 27	[6408-57-7]	61580	(162)	0.25
Acid Green 36	[1324-82-9]	61595	condense 4-biphenylamine with leucoquinizarin and quinizarin in the presence of boric acid, sulfonate and convert to the sodium salt	
Acid Green 37	[6424-94-8]	62515	condense cyclohexylamine with 5-nitro-2-anthraquinone sulfonic acid, brominate, condense with *p*-aminoacetanilide and convert to the sodium salt	
Acid Green 38	[6424-97-1]	62550	(182)	
Acid Green 40	[12219-87-3]			
Acid Green 41	[4430-16-4]	62560	(184)	2.50
Acid Green 42	[6425-06-5]	62575	condense *p*-toluidine with 1,4,6-trichloroanthraquinone; treat product with *p*-toluenethiol, sulfonate, and convert to the sodium salt	2.69
Acid Green 44	[6408-59-9]	61590	(165)	
Acid Green 71	[12219-91-9]			0.50
Acid Green 84	[12234-91-2]			1.70

Browns

Acid Brown 26	[6370-65-6]	65000	condense 1-aminoanthraquinone with 1-chloroanthraquinone, sulfonate with oleum and convert to the sodium salt	
Acid Brown 27	[1324-45-4]	66710	(185)	

Blacks

Acid Black 48	[1328-24-1]	65005	condense 1,4-diaminoanthraquinone with 1-amino-4-bromoanthraquinone in the presence of copper and sodium acetate, sulfonate with oleum, and convert to the sodium salt	
Acid Black 97	[6358-61-8]	65008		

260

These 1-amino-4-arylaminoanthraquinones are valuable as wool dyes because they can dye bright, level shades with very good fastness to light and washing. They are thus superior to earlier dyes even though their milling fastness is only 1–2.

Similarly, trifluoromethylaniline (58) and trifluoromethoxyaniline (59) have been employed to prepare dyes that produce a reddish-blue shade, also with very good fastness to light, water, perspiration, and washing.

A variety of substituted arylamines, even cyclohexyl amine, have been used with bromamine acid, in place of aniline, to prepare analogues in this series. Some of the less complex products include CI Acid Blue 62 (138) from cyclohexylamine, CI Acid Blue 41 (139) from *p*-amino-*N*-methylacetanilide, CI Acid Blue 51 (140) from *m*-aminobenzoic acid followed by esterification, CI Acid Blue 40 (141) from *p*-aminoacetanilide, and more recently an analogue from 2-methyl-4-*tert*-butylaniline (142) (60).

A similar class of blue to reddish-blue dyes that are fast to soda boiling and washing is comprised of 1-amino-4-*p*-acylamidoanilinoanthraquinone-2-sulfonic acids. These compounds contain an alkyl or cycloalkyl portion of the acyl group which has at least five carbon atoms, eg, hexahydrobenzoyl, undecoyl derivations of *p*-phenylenediamine.

Blue and green dyes for wool and nylon are derived by heating bromamine acid with various heterocyclic amines including 2-amino-3-methoxydibenzofuran (143) and aminocarbazoles (61), as well as diphenyl derivatives, such as 4-aminodiphenyl-4'-sulfocyclohexylamide (144) (62).

An outgrowth of studies on the condensation of bromamine acid with various arylamines has been the attempt to create an affinity in them for cotton and other cellulose fibers. The azo group is one functionality that has exhibited some promise. The olive and green wool dyes obtained by the condensation of bromamine acid with aminoazo compounds, such as (145), have some affinity for cotton.

Recently, compounds (146–147) were prepared by a similar but slightly more complex route (63–64). Dye (146) has been used on cotton and polyamide fiber and has demonstrated excellent fastness to wet processing, in particular to washing, water, alkali and rubbing, and perspiration, as well as good fastness to light. Dye (147) has been used on cellulose fibers to give fast blue shades.

Alterations to the 1-amino-4-anilinoanthraquinone-2-sulfonic acid, other than by using different amines for reaction at the 4-position, have been examined extensively. These are usually performed with the intention of achieving increased brightness, increased fastness and/or creating minor changes in the colors.

Additional sulfonation, eg, in the aryl substituent can be effected with moderately strong sulfuric acid. Precautions must be taken, however, to prevent oversulfonation as the reaction occurs quite readily. This procedure has been valuable in preparing fast blue and greenish-blue shades for polyamide fibers, dyes (148) and (149), respectively, and the highly reactive acid dye (150) (65–66).

Frequently, modifications to the anthraquinone nucleus will also alter the final character of a dye. These changes, however, must be made before the amine is condensed with the haloanthraquinone. For example, dyes that have a sulfonic acid group in position 6 or 7 of the anthraquinone, instead of in the aryl residue or position 2 (as bromamine acid derivatives) are prepared by condensing 4-halogen-1-hydroxy or aminoanthraquinone-6 (or 7)-sulfonic acids with aromatic amines. The resulting compounds display excellent leveling properties and greener shades than their isomeric

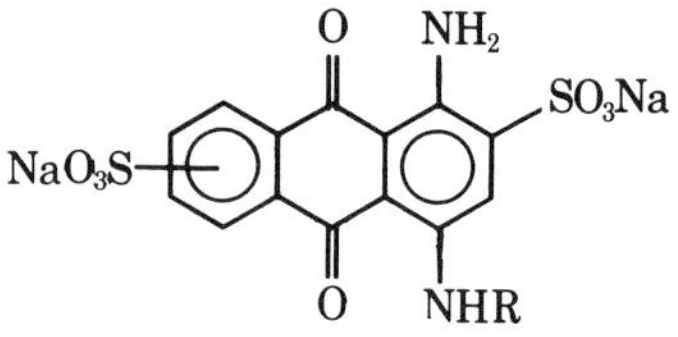

262 DYES, ANTHRAQUINONE

(151) 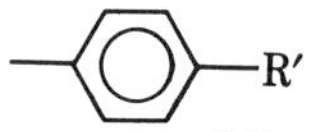 [structure: anthraquinone with NaO₃S, NH₂, SO₃Na, NHR, R = phenyl-R']

(152) 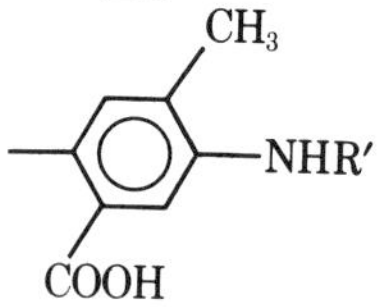 [structure: aryl group with CH₃, NHR', COOH]

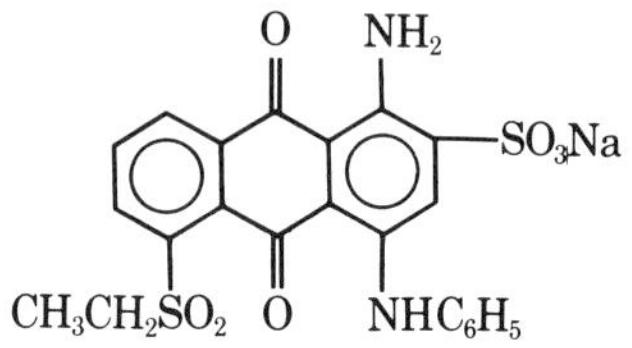

(153) [70321-20-9] [structure: anthraquinone with NH₂, SO₃Na, CH₃CH₂SO₂, NHC₆H₅]

analogues with the sulfonic acid group on the aryl group. Likewise, the blue and greenish-blue acid colors of dyes (151) (R' = acylamido, halogen, etc) (67) for wool and nylon and dyes (152) for cotton and wool are derived from 1-amino-4-bromoanthraquinone-2,6 (or 7)-disulfonic acid (68). They demonstrate a notable resistance to change of color in light.

When the sulfonic group in position 6 or 7 is a sulfonamide, the anthraquinone condenses with amines to produce blue or blue–green wool dyes with excellent lightfastness and, in addition, good leveling properties. An alkyl sulfone group (eg, $C_2H_5SO_2$—) in the 5-position of the anthraquinone nucleus such as CI Acid Blue 25 (137) generates greener and brighter shades (153) (69).

Sulfonic Acid Groups Elsewhere on the Dye Molecule. In some instances, such as CI Acid Blue 47 (154), there are no sulfonic acid groups on the anthraquinone nucleus, and the modification to the anthraquinone involves a simple aliphatic or aromatic substituent. This dye, eg, is produced by condensation of 1-amino-4-bromo-2-methylanthraquinone and p-toluidine followed by sulfonation of the aryl residue as described above. Similarly, CI Acid Blue 78 (155) is prepared by heating 1-amino-2,4-dibromoanthraquinone with toluidine in the presence of sodium acetate. This is followed by sulfonation with oleum.

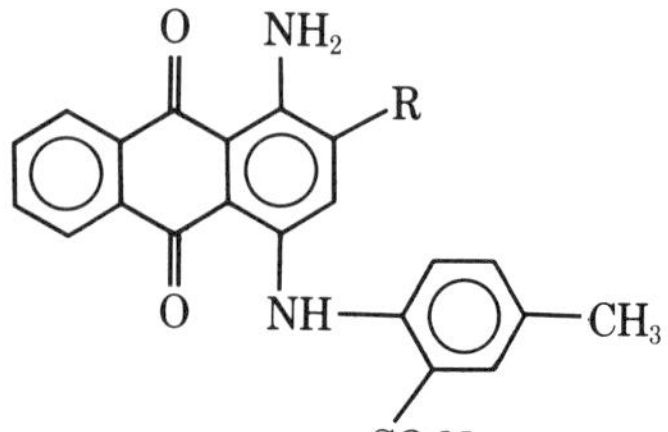

R
(154) CH₃ [4403-89-8]
(155) Br [6424-75-5]

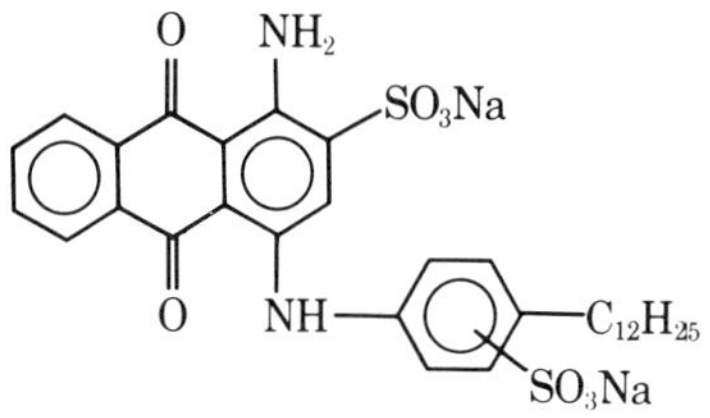

(156) [1324-53-4]

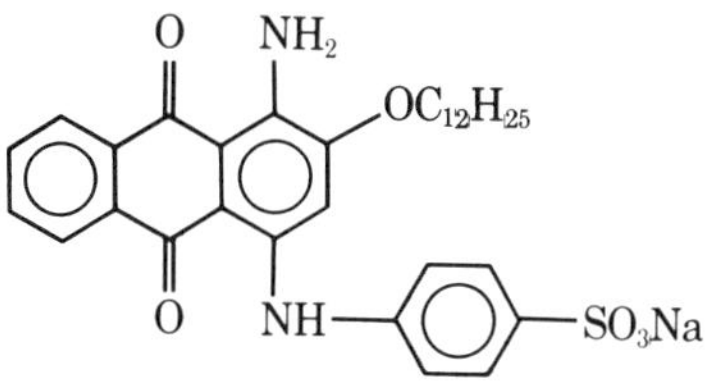

(157) [1324-52-3]

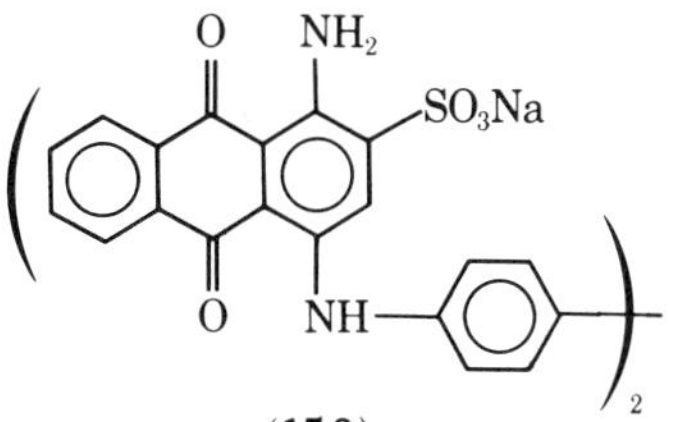

(158) [6408-47-5]

An hydroxyalkyl group on the 2-position or a hydroxy or alkoxy group in the 5- or 6-positions all improve leveling properties. In fact, one or more alkyl groups (C_4–C_{20}) may be introduced to an acid dye either directly to the anthraquinone or via the aromatic amine used in the condensation. CI Acid Blue 138 (156), and CI Acid Violet 51 (157) are two examples. Although the presence of the alkyl groups produce a marked influence on the fastness of these dyes to milling and potting, there is little effect on the shade of the dye.

Finally, oxidative coupling of dyes such as CI Acid Blue 25 (137) with manganese dioxide and sulfuric acid have been claimed to yield dimers of the type (158).

Colors of these wool dyes include the range of blue, violet, and brown, and their lightfastness is excellent.

Quinizarin, leucoquinizarin, and a variety of substituted hydroxy- and aminoanthraquinones have also served as starting materials for anthraquinonesulfonic acid dyes. These compounds allow more leeway in the functionality that can be introduced to the dye.

One of the simplest leucoquinizarin derivatives is CI Acid Violet 43 (159). It is synthesized by condensing equivalent amounts of leucoquinizarin and p-toluidine, followed by sulfonation.

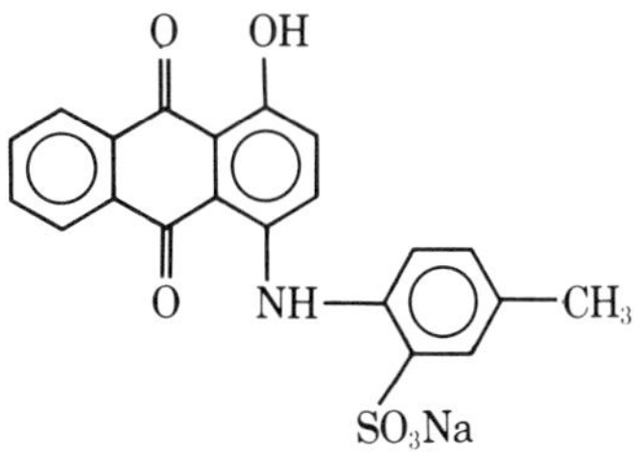

(159) [4430-18-6]

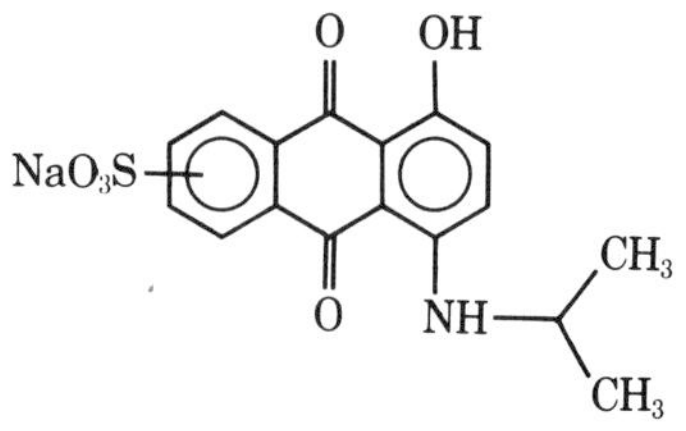

(160) [70969-39-0]

This dye colors wool and silk, from a neutral or acid bath, a bright, bluish violet. Its shade can be converted to a greenish-blue with excellent fastness properties by afterchroming.

Newer dyes prepared in this series by essentially the same route using different amines include (160). These are dyes for polyamide fibers and wool which produce fast violet shades (70).

Leucoquinizarin has also been condensed with two equivalents of a primary aromatic amine, then oxidized and finally sulfonated to yield 1,4-diarylaminoanthraquinones. This procedure has been employed to prepare the important wool dye CI Acid Green 25 (161) (71), which demonstrates a lightfastness of 6 but only fair acid and alkaline milling fastness, as well as its sister dye, CI Acid Green 27 (162), which also has a lightfastness of 6 and an alkaline milling-fastness of 5. Improvement in milling- and potting-fastness has been achieved by introducing dodecyl or dodecyloxy groups to the dye molecule.

Similarly, CI Acid Blue 80 (163) is derived from mesidine. This dye gives a brilliant blue color which is deeper than the previous two because of a bathochromic shift.

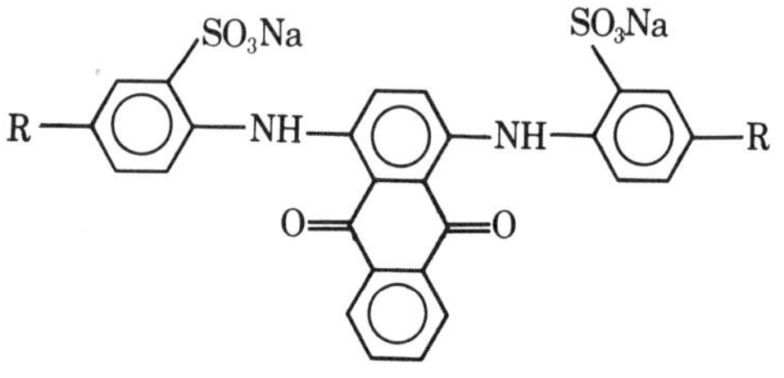

R
(161) CH$_3$ [4403-90-1]
(162) n-C$_4$H$_9$ [6408-57-7]

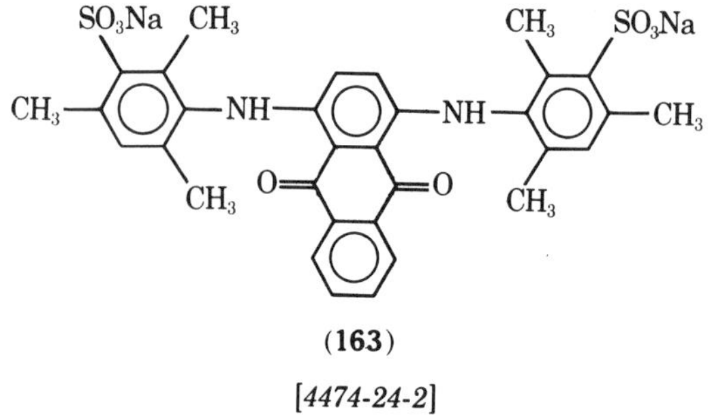

(163)
[4474-24-2]

Brightness of shade with outstanding fastness to washing, milling and light, in violet and blue dyes has been obtained by sulfonating the anthraquinone derivatives substituted by secondary amino groups which possess alkyl or aryl residues, such as 1,4-bis-*p*-butylanilinoanthraquinones.

Quinizarin, likewise, is condensed with two equivalents of an amine such as hydroxydimethylamine, in order to replace both its hydroxyl groups. The resultant bis(alkylamino)anthraquinone, in this case bis(hydroxydimethylamino)anthraquinone, is then sulfonated. This final product is a blue, lightfast, level dye (164).

(164)
[70321-45-8]

Quite like the examples given for leucoquinizarin, quinizarin has also been condensed with aromatic amines such as *p*-cyclohexylaniline and 1,2,3,4-tetrahydro-2-naphthylamine to produce CI Acid Green 44 (165) and CI Acid Blue 35 (166), respectively.

At least two routes have been found effective in synthesizing unsymmetrically substituted 1,4-aryl and/or alkylaminoanthraquinones. One method is the sequential treatment of leucoquinizarin or quinizarin with one equivalent each of two different primary amines. A second scheme proceeds via halogenation of 1-alkylaminoanthraquinones. The bromination is carried out in 96% sulfuric acid at 10°C, the bromine is then substituted by a second primary amine (different from the one already present), and the final step is sulfonation. Dyes (167) (72) and (168) (CI Acid Blue 27 (CI 61530)) are products of such routes.

The 1,4-diaminoanthraquinone nucleus is one of the least substituted starting materials used to prepare sulfonic acid dyes. It differs from the previous starting materials in that the reaction which generates the bis-alkylaminoanthraquinones is merely an alkylation. Sulfonation is carried out in the customary manner.

Water-soluble sulfonic esters of hydroxyalkylaminoanthraquinones are often useful for dyeing wool and nylon. Two such derivatives are acid dye (CI 61550) (169) and dye (170). They are prepared by sulfonation of the alkylation products of 1,4-diaminoanthraquinone with epichlorohydrin and ethylene oxide, respectively. Both are bright blue acid dyes with moderate fastness to light.

The 1,5- and 1,8-derivatives of anthraquinone have also been used extensively as starting materials for dyes. By means of a simple two step sequence (ie, reaction with toluidine followed by sulfonation) 1,5- and 1,8-dichloro- or dinitroanthraquinones can be used to prepare the bisarylanthraquinone disulfonic acid violets (CI 61710 and 61800) (171–172).

Two other dyes in this class are CI Acid Blue 43 (173) and CI Acid Blue 45 (174). Structure (173) is prepared by sulfonation with chlorosulfonic acid or 20% oleum, nitration, and reduction. It has demonstrated superior characteristics of fastness. The structure (174) is prepared by the same sequence, but with more extensive sulfonation.

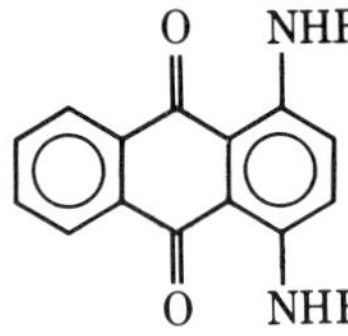

266 DYES, ANTHRAQUINONE

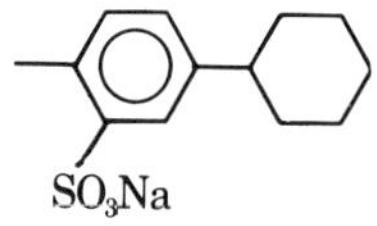

(Structures 165–170 showing 1,4-diaminoanthraquinone derivatives with substituents:)

(165) R = R′ = 4-cyclohexylphenyl-2-SO₃Na [6408-59-9]

(166) 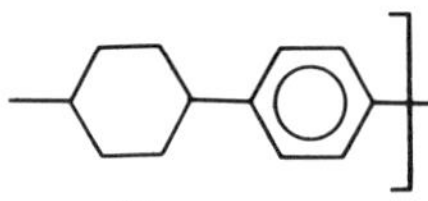 R = cyclohexyl-phenyl-SO₃Na [1324-61-4]

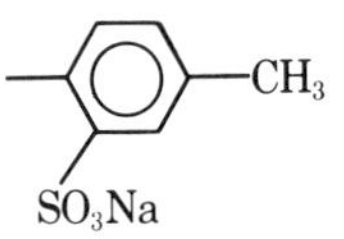

(167) R = 4-methylphenyl-3-SO₃Na

(168) R = 4-methylphenyl-3-SO₃Na, R′ = CH₃ [6408-51-1]

(169) R = R′ = CH₂CHCH₂Cl with OSO₃Na [6527-66-8]

(170) R = CH₂CH₂OH, R′ = CH₂CH₂OSO₃Na [70321-47-0]

R′ = CH₂CH₂CON(CH₃)₂ [70321-46-1]

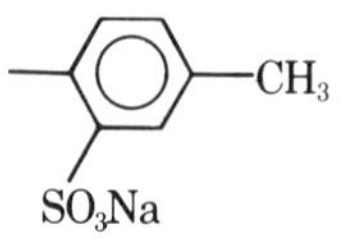

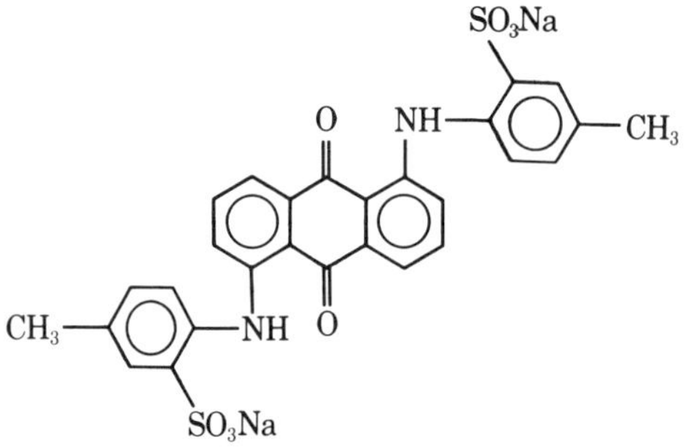

(171) [6408-63-5]

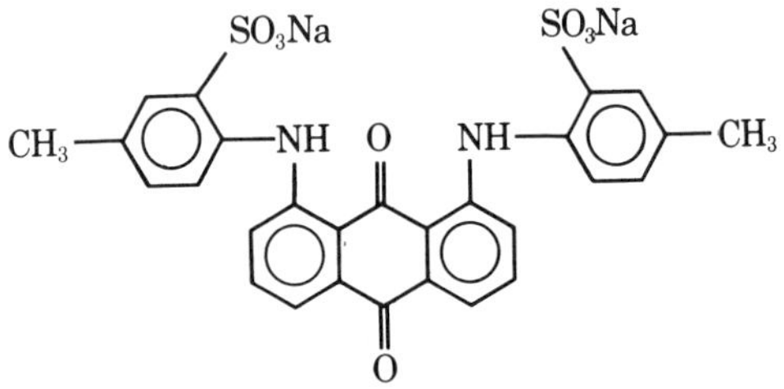

(172) [6408-68-0]

With slight variations on this same theme, dyes (175–176) have also been obtained (73–74). Dye (176) is especially good for dyeing nylon from a weak acid bath. As a bright greenish-blue shade, it exhibits excellent wet- and lightfastness.

Finally, extensive halogenation of 1,5- and 1,8-diaminoanthraquinone with sulfonyl chloride in o-dichlorobenzene will produce the tetrahalo derivative. Replacing two of the halogen atoms by p-toluidine and sulfonating yields CI Acid Blue 81 (177), for the chloro derivative and the CI acid dye 64510 (178) for the bromo analogue.

Polyfunctionalized anthraquinonesulfonic acids have also been prepared starting with (1) halogenated hydroxy and aminoanthraquinones, and (2) polyhydroxyanthraquinones.

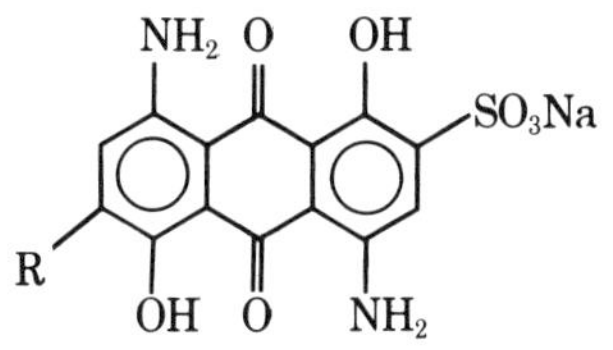

(173) H [2150-60-9]
(174) SO₃Na [2861-02-1]

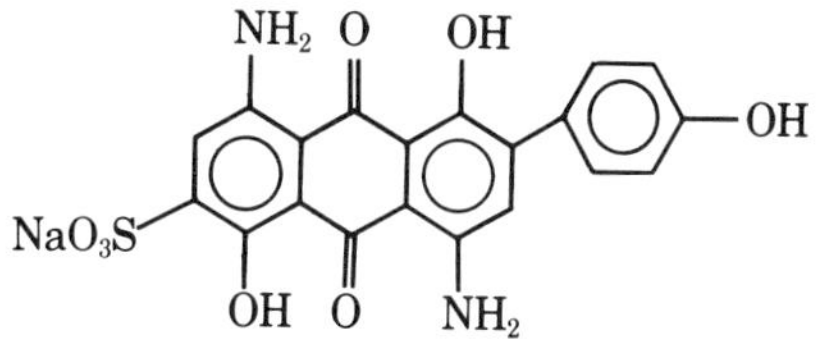

(175) [55694-11-6]

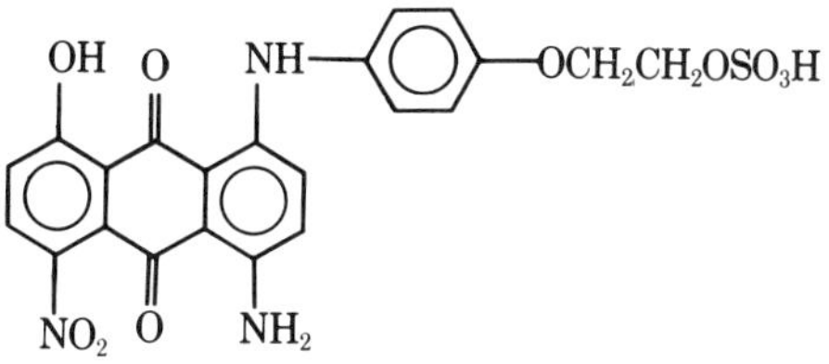

(176) [50586-86-2]

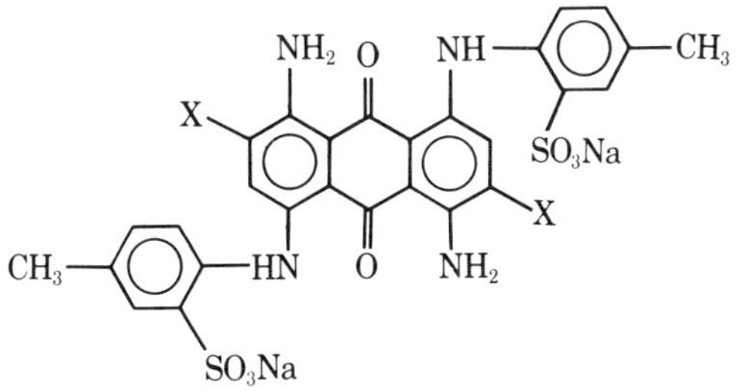

(177) Cl [8005-76-3]
(178) Br [8005-76-3]

1,4-Diamino-2,3-dichloroanthraquinone and 6-chloro- or 6,7-dichloroquinizarin are examples of type (1) starting materials. The trisulfonic acid of 1,4-diaminoanthraquinone, (179) is obtained by following the sequence of sulfonation and then treatment with sodium sulfite. The resultant dye is used on wool for clear blue shades of excellent fastness.

CI Acid Violet 41 (180) and CI Acid Violet 42 (181) are also prepared from 1,4-diamino-2,3-dichloroanthraquinone via nucleophilic substitution of phenol for the chloro atoms. The former is obtained by heating the anthraquinone compound with phenol in the presence of potassium carbonate at 165–198°C for 10 h. The latter evolves from heating the same two reagents with sodium sulfite, manganese dioxide, and water at 140°C in an autoclave.

The hydroxyl groups of polyhydroxyanthraquinones may be found in a variety of positions on the anthraquinone nucleus. A few examples of such compounds are given below to demonstrate their use in the preparation of acid dyes.

By heating 1,4,5-trihydroxyanthraquinone with p-toluidine, hydrogen chloride, boric acid, and zinc dust at 95°C, and sulfonating the product, CI Acid Green 38 (182) is obtained. In order to obtain unsymmetrically substituted dyes such as (183), the same starting materials, in this case the sulfonic acid of 1,4,5-trihydroxyanthraquinone, are treated sequentially with one equivalent of two different amines and the resultant diamino compound sulfonated on the aryl substituent, if desired (75).

The dihydroxy analogue of CI Acid Green 38 is derived from leuco-1,4,5,8-tetrahydroxyanthraquinone with two moles of p-toluidine in the presence of arsenic and boric acids. The sulfonated product, CI Acid Green 41 (184), has excellent lightfastness (7,8) and moderate fastness to milling (3).

One final group of important anthraquinonesulfonic acid dyes are the heterocyclic

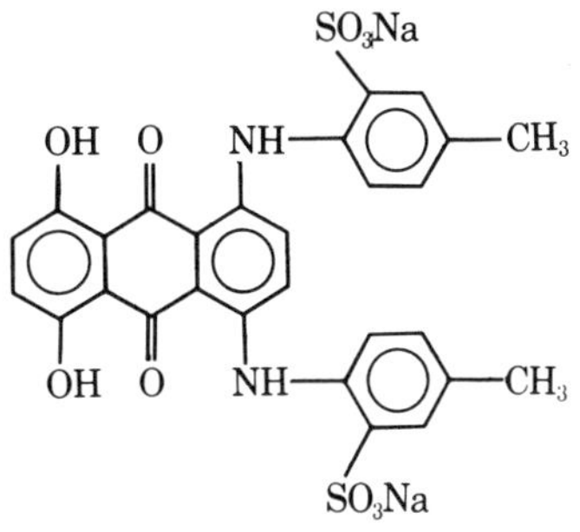

(184)

[4430-16-4]

derivatives. Although the acid and acid–mordant anthraquinone dyes produce principally blue, green, gray, and black shades, red and similar shades are available as the anthrone derivatives, and red–brown is found among the carbazoles.

When 1-benzamido-2-methyl-4β-naphthylaminoanthraquinone is dissolved in concentrated sulfuric acid, the product is a sulfonic acid of the carbazole type, CI Acid Brown 27 (185). It dyes wool in red-brown shades.

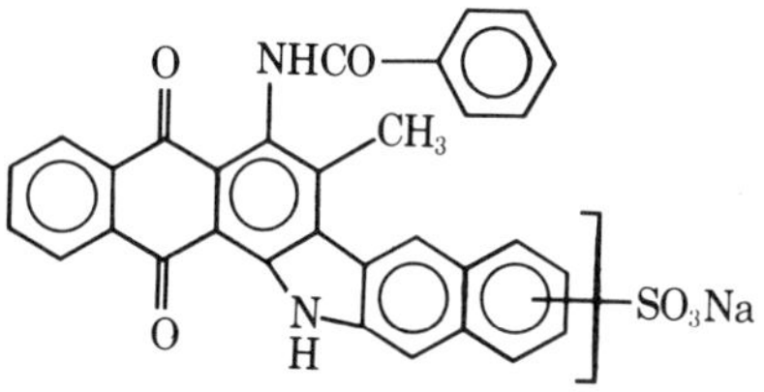

(185)

[1324-45-4]

The more valuable heterocyclics are 1,9-derivatives of anthrone. An important example is CI Acid Red 80 (186), which dyes red on wool and silk and demonstrates outstanding fastness to light and level dyeing. Its method of preparation involves cyclization of 1-(N-methyl)acetamido-4-bromoanthraquinone with boiling caustic soda solution, replacement of the bromine atom by p-toluidine and finally sulfonation. If p-toluidine is replaced by aniline or m-chloroaniline in the preparation, then the products are CI Acid Red 81 (187) or CI Acid Red 83 (188).

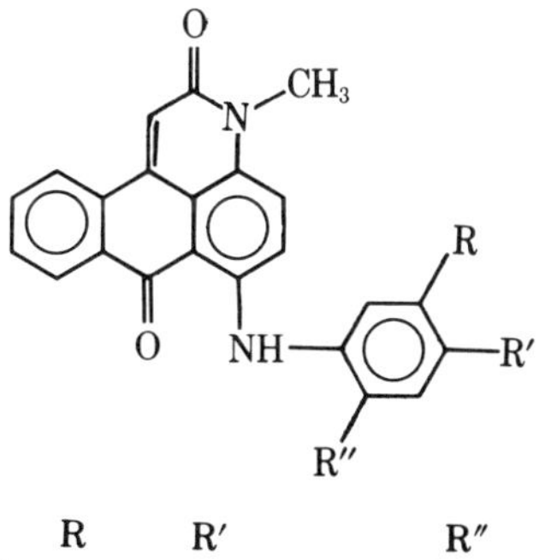

	R	R'	R''
(186)	H	CH$_3$	SO$_3$Na [4478-76-6]
(187)	H	SO$_3$Na	H [6846-33-9]
(188)	Cl	SO$_3$Na	H [6871-98-3]

The substitution of an aryl group for the N-methyl group in (186) causes improved fastness to perspiration.

CI Acid Violet 39 (189) is an example of a different type of anthrone. It contains a pyrimidone ring and is a reddish-violet shade with very good fastness. Synthesis involves condensation of 1-p-toluidino-4-methylaminoanthraquinone with urea in boiling phenol followed by sulfonation.

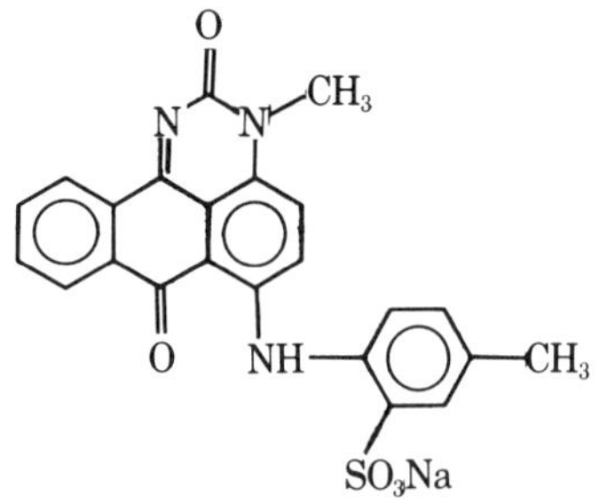

(189)
[6871-89-2]

Mordant Dyes

Because of their high cost, mordant dyes are no longer manufactured in large amounts. Nevertheless, they offer certain attractive features as dyes: a wide range of bright shades and excellent fastness properties.

Di- and trihydroxyanthraquinones are the older forms of mordant dyes. Having originally been obtained from natural sources, eg, madder root and morinda root, many of these were later produced synthetically from coal tar chemicals. Dyes such as these were used extensively in printing and exhibited fastness to light but inferior fastness to washing, wet treatments, and perspiration.

Mordant dyes are capable of combining with various metals to form complexes of different types. Although the salts of such multivalent metals as iron, copper, aluminum, and cobalt have been used for mordanting purposes, the perferred metal is chromium. Colors of the resulting dyes range from yellow through red, orange, brown, green, blue, violet, and black, depending upon the metal used and the structure of the anthraquinone compound.

Alizarin. The simplest and most important constituent of this group is 1,2-dihydroxyanthraquinone, commonly known as alizarin (190) (CI Mordant Red 11). Alizarin crystallizes from alcohol in brilliant brownish-yellow needles and sublimes at 110°C as orange-red needles, mp 290°C, bp 430°C. It is readily soluble in common organic solvents and dissolves in aqueous caustic soda producing a purple color.

This dye is produced commercially by heating the sodium salt of anthraquinone-2-sulfonic acid (silver salt) with aqueous soda and sodium nitrate (or chlorate) in an autoclave at 200°C. The synthetic material obtained however is a mixture of alizarin, flavopurpurin (1,2,6-trihydroxyanthraquinone), and anthrapurpurin (1,2,7-trihydroxyanthraquinone). (The latter two compounds are derived from anthraquinone disulfonic acid, a contaminant of the starting material, β-monosulfonic acid.)

Alizarin is a polygenetic dyestuff, consequently the color varies with the metal used as the mordant.

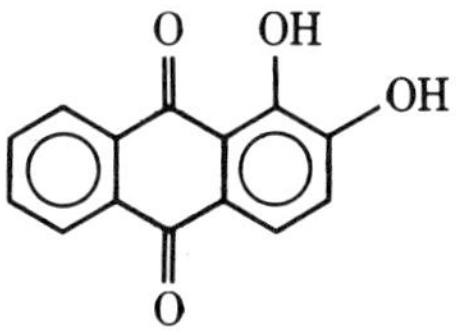

(190)
[72-48-0]

Mordant	Color	Mordant	Color
magnesium	violet	iron (ferrous)	black–violet
calcium	purple–red	iron (ferric)	brown–black
barium	purple–red	copper	brown–violet
strontium	red–violet	lead	purple–red
aluminum	rose–red	tin (stannous)	red
chromium	brown–violet	tin (stannic)	violet
		mercury	black–violet

The well-known Turkey Red dye is an aluminum mordant of alizarin used for dyeing and printing. It is prized for the beauty of its shade and its high fastness, but it is giving way to the azoic reds which are much simpler to produce.

A variety of derivatives have been prepared from alizarin and used as dyes. One such derivative is the β-sulfonic ester (191) derived from the reaction of the parent compound with sulfur trioxide and pyridine. This material has been found to be fast to rubbing when applied on cotton in conjunction with aluminum sulfate and calcium acetate.

3-Nitroalizarin (192) (CI Mordant Orange 14) formed by nitration in concentrated sulfuric acid in the presence of boric acid or 62% nitric acid in o-dichlorobenzene at 40°C has found limited use on aluminum mordant (76) for shades of orange.

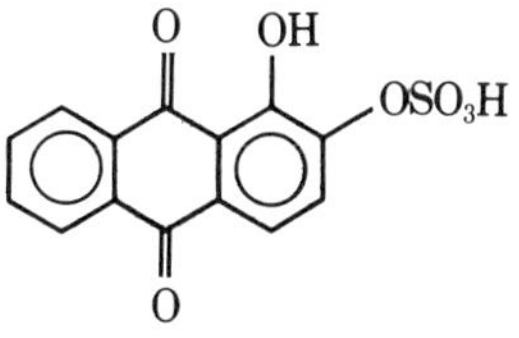

(191)
[70416-83-0]

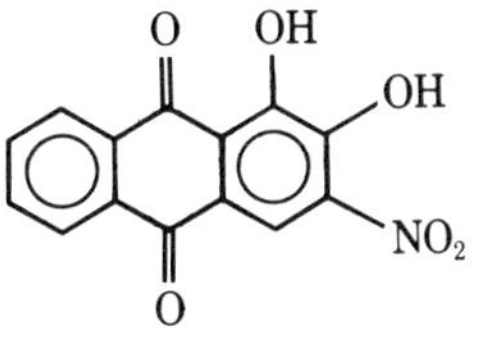

(192)
[568-94-3]

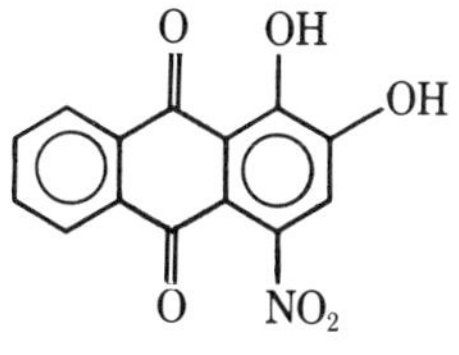

(193)
[2243-71-2]

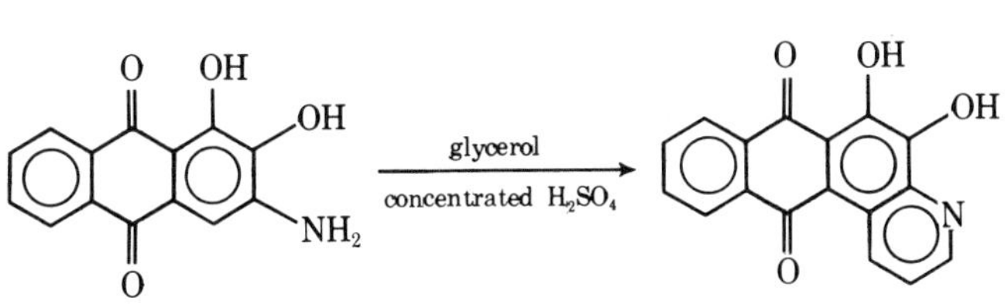

(194)
[568-02-5]

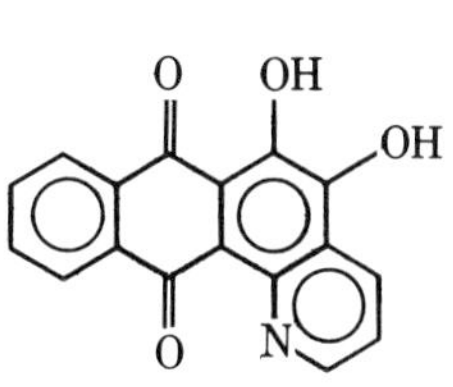

(195)
[10134-44-8]

4-Nitroalizarin (**193**) (CI Mordant Brown 44) is obtained by nitrating alizarin in sulfuric acid or oleum at −5 to −10°C. It is best prepared however, by nitration of alizarin dibenzoate, followed by alkaline hydrolysis (77).

An important pyridino-mordant dye (**194**) (CI 67410) is prepared by heating a mixture of 3-nitro- and 3-aminoalizarin with glycerol and 83% sulfuric acid at 110°C for 3 h. This dye gives reddish-blue color on chromium mordant (78).

Similarly, the Skraup reaction on 4-aminoalizarin produces a Mordant Dye 67405 (**195**) that when used on chromium mordant as a bisulfite paste yields a dull, bluish-green color (79).

Anthragallol, CI Mordant Brown 42 (**196**), has found practical utility as a fast brown on chromium mordant.

Oxidation of alizarin with manganese dioxide and sulfuric acid gives purpurin (Mordant Dye 58205) (**197**), which produces a scarlet color on aluminum mordant. It is a useful intermediate in the preparation of acid–mordant dyes.

When alizarin is nitrated with fuming nitric acid followed by boiling in water, 3-nitropurpurin is obtained (Mordant Dye 58215) (**198**) which gives red on aluminum mordant.

5-Hydroxyalizarin (CI Mordant Red 45) (**199**) is bordeaux on aluminum mordant. It can be derived by oxidation of alizarin with fuming sulfuric acid in the presence of boric acid, followed by hydrolysis (80).

Flavopurpurin (**200**) (CI Mordant Red 4) gives a dull red on chromium mordant. When it is sulfonated by oleum it produces reddish brown on chromium or aluminum mordant (**201**) (81).

When alizarin is treated with 80% fuming sulfuric acid at 30°C, it is not sulfonated, but two hydroxyl groups are added, giving CI Mordant Violet 26 (**202**), a violet dye on chromium mordant (82).

The action of manganese dioxide and sulfuric acid introduces an additional hydroxyl group to the quinalizarin structure. This dye, CI Mordant Blue 50 (**203**), gives a greenish-navy on chromium mordant (83).

By reaction of fuming sulfuric acid and sulfur with 1,5-dinitroanthraquinone,

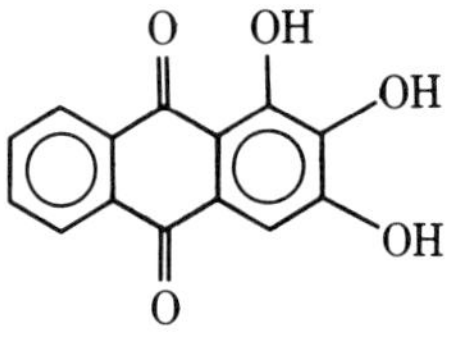

(**196**)
[602-64-2]

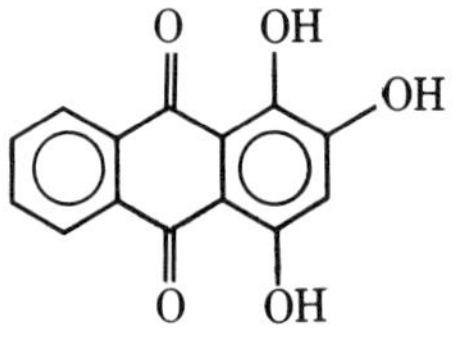

(**197**)
[81-54-9]

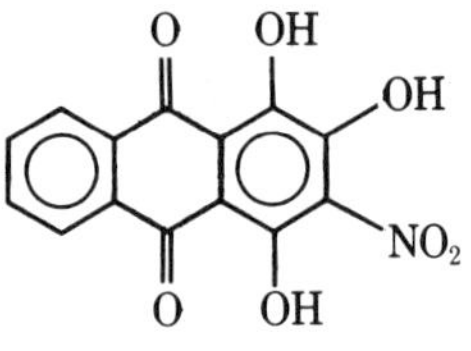

(**198**)
[6486-91-5]

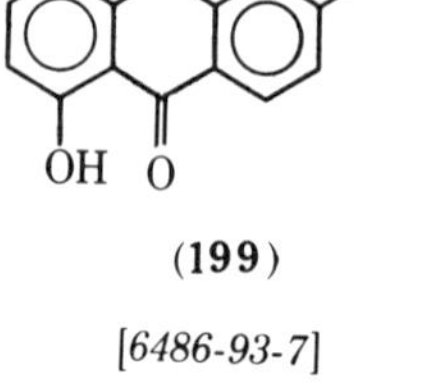

(**199**)
[6486-93-7]

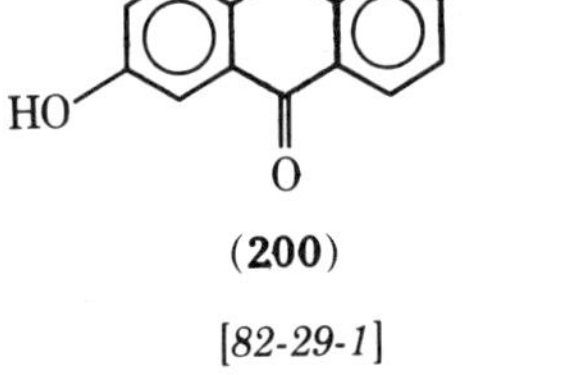

(**200**)
[82-29-1]

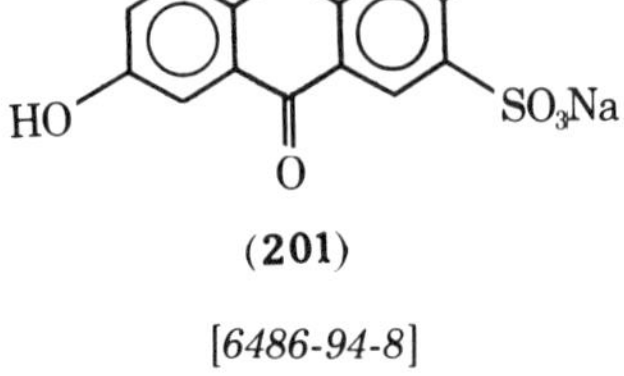

(**201**)
[6486-94-8]

or a mixture of the 1,5- and 1,8-compounds, a blue mordant dye (CI Mordant Blue 23) (**204**) is obtained. It produces a reddish-navy on chromium mordant (84).

1,2,4,5,6,8-Hexahydroxyanthraquinone (CI Mordant Blue 32) (**205**) produces a reddish-navy on chromium mordant and similarly, 1,2,4,5,7,8-hexahydroxyanthraquinone (Mordant Dye 58615) (**206**) dyes reddish blue on chromium mordant and violet on aluminum mordant. These are important dyes for coloring wool in all forms as well as for dyeing leather.

More complex substituted mordant dyes have been obtained, eg, in the manufacture of CI Mordant Blue 48 (**207**) where 1-amino-4-bromo-2-anthraquinonesulfonic acid is condensed with 5-(3-amino-4-methoxybenzylsulfonyl)salicylic acid and converted to the sodium or ammonium salt (85).

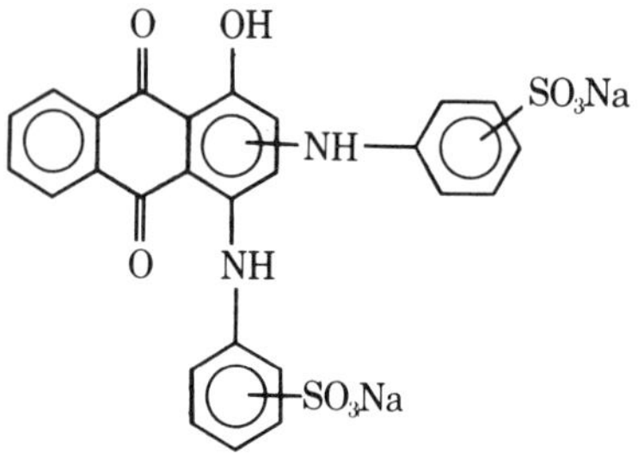

274 DYES, ANTHRAQUINONE

(208) [1324-21-6]

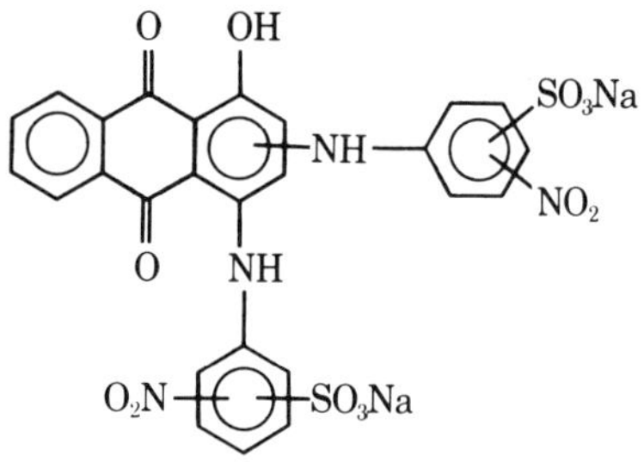

(209) [61968-77-2]

Commercially important mordant blacks are obtained by condensing purpurin with aniline in the presence of boric acid, sulfonating with sulfuric acid monohydrate and converting to the sodium salt. This gives CI Mordant Black 13 (208) (86). The same sequence using nitroaniline during the condensation gives CI Mordant Black 57 (209) (87).

Heat Transfer Dyes

Transfer printing refers to a process in which a thin film of volatile dyes applied to an inert substrate, such as paper, is transferred by dry heat to a suitable fabric held in contact with the paper. The heating and good contact required for the procedure are achieved with a hot iron, hot press or continuous calender. The fabrics colored are those that demonstrate an affinity for disperse dyes.

The heat transfer printing process is expected to encourage the present day trend towards the use of polyester as the dominant clothing fabric. Its advantages include the ability to reproduce intricate patterns with excellent clarity. On the other hand, limitations exist in the range of colors, and the degree of colorfastness exhibited on all fabrics, including polyester.

The chemistry of transfer printing dyes has received little attention and yet is associated with a serious concern of this new technology: the need to correlate sufficient volatility with good fastness properties. Research in this area of dyes found that in order to achieve adequate sublimation, the dye should have a molecular weight less than 350, nonionic polar groups should be kept to a minimum (eg, —NO$_2$, —CN, —SO$_2$R, —OH, —NH$_2$, —NHR), and with few exceptions no ionized groups should be present. Low molecular weight dyes such as (210), which prints yellow on polyester,

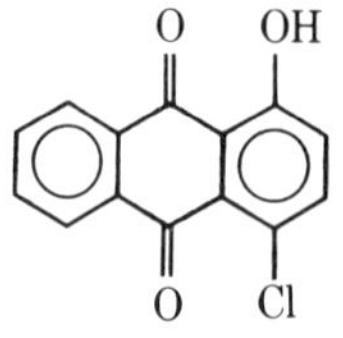

(210) [82-42-8]

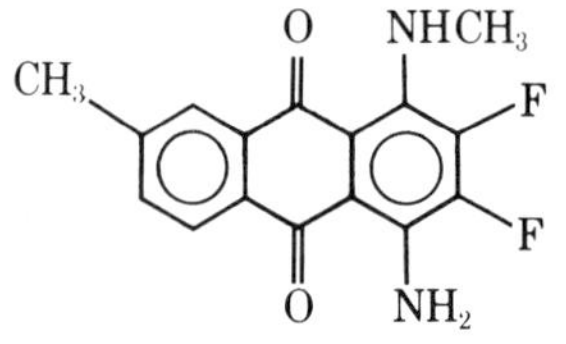

(211) [55447-77-3]

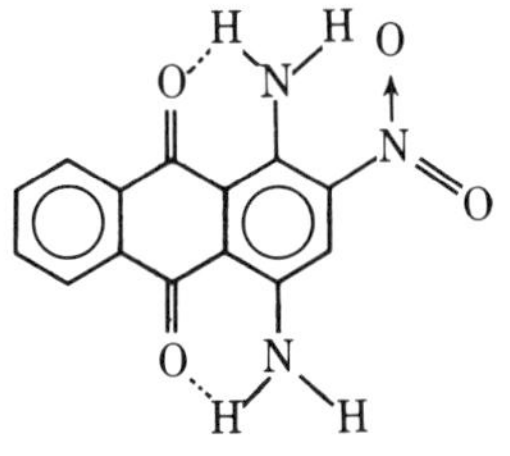

(212) [23677-62-5]

have found special applications commercially because of their highly volatile nature. This dye, eg, colors the fabric in only 15 s at 150°C.

Methods of improving volatility, other than lowering the molecular weight include steric crowding and hydrogen bonding. Compound (**211**) illustrates an anthraquinone dye whose lower melting point and increased volatility are owing to steric crowding. When electron acceptor groups are required in a chromagen, cyano groups and fluorine atoms have been used preferentially to nitro groups and chlorine atoms (88). These and similar choices are related to the color and fastness properties demonstrated by the dyes possessing these substitutions.

Intramolecular hydrogen bonding has been credited with maintaining volatility in dye molecules with polar group substitutions. The greenish-blue dye (**212**) (89), as well as various 1,4,5- and 1,4,5,8-hydroxy- and amino-substituted anthraquinones (90), such as CI Disperse Blue 56 (**213**), have found utility in the transfer printing process despite the fact that these groups via the delocalization of electron density are associated with the hydrogen bonds. The dyes actually being used in this process today are varied. A few examples are described below accompanied by their respective shortcomings.

Quinizarin (**214**), which meets the three requirements for adequate sublimation, has one of the least complex structures in this group of dyes (91). This dihydroxyanthraquinone has been used effectively for orange printing on polyamide fabric.

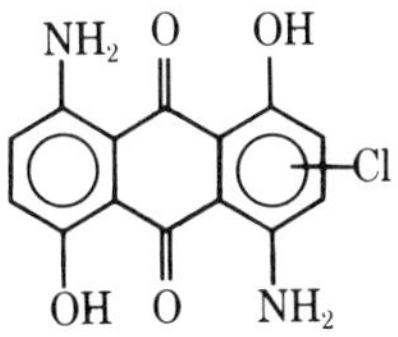

(**213**)
[12217-79-7]

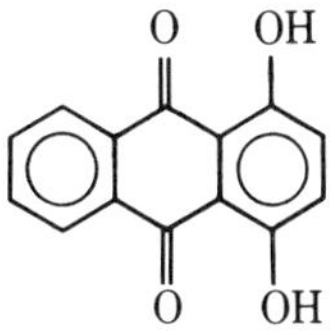

(**214**)
[81-64-1]

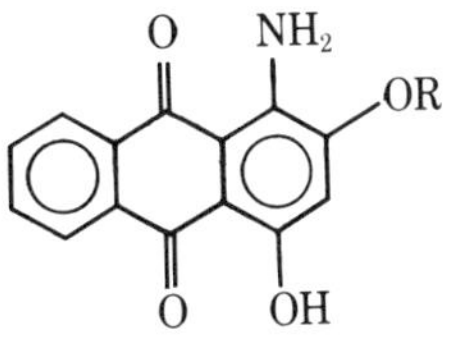

(**215**) CH$_3$ [2379-90-0]
(**216**) C$_6$H$_5$ [17418-58-5]

At the next higher level of complexity in structure are the monosubstituted 1,4-diamino- and 1,4-amino-hydroxyanthraquinones. The 2-alkoxy and 2-aryloxy compounds, such as (**215**) and (**216**), are good heat transfer dyes. The substitution in the 2-position of a nonpolar group maintains the subliming nature and produces pink to red dyes which have been used to impregnate cotton and polyester cotton fabrics (92).

The 1,4-bisalkylamino- and 1-amino-4-alkylaminoanthraquinones (**217–219**) have been popular for heat transferring, especially of blue to greenish-blue shades. Unfortunately, inadequate lightfastness has frequently been a problem with these derivatives.

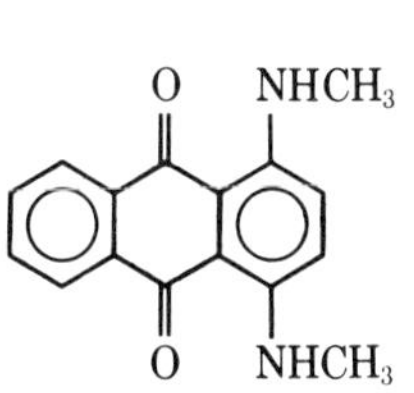

(**217**)
[2475-44-7]

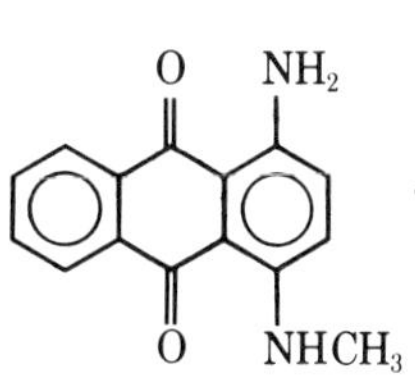

(**218**)
[1220-94-6]

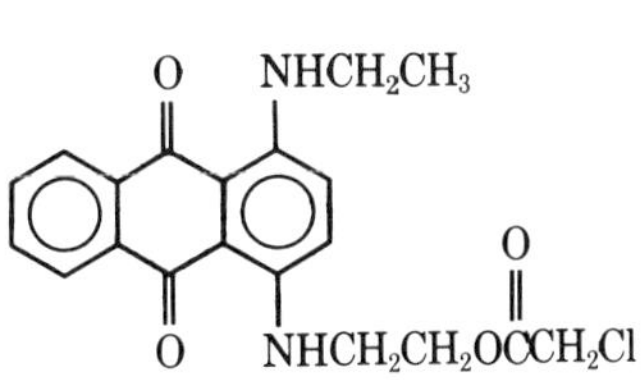

(**219**)
[61734-62-1]

276　DYES, ANTHRAQUINONE

Recently, however, research addressed to this problem has been of some assistance. It was discovered that the introduction of a cyano group in the 2- or 2,3-positions not only produced a useful bathochromic shift, but more importantly improved significantly the lightfastness of the dyes. Compounds such as (**220**) and (**221**) are now being used for transfer printing.

The choice of alkyl or aryl groups used to alkylate aminoanthraquinones also appears to be of some importance to these dyes. The branched chains such as isopropyl and isobutyl, eg, as in dye (**222**), have been popular in the alkylamino series, and the tolyl group, dye (**223**), has been used in the arylamino series.

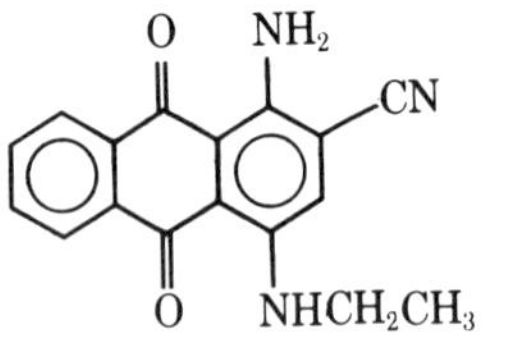

(**220**)
[62570-50-7]

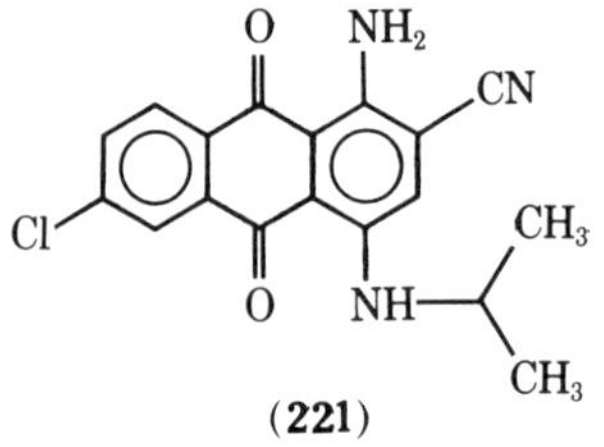

(**221**)
[63723-36-4]

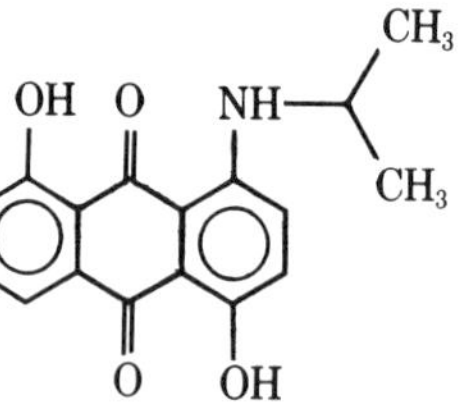

(**222**)
[70416-87-4]

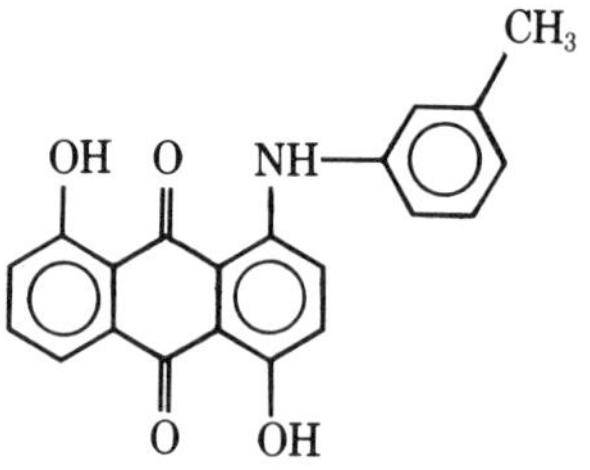

(**223**)
[51730-30-4]

Table 5.　United States Production of Anthraquinone Dyes from 1962–1972[a]

Year	Production, t	Value, $1000	Wt % of total dye production
1962	19,270	61,476	22.5
1963	18,570	61,084	20.1
1964	18,900	66,889	22.6
1965	21,620	72,664	23.0
1966	24,600	85,501	24.7
1967	23,450	86,671	25.1
1968	24,990	95,760	24.3
1969	23,510	98,476	21.6
1970	23,570	99,661	22.2
1971	21,410	108,252	19.4
1972	21,130	125,399	17.7

[a] Taken from the U.S. International Trade Commission Reports on Synthetic Organic Chemicals. After 1972 these reports no longer report data by chemical class.

Economic Aspects

Anthraquinone dyes and colorants make important contributions to a number of widely different usage groups including acid, direct, disperse, mordant, solvent, vat, pigment and reactive dyes as well as cosmetics. This importance despite the complexity of manufacture and consequently high cost, is attributed to the all-around fastness properties exhibited by many anthraquinones and unsurpassed by most other types of dyes. Although the high manufacturing costs have created significant competition from less expensive dyes particularly monoazo types, the anthraquinones continue to play an important role in the field of dyes for synthetic fibers. The percentage of total dye production and dollar value for these dyes are given in Table 5.

BIBLIOGRAPHY

"Anthraquinone and Related Quinonoid Dyes" in *ECT* 1st ed., Vol. 1, pp. 959–982, by Jesse Werner, General Aniline Works, Div. General Aniline & Film Corporation; "Anthraquinone Dyes" in *ECT* 2nd ed., Vol. 2, pp. 501–533, by A. J. Cofrancesco, General Aniline & Film Corporation.

1. U.S. Pat. 3,201,415 (Aug. 17, 1965), J. M. Straley and D. J. Wallace (to Eastman Kodak Company); U.S. Pat. 2,659,739 (Nov. 17, 1953), J. B. Dickey (to Eastman Kodak Company).
2. U.S. Pat. 3,249,626 (May 3, 1966), R. Neeff and O. Bayer (to Farbenfabriken Bayer Aktiengesellschaft).
3. U.S. Pat. 2,852,535 (Sept. 16, 1958), A. Peter and E. Wydler (to Sandoz A.G.).
4. U.S. Pat. 2,585,681 (Feb. 12, 1952), D. I. Randall and E. E. Renfrew (to General Aniline and Film Corp.); U.S. Pat. 3,989,449 (Nov. 2, 1976), L. Dixon (to L. B. Holliday & Co.).
5. Ger. Offen. 2,623,224 (Dec. 16, 1976), S. Koller (to Ciba-Geigy A.G.).
6. U.S. Pat. 3,929,401 (Dec. 30, 1975), F. Krock and R. Neeff (to Bayer Aktiengesellschaft); U.S. Pat. 3,920,702 (Nov. 18, 1975), F. Krock and R. Neeff (to Bayer Aktiengesellschaft); U.S. Pat. 3,968,131 (July 6, 1976), K. Maier (to BASF Aktiengesellschaft); U.S. Pat. 3,980,678 (Sept. 14, 1976), E. Yamada, and co-workers (to Sumitomo Chemical Co.).
7. U.S. Pat. 3,929,842 (Dec. 30, 1975), E. Yamada and M. Nishikuri (to Sumitomo Chemical Co.).
8. U.S. Pat. 4,035,395 (July 12, 1977), F. Krock (to Bayer A.G.).
9. U.S. Pat. 3,299,103 (Jan. 17, 1967), K. Maier (to BASF A.G.).
10. Ger. Offen. 2,644,145 (Apr. 21, 1977), W. Baumann (to Sandoz A.G.).
11. Ger. Offen. 2,454,327 (May 20, 1976), F. W. Kroeck (to Bayer A.G.).
12. U.S. Pat. 2,640,062 (May 26, 1953), G. W. Seymour and V. S. Salvin (to Celanese Corporation of America); U.S. Pat. 2,768,052 (Oct. 23, 1956), R. C. Johnson (to E. I. du Pont de Nemours & Co., Inc.); U.S. Pat. 2,844,598 (July 22, 1958), J. Gunthard (to Sandoz A.G.); U.S. Pat. 2,992,240 (July 11, 1961), F. Lodge (to ICI Ltd).
13. U.S. Pat. 4,041,052 (Aug. 9, 1977), H. Schwander and U. Karlen (to Ciba-Geigy A.G.).
14. U.S. Pat. 4,008,222 (Feb. 15, 1977), H. Machatzke (to Bayer A.G.).
15. U.S. Pat. 4,041,051 (Aug. 9, 1977), E. Yamada, H. Ito, and T. Akamatsu (to Sumitomo Chemical Co.).
16. U.S. Pat. 3,099,513 (July 30, 1963), G. Gehrke (to Bayer A.G.).
17. Ger. Offen. 2,634,169 (Feb. 17, 1977), U. Karlen (to Ciba-Geigy A.G.).
18. U.S. Pat. 3,072,683 (Jan. 8, 1963), J. M. Straley and R. Giles (to Eastman Kodak Co.); U.S. Pat. 3,240,551 (Mar. 15, 1966), J. W. Fitzpatrick, K. H. Lohmann, and O. Grosz (to Toms River Chemical Corp.).
19. U.S. Pat. 2,573,732 (Nov. 6, 1951), V. S. Salvin and J. R. Adams (to Celanese Corporation of America); U.S. Pat. 2,573,733 (Nov. 6, 1951), V. S. Salvin and J. R. Adams (to Celanese Corporation of America); U.S. Pat. 2,587,002 (Feb. 26, 1952), G. W. Seymour and V. S. Salvin (to Celanese Corporation of America).
20. U.S. Pat. 2,641,602 (June 9, 1953), J. M. Straley and J. B. Dickey (to Eastman Kodak Co.); U.S. Pat. 2,777,863 (Jan. 15, 1957), J. B. Dickey and E. B. Towne (to Eastman Kodak Co.); U.S. Pat. 3,154,568 (Oct. 27, 1964), R. A. Walker and V. S. Salvin (to Celanese Corporation of America).

21. U.S. Pats. 2,967,753, 2,967,754 (Jan. 10, 1961), P. Bitterli and J. Gunthard (to Sandoz A.G.).
22. U.S. Pat. 3,928,396 (Dec. 23, 1975), N. Kishi, Y. Ota, and N. Okuno (to Mitsubishi Chemical Industries Ltd.).
23. U.S. Pat. 3,184,455 (May 18, 1965), R. Neeff (to Bayer A.G.).
24. C. F. Allen, *J. Org. Chem.* **6,** 743 (1941).
25. *British Intelligence Objectives Subcommittee Report* **1493,** 8 (1946).
26. U.S. Pat. 3,195,973 (July 20, 1965), O. Fuchs and H. Rentel (to Farbwerke Hoechst A.G.).
27. U.S. Pat. 4,008,222 (Feb. 15, 1977), H. Machatzke (to Bayer A.G.).
28. U.S. Pat. 2,805,225 (Sept. 3, 1957), P. Grossmann and W. Jenny (to Ciba Ltd.).
29. U.S. Pat. 2,990,413 (June 27, 1961), G. Gehrke (to Bayer A.G.).
30. U.S. Pat. 3,992,421 (Nov. 16, 1976), R. Botros (to American Color and Chemical Corp.).
31. U.S. Pat. 2,628,963 (Feb. 17, 1953), J. F. Laucius and S. B. Speck (to E. I. du Pont de Nemours & Co., Inc.).
32. U.S. Pat. 2,753,356 (July 3, 1956), J. F. Laucius and S. B. Speck (to E. I. du Pont de Nemours & Co., Inc.).
33. U.S. Pat. 3,100,132 (Aug. 6, 1963), W. Jenny and R. Witwer (to Ciba Ltd.); U.S. Pat. 4,006,163 (Feb. 1, 1977), H. P. Kolliker (to Ciba-Geigy Co.).
34. U.S. Pat. 3,203,751 (Aug. 31, 1965), J. D. Hildreth (to Toms River Chemical Corp.).
35. U.S. Pat. 2,692,272 (Oct. 19, 1954), M. A. Perkins (to E. I. du Pont de Nemours & Co., Inc.).
36. H. Kaelliker and P. Caveng, *Chimia* **20,** 281 (1966).
37. U.S. Pat. 3,294,815 (Dec. 27, 1966), E. Hartwig and W. Braun (to BASF A.G.).
38. U.S. Pat. 3,254,078 (May 31, 1966), J. M. Straley and D. J. Wallace (to Eastman Kodak Co.).
39. U.S. Pat. 3,939,162 (Feb. 17, 1976), W. Elser and M. Ruske (to BASF A.G.).
40. U.S. Pat. 2,335,412 (Nov. 30, 1943), W. Hentrich and H. J. Engelbrecht (vested to Alien Property Custodian).
41. U.S. Pat. 1,004,107 (Sept. 26, 1912), P. Thomaschewski (to F. Bayer and Co.).
42. U.S. Pat. 2,042,757 (June 2, 1936), K. Zahn, H. Koch, and K. Weinard (to General Aniline Works, Inc.).
43. H. Labhart, *Helv. Chim. Acta* **40,** 1410 (1957).
44. U.S. Pat. 2,314,356 (Mar. 23, 1943), H. R. Lei and C. A. Young (to E. I. du Pont de Nemours & Co., Inc.).
45. U.S. Pat. 2,453,100 (Nov. 2, 1948), E. E. Renfrew (to General Aniline and Film Corp.); U.S. Pat. 2,453,104 (Nov. 2, 1948), C. G. Vogt (to General Aniline and Film Corp.).
46. *Field Information Agencies Technical Report 1313* **2,** 236 (Feb. 1, 1948).
47. Ger. Pat. 560,237 (Sept. 15, 1932), P. Nawiasky, B. Stein, and A. Vilsmeier (to I.G. Farbenindustrie).
48. U.S. Pat. 2,468,599 (Apr. 26, 1949), H. Z. Lecher, M. Scalera, and W. S. Forster (to American Cyanamid Co.); *Field Information Agencies Technical Report 1313,* **2,** 174 (1948).
49. U.S. Pat. 3,962,251 (June 8, 1976), W. Bachmann and T. Flatt (to Allied Chem. Co.).
50. *Field Information Agencies Technical Report 1313* **2,** 113 (1948).
51. U.S. Pat. 2,385,113 (Sept. 18, 1945), G. M. Smyth (to American Cyanamid Co.).
52. Ger. Offen. 2,632,653 (Feb. 10, 1977), H. Altermatt (to Ciba-Geigy A.G.).
53. U.S. Pat. 2,874,168 (Feb. 17, 1959), D. E. Graham and E. V. Hort (to General Aniline and Film Corp.).
54. U.S. Pat. 2,187,812 (Jan. 23, 1940), F. Baumann and H. W. Schwechten (to General Aniline Works).
55. U.S. Pat. 3,917,640 (Nov. 4, 1975), J. W. Fitzpatrick and H. N. Schmidt (to Toms River Chemical Corp.).
56. U.S. Pat. 2,302,729 (Nov. 24, 1942), M. Whelen (to E. I. du Pont de Nemours & Co., Inc.).
57. U.S. Pat. 2,468,599 (Apr. 26, 1949), H. Z. Lecher, M. Scalera, and W. S. Forster (to American Cyanamid Co.).
58. U.S. Pat. 2,439,626 (Apr. 23, 1948), W. Kern (to Soc. pour l'ind Chim. a Bale).
59. U.S. Pat. 4,035,397 (July 12, 1977), J. Pechmeze, P. Touratier, and L. Cabut.
60. Ger. Offen. 2,508,506 (Sept. 9, 1976), W. Harms and D. Huellserung (to Bayer A. G.).
61. U.S. Pat. 938,566 (1909), Fischer.
62. U.S. Pat. 2,233,496 (Mar. 4, 1941), F. B. Stilmar (to E. I. du Pont de Nemours & Co., Inc.).
63. U.S. Pat. 4,049,656 (Sept. 20, 1977), P. Hindermann and H. Meindl (to Ciba-Geigy A.G.).
64. Belg. Pat. 843,696 (Jan. 3, 1977), R. Mislin, W. Schoenauer, and K. Stecker (to Sandoz Ltd.).

65. Fr. Dem. 2,293,472 (July 2, 1976), J. Pechmez and P. Touratier (to Uegine Kuhlmann).
66. Belg. Pat. 831,283 (Jan. 12, 1976) A. Tertre (to ATSA).
67. Ger. Pat. 390,201 (Jan 4, 1924), (to Gesellschaft fur Chemische Industrie im Basel).
68. Swiss. Pat. 559,766 (Mar. 14, 1975), A. Buehler, H. Schuetz, and G. Hoelze (to Ciba-Geigy A.-G.).
69. U.S. Pat. 2,408,259 (Sept. 24, 1946), T. Holboro, W. Kern, and P. Sutter (to Society of Chemical Industry in Basle).
70. Ger. Offen. 2,640,195 (Mar. 24, 1977), P. Hindermann and J.-M. Adam (to Ciba-Geigy A.-G.).
71. U.S. Pat. 2,346,726, E. C. Baxbaum (to E. I. du Pont de Nemours & Co., Inc.).
72. U.S. Pat. 3,915,999 (Oct. 28, 1975), F. Glaser (to BASF A.G.).
73. U.S. Pat. 4,014,906 (Mar. 29, 1977), U. Karlen and H. Morawietz (to Ciba-Geigy Corp.).
74. U.S. Pat. 3,935,248 (Jan. 27, 1976), J. S. Mouson, D. Robert, and A. Ridyard (to ICI Ltd.).
75. U.S. Pat. 3,980,677 (Sept. 14, 1976), W. Hohmann, K. Wunderlich, and H. S. Bien (to Bayer A.G.).
76. U.S. Pat. 186,032 (Jan. 9, 1877), H. Caro.
77. Ger. Pat. 74,598 (Mar. 7, 1894), (to Bayer Co.); *Friedl* **3,** 264 (1890–1894).
78. *Friedl* **4,** 1134 (1895–1896).
79. *Friedl* **3,** 252 (1890–1894).
80. *Friedl* **9,** 691 (1908–1909).
81. *Ber. Dtsch. Chem. Ges.* **9,** 1678 (1876).
82. R. Schmidt, *J. Prackt. Chem.* **43,** 237 (1891).
83. L. Gattermann, *J. Prackt. Chem.* **43,** 250 (1891).
84. *Friedl* **4,** 283 (1895–1896).
85. *Friedl* **23,** 969 (1936).
86. *British Intelligence Objectives Subcommittee Report* **1484,** 26 (1946); *Field Information Agencies Technical Rev. Ger. Sci 1313* **2,** 210 (1948).
87. *British Intelligence Objectives Subcommittee Report* **1484,** 36 (1946); *Field Information Agencies Technical Rev. Ger. Sci 1313,* **2,** 236 (1948).
88. Jpn. Kokai 74 126,984 (Dec. 5, 1974), M. Miyatake, M. Anzai, and S. Ebashi (to Tottan Printing Company).
89. Brit. Pat. 1,403,619 (Aug. 20, 1975), J. F. Dawson and M. B. Masters (to Yorkshire Chemical Company).
90. Jpn Kokai 74 110,985 (Oct. 22, 1974), E. Yamada, M. Mishijuri, and S. Kawamura (to Sumitomo Chemical Company).
91. U.S. Pat. 4,029,468 (June 14, 1977), K. Maier (to BASF A.G.).
92. U.S. Pat. 3,915,628 (Oct. 28, 1975), W. Bossard (to Ciba-Geigy A.G.).

RACK HUN CHUNG
General Electric Co.

RUSSELL E. FARRIS
Sandoz Colors & Chemicals

DYES, APPLICATION AND EVALUATION

Application, 280
Theories of dyeing, 282
Dyeing machinery, 286
Preparation for dyeing, 294
Acid dyes, 297
Basic dyes, 301
Direct dyes 302
Disperse dyes, 305
Fiber-reactive dyes, 308
Sulfur dyes, 310
Insoluble azo dyes, 311
Vat dyes, 313
Miscellaneous types of dyes, 314
Dyeing of fiber blends, 316
Pigment dyeing, 319
Solvent dyeing, 319
Textile printing, 321
Paper coloring, 329
Leather dyeing, 331
Evaluation, 334
Fastness tests for textiles, 334
Application properties, 345
Qualitative and quantitative analyses, 346
Bibliography, 349

APPLICATION OF DYES

Dyeing may have been inspired by the accidental staining of primitive garments with food and drink or by the discoloration of sheep's fur with fruit juices while the animals grazed among wild shrubbery. The history of dye application was essentially unknown until about 1500 BC because earlier examples of dyed textiles survived only in favorable climatic conditions. However, ancient Egyptian, Babylonian, Hebrew, Indian, and Chinese examples of dyed textiles, as well as literary evidence of earlier dye application, are known. By the beginning of the Christian era, dye application had developed into a well-established technology as evidenced, eg, by the Egyptian *papyrus Leidensis* and *papyrus Holmiensis* which contain detailed descriptions of dyeing processes. Application of natural dyes to textiles developed steadily and became an important industrial craft, as evidenced by the existence of dyers' guilds or corporations as early as the 12th century AD in England. A radical departure from the craft occurred in the 19th century with the synthesis of the first synthetic dye. A few important dates in the history of this change include the syntheses of Mauveine (Mauve) [6373-22-4] by Perkins in 1856; Alizarine Red [72-48-0] by Gräbe and Lieberman in 1868 and Indigo [482-89-3] by von Baeyer in 1870; the development of azo dyes (qv) on fibers since 1880; the discovery of anthraquinone vat dyes (qv) by Bohn in 1901; the development of disperse dyes for cellulose acetate in 1922; and the development of fiber-reactive dyes which led to the introduction of a range of such dyes by ICI in 1956.

Myriads of dyes have been synthesized in laboratories around the world, a few hundred of which have found their way to large-scale production and industrial application. However, one can expect that this process will slow down in the near future as development of new dyes becomes more expensive, and as safety, health, and ecological restraints become more stringent.

Dyeing describes the imprintation of a new and often permanent color, especially by impregnating with a dye and is generally used in connection with textiles, paper, and leather. Printing may be considered as a special dyeing process by which the dye is applied in locally defined areas in the form of a thickened solution and then is fixed.

Generally dyes are dissolved or dispersed in a liquid medium before being applied to a substrate into which they are fixed by chemical or physical means, or both. Owing to its suitability, its availability, and its economy, water usually is the medium used in dye application; however, nonaqueous solvents have been studied extensively in recent years (see Dye carriers).

Textile substrates can be classified in three groups: cellulosic, protein, and synthetic polymer fibers. The first two are of hydrophilic nature and the last is of hydrophobic nature. Cellulosic fibers are vegetable fibers, such as cotton (qv), linen, jute, (see Fibers, vegetable), and regenerated cellulose fibers such as rayon (qv). Protein fibers comprise wool (qv), silk (qv), and various animal hair fibers. The synthetic-polymeric fibers can be classified according to their ionic properties: nonionic fibers are cellulose acetate and triacetate (qv), and conventional polyester fibers (qv); cationic fibers are polyamides (qv) and anionic fibers are polyacrylonitriles (see Acrylonitrile polymers).

Even and economical distribution of a small amount of dye (generally ≤3 wt % of goods) throughout the substrate and fixation of the dye are the keys to dyeing, ie, with regard to fastness to washing and to other deteriorating influences.

Production of dyeings of acceptable quality requires the use of many auxiliary products and chemicals. These include chemicals that improve fastness properties such as bleaching agents, wetting and penetrating agents, leveling and retarding agents, and lubricating agents. Other agents speed the dyeing process or are used for dispersion, oxidation, reduction, or removal of dyes from poorly dyed textiles.

Dyes of similar or identical chromophoric class are used for widely differing applications and, therefore, are classified according to their usage rather than their chemical constitution. As all dyes are water soluble at one stage of their application, it is also convenient to classify them according to their solubility groups; these can be permanent or temporary:

Permanent solubilizing group	*Types of dye*
—SO_3Na	acid, direct, mordant, fiber reactive
—$\overset{+}{N}H_3Cl^-$, —$\overset{+}{N}R_3Cl^-$	basic
—OH, —NH_2, —SO_2NH_2	disperse
Temporary solubilizing group	*Types of dye*
—ONa	vat, insoluble azo
—OSO_3Na	solubilized vat

Dyes with identical or similar solubilizing groups generally display similar dyeing behavior even though their main structure may vary substantially. Another important consideration of the use of a given dye for a specific application is the fastness properties obtained. Detailed information as to the application and fastness properties of commercial dyes is found in the pattern cards issued by their manufacturers.

The following classification of colorants for dyeing is used: acid, basic, direct, disperse, insoluble azo, sulfur, vat, fiber-reactive, miscellaneous dyes, and pigments. Acid dyes have free acid groups in the ionized state in the dyebath and are normally applied from an acidic dyebath on wool, silk, and polyamides. Basic dyes have cationic groups in the ionized state in the dyebath and are applied on acrylic and other fibers which have structural anionic groups. Direct dyes stain cellulosic fibers and do not require mordants. They are applied from salt-containing baths; the salt increases their affinity for the fiber. Disperse dyes are sparingly soluble in water and are applied as dispersions. Their transport to the fiber, however, is in monomolecularly dissolved form. Their greatest usage is on polyester fibers. Insoluble azo dyes are produced *in situ* by depositing in the fiber a temporarily solubilized phenolic compound and subsequently coupling the phenol with a diazonium salt. Fiber-reactive dyes are water soluble anionic dyes containing a reactive group that can form a covalent bond with a compatible group (typically with the hydroxyl groups of cellulose) on the substrate. Sulfur dyes, many compositions of which are still unknown, have affinity to cellulose in an alkaline reducing bath. After oxidation these dyes become insoluble and fixed on the fiber. Vat dyes are applied in a similar manner. They are insoluble in water and are applied as dispersions. They are then reduced in an alkaline bath, where they form the leuco compound which is water soluble and has affinity for cellulosic and other fibers. By oxidation they again become insoluble and are fixed to the substrate. Miscellaneous dyes include oxidation dyes and natural dyes both of which have limited commercial application. Pigments are not dyes. However, pigment-dyeing and pigment-printing are important commercial processes in which insoluble colorants are applied to textile materials and are attached by a physical process with binders (comparable to painting) (see Pigments; Printing processes).

Theories of Dyeing

The fact that dyes in solution in the dyebath leave the bath and are sorbed by the fiber material during the dyeing process has been of interest to chemists and physicists for many years. Theories of dyeing to describe the sorption (or desorption) process and the resulting dye-fiber bonds, using the tools of thermodynamics, have increased in complexity over the years. Furthermore, it must be recognized that the dyeing process is dynamic in nature. It normally entails changes in the dyeing temperature, the dye concentration, the bath composition, and the fiber structure. Although aspects of the dyeing process can be accurately described with modern theories and related models, a complete description of the dyeing process at the molecular level is still unavailable.

Affinity and diffusion are fundamental aspects of the dyeing process. The former describes the force by which the dye is attracted by the fiber, and the latter describes the speed with which it travels within the fiber from areas of higher concentration to areas of lower concentration. Alone or together with fiber characteristics, affinity and diffusion determine the speed of dyeing, its temperature dependency, the equilibrium exhaustion of the dye, and some of its fastness properties.

Two fundamental dyeing processes have been distinguished: the nonionic and the ionic dyeing processes. The former refers to the dyeing of synthetic fibers such as polyester, nylon and triacetate by disperse dyes. It is governed, as a first approximation, by diffusion behavior according to Fick's second law and by a Nernst-type isotherm. The dye-fiber bonds are probably hydrogen bonds, charge-transfer forces and van der Waals forces. The ionic dyeing process implies ionic groups in fibers and in dyes and the dyeing process involves ion-exchange mechanisms. Ion exchange is clearly involved in dyeing acrylic fibers with basic dyes and nylon with acid dyes as well as wool and silk with acid or direct dyes; the absorption isotherm in these cases is of the Langmuir type or may be a combination of Langmuir and Nernst types. More difficult to explain are dyeing mechanisms involving ionized dyes and substrates such as direct or vat dyes on cellulosic fibers, which contain few reactive groups. Dyeing isotherms are generally of the Langmuir type, and dye-fiber bonds are believed to be hydrogen bonds, ion-dipole interactions and van der Waals forces. Forces similar to these latter three are believed to attract reactive dyes to cellulosic fibers. The dye-fiber bond of properly fixed reactive dyes is a covalent bond and probably the best understood of all such bonds.

In recent years, dyeing theory also has concerned itself with optimal dyeing conditions to achieve level dyeing. Two fundamental approaches are possible. In the first, the dye is applied under carefully controlled conditions at a uniform rate of absorption so that level (uniform) dyeing is maintained from the beginning. Although the exhaustion proceeds at less than maximum rate, little or no time is necessary at the end of the exhaustion step to level out the dye. In the other approach, the dye is exhausted very rapidly, possibly under isothermic conditions at the maximum dyeing temperature, and the initially created unlevelness is then leveled out in the subsequent extended migration period. An important factor in this connection is the relative movement of textile material and of dye liquor. Several detailed and up-to-date treatises on the theory of dyeing have recently been published (1–5).

Dyeing Machinery

In the application of dyes, there have developed over the years three chief principles of dyeing textiles. (1) The dye liquor is moved as the material is held stationary. An illustration of this is raw-stock or package-machine dyeing. (2) The textile material is moved without mechanical movement of the liquor. Examples of this principle are jig dyeing and continuous dyeing which involves the padding of fabric. (3) A combination of (1) and (2) is exemplified by a Klauder-Weldon skein dye machine in which the dye liquor is pumped as the skeins are mechanically turned. Another example is a jet dyeing machine in which both the goods and the liquor are constantly moving. Dyeing equipment is listed under these three categories.

Transportation of Dye Liquor Through the Textiles (Raw Stock, Sliver, Yarn, or Cloth). Regardless of the form of the textile, the principle is generally the same. A large stainless steel kier, capable of withstanding sufficient pressure to reach a maximum operating temperature of 145°C, has one or more perforated spindles through which the dye liquor is pumped. Around this spindle, the textile is packed tightly in the form of a cake in a perforated basket, as yarn, or in a package as a sliver or two in a can, or as cloth around a beam barrel. The dye liquor is pumped through the textile and then flows to the bottom of the machine and into the return side of the pump (see Fig. 1). The stock or sliver machine pumps the liquor only in one direction (from the inside spindle to the outer shell) since the material does not have to be dyed as level as on package or beam machines.

284　DYES, APPLICATION AND EVALUATION

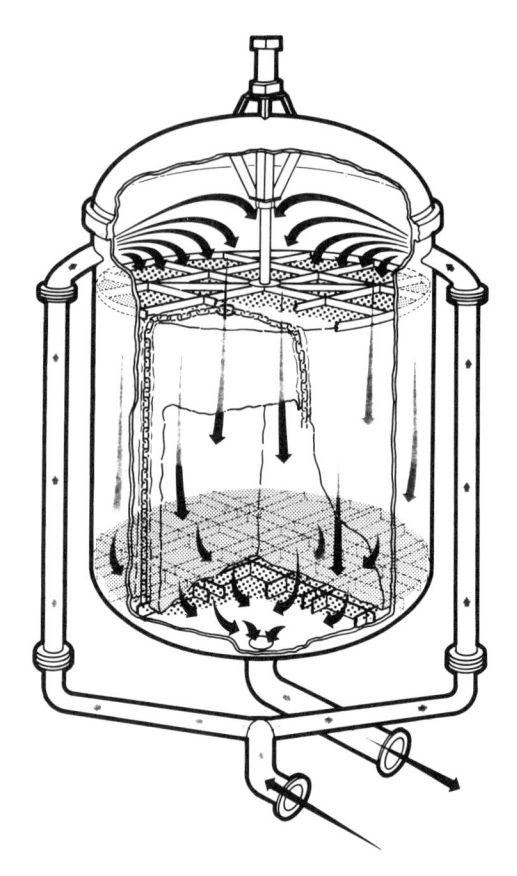

Figure 1. Schematic raw stock dye machine. Courtesy of Gaston County Dyeing Machine Company.

Package and beam machines (Figs. 2 and 3) have a unique automatic reversing mechanism which switches valves and changes flow direction on a preset schedule from inside to outside and from outside to inside. This ensures level dyeing on properly wound and loaded packages or beams.

The kiers are of varying size and production units and weigh 225–1400 kg; pump size and flow is adjusted proportionately. Most machines are highly versatile so as to accommodate the kind of textile being dyed; they are able to control rate of temperature rise, volume of liquor flow and time of dye application. Consider the following example: wool yarn is package-dyed at a maximum of 110°C and a flow of 33 L/(kg·min)(4 gal/lb·min)), however, the same machine dyeing 100% textured polyester may reach a temperature of 135°C and a flow of 75 L/(kg·min)(9 gal/lb·min)). Some machines can convert to dye stock, yarn packages, or beams of warp yarn.

In recent years, manufacturers of these machines have tried to save energy by reducing dyeing time while holding liquor ratio at a minimum. Some companies have been able to reduce the dyeing time of 100% textured polyester yarn from 60 to 15 min. Consequently, of course, production per machine has been increased greatly. When fiber blends are dyed, the time cycle is considerably increased since two different classes of dyes are applied. Fiber blends sometimes require two separate dyebaths with cleaning in between.

Figure 2. Beam dye machine. Courtesy of Parrott and Ballentine.

Transportation of the Textile Material With No Mechanical Movement of The Liquor. Probably the oldest example of dyeing in which the textile alone moves is the ancient box-type skein dye machine in which skeins of yarn were hung on wooden pegs with about three-fourths of the skein submerged in the dye liquor. To prevent stick marks and unlevelness, the sticks were turned by hand at frequent intervals.

Chain Warp Dyeing. This procedure is widely used in the dyeing of indigo, on warp yarns which are in the form of ropes or chains. In this procedure, several warp ropes are pulled through a series of tubs containing the dye liquor and gradually are dyed to the desired shade.

Jig Dyeing. Piece goods have been dyed for many years by means of a batch process on a dye jig (Fig. 4). Several hundred meters of cloth are wound around a beam roller on one side of a jig. The goods are run off the beam through a small-volume (400 L, ca 100 gal), V-shaped dyebath and are wound on the opposite beam. There are friction brakes on each beam which control tension. When the entire cloth has been wound on the opposite beam, the clutch is reversed and the cloth travels back in the opposite direction. The process of running the cloth from one beam to another and back again is called an end. The goods are run a sufficient number of ends until the desired shade is achieved.

Because contact with the dye liquor is for such a brief time, the cloth temperature is much lower than the dyebath. Any fibers or dyes requiring high temperatures (>80°C) should be dyed on an enclosed jig with open steam jets. Many improvements have been made to streamline performance and to reduce machine operation labor. Some of these are (*a*) tensionless jigs using variable speed electric motors with built-in drag for brakes; (*b*) automatic reversing equipment; and (*c*) automatic temperature and level controls.

These machines are used widely for goods that are easily creased, such as fabrics consisting of filament acetate, of heavy filament nylon, or of numbered cotton duck. They are also convenient for small dye lots and for sampling purposes.

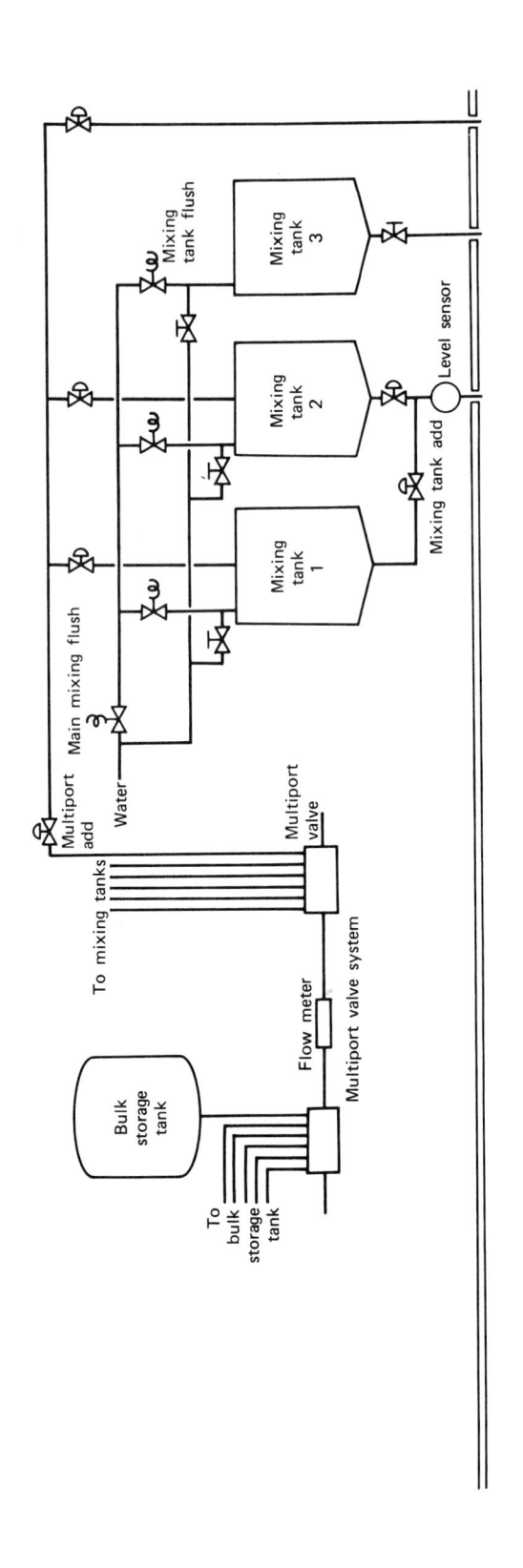

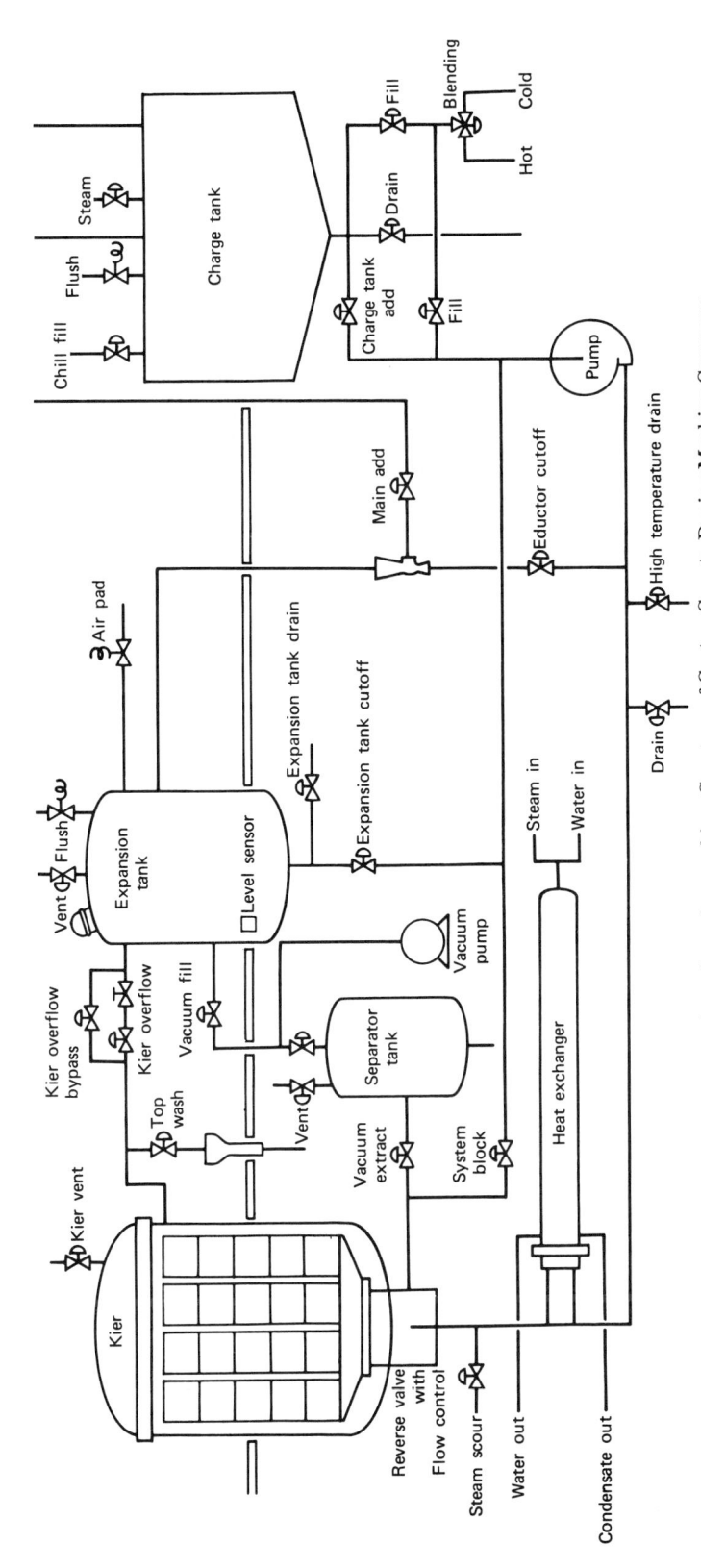

Figure 3. Schematic package dye machine. Courtesy of Gaston County Dyeing Machine Company.

Figure 4. Jig dye machine. Courtesy of Parrott and Ballentine.

Beck Dyeing. This is one of the oldest forms of mechanized piece goods dyeing. The machine (Fig. 5), in its basic form, consists of a shallow U-shaped box which has a gradual low curvature in the back and a rather high vertical rise in the front. About 2 m above the tub is a driven elliptical reel which transports the piece goods from the front side and allows them to slide down the back of the beck and return to the front.

Becks are generally constructed so as to run fabrics in either (*1*) open-width or

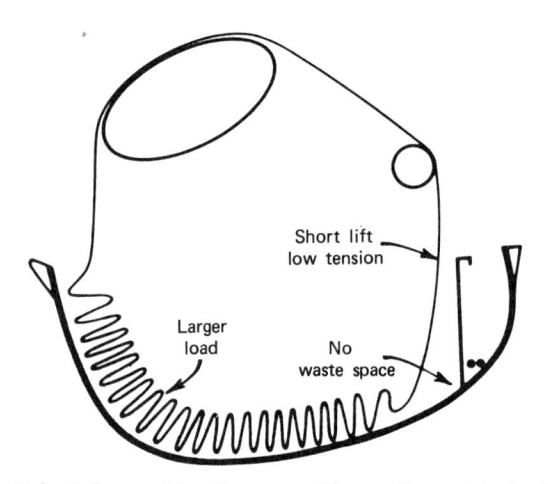

Figure 5. Schematic beck dye machine. Courtesy of Gaston County Dyeing Machine Company.

(2) rope forms. The open-width threading is normally used for heavyweight material, such as carpets, twills, and sateens, which would be damaged if crushed into a "rope" form. In the second form, the beck has a long bar running over the length of the tub and has pegs every 25–30 cm which keep strands separate and prevent them from tangling. When using becks of this type, the strands may be threaded as one long spiral rope which runs as a continuous belt. An alternative method involves threading one piece of cloth per peg division and sewing it together so that one has an endless belt between each set of pegs. All of the pieces in such an arrangement must be approximately the same length to achieve level dyeing.

The modern beck is constructed of stainless steel and may be built for atmospheric temperatures or it may be a heavy-walled, sealed unit suitable for high temperature/high pressure work. Sophisticated controls are added to monitor the entire dye cycle and ensure that dyeings are as consistent as possible.

Continuous Dyeing. Most woven goods, other than stretch fabrics, are continuously dyed provided each shade consists of sufficient yardage for economical operation. The amount of fabric required to justify continuous dyeing depends on the class or classes of dyestuff to be applied and on the available equipment. The equipment at hand may be a simple padder (Fig. 6) or a complete dye range (Fig. 7) 240 m in length.

If a padder is the sole application machine available, then one is limited to dyeing direct or to using fiber-reactive dyes—both by the pad-batch system. Most dye padders consist of a medium-density rubber roller across the width of which pressure can be applied. This roller presses against a stainless steel or a hard rubber roller. These rollers may be mounted either vertically or horizontally. The cloth is passed through a stainless steel pad box, down under a rod or a roller which is below the dye liquor level, between the squeeze rollers and is wound up to allow the dye to penetrate and to fix on the cloth. Wet add-on varies between 50 and 80%, depending on tightness of the squeeze rollers. The tightness of the squeeze is controlled by weight levers, hydraulic pressure, or compressed air.

Almost all of the modern continuous dye ranges (Fig. 7) are built containing the same components; the main difference is the unit that applies energy to the fabric.

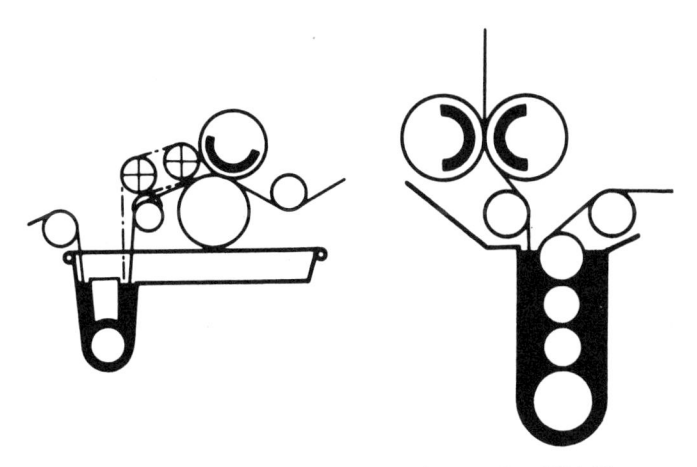

Figure 6. Dye padder. Courtesy of Zima Corporation, USA-Kuesters.

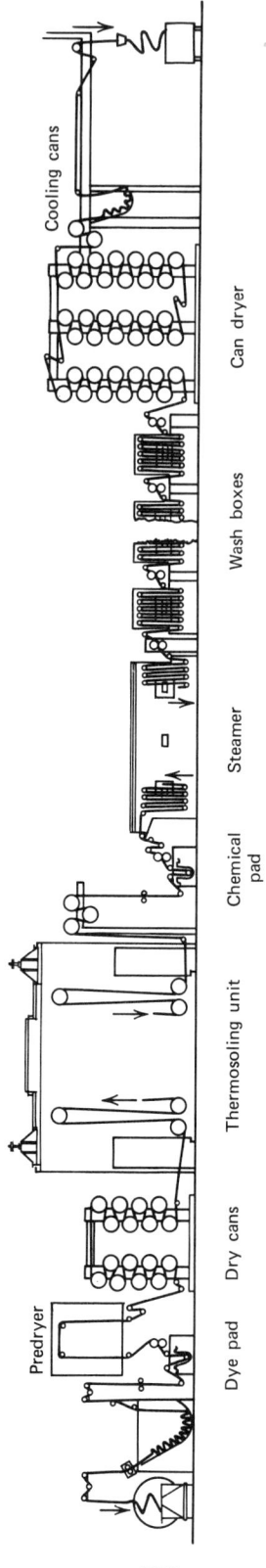

Figure 7. Pad-thermosol-chemical pad-steam dye range.

Examples are an electric infrared dryer, rather than a gas-fired unit, or a hot-air thermosol unit, rather than an oil-heated contact machine.

The infrared units reduce moisture content 20–30% and greatly minimize uneven migration of the dye on the wet goods when they go onto the drying cylinders. The dried cloth progresses into the thermosol unit where the dyestuff for the synthetic portion of the fabric is fixed. Goods then continue through the chemical pad for immersion into an alkaline (or reducing) solution, depending on whether fiber-reactive or vat dyes are applied to the cellulosic fibers. They then pass through eight to ten wash boxes, which contain various chemicals, depending on the class of dyestuff applied. Generally the sequence for vat dyes is (*1*) wash cold; (*2*) oxidize hot; (*3*) wash hot; (*4*) soap hot; (*5*) wash hot; and (*6*) dry (see Drying).

In the past, booster boxes, Williams units and molten-metal units were used in place of the steamer. However, the steamer is used almost universally in the modern dye range.

Machines Based on Movement of Both Dye Liquor and Material. As mentioned earlier, one example of a machine in which both yarn and dye liquor are moved is the Klauder-Weldon skein dye machine (Fig. 8); not only do the skeins turn, but the liquor is pumped in small streams over the yarn as the threads pass over the spindles. This process assures maximum uniformity and levelness.

Another extremely popular machine of this type is the jet dyeing machine (Fig. 9), which conserves energy by reducing the cloth to liquor ratio to 1:10 as compared to 1:20 for the beck. In this machine, the fabric which is in a rope form, is transported

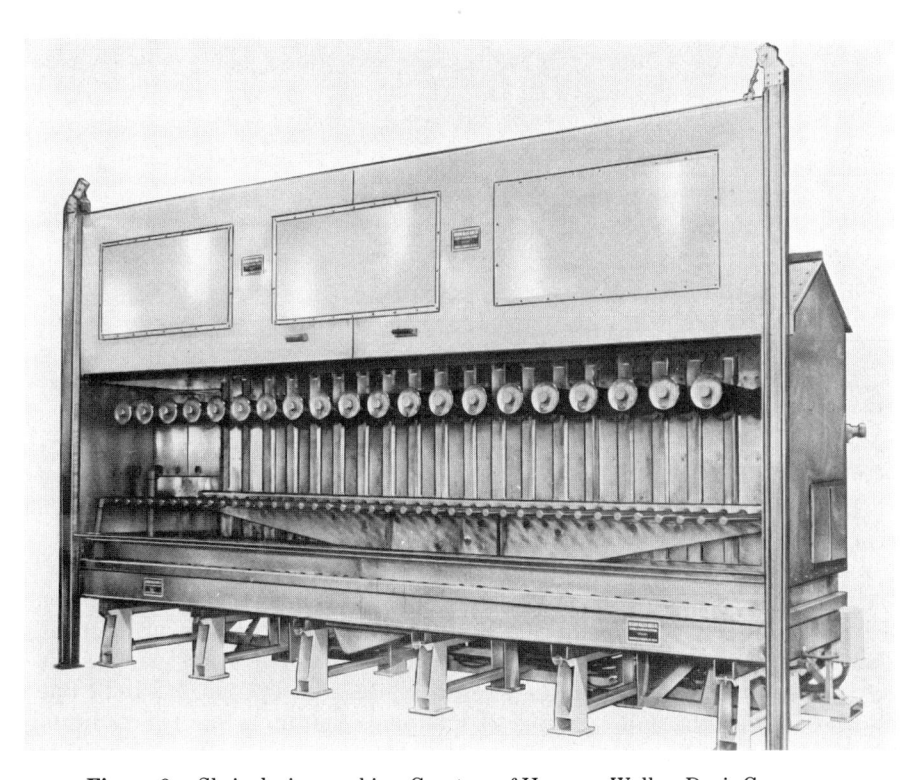

Figure 8. Skein dyeing machine. Courtesy of Hussong-Walker-Davis Company.

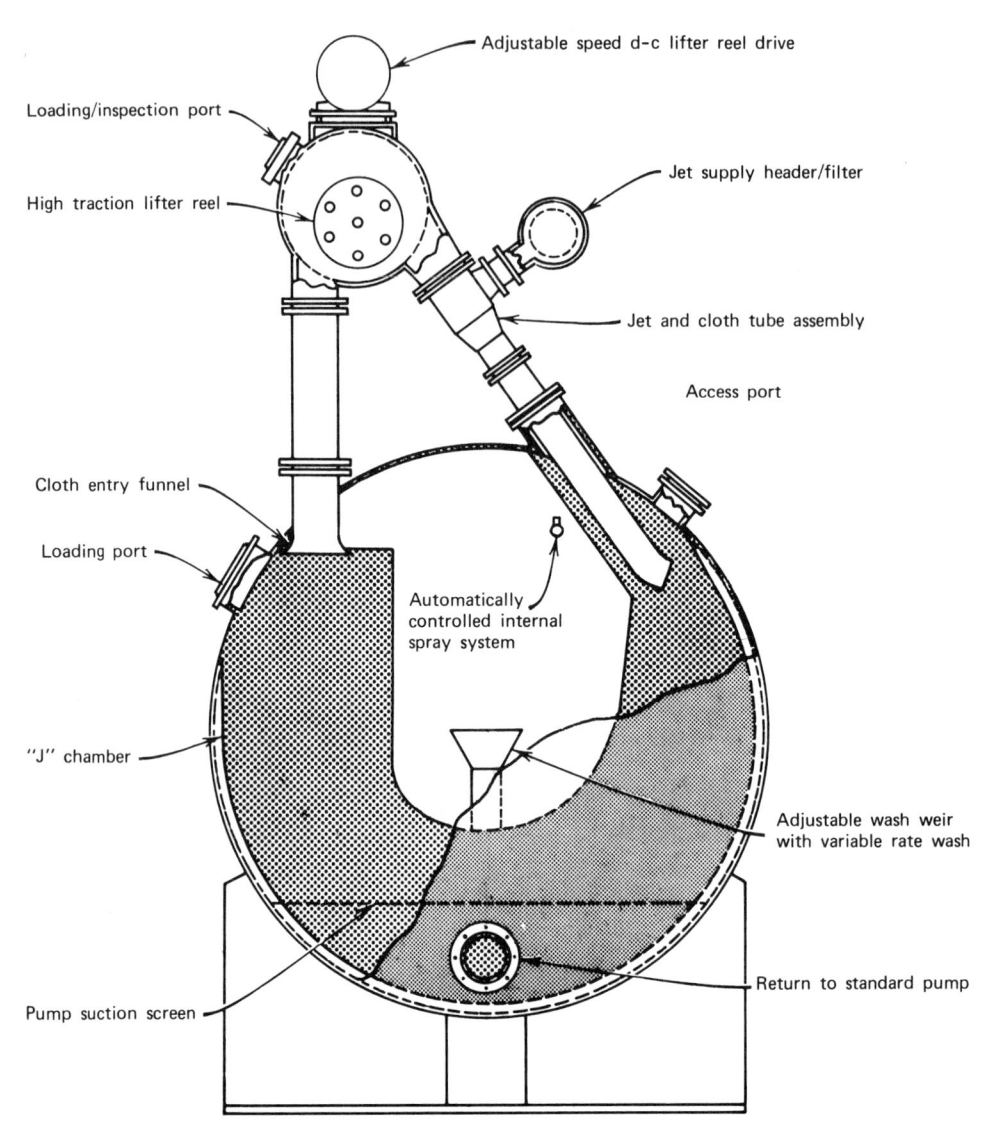

Figure 9. Jet dye machine (Mark V). Reel speeds up to 360 m/min and various automatic controls. Courtesy of Gaston County Dyeing Machine Company.

by movement of the dye liquor through a venturi jet. This method allows extremely gentle handling of the goods while it provides intimate contact between the dye liquor and each meter of material. The machine operates at 40–135°C.

In these times of extreme mindfulness of energy consumption, another new batch dyeing machine has entered the market, the Aqualuft (Fig. 10), which is manufactured by Gaston County Dyeing Machine Company. The Aqualuft dyes in an aqueous foam medium. The goods are lifted to the top of the machine by a reel and then pass through a venturi tube into which dye liquor foam is injected by a pump. It can be operated at 40–135°C. Since the unit operates with a foam medium rather than a liquid one, the cloth ratio runs as low as 1:6. This low ratio of total water consumed signifies savings in heating energy, effluent waste water, and total water required for dyeing.

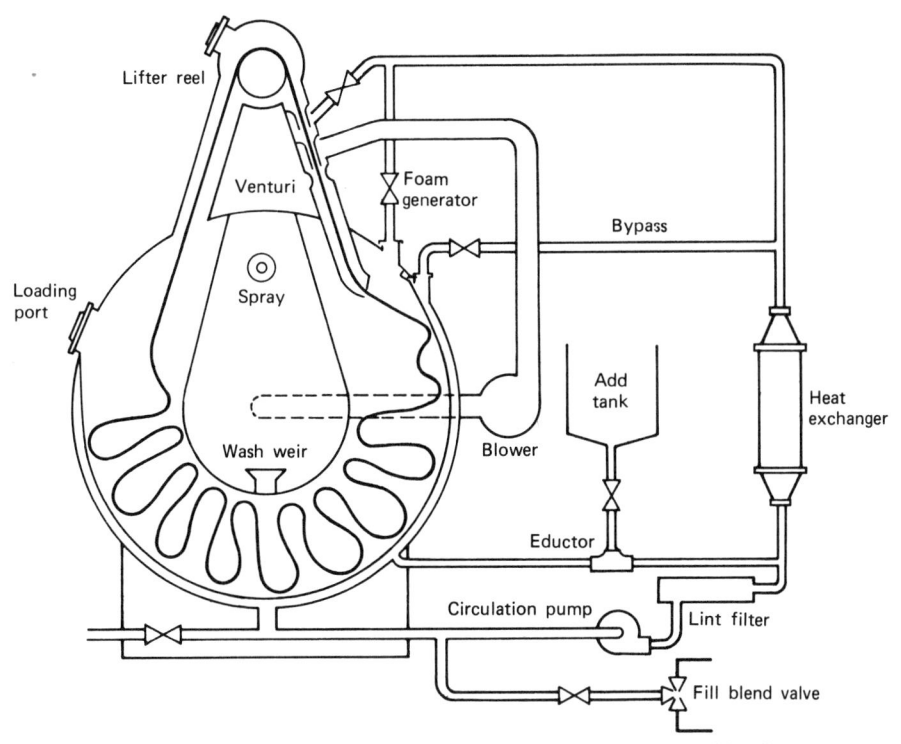

Figure 10. Aqualuft machine. Courtesy of Gaston County Dyeing Machine Company.

Control of Dyeing Equipment. Over the years, the dyer and machinery manufacturer have applied any mechanical or electrical equipment that would enable them, day after day, to produce repeatable dyeings of top quality. First, thermometers were installed in dye lines; these soon evolved into thermocouples with remote recording. Other improvements were soon developed, such as automatic four-way valves with variable-interval controls, flow controls, pressure recorders, hydraulic and air pressure sets on rollers, pH controls, etc (see Instrumentation and control).

In the late 1940s, chart and punch-card process controls were integrated in exhaust dyeing equipment, and the waves of improvements made the dyer's job more of a science and less of an art.

Today there are many completely automated computer-controlled exhaust dyehouses, especially those processing 100% synthetics. Some firms have a no-add procedure in the dyehouse by which the dyer loads the fabric or yarn, weighs the dye, punches a button and lets the computer take over the entire process. This procedure ensures a constant dyeing cycle and the only variables are the dye index of the fiber or the quality of the dyestuff.

In continuous dyeing, there is less automation because of the many variables and the rapidity of the dyeing process which require many adjustments during the period in which several thousand meters of textile are dyed. However, instrumental science has continued to advance rapidly so that at present there is at least one continuous range available which is entirely computer-controlled except for the make-up of the dye mix. This unit features punch-card computer control and closed-circuit television monitoring of all processes. Air flow, temperatures, chemical systems (reduction and oxidation), bath levels, and speeds are carefully regulated.

A recording spectrometer continually reads the reflectance of the shade. It does not adjust the dye mix but it enables the dyer to determine if any changes are dictated. Although such control equipment is expensive (>$1,000,000 per range), the quality of the dyeings and the amount of energy saved justify the capital outlay. The liquor flow per tub is kept at the lowest practical rate, and reduction and oxidation potentials are at minimums resulting in savings of energy through conservation of chemicals, steam, and hot water.

Preparation for Dyeing

The proper preparation of textile fibers, yarns, and fabrics is an essential and integral part of dyeing and of finishing. The purposes of the various preparative procedures are removal from the fibers of natural or applied impurities which might impede the penetration and leveling of the dyes on the fibers, conversion of the materials into proper physical condition and form for necessary mechanical operations, and in some cases, application of chemicals to the goods prior to dyeing. Improper finishing techniques, or the failure to recognize the physical and chemical properties of the treated fibers, may lead to a variety of problems during the dyeing cycle. The appearance of resist marks from silicate or lime-soap deposits, chafe marks from mechanical damage, poor absorbency, crease marks, and changes in the surface and dimensional characteristics of certain thermoplastic fabrics result from improper preparation.

The fibers may be classified into two general groups: natural and synthetic. Many of the synthetic fibers have structural and physical properties similar to those of the natural animal and vegetable fibers. Blends of various fibers require handling techniques that will protect all fabric's fibers from damage.

Acetate. Strong alkalies must be avoided in the scouring of acetate, since the surface of the cellulose acetate (qv) would be saponified by such treatment. Many fabrics tend to crease and therefore require open-width handling. Scouring is frequently carried out on a jig or beam using 1.0 g/L of surfactant and 0.5–1.0 g/L tetrasodium pyrophosphate for 30 min at 70–80°C.

Cellulose acetate having 92% or more of the hydroxyl groups acetylated is referred to as triacetate and is sold under the trade name of Arnel (Celanese Corporation of America) in the United States. This fiber is characteristically more resistant to alkali than the usual acetate and may be scoured, generally in open width, with aqueous solutions of a synthetic surfactant and soda ash.

Acrylics. Acrylics such as Acrilan (Monsanto Textiles Company), Creslan (American Cyanamid Company), and Orlon (E. I. du Pont de Nemours & Co., Inc.) fabrics, are scoured with a synthetic detergent at 45–65°C and are rinsed before further processing to remove tints, size, wax, grease, spinning oils, or other impurities which were applied or picked up during the weaving operation. Bleaching, when required, is usually accomplished by means of a sodium chlorite bleach (see Bleaching agents), a selected optical brightener (see Brighteners, fluorescent), or a suitable combination of the two. Acrylic-blend fabrics may require other bleaching agents if chlorine-sensitive fibers are present.

Most acrylic fabrics require a presetting in open width in boiling water to avoid dimensional stability problems during subsequent wet-processing steps.

Modacrylic fibers, such as Dynel (Union Carbide Chemicals Co.) and Verel (Tennessee Eastman Co.), contain less than 85 but at least 35% by weight acrylonitrile units (see Acrylic and modacrylic fibers). Fabrics made from these fibers are prepared for dyeing in a manner similar to that of the straight acrylics.

Cotton. Cotton (qv) fibers are coated with natural waxes and pectins, which are water-insoluble but removable by aqueous alkalies or by solvent-dewaxing processes. Removal of these impurities improves the absorbent properties of the natural cotton and its receptivity to dyestuffs. The whiteness of the cotton is improved and seed particles are destroyed by caustic boil-off and bleaching operations. Cotton may be made suitable for dyeing in a variety of forms, such as raw stock, yarn, or piece goods. Raw stock is normally dyed without thorough dewaxing, since the natural waxes aid in subsequent spinning operations. Surfactants are employed to aid the penetration of dyestuffs through the protective waxes. Flaws in the dye levelness are overcome by subsequent carding. Yarn is normally boiled to drive off volatiles prior to dyeing. Hard-twisted ply yarns are frequently given a kier-boiling prior to dyeing to improve levelness. Careful preparation of cotton piece goods is essential to achieve suitable dye penetration, fastness, and general appearance. Fabric construction dictates whether the fabrics will be processed in rope or open-width forms. Heavy piece goods, which are subject to rubs and crease marks, and certain other constructions must be handled in open width.

The first step in the preparation is singeing which removes lint and fuzz from the fabric surface. The goods are then impregnated with an enzyme solution to effect solubilization of the sizing. Enzyme solutions are not necessary when only poly(vinyl alcohol) sizing is used. However, most frequently, starch sizes are used as extenders for poly(vinyl alcohol) sizing and it is necessary to use enzyme solution. New developments in sizing have been made to utilize poly(vinyl alcohol) sizing, and subsequent scouring units are constructed to reclaim the sizing material for reuse. Desizing may be accomplished by pad-batch, jig, or pad-steam techniques. Material that must be handled in open width requires hot scouring on jigs or on open-width boil-off machines in 10–12° Twad caustic soda containing a surfactant. Materials which may be roped are scoured continuously with caustic or soda ash in rope-soaping machines. The goods are then rinsed, scoured with acetic acid, rinsed, and dried.

Before dyeing in light shades, the goods should be bleached with hydrogen peroxide and caustic soda to bleach the motes. This operation helps also in the removal of trace impurities which remain after boil-off. Although kiers are still used for boil-off and bleaching, they have been replaced largely by steam-heated J-boxes or steamers which allow continuous processing and reduction of processing time.

Mercerizing is accomplished by passing the cotton fabric through 15–30% caustic soda. Improved luster and increased dye affinity results. The fabric is normally held under tension during processing. In order to obtain cotton stretch fabrics, mercerization is carried out on fabric in a slack condition. Recently a mercerization technique using liquid ammonia, has been introduced.

Polyamides. Polyamides (qv) are thermoplastic fibers that retain the form produced by heat treatment. They are usually given an alkaline scour and then heat-set. The heat-setting treatment is conducted at ca 10°C above the subsequent wet-processing steps; this ensures good form retention after processing. Woven fabrics are usually heat-set on a contact heat-setting machine and nylon tricot is generally heat-set on a tenter frame having either Calrod units (heating element, General Electric Company) or steam chambers.

When bleaching is required, sodium chlorite is preferred as the bleaching agent. However, for economy of operation, bleaching is often accomplished with peracetic acid in the alkaline scouring operation. Cotton/nylon blends are frequently bleached first with hydrogen peroxide to remove motes, and sometimes are bleached further with sodium chlorite. Bleaching is followed with an antichlor treatment, involving chemicals such as 0.1% sodium bisulfite and 0.15% tetrasodium pyrophosphate.

Scouring may be conducted on jigs, boil-off machines, or kettles, depending upon fabric weight, construction, and crease tendency in the rope form. A combination of a synthetic detergent and soda ash is usually used and scouring is conducted at 85–100°C. Certain nylon blends may require less stringent conditions and the use of less alkaline builders, such as tetrasodium pyrophosphate.

Polyesters. Polyester fabrics, such as Dacron (E. I. du Pont de Nemours & Co., Inc.) Fortrel (Fiber Industries), Kodel (Eastman Chemical Products, Inc.), characteristically resist degradation by the usual processing chemicals (see Polyester fibers). A hot alkaline scour with a synthetic surfactant and with 1% soda ash or caustic soda is used to remove size, lubricants, and oils. Sodium hypochlorite is sometimes included in the alkaline scouring bath when bleaching is required. After bleaching, the polyester fabric is given a bisulfite rinse and, when required, a further scouring in a formulated oxalic acid bath to remove rust stains and mill dirt which is resistant to alkaline scouring.

Other fibers blended with polyesters in numerous blended fabrics require alternative methods of preparation. Generally, the scouring and bleaching procedures used for these blends are those employed for the blended fiber alone.

In order to remove long fibers, polyester fabrics are sometimes sheared prior to dyeing. Singeing of the polyester fabrics, which prevent pilling, is usually delayed until after dyeing to prevent spotty dye exhaustion in unevenly singed areas.

Polyester fabrics and many other hydrophobic synthetic fabrics require the application of an antistatic agent prior to printing to prevent the buildup of static charges at rapid printing speeds.

Silk. Silk (qv) in its raw state is coated with sericin. It is necessary to remove this gum in order to develop the silk luster and dyeability. Historically, 25–30% soap, owf (on weight of fiber), was used for degumming. Synthetic detergent systems, such as higher alcohol sulfates, and soda ash and boric acid, have replaced soap to a large extent. Buffered alkalies, especially polyphosphates, are frequently used with synthetic detergent systems. Strongly alkaline systems must be avoided to prevent attack on the proteinaceous silk fibers.

Spandex. Spandex elastic fibers which are chemically segmented polyurethanes are usually combined with other nonelastic fibers to produce fabrics with controlled elasticity (see Fibers, elastomeric; Urethane polymers). Processing chemicals must be carefully selected to protect all fibers present in the blend. Prior to scouring, spandex fabrics are normally steamed to relax uneven tensions placed on the fibers during weaving. Scouring, which is used to remove lubricants and sizing, is normally conducted with aqueous solutions of synthetic detergents and tetrasodium pyrophosphate, with aqueous emulsions of perchloroethylene or with mineral spirits and sodium pyrophosphate.

When bleaching of the spandex is required, a reductive bleach with sodium hydrosulfite and sodium metabisulfite is used. Cotton and spandex blends may require a hydrogen peroxide bleach at pH 9.0–9.5 prior to or instead of the normal reductive bleach. Chlorine-type bleaches which damage spandex are avoided.

Viscose Rayon. Viscose rayon (qv), because of its low wet strength, must be processed under minimum tension at all stages of preparation. Skeins contain few impurities and require only light scouring. Piece goods may contain starch or sizing compounds, which were applied prior to weaving, and may require an open-width enzyme desizing followed by scouring, either in open width or in rope form, depending upon fabric construction and weight.

Certain new types of staple rayon, known as high-wet-modulus rayons, are characterized by high strength and high wet modulus with low elongation. Fabrics prepared from this type of rayon are readily stabilized dimensionally by Sanforizing. In general, fabrics may be prepared for dyeing by scouring in mildly alkaline synthetic-surfactant solutions.

Wool. Raw wool (qv) must be cleaned before it can be efficiently carded, combed, otherwise processed, or dyed. Loose wool, as obtained from the sheep, contains 30–80% impurities consisting of wool grease, dried perspiration (suint), dirt, seeds, and burrs. Methods of scouring the wool vary widely, depending upon the type of wool and the amount and types of soil. The equipment for washing wool by the countercurrent method usually consists of four to five bowls in sequence, each ca a meter deep and fitted with 38-cm-deep perforated trays which support the wool. Forks keep the wool moving along toward heavy squeeze rollers which are located at the end of each bowl and are designed to squeeze out dirty liquor before the wool is passed on to the next bowl. Soap and soda ash, historically used for wool scouring, largely have been replaced by nonionic surfactants with or without soda ash. In order to prevent harshening of the wool, soda ash percentages are kept at a minimum. Scouring temperatures are normally 45–60°C, depending upon the system and type of wool to be cleaned.

Acid (Anionic) Dyes

The most common fiber types to be dyed with acid dyes are (in sequence of their importance): polyamide, wool, silk, modified acrylic, and polypropylene fibers as well as blends of the aforementioned fibers with other fibers, such as cotton, rayon, polyester, regular acrylic, etc. Approximately 80–85% of all acid dyes sold to the U.S. textile industry are used for dyeing nylon, 10–15% for wool and the balance for those fibers mentioned above. Acid dyes are organic sulfonic acids; the commercially available forms are usually their sodium salts, which exhibit good water solubility.

According to their structure, acid dyes belong to the following chemical groups: azo (qv); anthraquinone (qv); triphenylmethane (qv); pyrazolone; azine (qv); nitro; and quinoline (qv). Azo dyes represent the largest and most important group and are followed by anthraquinone and triarylmethane dyes. Of the other dye groups, very few products are of any commercial value.

Acid dyes can be divided into four groups:

(1) These are the level dyeing acid dyes with one sulfonic acid group. They offer excellent leveling, migration and, coverage of barré properties. Fastness to light is very good, while the wetfastness properties in heavier shades generally are only marginal. The latter can be improved with an aftertreatment of either tannic acid/tartar emetic or any other synthetic aftertreating agent. The dyes of this group should be used when wetfastness properties are of no major concern and when emphasis is put on good dyeing performance, such as coverage of barré. Typical representatives are CI Acid Yellow 49, CI Acid Red 337 and CI Acid Blue 40. They can be used for dyeing of apparel, knit goods, carpet, upholstery, etc.

(2) These are the neutral-dyeing acid dyestuffs. They are also monosulfonated and are very similar in their leveling, migration, and coverage of barré properties to group 1; however, because of their chemical structure (larger molecular size), these dyes exhibit superior wetfastness properties. Since these dyestuffs have excellent neutral affinity and good build up properties, they are suited especially for dyeing of medium to heavy shades. CI Acid Yellow 159 and CI Acid Red 299 are typical dyes of this class.

(3) These are the milling-type acid dyes. They are disulfonated dyes and provide dyeings with highest wetfastness properties. The leveling and migration properties of these products are much inferior than those of the monosulfonated dyes in groups 1 and 2. Coverage of barré is also very poor. CI Acid Yellow 79 is an example of this group. It is obvious that there are no acid dyes that combine all desirable properties. Those which offer the excellent dyeing characteristics such as good leveling, migration and coverage of barré, have only marginal wetfastness properties; those that provide high wetfastness do not level very well. The dyes in group 2 represent the best compromise. When the fabric is aftertreated with a synthetic aftertreating agent, wetfastness—in most instances—will be equal or similar to dyeings obtained from milling dyes.

(4) These are the premetallized dyes which also include mono- and disulfonated types. Premetallized dyes exhibit a rather high dyeing strike rate even at pH values of 7–8 and, therefore, are extremely difficult to dye level. Their wetfastness properties are either comparable or superior to milling-type acid dyes. CI Acid Yellow 151 is a representative of this group.

Dyeing Processes. The two major polyamide types commercially available today are nylon 6 and nylon 66. Nylon 6 represents a polycondensate of caprolactam, and nylon 66 is a polycondensation product of two individual components, adipic acid and hexamethylendiamine (see Polyamides). Both fiber types are very receptive to acid dyes under certain conditions.

A direct relationship exists between the chemical structure of an acid dye and its dyeing and wetfastness properties. The dyeing of polyamide fibers depends on a great number of variables. Nylon 6 has a greater affinity for dyes than does nylon 66 and this may affect fastness properties of the final dyeing. Heat history of the fiber influences dye affinity; for instance, heat used during the texturizing process of the fiber may sometimes vary. Barré develops as a result. Barré may also be caused by differences in knitting tension. Antisoil and antistatic fibers have become increasingly important and it is quite common that these modified fibers show a reduced dyeability. The dyeing process is influenced by a number of additional parameters, such as: (a) dyestuff selection, (b) type and quantity of auxiliaries, (c) pH, (d) temperature, and (e) time.

(a) Dyes are selected according to the intended use of the fabric to be dyed. The needed fastness properties and/or dyeing characteristics determine whether mono- or disulfonated acid dyes are used. It is not advisable to mix these two dyestuff types in a given formulation, especially not in medium to heavy shades. The monosulfonated dyes exhaust more rapidly and block the disulfonated dye from exhausting onto the fiber, especially if the depth of shade is close to the saturation point of the fiber. Attention also has to be paid to the K value of the individual dyes within a combination. K values give the dyer an indication of the sequence in which the dyes of a combination will exhaust onto the fiber. Since it is most desirable to use on-tone exhausting combinations, the K values of acid dyes in a formulation should not differ by more than ±0.5–1.0 unit.

(b) When a polyamide is dyed with acid dyes, dyeing assistants are usually employed. They can fulfill three functions: increase the leveling and migration properties of acid dyes, help to cover barré, and improve the compatibility of acid dyes in combinations.

Anionic products show a relatively high affinity for the nylon fiber at temperatures below the boil and temporarily block dyesites, slowing down the exhaustion of the dyes and thereby helping to achieve a level dyeing.

Nonionic-anionic auxiliaries not only show affinity for the fiber, but also form complexes with dyes. In this way, they slow down the exhaustion process and also achieve level dyeing by affecting the compatibility of the dyes. By proper use of such auxiliaries, normally incompatible combinations (dyes with substantially different K values) can become compatible (similar K′ values, ie, compatible in presence of the auxiliary) (6).

(c) The pH is one of the most important factors in dyeing polyamide with acid dyes. The dyestuff binding groups in a polyamide fiber are the amino end groups (NH_2—). To enable the dissolved and ionized dye to react with these amino groups, the latter must be activated. A very low pH of 3–3.5 provides many activated NH_2 groups and results in a rapid, almost instant, exhaust of all the dye present in the bath. In such a case leveling and migration would be rather poor. At the optimum pH, 85–95% of the dye in the bath is exhausted onto the fiber at the end of the dyeing cycle. The most effective pH can be calculated for certain acid dye combinations (6).

(d) The dyeing temperature for dyeing polyamide fibers with acid dyes should be as close to the boil as possible. The higher the final temperature is, the better are the chances of obtaining a good dyeing, since the temperature has a great influence on the leveling and migration properties of acid dyes. Best results in leveling out faulty lots are obtained if the goods are treated at elevated temperatures of 105–110°C.

Even more critical than the end temperature is the rate of temperature increase. Depending on the type of dyes used, exhaustion starts at 25–35°C when using level dyeing acid dyes, or only at 50–65°C when certain milling-type dyes are in the bath. If the correct pH has been chosen, the dyebath will be exhausted before the boil has been reached. This also means that it is not always necessary to start a dyeing at room temperature because there would be, in many instances, a considerable dead time before the actual exhaustion starts to take place.

(e) The dyeing time represents a considerable cost factor in any process and should be as short as possible without jeopardizing the quality of the dyeing. Common practice is to sample after 30–45 min at the boil. If dyed under optimum conditions, the dyeing time at the boil is reduced to 10–20 min.

Dyeing of Nylon Carpet. A great percentage of acid dyes used in the textile industry is used in dyeing of nylon and wool carpets. Nylon carpet fibers can be dyed in many different ways, eg, in raw stock form, where usually highest wetfastness dyes are applied, in yarn form and in piece form. The latter is divided into piece or beck dyeing and continuous dyeing. For the continuous dyeing process, the most popular machine is the Kuesters carpet-dyeing range, which consists of a wet-out padder, a dye applicator, a loop steamer, and 3–4 wash boxes. Usually a drying oven is used after a dyeing range.

Continuous dyeing of nylon carpet became of great interest when tufted wall-to-wall carpet gained in appeal. The ever-increasing demand for larger and larger yardages of one particular color led to the developments of continuous dyeing ranges which, theoretically, can dye 85×10^3 m^2 (1×10^5 yd^2) of carpet within 24 h with no shade difference from the first to the last yard. Besides the regular continuous dyeing method used for dyeing solid shades, many additional dye application techniques have been developed to create the very popular multicolored carpet styles. The TAK and Multi-TAK machines are the most important units which have been added to the Kuester dyeing range to make it more versatile. Both the TAK and Multi-TAK represent machines that enable the dyer to apply different dye liquors in the form of drops or lines or any variation thereof onto the carpet. Usually, the ground shade is applied first by either padding or use of the blade; then, before the carpet enters the steamer, additional color drops (2–4 different shades) are applied on top of the unfixed ground shade by either the TAK or Multi-TAK machine. These droplets form a variety of patterns depending upon the viscosity of the ground shade coloration, of the TAK droplets and/or of the machine settings. Some Kuester continuous dyeing machines also are set up in line with rotary print heads, thereby further widening the styling possibilities.

Dyeing of Wool. Wool (qv) has lost its former importance and represents, today, only a minor factor in fiber dyeing. Dyeing of wool may be carried out in the various stages of processing. Dyestuff selection depends upon the specific end uses of the fabric or yarn as well as on the stage of manufacture. Wool fabrics are dyed primarily by exhaust dyeing methods. Although continuous dyeing of wool has been discussed for some time, it has not become very important. Only a few units are running today which continuously dye wool top and raw stock. Wool is dyed as: (*1*) raw stock, (*2*) yarn, and (*3*) piece goods.

(*1*) When dyeing raw stock or top, the levelness of dyeing is of minor importance, but the highest possible wetfastness properties for further processing are required. This does not mean that unlevel dyeings will be accepted, but that a certain shade variation in the dyeings can be counterbalanced during the blending process. According to the requirements for the manufacturing processes, dye classes with highest wetfastness properties must be used, such as chrome and neutral premetallized and milling colors.

(*2*) For yarn dyeing, the dye selection is governed mainly by the end-use of the material. However, the subsequent manufacturing processes influence the selection of dyes as far as fastness properties are concerned.

(*3*) For dyeing on piece goods, the performance of the dyes in the dyebath is very important because a level-dyed piece of fabric is required above all other aspects. The dye selection for piece dyeing is, therefore, based on the performance of the dyes as well as on the fastness requirements for the specific end-uses.

Application of Acid Dyes to Silk. Because of economic reasons, very little pure silk (qv) is being processed these days. Blends of silk with other fibers (eg, polyester) are more common. For fastness reasons, fiber-reactive dyes are used quite frequently, however, when top wetfastness is not needed, selected acid and neutral premetallized dyes may be utilized. Dyeing is carried out at a pH of 5.0–5.5 with acetic acid and with 0.5% of an ethoxylated fatty acid derivative in the bath. Best exhaustion of acid dyes on silk is achieved at about 85°C.

Dyeing of Modified Acrylic and Polypropylene. These fibers in their modified acid-dyeable form are of very minor importance. Regarding acid-dyeable acrylic fibers, dyeing procedures are very specific for the few fiber types available and should be requested from the fiber producer. With acid-dyeable polypropylene, dyeing is usually carried out at the boil at pH 2.5–3.0 using formic acid and, if needed, sulfuric acid. It is important to select the proper dyes for modified fibers because most of them are rather poor in either wetfastness or lightfastness properties.

Basic (Cationic) Dyes

The first basic dye, mauve, was synthesized by Perkins in 1856; and shortly afterward fuchsine, methyl violet, aniline blue, and other classical basic dyes were developed. For years these dyestuffs were primarily used for dyeing brilliant shades on silk and cellulosic fibers that had been treated with a mordant; however, usage of cationic dyes for these fibers was limited due to very poor washfastness and lightfastness. Basic dyes did not achieve prominence until the polyacrylonitrile fibers were introduced in the early 1950s. Initially, acrylic fibers were very difficult to dye and, until acidic groups were incorporated as dye sites in the fiber, basic dyestuffs were of little interest. Cationic dyes are currently used in large quantities to dye acrylics (Orlon, Acrilan, Creslan, Zefran) and modified acrylics (Verel). Subsequent developments led to the introduction of acidic groups to polyester and polyamide fibers, further increasing the market for these dyes.

Dyeing of Acrylic Fibers. Basic dyes are water-soluble and dissociate into anions and colored cations. The cations have a strong affinity for the acidic group (sulfonic or carboxylic) and form salts. Because of these strong bonds, washfastness is usually outstanding, and lightfastness varies considerably, depending on the dyestuff. Basic dyes are usually applied in a batch process with skein, stock, and package dyeing more prevalent than piece dyeing on becks or jet machines.

The fundamental steps in the acrylic dyeing mechanism are: adsorption of the basic dyestuff at the fiber surface, which occurs only when the glass transition point of the fiber is exceeded; diffusion of the dye into the fiber as the fiber molecules acquire enough energy to move; formation of the dye-fiber bond; and migration of dyestuff from one dyesite to another or from within the fiber to the surface. The degree of migration varies considerably, depending on the dyestuff, but generally basic dyes do not migrate well.

Leveling of basic dyestuffs can be a major problem which often can be traced partially to differing exhaustion rates of individual dyestuffs which may occur in combination shades. One method for measurement of exhaustion rates is the time of half-dyeing or the time in minutes required for the fiber to absorb half as much dye as will be absorbed if dyed to equilibrium. Half-dyeing times do not adequately describe the behavior of basic dyestuffs in combinations where the individual dyestuffs could interfere with each other's exhaustion rate.

A method has been developed for determining compatability or K values (7). Dyeings of the basic dyestuffs, which are to be rated, are made in succession for a specified period of time with dyes having a known strike rate or K value. Five groups (K-1 through K-5) are used to classify dyes, with K-1 dyes being exhausted first when dyed in combination with dyes of any other group; dyes with a K-2 are taken up preferentially when combined with dyes of K-3 to K-5. For optimum leveling, the K values of basic dyes in a combination shade should be as close to each other as possible.

Every basic-dyeable fiber has a saturation value, ie, it has a given number of acidic dyesites which limit the quantity of basic dyestuff that can be fixed. The dyestuff also has a saturation factor, which is a measure of the relative molecular weight per cation in the dye. These two values determine the depth of shade obtainable with a given dye and fiber.

Acrylic fibers also vary considerably in their rate of dyeing, depending on whether they are of a dry-spun or a wet-spun fiber. Wet-spun fibers dye at a higher rate than dry-spun fibers. The saturation constant and dyeing rate of typical acrylic fibers are shown in Table 1.

To overcome the high affinity of some basic dyestuffs and to prevent unlevel dyeings, often a cationic retarder is used. Cationic retarders act as a colorless dyestuff and compete with the basic dyes for available dyesites. The optimum amount of retarder can be calculated (8).

Dyeing of Basic-Dyeable Polyester. Cationic-dyeable polyester has achieved a moderate degree of success because it lends itself to two- or three-color effects in blends with regular disperse-dyeable polyester, cellulosic fibers, or wool. Typical basic-dyeable polyester fibers are Dacron T-64 and T-92, Fortrel 402, and Trevira 640.

The dyeing rate of cationic-dyeable polyester is much lower than that of acrylic; however, an even greater difference exists in the diffusibility associated with each, which is estimated for the cationic-dyeable polyester to be only 10% of that for acrylics. This has to be overcome by dyeing at higher temperatures (110–130°C) and by using a carrier to improve penetration. Another use for the carrier is the prevention of cross-staining the disperse-dyeable portion of the blend by the basic dyestuffs.

Nonionic products must be present to act as antiprecipitants or to suspend particles resulting from the reaction between the cationic dyes and anionic dyestuffs when blends of basic-dyeable and disperse-dyeable polyester are dyed.

Direct Dyes

Direct dyes are water-soluble dyes which exhaust onto cellulosic fibers, such as cotton, rayon, and linen, from a salt bath. Because of this property, these dyes are called substantive dyes.

The first successful synthetic direct dye was Congo Red [573-58-0] (CI Direct Red 28), a diazo derivative prepared from benzidine by Boettiger in 1884. Despite its current use because of its economics and brightness, Congo Red is one of the more fugitive direct dyes.

Table 1. Relative Saturation Constant S_F and Dyeing Rate V of Various Acrylic Fibers

Fiber	S_F	V
Acrilan A-16	1.3	3.0
Acrilan B-90	0.9	1.1
Creslan 61	1.7	4.7
Creslan T-61	1.7	2.9
Orlon 42	2.1	1.8
Orlon 33	3.3	4.5
Orlon Sayelle	7.0	6.5

Nearly all direct dyes are azo products containing one or more sulfonic radicals which impart water solubility. A few contain carboxyl groups ortho to hydroxyl, a configuration which permits aftertreatment for improvement of fastness. Most direct dyes contain two or three azo groups, although monoazo and higher polyazo products are known. The following structural properties govern direct dyestuffs with substantivity or excellent affinity for cellulose fibers: elongated molecular structure with few side chains; several aromatic rings, which may be arranged in one place and which contain chemical groups that can form hydrogen bonds relatively easily; and few solubilizing chemical groups.

Direct dyestuffs form anions (negative charges) by dissociation. In aqueous solution, the molecules of direct dyestuffs are linked by hydrogen bonds, forming larger agglomerates or colloidal solutions. This reduces their solubility and promotes disposition on the fibers. The elongated structure of these agglomerates adapts them to the parallel arrangement of the cellulose molecule to which they are firmly linked by multiple hydrogen bonds.

Uses. The 1975 U.S. International Trade Commission report indicates that direct dyes are an important class of dyes manufactured in the United States: sales of direct dyes were fourth by volume and fifth by value among all dyes. Direct dyes are used extensively for coloring cotton and rayon textiles, on which the dyes give a full range of color. Also, direct dyes are used on paper and on many other fibers, such as linen, jute, hemp, silk, polyamide, and mixtures of other fibers.

An important consideration in the selection of direct dyes for a given purpose is their ease of application. It is for this reason, for example, that package dyes for domestic use contain direct dyes; such packages are usually mixtures of acid, direct, and other dyes. Garment dyeing of cellulosic fabrics is nearly always done with direct dyes.

Direct dyes, in addition to ease of application, also give excellent penetration, even coverage, level dyeing, and they do not generally impair luster of the material to be colored. The fastness properties without aftertreatment vary considerably. Therefore the end-use requirement dictates whether their application should be followed by an aftertreatment to improve either their wetfastness or lightfastness properties. Most manufacturers usually apply trade names to their direct dyes according to their lightfastness properties. The lightfastness varies from very poor to outstanding. More recently developed direct dyes are suitable for use on upholstery, drapery, automotive fabric and other fabrics where excellent fastness to light is important. Many direct dyes exhibit better lightfastness on rayon than on cotton.

Application. The amount of salt used in dyeing is governed by the liquor-to-fabric ratio, the total amount of dye, and the exhausting properties of the particular dyes. Direct dyes normally do not exhaust completely. Manufacturers' circulars and shade cards indicate the appropriate dye method, the general salt requirements and the optimum dyeing temperatures of specific dyes.

Individual direct dyes exhibit a distinctive equilibrium exhaustion, leveling capacity, and rate of absorption. The important factor regarding leveling is the equilibrium exhaustion. Generally dyes that exhaust to a high degree do not level well. Dyes which exhaust quickly and dye unlevel at first may still level during the dyeing process if their equilibrium exhaustion is relatively low. Corresponding information is found in most manufacturers' circulars.

When dyeing light shades or shades that contain dyes which do not migrate well, an effective leveling agent should be employed to obtain level dyeings. If chafe marks may be a problem, use of 1–2% of a nonionic softener during dyeing is recommended, ie, an effective dyebath lubricant.

In some cases, viscose fiber may still contain residual sulfur which may adversely affect the dyestuff (ie, reduction) during the dyeing process. A product, such as a combination of ammonium salts of aliphatic- and of aliphatic–aromatic sulfonic acids, is recommended at a concentration of 1–2% to prevent reduction of dyes in the dyebath.

Interest in blended fabrics of synthetic fibers and cellulosic fibers (ie, polyester/cotton, polyester/rayon, polyamide/rayon, etc) has prompted requests for direct dyes that are suitable for application at an acid pH and at temperatures above 100°C. The optimum dyeing condition for most direct dyes is at a neutral or alkaline pH. However, a limited selection of direct dyes suitable for acid pH application and for temperatures above 100°C may be obtained from manufacturers.

Aftertreatment. The wetfastness of many direct dyeings can be improved in various ways: aftertreatment with cation-active organic fixing agents; aftertreatment with formaldehyde; diazotization of the dyes on the fiber and development with amines, phenols, or naphthols; and aftertreatment with diazonium compounds. The lightfastness of some direct dyeings can be improved by aftertreatment with copper salts. Copper salts are used alone, in combination with chrome, or in combination with cation-active organic condensation products.

Cationic Fixing Agents. Most direct dyeings can be improved in wetfastness by aftertreatment in a fresh bath after dyeing has proceeded for 15–30 min at 25–60°C with 1–4% of a cationic aftertreating agent. The exact conditions are given by the manufacturers of specific products.

These aftertreating agents are generally of two kinds: those that are principally cation-active; and those that are cation-active resinous compounds. The specific product to be used should be screened for its effect on lightfastness and on shade change. Many products offered as dye fixatives impair the lightfastness; some are quite harmful. Often there might be an accompanying change in the shade of the dyeing. In addition, the products vary in their ability to be removed; thus if improperly removed or unevenly applied, any subsequent dye processing could produce unlevel dyeings.

Formaldehyde Aftertreatment. The fastness to bleeding and the washing characteristics of some direct dyes can be appreciably improved by aftertreatment with formaldehyde; little or no change in shade or lightfastness of the dyeing occurs. The addition of acetic acid is nearly always recommended in the aftertreatment.

Development of Diazo Dyes. Direct dyes that possess a free amino group in their composition can be diazotized after their application on the fiber and their subsequent coupling with a phenol, a naphthol, or an amine to obtain a final dyeing with one more azo groups in the molecule. As a consequence, the dyeing's fastness to washing increases. Direct dyeings rarely possess a fastness to washing equal to Class 3 (good); most of them are midway between Classes 1 (poor) and 2 (fair). Direct dyes that can be diazotized and coupled, however, can usually be placed in Class 3, meaning that they can withstand a wash test conducted at 50°C very well. A few even can be rated Class 3–4 in washing. Generally this class of direct dyes has poor to fair lightfastness. Therefore as demands for lightfastness increase, the use of this class of dye decreases.

The term development is applied to the coupling of a diazotized direct dye on the fiber with a suitable phenol, naphthol, or amine; the coupling components are termed developers. Development is accompanied by a change in shade which varies according to the dye from slight to very considerable, but the lightfastness changes little, if at all. Although they are commonly called D&D (direct developed) dyes, manufacturers use different brand names for each of them. Specific examples of dyes belonging to the diazo group are Primuline [6537-66-2] (CI Direct Yellow 59) and Phenazo Black BH [2429-73-4] (CI Direct Blue 2).

Since diazotized and developed dyeings in general are discharged in printing, they are used extensively where white and colored discharge effects are required on a solid ground color.

Aftertreatment With Diazonium Compounds. After they have been exhausted on the fiber, a limited number of direct dyes retain their ability to couple with diazonium compounds to form larger dye molecules. A concomitant increase in fastness to washing occurs which is comparable to the increase obtained with the diazo dyes upon development. The dyes suitable for this kind of aftertreatment are especially useful in the production of rich brown, olive, and black shades. However this method is not widely used in spite of its simplicity and effectiveness.

Only two diazo compounds are commonly employed for this type of aftertreatment, or development; they are diazotized *p*-nitroaniline and diazotized 2,5-dichloroaniline. These are sold in a convenient stabilized condition, ready for use, obviating the need for carrying out a diazotization; thus the fastness of the D&D dyes is attained in a one-step aftertreatment.

Aftertreatment With Copper Salts. About 10% of the ordinary direct dyes can be improved considerably with respect to fastness to light by an aftertreatment with copper salts. These are dyes that can chelate metal, ie, a metal atom can replace hydrogen(s) in the structure. In some instances, fastness to washing is appreciably improved. However, dyeings that are aftertreated with copper lose their improved lightfastness when copper is gradually removed from the fabric by repeated mild launderings.

A development in the field of direct dyes is aftertreatment with a mixture of a copper salt and an organic nitrogen-containing cation-active condensation product. Together the two considerably improve lightfastness and washfastness, as compared to the same type of improvements achieved with the copper aftertreating combination discussed above, or with the cation-active fixing agent alone. It is probable that chelate formation takes place between dye, copper, and cationic fixing agent. However, the more usual cation-active fixing agents in combination with a copper salt do not give an improvement in light- and washfastness which is equal to that obtained with the new type of fixative in combination with copper salt.

Disperse Dyes

Disperse dyes were introduced in the 1920s, and primarily were used for the dyeing of cellulose acetate fibers; however, later many of these dyes were found to be suitable for polyamide fibers. Subsequent synthetic fibers, such as triacetate and polyester, had different molecular structures which required new disperse dyes.

Hydrophobic fibers, such as acetate, triacetate, and polyester, show no affinity for ionic dyestuffs unless chemically modified to accept them. These fibers absorb nonionic organic colorants applied from a dispersion. Disperse dyes are colored organic substances with very low solubility in water and which, during the course of a dyeing, are transferred from dispersion via monomolecular solution to the fiber, ie, the dyestuff is believed to form a solid solution in the fiber, which acts as a solvent.

Fastness Properties. The fastness properties of the acetate dyes were not adequate on triacetate and polyester fibers, but newly developed disperse dyes can be dyed in a full range of shades which satisfy the fastness demands of the market. Fastness properties of disperse dyes vary widely, depending on the molecular structures of the dyestuff and of the fiber to which they are applied. When selecting dyestuffs to meet specific fastness requirements, special consideration must be given to:

Wetfastness. Generally speaking, the rule of "easy on-easy off" applies, ie, those products that exhaust easily will come off easily in a wash test.

Sublimation-Fastness. This property is especially important when selecting dyes for polyester and triacetate as they are heat-set at high temperatures (170–180°C) after dyeing to achieve dimensional stability.

Gas-Fastness. Certain disperse dyes, especially anthraquinone types, are sensitive to oxides of nitrogen and may change shade or fade drastically upon prolonged exposure. This phenomenon is more prevalent on acetate and polyamide than on triacetate or polyester. Gas-fading inhibitors, which have an affinity for the fiber and are attacked by the oxides of nitrogen, are often used to minimize this effect.

Lightfastness. This property is most important with respect to dyeing of upholstery or automotive fabrics. Reduced lightfastness is acceptable for other uses, eg, ladies' wear, which tends to be exposed to daylight less frequently before being replaced.

One very general classification of disperse dyes is determined according to the ease of diffusion of the dye into polyester fiber. The dyestuffs are then referred to in the trade as "low-, medium-, or high-energy." The differences from one disperse dye to another are due to molecular structure, size, or both.

Dyeing of Acetate. Very small quantities of acetate staple are dyed; however, large quantities of acetate filament are found in satin, taffeta, and tricot fabrics; these are usually dyed open-width on a jig owing to their inclination to crease or crack easily. A typical dyeing procedure on the jig involves addition of acetic acid and dispersing agent over two ends at 50°C. The disperse dye is added over two ends and the dyebath temperature is gradually raised to 80°C in 5°C increments with two passes at each temperature. The dyeing is completed after 30–60 min at 80°C.

Dyeing of Nylon. Nylon is dyed in numerous forms (nylon 6 and 66) on various pieces of equipment. By far, the major form of consumption of nylon is bulked continuous filament (BCF) for carpet yarns. Disperse dyes are applied on becks, skein dye machines, package dye machines, and by continuous methods. Disperse dyes are accepted widely as the dye of BCF yarns owing to their leveling characteristics as well as to their ability to cover dye affinity differences. Other methods of dyeing nylon include paddle wheels for hosiery and sweater bodies, and beam dyeing of tricot fabrics.

A typical procedure for any exhaust dyeing equipment includes setting the bath at 40°C with a pH of 6.0 for the dispersing agent and acetic acid. The goods are run for 10 min and the disperse dyes are added, then the bath is heated to 90–100°C over 30 min and is run for one hour, and lastly the goods are rinsed thoroughly and are dried.

Dyeing of Triacetate. Triacetate, which is completely acetylated cellulose, is a very hydrophobic fiber, as compared to secondary acetate, and consequently does not dye rapidly. The dyeing of triacetate is akin to the dyeing of polyester fibers, ie, it is necessary to increase the rate of diffusion of the disperse dye into the fiber by increasing the dyeing temperature to 110–130°C or using a dye accelerant or carrier, or both (see Dye carriers). Trichlorobenzene, butyl benzoate, methyl salicylate, and biphenyl are typical accelerants. Higher amounts (10–15% owf) are used for dyeing at the boil than at 110–130°C where 2–4% carrier (or in some cases, none) will suffice. Triacetate, like polyester, achieves dimensional stability by heat-setting which requires disperse dyes with good sublimation-fastness.

A typical dyeing procedure for triacetate includes setting a bath containing acetic acid, dispersing agent and carrier (amount based on dyeing temperature and liquor ratio) to pH 4.5–5.0; circulating 10 min and adding disperse dyestuffs; raising to 100°C or 110–120°C over 30–45 min; and running for 1–2 h, depending on the desired depth of shade. Triacetate may also be dyed continuously by the pad-dry-thermosol-scour method. The disperse dye is padded along with a suitable thickener and is dried. The goods are dried in a thermosol oven for 90 s at 190–200°C, rinsed, and scoured.

Dyeing of Polyester. Polyester is a polymer resulting from the ester exchange reaction of dimethyl terephthalate and a glycol, usually ethylene glycol. Typical polyester fibers include Dacron, Fortrel, Trevira, and Terylene.

Polyester was the fastest growing of all fibers during the 1960s and 1970s and this trend is not expected to change. Polyester is a very desirable fiber due to its economy, strength, and dimensional stability. Vast quantities of polyester filament are consumed in 100% polyester double knits and wovens, and an equally large amount of staple is consumed in blends with cotton, wool, and acrylic fibers.

Techniques for application of disperse dyes have advanced rapidly since the introduction of high temperature or pressure dyeing. Owing to its hydrophobic nature and very slow dyeing rate, considerable quantities of an accelerant or carrier were necessary to successfully dye polyester. High temperature or pressure dyeing becks have been succeeded or supplemented by jet dyeing machines and package-dyeing machines which give very rapid bath turnover. These developments have drastically reduced the time of dyeing as well as improved the quality of dyeing since many polyester fabrics prior to these developments were dyed with difficulty.

Dyeing Methods. Typical dyeing methods are:

Atmospheric Exhaust. This method is used primarily for dyeing of polyester blends when pressure equipment is not available. The bath is set at 50°C and at pH 4.5–5.0 (acetic acid), and dispersing agent and 10–15% of a suitable carrier are added. The temperature of the dyebath is raised over 45 min to 100°C and run for 1–2 h.

High Temperature Exhaust. This method of dyeing is most prevalent since it is much faster and more economical than atmospheric dyeing. The basic procedure remains the same whether dyeing proceeds on a pressure beck, a jet, a package-dyeing machine, or a high-pressure beam; however, no carrier or a much reduced amount of carrier is used. Higher rates of temperature rise may be utilized since most of these dyeing machines have circulating pumps producing relatively high liquor flow rates. Run times at the dyeing temperature may be reduced considerably since the dyeing rate and/or diffusion of the disperse dye takes place much more rapidly than at atmospheric pressure with a carrier. Optimized dyeing conditions can be calculated based on the knowledge of a few dye, fiber, and machinery parameters (9).

Continuous Dyeing. Millions of meters of polyester/cotton are processed annually by the thermofixation process. The disperse dye is padded onto the fabric along with a suitable thickener which prevents migration. The pad bath may contain vat or fiber-reactive dyestuffs for the cellulosic portion of the blend. The goods are pre-dried by ir heating and/or by steam-heated dry cans and then are thermosoled for 90 s at 180–200°C to fix the disperse dyestuff. The goods then pass through a reduction bath for vats or an alkali/salt bath for fiber-reactives, are steamed to fix the dyes for the cellulosic fibers, and the fabric then passes through a series of wash boxes and, lastly, is dried. The continuous method is the most economical method of dyeing large yardages as speeds in excess of 90 m/min are possible.

Dyeing of Acrylic Fibers. The use of disperse dyes for the dyeing of polyacrylonitrile fibers is limited. The dyes possess very limited wash- and lightfastness and are generally used only for pastel shades where leveling is the major consideration. Dyeing takes place in a bath at a pH of 5.0–6.0 for one hour at 100°C.

Fiber-Reactive Dyes

Fiber-reactive dyes are the newest major dyestuff group and have had their significant developments occur within the last 20 yrs. Fiber reactive dyes supplement vat, direct, and naphthol dyes, and are available in a full range of colors of outstanding brilliance and fastness properties. Fiber-reactive dyes can be applied to wool, silk, and polyamide; however, they are used primarily for cellulosic fibers. By proper selection of dyes, one can apply fiber-reactive dyes by almost all known batch processes, as well as semi-continuous and full-continuous dyeing methods.

A reactive dye contains a group that reacts with the hydroxyl (OH) group in cellulosic fibers and the amino (NH_2) groups in polyamide, silk, and wool. Wool also contains highly reactive SH groups. Cellulosic fibers react in an alkaline medium at pH 9–12. Wool, silk, and polyamide can be dyed in a weakly acid medium, and wool can also be dyed in a strongly acid bath at pH 2–3. The first rule of dyeing with reactive dyes can be derived from these pH effects: the less reactive the dye and the less reactive the displaceable group of the fiber, the higher the pH of the dyebath must be.

A large number of reactive systems have been described in patent literature; however, only a few have achieved any degree of prominence. The most important groups are shown below in Table 2.

Application to Cellulosic Fibers. The dyeing of cellulosic fibers with reactive dyes normally is limited to the alkaline pH range as the OH group of cellulose does not dissociate in acidic media, ie, the cellulose anion required for reaction with the dye is formed in the alkaline range:

$$\text{cell-OH} + \text{OH}^- \rightarrow \text{cell-O}^- + H_2O$$

When dyeing cellulosic fibers with reactive dyes in an alkaline medium, the cell-O$^-$ group is not the only reaction partner. Reactions are also possible with the OH$^-$ group of water, which implies that the reactive dyeing of cellulosic fibers is always accompanied by a partial hydrolysis of the dye being used.

Exhaust Dyeing. Application procedures may vary considerably as the different fiber-reactive groups have different dyeing properties which are based on their substantivity and reactivity.

The substantivity of a fiber-reactive dye largely determines the dyeing process

Table 2. Reactive Systems

Reactive group	Year introduced	Manufacturer	Trade name
dichlorotriazine	1956	ICI	Procion M
monochlorotriazine	1957	ICI	Procion H
		Ciba-Geigy	Cibacron
dialkylaminoethylsulfone	1958	Hoechst	Remazol
trichloropyrimidine	1960	Sandoz	Drimarene
		Ciba-Geigy	Reactone
dichloroquinoxaline	1961	Bayer	Levafix E
dichlorophthalazine	1963	Francolor	Elisiane
monochlorobenzothiazole			
methylsulfonylbenzothiazole	1967	Bayer	Levafix P
monochlorodifluoropyrimidine	1971	Bayer	Levafix EA/PA
		Sandoz	Drimarene R

to which it is suited. Reactive dyestuffs of medium-to-high substantivity are used for long-liquor-ratio dyeing, whereas dyes of medium-to-low substantivity are used for continuous dyeing and printing. It is desirable that there be a high degree of substantivity for exhaust dyeing, but it should not be so high that dyestuff hydrolysate, which occurs as a by-product during the dyeing process through reaction with water, cannot be removed easily from the fiber by rinsing and soaping. To obtain the necessary substantivity and to ensure that the dyestuff molecule is in a position that is favorable for reaction, sodium sulfate or sodium chloride is added to the dyebath in amounts ranging from 20–100 g/L. The substantivity of a fiber-reactive dye decreases with rising temperature and pH. Therefore, dyes of low reactivity need higher quantities of salt in exhaust dyeing to ensure sufficient yield.

The reactivity is determined largely by the chemical constitution of the reactive group. The degree of reactivity determines the type and quantity of alkali and the temperature required for the fixation of a dye, ie, it determines the application performance of a dye. Differences between dyes of different ranges and dyes of varying reactivity can also be found within one class of dyes which has the same reactive groups.

Although there are specific recommendations for each range of fiber-reactive dyes, there are similarities in all of the exhaust dyeing procedures. Usually dye and salt are added first (the salt may be added in portions) to exhaust the dyestuff, then alkali is added to produce the chemical bond with the fiber. Approximately 50–70% of the applied dye normally is fixed to the fiber. Typical alkalies are soda ash, sodium hydroxide, sodium bicarbonate, and trisodium phosphate. Often these are combined in various proportions to achieve the desired alkalinity. Depending on which reactive group is used, the dyeing temperature may vary from 30–80°C. Upon completion, the goods are given several rinses and are soaped hot to remove all unfixed dyestuff, ensuring maximum wetfastness.

Cold Pad-Batch Dyeing. In the cold pad-batch process, fiber-reactive dyes are padded together with a suitable alkali onto the cellulosic fiber and then are batched for 3–24 h at 30°C to achieve fixation. The dye is absorbed from the padding liquor without intermediate drying. The application method is an exhaustion process from a very short liquor ratio. The number of dyes suitable for the pad-batch process is larger than the number suitable for exhaust dyeing from a long liquor ratio since inadequate reactivity which may be associated with the former process can be balanced by using stronger alkalies. A major consideration in this process is the selection of dyes with good pad-bath stability in the presence of alkali, ie, those that do not hydrolyze. The precaution is necessary to prevent tailing, a gradual change in depth of dyeing from beginning to end.

Continuous Dyeing. Fiber-reactive dyes are suitable for various continuous dyeing methods.

Pad-Dry-Thermosol-Chemical Pad-Steam. This method is used widely because it is suitable for the average continuous dye range on which polyester/cotton is processed. The goods are padded with dyestuff, thickener, and wetting agent, and urea, dried, thermosoled at 190–195°C, padded with brine and alkali, steamed, rinsed, and soaped.

Pad-Dry-Thermosol. In this method, an alkali, such as sodium bicarbonate, is added to the pad bath, along with the dyestuff and urea, and fixation takes place with dry heat during drying and thermofixing. No chemical pad or steam is necessary. The goods are then rinsed and soaped to remove unfixed dyestuff.

Pad-Chemical Pad-Steam-"Wet-on-Wet". This procedure is used for terry toweling when intermediate drying is undesirable due to migration of dyestuff to the yarn tips. The fiber-reactive dye is padded on, and then brine and alkali are applied with an impregnated chemical pad. The goods are steamed to fix the dye; then they are rinsed and soaped.

Sulfur Dyes

The introduction of sulfur dyes, in 1873 generally is attributed to Croissant and Bretonniere who are said to have heated a mixture of sawdust and bran with sulfur, sodium sulfide, and caustic soda to form a product called Cachou de Laval [8001-76-1]. This dye had little commercial success. Vidal, in 1893, prepared a successful black dye from p-aminophenol, sodium sulfide, and sulfur. In 1900 Weinburg and Herz began working on a full line of sulfur dyes and Immedial Yellow [1326-40-5] was developed by Weinburg in 1903. These new dyes did not necessarily have to be oxidized with bichromate only (as was the case with some previous ones). Many new dyes continued to be discovered by fusing sulfur and alkali with a variety of amine, nitro, and phenol derivatives.

These early dyes were powders, varying in color among yellows, oranges, browns, bordeauxs, olives, and blacks. They were limited in brightness but offered good wet-fastness properties with limited lightfastness, depending greatly on the individual dye. Very little was known about the structure of the finished products except that they were complex, mixed products that produced a certain shade.

Gradually, as the dyestuff chemist's knowledge of the complex structure of sulfur dyes advanced, more dyes could be produced to meet specific requirements. They were supplied as dry powders and dissolved with sodium sulfide and alkali in a boiling bath for 10–15 min and then were applied to cotton. All of this changed in 1936 when Crist of the Southern Dyestuff Company announced the first prereduced liquid sulfur dyes for commercial use. This was a substantial step forward since the dyes were filtered and stabilized and could be applied from RT to boiling temperatures. Slowly, through intensified research efforts, chemical structures were identified and dyes producing brighter colors eventually were added to the available color palette. These included bright yellows, oranges, greens, and a red. The synthesis procedures for these dyes is called aqueous reflux sulfurization of dye intermediates and was discovered by Southern Dyestuff chemists.

Performance Characteristics. The fastness properties of sulfur dyes lie between those of diazotized direct dyes and vat dyes. They have better washfastness than diazotized or aftertreated direct dyes but are somewhat inferior in washfastness to vat dyes. However, as they lose depth, they do not stain other fibers. Lightfastness varies from mediocre to good. Chlorine fastness is rather poor in most cases, but there are a few which show fairly good fastness. Sulfur dyes in earlier days gave cotton an inabsorbent harsh hand (texture); however, with the use of chelating agents, ester softeners, and newer oxidizing agents, such as bromites, bromates, and iodates, much, but not all, of this problem has been overcome.

Some sulfur dyes in cellulosic fibers, especially blacks cause loss of tensile strength upon storage under conditions of high humidity and temperature. This problem has been largely minimized by proper oxidation, pH, buffering, and durable-press resin finishing.

Application. Sulfur dyes can be applied by exhaust dyeing on raw stock and on packages or on piece goods on the jig. However, continuous dyeing is the process by which most of the sulfur dyes are applied today. The prereduced dyes are padded and steamed for 30–45 s and then are washed, oxidized, soaped, washed, and dried.

Insoluble Azo Dyes (Naphthols)

The beginning of this class of dyes, which is used primarily for cellulosic fibers and for special purposes on acetate, polyester, nylon, and silk, had its start in 1880 when the English firm, Read, Holliday and Sons, found that a bright red shade could be produced from β-naphthol and diazotized β-naphthylamine. This was accomplished by applying the diazonium chloride salt of β-naphthylamine. This particular dyestuff was never of much importance by itself; however, it provided a process whereby a multitude of naphthol compounds reacted with an even greater number of diazotized organic amines on cotton. One of the first successful dyes was Para Red [6410-10-2], produced by Badische Anilin and Soda Fabriken who allowed β-naphthol to react with diazotized p-nitroaniline.

A whole new class mostly of reds and oranges was developed for cottons. Later more advanced naphthols and coupling agents were discovered; yellows, blues, and blacks appeared and most had versatile and good fastness properties. Soon they were known as the ice colors, since the diazotized amines had to be cooled in order to maintain their stability.

As knowledge and expertise in the use and manufacturing of naphthols and other dye components progressed, two developments occurred that greatly simplified the dyeing procedure. First, instead of having to diazotize and stabilize the bases, a method was discovered for the diazo-coupling component involving producing a complex metallic salt. Such products are more expensive, but they are convenient to use and can be prepared quickly. Second, stable alkaline solutions of most naphthols were prepared and made the volume user's job more convenient and quicker. No longer was it necessary to paste the powdered naphthol and ethanol or ethylene glycol monoethyl ether with strong caustic soda, and carefully add the mixture to a hot spring bath. With these improvements, growth of these dyestuffs increased substantially.

Application. As more and more naphthols were discovered, their varying affinity for the different fibers became apparent. As an example, the strike rate varied from 10% for Naphthol AS [92-77-3] to 98% for AS-S [2672-81-3]. As a rule of thumb, those naphthols exhibiting less than 25% substantivity are used for continuous dyeing. Since, in exhaust dyeing, the fiber is salt-rinsed to improve crock and washfastness, too much naphthol is lost with low substantivity naphthols. On the other hand, those naphthols with low affinity are highly suitable for continuous dyeing because they offer almost no substantivity in padding.

The reverse is true of substantive naphthols with "fast strike" for all types of exhaust dyeing, since they not only give better economy, but also give superior washfastness because they can be salt-rinsed before coupling to remove unfixed naphthol.

In continuous dyeing, the goods are padded through a hot solution (80°C) of the naphtholate which contains proper amounts of naphthol, solvent, caustic soda, and a stabilizing agent (usually a refined sulfonated castor oil). The goods with a wet pickup of 60–65% are evenly dried on controlled-heat-dry cans and then are cooled. They process through 2–3 tubs containing the cold (3–5°C) diazonium compound (salt or diazotized base) at a pH of 4.5–7.1, depending on the coupling characteristics of the diazotized product. The buffers that are used range from sodium acetate to zinc sulfate. After coupling for 30–45 s, goods proceed through 3 boxes of running cold rinses, followed by 2 boxes of hot (80°C) alkaline soaping. Afterwards, goods are rinsed hot and dried.

Generally, when exhaust dyeing, the naphtholate bath is made up as above with two exceptions. First, only medium-to-high substantivity naphthols are used for best fastness results. Second, the final bath temperature is adjusted to approximately 30°C, since maximum exhaustion occurs at this temperature. After exhausting the bath for 30 min, the bath temperature is dropped and the goods are rinsed with 15–30 g/L common salt at room temperature to remove unfixed naphtholate and excess caustic soda. The machine is charged with the cold developing bath, prepared as in the continuous dyeing bath, and coupling takes place for 20 min. Afterwards, the bath temperature is dropped further and followed by successive baths of cold rinsing, hot soaping, and hot rinsing.

It should be noted that naphtholated goods are especially sensitive to decomposition by light and should be protected from strong light before coupling.

In the period, 1930–1960, naphthols reigned supreme for both exhaust and continuous dyeing of bright shades on cellulosic fibers when fastness to light, bleaching, and washing was required. However, beginning in 1956 with the commercial introduction of fiber-reactive dyes, naphthols gradually began to lose importance because fiber-reactives offered greater convenience and ease of dyeing.

Additional further decrease of naphthol dyeing was due to the restrictions placed on the use and manufacture of many chemical intermediates by Government regulatory agencies, eg, OSHA. At present the number of naphthols and coupling agents, as well as the number of firms manufacturing these products, have shrunk substantially. However, for extremely bright red and cranberry shades, naphthols are the only dyes that give top all-around fastness, including fastness to chlorine bleaching.

Vat Dyes

The naturally occurring vat dyes in plants, such as woad and indigo [482-89-3], are known to have been used 4000 years ago (see Dyes, natural). Since antiquity, these, along with Tyrian purple [19201-53-7] extracted from shellfish, were the vat dyes applied by natural fermentation to cellulosic fibers. Vat dyes primarily are used to produce exceptionally fast colorations on any cellulosic fiber, chiefly cotton. However, under proper conditions, they will dye to some extent almost any fiber (eg, silk, wool, acrylic, polyamide, and polyester fibers) with varying degrees of fastness. Yield and fastness vary substantially, thus limiting considerably the practicability of commercial application. In 1897, von Baeyer first synthesized indigo, which was the start of rapid progress in the discovery of many vat dyes, such as Indanthrene which was synthesized four years later.

Vat dyes are complex organic molecules which are insoluble in water, but when their carbonyl groups are properly reduced in a solution of caustic soda and sodium hydrosulfite to the so-called leuco or soluble state, they exhibit an affinity for cellulosic fibers. Chemical oxidizing agents are usually used to hasten oxidation of the reduced dye within the fiber back to its insoluble form which is physically trapped. This results in shades usually of excellent wash-, chlorine-, and lightfastness. The dyes are sold as powders or pastes which form dispersions upon dilution with water.

Application. Exhaust. Vat dyes are usually applied by either continuous or exhaust procedures, and occasionally by the pad-jig method.

Exhaust dyeing, whether stock, yarn or fabric, is done by either circulating the dye liquor through textiles (eg, package or stock dyeing) or moving the textiles continuously in the dyebath (eg, skeins or jig dyeing).

In order to obtain best dyeing results, vat dyes are generally separated into four dyeing groups:

Group	Dying temp	Caustic soda or sodium hydrosulfite, g/L	Sodium chloride, g/L
1 (IK)	20–30°C	5.0	20.0
2 (IW)	45–50°C	4.0	10.0
3 (IN)	50–60°C	4.5	none
4 (IN—Spec)	50–60°C	special formulations	none

These are, of course, generalizations, as many dyes are more sensitive to alkali concentration than are others in the same class; therefore, one must consult the manufacturers for specific recommendations. Typically, the bath is brought to the dyeing temperature and 75% of the required caustic soda and sodium hydrosulfite is added. After 10 min, the rest of the caustic soda and sodium hydrosulfite is added along with the reduced dyestuff and dyeing then is continued for one hour.

Should the vat-pigmented procedure be preferred, as it produces more level results, the following procedure is used. Wet the stock, as with the reduced method; adjust the bath to proper dyeing temperature; add the diluted dye over a 10–15 min interval; circulate for 20 min. At this point, the predissolved caustic soda and sodium hydrosulfite are added and dyeing proceeds for one hour. If IK or IW dyes are used, a proper amount of common salt should be added in two portions after 15 min of operation. Following dyeing, the bath is dropped, the goods are washed, oxidized, rinsed, soaped, hot rinsed, and dried.

Continuous Dyeing. This is the dyeing method by which the vast majority of vat dyes are applied, amounting to approximately 3500 metric tons out of a total U.S. production of 5500 t (excluding indigo) of vat dyestuff. Fabrics continuously dyed range from heavy terry toweling to poplins to broadcloths. Fabric varies in fiber content from 100% cotton to blends, such as 65/35 polyester/cotton, 50/50 polyester/cotton, 80/20 polyester/cotton, 50/50 nylon/cotton, and 15/85 nylon/cotton (see Textiles).

Continuous dyeing of cottons began in 1929, when the Graniteville Company started the first complete-continuous vat-dye range in which the goods were padded through reduced vat dye and then quickly (before oxidation) transported into large boxes containing reduced dye, caustic soda, sodium hydrosulfite, and sodium chloride.

Later, the steamer process came into being with DuPont's patent of 1944. This is the system being used today in continuous dyeing. A schematic drawing of such a range is shown in Figure 7. The goods are padded, dried, padded with a solution containing sodium hydrosulfite, caustic soda and brine, steamed for 30–60 s, rinsed, oxidized, soaped, rinsed, and dried.

The greatest volume of continuously dyed cotton is in 65/35 or 50/50 polyester/cotton blends. Blends are padded through the pad containing vat and disperse dyes and then carefully dried to minimize migration, then they are placed in a thermosol oven where the disperse dyes are fixed on the polyester by dry heat (200–220°C), and finally are cooled before entering the chemical pad for usual development of the vat dyes.

Soluble Vats—Leuco Esters of Vat Dyes. The procedure for modifying vat dyes was discovered in 1921 by Bader and Sunder and involved the reaction of leucoindigo in pyridine solution with chlorosulfuric acid, then conversion of this sulfonated compound to the sodium salt by addition of the proper amount of caustic soda. This compound is water-soluble and slightly substantive to cellulose.

After the stabilized leuco dye is applied to the cellulose fibers, it may be oxidized to indigo by treatment in an acid bath of sulfuric acid and sodium nitrite. These reactions could be applied to many indigoid and anthraquinone derivative vat. With the advent of fiber blends, it became more economical to continuously dye pastel shades with pigments and heavier shades with disperse/vat dyes. Exhaust dyeing of leuco esters was replaced with conventional vat dyeing on advanced high-speed dyeing equipment.

Miscellaneous Dyes

Dyes Developed on the Fiber. *Phthalocyanine Development Dyes.* A technique used for creating bright blue and green shades on cellulosic fibers which has lost importance in recent years consists of application to the material of 3-amino-1-iminoisoindolamine compounds dissolved in selected solvents (generally by padding). In some cases these are polymeric and already contain complexed copper or nickel. In other cases the metal is applied from a separate bath in the form of a water-soluble complex. The dye is typically developed by a 3-min treatment at 140°C. Dyeings of fastness properties comparable to those of vat dyes are being produced in this manner. Such products are marketed as Phthalogen dyes by Bayer AG in Germany.

Onium Dyes. Onium dyes contain an alkyl halide group and are made soluble by creation of an onium salt. After padding the dyes onto cellulosic material, the dyes can be insolubilized and fixed by a heat treatment thereby creating dyeings with good fastness properties. Dyes of this type were marketed under the Alcian brand name by ICI in England.

Oxidation Dyes. *Aniline Black.* Oxidation dyes for cellulosic fibers, at one time widely used, have lost importance in recent years because of loss of strength of the textile material, and because of ecological reasons. They still play a role in textile printing because of their easy dischargeability. Aniline black [13007-86-8] is a full bluish-black which can be produced inexpensively in a continuous application. It is produced by oxidation of an aromatic amine, ie, aniline in the cellulosic fiber in an acid medium with an oxidizing agent and a catalyst. A typical padding formula consists of: 100 g/L aniline hydrochloride, 50 g/L potassium ferrocyanide, 40 g/L sodium chlorate, and 30 g/L tartaric acid.

The fabric is padded and dried, and steamed for 1–2 min at 100°C. Further oxidation occurs in an acidified sodium bichromate bath at room temperature to obtain a nongreening black. The fabric is then soaped.

One problem associated with the dyeing of aniline black is the unavoidable loss of tensile strength. Under carefully controlled conditions the loss can be kept at 10–20%. However, under practical conditions, it is frequently higher.

Diphenyl Black. Loss of tensile strength can be reduced by use of diphenyl black [101-54-2]. N-Phenyl-p-phenylenediamine is dissolved together with lactic and acetic acids and is applied in combination with sodium chlorate and a suitable catalyst (eg, ammonium vanadate). After padding and drying the goods are steamed and soaped. By using other organic amines, it is possible to produce brown shades in a similar manner.

Mordant Dyes. *Alizarine Red.* Alizarine Red [72-48-0] also known as Turkey Red, was of considerable importance before the arrival of insoluble azo dyes. It produces full red shades of excellent fastness properties. The cellulosic fiber is mordanted (generally several times) with a metal salt and an olive oil emulsion. Aluminum or calcium salts are used and the metal is believed to be fixed as fatty acid salt. The material is then treated with hydroxyanthraquinone compounds which are water-insoluble but which can form complexes with the fixed metal. Before synthetic alizarine red was produced in 1868, the basis of alizarine red was the natural dye madder.

Natural Dyes. Few, if any, natural dyes are still in commercial use today. However, they are being used in small amounts by artists and craftsmen (10). Before the advent of synthetic dyes they were used in often complicated and lengthy processes by commercial dyers (see Dyes, natural). Among the natural dyes of animal source are purple from the snail *Murex brandaris* and other snails and cochineal [1260-17-9] from the female insect of *Coccus cacti*. Well-known natural dyes from plants are indigo, madder [72-48-0], safflower, woad, and logwood. Mineral colorants such as chrome yellow [1344-37-2], ultramarine [1317-97-1], and others were applied as pigments, mainly in printing, or produced by chemical reaction on the fiber.

Mineral Dyes. Mineral dyes are popular because of their fastness properties, low cost, and the fact that some of them would render cellulosic materials fast to mildewing and rotting. This has been of particular, however, lessening interest for military applications. In the dyeing of mineral khaki, the fabric is padded with a mixture of ferrous and chromic acetates. The fabric is carefully dried and steamed for 4 min at 100–105°C.

It is then passed through a solution of sodium hydroxide and sodium carbonate to produce within the fiber the insoluble metal hydroxides, and it is then soaped. Since the advent of synthetic fibers and newer dyeing techniques, their use has rapidly declined.

Dyeing of Fiber Blends

Fiber blends combine the advantageous properties of two or more fibers into one fabric. The combination of the esthetics of a natural fiber (eg, hydrophilic properties, hand) and the physical properties of the synthetic fibers (eg, strength, abrasion resistance) is an important factor in the acceptance of fiber blends. Also the differences in dyeability between the many fibers on the market open a wide field of multicolored yarns and fabrics to the stylist.

Designs, which in the past could only be obtained by the cumbersome and time-consuming process of blending differently colored fibers into multicolored yarns and fabrics, can be dyed today by a relatively short procedure, if different fibers have been blended in an early stage of processing. This saves expense in inventories and shortens delivery times substantially. The multitude of fibers offered have resulted in an almost unlimited number of fiber blends in the field; usually two fibers are used together, but three-fiber blends are also relatively common. Of the two different fiber blends in use, the so-called intimate blend is obtained by mixing the fibers prior to spinning them, and blended fabrics can consist of one fiber in the warp and another in the filling or can be comprised of yarns of different fibers woven or knitted in patterns. The dyeing methods for both types of blends are very similar but an intimate blend does not show a shade difference between two fibers as clearly.

Fiber blends can be dyed into union shades or tone-on-tone. Often multicolor effects are obtained by coloring the individual components in different shades or by reserving one fiber. A complete reserving of a fiber is not possible in all cases.

Blends Containing Wool. *Wool/Cellulosic Fibers.* One of the oldest fiber blends in the textile market is the combination of wool and cotton or wool and viscose. Economy was the major reason for this blend. Selected direct dyes, which dyed both fibers from a neutral bath in a uniform shade, or a combination of neutral-dyeing acid dyes with direct dyes were used to dye this fiber blend. However, problems with cross-staining, shade reproducibility, and loss of strength of the wool fiber by boiling in a neutral dyebath (pH 6.5–7.0) led to development of the now common half wool-acid one-bath process. Selected direct and acid dyes are applied at pH 4.5–5.0 at 98–100°C. A phenolsulfonic acid condensation product is added as a reserving agent, to prevent the direct dyes from dyeing the wool under acid conditions. Smaller amounts (1–2% owg (on weight of goods)) are required for union shades and 3–4% owg are necessary to obtain contrasting bicolor effects. If optimum wetfastness properties are required, fiber-reactive dyes can be applied to both fibers by use of a two-bath process.

Wool/Nylon. Nylon has been blended with wool in order to give additional strength to the yarn or fabric. It is used mainly in the woolen industry for coats and jackets and, to a lesser extent, for socks and carpet yarns. Both fibers are dyed with the same products, however the fibers have different affinity to them. Generally level-dyeing acid dyes are applied. Disulfonic acid types are preferred for light to medium shades, because they dye both fibers more easily in the same depth as monosulfonic acid types. However, in most instances a reserving agent is needed for light to medium shades to balance the depth between nylon and wool. Without it, the nylon will dye more strongly.

For heavy shades, monosulfonic acid types are preferred to obtain the necessary build-up on the nylon. In case the nylon remains lighter than the wool, a small amount of disperse dye is added to the dyebath to build up the depth on this fiber.

Wool/nylon upholstery fabrics and carpet yarns require higher light- and wetfastness properties. Neutral premetallized dyes are used in these cases. However, they have a much higher affinity to the nylon than to the wool. Therefore, stronger retarding agents have to be employed, eg, phenolsulfonic acid condensation products. Higher amounts are required for light shades than for medium depth. Dark shades can be dyed without addition of a retarding agent.

Wool/Acrylic Fibers. This blend is being used for industrial and hand knitting yarns. The acrylic fiber is aesthetically similar to wool, increases the strength of the yarn and adds bulk to the goods. Special precautions are necessary since the two fibers are colored with dyes of opposite ionic type. Coprecipitation is prevented with the use of an antiprecipitant. Usually, level-dyeing acid dyes are used for the wool portion in combination with the cationic dyes for acrylic fiber.

Wool/Polyester Fibers. The 45/55 wool/polyester blend is the most common fiber combination in the worsted industry. Strength and excellent dimensional stability of the polyester fiber enables the creation of lightweight wear fabrics not obtainable before. Economy has modified the fiber ratio and 30/70 and 20/80 wool/polyester blends are as common as the classical 45/55 blend. Disperse dyes for polyester and acid or neutral premetallized dyes for wool are employed in a one-bath process. Should cationic dyes be used for the polyester portion, a one-bath procedure can only be employed for light to medium shades, whereas dark shades require a one-bath two-step process. Wool blends should not be dyed above 105°C so that deterioration of the fiber quality does not occur.

Blends Containing Cellulosic Fibers. *Cellulosic/Acrylic Fibers.* Commonly this blend is used in knitgoods, woven fabrics for slacks, drapery, and upholstery fabrics. Since anionic direct dyes are used for the cellulosic fiber and cationic dyes for the acrylics, a one-bath dyeing process is only suitable for light to medium shades. Heavy shades are dyed in a one-bath two-step process (acrylic fibers are dyed first, followed by the cellulosic fibers). Cotton/acrylic fiber blends are also used for high quality upholstery pile fabrics. Besides the one-bath exhaust dyeing procedure involving a very high ratio of liquor to fabric, a continuous pad-steam process is used to dye these fabrics. After padding, the goods are steamed for 7–15 min at 98–100°C. The material then has to be rinsed warm and cold before drying.

Cellulosic Fiber/Nylon Blends. These blends are used in fabrics for apparel, corduroy, and swim wear. If wetfastness requirements are relatively low, the nylon portion can be dyed with disperse dyes and the cellulosic fiber with direct dyes. A one bath procedure can be employed. For better wetfastness, the nylon portion is dyed with level dyeing acid colors together with the direct dyes in one bath at 95°C using a reserving agent to prevent the direct dyes from dyeing the nylon. An aftertreatment with a cationic fixative improves the wetfastness properties. For swim wear, the cotton portion is dyed with fiber-reactive dyes.

After rinsing hot and cold and soaping at the boil, the nylon portion is dyed with a phosphate buffer system. Selected acid and/or acid milling colors are applied. An aftertreatment with a phenolsulfonic acid condensation product will result in best wetfastness properties.

Cellulosic/Polyester Fibers. One of the most important fiber blends on the market is the mix of 35/65 or 50/50 cotton/polyester. High tenacity viscose fibers are sometimes used instead of cotton. Many of the apparel knitgoods consist of this fiber blend as do sheeting, shirting, and work-cloth fabrics.

Although the knitgoods are dyed in exhaust-dyeing procedures, most of the woven fabrics are dyed according to one of the continuous dyeing processes. The easiest way to dye cellulosic/polyester blends is a one-bath exhaust dyeing process with application of direct and disperse dyes. High temperature dyeing equipment is very frequently utilized for the dyeing of cotton/polyester blends. This requires a special selection of direct dyes that are stable to high temperature dyeing conditions. The dyeing is conducted for 30–45 min at 125–130°C. To improve wetfastness properties, the dyeings can be aftertreated with a cationic fixative, copper sulfate, or formaldehyde, depending on the class of direct dyes employed. Fiber-reactive dyes are frequently used for cotton/polyester blends. One-bath two-step or two-bath processes are applied, since most of the fiber-reactive dyes will be partly hydrolyzed under the high temperature dyeing conditions for polyester.

Continuous Dyeing. Large quantities of cotton/polyester fabrics are dyed continuously to promote shade consistency and economy. The processes vary with the class of dye applied to the cellulosic fiber: pigments, sulfur, fiber-reactive, or vat dyes depending on the fastness requirements for the finished goods. The simplest way is a pad-dry-cure process using pigments for pale shades on sheeting and on shirting fabrics.

For higher fastness requirements and for darker shades, combinations of disperse/vat and disperse/fiber-reactive dyes are applied. The most common procedures are: pad-thermosol–chemical pad–steam; pad-thermosol–chemical bath–steam; and pad-thermosol–chemical bath.

Blends Consisting of Synthetic Fibers. ***Polyester Fiber Blends.*** Disperse-dyeable and cationic-dyeable polyester fibers are frequently combined in apparel fabrics for styling purposes. Whereas the disperse-dyes will dye both fibers, but in different depths, selected cationic dyes will reserve the disperse-dyeable fiber completely, resulting in color/white effects.

Polyester Fiber/Nylon Blends. This fiber blend is used in apparel fabrics as well as in carpets. Disperse dyes will dye both fibers however they possess only marginal fastness properties on nylon. Therefore, it is important to select those disperse dyes that dye nylon least under the given circumstances. The nylon is dyed with acid dyes, selected according to the fastness requirements. The fiber blend is dyed in a one-bath process for one hour at 100°C or at 115–120°C. Cationic-dyeable polyester/nylon blends and blends of nylon with both polyester types are also in use. The three fiber blend is dyed according to a one-bath two-step process. The disperse-dyeable portion and nylon are dyed first in one bath, followed by the cationic-dyeable polyester.

Polyester Fiber/Acrylic Fiber Blends. This fiber blend is dyed in a similar fashion similar to that of the blends of the different polyester fibers. The selection of cationic dyes is substantially larger for the acrylic blend.

Nylon Blends. Differential-dyeing nylon types and cationic-dyeable nylon blends are used primarily in the carpet industry. The selection of cationic dyes for nylon is rather limited, most products have very poor fastness to light. These blends are dyed in a one-bath procedure at 95–100°C. Selected acid dyes are used for differential dyeing. Disperse dyes will dye all different types in the same depth.

Pigment Dyeing

The earliest known process of pigment dyeing involved rubbing variously colored clays directly into the fabric. The shade range was limited by the variations in the shades of clay owing to iron oxide content. In some cases, washfastness was improved by the crude application of casein or starch as a binder (see Pigments).

Modern pigment application had its practical beginning in 1942 with the introduction of the continuous application of a bath consisting of an oil-in-water emulsion of organic pigments with thermosetting and/or thermoplastic resinous binders. A catalyst may or may not be required, depending on the type of binder used. Light to medium depth shades are applied daily to thousands of meters of cotton and cotton/polyester blends, mostly in fabrics weighing 110–250 g/m^2. The dyeing pigments used are almost 100% organic, chiefly covering a range of azoics, carbon black, phthalocyanines, triarylmethanes, and dioxazine derivatives. They are finely dispersed (0.5–5.0 μm dia) and are attached by the resinous binders to the fiber surface. Many dyers do not look upon this form of pigment coloration of textiles as *dyeing*; nevertheless, millions of meters of fabric are dyed by this system each year. It has become of even greater importance with the growing concern for energy and pollution because of the following factors: (a) cold pad bath and no heat requirements beyond curing at 170–175°C; (b) no reducing or oxidizing chemicals required; and (c) less hot-soaping and rinsing required as for fiber-reactive, sulfur, or vat dyes.

The resultant shades generally have outstanding lightfastness and good washfastness. Of technical concern are abrasions near seams and sharp edges, which tend to become frosty in washing. Further, the crock- or rubbing-fastness must be carefully observed and controlled by the addition of more binder or by the reduction of the pigment concentration. Fabrics being used range from lightweight poplins and sheetings to corduroy.

Application. The application mix is made up by adding a suitable antimigrant in drying to three-fourths of the volume of ambient-temperature water. Afterwards, the binder is added with slow stirring, followed by the addition of the diluted pigment, then the catalyst (if required) is added, and the mix made to volume with cold water. The finished mix is let down into the padder of a regular dye range as shown in Fig. 6. The goods are padded at RT through this mix, passed through the infrared drying unit and onto pressure-controlled cans for drying, and then into a curing oven for 45–60 s at 170–175°C. Finally, the fabric is soaped, washed, and dried.

Specialty Uses. In addition to polyester/cellulosic and cellulosic fabrics, pigments may also be applied to 100% synthetic fibers of special construction for unique uses. Examples are 100% filament polyester for draperies and glass fabrics for bedspread, curtain, and upholstery uses. This type of fabric is usually run on a finishing frame by going through the padder, followed by an infrared unit with final drying and curing in the tenter housing. Quite often, these are pastel shades with little or no afterwash required.

Solvent Dyeing

Solvent dyeing generally refers to dyeing in nonaqueous media. In the early 1970s, solvent dyeing was expected to become the dyeing process of the future and was discussed and researched extensively (11). The primary interest was in the possibility of reduced water consumption and water effluent treatment costs; secondarily, improved productivity was expected. Although the argument for or against nonaqueous dyeing has not been finally settled by any means, it has at present been decided in favor of aqueous dyeing. For all practical purposes there is today no commercial solvent dyeing (except for a few very specialized applications). Among the arguments for aqueous dyeing are: water is a virtually harmless solvent; water consumption has a potential for substantial reduction by using more efficient processes, lower liquor ratios and water recirculation; and most plants have already made extensive investments in effluent treatment facilities.

Arguments against solvent dyeing are: very substantial new investments are required; nonaqueous solvents are potentially more harmful than water in various respects; nonaqueous solvents are substantially more expensive than water, requiring working in closed equipment and recirculation or recovery; working with nonaqueous solvents can lead to a problem of disposing of solvent-containing residues; potential use of millions of liters of nonaqueous solvents can lead to a substantial air pollution problem; and difficulties in dyeing of fiber blends.

Arguments for dyeing in nonaqueous solvents are: energy savings due to reduced energy requirements for heating and evaporating such solvents, and higher productivity due to increased dyeing speed without loss in quality of dyeing, in some cases.

The following specific advantages and disadvantages of solvent dyeing processes have been found.

Nonionic Dyeing Process (Disperse Dyes on Polyester). Processes have been developed to dye polyester in perchloroethylene or 1,1,1-trichloroethane with disperse dyes in an exhaust-dyeing process. The advantage of such processes is owing to a substantially higher dyeing speed, based on the faster diffusion of disperse dyes when applied from these solvents. At the same time, better leveling is obtained due to faster diffusion and a lower partition coefficient. Oligomers that are frequently a problem in aqueous polyester dyeing are dissolved in the solvent. However, virtually all commercial dyes have a substantial solubility in the proposed solvents and the yields in exhaust dyeing are generally very poor. The reason is believed to be in the comparable magnitude of the solubility parameters of polyester, disperse dye, and perchloroethylene (12). Proposals have been made for using fluoro- and fluorochloro-hydrocarbon solvents with lower solubility parameters than perchloroethylene. Another approach involves the synthesis of disperse dyes with lower solubility in perchloroethylene. These are dispersed in perchloroethylene in a manner comparable to conventional disperse dyes in water. Relatively good yields (70–90%) have been obtained from selected dyes. However, no commercial products of this type have appeared. Many of the advantages of such a process have been eroded by substantial reductions in dyeing time of aqueous dye cycles for polyester.

For the purpose of continuous dyeing of polyester, experimental disperse dyes of relatively high solubility in perchloroethylene have been selected. These can be applied from such solutions, dried and fixed conventionally by dry heat. However, substantial dye migration during drying was encountered and the process offers no technical or economical advantages over aqueous application. Furthermore, the important polyester/cellulosic blends are difficult to process in such a manner.

Ionic Dyeing Processes. Ionic dyeing processes, such as acid dyes on wool and nylon, basic dyes on acrylic fibers, and direct dyes on cellulosic fibers, require the presence of water or lower alcohols for the ionization of dye and fiber and fixation to take place. Furthermore, the solubility of these dyes in solvents like perchloroethylene is very low. Two approaches have been investigated to overcome the problems:

(1) Dyeing from emulsion, water soluble dyes. The dyes are dissolved in water. The water at ca 10–30% vol % of perchloroethylene is emulsified using a suitable emulsifier into the perchloroethylene.

(2) Dyeing from emulsion, water-insoluble dyes. The dyes are modified by masking the water-solubilizing groups to make them perchloroethylene soluble. They are dissolved in perchloroethylene and 10–30 vol % of water is emulsified into the perchloroethylene. During the dyeing process the dye reverts on the fiber surface to its original form and is fixed in the conventional way. Systematic investigation of such processes has revealed, however, that no technical advantages are gained from the use of the nonaqueous solvent over conventional aqueous dyeing.

Two processes that involve the use of alcohols as dye application media have been proposed. In the Remaflam process (13) dyes are padded onto fabrics from pad liquors containing substantial quantities of methanol. The fabric is dried by burning off the methanol in a specially constructed unit. Fixation of the dye is then by conventional (dry heat or steaming) means.

The other process involves ethylene glycol or similar compounds as a dye medium (14). The process is available to licensees from the U.S. Department of Commerce. Disperse and ionic dyes are dissolved in hot ethylene glycol and textile materials are dyed by immersion for short periods (10–30 s) in the solution at 110–150°C. The process is applicable, eg, to the dyeing of polyester and nylon with disperse dyes, wool and nylon with acid dyes, acrylic fibers with basic dyes, cellulosic fibers with direct dyes, as well as to application of finishing agents. Because of the relatively short immersion time, the process is particularly suited for continuous dyeing. The dyes are incompletely exhausted during the passage and need to be replenished. The ethylene glycol is rinsed from the fabric with water or methanol. At present, no commercial application of this process is known.

Textile Printing

The term textile printing is used to describe the production through a combination of various mechanical and chemical means, of colored designs or patterns on textile substrates. In printing on textiles, a localized dyeing process takes place, whereby in general the chemical and physical parameters of dyeing apply.

A stringent selection of dyes is necessary for textile printing, because the localized dyeing process is accomplished in an extremely short liquor ratio and with the assistance of a number of auxiliary chemicals which aid in dye fixation during the comparatively short fixation or dyeing time, but may adversely affect solubility or other properties of certain dyes. Furthermore, the dye solution or dispersion to be printed has to show certain viscosity characteristics so as to produce a defined pattern on the textile.

The process of textile print coloration can be divided into three steps. First, the colorant is applied as pigment dispersion, dye dispersion, or dye solution from a vehicle called print paste or printing ink, containing in addition to the colorant such solutions or dispersions of chemicals as may be required by the colorant or textile substrate to improve and assist in dye solubility, dispersion stability, pH, lubricity, hygroscopicity, rate of dye fixation to the substrate, and colorant-fiber bonding. The required viscosity characteristics of a print paste are achieved by addition of natural or synthetic thickening agents, or by use of emulsions. In most cases, after the print paste is deposited on the textile, the prints are dried and stored until further processing in dye fixation. However, for printing of carpeting materials, the drying step is generally omitted and the textile is exposed to the dye fixation step directly after printing. A dried print, as a rule, is not fast to washing or crocking, or may have inferior lightfastness and, with the exception of pigments, does not show the correct and final shade.

The second step in processing printed textiles is the fixation process. During the fixation process the printed textile material is exposed to heat, heat and steam, or chemical solutions at temperatures and lengths of time governed by the type of fiber processed and by the dye class itself. During this treatment, the dyes diffuse from the fiber surface into the fiber structure and undergo physical or chemical bonding to the fiber. In case of pigment prints, latexes applied simultaneously with the pigment develop their film-forming properties during a heat treatment and impart to the print the final desired fastness.

During the following after-scouring, the third step in the textile printing process, the prints are rinsed and scoured in a detergent solution in order to remove auxiliary chemicals, thickening agents, and portions of unfixed dyes remaining on the surface of the printed fibers. After the scouring operation, only the dye adhering firmly to the textile remains.

Frequently, the scouring operation includes the use of chemicals assisting in removal of dye or dye fragments and/or chemicals reacting with textile fibers, or with removed dye to prevent redeposition of dye during scouring. Pigment prints, in general, do not undergo a scouring operation.

A novel type of textile printing, transfer printing, employs the intermediate step of printing dye dispersions or dye solutions onto a temporary substrate, usually paper. From the paper, the dye is transferred to the textile by heat or heat and steam, while printed paper and textile are in close contact.

Dyes in Textile Printing. The selection of dyes or pigments for printing of a particular textile substrate is governed primarily by the ability of this substrate to accept the dye. Other considerations are the fastness specifications involved and styling requirements with regard to brilliancy or shades in the design. In some cases, for economic or technical reasons, a print is executed with two or more dyes from different dye classes printed along side each other, or even in combination with each other. The more important classes of dyes for textile printing and the fibers to which they are applied are discussed below.

Pigments. Pigment-printed textiles represent the highest percentage of all printed textiles. This is primarily due to the uncomplicated process and low cost of imparting colored patterns to textiles with pigment systems.

The water-insoluble pigment, used in most cases as an aqueous dispersion, has no affinity to textile fibers and is not able to enter into chemical or physical reactions with the fiber. For the bonding of pigments to the textile, a bonding agent, generally of a synthetic latex type, is incorporated in the print paste, which through its film-forming properties holds the embedded pigment firmly on the fiber surface.

Due to the possibility of fixing pigments to textiles without interaction with the fiber material, this class of colorants can be applied to a greater variety of fibers than any other coloring matter. This advantage largely accounts for the popularity of pigments in textile printing especially on fiber blends.

The binding agent or "low crock" is a dispersion or solution of such polymers as poly(acrylic acid) derivatives and butadiene–styrene copolymers. In most cases a cross-linking agent is applied in the same print paste, consisting of synthetic resin types, eg, melamine–formaldehyde derivatives, for added wash- and crock-fastness.

Both pigment dispersion and binding agent–cross-linker are incorporated into a clear emulsion or synthetic thickening solution. The clear concentrate, from which the clear is prepared, contains such necessary auxiliary chemicals as dispersing agents, emulsifiers, acid donors, lubricants, protective colloids, and synthetic thickening agents of the poly(acrylic acid) and maleic acid anhydride-copolymer type. For emulsion-clear printing, mostly oil-in-water emulsions are used. With the clear emulsion type, 10–70 wt % of mineral spirits are emulsified into the clear concentrate-water solution and for the clear all-aqueous synthetic-thickener type a solution of a suitable clear concentrate in water is prepared.

The previously common clear water-in-oil emulsion system has largely given way to the oil-in-water systems because of better fastness properties that can be achieved with the latter and also for ecological reasons, and more recently the all-aqueous clear systems have gained importance for reasons of cost and ecology (see Emulsions).

Pigment print systems have a number of inherent limitations, most notably the effect of the binding agent and cross-linking chemical on the hand and feel of the textile material; also sometimes insufficient fastness to crocking, abrasion, washing, and dry cleaning, noticed especially in heavier shades and on large coverage; and the sometimes objectionable property of pigments to camouflage fiber texture and luster. Pigment prints are fixed by heat treatment at 150–200°C, in 1–5 min. Afterscouring is generally not necessary.

Disperse Dyes. This dye class is used for printing of polyester, triacetate, acetate, and in certain cases of polyamide. Depending on the type of fiber and the available fixation equipment, there must be a careful selection of dyes and print paste composition. Disperse dyes are used in powder or paste form, or ready-to-prepare aqueous dispersions for incorporation into a thickener solution. For disperse dyes that show sensitivity to alkaline hydrolysis or reduction during fixation, an acid donor or acid, and, if necessary, a mild oxidizing agent, are added to the print paste.

For printing on polyester, the fixation conditions are more rigorous than on other disperse-dyeable fibers, owing to the slower diffusion of disperse dyes in polyester. For continuous fixation the prints are exposed at atmospheric pressure to superheated steam of 170–180°C, for 6–8 min. A carrier is added to the print paste for accelerated and full fixation.

In batch-type fixation, the prints are steamed in a pressurized autoclave with saturated or nearly saturated steam at 14–18 kPa (105–135 mm Hg) and 125–130°C, for 45–60 min. In this case, the amount of carrier can be reduced or completely omitted. Generally, a small amount of urea (5 wt % of print paste) is added to the print paste to aid in fixation.

Dry-heat fixation conditions of 170–215°C, for 1–8 min are less popular for printed fabrics, but are sometimes employed because of lack of other equipment.

Afterscouring of polyester generally includes a reduction clearing with sodium hydrosulfite and alkali at 60–80°C to remove any dye remaining on the fiber surface.

Printing on triacetate follows the same general rules as for polyester. For batch type pressure steaming, the steam pressure is reduced to 7–10 kPa (50–75 mm Hg), at 115–120°C.

Acetate requires a steam pressure of ca 3.5 kPa (25 mm Hg), 108°C for full fixation of disperse dyes. With selected disperse dyes of a higher rate of diffusion in acetate, in combination with a suitable carrier, continuous steam fixation under atmospheric pressure at 100–105°C during 20–30 min is also possible. A light scouring at 40–50°C, completes the operation.

On polyamide, disperse dyes have generally low wetfastness properties, making them unsuitable for printed textiles that require even moderate wash- or perspiration-fastness. Selected disperse dyes, however, are used in printing of carpeting materials where the wetfastness level is sufficient.

A new field for application of disperse dyes to, primarily, polyesters and polyamides is transfer printing. In this process, selected low-energy disperse dyes of lower fastness to sublimation are printed onto paper. This can be accomplished by printing aqueous, thickened dispersions of dyes, or by printing solutions of disperse dyes in thickened organic solvents. After the paper is dried, it is brought into close contact with the textile material and a temperature of 195–215°C is applied to the paper-textile sandwich for 20–60 s. At this temperature the dyes sublime and condense immediately on the fiber surface, and diffuse simultaneously into the fiber. This process does not require afterscouring.

Acid Dyes. These dyes have their greatest importance in printing of polyamide. The dyes are dissolved in hot water, which can, depending on the solubility of a specific dye, have an admixture of a dye-solvent such as thiodiethylene glycol or diethylene glycol monobutyl ether. Acid dyes in liquid form do not require dissolving, but can be added to the thickener solution directly. In case of print-dry operation, 3–5% of urea or thiourea is also added to the dye solution to act as humectant during fixation. Depending on the processing, additional print-paste additives may be used.

Print-Dry-Fixation Process. This process is usually employed on woven or knitted fabrics for apparel, swim wear, drapery, and upholstery materials. The print paste contains an acid donor, usually 1–3% ammonium sulfate, and in some cases a small amount of sodium chlorate to prevent reduction of the dye in fixation. Fixation is by continuous steam treatment at atmospheric pressure at 100–105°C, for 30 min, or in pressure steam of 3.5 kPa (ca 25 mm Hg) 108°C for one hour.

In afterscouring, most often a synthetic tanning agent is used to (*a*) prevent redeposition of dye on unprinted areas, and (*b*) to improve wetfastness properties of the print.

The print-fixation-rinse process, without drying between printing and fixation, is commonly used in printing of carpet materials. This process requires no or little humectant or dye solvent in the print paste, but due to the shorter steam fixation time of 5–8 min at 100–105°C, a stronger acid such as formic acid may be used to assure full fixation of the dye. Afterscour consists mainly of rinsing in water.

Printing of wool or silk with acid dyes is of minor importance. For these fibers the print paste is made with dye solvent, humectant (glycerine and urea), a suitable thickener and dilute organic acid. An oxidizing agent is also added. Fixation follows the procedure for polyamide with fully saturated steam.

Acid dyes can be printed on acetate, producing prints with very good wetfastness and exceptional brightness. The print paste contains a solvent, urea, and ammonium thiocyanate as fiber swelling agent to aid in diffusion of the dye. Again, fixation and scouring follow the procedures for polyamide.

Premetallized Dyes. This dye group is applied to the same textile fibers and with the same procedures as those with acid dyes. The premetallized dyes offer better fastness properties, but lack brilliancy of shade. Except in printing of carpeting, the neutral-dyeing types of premetallized dyes are applied generally without acid or acid donor.

Direct Dyes. A few selected direct dyes are used to complement the acid dyes in printing of polyamide. Printing of cellulosic fibers with direct dyes, as with acid dyes, has lost its importance owing to insufficient wetfastness properties of these dyes on cellulosic fibers, and cumbersome fixation and scouring procedures.

Reactive Dyes. This dye class represents, next to pigments, the main dye group for cellulosic fibers (ie, cotton and rayon). The dye molecule forms a chemical bond with the fiber, producing prints with exceptionally good wetfastness. Reactive dyes react with cellulose in the presence of alkali. However, on account of the different levels of reactivity and affinity of the dyes to cellulose, and depending on available equipment, a number of different fixation methods can be used. The selection of suitable thickeners is very important and usually a sodium alginate solution is the preferred thickening agent. The dye powder is dissolved with urea in water or, if suitable, sprinkled as dry powder into the thickener solution, where it dissolves during agitation. For steam and dry heat fixation, the reactive print pastes contain 5–30 wt % urea and 2–4 wt % soda ash or sodium bicarbonate. Fixation in steam of 100–105°C takes 2–8 min, in superheated steam of 150–180°C, 1–5 min, and in dry heat of 150–200°C, 1–5 min.

In so-called wet fixation methods, the print paste containing 5 wt % urea and no alkali is printed and dried. For fixation, the prints are passed through a solution of strong alkali, such as caustic soda, potassium carbonate, soda ash or combinations thereof, and salt, and the fixation takes place within 20–30 s or 6–24 h, depending on temperature and alkali concentration chosen.

A combination method is available for flash aging, where an alkali-free print is impregnated with a cold alkali solution and immediately passed through steam for $1/2$–2 min.

All reactive-dye prints require after fixation a vigorous rinsing and scouring at temperatures close to the boil, to remove unreacted or hydrolyzed dyestuff. This is necessary to obtain optimal fastness properties.

Basic (Cationic) Dyes. From this dye class a wide selection of shades can be produced to satisfy any demand for brightness or fastness of prints. Use of basic dyes is, today, confined mainly to polyacrylic textile fibers, acetate, and as complementary dyes for acid-modified polyester fibers that accept this class of dyes.

Basic dyes are used in powder or liquid form and are incorporated into the print paste by dissolving powder dyes first in water with the aid of a dye solvent (eg, thiodiethylene glycol) and an organic acid (eg, acetic acid) followed by addition of a suitable thickener. Liquid dyes can be added directly to the thickener solution. Maintaining an acid pH in the print paste, and during the drying and fixation steps, is important. This is accomplished by adding 0.5–1.0% of a nonvolatile organic acid (eg, citric acid) to the print paste. In most cases a fixation accelerator is used with the print paste.

During drying of prints, the temperatures should not exceed 125°C because some basic dyes are susceptible to shade changes at higher temperatures. Dye fixation on polyacrylic fibers and acetate can be by atmospheric steaming at 100–102°C, for 20–30 min or pressure steam of 3.5 kPa (ca 25 mm Hg) at 108°C, for 45–60 min.

With modified polyester only pressure steam will produce full fixation and color yield.

Afterscouring is by rinsing and detergent scour. A scouring auxiliary with affinity to the fibers can be used to prevent redeposition of rinsed out dye.

Vat Dyes. Applied to cellulosic fibers, vat dyes yield prints with excellent fastness properties. Their application can follow two different procedures.

All-In Method. The vat dye, in form of a dispersion of the water-insoluble dye is added to a thickener solution that also contains the necessary reducing agent such as sodium formaldehyde sulfoxylate, the alkali such as potassium carbonate, and glycerine as a hygroscopic material. After printing and drying at temperatures not exceeding 80°C, the prints are passed through steam at 100–102°C for 5–8 min. It is important that this steam is free of air. During the steam fixation the vat dye is reduced to its water-soluble leuco form and is absorbed by the fiber in this form.

Afterscouring consists of reoxidation of the leuco form to the insoluble vat dye, accomplished by addition of hydrogen peroxide or sodium dichromate to the water rinse, followed by boiling detergent solution.

Flash-Age-Fixation. The method of flash-age-fixation requires printing of a paste consisting of thickener solution and vat dye. The dried print is then impregnated with a solution of alkali, such as caustic soda and potassium carbonate, also containing a reducing agent such as sodium hydrosulfite and sodium formaldehyde sulfoxylate, and immediately passed for ½–5 min through air-free steam at 105–125°C. This is followed by oxidation and soaping.

Azoic Dyes. The azoic dyes are used primarily as complementary dyes alongside pigments or mordant dyes on cellulosic fibers. As commonly applied, they produce bright and dark shades of orange, red, maroon, navy blue, brown, and black.

The common method of application is to use both a naphthol derivative and a stabilized color base, usually in the form of a diazo imino compound in the same print paste. This mixture is soluble in dilute caustic soda and no coupling takes place at this stage. The dried prints are passed through steam at 100–105°C that contains acetic and/or formic acid vapor. As neutralization takes place on the print, the coupling occurs very rapidly and the insoluble azoic dye is formed.

Phthalocyanine Dyes. These dyes are synthetized as the metal complex on the textile fiber from, eg, phthalonitrile and metal salts. A print paste typically contains phthalonitrile dissolved in a suitable solvent and nickel or copper salts. During a heat or steam fixation of 3–5 min, the dye is formed. The color range is restricted to blue and green shades and can be influenced to some extent by the choice of metal salt. A hot acid bath during afterscouring completes the process.

Blends of Different Fibers. In certain cases it is desirable to print fiber blends with combinations of the appropriate dye classes, rather than with pigments. Only polyester/cellulose blends are of commercial importance and the following dye systems have been developed for them. The dyes of the different classes are contained in the same print paste and, therefore, are applied simultaneously in one print operation.

Disperse—Reactive Combinations. A careful selection of the disperse dyes is necessary with regard to their alkali sensitivity during fixation. Pressure-steam fixation is less suitable than super-heated steam or dry heat. Newer systems employ an afterscouring with strong alkalis to remove disperse-dye stain from the cellulosic portion.

Disperse—Pigment Combinations. These are applied from a print paste essentially similar in composition to a pigment print paste. However, fixation requires temperatures high enough to accomplish diffusion of the disperse dye into the polyester. The advantage of this system over a straight pigment print is better fastness to abrasion and washing.

Disperse—Vat Combinations. These require a two-step fixation. The disperse dye is fixed first, usually by dry heat, followed by impregnating of the textile with an alkali and reducing agent solution and short steam fixation for the vat dye. The selected disperse dyes fixed in the polyester fiber are not destroyed by the reducing agent, but disperse dye remaining on the cellulose is destroyed.

Methods of Printing. Textile materials can be printed at different steps of the textile manufacturing process. Woven fabrics comprise the largest percentage of printed goods. In recent years, knitted textile fabrics have considerably increased in importance. However, printing can also be done on yarns in skein form, or on warps being passed from a warp beam to another beam, or as yarn strands. Space printing is a process where a yarn, temporarily knitted into a loose fabric, is printed and then again de-knitted. Carpets can be printed in woven or tufted constructions. Vigoureux printing is the printing of woolen slubbing. Regardless of the state of the textile material, any printing process makes use of one of the following methods.

Roller Printing. The largest amount of printed textiles or transfer papers is produced by roller printing. However, this method of printing has been loosing ground steadily to other print methods.

The roller print machine consists of essentially a copper print roller which has the pattern or design to be printed engraved on its surface, and a cast iron cylinder of a diameter larger than the engraved print roller.

The print roller revolves in contact on one side with the cylinder and on the other side with a furnishing roller or furnishing brush, which in turn revolves in a "color box" (print paste reservoir) carrying the print paste to the engraved print roller. The fabric to be printed passes between the cylinder and the engraved print roller, absorbing print paste from the engraved areas of the print roller.

The fabric to be printed usually travels through the roller print machine superimposed on another absorbent fabric of slightly greater width, the "back gray" and on a rubber coated heavy fabric, the "blanket."

To confine the print paste to the engraved parts of the print roller, any excess that may have been applied by the furnishing roller is scraped off the unengraved, smooth surface of the print roller by a metal blade, the "doctor blade", prior to making contact with the fabric to be printed (see Coating processes).

Screen Printing. This printing process essentially consists of the transfer of print paste through a stencil to the substrate to be printed. The stencil is made of polyamide or polyester (also metal in some cases) bolting cloth stretched tightly over a metal frame. The pattern or design is produced in the stencil or screen by coating those parts that should not transfer the print paste with an impervious material. The design is transferred to a film and the film placed on the bolting cloth, which is coated with light-sensitive material. After light exposure, the unexposed parts, having been protected by the light-impervious areas of the film, are removed by rinsing and thus the pattern is created on the screen. To withstand the mechanical stress of prolonged printing, the light-sensitive coating has to be very strong or has to be reinforced with suitable screen lacquer.

For hand printing or silk screening, the fabric to be printed is stretched over long tables, which are covered with a blanket of layers of felt and back gray or felt- and rubber-coated material. The screen is placed on the textile material and the print paste is scraped across the screen by means of a squeegee, penetrating through the open areas in the screen onto the textile. Then the screen is lifted and moved along the table by hand or in a carriage to the next print position. A screen is used to deposit only one color, and for multi-color designs as many screens as there are colors in the design are necessary. It is important that the textile material does not move or shift during the printing process, therefore it is pinned to the back gray or glued to the blanket.

Mechanized screen printing is done on flat-bed screen printing machines. On these machines the screens remain stationary while the textile material is moved underneath the screens in intermittent steps. The textile material is glued to an endless blanket which serves as support for the textile material and which is cleaned with water once during each revolution. When the screens are in a lifted position, the blanket, and the textile material, with it, move forward. When they are in a lowered position the squeegees perform the printing. The movement of the squeegees can be controlled mechanically or electromagnetically (eg, in the case of metal squeegees).

A further development is the rotary screen-printing machine on which the screens have been fashioned into seamless perforated cylinders. The design is imparted to these rotary screens by methods similar to those used for flat screens; however, the screen is a preperforated metal sleeve.

Spray Printing. The newest method of application of print pastes to textiles is by a form of controlled spray process. In this process, a continuous stream of print paste droplets is directed towards the moving textile substrate. A pattern is created by diverting some droplets away from the textile, while others are allowed to reach the textile. The diversion of droplets can be accomplished by mechanical or by electrostatic means.

Styles of Printing. Most printing is done as direct or application printing. In this style the print pastes are applied to an uncolored fiber directly. In discharge printing the printing is done on textile material already uniformly dyed, whereby the dye is destroyed in a patterned fashion by such reducing agents as sodium formaldehyde sulfoxylate, zinc formaldehyde sulfoxylate or stannous chloride contained in the discharge-print paste. In addition to "white discharges," "color discharges" can be produced by combining a dye stable to the chosen discharging or reducing agent with the discharging agent in one print paste. For example, a cotton fabric dyed with a suitable direct dye can be discharged to white with sodium formaldehyde sulfoxylate. By adding a suitable vat dye to the print paste containing the reducing agent along with the necessary alkali, a color discharge can be produced.

In resist printing, print pastes are used that can inhibit the development or fixation of different dyes that are applied to the textile prior to or after printing. These resists can be of a chemical or mechanical nature, or combine both methods. For example, fiber-reactive dyes, which require alkali for their fixation, can be made resistant by printing a nonvolatile organic acid, such as tartaric acid, on the textile.

Paper Coloring

The major reasons for coloring paper (qv) are for aesthetic appearance and utilitarian purposes. Aesthetic appearance includes colored backgrounds for printed material, colored writing papers, colored household products to harmonize with interior decor, and many other diverse uses dictated by individual tastes. Utilitarian purposes include identification of multicopy forms, identification of manufacturer or marketer of specific materials or products, opaqueness or hiding power of packaged material, or to control consistency of paper manufactured from various colored raw materials.

Dyeing Processes. Paper may be colored by dyeing the fibers in a water suspension by batch or continuous methods. The classic process is by batch dyeing in the beater, pulper, or stock chest. Continuous dyeing of the fibers in a water suspension is adaptive to modern paper-machine processes with high production speeds in modern mills. Solutions of dyestuffs can be metered into the high density or low density pulp suspensions in continuous operation. Paper may also be colored by surface application of dyestuff solutions after the paper has been formed and dried or partially dried by utilizing size-press addition, calender staining, or coating operations on the paper machine. In addition, paper may be colored in off-machine processes by dip dyeings or absorption of dyestuff solution and subsequent drying, such as for decorative crepe papers.

Among the colorants that have been and are being used for the dyeing of paper are: natural inorganic pigments (ochre, sienna, etc); synthetic inorganic pigments (chromium oxides, iron oxides, carbon blacks, etc); natural organic colorants (indigo, alizarin, etc), and synthetic organic colorants. The last is the largest and most important group. For the dyeing of paper the following classes are most commonly used today:

Basic Dyestuffs. Basic dyestuffs are generally hydrochloride salts of color bases formed by dissociation partially of charged dyestuff ions (cations). Basic dyestuffs have a high affinity for mechanical pulps and unbleached pulps which have a large amount of acid groups in the fiber. The cationic dyestuff reacts with these acid groups to produce by salt formation very stable lakes that are insoluble in water. Bleached pulps cannot be dyed satisfactorily since they contain few acid groups after bleaching. Although most basic dyestuffs stain ligneous pulps, the addition of alum with or without sizing offers advantages depending on the depth of shade and type of paper required.

Basic dyestuffs usually have brilliant shades with high tinctorial strength, and on ligneous pulps, produce good fastness to water, steam, and calendering, and clear back water. Basic dyestuffs have poor lightfastness, and because of their poor affinity for bleached pulps, have a strong tendency to mottle and/or granite in blended furnishes with ligneous pulps.

Basic dyestuffs are usually used for dyeing of unbleached pulp or mechanical pulp such as wrapping paper, kraft paper, box board, news, and other inexpensive packaging papers. Their strong and brilliant shades also make them suitable for calender staining and surface coloring where lightfastness is not critical.

Acid Dyestuffs. Acid dyestuffs have no affinity for vegetable fibers, either as bleached or ligneous pulps. Although the dyestuff molecules penetrate into the capillaries of the fiber, there is no firm linkage since they are not bound chemically, nor is there any mutual attraction between the dyestuff and the fiber since they both have the same surface potential. This lack of affinity has the advantage that acid dyestuffs do not mottle in blended furnishes and produce level shades and even appearance when properly fixed by use of fixing agents, rosin size and papermakers' alum (aluminum sulfate). The acid dyestuffs are precipitated on the fibers by these materials, forming a linkage between the cationic-treated fibers and dyestuff molecules. Sizing improves the retention of lake particles by formation of aluminum resinate. Because of poor affinity and good solubility, acid dyestuffs have poor bleedfastness, and form colored backwater, and are therefore suitable for paper that does not require wetfastness, such as construction grades. Acid dyestuffs are most suitable for calender staining or surface coloring because of their solubility and brightness of shade.

Direct Dyestuffs. The difference in the affinity of direct dyestuffs for ligneous and bleached pulps is considerably less than basic dyestuffs, but direct dyes color bleached pulps more evenly because they are easily wetted and penetrate rapidly.

Direct dyestuffs generally have a high affinity for cellulose fibers and are therefore the most useful dyestuff type for unsized or neutral pH dyeings. Their bonding ability to nonligneous pulps and excellent fastness properties to light and bleeding make them very useful for all fine papers. The shades of direct dyestuffs are not as bright as those of acid or basic dyestuffs and in blended furnishes (bleached/ligneous pulps) mottling or graniting may occur.

Pigments. Synthetic organic pigments are replacing the use of some inorganic pigments for ecological reasons. The conditions of dyeing are quite different for pigments than those used for soluble dyestuffs. Pigments do not react chemically with the fiber, but are fixed physically and are dependent on filtration, adsorption, occlusion and flocculation. Paper dyeings with pigments have outstanding fastness properties, but poor affinity, low tinctorial strength, and two-sidedness problems limit their application to paper (see Pigments).

Fluorescent Whitening Agents. These whiteners (FWAs) are fluorescent substances which transform invisible ultraviolet light into visible blue light. Fluorescent whitening agents change the appearance of paper in two ways: by emitting light and therefore increasing the luminosity (brightness); and by changing the shade from yellowish-white to bluish-white.

FWAs are unfavorably influenced by factors such as low ultraviolet content light and high ultraviolet light adsorption of pulps, chemicals and fillers. FWAs can be used in the wet end or as a surface colorant or in combination to obtain the greatest efficiency or yield (see Brighteners, fluorescent).

Production of Colored Paper. All dyestuffs and pigments are taken up more readily by the fiber in the acid pH range. Aluminum sulfate (papermakers' alum) is more effective than acid in fixing acid and direct dyes or pigments. The aluminum ion forms salts with acid and direct dyes which do not readily dissolve, whereas with pigments the triple positive charge also ensures fixation on the negatively charged fiber. With bleached pulps, only direct dyestuffs may be used to color unsized or alkaline-sized papers. For household tissues of deep shades to be made at neutral pH (no alum), only a selection of direct dyestuffs with particularly high affinity can be used. Blended furnishes of pulp (bleached/ligneous) can be colored successfully with combinations of dyestuff classes such as basic and acid, direct and acid, and direct and basic dyes. Generally, the dyestuff with the greatest affinity should be added first and solutions of the cationic dyes cannot be mixed with solution of anionic dyes.

Batch Dyeing. Depending on operational conditions, the dyestuffs can be added to the pulper, beater, or stock chest. The advantages of this batch method are its simplicity and the fact that no additional equipment is necessary. Thorough mixing and optimum fixation on the fiber are ensured by adding the dyestuffs at an early stage of processing. However, the relatively long interval from addition to inspection of the dyed paper can lead to prolonged correction and color-change times.

The normal sequence of addition in the dyeing process is pulps, filler, dyestuffs, rosin size, and alum. The dyestuffs are either taken up by the fiber because of their affinity or they must be fixed on the fiber in a finely divided form by suitable fixing agents. Alum which is required to precipitate the rosin size has a strong precipitating and fixing effect on dyestuffs.

Continuous Dyeing. Continuous dyeing of the pulp by addition of dyestuff before the head box of the paper machine has been gaining in importance in recent years, because this method is very well adapted to the high production rates of modern paper machines. Compared to batch dyeing, the advantages of continuous dyeing are faster color changes owing to immediate response and less colored stock in the paper machine system. Continuous dyeing is even more important where the grade assortment includes white paper as well as colored paper. To utilize the advantages of continuous dyeing, the point of addition of the dyestuffs should be as close to the head box as possible. In actual practice this must be balanced against the dyestuff contact time and mixing requirements between dyestuffs and pulps and is dependent on the dyestuff affinity to the pulp. As a rule, only highly substantive dyestuffs can be utilized. To maximize dyestuff yield it is advisable to add the dyestuffs to the stock solution before dilution to facilitate adsorption of the dyestuff and to increase contact times and allow adequate mixing.

In the manufacture of colored papers, it is best to add the dyestuffs before addition of rosin size and alum. This is not always possible in continuous dyeing procedures where dyestuffs must be added to stock containing size and/or alum, and this may cause premature laking of the dyestuffs and subsequent loss of tinctorial strength and/or dullness of shade. The proper selection of dyestuffs can help to reduce these disadvantages.

The metering and addition of dyestuff solutions is critical to successful operation of continuous dyeing. It is also just as critical to meter other components of the process including pulp, broke, fillers, size, alum, fixing agents, retention aids, wet-strength resins, and other additions that affect dyeing. Exact and dependable operation, simple maintenance, and accurate control are essential requirements of a successful installation.

Leather Dyeing

The dyeing of leather (qv) presents some difficulties not encountered in the dyeing of textiles. Unlike textiles, leather is not a homogeneous product of definite composition whose chemical properties may be closely and accurately defined, but is rather a product derived from protein collagen (skin or hide substance) treated with one or more tanning agents. Although the chemical characteristics and properties of collagen have been accurately determined, the variety of materials that will convert it into leather leave the chemical composition of leather extremely indefinite. Not only may the compound used to convert hide substance into leather vary chemically over a wide range, but the quantities used, the method of application, and the physical condition of the hide prior to tanning or dyeing may vary, with each factor in turn affecting the dyeing properties of the resultant leather. Also, leather retains many of the properties originally associated with the parent substance, and these affect profoundly and, in many ways, limit the dyeing properties of the final product. Chief among these properties are sensitivity to extremes of pH, thermolability, and the tendency to combine with acidic or basic compounds.

Tanning. Leather may be produced (ie, the tanning of leather) by use of chrome, vegetable, aldehyde, syntan, oil, and many other tanning agents. Chrome leather accounts for the greater part of leather production, but is seldom employed as a self-tannage. Chrome is usually employed first, and then a vegetable, syntan, or resin "retannage" follows. Vegetable tannage is employed to a fairly large degree for specific types of leather. Syntans and aldehydes (eg, glutaraldehyde) are applied combined with chrome or vegetable agents, or as a retan or second tannage over either the chrome or vegetable agents.

Practically all chrome grain leather, used mainly for shoe or garment leathers, is given a retannage with a comparatively small amount of vegetable extracts such as quebracho, wattle, or other wood extracts, or syntans, resin tanning materials, or possibly combinations of two or all three. Such an anionic retannage not only imparts desirable characteristics to the leather, but also promotes more level and uniform dyeing. The use of an anionic retannage also tends to increase the penetration of the dyes into the leather, although full or complete penetration of dye is rarely accomplished in this manner. Larger percentages of dye are therefore required for such retanned leather to produce approximately the same full shades as on untreated leather.

Leather Dyes. The four main classes of dyes employed in the coloring of leather are the acid, acid/direct, direct, and basic types. On chrome leather, the direct dyes usually have greater affinity and produce fuller or heavier shades than do acid or chrome dyes. Acid/direct dyes as well as the metallized-type dyestuffs may be classified for the purpose of leather dyeing as the main types in use. Basic dyes color chrome leather weakly and unevenly, unless the leather is first mordanted or retanned with suitable materials, such as vegetable tannin, syntans, or previously applied acid and/or direct dyes. They may be used alone on vegetable-tanned leather to produce full shades or, as is done more frequently, following a preliminary coloring with acid or acid/direct dyes. In the latter case, basic dyes are used to impart fullness of shade with minimum coloring matter and cost.

Acid and direct dyes vary considerably in their penetrating properties on chrome leather. Although it is not possible to make a distinct division, the direct dyes as a class are more surface-dyeing than the acid dyes. Chemically, there appears to be a relationship between the number of sulfonic acid groups per dye molecule, the gross molecular weight, and the ability to penetrate into the interior sections of the leather. Thus, dyes with a low molecular weight and a relatively greater proportion of sulfonic acid groups penetrate more deeply than those with either a higher molecular weight or with fewer sulfonic acid groups per molecule. Since the direct dyes fall within the second category, they tend to dye only the surface of chrome leather.

Generally, in leather dyeing it is necessary to employ two or more dyestuffs to produce a given shade. Such blending is important, as the selected dyes should have approximately the same rates of diffusion; this is an important factor in the adjustment of shades. The performance of a series of dyes on textiles cannot be used as a guide for their use on leather. For most leather, only superficial dyeing or surface coloring is required. However, some shoe and garment leathers require a certain degree of penetration, and if this cannot be obtained by simply increasing the degree of neutralization, or the addition of alkali, it may be accomplished by blending surface-coloring dyes with penetrating-type dyes. This procedure is used to obtain enough penetration to resist or minimize subsequent buffing of the leather surface. If a proper selection of dyes is not made in these combinations, and if one dye tends to diffuse more rapidly than another, or is less firmly bound, the result is a two-layer effect at the expense of shade. The two-layer effect may develop during the drying process and cannot always be observed in the wet state of the dyed leather. Some penetrating dyes also migrate to the interior of the leather during the drying operation, and produce a change in shade on the surface.

In the case of almost all grain leather (except for special leathers, such as glazed calf, glazed kid, and all suede leather) the dye used is applied as "bottom" color and the final shade and finished product are obtained by the application of "top" coatings of pigment finishes. The greater opacity of pigments tends to fill and conceal the many imperfections (eg, those caused by insect bites, barbed-wire scratches, etc) that are normally present on average hides. An exception to this practice is the so-called aniline-finished leathers on which little or no pigment-type finishes are employed. In this type of leather, which is competing against the synthetic leathers, the greater transparency of the dyestuffs reveals the true grain-surface structure of the skin much more strikingly than do pigment finishes. A more rigorous selection of stock is necessary for this type of leather in order to avoid surface defects which are normally obscured by pigments (see Leather-like materials).

Leather Dyeing Methods and Equipment. The methods used in dyeing leather are quite simple and they obtain their names from the equipment employed, such as drum, wheel, paddle, brush, tray, or spray dyeing. Most leather is dyed in drums. Drum dyeing, is carried out in revolving cylinders ranging from ca 1–1.5 m in width and 2.4–3.7 m in dia. Dye and other solutions are fed into the rotating drum through a hollow axle or grudgeon. This method of dyeing is relatively rapid and efficient, as it permits the coloring of large amounts of leather in comparatively small amounts of solution, with attendant low utilization of dye and other materials. The amount of solution in drum dyeing ranges from 1–2 times the weight of wet leather or from 4–8 times the weight of dry leather. The usual practice is to employ the necessary quantities of dyestuff, which would range from 0.10–5.0 wt % of the blue (that is, wet) stock for grain leather. Dyeing is carried out at 46–52°C. The amounts of 85% formic acid used to exhaust the dyebaths are usually based on the ratio of one-half the amount of formic acid to that of dyestuff employed.

Dyeing can also be carried out in hide processors or automatic stainless-steel dyeing machines with Y-shaped cross-section to subdivide the interior into three loading sections (somewhat similar to a large-scale washing machine).

Spray dyeing also accounts for the coloring of large amounts of leather. Usually acid or basic dyes are selected for good solubility and level-dyeing properties. At present, proprietary ranges of premetallized or basic liquid dyes are available and used widely. A wetting agent and/or a solvent are usually incorporated in the dye solution to prevent droplet formation and to promote levelness and penetration after the solution has been deposited on the grain of the leather by the air-spray gun. This method

334 DYES, APPLICATION AND EVALUATION

is extensively used for shoe upper leather where only the grain side is sprayed and the flesh or reverse side is left uncolored with what is termed a natural back. This type of leather is mainly chrome-tanned cowhide that is retanned with relatively small amounts (2–5%) of vegetable tanning extracts such as quebracho, wattle, or sumac, sometimes combined with small percentages of syntans and/or resin tanning materials. The leather is then fat-liquored and processed in the normal manner. This method permits the tanning and storage of leather in the dry or natural state, and its rapid conversion into whatever shade is desired by spray-dye application. Spray dyeing is applicable to almost any type of leather where deep penetration is not required, and is the most economical method of applying a dye.

EVALUATION OF DYES

FASTNESS TESTS FOR TEXTILES

In an age of consumer activism and increasing international trade, test methods and procedures assume more and more importance. Generally accepted tests, instruments, and standards provide a common basis for the evaluation of quality aspects of dyed textiles.

In the development of test methods, an attempt is made to make them broad in scope and related to actual processing and use. Such tests are generally of an accelerated nature, ie, they reveal the fastness properties exhibited by the material over a lengthy period of time in a test of short duration.

Fastness of dyed textiles is evaluated in regard to natural destructive agents, such as daylight, weather, and atmospheric gases, as well as to various treatments the material is likely to undergo, such as washing, drycleaning, ironing, steaming, etc.

The international body active in test method development is the ISO (International Standards Organization). Fastness tests for dyed textiles fall under the jurisdiction of ISO Technical Committee 38 Textiles/Subcommittee 1 "Tests for Colored Textiles and Colorants." The United States is represented in ISO through the American National Standards Institute (ANSI). In the field of interest here, ANSI relies on the American Association of Textile Chemists and Colorists (AATCC) for test development and expert guidance. The AATCC currently has 15 active research committees concerned with colorfastness properties.

The purpose of these organizations in this area is solely to provide useful test methods. The setting of specifications on the basis of such test methods is a matter to be resolved between buyer and seller.

Colorfastness

Described below are the essential features of generally accepted tests (from the AATCC 1976 Technical Manual (15)) used in evaluating fastness properties of dyed, printed or otherwise colored textiles. For complete details and for sources of the specialized materials and equipment used in some of these test procedures, reference should be made to the current AATCC Technical Manual.

Acids and Alkalies (AATCC 6-1975) (ANSI L 14.2-1973/R 1960). This test method is designed to reveal the colorfastness of dyed or printed textiles to acid fumes, sizes, alkaline sizes, alkaline cleaning agents, and alkaline street dirt, by steeping the textile material in or spotting with the required solutions. The tested specimens are examined for changes in color.

Procedure. Acid tests: (*1*) The colored material is spotted with hydrochloric acid solution (100 mL 35 wt % HCl soln) at 21°C and allowed to dry at room temperature without rinsing; and (*2*) The colored material is spotted with 56% acetic acid and allowed to dry at room temperature without rinsing.

Alkaline tests: (*1*) The colored material is steeped for two minutes at 21°C in ammonium hydroxide solution (28 wt % NH_3) and dried at room temperature without rinsing; (*2*) The colored material is steeped for two minutes at 21°C in a 10% sodium carbonate solution and dried at room temperature without rinsing, (*3*) The colored material is hung in a 4-L bell jar placed on a glass plate over a 7.6-cm evaporating dish containing 10 mL of concentrated ammonium hydroxide (28 wt % NH_3). A 24-h exposure is made; and (*4*) The colored material is spotted with a freshly prepared calcium hydroxide paste, allowed to dry, and then brushed to remove the dry powder.

Evaluation and Classification. The effect on the color of the test specimens is expressed and defined by reference to the Gray Scale for Evaluating Change in Color. The results are reported as to class with respect to the specified test.

Colorfastness to Bleaching with Chlorine (AATCC 3-1975) (ANSI L14.57-1973/R 1963). This test method is used to determine the effect of sodium hypochlorite containing up to 0.3% available chlorine on dyed or printed cotton or linen textiles.

Procedure. The specimens are washed in sodium hypochlorite solutions under controlled conditions at pH 11.0 ± 0.2 and temperatures of 27 ± 3°C for 60 min.

Test	I	II	III	IV
available chlorine, %	0.01	0.1	0.2	0.3

Controls. The correct test performance is checked with the aid of two control dyeings: Control 1 is a dyeing on cotton cloth of 4% of Vat Violet VN [4424-87-7] (CI Vat Violet 13), 10% paste; and Control 2 is a dyeing on cotton of 40% Vat Brilliant Violet RK [3076-87-7] (CI Vat Violet 17), 10% paste.

Evaluation and Classification. Five classes of fastness are established with the help of the Gray Scale for Evaluating Change in Color. Control 2 fabric, when exposed to the four tests under satisfactory conditions will show the following classifications:

Test	I	II	III	IV
classification	4	4-3	3	3-2

Colorfastness to Bleaching With Peroxide (AATCC 101-1975) (ANSI L14.146-1973). This test is applicable to textiles of all kinds in all forms, with the exception of polyamides.

Procedure. A test specimen is placed between two 10.2 × 3.8 cm white test cloths, one of which is made of the same fabric as the specimen and the other of which varies as follows: If the first piece is wool, silk, linen, or rayon, the second piece should be cotton or multifiber fabric, if the first piece is cotton, or acetate, the second piece should be viscose rayon or multifiber fabric.

The colored swatch is placed between the two white cloths and sewn on all four sides, then subjected to the required test as given in Table 3. In tests, I, II, and III, the

336 DYES, APPLICATION AND EVALUATION

Table 3. Conditions in Testing for Colorfastness to Bleaching and Peroxide

Test conditions	Wool, I	Silk, II	Cotton, III	Cotton, IV
35% hydrogen peroxide, mL/L H_2O [a]	15.4	8.8	8.8	8.8
	(17.5 g)	(10.0 g)	(10.0 g)	(10.0 g)
sodium silicate, 1.41 sp gr (42° Bé) mL/L H_2O [a]		5.1	4.2	7.0
		(7.2 g)	(6.0 g)	(10.0 g)
sodium pyrophosphate, g/L H_2O [a]	5.0			
sodium hydroxide, g/L H_2O [a,b]			0.5	0.5
wetting agent, mL/L H_2O [a]			2.0	
pH, initial	9.0–9.5	10.5	10.5	10.5
time, h	2	1	2	1
liquor/cloth ratio	30:1	30:1	30:1	1:1

[a] Distilled water.
[b] Doubly sulfonated castor oil.

test swatch is rolled loosely and placed in a test tube with the appropriate solution for the indicated time. In test IV, the swatch is placed in saturated steam (99–101°C) for 1 h.

Evaluation and Classification. Staining and/or mark-off evaluated using either the AATCC Chromatic Transference Scale, or the Gray Scale for Evaluating Staining. Change of shade is evaluated by means of the Gray Scale for Evaluating Change in Color.

Carbonizing (AATCC 11-1975) (ANSI L14.3-1973/R 1960). Carbonizing is the process to which wool is subjected in order to remove impurities (such as burrs, bark, grass and cotton fibers) by soaking in a strongly acid solution. The stock or fabric is then dried and baked, after which the cellulosic material may be beaten or washed out. This test method reproduces any color change encountered in the carbonizing operation, and is applicable to dyed wool textiles.

Procedure. Four specimens of worsted test cloth (containing 12-effect floats of acetate, Acrilan 36, Arnel, cotton, Creslan 61, Dacron 54, Dacron 64, nylon 66, Orlon 75, silk, Verel A and viscose) are dyed. One is kept for comparison, and each of the other three is stirred and squeezed for 10 min at 24°C in 100 mL of sulfuric acid solution (50 g/L, or 6° Twad). The specimens are wrung to allow 60–80% pickup, dried at 80°C for 15 min, and then baked for 15 min at 110°C.

At this point, one sample is saved to show any change in shade while still strongly acid. A second sample is rinsed in running water for 30 min, and the third sample is rinsed twice in distilled water, then in a dilute soda ash solution, and finally in running water (rinsing is continued until the sample is neutral or slightly alkaline to litmus paper).

Evaluation and Classification. The test specimens are examined by comparison with dyed standards and classified as follows: *Class 1.* Poor fastness—a color which changes seriously in the acid bath and does not return to shade on rinsing; *Class 3.* Good fastness—a color which changes materially in the acid bath but returns to shade on rinsing, and *Class 5.* Excellent fastness—a color which is unaffected by the entire process of carbonization.

Evaluation may also be based on the Gray Scale for Evaluating Change in Color.

Standards. Standards for Classes 1, 3, and 5 are dyed with the following dyes:

Class 1. 2% Alizarine Red S (topchrome) [*130-22-3*] (CI 58005); *Class 3.* 2% Acid Cyanine (dyed acid) [*6448-97-1*] (CI 50230); and *Class 5.* 2% Alizarine Irisol R (dyed acid) [*4430-18-6*] (CI 60730).

Crocking (AATCC 8-1974/116-1974) (ANSI L14.72-1973/ANSI L14.212-1973). This test determines the fastness of a dyestuff to either wet or dry rubbing.

Procedure. The test specimen is placed on the base of the Crockmeter (Atlas Electric Devices Co., Chicago, Ill.) and a square of white test cloth is rubbed on the colored specimen by means of the Crockmeter finger (dry test cloth for dry crocking; test cloth wet out in distilled water for wet test; wet pickup, 65 ± 5%). The sample is rubbed twenty times; after this the staining of the white cloth is determined.

Classification and Evaluation. Staining is graded by use of the AATCC Chromatic Transference Scale or the Gray Scale for Evaluating Staining.

Degumming (AATCC 7-1975) (ANSI L14.4-1973). This test is applicable to dyed silk yarn which may be subjected to a degumming operation (boil-off in soap solution) during manufacture.

Procedure. The dyed yarn is braided with undyed gum-silk yarn and boiled for 2 h in a 0.5% soap solution at a constant 100:1 liquor ratio, then rinsed cold, and air-dried.

Evaluation and Classification. Alterations in color or staining of undyed silk are observed and rated by comparison with dyed standards or by use of the Gray Scale for Evaluating Staining or the AATCC Chromatic Transference Scale.

Standards. The standards are dyed as follows: *Class 3.* 5% Indigo MLB/6B Powder [*6417-56-7*] (CI 73075). *Class 5.* 5% Indigo MLB/4B Powder [*2475-31-2*] (CI 73065).

Drycleaning (AATCC 132-1976) (ANSI L14.241-1970). This test indicates the effect of repeated commercial drycleanings, using perchloroethylene as solvent, on dyed and printed fabrics. It is applicable for evaluating the colorfastness of all fibers, fabrics, and yarns which are intended for apparel or household use, and which are likely to be commercially drycleaned. The test is not intended to evaluate the durability of any finish, or applied flock, or metallic designs. Nor does it evaluate the resistance of colors to spot or stain-removal procedures (see Drycleaning).

Procedures. Samples (10 cm × 4 cm) are placed in a 10 cm × 10 cm bag of undyed cotton twill fabric together with 12 stainless steel disks 30 ± 2 mm diameter and 3 ± 0.5 mm thickness. Samples are run with 200 mL perchloroethylene for 30 min at 30 ± 2°C, after which they are blotted and air-dried.

Evaluation and Classification. Samples are evaluated for color change by comparison with the Gray Scale for Evaluating Change in Color.

Fulling (AATCC 2-1975). Fulling is a finishing process for woolen cloth involving the effect of heat, moisture, soap, and mechanical pounding. The purpose of this process is to shrink the cloth by felting action and thus make it firmer and thicker. The fulling test is devised for evaluating the colorfastness of dyed wool fabrics or yarns to mill-fulling.

Procedure. Two-gram dyed wool and two-gram white test cloth are formed into a bag, dyed sample inside. Then 6.3-mm steel balls are placed inside the bag. Tests are carried out under controlled conditions as shown in Table 4, in the Launder-Ometer using 20-cm steel tubes and 8 mL of solution. Yarns may be braided with white wool or other yarn.

Table 4. Test Conditions in the Fulling Test

Test	Temperature, °C	Soap conc, wt %	Na$_2$CO$_3$ conc, wt %	Time, h
I	32	3.75	1.5	0.5
II	38	3.75	1.5	1.5
III	43	3.75	1.5	4.5

Evaluation and Classification. The specimens are evaluated by running comparative tests against standard dyeings which represent minimum fastness for each class with respect to alteration of color staining. As an alternative, the Gray Scale for Evaluating Staining or the AATCC Chromatic Transference Scale may be used.

Standards. *Class 1.* 1% Alizarine Sky Blue B [6424-75-5] (CI 62105). *Class 3.* 2% Brilliant Milling Blue B [5863-46-7] (CI 42645), and *Class 5.* 2% Eriochrome Azurole B [1796-92-5] (CI 43830).

Dry Heat (Excluding Pressing) (AATCC 117-1976) (Based On ISO R 105/IV Part 2). This method assesses the resistance of the color of textiles of all kinds and forms to the action of dry heat, but excludes hot pressing.

Procedure. The material under test is sandwiched between a piece of undyed material of the same type as the test sample and a piece of multifiber test fabric No. 10A. The composite test sample is exposed for 30 s to one of the temperature conditions listed in Table 5 in a heating device such as the Scorch Tester (Atlas Electric Devices Co.).

Evaluation and Classification. The change in color of the test specimen is rated with the help of the Gray Scale for Evaluating Color Change. The staining of the undyed fabric(s) is evaluated using the Gray Scale for Staining or the AATCC Chromatic Transference Scale.

Hot Pressing (AATCC 133-1976). Colored samples are tested for dry, damp and wet pressing under controlled conditions.

Procedure. A 12 cm × 4 cm piece of material is exposed in a Scorch Tester. The bottom plate of the heating device is covered with asbestos sheeting, wool flannel, and dry, undyed cotton fabric. In case of dry pressing, the dry sample is placed on top of the cotton fabric. The top plate of the device is lowered for 15 s at either 110 ± 2°C, 150 ± 2°C, or 200 ± 2°C. In damp pressing, the sample is covered with a piece of un-

Table 5. Temperature Conditions for Composite Test Sample

Test	Temperature, °C
AATCC	
I	149 ± 2
II	163 ± 2
III	177 ± 2
IV	191 ± 2
V	205 ± 2
VI	219 ± 2
ISO	
I	150 ± 2
II	180 ± 2
III	210 ± 2

dyed cotton fabric that has been soaked in distilled water and squeezed to contain its own weight of water. In wet pressing, both the test sample and a piece of undyed cotton are soaked in distilled water and squeezed to contain their own weight in water. The wet test sample is placed on top of the dry cotton cloth and covered with the wet cotton cloth.

Evaluation and Classification. For determining change in color, samples are evaluated immediately and after the sample has been allowed to condition for 4 h at 20°C, using the Gray Scale for Evaluating Color Change, and staining is graded by comparison with the AATCC Chromatic Transference Scale or the Gray Scale for Evaluating Staining.

Lightfastness

The 1976 edition of the AATCC Technical Manual lists six tests for lightfastness: sunlight, daylight; carbon arc, alternate light and darkness; water-cooled xenon arc, continuous; and alternate light and darkness for both carbon arc and xenon arc.

Lightfastness (General Method) (AATCC 16-1974). Samples and standards are properly mounted, partially covered, and exposed for the appropriate number of hours under specified conditions.

Classification and Evaluation. Samples can be evaluated by three methods:

1. Based on AATCC blue-wool standards L2 through L9, as shown in Table 6. Lightfastness is evaluated in terms of color-change comparison with standard.

2. Based on standard sample. When an agreed-upon standard and sample are exposed until the standard shows a change equal to step 4 on the Gray Scale for Evaluating Color Change, the sample is considered satisfactory if no greater break appears, and unsatisfactory if a greater break is apparent.

3. AATCC Standard Fading Units. Samples may be classified according to the number of Standard Fading Units necessary to produce a change equal to step 4 on the Gray Scale for Evaluating Change in Color.

Table 6. Classification by Lightfastness

Less than	Equal	More than	Class
		L2	1
	L2	L3	2
L2		L3	2–3
	L3	L4	3
L3		L4	3–4
	L4	L5	4
L4		L5	4–5
	L5	L6	5
L5		L6	5–6
	L6	L7	6
L6		L7	6–7
	L7	L8	7
L7		L8	7–8
	L8	L9	8
L8		L9	8–9
	L9		9

Either the single step or the two step method can be used. In the single step method, the sample is faded to a step 4 break of the Gray Scale. In the two-step-method, a preliminary judgment is made when a step 4 break is obtained, then the sample is exposed further until a step 3 break is obtained. If the two results differ, both are reported.

Lightfastness-Carbon Arc (AATCC-16A-1974). Partially covered samples and standards are exposed and examined after periods as are necessary to produce a "just appreciable fading." Classification may be by any method given under General Method in this procedure.

Sunlight-Fastness (AATCC 16B-1974). Partially covered samples and standards are exposed in a glass-covered exposure cabinet and are allowed to remain in the cabinet only on sunny days during the hours between 9 am and 3 pm (Standard time); they are removed for inspection at frequent intervals. As each standard, beginning with L2, shows a change equal to step 4 on the Gray Scale for Evaluating Change in Color, all samples that show an equal change are removed from the cabinet.

Samples are classified as in the carbon-arc method, using AATCC standards or a standard sample. When desired, a standard agreed upon by the buyer and seller may be used in place of the AATCC standards. In this case, exposure is continued until the sample agreed upon shows a change equal to step 4 on the Gray Scale for Evaluating Change in Color.

Exposure time may be based on a mutually agreed upon number of langleys (4.184×10^4 J/m^2) of radiation. When a stated number of langleys are used, classification is based on the Gray Scale for Evaluating Change in Color. A sample that shows no greater change than step 4 is considered satisfactory. Classification may also be based on the required number of langleys (4.184×10^4 J/cm^2) necessary to cause a change equal to step 4 on the International Geometric Gray Scale. The report should include the method used as well as the date and location of testing.

Daylight Fastness (AATCC 16C-1974). Tests are conducted as for sunlight-fastness except that exposed samples are allowed to remain in the exposure cabinet 24 h a day, and are removed only for inspection. Classifications are as for sunlight-testing.

Lightfastness-Carbon Arc-Alternate Light-Darkness (16D-1974). This method requires a carbon-arc machine that is equipped with an atomizer, and which automatically controls the turning on and off of the lamp. A cam producing 1 h of darkness and 3.8 h of light is used. The apparatus is adjusted so that during the light-on period the black-panel temperature is $63 \pm 3°C$, and the rh is $35 \pm 5\%$. During the 1-h period the relatively humidity is $90 \pm 5\%$. Classification and evaluation are based on AATCC blue-wool standards or on a standard sample.

Lightfastness-Xenon Lamp-Continuous Light (AATCC 16E-1976). Samples are exposed in the same manner as in the carbon-arc machine with the black-panel temperature at $63 \pm 3°C$ and a rh of $30 \pm 5\%$. Evaluation and classification are based on AATCC blue-wool standards or on a standard sample.

Lightfastness-Xenon Lamp-Alternate Light-Darkness (AATCC 16F-1974). This method uses a cam as in Method 16D-1974 and follows the same procedure with the xenon tube replacing the carbon arc as a light source.

Carbon-Arc Lamp (Fade-Ometer). The Fade-Ometer is operated on ac at 15–17 A and 125–145 V across the arc. The arc is enclosed in a heat-resistant glass globe. The samples are mounted 25 cm from the arc and are on a frame that revolves about the arc 2 ± 1 times per minute. A blower unit in the base provides a flow of air through the machine allowing for temperature control, which is measured by use of a black-panel thermometer. The humidity in the chamber is automatically controlled.

Exposure Cabinet. The exposure cabinet for sunlight or daylight tests is constructed so as to face due south in the Northern Hemisphere, and due north in the Southern Hemisphere, with a slant approximately equal to the degree of latitude at which the cabinet is located. It allows free access of air to the samples and is covered with window glass at least 8 cm above the exposure area.

Black-Panel Thermometer. The unit consists of a bimetallic dial-type thermometer mounted on a suitable frame with the face (plate) of the frame and the stem of the thermometer sprayed with a heat-resistant black enamel.

Lightfastness Standards. Eight lightfastness standards L2 through L9 are available from AATCC. These standards are prepared by blending varying proportions of wool dyed with a very fugitive color, Eriochrome Azurol B [1796-92-5] (CI 43830), and wool dyed with a fast color, Indigosol Blue AGG [4086-05-9] (CI 73801), in such a manner that each numbered standard is approximately twice as fast as the preceding numbered standard.

Xenon-Arc Lamp. The apparatus uses a quartz-xenon burner tube which is water-cooled, with an inner Pyrex (Corning Glass Works) filter glass and an outer filter of clear glass. Samples are mounted at the required distance from the arc and are supported on a frame which revolves about the arc 2 ± 1 times per minute. Testing temperatures are measured by the black-panel thermometer, and moisture is added by controlled methods.

Alternate Wash-and-Light (AATCC 83-1974). This test is designed to provide a consumer end-use test which will simulate an average of five exposures to sunlight (duration, 6 h) followed by a washing without chlorine.

Procedure. A 6 cm × 15 cm specimen is exposed for twenty Standard Fading Units and is followed by a Wash IIIA.

Evaluation. The effect on the color is graded and expressed by comparison with the Gray Scale for Evaluating Change in Color.

Phototropism (AATCC 139-1975). This test is designed to show colors that are changed impermanently when exposed to strong light. This change takes place after a short exposure, but the color returns to its original shade when placed in the dark.

Procedure. Dyed samples 6 cm × 3 cm are exposed in the xenon arc fading apparatus for $1/25$ of the time necessary to obtain a step 4 Gray Scale Fade on an L-2 blue-wool standard. The sample is immediately inspected for a contrast between the original and the exposed portions. If the contrast is equal to or less than a step 4 of the Gray Scale for Evaluating Change in Color, the sample is considered not phototropic. If the contrast is greater than step 4, the specimen is left in the dark for one hour at $20 \pm 2°C$ and a rh of $65 \pm 2\%$. If the contrast has not disappeared, the sample is exposed for 15 s to steam at 101 kPa (1 atm) and reinspected. If the contrast disappears, the sample is considered to be phototropic. If the contrast remains, the sample is not considered phototropic and the contrast is due to low lightfastness. Fastness to light should be assessed in a parallel test.

Ozone in the Atmosphere Under Low Humidity (AATCC 109-1975) (ANSI L14174-1973). It has been found that ozone (qv) can destroy dyes applied to textiles. This test has been devised to expose and determine the fastness of dyed textiles to ozone at rh not exceeding 65%.

342   DYES, APPLICATION AND EVALUATION

Procedure. A specimen 10 cm × 6 cm and a swatch of Control No. 109 (available from Testfabrics, Inc., Middlesex, N.J.), are suspended in an exposure chamber at 18–28°C with rh <65%. Ozone should be present in concentrations that produce one cycle of fading in 1.5–6 h of test. Samples are exposed for the number of cycles necessary to cause a change of shade. A cycle is considered complete when the control matches the Standard for Fading (Testfabrics Inc.). Usually two cycles will cause a change in an ozone-sensitive dye.

Evaluation and Classification. Samples are graded by comparison with the Gray Scale for Evaluating Change in Color and the report will include the number of cycles to which the sample has been subjected.

Ozone in the Atmosphere Under High Humidity (AATCC 129-1975). This test is analogous to the previous test except that the test specimen is exposed in a chamber at 40 ± 5°C and 85 ± 5% rh, together with control sample No. 129.

Burned Gas Fumes (AATCC 23-1975) (ANSI L14.56-1973). This method assesses the resistance of colored textiles when exposed to atmospheric oxides of nitrogen as derived from the combustion of heating gas.

Procedure. Test specimens (5 cm × 10 cm) are exposed together with a strip of Control Sample No. 1 (Testfabrics Inc.), in a gas-fading apparatus (United States Testing Co., Hoboken, N.J.), until the control sample shows a change in color corresponding to that of the standard of fading. The gas-fading apparatus contains an open flame from combustion of illuminating gas or butane.

Evaluation. After each exposure cycle the specimens are compared with unexposed samples. The effect, if any, is classified with the aid of the Gray Scale for Evaluating Change in Color. The class number and the number of exposure cycles are reported.

Washfastness

Washing, Domestic; and Laundering, Commercial: Accelerated (AATCC 61-1975) (ANSI L14.81-1973). These accelerated laundering tests are designed for evaluating the washfastness of textiles. One 45-min test approximates the color loss and/or abrasive action of five average hand, commercial, or home launderings.

The conditions of the four different tests are given in Table 7.

Procedure. Samples (5 cm × 10 cm (Test IA) or 5 cm × 15 cm (all other tests)) to which a 5 cm × 5 cm piece of multifiber fabric has been sewn or stapled along one side are filled one each into 7.5 cm × 12.5 cm or 9 cm × 20 cm stainless-steel cylinders. The wash liquor and the steel balls as per Table 7 are added. The detergent used is AATCC Standard Detergent WOB. The containers are closed and subjected for 45

Table 7. Test Conditions

Test No.	Temperature, °C	Total liquor volume, mL	Detergent of total volume, %	Available chlorine of total volume, %	Number of steel balls	Time, min
IA	40	200	0.5		10	45
IIA	49	150	0.2		50	45
IIIA	71	50	0.2		100	45
IVA	71	50	0.2	0.015	100	45

min at the required temperature in a Laünder-Ometer (Atlas Electric Devices Inc.). At the end of the cycle, the containers are removed and the samples rinsed twice in 100 mL of water for 1 min at 40°C. They are scoured in 100 mL of 0.014% soln of acetic acid for 1 min at 27°C and rinsed again for 1 min at 27°C in water. The samples are then hydroextracted and dried by pressing with an iron at 135–150°C.

Evaluation. Test IA conforms to five careful hand launderings at 40°C. Test IIA conforms to five commercial launderings at 38°C or five home-machine launderings at medium or warm temperature setting. Test IIIA conforms to five commercial launderings at 49°C or five home launderings at 60°C. Test IVA conforms to five commercial launderings at 71°C or five home launderings at 60–66°C with chlorine present. Samples are evaluated for change in color with the corresponding Gray Scale and for staining with the aid of the AATCC Chromatic Transference Scale or the Gray Scale for Evaluating Staining.

Miscellaneous Fastness Properties

Perspiration (AATCC 15-1976). Perspiration may cause an undesirable change in shade or staining or both. This test is applicable to dyed, printed, or otherwise colored fabrics.

Procedure. A 5.7 cm × 5.7 cm swatch backed with No. 10 multifiber test cloth is immersed in the test solution for 15–30 min and squeezed to 2–2½ times its original weight. The samples are placed between glass or plastic plates and stacked in the AATCC Perspiration Tester (Atlas Electric Devices Inc.) or in the Perspirometer (Orange Machine and Mfg. Co., Point Pleasant, N.J.). A 3.63-kg weight is added to the Perspiration Tester or a 69-kPa (10-psi) pressure is adjusted on the Perspirometer. The loaded specimen unit is heated in an oven at 38 ± 1°C for ≥6 h.

The acid solution (pH 3.5) is made up of 10.00 g sodium chloride; 1.00 g of 85% lactic acid (USP); 1.00 g disodium hydrogen phosphate anhydrous, Na_2HPO_4; and 0.25 g histidine monohydrochloride; the solution is brought to one liter with distilled water.

Evaluation and Classification. Samples are evaluated for color change by comparison with the Gray Scale for Evaluating Change in Color. For evaluation of staining, either the AATCC Chromatic Transference Scale or the Gray Scale for Staining may be used.

Steam Pleating (AATCC 131-1974). This test is applicable to textiles intended to be used in pleated clothing. It is designed to measure the change in shade of dyed textiles that are subjected to a steam treatment, as in pleating.

Procedure. A 4 cm × 5 cm specimen is placed between two 4 cm × 5 cm pieces of undyed scoured fabric of the same type of fiber used in the test specimen and sewn along one side. The specimen is then mounted on a special Specimen Holder (Atlas Electric Devices Inc.) and steamed under one of the following three conditions:

Test	Time, min	Temperature, °C	Pressure, kPa
I mild	5	108	34
II intermediate	10	115	69
III severe	20	130	173

After exposure the samples are air dried below 60°C and conditioned for 4 h at 20 ± 2°C and 65 ± 2% rh.

Evaluation. The change in color of the test specimen is evaluated with the Gray Scale for Evaluating Color Change. The staining of the undyed fabric pieces is evaluated with the Gray Scale for Evaluating Staining. The type of test and the ratings figures are reported.

Stoving (AATCC 9-1975) (ANSI L14.9-1973). Stoving is a method for bleaching wool by subjecting it to sulfur dioxide fumes. This test is applicable to dyed yarns of all kinds which might be subjected to the sulfur dioxide stoving procedure.

Procedure. The dyed yarn is braided with undyed wool and undyed silk, soaked in a 0.5% soap solution, and hung in a bell jar with an excess amount of burning sulfur (flowers). The sample is allowed to remain in this atmosphere of sulfur dioxide fumes for 16 h. After thorough rinsing in cold water and extraction, the sample is dried and evaluated.

Evaluation and Classification. Samples are evaluated in comparison with dyed standards, or by comparison with the Gray Scale for Evaluating Change of Color. Staining of the undyed materials is evaluated with the AATCC Chromatic Transference Scale or the Gray Scale for Evaluating Staining.

Standards. Dyed on worsted wool yarn: *Class 1.* 1% Alizarine Yellow R [2243-76-7] (CI 14030); *Class 3.* 1% Naphthochrome Violet R [7452-51-9] (CI 43565); and *Class 5.* 1% Acid Alizarine Violet N [2092-55-9] (CI 15670).

Water Tests (AATCC 105-1975, ANSI L14.149-1973; AATCC 106-1975, ANSI L14.150-1973; AATCC 107-1975, ANSI L14.151-1973). Tests in fresh, sea, and chlorinated pool water are designed to measure the resistance in each of dyed, printed, or otherwise colored textile yarns and fabrics of all kinds.

Procedure. Distilled or demineralized water and artificial seawater are used because of the variables encountered in tap water or natural seawater. Seawater solution is made up of 30 g sodium chloride and 5 g magnesium chloride (anhydrous) brought to one liter with tap water. Chlorinated pool water solution is made up of 5 parts of a 1% soln of available chlorine per 10,000 parts of distilled water (5 ppm), adjusted to pH 8 with either sodium bicarbonate or acetic acid. Specimens for each type of water test are backed with multifiber test cloth.

For freshwater and seawater tests, the specimens are wet out in the suitable solution for 15 min, placed in the AATCC Perspiration Tester or Perspirometer at 69 kPa (10 psi) and placed in an oven at 38°C for 18 h, after which the specimens are removed and dried at room temperature without pressing.

Specimens for chlorinated-pool-water tests are kept in 200 mL of solution at 27°C in a closed jar for 4 h, during which time they are occasionally gently agitated. They are then placed in the Perspirometer and kept at 38° for 2 h.

Evaluation and Classification. All three water tests are graded for color change by comparison with the Gray Scale for Evaluating Change in Color; either the AATCC Chromatic Transference Scale or the Gray Scale for Evaluating Staining is used for rating the staining of the multifiber test cloth.

Water Spotting (AATCC 104-1975) (ANSI L14.148-1963/R-1969). This simple test indicates the fastness of textiles to water spotting. It does not show whether or not the discoloration is removable.

Procedure and Evaluation. A colored swatch is spotted with water. The color change is evaluated after 2 min, and again after drying, using the Gray Scale for Evaluating Change in Color.

APPLICATION PROPERTIES

Recently the AATCC has initiated a research program to develop test methods for the evaluation of application properties of dyes in dyeing processes. So far, three such test methods have been published and others are to follow shortly.

Compatibility of Basic Dyes for Acrylic Fibers (AATCC 141-1976). This test is intended to determine the behavior of a basic dye concerning compatibility when applied to acrylic fibers in the presence of other basic dyes. Compatible dyes according to this definition are those that exhaust at the same rate as the other dyes in the combination.

Procedure. The compatibility is determined in a so-called dip test where the dye under test is exhausted on several pieces of acrylic material in separate combinations with five basic dyes of established compatibility value. There are separate yellow and blue scales (AATCC) of five standard dyes each.

Evaluation. The compatibility value of the dye under test is that of the standard dye with which it produces on-tone dyeings throughout the sequence of dyeings. The compatibility value C ranges from 1 to 5 with half steps.

Dispersibility of Disperse Dyes: Filter Test (AATCC 146-1976). This test determines the dispersibility of disperse dyes as evaluated by filtering time and filter residue under standard conditions in aqueous media.

Procedure. Dye powder (2 g of standard and sample) is pasted with 180 mL water at 43–49°C. The pH is adjusted as shown in Table 8 and the volume brought to 200 mL. The dispersion is heated to 71°C and filtered in a prewarmed Buchner funnel through filter paper (as indicated in Table 8), under 56 ± 10 cm of vacuum. The time it takes for the dispersion to pass through the filter is recorded. The filter paper is dried for evaluation.

Evaluation. The residue on the filter paper is evaluated visually against the Filter Residue Scale from 5 to 1 (AATCC).

The results are rated according to filtering time as follows:

Class	Filtering time, s
A	0–24
B	25–49
C	50–74
D	75–120
E	>120

The type of test, the filtering-time class and the residue class are reported.

Table 8. Selection of Applicable Test

Test	Whatman filter combination	Application of dyes	pH of dispersion
I	#2 over #4	package dyeing of polyester	4.5– 5.0
II	#4 over #4	beck dyeing of polyester	4.5– 5.0
III	#4 over #4	dyeing of nylon carpet and apparel	7.0–10.0

Dyestuff Migration (AATCC 140-1976). The purpose of the test is to evaluate particulate migration of dyes that have been padded on a textile fabric during drying as well as the influence of auxiliaries on the migration. Such migration can be a substantial cause of unlevelness.

Procedure. The dye under test alone or together with auxiliaries is padded in a concentration desired on a 5 cm × 30 cm swatch of fabric. The swatch is immediately placed on a horizontal glass plate. A 9-cm dia watch glass is placed on a portion of the wet fabric. The fabric is dried at room temperature and the watch glass is removed.

Evaluation. The degree of particulate migration is estimated by comparison of the dye concentration in the area of the fabric which was covered by the watch glass to the dye concentration in the rest of the fabric. Three methods of evaluation are possible: (1) visual, by reference to the Gray Scale for Evaluating Change in Color; (2) by reflectance measurement and calculation of color difference; and (3) by quantitative extraction of the dye from disks of fabric of equal size from the covered and uncovered areas. Percent particulate migration (M_p) is calculated as follows:

$$M_p = 100 \frac{(1 - A_A)}{A_B}$$

where A_A is the absorbance at the analytical wavelength of the extract from the covered area. A_B is the absorbance at the analytical wavelength of the extract from the uncovered area.

QUALITATIVE AND QUANTITATIVE ANALYSES

A variety of analytical methods are discussed elsewhere in the Encyclopedia (see Analytical methods).

Paper and Thin Layer Chromatography

Both of these techniques are separation methods useful for dye identification. The dyes are extracted from fibers with suitable solvents (16) or dissolved if in powder form and applied to chromatographic paper strips or plates with the help of capillary pipettes. Paper strips or plates are then developed in a tank with a suitable eluent. These vary substantially according to the chemical nature of the dyes to be separated. Not only is it possible in this way to separate dye mixes into their component, but dye identification is feasible based on comparison with known dyes. In such cases, it is generally necessary to compare the dyes in at least two different eluent systems.

For semiquantitative or quantitative analysis, the separated dyes can be measured with a reflectance densitometer or they can be extracted from the plate or paper and measured by transmittance spectrophotometry.

High Pressure Liquid Chromatography

This modern version of the classical column chromatography technique is also used successfully for separation and quantitative analysis of dyes. It is generally faster than thin layer or paper chromatography; however, it requires considerably more expensive equipment. Uv and visible photometers or spectrophotometers are used as detectors and for quantitative purposes.

Spectroscopic Methods

Among these the most important in connection with dyes are vis, ir, and nmr spectroscopy.

Infrared Spectroscopy. This technique is not only very useful in identification of dyes by the so called fingerprint method, ie, comparison with dyes of known structure, but also to gain insight into the structural details of unknown dyes. Collections of ir spectra of commercial colorants are available (17), ir spectra of dyes are generally measured by the potassium bromide pellet method. The spectra are compared with spectra of known dyes or analyzed on the basis of group frequency correlations.

Nuclear Magnetic Resonance Spectroscopy. This technique is particularly useful in the structural elucidation of dyes as well as configurational analysis. Specialized versions of nmr such as ^{13}C nmr are also suitable for these purposes. Some of the dyes pose difficulties because they are insufficiently soluble in nmr solvents.

Visible Spectroscopy. Visible spectroscopy is the most commonly used technique for quantitative analysis of dye concentration. Visible spectroscopy can also be used for dye identification by the fingerprint method. Quantitative analysis is accomplished by transmittance spectrophotometry of dye solutions or by reflectance spectrophotometry of dyed textiles. Such analyses are currently performed by most dye manufacturers and many textile mills.

Solution Spectrophotometry. The relative strength of a dye (compared to a standard) or the dye concentration in solution can be determined accurately for most dyes by this method. Of greatest utility is a computer-interfaced recording visible spectrophotometer or a so called color-measuring instrument. Many suitable instruments are commercially available. Solution spectrophotometry is based on Beer's Law and the Lambert-Bouger Law concerning the relationships between absorption and concentration of the absorbing layer. It is important to prepare dye solutions that are stable and obey Beer's Law over the concentration intervals used (18–19). Measurements are generally taken at the wavelength of maximum absorption. The measurements are taken in such a way that the spectral internal transmittance $\tau_i(\lambda)$ is obtained. The spectral internal absorbance $A_i(\lambda)$ is then calculated as follows:

$$A_i(\lambda) = \log_{10} \frac{1}{\tau_i(\lambda)}$$

The spectral internal absorbance is linearly related to dye concentration. The specific spectral absorption coefficient $A_{c'}(\lambda)$, which is a constant for dyes obeying Beer's Law, is calculated:

$$a_{c'}(\lambda) = \frac{A_i(\lambda)}{lc'}$$

where l is the internal path length traversed by the light and c' is the dye concentration.

The relative strength of a dye sample can be calculated as a ratio of the specific absorption coefficients of standard and sample.

The calculation method is valid if standard and sample are chemically and physically identical and differ in relative strength only (see Color). The relative strength of dyes by the solution method is routinely determined with a standard deviation of approximately ±1.5% (20).

Relative strength determination by solution measurement is difficult in case of vat and sulfur dyes and leads to unreliable results in cases where some of the dye in the commercial product has no affinity for the fiber such as for certain direct and fiber-reactive dyes.

Reflectance Spectrophotometry. Because of discrepancies that can occur between strength and shade evaluations in solution and on textile substrates, the latter is often the preferred evaluation technique. In the case of dye manufacture, many dyes today are standardized in solution but there is always a final control step where dyeings are prepared. Historically, such dyeings have been evaluated visually for the relative strength and the shade of the dye under test on the substrate, compared to the standard. More and more attempts are being made to do such evaluations objectively. Instrumental evaluations of dye concentration or relative strength by reflectance measurement are generally based on the Kubelka-Munk relationship (21) between spectral reflectance and the Kubelka-Munk spectral absorption constant K and the spectral scattering constant S. This relationship is:

$$K/S = \frac{(1 - R_\infty)^2}{2 R_\infty}$$

where R_∞ is the spectral reflectance factor at complete opacity.

The scattering constant S of the textile material is assumed not to be influenced by the absorbed dye. In practice, the relationship between dye concentration and K/S value is generally not completely linear and comparisons should only be made of dyeings that differ by not more than about 30% in dye concentration. Many practical problems also have to do with the accuracy of the dyeing and the presentation of the dyed textile to the measuring instrument (22). The newest reflectance spectrophotometers on the market have measuring times for the complete visible spectrum of 2–8 s and it is practical to make repeat measurements for more reliable averages.

Standard deviations of relative dye strength determinations based on reflectance measurement of textiles tend to be 1.5–2 times larger than those from solution measurement.

Color Difference Evaluation and Color Sorting. Shade evaluation is comparable in importance to relative strength evaluation for dyes. This is of interest to both dye manufacturer and dye user for purposes of quality control. Objective evaluation of color differences is desirable because of the well-known variability of observers. A considerable number of color difference formulas that intend to transform the visually nonuniform CIE (International Commission on Illumination) tristimulus color space into a visually uniform space have been proposed over the years. Although many of them have proven to be of considerable practical value (Hunter Lab formula, Friele-MacAdam-Chickering (FMC) formula, Adams-Nickerson formula, etc), none has been found to be satisfactorily accurate for small color difference evaluation. Correlation coefficients for the correlation between average visually determined color difference values and those based on measurement and calculation with a formula are typically of a magnitude of approximately 0.7 or below.

Recently, in the interest of uniformity of international usage, the CIE has proposed two color difference formulas (CIELAB and CIELUV), one of which (CIELAB) is particularly suitable for application on textiles (23). (For an evaluation of the drawbacks of this formula see ref 24). This formula allows splitting of the total color difference (ΔE^*) into lightness difference (ΔL^*), hue difference (ΔH^*) and chroma difference (ΔC^*). The associated uniform color space, the $L^*a^*b^*$ space, and particularly the a*b* diagram are useful for illustrating relationships among colors (see Color).

A related application of color difference evaluation in textiles is color sorting. Most fabrics are dyed in pieces of 60–120 m length. Many such pieces may be dyed in a particular color in the same lot or in several lots over a period of time. The color of these pieces may vary slightly within a lot or from lot to lot for a number of reasons. For the garment manufacturer it is important to know which pieces can be grouped together for cutting so that the resulting garments look uniform in color. To accomplish this, individual pieces are grouped by a three-number code in a cinder block model. Assignment of colors to a particular cinder block is accomplished with the aid of color difference formulas (25).

Colorant Formulation. Historically, one of the most important tasks for the dyer is the formulation of dyes to match a given color standard (see Color). In recent years, art and experience are being aided more and more by computer formulation. Several commercial colorant formulation systems are on the market that consist of a color measuring instrument and a minicomputer with associated input and output units. The input information consists of reflectance data and concentration data of the dyes to be used as well as of the reflectance data of the standard to be matched. The reflectance data are converted to Kubelka-Munk data. Various algorithms have been proposed for the solution of the formulation problem (26), generally involving matrix algebraic operation. For various reasons the original calculated formula, when dyed, will not produce an exact match of the standard. Correction formulas can be used to improve the solution. Computer colorant formulation is particularly useful to determine the least expensive formulation from a group of dyes or to determine formulations exhibiting minimum metamerism (ie, a nearly spectrophotometric match). Computer colorant formulation systems require considerable technical experience for optimal usage, but offer potential for substantial savings (see Colorants).

BIBLIOGRAPHY

"Dyes, Application and Evaluation" in *ECT* 1st ed., Vol. 5, pp. 355–445, by P. J. Choquette, C. Z. Draves, Ludwig Fusser, O. F. Habel, J. S. Kirk, and F. A. Soderberg (Application), General Dyestuff Corporation; H. W. Steigler (Test Methods for Colorfastness of Textiles, under Evaluation), AATCC, Lowell Textile Institute, and F. C. Dexter (Spectrophotometric Curves, under Evaluation), Calco Chemical Division, American Cyanamid Division; "Dyes, Application and Evaluation" in *ECT* 2nd ed., Vol. 7, pp. 505–613, by G. M. Gantz and J. R. Ellis (eds.), J. J. Duncan, J. G. Freas, R. C. Hall, M. J. Landberge, J. Moss, F. O'Neal, W. N. Pardey, and T. Vogel, General Aniline & Film Corporation.

1. R. H. Peters, "The Physical Chemistry of Dyeing," in *Textile Chemistry*, Vol. III, Elsevier Scientific Publishing Co., Amsterdam, 1975.
2. I. D. Rattee and M. M. Breuer, *The Physical Chemistry of Dye Absorption,* Academic Press, London and New York, 1974.
3. C. L. Bird and W. S. Boston, eds., *The Theory of Colouration of Textiles,* The Dyers Company Publications Trust, London, 1975, distributed by the Society of Dyers and Colourists, Bradford, England.
4. "Practical Dyeing Problems—Analysis and Solution," *Proceedings of International Dyeing Symposium,* American Association of Textile Chemists and Colorists, 1977.
5. R. McGregor, *Diffusion and Sorption in Fibres and Films,* Vol. 1, Academic Press, London and New York, 1974.
6. Ref. 4, p. 154.
7. W. Beckmann, F. Hoffmann, and H. G. Otten, *J. Soc. Dyers Col.* **88,** 354 (1972).
8. W. Beckmann and O. Glenz, *Melliand Textilber* **49,** 1436 (1968).
9. W. Beckmann, *Text. Chem. Color.* **2,** 23 (1970).

10. R. J. Androsko, *Natural Dyes and Home Dyeing,* Dover Publications, Inc., New York, 1971.
11. "Textile Solvent Technology—Update '73," *Proceedings of Symposium, American Association of Textile Chemists and Colorists,* 1973.
12. Ref. 11, p. 163.
13. M. U. vonder Eltz and F. Schoen, *Textilveredlung* **9,** 3 (1974).
14. U.S. Pat. 4,056,354 (Nov. 1, 1977), A. G. Pittman, W. L. Wasley, and N. F. Getchell (to the United States of America as represented by the Secretary of Agriculture).
15. *1977 Technical Manual of the American Association of Textile Chemists and Colorists,* Vol. 53, 1977.
16. K. Venkataraman, ed., *The Analytical Chemistry of Synthetic Dyes,* Wiley-Interscience, New York, 1977.
17. *Sadtler Commercial Spectra,* Sadtler Research Laboratories, Inc., Philadelphia, Pa., 1973.
18. E. I. Stearns, *The Practice of Absorption Spectrophotometry,* Wiley-Interscience, New York, 1969.
19. R. G. Kuehni, *Text. Chem. Color.* **4,** 133 (1972).
20. C. D. Sweeney, *Text. Chem. Color.* **8,** 31 (1976).
21. D. B. Judd and G. Wyszecki, *Color in Business, Science and Industry,* 3rd ed., Wiley-Interscience, New York, 1975.
22. C. E. Garland, *Text. Chem. Color.* **6,** 104 (1974).
23. A. R. Robertson, *Color Res. App.* **2,** 7 (1977).
24. R. G. Kuehni, *J. Opt. Soc. Am.* **66,** 497 (1976).
25. F. T. Simon, *Die Farbe* **10,** 225 (1961).
26. R. G. Kuehni, *Computer Colorant Formulation,* Lexington Books, Lexington, Mass., 1975.

General References

W. F. Beeach, *Fiber Reactive Dyes,* Logos Press, London, England, 1970.
Dyeing of Paper, Bayer Farben Revue, Special Edition No. 4/1, 1974.
The Application of Vat Dyes, Monograph No. 2, American Association of Textile Chemists and Colorists, 1953.
R. W. Jones, *Sulfur Dyes, their Manufacture, Application, Fixation and Removal,* Martin Marietta Chemicals, Sodyeco Division, 1976.
L. Diserens, *The Chemical Technology of Dyeing and Printing,* Vol. I and II, Reinhold Publishing Corporation, New York, 1948.
R. Rys and H. Zollinger, *Fundamentals of the Chemistry and Application of Dyes,* John Wiley & Sons, Inc., New York, 1972.
E. R. Trotman, *Dyeing and Chemical Technology of Textile Fibers,* C. Griffin & Co., Ltd., London, 1964.
K. Venkataraman, *The Chemistry of Synthetic Dyes,* Vol. I–VI, Academic Press, New York, 1951–1978.
T. Vickerstaff, *The Physical Chemistry of Dyeing,* Oliver and Boyd, London, 1950.
F. Jacobs, *Textile Printing—Materials, Methods and Formulae,* Chartwell House, Inc., New York, 1952.
B. Knecht and G. C. Fothergill, *The Principles and Practice of Textile Printing,* C. Griffin & Co., Ltd., London, 1952.
K. Johnson, *Dyeing of Synthetic Fibers, Recent Developments,* Noyes Data Corp., Park Ridge, New Jersey, 1974.
H. Rath, *Lehrbuch der Textilchemie,* 3. Auflage, Springer Verlag, Berlin, 1972.

R. G. Kuehni
J. C. King
R. E. Phillips
W. Heise
F. Tweedle
P. Bass
C. B. Anderson
H. Teich
Mobay Chemical Corp.

DYES, NATURAL

In 1856 Perkin introduced the first synthetic dye, Mauve (see Azine dyes), to the English textile market and therewith not only founded the dyestuff industry but sounded the death knell for natural dyes. Since earliest times the use of color had been restricted to the natural products found in the environment—plants, some animals, and a few types of minerals. Within a few short years of Perkin's discovery a complete range of unnatural colors was available to the dyer, in most cases possessing fastness properties far superior to the natural product and, as the coal-tar industry grew, at a fraction of the cost. Today, with the exception of a few isolated primitive villages in the world, or for the handicraft worker who believes that nature knows best, these natural dyes are of little economic importance. However, although the use of natural colorants for textiles and fibers has declined, the use of nature-produced compounds for coloring foods and ingested products has recently increased. This has been prompted, in part, by government regulations, such as FDA's banning of such synthetic dyes as FD&C Reds 2 and 4 (see Colorants for foods, drugs, and cosmetics). With extensive testing of all synthetic products for carcinogenicity and mutagenicity, the food dye industry has returned to natural products as the source of its coloring bodies.

Natural dyes comprise those colorants (dyes and pigments) that are obtained from animal or vegetable matter without chemical processing. They are mainly mordant dyes, although some vat, solvent, pigment, direct, and acid types are known. There are no natural sulfur, disperse, azoic, or ingrain types. This article discusses the more important, formerly commercial dyes along structural lines: anthraquinones, naphthoquinones, indigoids, cartenoids, etc. Several general references (1–7) should be consulted if a greater interest in the historical and horticultural aspect of this subject is desired. An excellent reference for the chemical structures of natural dyes is ref. 8. For those interested in recent developments in the food and drug colorant industries, ref. 9 should be consulted. For all the voluminous literature now available to the synthetic dyestuff chemist, before 1862 all such literature would be concerned solely with natural dyes and their application. All knowledge to that time on unnatural dyes was contained in one small volume by F. Crace-Calvert (10).

Anthraquinones

Some of the most important natural dyes are based on the anthraquinone structure. They include both plant (madder, munjeet, emodin [518-82-1]) and animal (lac [6219-66-5], kermes [476-35-7], cochineal [1260-17-9]) types (see Dyes, anthraquinone).

Plant Anthraquinone Dyes. *Madder.* Madder, or now better known by its common chemical name, alizarin, is one of the most ancient of natural dyes. Reference is made to it in Egyptian, Arabic, Indian, Persian, Greek, Roman, and Germanic literature. Cloth dyed by madder has been found in predynastic Egyptian tombs. The approximately thirty-five species of madder plants belong to the order *Rubiaceae* which are found throughout Europe and Asia. The coloring material is concentrated in the roots of the plant, occurring to a maximum concentration of 4%. These perennial plants are allowed to remain in the ground for 18–28 mo at which time the roots are dug up, washed, dried, and finely ground. The primary use of madder is dyeing cotton Turkey

Red by mordanting with alum in the presence of lime. Madder also dyes wool with an alum or chrome cream of tartar mordant. It can be dyed from a single dye bath using sulfuric acid but the properties of the dyeings are not as fast as those produced on a mordant. Use of hard water or the addition of lime to the dyebath is essential. Madder also dyes silk when mordanted with basic aluminum sulfate; such dyeings have excellent fastness properties.

Although known for more than 6000 years, it was not until 1826 that the actual coloring matter was isolated from the root. Subsequent synthetic studies (11) arrived at 1,2-dihydroxyanthraquinone (1) as the structure of the major color-producing body; this was called alizarin, going back to the ancient name alizari or lizari for the root of the madder plant.

Alizarin was the first natural dyestuff to be produced synthetically in 1869 in von Baeyer's laboratory. Anthraquinone was dibrominated, then fused with alkali (12).

It was many years later that the true structure of this reaction series was elucidated: that the dibromo compound does not correspond to the dihydroxy product, but undergoes rearrangement during alkali fusion, probably involving fission of the quinone ring and ring closure to an alternative position. This method was not commercially acceptable, however, because of the expense of bromine and the technical inability at the time to carry out the caustic fusion on a large scale. Caro of Badische and Perkin in England independently discovered that anthraquinone could be heated with concentrated sulfuric acid to yield a water soluble sulfonic acid; this could then undergo caustic fusion and further air oxidation to yield alizarin (13).

In 1871 synthetic alizarin was made available commercially; shortly thereafter demand for the natural product ceased. Besides the two synthetic routes given above, alizarin is also made via several other routes (14–16):

Synthetic alizarin is superior to the naturally occurring type because it is pure 1,2-dihydroxyanthraquinone. The natural product contains mixtures of other isomers, natural dyes in their own right: purpurin (CI Natural Red 16, CI 75410) (**2**); purpuroxanthin (CI Natural Red 8, CI Natural Red 16, CI 75340) (**3**), and pseudopurpurin (CI Natural Red 14, CI 75420) (**4**). However, they do not contribute a synergestic effect to the properties and shade of pure alizarin.

The most important adduct of alizarin is the glycoside, ruberytheric acid [152-84-1] (**5**). The sugar constituent has been shown to be primeverose (17).

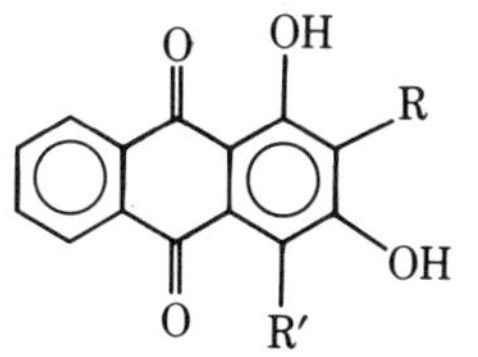

(**2**) R = H, R′ = OH
(**3**) R = R′ = H
(**4**) R = COOH, R′ = OH

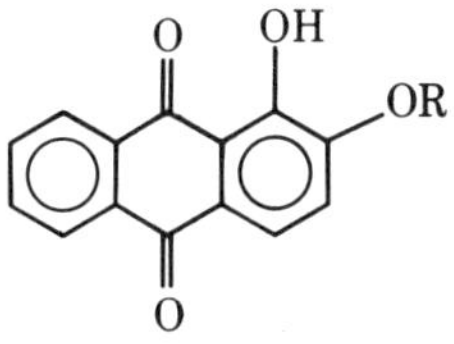

(**5**)

R = primeverosyl =

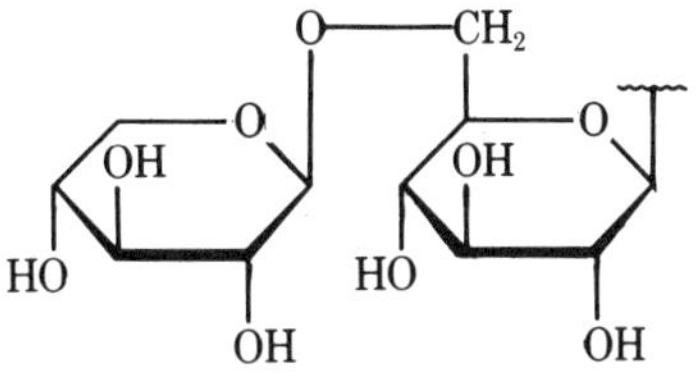

Animal Anthraquinone Dyes. *Kermes.* With the exception of lac dyes from ancient India, kermes or kermesic acid (**6**) is the oldest of all insect dyes. It is obtained from an oriental shield louse which lives on the leaves and stems of low, shrubby trees, the holm oak, *Quercus ilex* and the shrub oak, *Quercus coccifera*. The name kermes is derived from an Armenian word meaning little worm for which the later Latin equivalent was *vermiculus,* the basis of the English word vermillion, which is the color this insect dyes natural fibers. Mention is made throughout the ancient literature in reference to this color: Moses; Theophrastus, the Greek botanist; Pliny the Elder; the Roman emperor Diocletian. It was particularily important during the Golden Age of Venice, when its dyeings rivaled the fabled Tyrian royal purple, and the search for additional sources stimulated world exploration. The coloring matter is obtained from the female *Coccus ilices*. These insects have been collected historically by women, it being noted that they allowed their finger nails to reach a more than normal length in order to assist them in their work. One woman could pick about a kilogram of insects a day by candlelight, before daybreak, when the dew had not yet evaporated and the thorny holly or oak leaves were still soft.

The chemical structure of kermesic acid (6) was proposed by Dimroth (18) as:

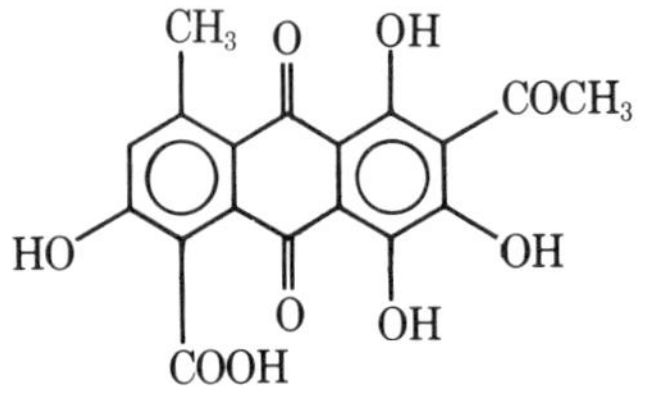

[Structure (6)]

(6)

It is slightly soluble in cold water; soluble in hot water giving a yellowish-red solution. On dissolving in concentrated sulfuric acid a violet red is obtained which turns blue on the addition of boric acid. The fame of its color is owing to the brilliant scarlet obtained when dyed on alum mordant.

Cochineal. Chemically, cochineal (7) is similar to Kermes, differing in the nature of one of the chemical constituents:

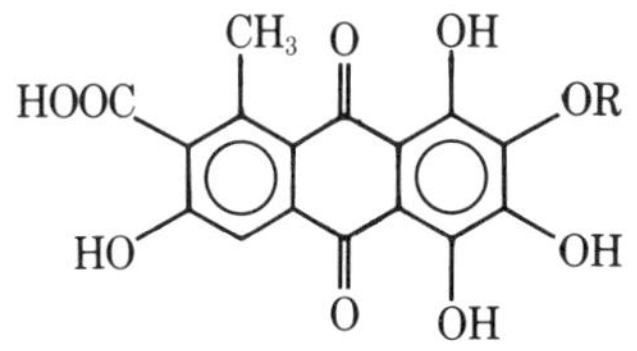

[Structure (7)]

(7)

R = α-D-glucopyranosyl,

[Glucopyranosyl structure]

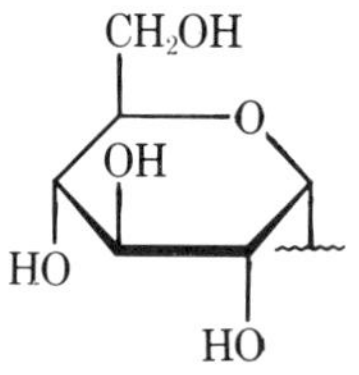

After the discovery of Mexico and the exploitive tactics of Cortez and the Spanish King, Charles V, cochineal replaced kermes as the major reddish orange on the continent; a monopoly that continued for over three hundred years. It was not until 1858 with the introduction of Aniline red that the Mexican-Spanish corner on the market was finally broken. Such a state secret was the source of this new world dye that even as late as 1725 it was commonly believed that cochineal was the seed of an American tropical plant. In actuality, cochineal is obtained from an insect of the same name which feeds on the cactus *Nopalea cochinellifera*. Even before the conquest of the Spaniards the value of the dye was legend. It is reported that the Aztecs used it as a medium of tribute; that in the 1480s Montezuma had extracted from the state of Huaxyacas (modern Oaxaca) a yearly tribute of twenty sacks of cochineal. Considering that ca 155,000 dried insects are required to produce one kilogram of cochineal dye, the value of such a tribute is enormous.

Cochineal insects require only three months to mature, but because of the rainy seasons in Mexico, labor-intensive operations are required. With the approach of the rainy season, branches of the cactus plants loaded with the young insects are cut off and stored indoors. The insects are carefully brushed off into bags or small wooden bowls. They are killed either by immersion in scalding water, heated ovens, or long exposure to the hot sun. The latter method yields the highest quality dye. The dried insects have the shape of irregular, fluted currantlike grains. These contain approximately 10% carminic acid (7).

The major use of cochineal was on tin-mordanted wool and alum- and tin-mordanted silk to produce brilliant scarlets. It is applied to wool from a bath containing an acid solution of stannous nitrate and tartar. Traces of copper and iron dull the dyeings. Fastness to light is good but that to washing is poor, ie, the dyeings become duller and bluer.

Of current interest, the alumina lake of cochineal, called carmine [1260-17-9], has found use in the food dye industry. It is permanently listed as approved by the FDA for ingested products. The lake is composed of not less than 50% carminic acid (7). Carmine is currently sold at $330/kg. Liquid formulations prepared by dissolving carmine in an ammonia–propylene glycol system are also commercially available. Recent uses have included coloring toothpaste, bakery goods, apple sauce, and pharmaceutical tablets (see Colorants for foods, drugs, and cosmetics).

Lac Dye. Lac is probably the most ancient of the animal dyes. It has been used in Southeast Asia and India since the beginning of recorded history. The word lac is derived from the Sanskirt *laksha* and is the same as the Hindu word *lakh* which means one hundred thousand, suggesting the countless thousands of insects which infest the trees. The dye is derived from the gum lac or viscous fluid made by the insect *Coccus laccae* which bore into the bark of trees and become enclosed in an exuding juice, which hardens into a resin (see Shellac; Resins, natural). The branches of all trees on which the lac insect reproduces are covered with this red–brown resin. This so-called stick lac is harvested and marketed as the crude product, containing about 10% coloring matter. A more concentrated form is obtained by extraction with hot sodium carbonate solution. The dye is precipitated by addition of alum or lime.

The dye yields scarlet and crimson shades which exhibit good fastness properties, especially to light and water. For a time lac was used in preference to cochineal in Europe in an attempt to break the Spanish monopoly, but because of the resinous and mineral impurities that are found in lac, its inferiority as a competitive dye caused its decline.

Structurally, the principal coloring body, laccaic acid (8), has been identified (19) as:

(8)

Table 1 summarizes many other naturally occurring dyes of anthraquinoid structure.

356 DYES, NATURAL

Table 1. Natural Dyes of Anthraquinoid Structure

Common name and structure no.	CAS Registry No.	CI Number	CI Name	Natural occurence	Reference
alizarin (1)	[72-48-0]	75330	Natural Red 6, 8, 9, 10, 11, 12	madder; chayroot	20 21
purpuroxanthin or xanthopurpurin (3)	[518-83-2]	75340	Natural Red 8, 16	madder; munjeet	22 23
rubiadin	[117-02-2]	75350	Natural Red 8, 16	madder; Galium	24 25
morindanigrin	[6219-65-4]	75360	Natural Red 19	Morinda umbellata	26
munjistin or purpuroxanthin-carboxylic acid	[478-06-8]	75370	Natural Red 16, 8	madder; munjeet	27 23
morindadiol	[1323-88-2]	75380	Natural Red 18	morinda root	28
soranjidiol	[518-73-0]	75390	Natural Red 18	morinda root	28
chrysophanic acid	[481-74-3]	75400	Natural Yellow 23	turkey rhubarb	29 30
purpurin (2)	[81-54-9]	75410	Natural Red 16, 8	munjeet; madder	31 32
pseudopurpurin (4)	[476-41-5]	75420	Natural Red 14, 9, 8	Galium; madder	33 34
morindon	[478-29-5]	75430	Natural Red 19, Natural Yellow 13	oungkouda	35 26
emodin or frangula-emodin	[518-82-1]	75440	Natural Green 2, Natural Yellow 14	purging buckthorn	36 37
laccaic acid (8)	[6219-66-5]	75450	Natural Red 25	Coccus laccae	38 39
kermesic acid (6)	[476-35-7]	75460	Natural Red 3	Coccus ilici	40 41
carminic acid or cochineal (7)	[1260-17-9]	75470	Natural Red 4	Coccus cacti	42 43

Naphthoquinone Types

Just as in the fused three-ring system of anthraquinones, in the two-ring naphthalene base system nature has developed enough conjugation to achieve color. Three major chemical systems yield natural dyeing materials: α-naphthoquinones, flavones, and anthocyanidins.

α-Naphthoquinones. The leaves of the tropical shrub henna, *Lawsonia inermis*, cultivated mainly in Egypt and India, upon extraction with hot water yield the yellow pigment lawsone which dyes wool and silk an orange shade. For many years a paste (a henna rinse) made from dried henna leaves has been used for tinting hair a reddish shade (see Hair preparations). Lawsone has been identified as 2-hydroxy-1,4-naphthoquinone (9). The isomeric 5-hydroxy compound (10), commonly called Juglone, Nucin, or Regianin, is found in the shells of unripe walnuts, usually in the colorless leuco form, α-Juglone (11). The shells are ground, fermented for two days, and extracted with hot water. In this processing the colorless hydro form undergoes air oxidation to yield the quinone structure that dyes natural fibers a yellow–brown hue.

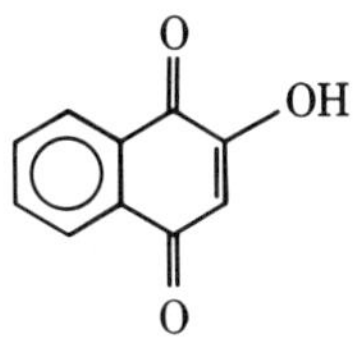

(9) 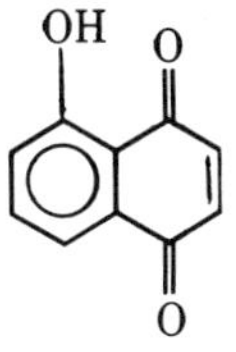 (10) 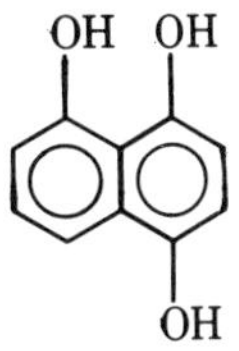 (11)

Lapachol (12) is found in Lapachol or Taigu wood from South America and in Bethabara wood from West Africa. The pigment forms a bright red, water-soluble sodium salt. It is isolated by extraction of the raw wood with cold sodium carbonate solution (1%); precipitation; extraction with ether. Until recently this dyestuff was much sought after for use in the manufacture of high quality bows and fishing rods.

Another natural dye, lomatiol [523-34-2] (13), which is very similar in structure to lapachol (the ω-hydroxy derivative), is found at the other end of the world. Lomatiol occurs in the Australian plant *Lomatia ilicifolia* as a yellow powder surrounding the nuclei seeds of the plant. Table 2 is a list of the natural dyes of α-Naphthoquinoid structure.

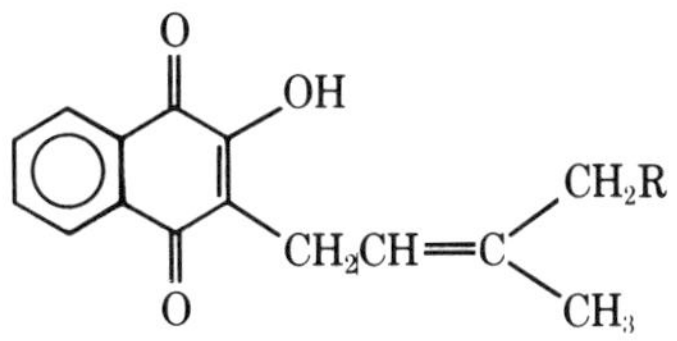

(12) R = H
(13) R = OH

Table 2. Natural Dyes of α-Naphthoquinoid Structure

Common name and structure no.	CAS Registry No.	CI Number	CI Name	Natural occurrence	Reference
lawsone (9)	[83-72-7]	75480	Natural Orange 6	*Lawsonia alba*	44
lapachol (12)	[84-79-7]	75490	Natural Yellow 16	taigu or lapachol wood	45
					46
juglone, nucin, or regianin (10)	[481-39-0]	75500	Natural Brown 7	walnut shells	47
deoxysantalin	[6771-96-6]	75510	Natural Red 22, 23	sanderswood, barwood	48
					49
alkannan	[517-90-8]	75520	Natural Red 20	*Anchusa tinctoria*	50
alkannin or anchusin	[23444-65-7]	75530[a]	Natural Red 20	*Anchusa tinctoria*	51
					52
shikonin or Tokyo violet	[517-89-5]	75535[a]		*Lithospermum erythrorhizon*	50

[a] Dyes CI 75530 and CI 75535 are enantiomers.

Flavones. A majority of the naturally occurring yellows have the flavone basic structure. Flavone [528-82-6] (**14**) (2-phenylbenzopyrone), a colorless material of mp 97°C is obtained from the primrose.

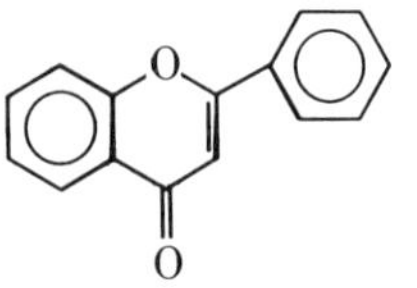

(14)

Most of the natural yellows are derivatives of hydroxyl and methoxyl substituted flavones or isoflavones (2-phenylchromones) eg, structure (**15**), which are present in the plant as glycosides or esters of tannic acid. Some of the more common naturally occuring yellows of this group include: luteolin (**16**) found in the herbaceous plant *Reseda luteola*, or dyer's weed, once cultivated extensively in England, France, Germany, and Austria. This was specified in most European government contracts for dyeing gold braid. When used on alum-mordanted silk the braid has the best fastness to light of any natural yellow. It has a long history: the Gauls used dyer's weed, and it is still in use in some parts of Europe for dyeing leather.

Morin (**17**) is found in the wood of the Osage Orange, *Maclura pomifera*, located throughout the United States. It is used in the United States on chrome-mordanted wool and nylon, tin-weighted silk and chromed leather. Another polyhydroxylated flavone is Quercetin (**18**). It is found primarily in the inner bark of Quercitron, *Quercus tinctoria*, an oak indigenous to Pennsylvania, Georgia, and the Carolinas. Lesser concentrations are present in horse chestnuts, onion skins, tea, and sumac. A red shade (CI Natural Red 1 [70247-65-5]) is obtained by rapidly extracting the bark with dilute ammonia, then boiling the extract with sulfuric acid. A yellow shade (CI Natural Yellow

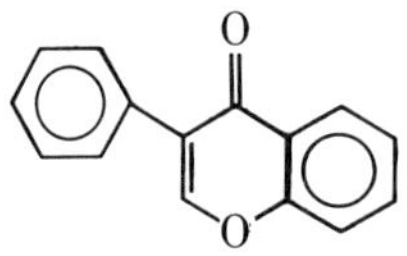

(15) [574-12-9]

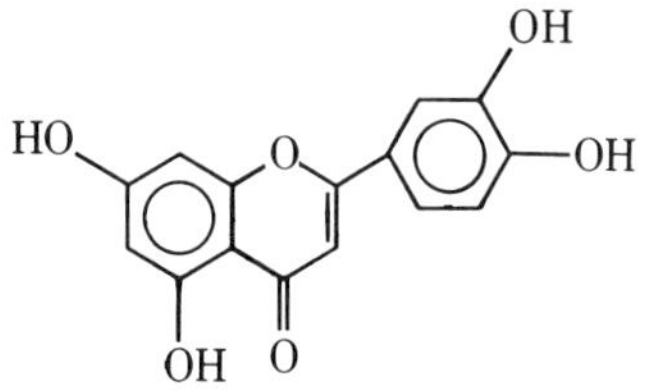

(16)

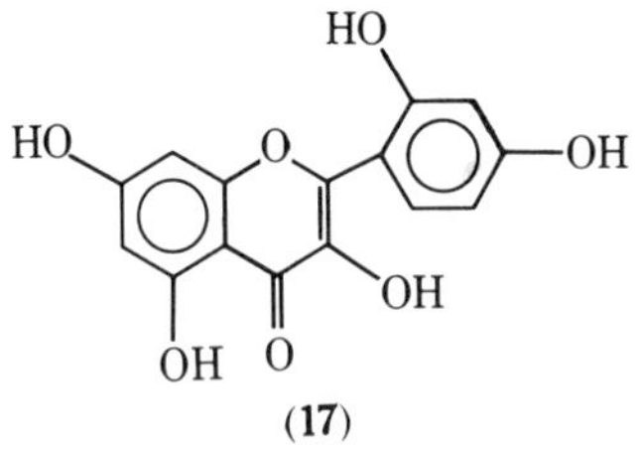

(17)

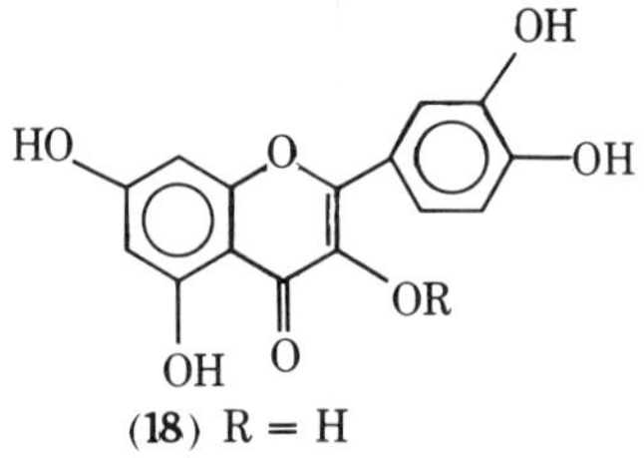

(18) R = H

(19) R = [rhamnose group with OH, CH₃, HO, HO]

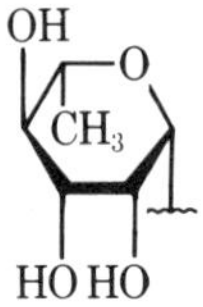

10 [117-39-5]) occurs when high pressure steam is used for extraction from the bark. This yellow is a mixture of Quercetin and its 3-rhamnoside derivative, Quercitrin (19). Many additional natural yellows have the flavonoid structure; they are summarized in Table 3.

Table 3. Natural Dyes of Flavonoid Structure

Common name and structure no.	CAS Registry No.	CI Number	CI Name	Natural occurence	Reference
pratol	[487-24-1]	75570	Natural Yellow 10	red clover	53
					54
apigenin	[520-36-5]	75580	Natural Yellow 1, 2	Greek chamomiles; weld	55
					56
luteolin (16)	[491-70-3]	75590	Natural Yellow 2	dyer's rocket, dyer's wood	57
					58
fukugetin	[6220-03-7]	75600		*Garcinia spicata*	59
					60
genistein or prunetol	[446-72-0]	75610		dyer's broom; greenwood	61
					62
fisetin	[528-48-3]	75620	Natural Brown 1	Venetian sumac	63
					64
datiscetin	[480-15-9]	75630	Natural Yellow 12	bastard hemp	65
					66
kaempferol, trifolitin, or indigo yellow	[520-18-3]	75640	Natural Yellow 13, 10	natal indigo; dyer's knotgrass	67
kaempferol 7-methyl ether	[569-92-6]	75650	Natural Green 2, Natural Yellow 13	sap green; Hungarian berries	66
morin (17)	[480-16-0]	75660	Natural Yellow 8, 11	dyer's mulberry; osage orange	68
					69
quercetin, meletin or sophoretin (18)	[117-39-5]	75670	Natural Yellow 10, 13, Natural Red 1	Persian berries; toon or Indian mahogany tree	70
					71
isorhamnetin	[480-19-3]	75680	Natural Yellow 10	red clover	72
					73
rhamnetin	[90-19-7]	75690	Natural Yellow 13	buckthorn	74
					73
xanthorhamnin	[1324-63-6]	75695	Natural Green 2, Natural Yellow 13	sap green; Hungarian berries	73
rhamnazin	[552-54-5]	75700	Natural Yellow 13	Persian berries; dyer's buckthorn	75
					76
quercimeritrin	[491-50-9]	75710	Natural Yellow 10	Indian cotton	77
					78
quercitrin (19)	[522-12-3]	75720	Natural Yellow 10	quercitron bark	79
					80
rutin	[153-18-4]	75730	Natural Yellow 10	Chinese or Avignon berries	81
					80
gossypetin	[489-35-0]	75750	Natural Brown 5, Natural Yellow 10	Indian cotton	82
					83

Dihydropyrans. Closely related in chemical structure to the above flavones are substituted dihydropyrans. The most important naturally occurring members are haematin [475-25-2] (**20**) and its leuco form, haematoxylin [517-28-2] (**21**). These are the principal coloring bodies of logwood [8005-33-2] (CI Natural Black 1, CI 75290) historically one of the most important natural dyes for dark shades of silk, wool, cotton, leather, wood, and animal bristles, hair, and fur. It is still used extensively in dyeing and tanning leather.

Logwood is the heartwood of *Haematoxylon campechianum*, a tree found in the moist climate of Mexico and Central America. It goes by a variety of names around the world: *bois de Campèche* in France (after the Provice of Campèche in Mexico where it was discovered); *blauholz* in Germany (because of its main use in producing blue dyeings); bloodwood, its common name in the 1700s, caused Linnaeus to apply the generic name *Haematoxylon* to the tree from which logwood is obtained.

The dyestuff is obtained from the wood by either the French or the American process. In the French process the logs, weighing several hundred kilograms, are reduced to chips and extracted with boiling water. In the American process this extraction is carried out under 103–207 kPa (15–30 psi) of steam. These weak extracts are then concentrated under vacuum or evaporation to obtain crystals or solid extract. During this concentration the hematoxylin is partially converted to its oxidized form. This rate of oxidation is increased with the addition of alkali. Logwood extracts may be unoxidized or oxidized to varying degrees, the haematin content varying from 20 to 100% of the total coloring matter present.

Brazilwood [600-76-0] (CI Natural Red 24; CI 75280) is closely related in chemical structure to Logwood. Again, a leuco form, brazilin [474-07-7] (**22**) and its oxidation product, brazilein [600-76-0] (**23**) have been identified.

(**20**) R = OH
(**22**) R = H

(**21**) R = OH
(**23**) R = H

This occurs in the wood of various species of *Caesalpina*, found in Asia and Central and South America. The Asian species was known in medieval times with the Portugese word *braza* referring to anything having a bright red color. Thus when Portugese explorers discovered that area of South America that had an abundance of trees which they knew produced brilliant red colors and they named the area Brazil. Although formerly used as a mordant dye on cotton, wool, and silk, this red dye has a very fugitive character and is no longer of economic importance.

Anthocyanidins (Flavylium Salts). Even the linguistic derivation of the name of this class of chemicals denotes the ancient connection with natural dyes: *antho* meaning flower and *kyanos* meaning blue in Greek. These are glycosides of hydroxylated 2-phenylbenzopyrylium salts (flavylium salts). The structure of the parent cation can be considered a resonance hybrid of oxonium forms (**24**) and (**25**), and carbenium forms (**26**) and (**27**) [14051-53-7].

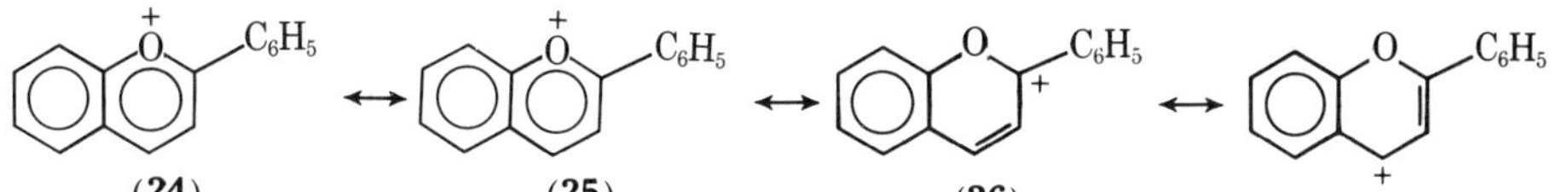

The anthocyanidins are usually isolated as the chlorides after hydrolytic fission of the glycoside with hydrochloric acid. The color developed by the plant is determined by the pH of the cell sap. For example, cyanin is red under acidic conditions (28), violet at neutral pH (29), and blue under alkaline conditions (30).

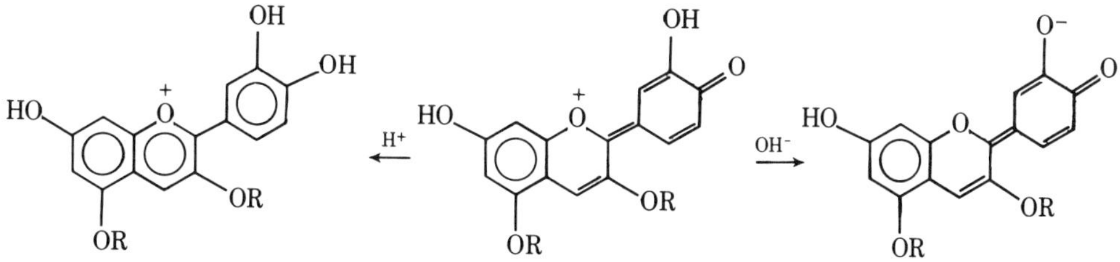

(28) [2611-67-8] red at pH 3

(29) [70116-96-0] violet at pH 7–8

(30) [70116-97-1] blue at pH 11

R = 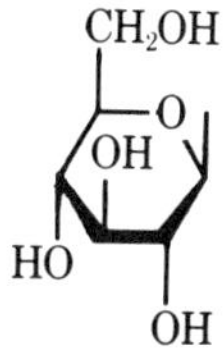 (β-D-glucopyranosyl)

Naturally occurring members of this class include: Carajurin [491-93-0] (31), obtained from the leaves of *Bignonia chica* in Central America and used by the Indians of that region to dye cotton and wool a brilliant red–orange. Awobanin [70116-95-9] (32) obtained from the flower Tsuyukusa, *Commelina communis* in Japan, and used in making awobana paper; it was formerly used in Japan for dyeing silk blue. Dracorhodin [643-56-1] (33), a resinous secretion found in the fruits of *Doemonorops propinqus* in Sumatra and Borneo. It is found in association with its bis derivative, Dracorubin [481-48-1] (34).

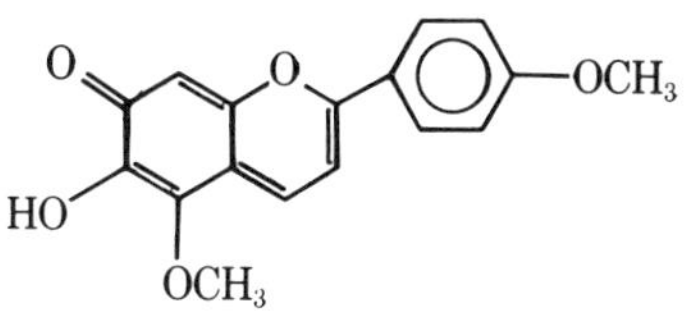

(31)

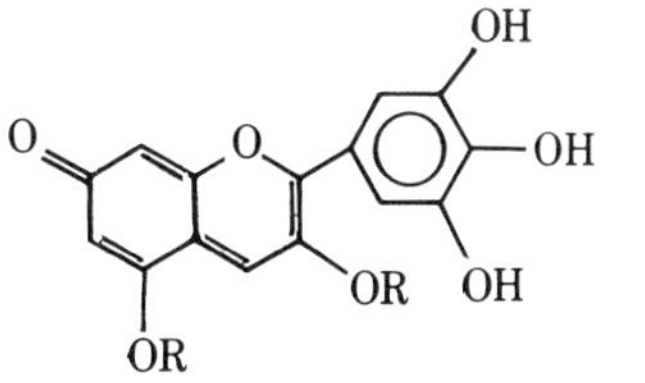

(32) R = $C_6H_{11}O_5$ (α-D-glucopyranosyl)

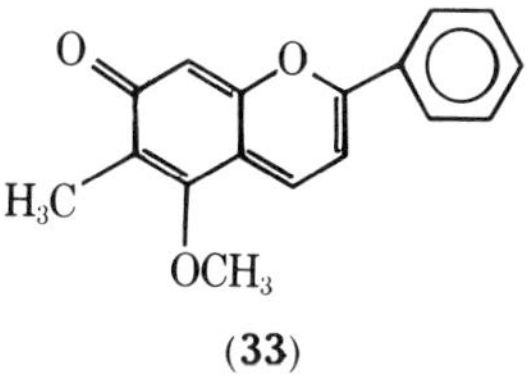

(33)

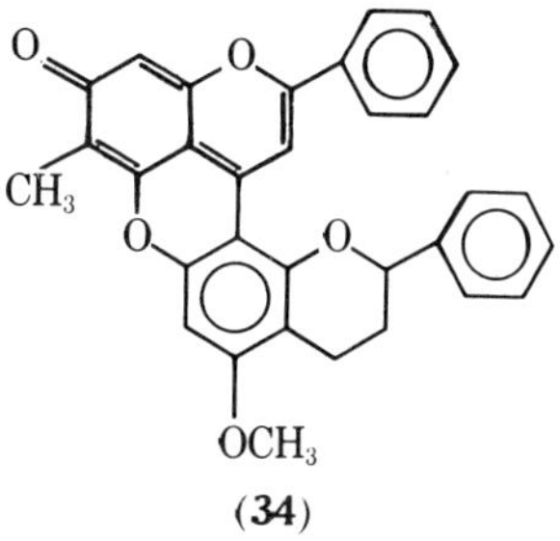

(34)

Together these two reds have the formidable name Dragon's Blood [9000-19-9] (CI Natural Red 31) which is used for coloring lacquers and varnishes and in the printing trade for preparing halftone plates for multicolor printing.

Natural dyes, particularly red shades, have become increasingly important for use in food coloring. With the FDA's withdrawal of FD&C Reds 2 and 4, new sources for food colors have been required. The initial approach by the food dye industry has been to return to natural dyes. Two natural products currently being produced commercially are of the anthocyanin class.

Cranberry juice concentrate is prepared from the juice of mature cranberries under low temperature and pressure. A powdered form is obtained by spray-drying a blend of this concentrate and malto-dextrin. Peonidin [134-01-0] (35) is the substituted anthocyanin primarily responsible for the color in cranberries.

Very similar in chemical structure is the extract (36) of grape skin, *enocionina*. This is obtained commercially by aqueous extraction of grape skins after pressing. In addition to the anthocyanin color body, tartaric acid, malic acid, tannins, and minerals are present in the concentrated extract. This is also available as a spray-dried powder. Both the cranberry and the grape extract have FDA approval as an additive for food use.

Alloxan Adduct. A yellow dye currently being used as a food colorant is the phosphate salt of riboflavin (37), vitamin B_2 (see Vitamins). This is found in varying amounts in all plant and animal cells, particularly good natural sources are milk, eggs, malted barley, yeast, liver, kidney, and heart. Excretion of riboflavin in the urine is responsible for the yellow color. Structurally, riboflavin is an alloxan derivative.

Besides the natural sources, riboflavin can be synthesized from 3,4-xylidine (38), ribose (39), and alloxan (40) (84).

Alloxan can be replaced by more readily available barbituric acid (84). In 1977 the price of riboflavin-5'-phosphate [146-17-8] was $110/kg.

Betanin

Another natural dye that is currently being used in food to take the place of the delisted FD&C Reds 2 and 4, is that obtained from red beet extracts, *Beta vulgarus*. This is available as beet juice concentrate, dehydrated beet root and spray dried extracts. Red beet root contains both red and yellow pigments of the class betaines; these are quaternary ammonium amino acids. The red colored bodies are betacyanine pigments of which the major constituent is betanin [7659-95-2] (41).

Indigoid Dyes

Two very important natural dyes have the indigoid structure. Indigo [482-89-3], the main product which gives this dyestuff class its name, is still one of the major dyes of the world, although now made via a synthetic route. Tyrian or Royal Purple [19201-53-7], although no longer in demand, was once the prize sought by the Caesars.

Tyrian Purple. Probably the most expensive dyestuff in history and the color with the most prestigeous associative value was the Royal Purple of the Caesars. Even its discovery has found a place in ancient mythology: the legend is that a sheep dog of the Greek hero Hercules, while attempting to bite into a shellfish, stained his jaws bright red. Hercules, noting the rarity of the hue, ordered a gown to be dyed with the newly discovered color.

The kingdom of Tyre did produce a purple dye as early as 1600 BC and distribution by the Phoenicians brought fame and fortune to the area until final conquest by the Arabs in 638 AD. The properties of this dyestuff, namely brilliance and nonfading, were so famous throughout the Greek and Roman worlds, and the source of supply so limited, that it was reserved only for the most powerful. Augustus Caesar decreed that only the Emperor and his household might wear "The Purple." Not surprisingly, the enemies of the Empire, particularly the early Christians, abhorred purple.

The source of this color is the pupura shellfish or *Murex brandaris* found in shallow waters throughout the Mediterranean. Each mollusk contains a few drops of glandular mucous; this fluid at first appears white, but on exposure to light changes to yellow–green and eventually violet or reddish-purple. Ancient recipes show that the shellfish were crushed, salted for three days, and boiled for about ten days. Different shades of purple could be obtained by mixing with honey or orseille. It was not until 1909 that the structure of the active dyestuff was finally elucidated (85). It is reported that it required 12,000 mollusks to obtain 1.4 g of dye on which the structural identification could be carried out. This was shown to be 6,6'-dibromoindigo [*19201-53-7*] (**42**).

Indigo. If any one color and fabric could be used to characterize taste or style in the United States, even reaching fad proportions in the mid 1970s, it would be blue jeans. A recent industry survey estimated world demand to be 13,000 metric tons per year (86).

The dyestuff behind this national color is indigo, now of course, a synthetic product (as CI Vat Blue 1; CI 73000) but originally a natural dye (CI Natural Blue 1; CI 75780) whose history started long ago in India, as the name denotes, long before Levi Strauss invented his jeans.

Indigo may be the oldest natural dye used by man; its method of preparation is described in ancient Sanskrit writings; Egyptian tombs have contained mummies wrapped in clothes dyed with indigo. The name itself is derived from the medieval name *indicum*, a Latin word originally used to cover all imports from India, but whose meaning gradually changed to specifically mean the blue vegetable dye whose importation greatly increased in Europe during the Middle Ages. The word *indicum*

(**42**)

replaced the Arabic word *al-nil* which meant blue; this linguistic root is still present in chemical terminology as the base for aniline, the principal building block of the synthetic dyestuff industry.

Natural indigo is obtained from the leaf of *Indigofera tinctoria*, a leguminose, widely distributed in Asia, Africa, and America. The dye is contained only in the leaf of the plant, unlike other vegetable dyes where the coloring matter is also found in the stalk, pods, and twigs. The shrublike plant, obtaining a height of 0.9–1.5 m, is harvested twice a year. Processing must be done rapidly; the plant is cut down early in the morning and as quickly as possible placed in vats and steeped in water for 9–14 h. The percentage of dye obtained is directly proportional to the freshness and moisture content of each leaf. The liquid is drawn off to beating vats, where it is air-oxidized by means of striking the slimy surface of the leaves with bamboo sticks. The liquid gradually changes from yellow–orange through dark green to blue as the air oxidation is completed. The precipitated crude indigo is allowed to settle for 2 h; the top water is drawn off and the sludge transferred to a large cauldron where it is heated to prevent further fermentation. When cool, the sludge is filtered through linen or cotton cloths onto trays where it dries to the consistency of a stiff paste. This is cut in bars or cakes for sale. Despite the crudeness of these operations, they may still be found today in the rural villages of India, the Philippines, and South America.

Not only was indigo used to dye fibers for garments, a powdered form has been used since antiquity as a paint and cosmetic. The Greeks and Romans developed a blue crayon which was widely used as an eye shadow. Until the invention of tempera paints in the Middle Ages, indigo was used as the basic blue in paint. It was even used medicinally, as an astringent. The Greek Dioscorides reported that it cleaned wounds and was used for ulcers and inflammations. As late as the 17th century it was prescribed for internal use.

The Venetians appear to be the first Europeans to use indigo obtained from India through their vast trading empire. However, its use did not spread immediately throughout Western Europe. A locally grown plant, *Isatis tinctoria*, known as woad had been used since antiquity for dyeing cloth blue. Although the chemical structure of the principal coloring body is the same for woad as Indian indigo, the woad industry was firmly entrenched in Europe and would not be replaced by the expensive import from the East until much cheaper sources from Central America were available two hundred years later. The threat to the woad industry by indigo was so great during the sixteenth and seventeenth centuries that England, France, and Germany passed laws prohibiting the importation of indigo under the belief that it was the devil's dye, being harmful, deceitful, and corrosive.

The true nature of indigo was kept secret for thousands of years. Like cochineal later, the dyers had no concept of the source of their dyestuff material. As late as 1705 some dyers believed indigo to be a mineral; a British patent was granted that year for obtaining the blue stone brought out of India from mines. It was not until the 1870s that a modern approach to structural identification of natural indigo was undertaken by A. von Baeyer in Germany. For many years the laboratories of von Baeyer in academia and Badische Anilin- und Soda-Fabrike in industry worked jointly on the indigo problem. The structure was finally elucidated by von Baeyer observing the interconversion of isatin (**43**), indoxyl (**44**), and indigo (**45**) (87):

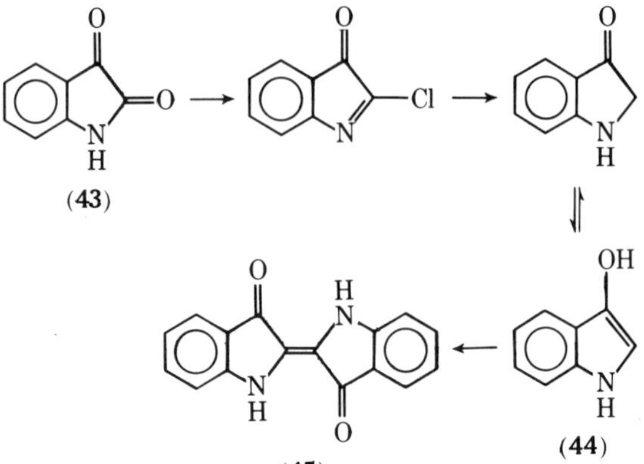

von Baeyer originally published the structural formula in the cis form; it was not until the x-ray studies of Reis and Schneider in 1928 that the trans configuration was proven (88).

A synthetic route was devised by von Baeyer to manufacture indigo. This process, starting with *o*-nitrobenzaldehyde, was sold to BASF with reduction to practice accomplished in 1897 (89).

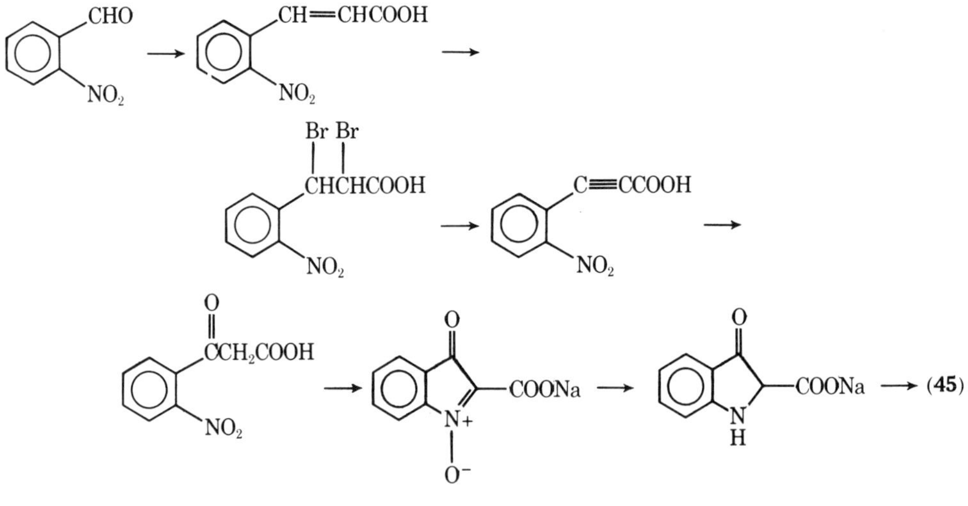

It was estimated that nearly £1,000,000 (in 1897, ca $5,000,000) was spent in the development of this process. Although this route did not achieve economic success, it did demonstrate the feasibility of manufacturing synthetic products that could compete with natural products. It also began a competition, primarily between Germany and England, for control of the multimillion dollar dyestuff market. The English textile market, geared for the natural product, demonstrated that the synthetic product did not have the same hue and dyeing characteristics as that produced by nature. They were correct. Natural indigo (**45**) also contains Indigotin [*482-89-3*] (CI 75780) (**46**),

Indirubin [479-41-4] (CI 75790) (**47**), Indigo Brown, Indigo Gluten, and Indigo Yellow [520-18-3] (CI 75640) (**48**).

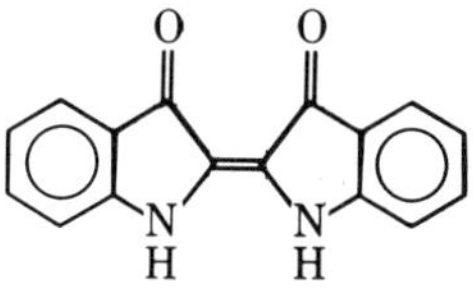

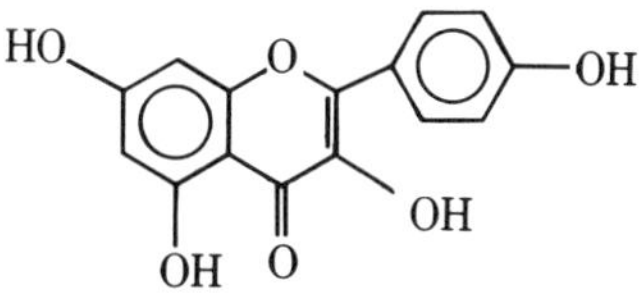

It did possess greater affinity for the fiber and better fastness properties. The Germans countered with a variety of recipes for adding additional ingredients to the dyebath so that their synthetic product would more closely approximate the natural one. Cost reduction programs for the natural product, and many new routes for the synthetic product were reported for a period of twenty years. As late as 1922 the English were still championing the natural product; however, the synthetic dyestuff industry was established.

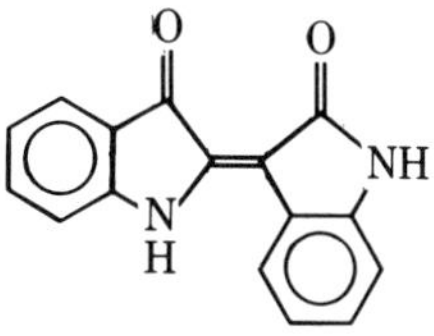

There are over thirty different synthetic routes to indigo (**45**) reported in the chemical literature. The first successful commercial process was developed by Heumann at BASF in the 1890s. It was in use at Ludwigshafen for thirty years.

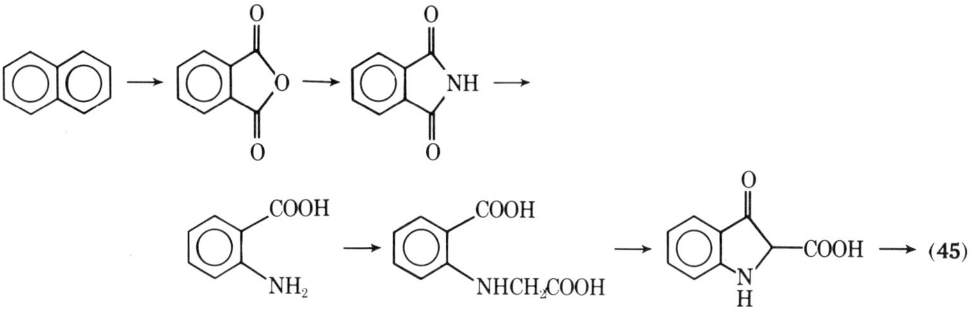

Another route, also introduced by Heumann, starts with the condensation of aniline with chloroacetic acid to give *N*-phenylglycine which on fusion with alkali gives indoxyl (**44**) which forms indigo (**45**) on air oxidation (90):

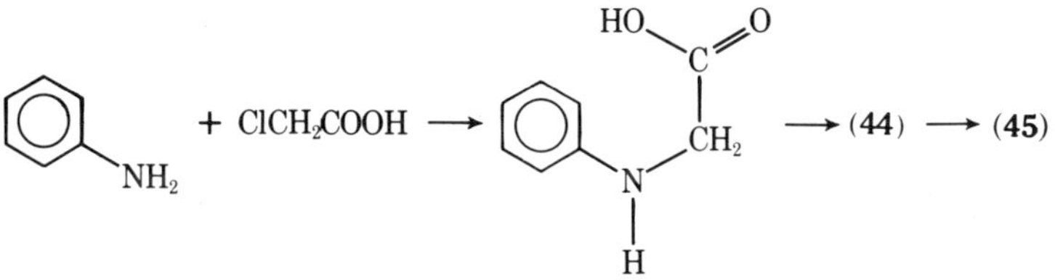

Variations on this scheme include the condensation of aniline, formaldehyde, and hydrocyanic acid to form *N*-phenylglycinonitrile, followed by alkaline hydrolysis to yield *N*-phenylglycine which gives indigo in 94% yields. This is the major commercial route used throughout the world. In an industry noted for its batch operations, this scheme was put on a continuous basis by BASF at Ludwigshafen (91).

Although the demand for indigo to color blue denim seems to be declining, most

Carotenoids

industry analysts feel that there will always be a comparatively high demand for denim work clothes. Even though the requirements may not stay at 13,000 t/yr, the product will remain one of the major industrial dyes.

Carotenoids

Although most color in the visible spectral range is generated by conjugated aromatic ring systems, another class of dyes found in nature obtains its color owing to the presence of long, conjugated double-bond chains (see Terpenoids). These dyes are the carotenoids; the class name being derived from the orange pigment found in carrots, carotene. They are also known as lipochromes because of their solubility and occurrence in fats.

Carotene. Carotene (Natural Yellow 26 [36-88-4]; Natural Brown 5; CI 75130) is widely distributed throughout the vegetable and animal kingdoms. Carotene was first isolated by Wackenroder in 1831, as ruby red crystals, from the common carrot; it was not until 1907 that Willstätter established an identification of carotene as a single product, $C_{40}H_{56}$, with mp 184°C. Kuhn (92), in the 1930s showed that the natural product contained three isomers: α (**49**), [432-70-2], mp 188°C; β (**50**) [7235-40-7], mp 184°C; and γ (**51**) [502-65-8], mp 178°C. His use of chromatographic absorption as a separative technique in this work received wide attention and established column chromatography as an analytical tool.

The β-isomeric form (**50**) is the most common structure; an average composition being 15% α, 85% β, and 0.1% γ. The major structural identification is based upon analysis of the products (α,α-dimethylglutaric acid (**52**), α,α-dimethylsuccinic acid (**53**), dimethylmalonic acid (**54**), and acetic acid (**55**)) resulting from oxidative cleavage reactions:

The γ-form is similar to Lycopene [502-65-8] (**56**), the red coloring matter found in tomatoes and other fruits:

There is also a close structural relationship between these two vegetable coloring dyes and vitamin A [11103-57-4] (57), the fat-soluble growth factor (see Vitamins).

(57)

One of the early clinical observations was that rats suffering from a deficiency of vitamin A could be cured by feeding with the extracts of green plants; a correlation between growth activity and the amount of carotene in the feed stock was shown. It was later established that the carotenoids are converted into vitamin A in the liver. The retina of the eye contains light sensitive pigments that are made up of proteins combined with carotenoids. In exposure to light the pigment bleaches, liberating the protein opsin and the carotenoid, vitamin A aldehyde. Carrots supply the starting materials for these chemical and biochemical transformations.

The carotenoids have found use as colorants in foods and ingested drugs. β-Carotene is permanently listed by the FDA as an FD&C approved colorant (see Colorants for foods, drugs, and cosmetics). At present it is available commercially as a 30% liquid suspension; a 24% semisolid suspension; a 10% beadlet-water dispersion, and a 3% emulsion.

Annatto. Another carotenoid, extract of annatto, is also used in food coloring. Currently its main use is in dairy products such as butter, margarine, and cheese. Annatto is obtained from the pulpy portion of the seeds of the plant *Bixa orellana*, found in India, Central America, and Brazil. After the seeds and pulp have been removed from the mature fruit, the residue is macerated in water and strained; the coloring matter is filtered, dried, and compressed into cakes. The chief ingredient of annatto is bixin [6983-79-5] (58).

In saponification, the methyl ester is hydrolyzed to give the diacid, nor-bixin. Annatto extract is available in aqueous solutions, oleaginous dispersions, and spray-dried powders.

Saffron. Saffron is obtained from the pistils of the *Crocus sativus*, a plant which flowers in the fall and is quite different from the common spring variety crocus. The name of the plant comes from the Arabic *za faran*, meaning yellow. During Greek and Roman times saffron enjoyed a variety of uses. Not only was it the principal yellow dye of the ancients, it was used extensively as a spice in cooking, as an ingredient in medicinal formulations, as a perfume in the public baths. Later, in the Middle Ages, it was found that when used as a mordant on paper with iron, a realistic imitation gold could be obtained; this made possible the beautiful illuminated manuscripts which the monasteries produced and which are still treasured art works. Like many ancient dyes, the natural source was limited; it has been estimated that it requires 140 stigmata to produce one gram of dyestuff. This dyestuff was even the subject of a war, the so-

(58)

called Saffron War of 1374. Because of its rarity, it was in great demand; Basle, Switzerland was the center of trade (the saffron blossom even became part of Basle's coat of arms). When a consignment of ca 360 kg of saffron from Italy was seized as booty, an army was sent out to recover the dyestuff and clear the roads for safe passage of commerce.

The major chemical constituent of saffron is crocin [42553-65-1] (**59**), the digentiobiose ester of crocetin [27876-94-4] (**60**).

The structural formulas (**56–57**) are based on the experimental observations that they take seven moles of hydrogen on catalytic hydrogenation; oxidation with chromic acid yields three to four moles of acetic acid per mole of pigment; and they are dibasic acids. The structures were confirmed by synthesis of perhydrocrocetin.

Chlorophyll

Unlike the other natural products discussed in this article, one might not consider chlorophyll in the same class as a dyestuff. This is probably because chlorophyll does not have affinity for the natural fibers, especially cotton and wool, which can be made into wearing apparel. However, if one overlooks this most common dye requirement, chlorophyll is indeed a dyestuff (CI Natural Green 3 [8049-84-1]; CI 75810). It is used extensively for coloring soaps, resins, inks, waxes (eg, candles); because it is physiologically harmless it is used in the coloring of edible fats and oils (eg, chewing gum, confectionery, egg white, gelatin), and for cosmetics, liniments, lotions, mouthwashes, and perfumes. In the usual sense of a dye, it is used to color leather, where it exhibits good penetrating power and is especially light stable.

Chemically pure chlorophyll is very difficult to prepare; therefore, the commercial product, like that found in nature, is a mixture along with several colored substances of the carotenoid family. The major components of the natural mixture have been designated chlorophyll a [479-61-8] (**61**) and chlorophyll b [519-62-0] (**62**), in a ratio of approximately 3 to 1, along with yellow, orange, and red colored bodies. The excellent hiding power of the green chlorophylls usually mask the rest whose presence is unknown until the chlorophyll is destroyed, as in the case of autumn leaves.

The elucidation of the chemical structure of chlorophyll has involved the work of a number of famous chemists, as far back as Berzelius in 1839. Willstätter received the Nobel Prize in 1915 for his work in this area (93). H. Fischer likewise became a Nobel laureate in 1930 for his synthesis of hemin [15489-47-1] (**63**), a related porphyrin

(94). He suggested the structures of what another Nobel laureate, R. B. Woodward (and seventeen associates), proved in a thirty-reaction sequence (95). The structures chlorophyll a (**61**), chlorophyll b (**62**), and hemin (**63**), are as follows:

(**61**) R = CH$_3$
(**62**) R = CHO

(**63**)

The fundamental ring system consists of four pyrrole rings linked at the α-positions by methine bridges. The chlorophylls are magnesium salt complexes, hydrogenated in the 7,8-position, with a 6,γ-ethanone ring. In chlorophyll b, an aldehyde group replaces the methyl group in position 3. There is a chemical resemblance to hemin (**58**), an iron salt complex from mammalian blood.

Commercial chlorophyll is obtained via a multistep process. The natural source of leaves (usually from stinging nettle, spinach, alfalfa, or corn) is extracted with petroleum ether (VM&P naphtha) to remove a majority of the carotenoids, fats, oils, and waxes; this is followed by another extraction with 80% acetone followed by filtration. Low boiling petroleum ether is added to this extract; the acetone is removed by washing with distilled water; there is another extraction with 85% methanol to remove most of the yellow pigments; the methanol is removed from the petroleum ether by extended washing with water. The chlorophyll precipitates, and it is isolated from the petroleum ether by centrifuging. The process, beginning again with acetone, is repeated five or more times. All this must be done in subdued light to prevent photodecomposition of the dissolved chlorophyll. Separation of chlorophyll a (**61**) from chlorophyll b (**62**) is accomplished by the chromatographic method of Winterstein and Stein (96) using a dried solution of the product in 1 part benzene and 10 parts petroleum ether with fine sugar (XXXXXX grade) as the absorbing medium. The separation is accomplished with Girard Reagent T (trimethylaminoacetylhydrazide chloride) which reacts preferentially with the formyl group of chlorophyll b, giving a water soluble product, whereas chlorophyll a remains unreacted and insoluble.

BIBLIOGRAPHY

"Chlorophyll" in *ECT* 1st ed., Vol. 3, pp. 871–882, by Paul Rothemund, Charles F. Kettering Foundation; "Natural Dyes" under "Dyes" in *ECT* 1st ed., Vol. 5, pp. 345–354, by G. E. Goheen and Jesse Werner, General Aniline & Film Corporation; "Plant Derivatives" under "Tints, Hair Dyes and Bleaches" in *ECT* 1st ed., Vol. 14, pp. 169–177, by Florence E. Wall, Consulting Chemist; "Chlorophyll" in *ECT* 2nd ed., Vol. 5, pp. 339–356, by P. Rothemund, Consulting Chemical Engineer; "Dyes, Natural" in *ECT* 2nd ed., Vol. 7, pp. 614–629, by A. J. Cofrancesco, General Aniline & Film Corporation.

372   DYES, NATURAL

1. *Colour Index,* 3rd ed. Vol. 3, Society of Dyers and Colourists, Bradford, English and American Association of Textile Chemists and Colorists, Research Triangle Park, North Carolina, 1971, pp. 3225–3255.
2. Ref. 1, Vol. 4, pp. 4623–4640.
3. W. F. Legget, *Ancient and Medieval Dyes,* Chemical Publishing Company, New York, 1944.
4. A. G. Perkin and A. E. Everest, *The Natural Organic Colouring Matters,* Longmans, Green and Co., New York, 1918.
5. F. Mayer, *The Chemistry of Natural Coloring Matters, ACS Monograph No. 89,* translated and revised by A. H. Cook, Reinhold Publishing Co., New York, 1943.
6. S. M. Edelstein and H. C. Borghetty, *The Plictho of Gioanventura Rosetti: Instruction in the Art of Dyers* (Translation of the first edition of 1548), MIT Press, Cambridge, Mass. and London, Eng., 1969.
7. S. P. Kierstead, *Natural Dyes,* Humphries, Boston, Mass., 1950.
8. T. S. Gore and co-authors, *Recent Progress in the Chemistry of Natural and Synthetic Colouring Matters and Related Fields,* Academic Press, New York, 1962.
9. D. E. Auslander and co-workers, *Drug and Cosmetic Industry* **121**(5), 36 (Nov. 1977); **121**(6), 55 (Dec. 1977).
10. F. Crace-Calvert, *Lectures on Coal-Tar Colours, and on Recent Improvements in Dyeing and Calico Printing Embodying Copious Notes Taken at the London Exhibition of 1862 with Numerous Dyed Patterns,*
11. C. Schunck, *Ann. Chem.* **66,** 176 (1847); F. Rochleder, *Ann. Chem.* **80,** 324 (1851).
12. C. Graebe and C. Liebermann, *Ber.* **1,** 49 (1868); *ibid.,* **2,** 14, 333 (1869).
13. L. F. Fieser and M. Fieser, *Organic Chemistry* 3rd ed., D. C. Heath and Co., Boston, Mass., 1956, pp. 902–903.
14. Brit. Pat. 293,328 (Feb. 12, 1884), C. Jenkins.
15. U.S. Pat. 1,744,815 (Jan. 28, 1930), J. Thomas and H. W. Hereward (Scottish Dyes, Ltd.).
16. Ger. Pat. 292,247 (Jan. 22, 1884), D. C. Millet.
17. D. Richter, *J. Chem. Soc.,* 1701 (1936); G. Zemplen and R. Bognar, *Ber.* **72,** 913 (1939).
18. O. Dimroth, *Ber.* **43,** 1387 (1911); O. Dimroth and W. Scheurer, *Ann. Chem.* **399,** 43 (1913); O. Dimroth and R. Fick, *Ann. Chem.* **411,** 315 (1916).
19. R. Schmidt, *Ber.* **20,** 1285 (1887); O. Dimroth and S. Goldschmidt, *Ann. Chem.* **399,** 62 (1913).
20. Ref. 4, pp. 23, 37, 52, 64.
21. Ref. 5, p. 117.
22. Ref. 4, pp. 29, 42.
23. Ref. 5, p. 120.
24. Ref. 4, p. 31.
25. Ref. 5, pp. 121, 125.
26. Ref. 5, p. 129.
27. Ref. 4, pp. 29, 32, 42.
28. Ref. 5, pp. 119, 129.
29. Ref. 4, pp. 53, 59. 71.
30. Ref. 5, p. 122.
31. Ref. 4, pp. 24, 42.
32. Ref. 5, p. 123.
33. Ref. 4, p. 27.
34. Ref. 5, p. 124.
35. Ref. 4, p. 46.
36. Ref. 4, pp. 55, 60, 62, 210.
37. Ref. 5, p. 130.
38. Ref. 4, p. 91.
39. Ref. 5, p. 144.
40. Ref. 4, p. 95.
41. Ref. 5, p. 141.
42. Ref. 4, pp. 77, 627.
43. Ref. 5, p. 137.
44. Ref. 5, p. 105.
45. Ref. 4, p. 102.

46. Ref. 5, pp. 109, 253.
47. Ref. 5, p. 106.
48. Ref. 4, p. 584.
49. Ref. 5, p. 147.
50. Ref. 5, p. 111.
51. Ref. 4, p. 73.
52. Ref. 5, p. 110.
53. Ref. 4, p. 203.
54. Ref. 5, p. 170.
55. Ref. 4, p. 154.
56. Ref. 5, p. 172.
57. Ref. 4, pp. 154, 159.
58. Ref. 5, p. 179.
59. Ref. 4, p. 160.
60. Ref. 5, p. 203.
61. Ref. 4, p. 159.
62. Ref. 5, p. 195.
63. Ref. 4, pp. 186, 461.
64. Ref. 5, p. 184.
65. Ref. 4, p. 620.
66. Ref. 5, p. 182.
67. Ref. 4, pp. 21, 60, 179, 212.
68. Ref. 4, pp. 174, 213, 219.
69. Ref. 5, p. 187.
70. Ref. 4, pp. 125, 175, 187, 201.
71. Ref. 5, p. 188.
72. Ref. 4, pp. 61, 203, 212.
73. Ref. 5, p. 189.
74. Ref. 4, pp. 207, 211.
75. Ref. 4, p. 207.
76. Ref. 5, p. 190.
77. Ref. 4, p. 227.
78. Ref. 5, p. 186.
79. Ref. 4, pp. 193, 200.
80. Ref. 5, p. 188.
81. Ref. 4, pp. 197, 200.
82. Ref. 4, p. 224.
83. Ref. 5, p. 192.
84. Ref. 13, pp. 1003–1004.
85. P. Friedlaender, *Ber.* **42**, 765 (1909).
86. *Chem. Week,* 19 (Mar. 3, 1976).
87. A. von Baeyer, *Ber.* **33**, 51 (1900).
88. Ref. 13, pp. 905–908.
89. Ger. Pat. 11,857 (Mar. 1886), (to Badische Anilin- und Soda-Fabrik); *Friedlander* **1**, 127 (1888).
90. K. Heumann, *Ber.* **23**, 3043 (1890); Ger. Pat. 54,626 (May 1890); *Friedlander* **2**, 100 (1891).
91. J. G. Kern and M. Stenerson, *Chem. Eng. News* **24**, 3164 (1946).
92. Ref. 13, pp. 951–952.
93. R. Willstätter and A. Stoll, *Investigations of Chlorophyll,* trans. by F. M. Schertz and A. R. Merz, Science Press, Lancaster, Pa., 1928.
94. H. Fischer and H. Orth, *Die Chemie des Pyrrols,* Vols. 1–3, Verlagsgesellschaft, Leipzig, Ger., 1934–1940.
95. R. B. Woodward and co-workers, *J. Am. Chem. Soc.* **82**, 3800 (1960).
96. A. Winterstein and G. Stein, *Z. Physiol. Chem.* **220**, 263 (1933).

RUSSELL E. FARRIS
Sandoz Colors & Chemicals

DYES, REACTIVE

Reactive dyes are colored compounds that contain functional groups capable of forming covalent bonds with active sites in fibers such as hydroxyl groups in cellulose, amino, thiol, and hydroxyl groups in wool or amino groups in polyamides. This bond formation between the functional group and the substrate results in high wetfastness properties. These dyes differ fundamentally from other types of dyes that owe their wetfastness to physical adsorption or mechanical retention (see Dyes and dye intermediates). The principal commercial applications of reactive dyes are in the dyeing of cellulose, wool, and nylon, either individually or as components of fiber blends. They have also found use in dyeing silk, hair, and leather (see Dyes, application and evaluation).

Development of Reactive Dyes

The first recorded chemical combination of a dye with cellulose was achieved in 1895 by Cross and Bevan (1). They used a complicated process involving the benzoylation of soda–cellulose, followed by nitration, reduction, diazotization, and finally coupling with phenolic compounds to form the covalently fixed azo dye. Several other attempts were subsequently made to attach dyes to cellulose by means of a covalent bond (2–6). The dominating feature of almost all of this early work was the use of severe conditions such as concentrated caustic soda to prepare soda–cellulose or the use of inert solvents. The use of mild conditions necessary to render the coloration process technically feasible and reasonably efficient was given little or no consideration.

During the 1930s fast dyeings on wool were produced using dyes containing chloroacetylamino groups (—NHCOCH$_2$Cl) (7), and β-chloroethylsulfamoyl groups (—SO$_2$NHCH$_2$CH$_2$Cl) (8). It was not realized at that time that the high wetfastness of these dyes was due to the reaction between the labile chlorine atom and the amino groups in the wool.

Dyes prepared from diazo components containing β-sulfatoethyl sulfone groups (—SO$_2$CH$_2$CH$_2$OSO$_3$H) (9), were also patented for application to cellulose acetate and nylon, since they contained no nuclear sulfonic acid groups. Hoechst (10) commenced research work on dyes containing vinyl sulfone and β-sulfatoethyl sulfone substituents, which led to the introduction of the Remalan group of reactive dyes (11). Although it was known that the high fastness of the Remalan dyes was associated with chemical bond formation between the dye and the fiber; the suitability of vinyl sulfone dyes as reactive dyes for cellulose was not realized until after the appearance of the Procion (ICI) reactive dyes.

Previous work on the reaction of soda cellulose with cyanuric chloride (5,12) led to a useful industrial method for the production of dyeings in which a covalent bond was formed between the dye and the fiber (13). This development resulted in the introduction of the first range of reactive dyes for cellulose marketed by ICI in 1956 as the Procion M Dyes. The initial members of this range were all highly reactive dichlorotriazinyl derivatives capable of reaction with cellulose under cold-dyeing conditions. They were quickly followed by the less reactive monochlorotriazinyl dyes requiring hot-dyeing conditions, which were marketed as the Procion H range (14). Direct dyes containing triazine residues patented previously by Ciba were used in the manufacture of a complete range of monochlorotriazinyl reactive dyes under the name Cibacron (15).

The success of the Procion dyes was followed by extensive research programs by all the major dyestuff manufacturers. As a result, a large number of reactive systems are mentioned in the patent literature (16–17). Consequently, reactive dyes containing 2,4,5-trichloropyrimidinyl groups were introduced under the trade names of Reactone (Ciba-Geigy) (18) and Drimarene (Sandoz) (19). In the early 1960s Bayer developed the Levafix P and E groups based upon 4,5-dichloro-6-methyl-2-methylsulfonylpyrimidine (1) (20) and 2,3-dichloroquinoxaline (2) (21) reactive groups, respectively; and BASF produced the Primazin P range with 4,5-dichloro-6-pyridazinone (3) (22) as the reactive species. Hoechst marketed their vinyl sulfone derivatives as the Remazol group (10). In addition, Bayer introduced similar dyes containing β-sulfatoethylsulfamoyl groups (—SO$_2$NHCH$_2$CH$_2$OSO$_3$H) (21), and BASF extended the Primazin range (23) to include dyes containing β-sulfatopropionamide groups (—NHCOCH$_2$CH$_2$OSO$_3$H), which combine with cellulose through nucleophilic addition reactions to form cellulose ether derivatives.

(1)
[17901-16-5]

(2)
[2213-63-0]

(3)
[932-22-9]

During the mid 1950s the major emphasis was on reactive dyes for cellulosic fibers, but 1959 heralded the introduction of the Procinyl (ICI) group of reactive disperse dyes designed specifically for application to nylon (24). Although reactive dyeing began with wool (7–8,11), it was not until 1963 that the first range of reactive dyes, specifically for application by exhaustion procedures, was marketed by ICI under the trade name Procilan (ICI). These dyes are generally 2:1 premetallized dyes containing reactive acrylamide groups which are not subject to hydrolysis in the dyebath (25–26). Reactive dyes for wool, however, gained importance with the introduction of the Lanasol (Ciba-Geigy) dyes containing the α-bromoacrylamide group (—NHCOC(Br)=CH$_2$) (27–28) which are noted for their bright shades, high reactivity and good wet-fastness properties. Similar properties are exhibited by the Drimalan F (Sandoz), Reactolan (Ciba-Geigy), and Verofix (Bayer) dyestuff groups containing a difluorochloropyrimidinyl group (29–31). The Hostalan dyes (Hoechst) that were developed to overcome the problems of unlevel dyeing (32–34) are the reaction products of Remazol (vinyl sulfone type) dyestuffs and N-methyltaurine (4).

CH$_3$NHCH$_2$CH$_2$SO$_3$H
(4)
[107-68-6]

Dichlorotriazinyl and monochlorotriazinyl dyes, although primarily designed for cellulose, have also been applied to wool. The Cibacrolan (Ciba-Geigy) group of monochlorotriazinyl dyes were introduced specifically for wool (35–36).

Tables 1 and 2 illustrate the various commercial types of reactive dyes developed

Table 1. Reactive Dyes for Cellulosic Fiber

Structure	Reactive group Name	Trade name	Manufacturer	Refs.
dye—NH—(dichlorotriazine ring with 2 Cl)	dichlorotriazine	Procion M	ICI	13,37–40
dye—NH—(triazine with Cl and X); X = NH₂, NHR, NHC₆H₅, OR, or SO₃H	monochlorotriazine	Procion H Procion H-E Cibacron Cibacron Pront Procion SP	ICI Ciba-Geigy Ciba-Geigy ICI	13,33,40–41
dye—NH—(pyrimidine with 3 Cl)	trichloropyrimidine	Reactone Drimarene X Drimarene Z	Ciba-Geigy Sandoz	18,33,42
dye—NH—(pyrimidine with SO₂CH₃, Cl, CH₃)		Levafix P	Bayer	20,42–43
dye—NH—(pyrimidine with F, Cl, F)	monochlorodifluoropyrimidine	Drimarene R Drimarene K Levafix EA Levafix PA	Sandoz Bayer	33,37,41
dye—NHC(O)—(pyrimidine with 2 Cl)	dichloropyrimidine	Reactofil	Ciba-Geigy	37
dye—NHC(O)—(pyridazine with 2 Cl)	dichloropyridazine	Solidazol	Cassella	33,42
dye—NHCCH₂CH₂—(pyridazinone with 2 Cl and O)	dichloropyridazinone	Primazin P	BASF	22,33,42
dye—NHC(O)—(quinoxaline with 2 Cl)	dichloroquinoxaline	Levafix E Cavalite	Bayer DuPont	14,21
dye—NHC(O)—(phthalazine with 2 Cl)	dichlorophthalazine	Solidazol Reatex Elisiane	Cassella Francolor	33,42
dye—NHO₂S—(benzothiazole with Cl)	chlorobenzothiazole	Reatex Elisiane	Francolor	39,42
dye—NHCCH=CH₂ (O)	acrylamide	Primazin	BASF	23,33

376

Table 1. (continued)

Reactive group Structure	Name	Trade name	Manufacturer	Refs.
dye—NCCH₂CH₂OSO₃H (with C=O and NH) dye—SO₂CH₂=CH₂ dye—SO₂CH₂CH₂OSO₃H	vinyl sulfone	Remazol Cavalite	Hoechst DuPont	10
dye—SO₂NH—CH₂CH₂OSO₃H	β-sulfatoethyl-sulfonamide	Levafix	Bayer	21
dye—NC—(ring)—SO₂CH₂CH₂Cl (with C=O, NH, Cl)	β-chloro-ethyl-sulfone	Solidazol N	Cassella	37
dye—NHCH₂OH	methylol	Calcobond	Cyanamid	42

for cellulosic fibers, and for wool and polyamide fibers respectively. The tables show the major reactive groups, the corresponding chemical structures, and the common trade names. Other trade names not included in the tables can be found in a summary by Siegel (20) and the *Colour Index* (55) (see Dyes and dye intermediates). Extensive patent literature and publications relating to reactive dyes have been reviewed by Davies (16) and Rosenthal (17) during the period 1967–1975.

Excellent monographs on reactive dyes have been provided (14,37). Comprehensive reviews on this subject have also been published (38–42,53,56–60) and a pertinent bibliography can be found in the *Colour Index* (55).

A reactive dye molecule may, for convenience, be regarded as a combination of the following units:

$$\text{dye—B—Y—X}$$

where dye is the chromophore (usually an azo, anthraquinone, or phthalocyanine residue); B is a bridging atom or group although this, in many cases, is part of the chromophoric system; Y is the unit carrying the reactive group (the activity of the reactive group depends to a large extent upon the nature of Y); and X is the group that reacts with the fiber. An additional water solubilizing group, which is not part of the chromophoric unit, may be found as part of the reactive group such as the sulfuric acid ester group of β-hydroxyethyl sulfone reactive dyes.

Chromophoric System. In principle, practically any desired chromophoric system can be combined with reactive groups to produce reactive dyes. The properties of the resulting dye, however, are affected by both of these groups. Proper combination is, therefore, needed to obtain dyes with good qualities such as high tinctorial strength, good solubility, good fastness properties, and economy.

Commonly used chromogens include azo, metallized-azo, anthraquinone, phthalocyanine, and metal-complex formazan derivatives. Azo compounds comprise the widest range of shade from greenish-yellow to black. For yellow dyes, coupling products of pyrazolones and aminopyrazoles are commonly used (61). Pyridone derivatives have gained much importance in the recent years as coupling components for yellow dyes (62). Brilliant red colors are usually based on aminohydroxynaphth-

378   DYES, REACTIVE

Table 2. Reactive Dyes for Wool and Polyamide Fibers

Structure	Reactive group	Name	Trade name	Manufacturer	Refs.
dye—NH—(triazine ring with Cl, X)	X = Cl X = NH$_2$, NHR, NHC$_6$H$_5$; X = OR	dichlorotriazine monochlorotriazine	Procion M Procion H Cibacron Cibacrolan Cibacron Pront	ICI ICI Ciba-Geigy Ciba-Geigy Ciba-Geigy	14–15,35,44
dye—NH—(pyrimidine ring with F, Cl, F)		monochlorodi- fluoropyrim- idine	Drimalan F Reactolan Verofix	Sandoz Ciba-Geigy Bayer	29–30,45–46
dye—NCCH$_2$Cl (with H, O)		ω-chloracetyl- amino	Cibalan Brilliant Drimalan	Ciba-Geigy Sandoz	44,47
dye—SO$_2$CH=CH$_2$ dye—SO$_2$CH$_2$CH$_2$OSO$_3$H		vinyl sulfone β-sulfatoethyl- sulfone	Remalan Remalan Remazolan	 Hoechst	10–11,43,48
(metal complex dye)—SO$_2$CH$_2$CH$_2$OSO$_3$H		β-sulfatoethyl- sulfone	Remalan Fast		
dye—CH$_2$NCCH$_2$CH$_2$SO$_2$CH$_2$CH$_2$OSO$_3$H		sulfatoethyl acrylamide sulfone	Lanafix	Sumitomo	33,38
dye—SO$_2$CH$_2$CH$_2$NCH$_2$CH$_2$SO$_3$H CH$_3$		methyltaurine- ethylsulfone	Hostalan	Hoechst	32–34
(metal complex dye)—NCCH=CH$_2$		acrylamide	Procilan	ICI	25–26,49–50
dye—NCC=CH$_2$ H Br		α-bromoacryl- amide	Lanasol	Ciba-Geigy	27,51–52
dye—CH$_2$NCC=CH$_2$ H Cl		α-chloroacryl- amide	Lanasyn Pure Blue FBL	Sandoz	33,42,53
(disperse dye)—NHCH$_2$CH—CH$_2$ \O/		epoxide	Procinyl^a	ICI	24,26,54
—CH$_2$CH—CH$_2$ OH Cl (and others)		3-chloro-2- hydroxypropyl			

^{a} Procinyl dyes used specifically for polyamide fibers.

alenedisulfonic acids (63). Chromium, copper and cobalt metal-complex azo dyes comprise the majority of metallized azo dyes. These dyes generally possess excellent lightfastness (64). Brilliant blue and green reactive dyes with high fastness to light are the main contribution of anthraquinone derivatives (65). Copper and nickel phthalocyanine reactive dyes give bright turquoise shades with good washfastness and satisfactory crocking fastness (66). Bright blue to green metal-complex dyes from formazan derivatives have been patented (67) and claimed to possess good fastness properties. Other groups mentioned in patent literature are also used as chromophoric systems for reactive dyes, but they have achieved little commercial importance up to the present.

The Bridging Group. The nature of the bridging group between the chromophore and the reactive group not only affects the shade, strength, and affinity of the dye, but can significantly affect its reactivity and the stability of the dye–fiber bond. Thus the use of oxygen or sulfur bridging groups is unsatisfactory, since they are readily hydrolyzed, and aliphatic or aromatic chains tend to lower the water solubility of the dye. In addition the length and flexibility of the bridging group has an affect on the degree of dye–fiber fixation. Therefore, amino and alkylamino groups are generally used as bridging groups in heterocyclic reactive dyes in view of the ease of synthesis, stability to hydrolysis, and minimum interference with solubility. In the case of dichloroquinoxaline (2) and dichlorophthalazine (5) reactive dyes that contain only one reactive chlorine group, the chromophore is attached to the reactive system through an amide bridging group. This bridging group is susceptible to acid hydrolysis, which can result in the rupture of the bond between the chromophore and the reactive system. Sulfonamide and amide bridging groups have been suggested as replacements for the sulfone group in the Remazol dyes. Although the stability of the dye–fiber bond is increased using these groups, a reduction in the reactivity of the dye is observed.

(5)
[4752-10-7]

The Reactive System. The combined unit Y–X in the above formula can be regarded as the reactive system. Generally, the reactive systems used in commercially available reactive dyes can be classified into two groups: (1) reactive systems based on nucleophilic substitution reactions, in which the mobile reactive group X is replaced by an attacking base, and (2) reactive systems based on nucleophilic addition reactions, in which a 1,2 trans addition of a nucleophile occurs across a polarized double bond. Examples of reactive systems (1) and (2) are halogeno-heterocyclic and vinyl sulfone systems, respectively. In some cases, both addition and substitution reactions are operable as in the α-bromoacrylamide types.

The reactivity of a reactive dye, therefore, is governed primarily by the chemical structure and arrangement of the reactive system (Y–X) and the variation of the chromophore to which the reactive system is attached.

Synthesis

Reactive dyes have a great advantage over direct dyes in that they do not depend on molecular complexity for adsorption. This allows a brighter spectrum of shades and more rapid diffusion characteristics compared with direct dyes.

The methods of synthesis of reactive dyes depend largely on the nature of the reactive component. Very thorough reviews have been presented of the methods of production of various types of reactive dyes (14,37). In general, the following methods are used: (*1*) prepare an azo dye containing a nucleophilic group, eg, —NH$_2$, —OH, and then condense the azo dye with a heterocyclic aromatic reactive system, eg, cyanuric chloride, tetrachloropyrimidine etc; (*2*) combine a diazotized aromatic amine with a coupling component containing a reactive system, or a diazonium compound containing a reactive system with a coupling component; (*3*) condense an aromatic compound containing a reactive system to an anthraquinone derivative, in particular, 1-amino-4-bromoanthraquinone-2-sulfonic acid (bromaminic acid); and (*4*) for phthalocyanine dyes which do not carry a suitable nucleophilic group for the condensation with the reactive system, it is necessary to introduce a bridging group, eg, a sulfonamide, to act as the nucleophile.

Reactive Dyes for Cellulosic Materials, Wool, and Nylon

Cellulosic Materials. Cellulose (qv) is a carbohydrate. It consists of (1 → 4)-β-D-glucan and is considered as a polyalcohol with three hydroxyl groups per unit of glucopyranose. As a consequence, any group that is capable of forming a covalent bond with alcoholic hydroxyl groups is a potential reactive system to be used in a dye molecule for cellulosic fiber. However, certain criteria must be met in order for this reactive system to be of practical use.

The most important criterion is that the dye can be applied in an aqueous medium. In this instance, there are two competing reactions. One is the reaction of the dye and the fiber, and the other is the reaction (hydrolysis) between the dye and water. In principle, it is possible to dye cellulosic fibers in an aqueous alkaline medium with water soluble reactive dyes (68–69). Under alkaline conditions, the hydroxyl groups in the cellulose molecules are partially ionized and can act as nucleophilic reagents. The reactive system of the dye molecule can then react either with cell-O$^-$ or OH$^-$. It has been shown (70–71) that the dissociation constant for cellulose is higher than that of water, thus cell-O$^-$ reacts faster than OH$^-$ with the reactive group of the dyestuff. This preference of dye–fiber interaction has been attributed to the higher nucleophilicity of the aliphatic hydroxyl groups of cellulose (72). In order for the hydroxyl anion of cellulose to attack the reactive system of the dye, they must be in proximity, ie, the dye molecules must be adsorbed onto the fiber. This property, known as substantivity, is the second fundamental condition that must be fulfilled before a dye–fiber interaction can take place.

Another important criterion is that the covalent bond formed between the dye and the fiber must be sufficiently stable to resist subsequent aftertreatments. In other words, the dye–fiber interaction must be relatively irreversible. In addition, the dye must have a suitable shelf-life, and be nontoxic and economical to produce.

Reactive systems can be classified into several groups depending upon the reaction mechanism (58):

Reactive Systems Based on Nucleophilic Substitution. This generally involves a base-catalyzed addition of a nucleophilic functional group of the cellulose to the electrophilic center of the reactive groups followed by the subsequent displacement of a suitable leaving group. This mechanism applies particularly to heterocyclic aromatic compounds containing labile leaving groups. Shown below is the reaction scheme of a dichlorotriazinyl functional group with cellulose (eg, Procion M dyes).

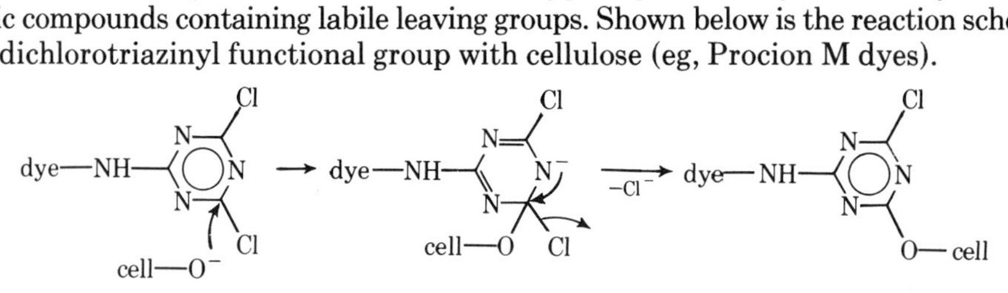

An excellent discussion has been presented of the mechanisms of the reaction of reactive dyes with cellulosic fiber (54). This nucleophilic substitution reaction mechanism also applies to other similar heterocyclic aromatic compounds such as monochloro-s-triazine (6) derivatives (Procion H and H-E type and Cibacron); 2,4,5-trihalopyrimidine (eg, (7)) (Drimarene K, R, X, and Z type, Reactone, Levafix EA and PA); 2,4-dichloropyrimidine (8) (Reactofil); 3,6-dichloropyridazine (9) (Solidazol); 1-H-4,5-dichloro-6-pyridazinone (3) (Primazin P); 2,3-dichloroquinoxaline (2) (Levafix E, Cavalite); 1,4-dichlorophthalazine (5) (Solidazol, Elisiane); and 2-chlorobenzothiazole (10) (Elisiane).

(6) [6153-86-2]
(7) X = Cl [5750-76-5]
(8) [3934-20-1]
(9) [141-30-0]
(10) [615-20-3]

The reactivity of the heterocyclic aromatic systems can be altered by making suitable substitutions in the ring: (1) the hetero-atoms in the aromatic ring are important in that they affect the electronegativity of the system; an increase in the number of hetero-atoms favors nucleophilic exchange, and consequently, the reactivity of a triazine derivative is higher than that of a pyrimidine derivative; and (2) replacing the halogen with an electron-attracting group will increase the reactivity, whereas an electron-donating group will decrease the reactivity; eg, replacement with an amine, such as in the case for Procion H and Cibacron dyes, reduces the reactivity.

An important substitution in the pyrimidine system is the replacement of the chlorine atom at the 5-position with an electronegative group such as $-NO_2$, $-CN$, $-COOH$, etc. ICI claimed that substituting a nitro group in the 5-position of the pyrimidine ring enhanced the reactivity to such an extent that the dye could be applied to cellulose under much milder conditions than trichloropyrimidyl derivatives (73). However, this was not suitable for practical application because the dye–fiber bond was weakened by the over-activation of the nitro group making the resulting dyeing deficient in washfastness. Dyes containing a cyano group in the 5-position possess high reactivity and afford satisfactory dyeings (13).

When a carboxyl group is substituted into the 5-position of the pyrimidine ring, the reactive system can be attached to the chromophore through an amide bridge. This results in another group of dyes, called Reactofil (74–75), which possess reactivity towards cellulose of the same order as that of a dichlorotriazinyl dye.

In general, the reactivity of N-heterocyclic reactive groups increases in the order of: chloropyrimidine < monochloro-s-triazine < dichloroquinoxaline < dichloro-s-triazine.

Nucleophilic substitution at a saturated carbon is demonstrated in β-substituted ethylamine derivatives. The reaction proceeds via an ethyleneimine intermediate that exhibits high reactivity towards nucleophilic substitution reactions. For cellulosic fiber, the leaving group at the β-position is generally a sulfato group. The neighboring group participation of the nitrogen is essential for the easy formation of the three-member ethyleneimine ring (21).

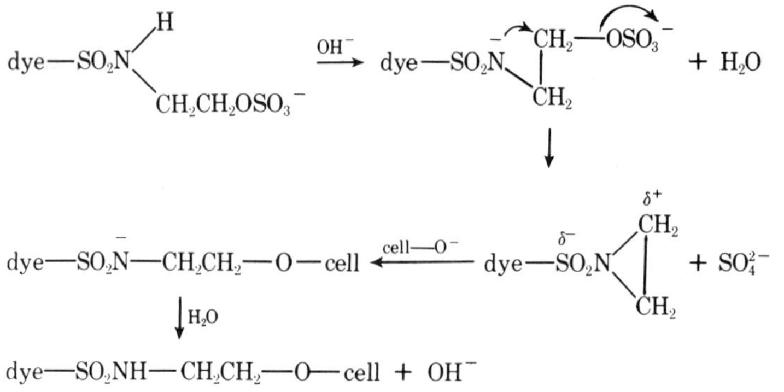

382 DYES, REACTIVE

$$\text{dye}-SO_2N\begin{array}{c}H\\CH_2CH_2OSO_3^-\end{array} \xrightarrow{OH^-} \text{dye}-SO_2N\begin{array}{c}^-CH_2-OSO_3^-\\|\\CH_2\end{array} + H_2O$$

$$\text{dye}-SO_2\overset{-}{N}-CH_2CH_2-O-\text{cell} \xleftarrow{\text{cell}-O^-} \text{dye}-SO_2N\begin{array}{c}\overset{\delta+}{CH_2}\\\overset{\delta-}{|}\\CH_2\end{array} + SO_4^{2-}$$

$$\downarrow H_2O$$

$$\text{dye}-SO_2NH-CH_2CH_2-O-\text{cell} + OH^-$$

Reactive Systems Based on Nucleophilic Addition. The mechanism of this reaction generally consists of two steps. The first step is a base-catalyzed elimination of a labile group (eg, —OSO₃H and Cl) generating a reactive double bond, followed by a 1,2 trans addition of the functional group of the cellulosic fiber to the double bond. Two commercially important groups of dyes react according to this mechanism: (1) Remazol, a vinyl sulfone derivative (dye—SO₂CH=CH₂) produced by Farbwerke Hoechst (10); and (2) Primazin, an acrylamide derivative (dye—NH—CO—CH=CH₂) produced by BASF (54) (see Acrylamide). Venkataraman and co-workers (76) have illustrated the reaction mechanism of cellulose with β-sulfatoethyl sulfone groups, the precursor of vinyl sulfone (Remazol), according to the following scheme.

$$\text{dye}-SO_2CH_2CH_2OSO_3Na + NaOH \rightarrow \text{dye}-SO_2CH=CH_2 + Na_2SO_4 + H_2O$$

$$\text{dye}-SO_2CH=CH_2 + \text{cell}-OH \xrightarrow{OH^-} \text{dye}-SO_2CH_2CH_2-O-\text{cell}$$

The electron-withdrawing nature of the bridge sulfonyl group activates the α-methylene hydrogen, thus facilitating the 1,2 elimination by a base. This property has the intrinsic disadvantage that the dye–fiber bond formed in dyeings can also be easily cleaved on subsequent aftertreatments. In fact, dyeings of vinyl sulfone dyes are less fast to boiling dilute sodium carbonate solution than are dyeings of chlorotriazinyl dyes. By substituting a vinyl sulfonamide group (—NH—SO₂CH=CH₂) at the bridge, the activating effect is reduced and the dyeings exhibit better alkali fastness (54).

By analogy, β-chloropropionylamides are the precursors of acrylamide. A base elimination of hydrogen chloride provides the reactive acrylamide form. However, this type of dyes (Primazin) has only limited use in dyeing cellulosic fibers and is mainly used for wool dyeing.

Reactive Systems Based on Both Nucleophilic Addition and Substitution. Several patents have been issued to Farbwerke Hoechst covering dyes with 2,2,3,3-tetrafluorocyclobutane-1-carboxamide [383-76-6] groups. The fluorine atoms on C-2 are activated by the carbonyl group on C-1 resulting in high reactivity of such dyes. The probable course of the reaction is shown below (14):

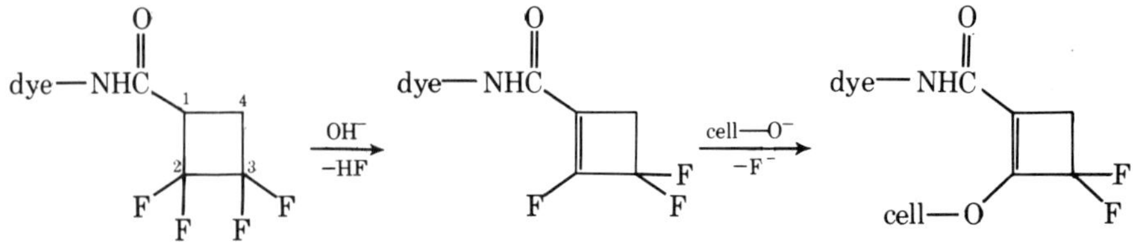

The first step is the base-catalyzed elimination of HF followed by nucleophilic substitution of the fluorine atom remaining on the double-bonded carbon by cellulose anions.

Dyes that React with Fibers Under Acid Conditions. All the reactive dyes mentioned above are applied to cellulosic fiber under alkaline conditions. The Calcobond dyes introduced by American Cyanamid Company are applied under acidic conditions. The reaction system is a methylolated nitrogen. The dye–fiber reaction is promoted by acid and heat and is believed to proceed according to the following mechanism (37):

$$\text{dye—NHCH}_2\text{OH} + \text{H}_3\text{O}^+ \rightleftarrows \text{dye—NHCH}_2^+ + 2\,\text{H}_2\text{O}$$

$$\text{dye—NHCH}_2^+ + \text{cell—OH} + \text{H}_2\text{O} \rightleftarrows \text{dye—NHCH}_2\text{—O—cell} + \text{H}_3\text{O}^+$$

Many dyes containing methylol groups attached to the chromophore in different ways are described in the relevant patent literature (77). The advantages mentioned for this type of dyes are: (1) they can be used together with disperse dyes, which are generally unstable in alkaline medium, to dye cotton and polyester blends; (2) the reactive system is similar to many textile-finishing agents and, therefore, can be applied together in a single operation. However, these dyes have failed to achieve much practical significance in dyeing of cellulosic fibers because they are incompatible with conventional reactive dyes and the cotton will lose up to 50% in tensile strength under the high temperature acidic fixation step (see also Amino resins).

Polyfunctional Fixing Agents That Form Covalent Bonds with Both the Dyestuff and the Substrate. Another approach to reactive dyeing is the linking of dyestuff molecules containing nucleophilic groups (eg, NH$_2$ and OH) with the nucleophilic hydroxyl anion of cellulose by means of a colorless polyfunctional component capable of reacting with both dye and fiber. Many different compounds have been disclosed in the patent literature as bridging compounds. Most of these suffer the disadvantage of having too low an affinity for cellulosic fiber and are only suitable for pad-dyeing and printing purposes. The most important polyfunctional reactive system is 1,3,5-triacryloyl-hexahydro-s-triazine (11) (triacryloformal). This forms the basis of the Basazol range of dyes (BASF). Hensel and Lützel (22) have postulated the following reaction scheme for structures (11) and (12):

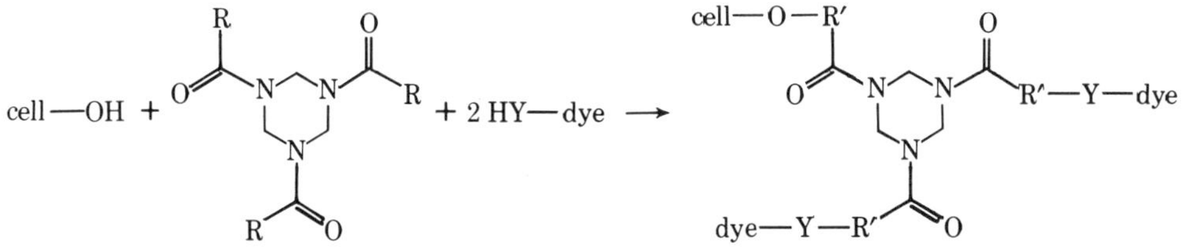

(11) R = CH=CH$_2$
[959-52-4]

(12) R = CH$_2$CH$_2$OH
[4002-62-4]

Y = nucleophilic group

R' = CH$_2$CH$_2$

The color yields of ordinary reactive dyes are usually not higher than 60–85% because of some hydrolysis in the dyebath. Color yields of about 90%, however, have been obtained with 1,3,5-triacryloformal (11) as the bridging compound. The reasons for this high yield are: (1) the carbonyl group has strong activating effect; and (2) contrary to the aromatic triazines, the activity of the substituents is not significantly altered after one or more of them has reacted. A good review of the general principle of dyestuff fixation by cross-linking agents has been presented (52).

Dyes Containing Several Reactive Groups. Conventional reactive dyes for cellulosic fibers suffer an obvious drawback in that only ca 70% of the dye is fixed onto the fiber, the remainder of the dye undergoes hydrolysis in the dyebath. There has been a continuous search for dyes that will exhibit high fixation efficiency. There are two possible methods to accomplish this goal: (1) use a reactive group that is not easily hydrolyzed; and (2) incorporate two or more reactive groups in a dye molecule to increase the chance of fixation. The latter method appears to have been the one mainly used in developing high fixation dyes. The high fixation dyes with multireactive groups are based upon statistical probability. That is, with two or more reactive groups there is better chance of fixation to the fiber. Furthermore, if one of the groups is hydrolyzed, the remaining groups may still react with the fiber. It must be stressed, however, that the presence of two or more reactive groups in a dye molecule does not necessarily provide high fixation. Other properties such as solubility, substantivity, and diffusibility are also determining factors.

The reactive groups of the dye can be linked as a reactive packet (17) with another group that does not carry a second chromophore (dye—NH—R_1—B—R_2; R_1 and R_2 = reactive groups; B = bridge). Another type of reactive group is a bridge that contains two chromophores (dye—NH—R_1—B—R_2—HN—dye).

Many patents have appeared covering dyes with two or more reactive groups. The favored group in such dyes is chlorotriazines (78). Two important groups of high fixation reactive dyes, Procion HE and Procion Supra, have achieved commercial importance. Procion Supra dyes (ICI) are intended primarily for use in textile printing applications but may also be employed in continuous dyeing processes.

Wool. The wool (qv) fiber contains many potential sites capable of forming covalent bonds with reactive dyes. Nevertheless, the incentive to invent reactive dyes for wool was not great because a wide variety of shades with high washfastness could be achieved with existing chrome dyes and milling acid dyes.

The reactive dyes first used on wool were designed primarily for application to cellulose. Difficulties were encountered with unlevel dyeing properties and hydrolysis of the dye during dyeing. These problems have now been largely overcome through development of reactive groups with little or no tendency to hydrolyze during the dyeing process, the use of special auxiliary products that minimize unlevel dyeing properties (53), and modification of the dyeing process.

In a review of the mechanism of reaction of proteins with reactive dyes, Shore (79) noted that the reaction with wool was more difficult to interpret than the reactive dye–cellulose systems (54,80) because of: (1) a greater variety of reactive groups in wool, the chemical reactions of which are complicated; (2) stronger and more localized interactions between the dye and fiber molecules; (3) an effect on dyeing behavior by the morphological structure of wool; and (4) susceptibility of the wool fiber degradation during dyeing. Many investigations during the last decade have also been directed towards the characterization of the main groups in the wool fiber that react with the dye (81–90). The reactive dye–wool systems can, however, be classified in the same general manner as the reactive dye–cellulose systems.

Reactive Systems Based on Nucleophilic Substitution. In an investigation of the rate of reaction of a dichlorotriazinyl dye with a series of simple model compounds containing the same functional groups as identifiable amino acids, the principal groups in proteins reacting with this particular dye were cysteine thiol groups, the primary amino groups of lysine and N-terminal amino acid residues, and the imidazole group of histidine (91). Several other studies (36,85,92) have also shown that one or more of these principal groups are capable of reaction with halogeno-heterocyclic dyes.

Several manufacturers have recently introduced reactive dyes based on 2,4-difluoro-5-chloropyrimidine, eg, Drimalan F (Sandoz), Reactolan (Ciba-Geigy), Verofix (Bayer), which exhibit a high degree of reactivity and resistance to hydrolysis. This property has been attributed to the capability of both fluorine atoms to react with nucleophiles (30).

Dyes containing chloroacetyl groups react with both the cysteine and histidine residues in wool (26). The reaction of a chloroacetyl dye with cysteine, histidine, and lysine depends upon the pH conditions employed (93–94). Several polymethine dyes (qv) containing N-chloroacetyl groups have been examined as fiber-reactive dyes for wool. The washfastness properties were improved significantly compared to the corresponding nonreactive dyes (95).

Reactive Systems Based on Nucleophilic Addition. Remalan dyes containing vinyl sulfone groups undergo considerable hydrolysis under neutral dyeing conditions (48). The Remazolan dyes were introduced to overcome this problem. They are applied to wool by a special method, which ensures complete reaction of the vinyl sulfone group with the fiber producing dyeings with excellent washfastness properties (47).

The Remazol dyes originally designed for cellulose can also be applied to wool (96–97). The unlevel dyeing properties of these conventional vinyl sulfone dyes led to the introduction of the Hostalan dyes (32–34). These dyes are prepared by reaction of vinyl sulfone dyes of the Remazol class with N-methyltaurine (4) under alkaline conditions. The reactive vinyl sulfone group is then gradually regenerated during the exhaustion dyeing process and level dyeing is attained by raising the dyebath temperature quickly to the boil. The reactions involved can be represented by the following scheme (98–99):

$$\text{dye}-SO_2CH=CH_2 + HN(CH_3)CH_2CH_2SO_3Na \xrightarrow{OH^-} \text{dye}-SO_2CH_2CH_2N(CH_3)CH_2CH_2SO_3Na \xrightarrow{H^+}$$

$$\text{dye}-SO_2CH_2CH_2\overset{+}{N}H(CH_3)CH_2CH_2SO_3^- \longrightarrow \text{dye}-SO_2CH=CH_2$$

Reactive dyes containing the acrylamide ($-NHCOCH=CH_2$) or β-chloropropionamide ($-NHCOCH_2CH_2Cl$) group exhibit lower reactivity for cellulose than the corresponding vinyl sulfone dyes. These dyes, therefore, can be applied advantageously to wool where a less reactive group is required to obtain adequate level dyeings. They also show practically no tendency to hydrolyze under normal wool dyeing conditions (26), unlike the triazine and vinyl sulfone types. This property is very desirable for satisfactory washfastness. The acrylamide reactive system forms the basis of the Procilan dyes (50) which exhibit good light- and washfastness properties, resembling those of the chrome dyes.

Reactive Systems Based on Both Nucleophilic Addition and Substitution. The Lanasol reactive dyes (Ciba-Geigy) containing α-bromoacrylamide groups were designed for application to wool. The chemistry of their reaction with wool has been detailed (51,100). Reaction of these dyes with the amino groups in wool proceeds through nucleophilic addition at the double bond, followed by elimination of HBr and formation of an aziridine compound through nucleophilic substitution. Subsequent hydrolysis and internal rearrangement, with ring opening, forms the stable compound as shown below:

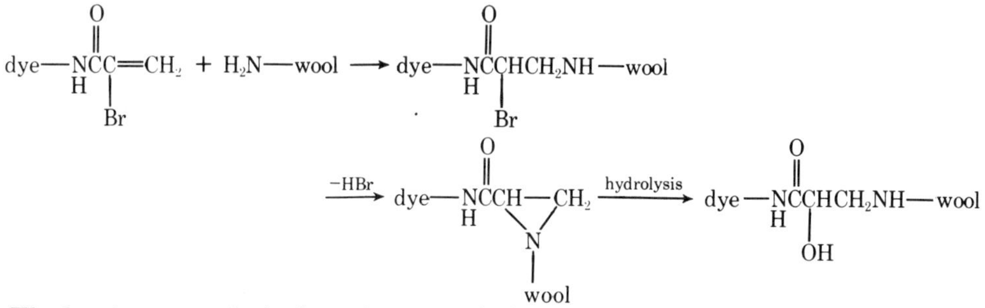

$$\text{dye—NHCOC(Br)=CH}_2 + \text{H}_2\text{N—wool} \rightarrow \text{dye—NHCOC(Br)HCH}_2\text{NH—wool}$$

$$\xrightarrow{-\text{HBr}} \text{dye—NHCOCH—CH}_2 \text{ (aziridine, N-wool)} \xrightarrow{\text{hydrolysis}} \text{dye—NHCOCH(OH)CH}_2\text{NH—wool}$$

The leaving group is the bromine atom (51) and the formation of the aziridine compound can also proceed by nucleophilic substitution of the bromine through the intermediate:

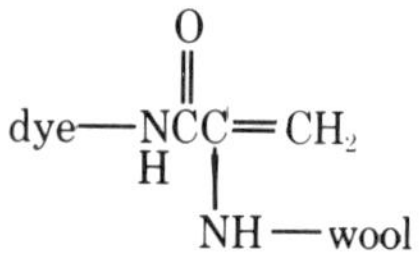

$$\text{dye—NHCOC(NH—wool)=CH}_2$$

Reactions Involving Disulfide Bonds. In recent years, dyes have been produced that, during or after application to the fiber, react covalently with themselves to form high molecular weight molecules.

Self-condensation dyes containing S-alkylthiosulfate (R—SSO$_3$H) and S-arylthiosulfate groups (Ar—S—SO$_3$H), known as Bunte salts (101), have been marketed as the Inthion (Hoechst) (102) and Dykolite (Southern Dyestuff Co.) (103–104) dyes, respectively, for application to cellulosic fibers.

An important member of the Inthion group, namely Wool Fast Turquoise Blue SW (—SO$_2$NHCH$_2$SSO$_3$H) (Hoechst) was shown to exhibit excellent wetfastness properties on wool (102). It has been suggested (105–106) that the Bunte salt dyes react with the cysteine thiol groups to form an asymmetrical disulfide, which then disproportionates to the insoluble disulfide dye as shown:

dye—SSO$_3$Na + wool—SH → dye—S—S—wool + NaHSO$_3$

2 (dye—S—S—wool) → dye—S—S—dye + wool—S—S—wool

Thus the improved wetfastness exhibited by these dyes could be due to the increased molecular size of the disulfide dye rather than covalent bond formation. It has also been suggested that dyes containing thiol groups form covalent bonds with wool (89,107–109).

Recently dyes containing β-isothioureidopropionamide (13) groups have been found suitable as reactive dyes for wool (110). A study of the mechanism of fixation on wool of several isothiouronium dyes, referred to as the Thiolan dyes, concluded

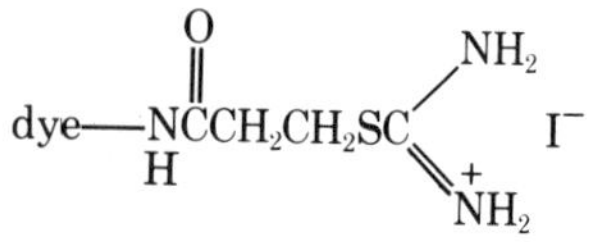

$$\text{dye—NHCOCH}_2\text{CH}_2\text{SC(NH}_2\text{)=NH}_2^+ \quad \text{I}^-$$

(13)

[56919-62-1]

that the high washfastness was due to insolubilization within the fiber rather than to mixed disulfide formation with cysteine (111). These dyes, therefore, can be regarded as condense dyes (112) rather than classical reactive dyes.

Reactions Involving Modified Wool. The introduction of shrink-resistant machine-washable wool necessitated the use of dyes that exhibit very high washfastness. This property could not be achieved satisfactorily using milling acid dyes, premetallized dyes or even chrome dyes. Fortunately, the newer reactive dyes for wool (eg, Lanasol, Drimalan, Verofix, Hostalan, etc) satisfied the stringent washfastness standards and, in addition, provided popular bright shades.

This modified wool is produced by a chlorine pretreatment followed by application of a resin to the fibers, for example the Chlorine-Hercosett process (113) (International Wool Secretariat), and the Dylan GRB and GRC processes (Precision Processes Textiles Ltd.).

The application of reactive dyes to Chlorine-Hercosett 57- (Hercules Powder Co.) treated wool has been studied in detail (99,114–116). Lewis and Seltzer (114) found that this resin-treated wool has a higher affinity for anionic dyes, owing to the high basicity of the resin, than either chlorinated or untreated wool.

Auxiliary leveling agents, such as Albegal B (Ciba-Geigy), Drimagen F (Sandoz) or Avolan RE (Bayer), which react with both the dye and the fiber, are recommended for use with the newer reactive dyes. These dyeing assistants reduce penetration of the dye into the fiber at low temperatures, but as the temperature is elevated, the auxiliary product–dye complex breaks down allowing the dye to react covalently with the fiber (31). The role of concentrated urea solutions in promoting the rate of reaction of reactive dyes with wool at low temperatures has been studied (117–120). Results of this work ultimately led to the development of a cold pad-batch process (121) which produces level, fast, bright dyeings on wool.

Reactive Dyes for Nylon. The level dyeing of nylon with high washfastness properties was initially unobtainable with conventional acid and disperse dyes, owing to variations in the physical and chemical characteristics of the fiber. The Procinyl reactive dyes were, therefore, developed. They are disperse dyes (devoid of solubilizing groups) containing a group capable of reaction with nylon. The patent literature indicates that the reactive group is not limited to one particular system and can include chlorotriazinyl, epoxy, 2-chloro-1-hydroxyethyl, β-chloroethylsulfamoyl, and aziridine groups. These dyes are stable in water and can be applied as conventional disperse dyes from weakly acid dyebaths. At the conclusion of dyeing, the dyebath is made alkaline in order to promote reaction of the dye and the fiber, resulting in a level dyeing with high fastness properties. Chemical combination of the Procinyl dyes with nylon has been demonstrated (24,122). It is suggested that reactivity is associated not only with the primary amino end groups of the fiber, but also with amide groups in the polymer chain. The possibility that dimerization or polymerization of the dye itself within the fiber has not been ruled out.

Recently the mechanism of fixation of disperse dyes containing a sulfonylazide ($-SO_2N_3$) or an arylazide (ArN_3) group on nylon has been examined (123). It was found that, under aqueous dyeing conditions, reaction occurs with the amino end groups of the polymer through a sulfonamide linkage. During heat-transfer printing, however, the azide group decomposes rapidly to a reactive nitrene species which undergoes insertion into carbon–hydrogen bonds of the polymer forming a nitrogen bridge (see Printing processes).

Analysis

The analysis of reactive dyes requires identification of the chromophore and the reactive system. The chromophore can be identified by the general methods adopted for any class of dyes (124). The identification of the reactive group after reaction with the fiber is a more difficult problem. The identification of reactive dyes on cellulosic fibers is described in refs. 125–126. These methods depend on the relative stability of the dye–fiber bond to acidic and alkaline hydrolysis. The extent of fixation of reactive dyes on wool fibers has been determined by solvent extraction techniques (127–130) (see Dyes, application and evaluation).

Dyestuffs, though not necessarily toxic, add organic carbon and color to waste water, which is objectionable to the public for esthetic reasons. The effective and economical removal of the various classes of dyes, including reactive dyes, from waste waters has been studied using techniques such as chemical coagulation or oxidation (131–132), activated carbon adsorption (133), polymeric adsorption/ion-exchange processes (134–135), high energy radiation (136), and electrolysis (137) (see Dyes and dye intermediates).

Economic Aspects

The annual production and sales figures for reactive dyes in the United States, compared to total dye production, over the period 1960 to 1976 are shown in Table 3.

A rapid growth of reactive dyes clearly occurred during the period 1960–1965 with production increasing by 545%. During the following six years growth continued, but at a slower rate of 234%. Reactive dye production has remained relatively constant since 1971 at approximately 1.2–1.4% of total dye production. A 29% decrease in production occurred in 1975, owing no doubt to the worldwide recession at that time. Sales of reactive dyes, however, increased by 27% during that year indicating that reduced production in the United States was counteracted by imports and overall reduction of inventories.

Table 3. Annual United States Production and Sales of Reactive Dyes [a]

Year	Production, t Total dyes	Fiber-reactive	% of total	Sales, t Total dyes	Fiber-reactive	% of total
1960	70,700	130	0.2	67,000	100	0.1
1965	94,000	720	0.8	86,200	710	0.8
1970	106,000	1,010	1.0	101,000	1,220	1.2
1971	111,000	1,680	1.5	105,000	1,590	1.5
1972	119,000	1,680	1.4	116,000	1,620	1.4
1973	129,000	1,680	1.3	121,000	1,560	1.3
1974	125,000	1,550	1.2	119,000	1,420	1.2
1975	93,500	1,100	1.2	94,700	1,400	1.5
1976	116,000	1,590	1.4	113,000	1,810	1.6

[a] Ref. 138.

In 1976 a total of 646 metric tons of reactive dyes was imported into the United States, representing 6.2% of the total dye imports. Approximately 97% of these imported reactive dyes were noncompetitive with reactive dyes manufactured in the United States.

In addition, reactive dyes have increased continuously in number since their inception in 1956. Currently, there are a total of 697 reactive dyes listed in the *Colour Index* (55). A breakdown by color is given in Table 4.

Reactive dyes have provided dyers with a full spectrum of shades. They possess high solubility, and good leveling power and are easy to apply. Reactive dyes are particularly compatible with environmental and effluent regulations because objectionable chemicals are not necessary for application. The low temperature method allows economical use of energy. However, the ideals of complete fixation and easy application represent the ultimate targets of dye makers.

Table 4. Reactive Dyes Listed in the *Colour Index*[a]

Color	1966	1971	1978
yellow	33	79	134
orange	25	60	86
red	54	115	176
violet	11	22	34
blue	41	101	175
green	0	18	23
brown	15	22	28
black	16	31	41
Total	*195*	*448*	*697*

[a] Ref. 55.

BIBLIOGRAPHY

"Dyes, Reactive" in *ECT* 2nd ed., Vol. 7, pp. 630–641, by D. W. Bannister, J. Elliott, and A. D. Olin, Toms River Chemical Corporation.

1. C. F. Cross and E. J. Bevan, *Res. Cellulose* **1**, 34 (1895).
2. G. Schroeter, *Ber.* **39**, 1559 (1906).
3. U.S. Pat. 1,567,731 (Dec. 29, 1925), F. Günther (to BASF).
4. D. H. Peacock, *J. Soc. Dyers Colour.* **42**, 53 (1926).
5. U.S. Pat. 1,886,482 (Nov. 8, 1932), R. Haller and A. Heckendorn (to Ciba); U.S. Pat. 2,025,660 (Dec. 24, 1935), R. Haller and A. Heckendorn (to Ciba).
6. J. D. Guthrie, *Am. Dyestuff Rep.* **41**, 13 (1952).
7. Brit. Pat. 341,461 (Jan. 16, 1931), (to I. G. Farbenindustrie).
8. Ger. Pat. 743,766 (June 8, 1944), H. Schweitzer and O. Bayer (to I. G. Farbenindustrie).
9. Brit. Pat. 587,467 (July 22, 19447), W. Müller and J. Scheidegger (to Ciba).
10. J. Heyna, *Angew. Chem. Int. Ed. Engl.* **2**, 20 (1963).
11. Brit. Pat. 712,037 (July 14, 1954), J. Heyna and W. Schumacher (to F. Hoechst); Brit. Pat. 733,471 (July 13, 1955), (to F. Hoechst); Brit. Pat. 740,533 (Nov. 16, 1955), (to F. Hoechst); Ger. Pat. 953,103 (Nov. 29, 1956), J. Heyna (to F. Hoechst).
12. J. Warren, J. D. Reid, and C. Hamalainen, *Text. Res. J.* **22**, 584 (1952).
13. W. E. Stephen, *Chimia* **19**, 261 (1965); Brit. Pats. 772,030 (Apr. 10, 1957), R. N. Heslop (to ICI); 774,925 (May 15, 1957), W. E. Stephen (to ICI); 781,930 (Aug. 28, 1957), R. N. Heslop (to ICI); 797,946 (July 9, 1958), F. R. Alsberg, and co-workers (to ICI); 798,121 (July 16, 1958), F. R. Alsberg, and co-workers (to ICI).

14. W. F. Beech, *Fibre—Reactive Dyes,* Logos Press, London, Eng., 1970, p. 13.
15. J. Wegmann, *J. Soc. Dyers Colour.* **76,** 205 (1960).
16. R. R. Davies, *Rev. Prog. Color. Relat. Top.* **3,** 73 (1972).
17. P. Rosenthal, *Rev. Prog. Color. Relat. Top.* **7,** 23 (1976).
18. H. Ackermann and P. Dussy, *Melliand Textilber.* **42,** 1167 (1961).
19. M. Capponi, *Am. Dyestuff Rep.* **53,** 913 (1964); Brit. Pat. 916,094 (Jan. 16, 1963), J. Benz and A. Schweizer (to Sandoz).
20. Brit. Pat. 1,120,761 (July 24, 1968), K. H. Schündehütte and K. Trautner (to Bayer); E. Siegel, *Chimia* **22** (Supplement), 100 (1968).
21. K. G. Kleb, E. Siegel, and K. Sasse, *Angew. Chem. Int. Ed. Engl.* **3,** 408 (1964).
22. H. R. Hensel and G. Lützel, *Angew. Chem. Int. Ed. Engl.* **4,** 312 (1965).
23. U. Baumgarte and F. Feichtmayr, *Melliand Textilber.* **44,** 267 (1963).
24. D. F. Scott and T. Vickerstaff, *J. Soc. Dyers Colour.* **76,** 104 (1960).
25. Ref. 14, p. 262.
26. A. N. Derbyshire and G. R. Tristram, *J. Soc. Dyers Colour.* **81,** 584 (1965).
27. *Ciba Rev.* **2,** 46 (1967); **4,** 41 (1967); **2,** 39 (1968).
28. D. Haigh, *Int. Dyer* **144,** 116 (1970).
29. Brit. Pats. 1,169,254 (Nov. 5, 1969), H. S. Bien and E. Klauke (to Bayer); 1,208,553 (Oct. 14, 1970), K. Gerlach (to Bayer).
30. D. Hildebrand and G. Meier, *Text. Prax.* **26,** 499, 577 (1971).
31. H. Egli, *Textilveredlung* **8,** 495 (1973).
32. F. Osterloh, *Text. Prax.* **26,** 164 (1971).
33. H. U. Von der Eltz, *Textilveredlung* **7,** 297 (1972).
34. Brit. Pat. 1,361,206 (July 24, 1974), (to F. Hoechst).
35. *Ciba Rev.* **6,** 50 (1963).
36. R. S. Asquith and D. K. Chan, *J. Soc. Dyers Colour.* **87,** 181 (1971).
37. D. Hildebrand, K-H. Schündehütte, and E. Siegel in K. Venkataraman, ed., *The Chemistry of Synthetic Dyes,* Vol. 6, Academic Press, Inc., New York and London, Eng., 1972.
38. P. Rys and H. Zollinger in C. L. Bird and W. S. Boston, eds., *The Theory of Coloration of Textiles,* The Dyers Company Publications Trust, White Rose Press Ltd., Mexborough and London, Eng., 1975, p. 326.
39. I. D. Rattee, *J. Soc. Dyers Colour.* **85,** 23 (1969).
40. R. L. M. Allen, *Colour Chemistry,* Appleton-Century-Crafts, New York, 1971, p. 201.
41. P. J. Dolby, *Text. Chem. Color.* **9,** 264 (1977); *Am. Dyestuff Rep.* **65**(3), 22 (1976).
42. P. Rys and H. Zollinger, *Text. Chem. Color.* **6,** 62 (1974).
43. N. Kollodzeiski, *Melliand Textilber.* **45,** 51 (1964).
44. H. Zollinger, *Rev. Text. Prog. U.S. Ed.* **11,** 218 (1959).
45. *Int. Dyer* **144,** 125 (1970).
46. F. Jones, *Rev. Text. Prog. London Ed.* **14,** 297 (1962).
47. F. Osterloh, *Melliand Textilber.* **49,** 1444 (1968).
48. F. Osterloh, *Melliand Textilber.* **41,** 1533 (1960).
49. Brit. Pat. 947,647 (Jan. 22, 1964), A. Blackhall (to ICI); Brit. Pat. 966,803 (Aug. 19, 1964), A. Blackhall (to ICI).
50. *Int. Dyer* **130,** 891 (1963); **131,** 31 (1964); **138,** 367 (1967); **144,** 116 (1970).
51. D. Mäusezahl, *Textilveredlung* **5,** 839 (1970).
52. J. Wegmann, *Melliand Textilber.* **49,** 687 (1968).
53. D. M. Lewis, *Wool Sci. Rev.* **49,** 13 (1974); **50,** 22 (1974).
54. O. A. Stamm, *J. Soc. Dyers Colour.* **80,** 416 (1960).
55. *Colour Index,* 3rd ed., Vol. 3, The Society of Dyers and Colourists, Bradford, Eng., and The American Association of Textile Chemists and Colorists, North Carolina, 1971, p. 3391; Vol. 5, 1975, p. 5249; Vol. 6, 1975, p. 6265.
56. E. N. Abrahart and co-workers, *Rep. Prog. Appl. Chem.* **56,** 141 (1971); **57,** 99 (1972); **58,** 113 (1973); E. N. Abrahart, *Dyes and their Intermediates,* 2nd ed., Chemical Publishing, New York, 1977, p. 185.
57. I. D. Rattee and M. M. Breuer, *The Physical Chemistry of Dye Adsorption,* Academic Press, Inc., New York and London, Eng., 1974, p. 224.
58. P. Rys and H. Zollinger, *Fundamentals of the Chemistry and Application of Dyes,* Wiley-Interscience, New York and London, Eng., 1972, p. 62.

59. J. A. Medley and K. L. Gardner, *Rev. Prog. Color. Relat. Top.* **3**, 67 (1972).
60. R. H. Peters, *Textile Chemistry,* Vol. 3, Elsevier Scientific Publishing Co., Inc., New York, 1975, p. 581.
61. Ger. Pats. 1,112,225 (Aug. 3, 1961), A. Fasciati, and co-workers, (to Ciba); 2,349,752 (Apr. 11, 1974), G. Heggar and G. Back (to Ciba-Geigy).
62. U.S. Pat. 3,847,894 (Nov. 23, 1972), D. R. A. Ridyard (to ICI); Brit. Pat. 1,368,824 (Oct. 2, 1974), P. W. Austin and D. S. Leitch (to ICI).
63. U.S. Pat. 2,892,829 (June 30, 1959), W. E. Stephen (to ICI).
64. Brit. Pats. 1,090,297 (Nov. 8, 1967), H. F. Andrew (to ICI); 1,138,833 (Jan. 1, 1969), P. A. Mack and R. Price (to ICI).
65. Ger. Pat. 1,148,341 (May 9, 1963), H. R. Schwander, J. P. Jung, and P. Hindermann (to Geigy).
66. Ger. Pat. 1,233,521 (Feb. 2, 1967), V. D. Poole (to ICI).
67. Brit. Pats. 1,191,741 (May 13, 1970) (to Geigy); 1,219,383 (Jan. 13, 1971) (to Geigy).
68. E. Bohnert and R. Weingarten, *Melliand Textilber.* **40**, 1036 (1959).
69. T. Vickerstaff, *J. Soc. Dyers Colour.* **73**, 237 (1957).
70. J. R. Aspland, A. Johnson, and R. H. Peters, *J. Soc. Dyers Colour.* **78**, 453 (1962); W. Ingamells, H. H. Sumner, and G. Williams, *J. Soc. Dyers Colour.* **78**, 454 (1962).
71. H. H. Sumner and T. Vickerstaff, *Melliand Textilber.* **42**, 1161 (1961).
72. D. Hildebrand, *Bayer Farben Rev.* (9), 29 (1964).
73. Brit. Pat. 891,601 (May 14, 1962), R. N. Heslop, R. Price, and V. D. Poole (to ICI).
74. P. Dussy, *Colourage* **17**, 31 (1970).
75. Brit. Pat. 1,051,684 (Dec. 14, 1966), H. Ackermann, H. Frei, and H. Meindl (to Geigy).
76. K. Venkataraman and co-workers, *Text. Res. J.* **40**, 392 (1970).
77. Brit. Pats. 922,403 (Apr. 3, 1963), (to American Cyanamid Co.); 1,006,131 (Sept. 29, 1965), (to American Cyanamid Co.); 1,100,599 (Jan. 24, 1968), (to American Cyanamid Co.).
78. Brit. Pat. 1,205,016 (Sept. 9, 1970), A. Crabtree and G. Griffiths (to ICI); Brit. Pat. 1,220,823 (Jan. 27, 1971), H. F. Andrew and G. Griffiths (to ICI); Brit. Pat. 1,231,411 (Sept. 4, 1969), R. N. Heslop (to ICI); U.S. Pat. 3,351,594 (Nov. 7, 1967), P. F. Clark and V. D. Poole (to ICI).
79. J. Shore, *J. Soc. Dyers Colour.* **84**, 408 (1968).
80. C. Preston and A. S. Fern, *Chimia* **15**, 117 (1961).
81. A. Schöberl, *Proc. Int. Wool Text. Res. Conf. Paris, (CIRTEL)* **3**, 301 (1965).
82. U. Baumgarte, *Melliand Textilber.* **43**, 1297 (1962).
83. I. W. Stapleton, *Text. J. Aust.* **43**, 61 (1968).
84. D. Hildebrand, *Text. Prax.* **25**, 292, 351, 428, 621 (1970).
85. E. Hille, *Text. Prax.* **17**, 171 (1962).
86. H. Zahn and P. F. Rouette, *Textilveredlung* **3**, 241 (1968).
87. J. F. Corbett, *Proc. Int. Wool Text. Res. Conf. Paris (CIRTEL)* **3**, 321 (1965).
88. H. Baumann, *Textilveredlung* **5**, 506 (1970); **8**, 145 (1973).
89. T. J. Abbott and co-workers, *J. Soc. Dyers Colour.* **91**, 133 (1975).
90. G. Reinert and co-workers, *Melliand Textilber.* **49**, 1313 (1968).
91. J. Shore, *J. Soc. Dyers Colour.* **84**, 413, 545 (1968).
92. A. D. Wirnik and M. A. Chekalin, *Text. Prax.* **17**, 577 (1962).
93. D. M. Lewis, I. D. Rattee, and C. B. Stevens, *Proc. Int. Wool Text. Res. Conf. Paris (CIRTEL)* **3**, 305 (1965).
94. D. M. Lewis, Ph.D. thesis, Leeds University, 1966.
95. D. J. Gale and J. F. K. Wilshire, *J. Soc. Dyers Colour.* **90**, 97 (1974).
96. H. U. Von der Eltz, *Text. Chem. Color.* **2**, 126 (1970).
97. H. U. Von der Eltz, *Melliand Textilber.* **52**, 687 (1971).
98. Farbwerke Hoechst, *Int. Dyer* **154**, 77 (1975).
99. R. R. D. Holt, *J. Soc. Dyers Colour.* **91**, 38 (1975).
100. D. Mosimann, *Text. Chem. Color.* **1**, 282 (1969).
101. C. D. Weston in K. Venkataraman, ed., *The Chemistry of Synthetic Dyes,* Vol. 7, Academic Press, Inc., New York, 1974, p. 35.
102. H. Luttringhaus, *Am. Dyestuff Rep.* **53**, 728 (1964).
103. C. D. Weston and W. S. Griffith, *Text. Chem. Colorist* **1**, 462 (1969).
104. U.S. Pat. 3,420,615 (Jan. 7, 1969), W. S. Griffith (to Martin-Marietta Corp.).
105. B. Milligan and J. M. Swan, *Text. Res. J.* **31**, 18 (1961).

106. F. Osterloh, *Milliand Textilber.* **44,** 57 (1963).
107. A. Robson and R. S. Stringer, *J. Soc. Dyers Colour.* **88,** 27 (1972).
108. R. S. Asquith and A. Puri, *J. Soc. Dyers Colour.* **87,** 116 (1971).
109. R. S. Asquith, P. Carthew, and T. T. Francis, *J. Soc. Dyers Colour.* **89,** 168 (1973).
110. G. B. Guise and I. W. Stapleton, *J. Soc. Dyers Colour.* **91,** 223 (1975).
111. G. B. Guise and I. W. Stapleton, *J. Soc. Dyers Colour.* **91,** 259 (1975).
112. Ref. 55, p. 3705.
113. H. D. Feldtman, J. R. McPhee, and W. V. Morgan, *Text. Manuf.* **93,** 122 (1967).
114. D. M. Lewis and I. Seltzer, *J. Soc. Dyers Colour.* **88,** 93 (1972).
115. K. R. F. Cockett and D. M. Lewis, *J. Soc. Dyers Colour.* **92,** 141 (1976).
116. E. Bellhouse, *J. Soc. Dyers Colour.* **91,** 33 (1975).
117. D. M. Lewis and I. Seltzer, *J. Soc. Dyers Colour.* **84,** 501 (1968).
118. K. R. F. Cockett, I. D. Rattee, and C. B. Stevens, *J. Soc. Dyers Colour.* **85,** 461 (1969).
119. R. S. Asquith and A. K. Booth, *Text. Res. J.* **40,** 410 (1970).
120. A. K. Gilchrist and I. D. Rattee, *Text. Chem. Color.* **5,** 105 (1973); **6,** 153 (1974).
121. J. D. M. Gibson, D. M. Lewis, and I. Seltzer, *J. Soc. Dyers Colour.* **86,** 298 (1970).
122. E. Elöd and U. Einsele, *Melliand Textilber.* **41,** 1377 (1960).
123. J. Griffiths and R. I. McDarmaid, *J. Soc. Dyers Colour.* **93,** 455 (1977); **94,** 65 (1978).
124. E. Clayton, *Identification of Dyes on Textile Fibres*, 2nd ed., The Society of Dyers and Colourists, Bradford, Eng., 1963.
125. A. Bode, *Melliand Textilber.* **40,** 1304 (1959).
126. F. Jordinson and R. Lockwood, *J. Soc. Dyers Colour.* **78,** 122 (1962); **80,** 202 (1964); **84,** 205 (1968); **86,** 524 (1970); **88,** 117 (1972); **90,** 23, 55 (1974).
127. K. R. F. Cockett, I. D. Rattee, and C. B. Stevens, *J. Soc. Dyers Colour.* **85,** 113 (1969).
128. J. R. Christoe and A. Datyner, *J. Soc. Dyers Colour.* **86,** 210 (1970).
129. R. S. Asquith, W-F. Kwok, and M. S. Otterburn, *Text. Res. J.* **48,** 1 (1978).
130. H. K. Rouette and co-workers, *Text. Res. J.* **41,** 518 (1971).
131. C. A. Rodman, *Text. Chem. Color.* **3,** 239 (1971).
132. R. H. Horning, *Text. Chem. Color.* **9,** 73 (1977).
133. P. B. Dejohn and R. A. Hutchins, *Text. Chem. Color.* **8,** 69 (1976).
134. S. L. Rock and B. W. Stevens, *Text. Chem. Color.* **7,** 169 (1975).
135. R. A. Montanaro and A. H. Noble, *The Textile Industry and the Environment—1973*, American Association of Textile Chemists and Colorists, North Carolina, 1973, p. 129.
136. F. N. Case, E. E. Ketchen, and T. A. Alspaugh, *Text. Chem. Color* **5,** 62 (1973).
137. U.S. Pat. 3,485,729 (Dec. 23, 1969), G. Hertz (to Crompton and Knowles Corp.).
138. *Synthetic Organic Chemicals—U.S. Production and Sales, 1960–1976*, U.S. Government Printing Office, Washington, D.C., 1961–1977; *Imports of Benzenoid Chemicals and Products—1976, U.S. International Trade Commission Publication 828*, U.S. Government Printing Office, Washington, D.C., Aug. 1977.

JOHN ELLIOTT
PATRICK P. YEUNG
Toms River Chemical Corporation

DYES, SENSITIZING

Spectral sensitizing dyes extend the wavelengths of light to which inorganic semiconductors (qv), organic semiconductors, and chemical reactions can be photosensitized. Spectral sensitizers are needed for the blue, green, and red portions of the visible spectrum as well as for the infrared (see Fig. 1). The sensitizing dyes can be ionic or nonionic, heterocyclic or nonheterocyclic, and in many cases exhibit more than one type of sensitization reaction. For example, spectral sensitization of silver halides is most common by conduction-band electron generation (n-type photoconduction), and trapped positive holes contribute less to the photoconductivity. However, changes in sensitizer structure can provide dyes that will sensitize with the photohole as the primary carrier (p-type photoconduction), and the photoelectrons are effectively trapped by the dyes (see Digital displays; Light-emitting diodes).

The detection of spectral sensitizing action often depends on amplification methods such as photographic or electrophotographic development or, alternatively, chemical analysis of reaction products. Separation of the photosensitization reaction from the detection step or the chemical reaction allows selection of the most effective spectral sensitizers. Prime considerations for spectral sensitizing dyes include: (1) the range of wavelengths needed for sensitization and (2) the absolute efficiency of the spectrally sensitized process. Because both sensitization wavelength and efficiency are important, optimum sensitizers vary considerably in their structures and properties.

Sensitization Wavelength and Efficiency

The wavelengths for which useful spectral sensitization can be achieved are best illustrated by the spectral sensitivity of commercial photographic plates and films. As shown in Figure 2, spectral sensitivity can be extended beyond the natural ultra-

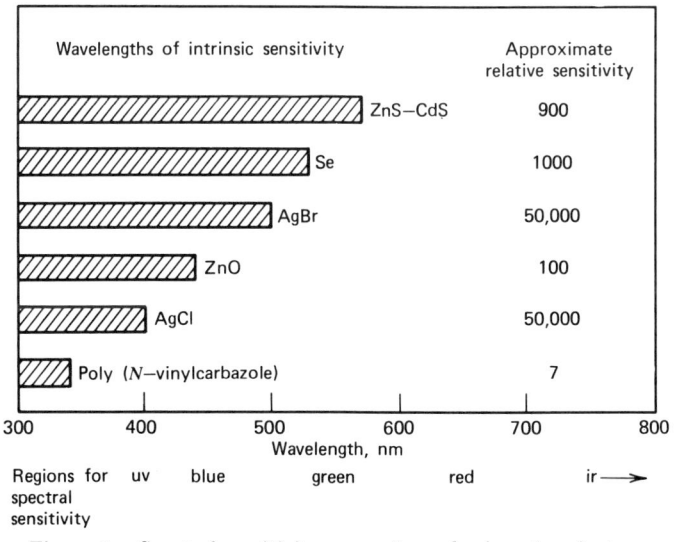

Figure 1. Spectral sensitivity ranges for undyed semiconductors.

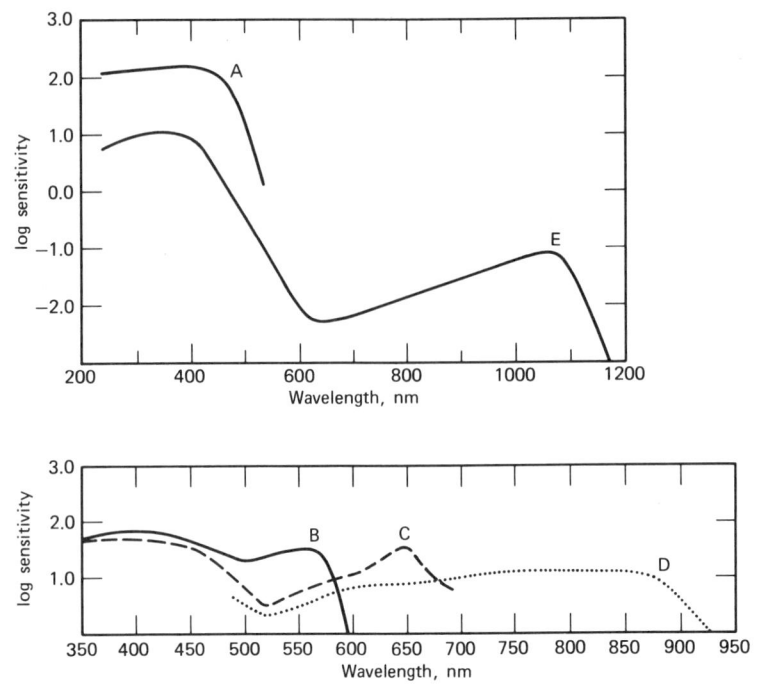

Figure 2. Spectral sensitivity curves for dyed silver halide plates and films.

violet and blue-light photosensitivity of silver bromide crystals (Fig. 2, A) to include green light (Fig. 2, B), red light (Fig. 2, C), and infrared wavelengths (Fig. 2, D and E). Some commercial products, like the Kodak spectroscopic film, type 103-H alpha, are sensitized to have a maximum sensitivity at specific wavelengths, such as the 645-nm line emitted by the sun's hydrogen atoms. Color films sensitive in the visible or infrared regions depend on three spectrally sensitized layers such as B, C, and D in the same film, but the separate amplification reaction produces the individual colored image dyes in each layer.

Undyed silver halides are intrinsically sensitive in the ultraviolet and blue regions of the spectrum (Fig. 1) and, unless competitive absorption occurs, dye-sensitized silver halides are sensitive to these same spectral regions as well as to the wavelengths of light absorbed by the dye. However, the absolute sensitivities in both the intrinsic (ultraviolet-blue) and spectral (dye-absorption) regions are affected by the chemical properties of the dyes. For example, easily oxidized dyes can increase the absolute intrinsic sensitivity of small silver halide grains (50–250 nm dia) by chemically trapping photoholes (photohole = absence of electron; dye + photohole = oxidized dye), thereby reducing inefficiency caused by photohole/photoelectron recombination. Dyes that are too easily reduced can decrease the absolute sensitivities of many silver halide films by being chemical traps for photoelectrons. As shown by the examples in Figure 2, even commercial products that employ the best possible spectral sensitizers for each wavelength region exhibit more than a tenfold variation in maximum sensitivity for the intrinsic (ultraviolet-blue) response of the silver halide. In cases of severe dye-induced sensitivity loss such as with the Kodak spectroscopic plate I-Z (Fig. 2, E), preexposure treatments termed hypersensitization (eg, bathing in dilute ammonia solution) are required to attain reasonable absolute sensitivities.

Structural Classes of Spectral Sensitizers

A useful classification of sensitizing dyes is the one adopted to describe patents in image technology. In Table 1, ITPAIS (Image Technology Patent Information System) dye classes are described and representative patent citations in the nearly complete ITPAIS file are listed as a function of major dye class. From these citations it is clear that preferred sensitizers for silver halides are polymethine dyes (cyanine, merocyanine, etc), whereas other semiconductors have more evenly distributed citations. Zinc oxide, for example, is frequently sensitized by xanthene (qv) or triarylmethane dyes (see Triphenylmethane dyes) as well as cyanines and merocyanines.

Spectral Sensitization of Silver Halides

The large number of patents for spectral sensitizers of silver halides indicates their extensive use in many photographic plates, films, and papers. Color films that give good color reproduction under a variety of illuminants (daylight, electronic flash, tungsten) and exposure times (short for electronic flash, long for available light) have required more extensive tailoring of spectral sensitizers than imaging systems based on other semiconductors. In addition, photographic products are often subjected to wide variations in temperature, humidity, and time before processing. Consequently, high priorities are also given to spectral sensitizers that do not degrade either the film or image quality, as well as provide efficient spectral sensitization at the desired wavelengths.

Important trends in the properties required of spectral sensitizers, as a function of general structural changes, are outlined below.

Spectral sensitizers (Fig. 3) used in practice may be illustrated using a few basic chromophores in the cyanine and merocyanine dye classes. The spectral sensitizers designated BN, GN, and RN spectrally sensitize the photoelectrons as primary carriers in silver halides. These are all essentially symmetrical cyanine and carbocyanine chromophores that can be adsorbed to commercial silver halides as ordered, narrow-absorbing arrays of molecules. The more efficient infrared sensitizers are derived from synmetrical dyes: IRN ($n = 1-3$). The spectral sensitizers like RP spectrally sensitize photoholes as primary image-forming carriers in silver halides. The merocyanine sensitizer MN and its derivatives ($n = 0$ and 2) sensitize by photoelectron production in silver halides.

The efficiency of spectral sensitization in photographic film (silver halide/gelatin) is a function of a dye's electrochemical reduction potential for spectral generation of photoelectron carriers and oxidation potential for spectral generation of photohole carriers. The relationships are diagrammed in Figure 4. Dyes with very negative reduction potentials are difficult to reduce and provide efficient spectral sensitizers where latent image is formed by photoelectron carriers. Dyes with high positive oxidation potentials are difficult to oxidize and provide efficient spectral sensitizers where latent image is formed by photohole carriers. In Figure 4 the symbols for the typical sensitizing chromophores (BN, GN, RP, BP, etc) shown in Figure 3 are included on the abscissa to indicate the usual electrochemical potential ranges exhibited by these various classes of dyes. Major changes in these potentials and the efficiency of spectral sensitization result from significant structural modifications of the heterocyclic groups in the dyes. Spectral sensitization by other classes of dyes on silver halides or other

Table 1. Spectral Sensitization and Dye Classification A. Patent Citations ITPAIS File[a]

ITPAIS classification	Silver halide	Supersensitization	ZnO	TiO$_2$	ZnS	Se	CdS, CdSe	Misc. inorganic semiconductor	Misc. organic semiconductors	Photodye-forming polymer or photopolymer
dyes, cyanine[b]	481	189	15	3	5	6	1	10	7	19
dyes, merocyanine[b]	184	35	17	2	2	2		3	7	16
all other polymethine dyes[b]	153	84	10	5	2	4		3	6	19
dyes, acridine	1		1	2	1	2			1	2
dyes, azine[c]			8	3	2	2			2	10
dyes, azo	2		2	1		3		1	5	
dyes, arylmethane	3		21	6	5	7		2	4	11
dyes, quinone type[d]	2		4	2	1	2		1	3	5
dyes, porphine[e]	1		5	4	1	6		1	3	4
dyes, xanthene[f]	8		32	11	8	9			7	13
dyes, pyrylium[g]	1								4	3

[a] ITPAIS, the Image Technology Patent Information System, developed and maintained by Eastman Kodak Company, Agfa-Gevaert (Antwerp/Leverkusen), and Fuji Photo Film Co., Ltd., encompasses selected patents and literature references related principally to the chemical aspects of image technology. The present file (Edition 12) contains approximately 20,000 indexed patents. Many literature references are now found in ITPAIS, selected from basic texts and reference works because the cited technology relates to specific patents that have been indexed. In addition, literature references are routinely selected for indexing from Research Disclosures. The present file contains more than 1,300 literature references. Search terms used for this table were very restrictive.
[b] Dyes, polymethine: used for dyes having at least one electron donor and one electron acceptor group linked by methine groups or aza analogues; allopolar cyanine, dye bases, complex cyanine, hemicyanine, merocyanine, oxonol, streptocyanine, styryl.
[c] Dyes, azine: used for azines, thiazines, oxazines.
[d] Dyes, quinone type: used for anthraquinones, indamines, indoanilines, indophenols, and miscellaneous quinones.
[e] Dyes, porphine derivative: used for chlorophyll, phthalocyanines, and hemin.
[f] Dyes, xanthene: used for eosin, fluorescein-type phthaleins, rhodamines, rose bengal.
[g] Dyes, pyrylium: also used for thiapyrylium and benzothiapyrylium.

Blue sensitizers (400–500 nm)

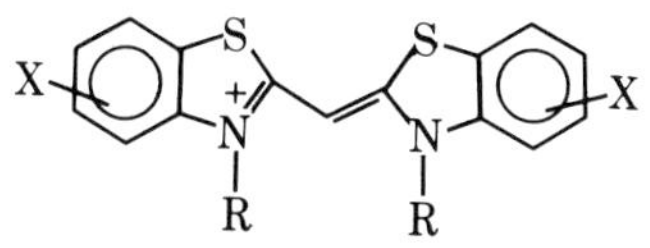

BN

Green sensitizers (500–600 nm)

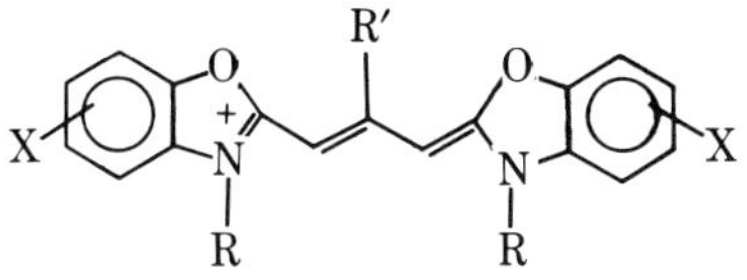

GN

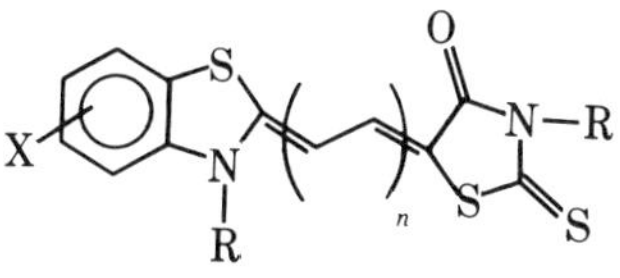

MN $n = 1$

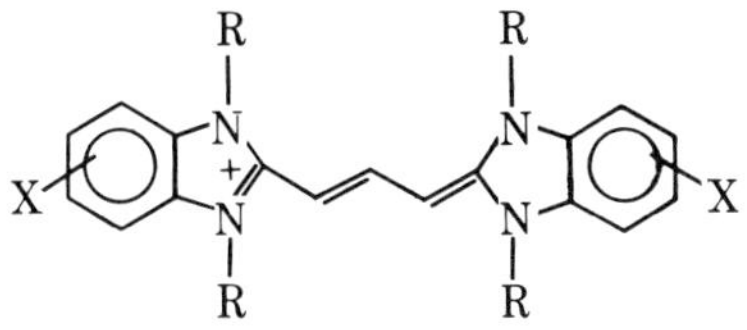

GN

Red sensitizers (600–700 nm)

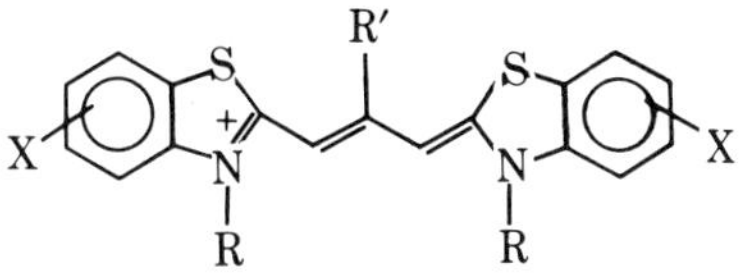

RN

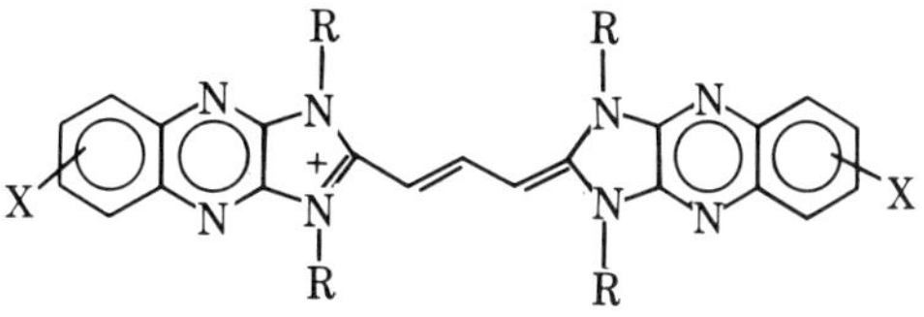

RP

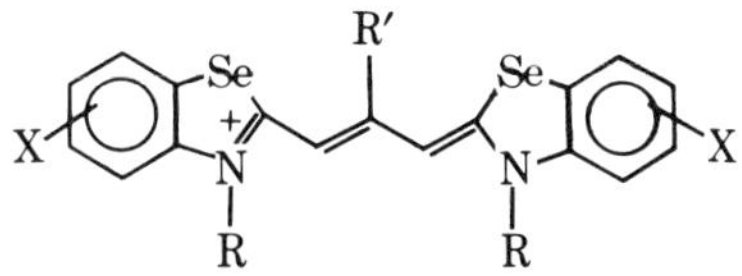

RN

Infrared sensitizers (>700 nm)

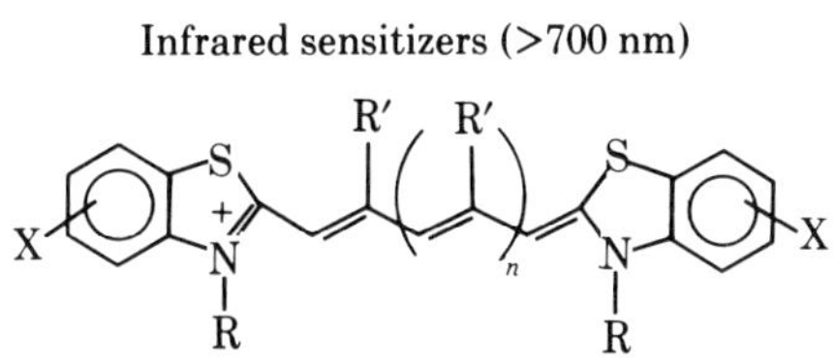

IRN $n = 1\text{-}3$

Figure 3. Spectral sensitizers.

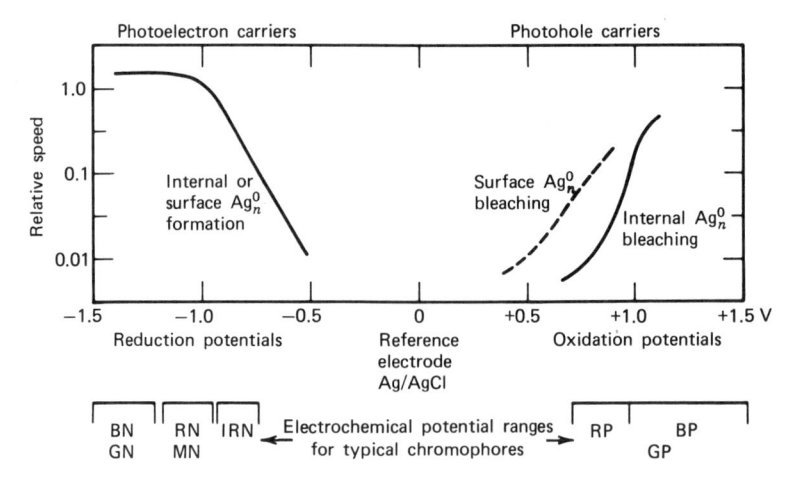

Figure 4. Spectral sensitization of silver bromides.

solid semiconductors may also be outlined in terms of relative energy levels defined by reduction or oxidation potentials (1–2). Further tailoring of the dye structures (to provide additional improvements in efficiency, spectral response at specific wavelengths, stronger adsorption, and improved solubility) is primarily accomplished by substituent modification. Substituents like methoxy, chloro, and phenyl on the aromatic rings in BN, GN, and RN shift both the solution absorption and spectral sensitization to longer wavelengths and reinforce the dyes' tendencies to form ordered, sharp-absorbing J-aggregates. In the carbocyanines from benzoxazole (GN) and benzothiazole-benzoselenazole (RN), simple out-of-plane substituents like ethyl or phenyl at position R' strongly increase J-aggregation.

Modification of dye solubility by substituent changes affects migration of spectral sensitizers between layers in a multilayer film and dye removal by photographic processing solutions. These changes are primarily in the substituents attached to heterocyclic nitrogen atoms. The structural types of these groups have been summarized in a review of the patent literature (3). Typical groups have included: ethyl and other alkyls; sulfoalkyl-like—$CH_2CH_2CH_2SO_3^-$; unsaturated groups (phenyl, allyl, —$CH_2CH=CHSO_3^-$); carboxyalkyl, phosphonoalkyl, and sulfatoalkyl-like —$CH_2CH_2CH_2OSO_3^-$. Merocyanines are typically insoluble in hydroxylic solvents, but soluble in pyridine. As an example, shown below are the solubilities for a carbocyanine chromophore where the first sulfopropyl group creates a less soluble zwitterionic dye and the second leads to a much more soluble anionic species.

	CAS Registry No.	Methanol solubility, mmol/L
R = R' = ethyl(bromide)	[18426-56-7]	9.5
R = ethyl, R' = sulfopropyl	[51859-32-6]	3.6
R = R' = sulfopropyl(H$^+$)	[23568-98-1]	25.5

Aggregation (self association) of spectral sensitizers into ordered arrays is of prime importance in silver halide films (2). The tendency of dyes to aggregate in ordered arrays is not only a function of the type of substituent but also the length of the chromophore and the surface of the solid substrate (4). H-aggregation broadens the spectral response of most sensitizers (relative to that for monomeric dyes). Panchromatic films can thus require only a single sensitizer if both the monomer and H-aggregates sensitize well. Infrared dyes like IRN (Fig. 3) absorb to silver halides in many forms and exhibit spectral responses over a broad range of wavelengths, owing to the presence of *trans*-monomeric dye, isomeric (*cis*) dye forms, and H-aggregates at much shorter wavelengths than the *trans*-monomeric dye. J-aggregation, on the other hand, leads to absorption bands at longer wavelength and narrower than the monomer (see comparison, Fig. 5b) and highly wavelength-selective spectral sensitization. The J-aggregates were first noted by Jelly, and the geometric arrangements of molecules needed for J-aggregate absorption are known for cyanine dyes (see Cyanine dyes). Since some J-aggregates are inefficient spectral sensitizers with low relative quantum yields for spectral sensitization, supersensitization of their response by small amounts of added compounds has been extensively studied (2,5–6). To improve the relative quantum yields, supersensitizers with suitable oxidation potentials (see Fig. 5a) are added at about one-tenth the concentration of J-aggregating dye. These supersensitizers separate the exciton in the aggregate by trapping the positive hole in the supersensitizer and freeing the photoelectron for sensitization.

Many lengthy reviews and several books cover specific aspects of spectral sensitizing dyes for silver halides. These include: synthetic methods (7–8), general sensitization mechanisms (1–2,9–10), electrochemical potentials and dye efficiency

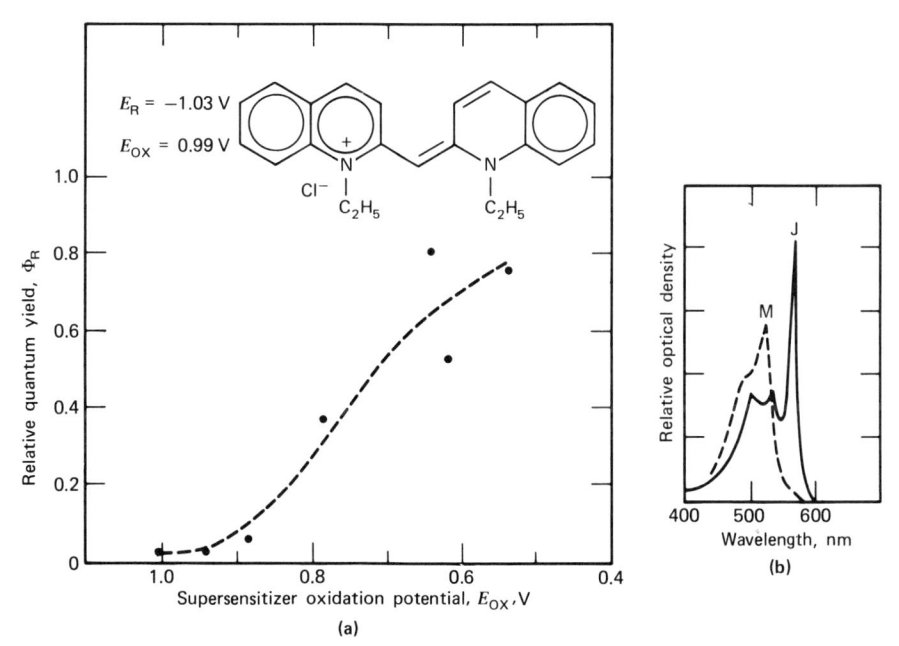

Figure 5. A J-aggregated spectral sensitizing dye. Structure in (**a**) is 1,1-diethyl-2,2-quinocyanine chloride [*2402-42-8*].

Table 2. Spectral Sensitizers for Inorganic and Organic Semiconductors[a]

Semiconductor	Sensitizer	CAS Registry No.
CdS (Cu doped)	thiacarbocyanine	[905-97-5]
	benzothiazolylrhodanines ($n = 1$)	
	rhodamine B	[81-88-9]
ZnO	thiacarbocyanine	
	erythrosin	[16423-68-0]
	eosin	[17372-87-1]
	phthalocyanine	[574-93-6]
	rhodamine B	
	rose bengal	[11121-48-5]
	methylene blue	[61-73-4]
	fluorescein	[2321-07-5]
TiO$_2$	benzothiazolylstyryl dye	
	thiacarbocyanine	
anthracene	rhodamine B	
poly(N-vinylcarbazole)	rose bengal	
	2,4,7-trinitrofluorenone	[129-79-3]
	pyrylium carbocyanines	
solid dye particles	phthalocyanines	
(photoelectrophoresis)	thioindigo	[522-75-8]
	flavanthrone	[475-71-8]
	quinacridone	[1047-16-1]
dye electrodes	benzothiazolylrhodanines ($n = 1$)	
(dopants = o-chloranil, tetracyanoethylene, iodine)	phthalocyanines	

Structural formulas for spectral sensitizers

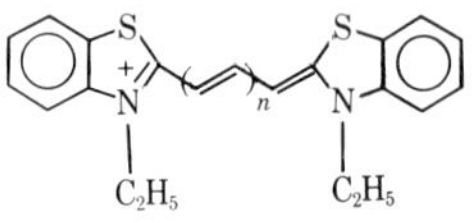

vinylogous thiacyanines
[thiacarbocyanine, $n = 1$]

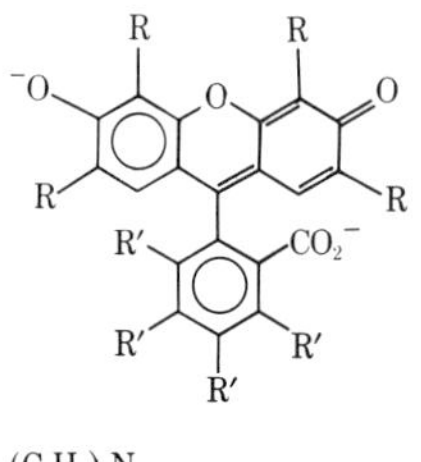

$R = R' = H$, fluorescein
$R = Br, R' = H$, eosin
$R = Br, R' = Cl$, phloxin [55280-65-4]
$R = I, R' = H$, erythrosin
$R = I, R' = Cl$, rose bengal

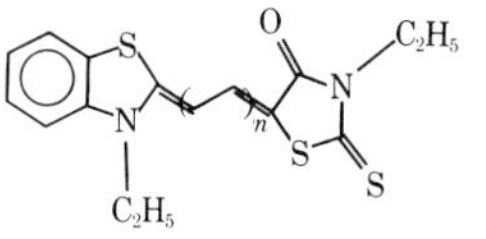

benzothiazolylrhodanines
[merocarbocyanine, $n = 1$
[3568-36-3]]

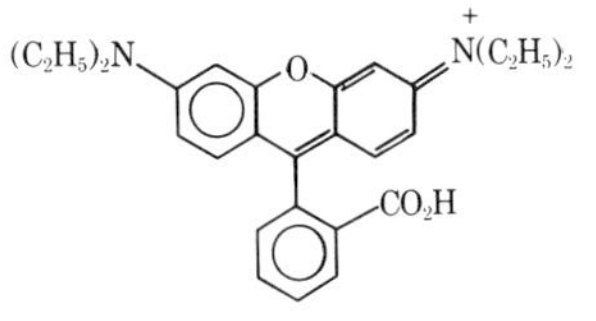

rhodamine B

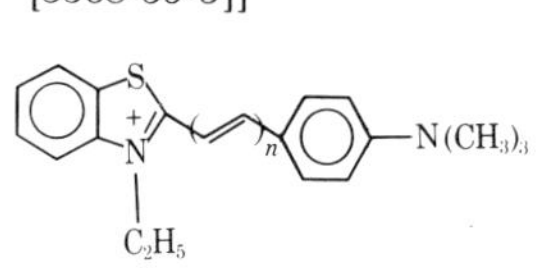

benzothiazolyl styryl ($n = 1$)
[3028-97-5]
(butadienyl, $n = 2$) [17818-85-8]

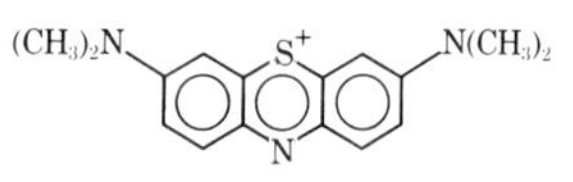

methylene blue

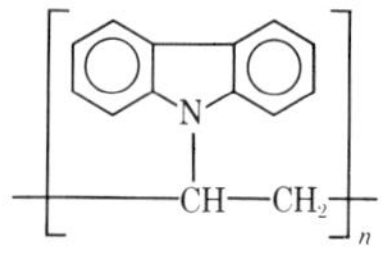

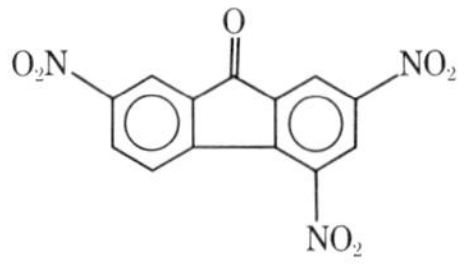

poly(N-vinylcarbazole):2,4,7-trinitrofluorenone
[39613-12-2]

Table 2. (continued)

Semiconductor	Sensitizer	CAS Registry No.
Cu-phthalocyanine [147-14-8]	flavanthrone	thioindigo

[a] Ref. 18.

(2,11–12), dye adsorption and aggregation (4), and supersensitization of J-aggregated (5–6) or monomeric dyes (13). For specific dye syntheses and properties of the major classes of silver halide spectral sensitizers, cyanine and related dyes, see Cyanine dyes.

Spectral Sensitization of Inorganic and Organic Solids

Many semiconductors other than silver halides exhibit spectrally sensitized photoconductivity. Inorganic semiconductors include selenium, germanium, CdS, HgO, HgI_2, ZnO, PbO, Cu_2O, thallium halides, and TiO_2. Reviews of the primary photoconductive mechanisms in these solids are available (14–17). As noted by the number of patent citations (Table 1) and the published literature, zinc oxide has been extensively investigated. Spectrally sensitized photoconduction has also been observed for organic semiconductors: anthracene, poly(N-vinylcarbazole), polyacetylene, copper phenylacetylide, phthalocyanines, and other solid dyes (14). Typical sensitizers for several semiconductors are listed in Table 2.

Zinc oxides can be spectrally sensitized by a variety of sensitizing dyes. The Dember effect of pressed ZnO powder was demonstrated early for cyanine dyes, eosin, erythrosin, and rhodamine B (19), and other dyes were also spectral sensitizers (20–22). The spectral sensitization is favored by strong dye adsorption (23–25). Commercial ZnO electrophotographic papers are reported to be sensitized with bromophenol blue, auramin, sodium fluorescein, rose bengal, and eosin (25) (see Electrophotography). Mechanistic studies dealing with dye-sensitized zinc oxide layers or crystals have used some of these sensitizers (rose bengal, eosin) (25–27). In addition, sensitization was observed for cyanine dyes (28), erythrosin (26), dihydroxy azo dyes (25), rhodamine B (25), and methylene blue (25). Adsorption of sensitizers like rhodamine B and dihydroxyazo dyes results from zinc salt formation with either chromophoric groups or nonchromophoric, charged substituents. Azo dyes without strong salt-forming ability (eg, 4,4′-azodiphenol) are not effective sensitizers (25). Halide ions significantly increase the spectral sensitization by rhodamine B (25). Electronegative organic molecules (eg, quinone) also function as cosensitizers for dyed ZnO (25) in analogy to the supersensitization of dyed silver halides.

Spectral sensitization of titanium dioxide is best accomplished by the cationic cyanine dyes shown in Figure 6. Dyed TiO_2 electrode systems have potential solar cell use and were successfully adapted for the photochemical decomposition of water to hydrogen and oxygen (29) (see Photovoltaic cells). Typical sensitizers include the vinylogous thiacyanines, styryl (and butadienyl) dyes, and complex cyanines (see Cyanine dyes). Nonionized merocyanine dyes and many zinc oxide sensitizers (rhodamine B, eosin, erythrosin) are not good sensitizers for TiO$_2$ emulsions (30). However, substitution of the cyanine, styryl, merocyanine and other chromophores with polyhydroxy-containing substituent groups improves the binding of the dyes to TiO$_2$ and their spectral sensitization (30). Coverage of the oxide surface with metal ions also improves the effectiveness of certain anionic dyes (29). Thin-film TiO$_2$ electrodes were sensitized by spectral sensitizers plus supersensitizers (29). For example, detectable photocurrents for oxacarbocyanine were improved by hydroquinone, and rhodamine B was made to sensitize using added 4,5-dihydroxy-m-benzenedisulfonic acid. Spectrally sensitized hydrogen generation was achieved in the region 500–630 nm using the TiO$_2$ films plus supersensitizers and rhodamine B (29).

Cadmium sulfide can be spectrally sensitized by cyanine, styryl, merocyanine, and xanthene dyes (28,31–33); examples are shown in Figure 7. In contrast to zinc oxide, which is sensitized by many dyes which desensitize silver halides, cadmium sulfide energy levels require dyes that are also good sensitizers for silver halide. In a comparison of dye energy levels for various photoconductors, vinylogous thiacyanines were noted to sensitize ZnO, AgBr, and CdS, but the silver halide desensitizer methylene blue sensitized only zinc oxide (28).

Dye-sensitized poly(N-vinylcarbazole) (PVK) has been investigated with respect to quantum efficiency of photohole generation by triarylmethane, pyrylium, cyanine,

	n = 1, styryl		n = 2, butadienyl	
		Wavelength		Wavelength
X	CAS Reg. No.	range, nm	CAS Reg. No.	range, nm
S	[3028-97-5]	420–670	[17818-85-8]	430–730
C(CH$_3$)$_2$	[60575-38-4]	420–650	[27996-72-1]	480–750
CH=CH	[117-92-0]	420–680	[50401-03-1]	440–650

Vinylogous thiacyanines

	CAS Reg. No.	Wavelength range, nm
n = 0	[2197-01-5]	380–480
n = 1	[905-97-5]	470–620
n = 2	[514-73-8]	560–710

Complex carbocyanines

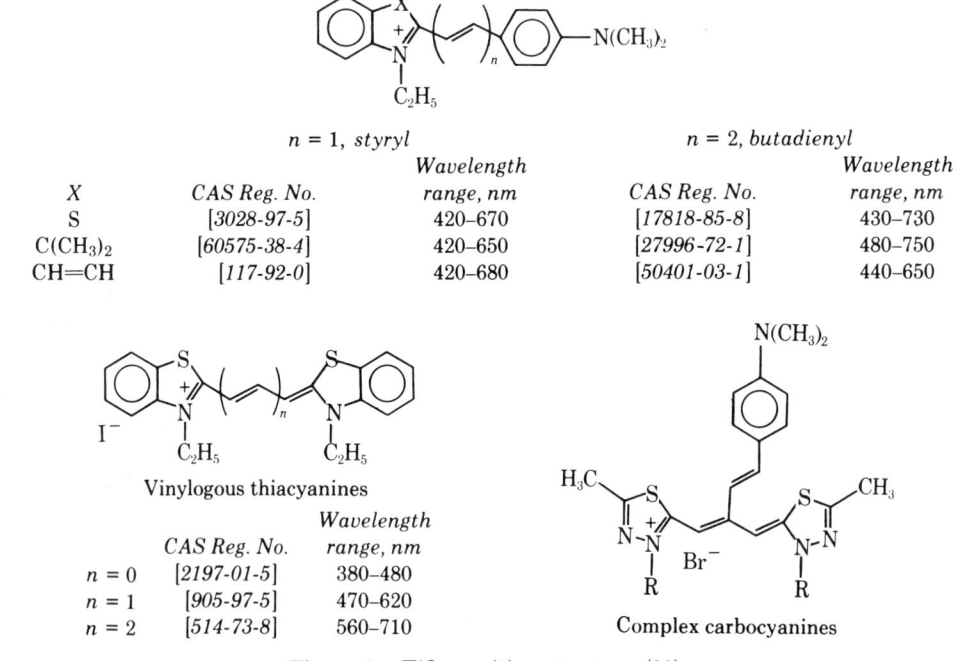

Figure 6. TiO$_2$ sensitizer structures (30).

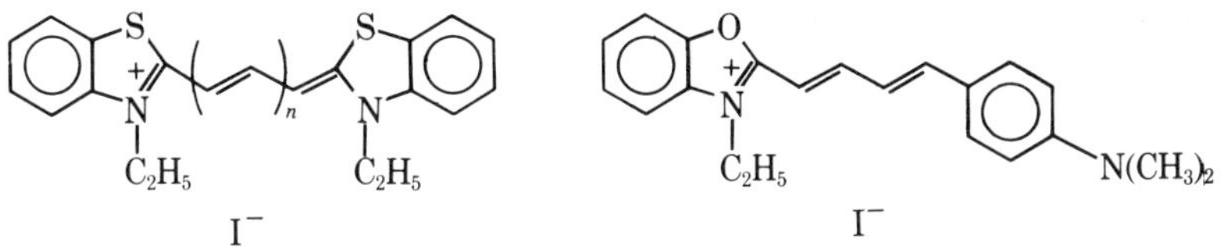

Cyanine	Carbocyanine		Butadienyl
n = 0	n = 1	λ_{max} photoconductivity	[69943-53-9]
[2197-01-5]	[905-97-5]	relative sensitivity	680 nm
530 nm	600 nm		6
40	62		

Figure 7. Examples of spectral sensitizers for Cu-doped CdS (31).

and carbocyanine dyes (34–36). The spectral sensitivity of a 1:1 poly(N-vinylcarbazole): 2,4,7-trinitrofluorenone mixture (Table 2) extends into the infrared (23–24, 36). As noted for the pyrylium sensitizers (34), the quantum yield shows a significant dependence on field strength.

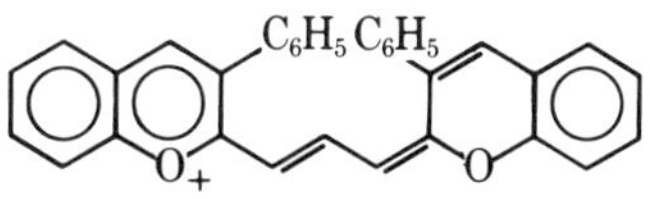

ClO$_4^-$ [42906-23-0]
pyrylium spectral sensitizer
for poly(N-vinylcarbazole), Φ = 0.33

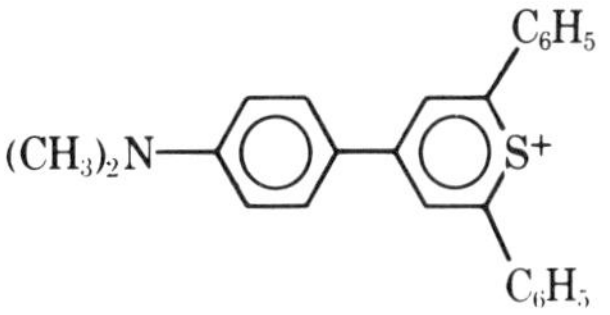

ClO$_4^-$ [14039-00-0]
or BF$_4^-$ [25966-12-5]
thiapyrylium spectral sensitizer
for poly(4,4'-isopropylidene-
diphenyl carbonate) cocrystalline
complex with the thiapyrylium dye

Arylalkane polycarbonates mixed with many organic photoconductors can be spectrally sensitized by thiapyrylium and other dyes, particularly when dye/polycarbonate cocrystalline complexes can be formed (35). Relative speeds are improved by vapor-treating the electrophotographic films before exposure to cause either H-aggregation or J-aggregation of the spectral sensitizers. For the 4-(dimethylaminophenyl)diphenylthiapyrylium perchlorate shown, the non-vapor-treated film shows a relative speed of 40 at 580 nm, whereas the vapor-treated coating gives a speed of 1800 at 685 nm (cocrystalline, aggregated dye).

Solid organic dyes may serve as photosensitive particles in electrophotographic systems (37–39) or as solid, spectrally absorbing electrodes in other research (16,40) or industrial applications (41). The solid photosensitive particles (Table 2) change electrical charge upon exposure and migrate to form an image in a process termed photoelectrophoresis (39). Since the effective spectral sensitivity of each particle coincides with its absorption, dye particles of different colors may be used to generate a colored image (37–39). More conventional uses of dye semiconductor properties include photoresistance cells, photoelectric cells, and spectrally sensitive detectors. The photoconductivity of dyes for potential application in these systems is significantly improved by doping the solid dyes (Table 2) (41).

Spectral Sensitization of Photoresists, Photopolymers, and Photopolymerization

Photoreactive polymers (qv) may be classed as negative working or positive working (42–43). Poly(vinyl cinnamate) (PVCN) is a commonly used negative photoresist which is spectrally sensitized to wavelengths longer than its intrinsic absorption (315 nm). Triplet sensitizers such as Michler's ketone, anthraquinone, and p-nitroaniline will photo-cross-link PVCN with similar quantum yields (44). N-Methyl-2-benzoyl-β-naphtholthiazoline (BN) is also cited as a triplet sensitizer, but the substituted pyrylium sensitizer TPP may cross-link by both triplet energy transfer and electron transfer mechanisms (43). Photo-cross-linkable polymers, derived from poly(1,2-dimethyl-5-vinylpyridinium methyl sulfate) or poly(1-methyl-4-vinylpyridinium p-toluenesulfonate), were spectrally sensitized efficiently in the region 270–540 nm by chromophores attached to the polymer backbone (45). Acrylamide was photopolymerized by decomposing diazonium compounds with azine dyes (methylene blue, azure A [8000-86-0]), acridine dyes (phenosafranine [81-93-6]), and xanthene dyes (rose bengal, pyronine G [92-32-0]) (46). The presence of an oxidizable activator like sodium p-toluenesulfinate was necessary for the spectral sensitization.

Poly(1,2-dimethyl-5-vinylpyridinium methyl sulfate) [27056-62-8] with chromophores (CH=CH—R) attached via the 2-methyl group

Certain positive-working photoresists depend on decomposition reactions of quinone diazides (44) or tetraarylborates (47). Triplet-state sensitizers do not sensitize the quinone diazide rearrangement to soluble carboxylate compounds (44). However, a wide variety of cationic dyes photosolubilize the tetraarylborate anion (47). Some of these are listed in Figure 8.

Dye-sensitized dichromated gelatins (48) have traditionally been treated with xanthene dyes and reducing agents to improve broadband exposure response. Recent studies have shown that thiazine dyes such as methylene blue and methylene green are better sensitizers. Other classes of dyes also sensitize: triphenylmethane (qv), azine (qv), indigoid (see Dyes, natural), disazo (see Azo dyes), and cyanine (qv) (in order of decreasing utility).

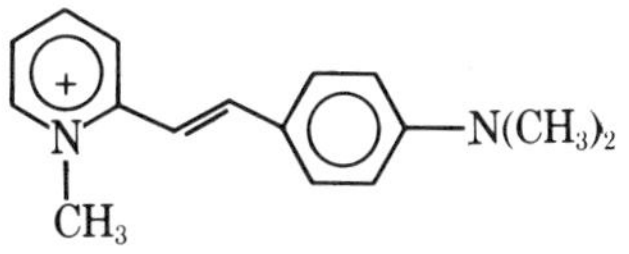

pyridinium styryl [2156-29-8]

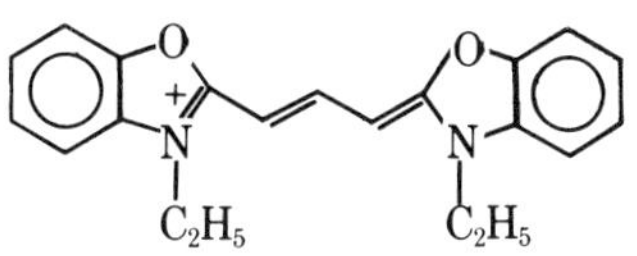

oxacarbocyanine [905-96-4]

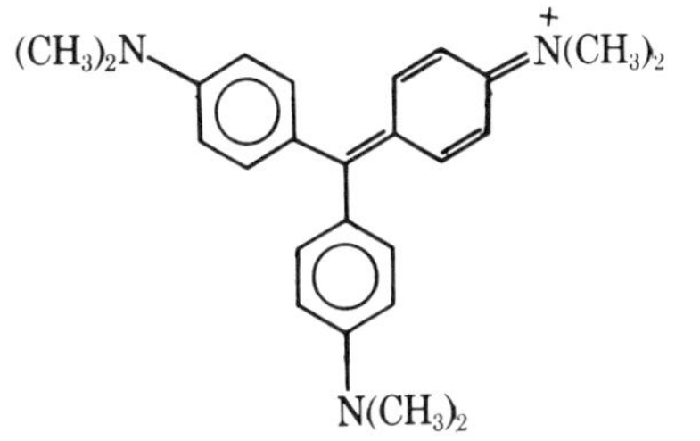

methyl violet [21325-96-2]

rhodamine B (Table 2)

methylene blue (Table 2)

Figure 8. Cations for tetraarylborates.

Dye-sensitized metal ion reductions in solution occur for those dyes that are photoreducible (49–51). Methylene blue is photoreduced to the leuco form by light plus ethylenediaminetetraacetic acid (electron donor). The leuco form acts to reduce metal ions. Ferric ion, eg, can be reduced in large amounts by dye sensitization, since metal ion reduction regenerates the dye. Other metals are also photoreducible: Ti(IV), Cu(II), Cr(VI)O$_4^{2-}$, Ag(I), Mo(VI)O$_4^{2-}$, Hg(II). Certain classes of dyes are effective: xanthenes such as fluorescein and halogenated derivatives, most thiazines (methylene blue, azure A), and certain 3,6-diaminoacridines. Azine dyes and azo dyes are not effective in solution photoreductions. Those dyes that can be photoreduced in solution through a long-lived excited state are sensitizers for photooxidation of *p*-diaminotoluene (52).

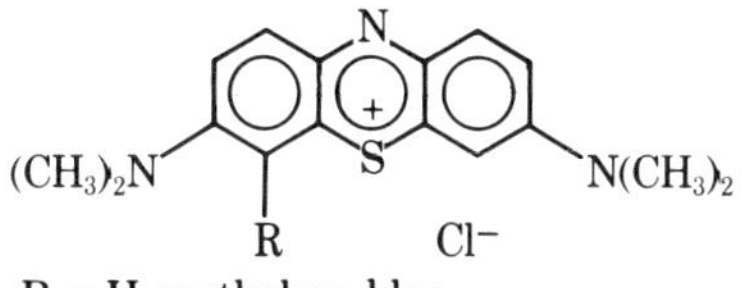

R = H, methylene blue
R = NO$_2$, methylene green [2679-01-8]

Dye-Sensitized Reactions in Solution: Singlet-Oxygen Generation and Photodynamic Reactions

Metastable singlet oxygen exists in two states $^1\Sigma_g$ and $^1\Delta_g$ which are 159 and 92 kJ (38 and 22 kcal) above the $^3\Sigma_g$ ground state, respectively (53–54). Because of the low energies involved, dyes of a wide variety of structures can sensitize the formation of singlet oxygen via the dye triplet state. The xanthene dyes (fluorescein derivatives, Table 2) are widely used as efficient sensitizers (55–56). Cyanine dyes (57), triphenylmethane dyes (56), porphins, and porphyrins containing transition metals are often poor sensitizers for these reactions. Triplet-sensitized photooxidation reactions are

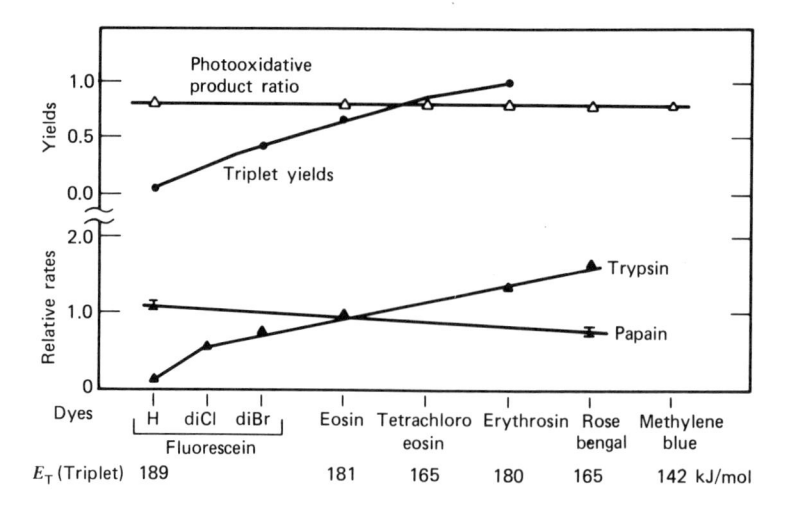

Figure 9. Photooxidation and photodynamic action. CAS Reg. No.: diCl [2320-96-9], diBr [31544-98-6], and tetrachloroeosin [26761-84-2]. Other CAS Reg. No. are shown in Table 2. To convert kJ to kcal, divide by 4.184.

classified as Type I or Type II (58–63). Experimental results that support a common species ($^1\Delta_g$) as the primary chemically reactive molecule in dye-sensitized photooxidations are shown in Figure 9.

Heterogeneous methods for producing singlet oxygen use sensitizers such as rose bengal (silica gel) (64), methylene blue (silica gel plus AlCl$_3$) (64), and rose bengal (polymer bound) (65). The polymer-bound rose bengal is marketed under the trade name Sensitox and is claimed to be as effective a sensitizer as solutions of rose bengal (65).

$$\text{H atom, or electron transfer} \xleftarrow[\text{substrate or solvent}]{\text{I}} {}^3\text{Dye} \xrightarrow[O_2]{\text{II}} {}^1O_2 + \text{Dye}$$
$$\searrow O_2^- + \text{Dye}^-$$

Sensitizing dyes, light, and oxygen can damage and inactivate virtually all classes of organisms through the photooxidation of proteins, polypeptides, individual aminoacids, lipids with allylic hydrogens, tocopherols (see Vitamins), sugars, and cellulose materials (54,66–69). This general type of photooxidation has been termed photodynamic action (68). Certain porphyrias in humans are characterized by a sensitivity to light through photosensitizing porphyrins deposited in the skin. Other sensitization reactions have medical value in the treatment of psoriasis and neonatal jaundice (70). The dependence of photodynamic action on dye structure is generally more marked than for simple singlet-oxygen reactions (63,71–72). Thiazine, porphyrin, and xanthene photosensitizers (methylene blue, rose bengal, phenosafranine, eosin, erythrosin, hematoporphyrin) may operate primarily through singlet oxygen (Type II reactions) in the photooxidation of methionine, whereas flavin and anthraquinone sensitizers tend to exhibit mainly Type I photooxidation (71). In a solid gelatin matrix where oxygen diffusion is limited, the photoinactivation of papain shows a very different dependence on xanthene dye structure than typical singlet-oxygen reactions (Fig. 9)

(63). High concentrations of amines (eg, $(C_2H_5)_3N$) in the presence of rose bengal can make Type I and Type II pathways competitive (72).

BIBLIOGRAPHY

1. H. Meier, *Spectral Sensitization,* Focal Press, New York, 1968.
2. W. West and P. B. Gilman in T. H. James, ed., *The Theory of the Photographic Process,* 4th ed., Macmillan, New York, 1977, p. 251; P. B. Gilman, *Pure Appl. Chem.* **49,** 357 (1977).
3. E. J. Poppe, *Z. Wiss. Photogr. Photophys. Photochem.* **63,** 149 (1969).
4. A. H. Herz, *Adv. Colloid Interface Sci.* **8,** 237 (1977); M. Kazmierezak and W. Markocki, *J. Signalaufzeichnungsmaterialien* **6,** 289 (1978).
5. W. West and P. B. Gilman, *Photogr. Sci. Eng.* **13,** 221 (1969).
6. J. E. Jones and P. B. Gilman, *Photogr. Sci. Eng.* **17,** 367 (1973); P. B. Gilman, *Photogr. Sci. Eng.* **18,** 418 (1974).
7. F. M. Hamer, *The Chemistry of Heterocyclic Compounds,* Vol. 18, A. Weissberger, ed., *The Cyanine Dyes and Related Compounds,* Wiley-Interscience, New York, 1964.
8. L. G. S. Brooker in C. E. K. Mees and T. H. James, eds., *The Theory of the Photographic Process,* 3rd ed., Macmillan, New York, 1966, p. 198.
9. R. C. Nelson, *J. Photogr. Sci.* **24,** 13 (1976); F. Dietz, *J. Signalaufzeichnungsmaterialien* **6,** 245 (1978).
10. H. Kuhn, D. Mobius, and H. Bucher, in A. Weissberger and B. W. Rossiter, eds., *Physical Methods of Chemistry,* Part IIIB, John Wiley & Sons, Inc., New York, 1972, Chapt. VIII.
11. R. W. Berriman and P. B. Gilman, *Photogr. Sci. Eng.* **17,** 235 (1973).
12. D. M. Sturmer, W. S. Gaugh, and B. J. Bruschi, *Photogr. Sci. Eng.* **18,** 49, 56 (1974).
13. O. Riester, *Proceedings of the 7th Conference on Scientific (Theoretical) and Applied Photography, Budapest, 1975; Photogr. Sci. Eng.* **18,** 295 (1974).
14. Ref. 1, pp. 101, 104, 111, 122.
15. H. Meier, *Photochem. Photobiol.* **16,** 219 (1972).
16. H. Gerischer, *Photochem. Photobiol.* **16,** 243 (1972).
17. R. M. Shaffert, *Electrophotography,* 2nd ed., John Wiley & Sons, Inc., New York, 1975, pp. 63–65, 351–358, 390.
18. H. Nakatsui and Y. Hishiki, *Photogr. Sci. Eng.* **22,** 18 (1978); R. H. Sprague and J. H. Keller, *Photogr. Sci. Eng.* **14,** 401 (1970); K. H. Hauffe, *Photogr. Sci. Eng.* **20,** 124 (1976); H. Meier, *Spectral Sensitization,* Focal Press, New York, 1968, pp. 104, 111, 122; R. M. Shaffert, *Electrophotography,* John Wiley & Sons, Inc., New York, 1975, pp. 63–65, 351–358, 390; V. Tulagin, *J. Opt. Soc. Am.* **59,** 328 (1969); H. Meier, W. Albrecht, and U. Tschirwitz, *Angew. Chem. Int. Ed.* **11,** 1051 (1972).
19. E. K. Putzieko and A. Terenin, *Zh. Fiz. Khim. SSSR* **23,** 676 (1949); *Chem. Abstr.* **43,** 7349 (1949).
20. C. J. Young and H. G. Greig, *RCA Rev.* **15,** 469 (1954).
21. H. Meier, *J. Phys. Chem.* **69,** 719 (1965).
22. A. Terenin and I. Akimov, *J. Phys. Chem.* **69,** 730 (1965).
23. H. Tsukahara, *Electrophotography (Japan)* **22,** 17 (1969).
24. W. F. Berg and K. Hauffe, eds., *Current Problems in Electrophotography,* Walter de Gruyter, Berlin, 1972, pp. 146, 178.
25. K. H. Hauffe, *Photogr. Sci. Eng.* **20,** 124 (1976).
26. W. Eckenbach, *J. Photogr. Sci.* **25,** 141 (1977).
27. L. I. Grossweiner and W. D. Brennan, *J. Photogr. Sci.* **17,** 189 (1969).
28. T. Tani, *Photogr. Sci. Eng.* **17,** 11 (1973).
29. P. D. Fleischauer and J. K. Allen, *J. Phys. Chem.* **82,** 432 (1978) and references therein.
30. R. H. Sprague and J. H. Keller, *Photogr. Sci. Eng.* **14,** 401 (1970); Ger. Offen. 1,912,102 (Sept. 18, 1969), P. B. Gilman and J. A. Haefner (to Eastman Kodak Co.).
31. H. Nakatsui and Y. Hishiki, *Photogr. Sci. Eng.* **22,** 18 (1978); **23,** 76 (1979).
32. J. W. Trusty and R. C. Nelson, *Photogr. Sci. Eng.* **16,** 421 (1972).
33. P. Yianoulis and R. C. Nelson, *Photogr. Sci. Eng.* **18,** 94 (1974); **22,** 268 (1978).
34. M. Ikeda and co-workers, *Photogr. Sci. Eng.* **19,** 60 (1975).
35. U.S. Pat. 3,615,414 (Oct. 26, 1971), W. A. Light (to Eastman Kodak Co.); P. Borsenberger and co-workers, *Third International Conference on Electrophotography,* Washington, D.C., November 15–18, 1977, p. 31.

36. K. Kriz, *Photogr. Sci. Eng.* **16,** 58 (1972); R. M. Schaffert, *IBM J. Res. Dev.* **15,** 75 (1971).
37. U.S. Pat. 2,940,847 (June 14, 1960), E. K. Kaprelian; U.S. Pat. 3,140,175 (July 7, 1964), E. K. Kaprelian; U.S. Pat. 3,143,508 (Aug. 4, 1964), E. K. Kaprelian.
38. U.S. Pat. 3,384,565 (May 21, 1968), V. Tulagin and L. M. Carreira (to Xerox Corp.); U.S. Pat. 3,384,566 (May 21, 1968), H. E. Clark (to Xerox Corp.).
39. V. Tulagin, *J. Opt. Soc. Am.* **59,** 328 (1969).
40. H. Meier, *Angew. Chem. Int. Ed.* **4,** 619 (1965).
41. H. Meier, W. Albrecht, and U. Tschirwitz, *Angew. Chem. Int. Ed.,* **11,** 1051 (1972).
42. J. L. R. Williams in E. Sélégny, ed., *Polyelectrolytes,* D. Reidel Publishing Co., Dordrecht-Holland, 1974, p. 507.
43. J. L. R. Williams and co-workers, *Pure Appl. Chem.* **49,** 523 (1977).
44. T. A. Shankoff and A. M. Trozzolo, *Photogr. Sci. Eng.* **19,** 173 (1975).
45. D. G. Borden and J. L. R. Williams, *Makromol. Chem.* **178,** 3035 (1977).
46. T. Yamase and co-workers, *Photogr. Sci. Eng.* **17,** 28, 268, (1973); **19,** 57 (1975).
47. D. G. Borden, *Photogr. Sci. Eng.* **16,** 300 (1972).
48. A. Graube, *Photogr. Sci. Eng.* **22,** 37 (1978).
49. G. K. Oster and G. Oster, *J. Am. Chem. Soc.* **81,** 5543 (1959).
50. F. Millich and G. Oster, *J. Am. Chem. Soc.* **81,** 1357 (1959).
51. U.S. Pat. 3,097,097 (July 9, 1963) G. K. Oster and G. Oster; D. F. Eaton, *Photogr. Sci. Eng.* **23,** 150 (1979).
52. G. Oster and co-workers, *J. Am. Chem. Soc.* **81,** 5095 (1959).
53. D. R. Kearns, *Chem. Rev.* **71,** 395 (1971).
54. C. S. Foote in W. A. Pryor, ed., *Free Radicals in Biology,* Vol. 2, Academic Press, New York, 1976, p. 85.
55. K. Gollnick and G. O. Schenck, *Pure Appl. Chem.* **9,** 507 (1964).
56. K. Gollnick, *Adv. Photochem.* **6,** 1 (1968).
57. G. W. Byers, S. Gross, and P. M. Henrichs, *Photochem. Photobiol.* **23,** 37 (1974).
58. K. Gollnick and co-workers, *Ann. N.Y. Acad. Sci.* **171,** 89 (1970).
59. V. Kasche and L. Lindquist, *J. Phys. Chem.* **68,** 817 (1964).
60. J. D. Spikes, *Radiat. Res.* 823 (1967).
61. D. R. Adams and F. Wilkinson, *J. Chem. Soc. Faraday Trans.* 2, 586 (1972).
62. B. Schnuriger and J. Bourdon, *Photochem. Photobiol.* **8,** 361 (1968).
63. J. Bourdon and M. Duranté, *Ann. N.Y. Acad. Sci.* **171,** 163 (1970).
64. R. Nilsson and D. R. Kearns, *Photochem. Photobiol.* **19,** 181 (1974).
65. D. C. Neckers, *Chem. Tech.* 108 (1978).
66. G. Laustriat and C. Hasselmann, *Photochem. Photobiol.* **22,** 295 (1975).
67. D. A. Lightner and C-S. Pak, *J. Org. Chem.* **40,** 2724 (1975); D. A. Lightner, *Photochem. Photobiol.* **26,** 427 (1977).
68. J. D. Spikes and B. W. Glad, *Photochem. Photobiol.* **3,** 471 (1964).
69. N. I. Krinsky, *Photophysiology* **3,** 123 (1968); M. M. Mathews-Roth and N. I. Krinsky, *Photochem. Photobiol.* **11,** 419 (1970).
70. R. J. Cremer, P. W. Perryman, and D. H. Richards, *Lancet i,* 1094 (1958); M. Gomirato-Sandrucci, N. Ansaldi, and M. L. Colombo, *Minerva Pediat.* **17,** 394 (1965); *Chem. Abstr.* **63,** 15194b (1965).
71. J. D. Spikes and M. L. MacKnight, *Ann. N.Y. Acad. Sci.* **171,** 149 (1970).
72. R. S. Davidson and K. R. Trethewey, *J. Chem. Soc. Perkin II* **178,** 173, 169 (1977); *J. Chem. Soc. Chem. Commun.* 674 (1975).

DAVID STURMER
Eastman Kodak Co.

E

ECONOMIC EVALUATION

Economic evaluation is the financial analysis of proposed courses of action. Company management must constantly evaluate new proposals. For production ventures the first test is to calculate capacity economics, ie, the return on investment for a plant of reasonable size, at a favorable location, operating at full capacity. If capacity economics do not show the proposed venture to yield a satisfactory return on investment under the most favorable conditions, then it must be rejected.

A venture that passes the capacity economics test is further scrutinized. For example, the marketing department determines whether the sales force can handle the product, and the purchasing department determines whether raw materials will be available. If no prohibitive flaws are found, profitability studies begin.

A profitability study produces a year-by-year forecast of the effect of the proposed venture on the profits and capital expenditures of the company. Many such studies may be required before a venture is commercialized. Before appropriating funds, management will demand profitability studies on all variations of the venture, eg, technology, site location, and financing. After the appropriation of funds, continual revision of the basic profitability study reflects such factors as changes in yields, capital estimates, and predicted sales volumes and prices.

Definition of Terms

In any chemical manufacturing operation, raw materials undergo chemical process conversions to a product in a production plant. This activity is supported by utilities which power and service the plant, and by personnel who operate and maintain the equipment. A sales force sells the product and a transportation system is required to move product to the customer.

Production Plant. A complex refers to a large group of chemical manufacturing facilities. Within the complex there may be several plants. Each plant, in turn, may be composed of two or more units or sections each of which conducts a single well-defined operation. Support facilities termed ancillaries, utilities, and offsites are also located within the complex or plant area. Figure 1 shows how a typical large complex might be arranged and subdivided.

Money spent to install and equip a production plant is called capital, and the facilities thus provided are termed assets. Capital assets generally are subdivided into fixed capital and other capital.

Fixed capital assets comprise production and supporting equipment; they may be subdivided:

(1) Battery limit or block limit capital refers to equipment within the immediate manufacturing area.

(2) Ancillary capital, or ancillaries, describes necessary facilities not physically located within the battery limit area of a unit, eg, storage tanks. Total process capital is the sum of battery limit and ancillary capital. It includes all facilities within a chemical complex or plant dedicated to a single process plant or unit.

(3) Offsite capital, or simply *offsites,* refers to general support facilities, such as office buildings, storehouses, maintenance shops, roads, land, waste disposal systems, steam and cooling water systems, electrical power distribution networks, cafeterias, and employee recreational facilities. If utility systems are defined separately, the term utilities and offsites is used, abbreviated U & O.

Total fixed capital is the sum of total process capital and offsite capital.

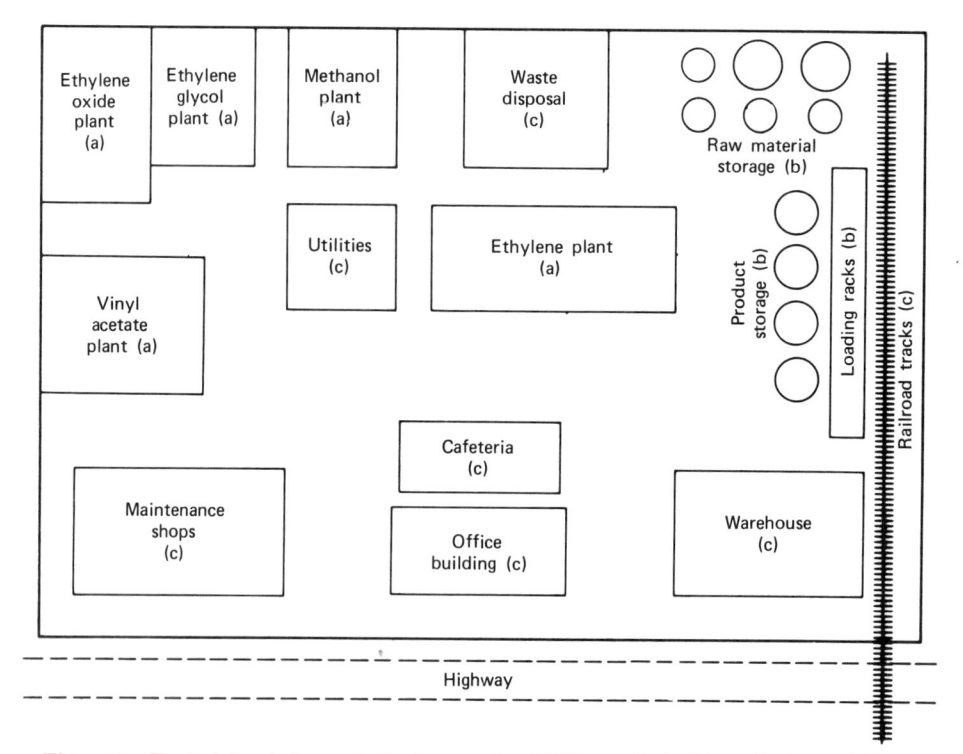

Figure 1. Typical chemical manufacturing complex. (a) Battery limit; (b) ancillaries; and (c) utilities and offsites.

In plant construction, expenditures are divided into capital and expense categories. Only the major expenditure, capital, is subject to depreciation (see below).

Other capital refers to less tangible facilities, or materials of relatively short life, which require the following commitment of capital funds.

(1) A license fee is the initial cash payment required to secure permission for use of necessary technology. Some contracts also require royalties.

(2) Catalysts with long lives are segregated from fixed capital for bookkeeping purposes. With lifetime less than one year, they may be charged as expense rather than capital.

(3) Working capital describes short-term assets, which can be readily converted to cash. Working capital and a related term, current assets, are defined below.

Total capital invested is the sum of total fixed capital and working capital (or current assets).

Production Cost. After a manufacturing facility has been built, continuing operating expenditures, such as what follows, are termed production costs, which are common to all chemical manufacturing.

Raw materials are substances which become, physically or chemically, part of the final product.

Catalysts are substances required to promote chemical reactions, but are not consumed and do not become part of the final product.

Chemicals are materials which are consumed in the process but, like catalysts, do not become part of the product.

Packaging supplies may or may not be required. If a chemical product is shipped in small quantities, container costs may be an important part of the total.

Royalties are sums paid regularly for the right to use technology that is patented or otherwise proprietary.

Utilities are facilities to provide power, heat, cooling, and other consumable functions.

By-product credit is used in manufacturing operations that produce by-products, which may be either undesirable or salable. The sales value of a by-product, less purification costs, results in a credit to production cost.

Labor costs include wages, salaries, and overtime paid to operating personnel.

Supervision or supervisory labor includes shift supervisors and overall superintendents.

Labor overhead represents costs related to payroll but not paid directly to employees. Included are payroll taxes, social security payments, pension plans paid for by the company, employee recreational subsidies, and health insurance premiums paid by the company. Labor overhead may be 20–40% of the direct payroll cost.

In addition to these costs related to a specific production operation, there are general plant costs:

Laboratory costs include analytical, testing, problem solving, and quality control functions.

Plant management includes supervisory and management personnel in overall control.

Staff support includes the technical specialists required in a complex modern industrial plant. Examples are accountants, personnel administrators, chemical, mechanical, and electrical engineers, secretaries, computer programmers, and purchasing agents.

Maintenance costs include both labor and material (see Maintenance).

Taxes, considered as part of production cost, are usually related to the value of plant assets, annual sales, or other standard. Income taxes are not included; these are discussed under Profit below.

Depreciation, in theory, is that portion of pretax earnings permitted to be set aside to replace aging plants. Only very rarely, however, are such funds actually accumulated in an escrow account. Thus depreciation becomes a paper cost deducted from pretax profit. Various methods of calculating the depreciation credit are discussed below.

Amortization, like depreciation, represents an income tax credit to replace obsolete facilities, but in general refers to intangible assets such as license fees or patents.

Total production cost is the sum of all the above costs, and is divided into variable and fixed costs.

Variable costs vary on an annual basis directly with volume or production. The variable cost per unit of product thus remains essentially constant; included are the costs of raw materials, catalysts, chemicals, packaging, royalties, utilities, and byproduct credits. In certain situations utilities may not be true variable costs; at least a portion of utility cost might better be classed as a fixed cost.

Fixed costs on an annual basis are constant. The annual cost is independent of production volume and thus unit cost is inversely proportional to annual production volume. Fixed costs include labor, supervision, labor overhead, laboratory operation, management, staff support, maintenance, taxes, and depreciation or amortization. Items such as laboratory costs, plant management and staff support, and property taxes are usually combined into plant overhead.

Company overhead includes headquarters and company management, sales, and perhaps research. These costs are not assignable to any one plant or product, and are often allocated as percentages of sales or capital assets.

Sales and Profits. The gross sales value of a product is what the customer pays for the product delivered.

The net sales value of a product is what the manufacturer receives for it as it leaves the plant site.

Profit is the money received from sale of products, which remains after production costs and prorated overhead have been paid. This is variously referred to as company profit, pretax profit, operating income, or net income. Income tax is not a factor at this point because these taxes traditionally are paid at the corporate, rather than the plant, level.

Profitability is a measure of relative economic desirability of an investment. Profit usually refers to dollar net gain realized annually from a venture, without reference to the investment, risk, or other cost. Profitability relates net gain or profit to investment.

Capacity Economics

Comparison of Ventures. Consider a proposal to produce at capacity 135,000 t/yr of a product expected to sell at 66¢/kg net sales return. Ventures A, B, and C, have been proposed, with capital investments and production costs as shown in Table 1. Venture C must be rejected, but A and B merit further consideration. Venture B requires 50 million dollars less capital than A, but in the long run the 14.5¢/kg production cost advantage of A over B would enable A to earn a greater profit than B in periods of low net sales returns. Details of such calculations are described in Table 1.

Calculation of Components. The proposal is to buy ethylene oxide and convert it to 89,500 t/yr of ethylene glycol.

Calculation of fixed capital, shown in Table 2, is 13.3 million dollars.

The plant production cost (see Table 3) is 44.9¢/kg at capacity or $40,250,000/yr, allowing for by-product credit. Before by-product credit, production cost is $45,500,000/yr.

The estimate of the working capital required by this plant when operating at capacity is given in Table 4.

The working capital represents a high percentage of the fixed capital because the

Table 1. Capacity Economics for Three Ventures

Factor	A	B	C
capacity, 1000 t/yr	135	135	135
fixed capital, millions of $	100	50	200
net sales return, ¢/kg	66.0	66.0	66.0
costs, ¢/kg			
utilities	2.9	7.3	8.8
raw materials	17.6	34.1	13.2
other variable costs	1.1	1.1	11.0
depreciation	7.3	3.7	14.7
other fixed costs	5.5	2.6	1.1
Total production cost	34.3	48.8	48.8
plant profit, ¢/kg	31.7	17.2	17.2
sales, administrative, and research, ¢/kg	3.3	3.3	3.3
profit before federal income tax, ¢/kg	28.4	13.9	13.9
profit after 48% federal income tax, ¢/kg	14.7	7.3	7.3
return after taxes on original fixed capital, %/yr	20	20	5

Table 2. Fixed Capital Calculation for the Hydration of Ethylene Oxide to 89,500 t/yr Ethylene Glycol[a]

Item	Capital, thousands of $
delivered equipment, E	1,002
installation, 47% E	841
instrumentation, 9% E	161
process piping, 50% E	901
electrical, 10% E	180
buildings, 15% E	269
Total process capital, P	4,154
utilities and offsites, 107% P	4,444
Total direct cost, D	8,598
engineering and supervision, 24% D	2,063
construction overhead, 6% D	516
contractor fee, 5% D	430
Total direct and indirect cost, T	11,607
contingency, 15% T	1,741
Total fixed capital	13,348

[a] Ref. 1.

414   ECONOMIC EVALUATION

Table 3. Production Cost Calculation for the Hydration of Ethylene Oxide to 89,500 t/yr of Ethylene Glycol[a]

Cost item	Units consumed per kg EG	Thousand units/yr	¢/unit	$1000/yr	¢/kg EG
variable costs					
ethylene oxide, kg	0.872	78,100	46.2	36,094	40.30
chemicals and supplies, ¢	0.04		0.04	39	0.04
utilities					
steam, kg	5.36	480,000	0.61	2,901	3.26
cooling water, L	113	102,000,000	0.00079	81	0.09
process water, L	2.3	212,000	0.0105	22	0.02
electricity, kW·h	0.029	2,560	1.75	45	0.04
gross variable cost				39,182	43.75
by-product credit					
diethylene glycol, kg	0.107[b]	9,540	44.0	4,200	4.71
triethylene glycol, kg	0.030[b]	2,730	38.5	1,050	1.17
Total credit				*5,250*	*5.88*
net variable cost				33,932	37.87
fixed costs					
labor,					
operators, 3 people per shift				250	
shift supervisors, 1 person per shift				100	
overall supervisors, 2 people				60	
direct labor				410	
labor overhead, 35% of direct labor				145	
Total labor				555	0.62
maintenance,					
3% of fixed capital				400	0.44
plant overhead,					
4.5% of fixed capital				603	0.68
depreciation, 10% of fixed capital				1,335	1.50
Total fixed cost				2,893	3.24
plant production cost				36,825	41.11
company overhead					
administration, research, sales: 5% of sales				2,350	
insurance: 1% of fixed capital				133	
distribution: 2% of sales				940	
Total				*3,423*	*3.83*
Total production cost				40,248	44.94

[a] Data derived from ref. 1, based on a production of 89,500 t/yr ethylene glycol (EG), 9540 t/yr diethylene glycol, and 2730 t/yr triethylene glycol; total fixed capital, $13,348,000.
[b] Units produced per kg ethylene glycol.

venture proposed involves only a simple processing step, ie, hydration, and therefore, the fixed capital is relatively low.

Based upon the above calculations, the capacity economics for a grass-roots plant to produce 89,500 t/yr of ethylene glycol from 78,100 t/yr of ethylene oxide are summarized in Table 5.

From Table 5 the relationship between return on investment at capacity and net sales return on ethylene glycol can be calculated:

Table 4. Estimate of Working Capital for the Production of 89,500 t/yr of Ethylene Glycol

Capital item	Days	$/day	Total, thousands of $
operating cash	20	124,700[a]	2,490
raw materials inventory	15	98,900[b]	1,480
product inventory	20	124,700	2,490
accounts receivable	30	124,700	3,720
accounts payable	30	98,900	(2,970)
working capital to support capacity operation			7,210

[a] Plant production cost before by-product credit.
[b] Raw materials cost before by-product credit.

Table 5. Capacity Economics for the Production of 89,500 t/yr of Ethylene Glycol[a]

	¢/kg	1000 $/yr
net sales, 89,500 t/yr	50	44,740
ethylene oxide, 78,100 t/yr at 46¢/kg	40.3	36,100
chemicals and utilities	3.4	3,080
by-product credit	5.9	(5,250)
net variable cost	37.8	33,930
plant fixed cost	3.3	2,890
plant production cost	41.1	36,820
plant profit	8.8	7,920
company overhead	3.8	3,420
profit before taxes	5.0	4,500
federal income tax 48%	2.4	2,160
profit after taxes	2.6	2,340
fixed capital		13,300
working capital		7,200
Total fixed and working capital[b]		20,500

[a] On-stream factor 0.90; Gulf coast location.
[b] Return after taxes: 11%.

net sales return, ¢/kg	44.9	47.1	49.3	51.5	53.7
return after taxes on total fixed and working capital, %/yr	0	5	10	15	20

These figures show that 2¢/kg in net sales return can correspond to an increase of 5% per year in return on investment at capacity.

Breakeven. The economics presented are for operation at full capacity, but from the data it is possible to determine at what percentage of capacity the profit becomes zero.

As the operating rate in the example of Table 5 drops, fixed costs remain constant at $6,310,000/yr. Available to pay for these is the difference between net sales return and variable costs, which in this example is:

$$(0.50 - 0.378) \text{ (plant production, 1000 t/yr)}$$

Hence the plant production rate at which profit becomes zero is given by:

416   ECONOMIC EVALUATION

$$\frac{\$6{,}310{,}000/\text{yr}}{(0.50 - 0.378)(1000)} = 51{,}700 \text{ t/yr}$$

This is 58% of full capacity.

Profitability Studies

  Defining the Proposed Venture. When calculating economics, possible repercussions of the proposed venture throughout the company must be considered. For example, a new product Y may reduce the sale of existing product X by replacing it in an important application. In this case the following reasoning should be used:

	Sales, million $		
	X	Y	Company
venture adopted	50	200	250
venture not adopted	100	0	100
effect			150

  A profitability study should look ahead at least fifteen years. For example, suppose that a company producing product D has substantial over-capacity: sales are 45,300 t/yr, but capacity is 135,000 t/yr. It is proposed to build a unit to convert, at capacity, 89,500 t/yr of D to its derivative E for which there is growing demand. In this case, the increased total demand for D is forecast to require expansion of the unit producing it within five years. Since this is the result of the operation of the E unit, the required expansion of the D unit must be considered a part of the E venture.

  Year-by-Year Economics. For each year in the life of the venture, net sales, production costs, after-tax profits, and expenditures of fixed and working capital are calculated.

  In the first year, significant costs chargeable specifically to the venture can be identified, such as research and development expenses. The last year is usually the one in which the major plant equipment is fully depreciated. Year-to-year economic data should be summarized in a cash-flow table (see Table 6). The cash flow in a given year equals the profit after taxes, plus the depreciation, minus the sum of the fixed and working capital invested in that year. The cumulative cash flow at the end of any year is the algebraic sum of all the earlier annual cash flows.

  Rating Ventures. The reader is cautioned not to depend on any one profitability criterion. Payout evaluations, eg, should not be independent of DCF considerations and the need for cash-flow generation.

  Payout Time. The payout or cash recovery time, is roughly the time required for the company to recover its original capital investment. More precisely, it is the length of time between the beginning of the venture and the date when the cumulative cash flow becomes positive. For example, the payout time for the venture described in Table 6 is six years.

  Year-by-Year Return on Investment. The profit after taxes, expressed as a percentage of gross fixed capital plus working capital, varies from year to year as shown in Table 7.

  Discounted Cash Flow Rate of Return. The discounted cash-flow (DCF) rate of return, also called the interest rate of return, the profitability index, and the rate of return by the investor's method, can be seen as the constant interest rate paid by a savings bank which accepts and returns cash on the same schedule as the venture. For example, the venture described in Table 6 might be compared to depositing $9,260,000

Table 6. Cash Flow Table of Year-by-Year Economic Data, in Thousands of Dollars

Year	Added net sales	Added profit after taxes	Added depreciation	New fixed capital	New working capital	Cash flow	Cumulative cash flow
1		(260)		9,000		(9,260)	(9,260)
2	5,000	310	600	6,000	200	(5,290)	(14,550)
3	10,000	2,050	1,000		400	2,650	(11,900)
4	14,000	3,300	1,000		600	3,700	(8,200)
5	15,000	3,200	1,000		300	3,900	(4,300)
6	15,000	3,200	1,000			4,200	(100)
7	15,000	3,200	1,000			4,200	4,100
8	14,000	2,700	1,000			3,700	7,800
9	14,000	2,700	1,000			3,700	11,500
10	14,000	2,700	1,000			3,700	15,200
11	13,000	2,200	1,000			3,200	18,400
12	13,000	2,200	1,000			3,200	21,600
13	13,000	2,200	1,000			3,200	24,800
14	12,000	1,700	1,000			2,700	27,500
15	12,000	1,700	1,000			2,700	30,200
16	12,000	1,700	1,000			2,700	32,900
Total	191,000	34,800	14,600	15,000	1,500	32,900	

Table 7. Year-by-Year Return on Investment

Year	%/yr	Year	%/yr
1		9	16
2	2	10	16
3	13	11	13
4	20	12	13
5	19	13	13
6	19	14	10
7	19	15	10
8	16	16	10

in a savings bank during year 1, depositing $5,290,000 in year 2, withdrawing $2,650,000 in year 3, withdrawing $3,700,000 in year 4, and so on. Annual withdrawals would continue through the sixteenth year but would then stop. At that time a final withdrawal of $1,500,000 working capital plus $400,000 salvage value (assumed equal to undepreciated fixed capital) would be made. In order to meet such a cash schedule, a savings bank would have to pay 20%/yr interest. The venture is, therefore, said to have a DCF rate of return of 20%/yr, a number unaffected by how the cash withdrawn is used.

The DCF calculation is a trial-and-error procedure. A DCF interest rate is selected, and the cash flow each year is multiplied by the discount factor for that year, as shown in Table 8.

The discount factor for the nth year is $1/(1+i)^n$ where i is the interest rate. For example, with 18% interest the discount factor for the tenth year is $1/(1.18)^{10} = 0.191$. Tabulations of discount factors are available in financial and mathematical handbooks. The calculation is repeated with different trial interest rates until the correct one is

418 ECONOMIC EVALUATION

Table 8. Calculation of DCF Interest Rate

Year	Cash flow, thousands of $	18% Interest Factor	18% Interest DCF	20% Interest Factor	20% Interest DCF
1	(9260)	0.847	(7843)	0.833	(7714)
2	(5290)	0.718	(3798)	0.694	(3671)
3	2650	0.609	1614	0.574	1534
4	3700	0.516	1909	0.482	1783
5	3900	0.437	1704	0.402	1568
6	4200	0.370	1554	0.335	1407
7	4200	0.314	1319	0.279	1172
8	3700	0.266	984	0.233	862
9	3700	0.225	833	0.194	719
10	3700	0.191	707	0.162	599
11	3200	0.162	518	0.135	432
12	3200	0.137	438	0.112	358
13	3200	0.116	371	0.093	298
14	2700	0.099	267	0.078	211
15	2700	0.084	227	0.065	176
16	4600	0.071	327	0.054	248
Total			*1131*		*(19)*

found, shown by the zero algebraic sum of the annual cash flows. In Table 8, 18% is not correct but 20% is. The DCF rate of return for the venture as described by Table 6 is, therefore, 20% per year.

If numerous DCF calculations are to be made, a computer should be programmed to do them; some pocket calculators are also so programmed (see Programmable pocket calculators).

Net Present Value. An alternative to the DCF calculation is to determine the net present value, or present worth, of the cash flows of the venture at an assumed interest rate. In Table 6, eg, the net present value of the cash flows from the venture at an interest rate of 18% would be 1.33 million dollars.

Rating the Proposal. The lower the payout time, the better. On the other hand, high values are desirable for year-by-year returns on investment, DCF interest rates, and net present values. It is difficult to be more definite about desirable values for economic criteria. A three-year payout time may be too long if the company needs cash urgently, but six years may be acceptable for a venture that leads the way toward a large and growing market. Generally speaking, a return below 5% on total fixed and working capital is unacceptable; returns of 25% or more are good unless the risk is unusually high.

In the case of DCF interest rate, 5% is too low. Above this, the acceptable DCF interest factor (or hurdle rate, as it sometimes is called) is a function of the cost of raising capital, the debt–equity ratio of the investor, the other proposed ventures available to the company, and the risk of the venture. Interest rates of 25% for DCF or greater should justify proposed ventures of normal risk.

Comparing Proposed Ventures. The best way to compare proposed ventures is to evaluate each and tabulate the results. It is important to do the evaluations over *the same span of time.* Suppose, eg, it is proposed to produce a raw material that at present is being purchased. Alternative A is to license a process. Alternative B is to develop a process that has shown promise in laboratory experiments. The comparison between A and B might be summarized as follows:

	A	B
date venture would begin	Jan. 1, 1985	Jan. 1, 1985
plant startup date	Jan. 1, 1988	Jan. 1, 1991
end of evaluation period	Jan. 1, 2006	Jan. 1, 2006
payout time, years	4	7
DCF interest rate, %	25	35

Either A or B is preferable to continuing to buy the raw material. Management must decide whether B's greater profitability justifies the three-year delay in payout.

EXAMPLE OF AN ACTUAL VENTURE

Suppose that a company produces ethylene oxide. The total production is sold to other companies. The company has already thoroughly evaluated a proposal to construct a new 227,000 t/yr ethylene oxide unit and has determined that the economics are attractive at the prices and volumes forecast.

Now consider a further alternative proposal not to sell 78,100 t of the ethylene oxide produced by the new unit but instead to hydrate it to 89,500 t/yr of ethylene glycol in a new unit and sell the remaining 149,000 t/yr of ethylene oxide to other companies. The question is whether this new proposal is profitable and should be included in the over-all project.

Sales and Raw Material Costs. The marketing department forecasts of net sales returns for ethylene glycol are given in Table 9 together with their previous forecasts for the net sales returns on ethylene oxide. (This venture will have to pay full net sales return for its ethylene oxide because the marketing department has previously stated that it can sell all the ethylene oxide if it is not converted to ethylene glycol.)

Capital Schedule. The new fixed capital is estimated at 13.3 million dollars, of which 7.8 million dollars is spent in the first year and 5.5 million dollars in the second. Table 10 summarizes the production cost schedule.

Working Capital. The working capital required to support capacity operation is 7.2 million dollars, as previously calculated. The value required in the early years of operation can be approximated by assuming that working capital is proportional to the plant production rate (Table 11).

Table 9. Net Sales Returns for Ethylene Glycol

Year of operation	Sales of ethylene glycol			Cost of ethylene oxide		
	Thousands of t	¢/kg[a]	Thousands of $	Thousands of t	¢/kg	Thousands of $
1	34	50.0	17,020	29	46.2	13,650
2	45	52.3	23,800	39	48.4	19,140
3	68	55.0	37,500	59	50.8	30,260
4	90	57.6	51,610	78	53.2	41,620
5	90	60.5	54,180	78	55.8	43,690
6	90	63.6	56,930	78	58.7	45,920
7	90	66.9	59,890	78	61.8	48,330
8	90	70.2	62,840	78	64.9	50,740
9	90	73.7	66,000	78	68.2	53,320
10	90	77.4	69,340	78	71.5	55,900

[a] Annual increase of 5%.

ECONOMIC EVALUATION

Table 10. Production Cost Schedule, Thousands of Dollars[a]

Cost item	First year	Second year[b]
capacity, %[c]	38	51
fixed cost at plant level	2,890	3,030
company overhead	3,420	3,590
net variable cost	12,890	18,170
total production costs	19,200	24,790

[a] Unless otherwise stated.
[b] A factor of 1.05 is included to allow for inflation and rising energy costs.
[c] Operation is at 76% of capacity in the third year and at 100% thereafter.

Table 11. Working Capital

		Working capital, thousands of $	
Year of operation	Capacity, %	Total	Added
1	38	2740	2740
2	51	3680	940
3	76	5480	1800
4	100	7210	1730

Startup Expenses. Although a new production unit is thoroughly tested before going into operation, nevertheless, minor and major problems can occur in the initial operation period. These entail startup expense in addition to normal production costs. In the above example, startup expense is estimated as three weeks of gross variable cost (before by-product allowance) at 50% of capacity:

$$\frac{21}{365} (39,180) (0.50) = \$1,130,000$$

Cash Flow and Profitability. Table 12 is obtained from the above calculations.

Estimating Fixed Capital

The cost of installing plant facilities for manufacturing a chemical product is termed fixed capital, and includes all tangible facilities, including allocated portions of general plant utilities and offsites. In addition to direct cost of equipment and installation, the costs of detailed engineering, project supervision, and contractor fee or profit are part of fixed capital.

Effect of Inflation. Cost estimates usually are based upon known past costs, adjusted for inflation. Plant construction-cost inflation may be greater than overall average inflation. Various indexes that track inflation are available and are updated regularly. Some are listed below.

Marshall & Swift (previously Marshall & Stevens) *Equipment Cost Index* (MSI) is based on a 47-industry average of manufacturing equipment and installation costs. Base year, with an index of 100, is 1926 (2).

Engineering News-Record Construction Index (ENR) gives costs of basic construction materials, such as concrete, lumber, and steel. Base year for 100 index is 1913 (3).

Table 12. Cash Flow for Ethylene Glycol Production by Hydration of Ethylene Oxide, Thousands of Dollars

Factor	1	2	3	4	5	6	7	8	9	10	11	12
fixed capital[a]	7,800	5,500	0	0	0	0	0	0	0	0	0	0
working capital[a]			2,740	940	1,800	1,730	0	0	0	0	0	0
capital invested[b]	7,800	5,500	2,740	940	1,800	1,730	0	0	0	0	0	0
net sales			17,020	23,800	37,500	51,610	54,180	56,930	59,890	62,840	66,000	69,340
startup costs			1,130									
plant costs plus company overhead			6,310	6,620	6,960	7,300	7,670	8,050	8,460	8,880	9,320	9,790
net variable cost			12,890	18,170	28,430	39,280	41,240	43,310	45,470	47,740	50,130	52,640
total product cost			19,200	24,790	35,390	46,580	48,910	51,360	53,930	56,620	59,450	62,430
profit before taxes			(3,310)	(990)	2,110	5,030	5,270	5,570	5,960	6,220	6,550	6,910
profit after taxes			(1,720)	(510)	1,100	2,620	2,740	2,900	3,100	3,230	3,410	3,590
depreciation			1,330	1,330	1,330	1,330	1,330	1,330	1,330	1,330	1,330	1,330
cash flow	(7,800)	(5,500)	(3,130)	(120)	630	2,220	4,070	4,230	4,430	4,560	4,740	4,920
cumulative cash flow[c]	(7,800)	(13,300)	(16,430)	(16,550)	(15,920)	(13,700)	(9,630)	(5,400)	(970)	3,590	8,330	13,250

[a] Return on total fixed and working capital at capacity, 13%/yr.
[b] Payout time, 9 years.
[c] DCF interest rate, 8%/yr; net present value, $4,560,000.

Chemical Engineering Plant Cost Index (CE) is based on cost of equipment, installation labor, buildings, engineering, and other components of chemical plant costs. Base point is average plant costs in 1957–1959 (4).

Nelson Refinery Index (Nelson) is oriented primarily toward construction practices typical of United States petroleum refineries, but may be applied to any similar plant. Base cost year is 1946 (5).

These indexes refer to construction cost in the United States. Others such as the *Cost Engineer Index* (U.K., 1958 = 100) and the *Dutch Composite Index* (The Netherlands, 1966 = 100) are applicable to other areas. Excellent comparisons are available (6–7). Table 13 compares major United States cost indexes.

Costs of similar plants built in different years are related by the ratio of construction cost indexes for those years. For example (based on Marshall-Swift Index):

$$\text{cost in 1976} = \text{cost in 1968} \times \frac{472}{273}$$

Slightly different results are obtained with different indexes. When estimating the future cost of a plant not yet built, projection of the cost index to a future value is necessary.

Effect of Plant Size. Large plants usually cost less than small plants per production unit (7–13). Total costs for plants of different sizes are related by the ratio of capacities, raised to a power less than 1.0.

$$\text{plant B cost} = \text{plant A cost} \left(\frac{\text{capacity B}}{\text{capacity A}}\right)^f$$

where f is a function of plant size, type, and location. For typical process industry plants of moderate size and complexity, of the overall average exponent is about 0.65.

Cost Estimates

Order of Magnitude. Initial screening of proposed projects requires an estimate, to within perhaps ±40% of the potential investment. This may be determined in two ways.

Previous Costs. Known costs of existing plants may be projected to future plants using similar technology, by the methods discussed above. Representative costs are available in the literature (7,9–11,13–14). Cost estimates may be obtained from engi-

Table 13. United States Process Plant Construction Cost Indexes

Year	MSI, 1926[a]	ENR, 1913[a]	CE, 1957–1959[a]	Nelson, 1946[a]
1968	273	1201	114	304
1970	303	1445	126	365
1972	332	1807	137	439
1974	398	2099	165	523
1976	472	2490	192	616
1978	545	2870	219	721

[a] Base year = 100.

neering contractors, and assessment data from process vendors and licensors often are available.

Cursory Process Design. If satisfactory data from existing or similar plants are unavailable, a cursory plant design based on proposed process is required:

(1) Prepare a flow sheet based on process chemistry from experimental and literature sources. Estimate heat effects, probable recycle streams, purification requirements, and environmental protection considerations.

(2) Calculate a material balance, including estimates of conversion, yield, and by-product formation as well as special solvents, chemicals, and catalysts.

(3) Estimate major equipment sizes. This should be related to cost functions such as power for turbines and motors, heat duty for furnaces, area for heat exchangers, and diameter and height for distillation towers. Accuracy is not necessary at this stage.

(4) Estimate major equipment costs. Cost data may be available from company or personal files, vendor information, or literature.

(5) Estimate the battery-limit capital cost by multiplying total delivered equipment cost by a suitable factor (8,12–13,16–20). Factors typical of United States Gulf Coast installations are listed below.

process type	fluid	fluid–solid	solid
factor	3.4–3.8	3.0–3.3	2.8–3.0

These factors do not include cost of land, ancillaries, and offsites. Lowest factors are used when equipment cost is relatively high.

(6) Estimate total fixed capital, ie, the cost of ancillaries, utilities, and offsites and battery-limit total. Typically ancillaries are about 10–12% of battery-limit capital; utilities and offsites comprise about 25% of total fixed capital. An additional 25% overall allowance usually is added for unforeseen contingencies.

Computer programs, which perform many of the above operations, some with equipment-cost data banks, are available.

Preliminary Cost Estimates. A more detailed estimate is useful for appropriation or budget authorization requests, and for choosing between alternative technologies whose overall economic positions are similar. Anticipated accuracy may be ±20%. The method is similar to that used for cursory process design above, but more detailed (7,9–10,12,16,19,21–22).

Definitive Cost Estimates. Detailed cost estimates are possible only after most of the engineering has been completed and material and labor requirements determined. Accuracy is ±5–10%.

Working capital can be estimated from inventory requirements, anticipated accounts receivable and payable, and short-term cash-flow operating expenses. A rough approximation is about 18% of total fixed capital. The sum of total fixed capital and working capital is the total capital invested.

Estimating Working Capital

Working capital represents assets committed to meet short-term operating costs, including (1) cash reserves, (2) inventories, (3) accounts receivable, and (4) accounts payable. In calculation of production costs, working capital is not depreciated (ie, no annual tax-deductible cost is charged) because the short-term value does not change and there is no obsolescence.

Cursory Estimate. Where an accurate estimate of working capital is not necessary, it may be related to fixed capital or product sales value.

Fixed Capital. Typical working capital requirements for large chemical plants are 10–20% of fixed capital investment. For small plants, or relatively simple plants that process expensive raw materials, or plants with seasonal business requiring large inventories of finished product, the percentage may reach as much as 100% of fixed capital. This is the least accurate method of estimating working capital and should be used with caution.

Annual Sales. Working capital may be estimated at 25–35% of annual net sales, subject to special requirements discussed above.

Detailed Estimate. More accurate estimates of working capital requirement are based on its individual components.

Cash Reserves. Cash needed for wages and salaries, fringe benefits, taxes, and other daily expenses including miscellaneous supplies may be estimated at 6–8% of plant production cost or 4–6% of annual net sales.

Inventories. The three kinds of inventory requirements to be considered are: (1) raw materials, typically 2–4 weeks' consumption valued at delivered cost; (2) intermediate or crude products, usually only about 1–2% of production cost; and (3) finished products, one month plant production cost.

Special requirements must also be considered. For example, if raw materials such as natural gas, or finished products (eg, carbon monoxide) are normally gaseous, very little may be kept in inventory. Conversely, much larger inventories may be necessary in remote locations dependent on intermittent transportation facilities, or when sales are highly seasonal (eg, fertilizers).

Accounts Receivable. Typical credit terms are based on a 30-day payment period, but the invoice is seldom sent on day of shipment. Depending on company procedures, the allowance for accounts receivable should include 30–45 days' product at net sales value.

The sum of cash reserves, inventories, and accounts receivable is current assets. Some companies prefer this to working capital in calculating profitability.

Accounts Payable. This credit to working capital represents raw materials received but not yet paid for. It may be estimated at 30 days' raw material cost.

$$\text{working capital} = \text{current assets} - \text{accounts payable}$$

Estimating Production Cost

Production cost is the continuing expenditure of money which must be made to permit a chemical manufacturing facility to operate. The elements of production costs are listed in Tables 14–15. Table 3 is an example of a production cost calculation.

Market Considerations

Good capital and production cost estimates are essential to good economic evaluations, and the percentage errors in these tend to be lower than those in the price and sales volume forecasts. Therefore, the price and sales volume forecasts should be closely scrutinized. For a better perspective, graphs or tabulations showing five-year forecasts of prices, sales volumes, and market shares should include current values and five-year histories.

Table 14. Elements of Production Costs

Variable costs	Remarks	Fixed costs	Remarks	Company overhead	Remarks
raw materials	based upon process stoichiometry, conversion, yield, and recovery efficiency	labor, operating	approximate labor requirements for continuous plant operations are given in Table 15	administration-research	2–4% of product sales value
catalysts	costs of regeneration or replacement are prorated over useful life	supervision, overhead	eg, unit superintendent, production engineer, fringe benefits, and payroll taxes, 20–40% of total direct labor costs	sales, marketing[d]	1–2% of product sales value
chemicals packaging supplies	delivered cost applied may include labor costs	maintenance	ranges from 2% of fixed capital to 5% for corrosive processes	distribution insurance	2% of product sales value 1–3% of total fixed capital annually
royalties	typically 2½–5% of product sales value	plant overhead[a]	3–5% of total fixed capital		
utilities	see refs. 3–6 for typical consumption	depreciation[b]	9.1% of total fixed capital may be charged as operating costs[c]		
by-product credit	total variable cost minus by-product credit gives the net variable cost				

[a] Factory general.
[b] Including amortization.
[c] Based on the weighted average standard life of a United States chemical plant of about 11 years.
[d] Specialty products may require separate sales force and technical service which may account for as much as 10% of sales in marketing costs.

Table 15. Equipment Controlled by One Operator per Shift

Equipment type	Large automated plants	Small batch plants
reaction	1 major reaction section	1–3 batch kettles
purification	1 train (3–5 operations) of distillation or extraction equipment	1–2 batch stills or extractors
compression	1 group of centrally located compressors	1–2 small, separated compressors
filtration	3–6 continuous filters	2–3 batch filters
tank farm and general	1–2 per major plant unit	1 per product

Prices. In periods of industry overcapacity, prices fall or at best rise more slowly than costs. In periods of high capacity utilization, prices increase. It is important in making a price forecast to take into account the economics of the best producer. The effect of increasing or decreasing costs of raw materials and energy has to be considered in price forecasts as well as new technology. When planning a new plant, the economic potential of new competing processes has to be weighed carefully, especially if they may be commercialized within a few years.

Sales Volume. The annual sales volume can be forecast by multiplying the total industry sales (merchant sales) by the percentage market share that the company expects. Alternatively, the major buyers in the field can be surveyed as to their needs.

BIBLIOGRAPHY

"Economic Evaluation" in *ECT* 2nd ed., Vol. 7, pp. 642–660, by Oliver Axtell, Celanese Chemical Company.

1. H. W. Scheeline, *Ethylene Glycols, Glycol Ethers, and Ethanolamines, Process Economics Program Report No. 70,* Stanford Research Institute, Menlo Park, Calif., Aug. 1971, updated Mar. 1978.
2. R. W. Stevens, *Chem. Eng.,* 124 (Nov. 1947).
3. *Eng. News Rec.,* 398 (Sept. 1, 1949).
4. T. H. Arnold and C. H. Chilton, *Chem. Eng.,* 143 (Feb. 18, 1963).
5. *Oil Gas J.,* 110 (Oct. 1, 1956).
6. O. P. Kharbanda, *Process Plant and Equipment Cost Estimation,* Sevak Publications, Bombay, India, 1977.
7. R. L. Johnson, *Chem. Eng.,* 146 (Mar. 10, 1969).
8. J. T. Gallagher, *Chem. Eng.,* 89 (Dec. 18, 1967).
9. K. M. Guthrie, *Chem. Eng.,* 140 (June 15, 1970).
10. C. E. Dryden and R. H. Furlow, *Chemical Engineering Costs,* 1966 ed., Ohio State University, Columbus, Ohio, 1966.
11. J. E. Haselbarth, *Chem. Eng.,* 214 (Dec. 4, 1967).
12. H. Popper, *Modern Cost—Engineering Techniques,* McGraw-Hill Book Company, New York, 1970.
13. F. C. Jelen, *Cost and Optimization Engineering,* McGraw-Hill Book Company, New York, 1970.
14. *Process Economics Program,* Stanford Research Institute, Menlo Park, Calif., various reports.
15. J. R. F. Alonso, *Chem. Eng.,* 86 (Dec. 13, 1971).
16. K. M. Guthrie, *Chem. Eng.,* 114 (Mar. 24, 1969).
17. F. A. Holland and co-workers, *Chem. Eng.,* 72 (Apr. 1, 1974).

18. R. H. Perry and C. H. Chilton, *Chemical Engineers Handbook,* 5th ed., McGraw-Hill Book Company, New York, 1973.
19. M. S. Peters and K. D. Timmerhaus, *Plant Design and Economics for Chemical Engineers,* 2nd ed., McGraw-Hill Book Company, New York, 1968.
20. H. C. Bauman, *Fundamentals of Cost Engineering in the Chemical Industry,* Reinhold Publishing Corporation, New York, 1964.
21. H. J. Lang, *Chem. Eng.,* 130 (Sept. 1947).
22. R. S. Aries and R. D. Newton, *Chemical Engineering Cost Estimation,* Mc-Graw-Hill Book Company, New York, 1955.

General References

Corporate investment strategy

H. Bierman, Jr., and S. Smidt, *The Capital Budgeting Decision of Investment Projects,* 4th ed., Macmillan Publishing Co., Inc., New York, Collier Macmillan Publishers, London, Eng., 1975.
E. I. du Pont de Nemours Education and Technology Division, *The DuPont Guide to Venture Analysis,* 1978.

Engineering economy

E. L. Grant and W. G. Ireson, *Principles of Engineering Economy,* 4th ed., The Ronald Press Company, New York, 1964.
G. W. Smith, *Engineering Economy: Analysis of Capital Expenditures,* 2nd ed., The Iowa State University Press, Ames, Iowa, 1973.

Economic evaluation in the chemical industry

J. Happel and D. G. Jordan, *Chemical Process Economics,* 2nd ed., Marcel Dekker, Inc., New York, 1975.
F. A. Holland, F. A. Watson, and J. K. Wilkinson, *Introduction to Process Economics,* John Wiley & Sons, Inc., New York, 1974.
H. Popper, *Modern Cost—Engineering Techniques,* McGraw-Hill Book Company, New York, 1970.
R. Woods, *Financial Decision Making in the Process Industries,* Prentice Hall, Englewood Cliffs, N.J., 1977.

Fixed capital estimation

O. P. Kharbanda, *Process Plant and Equipment Cost Estimation,* Sevak Publications, Bombay, India, 1977.
Equipment cost indexes are published frequently in *The Oil and Gas Journal,* Petroleum Publishing Company, Tulsa, Okla. and in *Chemical Engineering,* McGraw-Hill, New York.

Working capital estimation

T. B. Lyda, "How Much Working Capital will the New Project Need?" *Chem. Eng.,* 182 (Sept. 18, 1972).

Marketing research, sales forecasting and pricing

N. H. Giragosian, *Chemical Marketing Research,* Reinhold Publishing Corporation, New York, 1967.
D. D. Lee, *Industrial Marketing Research, Techniques and Practices,* Technomic Publishing Company, Inc., Westport, Conn., 1978.
J. B. Malloy, "Projecting Chemical Prices," *Chem. Ind.,* 942 (Dec. 7, 1974).

428 ECONOMIC EVALUATION

Profitability: effect of variables

L. R. Bechtel, "How Currency Exchange Affects Investments," *Hydrocarbon Process.*, 197 (May 1975).
A. B. Blake, "The Asset Disposal Decision," *Chemtech*, 282 (May 1975).
T. R. Brown, "Economic Evaluation of Future Equipment Needs," *Chem. Eng.*, 125 (Jan. 17, 1977).
D. S. Bruce, G. F. Tice, and E. C. Chang, "Phasing and Timing Alter Viability," *Oil Gas J.*, 88 (Mar. 21, 1977).
F. A. Holland and F. A. Watson, "Putting Inflation into Profitability Studies," *Chem. Eng.*, 87 (Feb. 14, 1977).
B. Reuben, "Economies of Scale: Diminishing Returns?" *Process Eng.*, 100 (Nov. 1974).
R. W. Schmenner, "Before You Build a Big Factory," *Harv. Bus. Rev.*, 100 (July–Aug. 1976).

Profitability: discounted cash flow interest rates

J. Dean, "Measuring the Productivity of Capital," *Harv. Bus. Rev.*, 120 (Jan.–Feb. 1954).
W. R. Hirschmann, "Continuous Discounting for Realistic Investment Analysis," *Chem. Eng.*, 211 (July 19, 1965).
W. R. Hirschmann, "Realistic Investment Analysis—II," *Chem. Eng.*, 132 (Aug. 8, 1965).
F. C. Jelen, "Methods for Economic Analysis," *Hydrocarbon Process.*, 161 (Sept. 1974).
F. C. Jelen, "Methods for Economic Analysis," *Hydrocarbon Process.*, 161 (Oct. 1974).
R. I. Reul, "Profitability Index for Investments," *Harv. Bus. Rev.*, 116 (July–Aug. 1957).
J. B. Weaver, "Profitability Measures," *Chem. Eng. News*, 94 (Sept. 1961).

Profitability: methods for calculating the discounted cash flow interest rate

O. Axtell, "What's Your Interest Rate?" *Hydrocarbon Process. Pet. Refiner*, 121 (Sept. 1961).
T. K. Y. Leung, "New Nomograph: Quick Route to Discounted Cash Flow," *Chem. Eng.*, 208 (June 1, 1970).
G. E. Mapstone, "Figure Discount Cash Flow Quickly," *Hydrocarbon Process.*, 99 (Dec. 1975).
N. H. Wild, "Program for Discounted-Cash-Flow Return on Investment," *Chem. Eng.*, 137 (May 9, 1977).

Profitability: other than interest rates

S. L. Andersen, "Venture Analysis: A Flexible Planning Tool," *Chem. Eng. Prog.*, 80 (Mar. 1961).
J. Happel, "The Venture Worth Method for Economic Balance in Chemical Engineering," *Chem. Eng. Prog.*, 533 (Dec. 1955).
F. C. Jelen, "Capitalized Cost as a Method for Comparison of Alternatives," *Chem. Eng. Prog.*, 413 (Oct. 1956).

Profitability: allowing for risk

H. M. Hawkins, "How to Evaluate Projects," *Chem. Eng. Prog.*, 58 (Dec. 1964).
D. B. Hertz, "Risk Analysis in Capital Investment," *Harv. Bus. Rev.*, 95 (Jan.–Feb. 1964).
F. A. Holland and F. A. Watson, "Project Risk, Inflation, and Profitability," *Chem. Eng.*, 133 (Mar. 14, 1977).
F. A. Holland, F. A. Watson, and J. K. Wilkinson, "Estimating Profitability When Uncertainties Exist," *Chem. Eng.*, 73 (Feb. 4, 1974).
W. D. Rowe, *An Anatomy of Risk*, John Wiley & Sons, Inc., New York, p. 488.
L. M. Serruya and Y. Allaire, "Statistical Treatment of Risk and Uncertainty in Capital Budgeting Decisions," *Bus. Q.*, 48 (Spring 1969).
G. Wells, "Simulation Strategy for Process Plant Projects," *Process Eng.*, 60 (Jan. 1975).

Profitability: critiques of various methods

S. B. Henrici, "Eyeing the ROI," *Harv. Bus. Rev.*, 88 (May–June 1968).
D. P. Herron, "Comparing Investment Evaluation Methods," *Chem. Eng.*, 125 (Jan. 30, 1967).
F. C. Jelen, "Pitfalls in Profitability Analysis," *Hydrocarbon Process.*, 111 (Jan. 1976).

E. M. Lerner and A. Rappaport, "Limit DCF in Capital Budgeting," *Harv. Bus. Rev.*, 133 (Sept.–Oct. 1968).

Research and development project evaluation

T. Garcia-Borras, "Research-Project Evaluations," *Hydrocarbon Process.*, 137 (Dec. 1976).
A. V. Perrella, "Pitfalls in Evaluating R & D," *Chem. Eng.*, 59 (Aug. 1, 1977).
O. Winter, "Preliminary Economic Evaluation of Chemical Processes at the Research Level," *Ind. Eng. Chem.*, 45 (Apr. 1969).

Economic evaluation: new ideas

Below are three journals which are entirely devoted to economic evaluation.
Engineering and Process Economics, Elsevier Scientific Publishing Company, Amsterdam, The Netherlands.
Engineering Economist, Engineering Economy Division, American Society for Engineering Education, Stevens Institute of Technology, Hoboken, N.J.
American Association of Cost Engineers Bulletin, Pratt Institute, Brooklyn, N.Y.

OLIVER AXTELL
J. M. ROBERTSON
Celanese Chemical Company

EGGS

The (hen's) egg is an important part of the human food supply. It is highly nutritious, contributing proteins, fats, vitamins, and minerals. Because of its unique composition, the egg has extremely useful functional properties when used by itself or as an ingredient in other foods. There are also nonfood uses for eggs and components of eggs.

Eggs are marketed in the shell and also as egg products in three different forms: liquid, frozen, and dried. Eggs are produced and utilized throughout the world. Table 1 indicates the amount of eggs and egg products consumed in the United States in recent years. There has been a trend toward reduced per capita consumption, and an increased usage of egg products.

Chemical Properties

The egg can be divided into three basic parts: ca 31% yolk, ca 58% white, and ca 11% shell. The mixture of white and yolk is called whole egg. Gross chemical composition of the egg contents is shown in Table 2.

Table 1. Per Capita Consumption of Shell and Processed Eggs, 1954–1978 [a]

Year	Egg consumption, number			Proportion of eggs processed, %
	Shell	Processed[b]	Total	
1954	351	25	376	6.6
1960	306	28	334	8.4
1966	283	30	313	9.6
1972	288	48	336	14.3
1975	262	40	302	13.2
1976	241	33	274	12.0
1977	235	37	272	13.6
1978	242	35	277	12.0

[a] Ref. 1.
[b] Shell egg equivalent.

Table 2. Chemical Composition of Egg Components

Component	Egg white, %	Egg yolk, %	Whole egg, %
water	88.1	48.8	74.6
solids	11.9	51.2	25.4
protein	10.1	16.4	12.1
lipids	trace	32.9	11.2
carbohydrates (both bound and free)	1.1	1.0	1.1
free glucose	0.4	0.2	0.3
ash	0.6	1.7	1.0

Shell. The egg shell is 94% calcium carbonate, with a small amount of magnesium carbonate (1%), and calcium phosphate (1%). A keratin-type protein, which is water insoluble, is found within the shell pores and in the outer cuticle, acting as a protective coating. The pores of the shells allow escape of both carbon dioxide and water during storage.

The shell is separated from the egg contents by two protein membranes. The air cell found at the large end of the egg is formed by separation of these membranes. Size of this air cell increases when eggs are stored because of water loss.

Egg White. Egg white consists of both thick and thin layers within the shell: outer-thin next to the shell, outer-thick, inner-thin, and inner-thick next to the yolk membrane. The thick layers have a much higher level of ovomucin, the least soluble of the egg white proteins. Otherwise, composition of the thick and thin portions are close to the same (2). Fibrous chalazae consisting of mucin are also contained within the egg white. They are attached to each side of the yolk and serve to anchor the yolk in the center of the egg.

Egg white contains mostly proteins with different physical and chemical characteristics. Table 3 gives the composition of egg white, showing the important proteins. Some proteins in egg white have certain biological activity that affords protection from microbiological growth. For example, lysozyme lyses certain microorganisms; conalbumin binds iron, retarding growth of bacteria which require iron; avidin binds the vitamin biotin; ovomucoid inhibits trypsin activity. All of these activities are destroyed when the egg white is cooked.

Table 3. Composition of Egg White[a]

Constituent	Amount, %	Isoelectric point	Mol wt	Unique properties
ovalbumin	54	4.6	45,000	denatures easily, has sulfhydryls
conalbumin	13	6.0	80,000	complexes iron, antimicrobial
ovomucoid	11	4.3	28,000	inhibits enzyme trypsin
lysozyme	3.5	10.7	14,600	enzyme for polysaccharides, antimicrobial
ovomucin	1.5	?		viscous, high sialic acid, reacts with viruses
flavoprotein-apoprotein	0.8	4.1	35,000	binds riboflavin
"proteinase inhibitor"	0.1	5.2		inhibits enzyme (bacterial proteinase)
avidin	0.05	9.5	53,000	binds biotin, antimicrobial
unidentified proteins	8	5.5, 7.5, 8.0, 9.0		mainly globulins
nonprotein	8			primarily half glucose and salts (poorly characterized)

[a] Ref. 3.

Yolk. The yolk is separated from the white by the vitelline membrane and is made up of layers which can be seen by careful examination. Egg yolk is a complex mixture of mostly lipids and proteins. Lipid components are: glycerides, 65.5%; phospholipids, 29.3%; and cholesterol, 5.2%. The phospholipids consist of 73% lecithin, 15% cephalin, and 12% other phospholipids. Of the fatty acids, 33% are saturated and 67% unsaturated, including 42% oleic acid and 7% linoleic acid. All of the lipids are associated with phosphoprotein.

Yolk can be separated into two fractions by using high speed centrifugation: granules and plasma (4). The granules contain a high percentage of high density lipoproteins (HDL) and lesser amounts of low density lipoproteins (LDL) and water soluble proteins (phosvitins). The plasma contains water soluble proteins (livetins) and finely dispersed LDL. Most of the glycerides reside in this LDL fraction. The glycerides apparently form the inner core of the LDL, which is surrounded by a phospholipid shell, with protein wrapped around the shell. Half of the water in egg yolk is bound to the proteins and lipoproteins and half is free.

pH. The pH of egg white is normally about 9.0. However, egg white in freshly laid eggs has a pH of about 7.6. pH increases as carbon dioxide is lost during storage. The pH of yolk is normally about 6.6, and in freshly laid eggs it is 6.0. The chemistry of eggs has recently been reviewed (5).

Nutritional Properties

Eggs are considered to be one of the most nutritious foods. Eggs have the highest quality protein of any food and are important as a source for iron, phosphorus, trace minerals, and certain vitamins. Lipids in eggs are easily digested, and the amount of unsaturated fatty acids is greater than in most animal products. A review of the nutritive value of eggs is given in ref. 6.

Nutritional composition of eggs is shown in Table 4. Most evidence indicates that nutritional properties are retained almost completely in all types of egg products (8).

Table 4. Nutritional Composition of Eggs[a]

Nutrient	Egg white	Egg yolk	Whole egg
Proximate			
water, g	88.07	48.76	74.57
food energy, kJ	204	1546	661
(kcal)	(49)	(369)	(158)
protein (N × 6.25), g	10.14	16.40	12.14
total lipid (fat), g	trace	32.93	11.15
carbohydrate, total, g	1.23	0.21	1.20
fiber, g	0	0	0
ash, g	0.56	1.70	0.94
Minerals			
calcium, mg	11	152	56
iron, mg	0.03	5.58	2.09
magnesium, mg	9	15	12
phosphorus, mg	11	508	180
potassium, mg	137	90	130
sodium, mg	152	49	138
zinc, mg	0.02	3.38	1.44
Vitamins			
ascorbic acid (vitamin C), mg	0	0	0
thiamine (vitamin B_1), mg	0.005	0.254	0.087
riboflavin (vitamin B_2), mg	0.285	0.436	0.301
niacin, mg	0.089	0.069	0.062
pantothenic acid (vitamin B_3), mg	0.241	4.429	1.727
pyridoxine (vitamin B_6), mg	0.003	0.310	0.120
folacin (vitamin B_c), µg	16	152	65
cyanocobalamin (vitamin B_{12}), µg	0.065	3.803	1.547
retinol (vitamin A), IU	0	1839	520
Lipids			
fatty acids			
saturated, total, g	0	9.89	3.35
14:0, g	0	0.09	0.03
16:0, g	0	7.27	2.46
18:0, g	0	2.53	0.86
monounsaturated, total, g	0	13.16	4.46
16:1, g	0	1.10	0.37
18:1, g	0	12.06	4.08
polyunsaturated, total, g	0	4.28	1.45
18:2, g	0	3.66	1.24
18:3, g	0	0.09	0.03
20:4, g	0	0.28	0.09
22:5, g	0	trace	trace
22:6, g	0	trace	trace
cholesterol, mg	0	1602	548
Amino acids			
tryptophan, g	0.156	0.241	0.194
threonine, g	0.451	0.890	0.596
isoleucine, g	0.618	0.939	0.759
leucine, g	0.883	1.396	1.066
lysine, g	0.625	1.110	0.820
methionine, g	0.394	0.417	0.392
cystine, g	0.251	0.291	0.289
phenylalanine, g	0.638	0.714	0.686
tyrosine, g	0.407	0.706	0.505
valine, g	0.759	1.000	0.874
arginine, g	0.592	1.136	0.777

Table 4 (*continued*)

Nutrient	Amount in 100 g of egg component		
	Egg white	Egg yolk	Whole egg
histidine, g	0.230	0.394	0.293
alanine, g	0.654	0.821	0.709
aspartic acid, g	0.897	1.370	1.204
glutamic acid, g	1.415	2.007	1.546
glycine, g	0.380	0.496	0.404
proline, g	0.381	0.682	0.482
serine, g	0.748	1.359	0.923

[a] Ref. 7.

Of the many components in eggs, cholesterol has recently received most attention. Considerable controversy surrounds the role of dietary cholesterol, its relationship to serum cholesterol, and its effect on arteriosclerosis. Actually, cholesterol is an important part of animal tissues and cells. For example, cholesterol is necessary in the production of vitamin D, certain hormones, and biosalts, which are essential for digestion of foods. Cholesterol is carried to tissue in the body by the blood. The body maintains a certain cholesterol level and can synthesize the amount it needs. When supplied through the diet, the body will synthesize a lesser amount. There has been contradictory evidence that high levels of serum cholesterol does or does not result in a higher risk of heart disease. However, as a precautionary measure, many physicians still advocate diets low in cholesterol for persons with higher than normal blood cholesterol and for those who may be prone to heart disease.

Functional Properties

Eggs function in many different ways to give food products in which they impart certain desirable characteristics. Functional properties are as follows:

Coagulation and Thickening Ability. Eggs have the ability to bind pieces of food together or to thicken foods such as in cakes, custards, omelets, and puddings. Egg proteins denature and coagulate during heating over a wide range of temperatures, 57–82°C (9). This is one of the most important properties and is the reason eggs can be prepared and used in so many different ways. The unique coagulating characteristics of egg white protein are demonstrated in the baking of an angel food cake. When the cake is baked some of the egg white protein begins to coagulate at a relatively low temperature, about 57°C, which sets the foam batter structure. This structure is elastic and all of the protein will not be completely coagulated until the cake has expanded and developed into its final set form at about 82°C. Many different ingredients and conditions have an effect on the temperature and time required for coagulation of egg protein, including salt, sugar and other carbohydrates, pH, and mechanical action.

Foam-Forming. Eggs have the ability to incorporate air and form a foam when beaten with a mechanical device, such as a wire wisk. Egg white can be easily whipped into very stable foams by itself or when mixed with other ingredients, such as sugar. With the formation of numerous bubbles, surface areas are formed within the foam and the native protein will unfold, spread, and denature at the liquid–air interface of the bubbles. Since this surface denaturation is irreversible, a netlike structure is formed to maintain the foam in a fairly stable form.

Whole egg and yolk will also foam at a somewhat lower rate than egg white, even though these materials contain large amounts of lipids. The lipids are in a highly emulsified state, but if the emulsion is broken, whole egg or yolk will lose its ability to foam. Free fat has an adverse effect on whipping properties of all egg products. The foam-forming power of eggs is quite sensitive and can be adversely affected by certain processing procedures: for example, heating liquid to too high a temperature, subjecting egg to excessive shear forces, and drying under conditions which release free fat.

Emulsifying Power. Eggs and egg products have good emulsifying characteristics, with egg yolk rated best, whole eggs next, and egg white somewhat less effective. The excellent emulsifying power is attributed to phospholipids, but other components are also involved. Some work has indicated that when lipids are extracted from whole egg the remaining portion has good emulsifying properties (10).

An excellent example of the emulsifying properties of egg is in the making of mayonnaise. Here egg is the only emulsifier, and it must support a stable emulsion containing a minimum of 65% vegetable oil. Emulsifying properties are also important in many baked items such as in cakes where fats and oils are present (see Emulsions).

Tenderizing. Eggs retard crystallization of sugar and contribute to smoothness, moistness, and certain desirable textural characteristics of candies and baked goods.

Moisture Retention. By binding ingredients, eg, in cakes, eggs offer a barrier against water evaporation. Thus they help to retard moisture loss during baking and storage.

Flavor. Eggs have a distinct flavor which makes them desirable for eating by themselves. They also contribute to the flavor and mouthfeel of baked goods and other food products in which they are used.

Color. Natural color in egg yolk comes from xanthophylls and other fat-soluble pigments (see Dyes, natural). Eggs contribute to color in products in which they are used, although this is seldom their primary function. Although egg color can be substituted by other coloring materials, there are some products in which U.S. Federal Standards permit only eggs as the coloring material, eg, in egg noodles.

Production and Marketing of Shell Eggs

Egg production has gone through considerable change. Thirty years ago most of the eggs were produced by small farm flocks, with the largest flocks consisting of only a few hundred birds. Since then flock size has grown steadily until today most eggs are produced from flocks of 30,000 birds or more. Several operations have more than one million birds. A review of egg production practices is given in ref. 11.

Eggs are graded and sorted according to size and quality factors, which include both shell quality and interior quality (12). At one time all eggs were candled by hand. The egg was placed before a candling light and given a quick twist. Appearance and motion of the yolk and size of the air cell gave an indication of the interior quality. Candling is also used for detecting and removing eggs with blood spots and those with checked or cracked shells and other obvious defects. Now eggs are flash candled on a continuous conveyor within a short time after being laid. Because of their freshness most of these eggs have uniformly high interior quality with the proportion of thick to thin egg white being relatively high. The USDA has an egg-grading program which is run on a voluntary basis in cooperation with each state. Most states also have their own egg-grading laws. Following are United States size or weight classifications for eggs:

Size or weight class	Minimum net weight per dozen, kg
jumbo	0.85
extra large	0.77
large	0.68
medium	0.60
small	0.51
pee wee	0.43

Eggs are downgraded according to the following conditions of the shell:

Dirty. The shell is unbroken and has adhering dirt or foreign material, prominent stains, or moderate stains covering more than $1/4$ of the shell surface.

Check. The individual egg has a broken shell or a crack in the shell, but with its shell membrane intact and its contents do not leak.

Leakers. The shell and membrane are broken so that the contents are leaking. USDA regulations prohibit use of this type of egg for human consumption.

Grade AA and A quality eggs, as described in Table 5, are generally recommended for most household uses, such as frying, poaching, and cooking in the shell. However, lower grades can usually be used for many cooking and baking purposes. Higher grade eggs have somewhat better functional properties such as foam-forming power (13).

The present handling and processing methods for shell eggs is now highly mechanized. This includes the steps of collecting, sorting, washing, drying, candling, grading, classifying by weight, and packing. The large end of the egg is generally spray-coated with mineral oil in order to retard escape of carbon dioxide and water, and thus retain quality for a longer period.

One of the quality changes noted in eggs during storage is the thinning of the thick whites, which results in a lower grade. This can be noted through a candling light by a more distinct appearance of the yolk and greater movement of the yolk. This can also be observed when the egg is broken out on a flat surface. The large percentage of thick whites in fresh eggs will cause the egg contents to stand high. Eggs of poorer quality will spread out, and the yolk will flatten. The Haugh unit is one way of expressing this quality factor and is determined by a formula which reflects the relationship between the egg white height and the weight of the egg. The USDA has a flock-certification grading program based on the broken-out appearance of the white and yolk.

Eggs can also lose quality by bacterial contamination, usually because of improper cleaning or washing procedures. Actually, eggs have several natural barriers against bacterial invasions such as the shell, membranes, and several antibacterial factors present in the egg white. However, when eggs are washed, if wash solution temperature is less than the temperature of the egg, wash solution can be pulled into the egg through the shell because of contraction of the egg contents. Eggs washed with water high in iron content have been found to be more susceptible to spoilage (14). Apparently, excess iron overcomes the ability of conalbumin, which binds iron, to inhibit bacterial growth. Almost all eggs are washed before they are packed. When done properly no bacterial contamination will occur. Washing is usually done in a continuous system with eggs being conveyed through a washing chamber where they are brushed and

Table 5. United States Standards for Quality of Individual Shell Eggs [a]

Quality factor	AA quality	A quality	B quality	C quality
shell	clean, unbroken, practically normal	clean, unbroken, practically normal	clean to slightly stained, unbroken, may be slightly abnormal	clean to moderately stained, unbroken, may be abnormal
air cell	3.2 mm or less in depth, practically regular	4.8 mm or less in depth, practically regular	9.5 mm or less in depth, may be free or bubbly	may be over 9.5 mm in depth, may be free or bubbly
white	clear, firm	clear, may be reasonably firm	clear, may be slightly weak	may be weak and watery
yolk	outline slightly defined, practically free from defects	outline may be fairly well defined, practically free from defects	outline may be well defined, may be slightly enlarged and flattened, may show definite but not serious defects	small blood clots or spots may be present, outline may be plainly visible, may be enlarged and flattened, may show clearly visible germ development but no blood, may show other serious defects

[a] Ref. 12.

flushed with wash solution. After washing the eggs are rinsed and sanitized with either a chlorine or iodine solution. Temperature of the washing solution must be 32°C or higher, or 11°C warmer than the egg (12). This assures that the wash solution will not be drawn through the shell.

Eggs are packed for either retail or commercial use. For retail use the usual pack is one dozen eggs per carton. For commercial use it is 30 dozen eggs per case. If eggs are to be broken for manufacture of egg products, they are sometimes placed on flats or trays, and these are stacked on racks which are moved to the point of use.

Eggs have become more and more uniform throughout the year and they are seldom stored for periods of more than two weeks. Several years ago when production was seasonal, some eggs were stored for long periods, up to a year. Storage temperature is usually around 10°C. Higher temperature permits more rapid decrease in quality, and at lower temperatures eggs will "sweat" when removed from the storage room.

Egg Products

There are three basic types of egg products: liquid, frozen, and dried. All of these have many uses, eg, in baked goods, bakery mixes, noodles, mayonnaise and salad dressing, candies, and ice cream. Several reviews have been published on the subject of processing and marketing of egg products (15–21).

Tremendous changes have taken place in the egg products industry through the past 20–30 years. Some of these changes are indicated below.

Government Regulations. The FDA issued a regulation in the 1960s requiring that all egg products must be pasteurized in a way that renders them salmonella-negative by a specified testing procedure. Compulsary inspection under USDA supervision began December 29, 1970. Before that time inspection was on a voluntary basis. USDA inspection was then specified by certain users of egg products and was required for purchases made by the military and by USDA for various government programs. USDA already had regulations setting forth minimum requirements under their voluntary inspection program. Through the years these requirements have been refined and have become more and more rigid. Specific pasteurization conditions have been set forth for every type of egg product, and sanitary requirements of equipment have been established through the use of E3A (Dairy and Food Industry Supply Association) standards (22).

Customer Specifications. Companies that purchase egg products have become very aware of the need for specifications. These include microbiological, chemical, and physical or functional specifications. Microbiological specifications usually include: total plate count; coliform; yeast and mold; and salmonella. At present many specifications also call for negative results on other types of pathogenic microorganisms, such as coagulase-positive staph, *E. coli,* and *Clostridium perfringens*.

Chemical specifications usually include: solids or moisture; fat; proteins. These have the purpose of determining whether the egg product contains the full amount of egg. Sometimes other specifications include color, salt, sugar, ash, acidity of ether extract, and pesticide residue. Standard test procedures are found in ref. 23.

Performance of the egg product in the customer's finished goods is probably the most important requirement of an egg product. There are certain laboratory tests that predict how the egg product will perform, and these tests are sometimes included in a specification. Physical and functional tests include: scrambled eggs and custards for flavor and coagulation ability; whip tests, indicating foaming ability and stability of foam; angel food cake, which shows general performance of egg white products; sponge cakes for general performance of whole egg and yolk products; and mayonnaise preparations for emulsifying ability.

Nutritional Labeling. Increased interest is being shown in the nutritional properties of egg products and how egg products will affect nutritional properties in foods in which they are used. Much of this has come about because of new labeling requirements.

Waste Disposal and Loss Control. There are three types of wastes in the egg products industry: fluid wastes to the sanitary sewer; solid wastes to some type of dump or landfill program; and emissions to the atmosphere, such as those from spray-dryer stack losses.

Fluid wastes going to the drain, which include a certain amount of egg, represent an economic loss and increase the cost of disposing of the waste. Some losses cannot be completely eliminated, but can usually be reduced. For example, eggs are sometimes lost in an egg washer because of cracked or checked shells or thin shells that are fragile. These losses can be minimized by proper maintenance and adjustment of the equipment.

Solid wastes include case materials and shells. Shells can be dried and recovered as a usable item. Since they contain a high amount of calcium carbonate, they can be recycled into the laying ration, replacing oyster shells and limestone (24).

Emissions into the atmosphere from spray dryers can be minimized by using appropriate collection systems. Cyclone collectors are adequate for some types of egg products, but not all. Cloth collectors or bag filters are usually necessary when drying egg white or products containing carbohydrates in order to achieve good collection efficiency. Wet scrubbers are sometimes used, but these present certain sanitation problems (see Air pollution control methods).

Egg Breaking. Egg breaking is the beginning step in the production of all types of egg products. Eggs are received at the plant in cases or on racks and are held in a tempering room at 10°C until broken. This temperature gives the optimum conditions for breaking and separating. At present almost all eggs are broken and separated by machine. A completely automated system is shown in Figure 1. Eggs are loaded onto the system, washed, rinsed, sanitized, flash-candled, and then fed to an egg breaking and separating machine. This machine cracks and opens the shell and separates yolks from whites. Each system can handle 18,000–25,000 eggs per hour. Three persons are required to operate this type of system: one to load; one to flash-candle eggs coming out of the washer; one to operate the egg breaking and separating machine. The cracker head with the empty shell and the separating cup with egg contents move in front of the operator of the machine. Separating cups with intact yolks move on and discharge the whites and yolks into separate collection trays. The operator must inspect all opened eggs and trip any separating cup where the yolk membrane has broken and the whites are mixed with the yolks; this material is diverted to a separate collection tray. The operator also inspects the whites after separation and removes any contaminating yolk. The whites flow on a tray in front of the machine where yolk particles can be scooped into the mix tray. The whites are tested for yolk content, which is usually held below 0.03% (25). The development of automatic egg breaking machines began in the late 1940s. The first models broke only 3600–5400 eggs per hour. This was actually quite an improvement over hand breaking and separating where, by using

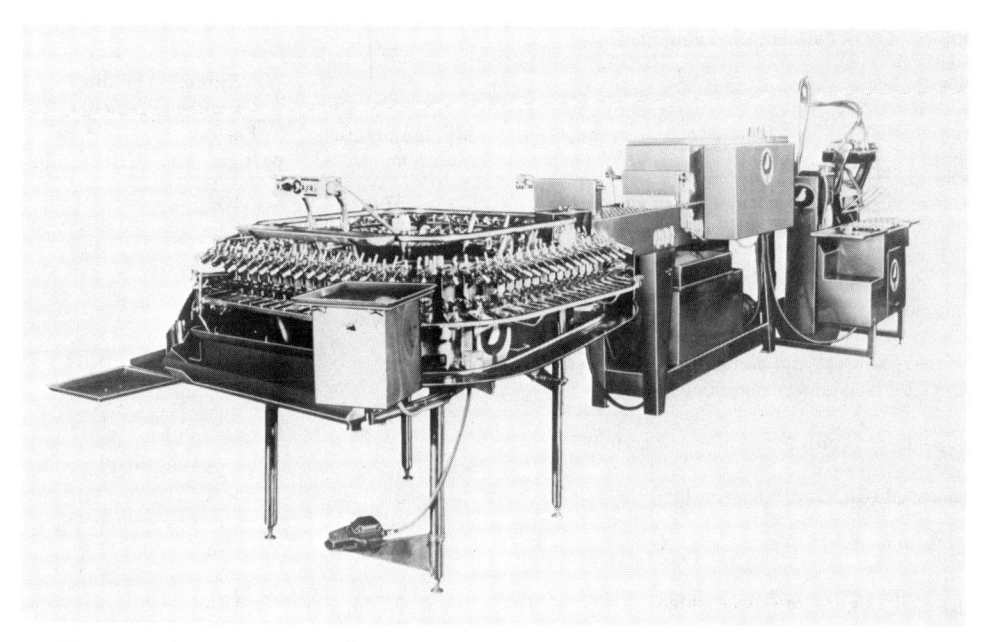

Figure 1. Automatic egg handling and breaking–separating system, Courtesy Henningsen Foods, Inc.

a hand breaking and separating device, one person could break and separate 1½–3 cases per hour.

Three streams flow from the egg breaking and separating machine: egg white, yolk, and mix. Each component is strained or clarified and thoroughly mixed before going to further processing. The mix is standardized to a solids level of 24.7–25.0% by adding yolk or whites. The yolk is standardized to a level above 43.0%, depending on specifications, by adding mix or whites. During separation from the white, some white adheres to the yolk, reducing its solids content. A white-free yolk would have about 51% solids, although United States standards permit solids of 43.0% minimum. The minimum is a mixture of 80% pure yolk and 20% white. Whole egg and yolk products are chilled to less than 5°C and egg whites to less than 7°C immediately after breaking.

Pasteurization. Since 1966 pasteurization has been required for all egg products in the United States. Table 6 shows the minimum pasteurization requirements by USDA. Except for egg white to be dried, it is necessary to pasteurize all liquid eggs whether they are for liquid usage or are to be frozen or dried. These pasteurization conditions are based on a bacterial kill equivalent to that obtained when heating whole egg liquid to 60°C for 3½ min. This has been established as effective for destruction of salmonella (27).

The purpose of pasteurization is to reduce or eliminate the possibility of having pathogenic microorganisms, such as salmonella, present in the finished egg product. Pasteurization must be accomplished without cooking the egg or damaging its sensitive functional properties. High temperature, short time (HTST) pasteurization equipment similar to that in the dairy industry is used for the pasteurizing liquid eggs. This

440 EGGS

Table 6. USDA Pasteurization Requirements [a]

Liquid egg product	Min temperature requirements, °C	Fastest particle, min	Average particle, min
albumen (egg white) (without use of chemicals)	57	1.75	3.5
	56	3.1	6.2
whole egg	60	1.75	3.5
whole egg blends (less than 2% added nonegg ingredients)	61	1.75	3.5
	60	3.1	6.2
fortified whole egg and blends (24–38% egg solids, 2–12% added nonegg ingredients)	62	1.75	3.5
	61	3.1	6.2
salt whole egg (with 2% or more salt added)	63	1.75	3.5
	62	3.1	6.2
sugar whole egg (2–12% sugar added)	61	1.75	3.5
	60	3.1	6.2
plain yolk	61	1.75	3.5
	60	3.1	6.2
sugar yolk (2% or more sugar added)	63	1.75	3.5
	62	3.1	6.2
salt yolk (2–12% salt added)	63	1.75	3.5
	62	3.1	6.2

[a] Ref. 26.

equipment must have all the standard components of an HTST pasteurizer, including constant-level tank, timing pump, heating plates, regeneration plates, holding tubes, flow-diversion valve, and cooling plates. Batch pasteurization of egg products is also permitted under certain conditions. For example, whole egg can be heated at 55.6°C for 35 min or 57.2°C for 15 min to accomplish pasteurization equivalent to heating the liquid in an HTST pasteurizer to 60°C for $3\frac{1}{2}$ min (28). All pasteurization equipment must meet E3A standards.

Several pasteurization methods are acceptable for egg white products. Since the natural pH of liquid egg whites is 9.0, a relatively low pasteurization temperature is quite effective. Other methods include: (1) Heat treatment of stabilized egg white at 60°C for $3\frac{1}{2}$ min, where the white is stabilized by adjusting to pH 7.0 with lactic acid and aluminum sulfate (29). At pH 7.0 stability of ovalbumin, lysozyme, ovomucoid, and ovomucin is greatly improved. The aluminum which is added complexes with conalbumin to give this protein much better heat stability. (2) Hydrogen peroxide can be added after heating plain egg white to 52°C followed by addition of catalase to break down the hydrogen peroxide. Here heating serves to inactivate the natural catalase in egg white before the hydrogen peroxide is introduced and also improves the bacterial killing power of hydrogen peroxide (30). (3) Heat treatment dried egg white at >54°C for a minimum of seven days. Combining liquid pasteurization with heat treatment of the dried product improves bacterial kill and gives greater assurance against presence of salmonella (31).

Liquid Egg Products. The most common liquid egg products are: white, yolk, and whole egg. Liquid eggs are generally consumed by large users, such as bakeries, who have the necessary handling equipment. They are transported in tank trucks holding approximately 20 metric tons. Temperature of egg yolk and whole egg must be held below 5°C and egg white must be below 7°C. For long distances it is necessary to refrigerate the transport tanks, eg, with liquid nitrogen. Portable refrigerated vats holding ca 500 kg are also used for transporting liquid eggs.

Specifications for liquid egg products depend on customer requirements. Typical specifications are as follows:

	Whites	*Yolks*	*Whole egg*
solids, % min	11.5	43.0	24.7
yolk content, % max	0.03		
fat, % min		25.0	10.0
protein, % min	9.0	13.0	11.3
standard bacterial plate count, max	25,000/g	25,000/g	25,000/g
coliform, max	10/g	10/g	10/g
salmonella	negative	negative	negative

Frozen Egg Products. The most common frozen egg products are:

Frozen Egg White. Because pasteurizing of egg white may have an adverse effect on whipping properties, a whipping aid is permitted as an optional ingredient in frozen egg white. For example, triethyl citrate is sometimes used (32). Stabilizers, such as gums, can also be added to increase viscosity and improve stability of the egg white foam (33).

Frozen Whole Egg. Pasteurized whole egg changes in consistency during freezing. When thawed it has a watery separated appearance, but after thorough stirring it appears to be uniform and smooth.

Frozen Fortified Whole Egg. This product is usually 70% whole egg and 30% egg yolk.

Frozen Plain Egg Yolk. Gelation of yolk takes place at $< -6°C$. When frozen egg yolk is thawed it has a gelled appearance and is difficult to handle. Special equipment is usually required and water is sometimes added. For this reason most frozen yolk products have ingredients, such as sugar or salt, added to reduce gelation and thus improve ease of handling.

Frozen Fortified Whole Egg with Syrup. Composition can vary, eg, one product contains 95% fortified whole egg and 5% corn syrup. Such a product has a relatively high viscosity with a smooth consistency. It has general use in the bakery.

Frozen Sugared Egg Yolk. The usual mixture is 90% egg yolk and 10% sugar. Advantage of this product over frozen plain egg yolk is that it has good fluidity when thawed.

Frozen Salted Egg Yolk. Although salt level can be 2–12%, it is usually 10%. This product is used almost exclusively in making mayonnaise and salad dressings.

Frozen Salted Whole Egg. Again, the usual salt level is 10%. This product is also used in mayonnaise and salad dressings.

Frozen eggs are packed in a number of different types of containers. The most common pack is in a metal can containing 13.6 kg of product. For institutional use plastic-coated fiber containers are sometimes used, and the package contains 13.6 kg or less. A blast freezer at −30 to −40°C is used for periods up to 72 h. For storage frozen eggs are held at −18 to −24°C.

Specifications for frozen egg whites, whole egg, and egg yolk are similar to those indicated above for liquid products. Specifications vary for frozen egg products with added carbohydrates and salt.

Dried Egg Products. Dried egg products are classified as follows:

Dried Egg White. Most dried egg white is made by spray-drying, which produces a powder form. Egg white is also pan-dried, where the product is separated into three different forms: flake, granular, and fines. The fines are usually milled to a powder. All egg white products have their natural glucose removed, which gives them excellent stability under almost any storage condition.

Glucose is removed from liquid egg white before drying by one of three methods: fermentation using a bacterial culture; fermentation using baker's yeast (*Saccharomyies cerevisiae*); oxidation to gluconic acid using a glucose-oxidase–catalase enzyme system (34). For the enzyme method oxygen is supplied by the addition of hydrogen peroxide. Reaction rate is controlled by the amount of enzyme, temperature, and rate of hydrogen peroxide addition. If glucose remained in egg white, the dried product would be very unstable. Reducing groups of the glucose combine with amino groups of the protein in a condensation reaction, followed by browning, and reduced stability (35).

Whipping aids help to preserve whipping properties of egg white during drying. Sodium lauryl sulfate is preferred by most users. Other approved whipping aids include: triethyl citrate (32), triacetin (36), and sodium desoxycholate (37). These are generally added at a level of less than 0.1% on a solids basis.

Functional properties of dried egg white products are similar to those of liquid and frozen egg white. Certain modifications in formulation and mixing techniques are sometimes helpful to give the best results. For example, in the preparation of angel food cakes, it is usually helpful to reduce the amount of acid salt, replace part of the granulated sugar with powdered sugar and part of the flour with wheat starch, and whip the meringue portion to a stiffer peak.

Dried Plain Whole Egg and Yolk. Included in this category are standard whole egg, standard yolk, stabilized (glucose-free) whole egg, stabilized (glucose-free) yolk, free-flowing whole egg, and free-flowing yolk.

Standard whole egg and yolk products are pasteurized and dried without further processing or inclusion of other ingredients or additives. Since the natural glucose is present, these are the least stable egg products, and it is recommended that they be held under refrigeration until used. Standard egg yolk has better storage life than standard whole egg probably because it has less glucose and a lower pH.

Glucose has been removed from stabilized whole egg and yolk before drying in order to give them better stability during storage (38). These products are recommended for use in mixes where shelf life at room temperature is important. Glucose is removed by either an enzyme procedure or a yeast fermentation. Procedures are similar to those used for egg white.

A free flowing, or anticaking agent is added to give dried whole egg and yolk improved flow characteristics. Sodium silicoaluminate added at a level up to 2% and silicon dioxide added at less than 1% are permitted for this purpose. Total of free flowing agent and moisture cannot exceed 5%. The free-flowing agent also improves dispersibility in water, making these products particularly well suited for use in mayonnaise and salad dressings.

All of the plain egg products have functional properties which compare favorably with their frozen or liquid counterparts. As a general rule in the production of many basic bakery items, these types of egg product can be dry-blended with other dry ingredients and water of reconstitution added with other liquids during the mixing stage.

These products cannot be whipped by themselves under ordinary whipping conditions. However, they can be used for making sponge cakes when appropriate emulsifiers are used.

Dried Blends of Whole Egg and Yolk with Carbohydrate. There are a number of different blends. The most common ones are: whole egg plus sucrose; whole egg plus corn syrup; whole egg plus yolk plus corn syrup; and yolk plus corn syrup. These are now the most widely used dried egg products for general bakery goods. Whole egg and yolk co-dried with carbohydrate at a level of about 10% on a liquid basis have good foaming properties. Corn syrup also gives to dried eggs anticaking characteristics, better flowability, and improved dispersibility in water. Dried blends of egg and carbohydrate function satisfactorily in both unemulsified and emulsified sponge cakes. They are also well suited for use in continuous mixing equipment.

Special Dried Egg Products. There are several other types of egg products where eggs are mixed and co-dried with other ingredients. One example is a scrambled egg mix which has been purchased by the United States government under USDA direction for various programs. This mix contains, on a solids basis, 51% whole egg, 30% skim milk, 15% vegetable oil, $1\frac{1}{2}$% salt, and $2\frac{1}{2}$% moisture.

The most important methods for producing dried egg products is by spray drying. It is essential that the dryer is of sanitary construction, meeting E3A standards. Several different types of spray dryers are used. Some dried egg products, such as egg white, are difficult to separate from the drying air and thus require collectors with high efficiency. Cloth filter collectors are commonly used. Liquid fed to the spray dryer is either cooled to below 5°C or preheated to above 59°C. Dried whole egg and yolk products are usually cooled to below 30°C as they are discharged from the dryer. This is done by exposure to cooling air or by contact with carbon dioxide (see Drying).

Finished dried egg products are generally packed in polyethylene-lined fiberdrums or boxes. Specifications for dried egg products vary depending on customer requirements. Examples of specifications are as follows:

	Spray-dried egg white	Spray-dried stabilized whole egg	Spray-dried stabilized egg yolk
moisture, % max	8.0	5.0	5.0
fat, % min		40	56
protein, % min	80.0	45.0	30.0
reducing sugar, % max	0.1	0.1	0.1
total bacterial plate count, max	10,000/g	25,000/g	25,000/g
coliform, max	10/g	10/g	10/g
yeast and mold, max	10/g	10/g	10/g
salmonella	negative	negative	negative

Other Special Egg Products. Several simulated egg products with low cholesterol have been developed for the purpose of replacing whole egg in the diet. These are in frozen, liquid, and dry forms. The base for these products is egg white, or egg white with some whole egg. Nonfat milk and vegetable oil are added to give a composition similar to whole egg. Certain ingredients are added to improve consistency when scrambled eggs are prepared, and nutrients are added to duplicate the nutritional properties of whole egg as closely as possible. Canadian regulations (39) specify that any simulated whole egg product must have the following nutritional properties:

B.22.032. No person shall sell any product simulating whole egg unless that product (*1*) is made from liquid, dried or frozen egg albumen or mixtures thereof; (*2*) has a protein rating of not less than 40 as determined by the official method; (*3*) notwithstanding sections *D.01.009* and *D.02.009,* contains per 100 g on a ready-to-use basis: not less than, (*a*) 50 mg calcium, (*b*) 2.3 mg iron, (*c*) 1.5 mg zinc, (*d*) 130 mg potassium, (*e*) 1000 IU (300 µg) vitamin A, (*f*) 0.10 mg thiamine, (*g*) 0.30 mg riboflavin, (*h*) 3.60 mg niacin, (*i*) 1.60 mg pantothenic acid, (*j*) 0.20 mg vitamin B$_6$, (*k*) 0.50 µg vitamin B$_{12}$, (*l*) 0.02 mg folic acid (folacin), and (*m*) 2.0 IU (1.34 mg) alpha tocopherol; and not more than 3 mg cholesterol; (*4*) has a calcium:phosphorus ratio of not less than one part calcium to four parts phosphorus; and (*5*) contains in the total fat of any fat or oil used not less than 40% *cis–cis*-methylene-interrupted (/═\/═\ ) polyunsaturated fatty acids and not more than 20% saturated fatty acids.

Some effort has been made to market shelled hard-cooked eggs. Equipment has been developed to prepare this type of product automatically and semiautomatically. Pickling is the most successful form of preservation. Several different types of pickling solutions have been used. Refrigeration alone for preservation of shelled hard-cooked eggs has presented many problems, eg, bacterial spoilage and syneresis.

The precooked long egg or egg roll has proven to be quite successful in both refrigerated and frozen forms. Processes have been developed to place the egg yolk in the center and the egg white on the outside so that when sliced the product has the appearance of a sliced hard-cooked egg. Gums and starches are added to egg white to help prevent syneresis (40).

There are several uses for eggs other than for human food, eg, as an adhesive; an ingredient in pet food, particularly for dogs and cats; a nutrient for culture media; and an ingredient in cosmetics and shampoos.

BIBLIOGRAPHY

"Eggs" in *ECT* 1st ed., Vol. 5, pp. 465–477, by Alexis L. Romanoff, Cornell University; "Eggs" in *ECT* 2nd ed., Vol. 7, pp. 661–676, by W. W. Marion, Iowa State University.

1. Statistical Reporting Service, USDA, Washington, D.C., data for 1954 to 1978.
2. R. H. Forsythe and D. H. Bergquist, *Poultry Sci.* **30,** 302 (1951).
3. R. E. Feeney, "Egg Proteins" in H. W. Schultz and A. F. Anglemeir, eds., *Symposium on Foods: Proteins and Their Reactions,* Avi Publishing Co., Inc., Westport, Conn., 1964.
4. J. R. Schultz and R. H. Forsythe, *Baker's Dig.* **41,** 56, 60 (1967).
5. W. D. Powrie, "Chemistry of Eggs and Egg Products" in W. J. Stadelman and O. J. Cotterill, eds., *Egg Science and Technology,* Avi Publishing Co., Inc., Westport, Conn., 1977, pp. 65–91.
6. F. Cook and G. M. Briggs, "Nutritive Value of Eggs" in ref. 5, pp. 93–108.
7. L. P. Posati and M. L. Orr, principal investigators, *Composition of Foods, Dairy and Egg Products, Raw-Processed-Dried, Agriculture Handbook No. 8-1,* Agriculture Research Service, United States Department of Agriculture, Washington, D.C., Nov. 1976.
8. G. J. Everson and H. J. Sounder. *J. Am. Diet. Assoc.* **33,** 1244 (1957).
9. S. R. Payawal, B. Lowe, and G. F. Stewart. *Food Res.* **11,** 246 (1946).
10. R. B. Chapin, *Some Factors Affecting the Emulsifying Properties of Hen's Eggs,* Ph.D. thesis, Iowa State College, Ames, Iowa, 1951.
11. M. H. Swanson, "Egg Production Practices" in ref. 5, pp. 7–28.
12. Regulations governing the grading of shell eggs and United States Standards, grades, and weight classes for shell eggs, *USDA 7 CFR,* Part 56, U.S. Government Printing Office, Washington, D.C., July 1, 1971.
13. H. H. Palmer, "Eggs" in P. C. Paul and H. H. Palmer, eds., *Food Theory and Applications,* John Wiley & Sons, Inc., New York, 1972, pp. 527–561.

14. J. A. Garibaldi and H. G. Baynes, *Poultry Sci.* **39,** 1517 (1960).
15. H. D. Lightbody and H. L. Fevold, *Adv. Food Res.* **1,** 149 (1948).
16. R. H. Forsythe. *Baker's Dig.* **38,** 52 (1964).
17. R. H. Forsythe, "The Science and Technology of Egg Products Manufacture in the United States" in T. C. Carter, ed., *Egg Quality: A Study of the Hen's Egg,* Oliver and Boyd, Ltd., Edinburgh, Scotland, 1968.
18. R. H. Forsythe, "Eggs" in S. A. Mat, Jr., ed., *Bakery Technology and Engineering,* 2nd ed., Avi Publishing Co., Westport, Conn., 1972.
19. D. H. Bergquist, "Eggs" in W. B. Van Arsdel, M. J. Copley, and A. I. Morgan, Jr., eds., *Food Dehydration,* 2nd ed., Vol. 22, Avi Publishing Co., Westport, Conn., 1973.
20. R. H. Forsythe and D. H. Bergquist, *Baker's Dig.* **47,** 84 (1973).
21. W. J. Stadelman and O. J. Cotterill, eds., *Egg Science and Technology,* Avi Publishing Co., Inc., Westport, Conn., 1977.
22. *Index of Published 3A and E3A Sanitary Standards and Accepted Practices,* Dairy and Food Industry Supply Association, Washington, D.C., Nov. 1, 1978.
23. *Official Methods of Analysis of the Association of Official Agricultural Chemists,* 12th ed., AOAC, Washington, D.C., 1975.
24. J. M. Vandepopuliere, H. V. Walton, and O. J. Cotterill, *Poultry Sci.* **54,** 131 (1975).
25. D. H. Bergquist and F. Wells. *Food Technol.* **10,** 48 (1956).
26. Regulations governing the inspection of eggs and egg products, *USDA 7CFR,* Part 59, U.S. Government Printing Office, Washington, D.C., June 30, 1975.
27. A. R. Winter, P. A. Greco, and G. F. Stewart, *Food Res.* **11,** 229 (1946).
28. A. W. Brant, G. W. Patterson, and R. E. Walters, *Poultry Sci.* **47,** 878 (1968).
29. F. E. Cunningham and H. Lineweaver, *Food Technol.* **19,** 1442 (1965).
30. U.S. Pat. 2,776,214 (Jan. 1, 1957), W. E. Lloyd and L. A. Harriman (to Armour and Co.).
31. U.S. Pat. 2,982,663 (May 2, 1961), D. H. Bergquist (to Henningsen Foods, Inc.).
32. U.S. Pat. 2,637,654 (May 5, 1953), H. L. Kothe (to Standard Brands, Inc.).
33. U.S. Pat. 3,219,457 (Nov. 23, 1965), H. F. Ziegler, Jr., and H. J. Buehler (to Anheuser-Busch, Inc.).
34. U.S. Pat. 2,482,724 (Sept. 20, 1949), D. L. Baker (to B. L. Sarett).
35. R. W. Kline and G. F. Stewart, *Ind. Eng. Chem.* **40,** 919 (1948).
36. U.S. Pat. 2,933,397 (Apr. 19, 1960), V. F. Maturi, L. Kogan, and N. G. Marotta (to Standard Brands, Inc.).
37. U.S. Pat. 2,881,077 (Apr. 7, 1959), L. Kline and A. D. Singleton (to National Dairy Products Corp.).
38. L. Kline and T. T. Sonoda, *Food Technol.* **5,** 90 (1951).
39. *Canadian Food and Drug Regulations, 73A-1,* April 28, 1977.
40. U.S. Pat. 3,510,315 (May 5, 1970), R. L. Hawley (to Ralston Purina Co.).

DWIGHT H. BERGQUIST
Henningsen Foods, Inc.

ELASTOMERS, SYNTHETIC

Survey, 446
Acrylic elastomers, 459
Butyl rubber, 470
Chlorosulfonated polyethylene, 484
Ethylene-propylene rubber, 492
Fluorinated elastomers, 500
Neoprene, 515
Nitrile rubber, 534
Polybutadiene, 546
Polyethers, 568
Polyisoprene, 582
Polypentenamers, 592
Styrene-butadiene rubber, 608
Thermoplastic elastomers, 626

SURVEY

The beginnings of the quest for synthetic rubber can be traced back to at least 1860 when Williams (1) demonstrated that natural rubber could be pyrolyzed into monomeric isoprene and oligomeric compounds. The general approach thereafter was to try to convert isoprene (**1**) or related conjugated dienes such as butadiene (**2**) into materials that resembled natural rubber (**3**) (see Rubber, natural).

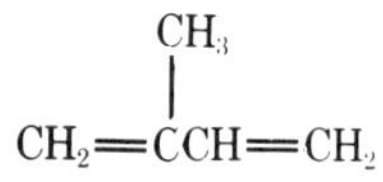

$$CH_2=\underset{\underset{CH_3}{|}}{C}CH=CH_2$$
(**1**)

$$CH_2=CHCH=CH_2$$
(**2**)

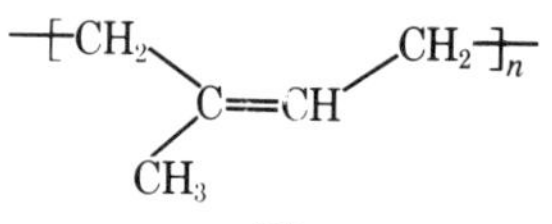

$$-[CH_2\underset{\underset{CH_3}{|}}{C}=CH-CH_2]_n-$$
(**3**)

Considerable progress was made utilizing a classical organic chemistry approach, but it was not until Staudinger's (2) concept of the macromolecule was accepted that a scientific foundation for present day developments was available. Polymer chemists can, to a considerable degree, now control the average chain length, distribution of chain lengths and isomeric unit structures in the polydiene macromolecule (3).

Furthermore, the chemical and stereochemical (4–6) structure requirements to provide elastomeric behavior over a specified temperature interval can now be predicted. Many elastomers have now been developed from monomers other than dienes (7–10). The production of synthetic rubber overtook that of natural rubber in 1960 (5). The articles following this brief survey describe in detail the unique properties, such as resistance to oil, aging, solvents, etc, that can be designed into a synthetic rubber through proper choice of structure and careful compounding. Moreover, block copolymer technology (11–13), which is also described, has permitted the development of the thermoplastic elastomers (qv). These materials can be processed like thermoplastics but have properties analogous to rubbers. In this case, "physical" cross-links are derived from the association of glassy or crystalline hard blocks and may be compared to the chemical cross-links between conventional thermoset elastomers (see Copolymers).

This article provides a discussion of rubber elasticity, a review of the classes, types, structure, and uses of synthetic elastomers and statistics concerning production, general markets and applications of synthetic elastomers (14–15).

Rubber Elasticity

A basic requirement for a rubbery material is that it must consist of long flexible chainlike molecules that can undergo rapid rotation as a result of thermal agitation. In the practical sense, this means a high molecular weight polymer that is well above its glass-transition temperature (T_g). Compilations of T_g values for various structures are available (16). Moreover, crystallinity must be low in the undeformed state of a single-phase rubber. A further requirement is that the long linear-chain molecules must be cross-linked by a few intermolecular bonds during processing to form an insoluble three-dimensional network. This was first demonstrated by Goodyear in 1839 when he cross-linked natural rubber by the reaction of the polydiene with sulfur. Such soft elastomeric solids are capable of undergoing large deformations of several hundred percent and of then recovering essentially completely when the stresses are removed. Details of the development of the theory of rubber elasticity have been discussed thoroughly in several books (17–20) and reviews (9,13,21–23) and can only be briefly outlined here.

Basically, the theories relate the change in molecular dimensions or chain vector lengths resulting from a deformation to the macroscopic strain. This is done by first introducing a relation between entropy and vector length for a single chain. A critical assumption is that the chain vector length changes on deformation in the same ratio as the corresponding dimensions in the bulk rubber (affine deformation). Next, the entropy of a network in the unstrained and strained state is derived, the difference being the entropy of deformation. Essentially constant volume is assumed. It is then possible to use standard thermodynamic procedures to obtain the work of deformation. By differentiation one can then calculate the forces required to produce a particular extent of strain. The work of deformation per unit volume (21) for the most general type of strain is given in equation 1:

$$W = \frac{1}{2} NkT(\lambda_1^2 + \lambda_2^2 + \lambda_3^2 - 3) \tag{1}$$

where λ_1, λ_2, and λ_3 are the three principal extension ratios of a cube of rubber, N is the number of network chains per unit volume, k is Boltzman's constant, and T is absolute temperature.

The case of simple extension, is shown in equation 2:

$$W = \frac{1}{2} NkT(\lambda^2 + 2\lambda - 3) \tag{2}$$

One can obtain the force per unit unstrained cross-sectional area by considering a specimen in the unstrained state to be in the form of a cube of unit edge length.

$$F = \frac{dW}{d\lambda} = NkT\left(\lambda - \frac{1}{\lambda^2}\right) \tag{3}$$

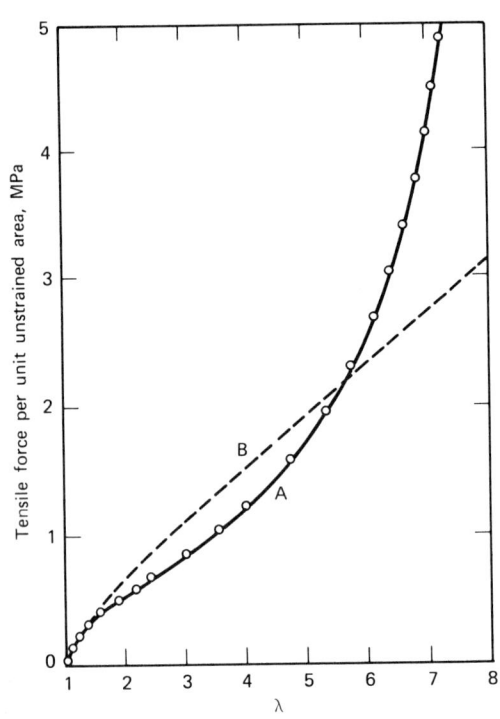

Figure 1. Force–extension curve for uniaxial extension (21). A, experimental; B, theoretical. To convert MPa to psi, multiply by 145. Courtesy of *Rubber Chemistry and Technology* and The Faraday Society.

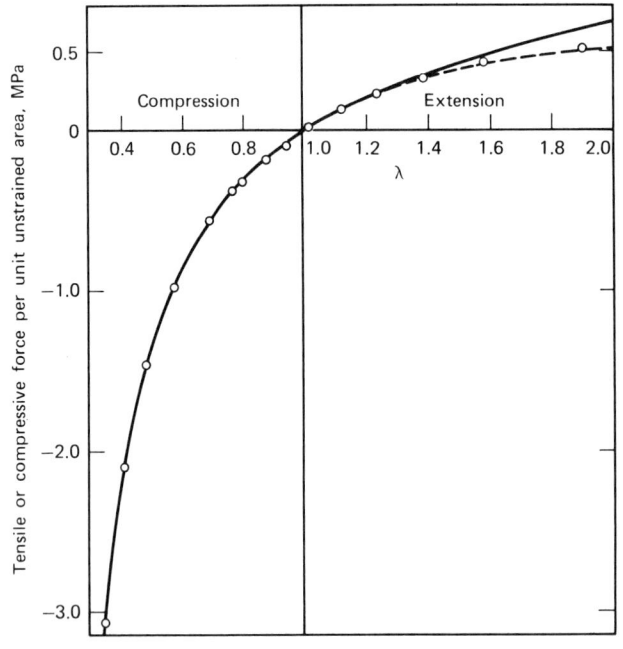

Figure 2. Uniaxial extension and compression (derived from equibiaxial extension) (21). Broken curve, experimental; full curve, theoretical. To convert MPa to psi, multiply by 145. Courtesy of *Rubber Chemistry and Technology* and The Faraday Society.

The quantity NkT in equation 1 can be shown to be equivalent to the shear modulus G (17,21). This can also be expressed in terms of a molecular weight between cross-links, M_c.

$$G = NkT = \frac{\rho RT}{M_c} \qquad (4)$$

where ρ = density of the rubber and R = the gas constant.

Analogous relationships of force to extension ratio can be calculated for other deformation modes such as uniaxial compression and simple shear, as well as simple extension.

The stress–strain relationships of the statistical theory of rubber elasticity are considered to define the properties of an ideal rubber. That is to say a material somewhat analogous to the ideal gas postulated in the kinetic theory of gases. It predicts the equilibrium gum vulcanizate stress–strain relations for any kind of strain will be defined by a single elastic constant (NkT). Secondly, it predicts that the rubbery modulus will be a function of the degree of cross-linking (cf. eq. 4).

The comparison of theory with experiment can be illustrated by the work of Treloar (24) for simple uniaxial extension, compression and shear strains as shown in Figures 1–3. The agreement in compression is particularly striking. In the more common case of simple extension (Fig. 1) the theory agrees to a first approximation up to several hundred percent where strain-induced crystallization effects or non-gaussian chain statistics, or both, are believed to become important. The significant deviations in the region from $\lambda = 1.5$ to $\lambda = 4$ have not been well explained in molecular terms. However, the data are usually better represented by a two-constant empirical Mooney equation.

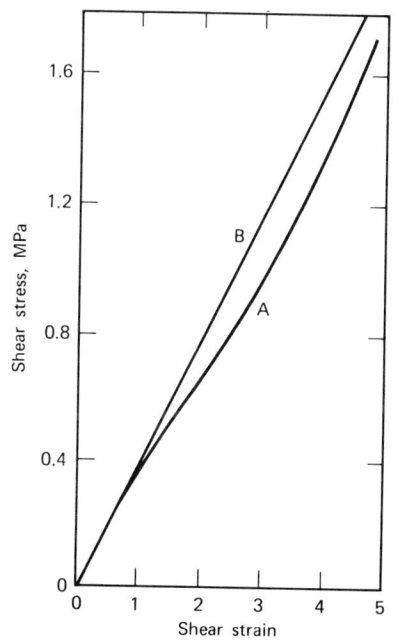

Figure 3. Stress–strain relation for simple shear (21). A, Experimental; B, theoretical. To convert MPa to psi, multiply by 145. Courtesy of *Rubber Chemistry and Technology* and The Faraday Society.

Table 1. Nomenclature and Structure of General Purpose Synthetic Rubbers

Name	CAS Registry Number	Chemical structure	Remarks
natural rubber (NR)	[9006-04-6]	>99% cis-1,4-polyisoprene	ca 6% nonhydrocarbon content
isoprene rubber (IR)	[9003-13-0]	$-(CH_2C(CH_3)=CHCH_2)_n-$ >97% cis-1,4-polyisoprene	Ziegler-Natta catalysis
butadiene rubber (BR)	[9003-17-2]	$-(CH_2C(CH_3)=CHCH_2)_n-$ (a) >97% cis-1,4-polybutadiene $-(CH_2CH=CHCH_2)_n-$ (b) ca 90% 1,4 with mixed cis/trans (c) ca 80% 1,4 with mixed cis/trans ca 20% 1,2 with mixed cis/trans	Ziegler-Natta catalysis anionic polymerization emulsion polymerization
chloroprene rubber (CR)	[9010-98-4]	ca 90% trans-1,4 $-(CH_2C(Cl)=CHCH_2)_n-$	emulsion polymerization
isobutylene–isoprene (IIR) rubber (butyl)	[29985-75-9]	$-(CH_2C(CH_3)_2)_m-(CH_2C(CH_3)=CHCH_2)_n-$ ca 98% ca 2%	cationic polymerization

nitrile–butadiene rubber (NBR)	[9003-18-3]	$-[CH_2CH=CHCH_2]_m-[CH_2CH]_n-$ $C{\equiv}N$	emulsion random copolymerization
styrene–butadiene rubber (SBR)	[9003-55-8]	$-[CH_2CH=CHCH_2]_m-[CH_2CH]_n-$ Ph	emulsion random copolymerization, also some anionic random copolymerization
ethylene–propylene copolymer (EPM)	[9010-79-1]	$-[CH_2CH_2]-[CH_2CH]_n-$ CH_3	Ziegler-Natta catalysis, random copolymerization
ethylene–propylene–diene terpolymer (EPDM)	[25038-36-2]	$-[CH_2CH_2]-[CH_2CH]-[CH_2CH]_n-$ CH_3 CH_2—CH=CH—CH_3	Ziegler-Natta catalysis, random copolymerization, ethylidene norbornene also used

$$F = 2\left(\lambda - \frac{1}{\lambda^2}\right)\left(C_1 + \frac{C_2}{\lambda}\right) \tag{5}$$

If C_2 equals zero, equation 5 reduces to the statistical theory with:

$$2\,C_1 = NkT \tag{6}$$

A plot of $F/2(\lambda - 1/\lambda^2)$ as a function of $1/\lambda$ often yields a straight line from which the values of C_1 and C_2 can be obtained (9,13,17–23). The empirical term C_2 is best understood as the deviation from ideality. It is very small for some highly flexible rubbers such as the siloxanes (22) (see Silicon compounds) and some networks formed by end-linking. Swelling of a network can sometimes decrease C_2 without sensibly changing C_1 (9,13,17–23). Many improvements to the theory have been proposed that are not discussed in this survey. Briefly, they incorporate corrections such as considering the ratio of the mean-square end-to-end distance of a network chain in the undistorted state to that for the chain free of network cross-links (25). The interested reader is directed to a particularly comprehensive recent review (13). Viscoelastic behavior is also discussed in detail in this review (13). Compounding technology is well covered in a recent monograph (26).

Classes of Synthetic Rubber

A brief historical perspective may be useful (4–5,7,15). The first important synthetic rubbers were based on 2,3-dimethylbutadiene which was initiated by sodium (see Initiators). These were produced commercially in Germany after the outbreak of World War I. They were not very good rubbers (perhaps owing to their high T_g) and production was discontinued after the war. The thiokol rubbers based on the interaction of glycols with sodium disulfides and polysulfides were discovered in 1927 (see Polymers containing sulfur). They are still useful in certain applications (eg, solvent resistance). Shortly thereafter, DuPont introduced chloroprene, which is a high *trans*-1,4-poly(2-chlorobutadiene) (4).

$$\underset{(4)}{\overset{Cl}{\underset{-\!\!\!+\!CH_2}{}}\!\!\!\!\!\!\!\!\!\!C\!=\!CH\!-\!CH_2\!\!\xrightarrow{}_{\!n}}$$

This rubber also displays good solvent resistance and was the first synthetic rubber vulcanizate to show good gum tensile strength. This is due to the fact that the high *trans*-1,4 structure permits crystallization to occur on stretching.

It is significant that the German chemical industry actually developed some butadiene–styrene emulsion copolymers (SBR) in the 1920s. In 1934, this technology was sufficiently advanced to allow production of Buna-N, a butadiene–acrylonitrile copolymer [9003-18-3]. Such systems are still the premier general purpose rubber for oil resistance.

An entirely different mechanism, namely cationic polymerization was utilized to develop Butyl rubber in 1937. This material is primarily polyisobutylene, with a small amount of isoprene copolymerized to provide sites for accelerated sulfur vulcanization. Butyl rubber is especially useful for its relatively low permeability to gases. Scientifically, it also demonstrates the concept of needing only limited functionality to achieve the essential network junctions.

Table 2. World Synthetic Rubber Production (Thousands of Metric Tons)[a]

Type of rubber	1973	1974	1975	1976
styrene–butadiene rubber, solid	3196	2974	2386	2824
styrene–butadiene rubber, latex	256	254	229	390
polybutadiene	796	769	666	795
polyisoprene	261	223	185	213
polychloroprene	313	318	291	326
ethylene–propylene	200	226	155	249
butyl	360	383	257	348
nitrile, solid	162	176	111	165
nitrile, latex	32	29	22	27
specialty[b]	73	74	63	76
Total synthetic[c]	5649[d]	5426[d]	4365[d]	5413[d]

[a] Ref. 15.
[b] Includes polyacrylates (solid and latex), chlorosulfonated polyethylene, chlorinated polyethylene, silicone, polysulfide, fluoroelastomers, epichlorohydrin and polyisobutylene. Excludes all polyurethane foams and elastomers.
[c] Excludes reclaimed rubber.
[d] Data do not include production from Communist countries.

Table 3. World Synthetic Rubber Production, % Distribution[a]

Geographic markets	1973	1974	1975	1976
North America	49.6	48.3	47.8	47.5
Western Europe	28.0	28.9	27.7	28.5
Asia, Africa, Oceania and Middle East	18.3	17.9	19.3	19.1
Latin America	4.1	4.9	5.2	4.9
Total world production	100.0	100.0	100.0	100.0

[a] Ref. 15.

One of the most important industrial achievements during World War II was the rapid development of the synthetic rubber industry in the United States. From a production of 8130 metric tons in 1941, an industrial mechanism capable of producing more than 1,000,000 t of rubber annually was achieved by the end of 1944 (15). The principal synthetic rubber developed was GR-S (Government Rubber-Synthetic) or SBR as it is now known. SBR is the largest volume synthetic rubber. The tremendous interest and importance in SBR spawned considerable research in the basic science of emulsion polymerization (3,5,7–8,10). The development of improved lower reaction temperature rubber polymerization processes and oil extension of very high molecular weight SBR followed.

The discovery of the Ziegler-Natta catalysts (qv) in the early 1950s revolutionized the synthetic rubber industry. These events have been lucidly discussed in recent books (4,10). For the first time an essentially exact duplicate of the natural rubber structure was achieved. Of perhaps equal importance, high cis-1,4-polybutadiene [9003-17-2] and totally or highly saturated rubbery copolymers based on ethylene and propylene (EPM and EPDM, respectively) were also prepared and subsequently commercialized. Related specialty rubbers such as the epichlorohydrin and propylene oxide-based rubbers were also developed via novel organometallic catalysis (see also below).

This review would certainly not be complete without some discussion of the

Table 4. Estimated Year End 1976 World Synthetic Rubber Capacity by Country, Region and Type, Thousands of Metric Tons[a]

Country	SBR solid	SBR latex	BR	IR	EP	Butyl	NBR solid	NBR latex	CR	Specialty[b]	Total synthetic
U.S.	1548	183	397	124	192	203	72	26	205	40	2990
Canada	145	10	65			42	10			na	272
Total North America	*1693*	*193*	*462*	*124*	*192*	*245*	*82*	*26*	*205*	*40*	*3262*
Argentina	55						2				57
Brazil	185	7	15				2	1			210
Mexico	65	5	20				2				92
Total Latin America	*305*	*12*	*35*				*6*	*1*			*359*
United Kingdom	240	53	40		20	45	10	7	25	na	440
Netherlands	140	7		75	40		5	3		na	270
Belgium	45		20			80					145
France	235	38	130	45	35	40	28	1	40	na	592
Federal Republic of Germany	185	59	70		20		25	6	60	na	425
Italy	270	25	45	30	50		12	3		na	435
Spain	55		10					3		na	68
Turkey	32		13								45
Total Western Europe	*1202*	*182*	*328*	*150*	*165*	*165*	*80*	*23*	*125*	*na*	*2420*
South Africa	40	4					1				45
Australia	50		20								70
India	30										30
Republic of Korea	25										25
Japan	690	61	210	85	40	50	35	8	77	na	1256
Total Asia, Africa, Oceania and Middle East	*835*	*65*	*230*	*85*	*40*	*50*	*36*	*8*	*77*	*na*	*1426*
Grand total excluding Communist countries	4035	452	1055	359	397	460	204	58	407	40	7467
Estimated total Communist countries	*1203*	*52*	*185*	*380*			*65*		*75*	*90*	*2050*
World total	5238	504	1240	739	397	460	269	58	482	130	9517

[a] Ref. 15.
[b] Includes polyacrylates (solid and latex), chlorosulfonated polyethylene, chlorinated polyethylene, silicone, polysulfide, fluoroelastomers, epichlorohydrin, and polyisobutylene. Excludes all polyurethane foams and elastomers. na = not available.

Table 5. Estimated Total Synthetic Rubber Shipments Within Geographical Regions, % Distribution by End Use [a,b]

End-use markets [c]	North America, %	Latin America, %	Western Europe, %
tire products	64.3	60.4	56.2
molded, extruded and calendered products	12.5		22.0
adhesives, cements, and sealants	2.9		0.6
latex foam products	3.5		
hose	2.8		0.9
footwear	2.7		3.2
belting	2.3		0.5
electrical	1.7		2.3
all other	7.3	39.6 [d]	14.3
	100.0	100.0	100.0

[a] Ref. 15.

[b] Includes dry styrene-butadiene rubber (emulsion and solution), styrene-butadiene latex, polybutadiene (solid and latex), polyisoprene (solid and latex), polychloroprene (solid and latex), ethylene-propylene, -butyl (solid and latex), nitrile (solid and latex), polyacrylates (solid and latex), chlorosulfonated polyethylene, chlorinated polyethylene, silicone, polysulfide, fluoroelastomers, epichlorohydrin, and polyisobutylene. Excludes all polyurethane foams and elastomers.

[c] *End-Use Market Definitions*

Tire products: includes all tire, tube and retread applications plus related products such as valve stems, innerliners.

Adhesives, cements, caulks, and sealants: elastomers used in conventional adhesives, caulking materials and sealants. Excludes such applications as pigment binders, nonwoven fabrics, textile scrims, paper saturation.

Belting: includes V-belts, conveyor belts, timing belts.

Electrical products: includes wire and cable insulation and jacketing materials, coatings for electrical insulation plus molded and extruded electrical products such as URD connectors, caps, cord protectors, ready-access terminals, harnesses.

Footwear: includes heels, soles, slab stock, boots, canvas, rubber soled footwear and related shoe products.

Hose: includes fabric and wire-reinforced constructions for automotive, industrial, and consumer applications.

Latex foam products: includes foamed products made from latex. Does not include elastomers consumed in blown or molded sponge.

Molded, extruded, and calendered products: covers all molded, extruded, and calendered sheet goods not specified elsewhere. Includes automotive, industrial and consumer products such as athletic goods, flooring, gaskets, household products, "O" rings, nonreinforced tubing, blown sponge, thread, rubber sundries.

All other: covers applications not elsewhere classified. Includes sales to compounders, resellers, brokers (ultimate usage is unknown to polymer producer), and miscellaneous categories such as fabric coating, fabric or paper saturation, asphalt binders, coatings, dipped goods, plastics modification.

[d] Includes all nontire applications for which separate data are not available by end use.

thermoplastic elastomers mentioned above (3–7,11–13). Basically, soft rubber-like regions are combined with microphase separated hard, glassy or crystalline domains either in a triblock or multiblock (segmented) copolymer, or a partially miscible blend. The soft regions provide the extensibility and the hard domains serve as physical network junctions and as microscopic reinforcing agents. The principal commercial examples of segmented systems are the thermoplastic polyurethanes and the polyester–polyethers (11–12). The triblocks are produced by the "living" anionic polymerization of styrene and the dienes (SDS) (11–12). The dienes may also be hydro-

Table 6. Specialty Cross-linked Rubbers

Name	Basic structure	Temperature use range, °C	Outstanding property	Some applications	Remarks
silicones (polydimethylsiloxane)	$-\text{SiO}-$ with CH_3 groups	−100 to 300	wide temperature range, resistance to aging, ozone, sunlight, very high gas permeability	seals, specialty molded and extruded goods, adhesives, biomedical	methyl vinyl siloxane added for improved cross-linking, fluoropropyl methyl grades available for polar solvent resistance
fluoroelastomers	$-CH_2CF_2\!+\!CF_2CF-$ with CF_3	−40 to 200 (continuous) and 300 (intermittent)	resistance to heat, oils chemicals	seals such as O-rings, corrosion-resistant coatings	many grades available which are optimized for application
polyacrylates	$-CH_2CH-$ with $C=O$, OR, $R = C_2H_5$, C_3H_8	−40 to 200	oil, oxygen, ozone and sunlight resistance	seals, hose	vinyl chloroethyl ether comonomer for vulcanization
epichlorohydrin	$-CH_2CHO-$ with CH_2, Cl	−18 to 150	resistance to oils, fuels, some flame resistance, low gas permeability	hose, tubing, coated fabrics, vibration isolators	available as a homopolymer or copolymer

Polymer	Structure	Temperature range (°C)	Properties	Applications	Comments

	—CH₂CH O—[—CH₂CH₂O—]
	    |
	    CH₂
	    |
	    Cl
		−40 to 135			wide distribution of chlorine types; total chlorine ca 25–45%

chlorosulfonated polyethylene	—CH₂CHCH CH₃—
	    | |
	    Cl SO₂
	        |
	        Cl
		−40 to 150	resistance to oils, ozone weathering, oxidizing chemicals	automotive hose, wire and cable linings for reservoirs	

chlorinated polyethylene	—CH₂CHCH₂—
	    |
	    Cl
		−40 to 150	resistance to oils, ozone chemicals	impact modifier, automotive applications	chlorine content variable

ethylene-acrylic	—CH₂CH₂—[—CH₂CH—]
	                |
	                C=O
	                |
	                OCH₃
		−40 to 175	resistant to ozone, weathering	seals, insulation vibration damping	only recently commercialized

propylene oxide	—CH₂CH O—
	    |
	    CH₃
		−60 to 150	low temperature properties	motor mounts	allyl glycidyl ether, comonomer for crosslinking

genated to produce more stable soft-center segments if 20–30% 1,2 structure is present. Higher 1,4 content can result in polyethylene-like crystallinity. The polyolefin elastomers are essentially physical blends of ethylene–propylene rubbers with isotactic crystalline polypropylene. The blend miscibility is optimized by either lightly cross-linking/grafting or through careful choice of the monomer sequence distribution in the rubber. This field has been very recently reviewed (27) (see also Olefin polymers).

Economic Aspects

The larger volume rubbers are often referred to as the general purpose (GP) rubbers. Similarly, the lower volume, higher priced systems are termed specialty rubbers. The accepted abbreviations and chemical structures for GP rubbers are shown in Table 1 (pp. 450–451).

World synthetic rubber production and the geographic distribution are in Table 2 and Table 3, respectively (p. 453). The data clearly show that SBR is still the most important synthetic elastomer, NBR and CR find much of their usage in oil- or solvent-resistant application.

Table 4 (p. 454) documents the estimated world synthetic rubber capacity for different elastomers at the end of 1976.

The major markets and their geographical distribution are shown in Table 5 (p. 455). Tire products are by far the most important application for synthetic rubber.

The cross-linked (thermoset) specialty rubbers are tabulated according to chemical structure, useful temperature range and special properties/applications in Table 6 (pp. 456–457). Although the specialty rubbers are more costly and hence smaller volume than the general purpose rubbers they fill the need for high performance materials in the rubber industry. Some are growing considerably faster than their GP counterparts.

BIBLIOGRAPHY

"Rubber, Synthetic" in *ECT* 1st ed., Vol. 11, pp. 827–852 by J. D. D'Ianni, The Goodyear Tire & Rubber Co., Akron, Ohio; "Elastomers, Synthetic" in *ECT* 2nd ed., Vol. 7, pp. 676–705 by William M. Saltman, The Goodyear Tire & Rubber Co.

1. G. Williams, *Proc. Roy. Soc.* **10,** 516 (1860).
2. H. Staudinger, *From Organic Chemistry to Macromolecules,* John Wiley & Sons, Inc., New York, 1970.
3. M. Morton, "Advances in Synthetic Rubber," in, E. B. Mano, ed., *Proceedings of the International Symposium on Macromolecules,* Elsevier, Amsterdam, 1975, p. 287.
4. W. M. Saltman, ed., *The Stereo Rubbers,* Wiley-Interscience, New York, 1977.
5. W. Cooper in A. Ledwith and A. M. North, eds., *Molecular Behavior and the Development of Polymeric Materials,* Chapman and Hall, Ltd., London, 1975, p. 150.
6. D. H. Richards, *Chem. Soc. Rev.* **6,** 235 (1977).
7. M. Morton, ed., *Rubber Technology,* 2nd ed., Van Nostrand Reinhold, New York, 1973.
8. G. S. Whitby, ed., *Synthetic Rubber,* Interscience, New York, 1954; F. R. Eirich, ed., *Science and Technology of Rubber,* Academic Press, New York, 1978.
9. A. S. Dunn, ed., *Rubber and Rubber Elasticity, J. Polym. Sci. Polym. Symp. 48,* Wiley-Interscience, New York, 1975.
10. J. P. Kennedy and E. Tornqvist, eds., *Polymer Chemistry of Synthetic Elastomers,* Wiley-Interscience, New York, 1968.
11. A. Noshay and J. E. McGrath, *Block Copolymers: Overview and Critical Survey,* Academic Press, New York, 1977.

12. J. E. McGrath, M. Matzner, L. M. Robeson, and A. Noshay, "Block and Graft Copolymers," in N. M. Bikales, ed., *Encyclopedia of Polymer Science and Technology,* Suppl. No. 2, Interscience Publishers, a division of John Wiley & Sons, Inc., New York, 1977, p. 129.
13. T. L. Smith in "Treatise on Materials Science and Technology," Vol 10 of J. M. Schultz, ed., *Properties of Solid Polymeric Materials, Part A,* Academic Press, New York, 1977, pp. 369–451.
14. E. R. Yescombe, *Plastics and Rubber World Sources of Information,* Applied Science Publishers Ltd., London, 1976.
15. *International Statistical Review of the Synthetic Rubber Industry, ISS NO 147-3662, Facts Book, 1977 Edition,* International Institute of Synthetic Rubber Producers, Inc., 45 Rockefeller Plaza, New York, N.Y. 10020.
16. J. Brandup and E. H. Immergut, eds. *Polymer Handbook,* 2nd ed., Wiley-Interscience, New York, 1975.
17. L. R. G. Treloar, *Physics of Rubber Elasticity,* 3rd ed., Oxford University Press, New York, 1975.
18. P. J. Flory, *Principles of Polymer Chemistry,* Cornell University Press, Ithaca, New York, 1953.
19. P. J. Flory, *Statistical Mechanics of Chain Molecules,* Wiley-Interscience, New York, 1969.
20. F. Bueche, *Physical Properties of Polymers,* Wiley-Interscience, New York, 1962.
21. L. R. G. Treloar, *Rubber Chem. Technol.* **47**(3), 626 (1974).
22. J. E. Mark, *Rubber Chem. Technol.* **48,** 495 (1975).
23. K. Dusek and W. Prins, *Adv. Polym. Sci.* **6,** 1 (1969).
24. L. R. G. Treloar, *Trans. Faraday Soc.* **40,** 59 (1944).
25. J. E. Mark, *Macromol. Rev.* **11,** 135 (1976).
26. R. Bobbit, ed. *The Vanderbilt Handbook,* Vanderbilt Rubber Co, Norwalk, 1978.
27. L. J. Fetters, D. Hansen, and J. E. McGrath, *Rubber Chem. Technol.,* in press, 1980.

JAMES E. MCGRATH
Virginia Polytechnic Institute and State University

ACRYLIC ELASTOMERS

Acrylic elastomers, in the current usage of the rubber industry, are rubbery polymers and copolymers in which esters of acrylic acid such as ethyl acrylate and butyl acrylate constitute a major portion. They exhibit heat and oil resistance which places them generally between nitrile rubber and silicone rubber in the spectrum of specialty elastomers. They make up a large part of the market in various automotive applications such as seals and gaskets because of their ability to retain properties in the presence of hot oils and other fluids, and to resist cracking or softening and other effects of exposure to heat and air. A closely related class of elastomers which fits the above definition has attained commercial importance recently. These are the ethylene–acrylic elastomers, copolymers of ethylene and methyl acrylate.

Acrylate elastomers originated in efforts in the United States to find substitutes for natural rubber before and during World War II. A series of patents and publications by workers at the U.S. Department of Agriculture Eastern Regional Laboratory in the early 1940s disclosed the discovery (1). The name Lactoprene was given to the new rubber (2), reflecting the thought that it could be made from lactic acid. The new product was suggested for all the applications of natural rubber, including tires, at a time when there was little definition of specific viscoelastic properties beyond the term rubbery. The first commercial acrylic elastomer was offered by the B. F. Goodrich Company in 1948 in two grades: Hycar PA, which is poly(ethyl acrylate) and Hycar PA-21 which was modified for easier vulcanization (see below). The latter is still available as Hycar 4021. In the early 1960s, other manufacturers entered the market and American Cyanamid introduced Cyanacryl acrylic elastomer. At the present time a series of Cyanacryl elastomers from American Cyanamid Company and Hycar elastomers from B. F. Goodrich Company supply the United States and export markets. Other domestic suppliers who entered the market in the 1960s have discontinued their polyacrylates. Some acrylic elastomers are also manufactured in Japan and in Europe.

Properties

Acrylic elastomers, as noted above, are polymers consisting largely of one or more esters of acrylic acid. The earliest commercial product was poly(ethyl acrylate) (1).

$$x\,CH_2{=}CHCOOR \longrightarrow {+}CH_2CH{+}_x\!-\!C({=}O)OR$$

(1)

The resulting structure of a saturated backbone with pendant ester groups is characteristic of all acrylic elastomers and accounts for their two most important properties. These are (1) resistance to aging in air at moderately elevated temperatures up to about 200°C, and (2) resistance to swelling, hardening, and other changes that take place in hot oils. The saturated backbone also makes them completely resistant to ozone, an important factor in weathering. Natural rubber, styrene–butadiene rubber (SBR), and acrylonitrile–butadiene (nitrile) rubber all have high unsaturation, even after vulcanization (see Rubber, natural). Consequently, they are more susceptible to oxidation, especially at high temperature, and attack by ozone than are the saturated polyacrylates. Resistance to hot oil depends on two separate effects. The first is the relative affinity of the rubber for oil. The hydrocarbon rubbers, natural and SBR, swell greatly, soften and lose strength. The polar groups in the nitrile and acrylic rubbers resist this effect to differing degrees, depending on the compositions of the individual elastomers. The other effect relates to the nature of the oil. Of specific concern are oils for high temperature service such as automotive engine lubricants and automatic transmission fluids. They contain additives that usually include sulfur and labile sulfur compounds. The additives cause continuing vulcanization (cross-linking) of unsaturated rubbers. Consequently, in spite of the fact that nitrile rubbers may swell no more than polyacrylates initially, they become hard and brittle in service and lose their sealing ability. The acrylic elastomers with their saturated backbone are not subject to this type of continuing vulcanization.

Low Temperature Properties. The principal disadvantage of poly(ethyl acrylate) in automotive applications is that it becomes stiff on cooling and brittle at a relatively high temperature. Typical brittle points of commercial poly(ethyl acrylates) are approximately −15°C, depending greatly on the test method. The demand for improved low temperature properties, ie, flexibility at lower temperatures has increased greatly. One means of achieving this goal is by utilizing longer-chain alkyl esters. Table 1 gives low temperature brittle points reported for homopolymers of straight-chain alkyl esters of acrylic acid (3). The brittle points were determined by flexing strips of polymer (about 2 mm thick) after two minutes' immersion in dry ice-cooled ethanol and noting the temperatures at which the samples broke. The values are similar but do not agree exactly with glass transition temperatures determined by other means. It is considered that the glass transition temperature relates to the internal mobility of the segments of the polymer chain, which is determined by the free volume of the polymer relative to the volume required for the motion of a segment (4). Increasing the length of the side chain increases the free volume of the system, thereby increasing the mobility. As a practical matter, the user is concerned with the stiffness or brittleness of a particular vulcanizate, which contains fillers and chemicals and is cross-linked, rather than with a theoretical glass transition temperature. Perhaps coincidentally, the values in Table 1 are in fair agreement with brittle points observed on such vulcanizates, at least for the lower members of the series.

Table 1. Brittle Points of Straight-Chain Polyacrylates[a]

Polyacrylate	CAS Registry No.	Brittle point, °C
methyl	[9003-21-8]	3
ethyl	[9003-32-1]	−24
n-butyl	[9003-49-0]	−44
n-hexyl	[27103-47-5]	−58
n-heptyl	[29500-72-9]	−60
n-octyl	[25266-13-1]	−65

[a] Ref. 3.

The problem of low temperature flexibility cannot be solved by simply utilizing a long-chain alkyl ester. As the size of the aliphatic ester increases, the affinity of the polymer for oils increases as well, resulting in excessive swelling of the polymer. For example, poly(n-butyl acrylate), which may have a brittle point below −40°C (depending on composition and conditions, as noted above) exhibits a volume change (swell) in ASTM No. 3 oil of well over 100%. Copolymerization of an oil-resistant monomer with a long-chain acrylate ester may solve the problem. Acrylonitrile, for example, has been used to improve the oil resistance of polyacrylates and other polymers. At the present time there is no significant use of such a copolymer in the United States although numerous literature references indicate that such procedures are employed in Japan and the USSR (see Acrylic ester polymers).

In the United States, alkoxyalkyl acrylates are used as part of the copolymer (see Copolymers). For instance, methoxyethyl acrylate (2) or ethoxyethyl acrylate (3) have alcohol portions whose free volumes approximate that of the four-carbon alcohol in n-butyl acrylate, but the oxygen atom in the chain reduces the affinity for oil.

$$CH_2=CHCOOCH_2CH_2OCH_3$$
(2)
$$CH_2=CHCOOCH_2CH_2OCH_2CH_3$$
(3)

For example, a copolymer can be obtained consisting of 50% butyl acrylate, 47% methoxyethyl acrylate, and 3% vinyl chloroacetate and exhibiting flexibility down to −41°C with a volume swell of 28% in No. 3 oil (5). Alternatively, a copolymer consisting of 24% butyl, 45% ethyl, and 28% methoxyethyl acrylate, with 3% vinyl chloroacetate, has a brittle point of −31°C and a swell in No. 3 oil of 18%. The compositions of the newer commercial products have not been disclosed, but the elastomers with greatly improved low temperature properties, that is, exhibiting flexibility to −25 or −40°C, are believed to involve copolymers generally similar to those described above.

Table 2 gives the typical vulcanizate properties of acrylic elastomers. These are generally for a vulcanizate of medium hardness that was cured for a short time in a mold and then postcured or tempered for several hours in an oven at 150–175°C. Such postcuring has been customary with acrylic elastomers with most of the available vulcanizing systems. More specific information regarding the effect of compounding

Table 2. Typical Properties of Acrylic Elastomers[a]

Property	Regular	Low temperature
Original		
100% modulus, MPa[b]	10.3	8.3
tensile strength, MPa[b]	15.2	10.3
elongation, %	215	150
hardness, durometer A	80	75
compression set, %		
70 h at 150°C	28	35
low temperature brittle point, °C	−15	−40
Air aging 168 h at 175°C		
100% modulus, MPa[b]	11.4	9.1
tensile strength, MPa[b]	12.9	9.3
elongation, %	180	130
hardness, durometer A	83	80
Fluid aging 70 h at 150°C		
ASTM Oil No. 1		
100% modulus, MPa[b]	9.7	6.9
tensile strength, MPa[b]	15.8	9.7
elongation, %	210	140
hardness, durometer A	81	74
volume change, %	+1	+2
ASTM Oil No. 3		
100% modulus, MPa[b]	9.7	6.2
tensile strength, MPa[b]	13.8	10.3
elongation, %	205	100
hardness, durometer A	72	63
volume change, %	+15	+27
Automatic transmission fluid		
100% modulus, MPa[b]	10.2	7.9
tensile strength, MPa[b]	15.5	6.8
elongation	170	95
hardness, durometer A	75	68
volume change, %	+5	+10

[a] A composite of data from suppliers' technical sales literature.
[b] To convert MPa to psi, multiply by 145.

variations and of exposure to different environments and conditions is available in suppliers' literature.

Vulcanization Sites. The simple homopolymer of ethyl acrylate, which was the first commercial product, could be vulcanized by strongly alkaline reagents such as sodium metasilicate pentahydrate. The cross-linking probably occurs as the result of a Claisen-type reaction between the carbethoxy groups in one polymer molecule and the α hydrogen atoms in an adjacent molecule (6). The reaction, however, suffers from a slow cure rate and limited versatility. Consequently, modified polyacrylates have been made containing a variety of reactive sites to facilitate vulcanization. Halogen groups were among the first to be investigated (7), and the first modified commercial acrylic elastomer was a copolymer of ethyl acrylate, about 95%, and 2-chloroethyl vinyl ether, about 5% (4). The side chain, with the chlorine atom activated by the ether linkage, provides a convenient cross-linking site.

$$x\ CH_2{=}CH(C{=}O)(OC_2H_5) + y\ CH_2{=}CH(OCH_2CH_2Cl) \longrightarrow (\!-\!CH_2CH(C{=}O)(OC_2H_5)\!-\!)_x(\!-\!CH_2CH(O)(CH_2CH_2Cl)\!-\!)_y$$

(4)

Vulcanization is accomplished by active polyamine systems such as triethylenetetramine with bisbenzothiazolyl disulfide (MBTS) or Trimene Base and sulfur. (Trimene Base is a product of indefinite structure obtained from the reaction of ethyl chloride, formaldehyde, and ammonia, produced by Uniroyal, Inc). These systems, and variations of them, were used for many years with acrylic elastomers. In spite of problems, such as volatility and limited storage life after compounding (bin stability), they did impart satisfactory properties which permitted the polyacrylates to attain their place in the array of specialty elastomers. Diamines, such as hexamethylenediamine carbamate in combination with an acid acceptor, eg, dibasic lead phosphite, are also effective in vulcanizing this type of acrylate. Another chlorine-containing comonomer with similar activity for cross-linking is 2-chloroethyl acrylate, often mentioned in the patent literature. Its copolymers are vulcanized with the same types of curing systems used with those of 2-chloroethyl vinyl ether.

By the early 1960s technological advances in automobiles, such as the trend to more powerful engines and smaller automatic transmissions, had increased operating temperatures and accelerated usage of acrylic elastomers in applications where nitrile rubber had formerly been adequate. A technological advance introduced with one of the new products at this time was a considerably more active vulcanization site provided by the use of vinyl chloroacetate (8) as a comonomer (5).

$$x\ CH_2{=}CH(C{=}O)(OR) + y\ CH_2{=}CH(O)(C{=}O)(CH_2Cl) \longrightarrow (\!-\!CH_2CH(C{=}O)(OR)\!-\!)_x(\!-\!CH_2CH(O)(C{=}O)(CH_2Cl)\!-\!)_y$$

(5)

The chlorine atom adjacent to the carbonyl group is more active by a factor of four or five than that in β-chloroethers and can be vulcanized with a variety of systems.

The previously described systems cure these polymers but are much too fast for the more active chlorine. They cause premature vulcanization or scorch during processing and are not practical. Ammonium salts of carboxylic acids, such as ammonium benzoate or ammonium adipate were the first recommended curatives (9), followed by the introduction of the popular soap–sulfur curing system (10). It consists of sulfur and a metallic carboxylate (a soap) such as sodium stearate, potassium stearate, sodium oleate, or similar materials. In none of the modern vulcanization systems for acrylates has the mechanism been elucidated, but it is apparent that in this case the soap acts as a polymer-soluble base. Variations in the soap may affect the base strength, the solubility, and the molar content for a given concentration. Thus the use of differing amounts of different soaps and varying sulfur levels (typically less than one part per hundred of rubber) provides considerable versatility to this type of cure system. Several other unsaturated monomers containing the α-chloroacetyl moiety on a side chain (11) impart cure activity similar to that of vinyl chloroacetate.

A somewhat different approach to incorporating an active chlorine vulcanization site was taken by workers at B. F. Goodrich Company (12). It employs vinylbenzyl chloride, where the activity of the chlorine atom is like that of the α-chloroacetyl group because it is vulcanized by the same agents.

In the 1960s, a nonhalogen cure site was introduced by the use of allyl glycidyl ether (6) as a comonomer (13) (see Allyl monomers and polymers).

$$CH_2=CHCH_2OCH_2CH-CH_2$$
$$\diagdown O \diagup$$

(6)

It provides a pendant epoxide group which can be cross-linked with systems similar to those used for the α-chloroacetyl sites, ie, ammonium salts, soap–sulfur, and others. Still another modification consists of the incorporation of a small amount of carboxylic acid such as methacrylic acid in the polymer, together with the usual chlorine-containing monomer. Such polymers can be cured with a quaternary ammonium salt (14), or by an alkali metal salt of a carboxylic acid or organophosphoric acid (15), in either case without sulfur.

An improved vulcanization system for the 2-chloroethyl vinyl ether site (16) consists of 2-mercaptoimidazoline and red lead or dibasic lead phosphite. It eliminates some of the problems of the older polyamine systems, thus extending the utility of the earlier polymer type.

The most widely used cure system for vulcanization at the present time appears to be the soap–sulfur system. The mercaptoimidazoline–red lead system is used to a lesser extent, whereas the ammonium salt system is still fairly widely used. Reference to the polyamine–sulfur system is often seen in Japanese and USSR literature.

Manufacture

Esters of acrylic acid, the raw material for acrylic elastomers, can be manufactured by a variety of methods (see Acrylic acid and derivatives). Acrylate esters polymerize readily, and for various purposes all the usual polymerization methods, ie, bulk, solution, emulsion, and suspension have been used. For elastomers, only emulsion and suspension polymerization are important.

Emulsion polymerization (17) is initiated by a water-soluble catalyst within the micelles (18) formed by an emulsifying agent (see Emulsions). The catalyst may be a persulfate salt or a redox system. After polymerization is complete, the excess monomer is steam-stripped and the polymer is coagulated by controlled addition to a salt solution; the crumb is collected on a screen, washed, and dried.

Suspension polymerization (19) utilizes agitation to produce monomer droplets suspended in the aqueous phase in the presence of a monomer-soluble catalyst such as benzoyl peroxide or azobisisobutyronitrile. A protective colloid such as bentonite, starch, or poly(vinyl alcohol) may be used to prevent the droplets from coalescing and keep the polymer particles separated. The droplets are ordinarily much larger than in an emulsion, averaging perhaps 0.5–2 mm in diameter for suspension polymerization. The polymerization is thus like many bulk polymerizations, occurring simultaneously in the small droplets. The water phase provides mobility, and important heat transfer and temperature control. After completion, the excess monomer is steam stripped, and the beads of polymer are collected and washed on a centrifuge or filter and dried on a vibrating screen or by means of an expeller-extruder.

Some of the effluents are similar from both processes. The steam-stripped monomer may be repurified or removed by means of a caustic scrubber. Since conversions are usually very high in practice, the latter method is often used. Traces of monomer in the dryer off-gases are probably in subtoxic concentration but have a characteristic sharp odor which is objectionable to many people; therefore, the off-gases may be incinerated. The aqueous effluent from the suspension polymerization contains a portion of the protective colloid used, as well as small amounts of organic material including traces of monomer and hydrolyzed fragments from the esters used. The aqueous effluent from the emulsion process includes, in addition to the same types of materials as from the suspension process, emulsifiers and salts from the coagulant bath. Treatment of effluents from rubber manufacture has been a serious concern for many years. However, since the volume of acrylic elastomers produced is negligible compared to the general purpose and larger specialty elastomers, effluents from acrylic elastomers do not present a problem.

Processing. Acrylic elastomers are processed in the same manner as other elastomers. A Banbury mixer is used on a large scale and mill mixing on a smaller scale or for occasional batches.

Parts can be shaped by compression, transfer, or injection molding; for the latter the more rapid vulcanization systems are used with the newer polymers (see Plastics technology). Cycles of 1–2 min above 200°C or 3–5 min at 165–175°C have been used for molding. Small diameter tubing for automotive use as well as jacketing for electrical wires can be formed by extrusion and cured in high pressure steam. Microwave curing is used with modern fast-curing acrylics (see Microwave technology). Alternatively, the compounded acrylate can be vulcanized in the form of a large cylinder from which the final parts, eg, transmission seals, are cut by means of a lathe. As esters, the acrylic elastomers are somewhat sensitive to moisture, but rubber-covered rolls can be produced in the same way as with other rubbers. Rolls of acrylic elastomer are used to apply hot paraffin to paper stock, lacquering sheet metal, and other applications involving solvents or high temperature.

The polyacrylate parts are usually subjected to a post-cure cycle after the primary cure. (This does not apply to rolls, which customarily need a cure of several hours.) Some of the newer vulcanization systems with fast-curing elastomers are capable of producing compression set values (the property most sensitive to extent of cure) in the primary-cure cycle comparable to values obtained with a postcure with the older polymers. Nevertheless, requirements for the automotive applications are so stringent that most parts are still postcured.

An interesting new processing technique for elastomers utilizes so-called powdered rubbers which are often more like beads or granules than powder (20). The suspension polymerization technique adapts readily to production of bead-form polymer (21), which may become an important form as powdered rubber technology advances.

Compounding. The distinctive compounding feature of acrylic elastomers is their vulcanization chemistry. For most applications they employ carbon black fillers, and require reinforcing grades for the development of good stress–strain properties. The most common filler is probably N550 black, formerly designated FEF (fast extruding furnace black). It exhibits a good balance of reinforcement, resistance to aging, and low compression set (see Carbon, carbon black). As in other elastomers, finer particles of carbon black produce higher tensile strength, modulus, and hardness, but at the same time tend to decrease resistance to aging in hot air and increase compression set. Mineral fillers (qv) are used to some extent, although acidic fillers are to be avoided because of their retarding effect on the vulcanization. An interesting use of mineral-filled stocks is in their production of color-coded replacement seals for automotive repair kits. Polyacrylates need less antioxidants than unsaturated rubbers, although many formulations include some antioxidants. The tertiary carbon atom in each ester group may be the site of oxidation. Antiozonants (see Antioxidants) are not required. Plasticizers are seldom used, since they tend to be fugitive at service temperatures, or by extraction in the oils. However, small concentrations of lubricants and other processing aids are often used. Stearic acid, often used as a lubricant in other elastomers, retards the cure in soap–sulfur cured polyacrylates, in effect by neutralizing some of the alkalinity provided by the soap. This effect should not be overlooked in compounding.

Economic Aspects

Reliable data on the production and consumption of acrylic elastomers are not available because their volume is a very small fraction of total rubber consumption. The total new rubber consumption in the United States for 1978 was estimated (22) to be 3,300,000 metric tons, of which about 2,500,000 t was synthetic. Consumption of acrylic elastomers in 1974 was probably 3000–6000 t (23) when total rubber consumption was 3,100,000 t (24). Although it is true that acrylic elastomers find use in many different applications, the major consumer is the automobile industry. As a result, the annual sales figures follow new car sales rather closely. The trend to smaller cars in the United States has apparently not resulted in reduced consumption of acrylics. On the contrary, the effort to produce lighter weight vehicles has led to smaller engines, transmissions, and engine compartments, which leads to higher operating temperatures and greater demand for the specific qualities offered by the polyacrylates. Prices of acrylic elastomers have remained remarkably stable over the years. The 1950 product was priced at $2.97/kg (5); the price for poly(ethyl acrylate) types in 1978 was ca $3.19/kg. The higher acrylates, those designed for better low temperature properties, are more expensive because of the higher cost of the ester raw materials.

Uses

The main uses of acrylic elastomers are in the automobile industry. Their service temperatures range from 150 to 200°C, and their resistance to attack by lubricating oils place them between the lower cost nitrile rubbers and the more expensive silicones and fluoroelastomers. Automatic transmissions of modern automobiles contain as many as sixteen elastomeric seals, most of which are polyacrylate. Other uses in the engine compartment include valve-stem oil deflectors, spark-plug boots, ignition-wire jacketing, and tubing of various types where exposure to heat and oil splashing are involved. Shaft seals, O-rings, grommets, packings, gaskets, and closed-cell sponge are other applications. Nonautomotive uses for polyacrylates include tank linings and roll coverings. The major markets involve carbon black-containing parts, and most of them are in enclosed devices. Nevertheless, the excellent resistance of polyacrylates to weathering, including complete resistance to ozone, suggest their use in outdoor applications where long life is desirable.

Ethylene–Acrylic Elastomers

Ethylene–acrylic elastomers, introduced in 1975 (25), utilize a different concept than the older acrylate elastomers and employ different chemistry for vulcanization.

The new ethylene–acrylic elastomer, Vamac, is a copolymer primarily of methyl acrylate and ethylene. A third monomer (7) provides pendant carboxyl groups for cross-linking with a diamine. The diamine usually recommended is hexamethylenediamine carbamate sold as Diak No. 1.

$$-(CH_2CH_2)_x- -(CHCH_2)_y- -(CRCHR')_z-$$
$$\qquad\qquad\qquad | \qquad\qquad |$$
$$\qquad\qquad\qquad C=O \qquad\quad C=O$$
$$\qquad\qquad\qquad | \qquad\qquad |$$
$$\qquad\qquad\qquad OCH_3 \qquad\quad OH$$

(7)

Like the traditional acrylic elastomers, the structure provides a fully saturated backbone which gives excellent resistance to aging, especially to ozone attack. The combination of oil resistance and low temperature flexibility is achieved by a somewhat different concept than in earlier polyacrylates. Previously, higher acrylic esters have been used to provide low temperature flexibility and comonomers (eg, acrylonitrile) or polar groupings in the ester alcohol (alkoxyalkyl acrylates) are used to maintain oil resistance. In the new polymer, the most highly oil-resistant alkyl acrylate, ie, methyl acrylate, is used, and flexibility is provided by ethylene as comonomer.

Typical properties of a commercial ethylene–acrylic elastomer are shown in Table 3 (26). A good balance of properties is indicated. The volume swell in ASTM No. 3 oil is higher than previously considered acceptable for polyacrylates, but the swell in some practical automotive fluids such as automatic transmission fluid, power-steering fluid, water and glycol–water mixtures is quite satisfactory, 15% or less, and the retention of properties after exposure to the fluids is good (27). The new elastomer is recommended for use at temperatures up to 200°C intermittent exposure. It is stated to have 1000 h life at 171°C, defined as the time at that temperature required to reduce its

Table 3. Typical Properties of Ethylene–Acrylic Elastomer, Press Cure 20 Min at 177°C[a]

Property	Value
Original	
100% modulus, MPa[b]	2.8
200% modulus, MPa[b]	6.4
tensile strength, MPa[b]	14.1
elongation, %	450
hardness, durometer A	63
Aged 7 d at 177°C[c]	
100% modulus, MPa[b]	4.5
tensile strength, MPa[b]	16.0
elongation, %	250
hardness, durometer A	68
Aged 3 d at 200°C[c]	
100% modulus, MPa[b]	4.5
tensile strength, MPa[b]	11.7
elongation, %	210
hardness, durometer A	72
Low temperature properties	
Clash-Berg torsional stiffness, T-10,000, °C	−26
brittle point, °C	−44
Volume increase in ASTM No. 3 Oil	
70 h at 149°C, %	57

[a] Ref. 26.
[b] To convert MPa to psi, multiply by 145.
[c] Oven.

breaking elongation to 50% absolute. The ethylene–acrylic elastomer is recommended for applications typical of acrylic elastomers such as seals, gaskets, O-rings, hose, and tubing. It has good electrical properties when loaded with mineral filler, thus making it useful for ignition-wire jacket, spark-plug boots and primary wire insulation where heat and oil resistance are also needed (see Insulation, electric). In addition, it is reported to have high vibrational damping characteristics (high tan δ), and to maintain them nearly constant over a wide temperature range (−30 to 160°C) and, at the same temperature, over a range of frequencies from 10 to 1000 Hz. This quality should make it useful in automotive damping applications.

Notice

The information and statements herein are believed to be reliable but are not to be construed as a warranty or representation for which we assume legal responsibility. Users should undertake sufficient verification and testing to determine the suitability for their own particular purpose of any information or products referred to herein. NO WARRANTY OF FITNESS FOR A PARTICULAR PURPOSE IS MADE.

Nothing herein is to be taken as permission, inducement, or recommendation to practice any patented invention without a license.

BIBLIOGRAPHY

1. W. C. Mast and co-workers, *Ind. Eng. Chem.* **36,** 1022 (1944); U.S. Pat. 2,509,513 (May 30, 1950), W. C. Mast and C. H. Fisher (to U.S. Department of Agriculture).
2. C. H. Fisher and co-workers, *Ind. Eng. Chem.* **36,** 1032 (1944).

3. C. E. Rehberg and C. H. Fisher, *J. Am. Chem. Soc.* **66,** 1203 (1948).
4. E. H. Riddle, *Monomeric Acrylic Esters,* Reinhold Publishing Corp., New York, 1954, pp. 61–62.
5. U.S. Pat. 3,488,331 (June 6, 1974), A. H. Jorgensen, Jr. (to B. F. Goodrich Company).
6. S. T. Semegen and J. H. Wakelin, *Rubber Age N.Y.* **71,** 57 (1952).
7. W. C. Mast, T. J. Dietz, and C. H. Fisher, *India Rubber World* **113,** 223 (1945).
8. U.S. Pat. 3,201,373 (Aug. 17, 1965), S. Kaizerman (to American Cyanamid Company).
9. U.S. Pat. 3,324,088 (June 6, 1967), T. F. Waldron (to American Cyanamid Company).
10. U.S. Pat. 3,458,461 (July 29, 1969), F. F. Mihal (to American Cyanamid Company).
11. U.S. Pat. 3,493,548 (Feb. 3, 1970), D. C. Chalmers (to Polymer Corporation, Ltd.); Brit. Pat. 1,169,955 (Nov. 12, 1969), U. Kuhlmann.
12. U.S. Pat. 3,763,119 (Oct. 2, 1973), R. D. DeMarco and H. A. Tucker (to B. F. Goodrich Company).
13. U.S. Pat. 3,335,118 (Aug. 8, 1967), G. A. Knavel and G. Rosen (to Thiokol Chemical Corporation).
14. U.S. Pat. 3,875,092 (Apr. 1, 1975), R. E. Morris (to B. F. Goodrich Company).
15. U.S. Pat. 3,912,672 (Oct. 14, 1975), R. E. Morris and H. A. Tucker (to B. F. Goodrich Company).
16. U.S. Pat. 3,335,117 (Aug. 8, 1967), M. B. Berenbaum and G. A. Knavel (to Thiokol Chemical Corporation).
17. W. C. Mast and C. H. Fisher, *Ind. Eng. Chem.* **41,** 790 (1949).
18. W. D. Harkins, *J. Polym. Sci.* **5,** 217 (1950).
19. Ref. 4, p. 54.
20. Some useful references among a large and growing literature on powdered rubber include N. Doak, *Eur. Rubber J.* **156,** 52 (Oct. 1974); M. E. Woods and W. Whittington, *Rubber Age,* 39 (Oct. 1975); J. P. Lehnen, *Eur. Rubber J.* 10 (Nov. 1977).
21. U.S. Pat. 3,976,611 (Aug. 24, 1976), R. R. Aloia (to American Cyanamid Company).
22. *Elastomerics,* 20 (Jan. 1978).
23. *Rubber Age* **107,** 38 (May 1975); D. Dworkin, *Rubber World,* 43 (Feb. 1975).
24. *Rubber Age* **107,** 12 (May 1975).
25. J. F. Hagman and co-workers, "Ethylene–Acrylic Elastomers—A New Class of Heat and Oil Resistant Rubber," presented to the Rubber Division, American Chemical Society, October, 1975.
26. J. F. Hagman, "Diamine Curing Systems" in *Vamac Ethylene–Acrylic Elastomer Product Bulletin EA 301,* E. I. du Pont de Nemours & Co., Inc., Wilmington, Del.
27. W. R. Abell, "Ethylene–Acrylic Elastomer for Demanding Automotive Applications," paper presented before Society of Automotive Engineers Passenger Car Meeting, Detroit, Mich., Sept. 26–30, 1977.

General References

C. H. Fisher, G. S. Whitby, and E. M. Beavers in G. S. Whitby, ed., *Synthetic Rubber,* John Wiley & Sons, Inc., New York 1954, pp. 900–910.
E. H. Riddle, *Monomeric Acrylic Esters,* Reinhold Publishing Corp., New York, 1954.
P. Fram in N. Bikales, ed., *Encyclopedia of Polymer Science and Technology,* Vol. 1, John Wiley & Sons, Inc., New York, 1964, p. 226.
T. M. Vial, *Rubber Chem. Technol.* **44,** 344 (1971).
Product bulletins on Cyanacryl acrylic elastomer from American Cyanamid Company, Bound Brook, N.J. 08805; Hycar acrylic elastomer from B. F. Goodrich Company, 6100 Oak Tree Blvd., Cleveland, Ohio 44131; Vamac ethylene–acrylic elastomer from E. I. du Pont de Nemours & Co., Inc., Wilmington, Del. 19898.

T. M. VIAL
American Cyanamid Company

BUTYL RUBBER

Commercial butyl rubber [9010-85-9], poly(methylpropene-co-2-methyl-1,3-butadiene) or poly(isobutylene-co-isoprene), is one of a family of copolymers discovered in 1937 (1–4) in the laboratories of the Standard Oil Development Company, now the Exxon Corporation. This discovery established the principle of limited chemical functionality for the preparation of vulcanizable elastomers. Under spur of the World War II shortage of natural rubber, commercial production began in late 1943, representing an outstanding engineering achievement considering the complexity of the process.

The first halogenated butyl rubber was discovered in the laboratories of the B. F. Goodrich Company (5–8). A brominated butyl rubber (Hycar 2202) was commercialized by Goodrich in small scale, batch preparation in 1954 but withdrawn from the market in 1969. A chlorinated butyl rubber (Chlorobutyl) was fully commercialized by Exxon in 1961 (9–10). In 1971, Polysar, Ltd. commercialized a brominated butyl rubber (Polysar X2).

These unusual products had considerable market impact. Before the advent of butyl rubber inner tubes, tires had to be reinflated frequently. With the gas impermeability provided by butyl rubber, the interval between pressure checks was extended significantly. Halogenated butyl rubber permitted the manufacture of tubeless tires having comparable qualities. Without doubt, the family of butyl rubbers contributed greatly to driving safety and improved tire durability and wearing characteristics, both in truck and passenger tires (see also Rubber, natural).

Physical Properties

The physical constants for butyl rubber have been tabulated (11–12). Values are given for unvulcanized and, in some cases, vulcanized and carbon black-filled polymer. Halogenated butyl rubbers have very similar characteristics.

The physical properties of butyl-type polymers responsible for their commercial success as elastomers are: low glass transition temperature, T_g ca $-70°C$ (11–12), high level of impermeability to common gases (13–14) including water vapor (15–16) over a broad temperature range, and high hysteresis (17) over a useful temperature range. No other elastomer displays this combination of characteristics.

An unusual structural feature may be responsible for these unique physical characteristics. Neglecting the small amount of isoprene residue present, the molecule has a crowded structure with two methyl branch groups on every other carbon atom. This crowded structure results in bond angle distortion (18), as shown below:

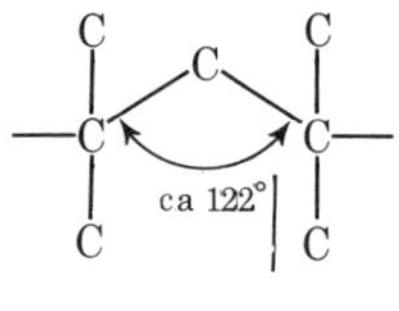

and an altered (compared to that for a normal tetrahedral angle, 109.5°) rotational potential-barrier profile. The resulting viscoelastic behavior can be interpreted in terms of a high monomeric friction coefficient and unusual WLF (Williams, Landel, Ferry) parameters (19). The behavior of butyl rubber at ambient conditions, placed on a generalized spectrum for elastomeric viscoelastic behavior, would suggest a much higher glass transition temperature than observed (20–21).

The distinctive mechanical properties of butyl rubber can best be illustrated by comparison with natural rubber. At $-75°C$ both are glassy hard; at room temperature natural rubber is resilient whereas butyl rubber displays very little resilience; at $100°C$ both are highly resilient.

These same structural features presumably are reponsible for butyl rubber's low thermal expansion coefficient (11–12) compared to other hydrocarbon elastomers. Bond angle distortion is the likely reason for butyl rubber's thermal degradation behavior (22), the relatively low heat of polymerization (23), and the low polymerization ceiling temperatures (24). Permeability is the product of diffusivity and solubility. The substantially saturated hydrocarbon nature of the polymer assures low solubility for common gases. The features outlined above assure low diffusivity. Tabulated values for permeability and diffusivity of many gases with butyl rubber have been published (13–16).

Elastomers are rarely used in gum form and many properties are determined by the kind and degree of cross-linking as effected by the curative system and the nature and amounts of the compounding ingredients (fillers, plasticizers, etc) employed; see refs. 25–31.

Chemical Properties

The isoprene in butyl rubber is linked predominantly by 1,4 addition at a level

$$y\ CH_3C(CH_3)=CH_2 + [CH_2=C(CH_3)CH=CH_2] \longrightarrow -(CH_2C(CH_3)=CHCH_2)_x-(CH_2-C(CH_3)_2)_y-$$

from about 0.5 to 2.5 moles per 100 moles of monomers, depending on grade; the residual olefin is in the trans configuration (32–33). The substantially saturated hydrocarbon nature of the polymer endows it with good chemical resistance, and good resistance to oxidative deterioration and to cracking by ozone.

The presence of hydrogen on carbons α to the double bond permits vulcanization by sulfur and accelerators (34). Other cross-linking systems can be employed usefully in certain applications (26–27). These are also the sites for initiation of autocatalytic oxidation by hydroperoxide formation (35–37). Butyl rubber vulcanizes more slowly than the highly unsaturated rubbers (in a given sulfur–accelerator recipe) and its vulcanizates display much greater resistance to oxidative deterioration. The radical reactions which ensue during oxidation lead to more chain cleavage than coupling and butyl rubber vulcanizates soften on oxidation. Both the low unsaturation and high hysteresis contribute to the ozone resistance of butyl (38).

In halogenated butyl rubber, allylic halogen is present and the double bond initially present in butyl rubber has been shifted (39–40).

$$-(C(CH_3)_2CH_2C(=CH_2)CHXCH_2CH_2C(CH_3)_2)- \qquad X = Br\ or\ Cl$$

Although this change has only little effect on many of the properties of identically compounded vulcanizates, it has important bearing on the vulcanization rate and the available routes to cross-linking (6–7,28–29). This structural feature provides the basis for substitution or elimination, leading to a highly reactive conjugated diene (41).

When commercial Chlorobutyl is vulcanized using a conventional butyl rubber sulfur–accelerator recipe, it is cured (ca 90% of maximum attainable cross-linking) in about one-fifth the time required for conventional butyl rubber. Brominated butyl rubber is inherently more reactive than chlorinated butyl rubber (28), but this can be moderated in commercial polymers by the presence of stabilizers that prevent premature dehydrohalogenation, cross-linking, and degradation in finishing and storage.

Halogenated butyl rubbers, but not butyl rubber itself, provide the cross-linking chemistry and kinetics necessary to obtain interfacial chemical bonding with highly unsaturated general purpose elastomers (6–7, 28–29,42–43). Compared to chlorinated butyl rubber, in some applications brominated butyl rubber offers the compounder a wider selection of curatives with which to attain the required level of adhesion to or covulcanization with the general purpose elastomers (43). With either commercial elastomer, as a practical matter, the curative system employed is of consequence (42–46).

For details on chemical properties as affected by compounding, etc, the reader is referred to the comprehensive literature (6,7,25–31,47).

Manufacturing and Processing

Virtually all of the world's production of butyl rubber is made by a precipitation (slurry) polymerization process in which isobutylene and a minor amount of isoprene are copolymerized in methyl chloride diluent at temperatures of -100 to $-90°C$ using aluminum chloride as catalyst. The polymer precipitates as it is formed as a finely divided, milky slurry. General descriptions of the process (25–26,48–56) and patent information (1–3,57–67) are found in an extensive literature. Alternative solution processes have also been described (68–72). Halogenated butyl rubbers are produced commercially by dissolving butyl rubber in hydrocarbon solvent and contacting the solution with gaseous or liquid elemental halogens.

Polymerization. Isobutylene (2-methylpropene) is the only four-carbon olefin that can be polymerized and copolymerized by cationic catalysis to the molecular weights required for elastomers. The polymerization is an extremely fast reaction even at $-100°C$. The typical number-average elastomeric molecule requires the successful completion of some 2000 or more monomer linkages (see Copolymers).

Catalysis. The catalysts employed in these polymerizations are Friedel-Crafts catalysts such as $AlCl_3$, BF_3, $TiCl_4$, and $Al(C_2H_5)Cl_2$. In many cases a cocatalyst is required to initiate polymerization and provide for efficient use of catalyst (54,73). The selection of catalyst and cocatalyst determines the nature of the negatively charged counter-ion, and thus the molecular weight of polymer produced under otherwise identical polymerization conditions.

Diluent. Although a diluent for the system may be chemically inert, it can have an effect other than dilution. The ionic reactions are affected by the dielectric constant and solvation power of the polymerization medium (74–75). Reported data suggest that methyl chloride participates chemically in the reaction (76). An important advantage of methyl chloride diluent is that aluminum chloride, an inexpensive and very effective catalyst, dissolves in it.

Temperature. The molecular weight of the polymer produced increases with decreasing polymerization temperature, other reaction conditions being the same. A plot of log $\overline{DP}$ versus T^{-1} is approximately linear and of positive slope over a broad temperature range (56,73–74).

Impurities. Impurities can lead to: (a) activation of catalyst (eg, H_2O in low $H_2O/AlCl_3$ molar ratio); (b) destruction of catalyst (eg, H_2O at higher $H_2O/AlCl_3$ ratios), (c) introduction of another chain-transfer reaction (eg, 2-butene or 2,2,4-trimethyl-1-pentene); (d) enhancement of the termination reaction (eg, 1-butene); and (e) interaction with carbonium ions to stop propagation (eg, dimethyl ether).

Impurities can be described as being catalyst activators (or cocatalyst), chain-transfer reagents (when their presence decreases the molecular weight of the polymer formed but not the catalyst activity), or poisons (when their presence decreases catalyst efficiency and activity). Some of these impurities have been evaluated in terms of poison and transfer coefficients (56,77). In general, most C_3 or higher olefinic and all oxygenated organic materials affect the reaction.

Comonomer. In contrast to the simple alkenes and cycloalkenes, conjugated dienes copolymerize with isobutylene at low temperatures. Their presence in the system depresses the molecular weight of the polymer. Monomer reactivity ratios for the system isobutylene–isoprene have been reported (78) ($r_1 = 2.5 \pm 0.5$, $r_2 = 0.4 \pm 0.1$) for use in the copolymerization equations (79). Isobutylene is the most reactive monomer. The product of the reactivity ratios is about unity, hence butyl rubber is a random copolymer.

Precipitation. As a consequence of precipitation polymerization, the polymer itself can participate in chemical reactions with entrapped monomers and catalyst under certain conditions. This can lead to chain branching and cross-linking, (80).

Detailed treatments of the polymerization chemistry are available (56,74,81).

Slurry Process. A simplified flow diagram of the slurry process for butyl rubber is presented in Figure 1. A number of commercial processes are available for production of the required high purity isobutylene (82–83). The most important is the sulfuric acid extraction process for recovery from dilute C_4 cracked petroleum streams. Significant quantities of isobutylene are also produced by dehydration of *tert*-butyl alcohol. Typical isobutylene, as supplied to the butyl rubber process, has a purity in the range of 95–99 wt % and includes varying amounts of propene, butane, 1-butene, 2-butene, isobutylene dimer, *tert*-butyl alcohol, and trace quantities of a variety of oxygen-containing compounds, depending upon the process employed. Additional purification is provided in a two-tower feed fractionation system.

As shown in Figure 1, a drying tower removes water and propene. 2-Butene and higher boiling compounds are rejected as a bottoms stream in a redistillation tower and returned to the isobutylene production plant. Isobutylene of 99.5% or higher purity is taken overhead and flows to a feed blend drum where it is mixed with fresh isoprene of >98.0% purity and a recycle stream of methyl chloride and unreacted isobutylene. 1-Butene cannot be efficiently separated from the feed isobutylene by distillation; its concentration in the process is controlled by purging a portion of the unreacted isobutylene in the recycle tower system. Depending upon the grade of butyl rubber to be produced, a typical feed blend contains from 25–35 wt % isobutylene, 0.4–1.25 wt % isoprene, with the balance methyl chloride. The feedstream is cooled to reaction

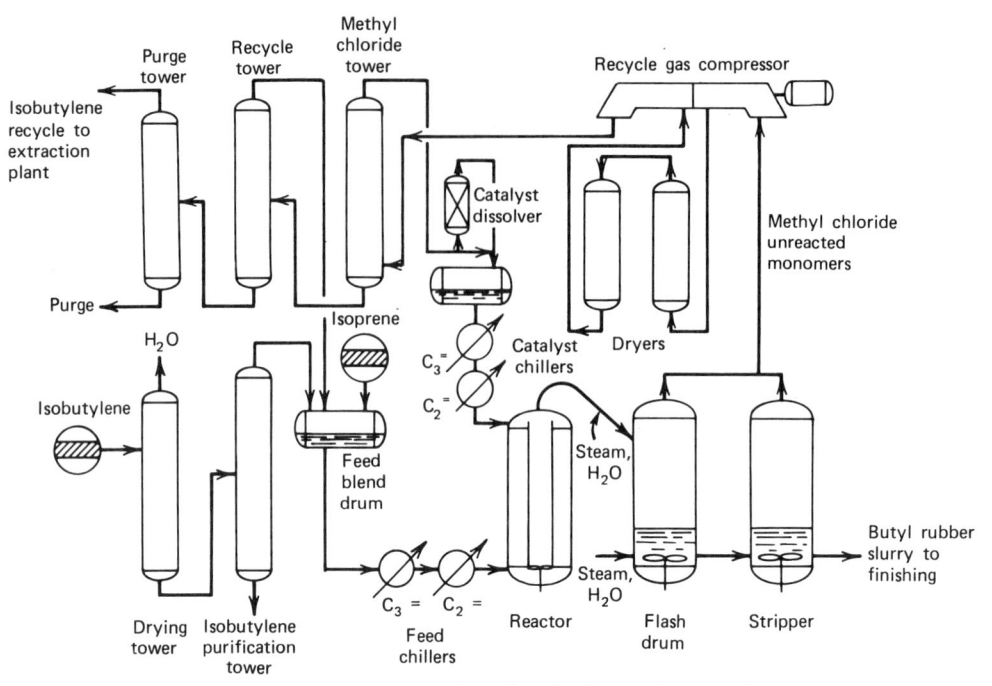

Figure 1. Schematic flow plan of slurry butyl rubber polymerization process.

temperature in a series of heat exchangers supplied with liquid propene or propane and ethylene as refrigerants.

Catalyst solution is prepared by passing purified liquid methyl chloride through a bed of granular aluminum chloride at 30–45°C. The resulting concentrated solution is diluted with additional methyl chloride and stored. Dilute catalyst solution can be chilled to −85 to −95°C and stored or the solution can be held at near ambient temperature and chilled just before use. Activity of the catalyst depends upon the amount of cocatalyst or coinitiators present and can be controlled (65,84). Typically from 2000–5000 kg of polymer are produced per kg of catalyst.

Chilled catalyst and feed mixture are continuously introduced to reactors constructed with a large central draft tube surrounded by concentric rows of heat transfer tubes. Materials of construction are $3\frac{1}{2}$ wt % Ni-steel, or other alloys suitable for low-temperature service. An axial flow pump at the bottom of the draft tube circulates slurry at high velocity to provide efficient mixing and good heat transfer. The copolymerization reaction is exothermic, releasing approximately 0.86 MJ/kg (370 Btu/lb) polymer. Heat is removed by transfer to boiling ethylene, supplied as liquid to jackets that enclose the tube section of the reactor. Reactor pressure ranges from 240–380 kPa (2.37–3.75 atm). The composition and molecular weight of polymer formed depend upon the concentrations of monomers in the reaction medium and the level of terminating and transfer agents present. These are a function of the feedstream composition and the extent to which monomers are converted. In practice, the principal operating variable is the flow rate of catalyst to the reactors. Typical residence time in the reactor is 30–60 min. Conversion of isobutylene ranges from 70–95% depending upon the grade of butyl rubber being produced.

Reactor operation is cyclical. Significant amounts of fouling occur during a run, depositing a film of rubber on heat-transfer surfaces and forming agglomerates that block reactor cooling tubes. Normally two or more reactors operate while others are washed with hot hexane or naphtha to remove deposited rubber. Typical runs vary from 24 to more than 48 h and depend upon production rate, feed purity, and the concentration of slurry and unreacted monomers in the reactor.

Polymer slurry containing 20–30 wt % butyl rubber continuously overflows from the reactor through a transfer line to a flash drum operating at 140–160 kPa (1.38–1.58 atm) and 65–75°C. Steam and hot water are mixed with the slurry in a nozzle as it enters the drum to vaporize methyl chloride and unreacted monomers which pass overhead and are recovered. The polymer agglomerates as coarse crumb-like particles in water. Metal stearates (eg, aluminum, zinc, or calcium stearate) are added and agitation is provided to control the crumb size. Other additives, such as antioxidants (qv), can also be introduced at this point. The slurry of polymer in water flows from the flash drum to a stripping vessel operated under high vacuum to remove remaining methyl chloride and unreacted monomers. Effective stripping of methyl chloride from both polymer crumb and water is particularly important. Methyl chloride is a toxic compound and any remaining in the polymer–water slurry would be lost from the process in the finishing operation. The stripped slurry is pumped to intermediate storage tanks and then to a finishing building for drying, baling, and packaging.

Vapors from the flash drums and strippers are compressed and cooled. Condensed water is separated and the vapor stream is dried by counter-current scrubbing with a glycol mixture followed by passage through beds of alumina. The dry gas stream containing less than 10 ppm (wt) water is further compressed and fractionated in a multitower system to recover a portion of pure methyl chloride for catalyst preparation and a recycle methyl chloride stream containing unreacted isobutylene. In a final tower, an isobutylene stream containing 1-butene is taken overhead and returned to the isobutylene production unit for repurification. A bottoms stream comprised of isobutylene, unreacted isoprene, and higher-boiling components formed in the reactor and fractionation towers is purged from the system and commonly used as a fuel component elsewhere in the chemical or refining complex.

In the finishing operation, butyl rubber slurry from intermediate storage is successively dewatered in a series of drying extruders. All butyl rubber plants use extruder lines supplied by Welding Engineers Inc. (King of Prussia, Pa.) or V. D. Anderson Co. (Strongsville, Ohio), but significant design modifications to this equipment have been developed by the butyl rubber producers. In a typical operation, slurry is introduced to a slurry feeder that produces a compressed crumb containing 20–25 wt % water. This is reduced to 5–10% in a dewatering extruder. Final drying to less than 0.5 wt % water is accomplished in a third extruder by allowing the compressed polymer melt to expand through a die to form an exploded crumb. This is conveyed by air in a transfer line to an enclosed fluidized-bed conveyor where water vapor is removed and the crumb is cooled to baling temperature. It is then compressed into 34- or 36.3-kg bales in baling presses, wrapped in polyethylene film, and packed into multiunit cardboard boxes or wooden crates containing ca 0.5 or 1.0 metric ton of product. Shipment to customers is by railroad car or truck.

The extremely low temperature demanded for butyl rubber polymerization requires an extensive compression–refrigeration system (see Refrigeration). Lowest temperature cooling services are provided by vaporization of liquid ethylene. Intermediate level refrigeration is supplied by propane or propene. The large power requirements of the refrigeration and recycle compressors and the extensive amounts of steam used in flash drums and fractionating towers make the butyl rubber process highly energy intensive compared to processes for other elastomers. Overall energy requirements range from 35 to 55 MJ/kg (15,000–24,000 Btu/lb) of butyl rubber produced, calculated in terms of equivalent fuel [at 43 MJ/kg (18,500 Btu/lb) of fuel heating value] required to produce the needed forms of energy. Energy requirements depend largely upon plant size, concentration of slurry produced in the reactors, and the effectiveness with which energy is utilized in the various processing steps.

Solution Process. Over the years, significant industrial activity has been devoted to the development of a hydrocarbon solution process for butyl rubber as evidenced by the patent literature (10,68–72). Such a process using hexane or pentane as solvent avoids the use of methyl chloride and would provide polymer directly in solution, an advantage when butyl rubber is to be halogenated to produce bromobutyl or Chlorobutyl. Up to 1978 there had been no commercial production of butyl rubber using solution polymerization technology. However, a pilot-scale development of a solution process has been in progress in the USSR, and a commercial facility is believed to be under construction.

Little is known at present of the USSR solution butyl rubber process beyond information disclosed in a recent patent (72). A hydrocarbon-soluble catalyst system is employed comprised of an aluminum alkyl halide catalyst with water or hydrogen sulfide as cocatalyst. Isopentane appears to be the polymerization solvent of choice, but other C_5 to C_7 hydrocarbons may be used. The comonomers dissolved in isopentane and the catalyst are chilled to below polymerization temperature and introduced simultaneously to a stirred continuous reactor. The polymerization is conducted in the range of -90 to $-50°C$. Ethylene is used as refrigerant in the reactor jacket.

The cold reactor effluent containing 10–12 wt % dissolved butyl rubber is contacted with a small quantity of methanol or ethanol to destroy the catalyst. The polymerizate is pumped through heat exchangers to cool the reactor feedstream and then flows to an aqueous degassing vessel in which steam and water are introduced to volatilize solvent and unreacted monomers. The resulting slurry is passed to a screen to remove water and then finished in extrusion drying equipment. The hydrocarbon solvent and unreacted monomer stream from the degassing vessel is condensed, washed with water, dried azeotropically, and then purified by distillation before being recycled to the reactor feed preparation section.

Halogenation. The two forms of halogenated butyl rubber produced in commercial quantities are bromobutyl and Chlorobutyl. The halogenation reaction is carried out in hydrocarbon solution using elemental chlorine or bromine (in a 1:1 molar ratio with enchained isoprene). The reactions are fast (Cl_2 faster than Br_2):

More than one halogen atom per olefin unit can be introduced (28,40). However, the reaction rates for excess halogen are lower and the reaction is complicated by chain fragmentation (28).

Hydrocarbon Solution. The hydrocarbon solution of butyl rubber having appropriate molecular weight and unsaturation is prepared either by direct dissolution of reactor slurry, referred to as a solvent replacement process (10, 85–86), or a process in which solid particles of finished butyl rubber are dissolved in hydrocarbon solvent under high agitation.

Solvent replacement. In two commercial plants halogenated butyl rubber is produced by integrating solvent replacement facilities. Cold butyl rubber slurry in methyl chloride from the polymerization reactor is passed to an agitated solution in a drum containing liquid hexane. Hot hexane vapors are introduced to flash overhead the methyl chloride diluent and unreacted monomers. Dissolution of the fine slurry particles occurs rapidly. The resulting solution is stripped to remove the last traces of methyl chloride and monomers, and brought to the desired concentration for halogenation by flash concentration. Hexane recovered from the flash concentration step is condensed and returned to the solution drum. Polymer solution is stored in heated, agitated tanks before halogenation. A combined vapor stream of unreacted monomers, methyl chloride, and hexane from the solution drum and stripping tower is compressed and fractionated to recover hexane for reuse. The remaining methyl chloride and unreacted monomers are sent to the diluent recovery section of the slurry butyl rubber process.

Bale or crumb dissolution. Alternatively, solutions of butyl can be prepared by dissolving solid particles of finished rubber in agitated dissolving vessels which do not require integration with the slurry butyl rubber plant. As in the solvent replacement process, hexane is the preferred solvent, primarily because it is easily removed after halogenation. The dissolution step can be carried out in batch dissolvers or in a multistage continuous system. Bales of finished butyl rubber having desired molecular weight and unsaturation are chopped or ground to small fragments and conveyed to the dissolving vessels, or crumb particles produced by the drying extruders in the finishing section of the slurry process can be used directly. Rubber particles must be introduced to the dissolving vessels without significant loss of solvent to the atmosphere or entrapment of air to avoid environmental and safety problems. The time required to reach complete solution depends upon temperature, size of particles, and the intensity of agitation, and typically ranges from 1 to 4 h. Solutions containing 15–20% polymer can be obtained, although energy requirements for mixing become large at the higher concentrations. Solutions from the dissolving vessels can be used directly for halogenation or stored.

The Halogenation Process. A simplified flow diagram for the halogenation process is given in Figure 2. Butyl rubber in solution is contacted with chlorine or bromine in a series of high-intensity mixing stages. In chlorination it is critically important that chlorine be introduced either as a vapor or as a dilute solution, since liquid chlorine reacts violently with hydrocarbon solutions of butyl rubber. The bromination reaction is slower and bromine may be used safely in liquid or vapor form. Hydrochloric or hydrobromic acid is generated during the halogenation step and must be neutralized. Commonly, a dilute aqueous caustic solution is used as neutralization agent and is intimately contacted with the solution under conditions of high agitation. The caustic phase may be settled and separated from the solution of halogenated butyl. Next,

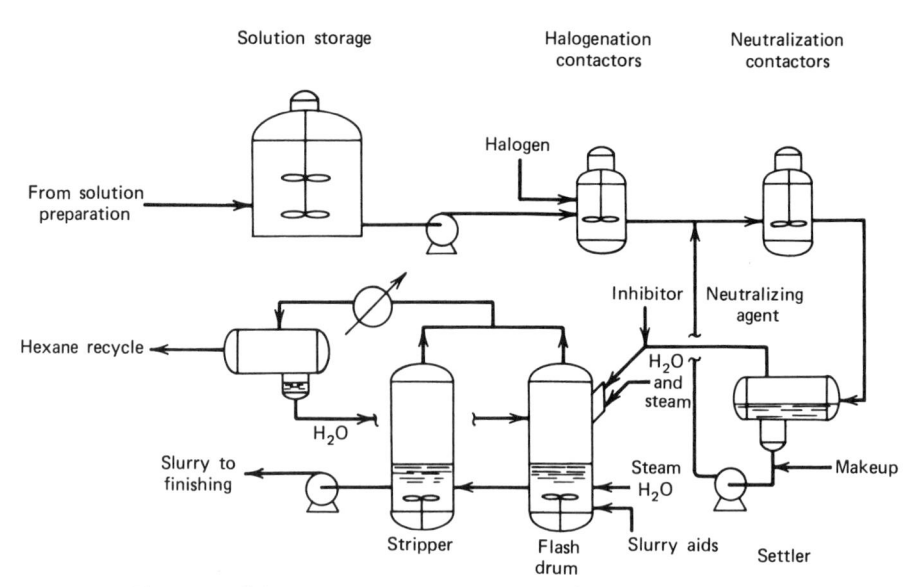

Figure 2. Schematic flow plan of halogenated butyl rubber process.

antioxidants, slurry aids, and stabilizers to protect against dehydrohalogenation are introduced and the solution is transferred to an agitated flash drum. Here the solution is contacted with steam and hot water to vaporize solvent and produce a slurry of crumb particles in water. The resulting slurry flows to a stripping vessel where the hexane content of the crumb is reduced to low levels. The finishing sequence is analogous to that described previously for butyl rubber.

Economic Aspects

Table 1 lists plant capacities, products, and location of producers of butyl rubber and halogenated butyl rubber elastomers. Butyl rubber is also produced in the USSR by the slurry process. The plants are believed to be located at Nizhnekamsk, Sumgait, and Yaroslavl, but the capacities are not known. Worldwide sales of butyl rubber and halogenated butyl rubber since 1962 (excluding those from the USSR plants) are shown in Figure 3. In 1977, butyl rubber polymer sales were equivalent to about an 80% utilization of production capacity.

The listed U.S. sales prices for the major grades of butyl rubber from 1963 to 1978 are plotted in Figure 4. The price increase since 1973 was caused by sharply escalating costs for energy and raw materials as the result of the large increase in world prices for crude oil. Prices for Chlorobutyl have been $0.014–0.018/kg higher than butyl rubber during this period. Chlorobutyl had a 1978 list price of $0.25/kg in the U.S.; bromobutyl (Polysar X-2) listed for $0.268/kg.

Specifications

The available grades of butyl rubber elastomers cover the viscosity average molecular weight range of about 325,000 to 500,000 and the unsaturation range of about 0.5–2.2 mol %. Chlorinated butyl rubber is offered in two grades, brominated butyl rubber in one. Grade lists and specifications are available from producers.

Table 1. Butyl and Halogenated Butyl Rubber Production Capacity

Producer	Products	Plant capacity, t/yr[c]
Exxon Chemical Company		
Baton Rouge, La.	Chlorobutyl	85,000
Baytown, Texas	butyl rubber	100,000
SOCABU, Notre Dame de Gravenchon, France[a]	butyl rubber	47,000
Fawley, England	butyl, Chlorobutyl rubbers	46,000
Japan Butyl Co., Ltd., Kawasaki, Japan[b]	butyl rubber	55,000
Polysar Limited		
Sarnia, Ontario, Canada	butyl, bromobutyl rubbers	42,000
Zwijndrecht, Belgium	butyl, cross-linked butyl rubbers	80,000
Cities Service Company		
Lake Charles, Louisiana	butyl, cross-linked butyl rubbers	38,500
Hardman Incorporated	liquid butyl rubber polymers	>450
Burke-Palmason Chemical Company	butyl rubber latex	2,000

[a] Part ownership by Companie Francaise de Raffinage.
[b] Jointly owned with Japan Synthetic Rubber Company, Ltd.
[c] Information supplied by producers.

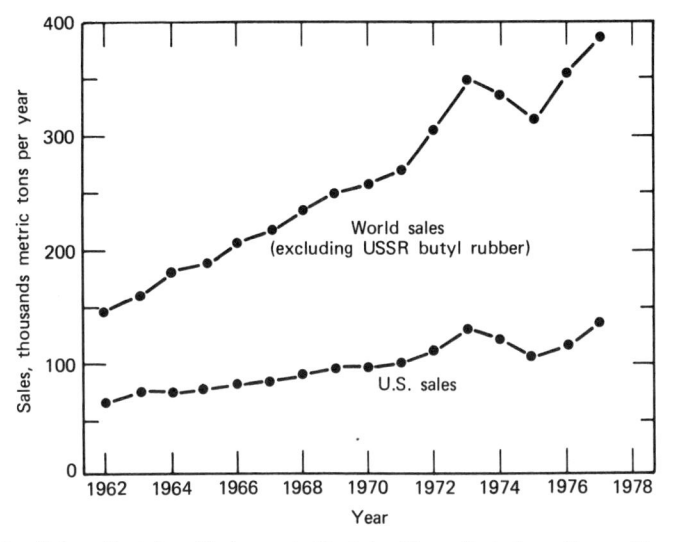

Figure 3. Sales of butyl and halogenated butyl rubbers. Data from Exxon Chemical Co.

Specifications are set for Mooney viscosity to reflect molecular weight, functionality level, volatiles and additives concentration. In addition, physical properties for standard vulcanizates may be specified. Results from scheduled laboratory tests are employed to monitor plant production for quality.

Health and Safety Factors

Table 2 lists the chemicals employed in the production of butyl rubber and halogenated butyl rubber that have personnel exposure limits established by OSHA or recommended by other groups.

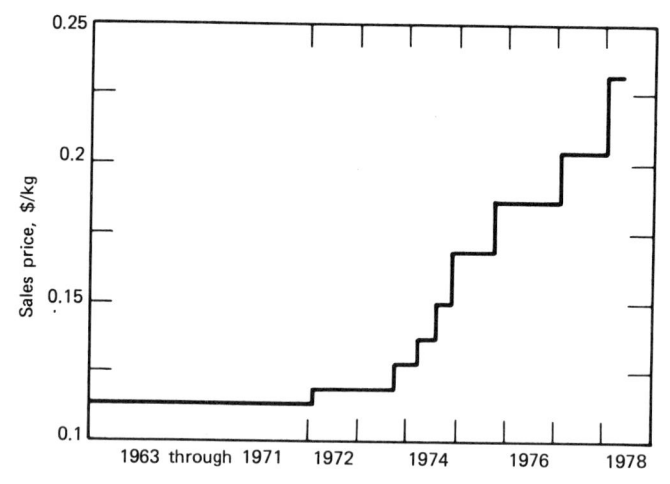

Figure 4. Butyl rubber sales price in United States. Data from Exxon Chemical Co.

Table 2. Exposure Limits[a] for Compounds Used in Butyl Rubber Manufacture

Compound	Exposure limit, ppm (vol)	Details
methyl chloride	TWA[b] = 100	OSHA
	max peak exposure = 300	for 5 min in a 3 h period
hexane	TWA = 500	OSHA
	TWA = 100	ACGIH (1976)
	STEL[c] = 125	ACGIH (1976)
	TWA = 100	with 15-min ceiling conc of 510 ppm (vol), NIOSH, 1977
chlorine	TWA = 1	OSHA
	STEL[c] = 3	ACGIH
bromine	TWA = 0.1	OSHA
	STEL[c] = 0.3	ACGIH

[a] As of April 1, 1978.
[b] Time-weighted average calculated for 8-h time period of exposure. Normal exposure based on 8-h shift, 40-h week.
[c] Short term exposure limit, ceiling that should not be exceeded during a 15 min exposure period with no more than four such exposures per day permitted.

Some grades of regular and halogenated butyl rubber have FDA approval for use in applications where they come directly or indirectly in contact with food. Molecular weight, functionality, and additive limitations are specified. The FDA regulations are published in the Federal Register and incorporated in the *Code of Federal Regulations*—Chapter 1, Title 21, Part 121 (Food Additives). Producers should be contacted for appropriate FDA grades. The FDA status of compounding ingredients employed in subsequent fabrication must also be considered over and beyond additives present in these grades as received.

Uses

The distribution of uses for butyl-type elastomers is given below (data from Exxon Chemical Co.).

Use	Percent
tire industry	85
mechanical goods	6
electrical applications	3
others	6

Butyl rubber is used mainly in the manufacture of inner tubes; halogenated butyl rubber for tubeless tire innerliners, heat-resistant inner tubes for heavy duty truck tires, and certain tire sidewall applications (stain barriers, composite sidewall compounds, etc). These uses take advantage of the impermeability of butyl rubber and its resistance to changes induced by environmental factors (aging, ozone cracking, etc). Some uses also take advantage of the damping characteristics of the polymer vulcanizates. Butyl rubber is also employed for wire insulation (see Insulation, electric), protective pipe or tank coatings, irrigation ditch or reservoir lining, collapsible shipping containers, and miscellaneous mechanical goods (hose, gaskets, etc).

Elastomers must be compounded to meet processing, property, and cost requirements for any particular application. Producers can supply suitable formulations for known applications and their technical service groups can assist in further development of these or in the development of compounds for new applications.

Derivatives

There are two other basic modifications of butyl rubber manufactured in low volume. A cross-linked terpolymer of isobutylene, isoprene, and divinylbenzene containing some unreacted substituted vinyl benzene appendages is available in two grades of differing degrees of cross-linking.

It is employed primarily in the manufacture of sealant tapes and caulking compounds (see Sealants). Because of the residual reactive functionality, it can be cross-linked by peroxides whereas conventional butyl rubber cannot (87). Cross-linked butyl rubbers produced by a Banbury mixer posttreatment of regular butyl rubber are also available.

The other modification is a low-molecular weight "liquid" butyl rubber manufactured by controlled molecular fragmentation of a standard butyl rubber. This, too, is available in two grades of about 20,000 and 30,000 viscosity-average molecular weight, respectively. The principal areas of use are in sealants, caulks, potting compounds (see Embedding), and coatings, where advantage can be taken of its relatively low viscosity in formulating high solids compounds which can be poured, sprayed, troweled, or spread.

In addition to these derivatives, butyl rubber is available in latex form. The latexes are manufactured by emulsifying solutions of standard butyl grades, and are available at 62% solids level. Major uses are in adhesives and coatings (see Latex technology).

BIBLIOGRAPHY

"Rubber, Halogenated Butyl" in *ECT* 1st ed., 2nd Suppl. Vol., pp. 716–734 by F. P. Baldwin and I. Kuntz, Esso Research and Engineering Company; "Elastomers, Synthetic" in *ECT* 2nd ed., Vol. 7, pp. 676–705 by William M. Saltman, The Goodyear Tire & Rubber Co.

1. U.S. Pat. 2,243,658 (May 27, 1941), R. M. Thomas and O. C. Slotterbeck (to Standard Oil Development Company).
2. U.S. Pat. 2,356,128 (Aug. 22, 1944), R. M. Thomas and W. J. Sparks (to JASCO).
3. U.S. Pat. 2,356,129 (Aug. 22, 1944), W. J. Sparks and R. M. Thomas (to JASCO).
4. R. M. Thomas, and co-workers, *Ind. Eng. Chem.* **32,** 1283 (1940).
5. U.S. Pat. 2,631,984 (March 17, 1953), R. A. Crawford and R. T. Morrissey (to B. F. Goodrich Co.).
6. R. T. Morrissey, *Rubber World* **138,** 725 (1955).
7. R. T. Morrissey, *Ind. Eng. Chem.* **47,** 1562 (1955).
8. U. S. Pat. 2,732,354 (Jan. 24, 1956), R. T. Morrissey and M. R. Frederick (to B. F. Goodrich Co.).
9. U.S. Pat. 2,964,489 (Dec. 13, 1960), F. P. Baldwin and R. M. Thomas (to Esso Research and Engineering Company).
10. U.S. Pat. 3,023,191 (Feb. 27, 1962), B. R. Tegge, F. P. Baldwin, and G. E. Serniuk (to Esso Research and Engineering Company).
11. J. Brandup and H. Immergut, eds., *Polymer Handbook,* Vol. 9, 2nd ed., John Wiley & Sons, Inc., New York, 1975.
12. L. A. Wood, *Rubber Chem. Technol.* **49,** 189 (1976).
13. V. Stannett in J. Crank and G. S. Park, eds., *Diffusion in Polymers,* Academic Press, New York 1968, Chapt. 2.
14. G. J. VanAmerongen, *Rubber Chem. Technol.* **37,** 1065 (1964).
15. J. A. Barrie, ref. 13, Chapt. 8.
16. Y. Iyengar, *J. Polym. Sci., Part B,* **3,** 663 (1965).
17. T. P. Yin and R. Pariser, *J. Appl. Polym. Sci.* **8,** 2427 (1964).
18. R. H. Boyd and S. M. Breitling, *Macromolecules* **5,** 1 (1972).
19. J. D. Ferry, *Viscoelastic Properties of Polymers,* 2nd ed., John Wiley & Sons, Inc., New York, 1961, p. 359.
20. *Ibid.,* p. 317.
21. M. L. Williams, R. F. Landel, and J. D. Ferry, *J. Am. Chem. Soc.* **77,** 3701 (1955).
22. S. Madorsky, *Thermal Degradation of Organic Polymers,* Interscience Publishers, Inc., a division of John Wiley & Sons, Inc. New York, 1964.
23. P. J. Flory, *Principles of Polymer Chemistry,* Cornell University Press, Ithaca, N.Y., 1953, p. 254.
24. P. H. Plesch, "Isobutene" in P. H. Plesch, ed., *The Chemistry of Cationic Polymerization,* Macmillan Co., New York, 1963, p. 146.
25. R. M. Thomas and W. J. Sparks in G. S. Whitby, ed., *Synthetic Rubber,* John Wiley & Sons, Inc., New York, 1954, Chapt. 24.
26. D. J. Buckley, *Encyclopedia of Polymer Science and Technology,* Vol. 2, Interscience Publishers, a division of John Wiley & Sons, Inc., New York, 1965, p. 754.
27. D. J. Buckley, *Rubber Chem. Technol.* **32,** 1475 (1959).
28. F. P. Baldwin and co-workers, *Rubber and Plast. Age* **42,** 500 (1960).
29. R. L. Zapp and P. Hous, "Butyl and Chlorobutyl Rubber" in M. Morton, ed., *Rubber Technology,* 2nd ed., Van Nostrand-Reinhold Co., New York, 1973, p. 249.
30. E. E. McSweeney and W. J. Mueller, *Rubber Age N.Y.* **76,** 247 (1954).

31. Bibliography No. 87, "Properties of Butyl and Nitrile Rubber," Rubber Division Library and Information Services, University of Akron, 1963.
32. J. Rehner, Jr., *Ind. Eng. Chem.* **36,** 46 (1944).
33. H. Y. Chen and J. E. Field, *J. Polym. Sci., Part B* **5,** 501 (1957).
34. F. P. Baldwin and co-workers, *Rubber Chem. Technol.* **43,** 522 (1970).
35. J. L. Bolland, *Trans. Faraday Soc.* **46,** 358 (1950).
36. D. Barnard and co-workers, "Oxidation of Olefins and Sulfides" in L. Bateman, ed., *The Chemistry and Physics of Rubber-Like Substances,* John Wiley & Sons, Inc., New York, 1963.
37. N. M. Emanuel, D. T. Denisov, and Z. K. Maizus in *Liquid Phase Oxidation of Hydrocarbons,* Plenum Press, New York, 1967.
38. M. Braden and A. N. Gent, *J. Polym. Sci.* **3**(7), 90 (1960).
39. I. G. McNeill, *Polymer* **4,** 15 (1963).
40. M. J. Melchior, to be published.
41. U.S. Pat. 3,816,371 (June 11, 1974), F. P. Baldwin and A. Malatesta (to Esso Research and Engineering Company).
42. R. L. Zapp, *Rubber Chem. Technol.* **46** 251 (1973).
43. J. Walker, R. H. Jones, and G. Feniak, Philadelphia Rubber Group Fall Technical Meeting, Sept. 22, 1972.
44. J. B. Gardiner, *Rubber Chem. Technol.* **43,** 370 (1970).
45. J. Walker, R. H. Jones, and G. Feniak, *Rubber Age N.Y.* **108,** 33 (1976).
46. R. F. Bauer and E. A. Dudley, *Rubber Chem. Technol.* **50,** 35 (1977).
47. Bibliography No. 77, "Aging of Butyl Rubber Vulcanizates," Rubber Division Library and Information Sources, University of Akron, 1967.
48. C. E. Schildknecht, *Vinyl and Related Polymers,* John Wiley & Sons, Inc., New York, 1952, p. 571.
49. R. J. Adams and E. J. Buckler, *Trans. Inst. Rubber Ind.* **29,** 17 (1953).
50. R. M. Thomas, *India Rubber World* **130,** 203 (1954).
51. C. E. Schildknecht, *High Polymers,* Vol. X, *Polymer Processes,* Interscience Publishers Inc., New York, 1956, p. 208.
52. J. Walker, *Rubber J.* **131,** 39 (1956).
53. R. A. Labine, *Chem. Eng.* **66**(24), 60 (1959).
54. H. S. Pylant, *Oil Gas J.* **61,** (Dec. 2, 1963).
55. R. Dolez, *Genie Chim.* **93**(2), 41 (1965).
56. J. P. Kennedy, "Polyisobutene and Butyl Rubber," in J. P. Kennedy and E. Tornqvist, eds., *Polymer Chemistry of Synthetic Elastomers,* Part I, Wiley-Interscience, New York, 1968, p. 311.
57. U.S. Pat. 2,399,672 (May 7, 1946), A. D. Green, E. T. Marshall, and S. Lane, (to Standard Oil Development Company).
58. U.S. Pat. 2,401,754 (June 11, 1946), A. D. Green (to Standard Oil Development Company).
59. U.S. Pat. 2,462,123 (Feb. 22, 1949), J. F. Nelson (to Standard Oil Development Company).
60. U.S. Pat. 2,463,866 (March 8, 1949), A. D. Green (to Standard Oil Development Company).
61. U.S. Pat. 2,474,592 (June 28, 1949), F. A. Palmer (to Standard Oil Development Company).
62. U.S. Pat. 2,523,289 (Sept. 26, 1950), P. K. Frolich (to JASCO).
63. U.S. Pat. 2,529,318 (Nov. 7, 1950), B. R. Tegge (to Standard Oil Development Company).
64. U.S. Pat. 2,530,129 (Nov. 14, 1950), J. H. McAteer and co-workers (to Standard Oil Development Company).
65. U.S. Pat. 2,581,147 (Jan. 1, 1952), H. G. Schutze (to Standard Oil Development Company).
66. U.S. Pat. 2,999,084 (Sept. 5, 1961), H. K. Arnold and E. R. Gurtler (to Esso Research and Engineering Company).
67. U.S. Pat. 3,005,808 (Oct. 24, 1961), R. T. Kelley, J. E. Walker, and B. R. Tegge (to Esso Research and Engineering Company).
68. U.S. Pat. 2,844,569 (July 22, 1958), A. D. Green and co-workers (to Esso Research and Engineering Company).
69. U.S. Pat. 2,927,912 (March 8, 1960), A. B. Small (to Esso Research and Engineering Company).
70. U.S. Pat. 3,361,725 (Jan. 2, 1968), P. T. Parker and J. A. Hanan (to Esso Research & Engineering Company).
71. S. Cesca and co-workers, *Rubber Chem. Technol.* **49,** 937 (1976).
72. Can. Pat. 1,019,095 (Oct. 11, 1977), N. V. Scherbakova and co-workers (to USSR). Also issued as East German Pat. 104,985 (April 5, 1974).

73. P. J. Flory, *Principles of Polymer Chemistry,* Cornell University Press, Ithaca, New York, 1953, p. 218.
74. P. H. Plesch, "Isobutylene" in P. H. Plesch, ed., *The Chemistry of Cationic Polymerization,* Macmillan Co., New York, 1963, p. 143.
75. M. Szwarc, *Makromol. Chem.* **89,** 44 (1965).
76. J. P. Kennedy and co-workers, *J. Polym. Sci. Part 4,* **1,** 331 (1963).
77. J. P. Kennedy and R. G. Squires, *J. Macromol. Sci.* **A1,** 805, 831, 847, 861, 961, 977, 995 (1967).
78. R. B. Cundall, "Co-polymerization" in P. H. Plesch, ed., *The Chemistry of Cationic Polymerization,* Macmillan Co., New York, 1963, p. 564
79. F. R. Mayo and C. Walling, *Chem. Rev.* **46,** 191 (1950).
80. W. A. Thaler and D. J. Buckley, Sr., *Rubber Chem. Technol.* **49,** 960 (1976).
81. J. P. Kennedy, *Cationic Polymerization of Olefins; A Critical Inventory,* John Wiley & Sons, Inc., New York, 1975, pp. 93–176.
82. S. R. Goel, J. C. Gupta, and H. K. Mulchandani, *Chem. Age India* **27**(8), 705 (1976).
83. G. T. Baumann and M. R. Smith, *Petrol. Ref. and Hydrocarbon Process.* **33**(5), 156 (1954).
84. U.S. Pat. 2,488,736 (Nov. 22, 1949), F. A. Palmer (to Standard Oil Development Company).
85. U.S. Pat. 2,940,960 (June 14, 1960), B. R. Tegge and co-workers (to Esso Research and Engineering Company).
86. U.S. Pat. 3,257,349 (June 21, 1966), J. A. Johnson, Jr. and E. D. Luallin (to Esso Research and Engineering Company).
87. J. Walker, G. J. Wilson, and K. J. Kumbhani, *J. Inst. Rubber Ind.* **8,** 64 (1974).

F. P. BALDWIN
R. H. SCHATZ
Exxon Chemical Co.

CHLOROSULFONATED POLYETHYLENE

Chlorosulfonated polyethylene is the term used to represent a group of synthetic elastomers derived from the reaction of a mixture of chlorine and sulfur dioxide on any of the various plastic polyethylenes (see Olefin polymers). The product of this reaction is a chemically modified form of the original polyethylene, and may contain 20–40% chlorine (mainly as secondary alkyl chlorides, RR'CHCl) and ca 1–2% sulfur, present mostly as secondary sulfonyl chloride groups (RR'CHSO$_2$Cl). Chlorosulfonated polyethylene is thus a saturated chlorohydrocarbon elastomer having sulfonyl chloride functions which are used as cross-linking or curing sites.

The useful processing properties of chlorosulfonated polyethylene are controlled by the molecular weight, molecular weight distribution, and degree of chain branching in the starting polyethylene. The useful chemical properties, such as resistance to ozone, oxygen, and other oxidizing chemicals, as well as solvent resistance, resistance to thermal degradation, and light discoloration, are believed to result from the stability of the carbon-to-secondary chlorine bond and the absence of any unsaturation along the polymer chain as by-products of either the chlorosulfonation or curing reactions.

The effects of the chlorosulfonation reaction on the properties of the polyethylene produced by free-radical polymerization was first studied by DuPont chemists in the late 1940s (1–3). Their efforts resulted in the first commercial chlorosulfonated polyethylene, a product based on a starting polymer having minimal chain branching and a molecular weight in the 20,000 range and containing approximately one chlorine for every seven carbon atoms and one sulfonyl chloride group for every 84 carbons. This product, called Hypalon S-2 was introduced in 1952 (4–5). A lower molecular weight, higher chlorine-content polymer, designed as a polymer base for solvent paints was introduced in 1958 under the name Hypalon-30.

In the late 1950s, the development of metal coordination catalysts for the polymerization of olefins led to the availability of linear polyethylene (high density) and ethylene–propylene copolymers, expanding the number of base polymers that could be chlorosulfonated. Research and development by DuPont (6–7) and Phillips (8) demonstrated that linear polyethylene gave chlorosulfonated products that were superior to earlier products both with regard to mechanical properties and rubber processability. Chlorosulfonated linear polyethylenes were made commercially available in 1959.

Chlorosulfonated polyethylenes (suitably compounded) provide elastomeric vulcanizates resistant to ozone and oxygen attack even under sunlight or ultraviolet radiation. They furthermore resist deterioration by heat, chemicals, and solvents. These properties have permitted the use of this class of specialty elastomers to grow to about 23,000 metric tons per year in 1977. Its major uses are in the areas of wire covering, hose, tubing, and sheet goods.

Physical Properties

Chlorosulfonated polyethylene, as sold by its manufacturer, is a raw synthetic rubber. To convert it to useful articles, other manufacturers must mix, or compound the polymer with selected fillers (qv), processing aids, stabilizing chemicals, and cross-linking agents or catalysts. The compound must then be shaped or molded and finally vulcanized. The properties of the finished product depend on the exact type and amount of chlorosulfonated polyethylene used, as well as the kind and amount of other agents. The physical properties of representative chlorosulfonated polyethylenes are summarized in Table 1.

The dynamic shear modulus of a 28% chlorine, 1.24% S, chlorosulfonated free-radical-based polyethylene ranges from 7 MPa to 2.1 GPa (1,000–300,000 psi) (10).

The working properties of chlorosulfonated polyethylenes are a function of the starting polyethylene and the chlorine content. As chlorine content increases from 0 to about 25%, the crystallizing tendency of the parent polymer decreases. At a chlorine content of ca 30% for free-radical-based polyethylene, and about 35% for linear polyethylene, the product possesses minimum stiffness. As the chlorine content is further increased, stiffness increases again and the polymers become more and more glassy in character, as well as more soluble in common solvents.

Chemical Properties

The chemical properties of chlorosulfonated polyethylene can be reliably predicted from the types of chemical functions present and the known chemistry of these functions in low molecular weight substances. Acid, ester, and amide derivatives have

Table 1. Physical Properties of Chlorosulfonated Polyethylenes[a]

Polyethylene base	Density Low		Density High[b]		
chlorine content, %	29	43	25	35	43
sulfur content, %	1.4	1.1	1.0	1.0	1.0
specific gravity	1.14	1.26	1.07	1.18	1.27
Mooney viscosity	30	30	40	30–95	77
processability[c]	fair	fair	excellent	excellent	excellent
solution properties					
Brookfield viscosity, mPa·s (= cP)					
25 wt % in toluene	1300	350			
5 wt % xylene	9	5	60	20 to 50	12
vulcanizate properties					
hardness, durometer A	45–95	60–95	40–95	40–95	60–95
abrasion resistance	very good	very good	excellent	excellent	excellent
tensile strength, MPa[d]					
gum stock	<8.3	<17.2	<27.6	<27.6	<24.6
black stock	<20.7	<24.1	<27.6	<27.6	<27.6

[a] Ref. 9.
[b] Linear.
[c] Includes mixing and capability to be molded, extruded, and calendered.
[d] To convert MPa to psi, multiply by 145.

been prepared and their infrared spectra studied (11). The characterization of sulfonic acid derivatives has been extended to derivatives of both free-radical-based and linear chlorosulfonated polyethylenes by comparison of the infrared spectra of the polymeric materials and model compounds (12). Detailed determination of the types of chlorine present in chlorosulfonated free radical polyethylene was made by examining the reaction kinetics and reaction products of the polymer with selected amines, zinc dust, and potassium iodide (13). The distribution of chlorine in the various possible functional types is shown in Table 2. The data show that a preponderance are secondary chlorines separated by four or more carbon atoms. The chlorine distribution suggests that good thermal stability can be expected from the polymers containing 25–35% chlorine.

Table 2. Chlorine Distribution in a Chlorosulfonated Free-Radical-Based Polyethylene

Chlorine type	Structure	Total chlorine, %
primary	RCH_2Cl	2.7
secondary	$RCHCl(CH_2)_nCH_2CHClR$[a]	71
	$RCHClCH_2CHClR$	
	$RCHClCHClR$	18
	$RCHClCH(SO_2Cl)R$	0.5
tertiary	R_3CCl	2.3–3.5
sulfonyl chloride	$RCH_2.SO_2Cl$	0.08
	R_2CHSO_2Cl	4.20

[a] $n \geq 2$.

Curing Chemistry. Early recommendations for cross-linking chlorosulfonated polyethylene polymers involved the use of divalent metal oxides to form metal sulfonate cross-links (2,14). Organic agents including diamines, alkanolamines, polyols, guanidines, thioureas, and many others were found to react readily with SO_2Cl groups to form vulcanized elastomers (15–16). Industrial practice has eliminated many of these choices and for the most part now uses one of the five systems (16) listed in Table 3. Differences between the curing of chlorinated polyethylene and chlorosulfonated polyethylene suggest that sulfonyl chloride groups play an activating role in the curing reactions. This is supported by the fact that vulcanizate properties are independent of polymer sulfur content over 0.6–1.8% sulfur. Loss of sulfur as SO_2 during the cure amounts to 40–60%. This would be expected to result in the development of unsaturation in the polymer. The outstanding ozone resistance of the vulcanizates, however, seems to deny the presence of even traces of unsaturation. A fully acceptable and complete mechanism for the curing of these polymers remains to be developed.

Resistance to Degradation. The generally good thermal stability of chlorosulfonated polyethylene reflects its resistance to attack by atmospheric oxygen. This property can be improved by various compounding additives, the most important of which is nickel dibutyldithiocarbamate (see Heat stabilizers). Recommendations for compounding to obtain maximum resistance to heat and weathering are available in the trade literature. A detailed study of the stress relaxation of chlorosulfonated polyethylene vulcanizates at 150°C (18) has shown that curing agents such as *m*-phenylenebis(maleimide) providing covalent cross-links give greater thermal stability than the conventional metal oxide–sulfur accelerator systems. In this study, comparison of intermittent and continuous stress–relaxation data demonstrated that the oxidative degradation occurred in the polymer chains rather than in the cross-links.

The oxidation resistance of chlorosulfonated polyethylenes extends to liquid–polymer interfaces. The polymer vulcanizates are resistant to attack by oxidizing acids such as dilute (95%) sulfur acid and dilute nitric acid (10–30%) as well as many other chemical fluids. Comparative qualitative data on the chemical and solvent resistance of 14 commercial elastomers, including chlorosulfonated polyethylene, to 500 chemical agents is presented in ref. 19.

The oxidation resistance of chlorosulfonated polyethylene also contributes to its resistance to ignition and burning. Determination of the flammability of Hypalon

Table 3. Practical Curing Systems for Chlorosulfonated Polyethylenes[a]

Design objective	Agents
carbon black-filled stocks	
general purpose and heat resistance	litharge (PbO) plus sulfur accelerator
lead-free stocks	magnesium oxide plus pentaerythritol, epoxy resin plus sulfur accelerator
white or colored stocks	
general purpose	magnesium oxide plus pentaerythritol
maximum color stability	high MgO plus sulfur accelerator
water resistance	tribasic lead maleate ($C_4H_2O_7Pb_4 \cdot H_2O$) plus accelerator and resin

[a] Ref. 17.

by the oxygen index method (20), shows that carbon black-loaded chlorosulfonated polyethylene has an oxygen index value of 32.4 vs 25.5 (poorer) for a whiting-filled stock (see Flame retardants).

Preparation and Manufacture

The preparation of chlorosulfonated polyethylene (21) by chlorosulfonation of hydrocarbons is represented by the following equation:

$$-(CH_2CH_2CH_2CH_2CH_2)_x- + 2\,Cl_2 + SO_2 \longrightarrow -(CHCH_2CH_2CH_2CH)_x- + 2\,HCl$$

with Cl and O=S=O / Cl substituents on the product.

The polymer is dissolved in carbon tetrachloride and traces of water are removed by azeotropic distillation (qv). A free-radical catalyst, usually 2,2′-azobisisobutyronitrile, is introduced (22). A mixture of chlorine and sulfur dioxide is bubbled into the polymer solution at the boiling temperature of the solvent and the reaction is continued until the desired amount of chlorine and sulfur dioxide has reacted. The by-product HCl and excess SO_2 are vented through a reflux condenser. The chlorosulfonated polymer solution is then refluxed for a short period to strip remaining HCl and SO_2. A polymer stabilizer, eg, an organic epoxide, is added and the polymer is isolated either as a crumblike particle by steam distilling the solvent (23) or as a reticulated film by direct evaporation on a double-drum dryer (24) (see Drying).

In manufacturing operations, the same batch process steps are used. The reaction vessel is a large volume glass-lined reactor capable of operating at 100–200 kPa (1–2 atm). By-product hydrogen chloride is scrubbed out with water in a countercurrent scrubber beyond the reflux condenser. The sulfur dioxide is separated and recycled to the feed gas streams. The aqueous HCl is neutralized with lime or sea shells ($CaCO_3$) for disposal as calcium chloride. The chlorosulfonated polyethylene is isolated as a polymer sheet on a double-drum dryer. The sheet is gathered into rope that is cut into short pieces called chips which are dusted lightly with talc to prevent agglomeration during storage.

The polymer chips are generally packaged in reinforced paper bags or corrugated paper boxes. The packaged product is shipped by box car or trucks.

Variations of the simple batch process described above are described in the patent literature. Chlorobenzene has been used as a solvent instead of carbon tetrachloride (25). Chlorosulfonation of suspensions or solutions of linear polyethylene, or both, in one or more stages has been described (7,26–28). Considerable effort has been extended to develop continuous processes, such as chlorosulfonation under pressure in a continuous tubular reactor using trichlorofluoromethane as the solvent (29), or chlorosulfonation of fused polyethylene in a screw extruder using sulfuryl chloride as the reagent (30). Staged, two-phase, countercurrent, continuous reactors for chlorosulfonation have been described (31).

Economic Aspects

The sole commercial producer of chlorosulfonated polyethylenes is the DuPont Company which markets a line of nine products under the name Hypalon Synthetic Rubber (9). Since the starting material for these polymers is primarily ethylene (qv), the materials costs are significantly related to petroleum prices. In addition, the nature of the manufacturing process (batch operation), the relatively low polymer concentration in the process stream, and the polymer isolation procedure (including CCl_4 removal from dryer vents) increase the costs.

Specifications, Standards, and Quality Control

Commercial chlorosulfonated polyethylenes are available under specifications as to chlorine and sulfur content, Mooney viscosity, and general physical appearance. The chemical composition during manufacture is determined by infrared analysis, but finished product is controlled by elemental analysis because the polymer stabilizer interferes with the ir method. The processing and use properties of the polymers are determined by standard methods for rubber testing as described in the ASTM methods manual (32).

Health, Safety, and Environmental Factors

The process for manufacturing chlorosulfonated polyethylene involves the well-recognized hazards of using chlorine, sulfur dioxide, and carbon tetrachloride. Safe handling procedures and toxicity limits for Cl_2 and SO_2 are well documented and their irritation potential provides significant warning if exposure occurs. Carbon tetrachloride exposure is more insidious and constant monitoring is necessary to maintain work area concentrations below the OSHA promulgated limit of 10 ppm in air (33).

The manufacturing waste products, ie, excess SO_2 and HCl are recycled and neutralized, respectively, with minimal environmental effect. Extensive efforts are necessary, however, to prevent significant loss of carbon tetrachloride vapors to the atmosphere during the polymer isolation step. Positive draft ventilation around the dryers and the use of condensers and carbon adsorbers to remove the solvent from this air stream has enabled operations to meet EPA requirements.

The polymer products are characterized by a low order of oral and skin irritation potential (33). They contain small but measurable amounts of carbon tetrachloride. Slow diffusion of the residual solvent occurs under normal storage. If ventilation is limited and large quantities of polymer are stored, monitoring the atmosphere of the warehouse for carbon tetrachloride is recommended.

Users of chlorosulfonated polyethylenes in rubber compounding and processing procedures are warned to provide good ventilation in the vicinity of processing machinery to avoid exposure to carbon tetrachloride and other vapors that may be generated from the polymer compounds at processing temperatures (34).

Uses

The current commercial polymers offering a range of chlorine contents in both free-radical-based and linear polyethylene-based materials are listed in Table 4.

The Hypalon 20 and 30 polymers are generally applied from compounded solutions in organic solvents and are used for coating fabrics, fluid-applied roofing, decorative and ozone-proof coatings for elastomeric articles, and as chemically resistant maintenance coatings (see Coatings).

Table 4. Uses of Commercial Chlorosulfonated Polyethylenes[a]

Name	Base[b]	Chlorine, %	Application or characteristics
solution coatings			
Hypalon-20	FRPE	29	good low temperature properties
Hypalon-30	FRPE	43	hard glossy films
Hypalon-48	LPE	43	hard glossy films
general purpose			
Hypalon-40[c]	LPE	35	used for belts, calendered
Hypalon-48S[d]	LPE	43	fabric coatings, wire and cable covers, hose, molded goods, and tank linings
Hypalon-45	LPE	25	used for pond and pit liners, roofing film, magnetic door seals, and gaskets

[a] Ref. 9.
[b] FRPE = free-radical polyethylene, LPE = linear polyethylene.
[c] Plus viscosity variants, Hypalon LD-999, Hypalon 405, and Hypalon 4083.
[d] Viscosity variant of Hypalon-48.

The Hypalon-40 and 48 polymers are generally compounded and applied using normal rubber mixing and shaping technology (see Rubber compounding). Uncured chlorosulfonated polyethylenes are more thermoplastic than other commonly used elastomers. They are generally tougher at room temperature, but soften rapidly as temperature increases. This property makes it necessary to warm stocks before extruding or calendering and requires good control of stock temperature during processing runs. Blends of chlorosulfonated polymers are often used to alleviate processing irregularities.

The Hypalon-40 series are usually compounded with curing agents, carbon black, or mineral fillers, etc, and are vulcanized to provide various forms and shapes. Hypalon-45, 48, and 48S are often compounded without curing agents and then calendered or extruded without a subsequent vulcanization step. The resultant products have good stress–strain properties in spite of being uncured. They find application in magnetic door closures, roofing film, and pit and pond liners (see Film and sheeting materials).

BIBLIOGRAPHY

1. U.S. Pat. 2,212,786 (Aug. 27, 1940), D. M. McQueen (to E. I. du Pont de Nemours & Co., Inc.).
2. U.S. Pat. 2,416,061 (Feb. 18, 1947), A. McAlevy, D. E. Strain, and F. S. Chance (to E. I. du Pont de Nemours & Co., Inc.).

3. U.S. Pat. 2,503,253 (Apr. 11, 1950), M. L. Ernsberger and P. S. Pinckney (to E. I. du Pont de Nemours & Co., Inc.).
4. R. E. Brooks, E. E. Strain, and A. McAlevy, *Rubber World* **127**, 791 (1953).
5. A. M. Neal, *Report 55-3,* Rubber Chemicals Div., E. I. du Pont de Nemours & Co., Inc., Wilmington, Del., Feb. 1953.
6. U.S. Pat. 2,982,759 (May 2, 1961), R. O. Heuse (to E. I. du Pont de Nemours & Co., Inc.).
7. U.S. Pat. 2,879,261 (Mar. 24, 1959), P. R. Johnson and M. A. Smook (to E. I. du Pont de Nemours & Co., Inc.).
8. U.S. Pat. 2,972,604 (Feb. 21, 1961), W. B. Reynolds and P. J. Canterino (to Phillips Petroleum Co.).
9. *Hypalon Synthetic Rubber Bulletin HP-210.1 (E-12336),* E. I. du Pont de Nemours & Co., Inc., Wilmington, Del., 1976.
10. S. F. Kurath, E. Passaglia, and R. Pariser, *J. Appl. Polym. Sci.* **1**, 150 (1959).
11. M. A. Smook, E. T. Pieske, and C. F. Hammer, *Ind. Eng. Chem.* **45**, 2731 (1953).
12. A. Nersasian and P. R. Johnson, *J. Appl. Polym. Sci.* **9**, 1653 (1965).
13. A. Nersasian and D. E. Andersen, *J. Appl. Polym. Sci.* **4**, 74 (1960).
14. M. A. Smook and co-workers, *Rubber World* **128**, 54 (1953).
15. W. F. Busse and M. A. Smook, *Rubber World* **128**, 348 (1953).
16. J. T. Maynard and P. R. Johnson, *Rubber Chem. Technol.* **36**, 963 (1963).
17. R. M. Straub, *General Compounding of HYPALON Synthetic Rubber, Bulletin A-63957,* E. I. du Pont de Nemours & Co., Inc., Wilmington, Del., 1969.
18. F. Haaf and P. R. Johnson, *Kautsh. Gummi Kunstst.* **22**, 4, 149 (1969).
19. Reprint, *The General Chemical Resistance of Various Elastomers,* originally published in *The 1970 Yearbook of the Los Angeles Rubber Group Inc.,* P.O. Box 2704, Terminal Annex, Los Angeles, Calif. 90054.
20. C. E. McCormack, *Rubber Age* **104**, 27 (1972).
21. U.S. Pat. 2,046,090 (June 30, 1936), C. F. Reed and C. Horn.
22. U.S. Pat. 2,640,048 (May 26, 1953), J. S. Beekly (to E. I. du Pont de Nemours & Co., Inc.).
23. U.S. Pat. 2,592,814 (Apr. 15, 1952), J. L. Ludlow (to E. I. du Pont de Nemours & Co., Inc.).
24. U.S. Pat. 2,923,979 (Feb. 9, 1960), J. Kalil (to E. I. du Pont de Nemours & Co., Inc.).
25. U.S.S.R. Pat. 134,848 (Dec. 15, 1960).
26. Ger. Pat. 1,068,012 (Oct. 29, 1959), (to Farbewerke Hoechst A.G.).
27. Brit. Pat. 842,763 (July 27, 1960), (to The Dow Chemical Co.).
28. U.S. Pat. 3,314,925 (Apr. 18, 1967), K. F. King (to E. I. du Pont de Nemours & Co., Inc.).
29. U.S. Pat. 3,296,222 (Jan. 3, 1967), S. Dixon and co-workers (to E. I. du Pont de Nemours & Co., Inc.).
30. U.S. Pat. 3,347,835 (Oct. 17, 1967), J. C. Lorenz (to E. I. du Pont de Nemours & Co., Inc.).
31. U.S. Pat. 3,542,747 (Nov. 24, 1970), R. E. Ennis and J. W. Scott (to E. I. du Pont de Nemours & Co., Inc.).
32. *Annual Standards (Pt. 37, Rubber Testing),* American Society for Testing Materials, Philadelphia, Pa.,
33. *CFR Title 29, Labor, Part 1910.1000,* U.S. Government Printing Office, Washington, D.C., July 1, 1978, Table Z-2, p. 640.
34. I. C. Dupuis, *Toxicity Information Related to Handling and Processing HYPALON, Bulletin E07672,* E. I. du Pont de Nemours & Co., Inc., Elastomer Chemicals Department, Wilmington, Del.

PAUL R. JOHNSON
E. I. du Pont de Nemours & Co., Inc.

ETHYLENE–PROPYLENE RUBBER

Ethylene–propylene rubber was first introduced in the United States in limited commercial quantities in 1962, and is now the fastest growing elastomer. The terpolymer modification, EPDM, includes, in addition to ethylene (E) and propylene (P), a small amount of a third monomer, a diene (D), to permit conventional sulfur vulcanization at pendant sites of unsaturation. Minor amounts of ethylene–propylene copolymer without diene unsaturation, called EPM, are also produced (see Copolymers).

In 1951 Ziegler in Germany discovered a new class of polymerization catalysts, comprising a transition metal halide associated with an organometallic reducing agent, typically an aluminum alkyl. These catalysts induced polymerization by an anionic mechanism. Ziegler catalysts (1) were first used commercially in the manufacture of linear, low pressure, high density polyethylene (see Ziegler-Natta catalysts).

In Italy, Natta showed that with selected catalysts of the Ziegler class ethylene and propylene copolymerized in an irregular or random way to yield an amorphous material with interesting elastic properties (2).

In 1963 Ziegler and Natta were jointly awarded the Nobel prize for chemistry in recognition of their discoveries which have led to several commercially important new plastics and elastomers including ethylene–propylene rubber.

The structure (1) of the regular, alternating amorphous copolymer of ethylene and propylene can be written:

$$CH_2{=}CH_2 + CH_3CH{=}CH_2 \longrightarrow {+}(CH_2CH_2\overset{\underset{\displaystyle CH_3}{|}}{C}HCH_2{)}_n{-}$$

(1)

This structure is remarkably similar to the structure of natural rubber, *cis*-1,4-polyisoprene (2):

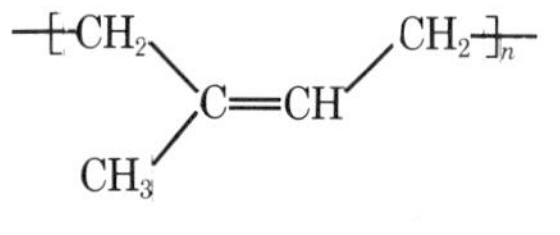

$$-{(}CH_2\underset{CH_3}{\diagdown}C{=}CH\diagup CH_2{)}_n-$$

(2)

It is not surprising, therefore, that the regular, alternating copolymer of ethylene and propylene is a decidedly rubbery material since its classical structure so closely approaches that of natural rubber (see Rubber).

Properties

Although the rubbery properties of the ethylene–propylene copolymers are exhibited over a broad range of composition, weight percentages of commercial products generally range from 50:50 to 75:25 ethylene/propylene.

In addition to the ethylene/propylene ratio, the average molecular weight of the rubber is controlled by selecting certain catalyst and polymerization variables. Whereas the polymer chemist generally measures average molecular weight in terms of intrinsic viscosity, the rubber compounder uses Mooney viscosity. The ethylene–propylene rubbers are controlled within the range of raw polymer Mooney viscosity that has been found to fit the various processing and applications requirements of the rubber industry and includes most other commercial synthetic rubbers. Mooney viscosity of EPM and EPDM is generally measured at four minutes after a one-minute warmup at 125°C. The measurement is expressed as ML 1 + 4 at 125°C.

The properties of a typical raw EPM are shown in Table 1; they are essentially the same for EPDM. Clearly, the low specific gravity of EPM and EPDM can be related to reduced unit costs in the fabrication of finished rubber products.

Polymer Unsaturation. The structure of EPM (1) shows it to be a saturated synthetic rubber. There are no double bonds in the polymer chain as there are in the case of natural rubber and most of the common commercial synthetic rubbers. The main chain unsaturation in these latter materials introduces points of weakness. When exposed to the degrading influences of light, heat, oxygen, and ozone, the unsaturated rubbers tend to degrade through mechanisms of chain scission and cross-linking at the points of carbon–carbon unsaturation. Since EPM does not contain any carbon–carbon unsaturation, it is inherently resistant to degradation by heat, light, oxygen, and, in particular, ozone.

The double bonds in natural rubber and the common polydiene synthetics are essential to their curing into useful rubber products using conventional chemical accelerators and sulfur. As a saturated elastomer, EPM cannot be cured or cross-linked using the long-established manufacturing practices and chemicals pertinent to the unsaturated rubbers. A more commercially attractive product is one that retains the outstanding performance features discussed earlier (eg, heat, oxygen, and ozone resistance) and includes some carbon–carbon unsaturation from a small amount of an appropriate diene monomer to accommodate it to conventional sulfur vulcanization chemistry.

Table 1. Properties of Raw Ethylene–Propylene Copolymers[a]

Property	Value
specific gravity	0.86–0.87
x-ray crystallinity	none
appearance	colorless
Mooney viscosity	varied
heat capacity, kJ/(kg·K)[b]	2.18
thermal conductivity, mW/(m·K)	355.2
thermal diffusivity, cm/s	1.9×10^{-3}
thermal coefficient of linear expansion per °C	1.8×10^{-4}
brittle point, °C	−95
glass transition temperature, °C[c]	−60
relative air permeability	
EPDM	100
natural rubber	100
IIR[d]	13
SBR[e]	65

[a] Ref. 3.
[b] To convert J to cal, divide by 4.184.
[c] Ref. 4.
[d] Isobutylene–isoprene rubber (5).
[e] Styrene–butadiene rubber (23% styrene).

Diene Monomers. The common diene monomers are isoprene and butadiene (qv). Efforts to introduce them into the EPM molecule were unsuccessful. However, search for appropriate dienes resulted in the discovery of, perhaps, fifty such chemicals. The commercially important ones are either nonconjugated straight-chain diolefins, or cyclic and bicyclic dienes.

These two classes have two features in common: (1) they are not conjugated, and (2) the activities of their two double bonds with respect to polymerization are different. The lowest molecular weight straight-chain diolefin that meets these requirements is 1,4-hexadiene:

$$CH_3CH=CHCH_2CH=CH_2$$

When this compound reacts with ethylene and propylene, the terminal double bond is active with respect to polymerization whereas the internal unsaturation is passive at this stage but remains in the resulting terpolymer as a substituent or pendant location for active sulfur vulcanization; poly(ethylene-co-propylene-co-1,4-hexadiene) [25038-36-2] (3) is an important commercial elastomer (6).

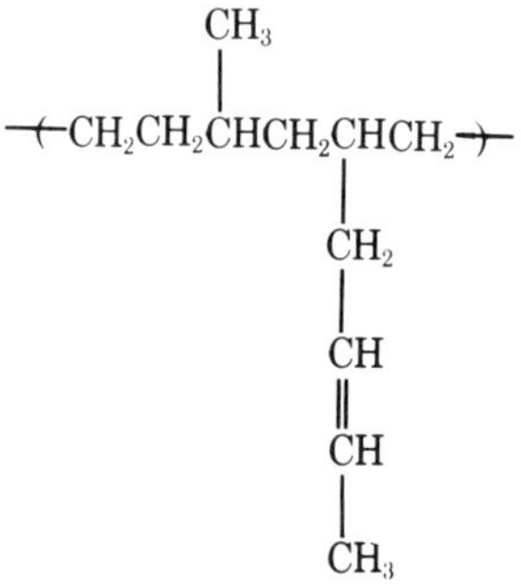

$$\begin{array}{c} CH_3 \\ | \\ -(CH_2CH_2CHCH_2CHCH_2)- \\ | \\ CH_2 \\ | \\ CH \\ \| \\ CH \\ | \\ CH_3 \end{array}$$

(3)

The bicyclic dienes used to introduce substituent unsaturation into ethylene–propylene rubber are typified by various derivatives of norbornene. Dicyclopentadiene (4) is used in certain EPDM grades that are made by a number of manufacturers (see Cyclopentadiene).

(4)

If enters the polymer readily with much higher polymerization efficiency than 1,4-hexadiene. The double bond in the bridged, or strained ring, is the more active with respect to polymerization and the five-membered ring with its double bond is left as a pendant substituent to the main polymer chain.

The most widely used diene in current commercial elastomer practice is ethylidenenorbornene, ENB (5), which gives poly(ethylene-co-propylene-co-ENB) [25038-36-2] (6) with EPDM.

[Structure (5): norbornene with =CHCH₃ substituent]

[Structure (6): —(CH₂CH₂CHCH₂—[ring])ₙ— with CH₃ groups]

Like the other bridged ring dienes, ENB shows a high rate of polymerization through the double bond in the bridged ring. The substituent internal double bond participates actively in sulfur cross-linking (7).

Both EPDM and EPM show outstanding resistance to heat, light, oxygen, and ozone because one double bond is lost when the diene enters the polymer and the remaining double bond is not in the polymer backbone but external to it.

The amount of diene that is incorporated into the EPDM molecule affects the curing rate of the rubber. Generally, from 4–5 wt % of diene gives a serviceable product.

Manufacture

The two principal raw materials for EPM and EPDM, ethylene (qv) and propylene (qv), both gases, are abundantly available at high purity. Propylene is commonly stored and transported as a liquid under pressure. Although ethylene can also be handled as a liquid, usually at cryogenic temperatures, it is generally transported in pipelines as a gas.

The manufacturing processes of EPDM are proprietary. A continuous solution process has been patented (8–9).

Ionic polymerizations using Ziegler-type catalysts are extremely sensitive to water and other polar materials, and dryness is essential. Water cannot be tolerated in any of the feed streams, although a few parts per million is a necessary and controllable maximum. The polymerization process is fully continuous. An inert hydrocarbon carrier solvent (eg, hexane) is used. This is dried in a column by azeotropic distillation (qv) and the dry solvent is fed to the polymerization vessels and used as a carrier for the catalysts to facilitate more precise control of these small streams. High-purity ethylene gas (from a pipeline supply) is dried by passing it through a molecular sieve (qv). High-purity propylene, from liquid storage, is similarly dried. Diene monomer, a liquid, is also sieve dried (see Drying agents).

The catalyst comprises two components (10). Vanadium oxytrichloride ($VOCl_3$), is a typical transition metal halide catalyst, whereas ethylaluminum sesquichloride [$(C_2H_5)_3Al_2Cl_3$], diethylaluminum chloride [$(C_2H_5)_2AlCl$], and diisobutylaluminum chloride [$(C_4H_9)_2AlCl$] are examples of typical metal alkyls (see Organometallics). Both types are dissolved in dry hexane to give precise flow control.

Dry hexane, ethylene, propylene, diene, and catalyst and cocatalyst solutions are continuously and proportionately fed to the first of a series of polymerization vessels. Polymerization of individual molecules, or chains, is extremely fast, and a few seconds at most, is the average life of a single growing polymer molecule from initiation to termination. The polymerization is highly exothermic, ca 2.559 MJ/kg (1100 Btu/lb). This heat must be removed, since the polymerization temperature (ca 35°C) has to be kept within narrow limits to ensure a product with the desired average molecular weight and distribution. Chilled jackets, draft tubes, and internal coils are used.

As the polymer molecules form and dissociate from the initiating catalyst, they remain in solution in the carrier hexane. The viscosity of the solution increases with increasing polymer concentration. The practical upper limit of solution viscosity is dictated by considerations of heat transfer, mass transfer, and fluid flow. At a rubber solids concentration of 8–10%, further increase in the solution viscosity is inadvisable and the polymerization is stopped.

The reactivity of ethylene is high whereas that of propylene is low. Various dienes have different polymerization reactivities (see Copolymers). The viscous rubber solution contains some unpolymerized propylene, more or less unpolymerized diene, and about 10% EPDM, all in homogeneous solution. This solution passes continuously into a flash tank where reduced pressure causes most of the unpolymerized monomers to leave as gases which are collected and recycled.

Catalyst residues, particularly vanadium, have to be removed as soluble salts in a washing and decantation operation. Vanadium residues in the finished product are kept to a few ppm. If oil-extended EPDM is the product, a metered flow of suitable oil is also added at this point.

The rubber is then separated from its solvent carrier by steam flocculation. The viscous cement is pumped through orifices into the vapor space of a violently agitated vessel partially full of boiling water. The hexane flashes off and, with water vapor, passes overhead to a condenser and to a decanter for recovery and reuse after drying. Residual unpolymerized ethylene and propylene appear at the hexane condenser as noncondensibles and are recovered for reuse after drying. The polymer, freed from its carrier solvent, falls into the water in the form of floc or crumb which looks very much like white popcorn.

The rubber floc or crumb, now a 3–4% slurry in hot water, is pumped over a shaker screen to remove excess water. The dewatered crumb is fed to the first stage of a mechanical-screw dewatering, drying press. Here, in an action similar to a rubber extruder, all but 3–6% of the water is expressed as the rubber is pushed through a perforated plate by the action of a screw. The cohesive, essentially dry rubber then passes into the second-stage press. This is similar to the first-stage dewatering machine, except that the mechanical action of the screw causes the rubber in the barrel to heat up to perhaps 150°C. As this rubber is extruded through the pelletizing perforated die plate at the end of the machine, the small amount of remaining water flashes as a vapor and dry rubber pellets are carried off on a cooling conveyor. The EPDM pellets are then continuously weighed, pressed into bales and packaged for storage and shipment.

Compounding. For curing EPDM, common rubber accelerators are used which are found in recipes for other synthetic rubbers. The choice of chemicals used in an EPDM vulcanizate depends on many factors, such as mixing equipment, mechanical properties, cost, safety, and compatibility.

A cure system consisting of tetramethylthiuram disulfide (TMTDS) activated with mercaptobenzothiazole (MBT) is basic, active, and safe. An even faster curing compound can be obtained by further activating the basic recipe with a dithiocarbamate such as zinc dibutyldithiocarbamate (ZDBDC) or ferric dimethyldithiocarbamate. Table 2 gives a typical cure system for EPDM–ENB, Table 3 gives the properties of the cured rubbers (see also Rubber compounding).

In general, EPDM compounds carry fairly high oil loadings. Indeed, EPDM polymer has a great capacity for extension with oil while retaining a good measure of its physical properties. Some producers add oil during manufacture. Oil-extended EPDM is made by first polymerizing a very high molecular weight, high Mooney viscosity polymer. Then, while the polymer is still in solution, the desired amount of oil is added. The oil-extended rubber is flash-flocculated, dried, and baled in the usual manner. The Mooney viscosity of the base polymer is reduced to a practical, workable level in the oil-extended EPDM.

Pigments, carbon blacks, and nonblacks, are added to every EPDM compound. The reinforcing, semireinforcing, easily processed and high abrasion blacks exhibit their typical behaviors. Similarly, the soft and hard clays, the reinforcing silica chemicals, and the calcium carbonates behave as expected (see Fillers).

Processing. The EPDM compounds can be successfully and economically mixed and processed on the machinery commonly found in various rubber fabricating plants. In mill mixing, EPDMs with lower Mooney viscosity in lightly loaded stocks are easier to handle. The high viscosity varieties of EPDM are tough on the mill but their processing improves as they readily accept high loading of oil and fillers. In Banbury mixing, the cycles and dump temperatures are about the same as would be used for SBR.

In general, EPDM compounds extrude readily on all commercial rubber extruders.

Table 2. Typical Cure Systems for EPDM-ENB, Parts by Weight

	A	B	C
component			
EPDM (ENB)	100	100	100
zinc oxide	5	5	5
FEF black	70	70	70
SRF black	35	35	35
naphthenic oil	100	100	100
paraffin wax	1	1	1
stearic acid	1	1	1
MBT	0.8	1.0	1.0
TMTDS	0.8	0.8	0.8
ROYLAC 133[a]	0.8		
tellurac[b]			0.8
tetrone A[c]		0.8	0.8
sulfur	0.8	0.8	0.8
compounded Mooney viscosity			
ML-4 at 100°C	28	27	26
Mooney scorch			
at 132°C			
scorch time, min	11¾	10½	9
cure rate, min	4	3½	3
at 121°C			
scorch time, min	32	18	17¼

[a] An activated dithiocarbamate accelerator.
[b] Tellurium diethyldithiocarbamate.
[c] Dipentamethylenethiuram tetrasulfide.

Table 3. Unaged Physical Properties of Cured EPDM–ENB[a]

Cure time, min	A	B	C
300% modulus, MPa[b]			
3	1.5	1.7	3.2
5	2.2	2.9	4.3
7.5	3.4	3.8	5.4
10	4.1	4.8	5.7
15	4.8	5.4	6.4
tensile strength, MPa[b]			
3	4.4	6.0	9.0
5	8.4	10.4	12.3
7.5	11.4	12.3	12.4
10	12.5	12.9	12.7
15	13.2	13.6	13.1
elongation, %			
3	860	900	830
5	970	940	780
7.5	860	830	680
10	810	740	640
15	760	690	600
hardness, Shore A			
3	43	48	50
5	48	50	51
7.5	50	52	53
10	50	53	54
15	50	55	55

[a] Cures at 205°C.
[b] To convert MPa to psi, multiply by 145.

This is important in wire applications and in automotive profile extrusions. Furthermore, EPDM compounds can be calendered both as unsupported sheeting and onto a cloth substrate (see Coating fabrics). The former is an important commercial application, since EPDM sheeting has been widely used for lining pits, ditches, and reservoirs and less widely as a roofing barrier (see Film and sheeting materials).

Highly durable, weather-resistant open- and closed-cell sponge can be made from EPDM. The various blowing agents commonly used for other elastomers are effective in EPDM (see also Foamed plastics).

The EPDM compounds can be cured on all of the common rubber-factory equipment. Press cures, transfer molding, steam cures, air cures, liquid metal cures, compression molding, and injection molding are all practical.

The EPDM compounds are nontacky. This sometimes causes building and cured adhesion problems in instances where it is necessary to join two stocks, as in the lap seams of a pit liner installation using EPDM sheeting, or in the building of an EPDM tire. These problems can be avoided by adding, where necessary, a few parts of a tackifier to the compound

Economic Aspects

Expressed as percentages of total annual rubber hydrocarbon consumption in the United States, EPDM has increased from nothing in 1964 to 2.2% in 1970 and

Table 4. United States EPDM Consumption in 1966–1975, Thousands of Metric Tons[a]

End use	1966	1969	1975
tires and related products	3.6	13.1	72.6
automotive	7.8	15.0	72.6
wire and cable	3.0	4.3	22.7
appliance, including hose	1.8	3.0	5.4
hose	2.7	6.4	11.4
belting	0.4	1.0	5.4
gaskets	0.4	2.0	13.6
rolls	0.5	0.5	4.5
proof goods	0.2	0.5	8.2
footwear	0.2	0.7	1.3
other	4.4	6.0	11.4
Total	25.0	52.5	229.1

[a] Ref. 4.

exceeded 10% in 1977. Table 4 illustrates the rapidly growing use of EPDM in all categories of rubber use.

Uses

In addition to the uses mentioned under Processing, an important application for EPDM is in blends with another rubber. Ozone resistance is thus provided without significant participation in co-cure, with the host rubber comprising the major portion of the blend. This technique has been applied in enhancing the ozone and weathering resistance of tire sidewalls and coverstrips. This use accounts for essentially all of the EPDM consumption in the tires and related products (see Table 4).

EPDM In Tires. At various times since its introduction EPDM has been hailed as the general purpose tire rubber of the future. A great amount of work has been done on the development of special grades of EPDM for the various components of passenger car tires, on the selection of compounding recipes adjusted to established tire fabrication practices, on the solution of building problems largely associated with the tack of EPDM compounds, and on the formulation of special treatments that would ensure an adequate level of adhesion between the cords and the carcass stocks (see Tire cord).

Although these problems have been solved, and it appears that an entirely serviceable all-EPDM tire can be built, there are no such tires currently being made. Economic factors favor the continued use of natural and general-purpose synthetic rubbers in tires.

BIBLIOGRAPHY

1. U.S. Pat. 3,113,115 (Dec. 3, 1963), K. Ziegler, H. Martin, and E. Holzkamp (to Karl Ziegler).
2. U.S. Pat. 3,300,459 (Jan. 24, 1967), G. Natta and G. Boschi (to Montecatini Edison S.p.A.).
3. G. Natta and co-workers, *Rubber Chem. Technol.* **36,** 1583 (1963).
4. Uniroyal Chemical, Division of Uniroyal, Inc., Naugatuck, Conn., unpublished work.
5. W. Hoffman, *Rubber Chem. Technol.* **37**(2), 164 (1964).

6. U.S. Pat. 2,933,480 (Apr. 19, 1960), W. F. Gresham and M. Hunt (to E. I. du Pont de Nemours & Co., Inc.).
7. U.S. Pat. 3,211,709 (Oct. 12, 1965), S. Adamek, E. A. Dudley, and R. T. Woodhams (to Dunlop Rubber Co., Ltd.) (assigned by mesne assignments to Hercules Powder Co. now Hercules, Inc.).
8. U.S. Pat. 3,341,503 (Sept. 12, 1967), J. L. Paige and S. M. DiPalma (to Uniroyal, Inc.).
9. P. J. Brennan, *Chem. Eng.* **72**, 94 (1965).
10. R. J. Kelly, *Ind. Eng. Chem. Prod. Res. Dev.* **1**, 210 (1962).

General Reference

F. P. Baldwin and G. Ver Strate, *Rubber Chem. Technol.* **45**(3), 709 (1972).

E. L. BORG
Uniroyal Chemical Co.

FLUORINATED ELASTOMERS

Fluorinated elastomers are a class of synthetic elastomers that are designed for demanding service applications in environments where combinations of extreme temperature ranges, chemicals, fluids, and/or fuels exist. The three basic fluorinated elastomer types are: fluorocarbons, fluorosilicones, and fluoroalkoxyphosphazenes (see also Fluorine compounds, organic).

A broad range of monomer and polymer developments was needed to achieve the balance of fluorinated elastomer properties that are known in industry. Military interest in the development of fuel and heat resistant elastomers for low temperature service initiated the development of fluorocarbon elastomers. In the 1950s the M. W. Kellogg Company in a joint project with the Quartermaster Corps, U.S. Army, and the 3M Company with the U.S. Air Force developed two commercial fluorocarbon elastomers: the copolymers vinylidene fluoride and chlorotrifluoroethylene, which became available from Kellogg in 1955 under the trade name of Kel-F (1–3); and a fluoroelastomer from 3M Company based on poly(1,1-dihydroperfluorobutyl acrylate) [424-64-6] sold as 3M Brand Fluororubber 1F4 in 1956 (4). The poor balance of acid, steam, and heat resistance of the latter elastomer limited its commercial use.

In 1957 the copolymers of vinylidene fluoride, $CF_2=CH_2$, and hexafluoropropylene, $CF_2=CFCF_3$, were developed on a commercial scale by DuPont (5–7) (Viton) and at 3M Company (Fluorel). In the 1960s terpolymers of vinylidene fluoride, hexafluoropropylene, and tetrafluoroethylene, $CF_2=CF_2$, were developed (8) and were commercialized by DuPont as Viton B. At about the same time, Montedison developed copolymers of vinylidene fluoride and 1-hydropentafluoropropylene as well as terpolymers of these monomers with tetrafluoroethylene marketed as Tecnoflon polymers (9–10). Other additions to the fluoroelastomer family have resulted from technological methods for development of elastomer chemical structure or new curing methods.

Elastomers have been developed for improved low temperature properties using perfluoromethyl vinyl ether, $CF_2\!\!=\!\!CFOCF_3$, as a comonomer with vinylidene fluoride and tetrafluoroethylene (11–12), with peroxide vulcanizable fluoroelastomers (13), and with totally perfluorinated monomers for improved solvent and heat resistance (14–17).

Copolymers of propylene and tetrafluoroethylene which are sold under the Aflas trade name by Asahi Glass Company, Japan, have been added to the fluoroelastomer family (18–22). A list of the principal commercial fluorocarbon elastomers is given in Table 1.

Table 1. Commercial Fluorocarbon Elastomers

Copolymer	CAS Registry No.	Trade names	Suppliers
poly(vinylidene fluoride-co-hexafluoropropylene)	[9011-17-0]	Fluorel	3M
		Viton A	DuPont
		Tecnoflon	Montedison
		Dai-El	Daikin
poly(vinylidene fluoride-co-hexafluoropropene-co-tetrafluoroethylene)	[25190-89-1]	Viton B	DuPont
		Dai-El G-501	Daikin
poly(vinylidene fluoride-co-hexafluoropropylene-co-tetrafluoroethylene) [25190-89-1] plus cure site monomer		Viton G (peroxide curable)	DuPont
poly(vinylidene fluoride-co-tetrafluoroethylene-co-perfluoromethyl vinyl ethyl) [56357-87-0] plus cure site monomer		Viton GLT (peroxide curable)	DuPont
poly(tetrafluoroethylene-co-perfluoromethyl vinyl ether) [26425-79-6] plus cure site monomer		Kalrez	DuPont
poly(tetrafluoroethylene-co-propylene)	[27029-05-6]	Aflas 100, 150	Asahi
poly(vinylidene fluoride-co-chlorotrifluoroethylene)	[9010-75-7]	Kel-F 3700	3M
poly(vinylidene fluoride-co-1-hydropentafluoropropylene)	[32552-63-9]	Tecnoflon SL	Montedison
poly(vinylidene fluoride-co-1-hydropentafluoropropylene-co-tetrafluoroethylene)	[29830-35-1]	Tecnoflon T	Montedison

Independent of the development of fluorocarbon elastomers for military applications, fluorosilicone elastomers were developed in the late 1950s at Dow Corning. They impart improved fluid resistance through the incorporation of fluorine, and they maintain the low temperature flexibility of the silicone elastomers. Fluoroalkylsiloxane research (23) resulted in the production of high molecular weight polymers (24–25) which were marketed as solvent resistant elastomers having excellent low temperature flexibility (26–28).

The fluorosilicone elastomers have repeating units of the general formula $CH_3SiOCH_2CH_2CF_3$ and contain small amounts of $CH_3SiOCH\!\!=\!\!CH_2$ used for cross-linking with peroxides (29–30). The fluorine content (30–40 wt %) is high enough to impart moderate oil and fuel resistance to vulcanizates which is sufficient for many uses but on balance does not equal the high fluid, compression set, and temperature resistance of fluorocarbon rubber. In the absence of significant chemical exposure, the useful temperature range of fluorosilicones is −54 to 260°C; where excellent low temperature flexibility and moderate fuel resistance is required, application in more specialized uses can be made. Since the commercial introduction of fluorosilicones in 1956 improvements have been made which include more consistent lot-to-lot properties, extension of the useful temperature range, improved fluid resistance and increased durability under static and dynamic stress.

The polyfluoroalkoxyphosphazenes are a recent commercial addition to the fluorinated elastomer family. Early work (31) culminated in studies of polymerization conditions conducive to the production of linear high molecular weight dichlorophosphazene polymers (32–34). These linear phosphazene polymers are substituted with fluorocarbon alcohols (35–40) as shown:

$$\begin{array}{c} Cl \\ | \\ \text{\textendash}[P=N]_n\text{\textendash} \\ | \\ Cl \end{array} + R_fCH_2ONa \longrightarrow \begin{array}{c} O\text{\textemdash}CH_2R_f \\ | \\ \text{\textendash}[P=N]_n\text{\textendash} \\ | \\ O \\ | \\ CH_2 \\ | \\ R_f \end{array}$$

$$R_f = CF_3, C_3F_7$$

Commercial interest in these elastomers was developed at Horizons Research. Commercial introduction was made by Firestone in 1975 (41). These elastomers are vulcanizable (by peroxide via incorporation of a cure-site monomer) and offer a unique set of low temperature properties similar to those of the fluorosilicones; however, with improved oil and solvent resistant properties from −55 to 175°C (41).

Properties

Table 2 lists the general physical properties of the main classes of fluoroelastomers. Table 3 summarizes general characteristics of vulcanizates prepared from commercially available fluoroelastomer gumstocks.

Table 2. Fluorine-Containing Elastomers

Type	Temperature use range, °C	Fluorine content, %	Valuable characteristics
fluorocarbon	−46 to 316	53–70	low compression set, high temperature stability, fuel oil, chemical resistance specific gravity, 1.80–1.86
fluorosilicone[a]	−54 to 232	30–40	low temperature flexibility softness specific gravity, 1.35–1.65
fluoroalkoxy-phosphazenes[b]	−54 to 232	30–40	low temperature flexibility fuel oil, chemical resistance

[a] Produced by Dow Corning (Silastic) and GE (FSE).
[b] Produced by Firestone (PNF).

Table 3. Fluorocarbon Elastomers Physical Property Ranges

Property	Value
tensile strength, MPa[a]	8.96–18.62
100% modulus, MPa[a]	2.07–15.17
elongation at break, %	100–500
hardness range	
Shore A–Shore D	50–55
compression set (method B 3.5 mm O-Ring), %	
70 h at 25°C	9–16
70 h at 200°C	10–30
1000 h at 200°C	50–70
specific gravity (gumstock)	1.80–1.86
General characteristics	
gas permeability	very low
flammability	self-extinguishing or nonburning (when properly formulated)
radiation resistance	good-to-fair
abrasion resistance	good and satisfactory for most uses
weatherability and ozone resistance	outstanding (unaffected after 200 h exposure to 150 ppm ozone)
low temperature flexibility	
Gehman T2	−8 to −18°C
(ASTM D T10	−16 to −23°C
1053) T100	−24 to −30°C
TR10 (ASTM D 1329)	−18 to −30°C
brittle point[b] (ASTM D 746)	−18 to −50°C

[a] To convert MPa to psi, multiply by 145.
[b] Highly dependent on grade of material used.

Thermal Stability. The retention of elongation after thermal aging of fluorocarbon elastomers is characteristic of their thermal stability. Figure 1 is a plot of percent retention of initial elongation vs days exposure to dry heat (150°C) for a number of oil-

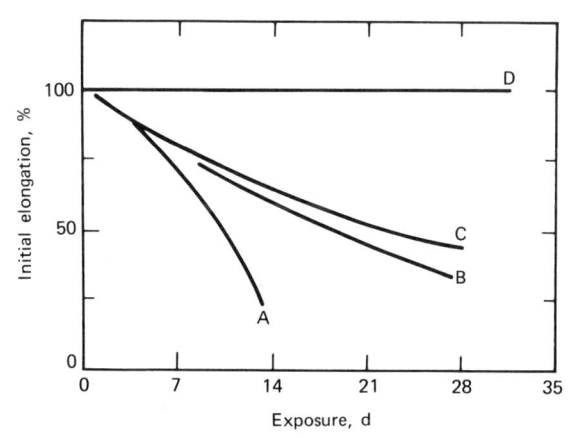

Figure 1. Comparison of elastomers of retention of elongation at 150°C. A, nitrile; B, EPDM (ethylene–propylene–diene monomer); C, acrylate; D, fluorocarbon.

resistant elastomers (42). As shown by Figure 1, fluorocarbon elastomers are far superior to any others. A more severe test at 205°C shows that a typical molded goods compound retains 95% of initial elongation after one year. Retention of tensile strength is another important characteristic of fluorocarbon elastomers. Figure 2 shows the results of long-term heat aging on a typical O-ring compound. Fifty percent of initial tensile strength is retained after a period of one year at 205°C, or after more than two months at 260°C.

Chemical Resistance. Table 4 summarizes the resistance of standard fluorocarbon elastomer compounds to various classes of chemicals.

Recent innovations in fluorocarbons elastomer development has led to more highly fluorinated materials which possess greater solvent resistance than standard resins. Included in this class of highly fluorinated materials is Fluorel LVS-76, Viton VTR-4590, and Kalrez. Figure 3 demonstrates the effect of high fluorine incorporation on improved volume swell resistance. The higher solvent-resistant materials are expected to find wide applicability in the petrochemical area.

Compression-Set Resistance. One property of fluorocarbon elastomers that makes them uniquely valuable to the sealing industry is their extreme resistance to compression set. Figure 4 plots set vs time for a compound prepared especially for compression-set resistance.

Manufacture and Processing

Manufacture of Fluorocarbon Elastomers. Elastomers listed in Table 1 have been prepared by high pressure, free-radical aqueous emulsion polymerization techniques (43–45). The initiators (qv) can be organic or inorganic peroxy compounds such as ammonium persulfate. The emulsifying agent is usually a fluorinated carboxylic acid soap, and the temperature and pressure of polymerization ranges from 80 to 125°C and 2.2 to 10.4 MPa (300–1500 psig). The molecular weights of the resultant polymers are controlled by the ratios of initiator to monomer or choice of chain-transfer reagents, or both. Typical chain-transfer reagents that can be employed are carbon tetrachloride,

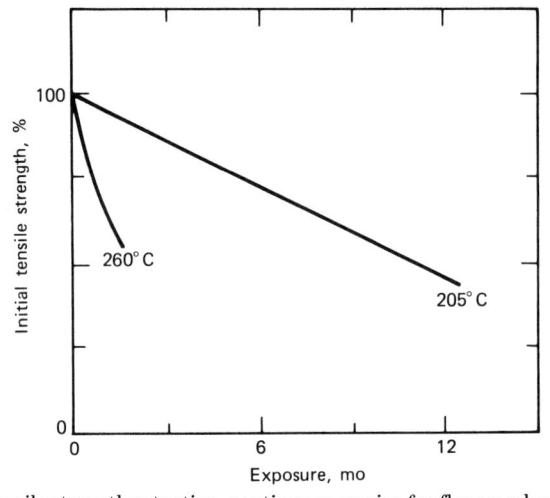

Figure 2. Tensile strength retention, continuous service for fluorocarbon elastomers.

Table 4. Fluid Resistance of Fluorocarbon Elastomers

excellent resistance[a]
- automotive fuels and oils
- hydrocarbon solvents
- aircraft fuels and oils
- hydraulic fluids
- certain chlorinated solvents

good to excellent resistance[b]
- highly aromatic solvents
- polar solvents
- water and salt solutions
- aqueous acids
- dilute alkaline solutions
- oxidative environments

not recommended[c]
- ammonia or amines
- strong caustic (50% sodium hydroxide above 70°C)
- certain polar solvents
  - (methyl ethyl ketone)
  - (methyl alcohol)

[a] May be used without reservation.
[b] Care must be taken in choice of proper gum and compound.
[c] Alternative materials should be considered.

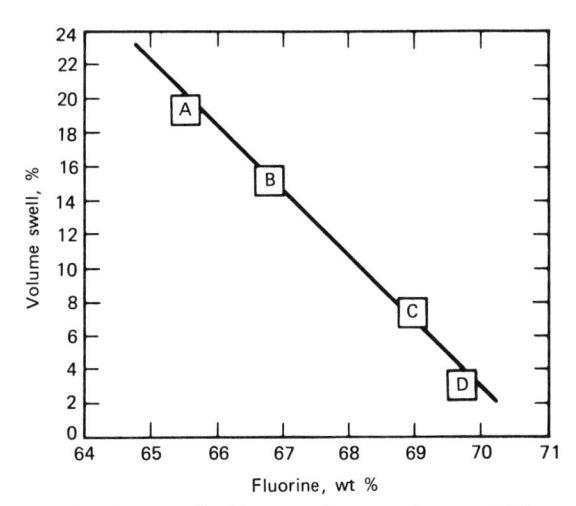

Figure 3. The percent volume swell of benzene for seven days at 21°C compared with the weight percent of fluorine on standard recommended compounds. A, copolymers of vinylidene fluoride–hexafluoropropylene (Fluorel, Viton, Diael, and Tecnoflon); B, terpolymers of vinylidene fluoride–hexafluoropropylene–tetrafluoroethylene (Viton B, Tecnoflon T, and Viton GH); C, Fluorel LVS-76 (Viton VTR-4590); and D, perfluoroelastomer (Kalrez).

methanol, acetone, diethyl malonate and dodecylmercaptans (46–48). A typical polymerization recipe is shown in Table 5.

The aqueous emulsion polymerization can be conducted by a batch or continuous

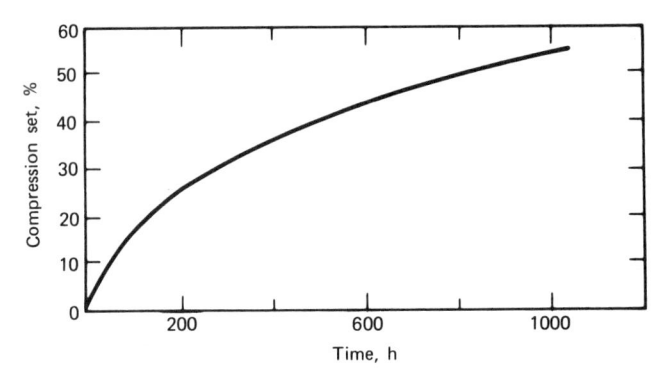

Figure 4. Compression set of a fluorocarbon elastomer, Compound I (see Table 8) at 200°C, 3.5 mm O-rings.

Table 5. Typical Fluoroelastomer Polymerization Recipe

Component	Amount, g
vinylidene fluoride	61
hexafluoropropylene	39
carbon tetrachloride	0.12
potassium persulfate	1.2
perfluorooctanoic acid	0.90
potassium phosphate dibasic	3.6
water	340

process (Fig. 5). In a simple batch process, all the ingredients are charged to the reactor, the temperature is raised, and the polymerization is run to completion. In a continuous process (49), feeding of the ingredients and removal of the polymer latex is continuous. The emulsion polymerization generally is not carried beyond 25% solids; however, it is possible to polymerize to stable latexes as high as 50% solids.

Discharge of latex from the reactor is controlled by a pressure control or relief valve. The polymer latex is then coagulated upon addition of salt or acid, or a combination of both. There are many satisfactory methods of effecting coagulation into a crumb. The crumb is washed, dewatered, and dried. Since most fluoroelastomers are sold with incorporated-cure systems, the final step in the process involves incorporation of the cure additives. This can be done on a two-roll mill, in a Banbury mixer, or in a mixing-extruder.

The manufacturing of the majority of fluoroelastomer gum includes the addition of an incorporated-cure system comprised of an organic phosphonium cure accelerator, such as triphenylbenzylphosphonium chloride; and a bisphenol cross-linking agent, such as hexafluoroisopropylidenediphenol (Table 6). These incorporated-cure systems offer improved compression-set performance, processing safety, faster cure cycles, and ensure quality control to the rubber molders, who need add only metallic oxides and reinforcing fillers for a complete formulation (50–55).

In addition to the incorporated-cure gums, there are also raw gums that contain no curatives, to which the rubber molder adds cure ingredients, such as diamines and peroxides (56), in addition to formulation (compound) ingredients.

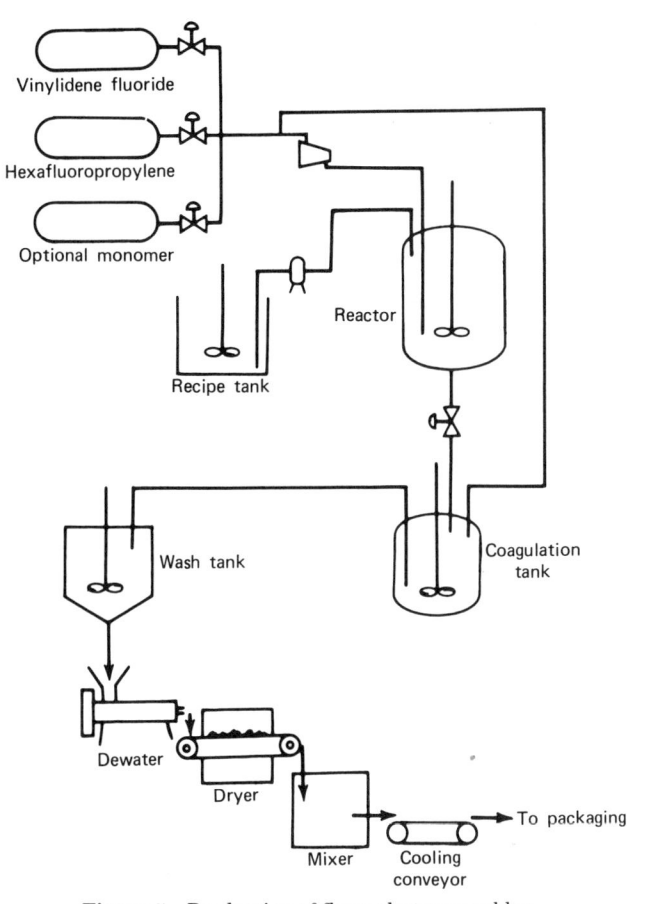

Figure 5. Production of fluoroelastomer rubber.

Manufacture of Fluorosilicones and Fluoroalkoxyphosphazene Elastomers. Like the fluorocarbon elastomers, fluorosilicones and fluoroalkoxyphosphazene elastomers are available in a range of viscosities, compounded masterbatches, and physical properties. For their specific manufacturing details, availability and uses, see Fluorine compounds, organic. The processing methods used with all the fluorine-containing elastomers are similar.

Processing of Fluorocarbon Elastomers. *Compounding.* Owing to the number of ingredients required in a conventional rubber recipe, fluoroelastomer compounding seems simple compared to what is typically encountered with other elastomers. However, the apparent simplicity of such formulations makes a selection of appropriate ingredients especially important if one is to obtain the excellent properties inherent in available gumstocks. A typical recipe is shown in Table 7.

When one has a firm idea of use requirements and rubber response to specific additives, a formulation may be selected. Uses generally fall into one of two classes: O-rings or molded goods.

O-Rings. In O-ring applications, the primary consideration is resistance to compression set. A fluoroelastomer gum is chosen for O-ring applications based on its molecular weight, cross-link density and cure system so that the best combination

Table 6. Typical Fluorocarbon Elastomer Incorporated-Cure Systems

Cure accelerator	Cross-linking agent
quaternary ammonium hydroxide salt tetrabutyl ammonium Captax, $\text{Captax}-S^- N^+(C_4H_9)_4$	hydroquinone, (para-dihydroxybenzene)
quaternary phosphonium salt triphenylbenzylphosphonium chloride, $(C_6H_5)_3\overset{+}{P}CH_2C_6H_5\ Cl^-$	hexafluoroisopropylidene-diphenol, $HO-C_6H_4-C(CF_3)_2-C_6H_4-OH$

Table 7. Typical Fluorocarbon Elastomer Compound

Component	Amount, phr[a]
rubber (may include curative)	100
inorganic base: magnesium oxide, calcium hydroxide	6–20
filler (reinforcing or nonreinforcing)	0–60
accelerators or curatives (if not included in base rubber, an example is Diak #1[b])	0–6
process aids	0–2

[a] Parts (wt) per hundred of rubber.
[b] Registered trademark of E. I. du Pont de Nemours & Co., Inc.

of processibility and use performance is given. An example formulation for such a use is Compound I in Table 8, based on Fluorel 2174. Long term compression-set resistance is described in Figure 6. Lower set values are achievable by use of higher molecular weight gumstocks at comparable cross-link densities. Compression-set resistance is also relatively sensitive to post-cure conditions. Compression-set resistance is very dependent on the cure system chosen. The phenolic cure system offers the lowest compression-set resistance available today, as shown in Table 9.

Molded Goods. In molded goods compounding, the most important physical property in the final vulcanizate is usually elongation to break, with compression set being a relatively minor consideration. Since complex shapes are often required, compound flow is also an important parameter. These objectives are generally best met by use of a gum with the lowest initial viscosity that is consistent with good physical properties and with a lower cross-link density to permit high elongation when compared to O-ring formulation.

Table 8. Fluorocarbon Elastomer O-Ring Applications

Typical formulation[a]	Amount, phr[b]
Fluorel 2174 (ML 1 + 10 at 120°C = 40)	100
MgO	3
Ca(OH)$_2$	6
MT Black	30
	139
physical properties[c] (3.5 mm cross section O-rings)	
tensile strength, MPa[d]	12.27
elongation at break, %	195
hardness, Shore A	75
compression set, method B for 70 h at 200°C, %	14
specific gravity	1.8

[a] Compound I. Qualifies for Mil-83248-1; AMS 7280A; and AMS 7276.
[b] Parts (wt) per hundred of rubber.
[c] Press cure: 5 min at 177°C; Post-cure: 24 h at 260°C
[d] To convert MPa to psi, multiply by 145.

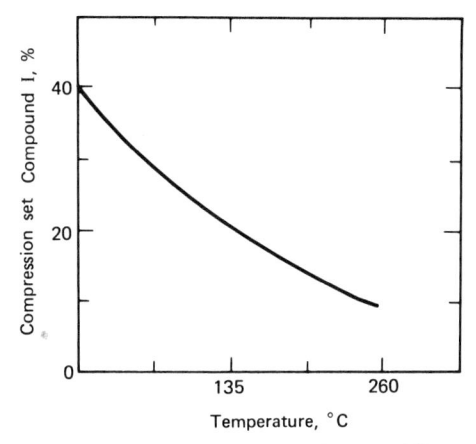

Figure 6. Effect of post cure temperature on compression set of Compound I (see Table 8) at 70 h at 200°C, 3.5 mm O-rings; post cure time of 24 h. Higher post cure temperatures lead to degradation of other properties such as elongation to break.

Comparison of starting viscosities and properties of a molded goods compound (see Table 10), with an O-ring compound (see Table 8) shows differences in elongation as a result of lower cross-link density. Even higher elongations are achievable with special formulations, as well as development of exceptionally high tear strengths in the press cure state.

The effects of specific fillers (qv) (black and nonblack) in molded goods applications is available (57) and is of special importance for water- and acid-resistant compounds. Table 11 illustrates the important effect of filler system technology on physical property retention in steam with phenolic and peroxide-curable raw gums. It is good policy to consult suppliers for specific recommendations to meet the balance of properties required.

Table 9. Effect of Cure System on Processing Safety and Compression-Set Resistance

Compound	Amount, phr[a] Amine	Bisphenol	Peroxide[b]
Fluorel 2178	100		
Fluorel 2174		100	
VT-R-4590[b]			100
Diak #3[c]	3		
MT Black	30	30	30
magnesium oxide	18	3	
calcium hydroxide		6	3
TAIC[d]			2
Luperco 101XL[e]			2
MS at 121°C[f]			
minimum	62	47	58
point rise (25 min)	2	1	10
compression set (3.5 mm O-rings), %			
air for 70 h at 200°C (ASTM D-395)	51	15	29

[a] Parts (wt) per hundred of rubber
[b] Data refer to a fluoroelastomer material offered by E. I. du Pont de Nemours & Co., Inc. and described in contribution No. 388 from Elastomer Chemicals Department.
[c] Registered trademark of E. I. du Pont de Nemours & Co., Inc.
[d] Triallylisocyanurate; available from American Cyanamid, Bound Brook, N.J.
[e] Registered trademark of Pennwalt Corporation, Buffalo, N.Y.
[f] Mooney scorch.

Processing. Fluoroelastomers can provide environmental resistance without sacrificing processability. Although this has not always been true, advances in cure system technology, coupled with appropriate means of gum preparation, have resulted in rubber stocks that are processible by any technique currently used in the rubber industry.

Formulation Parameters. Gum viscosity is of primary importance to the determination of processibility as this factor affects vulcanizate properties, especially compression set. Gums are available with ML 1 + 10 (Mooney viscosity, large rotor) (120°C) values of 30–160; a range of 30–60 is preferred for the optimum combination of flow and physical properties. Higher viscosities can cause excessive heat buildup during mixing without a compensatory gain in physical strength.

Formulation viscosity depends on gum viscosity and on filler selection (type and loading). A preferred range, as measured by MS-120°C (Mooney scorch), is 30–50.

Formulation stability and safety must also be considered when determining processibility as they are strongly affected by compounding ingredients and cure systems.

The data listed in Table 9 clearly show the effect that choice of cure system can have on scorch stability and processing minimums as determined by standard testing techniques (MS-121°C).

The most workable formulations are compounded with raw gums containing the phenolic or incorporated-cure systems. These rubbers offer the processor the best starting point for maximum processibility.

Table 10. Fluorocarbon Elastomer Molded Goods Applications

Typical formulation[a]	Amount, phr[b]
Fluorel 2176 (ML 1 + 10 at 120°C = 30)	100
Ca(OH)$_2$	6
MgO	3
MT black	30
Total	139
post-cured physical properties[c]	
tensile strength, MPa[d]	13.1
elongation at break, %	280
hardness, Shore A	73
compression set, for 70 h at 200°C (Method B, 1.27 cm disk), %	16
specific gravity	1.8

[a] Compound II. Qualifies for: all commercial uses, and Mil R-25897 Type II, Class 1.
[b] Parts (wt) per hundred of rubber.
[c] Press cure: 5 min at 177°C; post-cure: 24 h at 260°C.
[d] To convert MPa to psi, multiply by 145.

Table 11. Effect of Inorganic Base on Steam Resistance of Phenolic- and Peroxide-Curable Raw Gums

Compound	Amount, phr[a]		
Fluorel 2170	100	100	
Viton GH			100
MT black	30	30	30
magnesium oxide	3		
calcium hydroxide	6		2
Litharge		18	3
Luperco 101XL			2.5
triallylisocyanurate			2.5
Steam for 7 d at 121°C	Result		
tensile change, %	−64	−18	−10
elongation change, %	−35	+8	+8.5
hardness change	+1	+1	−5
volume swell, %	+10.8	+1.5	+1.0

[a] Parts (wt) per hundred of rubber.

Mixing. Fluoroelastomer formulations are processible by any standard technique. Open mill mixing is frequently used since most commercial gums mix well. Exceptions to this are very low viscosity gums which have a tendency to stick to the rolls, pelletized materials which are often difficult to band, and very high viscosity gums which are excessively tough (see Mixing and blending).

Internal mixing, commonly referred to as Banbury mixing, is widely used with fluoroelastomers. Gumstocks and compounds that are particularly successful fall in the viscosity ranges discussed earlier and use incorporated phenolic-type cure systems owing to their inherent processing safety. A typical Banbury mix cycle runs 6–8 min with a drop temperature of 120°C. The typical formulations in Tables 8 and 10 both are readily mixed in a Banbury.

Preforming. Extrusion preforming is easily accomplished if relatively cool barrel temperatures are used with either a screw or piston-type extruder (Barwell). Again, it is important that the gums be used in the appropriate viscosity ranges and that scorching be avoided.

Calendering operations are done routinely and warm rolls (40–90°C) are recommended for optimum sheet smoothness (see Coating processes). A process aid, such as low molecular weight polyethylene wax, is often used (see Abherents), and sheet thicknesses of 0.5–1.3 mm (20–50 mils) can normally be produced.

Molding. Compression molding is generally used when it is desirable to conserve material and when a molding operation is set up to allow preparation of large numbers of preforms with minimum labor costs (see Plastics technology). Flow requirements are minimal and high viscosity resins may be used.

Transfer molding minimizes preforming and is used for the production of very small parts; however, this technique often requires excessive amounts of material. Flow requirements can be quite high but fluoroelastomers are available that are effective in this application. Examples are Fluorel 2173 and 2176.

Injection molding is not widely used in the rubber industry owing to excessive material waste but is being broadly considered in the industry and gumstocks and techniques have been described in some depth (58) and are available.

All types of molding may be carried out at 149–199°C. Standard industry molding temperature is 177°C and this allows molding times of five minutes or less for most parts. A special accelerator (Fluorel 2172) is available from 3M Company which allows molders to carry out successful operations at a molding temperature of 150°C.

Post-Curing. Post-curing at elevated temperatures develops maximum tensile strength and compression-set resistance in fluoroelastomers. General post-cure conditions are 16–24 h at 233–260°C. Figure 4 shows the response of compression-set resistance to post-cure temperature.

Economic Factors

Annual worldwide fluorocarbon elastomer usage totals 2000–3000 metric tons. Approximately 60% of this usage is in the United States, 25% in Europe, and 10% in Japan.

Prices for fluorocarbon elastomers in 1978 were $23–220/kg and fluorosilicones (Silastic, Dow Corning; FSE, General Electric) were $44–220/kg. The PNF fluoroalkoxyphosphazenes (Firestone) in 1978 were $220/kg.

Specifications

Commercially available fluorocarbon elastomers meet automotive specifications in the HK section of ASTM D-2000 and SAE J-200. Fluorosilicone elastomers meet the FK portion of the same standards. ASTM D-1418 specifies designations of composition, eg, fluorocarbon elastomers are designated FKM, and fluorosilicones are designated FVMQ.

Available fluorocarbon elastomers offer the balance of those properties meeting the major O-ring specifications, such as AMS 7276, AMS 7280A, MIL 83248 Amendment 1 Type II, Class I and II.

Certain allowed grades and formulations of the fluorocarbon elastomers are qualified under FDA 121-2562 for use as rubber articles whose intended applications require repeated or continuous contact with food. Elastomer suppliers will provide assistance in formulating for specified uses.

Test Methods

The fluoroelastomer raw gums provided for rubber molding are tested for Mooney viscosity (ASTM D-1646) and for specific gravity (ASTM D-297). When compounded as described above, the stocks are tested for Mooney cure (ASTM D-1646), Mooney scorch (ASTM D-1646) and oscillating-disk rheometer cure rate (ASTM D-2705). The vulcanizates are evaluated regarding original physical properties (ASTM D-412 and D-1414), aged physical properties (ASTM D-573), compression set (ASTM D-395), and fluid aging (ASTM D-471). Low temperature properties described earlier are low temperature retraction TR10 (ASTM D-1329), torsional modulus (ASTM D-1053), and brittle point (ASTM D-746).

Health and Safety Factors

In general, under normal handling conditions, the fluoroelastomers have been found to be low in toxicity and irritation potential. The specific toxicological, health, and safe handling procedures are provided by the manufacturer of each fluoroelastomer product upon request (see also Fluorine compounds, organic).

Uses

About 40% of United States usage is industrial, half of which is directed to the automotive industry and the other half to pollution control. The industrial market typically consumes large numbers of O-rings and molded goods, such as diaphragms for pumps, valve seats, and hose. The automotive and related off-the-road equipment industry primarily uses O-rings for fuel sealing, bonded seals for shaft sealing, and needle tips for fuel and transmission control systems. Growth in this segment of the market is expected as automobile engine temperatures rise and fuels change to the more highly aromatic and alcohol-containing variety (see Gasoline). A primary need for this cost-conscious market segment is improved low temperature flexibility.

The pollution control industry uses large amounts of fluoroelastomer-coated cloth in relatively few, but very large, flue duct and expansion joint systems which conduct hot exhaust gases from thermal electric generating systems through scrubber systems prior to venting (see Coated fabrics).

The usage pattern in Europe is similar to that in the United States, except that use in petrochemical and hydraulic systems is more important, and pollution control is considerably less important.

In Japan usage is about 70% automotive, 15% in shipbuilding, and 8% for electrical applications.

The principal original use of fluoroelastomers was in the aircraft sealing industry; that market segment accounts for less than 10% of total fluorocarbon elastomer usage today.

BIBLIOGRAPHY

1. M. E. Conroy and co-workers, *Rubber Age* **76,** 543 (1955).
2. C. B. Griffis and J. C. Montermoso, *Rubber Age* **77,** 559 (1955).
3. W. W. Jackson and D. Hale, *Rubber Age* **77,** 865 (1955).
4. F. A. Bovey and co-workers, *J. Polym. Sci.* **15,** 520 (1955).
5. U.S. Pat. 3,051,677 (Jan. 2, 1962), D. R. Rexford (to E. I. du Pont de Nemours & Co., Inc.).
6. S. Dixon, D. R. Rexford, and J. S. Rugg, *Ind. Eng. Chem.* **49,** 1687 (1957).
7. J. S. Rugg and A. C. Stevenson, *Rubber Age* **82,** 102 (1957).
8. U.S. Pat. 2,968,649 (Jan. 17, 1961), J. P. Pailthrop and H. E. Schroeder (to E. I. du Pont de Nemours & Co., Inc.).
9. U.S. Pat. 3,331,823 (July 18, 1967), D. Sianesi, G. Bernardi, and A. Regio (to Montedison).
10. U.S. Pat. 3,335,106 (Aug. 8, 1967), D. Sianesi, G. C. Bernardi, and G. Diotalleri (to Montedison).
11. U.S. Pat. 3,235,537 (Feb. 15, 1966), J. R. Albin and G. A. Gallagher (to E. I. du Pont de Nemours & Co., Inc.).
12. Ger. Offen. 2,457,102 (Aug. 7, 1975), R. Baird and J. D. MacLachlan (to E. I. du Pont de Nemours & Co., Inc.).
13. U.S. Pat. 4,035,565 (July 12, 1977), D. Apotheker and P. J. Krusic (to E. I. du Pont de Nemours & Co., Inc.).
14. A. L. Barney, G. H. Kalb, and A. A. Kahn, *Rubber Chem. Technol.* **44,** 660 (1971).
15. A. L. Barney, W. J. Keller, and N. M. Van Gulick, *J. Polym. Sci. A-1* **8,** 1091 (1970).
16. G. H. Kalb, A. L. Barney, and A. A. Kahn, *Am. Chem. Soc. Dev. Polym. Chem.* **13,** 490 (1972).
17. S. M. Ogintz, *Lubric. Eng.* **34,** 327 (1978).
18. Y. Tabata, K. Ishigure, and H. Sobue, *J. Polym. Sci. A* **2,** 2235 (1964).
19. G. Kojima and Y. Tabata, *J. Macromol. Sci. Chem. A* **5**(6), 1087 (1971).
20. K. Ishigure, Y. Tabata, and K. Oshima, *Macromolecules* **6,** 584 (1973).
21. G. Kojima and Y. Tabata, *J. Macromol. Sci. Chem. A* **6**(3), 417 (1972).
22. G. Kojima, H. Kojima, and Y. Tabata, *Rubber Chem. Technol.* **50,** 403 (1977).
23. O. R. Pierce, E. J. McBee, and R. E. Cline, *J. Am. Chem. Soc.* **75,** 5618 (1953).
24. Can. Pat. 570,580 (Feb. 10, 1959), O. K. Johannson (to Dow Corning).
25. U.S. Pat. 3,002,951 (Oct. 30, 1961), O. K. Johannson (to Dow Corning).
26. Can. Pat. 586,871 (Nov. 10, 1959), E. O. Brown (to Dow Corning).
27. U.S. Pat. 3,179,619 (Apr. 20, 1965), E. O. Brown (to Dow Corning).
28. U.S. Pat. 3,006,878 (Oct. 31, 1961), I. D. Talcott (to Dow Corning).
29. O. R. Pierce and Y. K. Kim, *Rubber Chem. Technol.* **44,** 1350 (1971).
30. O. R. Pierce and Y. K. Kim, *J. Elastoplast.* **3,** 82 (1971).
31. H. N. Stokes, *Am. Chem. J.* **19,** 782 (1897).
32. H. R. Allcock and R. L. Kugel, *J. Am. Chem. Soc.* **87,** 4216 (1965).
33. H. R. Allcock, J. E. Gardner, and K. M. Smeltz, *Macromolecules* **8,** 36 (1975).
34. H. R. Allcock and G. Y. Moore, *Macromolecules* **8,** 377 (1975).
35. U.S. Pat. 3,702,833 (Nov. 14, 1972), S. H. Rose and K. A. Reynard (to Horizons Research).
36. U.S. Pat. 3,838,073 (Sept. 24, 1974), S. H. Rose and K. A. Reynard (to Horizons Research).
37. U.S. Pat. 3,700,629 (Oct. 24, 1972), K. A. Reynard and S. H. Rose (to Horizons Research).
38. U.S. Pat. 3,972,841 (Aug. 3, 1976), T. C. Cheng, G. S. Kyker, and T. A. Antkoroiak (to Firestone).
39. P. Touchet and P. E. Gatza, *J. Elastomers Plast.* **9,** 3 (1977).
40. R. E. Singler, N. S. Schneider, and G. L. Hagnauer, *Polym. Eng. Sci.* **15,** 321 (1975).
41. D. P. Tate, *J. Polym. Sci. Symp.* **48,** 33 (1974).
42. J. R. Dunn and H. A. Pfisterer, *Rubber Chem. Technol.* **48,** 356 (1976).
43. U.S. Pat. 3,051,677 (Aug. 28, 1962), D. R. Rexford (to E. I. du Pont de Nemours & Co., Inc.).
44. U.S. Pat. 3,053,818 (Sept. 11, 1962), F. J. Honn and S. M. Hoyt (to 3M Company).
45. U.S. Pat. 2,968,649 (Jan. 17, 1961), J. P. Pailthorp and H. E. Schroeder (to E. I. du Pont de Nemours & Co., Inc.).
46. U.S. Pat. 3,069,401 (Dec. 18, 1962), G. A. Gallagher (to E. I. du Pont de Nemours & Co., Inc.).
47. U.S. Pat. 3,080,347 (Mar. 5, 1963), C. L. Sandberg (to 3M Company).
48. U.S. Pat. 3,707,529 (Dec. 26, 1972), E. K. Gladding and J. C. Wyce (to E. I. du Pont de Nemours & Co., Inc.).
49. U.S. Pat. 3,845,024 (Oct. 29, 1974), S. D. Weaver (to E. I. du Pont de Nemours & Co., Inc.).

50. U.S. Pat. 3,655,727 (Apr. 11, 1972), K. U. Patel and J. E. Maier (to 3M Company).
51. U.S. Pat. 3,712,877 (Jan. 23, 1973), K. U. Patel and J. E. Maier (to 3M Company).
52. U.S. Pat. 3,752,787 (Aug. 14, 1973), M. R. deBrunner (to E. I. du Pont de Nemours & Co., Inc.).
53. U.S. Pat. 3,857,807 (Dec. 31, 1974), Y. Kometani and co-workers (to Daikin).
54. U.S. Pat. 3,864,298 (Feb. 4, 1975), Y. Kometani and co-workers (to Daikin).
55. U.S. Pat. 3,920,620 (Nov. 18, 1975), C. Ceccato, S. Geri, and L. Calombo (to Montedison).
56. J. E. Alexander and H. Omura, *Elastomerics* **2,** 19 (1978).
57. *FLUOREL 2170—Compounding with Various Fillers,* 3M Company Product Brochure, available from Commercial Chemicals Division, St. Paul, Minn. 55101.
58. B. H. Spoo, *Rubber Chem. Technol.* **49,** 1350 (1976).

General References

"Fluoropolymers" in K. J. L. Paciorek and L. A. Wall, eds., *High Polymers,* Vol. XXV, Wiley-Interscience, New York, 1972, pp. 291–313.
"Fluoroelastomers" in D. A. Stivers, ed., *The Vanderbilt Rubber Handbook,* R. T. Vanderbilt Co., New York, 1978, p. 244.
Rubber Technology, M. Morton, ed., Van Nostrand Reinhold Co., New York, 1973, pp. 407–439.
R. G. Arnold, A. L. Barney, and D. C. Thompson, *Rubber Chem. Technol.* **46,** 619 (1973).
J. C. Montermoso, *Rubber Chem. Technol.* **37,** 1521 (1961).

ARTHUR C. WEST
ALLEN G. HOLCOMB
3M Company

NEOPRENE

Neoprene [*31727-55-6*], or polychloroprene, was discovered by A. M. Collins in 1930 (1) during an investigation of the properties of 2-chloro-1,3-butadiene when this volatile liquid was fortuitously polymerized to a rubbery polymer that could be vulcanized or cured. Development of this research led to its initial manufacture and sale in 1931 (2) under the trade name DuPrene. As practical applications for the polymer multiplied, the generic name neoprene was applied to DuPrene and to all other polymers of chloroprene. Through the years, the use of these polymers has continued to increase to an annual worldwide consumption greater than 300,000 metric tons, or about 5% of the world market for elastomers. Neoprene has enjoyed this growth because it has filled much of the need for an elastomer resistant to oxidation, oil, and heat. Neoprene is now manufactured in the U. S., the Federal Republic of Germany, France, the USSR, Japan, and Northern Ireland.

The broad range of physical and chemical properties available in the family of chloroprene homo- and copolymers permits neoprenes to fulfill the requirements of many applications. This versatility arises from the chemistry of free-radical emulsion polymerization of chloroprene. All neoprene polymers are now prepared by free-radical emulsion polymerization (3–4).

Properties

Neoprene, as sold by its manufacturers, is a raw synthetic rubber. To convert it into useful objects it must be mixed or compounded with selected chemicals, fillers, and processing aids. The resulting "compound" is then shaped, or molded, and finally vulcanized. The properties of the finished product depend on the exact type and amount of neoprene and the compounding ingredients. Hence the physical and chemical properties are closely related to the compounding recipe (see Rubber compounding).

The basic physical properties of representative raw polymer, gum vulcanizate, and carbon black-filled vulcanizate are given in Table 1.

The dynamic mechanical properties of neoprene under slow deformation rates (shear and creep) are reported in refs. 6–7, the components of the complex dynamic shear modulus in refs. 8–9.

The elastomeric properties of polychloroprene are controlled by its microstructure; the processing properties are controlled by the distribution of molecular sizes and the number and distribution of long-chain branches and cross-links. The chemical properties (reactivity, aging, etc) are controlled by the microstructure and the inherent effects of the presence of one chlorine atom for every four carbon atoms. The microstructure in turn is controlled by the conditions of polymer preparation. Microstructure control and the effect of specific microstructures on polymer properties are of great technological importance.

Table 1. Physical Properties of Polychloroprene[a]

Properties	Raw polymer	Vulcanizates Gum	Vulcanizates Carbon black
density, g/cm^3	1.23	1.32	1.42
coeff. of vol. exp		610	
$\beta = 1/v \cdot \delta v/\delta T$, κ^{-1}	600×10^{-6}	720×10^{-6}	
Thermal properties			
glass transition temp, K	228	228	230
heat capacity, Cp, kJ/(kg·K)[b]	2.2	2.2	1.7–1.8
thermal conductivity, W/(m·K)	0.192	0.192	0.210
Electrical			
dielectric constant (1 kHz)		6.5–8.1	
dissipation factor (1 kHz)		0.031–0.086	
conductivity, pS/m		3 to 1400	
Mechanical			
ultimate elongation, %		800–1000	500–600
tensile strength, MPa[c]		25–38	21–30
Youngs modulus[d], MPa[c]		1.6	3–5
resilience[e], %		60—65	40–50

[a] Ref. 5.
[b] To convert J to cal, divide by 4.184.
[c] To convert MPa to psi, multiply by 145.
[d] Initial slope.
[e] Rebound.

Microstructure and Stereochemistry. Analysis of crystalline behavior and infrared absorption spectra showed (10–12) that polychloroprenes consist of linear sequences of predominantly *trans*-2-chloro-2-butenylene units (**1**). Small moieties of the structures formed by cis 1,4 polymerization (**2**), 1,2 polymerization (**3**), and by 3,4 polymerization (**4**) are also present. The amounts of (**2**), (**3**), and (**4**) present increase regularly with increasing polymerization temperature from a total of 5% at −40°C to about 30% at 100°C. The effect of these structural variations on the physical and chemical properties of the polymers was determined (10–12), and the microstructure studied with improved infrared techniques (13).

The sequence of isomers derived from head-to-tail (**5**), head-to-head (**6**), and tail-to-tail (**7**) monomer addition (14), was identified with the help of high-resolution nmr spectroscopy (see Analytical methods).

Structures (**6**) and (**7**) account for 10 to 15% of the sequence distribution, the values being in this range regardless of the cis-trans isomer proportions.

Further detail of polymer microstructure has been revealed by application of ^{13}C nmr techniques (15–16).

The stereochemistry of polychloroprene has been further defined by the unique synthesis of poly(*cis*-1,4-chloroprene) [*9010-98-4*] (17) and by the preparation of a stereoregular poly(*trans*-1,4-chloroprene) (18). Crystallinity, rate of crystallization, and melt transition temperatures are inverse functions of polymerization temperature because of increased introduction of structural irregularities as polymerization temperature increased.

The kinetics of polychloroprene crystallization have been studied using differential thermal analysis (19–20). The heat of fusion of the crystalline phase is approximately 96 kJ/kg (22.9 kcal/kg), and the activation energy for crystallization has been reported as 104 kJ/mol (24.9 kcal/mol) (21). Variations in molecular weight and degree of cross-linking have only slight effects on polychloroprene crystallization rate. Crystallization is most rapid at −12°C. The thaw temperature is about 15°C higher than the crystallization temperature (22).

Known values of amorphous polymer density (d = 1.23) and crystalline phase density (d = 1.35) permit calculation of the extent of crystallinity in a given polychloroprene. Polymer prepared at −40°C is about 38% crystalline (mp = 73°C); polymer prepared at +40°C is ca 12% crystalline (mp = 45°C).

Effect of Heat and Oxygen. Properly compounded polychloroprene vulcanizates have outstanding resistance to natural and accelerated aging if protected by an antioxidant (23). Practical and highly oxygen-resistant compounds of polychloroprene have been developed, and considerable information is available concerning compounding for specific durability properties (23–26). Hot-air aging of black polychloroprene stocks at 130°C results in a decrease of elongation to 100% in 14 d, whereas protected stocks may retain 200% elongation in the same exposure time. White stocks are far more sensitive. The effect of increasing temperature on reducing aging performance is, as one would expect, exponential. The question of how long a polychloroprene stock gives useful service depends on the environmental conditions (especially the temperature) and the specific compound recipe used. The stock may be useful for longer than 35 yr or for only a few h.

Investigations of the mechanism of oxygen attack on the polychloroprene molecule (27–28) propose a peroxy radical attack on α methylene hydrogens, followed by rearrangement:

$$\text{ROO·} + \left(\begin{array}{c}\text{C}=\text{CHCH}_2\\|\\\text{Cl}\end{array}\right) \longrightarrow \text{ROOH} + \left(\begin{array}{c}\text{C}=\text{CHCH·}\\|\\\text{Cl}\end{array}\right) \rightleftharpoons \left(\begin{array}{c}\text{ĊCH}=\text{CH}\\|\\\text{Cl}\end{array}\right)$$

Further peroxidation leads to β alkoxy peroxide radicals which decompose to carbonyl compounds (see also Hydrocarbon oxidation).

The thermal degradation of neoprene has been reviewed and studied (29) using modern thermal analytical techniques (tga and dta). Volatile degradation products liberated as temperature rises from 100 to 500°C include ethylene, a trace of chloroprene, and hydrogen chloride equivalent to at least 90% of the chlorine in the polymer. The degradation mechanism involves random, intramolecular evolution of HCl, leaving behind a predominance of triene structures in the polymer (30–31).

Flammability. Polychloroprene resists ignition (32–33). The rate of HCl evolution during forced ignition was measured (34); phosgene was not among the combustion products. Kindling temperature (900°C), vapor ignition temperature (250°C), and the effect of compounding agents on flammability have been examined in detail (35–38). The effect of composition variables on the rate of heat and smoke release from polychloroprene foam under forced ignition was recently described (39). Flame resistance of polychloroprene apparently arises from its low heat of combustion coupled with the evolution of nonflammable HCl in the early stages of decomposition, which chars the polymer.

Polymerization

Free-Radical Polymerization. The conjugated butadienyl structure of chloroprene is especially responsive to free-radical attack because of the influence of the electron-rich chlorine atom which facilitates free-radical addition to the monomer, permitting sequential reaction.

$$CH_2=\underset{Cl}{C}CH=CH_2 \xrightarrow{R\cdot} R CH_2\underset{Cl}{C}=CHCH_2\cdot$$

In fact, the chloroprene molecule is so much more reactive than other dienes or olefins that it prefers to self-polymerize. For this reason, most of the polymer produced commercially is homopolymer.

The polymerization of chloroprene is exothermic, with a heat release of 62.8–75.3 kJ/mol (15–18 kcal/mol) (40–41).

In the emulsion polymerization of chloroprene and other dienes, monomer droplets are dispersed in an aqueous phase by means of a suitable surface-active agent, generally at a pH of 10–12. Polymerization is initiated by the addition of a free-radical catalyst and the polymerization proceeds, preferably in an isothermal manner and generally at 20–50°C, until the desired conversion of monomer is achieved. (This is determined by measuring the specific gravity increase of the emulsion in relation to an empirical correlation of sp g vs conversion.) The polymerization is stopped by destroying the free radicals present by the addition of short-stopping and stabilizing reagents. After removing unpolymerized monomer, the polymer is isolated by destabilizing the colloidal system, separating the aqueous phase, and drying the polymer.

The properties of the finished polymer can be varied over a wide range by proper choice of (1) type and concentration of molecular weight modifier, (2) conversion of monomer, (3) presence of other monomers, and (4) temperature.

The effects of these independent variables are made more complex by the colloidal nature of the system, that is (1) types of surfactant present, (2) concentration of surfactants, (3) concentration of salts, (4) monomer droplet size, (5) polymer particle size, and (6) monomer-to-water ratio.

The details of the complex interrelationships of these independent variables are proprietary knowledge of the various polychloroprene producers.

The mechanism of the polymerization (42) of chloroprene is less well understood than the polymerization of styrene (qv) or 1,3-butadiene (qv). Carrying the emulsion polymerization to an excessively high conversion affects the molecular weight, molecular weight distribution, and processibility of the product polymer. The high polymerization rate of chloroprene imposes a serious disadvantage in this respect (43), giving rise to gel polymer at conversions as low as 30% (44–45) in the absence of molecular-weight modifiers or chain-transfer agents. This accounts for the fact that the first commercial polymers were peptizable sulfur copolymers rather than homopolymers. The discovery of mercaptans as chain-transfer agents permits soluble homopolymers to be made at conversions up to 65 to 70% (46). At this stage of the polymerization, the increased polymer–monomer ratio and the consequent polymer radical–polymer reaction lead to branched and cross-linked structures.

A variety of modifiers control the molecular weight of polychloroprene including alkyl mercaptans (47), dialkyl xanthogen disulfides (48), iodoform (49), benzyl (50) and alkyl (51) iodides, and trichloromethyl-substituted mercaptans (52). Of these, the alkyl mercaptans are the most frequently used; the mechanism of mercaptan chain transfer has been studied extensively (53–60).

Free-Radical Copolymerization. *Chloroprene–Sulfur Copolymers.* The need for molecular weight control to achieve polymer processability was initially resolved by the incorporation of sulfur into the polymer to provide cleavable moieties along the chain which would be attacked by thiuram disulfides much as polysulfide compounds are attacked by thiuram disulfide (61). Polymerization of chloroprene in the presence of sulfur produces a polymer containing linear di- or polysulfide segments (62–63). The desired molecular weight control is accomplished by the postpolymerization reaction with a thiuram disulfide or other thiophilic agents. This plasticization or peptization is carried out at a pH above seven. The complex mechanism of this reaction has been only partially elucidated (64–65). Sulfur-modified polychloroprenes (G polymers) have continued to form an important part of chloroprene polymer production.

Copolymerization with Vinyl and Diene Monomers. The comonomer most frequently used with chloroprene is 2,3-dichloro-1,3-butadiene, but over the years almost every diene and vinyl monomer has been explored for this purpose. The relative reactivity ratios of the more significant comonomers are listed in ref. 66. Comonomers that might improve polychloroprene properties include acrylonitrile (67–69), methacrylonitrile (69), 2-cyano-1,3-butadiene (70), 2,3-dichloro-1,3-butadiene (71–72), 1,2,3,4-tetrachlorobutadiene (73), α-chloroacrylonitrile (74), α,β-unsaturated acids, vinyl ethers, and esters (69), methyl vinyl ketone (69,75), and others (see Acrylonitrile; Copolymers).

ω-Polychloroprene. Under special conditions chloroprene polymerizes in the liquid or vapor state to form a hard, insoluble, granular mass known as ω or popcorn polychloroprene (76). In the presence of monomer, the polymerization appears to proceed without external initiation (77). This phenomenon is due to the presence of polymeric free radicals (seeds) whose termination rate is grossly reduced because of restricted mobility in the high-viscosity medium (78).

The formation of popcorn polymer, which occurs with most diene monomers, is observed in monomer storage vessels, transfer lines, and even in polymerization equipment. It can interrupt plant operations and represents a serious hazard.

Popcorn formation can be minimized by the presence of inhibitors such as nitric oxide (79), inorganic nitrites (80), anion-exchange resins in the nitrite form (81), N_2O_4 adducts of unsaturated organic compounds, and certain other free-radical inhibitors (82–87).

Polymerization Agents. Anionic surfactants of the rosin acid, fatty acid, alkyl sulfate, or alkyl arylsulfonate types are most frequently employed as emulsifiers and colloidal stabilizers in the polymerization of chloroprene (3,4,88–91). Cationic soaps, such as cetylpyridinium bromide or betaine as well as sulfate and sulfonate soaps, are used to obtain neutral or acidic emulsions (92–93). The choice of surfactant depends on the type of polymerization process and product desired as well as availability and other economic factors.

A wide variety of redox systems using inorganic and organic peroxy compounds serve as initiators (94–95), most commonly potassium persulfate or ammonium persulfate. Special initiators are sometimes needed to overcome the traces of antioxidants used in technical chloroprene to counteract adventitious exposure of the monomer to traces of oxygen (96). Included are acyclic azo compounds (97), nitrogen sulfide compounds such as heptasulfurimide (98), trialkyl boron compounds with cocatalysts (99), borazanes (100), and formamidine sulfinic acid (101) (see Initiators).

Short-stopping (termination) chemicals halt the polymerization at the desired monomer conversion. Most frequently used are thiurams (102), oxalates (103), aromatic amines and phenolic compounds (104), and N,N'-dialkyl hydroxylamines (105).

Manufacture

Neoprene polymers are mainly manufactured by batch emulsion polymerization, and the polymers are isolated by freezing-drying procedures. The polymerization of chloroprene involves the same steps as the emulsion polymerization of other diene monomers, namely: (1) emulsification, (2) initiation and catalysis, (3) heat transfer, (4) monomer conversion and molecular weight control, (5) short-stopping and stabilization, (6) monomer recovery, and (7) polymer isolation.

A typical production flow diagram is shown in Figure 1.

The emulsion polymerization of Neoprene GN (106–107), a general purpose polymer, is representative of the manufacture of many polychloroprene polymers and latexes.

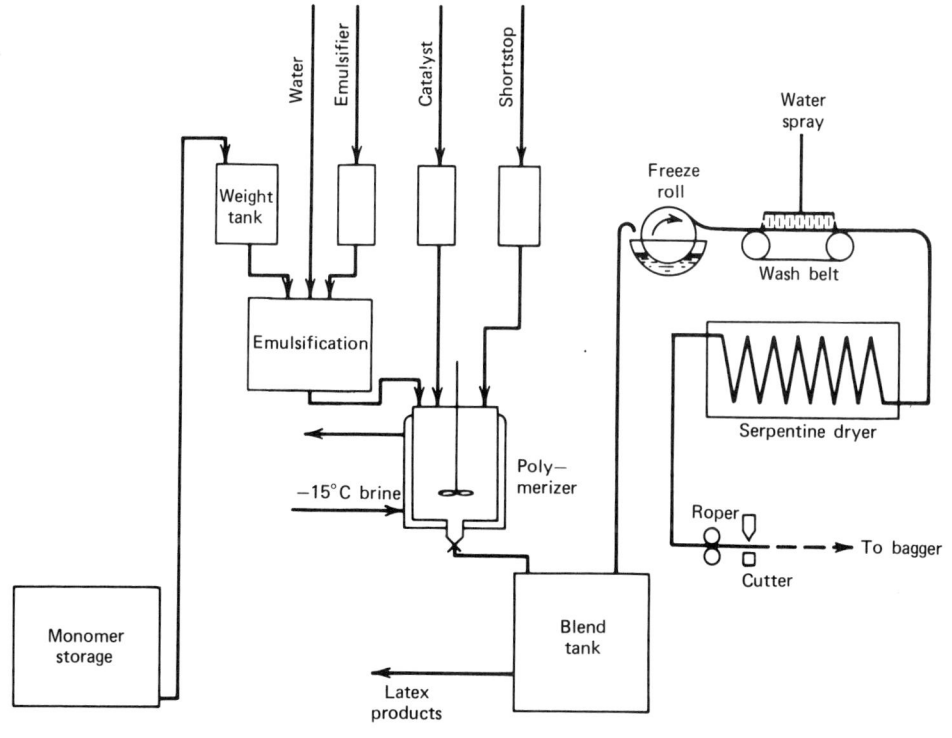

Figure 1. Schematic flow sheet for the polymerization and isolation of polychloroprene; the polymerization section is fully functional for the manufacture of polychloroprene latex products. Courtesy of *Rubber Chemistry and Technology.*

Appropriate amounts of sulfur and rosin are dissolved in chloroprene and the solution is emulsified with an aqueous phase containing sodium hydroxide and the sodium salt of a naphthalenesulfonic acid–formaldehyde condensation product as stabilizer. Sodium rosinate emulsifier is formed *in situ* on mixing the aqueous and organic phases. The two liquid phases are emulsified by recirculating through a centrifugal pump to give particles of about 3 μm in diameter. When emulsification is complete, the mixture is pumped to the polymerizer, a jacketed, glass-lined reaction kettle with a glass-coated agitator. An aqueous solution of potassium persulfate is added as required to initiate and maintain the polymerization. The temperature is kept at 40 ± 1°C by circulating brine through the kettle jacket and by changing agitator speed.

The monomer conversion is followed by measuring the specific gravity of the emulsion (108). Polymerization is stopped at 91% conversion (specific gravity 1.069) by adding an aqueous emulsion of a xylene solution of tetraethylthiuram disulfide, a plasticizer and stabilizer. The emulsion is passed through a steam stripper to recover unconverted monomer and then cooled to 20°C and aged at this temperature for about eight hours to solubilize the polymer (plasticization). After aging, the alkaline latex is acidified to pH 5.5–5.8. This arrests the plasticizing action of the thiuram disulfide, precipitates the rosin which is retained by the polymer, and prepares the latex for isolation of the polymer (109).

Neoprene is isolated from latex by the continuous coagulation of a polymer film on a freeze drum followed by washing and drying (110). The dry polymer is formed into ropelike sections and bagged. The success of this process depends upon having a latex that is completely coagulated within a few seconds at −10° to −15°C and that gives a film strong enough to withstand the stresses imposed upon it during washing and drying.

The acidified latex containing the rosin and an acid-stable dispersing agent is deposited as a frozen film on a roll cooled to about −15°C by circulating brine. The film is stripped, thawed and then washed and dried in air at ca 120°C. The dried film is discharged to a roper from which the rope is conveyed to the cutter and bagger.

Variations of this general polymerization process include continuous polymerization, improved process controls, alternative isolation procedures, and a multitude of recipe changes to produce subtle changes in process or product.

The most successful innovation has probably been the development of continuous polymerization (111–112). Its merits are significantly influenced by the objectives and economics of a specific manufacturing situation. In multiple product plants, batch operation seems preferable.

Film isolation can be speeded up and energy costs reduced. Among the improvements reported in patents are direct evaporation of the water from the polymer latex using a drum dryer (113) to produce a film, the use of additives (114–115), and direct infrared heating of wet film (116–117). An effort to separate polymer coagulum commercially from its aqueous phase in an extruder has been unsuccessful (118).

Compounding and Processing. Polychloroprene can be compounded in all types of rubber mixing equipment and shaped by molding, extrusion, and calendering. It can be vulcanized by practically all the methods used by the rubber and the wire and cable industries (23–25,119–123) (see Rubber compounding).

Curing Systems. For practical vulcanizates, metallic oxides, generally combinations of ZnO and specially calcined magnesia, must be used as a part of the curing systems. Zinc oxide alone gives a scorchy, flat curing stock with a low final state of cure. Magnesia alone retards the cure rate but, if sufficient time is allowed, gives a high state of cure. Oxides of lead can be used for stocks where superior water resistance is necessary. None of the other metal oxides provides the same desirable balance of processing safety and rate and state of cure while maintaining vulcanizate quality and age resistance.

The chloroprene–sulfur copolymers containing thiuram disulfide can be vulcanized by ZnO and MgO without added acceleration; however, the cure rate can be increased by an organic accelerator, ethylenethiourea being the most frequently used.

The mercaptan-modified (nonpeptizable) polychloroprenes cure very slowly with metal oxides alone. For a practical cure rate at temperatures of 140°C or higher, ethylenethiourea is used as accelerator, mostly alone or in combination with a thiazole or a thiuram disulfide. Combinations of a guanidine, thiuram monosulfide, and sulfur frequently offer excellent processing safety. A variety of di- and trialkyl thioureas are available and tetramethylthiourea has been found to balance processing safety with cure rate and state efficiently (124).

Antioxidants and Antiozonants. Antioxidants (qv) are essential in all neoprene compounds for good aging. For general purposes, N-phenyl-β-naphthylamine or N-phenyl-α-naphthylamine are most suitable where discoloration and possible staining are not a problem. In light-colored stocks, hindered phenols are often used. Antioxidants should be carefully selected in order to obtain maximum effectiveness and avoid interference with the curing process.

The resistance of polychloroprene vulcanizates to ozone is generally good, but superior protection requires the addition of antiozonants such as N,N'-alkyl -aryl-, or N,N' diaryl-p-phenylenediamines (see Antioxidants). If the amine function in the diaryl-p-phenylenediamine is hindered, the antiozonant does not reduce the storage stability of the compound (125). The p-phenylenediamine antiozonants are not desirable for light-colored compounds. For superior ozone resistance in such stocks, several proprietary compositions, generally unsaturated organic compounds of undisclosed structure, have been recommended (25).

In addition to curing agents and stabilizers, neoprene compounding includes a wide variety of fillers (qv), processing aids, and the like. Specific recommendations are available in the trade literature.

Vulcanization. The first polychloroprene (1930) was vulcanized after compounding with zinc oxide, sulfur, stearic acid, piperidiniumpentamethylenedithiocarbamate, and antioxidant, essentially a natural rubber curing recipe (see Rubber, natural). A somewhat later (1931) recipe used benzidine as the cross-linking agent in place of the sulfur and accelerator of the first formula, and demonstrated that the vulcanization chemistry of polychloroprene differs from that of other diene polymers.

The double bonds of polychloroprene are sufficiently deactivated by the electronegative chlorine atom so that direct vulcanization with sulfur is limited. The major cross-linking site is the tertiary allylic chloride structure (**3**) generated by 1,2 addition polymerization. The average allylic chloride content of polychloroprene (mercaptan modified) is about 1.5% of total chlorine present. This corresponds to an active cure site for every 67 monomer units.

Metal oxides are necessary, but not generally sufficient, vulcanization ingredients. The role of metal oxides involves an acid accepting function and mechanistically undefined actions, ZnO acts as a promoter and MgO as a retarder. Some insight into this complex situation is provided by a kinetic study (126) in which the vulcanization of polychloroprene with a metal oxide (5 ZnO, 4 MgO)–aldehyde–amine accelerator, using direct high-pressure-steam curing at temperatures from 150 to 200°C, was examined. The effect of temperature on the vulcanization rate constant showed an energy of activation of 126 kJ/mol (30 kcal/mol) for the curing reaction above 175°C, in the presence or absence of the organic accelerator. Below this temperature, in the presence of accelerator, the activation energy for vulcanization was only 42 kJ/mol (10 kcal/mol).

Regardless of the cross-linking role of metal oxides, their presence in polychloroprene polymer compounds is essential to controllable vulcanization and good aging of the vulcanizate.

The organic accelerators used in curing polychloroprene are believed to operate by either of two mechanisms. First, difunctional cross-linking agents such as diamines and bisphenols can be dialkylated by active chlorine polymer units, probably by 1,3 allylic shifts to yield stable cross-links (127):

$$2\ CH_2{=}CHCCl + H_2NRNH_2 \longrightarrow C{=}CHCH_2NHRNHCH_2CH{=}C + 2\ HCl\ (as\ MgCl_2\ or\ ZnCl_2)$$

$$\quad\ \ 1\ \ 2\ 3 \qquad\qquad\qquad\qquad\quad 3\ \ 2\ 1 \qquad\ \ 1\ 2\ \ 3$$

The $ZnCl_2$ formed is a strong Lewis acid and can accelerate the alkylation reaction. This mechanism is supported by the fact that difunctional curing agents do not cross-link polychloroprene that has been previously treated with piperidine or aniline to remove allylic chlorines.

The second mechanism has been proposed to explain the vulcanization of polychloroprene with substituted thioureas such as the commonly used ethylenethiourea (128).

The vulcanization of sulfur–chloroprene copolymers involves rearrangement of backbone polysulfide segments as well as reactions of allylic chlorine structures (129). Free-radical sulfur moieties such as RS· and RSS· are believed to arise from interaction of the polysulfide segments and the thiuram disulfide present in the polymer. These radicals react with each other or with the polymer backbone to yield sulfidic cross-links. Zinc oxide also reacts with the polysulfide units, reducing the number of sulfur atoms present per unit but increasing the number of polymer polysulfide units.

Vulcanization of polychloroprene with epoxy resins (130), dicumyl peroxide (131), thiosemicarbazide accelerators (132), agents absorbed on molecular sieves (133), aldehydes (134), and azidoformates (135) has been described. Tetramethylthiourea has been recommended as a safe processing curing agent for polychloroprenes (124). The curing of special polychloroprenes with tertiary amines has been disclosed (136).

Vulcanizate Properties. Compounding technology allows the preparation of vulcanizates meeting the engineering specifications of a wide variety of applications. In any of these compositions the inherent properties of neoprene, such as resistance to oils, greases, chemicals, and many solvents, and resistance to atmospheric degradation, and heat, flame, and fungus attack is present to a significant degree because neoprene polymer is the continuous phase in the composition. Any especially desired property can be optimized by appropriate variations in compounding to provide vulcanizates for any given application.

Neoprene vulcanizates show little, if any, change in property or appearance when exposed to alkalies, dilute mineral acids, or inorganic salt solutions (24,137–138). Chemicals of a highly oxidizing nature cause surface deterioration and loss of strength. Neoprene vulcanizates also withstand service with minimum swelling and little loss of strength in contact with aliphatic hydrocarbons, alcohols and many fluorinated liquids. Only limited life, however, can be expected from use in contact with chlorinated and aromatic hydrocarbons, organic esters, aromatic hydroxy compounds, and certain ketones. Comparative qualitative data on the chemical and solvent resistance of 14 commercial elastomers, including neoprene, to 500 chemical agents are presented in ref. 138.

Economic Aspects

The growth of production and use of chloroprene polymers over the years is reflected by the data in Table 2. In 1977, nine manufacturing locations with capacities ranging from 13,600 to 136,000 metric tons per year were operated to meet the annual demand of 320,000 t. DuPont operated plants at Louisville, Kentucky; LaPlace, Louisiana; and Maydown, North Ireland. Bayer A.G. operated one plant at Dormagen, FRG, and Distugil (Rhone-Poulenc) a plant at Champagnier, France. A Japanese firm, Denki Kagaku, operated a plant at Ohmi, Japan, in addition to a plant in Houston, Texas, formerly owned by Petrotex. A second plant in Japan was operated by Toyo Soda, and a third by Showa at Kawasaki. At least one substantial polychloroprene plant is operated near the city of Erevan, in the USSR.

The economics are based on the cost of the starting material. For many years, chloroprene monomer was made from acetylene. However, in the 1960s, the growth of polyethylene led to the relatively inexpensive by-product butadiene (qv) from the manufacture of ethylene. As a result, polychloroprene producers changed to butadiene processes, and polychloroprene prices remained at $0.88–1.10/kg until 1974. The rise in petroleum costs since that time has increased the price of butadiene and consequently polychloroprene prices are now in the range of $1.43 to 1.76/kg.

The production of raw materials for polychloroprene is energy intensive as shown

Table 2. World Consumption of Polychloroprene Polymer[a]

Year	Thousands of metric tons
1935	0.14
1940	2.6
1945	46
1950	51
1955	95
1960	135
1965	199
1970	254
1975	280
1977	320

[a] Ref. 42.

by the data of Table 3 where the energy requirements have been estimated for the raw material, conversion to chloroprene, and polymerization including isolation. These data provide strong support for butadiene as a raw material.

Analytical and Test Methods; Specifications

Modern techniques for characterization of polymers are routinely applied to neoprene. In conjunction with manufacturing operations, measurements of specific gravity determine monomer conversion, and determination of chlorine content and infrared measurements control product composition. Product specifications are based on appearance and volatile content (largely water); polymer processability is estimated by measuring raw polymer Mooney viscosity. In addition, scorch time (initial cure rate), cure time, and cure torque of a standard compound are measured in an oscillating disk rheometer (see Rheological measurements).

The quality of latex products is controlled by measurement of latex pH, emulsion viscosity, solids content, and chloroprene monomer content.

Sales specifications, generally based on the tests discussed above, are established by the individual polymer suppliers.

Storage and Shipment

Bagged neoprene is stored in air-conditioned warehouses maintained at temperatures below the freezing point of the polymer to prevent agglomeration of the cut polymer rope in the paper bags. Homopolymer types can be stored for many months, but sulfur copolymers change in processability and can be stored only for a few months.

Solid neoprene polymers in palletized paper bags, are shipped by truck or railroad box car. Neoprene latex products are shipped in alkali-resistant lined steel drums, or in stainless steel tank trucks or rail cars. Since freezing of latex must be avoided, because it may cause coagulation of the polymer particles, large lots of latex are often warmed to about 40°C before packaging and shipping. Except for drum lots, latex products are generally shipped soon after production.

Health, Safety, and Environmental Factors

The fire and health hazards of chloroprene monomer are of main concern in the manufacture of polychloroprene (see Chlorocarbons and chlorohydrocarbons, chloroprene). However, after steam stripping the monomer from the emulsion, the concentration of chloroprene in the system presents no acute hazards (139).

Table 3. Energy Requirement of Neoprene Polymer, MJ/kg [a]

	Starting material	
	Acetylene	Butadiene
raw material	202.3	119
conversion to chloroprene	27.7	22
polymerization and isolation	24.1	24.1
total energy requirement	254.1	165.1

[a] To convert MJ/kg to Btu/lb, multiply by 430.2; from ref. 42.

Neoprene latexes contain concentrations of chloroprene from a maximum of 0.5% to as little as 0.02%, depending on the specific latex type and details of product finishing. For toxicity and safe handling, see ref. 140.

The major effluents from a chloroprene polymerization plant are the gaseous vents from monomer transfers, the vapor exhausts from the polymer dryers, and the water washings from the polymer isolation lines. The gaseous exhausts are generally vented to the atmosphere in high-level stacks. The major organic components of the wash-belt effluent are salts of acetic acid. This stream is sent to sewage treatment plants where the BOD is reduced to acceptable levels.

Off-quality solid polymer is sold at reduced prices, but contaminated polymer and clean coagulum are buried in landfill operations.

Solid Neoprene Polymers

Over the years there has been constant modification and adaption of the basic polymer to meet the changing needs of the marketplace. The type originally introduced in 1932 was replaced in 1939 by the peptizable sulfur copolymers, stabilized with thiuram, which are still widely used because of their low nerve (elasticity or recovery in the uncured state), good building tack, high resilience, good tear strength, and resistance to flex cracking. The sulfurless, mercaptan-modified polymers, introduced in 1949, provided better storage stability, improved heat and compression set resistance, faster mixing, greater green strength, more latitude in processing safety and cure rate through the use of organic accelerators, and greater tensile strength in practical vulcanizates. In 1959, the tendency of W polymers to exhibit nerve (undesirable elastic memory) and die swell in processing operations was alleviated by the introduction of the WB polymers. These polymers contained a certain degree of pre-cross-linking to improve processibility (141). Further improvement in polychloroprene processibility and properties were made in 1970 with the introduction of the T polymers by DuPont.

The peptizable sulfur-copolymer types have less elastic memory than the nonpeptizable, mercaptan-modified polymers and therefore yield extrusions with low die swell and smooth sheets on calendering. Back-rinding (ie, internal pressure causing tear at joints) is less likely to occur during molding. These polymers have excellent tack and this combined with the low nerve results in compounds well-suited for application to fabric by frictioning (the feed roll runs faster than the fabric). In general, they flow together or "knit" well during molding.

The sulfur copolymers cure rapidly without added acceleration. Their vulcanizates are characterized by high resilience, dry surface, superior tear strength, and good adhesion to natural and styrene–butadiene rubber substrates.

The mercaptan modified polymers are generally off-white and do not contain added antioxidant. Their raw polymer stability is much superior to the sulfur copolymers and accounts for their constant viscosity, scorch, and cure rates for longer periods after manufacture. They resist breakdown on milling and mix faster with less heat development than the sulfur copolymers. Their extrusions resist distortion but are subject to surface roughness and die swell, unless their composition includes some precross-linking or their compounded stocks have a good balance of filler and plasticizer. Since these polymers require accelerators for acceptable curing, they offer greater latitude in processing safety and cure rate through choice of accelerator and variation of accelerator level. Their vulcanizates normally exhibit excellent tensile strength and compression set resistance, lower hardness and modulus at equivalent filler loadings, and longer heat-aging resistance vs the sulfur copolymers.

The adhesive-grade polychloroprenes are made in several viscosity grades, and although the viscosity is quite high, the polymer mills smoothly and dissolves readily in good solvents to yield smooth cements with polymer contents as high as 25%.

In 1977 there were 145 variations of solid polychloroprene polymers offered by the seven Western and Japanese producers. The general polymer types each producer offers are summarized in Table 4. Their 22 latex products are also indicated in this table.

Table 5 presents some specific applications of solid neoprene polymers.

Polychloroprene Latexes

Emulsion polymerization of chloroprene monomer produces a colloidal dispersion (or latex) of the polymer in which the polymer particle size is in the range of 5–200 mm. Neoprene latex has the general properties of natural rubber latex and can be used in the same applications. Because of its smaller particle size, it can also be used in novel applications where the natural rubber latex cannot be used (see Latex technology).

With the increased availability of synthetic surfactants and colloidal stabilizers, it has become possible to control the colloidal properties and thus the complex processes of polymer deposition and film formation. Since latex compounding and applications are almost all low-energy processes, expanding and new uses for polychloroprene latexes are anticipated.

As in the case of the solid polychloroprene polymers, the useful properties of polychloroprene latex products are controlled largely by the microstructure of the polymer, which in turn, is controlled by the polymerization temperature, the kind and amount of modifier present, and the extent of monomer conversion. High conversion polymers cannot be used for the processing of solid rubbers, but in latex technology, the gel structure of high conversion polymers does not interfere with processing the liquid dispersion, and in fact, the gel structure contributes to the physical properties

Table 4. Commercial Neoprene Polymer and Latexes, Numbers of Major Types and Grades by Various Producers[a]

Producer	Country	Trade name	Solid polymers Peptizable[b]	Solid polymers Nonpeptizable[c]	Solid polymers Adhesive	Latex Homopolymer	Latex Copolymer	Latex Specialties
Bayer	Fed. Rep. of Germany	Baypren	2	12	4	3	1	1
Denka	Japan	Denka	4	11	8	2	1	
Denka-U.S.	U.S.	neoprene	5	8		2		
Distugil	France	Butahlor	4	12	5	2		
DuPont	U.S.	neoprene	4	14	10	4	4	2
Showa	Japan	neoprene	4	14	10			
Toyo Soda	Japan	Skyprene	3	7	4			

[a] Ref. 42.
[b] Chloroprene–sulfur copolymers.
[c] Generally mercaptan-modified polychloroprenes.

Table 5. Applications of Solid Polychloroprene Polymers[a]

Use areas	Examples
automobile tires	in curb cushion strips and white sidewalls
cable sheathing and insulation	durable sheathed cables for mining, chemical, welding; household appliances; telephone drop wire
cellular rubber goods	weatherstripping for automobiles and buildings, skin diving suits, soft roll covers
conveyer belts	in mining, construction, and food industries
fabric coatings	for large flexible containers, protective clothing, camping equipment, waterproof goods
footwear	industrial shoes; soles and heels; waterproof gear
hoses	for chemical and oil industries, motor vehicles, construction, agriculture, and household applications
large sealing applications	dams and locks in waterways
linings and covers	for all tanks and pipelines where chemical and oil resistance is required
moldings and extrusions	strip stock for automobile and construction industries; load bearing pads; window channels; soil pipe gaskets; road sealing strips
power transmission belts	V-belts-automobiles; agricultural and industrial equipment
coating rolls	for printing, paper, fabric coating, foodstuffs, and other industries
roofing membranes	for flat roofs

[a] This table is a revision and update of Reference 25.

of the end product. The ratio of sol polymer to gel can be varied to give a number of subtle variations in a polychloroprene latex (142).

In latexes as in solid polymers, crystallinity arises from linearity in microstructure and is controlled by the polymerization temperature and the presence of comonomers and gel structure. The cohesive strength, modulus, tensile, hardness, and permanent set of latex polymer increase with increasing crystallinity but at the expense of a decrease in elongation, extensibility, and oil swell (142).

When the above property variations are combined with the changes in processing attributes created by the use of secondary surface-active additives, there is an infinite variety of combinations. In spite of the possibilities, it is surprising that among the 22 commercial latexes there are only about 15 chemical types.

All but three of the current commercial polychlorolatex products listed in Table 4 are anionic dispersions prepared from soaps formed *in situ* by the interaction of a rosin-derived acid and an alkali metal hydroxide. The special features of each product are best obtained from the producer's trade literature. Some of the applications of neoprene latexes (42) are: adhesives (contact bond, pressure sensitive, construction, cord laminating); asphalt modification; binders (asbestos fibers, paper fibers, curled hair, glass fibers); cement modification (aluminous, portland); coatings and proofing; dipped goods; extruded thread; impregnation; latex foam; mastics; saturants; and sealants.

Compounding and Processing. All of the anionic polychloroprene latexes can be compounded following the same general principles. Prior treatment of the compounding agents avoids flocculation of the latex colloid on mixing. Generally, it is preferable to predisperse water-insoluble liquids or solids using a surfactant that is also compatible with the latex colloid. Compounding agents may include metal oxides, antioxidants, curing chemicals, fillers, and plasticizers, depending on the use requirements.

All latex compounds, regardless of application, should contain at least 5 phr zinc oxide. Where extreme aging conditions are to be encountered, as much as 25 phr may be used. When polychloroprene is employed in conjunction with cellulosic or other acid-sensitive materials (ie, cotton, rayon, nylon, paper, etc), large amounts of zinc oxide should be used to minimize tendering of the substrate.

At least two phr of a good antioxidant, staining or nonstaining type depending on end use, should be employed.

Accelerators enhance the physical properties of polychloroprene films. Where tensile strength is of primary importance, tetraethylthiuram disulfide and water soluble sodium dibutyldithiocarbamate are suggested. For high modulus, thiocarbanilid is the preferred accelerator. Zinc dibutyldithiocarbamate and di-o-tolylguanidine are also frequently used.

Fillers such as clay, whiting, titanium dioxide, hydrated alumina, and fine silicas can be used to impart specific properties of polychloroprene film or to act as cost-reducing diluents. Carbon black provides color. Filler loadings of ten to fifty phr are frequently used in dipped goods and foams. As much as one hundred phr of filler, or more, may be employed in adhesive and coating compounds.

Mineral oils and light process oils are used as plasticizers at levels of from three to ten phr to provide soft flexible films. Ester plasticizers may be employed to obtain superior low temperature properties.

Anionic, amphoteric, and nonionic surfactants are used as both mechanical and chemical stabilizers. Natural and synthetic gums are used as thickeners.

In general, the polychloroprene latexes can be processed in the various ways and in the equipment originally designed for processing natural rubber and other elastomeric latexes. The requirements for each specific application, outside of compounding, are generally unique.

Handling and Safety. The effect of temperature on latexes must be taken into account when storing and handling polychloroprene latex. It becomes more viscous as temperature decreases, and may become practically nonfluid at about 7°C. Freezing the latex may result in irreversible coagulation. At high temperatures, chemical hydrolysis of active chlorine begins to occur more rapidly, resulting in a lowering of the latex pH. This reaction does not become significant until temperatures of 30 to 40°C are encountered.

The processing of polychloroprene latexes involves certain environmental considerations in view of EPA and OSHA regulations. For toxicity and safe handling, see ref. 140.

BIBLIOGRAPHY

"Neoprene" under "Rubber, Synthetic" in *ECT* 1st ed., Vol. 11, pp. 852–857, by H. W. Walker, E. I. du Pont de Nemours & Co., Inc.; "Neoprene" under "Elastomers, Synthetic" in *ECT* 2nd ed., Vol. 7, pp. 705–716 by C. A. Hargreaves, II, and D. C. Thompson, E. I. du Pont de Nemours & Co., Inc.

1. A. M. Collins, *Rubber Chem. Technol.* **46,** G48 (1973).
2. W. H. Carothers and co-workers, *J. Am. Chem. Soc.* **53,** 4202 (1931).
3. U.S. Pat. 1,967,861 (July 24, 1934), A. M. Collins (to E. I. du Pont de Nemours & Co., Inc.).
4. U.S. Pat. 1,967,865 (July 24, 1934), A. M. Collins (to E. I. du Pont de Nemours & Co., Inc.).

5. L. A. Wood, *Rubber Chem. Technol.* **49,** 196 (1976).
6. G. M. Martin, F. L. Roth, and R. D. Stiehler, *Trans. Inst. Rubber Ind.* **32,** 189 (1956).
7. L. A. Wood and F. L. Roth, *Rubber Chem. Technol.* **36,** 611 (1963).
8. J. H. Dillon, I. B. Prettyman, and G. L. Hall, *J. Appl. Phys.* **15,** 309 (1944).
9. T. P. Yin and R. Pariser, *J. Appl. Polym. Sci.* **7,** 667 (1963).
10. J. T. Maynard and W. E. Mochel, *J. Polym. Sci.* **13,** 235 (1954).
11. *Ibid.,* 251 (1954).
12. J. T. Maynard and W. E. Mochel, *J. Polym. Sci.* **18,** 227 (1955).
13. R. C. Ferguson, *Anal. Chem.* **36,** 2204 (1964).
14. R. C. Ferguson, *J. Polym. Sci. Part A,* **2,** 4735 (1964).
15. M. M. Coleman, D. L. Tabb, and E. G. Brame, *Rubber Chem. Technol.* **50,** 49 (1977).
16. E. G. Brame and A. A. Khan, *Rubber Chem. Technol.* **50,** 272 (1977).
17. C. A. Aufdermarsh and R. Pariser, *J. Polym. Sci. Part A,* **2,** 4727 (1964).
18. R. R. Garrett, C. A. Hargreaves, II, and D. N. Robinson, *J. Macromol. Sci. Chem.* A4 **8,** 1679 (1970); U.S. Pat. 3,660,368 (May 2, 1972), C. A. Hargreaves, II (to E. I. du Pont de Nemours & Co., Inc.).
19. B. Ya Teitelbaum and N. P. Anoshina, *Vysokomol. Soedin.* **7,** 978 (1965).
20. B. Ya Teitelbaum and co-workers, *Dokl. USSR Phys. Chem. Sec.* (English trans.) **150,** 463 (1963).
21. M. Hanok and I. N. Cooperman, *Proceedings International Rubber Conference Preprints Paper,* Washington, D.C., 1959, p. 582.
22. R. M. Murray and J. D. Detenber, *Rubber Chem. Technol.* **34,** 668 (1961).
23. R. M. Murray and D. C. Thompson, *The Neoprenes,* Elastomer Chemicals Department, E. I. du Pont de Nemours & Co., Inc., 1963.
24. N. L. Catton, *The Neoprenes,* Rubber Chemical Division, E. I. du Pont de Nemours & Co., Inc., Wilmington, Del., 1953.
25. S. Koch, ed., *Manual for the Rubber Industry,* Farbenfabriken Bayer A.G., Leverkusen, Germany, 1972, pp. 11–61.
26. E. Engelmann, *Institution of the Rubber Industry Journal* **3,** 77 (April, 1969).
27. H. C. Walter, G. J. Mantel, and E. K. Gladding, *Rubber Age N.Y.* **72,** 92 (1952).
28. H. C. Bailey, *Revue Generale du Caoutchouc Plastiques* **44,** 1495 (1967).
29. D. L. Gardner and I. C. McNiell, *Eur. Polym. J.* **7,** 569 (1971).
30. D. L. Gardner and I. C. McNiell, *Eur. Polym. J.* **7,** 593 (1971).
31. D. L. Gardner and I. C. McNiell, *Eur. Polym. J.* **7,** 603 (1971).
32. C. S. Williams, *Rubber "Flame Proofing,"* E. I. du Pont de Nemours & Co., Inc., Chemicals Division Report 38-8, May 1938.
33. *Flame Proofing,* E. I. du Pont de Nemours & Co. Inc., Rubber Chemicals Division, Report BL-180, Sept. 1944.
34. G. G. Skinner and J. H. McNeal, *Ind. Eng. Chem.* **40,** 2303 (1948).
35. D. C. Thompson, J. F. Hagman, and N. N. Mueller, *Rubber Age N.Y.* **83,** 819 (1958).
36. C. E. McCormack, *Rubber Age N.Y.* **104**(6), 27 (1972).
37. P. R. Johnson, *J. Appl. Polym. Sci.* **18,** 491 (1974).
38. P. R. Johnson, R. Pariser, and J. J. McEvoy, *Rubber Age* **107**(5), 29 (1975).
39. C. W. Stewart, Sr., R. L. Dawson, and P. R. Johnson, *Rubber Chem. Technol.* **48,** 139 (1975).
40. H. Walker as quoted by I. Williams, *Ind. Eng. Chem.* **31,** 1464 (1939).
41. S. Ekegren and co-workers, *Acta Chem. Scand.* **4,** 126 (1950).
42. P. R. Johnson, *Rubber Chem. Technol.* **40,** 650 (1976).
43. L. G. Melkonyan, *Arm. Khim Zh.* **21,** 187 (1968); Melkonyan and co-workers, *Arm. Khim Zh.* **22,** 873, 1062 (1969).
44. W. E. Mochel and J. H. Peterson, *J. Am. Chem. Soc.* **71,** 1426 (1949).
45. F. Hrabak and J. Zachoval, *J. Polym. Sci.* **52,** 131 (1961).
46. U.S. Pat. 2,234,215 (March 11, 1941), M. A. Youker (to E. I. du Pont de Nemours & Co., Inc.).
47. U.S. Pat. 2,426,854 (Sept. 2, 1947), P. O. Bare (to E. I. du Pont de Nemours & Co., Inc.).
48. Brit. Pat. 512,458 (Sept. 15, 1939), (to I. G. Farbenindustrie A.G.).
49. Brit. Pat. 858,841 (Jan. 18, 1961), (to Bayer Farbenfabriken A.G.).
50. U.S. Pat. 2,481,044 (Sept. 6, 1949), G. W. Scott (to E. I. du Pont de Nemours & Co., Inc.).
51. U.S. Pat. 2,518,573 (Aug. 15, 1950), G. W. Scott (to E. I. du Pont de Nemours & Co., Inc.).
52. U.S. Pat. 2,477,338 (Sept. 26, 1949), J. E. Kirby and W. E. Sharkey (to E. I. du Pont de Nemours & Co., Inc.).
53. W. E. Mochel and J. H. Peterson, *J. Am. Chem. Soc.* **71,** 1426 (1949).

54. J. W. McFarland and R. Pariser, *J. Appl. Polym. Sci.* **7**, 675 (1963).
55. M. Morton and I. Pirma, *J. Polym. Sci.* **19**, 563 (1956).
56. L. G. Melkonyan, R. V. Bagdasaryan, and V. Bunyatyants, *Arm. Khim. Zh.* **19**, 402 (1966); *Chem Abstr.* **66**, 3475W (1967).
57. L. G. Melkonyan, R. V. Bagdasaryan, and R. W. Karapetyan, *Arm. Khim. Zh.* **19**, 733 (1966); *Chem. Abstr.* **66**, 95944 K (1967).
58. R. A. Karapetyan, R. V. Bagdasaryan, and L. G. Melkonyan, *Arm. Khim. Zh.* **21**, 874 (1968); *Chem. Abstr.* **71** 4291j (1969).
59. R. A. Karapetyan, R. V. Bagdasaryan, and L. G. Melkonyan, *Arm. Khim. Zh.* **22**, 360 (1969); *Chem. Abstr.* **71**, 62020d (1969).
60. R. V. Bagdasaryan and L. G. Melkonyan, *Arm. Khim. Zh.* **20**, 401 (1967); *Chem. Abstr.* **68**, 13825b (1968).
61. U.S. Pat. 1,950,439 (March 13, 1934), W. H. Carothers and J. E. Kirby (to E. I. du Pont de Nemours & Co., Inc.).
62. W. E. Mockel, *J. Polym. Sci.* **8**, 583 (1952).
63. A. L. Klebanskii and co-workers, *J. Polym. Sci.* **30**, 763 (1958).
64. A. L. Klebanskii and co-workers, *Kauch. Rezina* **20**, 1 (1961); *Chem. Abstr.* **56**, 1574e (1962).
65. N. D. Zahdarov and co-workers, *Vysomolekul. Soeden.* **5**, 910 (1963); *Chem. Abstr.* **59**, 7737e (1963).
66. "Copolymerization" in G. E. Ham, ed., *High Polymers*, Vol. XVIII, Interscience Publishers, Inc., a division of John Wiley & Sons, Inc., New York, 1964, p. 720.
67. U.S. Pat. 2,395,649 (Feb. 26, 1946), F. C. Wagner (to E. I. du Pont de Nemours & Co., Inc.).
68. Brit. Pat. 858,444 (Jan. 11, 1961), (to Farbenfabriken Bayer A.G.).
69. U.S. Pat. 2,066,329 (Jan. 5, 1937), W. H. Carothers, A. M. Collins, and J. E. Kirby (to E. I. du Pont de Nemours & Co., Inc.).
70. U.S. Pat. 2,205,239 (June 18, 1940), A. S. Carter and F. W. Johnson (to E. I. du Pont de Nemours & Co.).
71. French Pat. 1,358,188 (April 10, 1964), R. Kobayashi (to Electro Chemical Industrial Company, Ltd.).
72. U.S. Pat. 1,965,369 (July 3, 1934), W. H. Carothers and G. J. Berchet (to E. I. du Pont de Nemours & Co.).
73. U.S. Pat. 3,058,960 (Oct. 16, 1962), C. A. Stewart, Jr. (to E. I. du Pont de Nemours & Company).
74. Brit. Pat. 858,841 (Jan. 18, 1961), (to Farbenfabriken Bayer A.G.).
75. Ger. Pat. 683,232 (Oct. 12, 1939), S. Kiesskalt and co-workers (to I. G. Farbenindustrie A.G.).
76. W. H. Carothers and co-workers, *J. Am. Chem. Soc.* **53**, 4203 (1931).
77. H. K. Banbrook, R. S. Lehrle, and J. C. Robb, *J. Polym. Sci. Part C*, **4**, 1165 (1964).
78. G. H. Miller, D. Chock, and E. P. Chock, *J. Polym. Sci. Part A*, **3**, 3353 (1965).
79. U.S. Pats. 2,942,037 and 2,942,038 (June 21, 1960), P. A. Jenkins (to Distillers Company, Ltd.).
80. U.S. Pat. 2,900,421 (Aug. 18, 1959), M. S. Kharasch and W. Nudenberg.
81. U.S. Pat. 2,842,602 (July 8, 1958), W. T. Norton (to E. I. du Pont de Nemours & Co., Inc.).
82. Brit. Pat. 878,118 (Sept. 27, 1961), (to Knapsack-Griesheim A.G.).
83. Ger. Pat. 1,114,806 (Oct. 12, 1961), K. Sennewald, W. Vogt, and H. Weiden (to Knapsack-Griesheim A.G.).
84. U.S. Pat. 3,417,154 (Dec. 17, 1968), H. E. Albert and P. G. Haines (to Pennsalt Chemicals Corp.).
85. U.S. Pat. 3,148,225 (Sept. 8, 1964), H. E. Albert (to Pennsalt Chemicals Corp.).
86. U.S. Pat. 2,770,657 (Nov. 13, 1956), J. R. Hively (to E. I. du Pont de Nemours & Co., Inc.).
87. U.S. Pat. 3,175,012 (March 23, 1965), G. P. Colbert (to E. I. du Pont de Nemours & Co., Inc).
88. M. Morton, J. A. Cala, and M. W. Altier, *J. Polym. Sci.* **19**, 547 (1956); M. Morton and I. Piirma, *J. Polym. Sci.* **19**, 563 (1956).
89. I. Williams and H. W. Walker, *Ind. Eng. Chem.* **25**, 1933.
90. D. E. Andersen and R. G. Arnold, *Ind. Eng. Chem.* **45**, 2727 (1953).
91. U.S. Pat. 3,074,899 (Jan. 22, 1963), and Ger. Pat. 1,104,701 (April 13, 1961), D. Rosahl and H. Esser (to Farbenfabriken Bayer A.G.).
92. U.S. Pat. 2,138,226 (Nov. 29, 1938), B. Dales and F. B. Downing (to E. I. du Pont de Nemours & Co.).
93. U.S. Pat. 2,263,322 (Nov. 11, 1941), H. W. Walker and F. N. Wilder (to E. I. du Pont de Nemours & Co. Inc.).

94. U.S. Pat. 1,967,861 (July 24, 1934), A. M. Collins (to E. I. du Pont de Nemours & Co., Inc.); U.S. Pat. 2,426,854 (Sept. 2, 1947), P.O. Bare (to E. I. du Pont de Nemours & Co., Inc.).
95. U.S. Pat. 2,569,480 (Oct. 2, 1951), E. J. Lorand (to Hercules Powder Co.).
96. U.S. Pat. 1,950,438 (March 13, 1934), W. H. Carothers, A. M. Collins, and J. E. Kirby (to E. I. du Pont de Nemours & Co. Inc.).
97. U.S. Pat. 2,707,180 (April 26, 1955), J. T. Maynard (to E. I. du Pont de Nemours & Co., Inc.).
98. U.S. Pat. 2,962,475 (Nov. 29, 1960), H. Malz and D. Rosahl (to Farbenfabriken Bayer A.G.).
99. Fr. Pat. 1,187,771 (July 12, 1971), R. Stroh, W. Sutter, and H. Haberland (to Farbenfabriken Bayer A.G.).
100. U.S. Pat. 3,236,823 (Feb. 22, 1966), G. Jennes, H. Sutter, and K. Nutzel (to Farbenfabriken Bayer A.G.).
101. U.S. Pat. 3,013,000 (Dec. 12, 1961), A. R. Heinz, D. Rosahl, and W. Graulich (to Farbenfabriken Bayer A.G.).
102. U.S. Pat. 2,259,122 (Oct. 14, 1941), H. W. Walker (to E. I. du Pont de Nemours & Co., Inc.).
103. Can. Pat. 661,997 (April 23, 1963), A. M. Hutchinson (to Distillers Company, Ltd.).
104. U.S. Pat. 2,576,009 (Nov. 11, 1951), J. R. Goertz (to E. I. du Pont de Nemours & Co., Inc.).
105. Ger. Pat. 1,138,944 (Oct. 31, 1962), F. P. Demme (to Pennsalt Chemical Co.).
106. U.S. Pat. 2,264,173 (Nov. 25, 1941), A. M. Collins (to E. I. du Pont de Nemours & Co., Inc.).
107. C. A. Hargreaves, II, "Elastomers by Radical and Redox Mechanisms: C. Neoprene," in J. P. Kennedy and E. G. M. Tornqvist, eds., *Polymer Chemistry of Synthetic Elastomers, Part 1,* Wiley-Interscience, a division of John Wiley & Sons, Inc., New York, 1968, pp. 227–252.
108. R. S. Barrows and G. W. Scott, *Ind. Eng. Chem.* **40,** 2193 (1948).
109. M. A. Youker, *Chem. Eng. Prog.* **41,** 391 (1947).
110. U.S. Pat. 2,187,146 (Jan. 16, 1940), W. S. Calcott and H. W. Starkweather (to E. I. du Pont de Nemours & Co., Inc.).
111. U.S. Pat. 2,384,277 (Sept. 4, 1945), W. S. Calcott and H. W. Starkweather (to E. I. du Pont de Nemours & Co., Inc.).
112. U.S. Pat. 2,831,842 (April 22, 1958), C. E. Aho (to E. I. du Pont de Nemours & Co., Inc).
113. U.S. Pat. 2,914,497 (Nov. 24, 1959), W. J. Keller (to E. I. du Pont de Nemours & Co., Inc.).
114. U.S. Pat. 3,365,518 (Jan. 23, 1968), R. W. Saville (to E. I. du Pont de Nemours & Co., Inc.).
115. U.S. Pat. 3,413,247 (Nov. 26, 1968), S. B. Schroeder (to E. I. du Pont de Nemours & Co., Inc.).
116. U.S. Pat. 3,054,192 (Sept. 18, 1962), D. Rosahl and co-workers (to Farbenfabriken Bayer A.G.).
117. U.S. Pat. 3,254,422 (June 7, 1966), C. J. Defeil (to E. I. du Pont de Nemours & Co., Inc.).
118. U.S. Pat. 2,371,722 (Oct. 24, 1940), F. W. Wanderer (to E. I. du Pont de Nemours & Co., Inc.).
119. D. B. Foreman in M. Morton, ed., *Rubber Technology* 2nd ed., Van Nostrand, New York, 1973, Chapt. XIII, p. 322.
120. C. A. Hargreaves, II and D. C. Thompson, "Chlorobutadiene Polymers," in N. M. Bikales, ed., "*Encyclopedia of Polymer Science and Technology,* Vol. 3, Interscience-John Wiley & Sons, Inc., New York, 1965, pp. 707–730.
121. D. C. Thompson, *Mechanical Molded Goods—Neoprene and Hypalon®*, E. I. du Pont de Nemours & Co., Inc., Wilmington, Del., 1955.
122. A. M. Neal and L. R. Mayo in G. G. Whitby, ed., *Synthetic Rubber,* John Wiley & Sons, Inc., New York, 1954, Chapt. 22, p. 767.
123. A. C. Stevenson in G. Alliger and I. J. Sjothun, eds., *Vulcanization of Elastomers*, Rheinhold, New York, 1963, p. 265.
124. D. H. Geschwind and W. F. Gruber, *Rubber Chem. Technol.* **44,** 1449 (1971).
125. D. H. Geschwind, W. F. Gruber, and J. Kalil, *Rubber Age* **99**(11), 69 (1967).
126. G. Bielstein, *Trans. Institution of the Rubber Ind.* **36,** 29 (1960); *Rubber Chem. Technol.* **34,** 319 (1961).
127. P. Kovacic, *Ind. Eng. Chem.* **47,** 1090 (1955) and references therein.
128. R. Pariser, *Kunstst.* **50,** 623 (1960).
129. N. D. Zakharov and co-workers, *Vysokomol. Soedin.* **5,** 910 (1963); English summary; *Chem. Abstr.* **59,** 7737 (1963).
130. N. D. Zakharov and G. A. Maiorov, *Sov. Rubber Technol.* (English trans.) **22,** 11 (1963).
131. L. D. Loan and J. Scanlon, *Rubber Plastics Age* **44,** 1315 (1963).
132. U.S. Pat. 3,022,275 (Feb. 20, 1962); F. Lober (to Farbenfabriken Bayer A.G.).
133. Can. Pats. 701,362 (Jan. 5, 1965) and 701,368 (Jan. 5, 1965), F. M. O'Connor (to Union Carbide Corp).

134. Y. Minoura and M. Tsukasa, *Rubber Chem. Technol.* **43,** 188 (1970).
135. O. S. Breslow, W. D. Willis, and L. O. Amberg, *Rubber Chem. Technol.* **43,** 605 (1970).
136. U.S. Pat. 3,686,156 (Aug. 22, 1972), J. F. Hagmann (to E. I. du Pont de Nemours & Co., Inc.).
137. *Bulletin E-02141, Engineering Properties of DuPont Neoprene,* E. I. du Pont de Nemours & Co., Inc., Elastomer Chemicals Department, Wilmington, Del.
138. Reprint "The General Chemical Resistance of Various Elastomers" originally published in the 1970 Yearbook of the Los Angeles Rubber Group, Inc., P.O. Box 2704, Terminal Annex, Los Angeles, Calif. 90054.
139. *Bulletin E-16572, Toxicity and Safe Handling Guidelines for DuPont Neoprene Solid Polymers,* E. I. du Pont de Nemours & Co., Inc., 1977.
140. *Bulletin E-07657, Toxicity and Safe Handling of Neoprene Latexes,* E. I. du Pont de Nemours & Co., Inc., 1977.
141. L. M. White, and co-workers, *Ind. Eng. Chem.* **37,** 770 (1945).
142. L. L. Harrel, Jr., in *Bulletin E-07659, Neoprene Latexes—Their Preparation and Characteristics,* E. I. du Pont de Nemours & Co., Inc., 1976.

General References

P. R. Johnson, "Polychloroprene Rubber," *Rubber Chem. Technol.* **40,** 650 (1976).

PAUL R. JOHNSON
E. I. du Pont de Nemours & Co., Inc.

NITRILE RUBBER

Nitrile rubbers, the amorphous copolymers in varying ratio of butadiene (qv) and acrylonitrile (qv), have been recognized as important specialty elastomers since their first disclosure in 1930 (1) and their importation into the United States in semi-commercial quantities from Germany in 1937. Importation of nitrile rubbers decreased when domestic manufacture began in 1940 and, by 1978, production by four principal domestic plants obviated the need for all but small quantities from foreign sources. Nitrile rubbers are classified as specialty rather than general purpose elastomers, as the vulcanized forms are used primarily for their oil, solvent and chemical resistance, even though they are capable of displaying excellent rubbery properties suitable for many applications at moderate cost. ASTM has recommended the designation NBR to identify the family of poly(acrylonitrile-*co*-1,3-butadiene) [*9003-18-3*] or nitrile–butadiene rubbers.

Properties

Nitrile rubber grades available in the marketplace have a 20–50% acrylonitrile content. When properly compounded and cured, the fuel and solvent resistance, abrasion resistance and resistance to gas permeation increase with increasing nitrile content; with decreasing nitrile content the low temperature properties and resilience improve. A second controlled variable in the manufacture of NBR is Mooney viscosity (2), a measure of average molecular weight ranging in commercial polymers from 30 to 90 arbitrary units. At the low end of this range processability is superior but dynamic mechanical properties may be poorer. Thus the manufacturer of finished rubber goods may select from among a large variety of NBR types to formulate rubber compounds meeting the combined requirements of processing during manufacture, stability under the intended environmental conditions and static or dynamic mechanical properties of the finished article.

The physical constants of unfilled gum nitrile rubber are listed in Table 1.

Oil and Solvent Resistance. Perhaps the most valuable property of NBR is its ability to be converted by conventional compounding, forming, and curing practices into parts that function reliably in service when in contact with fluids that swell or degrade other elastomers. Commercially established products for such service include seals, hoses, belts, wire and cable insulation, rolls, sponges and numerous molded or extruded mechanical items. The fluid resistance of compounds containing NBR, as measured by volume swelling in laboratory immersion (10)(eg, see Table 2) or by actual performance tests, is proportional to the acrylonitrile content, or the polarity imparted by the –CN groups on the polymer chains, and also is a function of the chemical nature of the oil or solvent. As a polar rubber, NBR is most resistant to hydrocarbons but is less resistant to polar fluids.

Volume swell alone is usually not the sole criterion for predicting performance of a compounded NBR; the effects of fluid exposure upon useful physical properties, such as tensile strength, elongation and hardness, are usually measured and specified for a particular application.

Resistance to Chemicals. In addition to physical action, such as absorption or extraction, reactive chemicals may be encountered that can chemically alter or degrade an NBR product. In general, the more polar the contacting chemical, the greater the degradation. Good resistance is exhibited when the contacting chemicals are water, inorganic salts, aliphatic hydrocarbons, soaps and most food products, but NBR is not recommended for contact with strong acids or alkalis, halogenated hydrocarbons, or polar organic compounds. Selecting a nitrile rubber for the formulation of hoses, seals, gaskets, and tank linings should involve laboratory testing of cured compounds against the chemical in question.

Low Temperature Properties. Although glass transition temperature, T_g, which is measured on raw NBR, predicts the approximate performance at reduced temperatures, practical measurements such as brittle point (ASTM D-746) and temperature retraction, TR-10 and TR-70 (ASTM D-1329), are generally employed in industry with cured compounds. Low temperature behavior is improved as the acrylonitrile content of NBR is reduced. Since oil resistance also declines with reduced nitrile level, the selection of an appropriate NBR type must often involve a compromise depending upon the given service condition.

Table 3 notes the changes in brittleness temperature and temperature retraction accompanying changing nitrile level in an unplasticized test compound.

Heat Resistance. Vulcanized NBR compounds are serviceable for continuous use up to 120°C in air and to 150°C in complete oil immersion and in the absence of air. For short exposure conditions, flexibility is retained at temperatures over 200°C. Manufacturers, therefore, are able to specify NBR for most automotive under-the-hood

Table 1. Nitrile Rubber Physical Constants

Property	Value	Reference
specific gravity		
20% AN[a]	0.95	
35% AN	0.99	
45% AN	1.02	
refractive index, n_D^{26}, 39% AN	1.5187	3
specific heat, J/(g·°C)[b], 35% AN	1.97	4
thermal conductivity, J/(cm·°C)[b]		
35% AN, 60°C	25.1×10^{-4}	5
35% AN, 140°C	18.4×10^{-4}	5
vol coeff thermal exp,		
26% AN, 20°C–120°C	130×10^{-3}	6
glass transition temperature, °C		
20% AN	−56	7
29% AN	−46	7
33% AN	−37	7
40% AN	−27	7
52% AN	−16	7
specific resistance, Ω/cm,		
0°C	5×10^{11}	8
20°C	3×10^{10}	8
dielectric strength, kV/mm		
27% AN	16	9
35% AN	21	9
dielectric constant 1 MHz		
27% AN, 0°C	4.0	9
27% AN, 20°C	5.5	9
27% AN, 50°C	11.0 (max)	9
27% AN, 100°C	9.0	9
40% AN, 0°C	4.0	9
40% AN, 20°C	4.8	9
40% AN, 65°C	14.0 (max)	9
40% AN, 100°C	12.0	9
volume resistivity, Ω·cm	10^9–10^{12}	9
power factor, 50 Hz		
27% AN −20°C	0.030 (max)	9
27% AN −5°C	0.001 (min)	9
35% AN −10°C	0.030 (max)	9
35% AN +5°C	0.002 (min)	1

[a] AN = acrylonitrile.
[b] To convert J to cal, divide by 4.184.

applications, eg, oil seals, water seals, gasoline hose tubes, emission control and air conditioner hoses, in place of more expensive specialty elastomers.

Oxygen and Ozone Resistance. Oxygen degrades NBR, as it does other unsaturated elastomers, in varying degree by embrittlement due to cross-linking. Ozone (qv), on the other hand, generates cracks in cured parts which are under stress; these cracks appear at right angles to the direction of stress. Resistance to oxygen may be improved by compounding with 2 to 3 parts of a commercial amine antioxidant, for example, Aminox (a polymeric amine) for black stocks, or a phenolic or an organic phosphite antioxidant, eg, Polygard (TNTP, Weston), for light-colored parts. To minimize ozone

Table 2. Volume Swell in Test Fluids vs Acrylonitrile Content[a]

NBR[b]	Acrylonitrile, %	Volume swell, % 70 h at 100°C in ASTM Oil no. 3[c]	70 h at RT In ASTM Fuel B[d]
18–80	22.0	32	52
ALT	26.0	30	47
BJ	29.0	21	40
BJLT	32.5	17	35
CLM	35.0	12	29
CJLT	39.0	7	23
D	44.5	4	5
Test compound[e]			
NBR[b]			100.0
SRF carbon black, N-774			70.0
zinc oxide			5.0
stearic acid			0.5
tetramethylthiuram monosulfide			0.2
spider sulfur			2.0
N-cyclohexyl-2-benzothiazolesulfenamide			1.0

[a] Ref. 11.
[b] Test carried out with Paracrils (Uniroyal).
[c] Aniline point 70°C.
[d] 70 vol % isooctane + 30 vol % toluene.
[e] Cured 4 min at 177°C.

attack, a combination of chemical antiozonant, typically a p-phenylenediamine derivative such as one of the Flexzone (Santoflex, Monsanto; Vulkanox, Bayer) series, at the level of 3 parts by wt in combination with about 3 parts of wax, such as Sunproof (Sunolite 240, Witco), is most beneficial (see Antioxidants and antiozonants).

Two other processes that impart ozone resistance and have assumed commercial importance are blending with poly(vinyl chloride) (PVC) and with EPDM, an ethylene propylene diene elastomer (see Ethylene–propylene rubber). For the former, several commercial blends, including Uniroyal's Paracril Ozo [52439-61-9] containing 30% of dispersed PVC, are available. In compounding, the stiffening influence of PVC is compensated by the use of ester plasticizers. Incorporation of 30% to 40% EPDM in a compounded NBR has been shown (12) to impart good ozone resistance. Often, compounding special accelerators (13) having a high affinity for the dispersed EPDM phase with the NBR–EPDM helps to achieve a co-cure with sulfur for best mechanical properties.

Permeability of Gases and Liquids. The polar nitrile groups in NBR strongly retard the diffusion of gases and liquids through hoses and diaphragms in direct proportion to the nitrile content (see also Barrier polymers). This polymer is often used to contain air, hydrogen, nitrogen and carbon dioxide and the medium-to-high nitrile types are used widely to contain both liquid and gaseous Freon refrigerants. In compounding, lamellar fillers reduce permeation, whereas polymeric plasticizers, in which the contacting medium is only sparingly soluble, should be used in preference to ester plasticizers (see Fillers).

538   ELASTOMERS, SYNTHETIC (NITRILE RUBBER)

Table 3. Low Temperature Properties

NBR[a]	Acrylonitrile, %	Brittle point, ASTM D-746, °C	Temperature retraction, ASTM D-1329, °C	
			TR-10	TR-70
18–80	22.0	−56	−39	−30
ALT	26.0	−51	−27	−20
BJ	29.0	−42	−27	−21
BJLT	32.5	−38	−26	−17
CLM	35.0	−29	−21	−17
CJLT	39.0	−21	−15	− 6
D	44.5	−11	−11	+ 4

Test compound (unplasticized)[b]

NBR[a]	100.0
SRF carbon black, N-774	70.0
zinc oxide	5.0
stearic acid	0.5
tetramethylthiuram monosulfide	0.2
spider sulfur	2.0
N-cyclohexyl-2-benzothiazolesulfenamide	1.0

[a] Tests carried out with Paracrils (Uniroyal).
[b] Cured 4 min at 177°C.

Mechanical Properties. Amorphous nitrile rubbers display inferior mechanical properties unless they are reinforced with appropriate carbon blacks or mineral pigments and unless they are vulcanized. In addition to tensile strength and modulus development, the reinforcing filler enhances other desirable properties such as abrasion resistance, tear strength, reduced compression set, controlled resilience and low water absorption. Generally, the carbon blacks are most widely used in industrial practice, with silicas and clays employed for light colors. For maximizing physical properties such as tensile strength, modulus, abrasion and tear resistance, the highly reinforcing carbon blacks, SAF (N-110) and ISAF (N-220), are employed; less reinforcing blacks, SRF (N-774) and MT (N-990), are useful for high elongation, low modulus, low compression set and low-cost compounding (see Carbon, carbon black).

Manufacture and Compounding

Emulsion Polymerization. The process of emulsion polymerization, which uses water as the carrier medium, has been developed extensively. Its commercial applications include production of SBR, styrene–butadiene rubber (qv) (14), as well as other rubbers, eg, polychloroprene (CR) (see Neoprene) and polyacrylates (see Acrylic elastomers) and NBR. Emulsion polymerization has technical and economic advantages over other possible polymerization methods such as bulk, solution or suspension techniques: (*1*) the conversion of monomers to copolymers is rapid, to the extent that the heat of the exothermic reaction may be removed and the temperature controlled within narrow limits; (*2*) controlled high molecular weights and hence good physical properties of vulcanizates may be achieved; (*3*) the product of the polymerization process, a high solids, low viscosity dispersion of rubber in water (synthetic latex), is readily transported and stored in process; and (*4*) the carrier medium, water, offers safety advantages compared with organic solvents and does not necessitate the expensive, energy-intensive recovery and purification operations associated with the use of solvents.

Nitrile rubbers generally are manufactured as either hot or cold types, as determined by the temperatures at which the polymerizations are controlled which, typically, are about 40°C and 5°C, respectively. The product advantages of lower content of branched and cross-linked polymer chains and of better processing are cited (15) as reasons for producing the cold varieties. Typical reactor recipes for the two types that are shown in Table 4 are similar to those developed for SBR under the auspices of the U.S. Rubber Reserve Board during and following World War II.

Shortstops are important to the production of quality-controlled nitrile rubbers as they terminate growing polymer chains and prevent further initiation of polymerization at some predetermined conversion of monomers. Stabilizers or antioxidants are added to the latex after polymerization (see Heat stabilizers).

1,3-Butadiene and acrylonitrile are the basic monomers that are combined in the water medium by a free-radical mechanism to form the high molecular weight polymer.

$$CH_2=CHCH=CH_2 \qquad CH_2=CHCN$$
$$\text{1,3-butadiene} \qquad \text{acrylonitrile}$$

The water not only serves as a reaction medium but also efficiently transfers the heat of polymerization from the reaction to the cooled reactor surfaces. The most commonly employed emulsifiers are fatty acid or rosin acid soaps (see Carboxylic acids) which are used frequently in combination with synthetic surfactants (qv). They stabilize the emulsion of monomers formed initially, provide the reaction sites or micelles, and support the dispersion of the polymer particles in the latex, thus preventing pre-agglomeration.

Cumene hydroperoxide and potassium persulfate initiators (qv) are used. Persulfates, which are relatively inactive at low temperatures, are preferred for the hot recipes; organic hydroperoxides are most satisfactory for the cold polymerizations. As shown in Table 4, cumene hydroperoxide functions as the oxidizing component in a redox catalyst system (17), and the reducing component is a complex formed from ferrous sulfate, ethylenediaminetetraacetic acid (EDTA) and sodium formaldehyde sulfoxylate (SFS).

Alkyl mercaptans, whose alkyl groups average about twelve carbon atoms, are the preferred ingredients used for modification or control of the average molecular weight of the finished, dry NBR product, as measured by the shearing disk Mooney viscosity method (2). This measurement is critical to the determination of the overall specifications for NBR related to processability during manufacture and compounding of finished products. As chain-transfer agents, the modifiers terminate the growth of some polymer chains and assume the role of free radicals to initiate other chains. The extent of termination is quantitatively related to the amount of mercaptan modifier in the reactor charge; for example, about 0.50 parts per 100 by wt of monomers of *tertiary*-dodecyl mercaptan would be sufficient to yield a Mooney viscosity in the desirable range for good processing at, eg, 80% monomer conversion.

Although not a part of the initial reactor charge, shortstops are vital to the production of specification NBR. These chemicals are needed to terminate the polymerization at a predetermined point, eg, 75–90%, for several reasons: the butadiene–

Table 4. Typical Reactor Charge Formulations[a]

Type	Hot	Cold
butadiene	variable[b]	
acrylonitrile	variable[b]	
water	180	180
soap	5.0[c]	4.6[c]
polymerized alkylnaphthylenesulfonic acid, sodium salt		0.20
KOH		0.03
KCl		0.30
ethylenediaminetetraacetic acid, sodium salt (EDTA)		0.02
ferrous sulfate		0.005
sodium formaldehyde sulfoxylate (SFS)		0.20
cumene hydroperoxide		0.20
potassium persulfate	0.3	
mercaptan modifier	variable[d]	

[a] Ref. 16.
[b] Total of monomers in charge equals 100: other ingredients are in parts per 100 by wt of monomers.
[c] May be fatty acid or disproportionated rosin acid soaps.
[d] Modifier adjusted to give required product molecular weight.

acrylonitrile ratio in NBR is a function not only of monomer charge levels but of monomer conversion levels either above or below a definite azeotropic ratio of ca 37% acrylonitrile. Very high conversion often results in branching or cross-linking; both are associated with poor processing. Also, owing to monomer depletion, the polymerization rate slows down in late stages to the point at which it is uneconomical to continue. Both inorganic and water-soluble organic reducing agents can function as shortstops, eg, sodium bisulfite and potassium dimethyl dithiocarbamate.

The general theory of the mechanism of emulsion polymerization was proposed by Harkins as the result of his studies that were sponsored under the U.S. Government's synthetic rubber program in 1942–1943 and were published in a series of papers (18) beginning in 1945. Harkins proposed that in the water phase free radicals initiate polymer nuclei from monomers that are solubilized in the interior of emulsifier micelles. The latter are colloidal aggregates of soap or other surface-active agents dispersed in the water. Emulsified droplets of monomers serve as reservoirs to keep the reaction sites supplied, but polymerization in the droplets is negligible. Finally, very small polymer particles, originating in the micelles and swollen with monomer, are the principal sites of the polymerization reaction, ie, chain growth and chain termination (see Emulsions; Latex technology).

The kinetics of emulsion polymerization were elucidated by Smith and Ewart (19) who demonstrated that the number of reaction sites increases with the emulsifier concentration and with the rate of formation of free radicals, but decreases with the growth rate of the free radicals. It was also shown that the polymer molecular weight is a function of the emulsifier concentration and the initiator level as well as the amount of modifier charged (19).

The two techniques that are practiced in the preparation of the NBR latex are batch, which is most suitable for short production runs, and continuous processing. In the batch process the measured amounts of ingredients (Table 4) are charged to an agitated, pressure-rated autoclave which is temperature-controlled by circulation of a coolant through a jacket or internal coils. Common vessel capacities range from

15–38 m³ (4,000–10,000 gal). The polymerization cycle usually requires a period of 5 to 12 h, depending on polymer type, reactor configuration, agitation and heat transfer capability. At the end of the reaction period, shortstop is added and the latex is transferred to a stripper for removal of unreacted monomers, and finally to an intermediate storage tank for testing. With the continuous method, the reaction ingredients are pumped through two or more stirred pressure vessels and the latex emerges from the last reactor in the series. As with the batch method, intermediate holding tanks may be employed for shortstopping, addition of stabilizers, stripping unreacted monomers and testing.

Alternative Polymerization Methods. NBR processing systems that have been considered as alternatives to the well-established emulsion approach include mass polymerization and solution or solvent polymerization, though economic factors have prevented commercial success. However, work on the suspension process using n-hexane as the suspension medium has continued in Japan by Furukawa and co-workers (20) who have produced an alternating copolymer of about 50% acrylonitrile, as opposed to the variable-ratio random copolymers available by emulsion techniques.

Conversion of NBR Latex to Dry Polymer. Low-boiling unreacted butadiene and higher boiling unreacted acrylonitrile are removed from short-stopped latex by vacuum-venting and steam distillation, respectively. These stripping operations, either batch or continuous, are significant users of electrical and thermal energy. The recovered monomers are condensed and stored for recycling to subsequent polymerizations in combination with fresh monomers.

Certain stabilizing chemicals are added to all grades of commercial nitrile elastomers to prevent their thermal or oxidative degradation during production operations, in storage as dry rubber, and to a limited extent, in the finished rubber products. Small amounts of these chemicals are incorporated into the polymer by adding them to the stripped latex prior to the subsequent finishing steps as emulsions or dispersions. Examples of suitable non-staining and non-discoloring stabilizing chemicals are 2,6-di(*tert*-butyl)-*p*-cresol and tri(nonylphenyl) phosphite (see Antioxidants). These materials are approved by the U.S. Food and Drug Administration for food use, and for materials intended to be in contact with foods. Approximately 1 to 2% of these chemicals in NBR gives adequate stabilization.

The stripped latex, properly stabilized, is tested against quality control specifications related to oil and solvent resistance. The most important of these specifications are Mooney viscosity, which is proportional to average molecular weight, and percent bound acrylonitrile. Subsequent operations are invariably continuous. Originally, the rubber is separated from latex at an alkaline pH, usually 8 to 10, by the addition of solutions of inorganic salts, eg, sodium chloride or aluminum sulfate, which are often used in conjunction with small amounts of mineral or weak organic acids.

The rubber crumb or curds, which occur as a slurry in water, may be filtered to remove most of the coagulating chemicals with the water and may be purified further by reslurrying in fresh water, filtering or screening again, followed by squeezing in an extruder. Drying of dewatered crumb may be accomplished in an extrusion dryer, a conventional apron, a fluid-bed hot air dryer, or a combination of these.

The dry polymer is usually compressed into bales, each weighing about 25 kg, wrapped in plastic film or packaged in paper bags and placed into shipping containers. Bale weights and packaging methods differ according to manufacturer's or customer's preference.

Compounding and Curing. In addition to reinforcing fillers, proper preparation for thermal curing also includes vulcanization agents. The most important of these are elemental sulfur and sulfur donors (see Rubber compounding). Organic peroxides are also used as cross-linking agents and they impart good high temperature resistance. Other chemicals that may be required to balance the vulcanization system are activators, such as zinc oxide and stearic acid, and accelerators which reduce the time and temperature requirements for sulfur cure. The latter include the very numerous thiazoles, sulfenamides, thioureas, guanidines, dithiocarbamates, thiuram disulfides, aldehyde-amine reaction products and combinations of these. A comprehensive review of sulfur curing systems (21) lists 80 common accelerators represented by over 1000 proprietary compositions.

Miscellaneous compounding ingredients include blowing agents for sponge; vinyl stabilizers for blends with PVC; bonding agents; peptizers; plasticizers; processing aids; tackifiers; colorants for mineral-filled compounds; ultraviolet absorbers and antirads.

The total compound of polymer, fillers, plasticizer and cure system may be mixed in conventional rubber equipment, such as open mills or internal mixers, then sheeted-off for molding, calendering or extrusion. Some precautions during these operations are often required to ensure good dispersion of ingredients and to prevent premature cure or scorching of stocks.

Any standard vulcanization technique used in the rubber industry may be used with nitrile compounds, including compression molding, transfer molding, injection

Table 5. Automotive Fuel Line Hose Tubing[a]

intermediate nitrile NBR[b]	100.0
zinc oxide	5.0
SRF carbon black, N-774	100.0
di(butoxyethoxyethyl) adipate (TP-95)	15.0
paraplex G-25	5.0
amine antioxidant (Aminox)	1.5
stearic acid	1.0
treated phthalic anhydride retarder (Esen)	0.5
tetramethylthiuram monosulfide (Monex)	1.5
sulfur	0.75
Total	230.25
Cured 20 min at 154°C	
modulus at 200%, MPa[c]	6.9
tensile strength, MPa[c]	14.4
elongation, %	430
hardness, durometer A	69
% volume swell, ASTM oil no. 3[d], 70 h at 100°C	8.0
% volume swell, ASTM Fuel B[e], 48 h at RT	22.7
fuel extraction on tubing, 24 h in Fuel B, g/7.8 × 10^{-3} m² surface area	0.38
precipitate in methanol	none
low temperature bending on tubing, 5 h at −40°C	pass
after immersion in ASTM no. 3 oil[d]	pass

[a] Ref. 22.
[b] Paracril BLT, 32.5% (Uniroyal).
[c] To convert MPa to psi, multiply by 145.
[d] Aniline point 70°C.
[e] 70 vol % isooctane + 30 vol % toluene.

molding, open-steam batch process for profile extrusions, open-steam continuous curing for insulated wire, dry-heat batch or continuous processes, or contact with liquid media or fluid beds. Less conventional methods involve microwave heating and electron beam radiation which are adaptable to continuous curing techniques (see Microwave technology).

Table 5 lists the components and key properties of a medium NBR compound, used for automotive fuel line hose, after vulcanization. Similar data can be obtained from manufacturers.

Economic Aspects

Table 6 gives production capacities for dry NBR.

Production of dry nitrile rubber in the United States (24–25) has been nearly constant in recent years and is marked by the recent withdrawal of two producers, Firestone, and Standard Brands (see Table 7).

Table 6. Production Capacities for Dry NBR in North America and Worldwide[a]

Producer and geographic area	Plant locations	Trade name	Capacity[b], metric tons
Producer			
B. F. Goodrich	Akron, Ohio	Hycar	33,000
	Louisville, Ky.		
Goodyear	Akron, Ohio	Chemigum	15,000
	Houston, Texas		
Uniroyal	Painesville, Ohio	Paracril	14,000
Copolymer	Baton Rouge, La.	Nysyn	5,000
Polysar	Sarnia, Ont., Can.	Krynac	10,000
		Total	77,000
Geographic area			
U.S.			67,000
Canada			10,000
Latin America			6,000
Western Europe			80,000
Japan			35,000
South Africa			1,000
Centrally-Planned-Economy Countries			65,000
		World total	264,000

[a] Ref. 23.
[b] North America data for 1977, worldwide data for 1976.

Table 7. Production of Dry Nitrile Rubber in the United States

Year	Production, metric tons
1958	33,000
1963	49,000
1968	72,000
1973	84,000
1976	62,000
1977	62,000

Approximately 90% of U.S. production is consumed domestically with the remainder exported. As of early 1978, the selling price range for NBR in the United States in truckload quantities was ca $1.52–1.72/kg, depending on and proportional to acrylonitrile content. During the same period preblended NBR/PVC grades sold for ca $1.52/kg.

Health and Safety Factors

Certain environmental restrictions on nitrile rubber manufacturing operations have been imposed by government agencies. The OSHA standard for worker exposure to butadiene monomer in air, 8 h time-weighted average, is 1000 ppm, and the maximum 15 min exposure is 1250 ppm. Acrylonitrile, believed to be carcinogenic, was regulated by an emergency temporary standard (26) at a level of 2 ppm time-weighted average and 10 ppm averaged over any 15 min period. A permanent standard for acrylonitrile in air has been issued (27). OSHA has also mandated practices to prevent skin and eye contact by workers.

Other regulations by state and federal environmental protection agencies apply to manufacturing operations and are site-specific for each plant location. Air and water effluents also are closely specified and controlled.

Uses

Uses of NBR products are listed in Table 8.

Table 8. Types of NBR Products and Extent of Use of Each[a]

Product	Extent of use, %
footwear	2.0
belts and belting	3.2
hose	7.5
o-rings, packings, gaskets	18.2
automotive, other mechanical goods	42.8
adhesives, sealants, coatings	3.2
rolls	2.4
coated fabric, film, sheeting	3.2
wire and cable insulation	1.0
blends (NBR/PVC, etc)	16.5
Total	100.0

[a] Ref. 28.

BIBLIOGRAPHY

"Rubber, Synthetic" in *ECT* 1st ed., Vol. 11, pp. 827–852 by J. D. D'Ianni, The Goodyear Tire & Rubber Co., Akron, Ohio; "Elastomers, Synthetic" in *ECT* 2nd ed., Vol. 7, pp. 676–705 by William M. Saltman, The Goodyear Tire & Rubber Co., Akron, Ohio.

1. Ger. Pat. 658,172 (Apr. 26, 1930), Brit. Pat. 360,821 (May 30, 1930), Fr. Pat. 710,901 (Feb. 4, 1931), U.S. Pat. 1,973,000 (Sept. 11, 1934), E. Konrad and E. Tschunkur (to I. G. Farbenindustrie A.G.).

2. *Method D-1646-74, Standard Test Method for Rubber from Natural or Synthetic Sources, Viscosity and Vulcanization Characteristics (Mooney Viscosity)*, American Society for Testing and Materials, Philadelphia, Pa.
3. R. H. Wiley and G. M. Brauer, *J. Polym. Sci.* **3**, 704 (1948).
4. N. Bekkedahl and R. B. Scott, *Rubber Chem. Technol.* **16**, 310 (1943).
5. J. Rehner, *J. Polym. Sci.* **2**, 263 (1947).
6. R. Vieweg and W. Schneider, *Kunststoffe* **31**, 215 (1941).
7. R. H. Wiley and G. M. Brauer, *Rubber Chem. Technol.* **22**, 402 (1949).
8. H. Roelig, *Rubber Chem. Technol.* **13**, 948 (1940).
9. W. J. Roff and co-workers, *Handbook of Common Polymers*, Butterworth & Co., London, 1974, p. 400.
10. *Method D-471-75, Standard Test Method for Rubber Properties, Effects of Liquids*, American Society for Testing and Materials, Philadelphia, Pa.
11. *Oil-Resistant Paracril Nitrile Rubber, Bulletin ASP-3440*, Uniroyal Chemical, Division of Uniroyal, Inc., Naugatuck, Conn.
12. M. S. Sutton, *Rubber World* **149**(5), 62 (Feb. 1964).
13. R. P. Mastromatteo, J. M. Mitchell, and T. J. Brett, Jr., *Rubber Chem. Technol.* **44**(4), (1971).
14. C. F. Fryling in G. S. Whitby, ed., *Synthetic Rubber*, John Wiley & Sons, Inc., New York, 1954, p. 224.
15. W. Hofmann, *Rubber Chem. Technol.* **37**(2), 81 (April–June 1964).
16. C. A. Uraneck and J. E. Burleigh, *J. Appl. Polym. Sci.* **12**, 1075 (1968).
17. V. C. Neklutin, C. B. Westerhoff and L. H. Howland, *Ind. Eng. Chem.* **43**(5), 1246 (1951).
18. W. D. Harkins, *J. Chem. Phys.* **13**, 381 (1945); **14**, 47 (1946); *J. Am. Chem. Soc.* **69**, 1428 (1947); and *J. Polym. Sci.* **5**(2), 217 (1950).
19. W. V. Smith and R. H. Ewart, *J. Chem. Phys.* **16**(6), 592 (1948).
20. J. Furukawa and Y. Iseda, *J. Polym. Sci. B7*, **47**, 561 (1969).
21. D. A. Smith, *Rubber J.* **149**(2), 55 (1967); **149**(3), 10 (1967).
22. *Uniroyal Chemical Bulletin ASP-3407*.
23. C. F. Ruebensaal, *The Rubber Industry Statistical Report*, IISRP, Uniroyal, Inc. (1977).
24. *Chem. Eng. News*, **52**(18), 12 (May 6, 1974).
25. Rubber Manufacturers Association Statistical Report, Synthetic Rubber Producers Report, New York, 1977.
26. *Fed. Regist.* **43**(11), 2600 (Jan. 17, 1978).
27. *Fed. Regist.* **43**(192), 45809 (Oct. 3, 1978).
28. D. Dworkin, *Rubber World* **171**(5), 43 (Feb. 1975).

General References

C. H. Lufter, "Vulcanization of Nitrile Rubbers," in G. Aliger and I. J. Sjothun, eds., *Vulcanization of Elastomers*, Reinhold, New York, 1964, p. 195.
"Nitrile Rubbers I," *J. Int. Rubber Ind.* **1**(5), 262 (1967); **1**(6), 319 (1967).
N. M. Bikales, ed., *Encyclopedia of Polymer Science and Technology*, Interscience Publishers, a division of John Wiley & Sons, Inc., New York, 1972 I.
W. Hofmann, "Nitrile Rubber," *Rubber Chem. Technol.* **37**(2), Pt. 2 (April–June 1964), a rubber review for 1963.
J. P. Morrill, "Nitrile and Polyacrylate Rubbers," in M. Morton, ed., *Rubber Technology*, 2nd ed., Van Nostrand Reinhold Co., New York, 1973, p. 302.
W. J. Roff, J. R. Scott, and J. Pacitti, "Nitrile Rubbers," in *Handbook of Common Polymers*, Butterworth & Co., London, 1971, Sec. 35, p. 396.
H. J. Stern, *Rubber, Natural and Synthetic*, 2nd ed., Maclaren & Sons, Ltd., London, 1967, p. 106.
H. E. Minnerly, "Nitrile Elastomers," in G. G. Winspear, ed., *Vanderbilt Rubber Handbook*, 1968, p. 99.
W. L. Semon, "Nitrile Rubber," in G. S. Whitby, ed., *Synthetic Rubber*, John Wiley & Sons, Inc., New York, 1954, p. 794.
C. E. Schildknecht, ed., *Polymerization Processes*, Vol. 29 of *High Polymers*, Wiley-Interscience, New York, 1977.

R. R. Barnhart, "Rubber Compounding," in A. Standen, ed., *ECT* 2nd ed., Vol. 17, John Wiley & Sons, Inc., New York, 1968, p. 543.

H. W. ROBINSON
Uniroyal, Inc.

POLYBUTADIENE

The history and development of polybutadiene rubbers closely parallels that of diolefin polymers in general. The highlights of the progress made in synthetic elastomers since their discovery have been reviewed in several articles (1–6).

Polybutadiene was first prepared and studied by Lebedev, a Russian, in 1910 using alkali metals for the initiator of diene polymerization (7–8). His work eventually led to the world's first industrial production facility for synthetic rubber.

Polybutadiene was produced in the USSR and Europe in the 1930s, but remained relatively unknown as an elastomer until the early 1950s, when solution diene polymerization using a coordination catalyst was discovered.

During World War I, Germany was looking for a substitute for natural rubber supplies from the Far East. In view of its plentiful supply, 1,3-butadiene [*106-99-0*], CH_2=CHCH=CH_2, became the logical monomer to study. Germany's interest in methyl rubber [*25034-65-5*], produced from 2,3-dimethyl-1,3-butadiene, and in polybutadiene produced with an alkali metal catalyst, gradually declined after World War I when natural rubber supplies were again accessible.

In the USSR, rod polymerization (9) of butadiene by metallic sodium was developed on a commercial scale and remained the mainstay of Soviet synthetic rubber production until the mid-1940s.

In the 1920s emulsion polymerization proved to be a feasible approach to the preparation of high molecular weight elastomers at high production rates. In the 1930s Germany concentrated on these emulsion rubbers. Buna S [*9003-55-8*], an emulsion styrene–butadiene copolymer (10), and Buna N [*9003-18-3*], an acrylonitrile–butadiene copolymer (11) were developed at this time, while activity in the United States centered on neoprene and butyl rubbers.

The advent of World War II again saw the withdrawal of natural rubber from the world market, and renewed research on synthetic rubbers. Both the United States and Germany turned to emulsion rubbers, produced by free-radical catalysis, to meet the wartime demand. Emulsion styrene–butadiene copolymers (formerly called GR-S) and acrylonitrile–butadiene copolymers (formerly called GR-N), produced in government owned and operated synthetic rubber plants (The Rubber Reserve Company), temporarily filled rubber needs.

Early work on the emulsion polymerization of butadiene led to a product inferior in properties to emulsion styrene–butadiene copolymers. Work on improving the emulsion polybutadiene system continued, however, in view of monomer abundance and low cost. The first commercially produced emulsion polybutadiene appeared on the market about 1950.

Another product of emulsion polybutadiene appearing about the same time was alfin rubber [9003-17-2], an elastomer synthesized with an alfin catalyst (sodium catalyst) (12–13). However, this polymer displayed poor processibility as a result of its high trans-1,4 content and molecular weight. Today alfin catalysts are mainly used for copolymerizing butadiene with styrene or isoprene.

Interest in synthetic polybutadiene was renewed owing to the discovery in the mid 1950s by Ziegler and co-workers (14) that ethylene could be polymerized to a high molecular weight polymer in the presence of a hydrocarbon solvent and a triethyl aluminum–titanium tetrachloride catalyst, using mild conditions of temperature and pressure. A myriad of synthetic elastomers soon followed.

An immediate offshoot of Ziegler's work was the synthesis of polyisoprene a synthetic natural rubber (15–17), in the mid 1950s followed by the synthesis of high-cis stereoregular polybutadiene (18). Commercial production of this polymer began in 1960. Only limited and expensive supplies of isoprene monomer were available, whereas butadiene monomer was plentiful and cheap.

The merits of polybutadiene, aside from cost, became quickly evident as it soon found worldwide acceptance in applications involving both tire and nontire uses. With its importance to the rubber industry, it ranks second in production only to emulsion SBR among all synthetic rubbers.

Butadiene Monomer. In the United States butadiene is used primarily in the production of general purpose rubbers (see Table 1). Tires and other fabricated rubber products account for 81% of butadiene's consumption (19).

In 1976 the butadiene demand in the United States was 1.7 million metric tons and by 1981 the demand is expected to increase to 2.0 million metric tons (20), an increase of nearly 4%.

Butadiene is produced commercially in the United States as a co-product in the manufacture of ethylene and propylene feedstocks and by dehydrogenation of butenes from refineries and butanes from natural gas. From the latter process, so-called primary butadiene is obtained. Since domestic consumption exceeds production, additional butadiene (co-product) is imported from Europe and Japan (see Butadiene).

Microstructure

The polymerization of butadiene is an example of an addition polymerization, wherein the repeating structural unit within the polymer chain has the same molecular weight as the entering monomer unit.

The configurations of polybutadiene are cis, trans, and vinyl. Since either or both of the double bonds in butadiene can be involved in the polymerization, the resulting

Table 1. General Purpose Rubbers from Butadiene

Rubber	CAS Registry No.	Production, %
styrene–butadiene (SBR)	[9003-55-8]	53
polybutadiene (PBR)	[9003-17-2]	18
neoprene	[9010-98-4]	7
nitrile (NBR)	[32240-92-9]	3
adiponitrile	[25249-61-0]	8
acrylonitrile–butadiene–styrene resins	[9003-56-9]	6
miscellaneous		5

polymer may have a variety of configurations. These result from the fact that the spatial arrangement of the methylene groups in polybutadiene allow for geometric isomerism to occur along the polymer chain. The different polymer structures of polybutadiene are given in Figure 1.

Participation of both double bonds in the polymerization process gives rise to a 1,4-addition, which may be cis-1,4- or trans-1,4-, depending on the disposition of groups about the polymer double bond. Participation of a single double bond results in a vinyl, or 1,2-addition. The three possible vinyl structures are (*1*) isotactic, where all vinyl groups have the same orientation; (*2*) syndiotactic, where vinyl groups alternate regularly; and (*3*) atactic, where isotactic and syndiotactic structures alternate randomly.

The four structurally different configurations of polybutadiene give rise to notably different behavior (21). Since the pure stereoisomers are highly regular, they are likewise highly crystalline. The cis-1,4 isomer crystallizes upon stretching over 200%. The trans-1,4 and vinyl isomers are crystalline without elongation.

The high-*cis*-1,4-polybutadiene is a soft, easily solubilized elastomer, which exhibits excellent dynamic properties, low hysteresis, and good abrasion resistance. A glass transition temperature of $-102°C$ (22) is reported for this rubber. This low glass temperature is believed to account for its excellent low temperature performance, high resilience, and abrasion resistance.

The *trans*-1,4-polybutadiene, in contrast, is a tough elastomer, having a reported glass transition temperature of $-107°C$ (23) and $-83°C$ (94% trans) (24). It is sparingly soluble in most solvents, in addition to its high hardness and thermoplasticity.

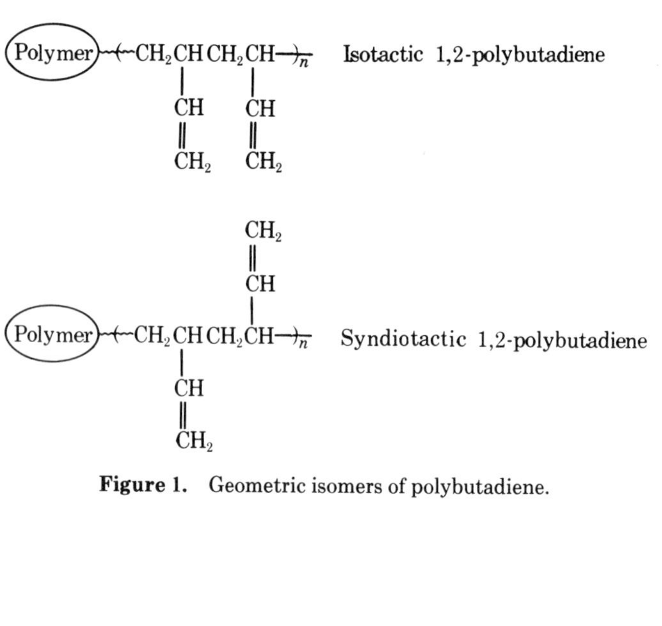

Figure 1. Geometric isomers of polybutadiene.

The 1,2-isotactic and 1,2-syndiotactic polybutadienes are rigid, crystalline materials, with poor solubility characteristics. A glass transition temperature value of −15°C (23) is reported for these rubbers. The atactic polymers are soft elastomers possessing poor recovery characteristics.

Another important factor influencing polymer properties is the interaction or relationship of a polymer chain unit of given microstructure with its neighboring units. An example of this is 1,2-polybutadiene, where both isotactic and syndiotactic configurations are possible.

The monomer unit may also be capable of adding unsymmetrically. In 1,2-addition the polymer chain is irregular, depending upon which end of the incoming monomer unit adds to the growing chain. In 1,4-addition the modes of addition are all equivalent.

Other polymer chain variations, such as connective linkages, may also lead to differences in polymer physical properties. Such linkages may actually form the basis for the differences in behavior between natural and synthetic polyisoprenes.

Polymerization Process

Most commercial processes employ solution polymerization. In general these systems are based upon organic lithium compounds or coordination catalyst containing metals in reduced valence states. On a laboratory scale these polymerizations are relatively straightforward. Special attention and care must be given to the elimination of the smallest traces of air, moisture, and deleterious impurities. The presence of even low levels may be not only detrimental to polymer conversion, but can also strongly influence the physical properties of the polymer.

Thus polymerizations are carried out using essentially pure dry monomer and solvents, such as aromatic, aliphatic, and alicyclic hydrocarbons. Catalysts are charged under some inert atmosphere, eg, nitrogen or argon. The polymerization is then carried out at a preselected temperature and time with adequate agitation of the reaction media. When the desired conversion of polymer is achieved, the catalyst is deactivated and the polymer stabilized against oxidation. The polymer is then recovered from the unreacted monomer and solvent, washed to remove catalyst residues, and dried.

Figure 2 is a flow sheet of the commercial production of polybutadiene. The polymerization is carried out continuously in a series of large reaction vessels, equipped with efficient agitation and cooled by either reflux or jacket cooling. The catalyst system is introduced into the monomer–solvent premix either as separate catalyst components or as a preformed catalyst complex.

Polymerization residence time can be adjusted through changes in the charge rate of the premix. Monomer concentration in the premix is determined by the heat exchanging capacity of the system and the ability of the polymerization process to transport the viscous cements. Monomer concentrations in aromatic solvents are generally 10–12 wt %; in aliphatic solvents they may be greater than 16 wt %.

After polymerization is complete, the viscous cement is simultaneously deactivated and stabilized with antioxidants, and then washed with water to remove catalytic residues. Solvent and unreacted monomer are steam distilled (steam stripping). More than 95% of the solvent and unreacted monomer are recovered after azeotropic drying and distillation and subsequently recycled.

By adding oil with adequate mixing to the polymer cement, polybutadiene can

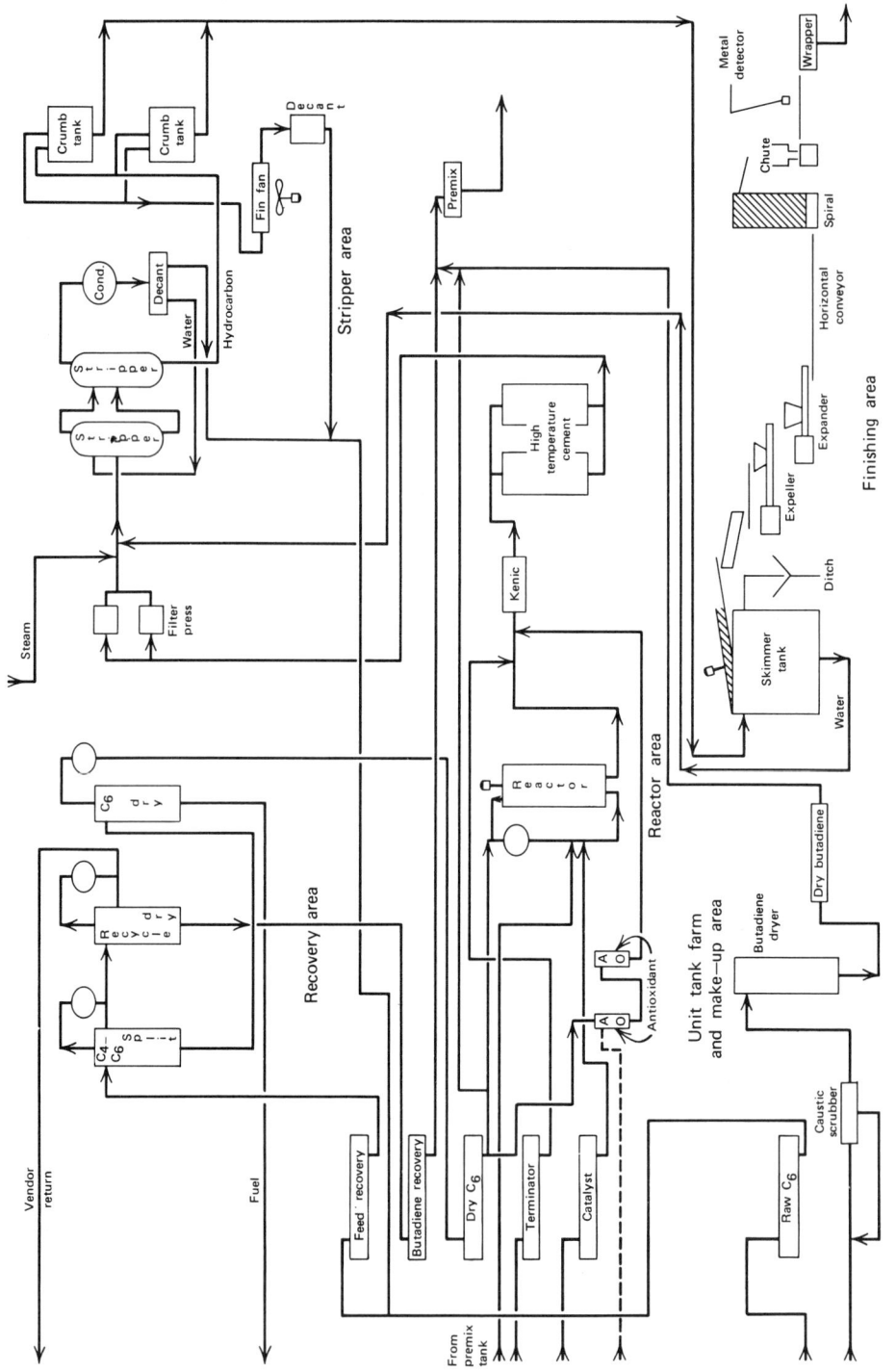

Figure 2. Flow sheet for the production of high-*cis*-1,4-polybutadiene in solution. Courtesy of The Goodyear Tire & Rubber Company.

be oil-extended with only minor process modifications. The uniform dispersion of the oil in the polymer is of utmost importance to the final product. It can then be recovered, washed, and dried in a manner similar to the oil-free polymer. Another modification is the addition of carbon black to the oil-extended solution of polybutadiene, forming a black masterbatch.

Cationic Polymerizations. In cationic polymerizations, the reaction is initiated by complex ions or ion pairs, produced by the interaction of Lewis acids with water, hydrogen halides, and other halogen containing compounds. When the Lewis acids are halides of aluminum, boron, titanium, or tin, a low solubility polybutadiene is produced, comprised of trans-1,4 and 1,2 units (25–26).

The polybutadienes formed by the cationic process are characterized by a high degree of cis–trans isomerization (27) and a low degree of saturation, estimated at 30–70% of theory (28). This gives rise to secondary reactions involving internal double bonds, resulting in a polymer having a high degree of intramolecular cyclization. Commercial applications of this technique for producing polybutadiene are not significant.

Free-Radical Polymerization. Free-radical polymerizations, whether performed in emulsion or solution, are initiated by an active radical R· formed by the decomposition of a peroxide, peroxysulfate, or similar free-radical forming reactions. The free radical R· takes part only in the initial step of polymer growth, eventually becoming far removed from the continuing site of polymerization.

Emulsion systems, in general, contain water, monomer(s), initiator, and an emulsifier. Polymerization is initiated when the monomer and free radical come into contact within a soap micelle in the aqueous phase.

The free-radical polymerization of butadiene is a function of the reaction temperature (29–31). As the polymerization temperature decreases, the cis-1,4 content also decreases, with no cis-1,4 content below −15°C. This same observation is made for polybutadiene segments of copolymers with styrene, acrylonitrile, and others.

Stereospecific Polymerization. The most outstanding characteristic of the Ziegler or Ziegler-Natta catalyst (qv) is its ability to produce highly stereoregular polymers. These catalysts are comprised, in general, of an organometallic compound and a transition metal compound from any Group IV–VIII, and may be homogeneous or heterogeneous. They may vary from anionic to cationic in behavior; however, the factors influencing this behavior are not always clear. Often more than two catalyst compounds must be added at very exacting ratios in order to achieve some particular catalyst activity or microstructure for the polybutadiene.

Much academic and industrial effort has centered on synthesizing the four stereoregular isomers of polybutadiene. Mostly it has been directed toward developing anionic catalyst systems for the synthesis of high (>90%) *cis*-1,4-polybutadiene, because of its excellent elastomeric properties.

Although a wide variety of possible catalysts for polymerizing butadiene has been reported, only catalysts containing titanium, cobalt, and nickel have been successfully developed commercially.

Catalysts

Alkali-Metal and Organoalkali-Metal Catalysts. The oldest ionic butadiene polymerization catalyst is metallic sodium. It was used in various modifications by the

Russians as early as 1910 and by the Germans during World War I to polymerize dimethylbutadiene. The early German Buna rubbers were polybutadienes prepared in the presence of sodium or potassium catalysts. Studies on the mechanism of butadiene polymerization by alkali metals, and particularly sodium (32–33), established the importance of organoalkali metal compounds in effecting catalytic activity and the anionic character of these polymerizations.

Many reagents serve as activators for sodium polymerizations, including CO_2, hydroxyl compounds, certain metal powders or oxides, and others. The sodium polymerization is highly exothermic and occurs in two stages (34), the first stage of polymerization being very slow and a second rapidly going to completion. Microstructure varies greatly depending upon the alkali metal used (see Table 2).

Similar results have also been observed for organoalkali-metal catalysts. The ability of the alkali metals to form 1,4-linkages decreases in the order of decreasing ionization potential of the alkali metal.

Lithium and its organic derivatives form 1,4-linkages, and in particular, cis-1,4-linkages (35%) with the greatest ease. All the other alkali metal catalysts give predominantly vinyl structure (45–65% vinyl).

Mechanistically, lithium metal first adds to a monomer unit, resulting in a dilithium species, which then proceeds to add other monomer units in a manner similar to organolithium compounds (38–40).

Butyllithium, developed commercially by the Firestone Tire and Rubber Company and the Phillips Petroleum Company, is the most commonly used organoalkali metal catalyst. Polymers prepared with butyllithium in heptane possess 40–50% trans, 40–50% cis, and 5–10% vinyl structures (41); this distribution is similar to that obtained in lithium metal polymerizations.

The polymer structure is independent of conversion, reaction temperature, and monomer concentration, but is dramatically affected by the introduction of electron donors, such as ethers, amines, and sulfides, which solvate the alkali metal (41–42). These additives substantially increase the vinyl structure.

Polymerization of butadiene by alkyl lithiums, proceeds via a living polymer, ie, no termination mechanisms are present, and the molecular weight of the polymers is proportional to the amount of butadiene monomer in the system, one polymer chain growing per molecule of monomeric alkyllithium. The polymers have a narrow molecular weight distribution, since all the growing chains have an equal probability of chain growth (43).

Table 2. Effect of Alkali Metal on Polybutadiene Microstructure, %

Metal	Ionization potential	cis-1,4	trans-1,4	1,2	Reference
Li	5.4	35	55	10	35
Na	5.1	23 (10)[a]	45 (25)[a]	32 (65)[a]	36
K	4.3	13 (15)[a]	49 (40)[a]	38 (45)[a]	36
Rb	4.2	7	31	62	37
Cs	3.9	6	35	59	37

[a] Values obtained in ref. 35.

With alkyl lithium catalysts the propagation process is similar to the initiation process, with monomer inserting between the lithium and the alkyl ions or the butadienyl residues of the ion, respectively, in the two processes.

Alfin catalysts (44–46) are related to the alkali metal catalysts. They are based on sodium and consist of a complex of NaCl, sodium alkoxide, and an alkenyl sodium, such as sodium isopropoxide and an allyl sodium. Alfin polybutadiene, however, has a molecular weight of up to several million, and is unprocessable by conventional techniques. Polymer microstructure is 68% trans-1,4, 17% cis-1,4, and 16% vinyl.

The molecular weight of alfin polymers can be modified to approximately 300,000 by the addition of modifiers, eg, 1,4-dihydrobenzene and 1,4-dihydronaphthalene (47). Molecular weight control occurs by hydrogen transfer.

Titanium-Based Catalysts. A majority of the early patents dealing with the titanium-based polymerization of butadiene reported cis-1,4 contents in the range of 60–70%, which proved to be unsatisfactory in rubber applications. In 1956 a high-cis-1,4-polybutadiene was prepared, using an $Al(C_2H_5)_3$–$TiCl_4$ catalyst.

The stereospecificity of the titanium-based catalysts depends largely upon the nature of the halogen atom attached to titanium. In the series $TiCl_4$, $TiBr_4$, and TiI_4, the cis-1,4 content increases 60–90% (48), proving that iodine is essential for the formation of high-cis-1,4-linkages.

Titanium-based catalysts are divided in four groups: (1) AlR_3 with $TiCl_4$, (2) AlR_3 with TiI_4, (3) ternary catalyst systems consisting of AlR_3 (or aluminum alkyl hydrides) with $TiCl_4$ and an iodine-containing compound, and (4) other organometallic compounds with TiI_4.

In the AlR_3 + $TiCl_4$ system, catalysts are prepared by the gradual addition of AlR_3 to $TiCl_4$. The most commonly used aluminum alkyls are triethyl and triisobutyl. Polymerization is performed in aromatic hydrocarbons, such as benzene or toluene. Aromatic solvents are essential for high polymerization rates and near quantitative yields of polymer, although mixtures with aliphatic solvents work to a lesser degree. Poorer catalytic activity, however, is observed in aliphatic solvents such as heptane.

In the preparation of $TiCl_4$ catalysts, minor changes in experimental conditions, eg, the Al–Ti molar ratio, can affect not only catalyst activity but also the nature of the final polymer (49).

$TiCl_4$ catalytic activity begins at an Al–Ti molar ratio of ca 0.5. At this ratio a powdery solid polymer with a low degree of unsaturation is obtained (50–51). This partially cyclized and cross-linked polymer is insoluble in benzene and other hydrocarbon solvents. Activity increases rapidly as the Al–Ti ratio approaches 1, where AlR_2Cl + β-$TiCl_3$ are the catalytic species in the system. The polymer at this ratio is an amorphous rubber, having a content of 75–78% cis-1,4, 21–23% trans-1,4, and 1–2% 1,2 (52). At an Al–Ti ratio of 2–2.2, tougher polymers, such as 40–42% cis-1,4, 56–58% trans-1,4, and 3–4% 1,2, are obtained. These polymers are comprised of a small high-cis (80%) extractable polymer portion and a large high-trans (90%) component. At a ratio of 3, only a small quantity of solid polymer is recoverable, in addition to a large amount of low molecular weight oligomers.

The AlR_3 + TiI_4 system, although not a significant titanium-based catalyst, produces a good yield of ca 80% cis-1,4-polybutadiene (53). The final product, however, is a highly gelled polymer.

The AlR_3 + TiI_4 system has been most extensively studied, since it produces a high (>90%) cis-1,4, linear, gel-free polybutadiene (48,54–57). This system has been successfully scaled up to a commercial production process. The catalyst becomes active at an Al–Ti molar ratio of 1, or slightly higher. As the Al–Ti ratio increases, a maximum is reached between 3 and 5, and decreasing slowly with increasing AlR$_3$.

The polymer cis content is not affected by the Al–Ti ratio but is a function of catalyst concentration. At low catalyst levels, better than 95% cis-1,4 polymer can be obtained. When using TiCl$_4$, the cis content does not depend on catalyst concentration. For the TiI$_4$ system, polymerization rate and conversion increase, whereas molecular weight decreases with increased catalyst concentration. The molecular weight is also a function of the polymerization temperature, monomer concentration, and ratio of catalyst components.

Titanium tetraiodide is insoluble in hydrocarbon media and, therefore, requires special treatment before catalyst make-up. In the absence of contamination from the environment, it can be ground dry or ball-milled with a solvent.

Titanium tetraiodide and AlR$_3$ form a heterogeneous catalyst. The solid and liquid portions of this catalyst have been separated but neither portion shows catalytic activity by itself (51,57).

Catalyst systems may also use mixed titanium halogen compounds such as TiI$_2$Cl$_2$ and TiICl$_3$ (51) which are both comparable to TiI$_4$ in catalytic activity.

Catalyst behavior similar to TiI$_4$ has also been reported for ternary systems of the following types: TiCl$_4$–I$_2$–Al(i-C$_4$H$_9$)$_3$ (57–59), TiCl$_4$–AlI$_3$–AlR$_3$, TiCl$_4$–AlR$_2$I–AlR$_3$ (60–61), and TiCl$_4$–AlI$_3$ combined with AlHCl$_2$·D, AlH$_2$Cl·D, or AlH$_3$·D, where D is an electron donor, such as (C$_2$H$_5$)$_2$O or N(CH$_3$)$_3$ (60,62). The addition of the donor molecule not only helps to stabilize the aluminum hydrides but also aids in solubilizing them in hydrocarbon solvent.

All titanium catalyst systems containing iodine behave similarly, since in all cases the final catalyst has iodine bound directly to the titanium (51–57).

Other organometallic compounds, besides aluminum alkyls, can be used in polymerizations as reducing agents for titanium, including organometallic compounds of lithium (63), sodium (64), zinc, and lead (65). Optimum activity in all of these systems is a function of catalyst–component ratios.

Systems containing LiAlH$_4$ in combination with TiCl$_2$I$_2$ or I$_2$–TiCl$_4$ have also been reported to yield 93–95% cis-1,4-polybutadiene (66).

Cobalt-Based Catalysts. A method of preparing high-cis-1,4-polybutadiene, using a catalyst consisting of CoCl$_2$ and (i-C$_4$H$_9$)$_3$Al, was patented (67) in 1956. Much of the preliminary polymerization work with cobalt involved heterogeneous catalysts containing alkyl aluminum halides and cobalt compounds, such as CoX$_2$ (X = halide), CoS, CoSO$_4$, CoCO$_3$, and Co$_3$(PO$_4$)$_2$ (68–69). Homogeneous catalysts are made from soluble Co(II) and Co(III) salts (octanoates, naphthenates, and acetylacetonates) and certain donor complexes such as CoCl$_2$·pyridine.

The soluble cobalt catalysts are charged at levels in the range of 0.01–0.1 millimoles per 100 parts (wt) monomer Co at Al–Co molar ratios between 100 and 1000. Heterogeneous catalysts, on the other hand, require ca 10 millimoles per 100 parts (wt) monomer Co at an Al–Co ratio of 5, which may be owing to the fact that heterogeneous cobalt catalysts are believed to form small quantities of soluble catalysts that actually initiate the polymerization. Only the soluble cobalt catalysts are of any practical importance.

Although CoCl$_2$ alone has been reported to polymerize butadiene to cis polymer

(70), most cobalt catalysts contain aluminum; the most important aluminum compounds are the dialkylaluminum chlorides (68,71–77).

The polymerization of butadiene in benzene with R_2AlCl–Co catalyst gives high conversions of >99% cis-1,4-polybutadiene. The molecular weight of this polymer varies between 10,000 and 1,000,000, depending upon experimental parameters. The high stereospecificity associated with this catalyst is limited to butadiene. With isoprene, polymers of mixed microstructures are prepared, having only 65% cis-1,4 content. Additional cobalt systems involving triethyldialuminum trichloride, $(C_2H_5)_3Al_2Cl_3$, $(C_2H_5)AlCl_2$, or mixtures have also been successfully used.

Much of the very early published work on cobalt catalysts is in disagreement, and many reported experiments were irreproducible. These discrepancies were later found to be the result of the presence of catalytic quantities of certain electron donors in these systems. Since the cobalt concentration used was very low, the efficiency of the catalysts was drastically affected by low levels of these donors. In fact, one pecularity of the cobalt system is that polymerization can occur at a rapid rate only in the presence of small quantities of water or certain other compounds.

Water is needed as an activator in reactions in benzene with the $(C_2H_5)_2AlCl$–$CoCl_2$·pyridine catalyst (78–79). In a thoroughly dry system no polymerization occurs after 24 hours reaction time at 6°C. When water or oxygen is added at 10 mol % with respect to the aluminum, rapid polymerization of high molecular weight, 98% cis-1,4-polybutadiene is obtained (80–81). In this system the strong Lewis acid, $O[(C_2H_5)AlCl]_2$, is believed to account for the observed activity.

Other examples of successful activators are $AlCl_3$, metallic Al, alcohols, reactive organic halides, halogens, hydrogen halides, and organic hydroperoxides (78,82). Activator concentration is generally 10–20 mol % with respect to the $(C_2H_5)_2AlCl$.

The number of polymer chains formed per cobalt atom in a polymerization system is very high. Since the molecular weight of the polymer changes very little with increasing conversion (83), chain transfer must be considered an important process in cobalt systems. Cobalt-based polybutadienes show significant branching with increased conversion, resulting in a very broad molecular weight distribution. The molecular weight initially increases rapidly with increased catalyst concentration, but gradually stabilizes, and in some instances, even decreases.

The polymer molecular weight can be regulated by the addition of certain chain-transfer reagents, including H_2 (84), ethylene, or propylene (up to 6 mol %), allene, methylallene, and certain unsaturated compounds (eg, cyclooctadiene).

Other methods of molecular weight control include changes in polymerization temperature, solvent composition, and monomer and catalyst concentration (85). High molecular weight polybutadiene can be produced by carrying out the usual cobalt polymerization, and then adding a reactive halide (eg, tert-butyl chloride) to link several long chains (86).

Polymerization with cobalt catalysts occurs at a very high rate in aromatic solvents, depending on the solvent used. Activity decreases in the order benzene > toluene > xylene > mesitylene. Mixed aromatic–aliphatic solvents have also been used, but activity depends upon composition (78). Aliphatic solvents alone give good polymerization rates (68,78), but polymers with lower cis-1,4 contents and molecular weights are obtained.

As indicated before, not all cobalt systems require water to achieve activity. The system CoX_2–AlX_3 (where X = Cl, Br, or I) is reported to give highly stereospecific 96% cis-1,4 polybutadiene in the absence of water and aluminum alkyl (87–90).

The system CoX_2–R_3Al (X = Cl, Br) (91–92) gives excellent activity at an Al–Co ratio of ca 1. At this ratio R_2AlCl is formed as an alkylation product of $CoCl_2$ with R_3Al. The R_2AlCl then reacts further with excess $CoCl_2$ to form the catalyst (93). At low Al–Co ratios, the catalyst, prepared in the absence of butadiene monomer and aged for a period of time, gives high-cis-1,4-polymer (91).

The $(i\text{-}C_4H_9)_3Al$–$CoCl_2$ pyridine system (78) at higher Al–Co ratios, on the other hand, gives no polymer in the absence of water, but high-cis-1,4-polymer at a H_2O–$(i\text{-}C_4H_9)_3Al$ ratio of one. Under these conditions, $O[Al(i\text{-}C_4H_9)_2]_2$, should be obtained. But an analogous compound $O[Al(C_2H_5)_2]_2$, obtained by the reaction of $(C_2H_5)_3Al$ with H_2O, in conjunction with $CoCl_3$·pyridine gives crystalline syndiotactic 1,2-polybutadiene (94).

Cobalt catalysts containing aloxanes, ie, aluminum alkyl compounds with Al—O—Al bonds, show polymerization activity (95). Since aloxanes are prepared by the reaction of aluminum alkyls with water, this system resembles cobalt systems where small amounts of water must be added to initiate polymerization. The catalyst formed with aloxanes give excellent yields of high-cis-1,4-polybutadiene.

Nickel-Based Catalysts

The study of nickel catalysis of butadiene polymerization has in recent years spurred not only considerable industrial interest as a commercial process but also academic interest concerning the mechanistic pathways of these reactions. A large number of nickel compounds and complexes, homogeneous or heterogeneous, give excellent yields of high-cis-1,4 (98%)-polybutadiene, either by themselves or in conjunction with other organometallic components, at relatively low reaction temperatures.

Catalytic activity has been observed in one-, two-, and three-component systems containing nickel. These systems can be divided into those based upon nickel salts and those based upon π-allyl nickel complexes. However, polymerization mechanisms proposed for both groups of catalysts include π-allylic complex formation as an intermediate.

In general, nickel salt catalyst systems consist of (1) an organic complex compound of nickel, (2) a Lewis-type acid (eg, boron trifluoride etherate) (96–99), and (3) an organometallic compound. The most notable example is the ternary system R_3Al–Ni octanoate–BF_3·etherate (96). The polybutadiene produced with this catalyst has a better than 98% cis-1,4-content and possesses a high molecular weight, a broad molecular weight distribution, and no substantial gel content. The catalyst efficiency for nickel is high, yielding 10^3–10^4 g of polybutadiene per gram of nickel. As in cobalt systems, chain transfer is the main mechanism of polymer chain termination. Polymerizations can be carried out in aromatic or aliphatic solvents.

Fluoride is essential in these systems for preparing high molecular weight, high-cis-polybutadiene, and many fluorine-containing organic and inorganic compounds, including HF (100), have been successfully used as the Lewis acid. Likewise, Li alkyls have been used to replace aluminum alkyls as the organometallic component (101–102).

All three components of these ternary systems are indispensable for catalytic activity. The order of addition of the components, preparation temperature, aging history, etc, greatly influence the activity of the resulting catalyst systems.

The second main group of nickel catalysts involve a unique class of hydrocarbon derivatives of nickel. They are classified as π-allyl complexes and involve the polycentric bonding of the delocalized π electrons of three carbon atoms with suitable bonding orbitals in the nickel atom (103–108).

The π-allylnickel halides have shown catalytic activity with butadiene. Polymer configuration depends upon the halide within the complex. With the chloride, a cis polymer is obtained, whereas the bromide and iodide give trans configurations (106). The trend to form the trans polymer apparently increases with decreasing electronegativity of the halogen used. Speculation as to the reason for this effect range from steric considerations to polar effects, although conclusive experimental evidence is lacking (109).

Various parameters affect the system. For example, in the presence of Lewis acids the π-allylnickel bromide gives a high-cis-polymer. When wet, π-allylnickel bromide yields a trans configuration, when dry, 60% cis forms (106). This is consistent with the observation that even small concentrations of any Lewis base favor formation of high-trans product. The choice of solvent can also influence activity and stereospecificity in a similar manner.

When the chlorine of π-allylnickel chloride is replaced by a halogen-substituted carboxylic acid, the resulting product polymerizes butadiene to high-*cis*-1,4-polybutadiene without the addition of cocatalyst (110–111).

The counterion induction effect in these complexes may be responsible for stereospecificity, rate, and yield (111–112). By varying the ratio of acid to nickel, a 50/50 cis–trans mixture of isomers was obtained (113), implying a type of alternating propagation mechanism in which successive insertions of cis and trans units in the growing chain occur. With a 1:1 molar ratio of acid to nickel the cis isomer was formed, at a ratio greater than 1:1 a near constant 50/50 cis–trans isomer was reported.

Many examples of polymerization using nickel in a reduced state are reported for binary catalyst systems. π-Complexes of cyclooctadienyl and bis(cyclooctadienyl)nickel(O) with metallic halides, acidic metal salts, and protonic acids have yielded highly stereospecific 1,4-polybutadienes (109,112,114–116).

Bis(cyclopentadienyl)nickel in association with various Lewis acids polymerizes butadiene to stereospecific products dependent upon the nature of the acid (114,116–117). A bimetallic complex is proposed in which the halogens act as bridging groups. Addition of butadiene to this complex is believed to result in a π-allyl system.

The reaction of π-crotylnickel halides with Lewis acid (B, Al, Ti, V, Mo, W halides, etc), yields products insoluble in hydrocarbons. These complexes initiate polymerization of butadiene to a high-cis-1,4 structure, regardless of the nature of the halogen used (117–121). The catalyst activity of these reaction products is higher than for the corresponding π-crotylnickel halides alone.

The π-allylic type bonds with nickel are not only instrumental in catalyst formation, but are believed to be the basis of the propagation mechanism in butadiene polymerizations (107), with butadiene π-bonded to nickel as an intermediate in the propagation step. The effect of solvent and counterions on this mechanism are by no means clear. It is known, however, that counterions are very important in determining the polymer microstructure as proved by the necessity of having fluoride or fluoride-containing ligands to prepare high-*cis*-polybutadiene. The electron affinity of the counterion can directly influence the d-electron density of the nickel; the more electron withdrawing the counterion, the higher the cis content of the polymer (122).

Other High-Cis-Producing Catalysts. Both cerium and uranium catalyst systems have been shown recently to give high-*cis*-1,4-polybutadiene (up to 99% cis).

A typical cerium catalyst system (123) consists of a cerium compound (eg, the octanoate salt), together with a mixture of an aluminum alkyl and an alkyl aluminum halide. The microstructure of the resulting polymers appears to be independent of the catalyst components' concentrations and the nature of the halogen attached to the aluminum.

Tris(allyl)uranium halides (124) and uranium alkoxides (125) have both been successfully used to prepare high molecular weight *cis*-1,4-polybutadiene (up to 99% cis). The microstructure of the polymer is again independent on which halogen is used. Activity of these catalysts can be enhanced by the addition of Lewis acids (eg, $TiCl_4$) (126).

Trans-1,4-Polybutadiene. Crystalline high-*trans*-1,4-polybutadienes (mp = 145°C) have been prepared with catalysts containing titanium, vanadium, chromium, rhodium, iridium, cobalt, and nickel. High-*trans*-1,4-polybutadiene was first prepared in 1955 (127) with a heterogeneous catalyst containing $(C_2H_5)_3Al$ and α-$TiCl_3$. Other heterogeneous catalysts, displaying similar activity, were prepared with $TiCl_4$, TiI_4, VCl_4, VCl_3, and $VOCl_3$. The highest content trans-1,4-polymers, however, were prepared with homogeneous vanadium catalysts (128).

Among the heterogeneous catalysts, VCl_3 shows the highest stereospecificity, giving 98–99% trans-1,4 content (129). The microstructure in this system is unaffected by the Al–V molar ratio; VCl_4 and $VOCl_3$ (130) give a 94–98% trans-1,4-polymer, containing a 10% amorphous fraction. Good polymerization activity is obtained only at relatively high vanadium halide concentrations, 1–2 wt % with respect to butadiene. Polymerizations are carried out at 20–50°C at Al–V molar ratios of 0.5–10.

Soluble vanadium catalysts can be prepared using an excess of $(C_2H_5)_2AlCl$ in conjunction with a soluble vanadium compounds, such as vanadium tri(acetylacetonate). The formation of low molecular weight polybutadiene with these catalysts can be avoided by adding the organoaluminum compound slowly to the vanadium compound during catalyst preparation (130).

The titanium halides are less stereospecific than the corresponding vanadium halides. The microstructure of the polybutadiene formed with the $(i\text{-}C_4H_9)_3Al$–$TiCl_4$ catalyst is a function of the Al–Ti molar ratio. At Al–Ti ratios below 1, a high-trans-1,4-polymer is obtained. At ratios greater than 1, a mixture of cis-1,4 (60–90%) and trans-1,4 structures are obtained (131–132).

The various forms of $TiCl_3$ (α, β, and γ) have also been prepared separately and used in conjunction with organoaluminum compounds as catalysts. The β-$TiCl_3$ gives predominantly cis-1,4-polymer with R_3Al or R_2AlCl, but trans-1,4-polymer with $RAlCl_2$ (131). The α and γ forms of $TiCl_3$ give low yields of high-trans-1,4-polymer with R_3Al (134).

In recent years, rhodium and nickel complexes have been used successfully to prepare 99–100% *trans*-1,4-polybutadiene in polar media, an example being aqueous emulsions (135–138). The products, in general, are low molecular weight polymers.

Crystalline trans-1,4-polybutadienes (97–99% trans) have also been prepared with nickel and cobalt catalysts. π-Allylnickel iodide polymerizes butadiene to a 95–97% trans-1,4-polymer (139). Likewise, soluble cobalt compounds and $(C_2H_5)_2AlCl$ with $N(CH_3)_3$ or $N(C_2H_5)_3$ modification result in trans-1,4-polymers (140).

1,2-Polybutadiene. The preparation of crystalline syndiotactic 1,2-polybutadiene (mp = 156°C) was first reported in a patent in 1955 (141), using homogeneous catalysts consisting of an aluminum alkyl, together with an acetylacetonate or alcoholate complex of titanium, vanadium, or chromium. Other halogen-free complexes of these metals likewise produced the syndiotactic polymer.

With halogen-free vanadium complexes, eg, vanadium tri(acetylacetonate), polymers having 78–86% vinyl content can be prepared. This vinyl content can be increased to greater than 90% by employing higher Al–V molar ratios (3–30) and/or by aging the catalyst (142).

Catalysts containing halogen-free titanium complexes, eg, $Ti(OR)_4$ and $Ti(NR_2)_4$, with an aluminum alkyl are slightly less stereospecific than the vanadium catalysts, giving an amorphous type polymer of greater than 85% 1,2 content (143). The absence of halogen appears to be a prerequisite to obtaining high vinyl polymer using vanadium and titanium catalysts.

Catalysts based upon molybdenum compounds, such as molybdenum tri(acetonylacetate), and an aluminum alkyl cocatalyst, give 80% syndiotactic 1,2-polybutadiene. Other cocatalysts, including organogallium compounds and lithium aluminum hydride give similar results (144–145).

Cobalt compounds, in the presence of a small amount of water and organoaluminum cocatalysts, such as R_3Al or $[Al(C_2H_5)_2]_2O$, give syndiotactic 1,2-polymer (146). This is in contrast to the previously discussed R_2AlCl–Co compound catalyst system, which gives cis-1,4-polymer. The explanation for the influence of halogen on microstructure is still uncertain.

Isotactic 1,2-polybutadiene (mp = 126°C) has been prepared with homogeneous catalysts obtained by the reaction of aluminum alkyl with several soluble compounds of chromium, such as the acetylacetonate, $Cr(CNR)_6$, chromium hexacarbonyl and others (147–149). Polymer conversion for these systems is low.

The microstructure of the polymer depends upon both the Al–Cr molar ratio and the time of aging. At lower Al–Cr ratios, the unaged catalyst gives microstructures (70–90% 1,2 structure), containing predominantly syndiotactic 1,2-polymer (148). At higher ratios the isotactic isomer predominates. Polymer conversion and microstructure are also dependent upon the nature of the chromium compound.

Atactic 1,2-polybutadiene can be prepared with alkali metal catalysts in the presence of electron donors, such as ethers or amines.

Processing and Curing

The processing characteristics of polybutadiene are influenced by the polymer microstructure, molecular weight, molecular weight distribution, and degree of branching (3–4,6,150–156). The polymer undergoes mastication, mixing, molding, and curing. A Banbury mixer and a roll mill are employed for mastication and mixing, a calender and extruder for molding.

Most polybutadienes are highly resistant to breakdown and have poor millbanding characteristics and rough extrusion appearance compared to SBR elastomers. The solution polybutadienes process satisfactorily when blended with other elastomers such as SBR. Emulsion polybutadiene processes better than solution polymer, but not as well as SBR, and is commonly blended with other elastomers for enhanced processing.

560 ELASTOMERS, SYNTHETIC (POLYBUTADIENE)

Certain chemical peptizers slightly increase breakdown and improve processing (3). A lower Mooney viscosity also improves processing, but may lead to cold-flow (4,6) problems. In addition, a broad molecular weight distribution and branching (152) both improve milling and extrusion behavior as compared to a linear polymer. Excessively high molecular weight polybutadiene tends to crumb on the roll mill.

Very high-*cis*-polybutadiene prepared with a uranium catalyst is reported (125) to have greatly improved millability, calenderability, tack, green strength, and adhesion to fabric. Such processability would eliminate the need for blending with other elastomers. The mill processabilities of polybutadienes comprised of variable microstructures and catalyst is presented in Table 3. The processability improves with increasing cis content. The high-cis-polymer also had the highest modulus (300%) and

Table 3. Processability of Poly(butadiene)s[a]

Catalyst	Composition,[b] %	25°C Raw polymer	25°C Black stock	40°C Raw polymer	40°C Black stock	60°C Raw polymer	60°C Black stock	75°C Raw polymer	75°C Black stock
uranium	98.5/1.0/0.5	good	good	good	good	good	good	good	fair
nickel	97.0/2.0/1.0	good	good	good	good	good	fair	fair	fair
cobalt	97.0/1.5/1.5	good	good	good	good	fair	fair	bad	bad
titanium	91.0/4.5/4.5	good	good	fair	fair	bad	bad	bad	bad
lithium	32.6/55.4/12.0	bad	bad	bad	bad	bad	bad	bad	bad

[a] Ref. 153.
[b] 1,4-*cis*/1,4-*trans*/1,2.

Table 4. Polybutadiene Curing Recipe

Ingredient	Parts by wt
Budene 1207	90.00
No. 1RSS	10.00
ISAF black	50.00
Circosol[a]	10.00
paraffin	2.00
Wingstay 100[b]	1.00
stearic acid	3.00
zinc oxide	3.00
Altax[c]	0.45
diphenylguanidine	0.75
sulfur	1.25

[a] Sun Oil Company.
[b] Goodyear Tire and Rubber Co.
[c] R. T. Vanderbilt Co.

Table 5. Polybutadiene Vulcanizate Properties

Property	Value
300% modulus, MPa[a]	707
tensile strength, MPa[a]	1666
elongation, %	540
Shore A hardness	59
tear strength, die C, MPa[a]	508

[a] To convert MPa to psi, multiply by 145.

tensile and tear strengths. Branched medium vinyl-polybutadienes (35–55% 1,2-content) process as well as blends of SBR and polybutadiene (153–157).

High-vinyl polybutadienes (>90% 1,2-content) are thermoplastics and are processed by the corresponding technology, eg, extrusion and blown-film methods (157).

Curing and compounding recipes are chosen by the individual manufacturer. Polybutadienes are cured employing conventional sulfur recipes or peroxide systems. Normally the polymer is blended with another elastomer, then mixed with filler (eg, HAF or ISAF blacks), typically an aromatic processing oil, wax, antioxidant, antiozonant, and curing ingredients at a later stage. For curing, thiazoles and sulfenamides are generally employed in combination with secondary accelerators or alone. Cure rates are close to SBR using similar loads of sulfur. A polybutadiene recipe for test curing employed by Goodyear (158) is given in Table 4. Typical vulcanizate's properties are given in Table 5.

The vulcanizate properties of a medium vinyl-polybutadiene–SBR blend, SBR, and an SBR–*cis*-polybutadiene blend are compared in Table 6. A test recipe for tire treads was employed, since the primary use of these blends is in tread stocks.

The trend in Table 6 shows the normal pattern based on polymer composition. The blend of SBR with *cis*-polybutadiene has significantly improved abrasion resistance and tear resistance, but the skid resistance is slightly decreased. The SBR–medium-vinyl-polybutadiene blend has enhanced blow-out resistance, improved abrasion, and decreased skid resistance. Elastomers with high mobility and low glass transition temperatures have high elasticity, very good abrasion, and poor skid and vice versa.

Medium-vinyl-polybutadiene may replace SBR partially or fully and has only recently become commercially important to the rubber industry. It could compensate

Table 6. Properties of Polybutadiene Blends[a]

	A	B	C
components, parts by wt			
45% vinyl PBD	68		
cis-4 1203			35
SBR 1712	32	100	65
styrene	7.5	23.5	15.3
vulcanizate properties			
300% modulus, MPa[b]	919	919	990
tensile strength, MPa[b]	2040	2090	1989
elongation, %	560	550	510
Shore A hardness	59	58	59
ΔT, °C at 15 minutes	42	42	39
Yerzley resilience, %	58	55	59
blow-out time, min	>60	17	15
Crescent tear 100°C, MPa[b]	374	374	444
abrasion index			
12875 km	106	100	137
wet skid index			
40 km/h	102	108	104

[a] Ref. 153.
[b] To convert MPa to psi, multiply by 145.

Table 7. Commercial Solution-Polymerized Polybutadienes

Producer	Location	Capacity, t/yr	Trade name	Type	Catalyst
North America					
American Synthetic Rubber Corp.	Louisville, Ky.	50,000	Cisdene	high-cis	titanium
Armstrong Rubber Co.	Borger, Tex.	18,000		high-cis	
Firestone Synthetic Rubber & Latex Co.	Lake Charles, La. Orange, Tex.	85,000	Diene	medium-cis	alkyllithium
General Tire & Rubber Co.	Borger, Tex.	37,000	Duragen	high-cis	
B. F. Goodrich Co., Chem. Div.	Orange Tex.	77,000	Ameripol CB	high-cis	cobalt
Goodyear Tire & Rubber Co.	Beaumont, Tex.	115,000	Budene	high-cis	
Phillips Chemical Co.	Borger, Tex.	15,000	Cis 4 Solprene	high-cis medium-cis	titanium
Texas-U.S. Chemical Co.	Port Neches, Tex.		Synpol E-BR	medium-cis	emulsion
Polysar, Ltd.	Sarnia, Ontario		Taktene	high-cis	
Latin America					
Coperco	Recife, Brazil	15,000	Coperflex	medium-cis	
Negromex S.A.	Salamonica, Mexico	20,000	Solprene	medium-cis	
Europe					
I.S.R. Ltd.	Grangemouth, Scotland	40,000	Unidene	medium-cis	alkyllithium
Petrochim N.V.	Antwerp, Belgium	20,000	Solprene	medium-cis	
Firestone France S.A.	Port Jerome, Fr.	50,000	Diene	medium-cis	
Michelin & Co.	Bassens, Fr.	50,000		high-cis	
Shell Chemie, S.A.	Marseille, Fr.	50,000	Cariflex BR		
Bunawerke Huels G.m.b.H.	Huls, Dermagen, FRG	70,000	Buna CB	high-cis high-cis	cobalt titanium
Anic S.p.A.	Ravenna, Italy	45,000	Europrene Cis	high-cis medium-cis	titanium alkyllithium
Calatrava	Santander, Spain	10,000	Solprene	medium-cis	
Potkim Kavcuk A.S.	Izmit, Turkey	13,000	Pet Cis		
Oceania					
Australian Synthetic Rubber Co., Ltd.	Melbourne, Australia	10,000	Austrapol CB	high-cis	
Phillips Australia Chem. Pty., Ltd.	Kurnell, Australia	10,000	Solprene	medium-cis	
Asahi Chemical Industry Co., Ltd.	Kawasaki, Japan	50,000	Tufdene	medium-cis	
Japan Elastomer Co.,Ltd.		10,000	Solprene	medium-cis	
Japan Synthetic Rubber Co., Ltd.	Yokkaichi, Japan	75,000	JSR Br	high-cis	
Nippon Zeon Co., Ltd.	Tokuyama, Japan	35,000	Nipol BR	high-cis	
Ube Industries, Ltd.		40,000	Ubepol BR		

a Estimated year end 1976 capacities.

for styrene, should a shortage of this monomer occur in the petrochemical industry. This would inevitably result in a price increase for butadiene monomer, since it would be in demand for both the existing conventional, as well as the new general purpose medium-vinyl-polybutadiene rubbers.

Economic Aspects and Uses

Both SBR and polybutadiene consumption are significantly affected by fluctuations in the tire industry and markets. Over 85% of polybutadiene production and 65% of total synthetic rubber consumption in the United States is used in the manufacture of passenger car, truck, and bus tires. Conventional passenger tires contain approximately 25% polybutadiene of the total blended rubber content.

The use of polybutadiene in the production of impact polystyrene has likewise grown in recent years. In 1974 approximately 42,700 metric tons of polybutadiene (ca 12% of domestic consumption) was used in this application. These polystyrene resins generally contain 3–5% polybutadiene. High impact polystyrenes may contain as high as 5–10% polybutadiene. End uses for these resins include toys, packaging, appliances, and many other applications.

In 1977 United States new rubber consumption reached 3,375,000 t, a 16% increase over 1976, a year of depressed markets. The share of synthetic rubber was 2,530,000 t, or 75% of the total, the remaining portion being natural rubber. In 1978 new rubber consumption decreased slightly to 3,300,000 t (2,480,000 t of synthetic rubber). Of this total production, 64.1% was for tire and tire-related uses. The long-range forecast for synthetic rubber consumption in North America in 1987 is 3,630,000 t (159). The relative market shares of natural and synthetic rubbers could change, if shifts occur in their market prices.

Table 7 gives manufacturers, location, and capacities of solution-polymerized polybutadiene. Figures for 1975 of world plant capacity for preparing solution polymerized polybutadiene indicate that the United States has the largest production capacity (29%), followed by Japan (21%), the USSR (14%), and France (11%).

Table 8 gives the world production of synthetic rubbers containing polybutadiene (160).

Table 8. United States and World Production of New Rubber, Thousands of Metric Tons

	1973 U.S.	1973 World	1974 U.S.	1974 World	1975 U.S.	1975 World	1976 U.S.	1976 World	1977[a] U.S.	1978[a] U.S.
styrene–butadiene rubber	1536	3452	1465	3228	1179	2615	1296	3214	1517	1500
polybutadiene	337	796	310	769	290	666	336	795	400	380
polyisoprene	119	211	93	223	61	185	62	213	75	70
polychloroprene		313	163	318	144	291	109	326	115	115
ethylene–propylene	120	200	126	226	84	155	114	249	140	140
butyl	160	360	164	383	80	257	122	348	145	135
nitrile	84	194	88	205	55	133	78	192	73	70
other	270		88		47		58		65	65

[a] Estimates.

The price of polybutadiene has been affected dramatically in recent years by shortages in monomer supply, and increasing feedstock and production costs. Polybutadiene prices declined from 1961 to 1972 from 66 ¢/kg to 48–55 ¢/kg, depending upon grade. From 1976 to 1978, the price rose from 48–55 ¢/kg to 86 ¢/kg, with plastics-grade polybutadiene at a 3.3–4.4 ¢/kg premium above the general purpose rubber.

Safety Factors and Testing

Polybutadiene rubbers, in dry form, are nontoxic. Compounding ingredients, however, may be slightly toxic or cause dermatitis in frequent contact. Special consideration must be given to ingredients added to polybutadienes used for food packaging materials (161).

When incinerated, rubber may emit toxic sulfide-containing vapors. Old tires, exposed to the atmosphere for a long time, may ignite spontaneously, caused by oxidation.

In the United States, rubber producers may face government regulations in the near future, that may make it mandatory for industries to change production processes that currently use aromatic solvents, particularly benzene.

Standard testing methods for the evaluation of solution polymerized polybutadiene and oil-extended polybutadiene are given in ANSI/ASTM D3189-73 and D3484-76, respectively (162). These methods specify the standard materials, test formulas, mixing procedures, and test methods for evaluation and production control of these rubbers.

BIBLIOGRAPHY

1. G. S. Whitby, ed., *Synthetic Rubber,* John Wiley & Sons, Inc., New York, 1952.
2. E. Tornqvist in J. P. Kennedy and E. Tornqvist, eds., *Polymer Chemistry of Synthetic Elastomers,* John Wiley & Sons, Inc., New York, 1968, Chapt. 2.
3. R. S. Hanmer and M. E. Railsback in M. Morton, ed., *Rubber Technology,* Van Nostrand-Reinhold, New York, 1973, pp. 199–219.
4. J. F. Svetlik in G. G. Winspear, ed., *Vanderbilt Rubber Handbook,* R. T. Vanderbilt Company, Inc., New York, 1968, pp. 89–98.
5. W. Cooper in W. M. Saltman, ed., *Stereo Rubbers,* John Wiley & Sons, Inc., New York, 1977, Chapt. 2.
6. B. D. Babitskii and V. A. Krol in I. V. Garmonov, ed., *Synthetic Rubber* (in Russian), Khimiya, Leningrad, U.S.S.R., 1976, pp. 76–99.
7. S. V. Lebedev, *Zh. Russ. Fiz. Khim. Ova.* **42,** 949 (1910).
8. *Ibid.,* **45,** 1249 (1913).
9. A. Talalay and M. Magat, *Synthetic Rubber from Alcohol,* Interscience Publishers, Inc., New York, 1945, p. 187.
10. Ger. Pats. 570,980 (Feb. 27, 1933), W. Bock and E. Tschunker (to I. G. Farbenindustrie); 558,890 (Jan. 9, 1927), M. Luther and C. Henck (to I. G. Farbenindustrie).
11. U.S. Pat. 1,973,000 (Sept. 11, 1934), E. Konrad and E. Tschunkur (to I. G. Farbenindustrie).
12. A. A. Morton, E. E. Magat, and R. L. Letsinger, *J. Am. Chem. Soc.* **69,** 950 (1947).
13. A. A. Morton, *Rubber Age* **72**(4), 473 (1953).
14. K. Ziegler and co-workers, *Angew. Chem.* **67,** 541 (1955).
15. Belg. Pat. 543,292 (Dec. 2, 1955), (to Goodrich-Gulf).
16. F. W. Stavely, *Ind. Eng. Chem.* **48,** 778 (1956).
17. S. E. Horne, Jr., *Ind. Eng. Chem.* **48,** 785 (1956).
18. *Wall St. J. East. Ed.,* (Apr. 26, 1956).

19. *Chem. Eng. News* **55**(46), 14 (1977).
20. K. Greene and R. E. Pennington, *Chem. Eng. Process.* **73**(7), 36 (1977).
21. G. H. Stempel, *Polymer Handbook*, 2nd ed., Vols. 1–5, John Wiley & Sons, Inc., New York, 1975.
22. G. S. Trick, *J. Appl. Polym. Sci.* **3**, 253 (1960).
23. W. S. Bahary, D. I. Sapper, and J. H. Lane, *Rubber Chem. Technol.* **40**, 1529 (1967).
24. T. S. Dainton and co-workers, *Polymer* **3**, 297 (1962).
25. W. S. Richardson, *J. Polym. Sci.* **13**, 325 (1954).
26. T. E. Ferington and A. V. Tobolsky, *J. Polym. Sci.* **31**, 25 (1958).
27. I. I. Boldyreva, B. A. Dolgoplosk, and Ye. N. Kropacheva, *Dokl. Akad. Nauk SSSR* **131**, 830 (1960).
28. V. Ya. Bogomol'niy, B. A. Dolgoplosk, and K. G. Miyesserov, *Vysokomol. Soldin. Ser. B* **10**, 370 (1968).
29. E. Hart and A. J. Meyer, *J. Am. Chem. Soc.* **71**, 1980 (1949).
30. I. Kolthoff, T. Lee, and T. Mairs, *J. Polym. Sci.* **2**, 199 (1947).
31. W. S. Richardson, *J. Polym. Sci.* **13**, 229 (1954).
32. K. Ziegler, H. Grimm, and R. Willer, *Ann. Chem.* **542**, 90 (1940).
33. A. Abkin and S. Medvedev, *Trans. Faraday Soc.* **32**, 286 (1936).
34. C. E. Marvel, W. J. Bailey, and G. E. Inskeep, *J. Polym. Sci.* **1**, 275 (1946).
35. H. E. Adams and co-workers, *Rubber Chem. Technol.* **45**, 1252 (1972).
36. R. V. Basova and co-workers, *Dokl. Akad. Nauk SSSR* **149**(5), 1067 (1963).
37. C. E. Bawn, *Rubber Plast. Age* **42**, 267 (1961).
38. M. Morton, *Rubber Plast. Age* **42**, 397 (1961).
39. S. Bywater, *Pure Appl. Chem.* **4**, 319 (1961).
40. C. E. Bawn and A. Ledwith, *Q. Rev. (London)* **16**, 361 (1962).
41. I. Kuntz and A. Gerber, *J. Polym. Sci.* **42**, 299 (1960).
42. V. A. Krapachev, B. A. Golgoplosk, and N. I. Nicolaev, *Dokl. Akad. Nauk SSSR* **115**, 516 (1957); *Rubber Chem. Technol.* **33**, 636 (1960).
43. P. J. Flory, *Principles of Polymer Chemistry*, Cornell University Press, Ithaca, N.Y., 1953, p. 338.
44. A. A. Morton, *Ind. Eng. Chem.* **42**, 1488 (1950).
45. A. A. Morton, J. Nedilow, and E. Schoenberg, *Rubber Chem. Technol.* **30**, 326 (1957).
46. A. A. Morton and E. J. Lanpher, *J. Polym. Sci.* **44**, 233 (1960).
47. *Chem. Eng. News* **47**, 46 (1969).
48. W. Franke, *Kautsch. Gummi Kunstst.* **11**, 254 (1958).
49. G. Natta, *Chim. Ind. (Milan)* **42**, 1207 (1960).
50. G. Natta and co-workers, *Chim. Ind.* **41**, 398 (1959).
51. P. H. Moyer and M. H. Lehr, *J. Polym. Sci. A* **3**, 217 (1965).
52. D. Morero and co-workers, *Chim. Ind.* **41**, 758 (1959).
53. Brit. Pat. 824,201 (Dec. 31, 1956), (to Chem. Werke Hüls).
54. I. Ja. Poddubnyi, V. A. Grechanovsky, and E. G. Ehrenburg, *Makromol. Chem.* **94**, 268 (1966).
55. W. Cooper and co-workers, *J. Polym. Sci.* **50**, 159 (1961).
56. Belg. Pat. 551,851 (Oct. 17, 1956), R. P. Zelinski and D. R. Smith (to Phillips Petroleum Co.).
57. W. M. Saltman and T. H. Link, *Ind. Eng. Chem. Prod. Res. Dev.* **3**(3), 199 (1964).
58. Brit. Pat. 938,089 (Dec. 2, 1960), (to Phillips Petroleum Co.).
59. Ger. Pat. 1,112,834 (1961), (to Firestone Tire and Rubber Co.).
60. W. Marconi and A. Mozzei, *J. Polym. Sci. A* **3**, 735 (1965).
61. W. Marconi and co-workers, *Chim. Ind.* **45**, 522 (1963).
62. M. H. Lehr and P. H. Moyer, *J. Polym. Sci. A* **3**, 753 (1965).
63. *Ibid.*, p. 231.
64. Brit. Pat. 920,244 (Oct. 3, 1960), (to Phillips Petroleum Co.).
65. Brit. Pat. 931,440 (Apr. 7, 1961), (to Phillips Petroleum Co.).
66. W. Marconi and co-workers, *Chim. Ind.* **46**, 245 (1967).
67. U.S. Pat. 2,977,349 (Nov. 7, 1956), C. E. Brockway and A. F. Ekar (to Goodrich Gulf Chem. Inc.).
68. C. Longiave, R. Castelli, and G. F. Croce, *Chim. Ind.* **43**, 625 (1961).
69. Ital. Pats. 592,477 (Dec. 6, 1957), 588,825 (Dec. 24, 1957), C. Longiave, G. T. Croce, and R. Castelli (to Montecatini S.p.A.).
70. W. S. Anderson, *J. Polym. Sci. A* **1**, 429 (1967).
71. A. Takahasi and S. Kombara, *J. Polym. Sci. B* **3**, 279 (1965).
72. Ital. Pat. 587,968 (Feb. 11, 1958), G. Marullo, and co-workers, (to Montecatini S.p.A.).

73. Ital. Pat. 594,618 (Apr. 24, 1958), G. Marullo, and co-workers, (to Montecatini S.p.A.).
74. Belg. Pat. 575,671 (Feb. 13, 1958), H. Tucker (to Goodrich-Gulf Chemicals).
75. Brit. Pat. 906,053 (Feb. 28, 1958), A. J. Canale, J. G. Balos, and C. H. Wilcoxen, Jr (to DeBataafsche Petr. Maat.).
76. Ital. Pat. 587,976 (Feb. 26, 1958), G. Natta, L. Porri, and L. Fiore (to Montecatini S.p.A.).
77. Brit. Pat. 776,326 (June 5, 1957) (to E. I. du Pont de Nemours & Co., Inc.).
78. M. Gippin, *Ind. Eng. Chem. Prod. Res. Dev.* **1**, 32 (1962).
79. *Ibid.*, **4**, 160 (1965).
80. A. I. Diaconescu and S. S. Medvedev, *J. Polym. Sci. A* **3**, 31 (1965).
81. W. M. Saltman and L. J. Kuzma, *Rubber Chem. Technol.* **46**, 1066 (1973).
82. V. N. Zgonnik and co-workers, *Vysokomol. Soldin.* **4**, 1000 (1962).
83. B. A. Dolgoplosk and co-workers, *Dokl. Akad. Nauk SSSR* **135**, 847 (1960).
84. C. Longiave, R. Castelli, and M. Terraris, *Chim. Ind.* **44**, 725 (1962).
85. V. N. Zgonnik and co-workers, *Vysokomol. Svedin.* **7**, 308 (1965).
86. F. Engel and co-workers, *Rubber Plast. Age* **45**, 1499 (1964).
87. H. Scott and co-workers, *J. Polym. Sci. A* **3**, 3233 (1964).
88. J. G. Balas, H. E. De la Mare, and D. D. Schissler, *J. Polym. Sci. A* **3**, 2243 (1965).
89. Ital. Pat. 679,829 (Dec. 12, 1962), G. Nata, A. Carbonara, and L. Porri (to Montecatini S.p.A.).
90. D. E. O'Reilly and co-workers, *J. Polym. Sci. A* **2**, 3257 (1964).
91. Brit. Pat. 859,698 (Dec. 4, 1958), (to Goodyear Tire & Rubber Co.).
92. Brit. Pat. 906,052 (Nov. 6, 1957), J. G. Balas, L. M. Porter, and R. J. Renolds (to De Bataafsche Petr. Maat.).
93. R. N. Kovalevskaya and co-workers, *Polym. Sci. USSR* **4**, 414 (1963).
94. C. Longiave and R. Castelli, *J. Polym. Sci. C* **4**, 387 (1963).
95. P. Racanelli and L. Porri, *Eur. Polym. J.* **6**, 751 (1970).
96. U.S. Pats. 3,170,904, 3,170,905, 3,170,906, 3,170,907 (Feb. 23, 1965), K. Ueda and co-workers (to Bridgestone Tire and Rubber Co.); 3,178,403 (Apr. 13, 1965), K. Ueda and co-workers (to Bridgestone Tire and Rubber Co.).
97. S. Kitagawa and Z. Harada, *Jpn. Chem. Q.* **IV-1**, 41 (1968).
98. Brit. Pats. 905,099, 906,334 (Dec. 31, 1959), (to Bridgestone Tire and Rubber Co.).
99. T. Matsumoto and A. Onishi, *Kogyo Kagaku Zasshi* **71**, 2059 (1968).
100. M. C. Throckmorton and F. S. Farson, *Rubber Chem. Technol.* **45**, 268 (1972).
101. C. Dixon and co-workers, *Eur. Polym. J.* **6**, 1359 (1970).
102. E. W. Duck and co-workers, *Eur. Polym. J.* **6**, 1359 (1970).
103. G. Wilke and co-workers, *Angew. Chem. Int. Ed.* **5**, 151 (1966).
104. G. Wilke, *Angew. Chem. Int. Ed.* **2**, 105 (1963).
105. L. Porri, M. C. Gallazzi, and G. Vitulli, *Pol. Lett.* **5**, 629 (1967).
106. L. Porri, G. Natta, and M. C. Gallazzi, *J. Polym. Sci. C* **16**, 2525 (1967).
107. F. Dawans and P. Teyssie, *Ind. Eng. Chem. Prod. Res. Dev.* **10**, 261 (1971).
108. E. O. Fischer and H. Werner, *Metal π-Complexes,* Elsevier Publishing Co., New York, 1966.
109. F. Dawans and P. Teyssie, *C. R. Acad. Sci. Paris* **263C**, 1512 (1966).
110. B. A. Dolgoplosk and co-workers, *Bull. Acad. Sci. USSR Chem. Ser.* **9**, 429 (1967).
111. F. Dawans and P. Teyssie, *J. Polym. Sci. B* **7**, 111 (1969).
112. J. P. Durand, F. Dawans, and P. Teyssie, *J. Polym. Sci. B* **5**, 785 (1967).
113. J. P. Durand and P. Teyssie, *J. Polym. Sci. B* **6**, 299 (1968).
114. F. Dawans and P. Teyssie, *J. Polym. Sci. B* **3**, 1045 (1965).
115. J. P. Durand, F. Dawans, and P. Teyssie, *J. Polym. Sci. B* **6**, 757 (1968).
116. F. Dawans and P. Teyssie, *C. R. Acad. Sci. Paris* **261C**, 4097 (1967).
117. E. I. Tinyakova and co-workers, *J. Polym. Sci. C,* **16**, 2625 (1967).
118. B. D. Babitskii and co-workers, *Dokl. Akad. Nauk SSSR* **161**, 282 (1965).
119. *Ibid.*, 583 (1965).
120. B. D. Babitskii and co-workers, *Vysokomol. Soedin.* **6**, 2202 (1964).
121. B. D. Babitskii and co-workers, *Izv. Akad. Nauk. SSSR,* 1507 (1965).
122. T. Matsumoto and J. Furukawa, *J. Macromol. Sci. Chem. A* **6**, 281 (1972).
123. M. C. Throckmorton, *Kautsch. Gummi Kunstst.* **22**, 293 (1969).
124. A. de Chirico and co-workers, *Makromol. Chem.* **175**, 2029 (1974).
125. M. Bruzzone and co-workers, *Rubber Chem. Technol.* **47**, 1175 (1974).
126. G. Lugli and co-workers, *Makromol. Chem.* **175**, 2021 (1974).

127. Ital. Pat. 545,952 (Mar. 12, 1955), G. Natta, L. Porri, and M. Mozzanti (to Montecatini S.p.A.).
128. G. Natta, L. Porri, and A. Carbonaro, *Rend. Accad. Naz. Lincei* **31**(8), 189 (1961).
129. G. Natta and co-workers, *Chim. Ind.* **40**, 362 (1958).
130. G. Natta, L. Porri, and A. Mazzei, *Chim. Ind.* **41**, 116 (1959).
131. G. Natta and co-workers, *Chim. Ind.* **41**, 398 (1959).
132. G. J. van Amerongen, *Adv. Chem. Ser.* **52**, 136 (1966).
133. Brit. Pat. 827,365 (Dec. 2, 1954, Apr. 21, 1955), S. E. Horne, F. Gibbs, and E. J. Carlson (to Goodrich-Gulf Chemicals, Inc.).
134. G. Natta and co-workers, *Gazz. Chim. Ital.* **89**, 761 (1959).
135. R. E. Rinehart and co-workers, *J. Am. Chem. Soc.* **83**, 4864 (1961).
136. V. A. Kormer and co-workers, *Am. Chem. Soc. Polym. Prepr.* **7**, 548 (1966).
137. V. A. Kormer, B. D. Babitsky, and M. I. Lobach, *Adv. Chem. Ser.* **91**, 306 (1969).
138. J. Zachoval and B. Veruovic, *J. Polym. Sci. B* **4**, 965 (1966).
139. W. Cooper and co-workers, *Adv. Chem. Ser.* **52**, 46 (1966).
140. L. Porri, G. Natta, and M. C. Gallazzi, *J. Polym. Sci. C* **16**, 2525 (1967).
141. Belg. Pat. 549,544 (July 15, 1955), G. Natta and L. Porri (to Montecatini S.p.A.).
142. G. Natta and co-workers, *Chim. Ind.* **41**, 526 (1959).
143. G. Natta, L. Porri, and A. Carbonaro, *Makromol. Chem.* **77**, 126 (1964).
144. G. Natta, *Nucleus (Paris)* **4**, 97, 211 (1963).
145. Ger. Pats. 1,144,925, 1,124,699 (Mar. 6, 1958), (to Phillips Petroleum Co.).
146. E. Susa, *J. Polym. Sci. C* **4**, 399 (1964).
147. Ital. Pat. 563,507 (Apr. 19, 1956), G. Natta, L. Porri, and A. Palvarini (to Montecatini S.p.A.).
148. G. Natta and co-workers, *Chim. Ind.* **41**, 1163 (1959).
149. G. Natta, *Mat. Plast.* **1**, 3 (1956).
150. K. Ninomiya and G. Yasuda, *Rubber Chem. Technol.* **42**, 714 (1969).
151. G. Alliger and F. C. Weissert, in J. P. Kennedy and E. Tornqvist, eds., *High Polymers*, Vol. 33, Wiley-Interscience, New York, 1968, Pt 1, Chapt. 3.
152. N. Tokita and I. Pliskin, *Rubber Chem. Technol.* **46**, 1166 (1973).
153. H. E. Railsback and N. A. Stumpe, Jr., *Rubber Age* **107**(12), 27 (1975).
154. E. W. Duck, *Eur. Polym. J.* **155**, 38 (1973).
155. J. R. Haws, L. L. Nash, and M. S. Wilt, *Rubber Ind.* **9**, 107 (1975).
156. K. H. Nordsiek, *Polym. Age* **4**, 332 (1973).
157. Y. Takeuchi, A. Sekimoto, and M. Abe, *ACS Symposium Series #4, A Symposium by the Division of Organic Coatings and Plastics Chemistry at the 167th Meeting of the ACS*, Los Angeles, Calif., Apr. 1–2, 1974.
158. *Elastomeric Materials*, The International Plastics Selector, Inc., San Diego, Calif., 1977.
159. *Chem. Market. Rev.*, 5 (Feb. 14, 1977).
160. Elast. **109**, 1, 27 (Jan. 1977).
161. N. I. Sax, *Dangerous Properties of Industrial Materials*, 4th ed., Reinhold Publishing Corp., New York, 1975.
162. *1977 Annual Book of ASTM Standards*, ASTM, Philadelphia, Pa.

L. J. Kuzma
W. J. Kelly
The Goodyear Tire & Rubber Company

POLYETHERS

For many years polyethers, such as $+CH_2CH(R)O+_n$, were considered to be excellent candidates for elastomers in the expectation that the oxygen atom would contribute greatly to chain flexibility and, thus, enhance elastomeric behavior (1). Poly(propylene oxide) [25322-69-4] with its low cohesive energy between chains should be a superior elastomer (1) but the known methods of polymerizing propylene oxide (qv) gave only low molecular weight, liquid polymers, not the desired high polymer. Such liquid propylene oxide polymers could be made largely hydroxyl-ended and then converted to a rubbery polyether urethane by reaction with a di- or polyisocyanate. Low density rubber foams can be made from the propylene oxide adduct of a polyol which is then chain-extended and cross-linked during fabrication with di- or polyisocyanate (2). However, this article discusses only polyether elastomers of sufficiently high molecular weight to be processed and fabricated by conventional rubber equipment and which are then cross-linked (ie, vulcanized) in a separate step. The preparation of such polyether elastomers required the development of new catalyst systems, specifically coordination catalysts, for polymerizing propylene oxide and other epoxides (see Polyethers articles for nonelastomeric polyethers from various monomers).

In 1957, some especially effective coordination catalysts for polymerizing epoxides to high polymers were discovered (3). Some of these new catalysts were the reaction products of organic compounds of aluminum, zinc, and magnesium with water. Unusually versatile was the combination of organoaluminum compounds with water and acetylacetone. These new catalysts led to the discovery and development of two classes of commercial polyether elastomers: (1) the epichlorohydrin (ECH) elastomers (see Chlorohydrins) (4–8), specifically the ECH homopolymer, abbreviated CO by ASTM (9), and the ECH-ethylene oxide (ECH–EO) copolymer [abbreviated ECO (9)]; and (2) the propylene oxide (PO) elastomers (10) (see Table 1). The epichlorohydrin elastomers were first introduced in 1965 under the trademark Hydrin, and later under the trademark Herclor. These elastomers are being made in quantities of thousands of metric tons per year.

In the early work on epoxide polymerization, high molecular weight, largely amorphous propylene oxide-unsaturated epoxide copolymers were made and their potential value as improved elastomers was recognized (10). Subsequently, properties of similar propylene oxide–unsaturated epoxide copolymer elastomers were reported (11). This new type of polyether elastomer (12–13) became commercially available under the trademark Parel in 1972. Parel 58 elastomer is a sulfur-curable copolymer of propylene oxide and allyl glycidyl ether (AGE) (see Allyl monomers and polymers).

The epichlorohydrin and propylene oxide elastomers are available only in dry form. At present the ECH elastomers are manufactured only by Hercules and Goodrich. The Hercules plant is located in Hattiesburg, Mississippi; the Goodrich Plant in Avon Lake, Ohio.

Structure

Epichlorohydrin Elastomers. In polyepichlorohydrin (1) the asymmetric monomer units can exist either in random stereosequences (atactic polymer) or in stereoregular

Table 1. Properties of Commercial Polyether Elastomers

	Epichlorohydrin elastomers		Propylene oxide elastomer
	ECH homopolymer	ECO	Parel 58
structure	$+CH_2CHO+_n$ 　　　$\|$ 　　CH_2Cl	$+CH_2CHOCH_2CH_2O+_n$ 　$\|$ CH_2Cl	$+CH_2CHO+_n +CH_2CHO+_{0.025n}$ 　　$\|$　　　　　　$\|$ 　　CH_3　　　　　CH_2 　　　　　　　　　$\|$ 　　　　　　　　　O 　　　　　　　　　$\|$ 　　　　　　　　$CH_2CH=CH_2$
name	chloromethyloxirane homopolymer	chloromethyloxirane copolymer with oxirane	methyloxirane copolymer with [(2-propenyloxy)methyl] oxirane
CAS Registry No.	[24969-06-0]	[24969-10-6]	[25104-27-2]
epichlorohydrin	100	68	
ethylene oxide, %	0	32	
chlorine, %	38.4	26	
specific gravity	1.36	1.27	1.01
Mooney viscosity of raw polymer at 100°C	48[a]	50–140	75
brittleness, temperature, °C	−18	−40	

[a] Reduced specific viscosity of about 1.4–1.6 (0.1%, α-chloronaphthalene at 100°C), corresponding to a weight average molecular weight of about 500,000 and a Mooney viscosity (ML-4 at 100°C) of ca 50.

sequences, such as isotactic polyepichlorohydrin where all the monomer units in a given chain have the same configuration.

$$\left(-CH_2-CH-O-\right)_{\overline{n}}$$
$$|$$
$$CH_2Cl$$

(1)

The units can be arranged head-to-head, tail-to-tail, and as is normally anticipated, head-to-tail. Commercial amorphous polyepichlorohydrin, Herclor H elastomer, is more than 97% head-to-tail (14). Stereoregular isotactic, crystalline polymer can be easily detected by conventional methods such as x-ray spectrometry or differential scanning calorimetry (dsc) or by isolating the crystalline fraction from the amorphous material by a low temperature crystallization from a solvent such as acetone (15) (see Polyethers). Commercial polyepichlorohydrins usually contain low levels of crystalline homopolymer since the amorphous rubber is preferred for most applications.

Propylene Oxide Elastomers. Propylene oxide homopolymer has, in principle, the same type of structural characteristics as polyepichlorohydrin. Many coordination catalysts give 20–30% head-to-head and tail-to-tail polymerization with propylene oxide (16). Presumably, the commercial product contains substantial amounts of such sequences, and by design, only small amounts of crystallinity. The presence of the unsaturated comonomer units also serves to reduce chain regularity and, hence, crystallinity.

Properties

The properties of epichlorohydrin and propylene oxide elastomers are listed in Table 1.

Epichlorohydrin Elastomers. The ECH homopolymers are essentially amorphous which contributes to their good processing and vulcanizate properties. The high chlorine content is responsible for their flame retardancy, which is somewhat lower for epichlorohydrin–ethylene oxide copolymer. A copolymer with a reduced specific viscosity of about 5.0 (about 100 Mooney) has a weight-average molecular weight of approximately 2.8 million. In general, even ECH–EO copolymers with very high Mooney viscosity exhibit fair processability with carbon black and mineral pigment filler.

The properties of the ECH homopolymer and copolymer vulcanizates are given in Table 2. These elastomers have generally good tensile properties, differing mainly in their resilience and low temperature brittleness. The homopolymer is a dead or low-resilience rubber at room temperature and, like most elastomers, its resilience improves with increasing temperature. At room temperature, ECO is highly resilient and its resilience is relatively unaffected by temperature, an unusual property for oil-resistant elastomers. Gum-vulcanizate tensile properties are generally poor.

The vulcanized homopolymer and ECO have equally excellent resistance to hot-oil and good resistance to solvents such as water and perchloroethylene (widely used in drycleaning), and to acids and bases. Heat-aging and ozone resistance are generally superior to neoprene and the nitrile rubbers. These elastomers, when properly stabi-

Table 2. Vulcanizate Properties of Polyether Elastomers

Property	Herclor H, homopolymer	Herclor C, copolymer	Parel 58
Mooney scorch at 121°C			
minutes to three-point rise			10
minutes to five-point rise	10	7	
100% modulus, MPa[a]	9.0	5.6	4.0
200% modulus, MPa[a]	15.5	11.4	8.5
300% modulus, MPa[a]		14.3	
tensile strength, MPa[a]	15.9	15.0	14.3
ultimate elongation, %	230	320	330
Shore A hardness	82	74	69
low temperature brittleness			
T_b, °C (ASTM D-746)	−18	−40	
low temperature stiffness			
$T_{10,000}$ (ASTM D-1053), °C			−90
Lupke rebound			
23°C, %	26	67	
100°C, %	75	63	
Bayshore resilience, %	9	36	48
Graves tear at 27°C			
(ASTM D-624), kN/m (lb/in.)	36 (204)	43 (247)	
volume change after immersion, %			
after 70 h			
ASTM oil No. 1 at 150°C		0	
ASTM oil No. 3 at 150°C	4	8	
ASTM fuel A, 23°C	0	1	50
ASTM fuel B, 23°C	8	11	140
H_2O at 23°C	0	5	4
H_2O at 100°C	+10	+10	5

[a] To convert MPa to psi, multiply by 145.

lized and vulcanized can be exposed for long periods of time to a temperature of 150°C. In long-term aging tests (17) the copolymer is reported to be only slightly inferior to heat-resistant EPDM compounds (see under Ethylene–propylene rubber) whereas the homopolymer is reported to be actually superior. The epichlorohydrin elastomers do not fail by embrittlement as most elastomers do on aging but rather gradually lose tensile strength, whereas elongation and hardness tend to level off.

Epichlorohydrin–ethylene oxide copolymer offers the oil resistance of nitrile rubbers (qv) together with the low temperature flexibility, flame resistance, and elastomeric properties similar to those of neoprene. Because of this excellent balance of properties combined with good heat-aging and ozone resistance, ECO has become the most widely used product of the ECH elastomer series. Both ECO and the homopolymer have good abrasion and tear resistance but generally poor electrical properties, as is characteristic of oil-resistant rubbers in general.

The homopolymer exhibits low air permeability (more than twofold lower than butyl rubber). It has excellent building tack, both dry and when freshened with a suitable solvent. The ethylene oxide copolymer is not quite as good in building tack but this may be because of its higher molecular weight rather than an inherent difference. The homopolymer also has somewhat better heat-aging resistance, ozone resistance, and flame resistance than ECO.

The ECO elastomer (Herclor C or Hydrin 200) is prepared with a one-to-one epichlorohydrin–ethylene oxide composition (2) (68 wt % epichlorohydrin) to obtain

$$+\!\!-\!\mathrm{CH_2}\!-\!\mathrm{CH}\!-\!\mathrm{O}\!-\!\mathrm{CH_2}\!-\!\mathrm{CH_2O}\!-\!\!\!\!\overline{}_n$$
$$\phantom{+\!\!-\!\mathrm{CH_2}\!-\!}|$$
$$\phantom{+\!\!-\!\mathrm{CH_2}\!-\!}\mathrm{CH_2Cl}$$

(2)

the optimum properties. This copolymer is completely amorphous and not sensitive to water. Copolymer can be made over the entire composition range. However, copolymers high in ethylene oxide typically either dissolve or swell in water after cross-linking, whereas copolymers high in epichlorohydrin can be crystalline because of isotactic sequences of epichlorohydrin and, thus, have less rubbery properties. Although these copolymers are intended to be uniform, the nature of the coordination catalysis often leads to some nonuniformity. Indeed, the 1:1 copolymer may even contain a small amount of water-soluble fraction which is very high in ethylene oxide. This type of copolymerization behavior (18) appears to be due to the presence of highly hindered sites in the coordination catalyst which permits easy access to ethylene oxide but not to the more hindered epichlorohydrin. The small amount of nonuniformity present, however, does not appear to have any known adverse effect on the use properties of these commercial products.

A valuable feature of the epichlorohydrin elastomers is their compatibility with each other. Blends of the two have intermediate properties offering wide latitude in the final composition.

The low temperature flexibility of the homopolymer can be improved by using plasticizers (19), blending with copolymer, or covulcanization with a diene rubber such as polybutadiene.

Propylene Oxide Elastomers. The vulcanizate properties of Parel 58 elastomer are shown in Table 2. The general tensile properties of black-filled vulcanizates are good and in the typical range of most large-volume elastomers. Detailed properties are reported in references 11 and 20. Abrasion resistance in an actual road tire test is poor (20). However, based on its favorable tear and resilience properties, this elastomer could be used as a tire casing rubber, although adhesion to other elastomers would have to be improved.

The outstanding properties of Parel 58 elastomer vulcanizates are high resilience, and excellent low temperature flexibility and flex life. In these properties they resemble natural rubber. In addition, Parel 58 elastomer offers moderate oil resistance and excellent resistance to ozone and heat, being superior to natural rubber in these characteristics. Gum vulcanizate tensile properties, however, are generally poor so that a reinforcing filler must be used. Good gum tensile properties can be obtained with a partly stereoregular, crystalline PO–AGE elastomer (21). However, such strain-crystallizable copolymers show larger energy losses at large strains than the strain-crystallizing neoprene and natural rubber.

Because of its effectiveness, cost and availability, allyl glycidyl ether is a desirable comonomer for sulfur vulcanization. A wide variety of unsaturated epoxide comonomers have been studied with propylene oxide (12–13).

Molecular Weight and Solution Behavior. *Epichlorohydrin Elastomers.* The weight-average ($\overline{M}_w$) molecular weight of amorphous polyepichlorohydrin can be determined from its dilute (0.1%) solution viscosity in α-chloronaphthalene at 100°C from the following relationship (15):

$$\log \frac{\eta_{sp}}{C} = \log [\eta] + 0.15 [\eta]C$$

$$[\eta] = 8.93 \times 10^{-5} \overline{M}_w^{0.73}$$

where C = concentration in g/100 mL, and $[\eta]$ = intrinsic viscosity.

Presumably, this relationship also gives an approximate molecular weight for ECO. Molecular weight distribution of these polymers is usually very broad, M_w/M_n, between 7 and 15; this is often typical for polymers prepared by coordination polymerization.

Other dilute solution studies of polyepichlorohydrin have been reported (22). Sedimentation constant (S) and intrinsic viscosity $[\eta]$ of benzene solutions are related to molecular weight (M) by the equations (23):

$$S = 0.145 \, M^{0.33} \text{ and } [\eta] = 0.155 \times 10^{-4} \, M^{0.89}$$

The unperturbed dimensions and the steric hindrance parameters of the homopolymer were determined from viscosity measurements in benzene (22). From measurements of homopolymer solubility in 39 solvents and intrinsic viscosity data in 19 solvents, the δd and δa parameters of the Hansen mode (24) were determined. The homopolymer is soluble in solvents (25) that have a solubility parameter (δ) between 8.6 and 9.8 and squared hydrogen-bonding parameter (γ^2) between 0.05 and 0.90 (26), or in solvents that have solubility parameters $\delta d \geq 7.9$, $\delta p \geq 5.5$, and $0.2 \leq \delta \leq 5.0$ (24).

Propylene Oxide Elastomers. Specific methods for determining the molecular weight of Parel 58 elastomer have not been reported. A dilute solution viscosity method (27) used for poly(propylene oxide) should be approximately applicable, ie, $[\eta] = 0.81 \times 10^{-4} \overline{M}_w^{0.85}$ where $[\eta]$ = intrinsic viscosity in benzene at 25°C and $\overline{M}_w$ = weight-average molecular weight. Solution property data have also been reported (28) for poly(propylene oxide). Based on intrinsic viscosity data and light scattering in hexane on fractions of polymer varying in degree of crystallinity, they show the following relationships:

$$\text{For hexane at 46°C, } [\eta] = 1.97 \times 10^{-4} \overline{M}_w^{0.67}$$

$$\text{For benzene at 25°C, } [\eta] = 1.12 \times 10^{-4} \overline{M}_v^{0.77}$$

The unperturbed dimensions of poly(propylene oxide) were studied from intrinsic viscosity and light scattering of fractions in a theta solvent, isooctane at 50°C (29). The value of $(\overline{r_0^2}/M)^{1/2}$ was 8.0 nm where $\overline{r_0^2}$ is the unperturbed mean square end-end distance of the polymer coils.

Polymer Preparation

Epichlorohydrin Elastomers. Epichlorohydrin self-polymerizes and copolymerizes with ethylene oxide by a coordination polymerization mechanism using an aluminum alkyl–water and aluminum alkyl–water–acetylacetone type catalyst systems (16), generally in hydrocarbon solvents. Solution polymerizations in benzene, toluene, or methylene chloride are generally used for both homopolymers and copolymers. Aliphatic hydrocarbons are generally not useful on a commercial scale because the polymer precipitates in a sticky form which is difficult to handle. Recently, a method of obtaining satisfactory slurries in aliphatic hydrocarbons has been reported (30). Bulk polymerization systems using special catalysts have been reported (31).

A mixture of high molecular weight, amorphous polyepichlorohydrin and crystalline, isotactic, high molecular weight polyepichlorohydrin can be prepared in ether diluent on a laboratory scale (15). The amorphous polymer is readily obtained by extraction with acetone. In commercial processes, the catalyst and diluent are designed to give a highly amorphous product and separation of crystalline material is not required.

In the coordination copolymerization of epichlorohydrin and ethylene oxide, ethylene oxide is about seven times more reactive than epichlorohydrin (18,32) which must be taken into account in order to prepare a uniform copolymer (33). Thus the monomer mixture needed to give the desired copolymer, ie, a 1:1 epichlorohydrin:ethylene oxide copolymer (68:32 wt %), requires an initial composition of 94% epichlorohydrin and 6% ethylene oxide. The monomer mixture is fed into the polymerization vessel at the rate at which the monomers are polymerized. Molecular weight can be controlled by the addition of carbonium ion precursors such as various organic halides, acid chlorides, anhydrides, etc (34).

After appropriate treatment to inactivate or remove the catalyst and, after the addition of a phenolic antioxidant, the polymer is isolated in dry form from solution by evaporation or steam stripping, as with other soluble polymers. Conditions of polymerization and isolation have to ensure a product essentially free of monomers and solvent.

The epichlorohydrin polymers have been prepared in latex form by emulsifying a solution of the polymer containing a small amount of nonsolvent to reduce viscosity, and then stripping the solvent (35). However, these elastomers are not commercially available in this form.

Propylene Oxide Elastomers. Propylene oxide is copolymerized with allyl glycidyl ether in a solution polymerization in an aliphatic, aromatic, or chlorinated hydrocarbon, in the presence of coordination catalysts such as the aluminum alkyl–water-acetylacetone (16), diethyl zinc-water (35–36), and complex cyanide catalysts (37), as well as others. The copolymer has the same composition as the monomer charge. Thus, complete conversion and a uniform copolymer is obtained. The commercial product, which is mostly amorphous, contains about 5% allyl glycidyl ether. Partly crystalline copolymer containing isotactic propylene oxide sequences are reported in the literature (18,38–41). Molecular weight is controlled by various carbenium ion precursors (34).

Poly(propylene oxide) is soluble in most organic solvents. The polymer is isolated in dry form as described above for the epichlorohydrin elastomers. Bulk polymerization is also feasible because of the perfect copolymerization. Specific approaches to low viscosity, slurry type polymerizations with isobutane as diluent have been reported in the patent literature (42–43).

Processing and Fabrication

Epichlorohydrin Elastomers. The epichlorohydrin elastomers can be processed and fabricated exceptionally well, and can be molded, calendered, and extruded. They are characterized by low mill shrinkage, excellent-to-good building tack, and excellent extrusion characteristics which require low extrusion temperatures and give very low die-swell. The highly fluid nature of these polymers at elevated temperatures gives superior moldings exhibiting excellent conformation, smooth surfaces, and negligible swell and shrinkage after vulcanization. These properties are valuable in the precision molding of intricate rubber parts.

The copolymer is available in a range of Mooney viscosity grades from 50 to 140. The lower viscosities exhibit the best processing and fabrication properties. These elastomers can be cured in open steam, and usually bond well to metals and many other rubbers.

The epichlorohydrin elastomers can be compounded by accepted techniques on conventional compounding equipment, and are easily processed in a Banbury mixer. Some difficulties have been experienced during the initial stages on two-roll milling prior to carbon black addition. Small amounts [0.75–2.0 parts by wt per 100 parts rubber (phr)] of antisticking agents such as stearic acid, sorbitan monostearate (Span 60, ICI America), zinc stearate, or a long-chain fatty amine provide excellent mill roll processability (see Abherents). Detailed processing suggestions are available (44–45).

As with natural rubber, the breakdown rate of the epichlorohydrin amorphous homopolymer is rapid and increases with decreasing mixing temperature. The copolymer, on the other hand, shows a much lower rate of breakdown at comparable temperatures which increases with increasing temperature (see Rubber compounding).

Propylene Oxide Elastomers. Parel 58 elastomer also processes well and can be fabricated very easily by molding, calendering, and extrusion. Because of its excellent molding qualities, intricate mold designs can be used in compression, transfer, or injection type curing equipment. It is readily compounded by accepted techniques on conventional compounding equipment. Mill compounding should be started on a cold mill in contrast to the ECH elastomers where a warm mill is preferred.

Compounding

Epichlorohydrin Elastomers. *Curing Systems.* Epichlorohydrin elastomers are cured or vulcanized (cross-linked) by adding reagents that can react difunctionally on the chloromethyl side chains of these polymers via nucleophilic type substitution reactions (see Table 3). In addition, a suitable acid acceptor is necessary for best vulcanization, generally to accept by-product HCl and sometimes to catalyze other aspects of the curing reaction. Acid acceptors such as calcium stearate, zinc stearate, basic lead carbonates, metal oxides such as lead oxides, and basic lead salts such as phthalates, stearates, etc, can be used under various conditions depending on the particular curative (7). Curatives include hexamethylenediamine or its carbamate (Diak No. 1), 2-mercaptoimidazoline (ethylenethiourea, NA-22), trimethylthiourea, finely divided urea, and even simple ammonium salts.

Table 3. Vulcanizate Formulations[a], Parts by Wt

Ingredient	Herclor H homopolymer[b]	Herclor C copolymer[b]	Parel 58[c]
polymer	100	100	100
FEF carbon black	50	50	
HAF carbon black			50
stearic acid	1.0	0.75	1.0
red lead	5.0	5.0	
nickel dibutyldithiocarbamate	1.0	1.0	1.0
2-mercaptoimidazoline (NA-22)	1.5	1.5	
zinc oxide			5.0
tetramethylthiuram disulfide			1.5
mercaptobenzothiazole			1.5
sulfur			1.25

[a] Cured 30 min at 160°C.
[b] From ref. 46.
[c] From ref. 47.

The NA-22 curing system has been the most widely used curative to date. It gives a good level of scorch resistance, about 7 to 8 minutes at 121°C, good initial properties, and a very good level of heat resistance. An acid acceptor stabilizer is required and red lead provides the best balance of properties. The NA-22 curing system responds to some nonlead-type stabilizers such as zinc oxide combined with pentaerythritol, which gives increased scorch resistance and good resistance to heat aging in air. Excellent physical properties can be obtained in silica-filled compositions such as Cab-O-Sil or Hi-Sil 233 by use of a combination of calcium stearate with NA-22 curative (48).

Trimethylthiourea is a slower curing agent than Diak No. 1 and gives higher scorch resistance at 121°C. Like Diak No. 1, trimethylthiourea requires an acid acceptor stabilizer for cure activation and maximum heat-aging resistance. Dibasic lead carbonate (white lead) provides the best balance of properties with this cross-linking agent.

The curing mechanism of epichlorohydrin elastomers with thioureas is believed to be similar to that proposed for neoprene (49) in which an intermediate isothiuronium salt is formed that reacts further with a metal oxide to form a sulfur cross-link and a urea. The intermediate isothiuronium salts have been prepared under conditions similar to or milder than vulcanization (50).

Sulfur markedly enhances tensile strength and elongation of carbon black and silica-loaded epichlorohydrin elastomer compounds cured with NA-22. Sulfur, in combination with NA-22 and an ionic-type secondary accelerator such as di-o-tolylguanidine, gives good scorch resistance, equally high levels of tensile strength, and somewhat higher elongation than obtained with NA-22 alone. Such systems are useful for covulcanization of blends of epichlorohydrin elastomers with sulfur, curable elastomers such as polybutadiene, or natural rubber.

Hexamethylenediamine gives relatively fast cures and may be incorporated on a two-roll mill, provided the stock is kept cool. Cure occurs in a few days at room temperature. Curing cements may be made from the epichlorohydrin elastomers as compound dispersions in a ketone or aromatic solvent by adding the diamine just before use. Such cements remain usable for at least one day at room temperature.

A simple sulfur-curing system (51) consists of sulfur, potassium sebacate, and poly(ethylene glycol). This system appears to function because the sebacate is a stronger base in the presence of poly(ethylene glycol) and generates S^- from S.

A special grade of the epichlorohydrin elastomers which can be cured with sulfur or peroxide was introduced in 1977 under the trade name Hydrin 400 (52). This product is reported to give faster cures, less mold fouling and sticking, and better cure compatibility with conventional unsaturated elastomers such as SBR or nitrile rubber. It behaves like an epichlorohydrin elastomer containing unsaturated groups such as made by copolymerization with unsaturated epoxides (53).

Stabilizers and Antioxidants. A phenolic such as Santonox R or Topanol CA is added to commercial polymers for protection during storage, processing, and cure. Additional antioxidants (qv) such as Agerite Resin, nickel dibutyldithiocarbamate, copper dimethyl dithiocarbamate, phenothiazine, 2-mercaptoimidazole, or various phenolics are often added to the compounding system along with at least one effective acid acceptor stabilizer in order to enhance aging resistance of the vulcanizate. The acceptor stabilizers also prevent mold sticking and improve cure and vulcanizate physical properties as well as heat-aging resistance. Certain lead compounds can sometimes cause mold fouling (see Heat stabilizers).

Vulcanizates of the epichlorohydrin elastomers often are somewhat corrosive to metals because of the chloride ion generated in the usual cure reaction (54). Certain additives, such as barium dinonylnaphthalene sulfonate (54), Vanplast 201, and o-toluic acid reduce corrosion and mold fouling in NA-22 curing systems containing lead stabilizers (55).

Fillers. The tensile strengths of unfilled epichlorohydrin elastomer vulcanizates, like those of most amorphous synthetic rubbers, are low (less than 3.5 MPa or 500 psi). These polyether rubbers are reinforced by fillers, including a wide variety of carbon blacks (FEF, MT, SRF, HAF, and ISAF). By varying type and the amount (30–120 phr) of filler, the hardness of the vulcanizate can be varied from 60 to 90 (Shore) and the tensile strength from 7 to 20 MPa (1000 to 2900 psi) (56). Silica fillers such as Cab-O-Sil give outstanding tensile and other properties, and such compositions have been studied for thread-type applications (48) (see Fillers; Carbon, carbon black).

Plasticizers. Certain esters and ethers such as di(n-octyl) phthalate (DOP), Skydrol 500A, dioctyl sebacate, dioctyl adipate, TP90B (high molecular weight polyether) and TP95 (high molecular weight polyether-ester) significantly reduce Mooney viscosity, improve scorch resistance, and extend the range of low-temperature flexibility of epichlorohydrin vulcanizates (19) (see Plasticizers).

Propylene Oxide Elastomers. *Curing Systems.* The propylene oxide elastomers are readily cured with a variety of sulfur accelerator systems. Typical formulas shown in Table 3.

Stabilizers. A small amount of phenolic antioxidant such as Santonox R or Topanol CA are added for protection during storage, processing, and cure. In addition, phenolics, amines, or nickel dibutyldithiocarbamate are also added during compounding to enhance ozone and high-temperature-aging resistance.

Fillers. A wide variety of both carbon black and silica-type fillers can be used with propylene oxide elastomers in much the same way as with other sulfur-curable elastomers. A silica filler such as Cab-O-Sil gives outstanding tensile strength and physical properties such as used in thread-type applications.

Economic Aspects

The 1978 price of epichlorohydrin elastomers was $3.33/kg homopolymer and $3.11/kg ECO. The comparatively high price is caused by the specialty, small-volume nature of these elastomers and the high cost of the monomers (ECH at about $1.05/kg and EO about 56¢/kg).

The 1978 price of Parel 58 elastomer was $3.20/kg. The somewhat high cost is due to the speciality, low volume nature of the polymer as well as the cost of monomers (PO is about 50¢/kg and AGE is about $4.4/kg).

Health and Safety Factors; Toxicity

Monomer. Many epoxides are suspected carcinogens. Specifically, in a recent NIOSH report (57), ECH has given positive results in animal tests, EO is in an indefinite category, PO is suspect, and AGE is not categorized. Thus, appropriate measures must be taken to obtain elastomers that are essentially free of monomers. Current OSHA TLV in ambient workroom air are (ppm, time-weighted average, unless otherwise noted, for a five-day, 40-h work week): ECH, 5; EO, 50; PO, 100; and AGE, 10 (ceiling concentration).

For ECH, NIOSH proposed a new TLV standard in September of 1976 of 0.5 ppm with a ceiling exposure of 5 ppm. This proposed standard has not been acted on as yet (July 1979). A recent report (58) notes that the level of ECH present in the ambient air during manufacture and storage and during processing in three customer plants was well below (<0.1 ppm) the proposed Federal standard of 0.5 ppm. The absence of ECH during storage and processing is believed to be due to the excellent adsorptive capacity of the carbon black additive used in most rubber formulations, and the reactivity of ECH with vulcanizing or stabilizer additives. However, to ensure maximum worker protection, proper ventilation should be provided at all times. A recent epidemiological study of workers at a Shell Chemical Plant exposed from 1948 to 1965 to ECH concentrations greater than Shell's current 1 ppm limit showed a lower mortality rate than the general population (59).

Curing Systems. The most widely used curative, ethylenethiourea (NA-22) has been reported to cause cancer and birth defects in laboratory animals (60). A recent epidemiological study (61) shows no evidence for ethylenethiourea causing cancer and birth defects in humans. It is recommended, however, that skin contact and breathing of dust of NA-22 be avoided (60).

Uses

Epichlorohydrin Elastomers. Epichlorohydrin elastomers are employed in many of the conventional specialty rubber applications such as seals and gaskets, diaphragms, hose, belting, wire and cable jackets, printing rolls and blankets, coatings, mechanical goods, and collapsible fuel containers (8) where the homopolymer and ECO or combinations improve oil resistance, high and low temperature properties, environmental resistance and fabricability over lower cost rubbers. Many specific applications are in the automotive area where better heat-, ozone-, and oil-resistant elastomers are needed. The resilience characteristics of the epichlorohydrin elastomers make them attractive candidates for sound and vibration damping applications. There are some specific applications in oil-well drilling equipment.

The homopolymer has great natural tack and solubility which is useful in adhesives. The low viscosity of solutions of the homopolymer permits formulations of high solids with low viscosity. It has almost unlimited compatibility with other adhesive ingredients in solvent systems, plus thermoplasticity for use in modified hot-melt systems. The solution properties of the homopolymer, which are important in adhesive applications (25), have been reported and applied to the footwear industry (62).

Epichlorohydrin elastomers are reported as additives to plastics such as PVC (63) and polysulfones to improve impact strength. Anion exchangers have been prepared from di- and polyamine cross-linked homopolymer (64). The copolymers, and some derivatives are reported for reverse osmosis membranes for desalinating water and other separation processes (65).

Propylene Oxide Elastomers. The most important application for propylene oxide elastomer is in motor-mounts because of its unusual heat resistance combined with its good rubber properties. Natural rubber cannot withstand the high under-the-hood temperatures as well.

Addition of propylene oxide rubber to tread rubbers improves fatigue life and resistance to heat and atmospheric aging (66). Studies have been reported on elastic fibers made from a vulcanized polymer of propylene oxide (67). Alloys of propylene oxide elastomers with polystyrene have been prepared by polymerizing styrene in the presence of these elastomers and found to give better weather-resistant, high-impact compositions than commercial high-impact polystyrene (17).

Epichlorohydrin Elastomer Derivatives

Extensive work has been directed toward the synthesis of derivatives from epichlorohydrin elastomers, generally via nucleophilic substitution reactions on the chloromethyl group. In such reactions base-induced degradation of the epichlorohydrin elastomers (68) can be an important side reaction in the presence of strong bases or poor nucleophiles. Side groups introduced into the epichlorohydrin elastomers include quaternary ammonium, amino, and diamino (69–79); imide (76); isothiuronium (50,80); carboxyl, carboxylate, sulfoxide, and phosphorus ester (81); thiosulfate (82); azide (83); thioether (84); aminothioethers (85); phosphinyl (86); hydroxyl (87); dithiocarbamate (88); and cinnamic ester (89–90). Epichlorohydrin–ethylene oxide copolymers with lower (2–35 mol %) ECH contents have been modified with groups such as α-pyrrolidate, glycolate, and amine at high substitutions (90–100%) without any change in intrinsic viscosity (32). Diverse applications have been reported for these modified polyethers, including flame retardants (86), photosensitive materials (89–90) and a variety of applications for water systems such as thickeners (69,71–73), flocculating agents (69), semipermeable membranes for reverse osmosis (75), drainage aids for paper manufacturing (77), de-emulsifiers (79), and shrinkproofing agents for wool (80).

Graft polymers of the epichlorohydrin elastomers with monomers that polymerize by a free-radical mechanism were prepared by a number of different routes, eg, by chain grafting (91), by a transition metal-labile ligand (such as carbon monoxide) complex system (92), and by a polymer hydroperoxide route (93). Vinyl chloride grafts on the homo- and copolymer were reported (63,94) as PVC additives to improve properties, especially impact strength (see Copolymers).

BIBLIOGRAPHY

1. C. C. Price, *Chemist* **38,** 131 (1961).
2. U.S. Pat. 2,866,774 (Dec. 30, 1958), C. C. Price (to University of Notre Dame).
3. E. J. Vandenberg, *J. Polym. Sci.* **47,** 486 (1960).
4. U.S. Pat. 3,158,580 (Nov. 24, 1964), E. J. Vandenberg (to Hercules Incorporated).
5. U.S. Pat. 3,158,581 (Nov. 24, 1964), E. J. Vandenberg (to Hercules Incorporated).
6. E. J. Vandenberg, *Hercules Chem.* **54,** 1 (Sept. 1965); *Rubber Plast. Age* **46,** 1139 (1965).
7. W. D. Willis and co-workers, *Rubber World* **153**(1), 88 (1965).
8. W. R. Leach, *Rubber World* **153**(2), 71 (1965).
9. *ASTM D1418, Standard Recommended Practice for Nomenclature of Rubbers and Rubber Latices,* American Society for Testing and Materials, Philadelphia, Pa.
10. E. J. Vandenberg and A. E. Robinson in E. J. Vandenberg, ed., *Polyethers,* American Chemical Society, Washington, D.C., 1975, p. 101.
11. E. E. Gruber and co-workers, *Ind. Chem. Prod. Res. Develop.* **3**(3), 194 (1964).
12. U.S. Pat. 3,728,320 (Apr. 17, 1973), E. J. Vandenberg (to Hercules Incorporated).
13. U.S. Pat. 3,728,321 (Apr. 17, 1973), E. J. Vandenberg (to Hercules Incorporated).
14. K. E. Steller in E. J. Vandenberg, ed., *Polyethers,* American Chemical Society, Washington, D.C., 1975, p. 136.
15. E. J. Vandenberg in W. J. Bailey, ed., *Macromolecular Syntheses,* Vol. 4, John Wiley & Sons, Inc., New York, 1972, p. 49.
16. E. J. Vandenberg, *J. Polym. Sci.* **7,** 525 (1969).
17. J. T. Oetzel, *Rubber World* **172,** 55 (April 1975).
18. E. J. Vandenberg, *Pure Appl. Chem.* **48,** 295 (1976).
19. *Bulletin ORH-7A, Plasticizers for Herclor C Elastomer* and *Bulletin ORH-16A, Plasticization of Herclor H Elastomer,* Hercules Incorporated, Wilmington, Del.
20. Jpn. Pat. 74 16,112, (Apr. 19, 1974) K. Shikata, S. Nakao, and M. Yoshinaga (to Tokuyama Soda Co., Ltd.).
21. W. Cooper, G. A. Pope, and G. Vaughn, *Eur. Polym. J.* **4,** 207 (1968).
22. H. Balcar and V. Bohackova, *Collect. Czech. Chem. Commun.* **42,** 2145 (1977).
23. L. S. Yasenkova and co-workers, *Vysokomol. Soedin. Ser. B* **13** 366 (1971).
24. C. M. Hansen, *J. Paint Technol.* **39,** 104 (1967).
25. J. Kozakewicz and P. Penczek, *Angew. Makromol. Chem.* **50,** 67 (1976).
26. E. P. Lieberman, *Off. Dig. Fed. Paint Var. Prod. Clubs* **34,** 32 (1962).
27. C. Shambelan, University of Pennsylvania, Ph.D. thesis, Feb. 1959.
28. G. Allen, C. Booth, and M. N. Jones, *Polymer* **5,** 195 (1964).
29. G. Allen, C. Booth, and C. Price, *Polymer* **8,** 397 (1967).
30. U.S. Pat. 3,634,303 (Jan. 11, 1972), E. J. Vandenberg (to Hercules Incorporated).
31. Ger. Offen. 1,941,690 (Apr. 2, 1970), T. Nakata and K. Kawamata (to Osaka Soda Co., Ltd.). See also U.S. Pat. 3,773,694 (Nov. 20, 1973).
32. T. N. Kuren'gina, L. V. Alferova, and V. A. Kropachev, *Vysokomol. Soedin. Ser. A* **11,** 1985 (1969).
33. C. A. Lukach and H. M. Spurlin in G. E. Ham, ed., *Copolymerization,* John Wiley & Sons, Inc., New York, 1964, p. 115.
34. U.S. Pat. 3,313,743 (Apr. 11, 1967), L. J. Filar and E. J. Vandenberg (to Hercules Incorporated).
35. U.S. Pat. 3,639,267 (Feb. 1, 1972), E. J. Vandenberg (to Hercules Incorporated).
36. R. Sakata and co-workers, *Macromol. Chem.* **40,** 64 (1960).
37. R. J. Herold in E. L. Wittbecker, ed., *Macromolecular Syntheses,* Vol. 5, John Wiley & Sons, Inc., New York, 1974, p. 9.
38. U.S. Pat 3,509,068 (Dec. 22, 1967); U.S. Pat. 3,649,561 (Apr. 28, 1972), J. Lal (to Goodyear Tire and Rubber Company).
39. Brit. Pat. 1,048,822 (Nov. 23, 1966), W. Cooper (to Dunlop Rubber Company).
40. U.S. Pat. 3,384,603 (May 21, 1968), G. Elfers (to Wyandotte Chemicals Corporation).
41. Ph. Teyssie, T. Oerhadi, and J. P. Broul in C. E. H. Bawn, ed., *Macromolecular Science,* Butterworth and Company, London, 1975, p. 216.
42. U.S. Pat. 3,957,697 (May 18, 1976), R. K. Schlatzer (to B. F. Goodrich Company).
43. U.S. Pat. 3,776,863 (Dec. 4, 1973), K. Shibatami and S. Nagata (to Kuraray Company, Ltd.).
44. *Bulletin ORH-2A, Suggestions for Processing Herclor C,* Hercules Incorporated, Wilmington, Del.
45. *Bulletin ORH-6A, Suggestions for Processing Herclor H,* Hercules Incorporated, Wilmington, Del.
46. *Bulletin ORH-23, Herclor Rubbers,* Hercules Incorporated, Wilmington, Del.

47. *Bulletin PRP, Parel 58 Elastomers,* Hercules Incorporated, Wilmington, Del.
48. E. J. Vandenberg, R. H. Ralston, and B. J. Kocher, *SRS/4, October 1969—Proceedings of the Fourth International Synthetic Rubber Symposium,* London, Sept. 30–Oct. 2, 1969.
49. R. M. Murray and D. C. Thompson, *The Neoprenes,* E. I. du Pont de Nemours & Co., Inc., Wilmington, Del., 1963.
50. U.S. Pat. 3,594,355 (July 20, 1971), E. J. Vandenberg and W. D. Willis (to Hercules Incorporated).
51. Y. Nakamura, K. Mori, and S. Kawamura, *Kobunshi* **33,** 591 (1976).
52. D. Walker, ed., *Rubber and Plastic News, RPN Technical Notebook,* Dec. 12, 1977, p. 20.
53. U.S. Pat. 3,158,591 (Nov. 24, 1964), E. J. Vandenberg (to Hercules Incorporated).
54. R. W. Turner, *Hercules Chem.* **58,** 14, (Apr. 1969).
55. *Bulletin ORH-20A, Corrosion Inhibitors in Herclor C Elastomer,* Hercules Incorporated, Wilmington, Del.
56. *Bulletin ORH-5C, Black Reinforced Herclor C Elastomer Compounds* and *Bulletin ORH-12A, Carbon Black Reinforcement of Herclor H Elastomer,* Hercules Incorporated, Wilmington, Del.
57. *Registry of Toxic Effects of Chemical Substances,* Reference Numbers: ECH, TX49000; EO, KX24500; PO, TZ29750; AGE, RR08750, National Institute of Occupational Safety and Health (NIOSH), U.S. Department of Health, Education, and Welfare, Rockville, Maryland, 1976.
58. *Bulletin ORH-24, Herclor Epichlorohydrin Elastomers, Definition of Hazards Associated with Handling and Processing,* Hercules Incorporated, Wilmington, Del., Nov. 1977.
59. *Chem. Week,* 36 (Oct. 5, 1977).
60. *Chemicals for Elastomers Bulletin 14A, NA-22 Handling Precautions and Toxicity,* E. I. du Pont de Nemours & Co., Inc., Wilmington, Del., May 1972.
61. D. M. Smith, *J. Soc. Occup. Med.* **26,** 92 (1976).
62. J. Kozakiewicz and P. Penczek, *Adhes. Age* **20**(7), 29 (1977).
63. F. Wollrab and co-workers, *Polym. Prepr.* **13**(1), 499 (1972).
64. E. E. Ergozhin, E. Zh. Menligaziev, and K. Kh. Tastanov, *Inst. Akad. Nauk. Kaz. SSR Ser. Khim* **27**(3), 57 (1977).
65. U.S. Pat. 3,567,631 (Mar. 2, 1971), C. A. Lukach and co-workers (to Hercules Incorporated); U.S. Pat. 3,567,630 (Mar. 2, 1971), C. A. Lukach (to Hercules Incorporated).
66. R. V. Boguslovskaya and co-workers, *Int. Polym. Sci. Technology,* **3**(9), T/93 (1976).
67. Fr. Pat. 1,440,449 (May 27, 1966) (to Dunlop Rubber Company).
68. E. J. Vandenberg, *J. Polym. Sci., Polym. Chem. Ed.* **10,** 2903 (1972).
69. U.S. Pat. 3,403,114 (Sept. 24, 1968), E. J. Vandenberg (to Hercules Incorporated).
70. E. Pulkkinen and T. Petaja, *Finn. Chem. Lett.* (1), 22 (1977).
71. U.S. Pat. 3,864,288 (Feb. 4, 1975), C. K. Riew and R. K. Schlatzer (to B. F. Goodrich Company).
72. C. K. Riew, *Polym. Prepr. Am. Chem. Soc. Div. Polym. Chem.* **14**(2), 940 (1973).
73. Ger. Offen. 2,540,310 (Apr. 8, 1976), R. K. Schlatzer, Jr., and H. A. Tucker (to B. F. Goodrich Company).
74. E. Bortel and R. Lamot, *Rocz. Chem.* **50,** 1765 (1976).
75. U.S. Pat. 4,005,012 (Jan. 25, 1977), W. Wrasidlo (to United States Department of the Interior).
76. T. V. Markman, N. A. Mukhitdinov, and M. A. Askarov, *Deposited Doc. 1974, VINITI 451-74; Chem. Abstr.* **86,** 121980d (1977).
77. U.S. Pat. 3,746,678 (July 17, 1973), C. R. Dick and E. L. Ward (to The Dow Chemical Company).
78. T. N. Kuren'gina, L. V. Alferova, and V. A. Kropachev, *Vysokomol. Soedin Ser. B* **15,** 885 (1973).
79. U.S. Pat. 3,591,520 (July 6, 1971), M. T. McDonald (to Nalco Chemical Company).
80. U.S. Pat. 3,694,258 (Sept. 26, 1972), E. J. Vandenberg and W. D. Willis (to Hercules Incorporated).
81. U.S. Pat. 3,417,036 (Dec. 17, 1967), E. J. Vandenberg (to Hercules Incorporated).
82. U.S. Pat. 3,706,706 (Dec. 19, 1972), E. J. Vandenberg (to Hercules Incorporated).
83. U.S. Pat. 3,645,917 (Feb. 29, 1972), E. J. Vandenberg (to Hercules Incorporated).
84. U.S. Pat. 3,417,060 (Dec. 17, 1968), D. S. Breslow (to Hercules Incorporated).
85. U.S. Pat. 3,804,900 (Apr. 16, 1974), R. A. Hickner and H. A. Farber (to The Dow Chemical Company).
86. U.S. Pat. 3,660,314 (May 2, 1972), E. J. Vandenberg (to Hercules Incorporated).
87. U.S. Pat. 3,415,902 (Dec. 10, 1967), R. A. Hickner and H. A. Farber (to The Dow Chemical Company).
88. T. Nakai and M. Okawara, *Bull. Chem. Soc. Jpn.* **41,** 707 (1968).
89. Ger. Offen. 2,124,686 (Dec. 2, 1971), H. Fukutomi and H. Ohotani (to Danippon Ink and Chemicals, Inc.).

90. Brit. Pat. 1,404,927 (Sept. 3, 1975), (to Mitsubishi Chemical Industries Co., Ltd.).
91. U.S. Pat. 3,632,840 (Jan. 4, 1972), E. J. Vandenberg (to Hercules Incorporated).
92. U.S. Pat. 3,627,839 (Dec. 14, 1971), E. J. Vandenberg (to Hercules Incorporated).
93. U.S. Pat. 3,546,321 (Dec. 8, 1970), H. Jabloner and E. J. Vandenberg (to Hercules Incorporated).
94. Ger. Offen. 2,147,290 (Apr. 13, 1972), F. Wollrab, F. Declerck, and P. Georlette (to Solvay et Cie).

E. J. VANDENBERG
Hercules Incorporated

POLYISOPRENE

Natural rubber is cis-1,4-polyisoprene [9003-31-0]. Worldwide production of synthetic cis-1,4-polyisoprene as a replacement for natural rubber is the end result of a long period of research by many people. The ASTM nomenclature for natural rubber is NR and for synthetic cis-1,4-polyisoprene it is IR. The IUPAC designation for both natural rubber and polyisoprene is poly-1-methylbutenylene; however, this nomenclature is rarely used in the rubber industry.

Natural rubber, *Hevea Brasilienses,* remained an enigma until the 19th century when serious efforts were made to understand, explain, and improve the natural elastomers. Faraday and Priestly conducted experiments with NR and Goodyear made his important discovery of vulcanization with it. Although the basic unit of natural rubber was postulated and proven to be isoprene and rubber then was proven to be polyisoprene, much doubt remained as to the arrangement of the polyisoprene units (see Rubber, natural). The configuration and arrangement of the polymer units was not elucidated as cis-1,4-polyisoprene until after the development of infrared, x-ray and other modern tools of analysis (1).

Synthetic polyisoprene was studied extensively during World War II when the supply of natural rubber was insufficient for military needs. However, the program was unsuccessful despite the numerous attempts that were made to polymerize isoprene to duplicate *Hevea.*

In 1954 the long-sought goal was reached. First, the BF Goodrich Company and shortly thereafter the Firestone Tire and Rubber Company announced their synthesis of cis-1,4-polyisoprene which essentially duplicated the structure and properties of natural rubber. Later disclosures showed that the two companies had used different catalyst systems to achieve the synthesis. BF Goodrich scientists used the newly discovered Ziegler polyethylene catalyst (2), whereas Firestone scientists used lithium metal (3). The rubber obtained with lithium catalysis had a cis-1,4 content of about 92%, but the polyisoprene from the Ziegler catalyst was closer to natural rubber with a cis-1,4 structure of about 98% (see Ziegler-Natta catalysts).

A commercial polyisoprene synthesized with the lithium catalyst was marketed by Shell Oil Co. in 1962. A rubber from a commercial process using the Ziegler catalyst was introduced by The Goodyear Tire & Rubber Company in 1963. In 1977 there were two producers of *cis*-polyisoprene in the United States, Goodrich and Goodyear, and one each in France, Italy, and the Netherlands, three in Japan, one in Rumania, and at least five in the USSR (4). Only Shell in the Netherlands is known to have used the lithium catalyst. All others are believed to use catalyst systems related to the Ziegler system.

Polymerization

Isoprene polymerizes to yield four different basic structures of polyisoprene (PI):

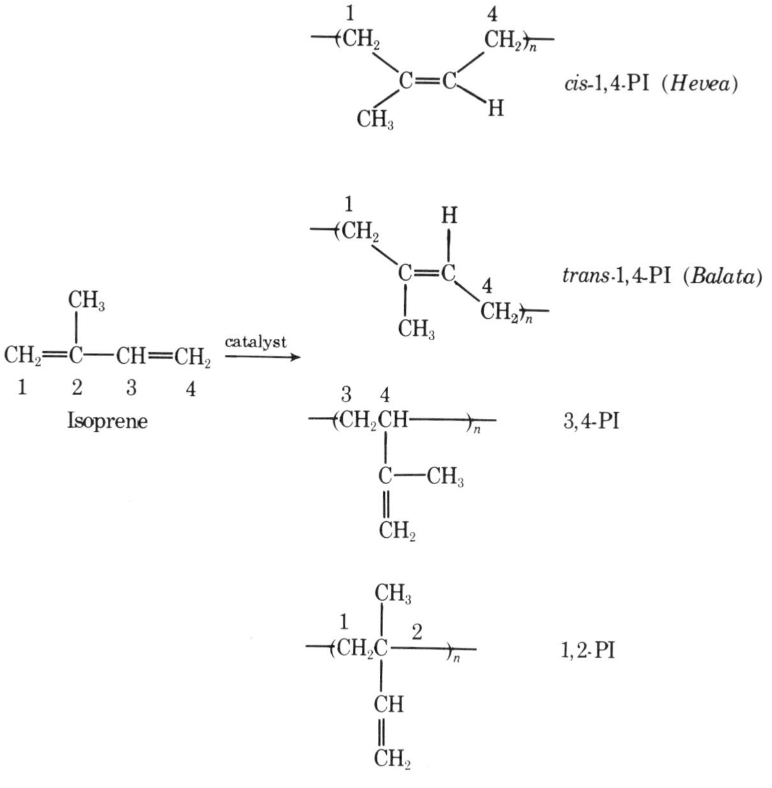

$$\text{CH}_2=\overset{\text{CH}_3}{\underset{2}{\text{C}}}-\underset{3}{\text{CH}}=\underset{4}{\text{CH}_2} \xrightarrow{\text{catalyst}}$$

Isoprene

cis-1,4-PI (*Hevea*)

trans-1,4-PI (*Balata*)

3,4-PI

1,2-PI

The poly cis-1,4, trans-1,4, and 3,4 structures can be made in high purity, but the poly-1,2 structure is obtained only in conjunction with the other three structures. Many catalyst systems yield polyisoprenes that are random copolymers of all four of the structural units. The structures can be readily analyzed by infrared and nmr techniques. The *cis*-1,4-polyisoprene structure is that of natural *Hevea* rubber but the trans-1,4 structure is that of naturally occurring *Balata*. These two diff   in their physical nature: *Hevea* is the familiar soft elastic rubber and *Balata* is a hard, crystalline polymer which has been used for golf ball covers. Only polymers of predominately cis-1,4 structure are used as general purpose rubbers. High *trans*-1,4-polyisoprene is offered commercially as a specialty product for the replacement of *Balata*.

Some typical values of the raw polymer properties for natural rubber and polyisoprene are shown in Table 1. The molecular weight [η] for some natural rubbers and for organolithium-catalyzed polyisoprene is very high initially. This frequently leads to false values for the Mooney viscosity. It is characteristic of all polyisoprenes that they undergo a reduction in molecular weight [η] during mechanical processing. Mooney viscosity can be used to follow molecular weight changes during processing after the initial mechanical breakdown step. The ratio of weight average molecular weight to number average molecular weight $M_w:M_n$, is a measure of molecular weight distribution or polydispersity—for a monodisperse polymer, the value is 1. Thus natural rubber has a wider molecular weight distribution than the synthetic polymers.

Ziegler Catalysts. The Ziegler coordination catalyst, which consists of an aluminum alkyl, R_3Al, and titanium tetrachloride, $TiCl_4$, polymerizes isoprene to stereoregular polyisoprene with 98% to 99% cis-1,4 structure, compared to natural rubber of 100% cis-1,4 structure. In the R_3Al, R may be an alkyl or an aromatic substituent (see Organometallics). High-purity isoprene that is free of polar compounds as well as acetylenes and cyclopentadiene is essential to the polymerization of isoprene. The polymerization is sensitive to the Ti–Al molar ratio: deviation from the optimum ratio, which is close to 1, results in greatly reduced catalyst efficiency, reduced molecular weights, and structure deviations. As the Ti–Al ratio is increased from about 1.2:1 to 2:1, trans structure is formed and eventually all trans-1,4 polymer results. Above ca 2:1 Ti–Al, cyclization occurs to give a resin. As the ratio is decreased from 0.8 to 0.5, the structure remains cis-1,4 but molecular weight decreases gradually and eventually only oily low molecular weight products are formed (6–7). Catalyst efficiency is reported to be improved by complexing the R_3Al with diphenyl ether (8), n-butyl ether (9), or tri-n-butyl amine (10). Best results are obtained when the catalyst is preformed at low temperatures, ie, below $-20°C$, by adding aluminum alkyl to the $TiCl_4$; gel content is also reduced (see below). At a Ti–Al ratio of 1, the heat of reaction of the $TiCl_4$ and $(i\text{-}C_4H_9)_3Al$ is estimated to be 251 kJ/mol (60 kcal/mol) (11).

A typical polymerization recipe is (12):

Component	Amount (parts by wt)
hexane	900
isoprene	100
$(i\text{-}C_4H_9)_3Al$	2.11
$TiCl_4$	(to give Ti–Al ratio of 0.8 to 1.1/1)
temperature, °C	0–50
time to 90% conversion, h	2–4

The polymerization may be carried out in aliphatic or aromatic hydrocarbon solvents and in chlorinated aromatic solvents such as chlorobenzene and dichlorobenzene. No polymerization occurs in solvents that can react with the catalyst, such as esters, ethers, amines, pyridines, alcohols, etc. The use of solvents such as carbon tetrachloride results in violent explosions: great caution is urged in using any but hydrocarbon solvents.

Use of high-purity isoprene and high-purity solvents produces a polymer molecular weight such that the polymer is useful for tires and other elastomer products. A change in the temperature of polymerization is a more effective method than a change in the amount of catalyst to achieve adjustments in molecular weight. The molecular weight changes inversely with temperature (12).

Table 1. Raw Polymer Properties[a]

	Hevea	Synthetic cis-1,4-polyisoprene	
catalyst		RLi	Ti–Al
cis-1,4 content, %	100	92	98
intrinsic viscosity [η], dL/g		ca 6.5	ca 3.5
Mooney viscosity at 100°C, ML-4			80
volatiles, %	1.0	0.1	0.1
ash, %	0.5–1.5	0.1	0.4
second-order transition, °C	−72	−70	−72
specific gravity	0.92	0.91	0.91
M_w/M_n	>3	2	2–3

[a] For additional physical constants see ref. 5.

Polymer molecular weight and molecular weight distribution can be measured accurately by gel permeation chromatography; however, the more convenient and widely used measurement of the molecular weight is intrinsic viscosity. In toluene at 30°C the intrinsic viscosity [η], expressed in dL/g, is related to molecular weight by the equation: $[\eta] = 1.9 \times 10^{-4} M^{0.745}$ (13). A drawback of both methods is the insoluble content (gel) of the elastomer prepared with the Ti–Al catalyst system. The gel content is higher in polyisoprene prepared in aliphatic solvents than in polymers prepared in aromatic ones, 20–25% vs 0–5%. The gel content is partially controlled by the order of addition of catalyst components (8) and/or by complexing the R_3Al with ethers (8–9) or amines (10). This gel is independent of conversion and is thought to originate from small amounts of 3,4 structure formed during the polymerization (12). Commercial polymers are characterized by a bulk (Mooney) viscosity measured with a shearing disk viscometer (14).

A closely related Ti–Al catalyst system is based on complexed aluminum hydride (see Hydrides) and on titanium tetrachloride (7,15). The system is free of the aluminum—carbon bond found in the usual R_3Al-type compounds. One commercial system uses poly(alkyliminoalanes), $+Al(H)N(R)+_n$ and $TiCl_4$. As in the $TiCl_4$–R_3Al system, close control of the Ti–Al ratio is essential (a ratio of 0.65 is optimum) for obtaining good catalyst efficiency and high molecular weights. Benzene is the preferred polymerization solvent in this system. Structurally, the polymer is the same and is in the same molecular weight range as polymer synthesized with the $TiCl_4$–R_3Al catalyst; however, the gel content is lower, 0–4% vs 20%.

Many other variations of the Ti–Al catalyst are reported in the literature for the polymerization of isoprene. There are also many reports on catalysts based on transition metal compounds other than titanium, none of which has reached the commercial stage. Both cis- and trans-1,4-polyisoprene can be prepared (7).

Organolithium Catalysts. The anionic polymerization of isoprene by organolithium catalyst in a hydrocarbon solvent also leads to a polymer of high cis-1,4 content (16). This is frequently reported as 92% cis-1,4 with about 6% 3,4 structure. The polymerization involves only two steps, initiation and propagation (17).

initiation: RLi + monomer (M) → RMLi

propagation: RMLi + n[M] → R(M)$_n$Li

A termination process will not occur in the absence of impurities. The polymer molecular weight can be estimated from the catalyst concentration and the amount of polymer formed, MW = K P/[RLi]. Very high molecular weight polyisoprene, [η] = 6–10, is formed. In contrast to the Ti–Al system, gel-free polymer results with the RLi catalyst. In toluene at 30°C the molecular weight is related to intrinsic viscosity by [η] = 2.00 × 10^{-4} $M^{0.728}$ (18).

n-Butyllithium and sec-butyllithium appear to be popular choices for the initiation of polymerization. Initiation with the sec-butyllithium is faster; however, propagation proceeds at the same rate regardless of the initiator (17) (see Initiators).

The polymerization rate is affected by choice of solvent. Polymerization rates are faster in aromatic solvents than in aliphatic ones. The rate of polymerization increases with temperature; an increase of fourfold has been reported for a 10°C temperature change (17). The active catalyst concentration has a pronounced effect on the rate of polymerization.

The microstructure of the polyisoprene is affected to a small degree by changes in the RLi concentration, by choice of aliphatic or aromatic solvent, by the temperature of polymerization and by the degree of monomer conversion. The high cis-1,4 content is favored by lower polymerization temperatures, low catalyst levels, and aliphatic solvent (16). The addition of polar materials, such as ethers or amines, has a pronounced effect on the 3,4 content of the polyisoprene. The addition of two moles of tetrahydrofuran per mole of RLi more than doubles the 3,4 content; an addition equal to 10% of the solvent raises the 3,4 content to over 60%.

When the RLi catalyzed polymerizations are run at room temperature and in a very pure system, chain transfer and termination steps do not occur, resulting in a so-called living polymer. Advantage can be taken of this characteristic to double polymer molecular weight by reaction with coupling agents (eg, methyltrichlorosilane, dimethyl phthalate), or to form star polymers by using polyfunctional coupling agents. Also, block polymers can be prepared by allowing the living polymer to react with an appropriate monomer such as styrene. The polymerization can be terminated with reactive shortstops that result in functional terminal groups, such as —OH and —CO_2H.

Other Catalysts. The polymerization of isoprene to the trans-1,4 structure is best carried out using a mixed VCl_3–$TiCl_3$–R_3Al catalyst (7). The catalyst is generally prepared in three stages which are: (1) reaction of R_3Al with $TiCl_4$ to give $TiCl_3$ followed by aging; (2) addition of VCl_4 to this mixture to give VCl_3, followed by aging; and (3) further acceleration by addition of R_3Al at time of usage. Polymerization is carried out in aliphatic solvents to give polymers of 98% trans-1,4-polyisoprene structure. Molecular weights are high, [η] = 4–5.

The resulting trans-1,4-polyisoprene is similar in properties to natural Balata. The crystalline melting temperatures for the synthetic α and β modifications are the same as those of the natural product and the rate of crystallization is equal to that of the natural material. However, the molecular weight of the trans polymer can be made higher than that of the Balata.

Isoprene can be polymerized to 99% 3,4 structure. The catalyst system most often used is based on $(RO)_4Ti$–R_3Al, where the Ti–Al ratio is ca 1:6. These have found no commercial use (19).

In an emulsion system isoprene polymerizes to an elastomer using free-radical initiation. The isoprene is emulsified in water by the action of soaps with an added initiator. Suitable free-radical initiators are potassium persulfate and a redox system such as cumene hydroperoxide–iron pyrophosphate (20). The polymer molecular weight is controlled through the addition of a mercaptan such as *tert*-dodecylmercaptan. Polyisoprene prepared in emulsion is of mixed structure and is not a replacement for natural rubber.

Isoprene may be polymerized with many other catalysts. These include (*1*) alfin catalyst—a mixture of allylsodium, sodium isopropoxide, and sodium chloride; (*2*) alkali metals; and (*3*) cationic catalysts such as $AlCl_3$. The cationic system is of interest for the copolymerization of 1–5% isoprene with isobutylene to form butyl rubber (qv) rather than for homopolymerization. The polyisoprenes prepared with the above catalyst systems have the following microstructures.

Catalyst	cis-1,4	trans-1,4	1,2	3,4	Ref.
free-radical	22%	69%	5%	24%	20
alfin	2	70	5	23	21
sodium	0	52	5	43	21
potassium	0	52	8	40	22
rubidium	3	50	8	39	22
cesium	8	67	8	17	22
cationic	0	91	4	6	21

These structures can be varied over a narrow range by choice of temperature, solvent, and catalyst concentration. None of these is of commercial importance.

Process. Using prepurified isoprene and solvents, the polymerization procedures are similar for both the Ti–Al and RLi systems. The basic steps are: (*1*) preparation and aging of the Ti–Al catalyst at a constant temperature in a small amount of the polymerization solvent (this step is not required in the RLi system); (*2*) polymerization in a single-stage or continuous reactor system; (*3*) charging the solvent and isoprene to the first reactor; (*4*) metering the catalyst into the first reactor; (*5*) shortstopping at desired conversion; (*6*) addition of antioxidant to the polymer solution; (*7*) extraction of the catalyst residues followed by steam–water coagulation of the polymer and stripping of the solvent and residual monomer; (*8*) dewatering of the polymer followed by mechanical drying; (*9*) compressing the crumb into approximately 35-kg bales and wrapping in film such as polyethylene; and (*10*) purification of recovered solvent and isoprene. All steps prior to shortstopping require an inert atmosphere; nitrogen is generally used. A complete flow sheet of a commercial plant has been published (23).

Compounding

cis-Polyisoprene is classified, together with SBR (see Styrene–butadiene rubber), polybutadiene (qv), and natural rubber, as a general purpose rubber. One characteristic of general purpose rubber is their extensive use in tires. The use properties result from the combination of elastomer with filler, softener, antioxidant, and vulcanizing agents. These are selected to achieve the desired product quality, and equally important, to allow satisfactory processing.

Since *cis*-polyisoprene has the same basic structure as natural rubber, techniques for the compounding and product application of natural rubber are directly applicable to *cis*-polyisoprene. *cis*-Polyisoprene and natural rubber have advantages over butadiene-derived rubbers, SBR, and polybutadiene, for applications requiring good tear strength, low hysteresis, good tensile strength in uncured compounds, good tensile strength at elevated temperature (100°C), and in applications where the construction of the article requires tack, or self-adhesion.

Commercial polyisoprene has certain advantages over natural rubber. These are good color, more uniform quality, low odor, faster breakdown and mixing, better extrusion and calendering, excellent mold flow, lower hysteresis, and controlled molecular weight.

Natural rubber has some advantages over *cis*-polyisoprene. It has somewhat higher tear strength, especially at elevated temperatures, higher tack, and higher tensile strength in high modulus compounds. In compounds containing reinforcing carbon black, natural rubber interacts with the carbon black to give stress-reinforcement beyond the yield point when the unvulcanized, or green, compound is subjected to tension. *cis*-Polyisoprene is essentially lacking this property of green strength.

Although the hydrocarbon portion of *cis*-polyisoprene and natural rubber are essentially identical, *cis*-polyisoprene is substantially 100% hydrocarbon, usually containing only an added antioxidant. Natural rubber contains 6–8% naturally occurring nonrubber materials. *cis*-Polyisoprene deviates from 100% cis-1,4 structure by 1–2%. It is free of the carbonyl groups attached to the natural rubber backbone. The presence of the nonrubber fraction in natural rubber is responsible for a number of the differences observed between *cis*-polyisoprene and natural rubber. These are: differences in vulcanization time, modulus, hysteresis or heat buildup, and hot tear resistance (24).

Green strength, however, is not changed by the addition of the nonrubber fraction from natural rubber to *cis*-polyisoprene (24). Failure to develop green strength is attributed to reduced rates of crystallization and reduced rubber carbon black interaction. Green strength is improved by the reaction of the *cis*-polyisoprene with maleic anhydride to introduce a low level of carbonyl groups along the hydrocarbon chain (25), or by adding a chemical promoter to the rubber as part of the mixing process (26). A somewhat different type of green strength is obtained by carrying out the polymerization of isoprene in the presence of ethylene (27). This gives a polymer with a greatly increased yield value and low elongation. Table 2 is a comparison of *cis*-polyisoprene and natural rubber in a carbon black-reinforced compound.

About 70% of the *cis*-polyisoprene is used with carbon black as the reinforcing filler (see Carbon, carbon black; Fillers). The remaining 30% is used with nonblack reinforcing filler, eg, silicas, calcium carbonate, soft or hard clay; or in gum applications. For each specific application, the individual fabricator selects the filler and adjusts the oil and vulcanization system to secure a compound suited to the particular processing equipment and to the vulcanizate properties required.

When the carbon black level is increased, Mooney viscosity of the compound, 300% modulus, hardness, abrasion resistance, and heat buildup (ΔT) increase. Tensile strength and tear resistance are, for most carbon blacks, optimum at 45–50 phr (parts per hundred of rubber).

When a nonreinforcing filler such as calcium carbonate or clay is substituted for the carbon black, tensile strength decreases with increased levels of filler, and hardness increases. Tear strength is relatively unaffected.

Adding oil to a *cis*-polyisoprene compound lowers processing temperatures, Mooney viscosity of the compound, hardness, tensile strength, tear strength, and abrasion resistance.

Table 2. A Comparison of *cis*-Polyisoprene with Natural Rubber

Rubber	Compound Mooney viscosity at 100°C, ML-4	Green strength TB-Ty, MPa[a]	Tack, MPa[a,b]	Cure time, 150°C/min	300% Modulus, MPa[a]	Tensile strength, MPa[a]	Elongation, %	Heat buildup, °C[c]	MOT tear, kN/m[a]
98% *cis*-1-4-polyisoprene	75	0.08	0.33	15	10.4	30	585	11	37
natural rubber	59	0.72	0.53	12	11.8	24.5	520	16	96
Recipe: 98% *cis*-polyisoprene or natural rubber							100		
antioxidant							1.0		
HAF black							50		
stearic acid							3.0		
zinc oxide							3.0		
naphthenic oil							15		
N-cyclohexyl-2-benzothiazol sulfenamide							1.7		
sulfur							1.7		

[a] TB = breaking stress; Ty = yield stress; to convert MPa to psi, multiply by 145.
[b] Ref. 28.
[c] Ref. 29.
[d] Ref. 30; to convert kN/m to dyn/cm, multiply by 10^6.

The accelerator system is chosen on the basis of processing time and temperature, the vulcanization time and temperature, and on the degree and type of cross-links desired in the finished product.

cis-Polyisoprene is frequently blended with natural rubber to take advantage of the better processing, slower vulcanization rate and uniformity of the *cis*-polyisoprene. It is blended with SBR and polybutadiene to gain the properties of the *cis*-polyisoprene together with improved resistance to breakdown, improved fatigue resistance, better abrasion resistance and better retention of properties imparted by the butadiene-based rubbers.

Uses

About 60% of synthetic *cis*-polyisoprene is used in tires. Following the pattern of natural rubber, this goes largely into truck tires, off-the-road tires, aircraft tires, and carcass and side-wall compounds of passenger car tires. Replacement of natural rubber with up to 20% of the 92% *cis*-polyisoprene and up to 40% of the 98% *cis*-polyisoprene is possible without significant differences in manufacturing or tire performance (31). In some cases, 100% replacement of natural rubber in passenger tires and truck tire treads has proven satisfactory. The remaining 40% of *cis*-polyisoprene goes into automotive bushings and motor mounts, belting, gaskets, footwear, battery separators, adhesives, and flooring. Applications using gum compounds are rubber sheeting, rubber thread, baby bottle nipples, and pharmaceutical sundries.

The high-purity, very light color and uniformity of the synthetic *cis*-polyisoprene, especially the 92% *cis*-polyisoprene, make it attractive for products based on chemical modification of rubber, eg, chlorinated rubber (see also Butyl rubber; Neoprene).

trans-1,4-Polyisoprene is used largely in golf ball covers. Other uses are hot-melt adhesives and orthopedic splints (see Prosthetic and biomedical devices) (32).

Economic Factors

Based on 1975 information, the total world capacity for the production of *cis*-polyisoprene is 735,000 metric tons (33). This is distributed as follows: United States, 123,000 t; Western Europe, 150,000 t; Japan, 85,000 t; and the USSR and Eastern Europe, 375,000 t. USSR capacity is expected to reach 1,000,000 t by 1980 (4).

The consumption of *cis*-polyisoprene in the non-Communist countries in 1976 was 253,000 t, or about 71% of the available capacity. The consumption of polyisoprene is expected to reach about 378,000 t in 1981 and 484,000 t in 1986. The corresponding consumption of natural rubber for 1976 was 2,702,000 t. This is expected to increase to 3,324,000 t by 1981, and 4,026,000 t by 1986 (34).

A shift in the pattern of tire production with increased production of radial tires in the United States will result in an increase in demand for the combination of natural rubber and *cis*-polyisoprene as a percentage of total tire rubber. The ability of the synthetic *cis*-polyisoprene to participate in this market will be determined, not by the technology of the synthetic rubber industry, but rather by the availability of cheaper by-product isoprene, and, ultimately the price of natural rubber.

The Institute of Synthetic Rubber Producers annually publishes statistics on the production of synthetic rubber and plant capacities (35), and at intervals, publishes a list of rubber producers, trade names, and the types of rubber that are available commercially (36).

BIBLIOGRAPHY

1. K. H. Meyer, *Natural and Synthetic High Polymers*, 2nd ed., Interscience Publishers, Inc., New York, 1950.
2. Belg. Pat. 543,292 (June 2, 1956), (to Goodrich-Gulf Chemical Co.); S. E. Horne, Jr., and co-workers, *Ind. Eng. Chem.*, **48**, 784 (1956).
3. F. W. Stavely and co-workers, *Ind. Eng. Chem.* **48**, 778 (1956).
4. *Rubber Trends* (75), 37 (Sept. 1977).
5. L. A. Wood, *Rubber Chem. Technol.* **49**, 189 (1976).
6. U.S. Pat. 3,728,325 (Apr. 17, 1973), E. J. Carlson and S. E. Horne, Jr. (to BF Goodrich).
7. W. M. Saltman, ed., *The Stereo Rubbers*, John Wiley & Sons, Inc., New York, 1977, Ch. 2.
8. E. Schoenberg, D. L. Chalfant, and R. H. Mayor, *Rubber Chem. Technol.* **37**, 103 (1964); Fr. Pat. 1,393,714 (March 26 1965), E. Schoenberg and D. L. Chalfant (to Goodyear Tire & Rubber Co.).
9. Can. Pat. 922,849 (March 13, 1956), L. C. Kreider (to B.F. Goodrich).
10. U.S. Pat. 3,165,503 (Jan. 12, 1965), H. Kahn and S. E. Horne, Jr. (to B.F. Goodrich).
11. I. V. Garmonov, *International Symposium on Isoprene Rubber, Moscow, U.S.S.R. Nov. 1972*; *Rubber Chem. Technol.* **46**, 555 (1973).
12. C. F. Gibbs and co-workers, *Kautschuk Gummi Kunst.* **13**, WT 336 (1960).
13. A. De Chirico, *Chem. Ind. (Milan)* **46**, 52 (1964).
14. *ASTM D-1646-74, 1977 Annual Book of ASTM Standards*, Part 37, American Society for Testing and Materials, Philadelphia, Pa.
15. A. Balducci and co-workers, *Rubber Chem. Technol.* **48**, 736 (1975).
16. H. E. Diem, H. Tucker, and C. F. Gibbs, *Rubber Chem. Technol.* **34**, 191 (1961).
17. J. P. Kennedy and E. G. M. Törnqvist, eds., *Polymer Chemistry of Synthetic Rubber*, Part II, Interscience Publishers, a division of John Wiley & Sons, Inc., New York, 1968, Ch. 6.
18. W. H. Beattie and C. Booth, *J. Appl. Polym. Sci.* **7**, 507 (1963).
19. G. Natta, L. Porri, and A. Carbonaro, *Makromol. Chem.* **77**, 126 (1964).
20. K. Sakota and co-workers, *J. Appl. Polym. Sci.* **20**, 2811 (1976).
21. W. S. Richardson and A. Scher, *J. Polym. Sci.* **10**, 353 (1953).
22. A. Tobolsky and C. E. Rogers, *J. Polym. Sci.* **40**, 73 (1959).
23. *Hydrocarbon Process.* **50**, 199 (Nov. 1971).
24. E. C. Gregg, Jr., and J. H. Macey, *Rubber Chem. Technol.* **46**, 47 (1973).
25. P. Luijk and J. M. Rellage, *Kautschuk Gummi Kunst.* **26**, 446 (1973).
26. J. R. Davies, and E. R. Rodger, *Internal Symposium on Isoprene Rubber, Moscow, U.S.S.R., Nov. 1972*; *Rubber Chem. Technol.* **46**, 574 (1973).
27. U.S. Pat. 3,684,785 (Aug. 15, 1972), H. Hasegawa and co-workers (to Nippon Zion).
28. J. R. Beatty, *Rubber Chem. Technol.* **42**, 1040 (1969).
29. *ASTM D-623-67 (reapproved 1972) Method A, 1977 Annual Book of ASTM Standards*, Part 37, American Society for Testing Materials, Philadelphia, Pa.
30. A. G. Vieth, *Rubber Chem. Technol.* **38**, 700 (1965).
31. J. E. Diamond, *International Symposium on Isoprene Rubber, Moscow, U.S.S.R., Nov., 1972*; *Rubber Chem. Technol.* **46**, 577 (1973); W. Kleemann and G. Erben, p. 578; A. L. Mekel, p. 578.
32. J. Peppeatt, *Rubber Chem. Technol.* **47**, 1309 (1974).
33. *Chem. Econ. Newsletter,* (May–June 1977).
34. *Chem. Age.*, 8, 9 (Oct. 15, 1976).
35. C. F. Ruebensaal, *The Rubber Industry Statistical Report and Changing Markets and Manufacturing Patterns in the Rubber Industry, 1977*, International Institute of Synthetic Rubber Producers, Inc., New York.
36. *The Elastomers Manual*, International Institute of Synthetic Rubber Producers, Inc., New York, 1977.

General References

G. S. Whitby, ed., *Synthetic Rubber*, John Wiley & Sons, Inc., New York, 1954.
M. Morton, ed., *Rubber Technology*, 2nd ed., Van Nostrand Reinhold Company, New York, 1973, Ch. 11.
W. M. Saltman, ed., *The Stereo Rubbers*, John Wiley & Sons, Inc., New York, 1977.

J. P. Kennedy and E. G. M. Törnqvist, eds., *Polymer Chemistry of Synthetic Elastomers,* Part I, Interscience Publishers, a division of John Wiley & Sons, Inc., New York, 1968, Ch. 2, Part II, 1969, Chs. 6 and 7.
J. C. W. Chien, ed., *Coordination Polymerization,* Academic Press, Inc., New York, 1975.
The Vanderbilt Rubber Handbook, R. T. Vanderbilt Company, Inc., New York, 1978.

HAROLD TUCKER
S. E. HORNE, JR.
BF Goodrich Company

POLYPENTENAMERS

Polypentenamer [25103-85-9] is obtained by the transition-metal-catalyzed polymerization of cyclopentene, and is one member of the homologous series of linear, unsaturated polymers designated *polyalkenamers*. The name polypentenylene has been used occasionally, in accordance with IUPAC-recommended nomenclature (1), but the designation polypentenamer has been most widely adopted for this polymer. Ring opening occurs via cleavage of the carbon-carbon double bond in cyclopentene, and thus the polymer may be represented structurally by bisalkylidene repeat units, but more commonly other equivalent representations have been employed:

$$\text{cyclopentene} \longrightarrow \begin{cases} =\!\!\text{CHCH}_2\text{CH}_2\text{CH}_2\text{CH}\!\!= \\ +\text{CH}=\!\text{CHCH}_2\text{CH}_2\text{CH}_2+ \\ +\text{CH}_2\text{CH}=\!\text{CHCH}_2\text{CH}_2+ \end{cases}$$

Cycloolefins such as cyclopentene possessing low ring-strain energy homopolymerize very poorly via addition across the double bond. Low molecular weight oligomers are normally obtained (2), although when ultra-high pressures are employed (eg, 6,580 MPa or 65,000 atm), cyclopentene can be converted in good yield to high molecular weight, semicrystalline saturated polymers (3).

A remarkable alternative polymerization process involving cleavage of the ring, catalyzed by an activated molybdena, was first described in a patent filed in 1957 (4). That process, although not very efficient, produced a high molecular weight rubbery polymer. Subsequent findings that certain Ziegler-Natta catalysts (qv) were surprisingly efficient in promoting this reaction (5), and that a unique new reaction of olefins termed *olefin metathesis* was implicated in cycloolefin polymerizations (6), have intensified interest in this fledgling field of polymer science.

Progress in this field has been documented in a succession of reviews concerning *trans*-polypentenamer [29300-20-7] (7–9). Extensive general reviews broadly cover cycloolefin polymerizations (10–11). Numerous other recent reviews deal with the basic chemistry, the olefin metathesis reaction (12–15).

Of the polyalkenamers, polypentenamer has generated the greatest interest for several reasons. The requisite monomer can be economically obtained from cyclopentadiene, an abundant petrochemical by-product. In addition, polypentenamer is a readily vulcanizable elastomer and possesses an intriguing combination of properties that indicate significant potential for use in tires and a variety of other applications.

Variations in polymerization techniques permit control of configuration of the double bond, and polymers ranging widely in structure have been examined. A virtually pure cis form is known and possesses excellent low temperature properties, but attention has been focused primarily on a readily prepared trans version which is of ca 85% trans structure.

This trans polymer has a glass transition temperature of ca −95°C, close to that of *cis*-1,4-polybutadiene, yet has a crystalline melting point just below room temperature. This fortuitous combination of properties results in a stress-crystallizing elastomer that is easily processed in conventional rubber-mixing equipment and exhibits good tensile, resilience, and abrasion characteristics. Although these properties were revealed by Natta and his colleagues during the early 1960s, the commercialization of polypentenamer has not as yet been achieved. Nevertheless, several corporations have pursued programs directed toward tire use evaluation, with Farbenfabriken Bayer A.G. having expressed an interest in manufacturing this polymer in the early 1970s.

Cyclopentene

Sources. The most commercially significant process for obtaining cyclopentene would utilize refinery C_5 fractions as starting material. Cyclopentene itself is a minor recoverable component of these streams, but much larger amounts of monomer can be obtained by recovery and hydrogenation of the more abundant component cyclopentadiene (see Cyclopentadiene). Numerous methods have been proposed for the hydrogenation step. Two particularly attractive routes utilize titanium (16) or nickel (17) catalysts. These processes are said to be at least 97% selective for the formation of cyclopentene under conditions where cyclopentadiene is quantitatively consumed.

According to a process developed jointly by BASF and Erdölchemie, which provides for the simultaneous recovery of isoprene and piperylene, cyclopentadiene is first isolated as dicyclopentadiene. This is then cracked, hydrogenated, and recombined with the initially separated C_5 fraction which contains some cyclopentene. Ultimate separation and recovery of monomers is based on an extractive distillation scheme (18) (see Azeotropic and extractive distillation).

The annual volume of dicyclopentadiene produced worldwide has grown rapidly, although in the United States, the production of dicyclopentadiene for chemical purposes was reported to have peaked in 1974 at ca 44,000 metric tons (19). This growth has been primarily in the form of surplus unrecovered dicyclopentadiene in refinery by-product streams. Recent economic developments have favored ethylene production based on the high temperature steam cracking of heavy-hydrocarbon feed streams. This process has already attained prominence in Europe and Japan. Substantial quantities of C_4 and C_5 dienes are major by-products of this process. Thus, a typical ethylene facility would produce (9–18) × 10^3 t of cyclopentadiene annually, recoverable as dicyclopentadiene (see Ethylene).

Properties. Cyclopentene is a colorless, pungent liquid under normal conditions. It is quite volatile and highly flammable. Its oral LD_{50} is 2.14 g/kg in rats (20). Cyclopentene in contact with air for extended periods forms a relatively stable hydroperoxide (21). Nevertheless, samples that have not been treated to remove peroxides should not be distilled to dryness. The physical properties of cyclopentene are summarized in Table 1.

Polypentenamer from Cyclopentene

Cyclopentene polymerizations have been conducted in bulk, but control of the reaction exotherm of ca 19 kJ/mol (4.5 kcal/mol) is best accomplished by solution polymerization methods if rapid rates of polymerization are anticipated. Suitable inert solvents include benzene, toluene, chlorobenzene, and methylene chloride. Polymerizations appear to be much slower in aliphatic solvents owing to reduced catalyst solubility. Usually reactions are carried out at RT or below, because polymer yields decrease markedly at elevated temperatures. This is a consequence of the rather low ceiling temperature for this polymerization.

Rather strict precautions for control of purity of monomer and solvent are required for optimum catalyst activity, as with other Ziegler-Natta polymerizations. Normally, allenes, acetylenes, and conjugated dienes which may be present must be removed. In addition, polar impurities such as oxygenated compounds cannot be tolerated, and these also must be eliminated. On the other hand, adventitious traces of polar contaminants such as water, or the easily-formed cyclopentene hydroperoxide, can serve rather erratically to enhance reactions by acting as catalyst modifiers.

Catalysts. Metathesis catalysts used for polymerizations vary widely in composition and reactivity toward different substrates. Relative reactivities of various catalysts toward terminal vs internal olefins, or moderately strained vs highly strained cycloolefins, are as unpredictable as are their stereospecificities. Metathesis polymerization catalysts universally utilize a transition metal compound, usually in combination with one or more cocatalysts.

Systems based on titanium, vanadium, zirconium, ruthenium, and iridium are known to polymerize highly strained cycloolefins, but their activity with cyclopentene is extremely low. Niobium, tantalum, and rhenium are somewhat more effective, but the preeminent catalysts for polymerizing cyclopentene are derived from tungsten

Table 1. Properties of Cyclopentene

Property	Value	Ref.
molecular weight	68.11	
freezing point, °C	−135.076	22
boiling point, °C	44.242	22
n_D^{20}	1.42246	22
d^{20}, g/mL	0.77199	22
C_p^{20}, J(g·°C)[a]	1.837	23
critical temperature, °C/MPa[b]	233/4.79	23
ring strain (estimated), kJ/mol[a]	20.5	24
	28.5	25

[a] To convert kJ to kcal, divide by 4.184.
[b] To convert MPa to atm., divide by 0.101.

and, to a lesser extent, molybdenum, as a result of their relatively high reactivity, availability, and relatively low cost.

Metathesis catalysts have been broadly categorized as being homogeneous or heterogeneous. The latter generally consist of transition-metal oxides or carbonyls deposited on active supports such as silica or alumina. However, they are normally active only at elevated temperatures unsuitable for cyclopentene polymerizations. Uncertainties regarding whether so-called homogeneous catalysts are indeed homogeneous have not been resolved. This distinction perhaps more aptly reflects the character of the individual catalyst components prior to reaction, rather than the state of the active species that is formed (see Catalysis).

Without the aid of cocatalysts, most transition metal compounds are generally inoperative for cyclopentene polymerizations, although slow reactions have been observed in some cases with WCl_6 alone (26). An interesting exception was found for the lower-valent tungsten complex, $C_6H_5WCl_3$, which by itself was quite active (27). Significant activity was also observed with certain tungsten–carbene complexes (28).

Two-Component Catalysts. Two-component systems were the first homogeneous catalysts to be reported. Efficient polymerizations of low-strain cycloolefins such as cyclopentene were first achieved in 1963 by use of tungsten and molybdenum halides in combinations with organoaluminum cocatalysts (5,29). Catalysts were described which were suitable for the preparation of both high cis and high trans versions of polypentenamer.

In the ensuing development of multicomponent catalyst systems, the transition metals have been employed in a wide variety of forms ranging from zero-valent coordination complexes to salts of the metals in their highest oxidation states. In conjunction with the transition metal compound, cocatalysts are used that are typically organometallic compounds or Lewis acids. A broad spectrum of substances has been shown to have some activity as cocatalysts, although aluminum derivatives typical of Ziegler-Natta systems have been most widely employed. Organotin derivatives appear to be somewhat less active.

The broad scope of two-component catalysts is evident in the partial listing of reported systems in Table 2. The subdivision according to stereospecificity is not unequivocal, since several combinations vary in this regard with polymerization conditions and catalyst-component ratio.

Individual features of some catalyst systems are noteworthy. Combinations that use aluminum halides (30–31) do not appear to involve reducing agents, although lower-valent forms of the transition metal are widely deemed essential for activity. The potential for steric control through choice of polymerization temperature may be a widespread phenomenon. Two rather typical systems based on WCl_6 in combination with tetraalkyltins (32) or trialkylaluminums (33) yield *cis*-polypentenamer of high steric purity at temperatures below $-30°C$, whereas at room temperature, products are obtained containing 75–85% trans structure. Steric control has also been obtained by variation of the Al/W ratio in a WF_6-based system (34).

Catalysts utilizing carbene compounds (35) and carbene precursors (36) are of considerable significance in view of recent evidence that metathesis reactions proceed via carbene–metal intermediates. Some of these catalysts are quite unique in their ability to polymerize cyclopentene at remarkably low concentrations of catalyst. With $TiCl_4$ as a cocatalyst, $(CO)_5W{=}C(OC_2H_5)R$ is effective at a cyclopentene–W ratio

Table 2. Representative Two-Component Polymerization Catalysts for Cyclopentene

Transition-metal compound	Cocatalyst	Ref.
trans-Directing catalysts		
WX_6	R_3Al, R_2AlX	5
WF_6	$RAlCl_2$ (Al/W > 2)	34
WCl_6	R_3SnH	39
WCl_6	$(C_6H_5)_3SnCl$	40
WCl_6	R_4Sn	41
WCl_6	$AlBr_3$	30
WCl_6	$C_6H_5CHN_2$	36
$NbCl_5$, $TaCl_5$	R_3Al	42
$TaCl_5$	CH_2CH_2O	43
$ReCl_5$	$(C_2H_5)_3Al$	44
$[(C_6H_5)_3P]_2(NO)_2MoCl_2$	$(CH_3)_3Al_2Cl_3$	45
$MoCl_3$ (stearate)$_2$	RMgX	46
$(CO)_5Mo{=}C(CH_3)OC_2H_5$	$C_2H_5AlCl_2$	35
$(\pi\text{-}CH_2{=}CHCH_2)_4W$	$AlBr_3$	47
$trans\text{-}Br(CO)_4W{\equiv}CCH_3$	$TiCl_4$	38
cis-Directing catalysts[a]		
WCl_6	$(C_2H_5)_4Sn$	32
WCl_6	$(C_2H_5)_3Al$[b]	33
WCl_6	$Na_3W(C_6H_5)_5$	11
WF_6	R_2AlCl (Al/W = 1)	34
$WCl_2\text{-}AlCl_3$	$(C_2H_5)_2AlCl$	48
MoX_5	R_3Al	49
$ReCl_5$	$(i\text{-}C_4H_9)_3Al$	50

[a] Low temperatures (−25 to −50°C) are generally required.
[b] Benzoyl peroxide is used additionally.

of 10^6 (37), and catalysts utilizing carbyne complexes such as $trans\text{-}Br(CO)_4W{\equiv}CCH_3$ are active at cyclopentene–W ratios of 10^9 (38).

Three-Component Catalysts. The activity of two-component polymerization catalyst systems is often markedly enhanced by the addition of a third component known as a modifier, promoter, or activator. The three-component approach (49) forms the basis for many excellent catalyst systems which are well suited for economical commercial processes. Oxygen-containing compounds such as alcohols, phenols, hydroperoxides, peroxides, molecular oxygen, and water were the first to be employed as modifiers. It was shown that for optimum activity, reaction of the modifier with the transition metal should occur prior to addition of the organometallic cocatalyst.

In a few instances, catalyst species resulting from the interaction of modifiers have been proposed. Modifiers such as alcohols react with transition-metal halides liberating HX and introducing alkoxy ligands on the metal. Epoxides apparently yield similarly substituted species, without the undesirable evolution of HX (34). β-Haloethanols yield alkoxy ligands bearing electronegative substituents on the β-carbon, resulting in improved catalyst stability and activity (34). Ethers have been used to improve catalyst stability when tetraalkyltin cocatalysts are employed. Stabilization apparently results from coordination of the ether with lower-valent transition metal species (47,51).

A seemingly endless variety of modifiers have been reported to be of value in enhancing the activity of catalyst systems based on tungsten and molybdenum halides and oxyhalides. In general, these systems yield predominantly trans polymers. In addition to the above-mentioned modifiers, the list includes carboxylic acids (52), acetals (53), alkyl hypohalites (54), nitroaromatics (55), and carbon dioxide (56). Oxygen and halogen (57), as well as quaternary amine salts (58), have been used with zero-valent tungsten complexes in combinations with organoaluminum compounds (see Organometallics).

Mechanism of Cyclopentene Polymerization

The Olefin Metathesis Reaction. Cyclopentene polymerization represents one application of a remarkable general reaction of olefins known as the olefin metathesis reaction. Pioneering studies at Phillips Petroleum demonstrated that acyclic olefins could be disproportionated at elevated temperatures over heterogeneous metathesis catalysts to mixtures of higher and lower molecular weight homologues (59–60). The development of high temperature heterogeneous disproportionation (61) and low temperature homogeneous polymerization processes occurred quite independently, until the common nature of these two developments was pointed out (6) and confirmed in sophisticated copolymerization studies (62) using ^{13}C-labeled cyclopentene. Since then, contributions from many sources toward understanding the metathesis reaction have aided greatly in the interpretation of polymerization phenomena.

Of primary significance was the discovery that this reaction proceeds via cleavage and reforming of carbon-to-carbon double bonds, without migration or exchange of protons (63–65). With acyclic olefins, redistribution of alkylidene moieties leads to a random product distribution at equilibrium (63). Attainment of thermodynamic equilibrium in the cis–trans isomer distribution normally accompanies redistribution, and cis-to-trans interconversions are inherent in the metathesis step.

$$2 \;\;>\!\!\overset{*}{C}\!\!\doteq\!\!C\!\!<\;\; \rightleftharpoons \;\; >\!\!\overset{*}{C}\!\!=\!\!\overset{*}{C}\!\!<\;\; + \;\;>\!\!C\!\!=\!\!C\!\!<$$

Elucidation of the mechanism of this reaction has posed a considerable challenge, and a variety of schemes have been proposed. One of the earliest schemes, which involved the combination of two olefins to give a quasicyclobutane intermediate, is illustrated for the reaction of 1-butene (66):

$$\begin{array}{c} CH_2{=}CHCH_2CH_3 \\ CH_2{=}CHCH_2CH_3 \end{array} \rightleftharpoons \begin{bmatrix} CH_2{\cdots}CHCH_2CH_3 \\ CH_2{\cdots}CHCH_2CH_3 \end{bmatrix} \rightleftharpoons \begin{array}{cc} CH_2 & CHCH_2CH_3 \\ \| & \| \\ CH_2 & CHCH_2CH_3 \end{array}$$

Views regarding the transition state in the metathesis reaction have evolved considerably. Currently, a fundamentally different mechanism, which is widely accepted, invokes the rearrangement of a complex containing a carbene–metal species and a coordinated olefin (67). This scheme is generally represented as proceeding through a four-membered metallocycle transition state:

$$\begin{array}{c} C{=}C^* \\ \vdots \\ C{=}M \end{array} \rightleftharpoons \begin{bmatrix} C{-}C^* \\ | \quad | \\ C{-}M \end{bmatrix} \rightleftharpoons \begin{array}{c} C^* \\ C \quad \| \\ \|\;{\cdots}\,M \\ C \end{array}$$

Several recent developments lend considerable support to this currently favored scheme. Carbene-bearing complexes such as diphenylcarbenetungsten pentacarbonyl have been found to undergo alkylidene exchange with acyclic olefins (68):

$$(CO)_5W=C(C_6H_5)_2 + \underset{H_3C}{\overset{H_3C}{>}}C{=}CH_2 \longrightarrow CH_2{=}C(C_6H_5)_2 + \text{Other products}$$

Carbene complexes are also effective catalysts for cycloolefin polymerizations (35), as well as for metathesis reactions (69). In addition, carbene precursors such as phenyldiazomethane are known to interact rapidly with WCl_6 to initiate polymerizations (36).

The relative merits of a variety of intricate arguments concerning the various mechanistic schemes proposed for this unique process have been critically assessed in several recent reviews (12–15,70).

Implications in Cyclopentene Polymerization. Several features of cyclopentene polymerizations are a direct consequence of olefin metathesis chemistry. Of prime importance is the fact that metathesis catalysts continue to react with all double bonds present in a given system throughout the course of polymerization reactions; thus, initially-formed polymer vinylene units continually undergo redistribution reactions. These reactions also consume acyclic olefins, which are usually present in traces adventitiously, but may be added intentionally for molecular weight control. The significant consequences for polymerizations which are carried to equilibrium are summarized below (6).

(1) The random incorporation of monoalkylidene units derived from acyclic olefins results in polymer chain ends. The number average degree of polymerization at equilibrium is determined by the ratio of reactants (monomer–olefin) consumed.

$$m\ RCH=CHR + n\ \bigcirc \longrightarrow m\ RCH{=}CHCH_2CH_2CH_2CH{\overline{=}_{n/m}}CHR$$

The use of large proportions of acyclic olefins can result in 1:1 codimers in good yield. This approach has formed the basis for linear diene syntheses, as in the reaction of ethylene with cyclopentene to yield 1,6-heptadiene (71):

$$\bigcirc + \underset{CH_2}{\overset{CH_2}{\|}} \longrightarrow \underset{CH_2}{\overset{CH_2}{\diagup}}$$

(2) The random coupling of bisalkylidene units [$=CHCH_2CH_2CH_2CH=$] explains the formation of macrocyclic rings as well as chains and thus, in the presence of acyclic olefins, ring-chain equilibrium results. The formation of macrocyclic oligomers in cyclopentene polymerizations has been observed (72). It was established in 1,5-cyclooctadiene polymerizations that ring formation occurs via intramolecular polymer reactions rather than by monomer-incorporation steps.

$$RCH{=}CHCH_2CH_2CH_2CH{=}CHR_{m+n} = RCH{=}CHCH_2CH_2CH_2CH{=}CHR_m + {=}CHCH_2CH_2CH_2CH{=}_n$$

At equilibrium, a most probable molecular weight distribution is expected in the chain fraction. In cyclopentene polymerizations employing trans-directing catalysts, the weight-to-number average molecular weight ratio ($\overline{M}_w/\overline{M}_n$) approached the expected value of 2 at high conversions (73).

(3) Monomer–polymer equilibrium is a significant aspect of cyclopentene polymerizations, and the reversible formation of cyclopentene from polypentenamer has been observed (72). This equilibrium is markedly influenced by the cis–trans content of the polymer as well as by the polymerization temperature. The equilibrium is shifted toward polymer by using: trans-directing catalysts, and low polymerization temperatures.

(4) Cis-to-trans interconversions proceed during polymerizations. Cyclopentene polymerizations tend to produce polymers having about 84% trans structure under conditions where equilibrium conversions are attained, excluding polymerizations conducted with cis-directing catalysts. The 84% trans structure has been independently found to be thermodynamically favored at room temperature, based on photochemical isomerization studies (74).

(5) Redistribution of monomer units during copolymerizations normally leads to the most probable sequence distributions at equilibrium. A broad distribution of monomer units was indicated qualitatively in cyclopentene–cyclooctene copolymers, as evidenced by the composition of macrocyclic oligomers which formed (75).

Reaction pathways that account for the above phenomena can be expressed in terms of the carbene–metal mechanism as follows (cis-to-trans interconversions accompany all steps):

Formation of chain ends:

$$\text{RCH=CHR} + \text{\ensuremath{\sim}\ensuremath{\sim}CH=M} \rightleftharpoons \text{\ensuremath{\sim}\ensuremath{\sim}CH} \parallel \text{RCH} + \text{CHR} \parallel \text{M}$$

Ring chain equilibrium:

$$\begin{array}{c}\text{\ensuremath{\sim}\ensuremath{\sim}CH=CH}\text{\textendash}(\text{C}_5\text{H}_8)_n\text{\textendash}\\ \text{M=CHCH}_2\text{CH}_2\text{CH}_2\text{\textendash}\end{array} \rightleftharpoons \begin{array}{c}\text{\ensuremath{\sim}\ensuremath{\sim}CH} \parallel \text{M} + \text{CH}\text{\textendash}(\text{C}_5\text{H}_8)_n\text{\textendash} \parallel \text{CHCH}_2\text{CH}_2\text{CH}_2\text{\textendash}\end{array} \quad (n \geq 2)$$

Monomer–polymer equilibrium:

cyclopentene + M=CH~~ ⇌ cyclopentene-M=CH~~

Redistribution of monomer sequences:

$$\begin{array}{c}\text{\ensuremath{\sim}A—A—CH=CH—A—A\ensuremath{\sim}}\\+\\ \text{\ensuremath{\sim}B—B—CH=M}\end{array} \rightleftharpoons \begin{array}{c}\text{\ensuremath{\sim}A—A—CH} \quad \text{CH—A—A\ensuremath{\sim}}\\ \parallel \quad + \quad \parallel \\ \text{\ensuremath{\sim}B—B—CH} \quad \text{M}\end{array}$$

The phenomena summarized above may not apply to polymerizations that terminate before attaining equilibrium, and kinetic control of the various processes can produce marked variances from these features. For example, olefins of different structures were found to vary widely in their effectiveness as molecular weight regulators under a given set of reaction conditions (34), indicating lack of attainment of equilibrium. The attainment of complete equilibrium is seldom achieved in practice, and kinetic effects often predominate.

Thermodynamics. The thermodynamic features that determine polymerizability for cyclic monomers are discussed in refs. 76 and 77. In ring-opening polymerizations, where the number and types of bonds do not change, the driving force is determined primarily by entropy and ring-strain energy. For small rings such as cyclopentene, polymerization is an antientropic process. Therefore, polymerizability is enthalpy-dependent and is derived from the release of ring strain. Calculated estimates of the low strain energy in cyclopentene have produced values of 20.5 kJ/mol (4.9 kcal/mol) (24) and 28.5 kJ/mol (6.8 kcal/mol) (25). These compare well with the calculated enthalpy for the hypothetical ring-opening polymerization of cyclopentane of −21.8 kJ/mol (−5.2 kcal/mol) (77).

The thermodynamics of cyclopentene polymerization are complicated by the fact that polymers of different cis content can be formed, and since these versions are not equal in stability, the equilibrium conversion state is influenced by the polymer microstructure, which in turn is determined by the polymerization catalyst. Equilibrium polymerization studies were first carried out with trans-directing catalysts (70), and monomer concentrations were determined as a function of temperature. From the relationship defining the dependence of $[M]_e$ on enthalpy ΔH_p and standard entropy ΔS_p^0 changes for ring-opening polymerizations:

$$RT \ln [M]_e = \Delta H_p - T \Delta S_p^0$$

values of $\Delta H_{p\text{-trans}}$ and $\Delta S_{p\text{-trans}}$ of −18.4 kJ/mol (−4.4 kcal/mol) and −62.3 kJ/(mol·K) (−14.9 kcal/(mol·°C)), respectively, were obtained. The ceiling temperature, above which high molecular weight trans polymer cannot be obtained, was extrapolated to be about 150°C.

Calorimetry (qv) was employed to measure the heat of polymerization of cyclopentene to a 65% trans polymer (78). The observed $\Delta H_{1,c}^0$ was ca −18.8 kJ/mol (−4.5 kcal/mol) at −7°C, and was estimated to be −20.5 kJ/mol (−4.9 kcal/mol) for polymerization to pure trans polymer.

A highly cis-directing catalyst based on WF_6 (34) gave equilibrium conversions of cyclopentene to a 97% cis polymer in bulk polymerizations at 0°C (74). The unique catalyst behavior prevented simultaneous attainment of cis–trans equilibrium, and allowed the monomer-cis polymer interconversion process to be examined. At 0°C, $[M]_e$ was ca 4 mol/L, and $\Delta H_{p\text{-cis}}$ and $\Delta S_{p\text{-cis}}^0$ were estimated to be −14.6 kJ/mol (−3.5 kcal/mol) and −66.5 kJ/(mol·K) (−15.9 kcal/(mol·°C)), respectively. These results were corroborated in an extensive study that utilized a cis-directing catalyst comprised of WCl_6 and tetraallylsilane (79). The ceiling temperature for the formation of cis-polypentenamer [38439-19-9] is 51°C.

The thermodynamically favored isomer composition for polypentenamer at 25°C has been determined to be ca 84% trans, based on photochemical isomerization studies (74).

Substituents on rings possessing little tendency to polymerize often render the monomer unpolymerizable, as with 2-methyltetrahydrofuran, and conflicting reports exist regarding the polymerizability of substituted cyclopentenes. 3-Methylcyclopentene is polymerizable, but reportedly not the 1- and 4-methyl isomers, nor 3-isopropylcyclopentene (34). In another study, the 3- and 4-methyl isomers of cyclopentene were both found polymerizable (80), but 1-methylcyclopentene's unreactivity was confirmed.

Other Routes to Polypentenamer

Two metathesis-based alternatives to cyclopentene polymerization have been considered. The polymerization of 1,6-cyclodecadiene (a dimer of cyclopentene) has been proposed as an indirect means of converting low cost ethylene and butadiene to polypentenamer (81):

$$C_2H_4 + 2\ C_4H_6 \rightarrow \text{[1,6-cyclodecadiene]} \rightarrow \text{[cyclodecene]} \rightarrow \text{[polypentenamer]}_n$$

However, this polymerization was found to be sluggish at low temperatures, and when elevated temperatures were employed, monomeric cyclopentene was obtained as the thermodynamically favored product.

The metathesis-induced conversion of α,ω-dienes to oligomeric polyalkenamers has been proposed as a general method. Thus a polybutenamer was obtained from 1,5-hexadiene (82). However, this method has as yet not been demonstrated for the preparation of polypentenamer.

$$(m + 1)\ CH_2{=}CH(CH_2)_n CH{=}CH_2 \rightarrow$$
$$CH_2{=}CH(CH_2)_n \text{+}CH{=}CH(CH_2)_n \text{+}_m CH{=}CH_2 + m\ CH_2{=}CH_2$$

A novel, radical-initiated polymerization involving the ring-opening of vinylcyclopropane was found to yield polypentenamer (83–84). However, this process is of little commercial value because of the low molecular weight of the product and the high cost of the monomer.

$$R\cdot + CH_2{=}CH\text{-}\triangleleft \rightarrow [RCH_2\dot{C}H\text{-}\triangleleft] \rightarrow RCH_2CH{=}CHCH_2CH_2\cdot$$

Properties of Polypentenamer

Characterization of Structure. The early characterization of polypentenamer was accomplished by ozonolysis (5). The reoccurrence of double bonds every five carbon atoms along the chain was established by the near-quantitative recovery of pentamethylenediol in reductive ozonolyses (62).

Spectroscopic examinations of polypentenamer have included infrared, ^1H and ^{13}C nmr techniques. In accordance with the metathesis mechanism of polymerization, vinyl groups are absent from the polymer. Microstructure determinations have relied primarily on infrared methods based on cis and trans vinylene absorptions at 13.7 nm and 10.35 nm. However, unduly high trans contents were often reported in early studies because of inaccuracies resulting from the use of polybutadiene absorptivities as standards. Recently, an improved method was developed utilizing absorptions at 7.1 and 10.35 nm; molar absorptivities of 152 and 9.4 L/(mol·cm) were obtained for these cis and trans bands, respectively (85).

^1H nmr spectra have been commonly employed to determine the amount of unsaturation present, but more specific information has been obtained from careful ^{13}C nmr studies (86). Quantitative Fourier-transformed pulsed spectra were obtained on a series of polypentenamers of 7–85% trans content. The α-carbons adjacent to cis and trans vinylene units gave well-resolved peaks which were used in these analyses. Further examination of the spectra revealed fine structure attributable to the isomer sequence distribution (86–87). ^{13}C-nmr studies have also allowed the observation of anomalous metathesis-induced changes in cis and trans sequence distributions during the course of polymerizations (88) (see Analytical methods).

Solution Properties. The effect of molecular weight on dilute solution properties in various solvents has been established for narrow molecular-weight-distribution polymer fractions of 80–85% trans structure (89). *trans*-Polypentenamers appear to exhibit higher intrinsic viscosities than cis polymers of comparable molecular weight, suggesting stiffer, more extended chains (90).

Solid-State Properties. The solid-state properties of polypentenamers are quite dependent on microstructure. Samples varying from ca 98% cis to 70% trans are amorphous under essentially all conditions. *cis*-Polypentenamer is quite resistant to crystallization, even at low temperatures. However, a sample possessing 99% cis content was induced to crystallize by being annealed at −75°C for one week (91), and gave a melting point of −41°C. Crystallographic data have not been obtained on this material.

For polymers containing 80–90% trans structure, crystallization occurs rapidly in unstressed samples at temperatures below 0°C. These polymers have been widely explored in evaluation programs because of their ability to crystallize readily under stress at ambient temperatures and above. As with natural rubber, this property is of technological value, since uncured stocks exhibit good tack and tensile strength and process well in mixing equipment because of the ability to develop high shear.

The rate of crystallization is markedly influenced by the trans content of polypentenamers (92), and by the temperature of crystallization (93). High rates of crystallization are often desired because of the good properties that result in many applications, but unfortunately, low temperature elasticity is severely impaired simultaneously. Thus a compromise in choice of microstructure is usually dictated.

In extensive x-ray crystallographic studies, the crystal structure of *trans*-polypentenamer (94–95) and trans-tactic copolymers of cyclopentene were determined (96). Two crystalline modifications were predicted for the homopolymer, but only one was observed. This orthorhombic unit cell has an identity period of 1.19 nm and contains two monomer units. Polymer chains are packed in a regular zig-zag fashion as with polyethylene, with the trans double bonds introducing only minor conformational defects. Random trans copolymers of cyclopentene with other cyclomonoolefins exhibit crystallinity in fibers, indicating isomorphic cocrystallization of dissimilar monomer units due to a common zig-zag conformation (96) (see also Olefin polymers).

Crystalline melting temperatures of *trans*-polypentenamers are dependent on isomer composition. Dilatometric methods were employed to measure melting temperatures for a series of polymers varying in trans content (93); the extrapolated T_m for a 100% trans polymer was 34°C. In studies employing differential scanning calorimetry (97), the equilibrium melting temperature for the pure trans polymer was extrapolated to be 40°C. Values for the heat of fusion of 8.08 and 12.0 kJ/mol (1930 and 2870 cal/mol) have been reported (96–97).

Self-reinforcement through the development of crystallinity during stretching of 85–89% *trans*-polypentenamers has been examined in some detail using cross-linked samples (98–99). The melting temperature of stretched samples increased with extension in accordance with Flory's equation (100). The dependence of melting temperature on elongation was found to be less than that of natural rubber because of the high heat of fusion of polypentenamer (99).

The glass transition temperature (T_g) is only slightly dependent on the microstructure of polypentenamer. In a detailed examination of the effect of structure and crystallinity on T_g, the T_g ranged from −106 to −91°C, as measured by differential scanning calorimetry, for polymers ranging from pure cis to 85% trans structure (97). The presence of crystallinity increased the T_g only slightly. In torsional pendulum experiments, the range of the T_g was found to be from −110 to −93°C for similar samples (101). Because of the low T_g of cis-polypentenamer and its exceptionally weak tendency to crystallize, its low temperature flexibility is excellent and is superior to that of other hydrocarbon elastomers (91).

Detailed calorimetric studies have been employed to measure the heat capacity and heat of fusion of polypentenamer as a function of polymer microstructure (102–103) (see Calorimetry).

Vulcanization. Because of the high level of unsaturation present, polypentenamers respond well to conventional cross-linking systems based on sulfur or sulfur-donor agents and accelerators. Several laboratories have examined the properties of vulcanized polypentenamer formulations in comparisons with those obtained from conventional elastomers, and the effects of variations in vulcanization parameters have been explored in some detail (11,34,92,104–108). The cross-linking process is highly efficient in polypentenamer (see Rubber compounding). In a detailed evaluation of cure parameters on vulcanizate properties, optimum results were obtained by using low levels of zinc oxide, stearic acid, and accelerator, and high cure temperatures of about 170°C were desirable (92). Cross-links appear to be quite stable, and cure reversion is minimal.

trans-Polypentenamers exhibit excellent mixing and processing characteristics, comparable to those of natural rubber (92). Compounding ingredients are rapidly and smoothly incorporated on the open mill and in internal mixers. Carbon black dispersion ratings as determined by microscopy are equal to those obtained with natural rubber mixes, and exceed those obtained with other general purpose elastomers. High molecular weight versions are said to incorporate very high levels of carbon black and oil without loss of desirable properties (11,105,107). Uncured stocks exhibit good tack and adhesion to fabric.

The properties of compounded polypentenamer stocks and their vulcanizates have prompted comparisons both with natural rubber (see Rubber, natural) and with polybutadiene (qv). The tensile strength (green strength) of uncured oil- and black-loaded stocks is high, and may exceed that of natural rubber stocks equally loaded in some cases. This characteristic is particularly attractive in the production of radial-ply tires, where green strength of carcass-coat stocks is important during tire construction. On the other hand, polypentenamer vulcanizates possess excellent resistance to abrasion, approaching that of 1,4-polybutadiene stocks. As a result, polypentenamer has been examined as a potential tire elastomer suitable in many areas of tire construction.

The resistance to aging of polypentenamer vulcanizates has been reported to be equal or superior to that of other tire elastomers. Retention of physical properties after exposure to hot air was better than that of polybutadiene and styrene–butadiene rubber (11), and resistance to crack formation on exposure to ozone was very good when only low levels of antiozonant were added (105).

Most polypentenamers have been of the high-trans variety. For these materials of ca 85–90% trans content, crystalline melting points were in the range of 15–20°C, and low-temperature hardening of vulcanizates has posed a limitation in the practical application of these materials. Polymers having a somewhat lower trans content near 80% provide substantially better low temperature properties (107). The loss in tack, green strength, and abrasion resistance that resulted with the decrease in trans content could be compensated for by using higher molecular weight polymers. However, the processability of these polymers was not measured.

cis-Polypentenamer has also been evaluated in carbon-black-loaded formulations (91). Because of its inability to crystallize readily, *cis*-polypentenamer does not mix easily with compounding ingredients, although satisfactory results have been obtained on a hot mill. Likewise, these stocks do not possess good green strength or tack. The uniqueness of this polymer is its excellent low-temperature flexibility, which even exceeds that of an 80% *cis*-polybutadiene. This is a consequence of the amorphous nature and low T_g of *cis*-polypentenamer.

Modification of Polypentenamer

Copolymerization. Metathesis polymerizations provide several pathways for the formation of random, block, or graft copolymers, but the properties of these materials have not been widely examined. In studies of the mechanism of cycloolefin polymerizations, a series of cyclopentene–cyclooctene copolymers were prepared (62,96). Partial ozonolysis of these copolymers revealed a distribution of C_{10}:C_{13}:C_{16} dimers that indicated a predominance of C_5–C_8 successions in the polymer as a measure of pronounced randomness (62). The x-ray spectra of cyclopentene–cycloheptene and cyclopentene–cyclooctene copolymers indicated the lack of crystallinity due to homopolymer sequences, but did reveal that random copolymers are isomorphic, and that linear trans monomer units can cocrystallize (96).

Copolymers of cyclopentene with bicyclic and polycyclic monomers have been described in the patent literature. Comonomers such as norbornadiene, vinylnorbornene, and dicyclopentadiene are used in amounts ranging from less than 1% to about 10% in order to improve processing and storage (cold-flow) properties of polypentenamer (109–112). In general, monomers possessing the highly reactive norbornene ring polymerize in nonrandom fashion, and therefore special procedures are required in order to obtain rubbery copolymers. These methods include sequential addition of monomers, whereby the more reactive monomer is added slowly to a polymerizing cyclopentene solution (113), and the use of organoaluminum iodides in combination with relatively large amounts of olefins as molecular-weight regulators (114).

An interesting copolymer of cyclopentene with the reaction product of 1,5-cyclooctadiene and hexachlorocyclopentadiene has been prepared (115). This material exhibits good flame resistance and resistance to swelling by oils as a consequence of its high chlorine content. In addition, it possesses good processability, and can be cured with sulfur-based cure systems (see Flame retardants).

Chemical Modification. Most chemical alterations of polypentenamer have been through reactions at the double bond. Addition of chlorine and bromine occurs smoothly in methylene chloride (116). The reaction product from the chlorination of *cis*-polypentenamer has a higher T_g (42°C) than that from the trans polymer (25°C). An unusual stereoregulation in the halogenation process was signified by the existence of crystallinity in these adducts (117).

The hydrogenation of polypentenamer has been accomplished using toluenesulfonyl hydrazide. As the extent of hydrogenation of the double bonds was increased, the percent crystallinity and melting temperature of the polymer increased (118). The dynamic mechanical properties of hydrogenated polypentenamer were found to resemble those of a linear polyethylene (119).

In a series of reactions, ionic derivatives of polypentenamer have been prepared. Hydroformylation was accomplished under mild conditions, yielding pendant aldehyde groups which were then converted to nitrile substituents via aldoxime intermediates (120) (see Oxo process). The addition of methyl thioglycolate to polypentenamer provided a route to a series of interesting materials possessing pendant ionic groups (121).

The noncatalyzed thermal rearrangement of *cis*-polypentenamer at 200–270°C results in cis-to-trans isomerization, accompanied by an unusual cyclization reaction (122):

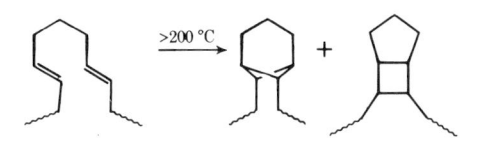

BIBLIOGRAPHY

1. J. Brandrup and E. H. Immergut, eds., *Polymer Handbook,* 2nd ed., John Wiley & Sons, Inc., New York, 1975.
2. F. Hofmann, *Chem. Ztg.* **57,** 5 (1933).
3. B. C. Anderson, C. L. Hoover, and O. Vogl, *Macromolecules* **2,** 686 (1969).
4. U.S. Pat. 3,074,918 (Jan. 22, 1963), H. S. Eleuterio (to E. l. du Pont de Nemours & Co., Inc.).
5. G. Natta, G. Dall'Asta, and G. Mazzanti, *Angew. Chem.* **76,** 765 (1964).
6. K. W. Scott and co-workers, *Adv. Chem. Ser.* **91,** 399 (1969).
7. A. J. Amass, *Br. Polym. J.* **4,** 327 (1972).
8. M. Dimonie, Cr. Oprescu, and Gh. Hubca, *Mater. Plast.* **10,** 417 (1973).
9. B. D. Babitskii and V. A. Kormer, *Sint. Kauch.,* 317 (1976).
10. N. Calderon, *J. Macromol. Sci. Rev. Macromol. Chem.* **7,** 105 (1972).
11. G. Dall'Asta, *Rubber Chem. Technol.* **47,** 511 (1974).
12. R. J. Haines and G. J. Leigh, *Chem. Soc. Rev.* **4,** 155 (1975); additional reviews are cited therein.
13. J. J. Rooney and A. Stewart, *Chem. Soc. Special Periodical Reports—Catalysis* **1,** 277 (1977).
14. T. J. Katz in F. G. A. Stone and R. West, eds., *Advances in Organometallic Chemistry,* Vol. 16, Academic Press, Inc., New York, 1977, pp. 283–313.
15. N. Calderon, J. P. Lawrence, and E. A. Ofstead in ref. 14, Vol. 17, 1979, pp. 449–492.
16. C. Lassau, D. V. Quang, and M. Hellin, *Hydrocarbon Process.,* 108 (Aug. 1973).
17. U.S. Pat. 3,784,481 (Jan. 8, 1974), C. Lassau, R. Stern, and L. Sajus (to Institut Français du Pétrole).
18. W. Graulich, paper presented at the 13th Annual Meeting of the International Institute of Synthetic Rubber Producers, Munich, FRG, 1972.
19. *Synthetic Organic Chemicals, Report of the U.S. International Trade Commission, 1974.*
20. *Toxic Substances List,* U.S. Dept. of Health, Education, and Welfare, Washington, D.C., 1973.
21. D. E. Van Sickle, F. R. Mayo, and R. M. Arluck, *J. Am. Chem. Soc.* **87,** 4832 (1965).
22. *Selected Values of Physical and Thermodynamic Properties of Hydrocarbons and Related Compounds,* published for the American Petroleum Institute, Carnegie Press, Pittsburgh, Pa., 1953.
23. R. W. Gallant, *Physical Properties of Hydrocarbons,* Vol. 2, Gulf Publishing Co., Houston, Tex., 1970.

24. J. D. Cox, *Tetrahedron* **19,** 1175 (1963).
25. P. v. R. Schleyer, J. E. Williams, and K. R. Blanchard, *J. Am. Chem. Soc.* **92,** 2377 (1970).
26. Ital. Pat. 784,307 (Dec. 2, 1967), G. Dall'Asta and G. Carella (to Montecatini Edison S.p.A.).
27. W. Grahlert and co-workers, *Plaste Kaut.* **22,** 229 (1975).
28. T. J. Katz, S. J. Lee, and N. Acton, *Tetrahedron Lett.*, 4247 (1976).
29. Fr. Pat. 1,394,380 (Mar. 12, 1965), G. Natta, G. Dall'Asta, and G. Mazzanti (to Montecatini Edison S.p.A.).
30. P. R. Marshall and B. J. Ridgewell, *Eur. Polym. J.* **5,** 29 (1969).
31. N. A. Sokolova and co-workers, *Vysokomol. Soed. Ser. B* **17,** 232 (1975).
32. G. Pampus and G. Lehnert, *Makromol. Chem.* **175,** 2705 (1974).
33. R. J. Minchak and H. Tucker, *Polym. Prepr. Am. Chem. Soc. Div. Polym. Chem.* **13,** 885 (1972).
34. P. Gunther and co-workers, *Angew. Makromol. Chem.* **14,** 87 (1970).
35. Fr. Pat. 2,110,598 (Apr. 6, 1972), Y. Chauvin, J. Soufflet, and D. C. Nhu-Hung Phung (to Institut Français du Pétrole).
36. B. A. Dolgoplosk and co-workers, *Eur. Polym. J.* **10,** 901 (1974).
37. Y. Chauvin, D. Commereuc, and D. Cruypelinck, *Makromol. Chem.* **177,** 2637 (1976).
38. E. O. Fischer and W. R. Wagner, *J. Orgmet. Chem.* **116,** C21, (1976).
39. Fr. Pat. 2,043,406 (Sept. 16, 1970), K. Nützel, K. Dinges, and F. Haas (to Farbenfabriken Bayer A.G.).
40. P. R. Hein, *J. Polym. Sci. Polym. Chem. Ed.* **11,** 163 (1973).
41. G. Pampus, G. Lehnert, and D. Maertens, *Polym. Prepr. Amer. Chem. Soc. Div. Polym. Chem.* **13,** 880 (1972).
42. Fr. Pat. 1,542,040 (Apr. 26, 1968), C. A. Uraneck and W. J. Trepka (to Phillips Petroleum Co.).
43. Fr. Pat. 2,012,675 (Mar. 20, 1970), N. Schön, G. Pampus, and J. Witte (to Farbenfabriken Bayer A.G.).
44. Ital. Pat. 859,384 (filed Mar. 28, 1969), G. Dall'Asta and P. Meneghini (to Montecatini Edison S.p.A.).
45. U.S. Pat. 3,558,518 (Jan. 26, 1971), E. A. Zuech (to Phillips Petroleum Co.).
46. Fr. Pat. 1,561,026 (Nov. 11, 1969), E. A. Zuech (to Phillips Petroleum Co.).
47. V. Kormer, *J. Polym. Sci. A1* **10,** 251 (1972).
48. U.S. Pat. 3,476,728 (Nov. 4, 1969), G. Natta, G. Dall'Asta, and G. Mazzanti (to Montecatini Edison S.p.A.).
49. Neth. Pat. 6,601,466 (Feb. 15, 1965), G. Natta and G. Carella (to Montecatini Edison, S.p.A.).
50. Fr. Pat. 2,025,142 (May 14, 1970), G. Günther, W. Oberkirch, and G. Pampus (to Farbenfabriken Bayer A.G.).
51. G. Lehnert and co-workers, *Makromol. Chem.* **175,** 2617 (1974).
52. Ger. Off. 2,051,798 (Apr. 27, 1972), F. Küpper, R. Streck, and K. Hummel (to Chemishe Werke Hüls).
53. Ger. Off. 2,006,767 (Aug. 19, 1971), N. Schön and co-workers (to Farbenfabriken Bayer A.G.).
54. U.S. Pat. 3,607,853 (Sept. 21, 1971), W. Oberkirch, P. Günther, and J. Witte (to Farbenfabriken Bayer A.G.).
55. Fr. Pat. 2,039,196 (Sept. 16, 1970), K. Nützel, F. Haas, and G. Marwede (to Farbenfabriken Bayer A.G.).
56. U.S. Pat. 3,780,009 (Dec. 18, 1973), G. Dall'Asta (to Montecatini Edison S.p.A.).
57. U.S. Pat. 3,597,403 (Aug. 3, 1971), E. A. Ofstead (to The Goodyear Tire & Rubber Co.).
58. Fr. Pat. 2,074,042 (June 22, 1971), H. W. Ruhle (to Esso Research and Engineering Co.).
59. Belg. Pat. 620,440 (Jan. 21, 1963), R. L. Banks (to Phillips Petroleum Co.).
60. R. L. Banks and G. C. Bailey, *Ind. Eng. Chem. Prod. Res. Dev.* **3,** 179 (1964).
61. G. C. Bailey, *Catal. Rev.* **3,** 37 (1969).
62. G. Dall'Asta and G. Motroni, *Eur. Polym. J.* **7,** 707 (1971); G. Dall'Asta, *Makromol. Chem.* **154,** 1 (1972).
63. N. Calderon and co-workers, *J. Am. Chem. Soc.* **90,** 4133 (1968).
64. J. C. Mol, J. A. Moulijn, and C. Boelhouwer, *Chem. Commun.*, 633 (1968).
65. A. Clark and C. Cook, *J. Catal.* **15,** 420 (1969).
66. C. P. C. Bradshaw, E. J. Howman, and L. Turner, *J. Catal.*, 269 (1967).
67. J. L. Hérisson and Y. Chauvin, *Macromol. Chem.* **141,** 161 (1970).
68. C. P. Casey and T. H. Burkhardt, *J. Am. Chem. Soc.* **96,** 7808 (1974).
69. W. R. Kroll and G. Doyle, *Chem. Commun.*, 839 (1971).

70. N. Calderon, E. A. Ofstead, and W. A. Judy, *Angew. Chem.* **88**, 433 (1976).
71. Fr. Pat. 1,511,381 (Jan. 26, 1968), G. C. Ray and D. L. Crain (to Phillips Petroleum Co.).
72. E. A. Ofstead and N. Calderon, *Makromol. Chem.* **154**, 21 (1972).
73. K. W. Scott, *Polym. Prepr. Am. Chem. Soc. Div. Polym. Chem.* **13**, 874 (1972).
74. E. A. Ofstead, paper presented at Division of Rubber Chemistry, Inc., Spring Meeting, Detroit, Mich., May, 1973.
75. J.-P. Arlie and co-workers, *Makromol. Chem.* **175**, 861 (1974).
76. F. S. Dainton and K. J. Ivin, *Q. Rev. Chem. Soc.* **12**, 61 (1958).
77. F. S. Dainton, T. R. E. Devlin, and P. A. Small, *Trans. Faraday Soc.* **51**, 1710 (1955).
78. D. Kranz and M. Beck, *Angew. Makromol. Chem.* **27**, 29 (1972).
79. I. A. Oreshkin and co-workers, *Eur. Polym. J.* **13**, 447 (1977).
80. E. A. Ofstead, unpublished results.
81. F. W. Küpper and R. Streck, *Makromol. Chem.* **175**, 2055 (1974).
82. G. Dall'Asta and co-workers, *Polym. Prepr. Am. Chem. Soc. Div. Polym. Chem.* **13**, 910 (1972).
83. T. Takahashi and I. Yamashita, *Bull. Chem. Soc. Jpn.* **37**, 131 (1964).
84. T. Takahashi and I. Yamashita, *J. Polym. Sci. Part B* **3**, 251 (1965).
85. C. Tosi, F. Ciampelli, and G. Dall'Asta, *J. Polym. Sci. Polym. Phys. Ed.* **11**, 529 (1973).
86. C. J. Carman and C. E. Wilkes, *Macromolecules* **7**, 40 (1974).
87. H. Y. Chen, *J. Polym. Sci. Polym. Lett. Ed.* **12**, 85 (1974).
88. K. J. Ivin, D. T. Laverty, and J. J. Rooney, *Makromol. Chem.* **179**, 253 (1978).
89. G. Gianotti, U. Bonicelli, and D. Borghi, *Makromol. Chem.* **166**, 235 (1973).
90. G. Dall'Asta, P. Meneghini, and U. Gennaro, *Makromol. Chem.* **154**, 279 (1972).
91. G. Dall'Asta and P. Scaglione, *Rubber Chem. Technol.* **42**, 1235 (1969).
92. F. Haas and D. Theisen, *Kaut. Gummi Kunstst.* **23**, 502 (1970).
93. A. Capizzi and G. Gianotti, *Makromol. Chem.* **157**, 723 (1972).
94. G. Natta and co-workers, *Makromol. Chem.* **91**, 87 (1966).
95. G. Natta and I. W. Bassi, *J. Polym. Sci. Part C* **16**, 2551 (1967).
96. G. Motroni, G. Dall'Asta, and I. W. Bassi, *Eur. Polym. J.* **9**, 257 (1973).
97. C. E. Wilkes, M. J. Pekló, and R. J. Minchak, *J. Polym. Sci. Symp.* **43**, 97 (1973).
98. G. Kraus and J. T. Gruver, *J. Polym. Sci. Part A-2* **10**, 2009 (1972).
99. U. Flisi and co-workers, *Eur. Polym. J.* **9**, 1187 (1973).
100. P. J. Flory, *J. Chem. Phys.* **15**, 397 (1947).
101. J. K. Gillham and J. A. Benci, *J. Appl. Polym. Sci.* **18**, 3775 (1974).
102. B. V. Lebedev, I. B. Rabinovich and V. Ya. Lityagova, *Dokl. Akad. Nauk SSSR* **237**, 877 (1977).
103. B. V. Lebedev and co-workers, *Vysokomol. Soed.* **A18**, 2444 (1976).
104. F. Haas and co-workers, *Rubber Chem. Technol.* **43**, 116 (1970).
105. H. J. Jahn and D. Theisen, *Ind. Gomma* **192**, 44 (1973).
106. W. Scheele and S. Fleige, *Kaut. Gummi Kunstst.* **27**, 125 (1974).
107. H. Tucker, R. J. Minchak, and J. H. Macey, *Polym. Eng. Sci.* **15**, 360 (1975).
108. B. D. Babitskii and co-workers, *Kauch. Rezina* **8**, 8 (1976).
109. U.S. Pat. 3,598,796 (Aug. 10, 1971), K. Nützel and co-workers (to Farbenfabriken Bayer A.G.).
110. U.S. Pat. 3,634,376 (Jan. 11, 1972), K. Nützel and F. Haas (to Farbenfabriken Bayer A.G.).
111. U.S. Pat. 3,778,420 (Dec. 11, 1973), J. D. Brown and C. A. Uraneck (to Phillips Petroleum Co.).
112. U.S. Pat. 3,781,257 (Dec. 25, 1973), G. Pampus, J. Witte, and G. Marwede (to Farbenfabriken Bayer A.G.).
113. Ger. Off. 2,437,104 (Feb. 20, 1975), H. Wakabayashi and co-workers (to Kanegafuchi Kagaku Kogyo).
114. Ger. Off. 2,451,848 (July 5, 1975), R. J. Minchak (to B. F. Goodrich Co.).
115. U.S. Pat. 3,634,374 (Jan. 11, 1972), A. J. Bell (to The Goodyear Tire & Rubber Co.).
116. G. Dall'Asta, P. Meneghini, and U. Gennaro, *Makromol. Chem.* **154**, 279 (1972).
117. G. Dall'Asta and co-workers, *Makromol. Chem.* **165**, 83 (1973).
118. K. Sanui, W. J. MacKnight, and R. W. Lenz, *J. Polym. Sci. Polym. Lett. Ed.* **11**, 427 (1973).
119. K. Sanui, W. J. MacKnight, and R. W. Lenz, *Macromolecules* **7**, 101 (1974).
120. *Ibid.*, 953 (1974).
121. K. Sanui, R. W. Lenz, and W. J. MacKnight, *J. Polym. Sci. Polym. Chem. Ed.* **12**, 1965 (1974).
122. M. A. Golub, *J. Polym. Sci. Polym. Lett. Ed.* **12**, 295 (1974).

EILERT A. OFSTEAD
The Goodyear Tire & Rubber Company

STYRENE–BUTADIENE RUBBER

In 1940 the United States government realized the deleterious effect that the spreading of war in the Far East would have on natural rubber supplies and consequently established a government corporation, The Rubber Reserve Company, which would accumulate and maintain a stockpile of natural rubber as well as launch a synthetic rubber research and development program. United States military demand for rubber increased markedly during World War II and inaugurated exchange of technical knowledge and coordinated research programs amongst interested United States rubber, oil, and chemical firms. Polybutadiene (qv) was rejected as an unsatisfactory substitute for natural rubber and a styrene–butadiene copolymer [9003-55-8] made in emulsion with a charge ratio of 25% styrene and 75% butadiene was chosen as the best possible general purpose rubber for development (see also Rubber, natural).

In mid 1942, production of GR-S (now SBR) began in a government plant; in 1945 production exceeded 830,000 metric tons. To achieve this goal, the United States government financed construction of fifteen SBR plants, two butyl rubber plants (see Butyl rubber), sixteen butadiene production facilities and five styrene plants. Between 1946 and 1955 these plants were sold to various private companies that have since maintained and improved most of them.

The history and development of large-scale production of Buna S (styrene–butadiene copolymer) in Germany has been reviewed; this program ran concurrently with the one conducted in the United States (1).

In 1948 D'Ianni (2) reported copolymerization of butadiene with styrene which involved a sodium-based alfin catalyst discovered by Morton (3) in 1945. Various efforts to commercialize this process have been unsuccessful (4).

Solution methods of polymerization have become popular recently and some of the resulting new rubbers are competing in cost and quality with emulsion SBR. Among these new polymers are various styrene–butadiene copolymers that are similar in properties to the emulsion type.

Copolymerization of butadiene with styrene via alkyllithium catalyst (5) was later found to be quite effective for tailoring the composition, structure, and properties of these copolymers. Similar copolymers were developed independently (6–7) in the United States and are now available under the trademarks Solprene (Phillips Petroleum Company), Kraton (Shell Chemical Company), and Stereon (Firestone Tire and Rubber Company).

It is possible to prepare solution copolymers of styrene and butadiene having a nonrandom, block structure. Block copolymers are molecules in which two or more chemically or structurally dissimilar segments are joined. Each segment or block usually is a long sequence of units of a single monomer, but it may also be a long sequence of randomly copolymerized units. Both types of block copolymers are known (see Copolymers).

Physical Properties

The glass transition temperature (T_g) is the temperature at which a soft viscous polymer changes to a hard glass. Thermodynamically, it is a second-order transition characterized by small changes in refractive index, specific volume, thermal expansion coefficient, sound velocity, etc. The T_g is dependent on structure and on order in the polymer.

For SBR copolymers prepared by emulsion polymerization at 50°C, the T_g is calculable from the bound styrene content (S = wt fraction of styrene) by:

$$T_g = (-85 + 135\ S)/(1 - 0.5\ S)$$

For a similar copolymer prepared at 5°C, T_g, is given by:

$$T_g = (-78 + 128\ S)/(1 - 0.5\ S)$$

In the case of block copolymers, two glass transitions are observed: the upper one corresponding to the glassy styrene blocks and the lower one corresponding to the rubber butadiene block (8). T_g was determined by differential thermal analysis for a series of solution copolymers (9), and values were reported from ca −90°C for very low styrene content to −36° for 51% styrene. The block-styrene contents were determined from the T_gs using a published equation.

Relationships between structural features and properties of the solution SBR copolymers have been studied (8,10–12). These results were simplified (10–11) by relating such properties as processibility, heat-buildup of vulcanizates in service, traction of tire treads, and abrasion resistance, to two measurable molecular parameters: macrostructure (molecular weight distribution and branching) and glass transition temperature.

Viscosity measurements in dilute solution are often used to estimate molecular weights of polymers using the Kuhn-Mark relation:

$$[\eta] = KM^a$$

where $[\eta]$ is the intrinsic viscosity (in dL/g), M is the polymer molecular weight, and K and a are experimentally determined constants. Values of K and a for SBR prepared at 50°C were 5.3×10^{-4} and 0.66, respectively, when the viscosity measurements were made in toluene at 30°C (see Rheological measurements).

Hot emulsion SBR polymers have molecular weight of $(1.5–4.0) \times 10^5$ (viscosity average) or $(0.3–1.0) \times 10^5$ (number average); cold rubbers, 2.8×10^5 (viscosity average), 5.0×10^5 (weight average) or $(1.1 - 2.6) \times 10^5$ (number average). Hot rubbers have lower molecular weight and more low molecular weight material as well as more chain-branching and cross-linking than cold rubbers. The solution SBR polymers have a narrower molecular weight range than emulsion polymers.

Because of its random sequence distribution, SBR polymers containing more than 10 mol % styrene do not crystallize even when stretched or cooled, and x-ray diffraction gives an amorphous halo. However, rubbers containing 5 mol % or less styrene can show some crystallization and show preferred orientation when stretched.

High molecular weight block SBR polymers are characterized by a microscopic two-phase structure that is formed by associations between blocks of the same kind of polymer (13). Elastomeric block polymers consist of an elastomer block that forms a continuous phase and a shorter block, which is usually glassy, that forms a dispersed phase. The unique physical properties of block copolymers arise from their two-phase structure, which is somewhat independent of polymer composition and the number of blocks per molecule. These unique properties include multiple transition temperatures, melt flow behavior resembling branched polymers and high tensile strength in the uncured state.

Additional physical properties for a typical SBR polymer, containing about 23.5% bound styrene content, are given in Table 1.

Raw Materials

Butadiene and styrene are the chief raw materials required to manufacture SBR. Others, which are required in smaller amounts, are the various emulsifiers, modifiers (eg, thiols), catalysts, initiators, shortstops, coagulating agents, antioxidants, and antiozonants.

Water is a major ingredient of the emulsion polymerization recipe as well as of the coagulation and product-washing operations. At one time water was considered a rather inconsequential component in the economics of the system; however, this can no longer be considered a valid presumption. The availability of sufficient quantities of water of adequate purity for polymerization and the subsequent treatment of the effluent water from the isolation operation prior to discharge, or prior to recycling, has become a significant factor in the overall economics of emulsion plants (see Water).

Similarly, solvent losses from solution polymerization processes have become very important considerations both from the economic viewpoint and the standpoint of tougher air quality standards. Air quality is now a very important industrial consideration for all synthetic rubber plants. Some organic solvents have fallen into serious disfavor as potential health hazards, whereas others merely tend to contribute to the hydrocarbon content of an industrial complex and thus generally affect the air quality of a given geographical area (see Air pollution; Air pollution control methods).

Butadiene. Butadiene production capacity in the United States is over 1,820,000 metric tons per year. About 60% of United States butadiene production comes from plants that dehydrogenate butane or butenes. The other 40% consists of coproduct from steam crackers, which produce primarily ethylene. Production from the latter source is expected to increase over the next few years as ethylene capacity increases. Butadiene for SBR amounts to about 55% of the total monomer production (see Butadiene).

Crude butadiene, produced either by dehydrogenation or obtained as by-product, must be extracted from other olefins or saturated hydrocarbons before it is used. Extractive solvents such as dimethylacetamide, dimethylformamide, acetonitrile, N-methyl pyrrolidinone, and furfural are used.

Dehydrogenation Processes. One-stage dehydrogenation (Houdry process) of n-butane is carried out at three companies. The Houdry process uses a catalyst of aluminum and chromium oxides and a 95+% n-butane feed. Butadiene is produced from n-butane in 57–63% yield. Process conditions can be altered so that the chief product is either butadiene or butenes or a mixture. One company uses a two-stage dehydrogenation process: in the first stage n-butenes are produced from a 98% n-butane feed, and in the second stage a compressed air–steam mixture is heated, mixed with n-butene, and passed over an oxidative dehydrogenation catalyst. C_4 components are recovered and the butadiene is extracted and purified.

The two other companies use n-butenes as the exclusive feed for butadiene. The n-butenes are obtained from a C_4 cut (from catalytic or steam-cracking off gases) from

Table 1. Poly(butadiene-co-styrene), SBR (About 23.5% Bound Styrene Content)[a]

Property	Unvulcanized	Pure-gum vulcanizate	Vulcanizates with ca 50 phr[b] of carbon black
density, g/mL	0.933 (0.9325–0.9335)	0.980 (0.940–1.000)	1.150
coefficient of volume expansion, $\beta = (1/V)(\delta V/\delta T)$, K^{-1}	660 × 10^{-6}	660 × 10^{-6}	530 × 10^{-6}
thermal			
glass transition temperature, K	209–214	221	221
heat capacity, C_p, kJ/(kg·K)[c]	1.89	1.83	1.50
$\delta C_p/\delta T$, kJ/(kg·K)[c]	3.2 × 10^{-3}		
thermal conductivity, W/(m·K)		0.190–0.250	0.300
heat of combustion, MJ/kg CO_2[c]	−56.5		
optical			
refractive index, n_D	1.5345 (1.534–1.535)		
dn_D/dT, K^{-1}	−37 × 10^{-5}		
electrical			
dielectric constant (1 kHz)	2.5	2.66	
dissipation factor	0.0009	0.0009	
mechanical			
compressibility, $B = -(1/V_0)(\delta V/\delta P)$, Pa^{-1} [d]	530	510	400
$\delta B/\delta P$, Pa^{-2} [d]	−2.7	−2.0	−1.8
bulk modulus (isothermal), GPa[e]	1.89	1.96	2.50
bulk wave velocity (longitudinal wave), km/s		1.485	1.510
strip (longitudinal wave) velocity, ν_1 (1 kHz), m/s		73	161
$\delta\nu_1/\delta T$, m/(s·K)		−0.2	
ultimate elongation, %		400–600	400–600
tensile strength, MPa[f]		1.4–3.0	17–28
initial slope of stress–strain curve, Young's modulus, E (60s), MPa[f]		1.6 (1.0–2.0)	3–6
shear modulus, G (60s), MPa[f]		0.53 (0.3–0.7)	2.0 (2.0–2.5)
shear compliance, J (60s), MPa^{-1} [f]		1.9	0.5
creep rate, $(1/J)(\delta J/\delta \log t)$, %/unit log t		7 (3–10)	12
complex dynamic shear modulus, G*, 1 Hz			
storage modulus, G', MPa[f]	0.66 (0.66–0.71)	0.76 (0.44–1.6)	8.7 (2.5–8.7)
loss modulus, G'', MPa[f]	0.087 (0.36–0.087)	0.083 (0.054–0.11)	1.9
loss tangent, G''/G'	0.13 (0.05–0.13)	0.11 (0.07–0.18)	0.22 (9.14–0.22)
resilience (rebound), %		65	40 (40–50)

[a] Ref. 14.
[b] phr = parts per hundred of rubber.
[c] To convert J to cal, divide by 4.184.
[d] To convert Pa to psi, divide by 6895.
[e] To convert GPa to psi, multiply by 145,000.
[f] To convert MPa to psi, multiply by 145.

612 ELASTOMERS, SYNTHETIC (SBR)

which isobutylene and butanes have been removed. Dehydrogenation of the n-butenes to butadiene is effected catalytically using Dow Type B catalyst (chromium oxide-stabilized calcium nickel phosphate) or Shell catalyst 205 (ferric oxide, chromium oxide, and potassium oxide).

By-Product Production of Butadiene. Cracking hydrocarbons, eg, naphtha fractions, gas oil, condensate, ethane–propane mixtures, and even crude oil, is widely used in the manufacturing of ethylene, and butadiene is one of the coproducts. Coproduct butadiene yield is dependent upon the severity of the cracking operation as well as the feed used. Presumably all United States ethylene producers now recover coproduct butadiene. For some solution SBR processes the by-product butadiene is separated as a hydrocarbon stream containing 30–60% butadiene. After removal of small amounts of poisons and the materials at the low and high boiling ends, the by-product stream can be used and further purification to the conventional polymerization grade concentration is unnecessary.

Heavy petroleum distillates can be coked to yield butadiene-containing gases and naphtha, gas oils, and petroleum coke.

Styrene. United States nameplate production capacity for styrene is about 3,200,000 t/yr, about 10% of which is used in SBR manufacture. Formerly, the principal use of styrene was SBR but this use is steadily declining in a relatively fixed tire market (see Styrene).

All United States styrene plants catalytically dehydrogenate high-purity ethylbenzene in the vapor phase to produce styrene. Yields of 90% are reported. The ethylbenzene is obtained mostly by alkylating benzene with ethylene; however, some is obtained by fractionating mixed xylene streams (20–30% ethylbenzene) that result from the catalytic reforming of naphtha (see Xylenes and ethylbenzene).

Styrene may also be a coproduct of propylene oxide manufacture. In this process ethylbenzene is oxidized to its hydroperoxide which reacts with propylene to yield propylene oxide and a coproduct, methyl phenyl carbinol. The carbinol is then dehydrated to styrene.

A third styrene production method economically extracts the styrene that is contained in pyrolysis gasoline, a by-product obtained when naphtha is cracked for olefin and for olefin manufacture (see Gasoline).

Raw materials for styrene manufacture seem to be available; however, since aromatics are an important ingredient for upgrading gasoline octane rating (especially no lead or low lead gasoline), it seems likely that spot shortages and gradual price increases can be expected (see Feedstocks).

Polymerization

The SBR chain contains styrene units as well as the three butadiene forms (cis-1,4, trans-1,4, and 1,2 or vinyl). These copolymers of styrene and butadiene may be *randomly* dispersed mixtures of the two monomers or *blocks* in which large segments of each kind follow one another, eg, SSSSBBBBBSSSBBBBB (S = styrene, B = butadiene), or grafts in which the segments dangle from the main chain.

Emulsion SBR normally contains about 23% styrene, which is randomly dispersed with butadiene in the polymer chains. The percentages of butadiene unit structures are about 18% cis, 65% trans, and 17% vinyl. SBR that is made in solution (see below) contains about the same amount of styrene as that made in emulsion and both the random and block copolymers have been produced in solution. They have lower trans, slightly lower vinyl, and higher cis contents than emulsion SBR.

The absence of spontaneous termination in alkyl lithium polymerization results in narrow chain-length distribution and few short chains. Since there is no transfer reaction, the polymer chains are linear. These linear polymers of narrow molecular weight distribution contrast with the branched, broader distribution polymers that are produced by free-radical emulsion polymerization. Chain branching and molecular-weight distribution can be effectively controlled in solution polymerization by properly utilizing the reactivity of the polymer anions (see below).

Styrene shortages and increasing styrene monomer costs have prompted the evaluation of medium-vinyl polybutadiene rubbers made in solution, as well as the utilization of emulsion SBRs with a reduced styrene content of 15% and lower, compared with the usual 23%.

Two methods of polymerization are in widespread use today and both are used for SBR. The emulsion process is used for standard SBR and the solution process for the other varieties.

Emulsion Polymerization. In emulsion polymerization, the monomer is emulsified in a medium such as water with the aid of emulsifying agents such as soaps and synthetic emulsifiers (see Emulsions; Surfactants). Initially the monomer is present in the form of emulsion droplets dispersed in the continuous aqueous phase. Many investigators formerly thought that the polymerization occurred within the emulsion droplets; however, subsequently it was shown that the emulsion droplets play only a minor role and that the polymerization in emulsion systems occurs primarily in the aqueous phase and not in the monomer droplets (see also Latex technology).

The classical micellar theory of emulsion polymerization as developed by Harkins (15–16), and described in mathematical detail by Smith-Ewart (17) has been undergoing a reappraisal. Considerable work published in the last ten to fifteen years has shown that the process of emulsion polymerization is more complicated and more subtle than previously believed.

The micellar theory generally was corroborated by investigations with sparingly soluble monomers like styrene and butadiene (18–23). Later, others (24–29) found that the rate of particle polymerization for styrene was not always constant, as predicted by Smith-Ewart kinetics, particularly at large particle sizes and high rates of initiation. Medvedev (30) dissented from the micellar theory, and proposed that the free radicals in the aqueous phase do not penetrate sufficiently into the latex particles, since they are restricted by the high internal viscosity of the polymer–monomer particles, and that polymerization, therefore, is a process occurring on the surface of the particles. Brodnyan (31) and co-workers concluded that, although there were only small kinetic differences between the two theories, the Medvedev mechanism may be more realistic because the changing particle size distributions indicate a shifting mechanism of polymerization as the surface-to-volume ratio of the particle changes. Roe (32) questioned the importance and role of micelles and concluded that monomer radicals are initiated and grow in the aqueous phase to assume the aspects of a polymer particle stabilized against flocculation by adsorbed emulsifier. Roe's calculations, which disregarded the micelles, result in the same mathematical expression for polymerization rate as the Smith-Ewart theory.

Robb (33) found that the number of particles per unit volume of latex varied with conversion during the steady-state period of emulsion polymerization of styrene. This, however, was inconsistent with the Smith-Ewart theory which required a constant number of particles during the same period. Other investigators have obtained good polymerization rates in the absence of surfactants (34–36).

A homogeneous nucleation theory has been developed (37–39) that allows for the exceptions to the micellar theory. According to this theory there is a homogeneous solution of monomer(s), stabilizer, and initiator in the aqueous medium at the beginning of polymerization. Polymerization begins with the formation of solvated oligomeric chains and these grow by free-radical combination of an initiator radical with a monomer molecule, followed by monomer addition, until at some critical chain length these oligomeric particles precipitate from solution and collapse to form a new phase of primary particles for propagation. The soluble oligomeric particles that subsequently are formed in solution either form more new particles or are captured by preexisting particles. Propagation is accomplished in the primary particles and the characteristics of propagation are governed by the type of monomers in the polymerization system their solubilities in water, the initiator concentration, the stabilizer type and the amount and conditions of agitation. The role of the surfactant, according to the homogeneous nucleation theory, is that of a stabilizer for the oligomeric particles that precipitate from solution to form primary particles. The surfactant is not required as a source of micelles or as an active and necessary component in the initiation process as required by the Harkins theory.

A theory has been proposed for emulsion polymerization that is based on a single internally consistent model that predicts several experimentally available data, such as the particle size, the conversion-time relationship, the dependence of particle size and molecular weight on conversion, and the influence of surfactant, initiator, and monomer (40). It differs from earlier theories in that the derived relationships are quantitative and contain no adjustable parameters. This excellent theory applies to a single monomer batch polymerization in which the initiator is soluble in the aqueous phase only, and its half-life is much longer than the duration of the conversion process, and the polymer is soluble in the monomer but insoluble in water.

In a homogeneous (solution or bulk) system containing free-radical initiators, the maximum degree of polymerization (the chain length) attainable is too low for elastomer use (molecular weights $<10^4$) unless the rate of polymerization is exceedingly slow. The low molecular weight of free radical, solution-polymerized polybutadiene is promoted by the relatively low rate of propagation to the high rate of termination: $k_p = 8$ L/(mol·s) and $k_t = 10^8$ L/(mol·s) plus a complicating cross-linking reaction (41). This limitation does not apply to emulsion systems for which both high propagation rate and high molecular weight are simultaneously possible, because the emulsion physically isolates each growing radical and prevents them from terminating one another. Other advantages for emulsion systems are: (1) the high transfer rate of the heat of polymerization through the aqueous phase (because of the considerable heat of diene polymerization (ca 75 kJ/mol or 18 kcal/mol), good heat transfer is necessary if temperature control is to be maintained) and (2) the easy removal of unreacted monomers and relatively constant fluidity despite the presence of high concentrations of high polymer.

The chain reaction in free-radical emulsion polymerization is initiated by radicals that are generated either by decomposition of peroxides or diazo compounds or by an oxidation-reduction reaction (a redox couple). Early free radical emulsion polymerizations, which were carried out at 50°C or higher, reflected the temperature dependency of the formation rate of free radicals and, therefore, of polymerizations. Subsequent research led to the discovery of oxidation-reduction reactions capable of generating radicals in sufficient numbers for adequate polymerization rates at temperatures as low as −40°C (see Initiators).

Commonly employed dissociative initiators are potassium peroxydisulfate, $K_2S_2O_8$, benzoyl peroxide and azobisisobutyronitrile (AIBN); cumene hydroperoxide and p-menthane hydroperoxide are oxidants for redox couples. The combination of potassium peroxydisulfate with a mercaptan, such as dodecyl mercaptan, is used to polymerize butadiene and styrene. In hot recipes the mercaptan furnishes free radicals, by its reaction with the peroxydisulfate, and it limits the molecular weight of the polymer by reacting with and terminating the growing polymer chain; the resulting thiyl radical initiates the growth of another chain. This use of mercaptan as a chain transfer agent or modifier is of great commercial importance in the manufacture of emulsion SBR and polybutadiene since the mercaptan allows control of the toughness of the product which otherwise may limit further processing.

A standard industrial polymerization recipe known as the mutual, standard, GR-S, or hot recipe is given in Table 2.

The conversion rate to polymer at 50°C is 5–6%/h. Polymerization is terminated at 70–75% conversion or less since higher conversions lead to polymers with inferior physical properties, presumably because of cross-linking in the latex particle to form microgel or highly branched structures. The termination is effected by the addition of a shortstop, such as hydroquinone (ca 0.1 part by weight), which reacts rapidly with radicals and oxidizing agents, thus destroying any remaining initiator and polymer free radicals. The unreacted monomers are then removed: first, the butadiene by flash distillation at atmospheric then at reduced pressure; second, the styrene by steam stripping in a column. A dispersion of antioxidant, such as a staining type, is added (1.25 parts) to protect the product (see Antioxidants). The latex is partially coagulated (creamed) by the addition of brine and then fully coagulated with dilute sulfuric acid or aluminum sulfate. The coagulated crumb is washed, dried, and baled for shipment. One of the first important improvements in this basic emulsion polymerization process was the adoption of continuous processing shown schematically in Figure 1.

The styrene, butadiene, soap solution, initiator, and activator solution (an auxiliary initiating agent) are pumped continuously from storage tanks into and through a series of agitated reactors and are maintained at the proper temperature and at such a rate that the desired degree of conversion is reached at the exit from the last reactor. Shortstop is then added, the latex is warmed by the addition of steam and the unreacted butadiene is flashed. Excess styrene is then steam-stripped and the latex is finished, often by blending with oil, and by creaming, coagulating, drying and baling (Fig. 1).

Cold Rubber. Another key improvement in SBR production was the use (about 1947) of more active radical initiating systems that permitted polymerization at 5°C with high rates of conversion. The cold SBR polymers produced at the lower temperature, but stopped at 60% conversion, were found to have properties superior to those of hot SBR. Two typical recipes for a cold SBR are given in Table 2.

At 5°C 60% conversion to polymer occurs in about 8–12 h. In the persulfate hot recipe, the mercaptan initiates and functions as a chain transfer agent, but in the cold

Table 2. Typical SBR Recipes

	Hot SBR 1000	Cold SBR 1500	Cold SBR 1502
composition, phm[a]			
butadiene	71	71	71
styrene	29	29	29
potassium peroxydisulfate	0.3		
p-menthane hydroperoxide (PMHP) (100%)		0.12	0.12
n-dodecylmercaptan (DDM)	0.5		
tert-dodecylmercaptan (TDM)		0.2	0.18
emulsifier makeup, phm[a]			
water (final monomer–water ratio adjusted to 1:2)	190	190	190
sodium stearate (soap flakes)	5		2.5
disproportionated rosin acid soap		4.5–5	2.55
trisodium phosphate dodecahydrate (buffer electrolyte)		0.50	
tripotassium phosphate (buffer electrolyte)			0.40
Tamol N[b] (secondary emulsifier)		0.02–0.1	0.02–0.1
Versene Fe-3[c] (iron complexing agent)		0.01	0.01
sodium dithionite (oxygen scavenger)		0.025	0.025
sulfoxylate activator makeup, phm[a]			
ferrous sulfate heptahydrate		0.04	0.04
Versene Fe-3[c]		0.05	0.05
sodium formaldehydesulfoxylate (SFS)		0.10	0.10
water		10	10
short stop makeup, phm[a]			
sodium dimethyldithiocarbamate (SDD)		0.10	0.15
sodium nitrite		0.02	0.02
sodium polysulfide		0.05	
hydroquinone	0.1		
water		8.0	8.0
polymerization conditions			
temperature, °C	50	5	5
final conversion, %	72	60–65	60–65
coagulation	salt–acid	salt–acid	salt–acid
antioxidant (about 1 part on rubber)	staining	staining	Agerite Geltrol[d] or Wingstay S[e] (nonstaining)
properties of coagulated dried rubber			
organic acid content, wt %	4–6	5–7	5–7
styrene content, wt %	24	24	24
Mooney viscosity at 100°C, ML-4, min	48	46–58	46–58

[a] Parts per hundred parts of monomer.
[b] Sodium salt of naphthalenesulfonic acid-formaldehyde condensate (W R Grace).
[c] Disodium salt of ethylenediaminetetracetic acid (EDTA) (The Dow Chemical Co.)
[d] Alkylated-arylated bisphenolic phosphite (R.T. Vanderbilt Co., Inc.).
[e] Styrenated phenol (Goodyear Tire & Rubber Co., Chemical Division).

redox recipes the mercaptan is not essential to initiation. The main difference lies in the initiator systems. The phosphates and ethylenediaminetetraacetic acid (EDTA) act as buffers and complex with ferrous ion, thereby maintaining a constant solubility and limiting its reactivity (oxidation potential). The reaction is very rapid at 0°C, the emulsion components are in separate phases and the reaction occurs only at the interface.

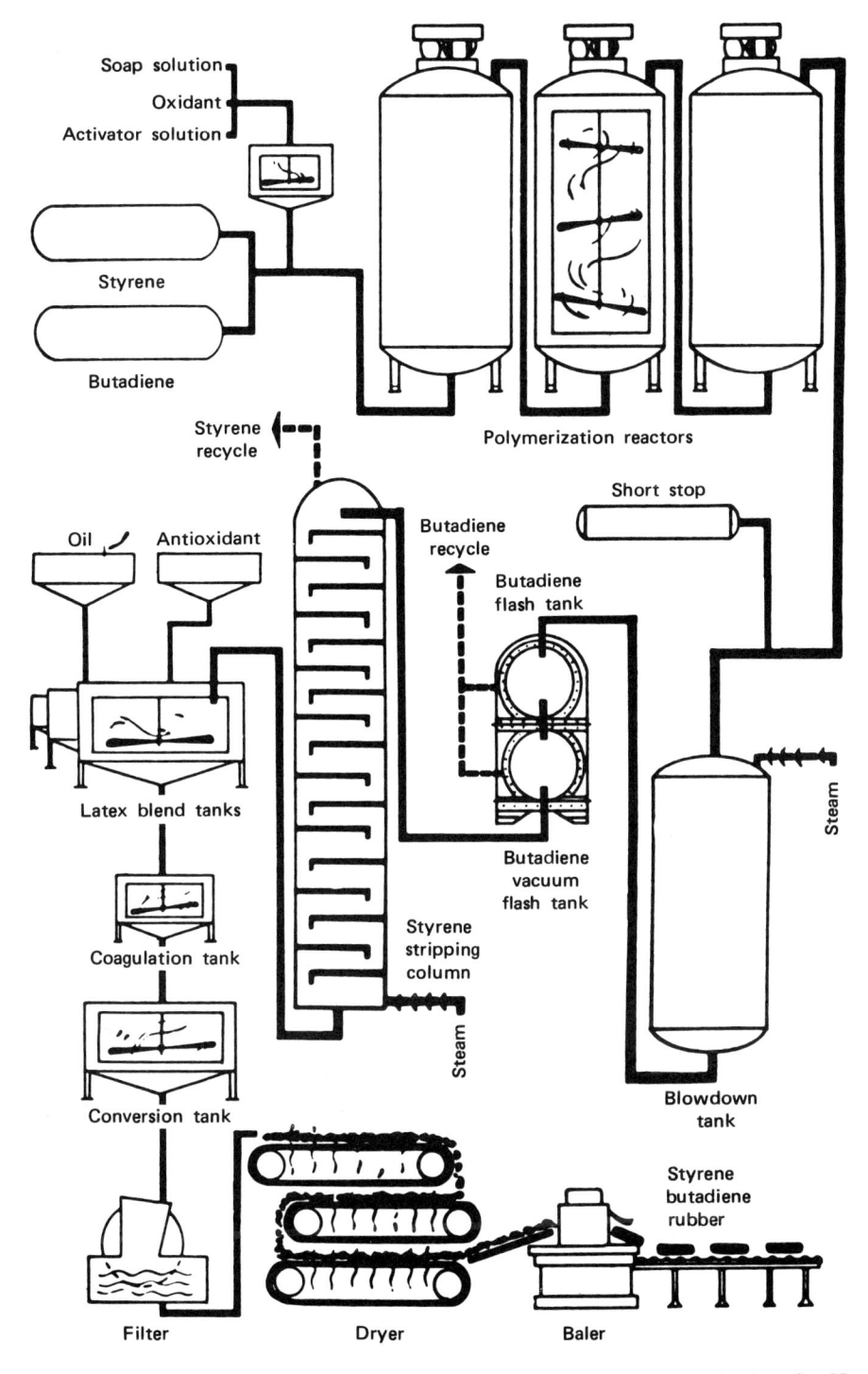

Figure 1. The production of styrene–butadiene rubber. Courtesy of Gulf Publishing Co., Houston, Tex. (42).

In many cold recipes, auxiliary reducing agents, such as sulfoxylates, are used as components of a redox cycle, ie, the ferric ion is reduced to the ferrous state. Reducing sugars are no longer in widespread use because of their cost and susceptibility to bacterial attack during storage.

A considerable improvement in the quality of SBR resulted from the availability of cold rubbers. It was found that difficult to process, gel-free elastomers with a higher than usual molecular weight could be modified by the addition of up to 50 phr (parts per hundred of rubber) of petroleum-base oils to permit handling in ordinary equipment. These extending oils improve processing characteristics but do not sacrifice useful physical properties. In the commercial polymerization process the oil is usually emulsified and blended with the latex before coagulation.

Recent trends in SBR production include specialized product grades designed for specific uses. The use of lighter-colored soaps, shortstops, antioxidants, and extending oils has permitted color control of SBR, which is important for many non-tire applications. Dithiocarbamates have been substituted for hydroquinone as shortstop except in some hot SBR where dark color is not objectionable. A shortstop such as sodium dimethyldithiocarbamate is more effective than hydroquinone in terminating radicals and destroying peroxides at the lower temperatures that are encountered in preparing the cold rubbers.

Masterbatches. Another improvement directed toward specific end uses is the preparation of carbon-black masterbatches of regular and oil-extended cold SBR. These are of interest to rubber manufacturers having limited mixing capacity and to those who wish to avoid factory handling of loose carbon blacks (see Carbon, carbon black).

Addition of carbon black in masterbatch manufacture is accomplished by slurrying the pigment in water which may contain an anionic surface-active agent. The SBR latex and slurry are then mixed with the emulsified extending oil and antioxidant. The resulting blend is coagulated rapidly, with vigorous agitation, by addition of an excess of acid or salt–acid mixture or acid plus a coagulating aid. The creaming step is usually omitted since it results in preferential precipitation of an unredispersible black with resultant poor qualities of finished product. Other methods for black masterbatching include blending the components in a steam jet and coagulation by addition of the latex–black blend to the salt–acid mixture.

Solution SBR. At present, successful stereospecific solution copolymerization has been carried out with only lithium and alkyl lithium catalysts. Coordination catalysts, such as are used for polybutadiene and polyisoprene, have not been satisfactory for styrene–butadiene copolymers (see Coordination compounds; Catalysis).

Several solution SBRs are offered commercially; two of them are random (Firestone's Stereon and Phillips' Solprene) and the others are block copolymers (Phillips and Shell). The random copolymers are rubbery and are similar to emulsion SBR yet with several improved properties. The block polymers tend to be thermoplastic and are not recommended for tire use. Both random products have narrower molecular weight distribution, less chain branching, higher cis content, lighter color and (usually) less nonrubber constituents than the emulsion SBRs. As a result, they are reported to have better abrasion resistance, better flex, higher resilience and lower heat build-up than the emulsion rubbers. Tensile, modulus, elongation and cost are comparable. The Solprenes are being offered in three oil-extended versions (with 37.5 phr aromatic or naphthenic oil and 50 phr naphthenic oil). Figure 2 shows a typical solution SBR flow diagram.

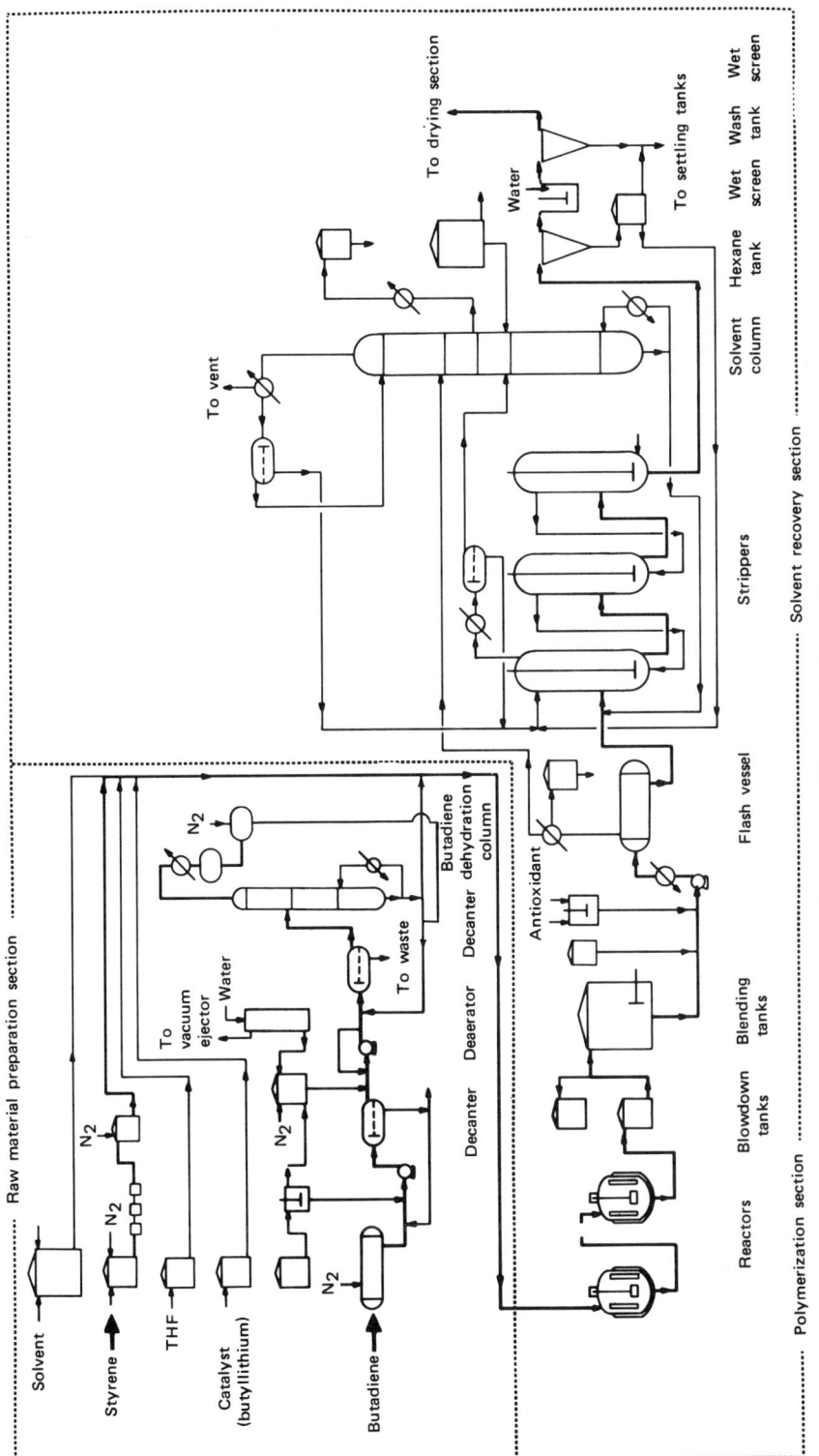

Figure 2. Styrene–butadiene rubber by continuous process.

Numerous aspects and especially the kinetics of alkyl lithium-initiated polymerization have been reviewed (43). These anionic or live polymerizations are capable of yielding two distinctive copolymers previously mentioned: block copolymers and uniformly sequenced copolymers, in addition to intermediate products. One interesting example of the latter is the radial teleblock polymers which have several polybutadiene chains extending from a central hub with a polystyrene block attached to the outward end of each polybutadiene segment.

Solution SBR polymerization readily lends itself to the design of many useful variations in the polymer macro- and microstructure (44). If a mixture of pure styrene and butadiene monomers is placed in a pure hydrocarbon solvent and initiated with n-butyllithium, a narrow molecular weight distribution butadiene–styrene tapered-block polymer is obtained. The addition of small amounts of ethers, tertiary amines, phosphites, surfactants, sulfides, and other modifiers suppresses or eliminates this tendency towards block polymerization. Several of these additives also change the stereochemistry of the diene polymerization. It is possible to obtain vinyl contents ranging from 9 to 90% or more (45–46).

Another method, which results in the elimination of the tendency to block formation without using modifiers and without changing stereochemistry, utilizes continuous addition of the desired mixture of monomers (either combined or separately) throughout the polymerization at a controlled rate (47–48). The monomer addition rate must be less than the rate of polymerization ensuring that the living polymer mixture never stores unreacted monomer for a sufficient time to permit long-chain block formation.

Still another means of controlling the solution polymerization involves the introduction of Na^+, K^+, Rb^+, or Cs^+ ions into the polymerization system (49). When ions of alkali metals other than Li^+ are present, styrene is incorporated early in the copolymerization and, in fact, can be incorporated at will depending on the ion concentration. Suitable choice of counter-ion combinations produces copolymers of variable or constant composition—with or without microstructure changes—depending on the conditions.

Linear polymers of narrow molecular-weight distribution exhibit Newtonian viscosity behavior and, consequently, cold flow in storage. This undesirable property can be eliminated by introducing a small degree of nonlinearity by adding a polyvinyl monomer, such as divinylbenzene, or a branching agent, such as tin tetrachloride (50–51). Increased breadth of molecular-weight distribution may also be obtained by mixing polymers of various chain lengths in solution.

Commercial production of solution styrene–butadiene copolymers requires more attention to detail than production of solution polybutadiene rubber. In addition to the high viscosity of the reaction mixture and the large heat of reaction, the large difference between the relative reactivity ratios of the two monomers is an important concern.

Various processes have been devised for producing copolymers that contain nonblock styrene during copolymerization. Composition is controlled and uniformly-spaced styrene units are produced in a continuous process in which the monomer ratio is maintained by the regulated addition of faster-polymerizing butadiene (52).

In another scheme the solvated monomer mixture and the catalyst are charged continuously into a reaction zone at 90–150°C, and the product mixture is withdrawn continuously into a finishing reactor or tubular zone (53). This very fast polymerization process produces such a small proportion of an end-block of styrene that copolymer properties are not adversely affected.

Compounding and Processing

Although synthetic rubber temporarily lost some of its market to natural rubber because of the increased manufacture of radial tires, it is expected that more synthetic rubber will be used in radials in the future and it should regain its former share of the tire rubber market.

Modern styrene–butadiene rubber no longer requires the repeated mixing cycles of the earlier synthetic rubber. Its extrusion properties are superior to those of natural rubber and its stocks have less tendency to scorch during processing. Although cold SBR is often preferable to hot SBR for optimum physical properties, use of hot SBR types can contribute to both processing and product improvements; they break down more readily, develop less heat, and accept more filler during processing. Blends for SBR and other rubbers, such as natural rubber or cis-polybutadiene also are made. Compounding recipes should be proportioned to balance the requirements for each type of rubber used.

All types of SBR use basic compounding recipes, as do other unsaturated hydrocarbon polymers that share the common ingredients of sulfur, accelerators, antioxidants, antiozonants, activators, fillers, and softeners or extenders. SBR requires less sulfur than natural rubber for curing. The usual range is 1.5–2.0 phr of sulfur; however, this range should be based on the rubber hydrocarbon only for oil-extended SBR. Because of their lower unsaturation, all styrene–butadiene rubbers are slower curing than natural rubber and require more acceleration.

Processing of SBR compounds is similar to that of natural and nonstyrene–butadiene rubbers. The ingredients are mixed in internal mixers or on mills and may be extruded, calendered, molded and cured in conventional equipment. Mixing procedures vary with the compound. In general, the rubber, zinc oxide, antioxidant, and stearic acid are mixed; then the carbon black is added in portions with the softener or oil (this may be considered a black masterbatch). It may be desirable at this point to dump the rubber into sheetform molds and cool the batch. The second phase includes adding the other ingredients, accelerator and sulfur being added last. Mixing is then continued until the sulfur is well dispersed.

Tire tread-wear and aging properties are superior to those of natural rubber. Building tack is still poor and dynamic properties are such that heavy-duty tires become too hot in use for satisfactory life or service. When reinforcing fillers such as carbon black or silica are not used, physical properties are poor. Cold SBR is preferable to hot SBR for maximum tensile properties and abrasion resistance. Hot SBR is no longer used for tire-tread compounds. In applications where lower physical properties can be tolerated hot SBR types may contribute processing advantages.

Present-day styrene–butadiene rubber often has better abrasion characteristics and better crack initiation resistance than natural rubber. The low unsaturation of styrene–butadiene rubbers ensures their good heat resistance and heat-aging qualities. SBR extrusions are smoother and maintain their form better than natural rubber does.

Mixing procedures for rubber stocks vary with different companies. Some plasticize the rubber in a separate operation from mixing before adding blacks and other ingredients. The procedure may be largely determined by the capacity and the mixing ability of the available equipment.

In working with a butadiene-rubber stock, the operator may add carbon black and other ingredients to make a masterbatch, thus eliminating the preliminary plasticizing step. If additional plasticization is required, the masterbatch may be remilled before its use in the finished batch, rather than plasticizing the rubber by itself. However, excessive remilling of SBR compounds may lead to gel formation and poorer extrusions.

The masterbatches of certain synthetic tread stocks must be remilled prior to the finishing step in which the vulcanizing ingredients are added. This is equally necessary for certain natural rubber tread stocks as well. Generally, the processing of synthetic tread stocks requires less milling time than that of natural rubber tread stocks because one step is eliminated.

The absence of green tack in SBR stocks necessitates application of a layer of natural rubber cement to ensure adhesion on all surfaces of the tread that are to be joined to other surfaces when the tire is built. This is done mechanically on the bottom side of the tread-tubing unit.

Bias and bias-belted tires are being displaced by the radial-ply types. Radial-ply tires are reported to be better than bias-belted tires because they give better tread wear, less rolling resistance, better ride at higher speeds, and lower running temperatures (54). The growth in use of radial tires has required larger proportions of natural rubber in carcasses and sidewalls in place of SBR to improve green tire strength. However, there is some indication that the poor green strength of SBR may be improved without loss of desirable properties and, if these more recent findings are proven successful in extensive tire testing, the increased use of SBR in radial tire construction may be possible (55–56).

The DuPont "Delphi Forecast" suggested that by 1980, dry blends of powdered rubber and compounding ingredients will be used for 20% of all domestic extruded rubber. This survey predicted that the rubber industry will be largely continuous in its operations and much more highly automated. New forms of materials, especially pellets and powders, will be displacing the familiar bale supplies of elastomers. Many fabricators, especially the smaller ones, will be buying their materials either fully or partially preformulated.

The block styrene–butadiene copolymers are thermoplastic elastomers that do not need a curing step and yield reprocessable scrap, which leads to a potentially lower production cost for the final product. Compounding and processing procedures for the radial block copolymers have been described (57).

The block styrene–butadiene copolymers are available in crumb form in the uncompounded state. Generally, they are compounded with fillers (qv), oils (cycloparaffin types) and sometimes other polymers, such as polystyrene. Polymer producers offer these compounded versions in pellet form for a variety of applications.

A number of processing methods are application to the block copolymers: solution or hot-melt techniques, extrusion, injection molding, blow molding, thermoforming, calendering, and Banbury or other high-shear mixing equipment (see Plastics technology).

Specifications and Standards

Commercial Grades. The International Institute of Synthetic Rubbers Producers, Inc., is now responsible for assigning numbers to various commercial grades of SBR and of butadiene polymers and latexes. The numbering system instituted under the government synthetic rubber program is still, in the main, adhered to by private industry, although there is a recent tendency for the companies to assign an ASTM code number or their own codes or trade names for newer products. The Institute numbering system is arranged as follows:

Series:
- 1000   hot polymers
- 1500   cold polymers
- 1600   cold black masterbatch with 14 phr or less oil
- 1700   cold oil masterbatch
- 1800   cold oil–black masterbatch with more than 14 phr oil
- 1900   miscellaneous dry polymer masterbatches
- 2000   hot latexes
- 2100   cold latexes

There are many types of SBR, but only a portion of these are available from any one manufacturer. Each producer will generally prefix the code number with a distinguishing trade name.

Economic Aspects and Uses

Passenger tires and tire products account for the major portion of SBR consumption, but extensive use of these synthetics occurs in the fabrication of a wide variety of products. About 65% of all SBR elastomer produced in the United States is used in the manufacture of passenger car tires. Two expanding markets for SBR are adhesives (qv) and chewing gum. A wide variety of SBRs are available for adhesive applications, and several of the crumb forms were designed specifically for the adhesives industry. The free-flowing crumb form eliminates the need for milling, cutting, or grinding equipment, and since these crumb particles retain the porous nature of the coagulated rubber, they can be dissolved in a solvent faster than milled or pelletized bale rubber. The crumb SBR polymers that are intended for adhesives applications are light colored and contain a nonstaining, nondiscoloring antioxidant.

The chewing-gum SBR polymers are food-grade rubbers that are designed specifically for use in gum base. These polymers are available in both bale and crumb forms and allow the gum manufacturers a wide latitude of compounding capabilities (see Gums).

The block styrenic copolymers are intended for applications in adhesives, caulks, sealants, coatings, food packaging, toys, tubing, sheeting, molding equipment, belting, shoe soles and heels, and miscellaneous uses. Their consumption increased to 24,000 metric tons by 1976, and they were typically priced at 21–26 ¢/kg, compared to 14–18 ¢/kg for typical random SBR copolymers (58). More than half of the total consumption of the styrenic block copolymers is for footwear. Kraton 1101 and 1102 rubbers are styrene–butadiene–styrene (SBS) block copolymers, whereas Kraton G, which has an ethylene–butylene midblock, is probably a styrene–diene–styrene copolymer in which the unsaturation of the diene component has been reduced by partial hydrogenation. This polymer has good oxidation resistance and is suitable for use in rubber wire and cable cover (see Insulation, electric).

The established pattern for SBR polymer consumption will undoubtedly continue and grow as the use of rubber products of all types increases. New synthetics such as the ethylene–propylene rubbers (qv) and others that are being developed may replace SBR in some applications, but SBR's current position as the most widely used synthetic will no doubt continue for some time.

A summary of the worldwide SBR producers and their respective capacities is available in the Stanford Research Institute *Chemical Economics Handbook* (59). The consumption for SBR in the United States and Canada is expected to climb from ca 1,234,000 t in 1976 to ca 1,829,000 t in 1987 (60).

BIBLIOGRAPHY

1. H. Logemann and G. Pampus, *Kaut. Gummi Kunstst.* **23**, 479 (1970).
2. J. D. D'Ianni, *Ind. Eng. Chem.* **40**, 253 (1948).
3. A. A. Morton, *Ind. Eng. Chem.* **42**, 1488 (1950).
4. *Chem. Eng. News*, 17 (Mar. 27, 1972).
5. N. N. Chesnokova and A. A. Korotkov, *Thesis of Reports to the Ninth Conference on the General Problems of Chemistry and Physics of High Molecular Compounds, Moscow, U.S.S.R., 1956*, p. 32.
6. W. W. Crouch, *Rubber Plast. Age* **42**, 276 (1961).
7. G. Alliger and F. C. Weissert, *Rev. Gen. Caoutchouc* **43**, 1321 (1966).
8. G. Kraus, C. W. Childers, and J. T. Gruver, *J. Appl. Polym. Sci.* **11**, 1581 (1967).
9. J. N. Anderson, F. C. Weissert, and C. J. Hunter, *Rubber Chem. Technol.* **42**, 918 (1969).
10. G. Alliger and F. C. Weissert, *Rev. Gen. Caoutchouc* **43**, 1321 (1966).
11. F. C. Weissert and B. L. Johnson, *Rubber Chem. Technol.* **40**, 590 (1967).
12. M. Hoffman, G. Pampus, and G. Marwede, *Kautschuk Gummi* **22**, 691 (1969).
13. R. Zelinski and C. W. Childers, *Rubber Chem. Technol.* **41**, 161 (1968).
14. L. A. Wood, *Rubber Chem. Technol.* **49**(2), 189 (1976).
15. W. D. Harkins, *J. Am. Chem. Soc.* **69**, 1428 (1947).
16. W. D. Harkins, *J. Polym. Sci.* **5**, 217 (1950).
17. W. V. Smith and R. H. Ewart, *J. Chem. Phys.* **16**, 592 (1948).
18. H. Gerrens, *Adv. Polym. Sci.* **1**, 234 (1959).
19. B. M. E. van der Hoff, *J. Phys. Chem.* **60**, 1250 (1956).
20. E. Bartholome and co-workers, *Z. Elektrochem.* **61**, 522 (1957).
21. M. Morton, P. P. Salatiello, and H. Landsfield, *J. Polym. Sci.* **8**, 111 (1952).
22. *Ibid.*, p. 215.
23. *Ibid.*, p. 279.
24. R. H. Ewart and C. I. Carr, *J. Phys. Chem.* **58**, 640 (1954).
25. J. W. Vanderhoff and co-workers, *J. Polym. Sci.* **20**, 225 (1956).
26. J. W. Vanderhoff and E. B. Bradford, *Tappi* **39**, 650 (1956).
27. C. P. Roe and P. D. Brass, *J. Polym. Sci.* **24**, 401 (1957).
28. W. H. Stockmeyer, *J. Polym. Sci.* **24**, 314 (1957).
29. B. M. E. van der Hoff, *J. Polym. Sci.* **33**, 487 (1958).
30. S. S. Medvedev, *International Symposium on Macromolecular Chemistry*, Pergamon Press, New York, 1959, pp. 174–190.
31. J. G. Brodnyan and co-workers, *J. Colloid. Sci.* **18**, 73 (1963).
32. C. P. Roe, *Ind. Eng. Chem.* **60**(9), 20 (1968).
33. I. D. Robb, *J. Polym. Sci. A-1* **7**, 417 (1969).
34. W. J. Priest, *J. Phys. Chem.* **56**, 1077 (1952).
35. R. H. Ottewill and J. N. Shaw, *Kolloid-Z.* **215**, 161 (1967); **218**, 34 (1967).
36. A. Kotera, K. Furusawa, and Y. Takeda, *Kolloid-Z.* **239**, 677 (1970).
37. R. M. Fitch, *Off. Dig. J. Paint Technol.* **37**, 32 (1965).
38. R. M. Fitch, *Br. Polym. J.* **5**, 467 (1973).

39. R. M. Fitch and C. H. Tsai in R. M. Fitch, ed., *Polymer Colloids,* Plenum Press, New York, 1971, pp. 73–116.
40. J. L. Gardon, *Rubber Chem. Technol.* **43,** 74 (1970).
41. D. H. Richards, *Chem. Soc. Rev.* **6,** 239 (1977).
42. *Pet. Refiner* **35**(12), 164 (1956).
43. H. L. Hseih and W. J. Glaze, *Rubber Chem. Technol.* **43,** 22 (1970).
44. R. N. Cooper and L. L. Nash, *Rubber Age* **104,** 55 (May 1972).
45. U.S. Pats. 2,975,160 (Mar. 14, 1961), R. P. Zelinski (to Phillips Petroleum Co.); 3,281,383 (Aug. 9, 1962), R. P. Zelinski and H. L. Hsieh (to Phillips Petroleum Co.).
46. Can. Pat. 867,450 (March 30, 1971), A. F. Halasa (to the Firestone Tire & Rubber Co.).
47. T. C. Bouton and S. Futamura, *Rubber Age,* 33 (Mar. 1974).
48. U.S. Pat. 3,094,512 (June 18, 1963), J. N. Short (to Phillips Petroleum Co.).
49. U.S. Pat. 3,294,768 (Dec. 27, 1966), C. F. Wofford (to Phillips Petroleum Co.).
50. U.S. Pat. 3,363,659 (Jan. 16, 1968), N. F. Keckler and B. L. Johnson (to the Firestone Tire & Rubber Co.).
51. C. A. Uraneck and J. N. Short, *J. Appl. Polym. Sci.* **14,** 1421 (1970).
52. Can. Pat. 769,096 (Oct. 10, 1967), (to the Firestone Tire & Rubber Co.).
53. Brit. Pat. 1,136,189 (Dec. 11, 1968), (to Shell International Research).
54. *Chem. Eng. News* **48,** 36 (Apr. 27, 1970).
55. R. G. Bauer, E. M. Friedman, and D. C. Rubio, *Am. Chem. Soc. Coat. Plast. Prepr.* **37**(1), 778 (Mar. 1977).
56. E. J. Buckler and co-workers, *paper presented at the International Institute of Synthetic Rubber Producers (IISRP) meeting in Monte Carlo, May 5, 1977.*
57. J. R. Haws and T. C. Middlebrook, *Rubber World* **167**(4), 27 (1973).
58. "Elastomers-Synthetic" in *Chemical Economics Handbook,* Stanford Research Institute, Menlo Park, Calif., 1977, 525.6621C.
59. "C_4 Hydrocarbons" in *Chemical Economics Handbook International,* Vol. II, Stanford Research Institute, Menlo Park, Calif., 1976, SBR Elastomer-1.
60. International Institute of Synthetic Rubber Producers, Inc. (IISRP) forecast, *Rubber Plast. News,* 32 (Feb. 21, 1977).

R. G. BAUER
The Goodyear Tire & Rubber Company

THERMOPLASTIC ELASTOMERS

Thermoplastic resins are polymeric structures that soften or melt at elevated temperatures, allowing them to be processed into fabricated products that, when cooled, recover the physical and chemical properties of the original resin. An elastomer, as defined by the ASTM, is a material that can be stretched at room temperature to twice its own length, held for 5 min, and upon release will return to within 10% of its original length over a similar period of time (1). In simpler terms, it is a material that, when stretched, snaps back into its original shape. Thermoplastic elastomers, often referred to as elastoplastics, are a versatile class of materials that combine many of the good properties of vulcanized elastomers with the processing characteristics of thermoplastics. Thus at normal temperatures they can be stretched with good elastic recovery. They have high tensile strength without curing or reinforcing fillers. Since they soften at higher temperatures, they may be injection-molded or extruded, and when cooled recover their physical properties. The materials are not chemically cross-linked, and scrap can be reprocessed. Their useful temperature range is limited by their softening points.

These qualities result from the unique physical network structures obtained from two mutually incompatible phases forced to coexist by chemical bonding in a specific relationship that allows the stiff molecular segments to serve as physical cross-links. Thus, a soft segment with a glass transition temperature (T_g) below room temperature is chemically linked at either end to a hard segment having a T_g above room temperature (the glass transition temperature is a second-order transition temperature at which the polymer changes from a hard, brittle glass to a rubbery or viscous polymer). Variations in physical properties may be obtained by changing the identity of the blocks, their linearity, the degree of polymerization of the individual blocks, or the ratio of one to another.

When compared to general purpose elastomers such as styrene–butadiene rubber (qv) (SBR), natural rubber (qv), and polybutadiene (qv), the thermoplastic elastomers must be classified as specialty elastomers with markets outside of tires, which consume 60% of the world's rubber production. As a frame of reference, 1976 was the year of highest world production in history of both synthetic and natural rubber at 11,085,000 metric tons. The share of 1976 consumption for synthetics was 67.5% (2). The amount for special synthetic elastomers including the thermoplastic elastomers was 101,000 t, or <1% of world consumption. However, the growth rate of the thermoplastic elastomers for the next five years is projected to be 15–20%; whereas, the projected growth rate of the general purpose elastomers is a more moderate 5%.

Of the three classes of thermoplastic elastomers to be discussed, the styrene–diene block copolymers are the largest volume (>50,000 t), the thermoplastic polyurethanes are next (>15,000 t), and copolyester ethers, the newest entry, are now >2000 t. The styrene-diene block copolymers cost $1.30–1.65/kg in 1978. The thermoplastic urethane and copolyester–ether elastomers are in the $2.90–4.20/kg range. Since the polymerization techniques, chemical compositions, properties, and uses differ greatly, the three copolymer types are discussed separately (see also Copolymers). Styrene–diene block copolymers have been commercial since 1965 and are available from Shell Oil Company and Phillips Petroleum Company. Thermoplastic polyurethane elastomers have been commercial for ca 20 yr and are available from Mobay Chemical Company, B. F. Goodrich Chemical Company, Upjohn Company, American Cyanamid Company, and

Uniroyal, Inc. The copolyester–ether thermoplastic elastomers were commercialized by DuPont in 1972.

Styrene–Diene Thermoplastic Block Copolymers

Synthesis. Preparation of styrene–diene block copolymers is achieved by forming a living polymer, a term coined to describe the product of a polymerization that has no termination or chain transfer reactions (3). This approach uses an anionic catalyst and permits the preparation of blocks having predictable molecular weights controlled by the ratio of monomer to initiator and narrow molecular weight distributions (see Initiators). The earliest discoveries of alkali metal-initiated polymerization of conjugated dienes occurred in 1911 (4–6). Subsequently, it was found that styrene could be polymerized with sodium and the characteristic reddish-yellow color of the styryl anion was observed (7). The exclusion of proton donors such as water or alcohols allows the polymerization to proceed without a termination step, and after the monomer is consumed, a second monomer may be added that polymerizes onto the first block resulting in two different chemically linked polymer chains. The preparation of thermoplastic elastomers from styrene and dienes (butadiene and isoprene) was first reported in patents (8–10). The patents describe ABA-type block copolymers in which A is a rigid block having a T_g above room temperature and B is a rubbery block having a T_g below room temperature. The importance of an abrupt change from the A to the B block is illustrated by the synthesis of tapered block copolymers (11–12).

The tapered or random segment of the chain gives some compatibility to the two phases and results in poorer physical properties. Star or radial block copolymers may be produced by using a coupling agent (divinylbenzene, silicon tetrachloride, or phosphorus trichloride) at the end of the polymerization to tie several chains together (13–15) as illustrated. The use of divinylbenzene as the coupling agent does not terminate the living chain.

$$4\ (S\text{-}B)Li + SiCl_4 \rightarrow 4\ LiCl + (S\text{-}B)_4Si$$

Physical and Chemical Properties. The difference between random and block styrene–butadiene copolymer [9003-55-8] is illustrated in a comparison of the glass transition temperatures as shown in Figure 1 in which the two T_gs for the block structure are clearly seen (16). The glassy end blocks, which are incompatible with the rubbery center blocks, serve as physical cross-links, as shown in Figure 2 (17). When the material is dissolved or heated above the T_g of the end blocks, the labile network disappears, and the rubber loses its strength. Precipitation or cooling of the copolymer restores the cross-links and the tensile strength of the rubber. The glassy domains, which can be seen by electron microscopy, have diameters of ca 30 nm. Since this distance is shorter than the wavelength of visible light, a transparent material results because visible light is not dispersed. Physical blends of the two homopolymers are not transparent. The domains deform during stretching and rupture occurs in the polystyrene phase. Typical molecular characteristics of a commercial SBS block copolymer produced by Shell under the trade name Kraton 1101 are shown in Table 1 (18).

Block copolymers of the AB- and BAB-type have poor tensile strength and low elongation, as shown in Table 2 (19). In AB and BAB block copolymers, there are free

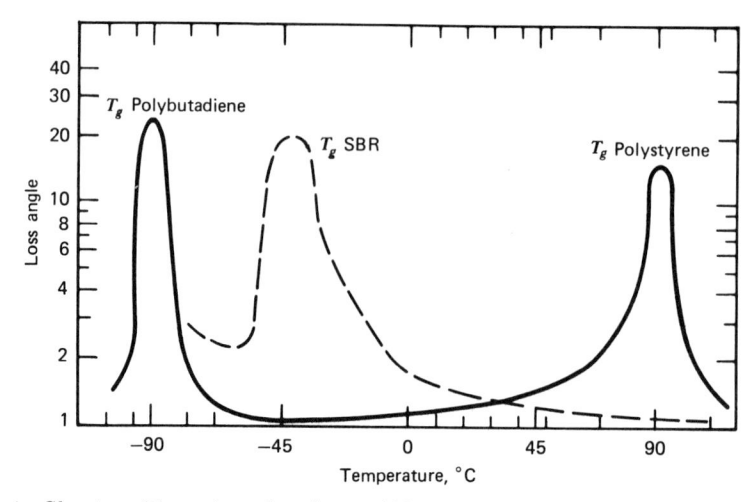

Figure 1. Glass transition points of random and block styrene–butadiene copolymers. —— Styrene–butadiene–styrene block copolymer; - - - styrene–butadiene random copolymer (SBR).

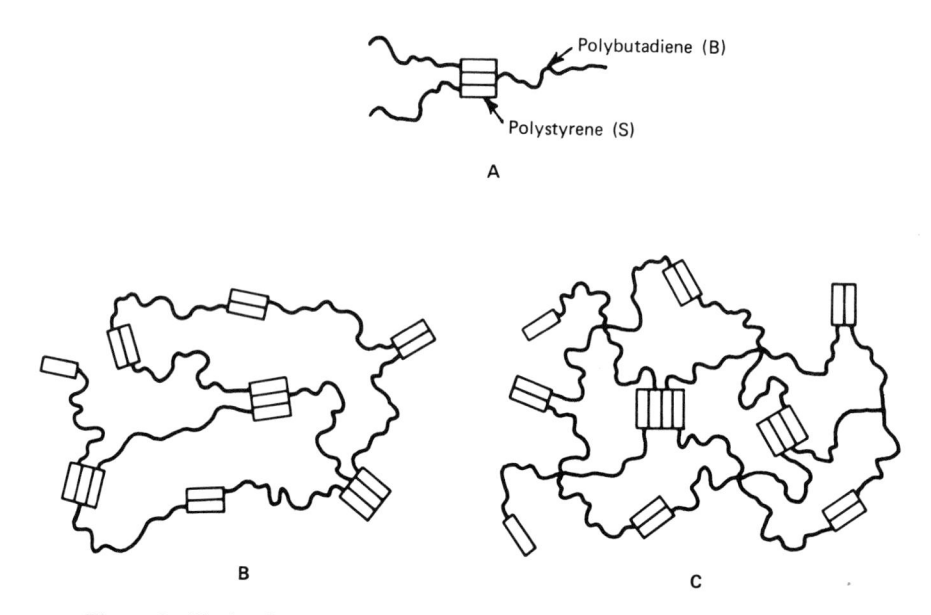

Figure 2. Block polymer structures. A, simple S-B; B, linear S-B-S; C, radial (S-B)$_n$ X.

ends not anchored in a glassy phase which do not participate in a network structure (Fig. 2).

The stress–strain curves of a series of styrene–butadiene–styrene (SBS) and styrene–isoprene–styrene [25038-32-8] (SIS) block copolymers made with sec-butyllithium are shown in Figures 3 and 4 (20). The numbers indicate the percent styrene in the copolymer. Stress–strain curves most closely resembling cross-linked elastomers are obtained from copolymers containing 30–50% styrene. Tensile strengths of 20.7–34.5 MPa (3000–5000 psi) with high elongations may be obtained from the raw,

Table 1. Molecular Characteristics of Kraton 1101 Copolymer

$\overline{M}_w$	1.02×10^5
$\overline{M}_n$	0.84×10^5
PS, wt %	33
molecular weight of each PS end block	1.7×10^4
molecular weight of the central PB block	6.8×10^4
configurational composition of the rubbery matrix:	
trans-1,4-PB, %	42
cis-1,4-PB, %	49
1,2-PB, %	9

Table 2. Comparison of SBS and BSB Polymers

Composition		
segmental molecular weights, $\times 10^{-3}$	10S-52B-10S	28B-20.5S-28B
styrene content, %	27.5	27
total molecular weight, $\times 10^{-3}$	73	76
Mechanical properties[a]		
100% modulus, MPa (psi)	1.66 (240)	0.48 (70)
extension at break, %	860	120
ultimate tensile strength, MPa (psi)	27.3 (3950)	0.48 (70)
Shore A hardness[b]	65	17

[a] Tensile specimens were cut from compression molded sheets using a die having a 2.5 cm constricted length. Extension rate was 5 cm/min.
[b] ASTM 1706-61.

Table 3. Typical Properties of ABA Thermoplastic Elastomers and Conventional Rubbers

	Kraton 1101[a,b]	Kraton 1107[a,c]	Natural rubber[d]	SBR rubber[d]
styrene, %	30	14		
tensile strength, MPa	31.8	21.4	20.8	14.5
(psi)	(4600)	(3090)	(3000)	(2100)
modulus at 300% ext,				
MPa	2.8	0.7	3.5	2.1
(psi)	(404)	(101)	(506)	(303)
elongation at break, %	880	1300	600	800
hardness, Shore A	71	37	55	45
specific gravity	0.94	0.92		

[a] Ref. 21.
[b] SBS (styrene–butadiene–styrene).
[c] SIS (styrene–isoprene–styrene).
[d] Ref. 22.

unfilled polymer. Shown in Table 3 is a comparison of mechanical properties of SBS block copolymers with vulcanized SBR and natural rubber illustrating the range inherent in the thermoplastic elastomers.

The chemical characteristics of the copolymers are determined by the nature of the components. Alteration of the chemical characteristics is achieved by altering one

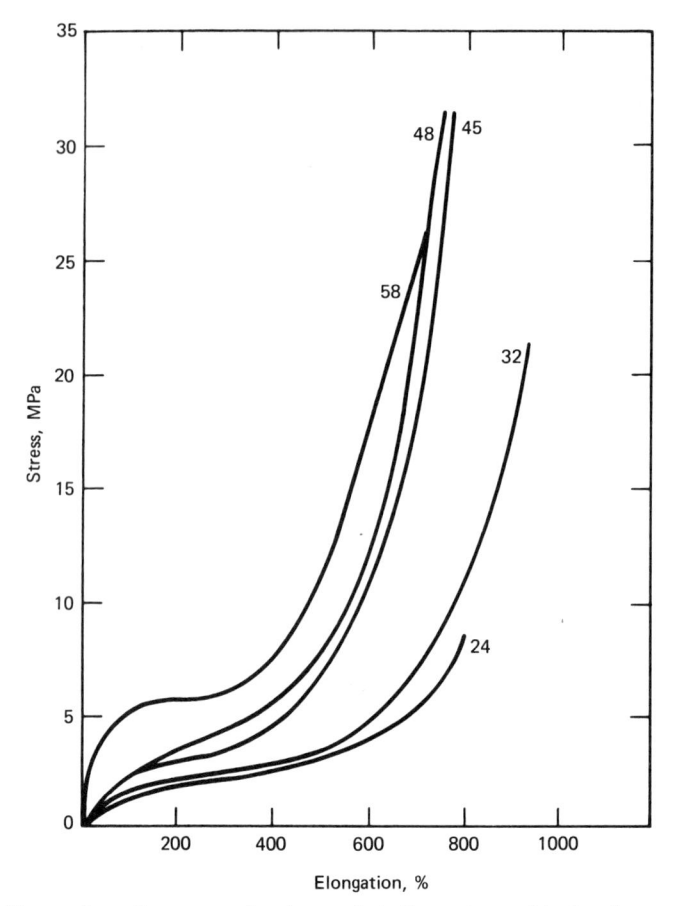

Figure 3. Stress–elongation curves for styrene-butadiene-styrene block polymers prepared with *sec*-butyllithium initiator; to convert MPa to psi, multiply by 145.

or more of the blocks. Unsaturation in the center blocks is responsible for lower oxidative stability, ozone resistance, and weather resistance. Saturation can be provided by hydrogenation of the center block.

The styrene–diene thermoplastic elastomers have excellent resistance to water, acids, and bases. Resistance to hydrocarbons, solvents, and oils is poor. The thermoplastic nature limits their utility to temperatures below 65°C depending on the stress. Elastic recovery, compression set, and creep properties are usually inferior to the chemically cross-linked elastomers.

Processing. The SBS elastomers may be processed by a wide variety of techniques including solution processing, extrusion, calendering, injection molding, blow molding, and vacuum forming. Standard rubber and plastics equipment is useful for processing the elastomers.

Suggested temperatures for extrusion are:

Feed zone	Intermediate zone	Final zone	Die	Stock
80–105°C	100–120°C	120–130°C	120–130°C	130–165°C

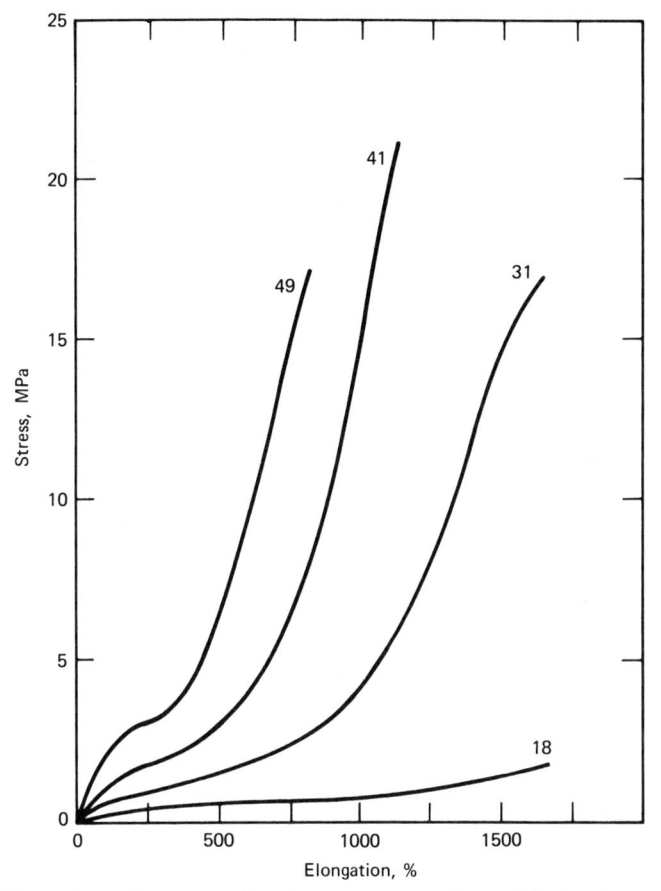

Figure 4. Stress–elongation curves for styrene–isoprene–styrene block polymers prepared with sec-butyllithium initiator; to convert MPa to psi, multiply by 145.

For injection molding of Shell Kraton G elastomers typical conditions are:

Barrel temperature, °C			Mold temperature, °C	Injection pressure, MPa (psi)	Back pressure, MPa (psi)	Cycle time, min.
Rear	Front	Nozzle				
200–225	215–250	230–260	50–65	69–103 (10,000–15,000)	0.7–3.4 (100–500)	0.5–1

A variety of fillers including clays, silicas, and calcium carbonate are used for increasing hardness. Naphthenic oil may be used as extenders, but aromatic oils are to be avoided since they tend to associate preferentially with the polystyrene segment which reduces strength (see Styrene–butadiene rubber).

Applications. Uses for the thermoplastic elastomers fall into two main sectors: primary raw materials for rubber products without vulcanization, and modifiers to upgrade the qualities of other rubbers and plastics.

The largest markets for the styrene–diene block copolymers are footwear, adhesives (qv), and mechanical goods. They are compounded and injection molded for use in canvas, noncanvas, and unit-sole footwear applications. The product has desirable properties of excellent wear-resistance, light weight, and good slip-resistance, and thus is expected to gain a major share of this market over the next five years. The SIS thermoplastic elastomers have properties that make them well-suited for pressure-sensitive and hot melt adhesives (qv). They have good peel strength (1.3 N/cm or 7.4 lbf/in. of width, 180°C) and high tack. They are soluble in common solvents and their thermoplastic character allows them to be formulated as hot melts. The inherent transparency of the block copolymers allows their use as adhesives in glass-to-glass laminates (see Laminated materials, glass).

Certain types of tackifying resins are used to increase the tack of the elastomers. The type of rubbery center block determines the most suitable resin tackifier with the saturated center blocks being the most difficult to tackify. Oils or plasticizers must be added in addition to tackifying resins. A variety of tackifying resins are available for the polyisoprene and polybutadiene center-block types such as rosin esters, pinene resins, and synthetic terpene resins (see Hydrocarbon resins). Increasing restrictions on the release of solvents to the atmosphere have provided impetus for the development of hot melt adhesives based on the styrene–diene thermoplastic elastomers. The SIS block copolymers are mixed with tackifying resins containing aromatic substituents or with aliphatic or alicyclic hydrocarbon resins having a softening point in the range 80–130°C. The mixture of tackifying resins, processing oil, and thermoplastic elastomer forming the hot melt pressure-sensitive adhesive is simply heated, applied to a substrate, and then cooled. The substrate may be paper, cloth, or films such as cellophane, vinyl resins, or polyethylene. Other industrial applications for styrene-diene thermoplastic elastomers include gasket compounds, hose and tubing, electrical cables, construction glazing, bitumen additive for roofing, packaging film, housewares, toys, and coatings.

Thermoplastic Urethane Elastomers

Thermoplastic polyurethane (TPU) elastomers are a special class of urethanes that can be processed as plastics and as cements for a wide range of applications (see Urethane polymers). These elastomers resulted from urethane polymer work in the 1930s by Professor Otto Bayer and co-workers at Farbenfabriken Bayer A.G. The polymers show the characteristic high tensile strength, and resistance to tear, abrasion, and ozone of urethanes without added reinforcement. Thermoplastic polyurethanes are formed by linking three basic components (23):

(1) A linear, hydroxyl-terminated polyol with mol wt of 500–3500. Polyol types are polyesters (qv) (adipate, azelate, isophthalate and caprolactone) and polyethers (qv) (polypropylene ether glycol, polytetramethylene ether glycol).

(2) A diisocyanate, which may be aromatic such as diphenylmethane-4,4'-diisocyanate (MDI), toluene-2,4-diisocyanate (TDI), or nonaromatic as dicyclohexylmethane-4,4'-diisocyanate (H_{12}MDI). The preferred diisocyanate is MDI. If a nonyellowing thermoplastic elastomer is required for coating applications, the nonaromatic H_{12}MDI is preferred (see Isocyanates).

(3) A low molecular weight glycol such as 1,4-butanediol, ethylene glycol, or 1,4-phenylene-bis-β-hydroxyethyl ether to serve as a chain extender.

In the preparation the linear polyol is combined with an excess of diisocyanate to give an isocyanate-terminated prepolymer which is stable for months and can be considered a living polymer. The next step involves reaction with the low molecular weight diol to form the higher molecular weight thermoplastic elastomer. The reactions can be represented schematically as follows (24).

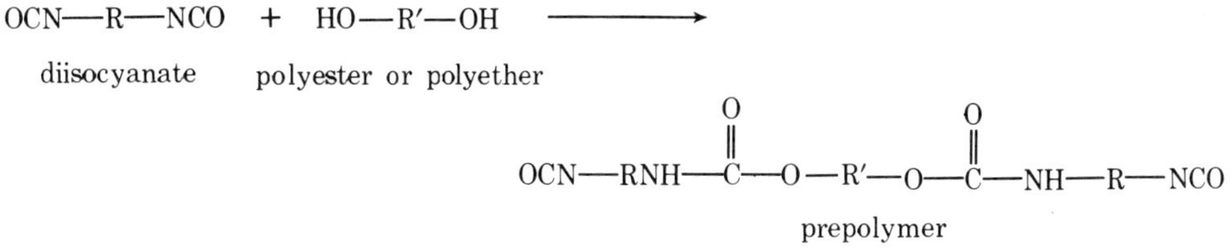

OCN—R—NCO + HO—R'—OH ⟶

diisocyanate   polyester or polyether

$$\text{OCN—RNH—}\overset{\overset{O}{\|}}{C}\text{—O—R'—O—}\overset{\overset{O}{\|}}{C}\text{—NH—R—NCO}$$

prepolymer

Chain extension with glycols takes place with the formation of urethane groups:

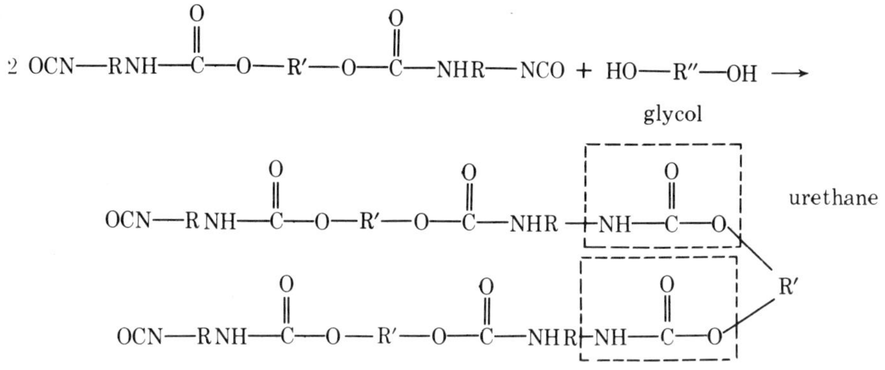

$$2\ \text{OCN—RNH—}\overset{\overset{O}{\|}}{C}\text{—O—R'—O—}\overset{\overset{O}{\|}}{C}\text{—NHR—NCO}\ +\ \text{HO—R''—OH}\ \rightarrow$$

glycol

$$\text{OCN—RNH—}\overset{\overset{O}{\|}}{C}\text{—O—R'—O—}\overset{\overset{O}{\|}}{C}\text{—NHR—NH—}\overset{\overset{O}{\|}}{C}\text{—O}$$

urethane

$$\text{OCN—RNH—}\overset{\overset{O}{\|}}{C}\text{—O—R'—O—}\overset{\overset{O}{\|}}{C}\text{—NHR—NH—}\overset{\overset{O}{\|}}{C}\text{—O}$$

R'

In a "one shot" technique, in which polyol, diisocyanate, and glycol are mixed in one step and permitted to react, a random distribution is obtained which is accompanied by slightly less useful physical properties.

Generally, polyester-based materials are selected for high strength, tear, chemical and heat resistance, and polyether-based materials are selected for low temperature flexibility, high humidity conditions, and resistance to attack by fungi and bacteria.

Morphology. To obtain good thermoplastic characteristics, an elastomer should be prepared with as high as possible a decomposition temperature and with the processing temperature maintained just slightly lower. Softening must occur before decomposition (25). The thermoplastic polyurethane (TPU) elastomer consists of linear primary chains made up of hard and soft segments joined end-to-end by covalent chemical linkages. The soft segment is the reaction product of the diisocyanate compound and the polyol. It tends to be low melting and largely amorphous. Good low-temperature flexibility is obtained with soft segments having low T_g values. The hard segments are the linear reaction products of the diisocyanate and the glycol. A high melting temperature (T_m) is desired for the hard segment to improve resistance to elevated temperatures. For practical purposes a single diisocyanate (usually MDI) is used and is common to both the hard and soft segments. The polyol and glycol are then selected on the basis of desired mechanical properties. There is a wide range of flexibility which includes polyol type and molecular weight, glycol extender composition and the mole ratios selected for combining with the diisocyanate. The compositional variables affect the degree of phase separation and hard-segment domain size. Characteristic domain dimensions are ca 5–10 nm (26). The hard segments tend to associate with each other through interchain urethane-urethane hydrogen bonding, and form domains in the mobile soft segment. A two-phase polymer system results. The hard segment domains tie the linear polymer chains to effect a cross-linked network which accounts for the elastomeric properties. The labile cross-links or tie points can be broken by heat. On cooling, the primary chains reform the network structure (27).

Strength and extensibility of TPU elastomers above room temperature decrease with increasing temperature because the soft domains flow at elevated temperatures. At 20–40°C there is increased toughness in the domains and strength increases dramatically owing to strain-induced crystallization (28). The thermal lability of hydrogen bonds in polyurethane elastomers was studied by ir analysis. At room temperature virtually all —NH— groups are hydrogen bonded. At temperatures above the glass transition temperature of the hard segment, the hydrogen bonds dissociate (29). The high strength of thermoplastic polyurethane elastomers cannot be attributed to hydrogen bonding directly but requires that hard segments be sufficiently long so that the domains are resistant to plastic flow.

Preparation, Properties, and Processing. In 1958, B. F. Goodrich Company described a new thermoplastic polyester–urethane elastomer for use without vulcanization (30). The thermoplastic, completely soluble, linear polymer was prepared from poly(tetramethylene adipate), 4,4'-bisisocyanatodiphenylmethane and 1,4-butanediol. Its average molecular weight is 36,000. Tensile strength at 24°C is >35 MPa (5000 psi) with elongation >500%. The preparation of thermoplastic polyurethanes is described in the patent literature (31–37).

In a typical preparation for a thermoplastic polyurethane, the molar amount of polyester (poly(tetramethylene adipate)) and glycol (1,4-butanediol) combined is substantially equivalent to the molar amount of the diisocyanate (diphenylmethane-4,4'-diisocyanate) (31). The product has the following physical properties: tensile strength, 42 MPa (6100 psi); elongation, 650%; 300% modulus, 7 MPa (1000 psi); and Shore A hardness 85. The polymers may also be prepared in dimethylformamide (DMF) solution at 30% solids under random polymerization conditions (38). The structure of the essentially linear thermoplastic polyurethane from poly(tetramethylene adipate), 1,4-butanediol and diphenylmethane-4,4'-diisocyanate is shown below.

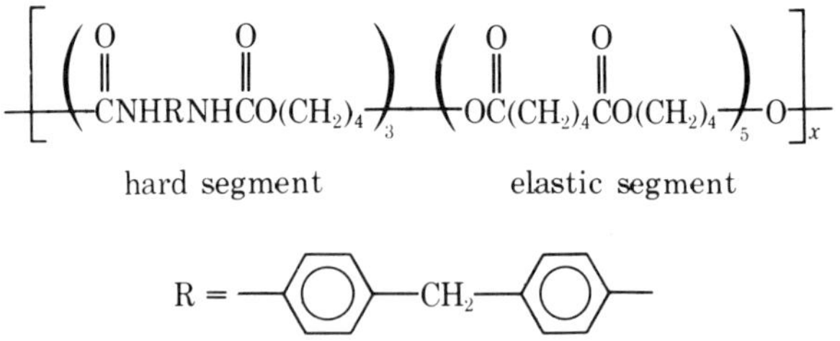

$$\left[\left(\begin{matrix}O\\\|\\-CNHRNHCO(CH_2)_4\end{matrix}\right)_3\left(\begin{matrix}O\quad O\\\|\quad\|\\-OC(CH_2)_4CO(CH_2)_4\end{matrix}\right)_5-O\right]_x$$

hard segment   elastic segment

$$R = -\bigcirc\!\!-CH_2-\!\!\bigcirc-$$

The tensile strength and 300% modulus values increase with polymer mol wt, reaching a plateau at $\overline{M}_n$ 36,000–40,000. The flex life of samples decreases with increasing $\overline{M}_n$. Abrasion resistance increases with polymer $\overline{M}_n$. Tear resistance is independent of mol wt. The polymers show good resistance to gamma irradiation (39).

Commercial thermoplastic urethane elastomers are formulated to meet specific product needs and are processable by injection molding, extrusion, calendering, blow molding and vacuum forming. Since urethane elastomers are susceptible to moisture pickup during storage they should be dried for one hour at 100–110°C just prior to use. They are supplied in a range of hardnesses from S_A 80 to S_D 55 (40–42) (see Hardness). Lower hardness thermoplastics can be made, but they are more difficult to process. Suggested operating conditions for injection molding, extrusion and calendering follow (43–46).

Injection molding conditions for Cyanaprene thermoplastic polyurethane elastomers (S_A 80 to S_D 57) are shown below:

Temperature, °C				Pressure, MPa (psi)		Cycle time, min
Rear	Front	Nozzle	Mold	Injection	Back	
185–205	195–215	200–215	25–50	85 (12,300)	7–14 (1000–2000)	0.5–1.0

After injection molding the elastomer can be cured in an oven for 8–24 h at 110–120°C. At room temperature, maximum tensile strength will be developed after 2 wk. Regrind may be used in concentrations up to 50% in virgin material.

Extrusion conditions for Texin thermoplastic polyurethane elastomers (S_A 80 to S_D 55) (pressure 5–28 MPa, 700–4000 psi) are shown below:

	Zone				melt	die
	1	2	3	4		
temperature, °C	94–125	120–150	135–150	150–175	150–185	105–165

Operating conditions for inclined "Z" calender, stock temperature, 160–175°C are shown below:

	1	2	3	4
roll speeds, m/min	0.9–1.7	1.0–1.8	1.2–2.4	2.2–4.3
roll temperature, °C	145–175	145–185	140–185	135–170

The thermoplastic urethanes are compatible with many other polymers (eg, ABS, polystyrene, Kratons, and Surlyn ionomer) in polyblends. The polyblend technique can be used to improve properties and lower the cost of the polyurethane compound.

Applications. Since the urethane elastoplastics incorporate exceptional resistance to abrasion, fuel and oils, and high tensile, tear- and load-bearing properties, and are available in a broad durometer range, they are candidates for demanding applications in such areas as automotive, sporting, general mechanical goods, fabric coatings, and biomedical applications such as intra-aortic balloons (see Prosthetic and biomedical devices). Specially prepared thermoplastic urethanes have acceptability under FDA regulations 121.2520, 121.2522, and 121.2562 for food-contact uses. These materials are also acceptable by the USDA for use in processing areas for contact with meat or poultry food products prepared under FDA inspection (see Meat products). In the magnetic tape (qv) market thermoplastic polyurethanes are used as the binder to adhere metallic oxides to polyester film in audio, video, and computer tapes.

Thermoplastic Copolyester–Ether Elastomers

Segmented copolyester–ethers represent a novel family of commercial thermoplastic elastomers derived from terephthalic acid (T), polytetramethylene ether glycol (PTMEG), and 1,4-butanediol (47–49). They offer an unusual combination of easy processing and high performance under environmental extremes (see Polyesters; Polyethers).

Synthesis. The polyester-ether copolymers are prepared by titanate ester (tetrabutyl titanate)-catalyzed melt transesterification of a mixture of dimethyl terephthalate, polyether glycol, and excess 1,4-butanediol (50). The reaction sequence is as follows:

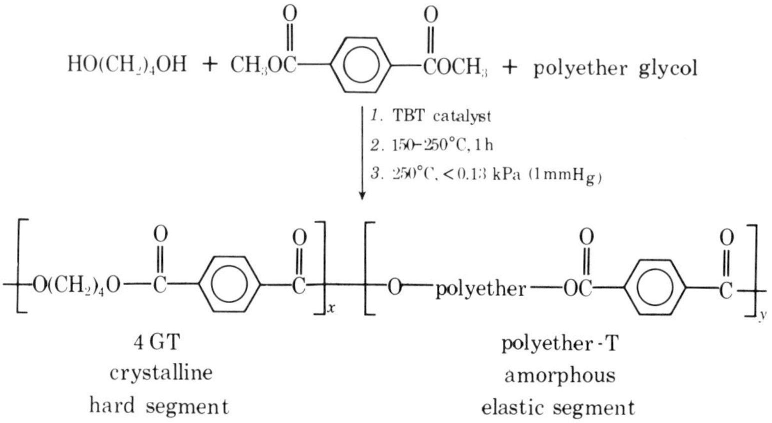

$$HO(CH_2)_4OH + CH_3OC\text{-}C_6H_4\text{-}COCH_3 + \text{polyether glycol}$$

1. TBT catalyst
2. 150–250°C, 1 h
3. 250°C, <0.13 kPa (1 mmHg)

$$\left[\text{-}O(CH_2)_4O\text{-}C(O)\text{-}C_6H_4\text{-}C(O)\text{-}\right]_x \left[\text{-}O\text{-polyether-}OC(O)\text{-}C_6H_4\text{-}C(O)\text{-}\right]_y$$

4 GT	polyether-T
crystalline	amorphous
hard segment	elastic segment

The stoichiometry favors the formation of long sequences of tetramethylene terephthalate (4GT) units resulting in block copolymers consisting of crystallizable 4GT hard segments and amorphous elastomeric poly(alkylene ether terephthalate) soft segments (PTMEG-T). A number average molecular weight in the range of about 25,000–30,000 is required for thermoplastic elastomer properties. The code 4GT refers to the four carbon glycol, 1,4-butanediol (4G), which is esterified with terephthalic acid (T). The polymers exhibit a continuous two-phase domain structure where the crystalline short-chain polyester hard segments contribute to strength and the amorphous soft polyether ester segments account for the flexible elastomeric nature. The system undergoes phase separation by crystallization of the 4GT hard segments. A thermally reversible three-dimensional network is formed resembling conventionally cross-linked elastomers.

The differential scanning calorimeter (dsc) thermogram of a copolyester ether containing 42% PTMEG-T soft segment and 58% 4GT hard segment shows the presence of two transitions: the glass transition of the amorphous phase at about −50°C and the melting point of the crystalline phase at about 200°C (Fig. 5). This combination of a low glass transition temperature and high melting point makes the polymers suitable for applications at temperatures far above the upper limit of styrene–diene and polyurethane elastomers.

The 4GT sequences were found to crystallize as lamellar domains of ca 10-nm

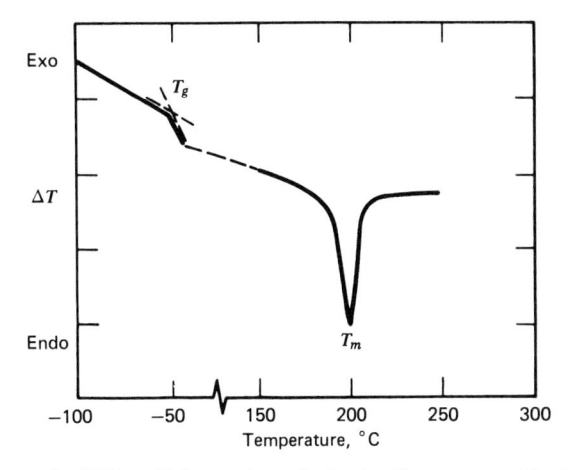

Figure 5. Differential scanning calorimetry thermogram, 58 wt % 4GT.

width and up to several hundred nanometers in length. These crystalline regions are not discrete as are the polystyrene domains in the SBS triblock polymers but are continuous and highly interconnected (51). In essence, a continuous crystalline network is superimposed on a continuous amorphous network. When stress is applied, it is transmitted initially through the crystalline phase and gradually transferred to the elastomeric portions of the polymer network. The failure of this phase ultimately results in rupture (52–53).

Physical Properties. The physical properties can be varied by changing the concentration of hard and soft segments. The desirable amount of 4GT hard segments is 33–76% (Table 4).

Polymers containing less than 30% 4GT hard segments have relatively low strength. The soft segment (PTMEG-T) is a soft gum that flows under stress. Medium-hardness copolyesters containing about 55–60% 4GT hard segment offer the best compromise between low and high temperature performance.

Table 4. Typical Properties of Segmented Polyether Esters

4GT hard segment, %	33	58	76
polymer melt temperature (by dsc), °C	176	202	212
specific gravity	1.15	1.20	1.22
durometer hardness	92A	55D	63D
tensile strength, MPa (psi)	39.3 (5700)	44.1 (6400)	47.5 (6900)
elongation at break, %	810	760	510
10% modulus, MPa (psi)	3.58 (520)	10.0 (1450)	16.9 (2450)
Bashore resilience, %	60	53	40
flexural modulus, MPa (psi)	44.8 (6500)	206 (30,000)	496 (72,000)
brittle point, °C	< −70	< −70	< −70
Izod impact resistance, J/cm (ft.-lb./in.)			
22 °C	10.7 (ca 20)	10.7 (ca 20)	10.6 (ca 20)
−40°C	10.7 (ca 20)	10.7 (ca 20)	0.32 (ca 0.6)
oil swell (ASTM NO. 3 oil, 7 days at 100°C), % vol increase	22.0	12.2	6.6

Table 5. Suggested Settings for Processing Hytrel Polymers

Injection molding		Extrusion	
cylinder temperature, °C		barrel temperature, °C	
rear	141–210	zone 1	155–210
center	146–210	zone 2	170–215
front	163–213	zone 3	170–225
nozzle	163–213	adapter and neck, °C	170–225
melt temperature, °C	171–238	die, °C	170–230
mold temperature, °C	21	melt, °C	170–235
cycle, min	0.5–3		
injection pressure,			
MPa	42–100		
(psi)	(6100–14,500)		
back pressure,			
MPa	3.4–5.5		
(psi)	(500–800)		

Tensile strength depends upon the chemical structure, copolymer composition, chain length, and crystalline morphology. The tear strength appears to be a function of polyether structure and concentration. Increasing the polyether content results in more elastic polymers of lower modulus. Hardness is dependent upon polyester concentration. Low temperature properties are dependent upon concentrations of 4GT and polyether units. Melting points appear to be a function of the 4GT block lengths.

The properties can be modified by incorporation of various conventional fillers (qv) such as carbon black, clays, and chopped fiberglass. These additives increase the modulus at various elongations.

Processing. The thermoplastic copolyester-ether elastomers commercialized as Hytrel by DuPont can be processed by injection, blow, compression, transfer, or rotational molding. They can be readily calendered and extruded (54–56). The polymers have a relatively low melt viscosity, good melt stability, and rapidly harden from the molten state. The excellent melt stability enables use of up to 50% regrind mixed with virgin polymer.

Injection molding is performed in standard machines using general purpose screws. No special techniques are needed in extruding these polymers. The high modulus facilitates fabrication of tubing and crossheading of covers on hose and cable jackets. Prototype parts can also be prepared by machining blocks prepared by melt casting (57–58).

Suggested settings for processing Hytrel polymers are listed in Table 5 (59–60).

Applications. The copolyester–ether elastomers have superior tensile and tear strength, flex life, and abrasion resistance, combined with low oil-swell, and good electrical resistance. Some of the many uses include as a replacement for cured rubber and rubber-metal parts with a one-component elastomer unit. Applications include tube, or tube and jacket of hydraulic hose; electrical cable connectors; airbrake tubing for tractor trailers and sporting goods; rotationally molded tires for garden tractors; blends with PVC for use in wire and cable jackets; and a variety of automotive, machinery, and construction equipment parts.

BIBLIOGRAPHY

1. *ASTM Special Technical Bulletin,* No. 184.
2. *Rubber World* **177**(4), 34 (1978).
3. M. Swarc, *Nature* **178**, 1168 (1956).
4. C. Harries, *Ann.* **383**, 184 (1911).
5. U.S. Pat. 1,058,056 (Apr. 8, 1913), C. Harries (to Farbenfabriken Bayer).
6. Brit. Pat. 24,790 (Oct. 25, 1911), F. E. Matthews and E. H. Strange.
7. W. Schlenk and co-workers, *Ber.* **47**, 473 (1914).
8. U.S. Pat. 3,265,766 (Aug. 9, 1965), G. Holden and R. Milkovich (to Shell Oil Co.).
9. U.S. Pat. 3, 231, 635 (Jan. 25, 1966), G. Holden and R. Milkovich (to Shell Oil Co.).
10. U.S. Pat 3,287,333 (Nov. 22, 1966), R. P. Zelinski (to Phillips Petroleum Co.).
11. R. E. Cunningham, M. Auerbach, and W. J. Floyd, *J. Appl. Polym. Sci.* **16**, 163 (1972).
12. U.S. Pat. 2,975,160 (Mar. 14, 1961), R. P. Zelinski (to Phillips Petroleum Co.).
13. U.S. Pat. 3,949,020 (Apr. 6, 1976), R. T. Prudence (to Goodyear Tire & Rubber Co.).
14. R. L. Mawes, *J. Inst. Rubber Ind.* **6**(3), 121 (1972).
15. M. Morton and co-workers, *J. Polym. Sci.* **57**, 471 (1962).
16. B. H. Sorenson, *Plastic News* 11 (Feb. 1976).
17. J. R. Haws and T. C. Middlebrook, *Rubber World* **167**(4), 27 (Jan. 1973).
18. E. Pedemonte and co-workers, *Polymer* **16**, 531 (1975).
19. W. R. Hendricks and R. J. Enders in M. Morton, ed., *Rubber Technology,* Van Nostrand Reinhold Co., New York, 1973, p. 515.
20. R. E. Cunningham and M. R. Treiber, *J. Appl. Polym. Sci.* **12**, 23 (1968).
21. *Shell Chemical Company Tech Bulletin SC-68-77,* March 1977.
22. S. C. Wells, *Mach. Des.* **44**, 52 (Dec. 1972).
23. C. S. Schollenberger and L. E. Hewitt, *Soc. Plast. Eng.* **23**, 48 (1977).
24. T. H. Rogers and co-workers, *J. Elastomers Plast.* **8**, 116 (1976).
25. E. Mueller and co-workers, *Rubber Chem. Technol.* **26**, 493 (1953).
26. R. Bonart, *J. Macromol. Sci., Phys.* **B2**(1), 115 (1968).
27. T. L. Smith, *J. Polym. Sci., Polym. Phys. Ed.* **12**, 9, 1825 (1974).
28. G. L. Wilkes and R. Wildnauer, *J. Appl. Phys.* **46**, 10, 4148 (1975).
29. R. W. Seymour and S. L. Cooper, *Macromolecules* **6**, 48 (1973).
30. C. S. Schollenberger, H. Scott, and G. R. Moore, *Rubber World* **137**, 549 (1959).
31. U.S. Pat. 2,871,218 (Jan. 27, 1959), C. S. Schollenberger (to B. F. Goodrich Co.).
32. U.S. Pat. 3,015,650 (Jan. 2, 1962), C. S. Schollenberger (to B. F. Goodrich Co.).
33. U.S. Pat. 3,214,411 (Oct. 26, 1965), J. H. Saunders and K. A. Pigott (to Mobay Chemical Co.).
34. U.S. Pat. 3,233,025 (Feb. 1, 1966), B. F. Frye, K. A. Pigott, and J. H. Saunders (to Mobay Chemical Co.).
35. U.S. Pat. 3,632,845 (Jan. 4, 1972), J. E. Brownsword (to Goodyear Tire & Rubber Co.).
36. U.S. Pat. 3,446,781 (May 27, 1969), J. E. Brownsword (to Goodyear Tire & Rubber Co.).
37. U.S. Pat. 3,769,245 (Oct. 30, 1973), F. D. Stewart and C. S. Schollenberger (to B. F. Goodrich Co.).
38. C. S. Schollenberger and K. Dinbergs, *J. Elastoplast.* **5**, 222 (1973).
39. C. S. Schollenberger and co-workers, *Rubber World* **142**(6), 81 (1960).
40. K. A. Pigott and co-workers, *Mod. Plast.* **40**(4), 117 (1962).
41. K. A. Pigott and co-workers, *Soc. Plast. Eng. J.* 1281 (Dec. 1963).
42. K. A. Pigott and co-workers, *Ind. Eng. Chem. Prod. Res. Dev.* **1**(1), 28 (1962).
43. *Cyanaprene Bulletin, 5-26082500,* American Cyanamid Company, Jan., 1975.
44. *Texin Bulletin,* Mobay Chemical Company, 1969.
45. *Estane Bulletin, Polyurethanes, TSR 76-01,* B. F. Goodrich Co., 1976.
46. *Estane Polyurethanes, Bulletin TSR 72-03,* B. F. Goodrich Co., 1977.
47. M. Brown, *Rubber Ind.* **9**(3), 102 (1975).
48. R. P. Kane, *Paper No. 3, American Chemical Society Meeting No 172,* Rubber Division, San Francisco, Sept., 1976.
49. G. K. Hoeschele, *Polym. Eng. Sci.* **14**(12), 848 (1974).
50. J. R. Wolfe, Jr., *Rubber Chem. Technol.* **50**(4), 688 (1977).
51. R. J. Cella, *J. Polym. Sci., Part C* **42**, 727 (1973).
52. J. R. Knox, *Polymer Age,* 357 (Oct. 1973).

53. A. Lilaonitkul and S. L. Cooper, *Rubber Chem. Technol.* **50**(1), 1 (1977).
54. U.S. Pat. 3,651,014 (Mar. 21, 1972), W. K. Witsiepe (to E. I. du Pont de Nemours & Co., Inc.).
55. U.S. Pat. 3,763,019 (Oct. 2, 1973), W. K. Witsiepe (to E. I. du Pont de Nemours & Co., Inc.).
56. U.S. Pat. 3,766,146 (Oct. 16, 1973), W. K. Witsiepe (to E. I. du Pont de Nemours & Co., Inc.).
57. E. J. Roddy, *Paper No. 5, American Chemical Society, Minneapolis Meeting*, Rubber Division, 1976.
58. A. L. Goodman and co-workers, *Paper No. 26, American Chemical Society, Minneapolis Meeting*, Rubber Division, 1976.
59. *"Hytrel" Injection Molding*, DuPont Bulletin A87314, 1976.
60. M. Brown, *Extrusion of Hytrel*, DuPont Bulletin Hyt-403A, 1967.

General References

H. L. Stephens, *Prog. Rubber Technol.* **37**, 65 (1974).

A. Noshay and J. E. McGrath, *Block Copolymers: Overview and Critical Survey*, Academic Press, New York, 1976.

J. V. Dawkins in D. C. Allport and W. H. Janes, eds., *Block Copolymers*, John Wiley & Sons, Inc., New York, 1973.

"High Polymers" in J. P. Kennedy and E. G. M. Tornqvist, eds., *Polymer Chemistry of Synthetic Elastomers*, Part I, Wiley-Interscience, a division of John Wiley & Sons, Inc., 1968.

W. M. Saltman, ed., *The Stereo Rubbers*, John Wiley & Sons, Inc., New York, 1977.

J. H. Saunders and K. C. Frisch, *Polyurethanes, Chemistry and Technology*, Parts I and II, Wiley-Interscience, a division of John Wiley & Sons, Inc., New York, 1962, 1964.

P. Wright and A. P. C. Cumming, *Solid Polyurethane Elastomers*, Gordon and Breach Science Publishers, New York, 1969, Chapt. 8.

S. L. Cooper, J. C. West, and R. W. Seymour, "Polyurethane Block Polymers" in *Encyclopedia of Polymer Science and Technology*, Supplement Vol. 1, Wiley-Interscience, a division of John Wiley & Sons, Inc., 1976, pp. 521–543.

A. F. FINELLI
R. A. MARSHALL
D. A. CHUNG
The Goodyear Tire & Rubber Co.

ELECTRICAL CONNECTORS

Electrical connectors are mechanical devices that connect wires, cables, printed circuit boards, and electronic components to each other and to related equipment. Connector designs include miniature units for microelectronic applications, specialized cable, rack and panel designs for incorporating combinations of a-c, d-c, and radio-frequency conducting contacts, and high-current connectors for industrial application and for transmission and distribution of electrical power in overhead and underground networks. Further categorization of connectors can be made according to: their application, whether they permanently join conductors and components or permit separation and rejoining; the means used to effect connection, whether by fusion (welding, soldering) or by pressure (the values of which can be small or can be great enough to severely deform metal); the distribution type, whether of power or of low (signal) levels of current; and the conductor size. The term electrical contact describes the junction between two or more current-carrying members that provide electrical continuity at their interfaces. Connector contacts ordinarily remain stationary in active circuits, eg, they are not mated or separated. Components having electrical contacts other than connectors include circuit breakers, switches, relays, and contactors that are designed to interrupt or to establish current flow in active circuits, and slip rings and brushes which transmit current from a stationary to a moving frame of reference.

Many connectors for single conductors have an insulating sleeve, and almost all connectors that join two or more conductors have a plastic body, or dielectric, which separates the contact elements. Metal or plastic shells with mechanical aids, such as screws, levers, and other coupling devices to facilitate joining and separation of the contacts, also may surround a connector. The shell may have mounting features for securing the connector to a chassis, supports for the wires and cables, and polarizing keys for prevention of improper mating.

Connector Configurations

Electronic Connectors. The complexity and size of many electronic systems necessitate construction from relatively small building blocks which are then assembled with connectors. An electronic connector is a separable electrical connector used in telecommunications apparatus, computers, and in signal transmission and current transmission $\leq 5A$. Electronic connectors are favored over permanent or hard-wired connections because they facilitate the manufacture of electronic systems; also, they permit assemblies to be easily demounted and reconnected when inspection, replacement, or addition of new parts is called for.

The emergence of integrated circuit technology, which was characterized by the development of direct linkage of many circuit components or linkage via conductive paths on tiny ceramic substrates, was thought to threaten the growth of the electronic connector market. However, the explosive growth of electronics in most areas of manufacturing, business, and consumer products created new markets with resultant expansion of the connector field.

Electronic connectors may connect internally or externally. Internal connections may be between a component and a printed circuit board or wire (Fig. 1a); a printed

642 ELECTRICAL CONNECTORS

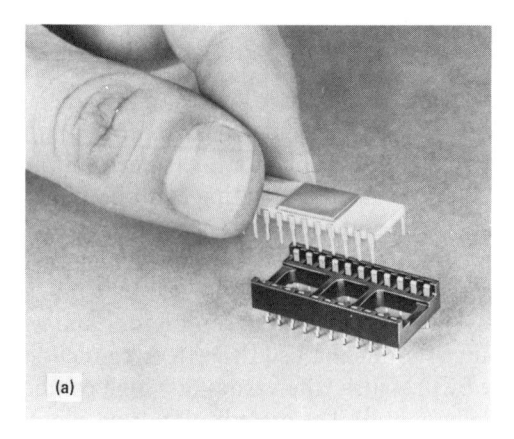

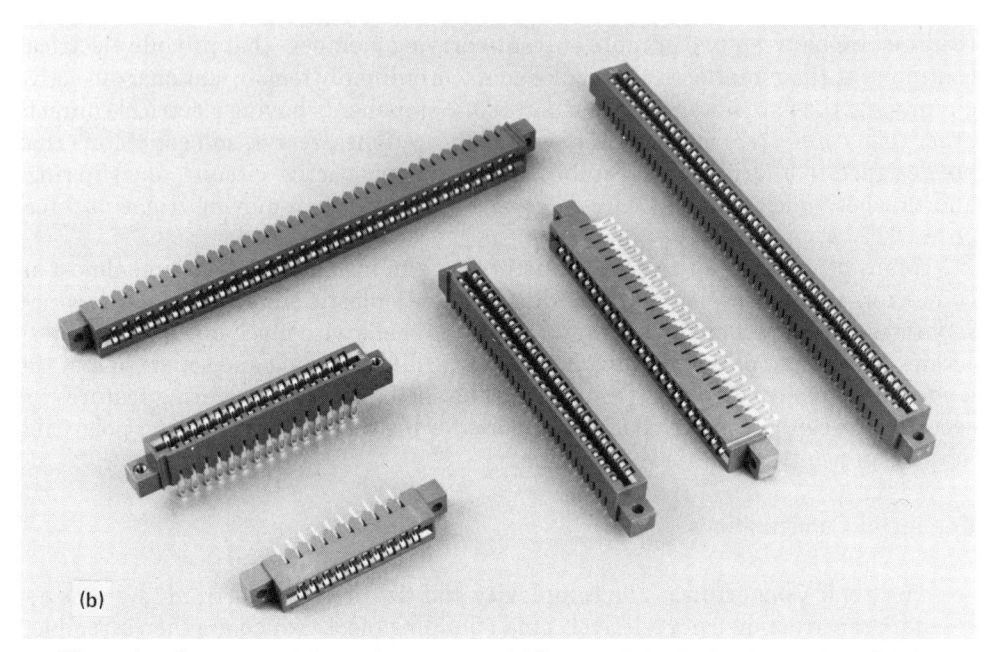

Figure 1. Some types of electronic connectors. (**a**) Receptacle for dual-in-line package (DIP) semiconductor integrated circuit. (**b**) Connectors for printed circuit boards with edge contacts; (two-piece connectors have male and female connector halves, one of which is attached to the printed circuit board, usually by soldering.). (**c**) Rectangular connector for chassis mounting. (**d**) Circular connector for cable. **a,d**: Courtesy of Burndy Corporation; **b,c**: courtesy of AMP Incorporated.

circuit board and a wire or another printed circuit board which is in a chassis (Fig. 1**b**); and between chassis in the same cabinet (Fig. 1**c**). External connectors join separate pieces of equipment (Fig. 1**d**).

Another type of electronic connector joins coaxial conductors. These have a solid or stranded center-conductor surrounded by a dielectric. The dielectric is covered with a conductive shield made of metal braid or tape and with a layer of insulation. Coaxial cable connectors terminate the center-conductor and the shield. These are used primarily in radio frequency circuits.

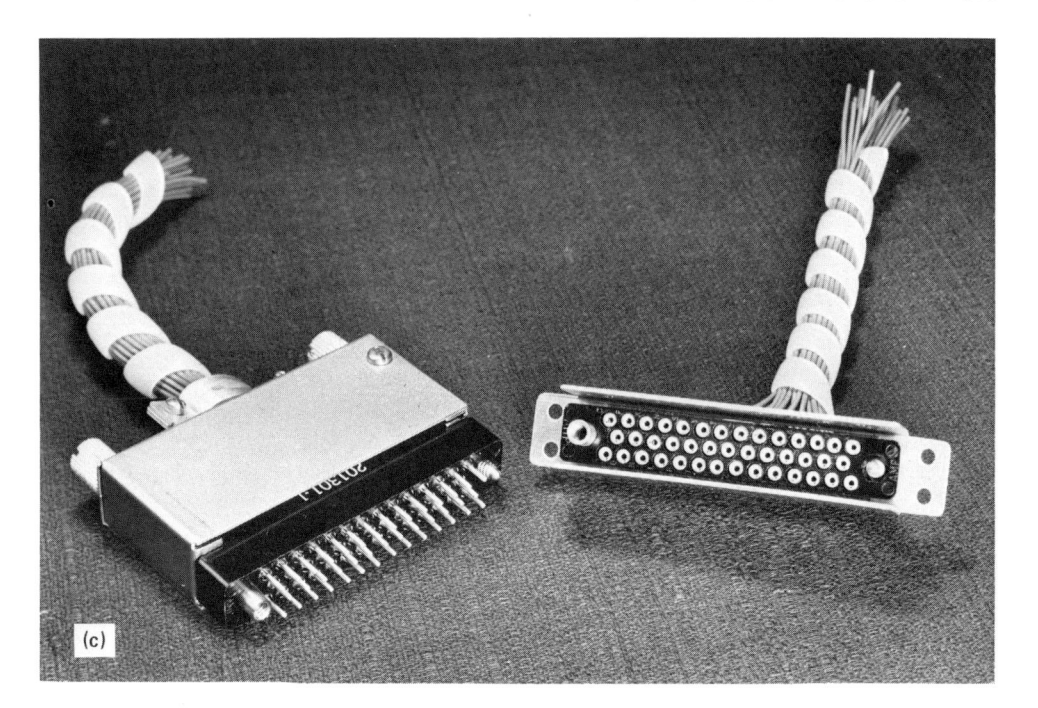

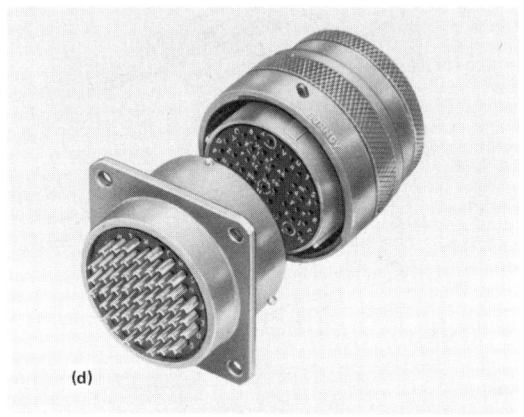

Figure 1. (*continued*)

The contacts of an electronic connector have spring elements which press the mating surfaces together with a predetermined force, usually in the range of 0.25–5 N (0.056–1.12 lbf); this range depends on the connector design and the materials from which the contact is made. Figure 2 illustrates typical spring designs. Mating of the connector is usually effected by sliding the surfaces together. Less frequently, the contacts are butted after they have been positioned in zero insertion force (ZIF) connectors. Butting requires separation of contact springs by means of a cam and lever or their equivalents. Butting is preferred to sliding the halves together as a method of mating connectors having more than 100 contacts because it requires smaller friction forces. Butting not only reduces mechanical wear, especially when soft metal finishes such as tin plate (1–2) are used as the contact material, but facilitates connection to the fragile legs of some components (Fig. 1a).

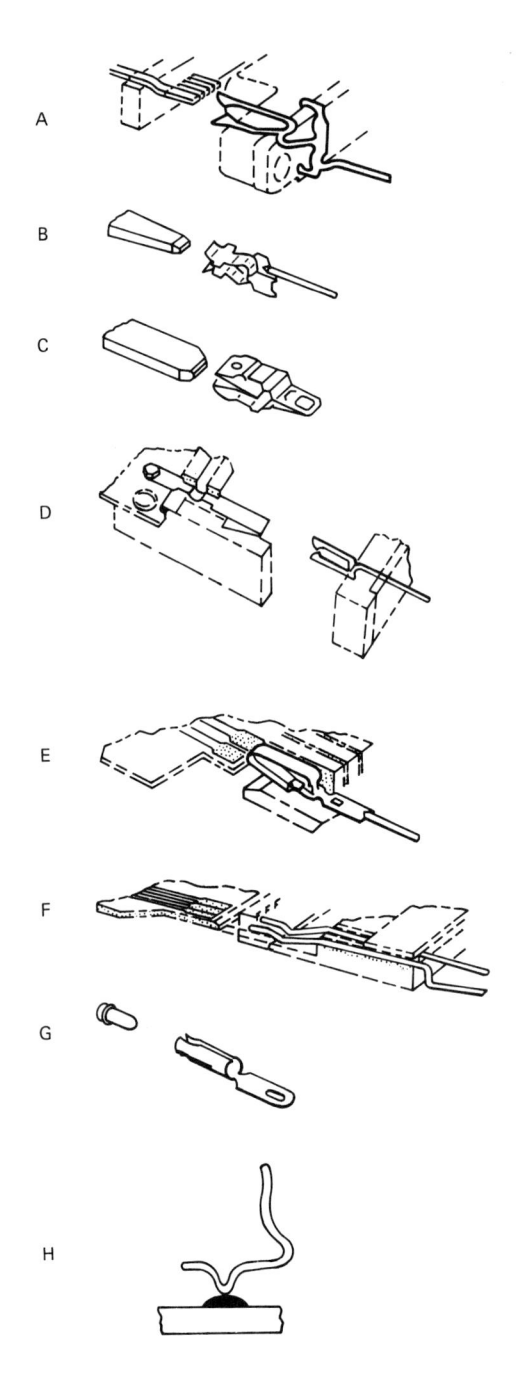

Figure 2. Contact spring configurations for separable electronic connectors. Examples A–G are engaged by sliding the spring contact and its mating member together. A–D, Blade-fork contacts; E, folded cantilever contact for printed circuit boards; F, straight cantilever contact for printed circuit boards; G, pin-socket contact; H, butting contact. A–G. Courtesy of Bell Laboratories.

Another design which is purported to provide unusually low mating forces employs bundles of wires in both halves of the connector that intermesh, like two hair brushes, when the parts are connected (3).

Joining to Electronic Connectors. The most widely used techniques for the termination of wires to separate contacts are the soldering (see Solders), welding (qv), crimping, solderless wrapping, and slotted-beam methods. Except for crimping and welding, it is usually possible to replace wires to a contact a limited number of times if repair or wiring changes are necessary.

Soldering materials are alloys that are composed primarily of tin and lead, and have low melting temperatures relative to the conductor metals which are being soldered. Welding requires sufficiently high temperatures for the fusion of metals.

Crimping is the compression of the back end of the separable contact, a tube, onto the wire conductor with a special tool that severely deforms both wire and barrel. This technique is suitable for joining both solid and stranded wires to connectors. Figure 3 illustrates typical crimp connections. Once the tool is removed the wire remains under the radial compression that is provided by the barrel. The force required to pull the copper wire from a crimped contact is approximately the same as its breaking strength. A soft metal, such as brass, is better for crimping than one having considerable springback, such as phosphor bronze (see Copper alloys).

In the solderless wrap (Fig. 4) or wire-wrap connection, a wire conductor is coiled around the back end of the separable contact, which has a square or rectangular cross section (4). The corners of the solderless wrap post and the areas of the wire that are in contact with it are severely deformed. In a properly made wrap, the force required

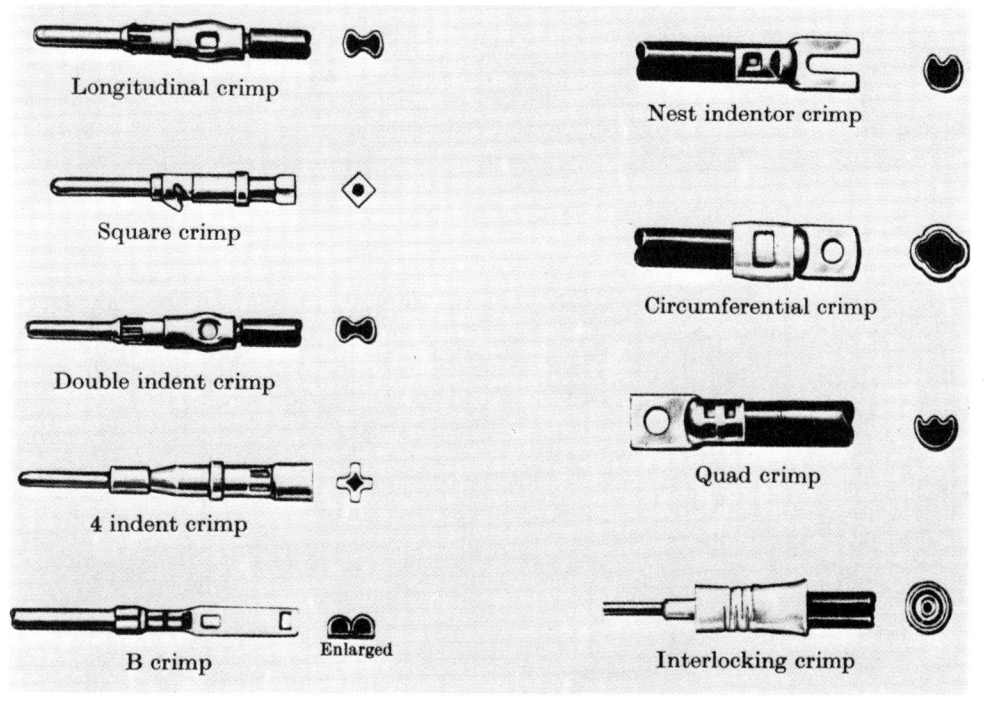

Figure 3. Typical crimp contacts. Insets are cross-sections of crimp indentations. Courtesy of Burndy Corporation.

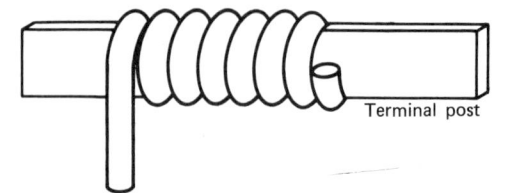

Figure 4. Solderless wrap connection.

to slide the wire along the post exceeds the breaking strength of the wire. The method is suitable only for solid wire, and special tools are used to make this connection.

Slotted beam or U-contacts describe a versatile design for the termination of solid wire and require that the wire be pushed into a narrow slot between two moderately rigid tines, or beams, at the back end of the separable contact (Fig. 5). The edges of the beams displace the insulation, squeeze the wire, and keep it in compression for the life of the connection. This termination method was developed for terminating conductors in a gang using flexible flat cables with round conductors (5).

Methods used to secure a wire to the back end of a separable contact include the taper pin and the solderless clip. The former is a cylindrical tapered body with a hollow end into which a wire is crimped; the front of the taper pin is forced into the back end of the connector contact which has conforming shape. The solderless clip (Fig. 6) has a spring which traps the solid or stranded wire against a post at the back end of the separable contact; the clip encircles both the wire and the post (6).

Splicing Connectors. Splicing connectors are used to permanently join wire to wire. Some are simple sleeve barrels that are crimped to bare wire; others are preinsulated with the crimp made by compressing the sleeve with its positioned insulation onto wires which may or may not have insulation. Two types of splicing connectors, which require insulation displacement and are used in the telecommunications industry, are illustrated in Figures 7 and 8. The connector in Figure 7 has a spring-tempered phosphor bronze liner that pierces the insulation and contacts the wire and

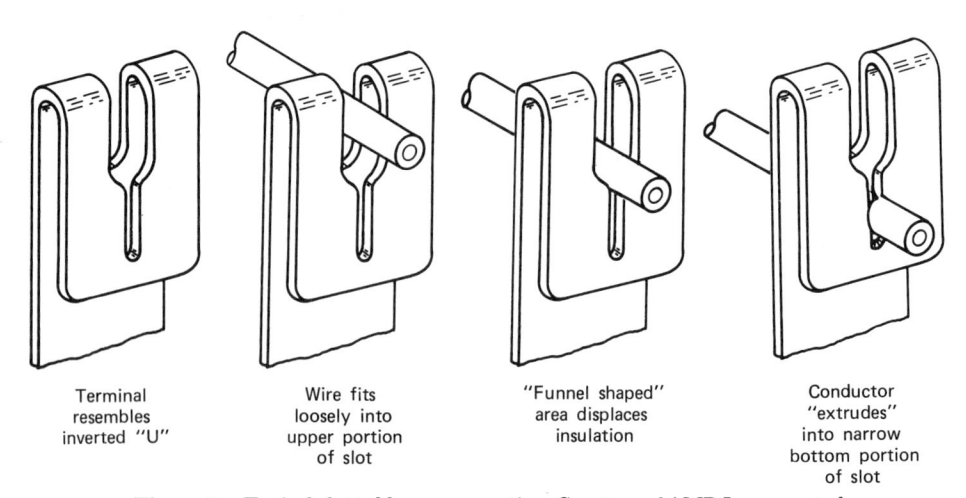

Terminal resembles inverted "U" | Wire fits loosely into upper portion of slot | "Funnel shaped" area displaces insulation | Conductor "extrudes" into narrow bottom portion of slot

Figure 5. Typical slotted beam connection. Courtesy of AMP Incorporated.

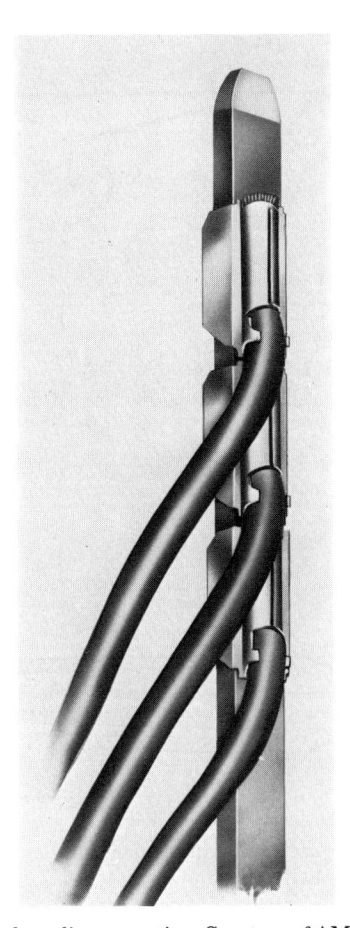

Figure 6. Solderless clip connection. Courtesy of AMP Incorporated.

an outer shell of soft brass (7). This connector is suitable for paper-insulated solid conductors. The slotted-beam connection principle is exemplified in the connector shown in Figure 8. This connector gang-terminated 25 pairs of wires with either paper or plastic insulation (8). The conductor can be aluminum or copper and its diameter ranges from 0.40–0.81 mm (20–26 American Wire Gage). A grease sealant is used to provide moisture resistance.

Terminals. Terminals are connectors with which individual wires are designed to be screwed down at their separable ends, and to which conductors are permanently joined at the back end, usually by crimping. Figure 9 illustrates common terminal configurations. Either ring or open tongue configurations provide a terminal for the screw connection.

Utility and Industrial Connectors. Connectors used in power distribution systems are nearly always of the permanent type and are usually made for single conductors. Sleeve barrels are used to splice cable by crimping (Fig. 10a). Insulating covers may be applied to the connector after the joint is made between the connector and the insulated cable. Clamp-type connectors are used with one or more clamping bolts and may have additional resilience when used with dished washers made of spring steel

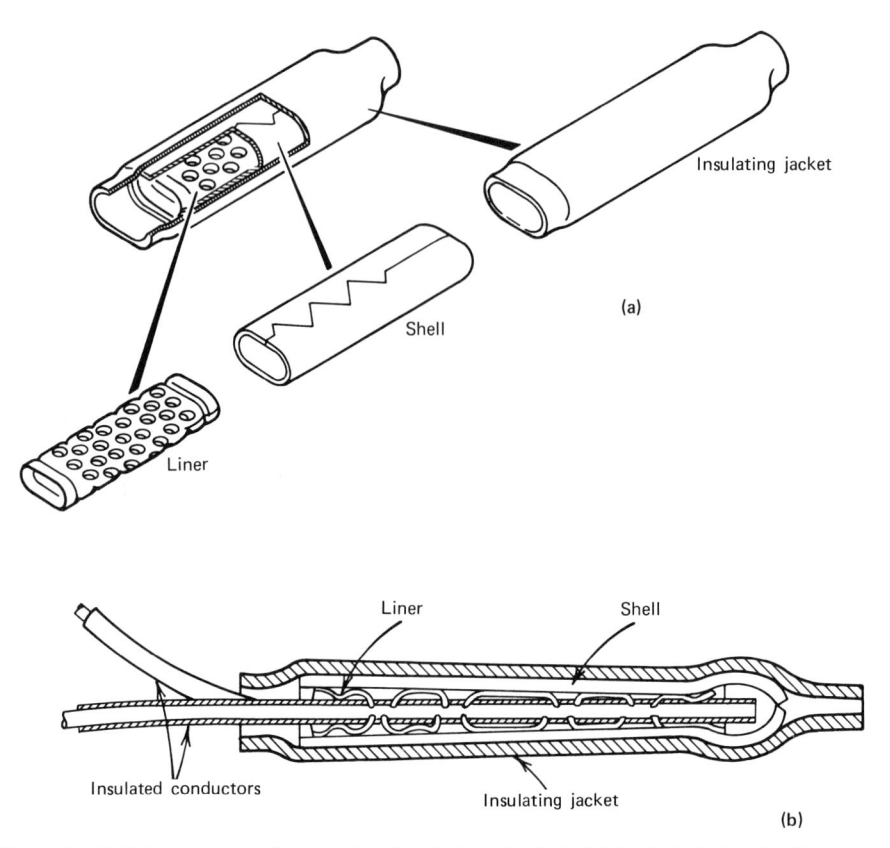

Figure 7. Splicing connector for two wires (insulation piercing). (**a**) Exploded view. (**b**) Cross section after pressing on wires. Courtesy of Bell Laboratories.

(Belleville washers). Figure 10**b** illustrates a typical clamp connector in a terminal configuration which is used for insulated conductors in building construction applications. Most crimp power connectors for aluminum and copper cable are made of aluminum. Those that are fabricated from copper alloys are for copper cable only. Clamp connectors are made of copper and aluminum alloys.

Methods of Application. Attachment of separable contacts to conductors may be achieved with automated machinery or with specialized hand tools. The automated machinery method is popular in large-volume original equipment markets for products such as strip terminals, rack and panel connectors, and printed-circuit-board connectors. This machinery applies contact components to conductors, but the insertion of the conductors and the applied contacts into connector housings is usually performed manually. The cases in which the desired connection is not amenable to automation require the use of hand-tooled application. Products included in this category are heavy duty terminals and splices, connectors to which conductors are terminated by soldering and small wire terminals. Hand-tool application is typically associated with utility, construction, maintenance and repair, and small-volume original equipment markets. The largest connectors, which are employed by utilities for construction and servicing of power distribution lines, require installation tools powered by hydraulic means, compressed air, or small explosive charges.

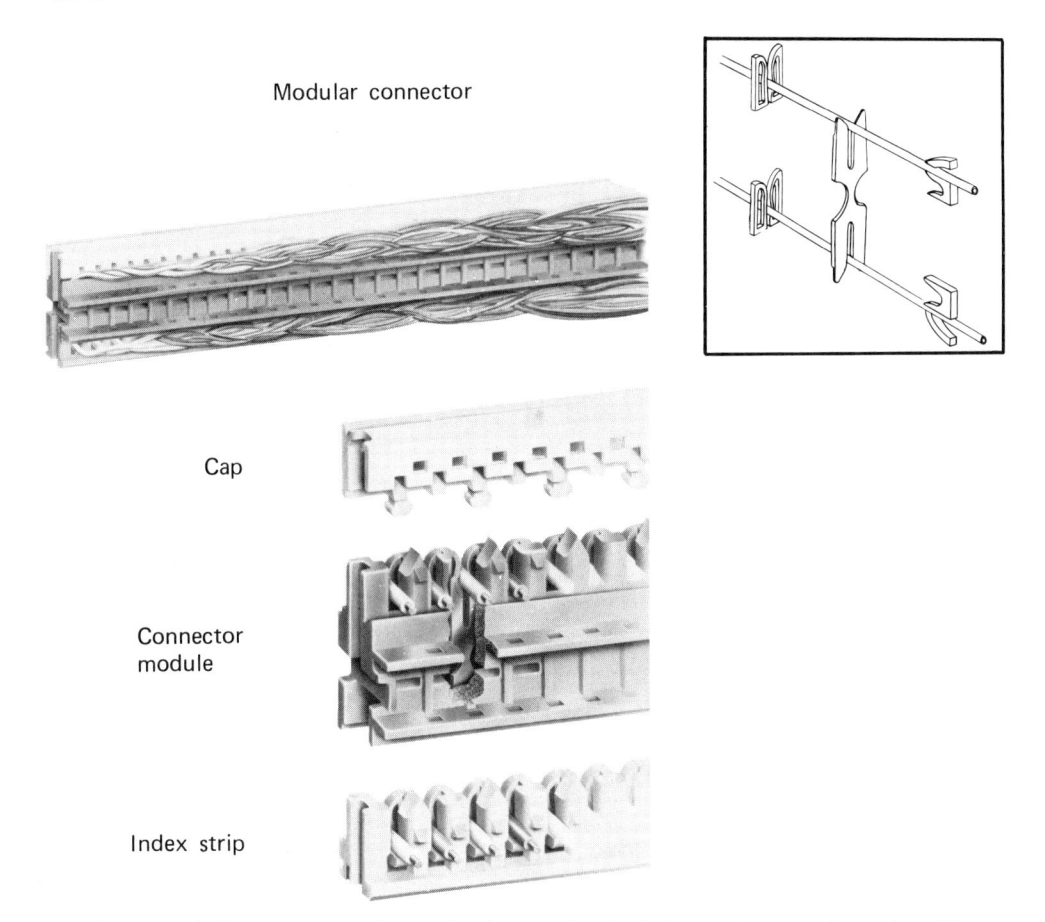

Figure 8. Splicing connector for terminating 25 pairs of telephone wire; inset shows slotted beam insulation piercing contact. Courtesy of Bell Laboratories.

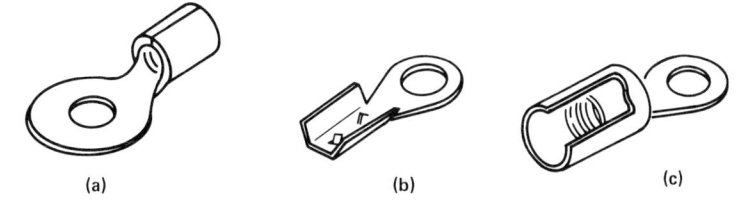

Figure 9. Some crimp terminal configurations. (**a**) Straight barrel, 90° tongue. Wire without insulation is crimped in barrel. (**b**) Open barrel with insulation-piercing lances. (**c**) Nylon or poly(vinyl chloride) preinsulated terminal accommodating and supporting wire insulation. Wire without end insulation is inserted in terminal and is crimped. The terminal sleeve is not broken but conforms to the shape of the crimp indent.

Contact Principles

The design of electrical contacts and the selection of contact materials are based on a body of interrelated physical, metallurgical, and chemical principles that have

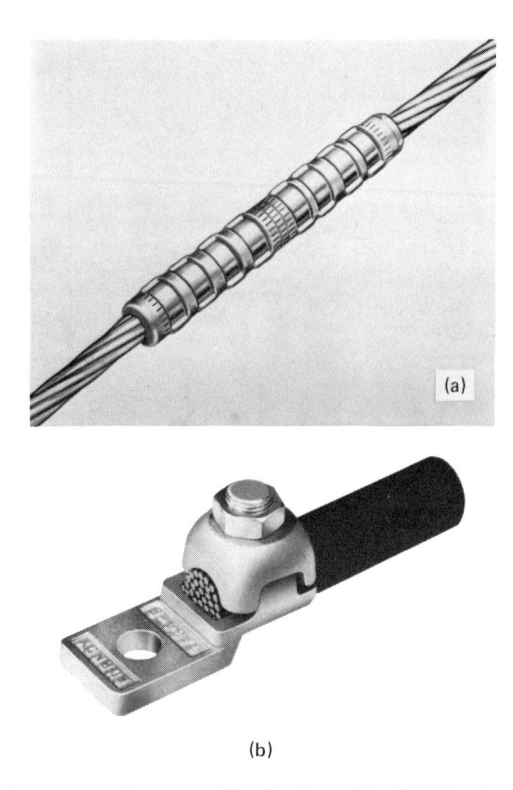

Figure 10. Typical connectors for power distribution cable. (**a**) Aluminum sleeve barrel for uninsulated cable. (**b**) Clamp-type terminal connector for insulated cable used in a building construction. Courtesy of Burndy Corporation.

developed since the beginning of inquiry into the nature of electricity. However, it was not until the early 1920s, when Ragnar Holm pioneered studies on the origin of resistance to current flow at the surfaces of touching metallic bodies, that contact science emerged as a specialized field of physics and materials technology.

Nature of Mechanical Contact. The surfaces of solids are irregular on a microscopic scale. Even nominally plane, smooth surfaces have a large-scale waviness on which is superimposed a roughness with peak-to-valley distances of several micrometers. When two metallic bodies are placed in contact at a light load, they touch at only a few small spots, or asperities. As the load is increased, more and more asperities come into contact and the surfaces move together. The true area of contact depends, therefore, on normal load and on the hardness of the metal. The real area of contact is only a fraction of the apparent area in most cases, except at very high loads. For example, the ratio of real to apparent contact areas of finely lapped steel flats having an apparent area of one cm^2 is ca 10^{-4} at a force of 10 N (2.25 lbf).

Nature of Electrical Contact. If metallic surfaces are covered by a nonconducting layer, such as an oxide or a sulfide tarnish film, the area of metallic contact will be zero provided that the film is unbroken. Significant current will not flow between such surfaces except when the film is less than 2–3 nm in thickness. At or less than these thicknesses, electron tunneling (tunnel conduction), which is voltage independent, occurs by a wave-mechanical effect analogous to the transmission of light through metal foil of thickness comparable to the wave length (9). If the nonconductive layer on a surface is discontinuous or is punctured, the mechanical load is borne by both film and metal. Current then flows through the metallic spots, called a spots (10). The lines of electric flow converge at these spots, as illustrated schematically in Figure 11. Constriction resistance, the increase of resistance beyond that of a continuous solid, ie, not having an interface (11), originates at this convergence.

In the simplest case, for a single circular contact spot with identical metals having a uniform film, contact resistance R has the relationship:

$$R = R_c + R_f$$

$$R = \frac{\rho}{2a} + \frac{\sigma}{\pi a^2}$$

where R_c = constriction resistance, R_f = film resistance, a = radius of the a spot, ρ = bulk resistivity of the contact metal, σ = film resistance (ohms/unit area). The radius of the a spot can be calculated from the hardness of the metal and the applied load.

Since a length of metal associated with the connector contact is ordinarily in the path between the contact end to which a wire is terminated and the contact interface, its resistance (bulk resistance) must be added to contact resistance when considering the connector as a circuit element. This overall resistance is sometimes erroneously called contact resistance.

Resistance Heating of Contacts. The contact material, contact area, and its heat dissipating ability, as well as the heat dissipating ability of the structure to which it is attached, limit the amount of current which a contact can transport. Excessive current will heat and soften the metal contact, and this softening will result in an increase in the surface area of the contact and a corresponding reduction in contact resistance.

At higher currents, the mating junction will melt. Both softening and melting occur at characteristic voltages. Typical values are shown in Table 1.

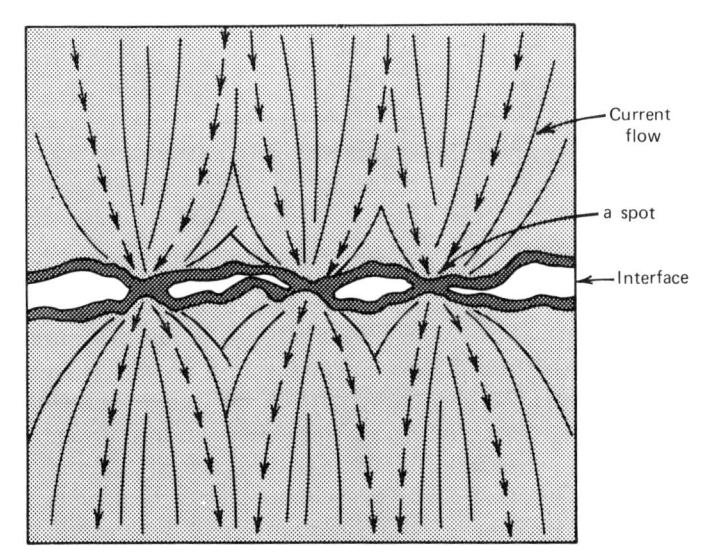

Figure 11. Microscopic view of contact interface (schematic); constriction resistance originates in the constriction of current flow through the touching metallic junctions of the mating surfaces.

652 ELECTRICAL CONNECTORS

Table 1. Some Softening and Melting Voltages[a]

Metal	Softening °C	Softening V	Melting °C	Melting V
Sn	100	0.07	232	0.13
Au	100	0.08	1063	0.43
Ag	150–200	0.09	968	0.37
Al	150	0.1	660	0.3
Cu	190	0.12	1083	0.43
Ni	520	0.22	1453	0.53
W	1000	0.6	3380	1.1

[a] Ref. 12.

Voltage Breakdown of Films. If a significant voltage can be passed by a circuit across a film-covered contact, the film, depending on its thickness and composition, may break down electrically. This action, called the coherer effect or fritting (13), results in the formation of minute metallic conductive paths through the film. Puncturing is of the order of 0.1 V/nm of film. The potential drop across the film after puncturing is the melting point voltage.

An Overview. Metallic contact between surfaces of separable connectors usually is obtained either by using noble metals, which are essentially film-free, or by designing the contact so that any films that are present are broken before or as the surfaces are brought together. For example, noble metal coatings on base metal substrates having high conductivity, such as copper alloys, are used for small multicontact connectors which can provide small mechanical loads normal to the surface. Alternatively, soft metal contacts of, for example, tin and tin-lead alloy platings can be used since films on their surfaces are easily disrupted on closure. In the latter case, mechanical wear with sliding connector designs may severely limit the numbers of insertions and withdrawals that are possible. In the case of power connectors, severe deformation of the aluminum or copper surface is obtained by crimping or clamping which facilitates disruption of the oxides on both the connector and conductor surfaces. Tin plate is used widely on aluminum connectors, and wire brushing of the cable (especially if it is made of aluminum) and joint aids (described later) increase the areas of metallic contact and thereby lower contact resistance. A large part of practical connector engineering is devoted to designing metallic contacts that need minimal maintenance, especially for those intended to serve in chemically agressive atmospheres and in high temperature applications.

The contact resistance of any electrical connector in a circuit must be stable and low for proper functioning of that circuit. Low voltage circuits, which are common in modern electronic systems, have open circuit voltages of not more than a few volts, insufficient levels for the fritting of films that grow on base metals from environmental exposure. Metallic bridges established by fritting are fragile, and the voltage drop across fritted surfaces may permit heating with consequent degradation of the contact. It is, therefore, generally undesirable to rely on fritting as a method that may establish current flow in a connector. Low contact resistance should be achieved with the use of noble metal contacts or of base metals concurrently with methods that mechanically perforate any insulating films which are present.

Typical separable electronic connectors have contact resistances that range from several milliohms to tenths of ohms. Utility–industrial connectors, which are established by very high pressure connection and are exemplified by a bolted joint, have contact resistances of microohms. Crimping, solderless wrapping, and other methods used to establish connection of wires and other conductors to electronic connectors have associated contact resistances of ca 10–100 $\mu\Omega$. The bulk resistance of the conducting elements, which include connector parts such as the spring, barrel, and pin, is usually several times greater than the contact resistances of the separable contact interface and the terminal ends. The overall resistance of electronic connectors generally ranges from 3–30 mΩ.

Materials and Processing

Contact Substrates. Some of the substrate metals used in connectors have already been mentioned. The substrate must be able to be terminated readily as well as be a good electrical conductor. In electronic connectors the substrate may serve as a spring element. The most widely used spring materials for connectors are the copper alloys: 98.1 wt % Cu; 1.9 wt % Be (Copper Development Association (CA) 172); 94.8 wt % Cu, 5.0 wt Sn, 0.19 wt % P (phosphor bronze, CA 510); and 88.2 wt % Cu, 9.5 wt % Ni, 2.3 wt % Sn (CA 725) (see Copper alloys). Sometimes springs made of metals that have poor conductivity, such as stainless steel, are used as inserts in connector barrels of brass or similar metals which are inexpensive and can be terminated easily.

The contact ends of printed circuit boards are copper. Alloys of nickel and iron are used as substrates in hermetic connectors in which glass is the dielectric material. Terminals are fabricated from brass or copper; from nickel, for high temperature applications; from aluminum, when aluminum conductors are used; and from steel when high strength is required. Because steel has poor corrosion resistance, it is always plated with a protective metal, such as tin.

Contact Finishes. Base metal substrates quickly develop oxides, tarnish, or corrosion films, in humid, polluted atmospheres. Since these films may prevent adequate metal-to-metal contact when the connector or connector–conductor surfaces are mated, coatings of other metals commonly are used to obtain corrosion resistance, to provide conductivity, or to facilitate termination to conductors by soldering, wire wrapping, or by other means. Application of finishes is achieved by electroplating (qv), cladding, and by hot-dipping when low melting metals such as tin are used (see Metal surface treatments). Selective application of a contact finish to portions of the substrate can be accomplished using any of these coating techniques. The major noble contact finishes are gold, palladium, and rhodium. The nonnoble finishes are tin, silver, nickel, and indium. Alloys of most of these metals also are widely used.

Requirements for Noble Metal Electroplated Finishes. Because it is chemically stable, gold is the noble metal most extensively used in the contacts in electronic connectors; multiple metric ton quantities (hundreds of thousands of troy ounces) are consumed annually for this purpose. The high cost of noble metals requires that they be sparingly, however effectively, used. The chief requirement of a plated finish is sufficient thickness. Gold coatings in thicknesses of 0.5–5 μm are used; the greater thicknesses are required for critical applications and for wear resistance when large numbers of engagements of the connector are specified. Rhodium ordinarily is plated from only 0.5–1 μm as thicker coatings have a tendency to crack spontaneously. The thickness of palladium electrodeposits parallels those of gold.

Porosity ranks next to thickness in importance, especially when the finishes must be operable in environments that are polluted or that otherwise promote tarnish or corrosion. Pores are openings in the surface that extend to the substrate or underplate. They can be intrinsic in the coating (14), or can be produced by mechanical wear or by forming operations which are involved in manufacturing the contact and follow plating. In some environments the substrate metal can tarnish or corrode around the coating pores and can produce localized areas of insulating films which cause contact resistance to increase. Porosity is less important for connectors that operate indoors at moderate to low relative humidities and in the absence of corrosive pollutants (15).

Hard deposits enhance a connector's resistance to mechanical wear. Rhodium, palladium, and gold electrodeposits containing 0.1–0.5% cobalt or nickel are more wear resistant than pure gold finishes (16). Deposits must be solderable if their associated terminations are to be soldered. Gold deposits thicker than 2.5 μm form brittle gold-tin intermetallics (17) during soldering and, therefore, give weak joints. Contaminated surfaces are also difficult to solder.

Underplatings normally are used for noble metal electrodeposits. Zinc from brass rapidly diffuses through gold plate and forms insulating films on the plated surface. Its rate of diffusion is reduced by copper or nickel underplatings (18). Diffusion of copper through gold plate, especially at temperatures above 100°C, can be troublesome, and nickel underplate is used to retard the diffusion rate (18). Copper and nickel underplatings are usually used to lower porosity in a noble metal deposit (14). The effective hardness of a multilayer coating is increased with hard underplates. The wear resistance of gold deposits can be improved by nickel underplating (19). Pure electroplated nickel is ductile and harder than most substrate metals, and the thickness of the nickel underplate should approximate or exceed the thickness of the gold layer in order to significantly reduce wear. Nickel underplate is used in place of rhodium, which is brittle, to minimize cracking. The corrosion tendency of contacts that have porous noble metal finishes is lowered by using nonreactive underplatings, such as 65 wt % Sn, 35 wt % nickel (20).

Requirements for Clad Inlay Noble Contact Materials. Insertion of a strip of metal into a groove in a base metal substrate which is then metallurgically bonded to the substrate by rolling at high pressure (Fig. 12) is referred to as cladding (21). The metals, of course, must be ductile, and some copper alloys, such as CA 725 and CA 150, which are used in connector springs, are especially suitable for this purpose. Pure gold, pure palladium, and alloys of gold with silver or nickel, such as 75 wt % Au and 25 wt % Ag,

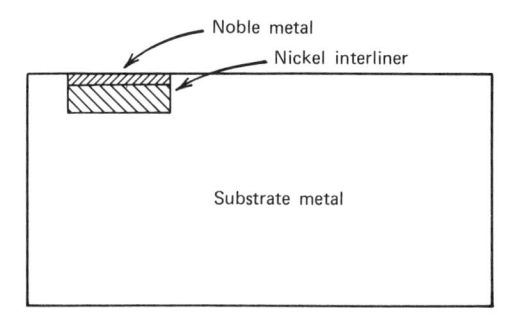

Figure 12. Clad inlay contact material (schematic).

and 96 wt % Au and 4 wt % Ni, are widely used clad inlays for electronic connector contacts. Since intermediate heat treatments as well as several passes through the rolls are usually necessary in order to achieve the desired thickness reduction, a nickel interliner is ordinarily used as a diffusion barrier. The same thickness and porosity requirements for electrodeposits apply to clad noble metal finishes.

Cladding may be less expensive than selective electrodeposition when coatings greater than 2.5–5 μm thick of gold are required but may be more expensive than electrodeposition with thinner coatings. Selective techniques are most easily used with sheet metal substrates that are to be machine stamped and formed into contacts. Clad noble metals are considerably more ductile (and less hard) than comparable electrodeposits and, therefore, are better suited to forming operations. Contacts that are made into separate parts from rod by screw machining are usually coated on all exposed surfaces by barrel electroplating.

Requirements for Noble Metal Weldments. Noble metal contact buttons, 25–75 μm thick, are made by resistance welding (qv) a rod of the material to the substrate, which usually is a contact spring, and by cutting the rod and forming the button to the desired shape. Pure gold and gold-silver alloys are the most commonly used metals.

Requirements for Base Metal Finishes. The low cost of base metal finishes obviates their selective coating. Electrodeposition is used for 0.5–5 μm thick coatings of tin and tin-lead alloy (usually about 50 wt % Sn, 50 wt % Pb) on electronic connector contacts, on contacts on the edges of printed circuit boards and on terminals. Aluminum connectors that have utility-industrial applications are more thickly coated, and hot dipping in molten tin is common.

Whiskers are filamentary growths a few micrometers in diameter that occur spontaneously from tin and other low melting pure metals in response to lattice strains. Whiskers can attain lengths of several millimeters and may short-circuit adjacent contacts. The presence of whiskers is unimportant if currents are in the ampere range, because bridging is instantly cleared by melting. Whisker formation can be retarded by combining small amounts of alloying elements, such as lead or copper, with tin coatings or by fusing the coatings after plating (22).

Conductive Elastomers. Conductive elastomers (23), rubbers that are made conductive by molding metal or carbon powders in them, have characteristics of both a contact material and a spring. Silicone rubbers, neoprene, polyurethane, and other elastomers have been used; however, silicones are the most popular because they have a low compressive set and operate over a wide temperature range, from ca −65 to 200°C (see Polymers, conductive; Antistatic agents). Particle loadings are high, eg, 70 wt % or more for metals, because there must be particle contact through the body for the elastomer to be conductive. Silver is used in those contacts that must be highly conductive and other metals are used in systems where higher resistance can be tolerated.

However, conductive elastomers have only ca $\leq 10^{-3}$ of the conductivity of solid metals. Also, their contact resistance changes with time when they are compressed. Therefore, they are not used where significant currents must be carried or when low or stable resistance is required. Typical applications, which require a high density of contacts and easy disassembly for servicing, include connection between liquid crystal display panels (see Liquid crystals; Digital displays) and between printed circuit boards in watches. Another type of elastomeric contact has a nonconducting silicone rubber core around which is wrapped metalized contacts which are separated from each other by insulating areas (24).

Contact Lubricants. Debilitating wear of separable connector contacts, which may occur if the metallic coating is thin or if forces normal to the contact surfaces are high; can be minimized by coating the contacts with thin films of organic lubricants (16). Viscous mineral oils and poly(phenyl ethers), soft microcrystalline waxes, and petrolatum have been used on electronic connectors. Although most of these lubricants are insulators, the few asperities of the metal pieces that make contact through the film provide low contact resistance, which is indistinguishable from that of unlubricated contacts. Graphite can be burnished on the surface of gold electrodeposits as a contact lubricant, but the contact resistance of such a system tends to be variable at low contact pressures. Some lubricants have corrosion inhibiting or antitarnishing properties for base metals and for porous noble metal finishes (16).

Joint Aids. Good metallic contact with aluminum connectors or aluminum conductors, including those with severly deformed surfaces, is difficult if aluminum oxide is present. It is believed that contact occurs by extrusion of the substrate metal during deformation through minute cracks produced in the oxide; and that some degree of geometric matching of cracks is necessary (25) when both of the surfaces are aluminum. It has been found that finely divided zinc incorporated in grease can be coated on the interior surfaces of an aluminum connector to provide lower initial contact resistance and better long term resistance stability (26). The grease is also able to retard ingress of air and water which can seriously degrade reliability of aluminum connectors, especially in exposed out-of-doors service in power distribution networks. Other soft metal coatings are effective, such as indium which is used (27) as a plating on copper alloy connectors intended for permanent joints to aluminum communications cables having slotted-beam connectors.

Insulators

Molded plastics serve as insulators for multi-contact connectors and glass is the insulating material used in hermetic connectors intended for bulk-head mounting. Ceramics (qv) are employed in some high voltage power connectors. Hard rubber shells insulate connectors that serve underground power distribution cables (see Insulation, electric). Nylon and poly(vinyl chloride) sleeving are used with preinsulated terminals. A wide variety of plastics are employed in electronic connector bodies depending on their size, the strength requirements, the complexity of the design, and the service environment (28). Table 2 lists some properties of those plastics used for electronic connector bodies. The first six materials are thermoplastic and the remainder are thermosetting. Thermoplastic materials are preferred because they are easy to manufacture. Many of these plastics are made with fillers (qv) which enhance various properties, such as glass fibers for strength.

Reliability and Testing

Connectors must be as reliable as any component in the circuits that they serve. Reliability requirements of each connection are particularly stringent in complex apparatus or where signal and power must be carried long distances along a séries of connector contacts. The consequences of failure include not only loss of service in a

Table 2. Some Plastics Used in Electronic Connector Bodies[a]

Material	Heat resistance, °C	Dielectric strength, MV/m	Dielectric constant	Water absorption, wt %	Flammability	Tensile strength, MPa[b]	Relative cost
nylons	80–150	15–20	3.6–4	1.5–8.5	self ext[c]	60–80	0.90
poly(phenylene oxide)	130–150	40	2.6–2.9	0.06	slow burn to self ext[c] non-drip	100–120	1.35
polysulfone	150–170	17	3.1	0.2	self ext[c]	70	1.20
polycarbonate	100–130	18	3.3–3.8	0.1	self ext[c]	55–65	0.95
acrylonitrile butadiene-styrene (ABS)	70–100	14–18	2.9–3.2	0.2–0.4	slow burn	35–60	0.40
polyester	120–140	13–19	3.1–3.3	0.1–0.3	slow burn	55–100	0.75
diallyl phthalate	170–230	15.5–18	3.2–6.0	0.2	self ext to[c] non-burning	40–75	1.00
phenolics	170–290	8–16	5.4–13.2	0.27	self ext[c]	35–55, 35–125 (glass filled)	0.25
melamine–formaldehyde	200	7–12	7.5–10.0	0.1–0.2	self ext[c]	35–70	
alkyd	230	10–21	4.5–6.5	0.1–0.2	self ext[c] and non-burn	20–60	0.35–0.70
epoxy	150–260	12–16	4.4–5.0	0.1–0.5	slow burn to self ext[c]	70–210	0.60

[a] Ref. 29.
[b] To convert MPa to psi, multiply by 145.
[c] Ext = extinguished.

telephone switching system or communication network, but the great expense of locating the failure, exposing the connection and making a replacement. It has been estimated (8) that the average cost of replacement of a single wire splice connection in telephone cable is in excess of $100; the cost of a new connector is only a few cents. It is not surprising, therefore, that considerable effort has been directed to determining the causes of connection failures and to learning how to minimize the likelihood of their occurrence. Acceptable failure rates range from <1 in 10^9 operating hours for contacts in air-frame (30) electrical systems and in some telecommunications equipment, to 100–1000 in 10^9 operating hours in instruments, to even larger rates for contacts in many consumer products. A failure is defined as exceedance of contact resistance, which can be as little as twice the initial contact resistance, that causes circuit malfunction. The required lifetimes of connectors may be ≥40 yr, although most required application times are shorter (see Materials reliability).

Mechanisms of Failure. The causes of connector contact failure can be of a thermal, chemical or mechanical nature, in addition to misapplication and physical abuse.

Thermal. If a connector is in an environment hotter than that for which it was designed, or if it undergoes significant resistive heating due to excessive current flow, the following can occur and eventually may cause failure: (*a*) stress relaxation of the contact spring, or creep of the deformed metal, as with crimp barrels, which are responsible for maintaining pressure contact; (b) acceleration of chemical reactions that cause failure. For example, rapid thickening of the oxide on base contact surfaces may occur even when they are mated if the contact is not gas tight, eg, if the contact is unable to resist encroachment of the atmosphere at the mating site; (c) accelerated diffusion of substrate metals or of base hardener and impurity metals through the surface of noble metal platings and the formation of insulating films; and (d) volatilization of connector contact lubricants.

Alternating temperature cycles may be more damaging than a constant elevated temperature because differences in the amount of expansion and contraction of the members supporting the contacts can cause breakage of the metallic junctions (a spots). When there is an insulating film on the surface between the a spots as with base metals, metallic contact cannot be maintained owing to the movements. Heat cycling forms the basis of much connector testing, eg, 500 cycles of heating by current flow to 100°C above ambient followed by cooling to room temperature for aluminum conductor overhead power distribution connectors (26). A connector is acceptable if this stress does not cause significant change in overall resistance.

It is not practical to determine the contact resistance in power connectors. The resistance of the connection of a specified length of conductor on each side of the connector is measured and is called the overall resistance or the connection resistance. One industry specification (31) defines the included lengths and requires the stability of the connection resistance to be within ±5% of its average value throughout the heat-cycle test.

Chemical. Contacts made of base metals and contacts coated with noble metals and having pores may be subject to chemical reactions with air pollutants which cause formation of insulating films. Above a critical humidity, usually 70%, galvanic corrosion occurs at pore sites if the substrate is corrodible. Tarnish films, which may form if copper or silver reacts with elemental sulfur or with H_2S, may spread (32) from the pore sites to the remaining surface of the noble metal. Soils such as hygroscopic dust contamination and fingerprints lower critical humidity. SO_2 and HCl are chemically aggressive pollutants at parts per billion levels.

Another humidity related degradation is silver migration from silver conductors (33): the transport of silver from one conductor to another through or on the surface of the insulator under the influence of a d-c electrical field. The positively charged conductor loses silver by oxidation in the form of positive ions, which then move through moisture paths to the cathode where they are reduced to metallic silver. The silver may grow in dendritic form and may eventually short-circuit the conductors. Current leakage between adjacent conductors through adsorbed films of moisture may occur, especially if ionic contaminants are present.

Chemical degradation can be avoided by using closed structures, which may have protective covers or which may be fully hermetic, as well as by using unreactive metals. Barrier coatings are effective in some cases, and polyethylene–polybutene grease (8) is used in some splicing connectors for telecommunications cable.

Many techniques have been developed for the accelerated testing of connector contacts. These include elevated relative humidity exposure, cycling temperature–humidity procedures, and aging in chambers containing simulated atmospheres which include individual or combined pollutants. The International Electrotechnical Commission has published a standard method for testing connectors which involves exposing the connectors for 4, 10, or 21 d in a chamber to a flowing stream of air containing 25 ppm of SO_2 at 75% rh and 25°C (15). Because of the complexity of environmental effects, age acceleration factors with respect to real conditions are generally unknown for tests in current use. Nevertheless, they serve as a design aid for estimating the relative merits of candidate connector contact materials and contact structures and for qualifying products.

Mechanical. Premature wearout or loss of contact metal during engagement and separation can result in loss of tolerances, reduced spring forces, formation of loose metallic wear debris, which may short-circuit contacts, and development of porosity in noble metal contacts. Underplatings, contact lubricants and hard materials reduce mechanical wear.

Fretting corrosion (34–35) can lead to high contact resistance of base metal contacts, such as tin plate in electronic connectors. Small cyclical displacements of the connector halves occur due to external vibration or to differential thermal expansion and contraction of the mating contacts. The wear debris that is formed remains in the contact zone. The accumulation of oxide debris in the contact region leads to increased contact resistance. Solutions to this problem are structures that do not permit movement of contact surfaces with respect to each other or use of gold as a contact finish.

Contacts containing platinum metals, such as palladium and rhodium, also may be subject to resistance problems due to the formation of so-called friction polymer (36) on their surfaces. This material is an insulating contaminant which originates in organic materials in the vicinity of the contact, such as organic vapors that adsorb on the surface, and subsequently polymerize to tough solids as a result of small mechanical displacements of the contacting surfaces.

Economic Aspects

The largest markets for connectors are in the United States and Japan. The Federal Republic of Germany, the United Kingdom, and France follow with similarly sized markets. The connector market (37) can be categorized as follows, with the estimated 1978 sales in the United States as a percent of total: military and aerospace (20%), electronic data processing (20%), communications (10%), instruments (10%), consumer including automotive (10%), utility (10%), building and other construction (10%), maintenance and repair (10%). The total U.S. market in 1978 for electronic connectors was approximately $850,000,000 and $600,000,000 for small splicing connectors, terminals, and utility–industrial connectors. About 20% of the 1978 electronic connector market was represented by consumers, largely in the telecommunications and automotive industries, who manufactured many of their own connectors. The growth of sales was ca 10%/yr from 1970 to 1978.

The per contact cost of electronic connectors ranges from less than one cent to one dollar or more, depending on the sophistication of the connector, the number of contacts in it, and the materials that are used. The average cost is five to six cents. Fiber optics (qv) technology will have an impact on conventional (metallic) electronic connectors. Transmission of information by light through optical fibers made of glass or plastic is potentially less expensive than transmission of electric signals through wire. The economic value of fiber optics is particularly pertinent with regard to certain applications in the telecommunications, computer, and military-aerospace fields.

660 ELECTRICAL CONNECTORS

Optical connectors are used in the source-to-cable, cable-to-cable, and cable-to-detector areas. The design and manufacture of reliable, low loss, easily installed optical connectors has been a key problem in the development of fiber optic systems. Future improvements in optical connectors and optical fibers may result in significant displacement of electrical connectors and wiring where information, rather than power, is being transmitted.

BIBLIOGRAPHY

1. M. D. Lazar, S. M. Garte, and R. P. Diehl, *IEEE Intercon '73*, New York, March 30, 1973.
2. M. Antler, W. G. Graddick, and H. G. Tompkins, *IEEE Trans. Parts, Hybrids, Packag.* **PHP-11,** 35 (1975).
3. R. S. Nelson, *Multiwire Brush Contact Connector, Report SAND 75-0245*, Sandia Laboratories, Albuquerque, N. M.
4. R. F. Mallina, *Bell Syst. Tech. J.* **32,** 525 (1953).
5. J. O. Knudson, *Proceedings Sixth Annual Connector Symposium*, Electronic Connector Study Group, Cherry Hill, N.J., 1973, pp. 207–219.
6. H. B. Brown, *Proceedings Institute of Printed Circuits Conference*, Orlando, Fla. 1977.
7. H. J. Graff, J. M. Peacock, and J. J. Zalmans, *Bell Syst. Tech. J.* **40,** 131 (1963).
8. D. R. Frey and D. C. Borden, *Proceedings Second International Symposium on Subscriber Loops and Services*, London, England, May, 1976, pp. 68–72 (IEEE Conference Publication No. 37).
9. R. Holm, *Electric Contacts*, 4th ed., Springer-Verlag, New York, 1967, pp. 118–134.
10. Ref. 9, p. 8.
11. Ref. 9, pp. 9–26, 124–125.
12. Ref. 9, pp. 87–92, 436–438.
13. Ref. 9, pp. 135–152.
14. M. Clarke in R. Sard, H. Leidheiser, Jr., and F. Ogburn, eds., *Properties of Electrodeposits, Their Measurement and Significance*, The Electrochemical Society, Princeton, N.J., Chapt. 8, pp. 122–141.
15. M. Antler and J. J. Dunbar, *IEEE Trans. on Components, Hybrids, and Manufacturing Technology* **CHMT-1** (1), 17 (1978).
16. M. Antler, *IEEE Trans. Parts, Hybrids, Packag.* **PHP-9,** 4 (1973).
17. C. J. Thwaites in F. H. Reid and W. Goldie, eds., *Gold Plating Technology*, Electrochemical Publications, Ltd., Ayr, Scotland, 1974, Chapt. 19, pp. 225–245.
18. M. Antler in ref. 17, Chapt. 36, pp. 478–494.
19. M. Antler and M. H. Drozdowicz, *Bell Syst. Tech. J.* **58,** 323 (1979).
20. M. Antler, M. H. Drozdowicz, and C. F. Hornig., *J. Electrochem. Soc.* **124,** 1069 (1977).
21. R. J. Russell, *Proceedings Seventh Annual Connector Symposium*, Electronic Connector Study Group, Inc., Cherry Hill, N. J. 1974, pp. 79–94.
22. A. Mendizza and P. C. Milner "Materials Technology", in *Physical Design of Electronic Systems*, vol. II, Prentice-Hall, Inc., Englewood Cliffs, N.J., 1970, Chapt. 5, p. 247.
23. T. P. Piccirillo and co-workers, *Proceedings Holm Seminar on Electrical Contacts*, Ill. Inst. Tech., Chicago, Ill., 1976, pp. 71–78.
24. C. Curry, *Proceedings Seventh Annual Connector Symposium*, Electronic Connector Study Group, Inc., Cherry Hill, N.J., 1973, pp. 211–216.
25. J. R. Osias, and J. H. Tripp, *Wear* **9,** 388 (1966).
26. I. F. Matthysse, *Basic Connection Principles*, 2nd ed., Burndy Corp., Norwalk, Conn., 1965.
27. R. W. Barnard and J. P. Pasternak, *Proceedings Holm Seminar on Electric Contact Phenomena*, Ill. Inst. Tech., Chicago, Ill., 1968; pp. 27–41.
28. R. Saunders, C. A. Harper, in *Handbook of Electronic Packaging* McGraw Hill, New York, 1969, Chapts. 6–7.
29. N. Spatz, *Machine Design* **50** (2), 112 (Jan. 26, 1978).
30. J. E. Atkinson, *IEEE Trans. Reliab.*, 8 (June, 1964).
31. *EEI-NEMA Standards for Connectors for Use Between Aluminum or Aluminum-Copper Overhead Conductors, EEI Pub. No. TDJ-162*, Edison Electric Institute, 90 Park Ave., New York, N.Y., 10016, Aug. 1973.

32. Ref. 22, p. 237.
33. Ref. 22, p. 249.
34. R. B. Waterhouse, *Fretting Corrosion*, Pergamon, New York, 1972, pp. 60–62.
35. E. M. Bock and J. H. Whitley, *Proceedings Holm Seminar on Electrical Contacts*, Ill. Inst. Tech., Chicago, 1974, pp. 128–138.
36. H. W. Hermance and co-workers *Bell Syst. Tech. J.* **37,** 739 (1958).
37. "Spotlight on Connectors," supplement to *Electronic News*, Oct. 17, 1977, 60 pages.

General References

R. Holm, *Electric Contacts Handbook*, 4th ed., Springer-Verlag, New York, 1967; comprehensive treatment of electric contact theory.
G. L. Ginsberg, ed., *Connectors and Interconnections Handbook*, Basic Technology, vol. 1, The Electronic Connector Study Group, Inc., P.O. Box 1428, Camden, N.J. 08101; many articles on electronic connectors; emphasis on connector design and applications.
F. H. Reid and W. Goldie, eds., *Gold Plating Technology*, Electrochemical Publications, Ltd., Ayr., Scotland, 1974; available from American Electroplaters' Soc., Winter Park, Fla., 32787; comprehensive treatment of gold, the most important contact materials for electronic connectors (plating processes, deposit properties, applications including connectors).
Physical Design of Electronic Systems, Integrated Device and Connection Technology, vol. III, Prentice Hall, Englewood Cliffs, N.J., 1971; part III of this volume (on "Contact and Connection Technology") gives Bell System point of view which emphasizes telecommunications applications.
C. A. Harper, ed., *Handbook of Electronic Packaging*, McGraw-Hill, New York, 1969; authoritative treatise, includes chapters on connectors and interconnecting devices, packaging, bonding techniques, and wire and cable specifications.

Comprehensive Bibliographies

The American Society of Testing and Materials published a *Bibliography on Electrical Contacts, Circuit Breakers and Phenomena* with supplements covering the world literature through 1965. The Institute of Electrical and Electronics Engineers' publication 70M68PMP continues this bibliography through 1969. The *Proceedings of the Holm Seminars on Electrical Contact Phenomena* (called the *Holm Conference on Electrical Contacts* since 1977) of the Illinois Institute of Technology, Chicago, Ill. includes additional supplements as follows: 1970–1971 Supplement with the 1971 *Proceedings;* 1972 Supplement in the 1972 *Proceedings* (the Seminar was concurrent with the Sixth International Conference on Electrical Contact Phenomena); 1973–1975 Supplement with the 1975 *Proceedings*; and 1976–1977 Supplement with the 1977 *Proceedings*.

Trade Publications

There are several trade publications serving the connector and related fields, including *Electronic Packaging and Production,* Milton S. Kiver Publications, Inc., Chicago, Ill. 60606, and *Electronic News,* Fairchild Publications, New York.

Conferences

The Annual Connector Symposia (conducted by The Electronic Connector Study Group, Inc., P.O. Box 1428, Camden, N.J. 08101) emphasize developments in design. The Annual Holm, Conferences on Electric Contacts of the Illinois Institute of Technology (Dept. of Electrical Engineering, Chicago, Ill.) attract papers in all areas of contact science and technology including connectors. Both organizations issue *Proceedings*.

MORTON ANTLER
Bell Telephone Laboratories

ELECTRIC INSULATORS. See Insulators, electric.

ELECTROANALYTICAL METHODS. See Analytical methods.

662 ELECTROCHEMICAL MACHINING

ELECTROCHEMICAL MACHINING. See Electrolytic machining methods.

ELECTROCHEMICAL PROCESSING

Introduction, 662
Inorganic, 671
Organic, 696

INTRODUCTION

Electrochemical processes involve the interconversion of electrical and chemical energy by means of a reaction at an electrode. Electrical charge may be fed to an electrolysis cell to induce chemical reactions (synthesis, metal winning (see Extractive metallurgy) or refining, etc), or chemical reactions may be run in a cell to generate electricity (batteries, accumulators, fuel cells) (see Batteries). Since the electrode reaction occurs at a surface, electrochemical techniques may also be used for surface treatment (electroplating (qv), electropolishing, anodizing) (see Metal surface treatments) or machining (see Electrolytic machining methods).

Electrosynthesis was first carried out by Davy in 1807 for the production of sodium and potassium, and in 1833 by Faraday who performed the first known example of the Kolbe reaction. Today this reaction remains one of the most useful in organic electrosynthesis. Generally speaking, all chemical reactions may be performed electrochemically and there are often great advantages to be obtained in doing so: electrochemical reactions allow control of selectivity and reaction rate through the electrode potential; are inherently pollution-free; have high thermodynamic efficiency; make possible reactions at ambient temperature; can often reduce the number of reaction steps; can often use cheaper starting materials; by means of electrolytic regeneration can use catalytic quantities of chemical oxidizing or reducing agents; and can often perform a synthesis electrogeneratively (1).

However, disadvantages are: electrochemical engineering and technology are far less developed than chemical engineering; many reaction variables may be involved with complex interdependences; long term stability of process components is often poor; and work-up of product is often costly.

The Electrochemical Cell

An electrochemical cell consists of at least two electrodes (anode and cathode) that dip into an electrolyte contained in a cell or reactor housing. The cell may be constructed so that the electrolytes at the anode and cathode are separated (anolyte and catholyte). One of the first cells studied was the Daniell cell and it is used as the basis of many introductions to the thermodynamics of electrochemical processes (2). It consists of a zinc anode and copper cathode dipped into solutions of their sulfates. A porous separator prevents mixing of the two solutions. Symbolically the cell is written:

$$Cu|Cu^{2+}\|Zn^{2+}|Zn$$

The single vertical line represents the phase boundary and the double broken line represents a liquid junction between the anolyte and catholyte (2). The reactions taking place at the electrodes are:

$$Zn^{2+} + 2e \rightleftarrows Zn \quad (Zn - 2e \rightleftarrows Zn^{2+}) \quad E^0 = -0.7628 \text{ V} \quad (1)$$
$$Cu^{2+} + 2e \rightleftarrows Cu \quad\quad\quad\quad\quad\quad\quad\quad E^0 = +0.3402 \text{ V} \quad (2)$$

By convention, they are written as reductions (3). The overall cell reaction is the sum of a cathodic and an anodic reaction:

$$Cu^{2+} + Zn \rightleftarrows Cu + Zn^{2+} \quad (3)$$

for which the change of Gibbs free energy ($\Delta G°$) in this case is -212.36 kJ/mol (-50.73 kcal/mol). When the two electrodes are connected externally through a load (high resistance is assumed so that the reaction proceeds reversibly), nF coulombs of charge per mole flow from the cathode to the anode through the potential difference E_{rev} (n is the number of electrons per molecule transferred, F is the Faraday constant and E_{rev} is the reversible potential of the cell). The ΔG is the free energy of the system exchanged with the surroundings minus the $P\Delta V$ work. Since the system is designed so that no other work is performed besides electrical work:

$$\Delta G = -nFE_{rev} \quad (4)$$

Potential is the energy scale in electrochemical processes just as temperature serves for chemical reactions, and wavelength for spectroscopic and photochemical methods.

The reversible cell potential may be calculated from the ΔG. For electricity sources, the reversible cell potential is the maximum voltage available from the element, whereas for synthesis its significance is the minimum voltage required to bring about the reaction. The voltage efficiency (ξ) for synthesis may thus be defined:

$$\xi = E_{rev}/E_{cell} = -\Delta G/nFE_{cell} \quad (5)$$

The actual cell potential deviates from the reversible potential owing to potential drop in the electrolyte and the phenomenon of overvoltage, which are important when the cell current is not zero, as is the case in processing applications.

The concentration dependence of the reversible potential is given by the Nernst equation which is derived from equation 4:

$$E_{rev} = E° + \frac{RT}{nF} \ln \prod a_i^{v_i} \quad (6)$$

where $E°$ ($= -\Delta G°/nF$) is the standard potential, ie, the cell potential with all species in their standard states, a_i is the activity of the ith species and v_i is its stoichiometric coefficient. The stoichiometric coefficient is positive for the reactants and negative for the products. The product Π is taken for all species involved in the electrode reaction. Thus for the Daniell cell:

$$E_{Daniell} = -0.7628 - (-0.3402) + \frac{RT}{2F} \ln \frac{a_{Cu^{2+}}}{a_{Zn^{2+}}} \quad (7)$$

It has been found useful to define a single or standard electrode potential as its potential versus the normal hydrogen electrode (NHE). The potential of the normal hydrogen electrode ($H^+ + e \rightleftarrows 1/2\ H_2$) is by convention taken to be 0.000 V. Single electrode potentials may therefore be obtained by measuring the potential of the cell:

$$Pt|H_2|H^+\|X|M \quad \text{where} \quad X|M \quad \text{represents the electrode of interest}$$

For example, to obtain the potential of the copper electrode, the potential of the cell $Pt|H_2|H^+\|Cu^{2+}|Cu$ should be measured. The cell has been written here according to the IUPAC convention with the NHE on the left (3). Tables listing standard electrode potentials are available (4). By using the Nernst equation, the reversible potential for a particular set of conditions may be calculated. The cell potential is given by the difference of the two separate electrode potentials. For organic processes, which are usually irreversible, it is often useful to calculate a standard (reversible) electrode potential from ΔG using equation 4.

Rate of Electrode Reactions

In 1905, Tafel found empirically that the rate of an electrode reaction could be expressed by an equation of the form

$$E = a - b \ln j \tag{8}$$

The electrode reaction rate, observable through the current density j, is exponentially related to potential. Later, Butler and Volmer using the concept of a reaction activation energy derived the expression:

$$j = j_0 \left[\exp\left(\frac{\alpha n F \eta}{RT}\right) - \exp\left(-(1-\alpha)\frac{nF\eta}{RT}\right) \right] = j_a + j_c \tag{9}$$

giving the net current as the algebraic sum of the forward (cathodic) j_c and backward (anodic) j_a currents ie, $j = j_c + j_a$:

$$O + ne \underset{j_a}{\overset{j_c}{\rightleftharpoons}} R \tag{10}$$

The term j_o, called the exchange current density, is the magnitude of j_c or j_a at zero overpotential $\eta (\eta = E - E_{rev})$, and α, the charge transfer coefficient, is a symmetry factor for the activation barrier. The exchange current density is thus a measure of rate and it may be related to the electrode reaction rate constant (5). For example, for the redox reaction in equation 10 one obtains:

$$j_o = k_{sh}(c_O)^{1-\alpha}(c_R)^\alpha \tag{11}$$

where c_O and c_R are the concentrations of O and R, respectively and k_{sh} is the rate of the electrode reaction at the standard electrode potential $E°$. The exchange current density may be obtained from a Tafel plot, ie, a plot of $\ln j$ against E (Fig. 1), by extrapolating the linear section to the reversible potential. For irreversible reactions, a pseudo j_o may be obtained by extrapolating the plot back to the E_{rev} calculated from ΔG (see also Corrosion).

Factors Limiting the Overall Reaction Rate. The Tafel plot shows that the current (proportional to reaction rate) obeys an exponential law over only a limited range of overpotential. At high overpotentials, the current takes a limiting value. Thus only at low overpotentials is the process charge transfer (or activation) controlled. Electrochemical reactions are, however, made up of several consecutive steps. Those preceding the electrode reaction are (6–7):

Step 1. Transport of the electroactive species from the bulk solution to the diffusion layer.

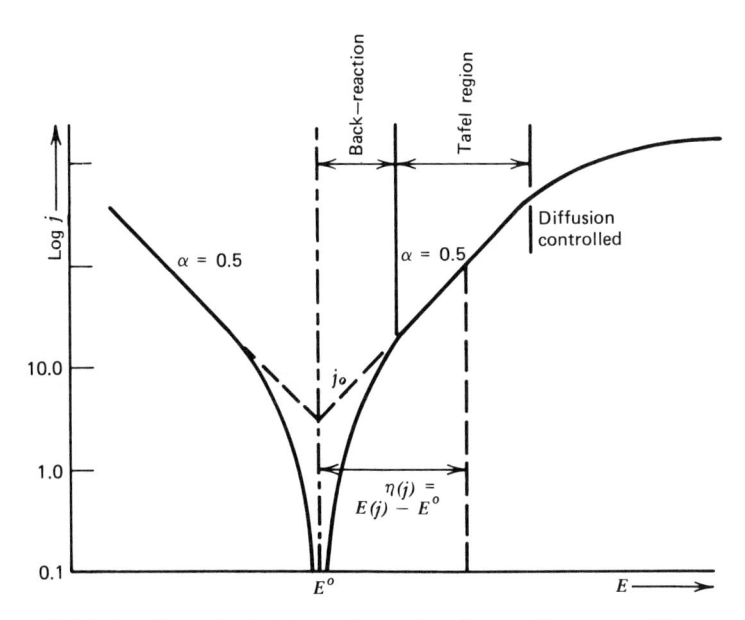

Figure 1. Tafel plot of hypothetical electrode reaction; slope and intercept of linear portion of log(j) against E give the values of the kinetic parameters α and j_o.

Step 2. Diffusion and migration through the diffusion layer to the electrode surface.

In addition, chemical reactions involving electroactive species (8), eg,

$$A + O \underset{\overleftarrow{k}}{\overset{\overrightarrow{k}}{\rightleftarrows}} B \tag{12}$$

(where O is an electroactive species partaking in equation 10) can also influence the rate of transport to or from the electrode. An example is the manufacture of chlorates. This takes place by a slow chemical reaction between electrogenerated HOCl and OCl⁻ and is the rate determining step. In this case, the homogeneous kinetics has a profound influence on the design of the cell unit.

Lastly, adsorption and desorption must not be neglected since they can greatly alter the surface concentration of species and the rate of the electron transfer.

These steps must also be considered for the processes following the electrode reaction (charge transfer) in addition to possible changes of phase, eg, gas evolution or crystallization. Any one of these, including the charge transfer itself, may be rate determining.

Generally, the bulk solution is considered to be well mixed; therefore step 1 may be neglected as the rate determining process. (This may not be valid for large conversions of the electroactive species in a single pass of the electrolytic reactor.) However, the degree of stirring does determine the thickness of the diffusion layer and therefore the rate of step 2. The diffusion-limiting current density (limiting implies zero concentration of electroactive species at the electrode surface) determined by step 2 and neglecting migration is:

$$j = nFDc/\delta = nFck \tag{13}$$

where F is the Faraday constant, c is the concentration of the electroactive species with diffusion coefficient D, k is the mass transfer coefficient, and δ is the thickness of the diffusion layer. Thus c/δ is the concentration gradient at the electrode surface. If the process is controlled, however, by the rate of a chemical reaction, then the flux of electroactive species O to the electrode is given predominantly by the rate of dissociation of B in a reaction layer. The current density is then given by an expression equivalent to equation 13 but with the diffusion layer thickness replaced by the reaction layer thickness μ. For the case of the simple preceding chemical reaction given by equation 12 $\mu = \sqrt{D/k}$ and $i = nFC\sqrt{k/D}$, ie, when $\mu < \delta$ (6). However, the chemical kinetics may be much more complex than the example given here.

In order to maximize the productivity, usually an industrial process is performed with as high a mass transfer rate to the electrode as possible. This is determined often by the hydrodynamic conditions present. Many correlations for these exist in the literature (9); some of the important ones are shown in Table 1.

Productivity

The concentration of the electroactive species during a batch electrolysis of volume V of feed is given by:

$$c = c_o \exp(-kAt/V) \tag{14}$$

where c_o is the initial concentration, A is the electrode area, t is the time and, k is the mass transfer coefficient ($k = D/\delta$). Thus for the batch processing of 1 t/h feedstock (mol wt 100 g/M) to a conversion of 99 % with an initial concentration of 1 kM/m^3 and a mass transfer coefficient of 10^{-3} cm/s, an electrode area of 1279 m^2 is required. Large electrode areas are therefore typical of electrochemical processing, and it is usually desirable to have a large electrode while maintaining a compact form (14). The success of this exercise is determined by the specific electrode area A_s defined as the electrode area divided by the volume containing the electrode. A comparison of some well-established modern cells is shown in Table 2.

Table 1. Mass Transfer Correlations of Various Hydrodynamic Regimes in Electrolytic Cells

parallel-plate reactor with turbulent flow[a]	$Sh = kd_h/D = 0.023Re^{0.8}Sc^{0.33}$
parallel-plate reactor with turbulence promoters[b]	$Sh = k \cdot d_h/D = 0.272Re^{0.631}Sc^{0.33}(D_e/\Delta l)^{0.357}$
rotating cylinder[c]	$Sh = k \cdot R/D = 0.0642\nu^{0.3}Sc^{-0.644}(R \cdot u_r)^{0.7}/D$
packed bed[d]	$Sh = k \cdot d_p/D = 1.44Re_p^{0.58}Sc^{0.33}$
natural convection at vertical electrodes[e] ($10^4 < GrSc < 10^{12}$)	$Sh = k \cdot L/D = 0.66(Gr \cdot Sc)^{0.25}$

[a] Ref. 10.
[b] Ref. 11.
[c] Ref. 7.
[d] Ref. 12.
[e] Ref. 13.

Table 2. Specific Electrode Areas of Electrolysis Cells

Cell	A_s, cm^{-1}
Hooker chlorine cell	0.037
rotating or wiped electrode	0.1
tank electrolyzer	0.13
filter press	0.3–1.7
capillary gap	1.0–5.0
packed bed	10.0–50.0
Swiss-roll	20.0–50.0
fluidized bed	20.0–100.0

Current and Potential Distribution

An electrosynthesis is particularly easy to control simply through the selection of the electrode potential. The full advantage may be taken only when the potential is uniform over the whole electrode area. Planar, two-dimensional electrodes may have a nonuniform potential distribution owing to: nonuniform current density distribution resulting from concentration changes or geometric factors; or finite electrode resistance (15–16). Similarly, three-dimensional electrodes (porous or particulate, etc) suffer from nonuniform electrode potential distribution owing mainly to the potential drop in the electrolyte phase in the pores (17). The potential distribution (18–19) is given by the solution of the Laplace equations for the two phases:

$$\nabla^2 \psi_s = 0; \quad \nabla^2 \psi_e = 0 \tag{15}$$

where ψ_s and ψ_e are the potentials in the solution and electrode phases, respectively. Often the potential variation in the electrode phase is negligible. The problem is usually subdivided into various degrees of rigor according to the following approximations: primary current density (cd) distribution is obtained when the activation and concentration overvoltages are negligible, ie, it is dependent simply on the geometry of the system; secondary cd distribution is obtained when the activation overvoltage is included in the derivation; and ternary cd distribution is obtained for a full derivation with geometric factors and activation and concentration overvoltage influences considered.

Generally a good cd distribution is obtained at large Wagner numbers Wa = $\kappa(d\eta/di)/l$, where κ is the conductivity of the electrolyte $(\Omega \cdot \text{cm})^{-1}$ and l is a characteristic length.

Energy Balance

The heat produced by an electrochemical reaction is determined by the enthalpy change accompanying the reaction and the electrical energy input to the cells (20–21). The heat balance for the electrolysis system is thus:

$$h_2 - h_1 + IE - Q = 0 \tag{16}$$

where h_1 and h_2 are the enthalpy fluxes for reactants and products, respectively (kJ/s), IE is the electrolysis power input (kW = kJ/s) and Q is the heat flux to the environment. For isothermal operation, Q is the heat that must be dissipated in a heat exchanger.

Optimization

The optimum operating conditions of an electrochemical process may be equated to the minimum total cost for a given production rate (22–24) or to the maximum return on invested capital (25). The total cost is the sum of the investment and running cost. The investment cost may be broken down into the contributions from: items that are functions of the system-independent variables, eg, the electrolysis cell—including construction and material contributions, pumps for circulation of the electrolyte, piping and fittings, and buildings; and those that are dependent only on the plant capacity, eg, power supplies, bus-bars, and switching gear.

The running costs may likewise be divided between: contributions that are functions of the independent variables, eg, electricity for electrolysis: a function of the decomposition potential, overvoltage and potential loss in the rectifiers, bus-bars, and electrolyte; electricity consumed by the pumps for electrolyte circulation; (cooling water, N_2, steam); maintenance, overheads, etc; and those that are dependent only on the plant capacity, eg, feedstock, electrolyte, and solvents.

There are also several different operating modes, eg, batch or continuous and direct or indirect, that should be considered in the optimization as well as the possibility of different electrochemical routes, eg, oxidation or reduction to the desired product. The search for a global minimum of the total cost is further complicated by the many interrelationships and constraints on parameters. Usually only a partial optimization is attempted. Examples are: Optimization of the current density for minimum electrolysis power plus electrode costs (22), which leads to:

$$j_{opt} = \sqrt{a/bR} \qquad (17)$$

where a is the specific electrode cost, b is the unit cost of electric energy, and R is the cell resistance. The optimization of the electrolyte flow velocity for minimum electrode and pumping costs (26–27). The optimization of the cell dimensions from a consideration of the material and construction costs (28). However, nonelectrochemical steps, eg, product work-up, often play a very decisive role in the profitability of the process (29).

Process Trends

The optimum current density given by equation 17 usually gives a cell potential that is much greater than the thermodynamic reversible value, owing to the large overvoltage and IR voltage drops that are obtained. Higher efficiency may be realized if the overvoltage is minimized by giving the electrodes catalytic activity or by distributing the electrode current over three dimensions (higher specific electrode area). Two of the most important areas of development are, therefore, improving electrode materials through the use of catalytic coatings, eg, DSA (Dimensionally Stable Anodes, Diamond Shamrock) electrodes (30) (see Metal anodes), and the development of compact cells with large electrode areas (31). The latter development not only facilitates lower overvoltages but enables electrochemical processing to be applied to very dilute solutions such as encountered in wastewater streams.

Current trends can be followed in the reviews that appear annually in the *Journal of the Electrochemical Society*.

Nomenclature

a	= activity, Tafel coefficient, specific electrode cost
A	= electrode area
A_s	= specific electrode area (electrode area divided by cell volume)
b	= Tafel coefficient, unit cost of electricity
c	= bulk concentration of electroactive species
D	= diffusion coefficient
D_e	= equivalent diameter
d_h	= hydraulic diameter
d_p	= particle diameter
$E°$	= standard potential (V)
F	= faraday constant (96,485 C/M)
ΔG	= change in Gibbs free energy
$Gr = g(\rho_b - \rho_s)L^3/\nu^2\rho_b$	= Grashof number
g	= gravitation constant
h	= enthalpy flux
j	= current density
IE	= electrolysis power input
k	= mass transfer coefficient; rate
L	= electrode length
Δl	= promoter spacing
NHE	= normal hydrogen electrode
O	= oxidant
Q	= heat flux
R	= cylinder diameter; resistance, reductant
$Re = ud_h/\nu$	= Reynolds number
$Re_p = ud_p/\nu$	= Reynolds number based on particle diameter
$Sc = \nu/D$	= Schmidt number
Sh	= Sherwood number
u	= electrolyte flow velocity
u_r	= peripheral velocity of rotating electrode
Wa	= Wagner number
α	= charge transfer coefficient
δ	= thickness of the diffusion layer
η	= overpotential
κ	= electrolyte conductivity
μ	= thickness of reaction layer
ν	= kinematic viscosity of electrolyte; stoichiometric coefficient
ξ	= voltage efficiency of reversible cell
ψ	= potential
∇^2	= Laplace operator

Subscripts

a	= anodic
c	= cathodic
e	= electrode
rev	= reversible
s	= solution
sh	= at standard electrode potential
L	= limiting

BIBLIOGRAPHY

"Electrochemistry" in *ECT* 1st ed., Vol. 5, pp. 495–549, by G. C. Akerlof, Mellon Institute of Industrial Research; "Electrochemistry" in *ECT* 2nd ed., Vol. 7, pp. 784–841, by G. C. Akerlof, Aerospace Department, Princeton University.

1. S. H. Langer and H. P. Landi, *J. Am. Chem. Soc.* **85**, 3043 (1963).
2. K. J. Vetter, *Electrochemical Kinetics,* Academic Press, New York, 1967, Chapt. 1.
3. R. Parsons, *Pure Appl. Chem.* **37**, 504 (1974).
4. G. Milazzo and S. Caroli, *Tables of Standard Electrode Potentials,* John Wiley & Sons, Inc., Chichester, 1978.
5. Z. Galus, *Fundamentals of Electrochemical Analysis,* Ellis Horwood Ltd., Chichester, 1976, Chapt. 3; H. R. Thirsk and J. A. Harrison, *A Guide to the Study of Electrode Kinetics,* Academic Press, London, 1972.
6. Z. Galus, *Fundamentals of Electrochemical Analysis,* Ellis Horwood Ltd., Chichester, 1976, Chapt. 5.
7. C. W. Tobias, M. Eisenberg, and C. W. Wilke, *J. Electrochem. Soc.* **99**, 395C (1952).
8. Ref. 6, Chapts. 8–13, 15.
9. J. R. Selman and C. W. Tobias, *Adv. Chem. Eng.* **10**, 211 (1978).
10. D. J. Pickett and K. L. Ong, *Electrochim. Acta* **19**, 875 (1974).
11. F. B. Leitz and L. Marincic, *J. Appl. Electrochem.* **7**, 473 (1977).
12. K. R. Jolls and T. J. Hanratty, *AIChE J.* **15**, 199 (1969).
13. N. Ibl, *Electrochim. Acta* **1**, 117 (1959).
14. D. J. Pickett, *Electrochemical Reactor Design,* Elsevier Scientific Publishing Co., Amsterdam, 1977, Section 5.3.
15. J. Newman, *Ind. Eng. Chem.* **60**(4), 12 (1968).
16. N. Ibl, *Technique de l'ingénieur, D902,* Paris (1976).
17. C. W. Tobias and R. Wijsamn, *J. Electrochem. Soc.* **100**, 459 (1953).
18. P. M. Robertson, *Electrochim. Acta* **22**, 411 (1977).
19. J. S. Newman and C. W. Tobias, *J. Electrochem. Soc.* **109**, 1183 (1962).
20. Ref. 14, Section 1.8.
21. T. R. Beck, *Improvements in Energy Efficiency of Industrial Electrochemical Processes,* Report ANL-OEPM-77-2, Electrochemical Technology Corp., 1977.
22. N. Ibl and E. Adam, *Chem. Ing. Tech.* **37**, 573 (1965).
23. N. Ibl and E. Schalch, *Chem. Ing. Tech.* **41**, 208 (1969).
24. N. Ibl and P. M. Robertson, *Electrochim. Acta* **18**, 897 (1973).
25. T. R. Beck in E. Yeager and A. J. Salkind, eds., *Techniques of Electrochemistry,* Vol. 3, Wiley Interscience, a division of John Wiley & Sons, Inc., New York, 1978.
26. P. M. Robertson and N. Ibl, *J. Appl. Electrochem.* **7**, 323 (1977).
27. A. Storck and F. Coeuret, *Ing. Eng. Chem. Process Des. Dev.* **17**(1), 99 (1978).
28. R. Nadebaum, PhD Thesis, University of Waterloo, Canada, 1974.
29. N. Ibe and co-workers, *AIChE., Symp. Ser.* **75**(185), 45 (1979).
30. J. W. Kühn von Burgsdorff, *Chem. Ing. Tech.* **49**, 294 (1977).
31. A. T. Kuhn and R. W. Houghton, *Topics in Pure and Applied Electrochemistry,* SAEST, Karaikudi, India, 1975.

General References

J. O'M. Bockris and A. K. N. Reddy, *Modern Electrochemistry,* Plenum Press, New York, 1970.

K. J. Vetter, *Electrochemical Kinetics,* Academic Press, New York, 1967.

P. Delahey, *New Instrumental Methods in Electrochemistry,* Interscience Publishers, Inc., a division of John Wiley & Sons, Inc., New York, 1965.

D. J. G. Ives and G. J. Janz, *Reference Electrodes,* Academic Press, New York, 1961.

Z. Galus, *Fundamentals of Electrochemical Analysis,* Ellis Horwood Ltd., Chichester, 1976.

H. R. Thirsk and J. A. Harrison, *A Guide to the Study of Electrode Kinetics,* Academic Press, London, 1972.

D. J. Pickett, *Electrochemical Reactor Design,* Elsevier Scientific Publishing Company, Amsterdam, 1977.

C. L. Mantell, *Electrochemical Engineering,* McGraw-Hill Book Co., New York, 1960.

P. Gallone, *Trattato di Ingegneria Elettrochimica,* Tamburini Editore, Milan, 1973.

A. Schmidt, *Angewandte Elektrochemie,* Verlag Chemie, Weinheim, 1976.

A. T. Kuhn, *Industrial Electrochemical Processes,* Elsevier Publishing Co., Amsterdam, 1971.

E. Heitz and G. Kreysa, *Grundlagen der Technischen Elektrochemie*, Verlag Chemie, Weinheim 1977.
J. Electrochem. Soc., annual reviews.

PETER M. ROBERTSON
Eidgenössische Technische Hochschule

INORGANIC

Electrochemical processes involve the transfer of electrons between an electrode and a substrate in solution. The energy required is in the range 0–3.5 eV, and depends on the electrode material and the substrate. This is a moderate energy input in comparison to photochemical or radiation activation methods. Nevertheless, it is sufficient to produce the strongest oxidizing and reducing agents known, ie, F_2 and solvated electrons, respectively. It would indeed be difficult to develop manufacturing processes for chlorine, sodium, and fluorine that improve on the electrochemical ones (see Alkali and chlorine products; Chemicals from brine; Fluorine).

The importance of the electrochemical processing of inorganic chemicals for the production of either elements or compounds is easily appreciated from the fact that in the U.S. they consume 6% of all the electricity generated or about 16% of the electricity used by industry. Table 1 lists the most important processes that are now practiced, however, other industrial operations that involve an electrode reaction, eg, metal refining (see Aluminum; Extractive metallurgy), energy sources (batteries, fuel cells (see Batteries)), electroplating (qv), electropolishing, anodizing (see Metal surface treatments), electrochemical machining (see Electrolytic machining methods) are not discussed in this article.

Hardware for Electrochemical Processing

The distinguishing feature of an electrochemical process is the electrolysis cell and its power supply. Like many chemical processes, the feedstock must be made up by the addition of solvents/electrolytes and the product must be extracted or worked-up. This often involves much additional equipment as is demonstrated by the flow sheet shown in Figure 1 of a chlor-alkali process. However, only the electrochemical cell and its components are discussed, not the unit operations involved (see Alkali and chlorine products).

In an electrochemical cell, feedstock is transported to the electrode/electrolyte interface. The design of an electrolysis cell is therefore based, among other factors, on optimization of: transport of electroactive species from the cell volume to the electrode surface; materials and topography of the electrode; and possible need to separate the products or reactants of the anode and cathode reactions. The main design possibilities of electrolysis cells are summarized in Table 2.

Industrial cells for inorganic processing seldom employ pumps to provide convection. Usually transport of the electroactive species to the working electrode is en-

Table 1. 1977 U.S. Production Statistics for Electrolytically Manufactured Elements and Compounds, Thousands of Metric Tons[a]

Substance	Amount
Metals	
aluminum[b]	4500
beryllium[b]	na
boron[b]	ca 0.007
cadmium	ca 2
chromium	ca 132
cobalt	0.22 (est)
copper	1500
lithium[b]	na
magnesium[b]	115
manganese	na
nickel	60
sodium[b]	157
titanium[b]	ca 41
zinc	524
Chemicals	
chlorine/caustic	10,663
fluorine	14 (est)
hydrogen/oxygen	na
lithium carbonate	21.3
manganese dioxide	21
perchlorates	na
permanganates	15
persulfates	na
potash/caustic	309
soda ash	8000
sodium chlorate	218.5

[a] Based on information from the U.S. Bureau of Mines and ref. 1.
[b] By fused-salt processes.

hanced simply by the gas generation within the cell. Examples are the cells for fluorine (Fig. 2), chlorate (Figs. 3 and 4) and aluminum production (Fig. 5). Gas generation at the counter-electrode in the Krebs (Fig. 3) and Diamond Shamrock/Huron Chemicals (Fig. 4) chlorate cells are employed not only for local stirring within the electrolysis compartment but also to maintain circulation to the bulk reaction volume. The Alcoa cell for the production of aluminum by the electrolysis of $AlCl_3$ (Fig. 5) also relies on gas pumping of the electrolyte. In this case, the product of the secondary electrode reaction (chlorine) can exit only from one side of the electrolysis chamber. A clockwise flow of electrolyte is thereby induced.

The electrodes shown in Figures 2–6 illustrate monopolar and bipolar types. The fluorine cell (Fig. 2) contains only monopolar electrodes. They are the inner carbon anodes and the steel tank cathode. Monopolar electrodes have a direct electrical connection with the external power supply. The total active area of a monopolar electrode has a nominally uniform potential with respect to the electrolyte. A disadvantage is the resulting low cell voltage. Cells are therefore more often designed with a stack of bipolar electrodes between monopolar feeder electrodes. Bipolar electrodes have no electrical connection to the power supply. Instead they are polarized by the potential gradients in the cell. The bipolar electrode is polarized anodically on the side

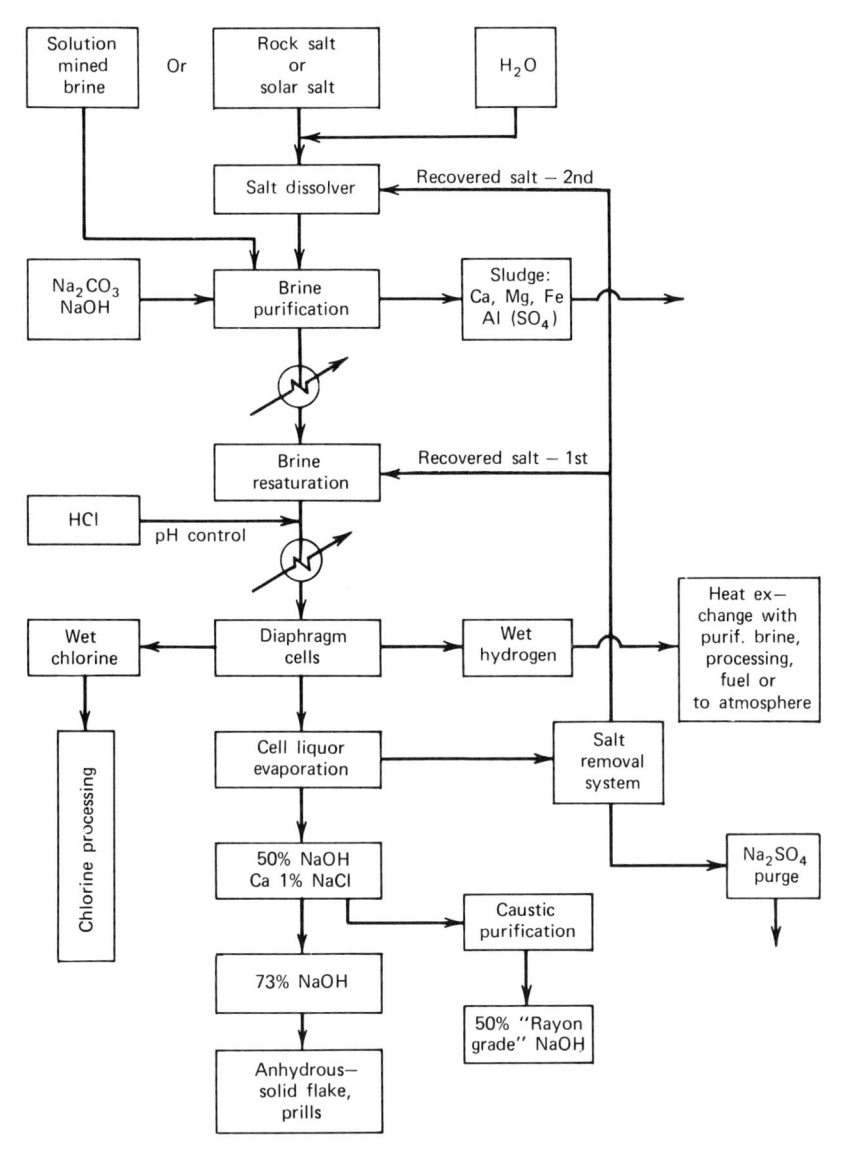

Figure 1. Flow sheet for chlor-alkali electrolysis by the Hooker diaphragm process.

facing the negative feeder electrode and cathodically on its reverse side. The cell voltage is then multiplied by the number of electrode pairs. This allows the use of higher voltage and correspondingly higher-efficiency rectifier stacks. Bipolar electrodes are used in the water electrolysis cell (Fig. 6) and the Alcoa aluminum cell (Fig. 5). The flow of electricity and ionic charge are colinear. The bipolar electrodes in the chlorate cell shown in Figure 4 are of an alternating arrangement. Here, the electricity flow in the electrodes is perpendicular to the ionic flux between the electrodes.

Often mixing of the products or reactants of the working and secondary electrodes is undesirable. Obvious examples are the electrolysis of water to oxygen and hydrogen or the electrolysis of brine to chlorine and hydrogen. In such cases, the cell is built with

674　ELECTROCHEMICAL PROCESSING (INORGANIC)

Table 2. Cell Types Used for Inorganic Electrochemical Processing

Cell type	Example of application
One-compartment cells	
inert vessel with immersed electrodes	
monopolar electrodes	chlorates, perchlorates, MnO$_2$, etc
bipolar electrodes	chlorates, metal refining
vessel is one electrode (monopolar electrodes)	chlorates, fluorine
horizontal liquid-metal cathode	aluminum, chlorine/caustic
Two-compartment cells	
diaphragm cells	
monopolar electrodes	chlorine/caustic
bipolar electrodes	
filter press	
bipolar electrodes	hydrogen/oxygen

a membrane or diaphragm to separate the anode and cathode compartments. This is exemplified in Figure 6 which shows a Brown-Boveri water electrolysis cell. For the separation of products not dissolved in the electrolyte (eg, gas or fluid phases that are immiscible with the electrolyte phase) porous (asbestos) or gauze diaphragms often suffice. However, for the selective separation of ionic species between compartments, an ion-exchange membrane is normally employed.

There is a growing tendency to incorporate turbulence promoters in the electrolysis compartment of cells with forced convection. The turbulence promoters take the form of nets (2), inert beads (3), strips, or rods perpendicular to the flow (4–5) or even forms machined on the electrode itself (eg, grooves). Turbulence promoters are especially useful at low flow rates (Reynolds numbers that correspond to laminar flow in the equivalent open channel) by providing an increased and uniform mass transfer rate over the whole electrode. The productivity of the cell may be thus increased.

The electrode types currently encountered in technical processes are given in Table 3 together with examples of their application. Electrodes are the most delicate components in electrolysis cells. Many of the so-called inert electrodes, notably graphite and carbon, corrode significantly and as a result have relatively short lifetimes (eg, 60 d in a Cl$_2$ diaphragm cell). Recently, new electrode types that are stable for several years have been offered. The most important are the dimensionally stable anode (DSA) types (7). These are produced by baking a catalyst layer onto a valve metal support (Ti, Ta). The catalyst may be a noble metal such as Pt or an oxide thereof (PtO$_2$, RuO$_2$) or a proprietary mixture of catalyst, binder, and stabilizer. Activated electrodes offer superior corrosion resistance and a lower reaction overvoltage. Reactions therefore proceed at lower absolute potentials, and thereby consume less power (see Metal anodes).

Production Conditions

Electrolysis of Chloride Solutions. Chloride may be oxidized electrochemically to chlorine or hypochlorite and chlorate and perchlorate. The distribution of products

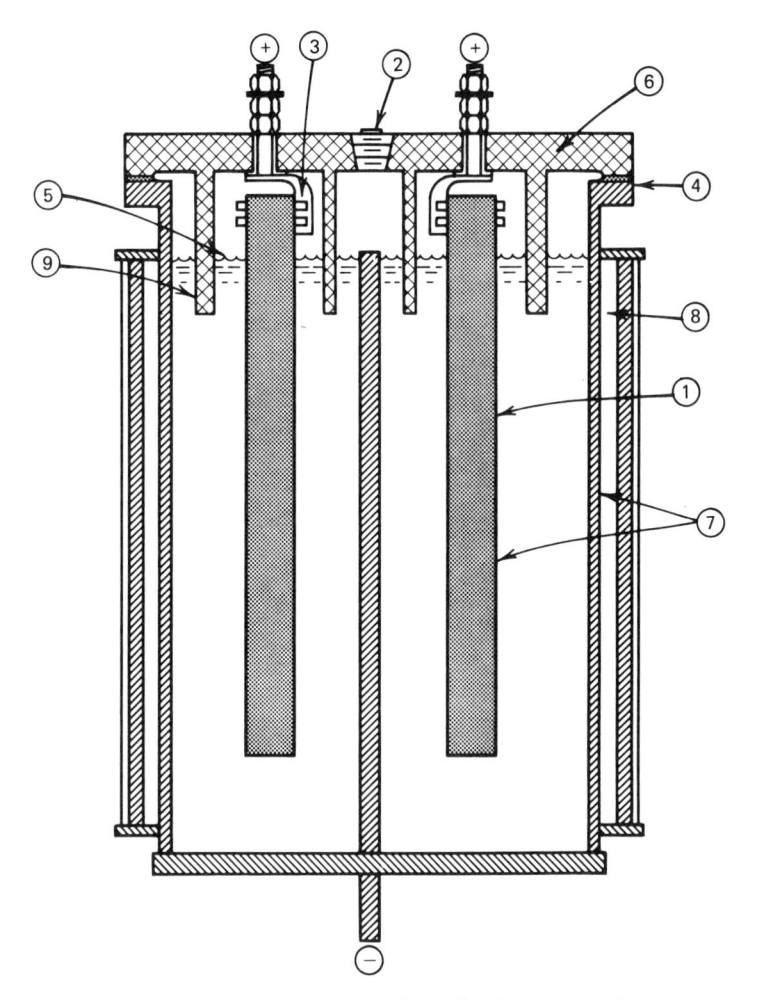

Figure 2. Allied Chemical Co. cell for the production of fluorine. 1, anode; 2, topping-up port; 3, anode support; 4, rubber gasket; 5, electrolyte level; 6, cell top; 7, cathode; 8, cooling jacket; 9, gas separation skirt.

of the first oxidation step:

$$2\,Cl^- - 2\,e \rightarrow Cl_2 \qquad (1)$$

is very pH-dependent and is determined by the equilibria (8–9):

$$Cl_2 + H_2O \rightleftarrows ClO^- + Cl^- + 2\,H^+ \quad K = 3.88 \times 10^{-4} \qquad (2)$$

$$ClO^- + H^+ \rightleftarrows HOCl \quad K = 3.13 \times 10^7 \qquad (3)$$

The distribution of species for a total chlorine concentration of 1 M is shown in Figure 7. Chlorine is the major product at < pH 5, and predominates above pH 10. The pH limits for the manufacture of chlorine and hypochlorite are therefore defined. In the intermediate pH range in which both OCl$^-$ and HOCl are present in solution, chlorate is produced through the slow reaction (10):

$$2\,HOCl + ClO^- \rightarrow ClO_3^- + 2\,Cl^- + 2\,H^+ \qquad (4)$$

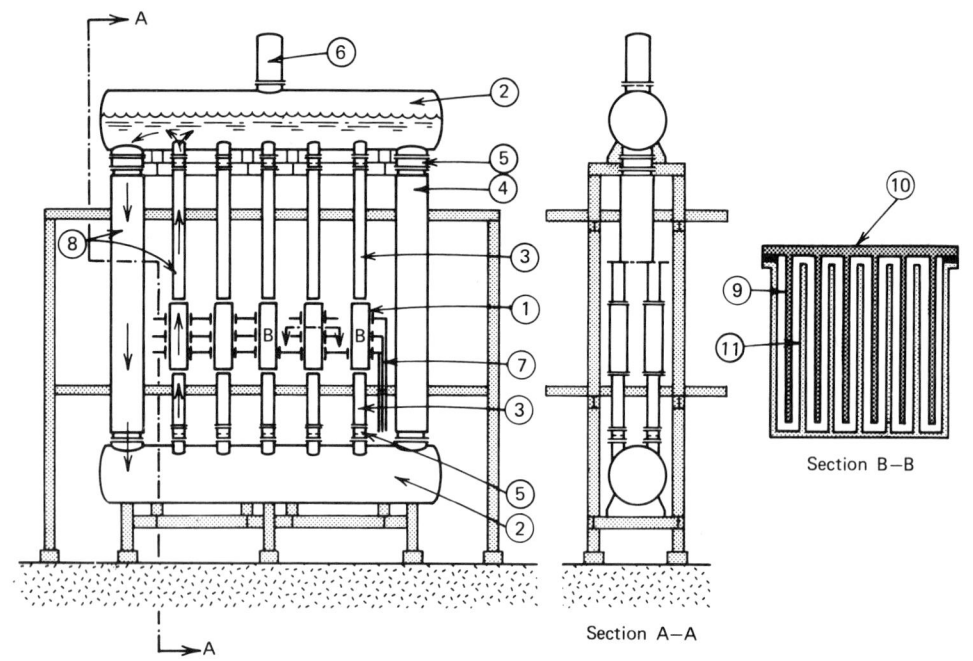

Figure 3. Krebs (Zürich) cell for the production of chlorate; tank and pipes are titanium or steel polytetrafluorethylene. 1, electrolysis cells; 2, reaction tanks; 3, risers; 4, downcomers with cooling jacket; 5, expansion joints; 6, hydrogen demister; 7, bus bars; 8, circulation of electrolyte through gas evolution at the counter electrode; 9, anodes (activated titanium); 10, titanium flange onto which the anodes are bolted; 11, steel cathodes welded directly onto the cell housing.

Table 3. Electrode Types Used in Inorganic Electrochemical Processing [a]

Classification	Example	Application
Functional		
inert	Pt anode	persulfates
active, consumable	C anode	aluminum
base for deposition	Hg cathode	chlorine/caustic
Geometric		
planar	DSA sheet anode	chlorine/caustic
mesh/expanded metal	steel net cathode	chlorine/caustic
porous	graphite	wastewater treatment
particulate	copper beads	metal deposition

[a] Based on classification of Beck (6).

It is also possible to obtain chlorate electrochemically by the further oxidation of OCl⁻ at higher anodic potentials:

$$6\,ClO^- + 3\,H_2O \rightarrow 2\,ClO_3^- + 4\,Cl^- + 6\,H^+ + 1\tfrac{1}{2}\,O_2 + 6\,e \tag{5}$$

but the reaction effectively gives a low current efficiency and is therefore undesirable (11–12).

The oxidation of chlorate to perchlorate is a purely electrochemical step and is pH independent:

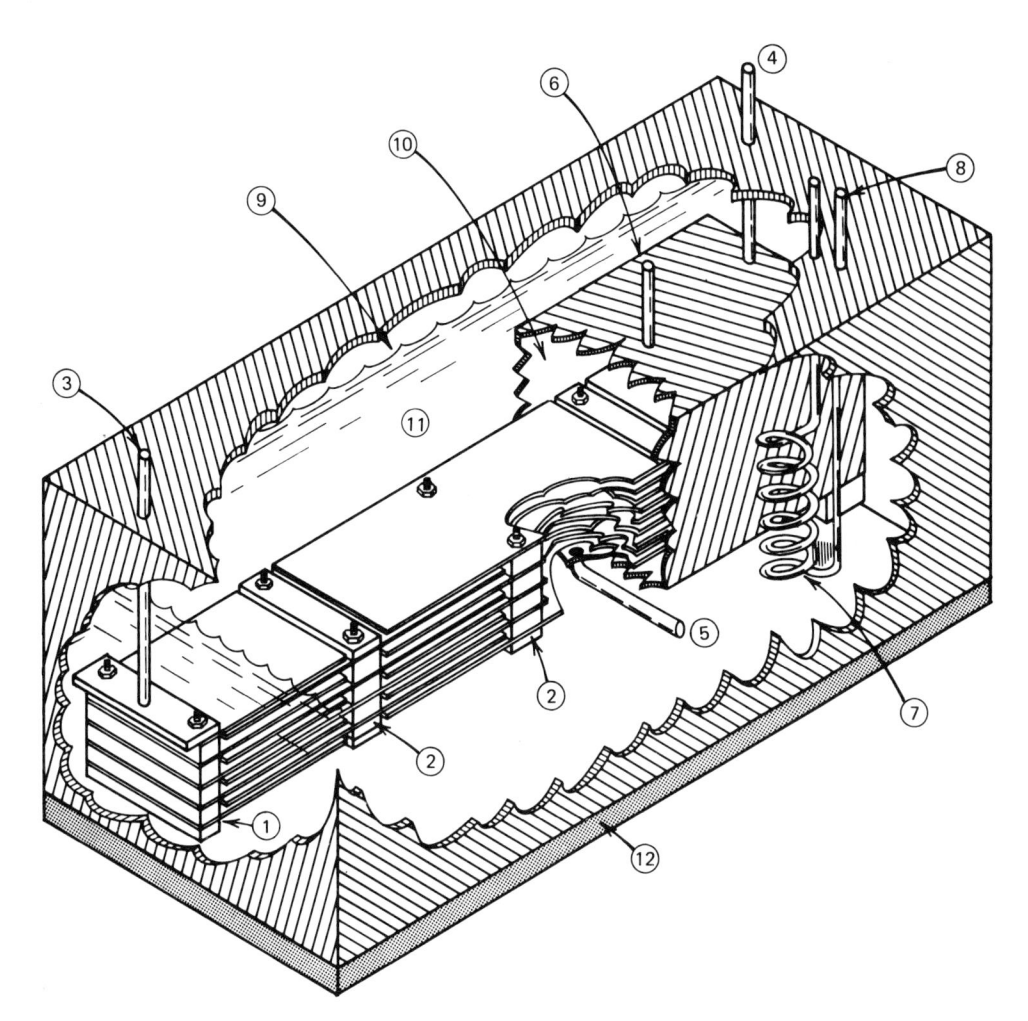

Figure 4. Diamond Shamrock/Huron Chemicals cell for the production of chlorates. 1, cathode feeder electrodes; 2, bipolar electrode units; 3, cathode current feeder; 4, anode current feeder; 5, electrolyte input to electrolysis chamber; 6, electrolyte exit from electrolysis chamber; 7, cooling coil; 8, cooling water connections; 9, electrolyte level; 10, electrolysis chamber; 11, chemical reaction chamber.

$$H_2O + ClO_3^- \rightarrow ClO_4^- + 2 H^+ + 2 e \qquad (6)$$

Perchlorate is usually not synthesized directly from chloride owing to the low current efficiency that would occur as a result of the simultaneous reaction of equation 5. The electrode materials used for chlorate production (eg, graphite) are in any case unsuitable for perchlorate synthesis. The reaction is therefore carried out in its own special reactor with platinum or PbO_2 anodes.

The electrolysis of brine is one of the oldest and certainly one of the most important and widespread industrial electrochemical processes (13–18). The overall reaction is:

$$2\ NaCl + 2\ H_2O \xrightarrow{\text{electrical energy}} 2\ NaOH + Cl_2 + H_2 \qquad (7)$$

There are three main technologies available for carrying out this process: diaphragm

678 ELECTROCHEMICAL PROCESSING (INORGANIC)

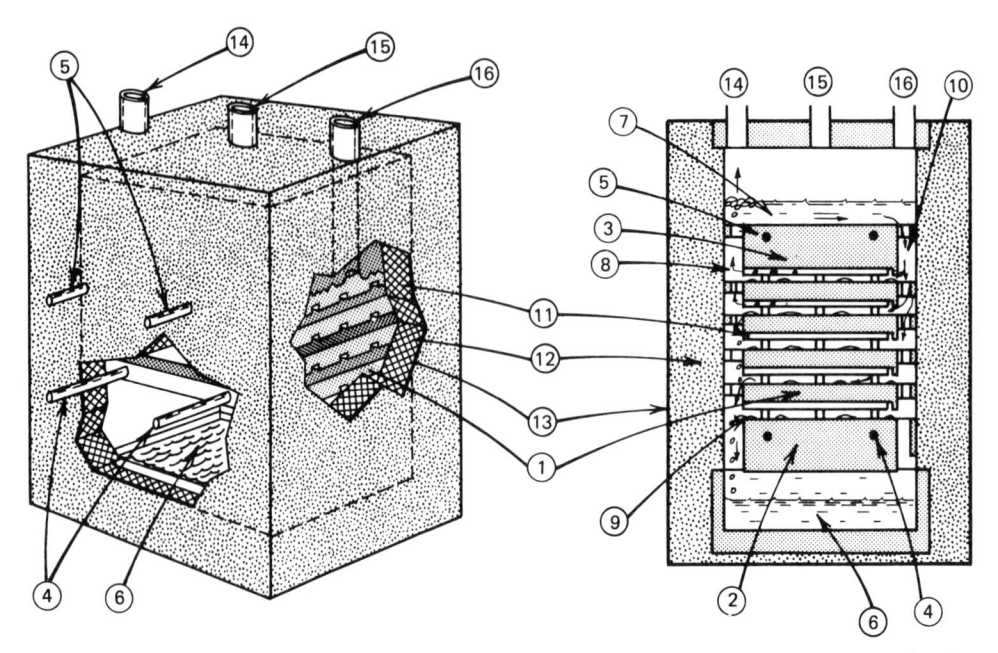

Figure 5. The Alcoa cell for the production of aluminum by the electrolysis of AlCl$_3$. 1, bipolar electrode (top is cathode surface); 2, cathode feeder electrode; 3, anode feeder electrode; 4, cathode current feeder; 5, anode current feeder; 6, pool of molten aluminum; 7, AlCl$_3$ melt; 8, path taken by anode product chlorine on its way to exit 14, this generates a clockwise circulation electrolyte; 9, aluminum transported to collection pool (6); 10, clockwise circulation of electrolyte; 11, channels cut in anode (open only on left side); direct gas to exit on left and thereby create circulation; 12, fire-resistant lining; 13, steel casing; 14, chlorine exit; 15, electrolyte topping-up port; 16, aluminum extraction port (tube inserted and metal sucked out).

cells, mercury cells (diaphragmless), and membrane cells. The latter is a recent development and should make a great contribution to future production.

Diaphragm Cell Technology. The complete reaction represented by equation 7 occurs in the diaphragm cell itself. At the anode, which is either carbon or a DSA material, the brine is oxidized to chlorine:

$$2\,Cl^- - 2\,e \rightarrow Cl_2 \qquad (8)$$

and at the cathode, which is usually steel, water is reduced to hydrogen:

$$2\,H_2O + 2\,e \rightarrow H_2 + 2\,OH^- \qquad (9)$$

A diaphragm (eg, asbestos (qv)) is formed on the cathode to separate the hydrogen and oxygen streams and allow close spacing of the anode and cathode. Since the diaphragm does not completely prevent electrolyte flow between the cell compartments, the highest practicable hydroxide concentration attainable is about 10%. The caustic stream must, therefore, be worked-up extensively for most purposes, which is a disadvantage of the diaphragm process. Operating experience also shows that asbestos diaphragms have a very limited life (ca 6 mo) and considerable maintenance is thus required. Recent improvements of the process are the incorporation of DSA electrodes and better diaphragm formulations (asbestos/plastic composites). The use of DSA anodes gives far less waste in comparison to graphite (which corrodes), and therefore,

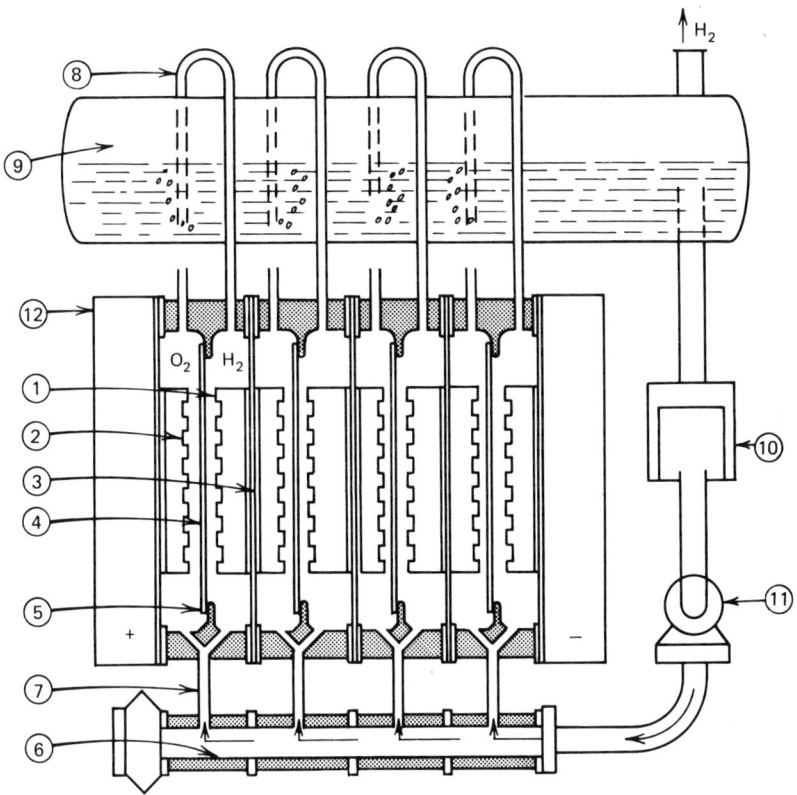

Figure 6. Section through a BBC (Brown-Boveri cell) for water electrolysis. 1, cathode; 2, anode; 3, center electrode; 4, diaphragm; 5, cell frame; 6, manifold; 7, feedpipe; 8, overflow pipe; 9, gas separator; 10, filter; 11, pump; 12, press plate.

leads to a cleaner operating environment. The operating conditions of diaphragm cell processes are given in Table 4.

Mercury Cells. In the mercury cell, a cathode material with a high hydrogen over voltage, mercury, is employed to suppress hydrogen evolution and permit sodium deposition:

$$Na^+ + e \rightarrow Hg/Na \qquad (10)$$

Since no hydrogen is generated, it is possible to construct this cell without a diaphragm. The cell is built around a long, gently sloping trough, down which there is a continuous flow of mercury (cathode). The anodes (graphite was the original anode material but it is superseded by DSA materials) are held about 5 mm above the mercury. The sodium amalgam leaving the trough passes through a decomposer containing graphite where water is reduced to hydrogen and caustic:

$$2\ Hg/Na + 2\ H_2O \rightarrow 2\ NaOH + H_2 \qquad (11)$$

A concentration of 50% NaOH is produced in this step. This highly concentrated solution requires little work-up, but it is achieved at the expense of a higher operating cell voltage and the possibility of mercury contamination of the environment. Although

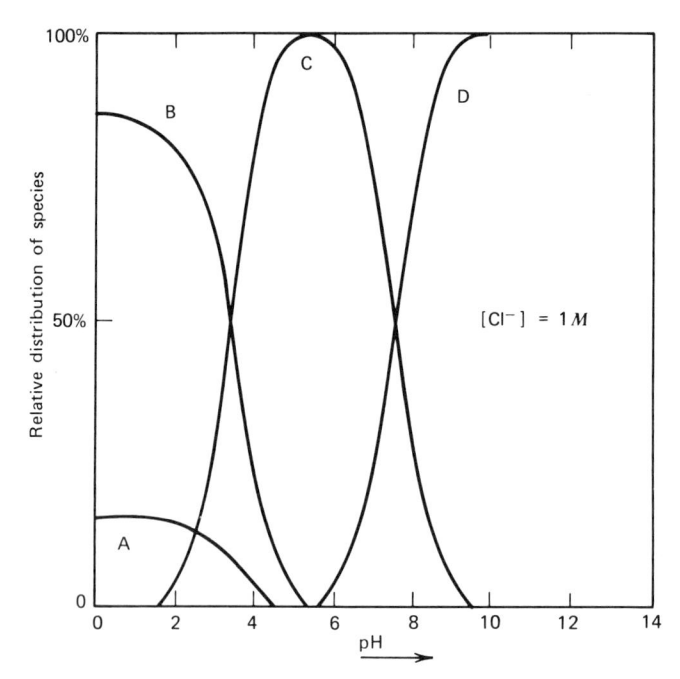

Figure 7. Distribution of Cl_2 in electrolysis of a 1 M chloride solution. A, Cl_3^-; B, Cl_2; C, HOCl; D, ClO^-.

the latter problem has been solved (19–20), it often still weighs heavily against the choice of mercury cells in new plants. The operating conditions of mercury cell processes are given in Table 5.

Membrane Cells. A disadvantage of the diaphragm cell is the relatively low NaOH concentration that may be reached. This is due to the finite electrolyte flow rate through the diaphragm with its lack of ion selectivity. Until recently there were no cation-exchange membranes available that were selective enough to allow a standing OH^- concentration difference of 10 M. Membranes based on perfluorosulfonic acids (DuPont (21)) and perfluorocarboxylic acids (Asahi Chemical Co. (22) and Asahi Glass (23)) do, however, have the necessary selectivity. The current efficiencies of some of these membranes are shown in Figure 8. Diamond Shamrock (25), Hooker (26) and Asahi Chemical (27) are now marketing large membrane cells (see Fluorine compounds, organic). For several years Ionics has been marketing membrane cells (Chloromat) that are not intended for very large tonnages.

Chlorates. The manufacture of chlorates is carried out by the electrolysis of brine at pH 6.5–7.4 with graphite or DSA anodes and steel cathodes (28–29). The formation of chlorate occurs by a purely chemical reaction between OCl^- and HOCl (eq. 4). Since this volume reaction is slow, a sufficient hold-up time must be incorporated in the circuit to allow the reaction to proceed to completion. This is achieved in simple tank electrodes by using a low specific electrode area (30). Alternatively, a compact electrolysis cell together with a separate chemical reactor, as in the Krebs cell (31), may provide the same results. Cells for chlorate synthesis are usually diaphragmless. In order to minimize back reduction of electrolysis products at the counter-electrode, sodium chromate is added which forms a membrane film over the cathode and effec-

Table 4. Operating Data of Diaphragm Cells

	Roof electrodes	Hooker Type S-4	Hooker Type H-2	Hooker Type H-4	Solvay Type DBT 26	Diamond Type DS 85	De Nora-Diamond Type Elincor 46	De Nora-PPG Type Glanor 1144	Krebs-Kosmo DZ 2	Nisso Engng. Type BW 60
current, kA	24	55	80	150	70	150	70–80	80	6	60
current density, kA/m^2	1.4–1.6	1.5–1.6	2.45	2.35	2.64	2.74	2.45	2.19	2.5	2.5
current efficiency, %	88–90	96.2	97	95.8	97.5	96.5	96.5	97	96	95
cell voltage, V	3.9–4.1	3.9–4.0	3.9	4.06	3.88	4.15	4.06	3.75[b]	3.5[b]	3.75[b]
energy consumption[a], kW·h/kg	3.4	3.1	3.05	3.2	3.0	3.25	3.17	2.92	2.75	2.98
temperature, °C							96	65–70		
anode material	graphite	graphite	Ti	Ti	Ti	Ti	Ti	Ti	Ti	Ti
conc. of electrolyte:										
NaOH, g/L	135	140	140	160	130	130–150	130–140	130–150	140	140
NaCl, g/L	140	185	180			175–210	180–210	180–210	180	185
cell capacity, t Cl$_2$/d	0.68	1.68	2.47	4.56	2.17	4.95	2.15	27.1	9.15	9.07
diaphragm lifetime, mo		3–4	6–11	7–13		12–24	4	15–18	12	8

[a] Calculated from the average values of potential and current efficiency.
[b] Single cell potential.

Table 5. Operating Data of Mercury Cells

	Uhde 190-75/100G	Uhde 350-100/130M	De Nora 20 H6	De Nora 33 M2	Solvay MAT 17	Krebs-Kosmo 23.2-70	Krebs (BASF) 2 T-200-8	Olin E-812	Kureha UC-26	Asahi-Glass
current, kA	132-144	345-450	300-330	500-600	170	280-300	200	260-288	320	60-70
electrode area[a], m^2	18.7 (A)	33.8 (A)	32.3 (A)	36.1 (A)	16.7 (A)	23.2 (C)	25 (C)	28.8 (C)	32	7.2
current density, anode, kA/m^2	7.7-10.3	10.2-13.3	9.3	13.85	ca 10.7	12	ca 8.4	ca 9.5	10	8.4
cathode, kA/m^2	7.5-10.0	10.0-13.0	8.8	13.3	10.2		8	9.1	ca 9.5	7.3
current efficiency, %	95	96	95-97	97-98	97	96-97	96	95	95-96	95
cell voltage[b], V	4.06	4.08	4.3	4.0	4.3	4.25	4.38	4.3	4.6	4.54
energy consumption[c], kW·h/kg Cl$_2$	3.23	3.21	3.36	3.1	3.35	3.28	3.45	3.42	3.64	3.61
anode material	graphite	Ti	graphite	Ti	Ti	Ti	graphite	graphite	graphite	graphite
no. of anodes per cell	120	54	120	66		36	144	96	60	24
cell inclination, mm/m	15	15-20	10	15		18		15	25	
mercury per cell, t	1.84	4.6	3.28	4.00		2.75	2.8	3.8	3.2	0.95
mercury consumption, kg/t Cl	0.002-0.025		0.1-0.15	0.05-0.075					0.13-0.15	0.2
graphite consumption, kg/t Cl$_2$	2.1		2.5						2.3-2.6	
cell capacity[b], t Cl$_2$/d	4.1	10.7	9.3	15.5	5.3	8.7	6.2	8.0	9.8	1.8

[a] A = anode, C = cathode.
[b] For normal loading.
[c] Calculated from the average values of potential and current efficiency.

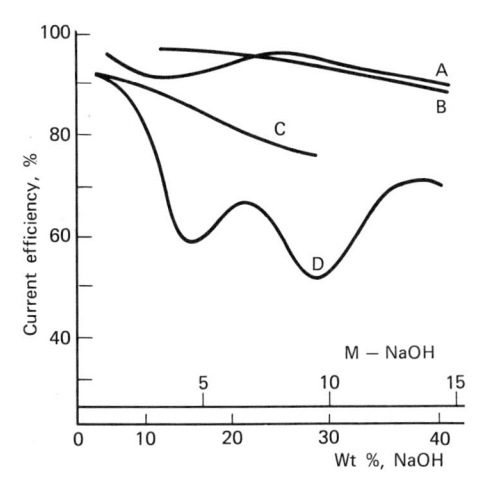

Figure 8. Efficiencies of membranes for chlor-alkali production. A. 85°C 22% NaCl in feed, 3.1 kA/m², Nafion 227 membrane, E.I. du Pont de Nemours and Co., Inc., 1976–present (24); B. 90°C 15.7% NaCl in feed, 5 kA/m² Asahi membrane, Asahi Chemical Industry Co., 1978–present; C. 85°C, 22% NaCl, 3.1 kA/m², Nafion 315 membrane, E.I. du Pont de Nemours and Co., Inc., 1973–1976 (24); D. 85°C, 22% NaCl, 3.1 kA/m², Nafion 427 membrane, E.I. du Pont de Nemours and Co., Inc., 1971–1974 (24).

tively prevents diffusion of anions to the cathode. In the Krebs cell (Fig. 3), the electrolysis cell itself is compact and a fairly large chemical reactor vessel must be provided for the reaction, 50 L/kA of cell current. Several of these units are cascaded to give a high conversion. The individual cells themselves operate at a steady state with a relatively low conversion factor. The operating conditions for chlorate synthesis are given in Table 6. The main use for sodium chlorate is as a precursor for chlorine dioxide, which is used as a bleach in the pulp and paper industry (see Chlorine oxygen acids and salts).

Perchlorates. Chloride solutions may be electrolyzed directly to perchlorate; however, only low yields are obtained (32–34). The normal procedure is to electrolyze a concentrated solution (300–600 g/L) of the chlorate. The process is less sensitive than chlorate manufacture to operating conditions and it is often carried out in relatively primitive tank electrolyzers. The choice of the anode material is limited to massive platinum or platinized titanium or PbO_2 (35). The current efficiencies with PbO_2 electrodes are about the same as with platinum. A comparison of the operating features of perchlorate cells is given in the article Chlorine oxygen acids and salts, perchlorates.

The world production of perchlorates is estimated to be about 100,000 metric tons per year. It is used mainly for the production of the rocket fuel, ammonium perchlorate.

Manganese Dioxide. Although most battery needs are met with selected MnO_2 ores, there are special applications, such as high capacity cells, that require the very pure product that is provided by electrolytically produced manganese dioxide (36). The production method involves the leaching of reduced manganese ore (MnO) with H_2SO_4 (or spent electrolyte) and anodic deposition of MnO_2 on a graphite or titanium anode at 90°C. The reaction is shown below.

$$\text{Anode:} \quad 2\,Mn^{2+} \rightarrow 2\,Mn^{3+} + 2\,e \tag{12}$$

$$2\,Mn^{3+} \rightleftharpoons Mn^{4+} + Mn^{2+} \tag{13}$$

Table 6. Operating Data of Chlorate Cells

	Western Electrochemical Co.	Krebs old	Krebs new	Daiki Engineering Co.	Corbin (Péchiney)	Pepcon (Uhde)	Lurgi	Chemetics[a] (Chemech)	Krebs-Kosmo	Krebs Zürich
current, kA	5	12	6	3	3			8		3
electrode area, m^2	5		2	5	5			10		20
current density, anode, kA/m^2	0.1–1.0		3	2	0.2–0.6	1.25–1.55	3	0.8	3.5	
current density, A/L	2–14	8	9	6	2–10					
current efficiency, %	75	82–88	94–96	96	82–90	88	92	94	93–95	95
cell potential, V	2.8–3.5	3.2	3.3–3.7	3.45	2.9–4.0	3.5–3.6	3.3–3.4	3.0	3.2	3.0–3.3
energy consumption, kW·h/kg[b]	6.4	5.7	5.57	5.4	6.1	6.1	5.5	4.9	5.1	ca 5
temperature, °C	30–40	44	60	75–80	30–40	60	80	40–50	60	80
pH			6.4–6.5	6.2				6.5		
anode material	graphite	graphite	Pt-Ir/Ti	Pt/Ti	graphite	PbO$_2$/graphite	Pd-Ir/Ti	graphite	Pt/Ti	Pt/Ti
anode consumption, kg/t NaClO$_3$	4–8				4–8					
end conc. of electrolyte, g NaClO$_3$/L		400	550–600	350		750	400	560–600	400–600	

[a] Bipolar cell.
[b] Calculated from the average values of potential and current efficiency.

$$Mn^{4+} + 2\,H_2O \rightarrow MnO_2 + 4\,H^+ \tag{14}$$

$$Cathode: \quad 2\,H^+ + 2\,e \rightarrow H_2 \tag{15}$$

The current density is normally 6–11 A/cm^2, and the reaction has a current efficiency of 80–90%.

Improvements actively being investigated range from the use of higher current densities through anode pretreatment (in an F$^-$/H$_3$PO$_4$/HNO$_3$ bath) (37), or the electrolysis of Mn(NO$_3$)$_2$ melts (38), to improvements of the deposit quality by the electrolysis of manganese oxide suspensions (39).

World production of MnO$_2$ by electrolysis in 1976 was estimated to be about 83,000 t of which Japanese manufacturers contributed 50%. Production capacity is expanding. Both Mitsui (40) and Tekhosha (41) have completed new plants of 12,000 t/yr capacity in Ireland (1976) and Greece (1975), respectively (see Manganese compounds).

Water Electrolysis. Usually on-site hydrogen/oxygen generation is performed on a relatively modest scale in plants for fat hardening, synthetic jewel manufacture (see Gems, synthetic), or in metallurgical processes, etc (42). However, where a source of cheap electricity is available, electrolytic hydrogen may even provide the feedstock for an ammonia plant, as is the case at Aswan, Egypt, where 56 Brown-Boveri EBK 385-70 electrolyzers produce hydrogen at the rate of 210 m^3/min (43).

The electrolysis cells for hydrogen/oxygen generation are usually filter-press units incorporating bipolar electrodes and asbestos diaphragms to separate the hydrogen and oxygen streams. Since the electrolysis is carried out in alkaline electrolyte, the use of an inexpensive anode such as Ni is facilitated. Owing to its high conductivity, 25% KOH is the universally favored electrolyte. A schematic cross-section through a typical cell is shown in Figure 6. The reversible potential for the overall process

$$2\,H_2O \rightarrow 2\,H_2 + O_2 \tag{16}$$

based on,

$$E_r = E^\circ - \frac{RT}{4F} 2.303 \log (P_{O_2} \cdot P_{H_2}^2) \tag{17}$$

is 1.23 V at normal temperature and pressure. In practice, however, no gas evolution is observed below 1.65 V and the actual operating voltage is 1.8–2.6 V. The energy efficiency of the process is therefore of the order of 45–65%. Equation 17 also shows that the magnitude of the reversible potential of the process increases by about 90 mV for a tenfold increase in operating pressure. High pressure gas may therefore be generated at the cost of only a minimal increase of electrolysis voltage. Several of the filter-press electrolyzers take advantage of this fact and are constructed for operation at 3.04 MPa (30 atm) (44–45). The need for compressors in the gas line is thereby eliminated.

Fluorine. Fluorine (qv) is the most reactive product of all electrochemical processes (46). It was first prepared by Moissan in 1886 but has only been manufactured on a technical scale since the 1940s, mainly for use in uranium enrichment plants (47). The process, which uses KF/HF electrolytes, may be categorized according to electrolyte composition and operating temperature. The original Moissan process involved the electrolysis of HF containing a small amount of KF (composition KF:12 HF) at −80°C (48). The low temperature is necessary with this electrolyte composition to reduce the HF vapor pressure to an acceptable level. However, corrosion of the cell materials and electrodes is a serious problem.

To overcome the high corrosion rates, Argo and co-workers in 1919 proposed the use of an electrolyte with a low HF content: KF:HF (49). The melting point is high and the electrolysis must be carried out at 250°C. Although corrosion resistance was satisfactorily solved using Cu, Mg, or Ag cell materials, a disadvantage was the necessity to use a highly anhydrous electrolyte and the tendency to form a passivating layer of carbon monofluoride on the anode (50).

Manufacture of fluorine is now based on the medium temperature (65–75°C) electrolysis of KF.2HF. Several 1–2 kA cells operating with this electrolyte were developed in the large wartime effort that started in 1943. A design from Hooker Chemical Corp. (51) was chosen by the Atomic Energy Commission (now DOE) and resulted in the type "C" and modern type "E" cell of 6 kA capacity (52).

Anodes for medium-temperature cells are preferably ungraphitized carbon (53) or the porous carbon used by ICI (54). It has been claimed that with a porous carbon anode a current density (cd) of up to 1080 A/m^2 is feasible without adverse polarization (passivation) effects, much more than the 270 A/m^2 for normal carbon anodes. Cathodes are normally steel. Considerable expertise is involved in providing a lasting electrical contact between the anode and the current feeder owing to creepage of the electrolyte or F$_2$ into the joint and the ensuing corrosion. A typical cell for the production of fluorine is shown in Figure 2.

Permanganate. Potassium permanganate is prepared from manganese dioxide or directly from manganese metal or alloys (55). The synthesis of permanganate from manganese dioxide first involves the preparation of potassium manganate by roasting manganese dioxide with potassium hydroxide in air:

$$MnO_2 + 2\ KOH + \tfrac{1}{2}\ O_2 \rightarrow K_2MnO_4 + H_2O \quad (18)$$

The manganate is dissolved, filtered and then oxidized to permanganate at a concentration of 120–180 g/L with 80–120 g/L KOH in an electrolytic cell. The electrolyzers used by the Carus Chemical Company employ bipolar steel electrodes. The cathode side is specially formed so that its active area is only $\tfrac{1}{150}$th of that of the anode. The back reduction of permanganate is presumably hindered by this means (56). The electrolysis practiced by Boots Pure Drug Co (57) uses nickel anodes and mild steel cathodes in cells arranged in cascade with interstage product crystallization (see Manganese compounds). The preferred method in the USSR appears to be a direct synthesis from manganese alloys. The electrode reactions are:

$$Anode: \quad Mn + KOH + 3\ H_2O \rightarrow KMnO_4 + 7\ H^+ + 7\ e \quad (19)$$

$$Cathode: \quad 7\ H^+ + 7\ e \rightarrow 3.5\ H_2 \quad (20)$$

The process introduced in 1957 at the Rustavi Chemical Combine used cast ferromanganese anodes and a KOH/K$_2$CO$_3$ electrolyte. The purpose of the K$_2$CO$_3$ is to prevent the formation of an anode mud which results from the dissolution of iron in the alloy.

Recently a packed bipolar bed cell was piloted at Rustavi (58). The cell dimensions are 0.5 m dia and 0.5 m height and it was filled with ferromanganese nodules. Energizing the bed with a potential of 90 V gave a cell current of 6.5 kA. Although ton quantities have been prepared, the bipolar cell is not believed to be in commercial use for the production of permanganate.

Peroxidisulfates. The electrolytic production of peroxidisulfates was once important for the manufacture of hydrogen peroxide (59). In 1956 the annual world production was 40,000 t of which more than 80% was obtained through electrolytically synthesized peroxidisulfates. The generation of peroxidisulfates is carried out by electrolyzing bisulfates or sulfates in acidic solution at a platinum anode:

$$2\ (NH_4)HSO_4 \rightarrow (NH_4)_2S_2O_8 + 2\ H^+ + 2\ e \qquad (21)$$

Hydrogen peroxide is produced by hydrolysis:

$$(NH_4)_2S_2O_8 + 2\ H_2O \rightarrow 2\ (NH_4)HSO_4 + H_2O_2 \qquad (22)$$

The development of the autoxidation of alkyl anthraquinones in 1953 by Dupont led to a rapid increase in the production of H_2O_2 but a sharp decline in the importance of the electrolytic route. In 1975, the annual world production of H_2O_2 was 400,000 metric tons, of which only a small amount was electrolytically produced.

The electrolysis plants that are still operating have been subjected to considerable improvement. Descriptions of modern plants have been given (60–61). Platinum, the universal anode material, is used in the form of platinized titanium, massive platinum, or platinum foil-clad titanium. Tank, cylindrical, and filter-press cells have all been employed. The introduction of additives such as HF, Na_2SiF_6, H_2SiF_6, and NH_4SCN in small amounts can improve the current efficiency by up to 20% (62). The electrolysis conditions of some reported processes are given in Table 7.

Electrowinning of Metals

The metals that are currently produced by electrolysis (67) are shown in Table 1. Fused salt processes are used when the reactivity of the metal does not allow deposition from aqueous solutions.

Electrowinning from Aqueous Solutions. The steps involved in the electrowinning of metals from aqueous solutions are: (a) conversion of the ore to an acid-soluble form (usually to the oxide by roasting); (b) leaching in acid (sulfuric acid); (c) purification and possibly concentration; and (d) electrolysis. The acid used in the leaching step

Table 7. Conditions for the Electrosynthesis of Peroxidisulfates

electrolyte	H_2SO_4/Na_2SO_4 mole ratio 1:1–2	H_2SO_4, 500–520 g/L	H_2SO_4, 50–300 g/L Na_2SO_4 or $(NH_4)_2SO_4$, 50–300 g/L
promoter, g/L		HCl, 0.03 NH_4SCN, 0.3	
product concentration	$Na_2S_2O_8$, 22–32.7%	$H_2S_2O_8$, 300 g/L	$Na_2S_2O_8$ or $(NH_4)_2S_2O_8$, 27–33%
current density, kA/m²	5–20	anode: 0.05 cathode: 0.01	anode: 8
cell voltage, V	4.6–5.4		3.6–3.9
current efficiency, %		75	78
temperature, °C	10–30	10–15	
anode		Pt-clad Ti	massive
references	63	64	61, 65–66

is regenerated during the electrolysis and is therefore recycled. Theoretically, there is no net consumption of leaching acid. In many cases, metals more noble than the product are present in the leached ore as impurities. Often these metals give low overvoltages for hydrogen evolution. Their codeposition apart from contaminating the product could also result in considerable hydrogen evolution and a loss of current efficiency. Therefore, it is important to remove such impurities before the electrolysis (step c). This is especially important for metals such as Zn and Mn that have deposition potentials close to hydrogen evolution.

The electrowinning of metals from aqueous solutions is generally carried out in tank electrolyzers under similar conditions. The main process conditions are given in Table 8.

Electrowinning From Fused Salts. *Aluminum.* Aluminum (qv) is produced world-wide by the Bayer-Hall-Héroult process. It consists of the electrolysis of alumina dissolved in molten cryolite (68). The overall reactions are:

$$2 Al_2O_3 + 6 C \rightarrow 6 CO + 4 Al \tag{23}$$

$$2 Al_2O_3 + 3 C \rightarrow 3 CO_2 + 4 Al \tag{24}$$

Oxygen is not evolved at the anode, instead the graphite is oxidized to CO and CO_2. The cell gas contains ca 70–90% CO_2 although the thermodynamics favor CO production (eq. 23). The cells are constructed from large steel tanks lined with refractory insulating bricks with an outer covering of graphite which acts as current feeder to the molten aluminum cathode pool. There are two anode types in use today. The prebaked (prefabricated) anode is a large carbon block. As equations 23 and 24 show, graphite is consumed and the anode blocks must be periodically renewed. The second type of anode (Soderberg anode) is continuously formed by addition of a carbon/pitch paste to a mold held above the melt. The paste is added at the rate of loss of graphite. The whole mass moves slowly down toward the melt and is baked *in situ* (see Carbon).

Table 8. Production of Metals by Electrolysis of Aqueous Solutions

Metal	Anode	Cathode	Electrolyte	Temperature, °C	Cell voltage, V	Cathode current density, kA/m^2	Energy requirement, kW·h/kg
Zn	Pb–Ag	Al	100–200 g/L Zn as sulfate	35	3.5	0.3	3.3
Cu	Pb–Sb (6–15%)–Ag (0–1%)	Cu	20–70 g/L Cu 20–70 g/L H$_2$SO$_4$	30–35	2.2	0.1	2.2
Mn	Pb–Ag (1%)	Hastalloy 316 stainless steel or Ti	30–40 g/L Mn as sulfate 125–150 g/L ammonium sulfate 0.1 g/L SO$_2$ (added to prevent MnO$_2$ deposition)		5.1	0.4–0.6	8–9
Co	Pb–Sb–Ag	stainless steel	15–50 g/L Co	50–65	5.0	0.035–0.04	4.5

A new development has been pioneered by Alcoa (69) based on the electrolysis of a molten AlCl₃–LiCl–NaCl electrolyte in a monopolar or bipolar cell with graphite anodes and cathodes (Fig. 5). The chlorine produced at the anode is used to generate the AlCl₃ from alumina and petroleum coke at 700–900°C. The overall reactions are:

$$Al_2O_3 + 3\,C + 3\,Cl_2 \rightarrow 2\,AlCl_3 + 3\,CO \quad (25)$$

$$\text{Cell reaction:} \quad 2\,AlCl_3 \rightarrow 2\,Al + 3\,Cl_2 \quad (26)$$

The Alcoa process consumes less energy than the conventional Hall-Héroult process. Moreover, the bipolar cells give a greater space-time yield.

Sodium. Sodium is manufactured by the electrolysis of fused NaCl (42 wt %) — CaCl₂ (58 wt %) in a Downs cell (70). The purpose of the added CaCl₂ is to lower the melting point of the electrolyte. With the composition shown, an operating temperature of 580°C is obtained. The Downs cell consists of several groups of concentrically arranged anodes and cathodes. A metal diaphragm is placed between each anode-cathode pair to prevent contact of the chlorine and sodium streams which rise to the top of the cell. The chlorine is collected in a hood and the liquid sodium is channeled to a riser pipe. The collected metal is allowed to cool to 110° in order to precipitate the calcium metal which is electrodeposited with sodium. The filtered sodium contains less than 0.04 wt % calcium (see Sodium).

Magnesium. Magnesium is produced by the electrolysis of a molten electrolyte of composition (71); 20 wt % MgCl₂, 20 wt % CaCl₂, 60 wt % NaCl at 700–720°C in Dow electrolysis cells. The cells contain conical graphite anodes and cathodes arranged concentrically. The operation of the Dow cell is very similar to the Downs sodium cell. The magnesium rises and is diverted to a collection trough. Chlorine is also collected in a hood covering the cell. Details of the fused-salt metal winning processes are given in Table 9 (see Magnesium; Chemicals from brine).

Table 9. Production of Metals by Electrolysis of Fused Salts

Metal process	Anode	Cathode	Melt	Temperature, °C	Cell voltage, V	Cathode current density, kA/m²	Energy consumption, kW·h/kg
Al/Hall-Héroult	Prebaked graphite	Al	Cryolite, 80–85%	940–980	4.1	3–3.5	
			CuF₂, 5–7%				13–15
			AlF₃, 5–7%				
	Soderberg graphite	Al	Al₂O₃, 2–8%				
Al/Alcoa	graphite	graphite	NaCl, 50%	700	ca 3.3	8–23	9.9
			LiCl, 45%				
			AlCl₃, 5%				
Na	graphite	steel	CaCl₂, 58%	580 ± 10	5.7–7	9.7	
			NaCl, 42%				
Mg	graphite	steel	MgCl₂, 20%	700–720			18.5
			CaCl₂, 20%				
			NaCl, 60%				

Electrochemical Processing of Wastewaters

Electrodeposition is a well-established process in the photographic industry for silver recovery from bleach and fixing solutions (72). However, as a result of the development of new cell designs, electrochemical methods are also applied to other less expensive metals at much lower concentrations. As shown by equation 14 (p. 664), the processing of wastes to a low end concentration of electroactive species often combined with very high throughput rates requires a very large electrode area. Several novel electrodes and cells have appeared recently that are capable of providing the required electrode area and are, therefore, ideal for wastewater processing. These designs include the use of porous electrodes (73), fixed-bed electrodes (74–76) and fluidized-bed electrodes (77), and rolled-up electrode sandwich constructions such as the Swiss-roll cell (78–80) and the Dupont ESE Jelly-roll cell (81). Some of these new designs are just starting to appear on the market. Akzo is offering a fluidized-bed cell (Fig. 9) for wastewater treatment (82), which will no doubt have an impact on the market.

In addition to metal recovery, which takes place cathodically, some waste may also be treated electrochemically by an oxidative process. Electrolytically generated hypochlorite may be used for the oxidative destruction of CN^- or the sterilization of

Figure 9. An AK20 fluidized cell for wastewater treatment.

domestic wastes. Thus on-site hypochlorite generation is becoming increasingly popular for marine sanitation applications (Letrasan), and swimming pool sterilization as well as for municipal waste treatment works (83–85). A summary of electrochemical waste treatment is given in Table 10 (see Wastes; Recycling).

In addition to the active electrolytic processes described here, well-established techniques such as electrodialysis (qv) and electroflotation (see Electrodecantation) are also employed.

Safety and Environmental Aspects

Hazards Associated With the Mixing of Anode and Cathode Gases.

The electrolysis of aqueous solutions often leads to the formation of gaseous products at both the anode and the cathode. The main examples are the H_2/Cl_2 and H_2/O_2 products of the chlor-alkali and water electrolysis plants. The electrode reactions:

$$\text{Anode:} \quad 2\,Cl^- \rightarrow Cl_2 + 2\,e \tag{27}$$

$$\text{or} \quad 4\,OH^- \rightarrow O_2 + 2\,H_2O + 4\,e \tag{28}$$

$$\text{Cathode:} \quad 2\,H^+ + 2\,e \rightarrow H_2 \tag{29}$$

are separated by a diaphragm to obtain pure products uncontaminated by the other gas. Mixing of these gases is potentially dangerous since the nonexplosive mixture range is very narrow. This is clearly seen in Figure 10, which shows the explosion and detonation limits for $Cl_2/H_2/O_2$ mixtures (91). In chlor-alkali production the events that lead to a dangerous mixture are dependent on the cell type (90). In the case of the diaphragm cell, the following problems can lead to H_2/Cl_2 mixtures: seepage of hydrogen through the diaphragm into the chlorine stream (normally amounts to about 0.3–0.5% H_2 in Cl_2) by perforation, thin spots, or imbedded metal particles; and low brine levels which leaves part of the diaphragm above the liquid level. Hydrogen, which diffuses about 35 times faster than chlorine, enters the anode compartment. A low brine level can occur because of boiling in the cell (a growing problem with larger cells), which gives a false reading on the manometer.

In the case of mercury cells the following situations would lead to a Cl_2/H_2 hazard: loss of mercury from the cell (pump failure) and evolution of hydrogen from the current feeders; contamination of the mercury by heavy metal impurities in the brine or the carbon anodes (which decompose); drop of d-c voltage which, as a result of incomplete polarization, may result in hydrogen generation; anode breakage and subsequent contact with the mercury results in such a large current density that hydrogen is evolved; too high acidity ($< pH\ 2$) favoring hydrogen generation; too low a brine concentration resulting in electrolysis above the limiting current; and insufficient decomposition of the amalgam (too little water in the decomposer) >0.6 wt % Na.

Processes that use diaphragmless cells, eg, for chlorate, permanganate, manganese dioxide, etc, must be carefully controlled to ensure that the current efficiency is high enough and that the percentage of by-product oxygen is low. Oxygen/hydrogen mixtures have a much narrower nonexplosive range than chlorine/hydrogen mixtures. Processes in which the possibility of H_2/O_2 mixing exists should therefore be meticulously monitored and controlled.

Table 10. Electrochemical Treatment of Wastewater

Pollutant (desalination process)	Concentration[a] allowed in U.S., ppm[b]	Source and concentration, ppm[a]	Nature of electrochemical treatment (cell electrode type)	Concentration, ppm initial	Concentration, ppm final	References
Ag $\left(Ag^+ \xrightarrow{e} Ag\right)$	0.05	50–250 photographic fixers,	rotating disk electrode tank electrolyzers	5000	500	72
			Dupont ESE cell	234	10	81
			Swiss roll			79
Cu $\left(Cu^{2+} \xrightarrow{2e} Cu\right)$	1.0	20–120 metal plating rinse water,	packed bed	6.3	0.15	74
			fluidized bed	2000		77
			fluidized bed (Akzo)	100	1	82
			Swiss roll	630	0.63	78
			Dupont ESE cell	20	0.6	81
			ECO-cell (rotating cylinder)	130	2	87
Fe $\left(Fe^{2+} \xrightarrow{-e} Fe^{3+}\right)$	0.3	100–150 coal mines,	packed bed	85% conversion		75
				700	0.1	76
Hg $\left(Hg^{2+} \xrightarrow{2e} Hg\right)$		chem. plant	Dupont ESE cell	30	0.3	83
		2.5–10 chlor-alkali,	porous electrode	3	0.03	81
			Swiss roll	140	0.2	79
			fluidized bed (Akzo)	5	0.05	82
Cr (ppt as insol. Pb salt) $\left(Cr(VI) \xrightarrow{3e} Cr(III)\right)$	0.05	2.5–70 metal plating,	Andeco cell (sacrificial lead anode)	200	0.05	88
			Swiss roll	100	0.05	80
SO$_2$ $SO_2 \xrightarrow{ox} H_2SO_4$		paper-pulp	United Technologies (absorption on C slurry then electrolytic oxidation)	90–100% removal		89
organic/domestic waste (oxidation with electrogenerated hypochlorite)		domestic	Pacific Engineering Pepcon cells			83
			Diamond Shamrock Sanilec and Letrasan cells			84
			Ionics Chloromat cells			85

[a] Ref. 86.
[b] Refer to the latest regulations

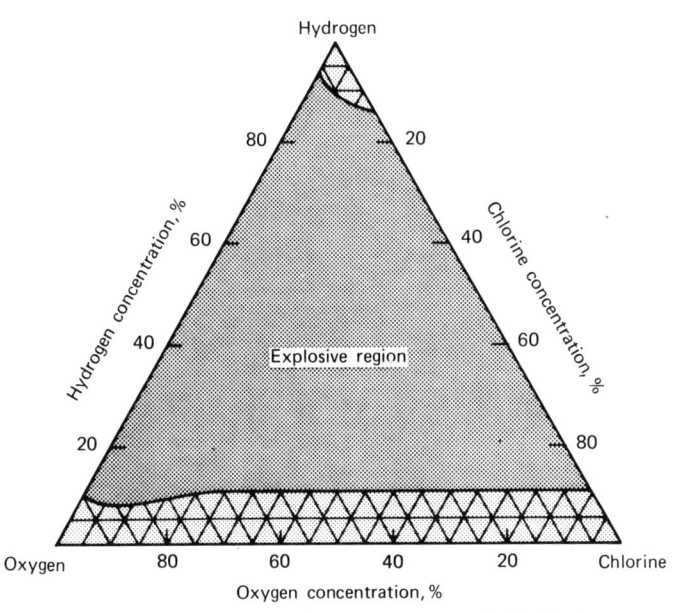

Figure 10. Hydrogen–chlorine–oxygen explosive limits.

The Mercury Problem. Mercury is regarded as one of the most toxic heavy metals. Legislated limits for mercury in wastewaters have therefore been set at very low levels (19–20). In some countries a value of 10 ppb, which is little above the detection limit, has been enforced. Only since about 1970 have the losses from mercury cell plants been regarded as a serious hazard to the environment.

Losses from mercury cell operation result from brine treatment, drainage, thick mercury treatment, emissions from the cell room area and entrainment with caustic and hydrogen. Mercury may be removed from liquids by either reduction and filtration or oxidation (with hypochlorite) and absorption by an ion-exchange resin. Good results have also been obtained by absorption in beds of active charcoal. The frequency of opening the cells has dropped owing to the installation of DSA electrodes, which require little maintenance, and has lowered the ventilation losses to the cell room to negligible levels. Modern mercury cell plants have an emission of only 1 g Hg/t Cl.

Fluoride From Aluminum Smelters. The operation of aluminum electrolysis at 950°C results in considerable evaporation and generation of waste gases containing fluorine in various forms. This emission is particularly undesirable due to the high toxicity of fluorine to livestock and plants. A bibliography of fluorine emission problems, controls and countermeasures is available (92). The elimination of fluorine (or fluorides) from the waste and ventilation gases of an aluminum factory is a considerable problem owing to the waste quantity involved: up to 2.5×10^6 m^3/t Al. Use of a Soderberg furnace allows the cell gas to be collected and purified, usually by combustion, with a cyclone and electrostatic precipitators for dust removal, and spray scrubbers or filter-plate columns for HF removal. Purification of the air vented from the plant is usually accomplished in a spray chamber mounted on the roof. Typical quality values for the air expelled are 10 and 2 mg/m^3 for the cell and vent gas, respectively. However, the total emitted F$^-$ can be considerable, and 680 kg/d is certainly not uncommon (93).

BIBLIOGRAPHY

"Electrochemistry" in *ECT* 1st ed., Vol. 5, pp. 495–549, by G. C. Akerlof, Mellon Institute of Industrial Research; "Electrochemistry" in *ECT* 2nd ed., Vol. 7, pp. 784–841, by G. C. Akerlof, Aerospace Department, Princeton University.

1. C. J. Harke and J. Renner, *J. Electrochem. Soc.* **125**, 455C (1978).
2. F. Schwager, P. M. Robertson, and N. Ibl, *Abstract 32, Extended Abstracts,* ISE Meeting, Zürich, Switzerland, 1976.
3. A. Storck, F. Vergnes, and P. LeGoff, *Powder Technol.* **12**, 215 (1975).
4. F. B. Leitz and L. Marincic, *J. Appl. Electrochem.* **7**, 473 (1977).
5. A. Storck and F. Coeuret, *Electrochim. Acta* **22**, 1155 (1977).
6. T. R. Beck in E. Yeager and A. J. Salkind eds., *Techniques of Electrochemistry,* Wiley Interscience, a division of John Wiley & Sons, Inc., New York, 1978.
7. J. W. Kühn von Bürgsdorff, *Chem. Ing. Tech.* **49**, 294 (1977).
8. R. E. Connick and Yuan-tsan Chia, *J. Am. Chem. Soc.* **81**, 1280 (1959).
9. W. Latimer, *Oxidation Potentials,* Prentice-Hall, New York, 1952.
10. D. V. Kokoulind and L. I. Krishtalik, *Soviet Electrochem.* **7**, 325 (1971).
11. N. Ibl and D. Landolt, *J. Electrochem. Soc.* **115**, 713 (1968).
12. D. Landolt and N. Ibl, *Electrochim. Acta* **15**, 1165 (1970).
13. A. T. Kuhn, *Industrial Electrochemical Processes,* Elsevier Publishing Co., Amsterdam, 1971, Chapt. 3.
14. J. S. Sconce, ed., *Chlorine,* ACS Monograph, Reinhold, New York, 1962.
15. *Pamphlet 10,* The Chlorine Institute, New York, 1976.
16. T. C. Jeffrey, P. A. Danna, and H. S. Holden, eds., *Chlorine Bicentennial Symposium,* The Electrochemical Society Softbound Series, Princeton, N.J., 1974.
17. G. Guglielmi, *Int. Chem. E. Symposium Series* **50**, D1 (1977).
18. K. Hass and P. Schmittinger, *Electrochim. Acta* **21**, 1115 (1976).
19. W. C. Gardiner in T. R. Beck and co-eds., *Electrochemical Contributions to Environmental Protection,* The Electrochemistry Society Softbound Symposium Series, Princeton, N.J., 1972, p. 16.
20. A. Schmidt, *Angewandte Elektrochemie,* Verlag Chemie, Weinheim, 1976, p. 162.
21. U.S. Pat. 3,770,567 (June 6, 1971), W. Gratt (to E. I. du Pont de Nemours & Co., Inc.).
22. Jpn. Kokai 78 73,484 (Nov. 15, 1976), M. Hamada and co-workers (to Asahi Chemical Industry Co., Ltd.).
23. Ger. Offen. 2,638,791 (March 3, 1977), H. Ukihashi (to Asahi Glass Co., Ltd.).
24. C. J. Hora and D. E. Malone, *Electrochemical Society Meeting,* Atlanta, Georgia, Oct. 1977.
25. V. H. Thomas and K. J. O'Leary, ref. 16, p. 218.
26. W. C. Gardiner, *J. Electrochem. Soc.* **125**, 22C (1978).
27. N. R. Iammartino, *Chem. Eng.* **83**(13), 86 (1976).
28. J. C. Schumacher, *J. Electrochem. Soc.* **116**, 68C (1969).
29. Ref. 13, p. 92.
30. U.S. Pat. 3,819,503 (June 25, 1974), H. V. Casson, J. S. Bennett, and R. E. Loftfield (to the Diamond Shamrock Corp. and Huron Chemicals).
31. J. Fleck, *Chem. Ing. Tech.* **43**, 173 (1971).
32. J. C. Schumacher, *ACS Monograph 146,* Reinhold, New York, 1960.
33. A. Legendre, *Chem. Ing. Tech.* **34**, 379 (1972).
34. E. Hausmann and E. Kramer, *Chem. Ing. Tech.* **43**, 170 (1971).
35. Ref. 13, pp. 525–533.
36. Ref. 13, p. 503.
37. U.S. Pat. 3,841,978 (Oct. 15, 1974), San-Chéng Lai (to the Kerr-McGee Chemical Corp.).
38. U.S. Pat. 4,048,028 (Sept. 13, 1977), S. Senderoff (to Union Carbide Corp.).
39. Jpn. Kokai 76 104,499 (Sept. 16, 1976), K. Nakamura.
40. *Chem. Age* 10 (April 16, 1976).
41. *Eur. Chem. News,* 24 (Sept. 27, 1974).
42. D. H. Smith in ref. 13, Chapt. 4.
43. M. Aeshlimann, *BBC Braun Boveri,* Publication CH-IS 411290E, Baden, Switzerland.
44. A. E. Zdansky, *Dechema Monograph* **33**, 92 (1959).
45. J. T. Anderson, *Chem. Process. London* **12**, S13 (1966).
46. A. J. Rudge in ref. 13, Chapt. 1.

47. R. J. Ring and D. Royston, *Australian Atomic Energy Commission Report,* AAEC/E281, 1973.
48. H. Moissan, *C. R. Seances. Acad. Agric. Fr.* **102**, 1543 (1886).
49. W. L. Argo and co-workers, *Trans. Am. Electrochem. Soc.* **35**, 335 (1919).
50. H. R. Newmark and J. M. Siegmund, *Trans. Electrochem. Soc.* **91**, 367 (1947).
51. R. L. Murray, S. G. Osborne, and M. S. Kircher, *Ind. Eng. Chem.* **39**, 249 (1947).
52. J. Dykstra and co-workers, *Ind. Eng. Chem.* **47**, 883 (1955).
53. J. Dykstra and W. C. Paris, *USAEC Report K-1428,* 1959.
54. Brit. Pat. 642,812 (Sept. 13, 1950), W. N. Howell and H. Hill (to Imperial Chemical Industries Ltd.).
55. Ref. 13, p. 504.
56. U.S. Pats. 2,843,537 (July 15, 1958), and 2,908,620 (Oct. 13, 1959), M. B. Carus (to Carus Chem. Co.).
57. *Br. Chem. Eng.* **9**, 383 (1964).
58. R. I. Agladze and G. R. Agladze, *Elektrokhimiya* **13**, 622 (1977).
59. P. W. Sherwood, *Chem. Ing. Tech.* **32**, 459 (1960).
60. W. Thiele and H. Matschiner, *Chem. Technol.* **29**, 148 (1977).
61. M. Schleiff, W. Thiele, and H. Matschiner, *Chem. Technol.* **29**, 679 (1977).
62. G. A. Seryshev and co-workers, *Zh. Prikl. Khim.* **49**, 1193 (1976).
63. Czech. Pat. 145,656 (July 29, 1972), J. Balej.
64. S. S. Markov and co-workers, *Zh. Prikl. Khim.* **49**, 1632 (1976).
65. East Ger. Pat. 99,548 (Aug. 12, 1973), W. Thiele.
66. East Ger. Pat. 119,197 (Apr. 12, 1976), W. Thiele and K. Wildner.
67. M. Barbier in R. P. Ovellette, J. A. King, and P. N. Cheremisinoff, eds., *Electrotechnology,* Vol. 1, Ann Arbor Science Publishing, Inc., Ann Arbor, Mich., 1978.
68. K. Grjotheim and co-workers, *Aluminum Electrolysis,* Aluminum Verlag, Dusseldorf, 1977.
69. J. G. Peacey and W. G. Davenport, *J. Met.* 25 (July, 1974).
70. M. Sittig, *Sodium: Its Manufacture and Uses,* Reinhold Publishing Co., New York, 1956.
71. A. S. Emery, *Principles of Magnesium Technology,* Pergamon Press, Oxford, 1966.
72. M. L. Schreiber, *J. SMPTE,* **74**, 505 (1965).
73. G. A. Carlson and E. E. Ester, in ref. 17, p. 159.
74. A. K. P. Chu, M. Fleischmann, and G. J. Hills, *J. Appl. Electrochem.* **4**, 323 (1974).
75. N. B. Franco and R. A. Balonkus, *Report from NTIS, no. PB232764,* 1974.
76. G. B. Adams, R. P. Hollandsworth, and D. N. Bennion, *J. Electrochem. Soc.* **122**, 1043 (1975).
77. S. Germain and F. Goodridge, *Electrochim. Acta* **21**, 545 (1976).
78. P. M. Robertson and N. Ibl, *J. Appl. Electrochem.* **7**, 323 (1977).
79. P. M. Robertson and co-workers, *Chem. Ind.,* 459 (July 1, 1978).
80. P. M. Robertson, F. Schwager, and N. Ibl, *J. Electroanal. Chem.* **4**, 323 (1974).
81. U.S. Pat. 3,859,195 (Jan. 7, 1975), J. M. Williams (to E. I. du Pont de Nemours & Co., Inc.).
82. G. van der Heiden, C. M. S. Raats, and H. F. Boon, *Chem. Ind.* 465 (July 1, 1978).
83. R. C. Rhees in ref. 19, p. 147.
84. Diamond Shamrock sales literature on Sanilec systems, Publication E-SC-23, 1976.
85. D. J. Vaughan, *Dupont Innovation* **4**(3), 10 (1973).
86. "Wastewater Treatment Technology," *Report from NTIS, no. PB216162,* Feb. 1973.
87. F. S. Holland, *Chem. Ind.,* 453 (July 1, 1978).
88. U.S. Pats. 3,766,037 (Oct. 16, 1973), and 3,926,754 (Dec. 16, 1974), Sung Ki Lee (to Andeco Inc.).
89. *Environ. Sci. Technol.* **11**, 223 (1977).
90. R. F. Schwab and W. H. Doyle, *Electrochem. Techn.* **5**, 228 (1967).
91. E. W. Lindeijer, *Rec. Trav. Chim.* **56**, 105 (1937).
92. "Air Pollution Aspects of Emission Sources: Primary Aluminum Production. A Bibliography with Abstracts," *Report from NTIS, no. PB 224867,* Environmental Protection Agency, June, 1973.
93. C. G. Dobbs, *Abstract 47061* in section "Emission Sources," ref. 92.

PETER M. ROBERTSON
Eidgenössiche Technische Hochschule

ORGANIC

Synthesis of organic chemicals by electrolysis is a very old science but a relatively new technology. Electrolytic oxidation, reduction, and coupling of organic molecules have been carried out in the laboratory for more than a century, but only in the past 10–15 years have organic chemicals been produced on an industrial scale by electrochemical means. Factors stimulating recent interest in this area are: (*1*) development of stable, noncorroding anodes, (*2*) availability of durable, ion-permselective membranes, (*3*) significant improvements in design and construction of electrolytic cells, and (*4*) a better understanding of and some novel solutions to mass transfer problems in electrolysis systems.

Comprehensive monographs dealing with electro-organic chemistry have been published (1–4). Several review articles devoted primarily to laboratory electrosynthesis of organic chemicals have appeared (5–6), and other authors have concentrated on commercial applications (7–9). An extensive bibliography covering the entire field of electro-organic synthesis from 1801–1975 has been assembled (10).

Electrochemical synthesis of organic chemicals may offer economic advantages over the more conventional chemical processing schemes for one or more of the following reasons: lower-cost raw materials; fewer reaction steps; higher selectivities; environmentally troublesome by-products are avoided; and product recovery and purification is simplified. Against these must be weighed the cost of the required electrical energy and relatively high electrolysis system capital cost. However, in many instances the electrochemical route provides a means of selectively directing the reaction via potentiostatic control to an end product not readily obtained by the usual oxidation–reduction agents.

Hardware for Electro-Organic Processing

Cells. Electrochemical cells for the production of organic compounds are as diverse in design and construction as are chemical reactors for nonelectrolytic processes. Cell materials and geometry must be tailored to meet specific process and economic considerations and the interrelationship between these factors may differ greatly from one organic system to the next. The literature on electro-organic processing does permit generalizations on cell components and configuration.

Electrodes. The combination of electrodes and electrolyte represents the heart of an electro-organic process. These two factors work together to establish the environment in which the electrochemical reactions occur and changes in either can markedly alter conversion and product selectivity. The search for the optimal electrode/electrolyte combination is the primary task of electro-organic research and determines to a large extent the success of a system.

The selection of the cathode material depends on whether that electrode is to be the working electrode or the counter-electrode for some anodic process of interest. In the case of electrochemical reductions and reductive dimerizations, the cathode material may be critical in directing the electrochemistry toward the desired reaction products. Frequently, high hydrogen-overvoltage materials, such as lead, mercury, tin, zinc, cadmium, and graphite, are required for satisfactory product selectivity and current efficiency in these cathodic processes. Cell design problems associated with mercury have been circumvented in some cases by use of amalgamated base metals (11). Lead may be hardened by alloying it with a few percent of antimony or calcium to provide a more satisfactory engineering material. Electroplated or hot-dipped coatings of cadmium and tin have been employed to reduce the electrode cost and improve the structural strength and rigidity relative to the use of the pure metals. For electro-organic processes in which the reaction of interest takes place at the anode, the cathode may serve merely to complete the electrochemical circuit via the reduction of water to form hydrogen gas. In these situations carbon steel is commonly employed for the cathode because of its low cost and good structural properties.

The quality of the cathode surface may be as important in synthesis performance as the selection of cathode material. Numerous instances have been cited in the literature in which trace quantities of metallic or organic impurities have deposited on the cathode surface upon continued use to detrimentally affect the electrode process. Similarly, minor amounts of extraneous metals in plated coatings have been shown to exhibit a strong influence on reaction selectivity. Stringent specifications on both raw material and electrode compositions are often necessary for sustained high process performance.

The choice of anode material for electro-organic processing is frequently more difficult than the selection of cathode material because of the tendency of metals to anodically dissolve. Platinum, platinized-titanium, or platinized-tantalum are often used in the laboratory, but the corrosion rates of platinum and other noble metals may be too great for economical application in commercial processes, particularly in acidic media. A comprehensive survey of materials found suitable for large-scale electrochemical processing has been given (12). Lead, alloyed with silver, antimony, thallium, or other metals, has been suggested for use with electrolytes of dilute sulfuric acid (13), and iron, carbon steel, and magnetite have been proposed for neutral alkali-phosphate electrolytes (14–15).

Various types of graphite have been used for anodes, particularly in systems employing halide anions. Lead dioxide deposited on graphite, titanium, or tantalum has found some application as an anode in electro-organic processing (16–17). For oxygen-evolving electrolysis in strongly alkaline environments mild steel, nickel, and stainless steel are generally suitable anode materials.

A significant recent advance in anode technology has been the development of the DSA (dimensionally stable anode) (18–19) (see Metal anodes). These materials have excellent stability, with a demonstrated caustic–chlorine service life in excess of five years. One form of DSA developed for chlorine-evolving service is a very thin mixed coating of Ru_2O_3–TiO_2 on a titanium substrate (20). An anode for oxygen evolution has been proposed employing a mixed Ta_2O_5–IrO_2 on an electroconductive base metal (21). Many other DSA coatings consisting of combinations of noble-metal oxides and valve-metal oxides have been described in the patent literature (22).

Electrolytes. Desired properties of electrolytes for electro-organic processes are: high electrical conductivity; low raw-material cost; some degree of solvency for the reactants; and compatibility with the counter-electrode. Good electrical conductivity, although of little consequence in laboratory synthesis, may prove extremely important in providing acceptable electrical power cost on a commercial scale. Exotic electrolytes are also undesirable for economic reasons. High solubility of the reactants in the electrolyte has been suggested as a requirement for electro-organic processes, but commercial successes using emulsions of organic reactants in aqueous supporting electrolyte has tempered this demand. Compatibility of the electrolyte with the counter-electrode permits the electrolysis to be conducted in the absence of a diaphragm.

Both aqueous and nonaqueous electrolytes are employed in electro-organic processes. Aqueous systems are generally preferred if water does not interfere with the desired electrode reaction. Electrolytes frequently mentioned are sulfuric acid, perchloric acid, alkali hydroxides, quaternary ammonium hydroxides, and alkali or quaternary ammonium salts of a wide variety of organic and inorganic acids. Since many organic species have limited solubility in aqueous media, solvents may be added to allow increased concentrations of the reactant in the electrolyte. Examples of some commonly employed solvents are methanol, ethanol, isopropyl alcohol, acetonitrile, dimethylformamide, and dimethyl sulfoxide.

Nonaqueous electrolytes are used when potentials beyond those required to discharge water are necessary in order to carry out the desired electro-organic reaction or when the reactants or products are unstable in the presence of water. Typical systems in use are anhydrous solutions of quaternary ammonium or alkali halides or perchlorates in the above-listed polar solvents and in blends of these with various nonpolar solvents (23).

Diaphragms. The use of a diaphragm is necessary in many electro-organic processes to segregate the anolyte and catholyte zones to prevent oxidation of a catholyte reactant or product, or reduction of one of the anolyte components, or to obviate the often troublesome search for a stable anode material. If the desired cathode reaction requires an electrolyte that is corrosive to all the common anode materials, the cell may be divided with a diaphragm, allowing the use of dilute sulfuric acid and Pb or PbO_2 as the anolyte–anode combination. The desired properties of an electrolysis diaphragm are: low electrical resistance; good chemical and physical stability; high resistance to diffusion of anolyte and catholyte constituents between compartments, except for transport of the desired current-carrying ions; and low cost.

The two basic types of diaphragms employed in electrolytic cells are porous media and ion-exchange membranes. Porous diaphragms pose a barrier to the transport of molecular and ionic species between the electrolyte compartments by limiting convective flow and molecular diffusion. Ion-exchange (or ion-permselective) membranes provide a high degree of mobility for either cations or anions while offering a very effective barrier to neutral molecules and counter-ions.

In laboratory studies of electro-organic processes, porous diaphragms of sintered glass or unglazed porcelain are frequently used, but these are usually unsuitable for industrial operations because of their fragile nature. A wide variety of woven and nonwoven fabrics have been used as electrochemical cell diaphragms. These are offered by filter-cloth vendors in fiberglass, nylon, polyethylene, polypropylene, poly(vinyl chloride) and polytetrafluoroethylene (PTFE). Asbestos mats have been employed for years in caustic–chlorine cells. Microporous plastic sheets of polyethylene, PVC, PTFE, and rubber have also been suggested.

Ion-exchange membranes are thin sheets of polymeric material containing pendant groups capable of selective exchange reactions with either cations or anions. In general, the polymer contains either sulfonic acid or carboxylic acid groups for exchange of cations or is aminated to produce anion-exchange capabilities. Commercially available ion-exchange membranes are 0.1–1.0 mm in thickness with electrical resistances of 1.5–20 $\Omega \cdot cm^2$ (measured in 0.5 N NaCl at 25°C). Reviews of properties and applications of commercially available ion-exchange membranes have been published (24–26).

Some of the polymer formulations employed for ion-exchange membranes in electrolytic processes are: (*1*) polystyrene cross-linked with divinylbenzene and sulfonated to provide cation-exchange sites (see Styrene plastics), eg, Ionics CR 61 (Ionics, Inc.); (*2*) a sulfonated fluorocarbon copolymer marketed under the trademark Nafion by DuPont; and (*3*) perfluorocarboxylic acid polymers developed by Asahi Glass and by Asahi Chemical (see Fluorine compounds, organic). The Ionics membrane has been used commercially by Monsanto in their electrochemical process for production of adiponitrile, and the Nafion membrane has been proposed for electroreductive coupling of acetone to give pinacol (27). A new polymeric membrane based on polybenzimidazolone has recently been developed by Teijin of Japan for electrochemical applications. The efforts in the caustic–chlorine industry to replace asbestos diaphragms with hydraulically impermeable ion-exchange membranes has spurred a large number of new diaphragm candidates (28) (see Membrane technology; Ion exchange).

Cell Designs. A broad range of electrochemical cell designs has been proposed and several comprehensive reviews have been published (29–32). Although the characteristics of the cell itself has received much attention in the literature, it is important to recognize that it is the cost of the entire electrolysis system with its associated rectifiers, busbars, tanks, pumps, heat exchangers and other auxiliaries that sets the total capital burden. The cells may represent a relatively small fraction of the total cost of the electrolysis system. By the same token, electrolysis requires added pumping power, rectifier losses, and in some instances, refrigeration power; an example of the significance of these other power loads is the UCB–MCI electrolytic process for adiponitrile, in which the cell power is 75% of the total electrical requirements (33).

The cell types may be classified as follows: (*1*) plate-and-frame cells; (*2*) planar or cylindrical electrodes in tanks; and (*3*) particulate electrode cells, either fixed-bed or fluidized-bed types.

The plate-and-frame design is probably encountered most frequently in electro-organic processing. It consists of a stack of parallel planar electrodes separated by insulating frames, which form compartments for electrolyte. In some cases the electrodes are bipolar, such that one surface serves as the anode for one electrolyte compartment, and the opposing surface acts as the cathode in the adjacent compartment. Diaphragms may be used to separate anolyte and catholyte zones. A number of individual cells may be placed together in a filter press to form a multicell unit. An example of a plate-and-frame design used by Monsanto for commercial production of adiponitrile from acrylonitrile (qv) is shown in Figure 1a (34). In this design a lead cathode is affixed to one side of a thick polypropylene sheet and a lead alloy anode is attached to the other side with internal lead lugs providing the electrical connection between adjacent cells in bipolar arrangement. Ion-exchange membranes in polypropylene supporting frames provide separation of anolyte and catholyte compartments. A number of other versions of plate-and-frame cells have been described (35–37).

Several variations of cell designs based on planar electrodes immersed in hydraulically sealed tanks have been proposed for commercial electro-organic processes. Such a cell designed for electrochemical production of propylene oxides (qv) is shown in Figure 1b (38). It consists of a bank of hollow porous-graphite anodes, through which the propylene is introduced, and steel screen–wire cathodes with an asbestos dia-

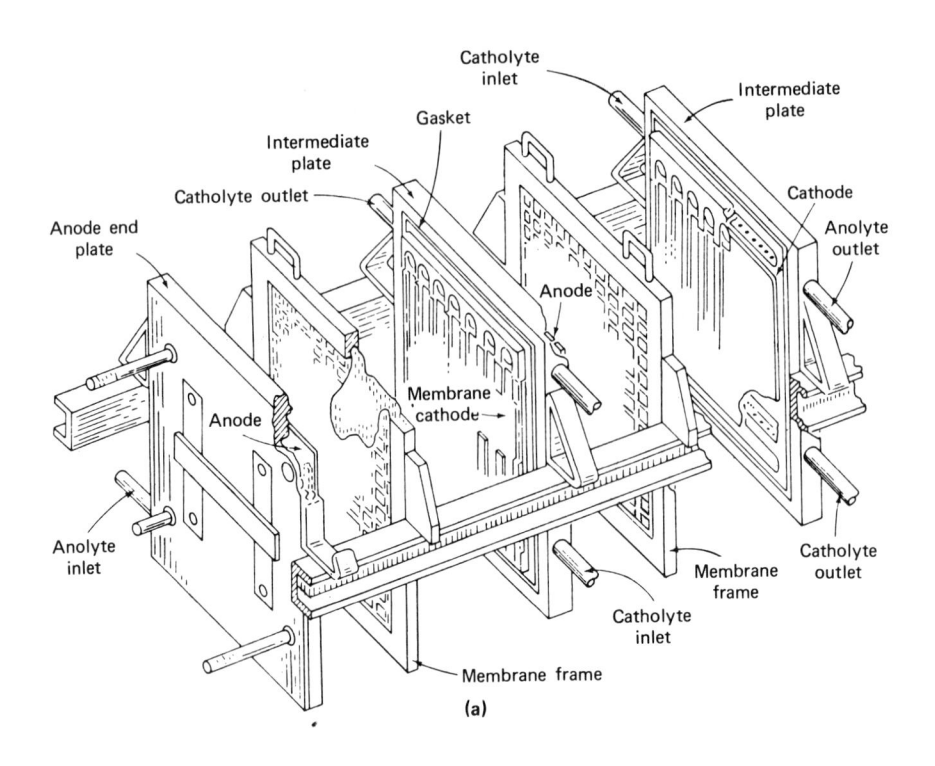

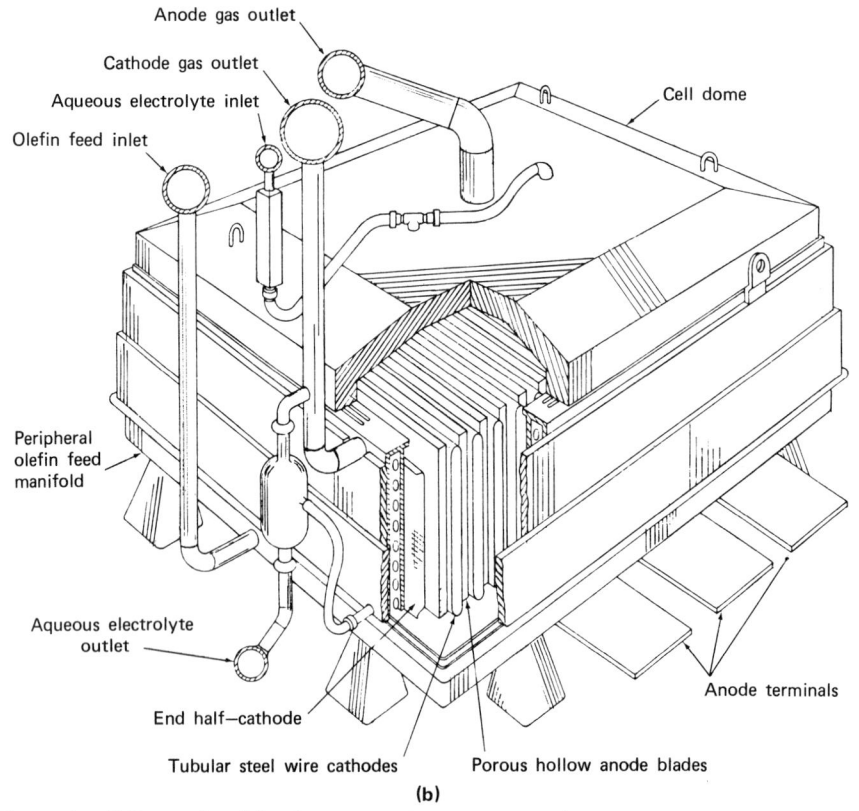

Figure 1. Cells employed in electro-organic processing: (**a**) Monsanto cell for adiponitrile; (**b**) LeDuc cell for olefin oxidation; (**c**) Beck-Guthke capillary-gap cell; (**d**) Fleischmann-Jansson Pump cell; (**e**) Nalco cell for tetramethyllead; (**f**) Backhurst fluidized-bed cell.

(c)

(d)

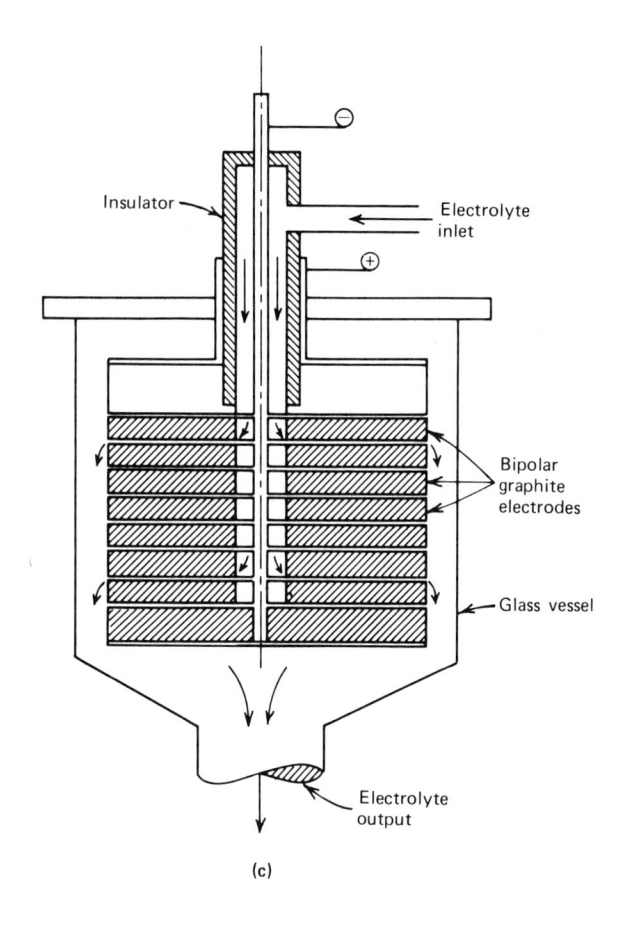

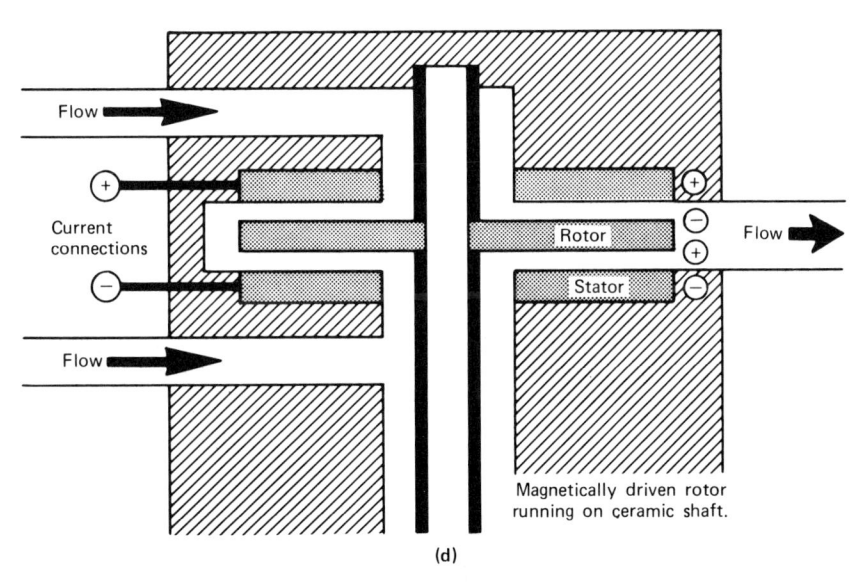

Figure 1. (*continued*)

702 ELECTROCHEMICAL PROCESSING (ORGANIC)

(e)

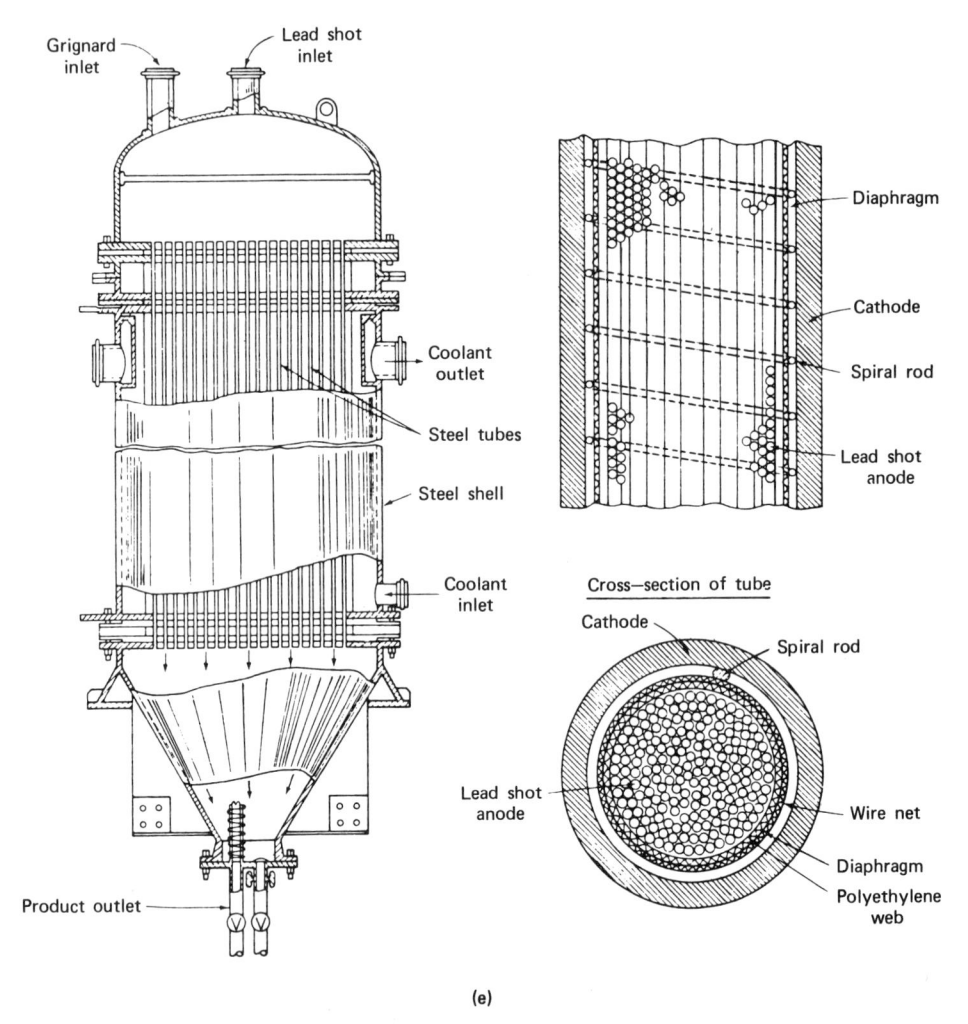

Figure 1. (*continued*)

phragm deposited on the exterior of the cathode screens. A plate-in-tank design employing solid block electrodes with hydraulic flow distribution grids in the electrolyte path for use with emulsion systems has also been described for electro-organic synthesis (39).

Another variation of the planar electrodes in a tank is the capillary-gap cell of Beck and Guthke (16). This cell (Fig. 1c) is a stack of cylindrical, bipolar graphite plates, 100 cm^2 in area and 1 cm in thickness. The plates are separated by thin (125-μm) plastic strips forming very narrow electrolyte passages, which are particularly suitable for use with low-conductivity electro-organic systems. The anode surfaces of the graphite plates are generally covered with a 200-μm layer of lead dioxide for improved durability. Commercial experience in organic synthesis with a later version of the capillary-gap cell has been described (40). Close spacing of electrodes is also featured in a novel design termed the Swiss-Roll cell, which consists of a rolled-up electrode and membrane sandwich installed in a cylindrical housing (41).

[Figure with labels: Electrolyte outlet, Porous plastic, Glass cylinder, Fluidized cathode bed, Fluidized anode bed, Cylindrical metal gauze cathode, Cylindrical metal gauze anode, Porous disk, To standard calomel electrode, Electrolyte inlet]

(f)

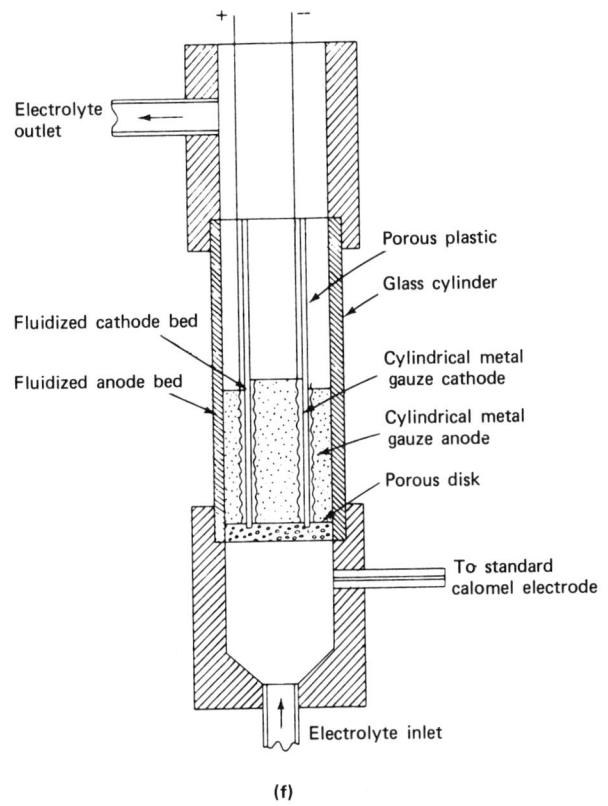

Figure 1. (*continued*)

A cell utilizing rotating cylindrical electrodes suspended in a tank has been employed in a wide range of electro-organic syntheses at the Central Electrochemical Research Institute in India (42). This design, which represents a scale-up of the rotating disk electrode frequently used in laboratory electrochemistry studies, is claimed to provide excellent mass transfer rates and, therefore, to be operable at high current densities. A more sophisticated variation of the rotating disk cell is the bipolar pump cell (Fig. 1d). It consists of a rotating annular disk electrode positioned close to conforming stator electrodes, such that the cell acts as its own pump. With supplemental pumping by an external device, the pressure distribution, residence time and mass transfer can be varied with a large degree of independence through adjustment of the rotation speed of the electrodes and the electrolyte flow rate (43).

Particulate-electrode cell designs have received much recent attention as part of a search for cells with improved space-time yields and superior mass-transfer characteristics. In fixed beds, the electrode particles are usually spherical in shape, but designs employing stacked rods, rings and fiber mats are also described (44). One fixed-bed design used commercially by Nalco Chemical for the production of tetraalkyl lead compounds is shown in Figure 1e (45). The cell consists of a series of 50 mm dia by 75 cm steel tubes, filled with lead pellets. A polypropylene screen lines the inside of the tube to electrically isolate the lead shot from the steel. Coolant flows through the zone outside the tubes for temperature control. An electrolyte composed of an ether solution of Grignard reagent and excess alkyl halide is passed through the tubes and the lead shot is anodically converted to tetraalkyllead, with the steel tubes serving as cathodes. A later version of the Nalco cell uses a porous ceramic lining in place of the polypropylene screen (46). A cell employing a bed of lead shot for cathodic removal of trace amounts of deleterious metal ions has been employed commercially by Monsanto (47).

In recent years a number of cell designs have appeared in which one or both electrodes are in the form of suspended particles. One of the first designs of this type employed 75–1000-μm metal or metal-coated particles, which were fluidized to 10–20% bed expansion by upward flow of electrolyte (Fig. 1f) (48). Spheres of copper-coated glass or polystyrene were found to be particularly suitable. A plate or rod immersed in the fluidized bed served to feed current to the ebullient mass. A commercial-scale unit employing a series of particulate electrode cells has been described (49).

Scale-Up Considerations. In the scale-up of an electro-organic reaction system from the bench-top experiment to a commercial prototype, many of the engineering considerations are the same as for a conventional chemical process. Assuming an electrode–electrolyte combination has been identified in the laboratory that gives satisfactory product selectivity and current efficiency, the following factors generally need to be addressed during scale-up: electrode and bulk-electrolyte kinetics; mass transfer in the cell; heat removal; cell productivity and cost; current distribution; voltage minimization; and materials of construction (50). Electro-organic systems often pose unique problems in mass transfer because of the low solubility of the reactants in the electrolyte. Similarly, heat removal and current distribution may require special attention with the lower conductivity electrolytes frequently employed.

The enhancement of mass-transfer rates in electrochemical cells has received much attention in recent years (29,51). Some of the approaches being used in electro-organic systems are: (1) high velocity flow via external circulating means with open channels (52–54), or with nonconductive turbulence promotors (55), or with mesh electrodes (37); (2) rotation of cylindrical or disk electrodes (43,56); (3) turbulence promotion via gases (57–58); (4) vibration of electrode (16); (5) fluidized electrode particles (59–60); and (6) fluidized nonconductive particles (61). Since the trend in commercial cell designs is to use parallel plate electrodes, either as a plate-and-frame assembly or capillary-gap device, forced circulation of electrolyte is probably the most common means of providing satisfactory mass-transfer rates. For example, in the Monsanto cell for electro-reductive coupling of acrylonitrile to adiponitrile, electrolyte velocities of 2 m/s are used in an open electrolyte channel 3 mm wide, corresponding to a Reynolds number of ca 6000 (34). With the use of turbulence-promoting spacers, superficial electrolyte velocities of 0.2–0.6 m/s are reported to give comparable results for this same electrode reaction.

Economic Optimization. The commercialization of an electrochemical process for large-scale production of organic chemicals depends primarily on three economic criteria: (1) product selectivity, (2) electrical power usage, and (3) electrolysis system capital. The attainment of a high yield of the desired product from the basic raw materials is, of course, a goal in all chemical processes. An important feature of many electro-organic processes is that lower-cost feedstocks may be employed than with conventional processing. The electrical power usage is a function of the electron change for the synthesis reaction, the cell voltage, and the current efficiency of the process. From Faraday's law, an electro-organic reaction involving a two-electron transfer can be calculated to yield a product with a molecular weight of 100 in a 5-V cell requiring 2.68 kW·h/kg of product at 100% current efficiency. Typically, power usages for organic products range from 3–6 kW·h/kg (see under Processes). The capital requirements for electro-chemical processing has received little attention, greatly handicapping realistic assessment of electro-organic processes. A study by Keating and Sutlic of DuPont suggests that the installed cost of an electro-organic synthesis plant in 1978 would be ca $11,000–15,000/m^2 ($1000–1400/ft^2) of active electrode surface area (37).

Power usage and electrolysis system capital are strongly dependent on the current density employed in the process. Although the number of cells required for a given production rate varies inversely with current load per cell, the voltage increases with current density for a specified cell design. Since higher cell voltages translate to greater electrical power inputs, generally the cost of both rectifier and process coolers will increase as well. The selection of the optimum current density at which to design the commercial facility is, therefore, an important economic consideration. An example of a current density optimization is shown in Table 1 for a hypothetical organic material produced at a rate of 10,000 metric tons per year. It is noteworthy that at high current densities the costs of rectifiers and heat exchangers may significantly outweigh the cell costs. The plot of total production cost, including a 30% before-tax return-on-investment, is shown in Figure 2, from which the minimum cost is seen to occur at a current density of about 4 kA/cell. In addition to the cell costs, a multitude of factors can influence the current density optimization, particularly the power cost, the required rate of return on capital, the cost of membranes and the method of heat removal. Thus the optimization can be complex and reliably predicted only with considerable pilot-plant or commercial experience.

Table 1. Current Density Optimization, 10,000 Metric Tons per Year Organic Product[a], 1979 Cost Basis

Factor					Unit cost
cell current, A	1000	2000	4000	6000	
current density, A/dm^2	15	30	60	90	
voltage/cell, V	5.1	6.4	9.0	11.5	
number of cells	840	420	210	140	
equipment cost, 10^3$					
cells	1680	840	420	280	$2000/cell
rectifiers	210	270	380	480	$50/kW
heat exchangers	190	240	340	430	
pumps and tanks	350	250	160	110	
Total equipment cost	2430	1600	1300	1300	
electrolysis capital, 10^3$	9130	5100	3270	2760	
operating costs, 10^3$/yr					
electricity	699	877	1234	1577	2¢/kW·h
cooling water	38	48	68	87	1.3¢/m^3
membranes	168	84	42	28	$100 each
cell maintenance	84	84	84	84	10¢/(cell-A)
other maintenance	108	88	75	70	5% processing equipment
cell labor	320	160	80	80	$20,000/operation
Total operating costs	1417	1341	1583	1926	
capital charges at 30%	2739	1530	981	828	
production cost, 10^3$/yr	4156	2871	2564	2754	

[a] Equivalent weight = 50, current efficiency = 80%, assumed current-voltage relationship and selectivity independent of current density.

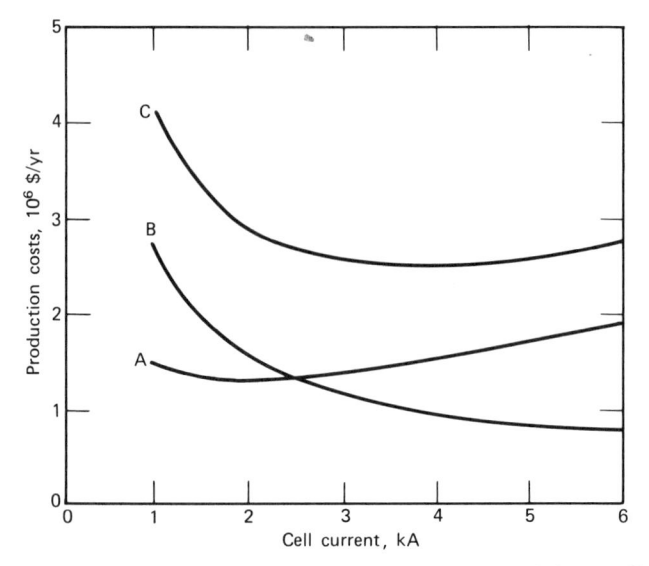

Figure 2. Current density optimization. A, Operating cost; B, capital charges; C, total production cost.

Industrial Electro-Organic Processes

Table 2 gives a summary of some of the principal electro-organic processes that have been operated on an industrial scale, or at least carried through the pilot-plant stage for certain products (1–7). One of the first large-scale applications was a plant built in 1937 by Atlas Powder Company (now ICI Americas, Inc.) to produce 1400 metric tons per year of sorbitol and mannitol by cathodic reduction of glucose (62) (see Alcohols, polyhydric). This process was of limited commercial importance, and was displaced after a few years by a more economical route using high pressure catalytic hydrogenation. The emergence of the next truly large-scale electro-organic process was in 1964 with the startup by Nalco at Freeport, Texas, of a facility to produce 15,000–18,000 t/yr of tetraalkyllead compounds. Following close behind Nalco, a plant to manufacture 15,000 t/yr of adiponitrile by electroreductive coupling of acrylonitrile was brought on-stream by Monsanto at Decatur, Alabama, in 1965.

Although the scale of operation of other known electro-organic processes is considerably below those of the Nalco and Monsanto processes, several are important in providing routes to organic compounds that are difficult to prepare selectively by more conventional means. As an example, dihydrophthalic acids can be produced in yields exceeding 90% by reduction of phthalic acid at lead and mercury cathodes, whereas catalytic hydrogenation leads to a mixture of the di-, tetra-, and hexahydrophthalic acids (76–77). Similarly, the electroreductive coupling of acetone to pinacol has received considerable attention over the years as a highly selective route to this intermediate, useful in production of pharmaceuticals and pesticides (78–79). An anodic process of commercial importance is the Kolbe synthesis of dimethyl sebacate from monomethyl adipate (70–72), which has been conducted on an industrial scale in the USSR. Electrochemical fluorination of organic chemicals has been practiced commercially by the 3M Company for a number of years (73), and more recently, ex-

Table 2. Electro-Organic Processes Operated on a Commercial or Pilot-Plant Scale

Product	CAS Registry No.	Structure No.	Reactants	Operating scale	Cell type[a]	Anode material	Cathode material	Diaphragm material[b]	Current density, A/dm^2	Cell voltage, V	Product selection, %	Current efficiency, %	Power usage, kW·h/kg	Ref.
sorbitol Atlas[c]	[50-70-4]	(1)	glucose	commercial	Tank	Pb	Hg–Pb	ceramic	4	20				62
tetraalkyllead Nalco			Pb, RCl, RMgCl	commercial	Bed	Pb	steel	polyethylene cloth	1.5–3	15–30	96–99	175	4–8	63
Ziegler			Pb–KAlR	pilot-plant	Tank	Pb	Hg	none	50	2.5	100	100	1	64
adiponitrile Monsanto	[111-69-3]	(2)	acrylonitrile	commercial	P–F	Pb alloy	Pb	IX	20–100	6–12	90–93	90–92	3–6	65
Asahi			acrylonitrile	commercial	P–F	Pb–Sb	Pb	IX	20	6–7	91	88–89	4	66
BASF			acrylonitrile	pilot-plant	CG	PbO$_2$–C	C	none	7–10	4–5	82–92	80–83	3	16
Rhone-Poulenc			acrylonitrile	laboratory	P–F	Pb–Sb	Pb	none	5–8	5–8	84–91	85–90	3–5	67
UCB-MCI[d]			acrylonitrile	pilot-plant	Tank	magnetite	C	none	8	4–4.5	85–90	>80	3	68
Phillips Petroleum			acrylonitrile	laboratory	P–F	Pb	Pb	none	20	4	91	90	3	
dimethyl sebacate USSR	[106-79-6]	(3)	monomethyl adipate	commercial	Tank	Pt–Ti	steel	none	6	12	82	58	6	70
BASF			monomethyl adipate	pilot-plant	CG	Pt	Ni	none	25	12–15	80–85	60–70	5–6	71
Asahi			monomethyl adipate	pilot-plant	P–F	Pt–Ti	steel	none	20	13–16	90–92	86–90	4	72

Table 2. (continued)

Product	CAS Registry No.	Structure No.	Reactants	Operating scale	Cell type[a]	Anode material	Cathode material	Diaphragm material[b]	Current density, A/dm²	Cell voltage, V	Product selection, %	Current efficiency, %	Power usage, kW·h/kg	Ref.
fluorinated organics 3M			acid fluorides	commercial	Tank	Ni	Fe	none	2	5–8				73
Phillips Petroleum			various hydrocarbons	pilot-plant	Tank	C	Ni	none	6–30	6–9				74
propylene oxide Kellogg	[75-56-9]	(4)	propylene	pilot-plant	Tank	C	steel	asbestos	80–110	3.5–4	82–86	70–80	5	38
Bayer			propylene	pilot-plant	Tank	C	steel	acrylic	2–5	3–5	90	88	4–5	75
dihydrophthalic acid California Research	[22919-28-4]	(5)	phthalic acid	pilot-plant	Tank	Pb	Hg	ceramic	3–40	4–11	98	90–95	1.6–4	76
BASF			phthalic acid	commercial	P–F	PbO₂	Pb	IX	10	9	94	75	4	77
pinacol Diamond Shamrock	[76-09-5]	(6)	acetone	pilot-plant	P–F	DSA	Pb	IX	16	5–6		50–60	4–5	78
BASF			acetone	pilot-plant	CG	PbO₂–C	C	none	10	7–10	90	50	9–15	79
tetraalkylthiuramdisulfide DuPont		(7)	dialkyldithiocarbamate	pilot-plant	P–F	Pt	steel	IX	48–65	6–7	ca 100	91–95	1.4–1.7	80

[a] Cell type: P–F = plate-and-frame; CG = capillary gap.
[b] Diaphragm material: IX = ion-exchange membrane.
[c] Now ICI Americas, Inc.
[d] UCB, S.A. (Belgium) and Ministry of Chemical Industry USSR.

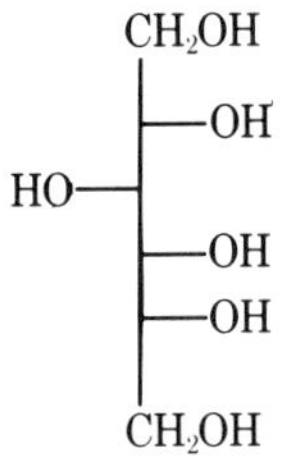

(1) HOCH₂-CH(OH)-CH(OH)(HO-)-CH(OH)-CH(OH)-CH₂OH [hexitol with structure shown]

(2) NC(CH₂)₄CN

(3) CH₃OCO(CH₂)₈COOCH₃

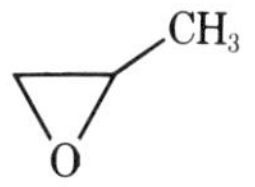

(4) methyloxirane (propylene oxide)

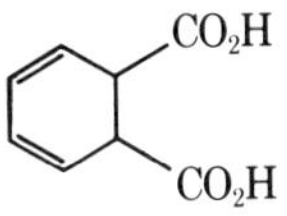

(5) cyclohexadiene-1,2-dicarboxylic acid

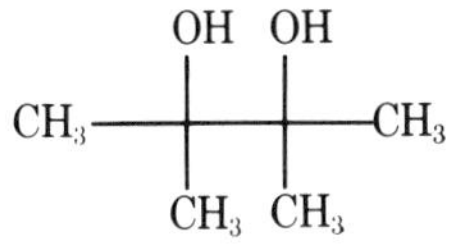

(6) CH₃-C(OH)(CH₃)-C(OH)(CH₃)-CH₃

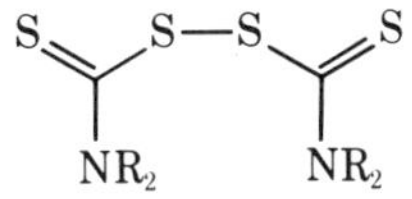

(7) R₂N-C(=S)-S-S-C(=S)-NR₂

tensive laboratory studies in this area have been described by Phillips Petroleum Company (74) (see Fluorine compounds, organic).

An electrochemical route to propylene oxide (qv) has received considerable attention, but as yet has not been commercialized. The chemistry is related to one of the early processes for producing propylene oxide involving the reaction of propylene with chlorine and water to give the chlorohydrin followed by saponification with lime to form propylene oxide and by-product calcium chloride (see Chlorohydrins). The electrochemical route involves passing gaseous propylene (qv) through the anode compartment of a divided cell in which both anolyte and catholyte are a dilute solution of sodium chloride. Hypochlorous acid produced in the anode compartment combines with the propylene to form propylene chlorohydrin, which diffuses through the porous diaphragm and is decomposed in the alkaline environment generated by the cathodic reduction of water. The principal reactions are:

anode:

$$2\,Cl^- \longrightarrow Cl_2 + 2e$$

$$Cl_2 + H_2O \longrightarrow HOCl + H^+ + Cl^-$$

$$HOCl + CH_2{=}CHCH_3 \longrightarrow HOCH_2CHClCH_3 + ClCH_2CHOHCH_3$$

cathode:

$$2e + 2H_2O \longrightarrow 2OH^- + H_2$$

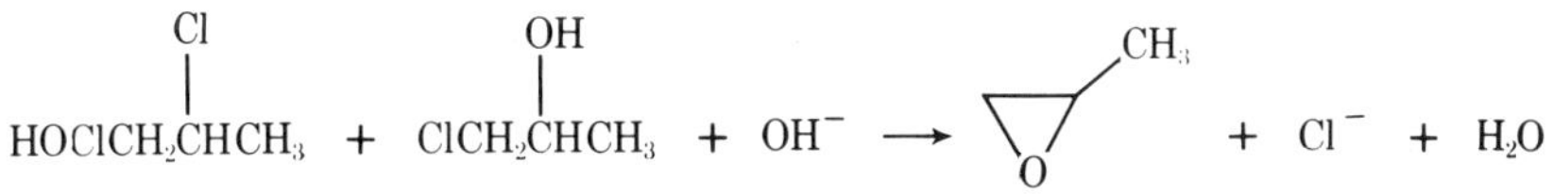

$$HOCH_2CHClCH_3 + ClCH_2CHOHCH_3 + OH^- \longrightarrow \text{propylene oxide} + Cl^- + H_2O$$

overall:

$$CH_2{=}CHCH_3 + H_2O \longrightarrow \text{propylene oxide} + H_2$$

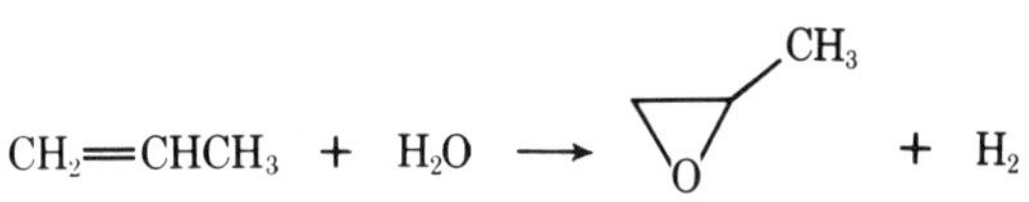

A cell designed specifically for olefin oxidation is shown in Figure 1b. The chief advantage of the electrochemical route to propylene oxide is elimination of the need for chlorine and lime, as well as avoidance of the calcium chloride disposal problem. An indirect electrochemical approach meeting these same objectives employs the chlorine produced at the anode of a membrane cell for preparing the propylene chlorohydrin external to the electrolysis system. The caustic made at the cathode is used to convert the chlorohydrin to propylene oxide, reforming a NaCl solution which is recycled. Attractive economics are claimed for this combined chlor–alkali electrolysis and propylene oxide manufacture (81).

Electrohydrodimerization of Acrylonitrile to Adiponitrile. Adiponitrile is commercially important as an intermediate in production of hexamethylenediamine, which in combination with adipic acid is employed in the manufacture of nylon-6,6 polymer (see Adipic acid; Polyamides). The total worldwide production of adiponitrile in 1973 has been estimated at 800,000 metric tons (82). In the early 1960s the predominant commercial routes to adiponitrile involved either reaction between adipic acid and ammonia under dehydration conditions or the chlorination of butadiene to dichlorobutenes, followed by combination with sodium cyanide to yield dicyanobutenes and hydrogenation of the double bond. The successful, commercially attractive electroreductive coupling of acrylonitrile at mercury and lead cathodes to form adiponitrile at high material and current efficiencies was first reported in 1963 by Baizer (83). The high selectivities resulted from the use of solutions of certain quaternary ammonium salts, such as tetraethylammonium p-toluenesulfonate, as the supporting electrolyte. As a major producer of acrylonitrile and nylon-6,6, this electrochemical route to adiponitrile offered Monsanto significant economic advantage over the adipic acid-based route. Following intensive process development efforts, a facility was started up by Monsanto in 1965 at Decatur, Alabama, to produce 15,000 t/yr of adiponitrile via this technology (65).

The electrode reactions involved in the electrohydrodimerization (EHD) are as follows:

cathode:
$$2\ CH_2{=}CHCN + 2\ H_2O + 2\ e \rightarrow NC(CH_2)_4CN + 2\ OH^-$$

anode:
$$H_2O \rightarrow 2\ H^+ + \tfrac{1}{2}\ O_2 + 2\ e$$

The hydrogen ions formed at the anode pass through the membrane and combine with hydroxyl ions produced at the cathode to give the overall reaction:

$$2\ CH_2{=}CHCN + H_2O \rightarrow NC(CH_2)_4CN + \tfrac{1}{2}\ O_2$$

Although adiponitrile yields are generally around 90%, some by-products are formed including propionitrile, 1,3,5-tricyanohexane (acrylonitrile hydrotrimer), hydroxypropionitrile, and bis(cyanoethyl) ether.

A schematic diagram of the original Monsanto adiponitrile process is given in Figure 3. The electrolysis section consists of separate catholyte and anolyte circulation systems with segregation of the two electrolytes within the cell provided by a cation-permselective membrane. The catholyte, an aqueous solution of acrylonitrile, quaternary ammonium salt and reaction products, is circulated between the membrane and the lead cathode surface at velocities of 1–2 m/s to minimize polarization effects

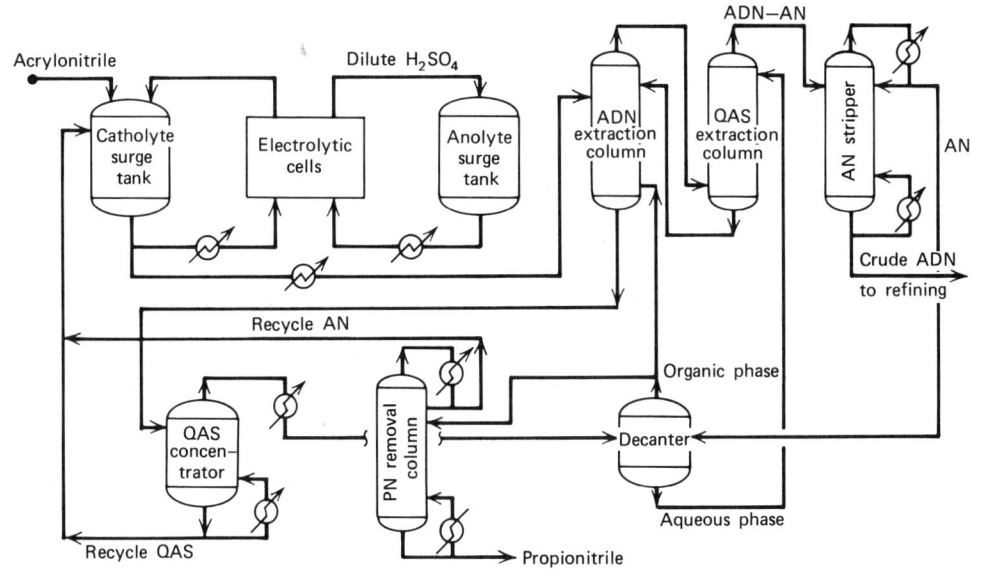

Figure 3. Monsanto adiponitrile process. QAS, quaternary ammonium salt; ADN, adiponitrile; PN, propionitrile; AN, acrylonitrile.

and limit the catholyte temperature rise. The anolyte is a dilute solution of sulfuric acid (see Table 2). Anode and cathode sheets are mounted on opposing sides of polypropylene plates containing internal lead connecting lugs. The membranes are housed in plastic frames which form the anolyte and catholyte compartments when closed between the polypropylene plates holding the electrodes (Fig. 1a). Assembly of an alternating sequence of these plates and frames in a press provides a bank of bipolar cells, such that electrical power need be introduced only on the two end electrodes. Typically, 20–40 cells are combined in a press to form a cell unit. Catholyte and anolyte are introduced in parallel flow through distribution plenums formed in the polypropylene plates. Heat generated in the electrolyte stream by virtue of the electrolyte ir drop is removed in heat exchangers external to the cells.

A slip-stream withdrawn from the circulating catholyte is cooled and countercurrently extracted in a multistage contactor with additional acrylonitrile for recovery of the adiponitrile. The adiponitrile-free catholyte is then evaporated to remove water entering the catholyte through the membranes via electro-osmosis from the anolyte and returned to the catholyte loop. The acrylonitrile–adiponitrile extract is countercurrently contacted with water to recover quaternary ammonium salt from the organic phase. The acrylonitrile is then stripped and the crude adiponitrile is further purified by distillation for subsequent hydrogenation to hexamethylenediamine. By-product propionitrile is removed from the recycle acrylonitrile in a separate distillation column.

The Monsanto plant at Decatur, Alabama, has been improved and expanded several-fold since start-up in 1965 (see Fig. 4). A facility to produce 90,000 metric tons per year of adiponitrile using the Monsanto EHD technology was scheduled for start-up in 1978 at Teesside in northeastern England by Polyamide Intermediates Ltd., a joint venture of Monsanto and Montedison, S.p.A.

Figure 4. Monsanto adiponitrile synthesis area, Decatur, Alabama.

A later version of the Monsanto EHD process has been described involving an aqueous electrolyte of about 10% disodium phosphate together with a relatively small amount of an acrylonitrile–adiponitrile organic phase, electrolyzed in an undivided cell with a cadmium cathode and carbon steel anode (84). Tributylethylammonium salts are added at concentrations of 2–10 mM for enhancing the adiponitrile selectivity and ethylenediaminetetraacetic acid is introduced to sustain the cathode current efficiency. The absence of a membrane and the high electrical conductivity of the electrolyte provides improved power efficiency over the divided-cell system; at 55°C, a current density of 20 A/dm^2 and an electrode gap at 2.5 mm, the cell operates at ca 4 V. With the claimed 87–89% adiponitrile selectivities, this is equivalent to a power usage of 2.6 kW·h/kg adiponitrile.

A number of other variations of the electroreductive coupling of acrylonitrile to adiponitrile have been reported, as indicated in Table 2. The only other process known to have been commercialized is that of Asahi Chemical Industry Company, Ltd., which began operation of a plant to produce 20,000 t/yr of adiponitrile in Japan in 1971 (66). The Asahi process employs a dual-compartment cell, divided by a cation-exchange membrane and the catholyte is an emulsion of acrylonitrile in an aqueous solution of tetraethylammonium sulfate. The chief distinction from the Monsanto process is that the acrylonitrile solubility in the supporting electrolyte is relatively low (3–7%). An acrylonitrile supply is maintained within the cell by feeding an emulsion of both organic and aqueous phases (see Table 2). Dilute sulfuric acid is used as anolyte. The cell is similar to an electrodialysis device in that it is a stack consisting of a repeating sequence of: lead cathode sheet, plastic catholyte spacer, cation-exchange membrane, anolyte spacer, and lead-alloy anode sheet (36). Raw material and steam consumption for the Asahi process per kilogram of adiponitrile are reported to be 1.1 kg acrylonitrile and 5 kg steam (85).

Another version of the EHD process has been carried through the pilot-plant stage by Union Chimique-Chemische Bedrijven (UCB), in a joint development with the Ministry of Chemical Industry for the USSR (68). The UCB–MCI process employs an emulsion of acrylonitrile in an aqueous potassium phosphate electrolyte in the absence of a membrane (see Table 2). Electrolysis is conducted at an operating temperature of 20°C. An acrylontrile usage of 1.18 kg per kg adiponitrile is claimed and electrical power consumption is stated to be 2.83 kW·h/kg for electrolysis and 0.95 kW·h/kg for refrigeration (33).

Electrosynthesis of Tetraalkyllead. Tetraalkyllead compounds have been used since the early 1920s as additives to gasoline to reduce the burning rate and prevent detonation in internal combustion engines (see Gasoline). Although tetraethyllead (TEL) was the first of the group to find commercial application, and still is predominant, the tetramethyl- and tetra(methyl–ethyl) compounds were later introduced as antiknock agents. The use of lead compounds in gasoline has been under strong environmental pressures in the United States since the late 1960s, causing production in the United States and Canada to peak in the early 1970s and fall to ca 400,000 t/yr in 1977 (86). Production outside the United States in 1977 was on the order of 300,000 t/yr. Although a further decline in domestic demand is predicted with the pending EPA limits on lead concentrations in gasoline of 130 mg/L (0.5 g/gal) in 1979, foreign production is expected to show some growth with a worldwide total of 500,000 t/yr forecast in 1980.

The first process used for the manufacture of tetraethyllead involved a batch reaction between sodium–lead alloy and ethyl chloride. This basic approach, with improvements and variations, was the predominant commercial route to antiknock compounds from the early 1920s to the early 1960s. In 1960 manufacture of tetramethyllead (TML) was begun by Ethyl Corporation using the combination of methyl chloride and sodium–lead alloy. In 1964 Nalco Chemical Company started up a facility at Freeport, Texas, employing electrochemical synthesis of TML or TEL from Grignard reagent (qv) and lead shot. The plant's production capacity was rated at 15,000 t/yr tetramethyllead or 18,000 t/yr tetraethyllead (63). However, initial production was exclusively for tetramethyllead.

The original Nalco electrolysis system employed ten 30-m^3 (ca 8000-gal) vessels containing a bank of steel tubes with porous diaphragm interior liners (see Table 2). Lead shot placed inside the lined tubes served as a consumable anode, and the inner surface of the steel tubes acted as the cathode (Fig. 1e). An ether solution of Grignard reagent and excess alkyl halide was passed through the tubes at 40–50°C. Heat removal was provided by circulation of a refrigerant on the exterior of the steel tubes. If the Grignard reagent is assumed to dissociate into a cation MgCl$^+$ and an anion R$^-$, the electrode reactions may be written as follows:

anode:
$$4 \text{ R}^- + \text{Pb} \rightarrow \text{R}_4\text{Pb} + 4\ e$$

cathode:
$$4 \text{ MgCl}^+ + 4\ e \rightarrow 2 \text{ Mg} + 2 \text{ MgCl}_2$$

Free magnesium is not actually deposited at the cathode, but rather combines with the excess alkyl chloride present to regenerate Grignard reagent. The overall reaction including the initial feed preparation is:

$$2 \text{ Mg} + 4 \text{ RCl} + \text{Pb} \rightarrow \text{R}_4\text{Pb} + 2 \text{ MgCl}_2$$

Proper control of the concentration of alkyl chloride in the electrolyte is critical to maximizing the yield of tetraalkyllead. If the concentration becomes too high, loss of Grignard reagent is incurred via a Wurtz-Fittig side reaction:

$$R-Cl + RMgCl \rightarrow R-R + MgCl_2$$

On the other hand, low levels of alkyl halide lead to the formation of free magnesium at the cathode, which can readily bridge the gap to the anode shot and short-circuit the cell. Under optimal process conditions, yields of up to 96% on the Grignard reagent are reported (87).

An improved version of the Nalco cell for production of tetraalkyllead has been described in which the lead shot fills virtually the entire vessel, rather than being inside tubes placed in the vessel (88). The cathodes consist of a number of solid rectangular steel blades vertically positioned within the shot bed. These are covered first with a steel wire mesh, which serves to increase the current capacity of the cathode and provide an electrolyte flow path, and then with a fiberglass cloth to electrically insulate the cathode assembly from the lead anode. This cell is claimed to offer a greater productivity than the original Nalco design by virtue of its increased effective electrode area. Also, since individual cathode elements can conveniently be fused, a cathode developing a short circuit can be independently taken off-line without shutting down the entire cell vessel.

A schematic flow sheet of the Nalco tetramethyllead process is shown in Figure 5 (89). The Grignard reagent is prepared from magnesium metal turnings and methyl chloride in an ether solvent, such as the diethyl ether of tetraethylene glycol (see Grignard reaction). The reaction is carried out at 35–40°C and an absolute pressure of ca 200 kPa (2 atm), to give methylmagnesium chloride in greater than 98% yield. This solution, containing excess methyl chloride is fed to the electrolysis cells described previously, where conversion of the Grignard reagent is nearly quantitative. The cell

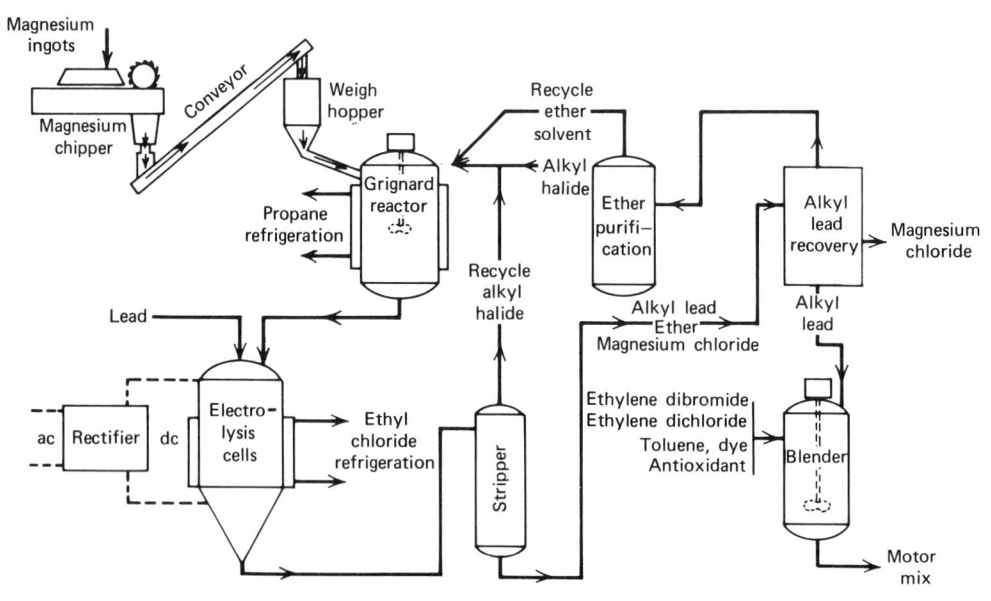

Figure 5. Nalco tetramethyllead process. Courtesy of Gulf Publishing Company (89).

effluent passes to a column where unreacted methyl chloride is removed with an inert gas and reabsorbed in the mixed ethers for recycle. The liquid bottoms from the stripper are extracted with water in a series of stages to separate the tetramethyllead, magnesium chloride, and mixed ethers. The ethers are purified by distillation and recycled. The operation for production of tetraethyllead is similar, with substitution of ethyl chloride for methyl chloride as the make-up alkyl halide.

Since start-up, the number of cells in Nalco's Freeport Plant has been increased from 10 to 14 and the productivity per cell has been raised substantially. Improvements have also been made in the yield based on magnesium, the electrolysis power efficiency and the energy utilization in the chemical processing areas. The Grignard preparation, which was a batch operation in the original commercial facility, has been converted to a continuous process.

Another electrochemical process for tetraethyllead was developed by K. Ziegler in the early 1960s but no commercial applications are known (64). In the Ziegler process molten potassium–aluminumethyl complex ($KAl(C_2H_5)_4$) is electrolyzed to yield TEL by the following equation:

$$4\ KAl(C_2H_5)_4 + Pb \rightarrow Pb(C_2H_5)_4 + 4\ K + 4\ Al(C_2H_5)_3$$

The potassium generated at the cathode forms an amalgam, which reacts with the triethylaluminum to regenerate three fourths of the original potassium–aluminum complex:

$$4\ Al(C_2H_5)_3 \rightarrow 3\ KAl(C_2H_5)_4 + Al$$

Thus the products from the cell are an organic phase consisting essentially of tetraethyllead and a potassium amalgam phase containing a suspension of finely divided aluminum. Yield of TEL is virtually quantitative. Electrolysis is conducted at 100–110°C. The amalgam phase from the cell is processed in a rather complex sequence of chemical and electrochemical steps, the overall result of which is to combine ethylene and hydrogen with the potassium and aluminum in the mercury to produce the $KAl(C_2H_5)_4$ required as feed to the cell.

Kolbe Synthesis of Sebacic Acid. Sebacic acid is an important intermediate used in the manufacture of polyamide resins, and its esters are applied in the manufacture of lubricating oils and plasticizers. Current worldwide demand is estimated to be ca 20,000 t/yr, virtually all of which is produced by alkaline hydrolysis of castor oil (qv). In the search for an alternative route to sebacic acid, independent of the sharp cycling of the castor oil supply, interest has recently focused on the Kolbe synthesis of dimethyl sebacate from monomethyl adipate. Sebacic acid is readily produced from the dimethyl ester by hydrolysis.

The Kolbe reaction, discovered in 1849, involves the anodic dimerization of an n-carbon atom carboxylic acid salt to give a hydrocarbon containing $2n - 2$ carbon atoms plus two moles of carbon dioxide. If instead of a monobasic acid the half-ester of a dibasic acid is employed, the product is the diester of a dibasic acid with $2n - 2$ carbon atoms. Thus anodic oxidation of the sodium salt of monomethyl adipate in methanol solution at a platinum anode gives the following:

$$2\ CH_3OOC(CH_2)_4COONa \rightarrow CH_3OOC(CH_2)_8COOCH_3 + 2\ Na^+ + 2\ CO_2 + 2\ e$$

  sodium monomethyl   dimethyl
  adipate   sebacate

In laboratory studies material yields of 80–90% of theoretical and current efficiencies of 50–90% have have reported (70–72). A comparison of processing conditions used by several different investigators is given in Table 2. Pilot-plant studies have been conducted by Asahi Chemical Industry, BASF, and a group in the USSR.

The USSR process is reported (90) to have been operated on a commercial scale of 1500 t/yr. The electrolyte composition was about 30% monomethyl adipate, 10% sodium salt of monomethyl adipate, 5% water, and the balance methanol. The cell operated without a diaphragm at a 5 mm electrode gap. Material yield in the pilot plant was 83% and current efficiency was 58% (see Table 2).

A schematic diagram of the plant is shown in Figure 6. Monomethyl adipate is dissolved in a mixture of recycle methanol and recycled electrolyte, and neutralized with sodium carbonate to give the desired ratio of salt to free acid. This solution is electrolyzed as described previously. Forced circulation of electrolyte is provided to prevent saponification of the esters in the alkaline zone near the cathode surface. The cell effluent passes to a distillation column for removal of the methanol overhead, and the bottoms are decanted to yield a crude sebacate ester upper layer and a lower layer containing water and unreacted sodium monomethyl adipate. Part of the water is removed from this latter stream by evaporation, prior to its return to the electrolyte make-up tank. Platinum lost from the anode during electrolysis collects as a layer of suspended solids at the decanter interface and is removed for metal recovery. The crude dimethyl sebacate may be distilled to produce a high-purity ester or it may be saponified with sodium hydroxide or sulfuric acid to yield sebacic acid. With an assumed yield of sebacic acid from dimethyl sebacate of 96%, the material usages per kilogram of sebacic acid produced are estimated to be 2.0 kg monomethyl adipate and 9.6 g Pt (2.0 g nonrecoverable).

Asahi Chemical Industry has developed a version of the Kolbe process for dimethyl sebacate with both material and current yields in the range of 90–92% (72). The Asahi electrolysis cell employs parallel-plate electrodes positioned 2–3 mm apart by a polyethylene spacer. The electrolyte is typically a methanol solution containing 20 wt % of monomethyl adipate, neutralized to an extent of 30 wt %, and 20 wt % of dimethyl sebacate. Electrolysis is carried out at 55°C with a flow velocity of electrolyte through the cell of 2.5–4 m/s. Asahi announced the construction of a 0.5 t/d pilot plant for electrochemical production of sebacic acid in 1973 (91).

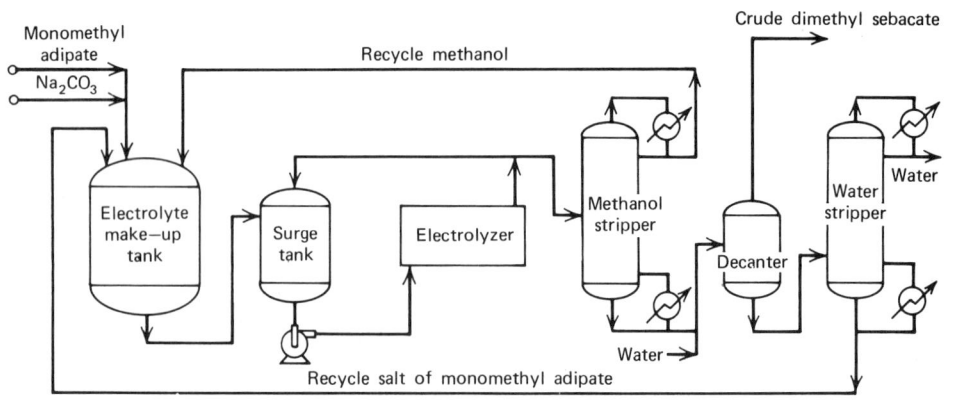

Figure 6. USSR's dimethyl sebacate process.

Manufacture of sebacic acid via the Kolbe synthesis has been studied by investigators at BASF employing a capillary-gap cell (92). This cell consists of a bipolar stack of 10-mm thick graphite disks having an outside diameter of 117 mm and a 30 mm central bore for electrolyte feed. The disks are positioned at a 0.5 mm electrode gap by radial polypropylene strips. A 40 μm platinum foil is bonded to the anode side of each disk and a 1-mm sheet of steel is provided on the cathode face. The electrolyte is a 44 wt % solution of monomethyl adipate, with 5% of the acidity neutralized via sodium methylate solution. Electrolysis is carried out at 42°C with an electrolyte flow velocity of 41 cm/s at the inlet to the annular cell compartment and 11 cm/s at the exit. A yield of dimethyl sebacate of 82% is obtained. Similar performance has been reported by BASF using a vibrating-screen cell (71).

Electrochemical Fluorination. Perfluorinated organic compounds are important industrial surfactants and textile treating agents (see Fluorine compounds, organic). Ammonium perfluorooctanoate has been employed as an emulsifying agent in the polymerization of tetrafluoroethylene (93). Other perfluorocompounds have been used to prepare a fire-fighting foam which has proven effective in smothering petroleum fires (94) (see Plant safety). A polymer formed from dihydroperfluorooctyl acrylate has been reported to provide dirt repellency in a formulation applied to textile fabrics (95) (see Textiles).

The discovery that a number of organic compounds could be fluorinated by electrolysis in a solution of anhydrous hydrogen fluoride was made by Simons around 1940 (96). In the Simons process, fluorination of the dissolved organics takes place at a nickel anode, generally without generation of free fluorine, while hydrogen is evolved at the iron cathode of a diaphragmless cell. Alkali fluorides may be added for improved electrolyte conductivity. Typically, the electrolysis is carried out at low temperatures (0–20°C), low current densities (1–2 A/dm^2) and operating cell voltages of 5–6 V. Generally, a large number of products are formed and the yields and current efficiencies are poor by commercial standards. Electrochemical fluorinations of a wide range of compounds including hydrocarbons, alcohols, ketones, carboxylic acids and amines were studied (97). In recent years extension of this approach has been undertaken by many other investigators and several comprehensive reviews of the field have been published (98–100).

The first commercial production of electrofluorinated organics was carried out by 3M Company at Hastings, Minnesota, in 1951 (101). This facility employed a 10 kA cell and had a capacity of about 115 kg/d of fluorocarbon products. The cell employed alternating planar nickel and iron sheets at 12.5 mm spacing, immersed in a cylindrical steel vessel 180 cm high by 120 cm dia. Cooling was generally provided by coils of tubing within the cell vessel. Vapors consisting of hydrogen, hydrogen fluoride and light fluorocarbons were passed through a condenser to remove the HF for return to the cell. This cell was used to produce a wide variety of fluorochemicals ranging from the simplest fluorocarbon, CF_4, to relatively high molecular weight perfluorinated tertiary amines and organic acids.

The electrofluorinated organics of principal industrial importance are perfluorooctanoic acid, $CF_3(CF_2)_6COOH$, and perfluorooctanesulfonic acid, $CF_3(CF_2)_7SO_3H$. Typical conditions for preparation of the latter compound involve batch electrolysis of a 10 wt % solution of n-octanesulfonyl chloride in anhydrous liquid hydrogen fluoride at 101 kPa (1 atm) and 17–19°C, employing an anode current density of 2 A/dm^2. An insoluble liquid phase is formed during electrolysis and is readily withdrawn from the bottom of the cell; distillation of this material gives perfluorooctanesulfonyl fluoride in 30–35% molar yields. The potassium salt may be formed by hydrolysis of the sulfonyl fluoride with KOH and the corresponding acid is prepared by distilling from a mixture of the salt in 100% H_2SO_4 (102).

An improved electrofluorination system of Phillips Petroleum Company has received extensive study. Instead of the flat nickel anode of the 3M cell, a porous carbon anode is employed, and the organic feedstock is introduced through the pores (74). By judicious control of the organic feed-rate and anode current, varying degrees of fluorination can be achieved within the porous anode. Typically, a mixture of potassium fluoride and hydrogen fluoride (1:2 mole ratio) is employed as the electrolyte at operating temperatures of 60–105°C. Current densities of 6–30 A/dm^2 are used and cell voltages are generally 4–12 V. Current efficiencies to fluorinated products of 80–100% are reported using light hydrocarbon feedstocks.

BIBLIOGRAPHY

1. M. M. Baizer, ed., *Organic Electrochemistry: An Introduction and a Guide,* Marcel Dekker, Inc., New York, 1973.
2. N. L. Weinberg, ed., *Techniques of Electro-Organic Synthesis,* John Wiley & Sons, Inc., New York, 1974.
3. A. P. Tomilov and co-workers, *Electrochemistry of Organic Compounds,* translated by J. Schmorak, Halsted Press, a division of John Wiley & Sons, Inc., New York, 1972.
4. A. N. Frumkin and A. B. Ershler, eds., *Progress in Electrochemistry of Organic Compounds,* translated by D. E. Hayler, translation edited by P. Zuman, Plenum Press, London, Eng., 1971.
5. S. Wawzonek, *Int. J. Methods Synth. Org. Chem.,* (6), 285 (June, 1971).
6. M. Ya. Fioshin, *Sov. Electrochem.* **13**(1), 1 (1977).
7. D. E. Danly, *Hydrocarbon Process.* **48**(6), 159 (June 1969).
8. A. T. Kuhn, *Br. Chem. Eng.* **16**(2–3), 149 (1971).
9. J. L. Fitzjohn, *Chem. Eng. Prog.* **71**(2), 85 (1975).
10. S. Swann and R. Alkire, *Electro-Organic Synthesis Technology, AIChE Symposium Series No. 185,* Vol. 75, 1979, pp. 61–63.
11. U.S. Pat. 3,471,381 (Oct. 7, 1969), H. Suter and co-workers (to Badische Anilin- und Soda-Fabrik).
12. K. B. Keating, "Electrode Selection for Electrochemical Processes," paper presented at 78th National AIChE meeting, Salt Lake City, Utah, Aug. 1974.
13. G. Z. Kiryakov, N. F. Razina, and Y. D. Dunsev, *Chemical Institute of the Academy of Sciences of the Kazakh S.S.R.* **6,** 4 (1960); W. Ger. Pat. 2,036,881 (July 24, 1970), (to Ionics, Inc.).
14. U. S. Pat. 3,616,321 (Oct. 26, 1971), A. Verheyden and J. Walravens (to UCB S.A.).
15. U.S. Pat. 3,897,318 (July 29, 1975), W. A. Heckle and D. L. Sadler (to Monsanto Co.).
16. F. Beck and H. Guthke, *Chem. Ing. Technol.* **41,** 943 (1969).
17. D. W. Wabner and co-workers, *Z. Naturforsch.* **31b,** 39 (1976).
18. L. M. Yakimenko, *Sov. Chem. Ind.,* (3) 174 (1973).
19. J.-W. Kuhn-von Burgsdorff, *Chem. Ing. Technol.* **49,** 294 (1977).
20. U.S. Pat. 3,846,273 (Nov. 5, 1974), G. Bianchi and co-workers (to Electronor Corp.).
21. Can. Pat. 989,773 (May 25, 1976), O. de Nora and co-workers (to Electronor Corp.).
22. J. A. M. Le Duc, "Industrial Electrodes 1976," *Electrochemical Industry and Technology,* Vol. 3, International Electrochemical Institute, Millburn, N.J., 1976.
23. C. K. Mann in A. J. Bard, ed., *Advances in Electroanalytical Chemistry,* Vol. 3, Marcel Dekker, New York, 1969, p. 57.
24. A. T. Kuhn, ed., *Industrial Electrochemical Processes,* Elsevier Scientific Publishing Co., New York, 1971, pp. 590–591.
25. G. Richter, *Int. Chem. Eng.* **17,** 57 (1977); *Chem. Ing. Technol.* **47,** 909 (1975).
26. W. Pusch, *Chem. Ing. Technol.* **47,** 914 (1975).
27. U.S. Pat. 3,992,269 (Nov. 16, 1976), T. T. Sugano and co-workers (to Diamond Shamrock Corp.).
28. *Chem. Eng.* **84** (Feb. 18, 1974); *Chem. Week,* 33 (Mar. 24, 1976); *Chem. Week,* 53 (Mar. 15, 1978).

29. A. T. Kuhn and R. W. Houghton in S. K. Rangarajan, ed., *Topics in Pure and Applied Electrochemistry*, SAEST, Karaikudi, India, 1975, pp. 133–166.
30. R. W. Houghton and A. T. Kuhn, *J. Appl. Electrochem.* **4,** 173 (1974).
31. A. P. Tomilov and M. Ya. Fioshin, *Br. Chem. Eng.* **16,** 154 (1971).
32. P. Gallone, *Electrochim. Acta* **22,** 913 (1977).
33. *Hydrocarbon Process.* **56**(11), 127 (1977).
34. Belg. Pat. 699,284 (May 31, 1967), D. E. Danly and R. W. McWhorter (to Monsanto Co.).
35. U.S. Pat. 3,119,760 (Jan. 28, 1964), R. W. Foreman and F. Veatch (to Standard Oil Co. of Ohio).
36. U.S. Pat. 3,657,099 (Apr. 18, 1972), M. Seko and co-workers (to Asahi Kasei).
37. K. B. Keating and V. D. Sutlic, in ref. 10, pp. 76–88.
38. U.S. Pat. 3,342,717 (Sept. 19, 1967), J. A. M. Le Duc (to Pullman, Inc.).
39. U.S. Pat. 3,575,839 (Apr. 20, 1971), M. A. Melnikov-Eikhenvald and co-workers.
40. F. Wenisch and co-workers, in ref. 10, pp. 14–18.
41. P. M. Robertson and N. Ibl, *J. Appl. Electrochem.* **7,** 323 (1977).
42. S. Garesan and co-workers, *Chem. Age India* **13,** 346 (1964).
43. R. E. W. Janssen and R. J. Marshall, *The Chemical Engineer,* 769 (Nov.–Dec. 1976).
44. F. Goodridge, C. J. H. King, and A. R. Wright, *Electrochim. Acta* **22,** 347 (1977).
45. U.S. Pat. 3,391,067 (July 2, 1968), D. A. Braithwaite (to Nalco Chemical Co.).
46. U.S. Pat. 3,479,274 (Nov. 18, 1969), L. L. Bott (to Nalco Chemical Co.).
47. U.S. Pat. 3,457,152 (July 22, 1969), J. N. Maloney, Jr., C. R. Campbell, and R. Johnson (to Monsanto Co.).
48. J. R. Backhurst and co-workers, *J. Electrochem. Soc.* **116,** 1600 (1969).
49. U.S. Pat. 4,019,968 (Apr. 26, 1977), P. M. Spaziante and C. Traini (to Parel S.A.).
50. R. B. MacMullin, *Electrochem. Tech.* **2,** 106 (Mar.–Apr. 1964).
51. J. Newman, *Ind. Eng. Chem.* **60,** 12 (Apr. 1968).
52. C. W. Tobias and R. G. Hickman, *Z. Phys. Chem.* **229,** 145 (1965).
53. G. A. Ashworth and R. E. W. Janssen, *Electrochim. Acta* **22,** 1295 (1977).
54. R. Dworak and H. Wendt, *Ber. Bunsenges. Phys. Chem.* **80,** 77 (1976).
55. F. B. Leitz and L. Marincic, *J. Appl. Electrochem.* **7,** 473 (1977).
56. M. Eisenberg, C. W. Tobias, and C. R. Wilke, *J. Electrochem. Soc.* **101,** 306 (1954).
57. A. R. Despic, "Some New Methods of Intensifying Mass Transfer in Industrial Electrolysis," paper presented at the Chicago meeting of the Electrochemical Society, May 1973.
58. N. Ibl, *Chem. Ing. Technol.* **43,** 202 (1971).
59. G. Kreysa, S. Pionteck, and E. Heitz, *J. Appl. Electrochem.* **5,** 305 (1975).
60. M. Fleischman and R. E. W. Janssen, *Chem. Ing. Technol.* **49,** 283 (1977).
61. P. LeGoff and co-workers, *Ind. Eng. Chem.* **61**(10), 8 (1969).
62. R. L. Taylor, *Chem. Met. Eng.* **44,** 588 (1937).
63. L. L. Bott, *Hydrocarbon Process.* **44**(1), 115 (1965).
64. U.S. Pat. 3,372,097 (Mar. 5, 1968), K. Ziegler and H. Lehmkuhl (to Karl Ziegler).
65. J. H. Prescott, *Chem. Eng.,* 238 (Nov. 8, 1965).
66. M. Seko, *Chem. Econ. Eng. Rev.* **7,** 20 (Oct. 1975).
67. U. S. Pat. 3,630,861 (Dec. 28, 1971), J. Bizot, G. Bourat, and D. Michelet (to Rhone Poulenc S.A.).
68. C. Van Eygen and co-workers, *Chem. Ind.* **104**(1), 71 (1971).
69. W. V. Childs and H. C. Walters, in ref. 10, pp. 19–25; U.S. Pat. 3,689,382 (Sep. 5, 1972), H. M. Fox and F. N. Ruehlen (to Phillips Petroleum Co.).
70. E. P. Kovsman, G. N. Freydlin, and Yu. M. Tyurin, *Sov. Chem. Ind.* (1), 13 (Jan. 1973).
71. U.S. Pat. 3,652,430 (Mar. 28, 1972), F. Beck and co-workers (to Badische Anilin- und Soda-Fabrik A.G.).
72. U.S. Pat. 3,896,011 (July 22, 1975), T. Isoya, R. Kakuta, and C. Kawamura (to Asahi Kasei).
73. Brit. Pat. 686,678 (Jan. 28, 1953); Brit. Pat. 709,588 (May 26, 1954); Brit. Pat. 763,673 (Dec. 12, 1956), (to Minnesota Mining & Manufacturing Co.); U.S. Pat. 2,717,871 (Sep. 13, 1955), H. M. Scholberg and H. G. Brice (to Minnesota Mining & Manufacturing Co.).
74. U.S. Pat. 3,511,760 (May 12, 1970), H. M. Fox and F. N. Ruehlen (to Phillips Petroleum Co.).
75. Brit. Pat. 1,090,006 (Nov. 8, 1967), W. Kronig and J. Grolig (to Farben Fabriken Bayer A.G.).
76. P. C. Condit, *Ind. Eng. Chem.* **48,** 1252 (1956).
77. U.S. Pat. 3,471,381 (Oct. 7, 1969), H. Suter and co-workers (to Badische Anilin- und Soda-Fabrik A.G.).

78. N. Shuster, A. D. Bakinsky, and K. J. O'Leary, paper presented at the Electrochemical Society meeting, Washington, D.C., May 2–7, 1976.
79. U.S. Pat. 3,899,401 (Aug. 12, 1975), H. Nohe and F. Beck (to Badische Anilin- und Soda-Fabrik A.G.).
80. L. H. Cutler, in ref. 10, pp. 103–107.
81. K. H. Simmrock, in ref. 10, pp. 89–102.
82. V. D. Luedke in J. J. McKetta, ed., *Encyclopedia of Chemical Processing and Design,* Vol. 2, Marcel Dekker, Inc., New York, 1977, pp. 146–162.
83. M. M. Baizer, *J. Electrochem. Soc.* 111, 215 (1964).
84. U.S. Pats. 3,830,712 (Aug. 20, 1974), C. R. Campbell and D. E. Danly; 3,897,318 (July 29, 1975), W. A. Heckle and D. L. Sadler; 3,898,140 (Aug. 5, 1975), J. H. Lester, Jr., and J. S. Stewart (to Monsanto Co.).
85. *Hydrocarbon Process.* 56, 126 (Nov., 1977).
86. *Chem. Week,* 21 (Mar. 29, 1978).
87. U.S. Pat. 3,380,900 (Apr. 30, 1968), D. G. Braithwaite and L. L. Bott (to Nalco Chemical Co.).
88. U.S. Pat. 3,573,178 (Mar. 30, 1971), G. E. Blackmar (to Nalco Chemical Co.).
89. L. L. Bott, *Hydrocarbon Process,* 44(1), 115 (1965).
90. M. M. Baizer, private communication.
91. *Chem. Econ. Eng. Rev.,* 58 (May 1973).
92. U.S. Pat. 3,787,299 (Jan. 22, 1974), F. Beck, J. Haufe, and H. Nohe (to Badische Anilin- und Soda-Fabrik A.G.).
93. Brit. Pat. 1,071,992 (June 14, 1967), D. L. Schindler (to E. I. du Pont de Nemours & Co., Inc.).
94. U.S. Pat. 3,258,423 (June 28, 1966), R. L. Tuve and E. J. Jablonski (to Secretary of U.S. Navy).
95. E. L. Grajeck and W. H. Peterson, *Text. Res. J.* 32, 320 (1962).
96. J. H. Simons in J. H. Simons, ed., *Fluorine Chemistry,* Vol. I, Academic Press, Inc., New York, 1950, p. 414.
97. J. H. Simons and co-workers, *J. Electrochem. Soc.* 95(2), 47 (1949).
98. J. Burdon and J. C. Tatlow in M. Stacey, J. C. Tatlow, and A. G. Sharpe, eds., *Advances in Fluorine Chemistry,* Vol. I, Butterworth's Scientific Publications, London, Eng., 1690, p. 129.
99. S. Nagase in P. Tarrant ed., *Fluorine Chemistry Reviews,* Vol. I, Edward Arnold Ltd., London Eng.; Marcel Dekker Inc., New York, 1967, p. 77.
100. A. J. Rudge in A. T. Kuhn, ed., *Industrial Electrochemical Processes,* Elsevier Publishing Co., London, Eng., 1971, p. 71.
101. J. H. Simons and T. J. Brice in J. H. Simons, ed., *Fluorine Chemistry,* Vol. II, Academic Press, Inc., New York, 1954, p. 340.
102. U.S. Pat. 2,732,398 (Jan. 24, 1956), T. J. Brice and P. W. Trott (to 3M Co.).

DONALD DANLY
Monsanto Chemical Intermediates Co.

ELECTROCHEMISTRY. See Electrochemical processing.

ELECTRODECANTATION

Electrodecantation (also called electrogravitational separation and electrophoretic convection) is based on a stratification phenomenon that may take place when colloidal dispersions are subjected to an electrical field between vertical membranes permeable to the electrical current and impermeable to the colloid. Semicolloids and electrolytes also stratify in a similar manner on selectively permeable membranes; however, the basic phenomenon differs somewhat in the mechanism of the formation of the stratifying layers on the membranes. With colloids, the charged particles migrate under the influence of the electrical field toward one of the poles (see Electrodialysis) and are retained and accumulated on the membranes. Under certain conditions, the thin concentrated layer on the membrane surface moves up or down between the surrounding liquid boundary and membrane interface, according to its relative density; at the opposite membrane the movement of the dilute layer takes place in the other direction. As shown in Figure 1, the stratified layers formed can be separated by decantation (1–2) (see also Gravity separation).

This phenomenon was first observed by Pauli in experiments on electrodialysis and had been used by him and his co-workers for studying the structure and composition of a great variety of colloidal dispersions including biological liquids and proteins (3–5). The separation into stratified layers was found to be independent of the particle size. The behavior of the material retained on the membrane depends upon the migration velocity of the particles in the electrical field vel_e, and the velocity of gravitational movement of the enriched layer vel_g. If vel_g is greater than vel_e, stratification will take place continuously, but if vel_e is greater, a deposit will accumulate on the membranes. If this deposit were allowed to remain in contact with the membrane, the particles would be drawn together and lose liquid by electroosmosis, leading to coagulation. This could be prevented by reversing the polarity of the electrodes periodically, when the deposit moves away from the membrane. Being thus freed from the frictional resistance, the deposit immediately stratifies. Colloids in which the particles have only a slight density difference from the continuous phase can be readily stratified by this method. In order to achieve continuous removal of the stratified layers, the potential gradient and the period for current reversal are of importance. Working conditions depend on the concentration of the colloid to be separated, the relative density of the disperse phase, the migration velocity, and the flow property of the layer retained by

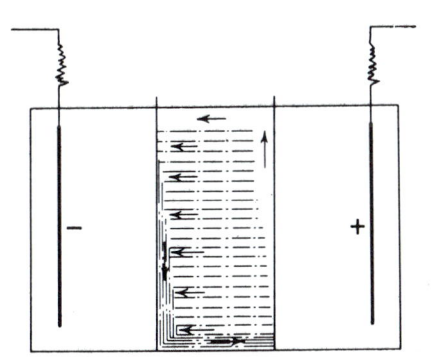

Figure 1. Mechanism of electrodecantation.

the membrane (6). The greater the difference in density between the colloid and the disperse phase, the faster a separation can be accomplished. The interval between reversal may vary from a few minutes to half an hour. For colloidal particles, neutral membranes are satisfactory, preferably with a smooth surface such as regenerated cellulose from cellulose nitrate or cellulose xanthate (cellophane).

Proteins and biological sera have been separated and also fractionated at isoelectric points of the constituents, such as from nonmigrating globulins (7). Tetanus and diphtheria toxins were freed from the protein fraction at the isoelectric points of the toxins (8). The electrogravitational effects for separating electrolytes on permselective ion-exchange membranes, which are used also for separating hemicolloids and colloidal electrolytes, have been investigated (9); ligninsulfonic acid has been separated by stratification (10).

For preparatory laboratory use, a simple form of equipment was suggested by Pauli (5). A multiple-membrane laboratory unit made with transparent rings is shown in Figure 2. This unit has improved efficiency because the rate of separation is directly proportionate to the membrane surface area. The two electrodes A and B are insulated from the frame F. The membranes C separate the electrode compartments from the working cells and extend over the whole cross section, whereas the membranes G in the working cells terminate several millimeters above the bottom and below the top of the cell. In this manner, the layers formed and stratified on all the membranes can be withdrawn through the center manifold H. The unit is gravity-fed continuously from a container D; the rate being controlled by a suitable valve I. The liquid from the container D is distributed to the spaces between the membranes through manifolds

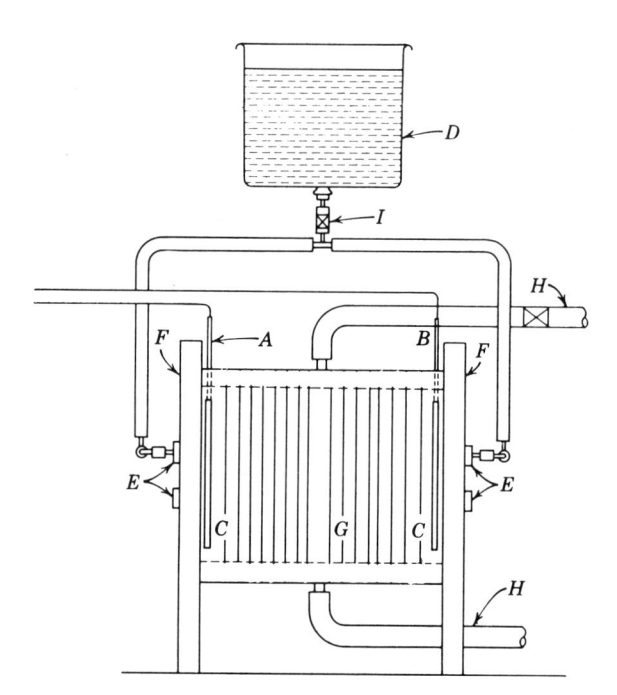

Figure 2. Diagrammatic view of a laboratory electrodecantation cell. Courtesy of Brosites Machine Co.

E (1). The colloid is fed into the cells and is separated in layers and removed continuously. To eliminate the heating effect during the separating of biological liquids, Polson (11) introduced intermediate cells, between the membrane compartments, through which a cooling solution is circulated. Through manifolds E, the membranes, as well as the rings, have interconnecting perforations, punctured to connect with the cells. The colloid is distributed evenly and fed into the cells, separated in layers, and removed continuously. An innovation uses multimembrane cells which are separated from each other by cells filled with circulating buffer solutions in cooling tubes (11). These units, working with controlled pH and temperature, have been widely used for studying separation from biological liquids.

Figure 3 is a schematic reproduction of Polson's unit (11) which can be used vertically or on an angle. The cells with multiple membranes M_1, M_2, and M_3 are filled with the solution to be separated. The cells C that hold the buffer solution are placed between the working cells. The electrode cells can have vertical or angular arrangements of electrodes. Usually carbon electrodes are used, and the electrode cells are filled with precooled buffer. The buffer solution is circulated within the C cells. The solution to be separated in the multimembrane cells is usually precooled.

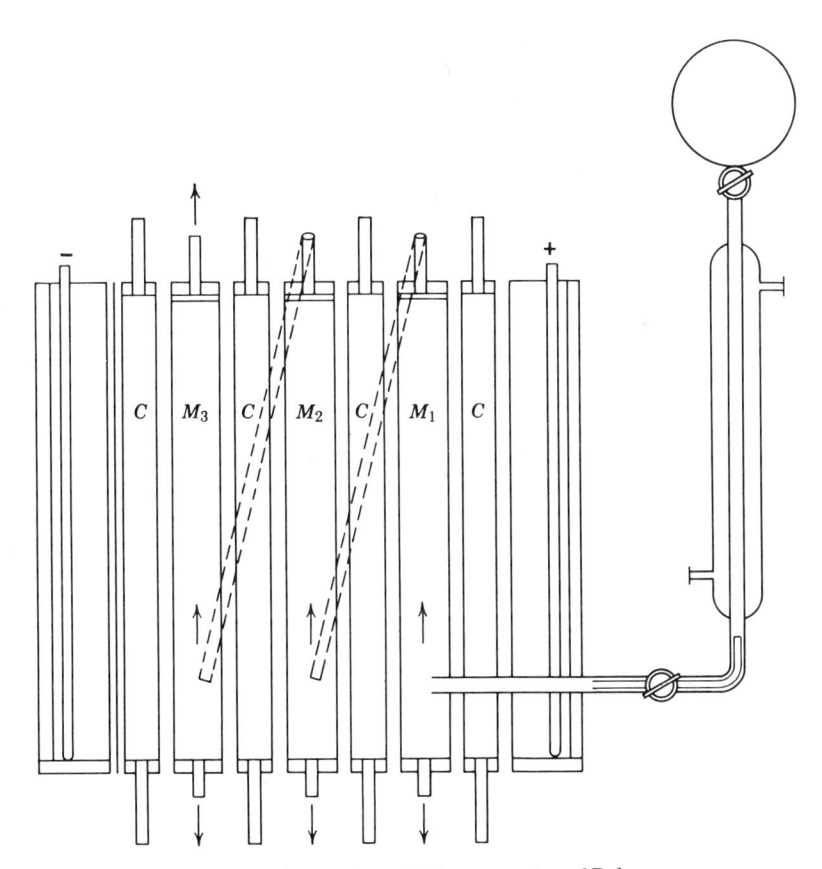

Figure 3. Electrodecantation apparatus of Polson.

Research and Development

Electrodecantation of biological systems has been the subject of frequent publications. It has been used for isolation of immunoglobulin antibody activity in relation to the hepatitis-associated antigen (12). The active immunoglobulin hepatitis-associated antigen (HAA) was separated by an electrodecantation method (13). Various isoelectric points were obtained by means of convectional electrophoresis on modified electrodecantation fractions of human gamma globulin (13) (see Analytical methods).

The sedimentation of albumin was detected via electrodecantation of serum proteins (14). The sedimentation occurred in discontinuous experiments through the influence of gravity as well as electricity. In continuous single-stage experiments, the separation of the globulin from other serum proteins could be obtained from solution only 50% of the time owing to the homogeneous distribution of globulin throughout the solution. Only the upper half of the cell had solution free from albumin and other proteins. In experiments with continuously operated triangular cells, the process was optimized to give globulin of 97.5% purity in a yield of 80% serum flow with a rate of 0.5 L/h in a block composed of 40 cells.

Enzyme Purification by Electrodecantation. Electrodecantation has been applied to the initial fractionation from crude material of pig kidney β-D-glucosidase and sheep testicular β-N-acetylglucosaminidase (15). Electrodecantations of pig kidney extract and sheep testicular extract were carried out at pH 4.9 and pH 6.3, respectively. A six to tenfold increase in specific activity could be obtained with good recoveries after a single cycle of electrodecantation. The technique has also been used to purify further an extracellular *Bacillus subtilis* protease preparation. Attempts to use electrodecantation for the concentration of very dilute enzyme solutions resulted in considerable loss of activity. The limitations and potential use of the technique in laboratory-scale enzyme preparation are discussed in ref. 15.

Poly(ethylene glycol) (PEG, mol wt 6000) has been used on an industrial scale for isolation of some important compounds of human plasma (16). Unlike the majority of protein fractionation agents, PEG separates protein compounds from a natural mixture by an exclusion mechanism. The mechanism involves the insolubilization of individual compounds by so-called superconcentration of the protein compounds in the inter-PEG spaces by a displacement mechanism (see Blood fractionation).

Concentration of Macromolecules by Electrophoresis Sedimentation. An electrophoresis-sedimentation apparatus used for the concentration of macromolecules differs from electrodecantation instruments in that it employs horizontal electrodes and semipermeable membranes of reduced areas in order to minimize the adsorption of macromolecules to the membranes (17). Bovine serum albumin (bovine and human hemoglobulin), γ-globulin, concanavalin A, very low density-, low density-, and high density-serum lipoproteins, and cytoplasmic RNA were concentrated ten to fiftyfold in relatively short times with overall recoveries of 86–96%.

Industrial Use

The first commercial application of electrodecantation has been in the concentration of rubber latex. The basic principles used have been adopted for most latex processes, eg, multimembranes, the continuous process, and the periodic reversal of polarity (18). Improvements in the placement of the membranes have been reported (19) (see Rubber, natural).

The first commercial application and design was identical to the design in Figure 2, in which 100 membranes were placed vertically, 3–10 mm apart. Using this process, the Dunlop Rubber Company in Malaya has produced hundreds of thousands of liters of latex for many years (20). The power consumption depends on the conductivity of the solution, and the number of membranes between the electrodes.

In a process comparable to that of Dunlop, an aqueous polytetrafluoroethylene dispersion of 4 wt % (including a 0.2 wt % sulfonic acid detergent, primarily sodium laurylsulfonate) was concentrated to 55–74 wt % polymer through electrodecantation (21).

Polytetrafluoroethylene dispersions of 3–30 wt % solids have concentrated to 50–60 wt % using graphite electrodes at 50 A/m^2 maximum current density to prevent formation of deposits on the membranes (22). The period of intermittent reversal of polarity (4–60 min) depends on the distance between the membranes (see Fluorine compounds, organic).

Electrodecantation has been used for concentrating electrolytic coating baths (23). After rinsing electrocoated surfaces, the residual rinsing fluid can be concentrated by use of electrodecantation with permselective membranes.

BIBLIOGRAPHY

"Electrodecantation" in *ECT* 1st ed., Vol. 5, pp. 549–551, by P. Stamberger, Technical Consultant; "Electrodecantation" in *ECT* 2nd ed., Vol. 7, pp. 841–846, by P. Stamberger, Technical Consultant.

1. F. Blank and E. Valko, *Biochem. Z.* **195,** 220 (1928).
2. E. J. W. Verwey and K. Kruyt, *Z. Phys. Chem. A* **167,** 149 (1933).
3. W. Pauli, *Anz. Akad. Wiss. Wien. Math Naturwiss. Kl.* **60,** 10 (1923).
4. W. Pauli, *Trans. Faraday Society* **31,** 19 (1933).
5. W. Pauli, *Helv. Chim. Acta* **25,** 155 (1942).
6. P. Stamberger, *J. Colloid Sci.* **1,** 93 (1946).
7. H. Gutfreund, *Biochem. J.* **37,** 186 (1943).
8. J. F. Largier, *Biochem. Biophysics Acta* **16,** 291 (1955); **21,** 433 (1956); *J. Immunologie* **79,** 181 (1957).
9. V. J. Frilette, *J. Phys. Chem.* **61,** 168 (1957).
10. W. O. Jean and A. J. Goring, *Tappi* **47,** 165 (1946).
11. U.S. Pat. 2,801,962 (Aug. 6, 1957), A. Polson; Brit. Pat. 726,183 (Mar. 16, 1955), A. Polson.
12. V. A. Burlev, I. F. Kiryukhin, and T. V. Golosova, *Lab. Delo* (10), 588 (1974); *Chem. Abstr.* **82,** 71401 (1975).
13. I. F. Kiryukhin and co-workers, *Vopr. Med. Khim.* **15c,** 610 (1969); *Chem. Abstr.* **72,** 74825 (1970).
14. F. Lappe, *Hoppe Seyler's Z. Physiol Chem.* (10), 353 (1975).
15. B. G. Winchester, H. Caffrey, and D. Robinson, *Biochem. J.* **121,** 161 (1971).
16. A. Polson and C. Huiz-Bravo, *Vox. Sang.* **23,** 107 (1972); *Chem. Abstr.* **77,** 85129 (1972).
17. I. Posner, *Anal. Biochem.* **70,** 187 (1976).
18. Brit. Pats. 505,732; 505,753 (Sept. 12, 1938), P. Stamberger and E. Schmidt (to Dunlop Plantation Ltd.); U.S. Pat. 2,247,065 (June 24, 1951), W. Pauli and P. Stamberger.
19. Brit. Pats. 541,304; 541,305 (Nov. 21, 1941), A. E. Murphy, J. F. Paton, and J. Ansell (to Dunlop Rubber Co., Ltd.); U.S. Pat. 2,331,494 (Oct. 12, 1943), A. E. Murphy, J. F. Paton, and J. Ansell (to Dunlop Rubber Co., Ltd.).
20. A. E. Murphy, *Trans. Inst. Rubber Ind.* **18,** 173 (1971).
21. Brit. Pat. 642,025 (Aug. 23, 1950), (to E. I. du Pont de Nemours & Co., Inc.).
22. A. Ferse, *Paste und Kautschuk* **18,** 35 (1971).
23. Ger. Offen. 2,136,773 (Mar. 9, 1972), A. B. Coke.

General Reference

M. Bier, *Electrophoresis*, Vol. I, Academic Press, Inc., New York, 1958, pp. 270–310.

PAUL STAMBERGER
Crusader Chemical Co., Inc.

ELECTRODIALYSIS

Electrodialysis is a process for moving ions across a membrane from one solution to another under the influence of a direct electric current (see also Dialysis). Classically the process was carried out in three-compartment electrolytic cells in which the compartments were separated from each other by essentially nonselective membranes. The end-compartments contained electrodes. In 1940 a multicompartment electrodialysis process using ion-selective membranes (Fig. 1) was suggested in which membranes selective to anions A alternate with membranes selective to cations C (1). When a d-c potential is applied, cations M^+ tend to move toward the negatively charged cathode. They are able to permeate the cation-selective membranes but not the anion-selective membranes if the latter are perfectly selective. Similarly, anions X^- tend to move toward the positively charged anode. They are able to permeate the anion-selective membranes but not the cation-selective membranes (again if the latter are perfectly selective). As a result, the odd-numbered compartments in the figure become depleted in electrolyte and the even-numbered compartments enriched. The membranes available in 1940 were not commercially practical; those that had high selectivity also had high electrical resistance; those that had low electrical resistance also had low selectivity. None was sufficiently mechanically strong or chemically stable for practical purposes.

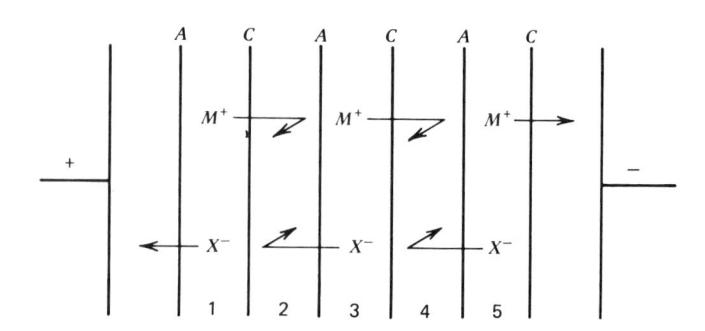

Figure 1. Principle of multicompartment electrodialysis (see text).

Membranes

In 1950 ion-selective membranes having high selectivity, low electrical resistance, good mechanical strength, and good chemical stability were described (2). These were essentially insoluble, synthetic, polymeric organic ion-exchange resins in sheet form. The chemical structures of typical modern membranes of this type are shown schematically in Figure 2. The membranes illustrated consist in the case of the cation-exchange resin (cation-selective membrane) of polystyrene having negatively charged sulfonate groups chemically bonded to most of the phenyl groups in the polystyrene. The negative charges of the sulfonate groups are electrically balanced by positively charged cations (counterions). Sulfonated polystyrene swells greatly in water. The amount of swelling is controlled typically by including cross-linking agents in the polymer (shown as divinylbenzene in Fig. 2); by incorporating electrically neutral polymers; or by having substituents (such as fluorine) on the polymer that decrease the affinity for water. The positively charged counterions are appreciably dissociated from the bound, negatively charged groups into the imbibed water in which they are mobile. They may be exchanged for other cations from an ambient solution, maintaining the electrical neutrality of the membrane. The high concentration of counterions in ion-exchange resins is responsible for the low electrical resistance of the membrane. (The concentration of counterions is typically >1 mequiv/cm^3 of ion-exchange membrane.) The high concentration of bound, negatively charged groups tends to exclude mobile, negatively charged ions (co-ions) from an ambient solution and is responsible for the high ion selectivity of the membranes (see Ion exchange; Ion selective electrodes; Membrane technology).

The anion-selective membranes represented in Figure 2 consist of cross-linked polystyrene having positively-charged quaternary ammonium groups chemically bonded to most of the phenyl groups in the polystyrene. In this case the counterions are negatively charged and are the principal carriers of the electric current.

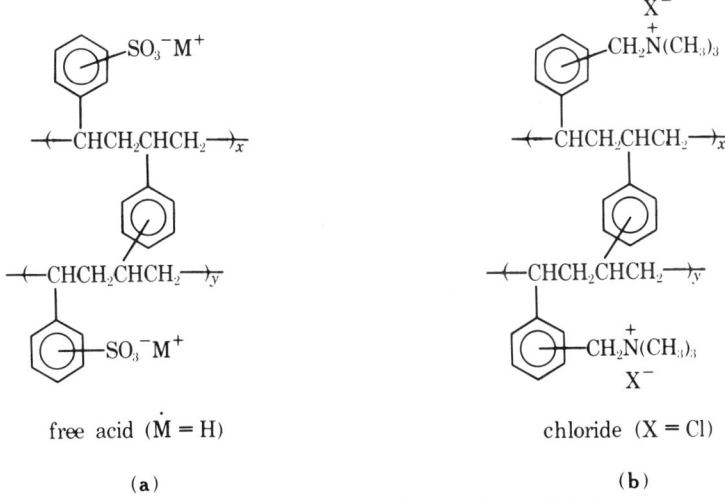

Figure 2. Schematic representation of ion-exchange resins. (a) Cation-exchange resin; (b) anion-exchange resin.

Commercially available membranes are usually reinforced with woven, synthetic fabrics to improve the mechanical properties. Several hundred thousand square meters of ion-selective membranes per year are now produced, the mechanical and electrochemical properties varied by the manufacturers to suit the end applications. The electrochemical properties of most importance for electrodialysis are (1) the electrical resistance per unit area of membrane, (2) the ion transport number (related to current efficiency), (3) the electrical water transport (related to process efficiency), and (4) the back-diffusion (also related to process efficiency).

Commercial ion-selective membranes have thicknesses of ca 0.15–0.6 mm and electrical resistances of ca 3–20 Ωcm^2 at RT when in equilibrium with 0.5 N sodium chloride. The electrical resistances are somewhat higher in more dilute solutions since co-ions are more effectively excluded from the membrane by the ion-selective resin. The electrical resistance decreases with increasing temperature at a rate of ca $-1.8\%/°C$. The electrical resistance of an electrodialysis apparatus depends in large part on the electrical resistances of the membranes in the case of electrolyte solutions having concentrations in excess of about 0.1 N. For more dilute solutions the resistance of the apparatus tends to be dominated by the resistance contributed by the solution.

The ion transport number is defined as the fraction of current carried through the membrane by counterions. If the concentration of fixed charges in the membrane is high compared to the concentration of the ambient solution, then the mobile ions in the ion-selective membrane will be mostly counterions and co-ions will be effectively excluded. The ion transport number will then approach 1. Commercial membranes have ion transport numbers in dilute solutions of ca 0.85–0.95. The relationship between ion transport number and current efficiency is shown in Figure 3 where $\bar{t}_-^A$ is the fraction of current carried by the counterions (anions) through the anion-selective membrane and $\bar{t}_+^C$ is the fraction of current carried by the counterions (cations) through the cation-selective membrane. The remainder of the current $(1 - \bar{t}_-^A) = \bar{t}_+^A$ in the case of the anion-selective membranes and $(1 - \bar{t}_+^C) = \bar{t}_-^C$ (in the case of the cation-selective membranes) is carried by co-ions and constitutes an electrical inefficiency. The net transport of electrolyte is then given by:

$$\bar{t}_-^A - (1 - \bar{t}_+^C) = \bar{t}_-^A + \bar{t}_+^C - 1 = \bar{t}_-^A - \bar{t}_-^C$$

The electrical water transport is defined as water that accompanies the electrical transport of ions through the membranes. Owing to the high concentration of counterions in ion-selective membranes, the transport of ions and water are closely coupled,

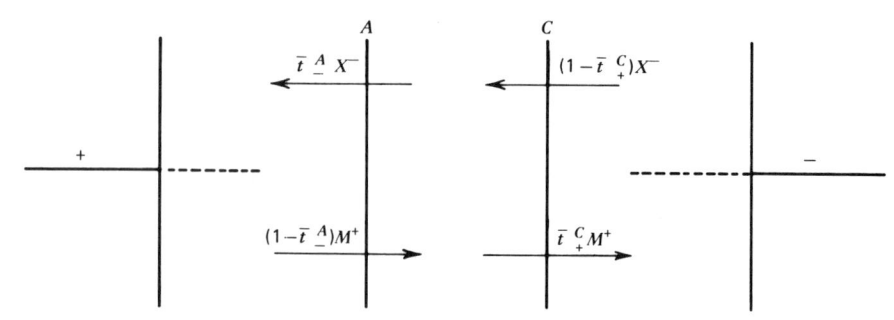

Figure 3. Relationship between current efficiency and ion transport number.

ie, the internal solution tends to move in a pistonlike flow. The electrical water transport is ca 100–200 cm^3 per equivalent of ions transferred, toward the low end of the range for anion-selective membranes and toward the high end for cation-selective membranes. In the case of dilute solutions such water transport is not of significant engineering importance. However, in more concentrated solutions, such as seawater, the water transport can be a significant fraction of the volume of solution electrodialyzed and, therefore, constitute a process inefficiency. Generally, membranes with low water contents and high concentrations of fixed, charged groups have low electrical water transport. Such membranes also tend to have high ion transport numbers and are therefore preferred for concentrated electrolytes.

Back-diffusion is the transport of co-ions (and an equivalent number of counterions) under the influence of the concentration gradients developed between enriched and depleted compartments during electrodialysis. Such back-diffusion counteracts the electrical transport of ions and hence causes decrease in process efficiency. Back-diffusion depends on the concentration difference across the membrane and the selectivity of the membrane; the greater the concentration difference and the lower the selectivity the greater will be the back-diffusion. Designers of electrodialysis apparatus, therefore, try to minimize concentration differences across membranes and utilize highly selective membranes. Back-diffusion between sodium chloride solutions of zero and one normal is generally less than ca 2×10^{-6} mequiv/(s·cm^2).

Apparatus

Electrodialysis apparatus is fundamentally an array of alternating anion-selective and cation-selective membranes terminated by electrodes. The membranes are separated from each other by gaskets which form fluid compartments. Compartments that have anion-selective membranes on the side facing the (positively charged) anode are electrolyte-depletion compartments (also called demineralizing, diluting, diluate, or dilute compartments). The remaining compartments are electrolyte-enrichment compartments (also called concentrating, concentrate, or brine compartments). The enrichment and depletion compartments also alternate through the array. Holes in the gaskets and membranes register with each other to provide two pairs of internal hydraulic manifolds to carry fluid into and out of the compartments, one pair communicating with the depletion compartments and the other with the enrichment compartments. Much effort has been expended on the design of the entrance and exit channels from the manifolds to the compartments to prevent unwanted cross-leak of fluid intended for one class of compartment into the other class. This effort has been made increasingly difficult by the trend to thinner membranes and gaskets (the latter determining membrane-spacing and the thickness of the fluid compartments) to reduce energy consumption. A contiguous group of two membranes and their associated two fluid compartments is called a cell pair. A group of cell pairs and their associated end electrodes are called a stack or a pack. Generally one hundred to several hundred cell pairs are arranged in a single stack, the choice being made on the basis of electrodialysis capacity desired, the uniformity of flow distribution achieved among the several compartments of the same class in a stack and the maximum total direct current potential desired. One or more stacks may be arranged in a press, designed to compress the membranes and gaskets against the force of fluid flowing through the compartments thereby preventing fluid leaks to the outside and internal cross-leaks between compartments. For small presses such compression is usually provided by tie-rods; for larger presses hydraulic rams are frequently used.

As mentioned previously, commercial membranes typically have thicknesses of ca 0.15–0.6 mm. The compartments between the membranes typically have thicknesses of ca 0.5–2 mm. The thickness of a cell pair is therefore ca 1.3–5.2 mm, commonly about 3.2 mm. One hundred cell pairs of the latter wire have a combined thickness of ca 320 mm. The effective area of a cell pair for current conduction is generally ca 0.2–2 m^2.

In concentrated electrolytes the electric current applied to a stack is limited by economic considerations, the higher the current I, the greater the power consumption W in accordance with: $W = I^2 R_s$, where R_s is the electrical resistance of the stack. In relatively dilute electrolytes the electric current that can be applied is limited by the ability of ions to diffuse to the membranes. This is illustrated in Figure 4 for the case of an anion-selective membrane. When direct electric current is passed, a fraction ($\bar{t}_-^A$, generally 0.85–0.95) is carried by anions passing out of the membrane-solution interface region and through the membrane. In the bulk solution a fraction (t_-^s) of the current is carried by anions passing into the interfacial region. Generally t_-^s is significantly less than $\bar{t}_-^A$. For example, in the case of sodium chloride, t_-^s is ca 0.6. As the electric current continues to pass, the interfacial region becomes depleted in electrolyte. The difference between the quantity of anions transferred out of the interfacial region and those transferred in must be made up by convection and diffusion. If the interfacial region is defined as the region of streamline flow, then diffusion is the only mechanism (other than conduction) available to bring anions into the region. At steady state the concentration of electrolyte in the solution at the membrane surface must be reduced sufficiently from the bulk value to provide the concentration gradient necessary to bring in by diffusion the difference between the anions carried out of and into the interfacial regions by the electric current. This may be expressed by:

$$i\bar{t}_-^A = it_-^s + FD(c_o - c)/\delta$$

where Faraday's constant F is the quantity of electric current required to transfer one equivalent of ions through a perfectly selective membrane, 96,500 C or 26.8 A·h; i is the current density, A/cm^2; D is the diffusion constant for the electrolyte; c_o, equiv/L, is the concentration of the electrolyte in the bulk (nonstreamline region); c, equiv/L is the concentration of electrolyte in solution at the membrane surface; and δ is the

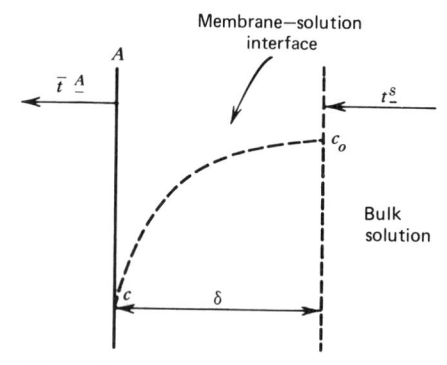

Figure 4. Current limitation controlled by diffusion to membrane.

thickness of the interfacial (streamline flow) region. This expression may be rearranged to:

$$i = \frac{DF(c_o - c)}{\delta(\bar{t}_-^A - t_-^s)}$$

Hence in any given situation the maximum current density which can be carried by a combination of electrical conduction and diffusion occurs when c (the concentration in solution at the membrane surface) approaches zero, ie:

$$i_{max} = \frac{FDc_o}{\delta(\bar{t}_-^A - t_-^s)}$$

For typical electrodialysis apparatus δ appears to be of the order of 5×10^{-3} cm, and $\bar{t}_-^A$ is about 0.9. For dilute sodium chloride solutions D is ca 1.5×10^{-5} cm^2/s, and t_-^s about 0.6. Under these circumstances the ratio i_{max}/c_o is ca 1 A·L/equiv·cm^2, ie, i_{max} is ca 0.1 A/cm^2 when c_o is ca 0.1 N. Increasing temperature increases D by ca 1.8%/°C, and decreasing δ increases i_{max}. Therefore, designers usually include some kind of structure in the electrodialysis compartments to break up the streamline interfacial region and bring electrolyte as close as possible to the membrane surface by convection. The maximum current also increases as the difference $(\bar{t}_-^A - t_-^s)$ decreases. As $\bar{t}_-^A$ approaches t_-^s, the diffusion-controlled limitation on current density vanishes and the current density may be increased without limit. However, power consumption increases as i^2R and puts an economic limitation on current density. When $\bar{t}_-^A$ approaches t_-^s, the anion-membrane loses its selectivity and becomes essentially a neutral membrane. The net transport of electrolyte is given by $(\bar{t}_-^A - \bar{t}_-^C)$. Thus for many electrolytes, an electrodialysis apparatus utilizing alternating neutral and cation-selective membranes exhibits substantial depletion and enrichment (3). For example, in the case of sodium chloride solutions the net transport of electrolyte is 0.8 (current efficiency, 80%) when the ion transport numbers of the anion- and cation-selective membranes are each 0.9. If the anion membrane is replaced by a neutral membrane, the net transport is 0.5 (current efficiency, 50%) when the concentration difference between enriched and depleted compartments is small.

Polarization. When the applied current density equals i_{max}, the anion-membrane (and the apparatus) are said to be concentration polarized or simply polarized. At i_{max}, the fluid at the surface of the membrane is depleted of electrolyte essentially completely and the electrical resistance of the apparatus increases substantially even though the bulk solution in the depletion compartments still may contain an appreciable concentration of electrolyte. This is a sign of polarization.

At the polarization current density, ions resulting from the dissociation of water have concentrations comparable to the concentration of electrolyte at the surface of the membrane. A significant fraction of the current through the anion-selective membrane is then carried by hydroxide ions into the enrichment-compartments. Hydrogen ions are carried into the bulk solution in the depletion-compartments. Changes in the pH of the enrichment- and depletion-compartments are another sign of polarization.

If i_{max} is exceeded substantially, electrolytes that are insoluble at high pHs (such as calcium carbonate and magnesium hydroxide) may precipitate at the interface between the anion selective membrane and the enriched solution. This is a third sign of polarization.

In sodium chloride solutions the ion transport number for Na⁺ is about 0.4 compared to about 0.6 for Cl⁻. It would be predicted that a cation-selective membrane should polarize at lower current densities than an anion-selective membrane. Careful measurements show that cation membranes polarize at lower current densities. However the effects on electrical resistance and pH are not as significant as those found when anion-selective membranes polarize. Such differences in behavior have not been satisfactorily explained (4).

Many fluids of natural origin contain detectable quantities of high molecular weight organic anions (humic, fulvic, and tannic acids are good examples) which can be carried to and deposited on anion-selective membranes. Such deposits can behave as thin films partially selective to cations (5). The interfaces between such films and the underlying anion selective membranes will then act as very thin stagnant depletion compartments and the anion-selective membranes may exhibit polarization at current densities that are much lower than would be expected for new membranes in the absence of such anions.

Performance. The performance of an electrodialysis stack may be estimated by considering the material balance around the stack:

$$\Delta(vc) = v_i c_i - v_o c_o = \frac{\bar{i} E A_p N_p}{F}$$

where v_i and v_o are the flow rates in L/s into and out of the depletion-compartments of the stack, respectively; c_i and c_o are the concentrations in equiv/L into and out of the depletion-compartments, respectively; $\bar{i}$ is the average current density in A/cm²; E is the net ion transfer $(\bar{t}_-^A - \bar{t}_-^C)$; A_p is the current-carrying area per cell pair, cm²; N_p is the number of cell pairs in the stack; and F is Faraday's constant. Typically a stack is designed so that $v_o c_o$ is ca 50% of $v_i c_i$. This is because $\bar{i}_{max}$ is determined by c_o. When c_o is very small compared to c_i then $\bar{i}$ may be uneconomically low. (Generally, as a matter of conservative engineering practice, the stack is designed so that $\bar{i}$ is appreciably lower than $\bar{i}_{max}$.) The power consumption W may be estimated from $\bar{i}^2 R_p A_p N_p$ where R_p is the electrical resistance of a cell pair per unit area (Ωcm²). R_p is given by:

$$R_p = R_d + R_e + R_C + R_A + R_{df} + R_T$$

where: R_d is the average resistance of a depletion-compartment; R_e is the average resistance of an enrichment-compartment; R_C is the average resistance of the cation-selective membranes; R_A is the average resistance of the anion-selective membranes; R_{df} is the resistance contributed by depletion in the membrane-solution interface; and R_T is the resistance offered by the concentration potential between the enrichment and depletion compartments (a measure of the thermodynamic work required to achieve the concentration difference).

Generally these individual contributions to R_p are not separately measured or calculated. Instead R_p itself is correlated with $\bar{i}$, c_o and c_i. For dilute solutions the following correlation may be used (6):

$$R_p = b + \frac{a}{c_a}$$

where a and b are empirically determined constants and c_a is the average concentration defined as:

$$c_a = \frac{2 c_{le} c_{ld}}{c_{le} + c_{ld}}$$

where

$$c_{le} = \frac{c_{ei} - c_{eo}}{\ln(c_{ei}/c_{eo})}$$

$$c_{ld} = \frac{c_{di} - c_{do}}{\ln(c_{di}/c_{do})}$$

The subscripts have the following significance: d indicates depletion compartment; e indicates enrichment compartment; i indicates inlet concentration; and o indicates outlet concentration.

Typically a is ca 1–2.5 $\Omega \cdot$L/(equiv$\cdot$cm^2) and b is ca 20–40 Ωcm^2, both at room temperature. The values of a and b within the ranges depend upon electrolyte composition and upon design of the electrodialysis apparatus. As stated above, R_p decreases with increasing temperature at ca 1.8%/°C. Typical d-c voltages are ca 0.5–1 volt per cell pair. Assuming an overall current efficiency of 90% this implies a d-c electrical energy requirement of ca 15–30 W$\cdot$h/equiv of electrolyte transferred. To this must be added energy losses during conversion of ac to dc and the energy consumption of auxiliary equipment such as pumps and instrumentation; these losses are usually one third of the d-c energy consumption. In the case of relatively concentrated electrolytes, for economic reasons designers tend to choose d-c voltages per cell pair from the low end of the range mentioned above.

Uses

High Purity Water. Electrodialysis has been used to some extent, in conjunction with other processes to produce pure water for use in high pressure boilers and in electronics and pharmaceuticals manufacture (7–9). Usually, electrodialysis is used to reduce the electrolyte concentration of the water processed to ca 1 mequiv/L. Further deionization is then accomplished by chemically-regenerated ion exchange; organic matter is removed by activated carbon; and particulate matter is removed by filtration through microporous membranes. This application is expected to increase in importance.

Brackish Water. Production of potable water from brackish water has been one of the principal applications of electrodialysis. Equipment used primarily for the brackish-potable conversion is shown in Figure 5. Several hundred plants have been installed, ranging in capacity from a few to more than 10,000 m^3/d. The term brackish water is used to define a water that is more saline than potable water but appreciably less concentrated than seawater. Brackish waters readily available around the world have concentrations of electrolyte of ca 20–200 mequiv/L. (Potable water generally has <10 mequiv/L; seawater has ca 600 mequiv/L.) A typical application would be demineralization from 0.05 N to 0.005 N. In this case, at one volt per cell pair, using the guidelines mentioned above, d-c energy consumption would be expected to be ca 1.35 kW$\cdot$h/m^3 of potable water. Auxiliaries would add about 0.45 kW$\cdot$h. The specific production rates are ca 0.5 m^3/(d$\cdot$m^2) of effective cell-pair area. In 1978 the installed cost per square meter was about $100 for large plants, including the immediate electrodialysis auxiliaries such as a-c/d-c conversion equipment, pumps, and instrumentation, but not including the costs of land, buildings, brackish water collection and pretreatment, and potable water storage and distribution. (Small plants may cost two

Figure 5. Aquamite XX equipment used primarily for production of potable water from brackish water.

to four times as much.) Pretreatment generally consists of coagulation and/or filtration to remove suspended particulate matter, less frequently activated carbon to remove colorants.

Direct current enters an electrodialysis stack at one end through a positively charged anode, generally a refractory metal such as titanium, niobium, zirconium, or tantalum coated with a thin film of a noble metal or noble metal oxide (see Metal anodes). The transition from an electron current in the anode to an ionic current in electrolyte bathing the anode produces hydrogen ions, and oxygen from the salts in the brackish water. Because hundreds of cell pairs are usually used in an electrodialysis stack, the number of equivalents of hydrogen ions and oxygen is generally $\ll 1\%$ of the number of equivalents of electrolyte removed by the membranes. The anodes are usually flushed with some of the brackish water to carry away the anode products. The electric current leaves the stack at the other end through a negatively charged cathode, generally a metal such as nickel or a nickel alloy. At the cathode, the electric current produces hydrogen gas and hydroxide ions, again in very small quantities compared to the amount of electrolyte removed through the membranes. However, the hydroxide ions can cause precipitation of calcium carbonate or magnesium hydroxide, or both, from ions in the brackish water.

In some electrodialysis plants, strong mineral acids such as sulfuric or hydrochloric are added to the brackish water which flushes the cathodes to prevent precipitation. These acids are an additional cost and an inconvenience particularly for small plants.

One manufacturer has solved this problem by designing a reversing-type electrodialysis stack, in which each electrode is alternately anodic and cathodic (10). Acid is not added to the cathode. After passing current through the stack for a period of ca 15–30 min, the direction of electric current is reversed for a similar period. The electrode that was cathodic in the first half of the cycle and precipitated some alkali-insoluble salts becomes anodic in the second half. Hydrogen ions generated during the anodic half of the cycle dissolve and separate insoluble salts from the electrode. In the electrodialysis stack, compartments that were enriching during one half of the cycle are depleting during the other half. Appropriately placed automatic valves interchange the entering and exiting fluid streams at the beginning of each half cycle. A few hundred of such reversing-type plants have been in operation successfully for several years. Capacities are a few to ca 10,000 m^3/d. The periodic reversal also appears to remove foreign substances from the electrodialysis stack *per se*. Such materials may include: alkali insoluble salts such as calcium carbonate produced by marginal concentration polarization at the interface between the depleting fluid and the anion-selective membrane; poorly soluble salts such as calcium sulfate, precipitated if their solubility limits are exceeded in the enrichment compartments; and high molecular weight organic anions such as humic and fulvic acids which, if present, may tend to be absorbed in the anion selective-membranes.

Seawater. Production of salt from seawater is the second principal application of electrodialysis. The number of plants is small, approximately ten in all but each plant is large in terms of membrane area installed. The combined production capacity is in excess of one million metric tons of crystalline sodium chloride per year. The membranes are thin and have low water transport and high ion transport numbers for monovalent ions in concentrated brine solutions. Generally, sea water is filtered, warmed with waste heat, and passed slowly through the depletion compartments of an electrodialysis stack. The liquid in the enrichment compartments is produced by the water and ion transport through the membranes; there is no separate feed of seawater to the enrichment compartments. Under these circumstances, the concentration in the enrichment compartments can reach ca 20 wt % of solids. The resulting brine is then processed to solid, crystalline sodium chloride in evaporator crystallizers (see Evaporation). For economic reasons, the seawater is depleted by only ca 50% of its salt (see also Chemicals from brine).

A few electrodialysis plants successfully produce potable water instead of brine from seawater. Recent technological innovations suggest that this application will become more popular (11–14). These innovations include the use of larger, thinner membranes having low electrical resistance, high ion transport numbers, and low water transport; thinner compartments between the membranes (≤0.5 mm); and operation at elevated temperatures (60–80°C). All of these advances lead to reduced energy consumption (ca 5–10 kW·h/m^3 of potable water) even at the high current densities required to achieve practical specific production rates. The elevated temperatures may be obtained from waste heat, solar heating (see Solar energy), or from recuperative heat exchangers driven by the heat released by the thermodynamic inefficiency of the electrodialysis process (see Water, desalination).

Cheese Whey. Whey is the serum remaining from the production of cheese from milk (see Milk products). It contains about half of the solids originally present in milk (lactose, lactalbumin, lactoglobulin, minerals) plus some of the salt and/or acid used in the production of cheese. The protein in the whey has high nutritional value but its uses are limited by the high electrolyte content. Electrodialysis is used to remove most of the electrolytes (15). Generally, the resulting product is concentrated and spray-dried. In 1978 world production of electrodialyzed whey was estimated to be >100,000 metric tons on a dry basis. Ion exchange is also used to de-ionize whey. The advantages of electrodialysis in this application are: (1) there is less denaturation of the protein; (2) it can easily be made a continuous process; and (3) there is less loss of biodegradable substances and, therefore, less environmental pollution.

Future Applications

In addition to the predicted substantial increase in production of potable water from seawater, electrodialysis should find several new applications in the future including recycling of cooling tower blow-down; electrowinning of metals; and generation of electricity from the mixing of waters of different salinity.

In many inland regions of the United States evaporative cooling towers are used to dissipate waste heat from steam electric power plants, either owing to shortages of cooling water or to the desire to avoid thermal pollution of the available receiving bodies of water. However, such evaporative coolers concentrate the minerals in the make-up water fed to the coolers. As the concentration increases, corrosion of the materials of construction of the coolers may increase and substantial amounts of minerals may be carried out of the tower as spray, sometimes constituting a pollution problem. Therefore, part of the recirculated water is bled (blown-down) from the cooling tower. In some cases there is also no suitable natural receiving body of water for such blow-down which is then instead sent to man-made natural-evaporation basins. Such basins are generally lined to prevent salt pollution to underground aquifers and can constitute a substantial fraction of the cost of a power plant. Electrodialysis can be used directly on the relatively hot blow-down from the tower to produce an enriched stream for disposal and a depleted stream for recycle to the tower. In this way the salts that must be removed from the recirculating cooling water can be obtained in a much more concentrated solution, enabling the evaporation basin to be much smaller (see Water, industrial water treatment).

As the shallow, comparatively rich bodies of minerals containing copper, nickel, zinc, cobalt, and similar metals are depleted, deeper, less rich bodies will be increasingly used. Deep bodies may be mined by *in situ* leaching. Even shallow bodies of comparatively low metal content may be processed by leaching. (Included among such shallow bodies are the dumps of processed ore from conventional metal-winning processes.) The metal-bearing leach liquors may often be too dilute and voluminous for convenient processing by conventional metal-winning techniques. When this occurs electrodialysis can be used to obtain a smaller volume of liquor enriched in the valuable metal (see Extractive metallurgy).

The production of one cubic meter of potable water from a large volume of seawater requires a minimum thermodynamic energy of ca 0.75 kW·h. The mixing of such a volume of potable water with seawater will, therefore, dissipate the same quantity of free energy. This is equivalent to the potential energy of 1 m^3 of water falling a distance of ca 275 m. Thus the mixing of a fresh water river with seawater probably dissipates more potential energy than that of the hydroelectric projects that could be developed on the river. Part of this energy can be extracted in a reverse electrodialysis apparatus in which fresh water is passed through the enriching compartments and saline water through the depletion compartments. Mixing takes place by ions migrating through the ion-selective membranes, such migration generating an electric current which may be drawn from the end electrodes. The membranes must be highly selective and have low electrical resistance, and the electrodialysis apparatus must have low cost and accept turbid water. However the total potential energy available worldwide is very large and renewable and is apparently available without serious environmental pollution problems. A concerted effort could probably develop this source. The first applications should probably be on the Great Salt Lake or other bodies of water nearly saturated in salts. Such bodies may be regarded in this connection as solar energy (qv) collectors.

Nomenclature

A_p = current-carrying area per cell pair, cm^2
c = concentration, eq/L
D = diffusion constant, cm^2/s
E = net ion transfer or transport, $\bar{t}_-^A - \bar{t}_-^C$
F = Faraday's constant, 96.5 kA·s
i = current density, A/cm^2
I = current, A
N_p = number of cell pairs in a stack
R_p = electrical resistance of cell pair per unit area, Ω·cm^2
R_s = electrical resistance of stack, Ω
$\bar{t}_-^A$ = fraction of current carried by anions
$\bar{t}_+^C$ = fraction of current carried by cations
v = flow rate, L/s
W = power consumption, W
ρ = thickness of interfacial (steamline flow) region, cm

BIBLIOGRAPHY

"Electrodialysis" in *ECT* 1st ed., Vol. 5, pp. 20–26, by P. Stamberger, Technical Consultant; "Electrodialysis" in *ECT* 2nd ed., Vol. 7, pp. 846–865, by W. K. W. Chen, Celanese Plastics Company.

1. K. H. Meyer and W. Strauss, *Helv. Chim. Acta* **23**, 795 (1940).
2. W. Juda and W. A. McRae, *J. Am. Chem. Soc.* **72**, 1044 (1950).
3. U.S. Pat. 2,872,407 (Feb. 3, 1959), P. Kollsman.
4. J. C. R. Turner, *Proc. 6th Int. Symp. Fresh Water from the Sea* **3**, 125 (1978).
5. G. Grossman and A. A. Sonin, *Desalination* **10**, 157 (1972).
6. E. A. Mason and T. A. Kirkham, *Chem. Eng. Prog. Symp. Ser.* **55**, 173 (1959).
7. W. E. Katz, *Proc. Am. Power Conf.* **33**, 830 (1971).
8. *Ind. Water Eng.*, (June–July 1971).
9. C. R. Zmolek, *Ind. Water Eng.*, (Dec. 1977).
10. W. E. Katz, *Int. Congr. on Desalination and Water Reuse, Tokyo, Japan, Dec. 1977.*
11. H. Behret, H. Binder, and A. Köhling, *Proc. 6th Int. Symp. Fresh Water from the Sea* **3**, 15 (1978).
12. R. Eggersdorfer, H. J. Hampel, and K. H. Scherer, *ibid.*, p. 57.
13. T. Kawahara, K. Asaka, and K. Suzuki, *ibid.*, p. 95.
14. W. A. McRae and co-workers, *ibid.*, p. 101.
15. L. E. Slater, *Food Eng.* (Mar. 1976).

General References

J. R. Wilson, ed., *Demineralization by Electrodialysis,* Butterworths, London, Eng., 1960.
L. H. Shaffer and M. S. Mintz in K. S. Spiegler, ed., *Principles of Desalination,* Academic Press, Inc., New York, 1966, pp. 199–289.
A. A. Sonin and R. F. Probstein, *Desalination* 5, 291 (1968).

WAYNE A. McRAE
Research Ionics, Inc.

ELECTROLESS PLATING

Electroless plating is the controlled autocatalytic deposition of a continuous film by the interaction in solution of a metal salt and a chemical reducing agent. This definition excludes such processes as conventional electroplating (qv), which uses externally supplied electrons as the reducing agent; uncontrolled precipitation of metal powders; immersion films produced by displacement reactions; thermal decompositions; and the various types of sputtering and evaporation techniques (see Film deposition techniques). Electroless deposition can give films of metals, alloys, compounds, and composites on both conductive and nonconductive surfaces (1–5).

Electroless plating is almost as old as modern electroplating (6–7). The first description by von Liebig in 1835 was the reduction of silver salts by reducing aldehydes. Despite its early start, progress in this field remained slow until World War II.

One significant problem was the need for very high purity chemicals. This requirement is very important as the surfaces being plated must maintain autocatalytic activity. Many substances react with and poison the catalytic surface even if they are present in trace amounts. Impure chemicals can also cause the opposite problem, overactivity. Colloidal materials and fine particles can promote uncontrolled autocatalytic reactions owing to their extremely high surface areas. Another reason for lack of wide interest was cost. Chemical reducing agents are much more expensive than electricity. If the deposits are not significantly different, there is no reason to use the more expensive process.

The growth of electroless plating after World War II is directly traceable to three developments: (*1*) the discovery that some alloys produced by electroless deposition, notably nickel–phosphorus, have unique properties that allow them to compete directly with conventional electroplating on metals; (*2*) the growth of the electronics industry, especially printed circuits; and (*3*) the large-scale introduction of plastics into everyday life. This gave a large potential market for any processes that could help simulate the appearance and properties of metal on nonconductive plastics.

Modern electroless plating began in 1944 when Brenner and Riddel rediscovered that hypophosphite could cause nickel deposition (1,7–9). Their subsequent work led to the first patents on commercially usable electroless nickel solutions. Many others did extensive work on electroless nickels during the 1950s and early 1960s.

Although these solutions were very useful for coating metals, they could not be used on most plastics because their operating temperature was 90–100°C. The first electroless nickel solution capable of wide use on plastics was introduced in 1966 (10). It can be operated at room temperature and is extremely stable to autodecomposition.

Electroless copper solutions underwent a similar development during the same period (11). Early printed-circuit boards used mechanically attached eyelets to provide electrical conductivity between the copper sheathing laminated on the two sides. Electroless copper plating first replaced the mechanical eyelets, and later became good enough to plate some or all of an unclad board in the semiadditive and additive processes.

Theory

The theory and practice of electroless plating parallels that of electrolytic plating. The main difference is that the electrons used for reduction are supplied by a chemical reducing agent present in solution. In electroplating, these electrons are supplied by an external source such as a battery or generator. This means that electroless solutions are not thermodynamically stable because the reducing agent and the metal salt are always present and ready to react.

The ideal electroless solution would deposit metal only on an immersed article, never as a film on the sides of the tank or as a fine powder. Room temperature electroless nickel baths now closely approach this ideal. Electroless copper plating has been greatly improved in the past ten years and is beginning to approach this stability when carefully controlled. The reason for these goals is clearly economic. Electroless plating costs more per weight of deposited metal than electrolytic plating. Any metal that plates in an unwanted location greatly increases the costs and it must be removed because it can promote decomposition of bath components. If complete bath decomposition does occur, the whole solution must be disposed of and a fresh bath prepared.

Any metal that can be electroplated can theoretically be deposited by electroless plating. Most baths reported have remained laboratory curiosities. Only a few metals—nickel, copper, gold, and silver—are used on any significant scale.

Electroless solutions contain a metal salt, a reducing agent, a pH adjuster or buffer, a complexing agent, and one or more additives to control stability, film properties, deposition rates, etc. The metal salt and reducing agent must be replenished at periodic intervals because they are consumed. Buffers, complexers, and other additives are necessary only for compensation of drag-out losses. Of the large number of potential reducing agents, only formaldehyde, hypophosphite, and to a much lesser extent, organoboron compounds are generally used. The latter two reducing agents give phosphorus- or boron-containing alloys. The detailed theory of the process has been discussed thoroughly in a number of works (3,11–14).

A major advantage of electroless solutions is their ability to give conductive metallic films on properly prepared nonconductors, and their ability to uniformly coat any platable object. No specially shaped anodes, shields, or complex rack designs are necessary. The most complex geometric shapes receive a very uniform plated film. Film thicknesses range from <0.1 μm, where only conductivity or reflectivity is wanted, to ≥1 mm for functional applications.

Equipment

Plating Tanks. Most metal tanks, except passivated stainless steel, cannot be used to hold electroless plating baths because the metal initiates electroless plating on itself. Tanks are preferably plastic or lined with a plastic or rubber coating. The tank linings must be stripped of metal deposits with acid at periodic intervals.

Some hot nickel and flash electroless copper solutions are plated to the point of exhaustion and then discarded. Most baths are now formulated to give bath lives of ≥6 turnovers of the bath constituents. All regenerable solutions should be filtered to remove particulates that can cause deposit roughness and bath instability.

Replenishment should be done with care as massive additions of chemicals can cause decomposition. Hot nickel solutions are often cooled to 60°C before addition of concentrated replenishing solutions. Maximum stability of electroless copper baths is obtained when continuous replenishment is practiced. A number of machines are available that continuously analyze copper baths and make additions. Room temperature nickel processes are much less demanding in most respects.

Passivated stainless steel or quartz heaters can be used but may become plated owing to local overheating of the solution. The preferred methods are a double-jacket heated tank or nonmetallic heat exchangers (see Heat exchange technology).

Rack Design and Loading Factors. Tank and rack design are less important than tank loadings and rack coatings, especially when plating nonconductors. The parts may be closely racked without harm if reracking is done after electroless plating; otherwise racking should be the same as for electrolytic plating. However, more racks may be placed in a given electroless tank than in a similar electrolytic tank. The critical point is not to overload the plating bath. Electroless baths rapidly become overactive and decompose when too much surface area is being plated at once. Most vendors recommend a maximum loading of ≤245 cm^2/L (1 ft^2/gal).

Two sets of racks were originally used when plating nonconductors, with reracking after the electroless bath, since all rack surfaces were plated. The racks were stripped and re-used for each cycle. Racks for plating without reracking have standard copper rack splines with stainless contacts (15). They are coated with a special poly(vinyl chloride) plastisol which absorbs large amounts of chromic acid from the etchant used in plating on plastics. This chromic acid cannot be completely removed during the subsequent rinsing and neutralization steps. This inhibits deposition of electroless nickel and electroless copper on its surface.

Safety and Waste Disposal

Most solutions are not particularly hazardous although they may be highly acidic or basic. All reducing agents (solids or liquids) should be stored separately from oxidizing substances such as chromic acid, in case of spills.

Plastic etchants containing sulfuric acid, chromic acid, or both, should be treated and neutralized in the same way as chromium plating baths. Electroless silver solutions must be neutralized before discarding because a dried silver–amine complex can explode (see Explosives). Waste treatment should always conform with current rules. The electroless baths, especially copper, may be difficult to treat owing to the presence of strong complexers. The vendors' recommendations should be followed for specific cases.

Plating on Metals

The first large-scale process was the Kanigen electroless nickel process from General American Transportation Co. This hot nickel process uses a hypophosphite reducing agent. Properties of electroless nickel deposits vary greatly depending on the reducing agent used. Hydrazine gives a practically pure nickel, and the organoboron reducing agents give very hard nickel–boron alloys. The most widely used hypophosphite baths deposit a range of nickel–phosphorus (1–15 wt % P) alloys with unique properties. Acidic baths (pH 4.0–5.5) are preferred, but alkaline baths (pH 8–10) are also used. Operating temperatures are 70–95°C. Table 1 shows some typical formulations. A large number of commercial baths are available with properties tailored to suit specific applications (17).

The engineering properties of electroless nickel have been summarized (18). The Ni–P alloy has good corrosion resistance, lubricity, and especially hardness. It can be heat-treated to a hardness equivalent to electrolytic hard chromium (Fig. 1), and its lubricity is comparable. Thus it is not surprising that the main applications for electroless nickel are in replacement of hard chromium.

The advantages over hard chromium include safety of use, ease of waste treatment, plating rates of as much as 40 μm/h, low porosity films, and the ability to uniformly coat any geometric shape without burning or use of special anodes. The uniformity of coating thickness can also minimize expensive machining after plating, and can be used to salvage parts that have been overmachined. Electroless nickel has superior corrosion and erosion resistance as compared to electrolytic nickel. It is used extensively on molds, pistons, pump parts, oil field equipment, dies, compressors, tanks, and piping.

The market size for electroless nickel solutions is not known with any certainty. The best guesses are based on hypophosphite production, as virtually all is used for electroless nickels. An estimate for 1977 usage is 900 metric tons of nickel, of which 85% may be for plating on metals and 15% for plating on nonconductors. However, in terms of plated area, plating of nonconductors is at least ten times as great owing to the much lower thicknesses used.

Electroless nickel can be used to plate aluminum. The adhesion is often poor unless the aluminum is etched to remove oxides. The best method is to use an intermediate zincate deposit. Adhesion can vary widely depending on the aluminum alloy used.

Table 1. Typical Formulations for Electroless Nickel on Metal

	A[a]	B[a]	C[b]
nickel chloride, g/L	30	30	21
sodium hypophosphite, g/L	10	10	24
ammonium chloride, g/L	50		
sodium citrate, g/L	100		
sodium glycolate, g/L		50	12
pH	8.5–10	4–6	5.7
temperature, °C	90–95	88–99	97–98
plating rate, μm/h	10	12.5	40

[a] Ref. 9.
[b] Ref. 16.

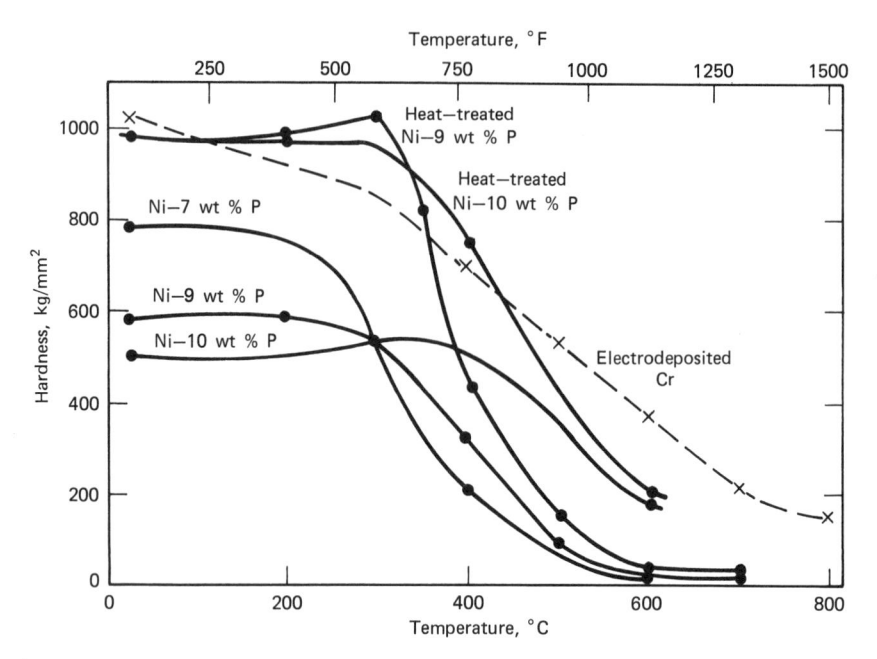

Figure 1. Hardness of electroless nickel–phosphorus alloys compared with electrolytic chromium at various temperatures. Heat-treating was at 400°C (18).

Electroless nickel or nickel–lead alloys can improve the solderability and braisability of aluminum even when a continuous film is not present (19). Electroless nickel systems based on dimethylamineborane reducing agents are used to coat aluminum contacts and semiconductors in the electronics industry (20) (see Electrical connectors).

Preparation of the Substrate. All parts must be cleaned thoroughly before plating. Any traces of dirt or oxide will either prevent deposition or lead to adhesion loss. Nickel can be spontaneously deposited from hot electroless solutions on most common metals, including mild steel, beryllium–copper, aluminum, stainless steel, and titanium. Generally, special etchants are used for difficult materials such as titanium and some stainless steels. Lead and a few other materials cannot be plated directly because they poison the electroless reaction. A thin electrolytic layer may be used to mask such surfaces to make them platable. Where initiation is slow or difficult, such as with copper substrates, it can be started by briefly contacting the inactive part with one that is actively plating, by giving it a brief cathodic pulse of current, by contact with a dissimilar metal, such as iron, or by prior immersion in a dilute solution of a precious metal such as platinum or palladium.

The parts are allowed to plate until the desired thickness is reached. This point can be monitored by knowledge of the bath operating rate, or by direct thickness or weight measurement of standard test parts.

Plating on Nonconductors

All nonconductors can be electrolessly plated but only a few can be plated to give good adhesion and appearance. A highly active, unstabilized electroless bath will coat any object it contacts, including its container. This process is called encapsulation because there is little or no adhesion between the metal deposit and the substrate (see also Microencapsulation).

Highly active baths are used rarely in printed circuits or plastics plating. The newer, more stable baths will not plate spontaneously on any nonconductor. The surfaces must first be coated with a catalyst to initiate the electroless deposition process. However, activation alone is not sufficient, for this will give encapsulation. There must be a way to form an adherent bond between the metallic film and the nonconductive substrate.

Most molded plastics have a very smooth, hydrophobic surface which must be modified. Formerly, the surface layer was abraded by drilling or punching, as in a printed circuit with through-holes, and by sanding or tumbling the part. At present, chemical etchants are used to oxidize and roughen the surface. The resultant hydrophilic surface promotes good metal to plastic adhesion.

Plating of Plastics. This market began in 1963 on a very small scale using electroless copper. It was soon discovered that ABS (acrylonitrile–butadiene–styrene) engineering thermoplastic was the easiest plastic to plate. Plated ABS has about 90% of the market (see Fig. 2), most of the remainder is polypropylene and modified poly(phenylene oxide).

Over 10^7 m^2 (1.1×10^8 ft^2) of plastics are plated in fewer than 75 United States plants (1977 estimate) with electroless chemical costs of $10–12,000,000. Individual plants process from 4.6×10^3 to $>1.3 \times 10^6$ m^2/yr (50,000–>15,000,000 ft^2/yr) of plastic surface area. Automotive items make up over 60% of the market on a plated area basis; the remainder is hardware, plumbing, and decorative items.

Electroless films have two functions: (1) they provide an electrically conductive layer which allows further coating by electroplating; and (2) they provide a secure bond between the plastic and the electroplated layer. A number of ASTM standards have been issued that cover thickness, adhesion, and thermal-cycling resistance of the total plated film (Table 2). In addition, the principal auto producers have their own sets of standards. Adherence to these standards necessitates close cooperation between resin suppliers and molders, part designers, platers, and users.

Plated plastics have several disadvantages. Plating normally lowers impact strength. The coefficient of thermal expansion is much higher for plastics than for metals, so stress build-up and adhesion loss can occur on severe thermal cycling. Blistering can occur during corrosion. The relatively low heat distortion temperature of most plated plastics can also limit applications. Molding cycles are relatively long.

One of the most important advantages is that weight savings can be as much as 60% as compared to an equivalent all-metal part. The molded plastic parts need no buffing or other finishing step before plating. Plated plastics have improved tensile strength, elasticity, flexural strength, a reduced total coefficient of thermal expansion, and improved abrasion and weathering resistance. Whereas a metal part may completely corrode away and fail in service, only the surface of a plated plastic can be corroded (21–22). This presupposes that the part design was adequate, the proper plating-grade resin was selected, and that the molding conditions were correct. For example, injection-molding that is done too quickly at too low a temperature is a major cause of excessive stress leading to thermocycle failures. All of the principal suppliers of plating-grade resins (Borg-Warner, General Electric, Monsanto, Uniroyal) offer detailed help on these problems.

Figure 2. Jeep grilles of ABS plastic. The grilles are 1.68 m by 0.53 m, have more than 2.3 m² of surface area, and weigh 3.18 kg. (**a**) The truck grilles have been plated with electroless copper. (**b**) The grilles have been plated with the complete automotive specification, which includes a final layer of 0.25 μm electrolytic chromium. Courtesy of J. Schrantz, *Industrial Finishing Magazine*.

Table 2. ASTM Standards for Plating Nonconductors

B532-70	recommended practice for evaluation of the appearance of plated plastic surfaces
B533-70	measurement of peel strength of metal-plated plastics
B553-71	recommended practice for thermal cycling test for evaluation of electroplated plastics
B554-71	recommended practice for measurement of thickness of metallic coatings on nonmetallic substrates
B571-72	adhesion of metallic coatings
B604-75	decorative electroplated coatings for plastics

Process Details. A typical process cycle for plating on plastics is shown in Table 3. The cost of chemicals is ca $0.65–1.30/m² (6–12¢/ft²). Successful electroless plating depends on the optimized interaction of five separate complex chemical solutions (23–24):

Etching is necessary to give good metal to plastic adhesion. Solutions of strongly oxidizing chromic acid–sulfuric acid–water or chromic acid–water are operated near

Table 3. Plating on Plastics[a]

Step	Solution	Temperature, °C	Time, min
etchant	CrO_3–H_2SO_4	60	5–10
neutralizer	mild Cr^{6+} reducer	25	½–2
catalyst	$PdCl_2$–$SnCl_2$–HCl	25	1–5
accelerator	mild acid or alkali	50	1–2
electroless nickel bath	nickel salts, complexer, buffer, hypophosphite; pH 8–10	25	5–10
or			
electroless copper bath	copper salts, complexer, buffer, formaldehyde; pH 12–13	25–40	5–10

[a] As in all plating processes, thorough rinsing is necessary after each step. This is a typical ABS plating cycle; for other plastics plating differs mainly in the pretreatment or etchant and neutralizer steps. The electroless metal deposit is typically 0.15–0.5 µm thick; electrolytic plating follows.

the point of mutual saturation. The etchant both physically roughens the surface and chemically modifies it to give a very hydrophilic surface. Metal adhesion occurs owing to the combined effects of chemical bonding and mechanical locking to the roughened surface.

Neutralizing removes the large amount of hexavalent chromium from the surface of the part. Hexavalent chromium shortens the life of the catalyst, and trace amounts completely inhibit electroless nickel deposition. The neutralizer is usually a mildly acidic or basic reducing agent, but other types of neutralizers are available especially for substrates that are difficult to plate.

Catalysis is done by a stabilized reaction product of stannous chloride and palladium chloride which is sold dry (25) or as a concentrated solution in hydrochloric acid (26–27). The older two-step process consisted of separate hydrochloric acid solutions of stannous chloride and palladium chloride but is now rarely used except in special applications. Catalyst absorption is typically 1–5 µg Pd/cm². Other precious metals can be used, but they are not as cost-effective.

Acceleration modifies the surface layer of palladium nuclei, and stannous and stannic hydrous oxides and oxychlorides. It can take place in any acid or alkaline solution in which excess tin is appreciably soluble and catalytic palladium nuclei become exposed.

Electroless metal is provided by either copper or nickel; a number of papers have indicated that copper may be better for corrosion resistance (22,28–29). A nickel–copper–phosphorus alloy of low copper content is also available (30). Although the electroless copper plating systems are similar to those used in printed-circuit processing, electroless nickel processes differ greatly from those used in metal plating and are not interchangeable. Room temperature nickel solutions are mildly alkaline and give a low and uniform deposition rate.

Electroless nickel systems are easily controlled and replenished; bath lives of several years are common. Electroless copper solutions generally have much shorter bath lives unless they are very carefully controlled. As only thin metal layers are deposited, the drag-out losses can be comparable to the actual usage of metal in plating. This drag-out helps bath stability by keeping the reaction products from building up.

A plating time of 5–20 min gives a 0.15–0.5-μm film that allows electroplating of the part as though it were totally metallic. Precautions must be taken to control the initial current during electroplating or the thin film may overheat and burn off. A nickel or copper strike is usually applied first, followed by any desired decorative or functional coatings. Recommended minimum electroplate thicknesses for a bright chromium finish are: bright acid copper, 15 μm; semibright nickel, 20–30 μm or bright nickel, 7–15 μm; and chromium, 0.25 μm (ASTM B604-75). Special stop-off paints are commonly applied to the backs of large auto parts, such as grills, to confine metal plating to appearance surfaces only.

Printed Circuits. This is by far the most diverse field of electroless plating. Numerous variations exist in solutions, processes, and techniques, both in laboratory and commercial form, and are used to create a great variety of products (20,31). All have the basic purpose of producing a layer of highly conductive copper in specified areas. Modern electroless copper films have a ductility and conductivity identical to that of electrolytic copper (32). The three basic classes of baths, shown below, differ mainly in stability, ease of regeneration and control, operating temperature, cost, and deposition rate.

Copper bath	Thickness desired	Processing type
flash	0.5–1 μm	subtractive
fast	1–2.5 μm	semiadditive
heavy build	25–40 μm	additive

Note that processing type above refers to how the copper circuits are fabricated (see below).

The total market for printed-circuit electroless-plating chemicals is estimated at $12–15,000,000 in 1977. This does not include $6–8,000,000 for copper etchants. Over 7×10^6 m^2 (75×10^6 ft^2) of printed-circuit boards were plated in 1977. Subtractive processes accounted for 90% of the total; the rest was equally split between semiadditive and additive processes. Printed-circuit boards are composed of epoxies, phenolics, and other heat-stable dielectrics; flexibile films of polyester and polyimide are also plated. The ratio of rigid to flexible surface areas plated is about 10:1. They are used in communications, instruments, controls, consumer markets, military, aerospace, and business applications (see also Integrated circuits).

The three major processing techniques are briefly summarized in Table 4. Additive and semiadditive processing can give material savings and higher circuit densities, whereas subtractive processing is technologically easier.

Subtractive processing was the first method to be used. Copper foil (31 mg/cm^2 or 1 oz/ft^2) is laminated to both sides of a flat board. Through-holes are punched or drilled. The board is cleaned, catalyzed, and placed in a flash electroless copper bath. A thin conductive layer of copper deposits on the entire board, including the nonconductive holes. The board is electrolytically plated to give 25-μm copper deposits in the through-holes. Further processing includes positive masking (see Photoreactive polymers) followed by etching away all unwanted copper to leave the desired circuit. As much as 80% of the original copper can be wasted. The electroless copper step replaced mechanical implantation of copper eyelets. Uniform thickness, conductivity, and adhesion of the copper are critical.

Semiadditive processing was developed to reduce copper waste. Thin 50-μm copper foil (4.5 mg/cm^2 or 0.15 oz/ft^2) laminates are used. Hole-forming, catalysis, and

Table 4. Printed-Circuit Processing Techniques, Representative Cycles[a]

Subtractive process	Semiadditive process	Additive process
copper-clad board	thin copper-clad board	unclad board
form holes	form holes	adhesive coat
catalyze	catalyze	form holes
accelerate	accelerate	etch surface
electroless copper	electroless copper	catalyze
positive resist coat	negative resist coat	negative resist coat
dissimilar metal electroplate	electrolytic copper	accelerate
etch copper	dissimilar metal electroplate	electroless copper
	strip resist	
	etch copper	

[a] Further processing follows in each case.

electroless copper plating are done as for subtractive circuitry. A strippable reverse-resist coating is then applied, copper is electroplated to the desired thickness, the resist is removed, and the whole board is etched. The original thin copper layer is quickly removed to leave the desired circuit. Copper waste may be less than 50%.

The most important of the numerous variants is the CC-4 process (33–34). An adhesive layer consisting of an epoxy paint containing a polybutadiene or poly(acrylonitrile–butadiene) emulsion is coated onto an unclad circuit board. This adhesive can be etched with chromic–sulfuric acid, catalyzed, and plated in a manner similar to an ABS plastic. It is cheaper than a thin-foil laminate, and pinholes and other imperfections are reduced.

A second type is the Pladd II system (35). A specially anodized aluminum foil is laminated to the printed circuit board. The aluminum is completely etched away just before use to give a microscopically rough surface which can be catalyzed and electrolessly plated in a conventional way. An electroless nickel layer can sometimes be used to provide the initial thin conductive film, eliminating the electroless copper step.

Another method is the swell-and-etch process. Special epoxy laminates with resin-rich surfaces are immersed in a strong organic solvent to make them easily etchable with chromic-sulfuric acid; further processing is the same as that described above.

Additive processing is the most technologically demanding. The copper circuit is formed directly on the board without a continuous copper film. Heavy-build electroless coppers are used to build up the final thickness of the entire circuit.

The CC-4, the Pladd II, and the swell-and-etch processes can all be used. The plain circuit board is etched, catalyzed, dried, and coated with a negative resist. The patterned board is activated and plated with electroless copper to the final circuit thickness of 25–50 µm. Electrolytic copper plating cannot be used because there is no continuous conductive film linking the discrete circuit paths.

Printed-Circuit Etchants. The two types of etchants used in printed-circuit board processing are often confused. The etchants used in the CC-4 and the swell-and-etch process are saturated chromic acid–sulfuric acid solutions of the type used for plating on plastics. They are used only to etch a portion of the organic component of the board and have a direct contribution to electroless plating since they improve adhesion and performance.

Copper etchants do not directly influence the electroless plating process. They are used merely to remove unwanted copper and should not affect the deposit properties. Many types have been used commercially, including chromic–sulfuric acid, ferric chloride, and ammonium persulfate. The costs of waste treatment and disposal have led to disuse of these throw-away systems. Newer types of regenerable etchants predominate, including cupric chloride, stabilized peroxide, and proprietary alkaline baths.

Emerging Printed-Circuit Technologies. A large number of newer or specialized processes are known. They include laminates containing dispersed catalyst to improve copper coverage in through-holes; catalyzed adhesives; two-step palladium–tin catalysts for maskless photo-forming; nonnoble metal catalysts; photoactivated nonnoble metal catalysts; and formaldehyde-free electroless copper systems.

Plating on Glass. Mirror production is the major application for electroless silver. The glass support is cleaned, catalyzed using a two-step catalyst, and coated on one side with an opaque silver film. A large number of recipes exist for the silver bath. Most are based on ammoniacal silver nitrate, with a reducing agent of sugar, formaldehyde, hydrazine, dimethylamineborane, etc (3,13,36). These baths are generally unstable and are used until exhausted.

Architectural Glass. The other important commercial glass-plating application is for production of architectural reflective glasses. Translucent metal films are used for decoration and for reduction of environmental heat gain. Electroless plating is used by one producer for this type of product (37).

These glass sheets are available in a wide range of optical densities and colors. The glass sheets are cleaned, coated with a two-step catalyst, and sprayed with a highly active electroless bath that deposits a very thin translucent colored film. The deposits are not thick enough to be weather- or wear-resistant so only the inner side of the glass is coated. Electroless gold, silver, nickel, and copper are used singly or in combination to give the desired effects.

Ceramic Plating. Ceramic plating is similar to glass plating in that a two-step catalyst is used. The surface is often mechanically roughened or chemically etched to improve the metal adhesion. Ceramic resistors are usually coated with electroless nickel although ceramic circuit boards use electroless copper.

Composite Plating. An electroless nickel matrix can be used to securely bond diamonds to cutting tools. The NYE–CARB process gives a silicon carbide-electroless nickel composite which has extremely high abrasion resistance (40) (see Composite materials).

Other Applications

Electroless Gold and Palladium. There are few distinct applications for electroless gold, though immersion (replacement) gold baths are commonly used. A number of baths have been formulated that deposit gold or palladium (3–4,38). They are used mainly in specialty applications in the electronics industry where the need for pore-free or discontinuous coatings justifies the high price. A new development is the use of nickel–molybdenum–boron and nickel–tungsten–boron alloys as partial or complete replacements for gold in electronic applications (39) (see Electrical connectors).

Miscellaneous. The patent literature contains many unusual baths and applications, most of which have never been commercialized. One series of patents covers extensive experiments in the use of electroless nickel coatings to stabilize friable strata *in situ* in oil wells. Another uses a coating of electroless nickel or nickel phosphide to stabilize elemental phosphorus during storage and shipment.

A recent patent describes the use of electroless solutions in dentistry for improving the adhesion of fillings (41) (see Dental materials). Many cobalt–phosphorus and nickel–cobalt–phosphorus alloys have good magnetic properties and have been proposed for use in computer memories and magnetic recording tape (2,42) (see Magnetic tape). Objects such as leaves, insects, and seashells can be coated by encapsulation (13) (see Microencapsulation). Electroless platinum can be plated on ion-exchange membranes for fuel-cell applications (see Batteries). Polymer microcapsules have been electrolessly plated to give products of controlled permeability. Composites of electroless nickel with Teflon (polytetrafluoroethylene) have been prepared. Thin electroless copper coatings have been applied to nitrocellulose propellant compositions to allow ignition by low voltage direct current (see Explosives). Electroless nickel and copper have been used to uniformly coat glass microspheres for tests of laser-induced hydrogen fusion (2,13,42–43) (see Fusion energy).

BIBLIOGRAPHY

1. "Symposium on Electroless Plating," *Am. Soc. Test. Mater. Spec. Tech. Publ.* **265**, (1959).
2. J. MacDermott, *Plating of Plastics with Metals, Chemical Technology Review No. 27*, Noyes Data Corp., Park Ridge, N.J., 1974.
3. W. Goldie, *Metallic Coating of Plastics*, 2 vols., Electrochemical Publications Ltd., Middlesex, England, 1968.
4. F. Pearlstein, and H. E. Bellis, *Electroless Plating of Metals*, American Electroplaters Society, Winter Park, Fla., 1972.
5. W. Mindt, *J. Electrochem. Soc.* **117**, 615 (1970); **118**, 93 (1971).
6. "The Industrial Revolution" in C. Singer and co-eds., *History of Technology*, Vol. 4, Oxford University Press, New York, 1958.
7. A. Wurtz, *C. R. Acad. Sci. Paris* **18**, 702 (1844); **21**, 149 (1845).
8. U.S. Pat. 1,207,218 (Dec. 5, 1916), F. G. Roux (to Societe L'Aluminium Francais); P. Breteau, *Bull. Soc. Chim.* **9**, 515 (1911).
9. U.S. Pat. 2,532,283 (Dec. 5, 1950), A. Brenner and G. Riddell (to U.S. Government).
10. U.S. Pat. 4,061,802 (Dec. 6, 1977), F. E. Costello.
11. E. B. Saubestre, *Plating* **59**, 563 (1972).
12. K. M. Gorbunova and A. A. Nikiforova, *Physiochemical Principles of (Chemical) Nickel Plating*, Academy of Sciences, Moscow, U.S.S.R., 1960 (Engl. transl. available from U.S. Government Office of Technical Services, Washington, D.C., 1963.
13. B. Y. Kaznachei, *Itogi Nanki Electrokhim.* 196 (1965); *Chem. Abstr.* **68**, 89414n (1968).
14. P. Cavallotti and G. Salvago, *Electrochim. Metal.* **3**, 23, 239 (1968).
15. D. A. Thompson, *Rack Design Considerations*, American Society of Electroplated Plastics, Washington, D.C., 1973.
16. U.S. Pat. 2,999,770 (Sept. 12, 1961), G. Gutzeit and A. Krieg (to General American Transportation Corp.).
17. *Metal Finishing Guidebook & Directory*, Metals and Plastics Publications, Inc., Hackensack, N.J., updated volume yearly.
18. *The Engineering Properties of Electroless Nickel Deposits*, International Nickel Co., New York.
19. U.S. Pat. 4,028,200 (June 7, 1977), K. Dockus (to Borg-Warner Corp.).
20. N. Feldstein, *Plating* **61**, 146 (1974).
21. R. Harris, *Appliance* **30**, 3 (1973).
22. R. G. Wedel, *Int. Metal Rev.* **217**, 97 (1977).
23. G. A. Krulik, *J. Chem. Ed.* **55**, 361 (1978).

24. G. A. Krulik, "The Catalytic Process in Electroless Plating," *63rd American Electroplaters Society Technical Conference, Denver, Col., 1976.*
25. Fr. Pat. 2,310,158 (May 5, 1976), G. A. Krulik and M. N. Jameson (to Borg Warner Corp.).
26. U.S. Pat. 3,011,920 (Dec. 5, 1961), C. R. Shipley (to Shipley Co.).
27. U.S. Pat. 3,682,671 (Aug. 8, 1972), R. J. Zeblisky (to Photocircuits Co.).
28. G. A. DiBari, *Plating* **60,** 1252 (1973).
29. G. L. Mallory, *Plating* **61,** 1005 (1974).
30. U.S. Pat. 3,764,352 (Oct. 9, 1973), M. Gulla (to Shipley Co.).
31. C. F. Coombs, *Printed Circuits Handbook,* McGraw-Hill, New York, 1967.
32. J. J. Grunwald, H. Rhodenizer, and L. Slominski, *Plating* **57,** 1004 (1970).
33. U.S. Pat. 3,625,758 (Dec. 7, 1971), F. T. Stahl and co-workers (to Photocircuits Corp.).
34. G. Messner and T. J. Sarnowski, "Additive Processes and Their Economics," *First National Plated Printed Circuits Conference, New York, Aug. 26–27, 1974.*
35. B. E. Norsworthy, *A Simplified System for Producing Additive Printed Circuits,* MacDermid, Inc., Waterbury, Conn.
36. F. Pearlstein and R. F. Weightman, *Plating* **61,** 154 (1974).
37. U.S. Pat. 3,671,291 (June 20, 1972), R. G. Miller and R. L. Cavitt (to PPG Industries, Inc.).
38. R. Sard, Y. Okinaka, and J. R. Rushton, *Plating* **58,** 893 (1971).
39. U.S. Pat. 4,019,910 (Apr. 26, 1977), G. O. Mallory (to The Richardson Chemical Co.).
40. U.S. Pat. 3,723,078 (Mar. 27, 1973), K. Parker (to General American Transportation Corp.).
41. U.S. Pat. 3,995,371 (Dec. 7, 1976), T. J. O'Keefe (to the University of Missouri).
42. A. F. Schmeckenbecker, *J. Electrochem. Soc.* **113,** 778 (1966).
43. A. Mayer and D. S. Catlett, *Plating* **65,** 42 (1978).

General References

Ref. 2 for U.S. patent literature 1966–1973.
Refs. 3, 11–14, and 22 for especially complete bibliographies.

Sources of Further Information

American Society of Electroplated Plastics, Inc., 1000 Vermont Ave., N.W., Washington, D.C. 20005.
The Institute for Printed Circuits, 1717 Howard St., Evanston, Ill., 60202.
The American Electroplaters Society, 1201 Louisiana Ave., Winter Park, Fla. 32789.
Plus any supplier of electroless processes (see ref. 17).

GERALD A. KRULIK
Borg-Warner Corporation

ELECTROLYTIC MACHINING METHODS

A tool composed of a harder material is used to shape a given metallic sample of material mechanically according to a specified pattern. As in the aerospace industry, the ever-increasing development of high strength, high temperature alloys places extreme demands on the hardness of the material from which a tool may be fashioned. Consequently, one is forced to employ grinding techniques that are not only laboriously slow and highly expensive but also severely limited with respect to the intricacy of the part to be machined. It, therefore, becomes necessary to find new methods for machining these new, extremely hard alloys.

One of the promising novel machining methods is the atom-by-atom removal of a metal by anodic corrosion which has been called electrochemical machining (ECM) (1–5). Because the metal is removed atom-by-atom, ECM affords the opportunity to machine a given workpiece without work-hardening, burring or smearing the metal, without regard to the hardness of the metal being cut, and virtually without tool wear.

In a chemical reaction, electron transfer between the reacting species occurs by the oxidation (loss of electrons) and the reduction (gain of electrons) taking place at the same site, and the energy liberated appears as heat and possibly light. For an electrochemical reaction to occur, the oxidation must take place at a site remote from the reduction. This situation is accomplished by interposing an electrolyte between the conductor (anode) at which oxidation occurs and that (cathode) at which reduction occurs (see Electrochemical processing).

When iron (qv) dissolves, ferrous ions enter the electrolyte by donating electrons to the anode. Eventually, the anode becomes so negative that further dissolution of iron is prevented by the electrostatic attraction between negatively charged anode and the positive ions in solution. If a sink for the electrons on the anode is provided by a battery that is connected in an external circuit between the anode and cathode, continuous dissolution of the iron anode can be obtained. Current is carried in the circuit by electrons (electronic conduction), and internally by ions (electrolytic conduction) through the electrolyte. The dissolution of metal at an anode driven by an external source of current, such as a battery or rectifier, is termed anodic corrosion (6) (see Corrosion; Metal anodes).

Principles of the ECM Process

In the ECM operation, the metal workpiece to be shaped is the anode, and the tool that produces the shaping is the cathode. The electrodes are connected to a low voltage source of direct current (dc). The anode and cathode are held in position by a properly designed fixture, and a solution of a strong electrolyte is pumped between the two electrodes. If there are no side reactions, the passage of each Faraday of electrical charge (96,500 C) results in the dissolution of an equivalent weight of metal. For the ECM process to be commercially competitive with conventional machining methods with a metal removal rate of ca 0.3 cm^3/s, the electrical charge must be passed at a rapid rate requiring the use of very high current densities (50–500 A/cm^2). In many cases, the upper limit to the current density that can be attained is determined by the availability of high capacity rectifiers. Voltages applied between the anode and cathode are normally 5–25 V.

The Solution Gap. At these high current densities, the solution path through which the current flows must be kept small to reduce an otherwise intolerably high IR loss. The typical gap between anode and cathode for acceptable electrochemical machining is 0.05–9.3 mm. Because metal is removed during the ECM process, the gap widens during the machining operation. The current density (ie, the metal removal rate) falls to low values unless relative motion between anode and cathode (ie, the feed rate) is maintained by a mechanically driven feed system. Either the anode is fixed and the cathode is moved or *vice versa*. Should the feed rate, for any reason, become too large for a given metal removal rate, a possible short circuit between the electrodes results, causing catastrophic damage to the tool and workpiece by spark erosion, thermal melting, welding, and tearing of the metal. Therefore, it is necessary to incorporate a protective device based, usually, on the detection of an increasing rate of current (7) which shuts down the current within microseconds when the gap becomes smaller than a specified, safe distance. Because there is a balance established between opposing forces of feed rate, which tends to narrow the gap, and of current density, which tends to widen it, an equilibrium or steady-state gap distance between workpiece and tool is reached quickly for a given feed rate and current density within certain boundary values (8–10).

Electrolytes. The current is conducted across the gap by the ions of a suitable electrolyte. Solutions of acids (HCl, H_2SO_4) have been used, but these electrolytes are very corrosive to the structural material of the machine (11). In addition, metal dissolved from the anode can be plated onto the cathode, thus causing the shape of the tool to change, and producing an unwanted change in shape of the workpiece. Alkaline electrolytes (NaOH, KOH) form protective anodic films on the workpiece surface and prevent metal dissolution. An exception is tungsten (qv), which can be machined only in strong alkali (7,12). The only practical electrolytes are neutral salt solutions in which a metal ion precipitates as a sludge. The most universally used salt is NaCl because of its availability and low cost (13). To reduce the IR drop in the gap, strong NaCl solutions of 5–25 wt % must be employed (14–15).

Any parameter that changes the conductivity of the electrolyte changes the current density, and hence, the metal removal rate.

Temperature Effects. If the electrolyte were to remain stagnant in the gap, enough heat would be generated by the large power or I^2R (W) loss at high current densities to vaporize the solution and thus bring the ECM process to a halt. This is prevented by pumping the electrolyte through the gap with a high pressure pump at a rate of 15–150 L/s and a pressure of 0.7–2.7 MPa (7–27 atm). An expression for the change in temperature, ΔT, of the solution entering and leaving the gap was derived for a cavity-sinking operation (8). Experimentally, measurements of ΔT are as high as 45°C (16). As much as a 100% change in the density and a 50% change in viscosity of a 10% solution of NaCl can be produced by such values of ΔT. Unless ΔT is kept below 10°C, uneven metal removal at hot spots in the gap may ruin the integrity of the machined part (14,17). Usually, a heat exchanger is incorporated in the flow line to control the temperature of the intake solution within ±1°C (see Heat exchange technology). Although ΔT can be lowered by increasing the solution flow rate, a point is reached above which cavitation of the fluid occurs, producing striations and grooves in the metal surface. The flow patterns in the gap were studied by addition of ink to the solution (9). These studies revealed that increasing the back pressure at the exit of the gap reduced the range of cavitation. Therefore, with proper design, cavitation effects can be avoided in the high-flow ECM system.

Corrosion Products. Even if temperature effects could be ignored, corrosion products from the metal dissolution process at the anode would accumulate in the gap; and the IR drop across the solution path would become so large that metal removal would stop. The second reason for a high solution flow rate is to sweep corrosion products out of the gap. As long as the corrosion products remain in a uniform colloidal suspension, good ECM finishes are produced (11); but if the solution contains particulate matter (lumps of coagulated hydroxides, nonmetallic inclusions, metallic grains), which lower the electrolyte conductivity (18) and increase the viscosity, the resulting uneven flow rates and nonuniform current distribution produce poor surface finishes. Customarily, a filtering unit is added to the flow line to remove the interfering particulates (13,15,19). Other sludge removal methods, such as centrifugal separation (qv), use of settling tanks, and frothing and flotation (qv) methods, have been employed (19).

Pressure Effects. At the cathode (usually made of brass, copper, or tungsten–copper (see Copper alloys)) where the reduction of water to gaseous hydrogen takes place ($2 H_2O + 2 e \rightarrow H_2 + 2 OH^-$), bubbles of H_2 are injected into the stream of electrolyte. These bubbles decrease the conductivity of the electrolyte and produce nonuniform current distribution (20). Photographic studies of gaps have been performed in flow cells made of clear plastic (9,21–25). A two-phase region known as the bubble layer (24) was observed to increase in thickness along the direction of solution flow at the cathode. It was reported that both the bubble size and the thickness of the bubble layer diminished with increasing flow rate as expected from the required increase in hydrostatic pressure. To withstand the high solution pressures, an ECM machine must be ruggedly built.

Types of ECM Operations

In the external shaping operation diagrammed in Figure 1a the cathode tool is shaped to fit the desired contours of the finished workpiece (a turbine blade in this case). As shown, two cathodes may be used to shape both sides of the blade simultaneously. Turbine blades that require 1–2 h to finish by conventional methods are machined by ECM to tolerances of 0.008 cm in 5–10 min (26). Since no pressure is applied to the workpiece, thin foils or honeycomb structures may be machined as easily as heavy castings without warping, bowing, twisting, burring, or smearing of edges. In addition, the anodic corrosion process does not induce mechanical stresses in the metal lattice, or thermal modification of metallurgical structures. It is easy to machine both extremely hard and soft materials. In certain alloys, however, the fatigue properties of the metal may be lowered (27–29), but the fatigue strength may be restored by vapor-blasting or shot-peening the machined surface. Embrittlement of the workpiece by hydrogen is not a concern in ECM because H_2 is liberated only at the cathode. There is no tool wear in ECM since there is no physical contact between tools and workpiece. Parts may be reproduced indefinitely as long as the tool is not damaged by sparking and short circuits. In order to obtain acceptable surface finishes (0.5–1.0 μm), it is essential to have uniform flow of electrolyte between the anode and cathode by proper design of the tool and the ECM fixture (electrochemical cell). Although cathode and fixture design requires considerable skill and time (ie, it may be expensive), these initial tooling costs can be recovered quickly through the high efficiency of the ECM process and the virtual lack of tool wear, if the replication of the part to be machined is high.

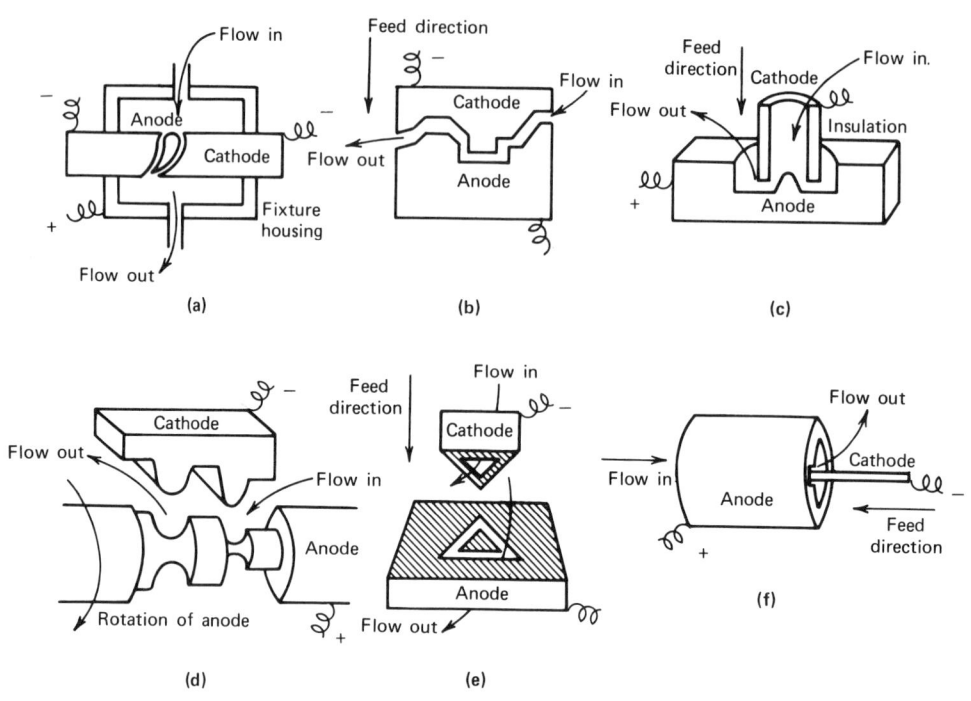

Figure 1. Various ECM operations. (a) External shaping; (b) cavity-sinking; (c) plunge-cutting; (d) turning; (e) trepanning; (f) internal grooving.

Figure 1b shows the ECM operation of cavity (die) sinking. The cathode is fashioned according to the shape of the cavity to be machined in a block of metal. A uniform current density is attained across the face of the cathode tool after it has completely entered the workpiece. A complex cavity in fully hardened steel, generally requiring many conventional machining operations, can be fashioned by ECM with good surface finish in just one operation.

Another important ECM operation is plunge-cutting, the essential parts of which are sketched in Figure 1c. In this hole-cutting operation, the electrolyte is pumped through a hollow cathode that has a shape corresponding to that of the hole to be machined in the workpiece. The exterior walls of the hollow tool are coated with an electrically insulating film (epoxy coating) so that the metal is removed only at those sites on the workpiece opposite the face of the tool. As noted in Figure 1c, a small metal slug is trepanned in the center of the plunge-cut. There is danger that this slug may hinge at the end of the cut, causing a short circuit by falling against the side of the cathode. To avoid this damaging action, a false or backup workpiece is attached to the bottom of the anode (30). Although there is usually some tapering of a plunge-cut hole, the taper may be minimized somewhat by a compensating design of the cathode.

In Figure 1d the symmetrical contouring of a rotating rod anode by a fixed cathode of proper design is accomplished by the turning (lathe) operation.

Many complex shapes of uniform thickness and good surface finish can be machined with a high degree of replication from a metal slab in the trepanning operation pictured in Figure 1e.

In the internal grooving operation shown in Figure 1f, a thin or pointed cathode, with all sides except the face insulated with epoxy coating, is passed down the inside wall of a hollow cylinder to form a groove (eg, a key slot). If the cylinder is rotated, a helical groove is obtained. In this operation as well as in the others, it is important to obtain proper alignment of the tool with the workpiece and to maintain this alignment throughout the machining operation to avoid loss of dimensional and geometrical control.

The wire-cutting operation shown in Figure 2 reduces the amount of high-current electricity needed to rough out large volumes of metal by ECM. The cathode is either a wire over which the electrolyte is sprayed, or a thin hollow tube through which the solution is pumped and exits through a series of holes on the cutting face of the tube. Sections of metal are removed by passing the wire through the metal. Tube-end cutting and metal slicing are possible.

To remove burrs from a part, a deburring machine shown schematically in Figure 3 is employed. In contrast to other ECM machines, the deburring operation uses fixed electrodes held at gap spaces wider than other operations. Usually, the solution flow rate is lower.

With electrochemical grinding (ECG), the cathode is an electrically conducting grinding wheel mounted on an insulated drive shaft. A jet of electrolyte is directed against the wheel. On the face of the wheel, diamond bort is embedded and these diamonds maintain a constant gap space between the wheel and the work (see Carbon, diamond). Nearly 100% of the metal is removed electrochemically. As a result, only light pressure is applied to the wheel whose life is thereby increased by more than an

Figure 2. Demonstration of wire-cutting operation. Courtesy of the Society of Automotive Engineers, Inc.

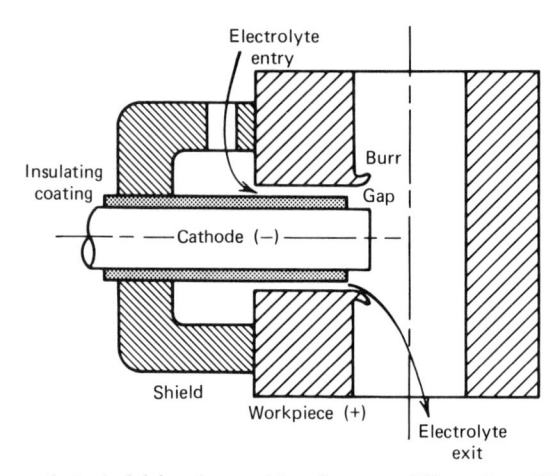

Figure 3. Diagram of a typical deburring machine. Courtesy of Chem-Form Division, KMS Industries, Inc.

order of magnitude. Delicate structures such as metal honeycombs and felts can be machined without burring or warping.

Modern Development of the ECM Process

In most of the early work, solutions of NaCl were used as the electrolyte, but relatively large overcuts were obtained with rounding of corners and edges and tapering of holes. Even large overcuts could be tolerated if they were uniform and predictable because compensating corrections could be made in the design of the tool to overcome the lack of control of dimensions and geometry. It appears, however, that serious limitations to the amount of agreement between the predicted and the actual overcut exist because the uniformity of the cut decreases as the overcut becomes larger. By considering the electrolyte merely as another ohmic element of the ECM system, most investigators fail to take adequate account of the electrochemistry involved in the derivation of overcut equations from a theoretical analysis of the ECM process (9,10,21,31–32). However, with equations that are understandably complex, a reasonable mathematical model has been established (33). Some success with empirical methods such as the cosine method and simulation operations for predicting overcut has been reported (34–35). Until the electrochemical properties of the anodic film and electrolyte flow effects can be factored into the mathematical expression, a useful solution of this central problem will not be obtained. As yet this formidable task has not been accomplished.

Because NaCl solutions are highly corrosive and support strong stray currents, addition agents have been added to the NaCl electrolytes to suppress these effects. Nitrates, nitrites, amines, polyalcohols, etc, have been employed as corrosion inhibitors to protect the machinery (19,36), and anodic film-forming agents have been used to suppress metal removal by stray currents. Although these additives improve the overcut characteristics of NaCl electrolytes, all of the problems are not eliminated. In addition, the cost of the electrolyte was drastically increased (by 2500–10,000%) (see Corrosion and corrosion inhibitors).

Electrolyte Development. In the attempted application of ECM machines using the NaCl electrolytes for everyday operations in the automotive industry, dimensions could not be held, tolerances could not be met, and because of unwanted metal removal at points remote from the cutting face of the cathode (stray or wild cutting), the ECM operation could not be made compatible with other manufacturing operations and the workpiece integrity was destroyed. Consequently, a search for a completely new type of ECM electrolyte that would provide its own inherent protection of sites against unwanted metal removal is most desirable.

From an investigation of the ECM properties of several hundred electrolytes and combinations of electrolytes, it was reported that solutions of sodium chlorate, $NaClO_3$, provided a superior ECM electrolyte with respect to high metal removal rates, excellent surface finish, and the desired control of dimensions and geometry for the machining of iron, steel, and ferrous alloys (37) (see Metal surface treatments).

As an example of the type of dimensional control obtained, notches that were machined in fully hardened 5160H steel bars with NaCl (bottom) and $NaClO_3$ (top) electrolytes are shown in Figure 4.

At all concentrations, the metal removal rate for steel in $NaClO_3$ solutions is about the same as for that in NaCl up to the solubility limit for NaCl. Since $NaClO_3$ is more soluble than NaCl, the metal removal rate increases further with increasing concentrations of $NaClO_3$ up to ca 600 g/L. The maximum cutting rate in $NaClO_3$ is twice the maximum rate in NaCl, and the machined surface still retains the mirror-bright surface at these high cutting rates.

When wet or in 65–75% rh conditions, $NaClO_3$ (as well as $NaNO_3$) does not present a fire hazard, but when dry and in the presence of finely divided organic materials, combustion danger is severe (38) (see Chlorine oxygen acids and salts).

Figure 4. Notches cut in steel samples by ECM in $NaClO_3$ (top) and NaCl (bottom). Courtesy of The Electrochemical Society, Inc.

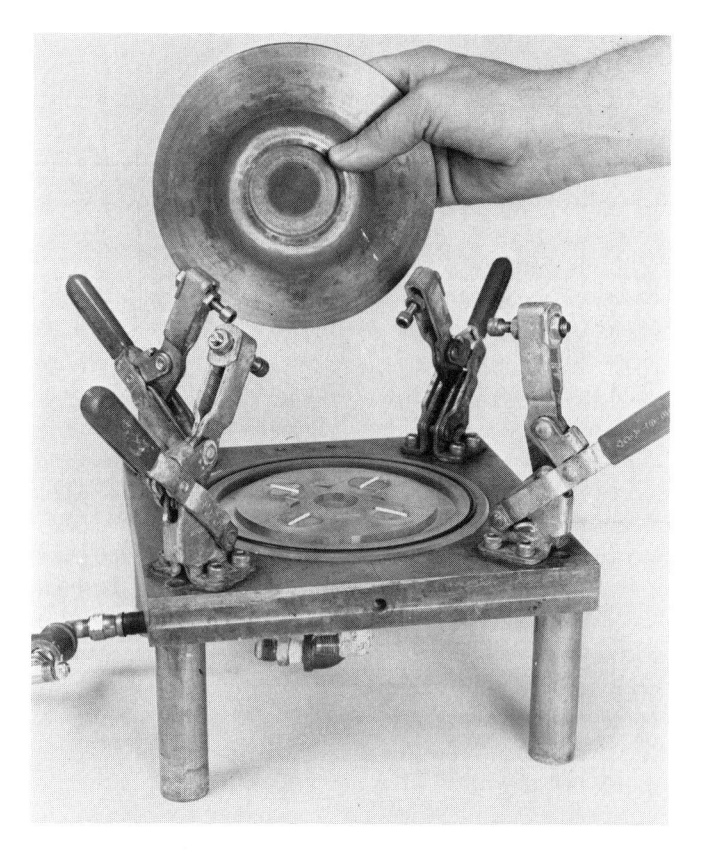

Figure 5. Embossing fixture for hydrostatic bearing pads showing cathodes, clamp lid, and clamps.

Fundamental Investigations. Investigations of iron and steel anodes in various electrolytes with constant potential polarization studies (39–42), constant-current film-stripping measurements (39–40), current-efficiency determinations (43–45), and high-rate flow cell results (45–47) showed that the electrochemical machining characteristics are strongly dependent on the electrochemical properties of the anodic films formed on the metal surface of the workpiece in a given electrolyte (48). A good ECM electrolyte is one in which a potential-dependent, protective film is formed on the metal surface of the workpiece. In the high current density region at sites on the workpiece directly opposite the cutting face of the cathode tool (at the bottom of a hole), metal is removed rapidly because a protective film is not present. Along the sides of the hole in the low current density region remote from the cutting face of the tool, the metal surface is protected by an oxide film and little or no metal is removed. For the machining of iron and ferrous alloys by ECM, this good dimensional control is obtained with $NaClO_3$ electrolytes (40–41,43).

A phenomenological model of the potential-dependent protective film has been proposed (43–44,49–50). This film is composed of hydrated ferric oxide units hydrogen-bonded together to form a three-dimensional matrix. At low current densities, this matrix covers the anode surface and hinders metal dissolution; at high current

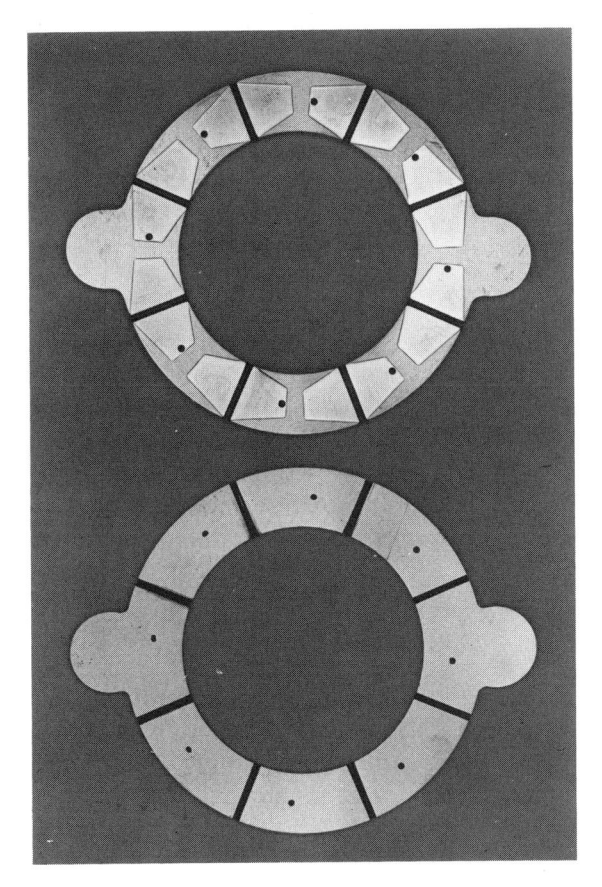

Figure 6. Hydrostatic bearing pads machined in NaClO$_3$ solution (top) from blank (bottom) by embossing type operation.

densities an ion exchange takes place between anions adsorbed from solution on the film and the oxide ions of the film to form soluble products which dissolve away, allowing metal to dissolve at high rates. The electrochemical properties of the film also determine whether the machined surface is electropolished, matte, or roughened (43,49–50).

Apparently, there is no universal electrolyte in which all metals can be electrochemically machined equally well. Since the end product of the ECM operation depends strongly on the nature and properties of the anodic film formed on the metal in a given electrolyte, and since these properties of the anodic film are different from metal to metal in a given electrolyte (45,51), obtaining a commercially acceptable product from ECM depends greatly on the choice of electrolyte.

Innovations in ECM Operations

Static Fixture Finishing and Sizing. One of the new concepts in the development of the ECM process is the use of ECM fixtures as an accurate sizing device for finishing surfaces. In such a system, there is no large, expensive ECM machine involved.

Figure 7. The fixture for pattern ECM with anode holder and anode sample panel at left and machined cathode at right. Courtesy of The Electrochemical Society, Inc.

Embossing. It is possible to design an arrangement of ECM fixtures such that embossed-type patterns may be applied to finely finished steel surfaces because of the low overcut nature of NaClO$_3$ solutions. An example is an application employing the fixture shown in Figure 5. Oil pads of 1.3–12.7 μm depth have been machined in bearings with excellent lineage, geometry, and depth control as exemplified by the hydrostatic bearing pictured in Figure 6. With ECM machining time of only 1–2 s for the production of hydrostatic bearing pads, a method for industrial production becomes entirely feasible where previously none existed.

In yet another type of embossing, in a process termed Intricate Pattern ECM, a half-tone type pattern of 12.6 lines/cm is embossed into steel (52). The resultant 50 μm depth is obtained in 15 s. The fixture used is shown in Figure 7, and the end product in Figure 8.

Broaching. One of the limitations of the ECM process is the inability of the system to machine large areas. At an average current density of 39 A/cm^2, the usual ECM machine (10 kA) is capable of machining only 258 cm^2 at one time. The area limitation can be overcome in many cases by using a moving, limited-area cathode that removes metal as it travels past the anode. If the geometry of the anodic workpiece is regular, it is possible to produce the part with an electrolytic grinder, an electrolytic lathe, or an electrolytic broaching machine.

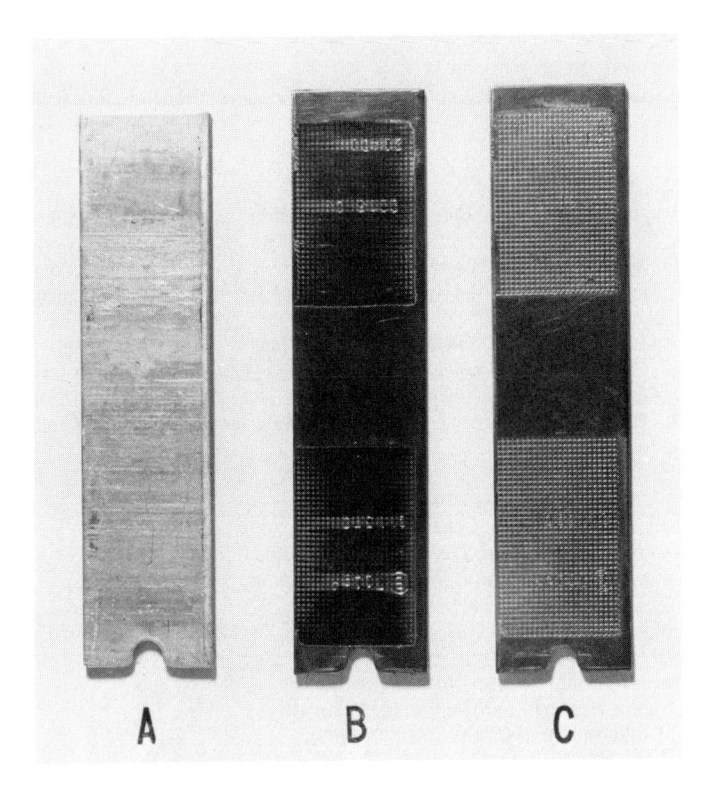

Figure 8. The test panels of mild steel, 20 mm × 90 mm × 1 mm: A, the blank; B, a washed out pattern because of wide gap space or excessively high applied potential; and C, a sharp pattern obtained at a gap of 0.20 mm, at a preset applied voltage of 5 V, and a solution flow rate of 2 L/min. Courtesy of The Electrochemical Society, Inc.

In principle, the ECM broaching operation is one in which the inside diameter of an anodic workpiece is machined by a cathode passing through the anode. By reversing the electrical polarity and by any necessary fixture component arrangement, the outside diameter of the workpiece may be machined. Although most interest is directed towards machining of the ID or OD of workpieces having a right circular cylinder geometry, the ECM broaching of other right cylinder geometries (oval, square, rectangle, etc) can be carried out with the proper fixturing and cathode design.

The broaching machine can accept a part for which the length of required stroke may be 1–50 cm, and has the capability of directing flow of the electrolyte from either the top or bottom of the fixture. By using a variable drive unit, the broach has a stroke travel speed ranging from 0.13 to 2.54 cm/s, thus a cathodic "window" may be moved through an anodic area 50 cm long in 20 s.

Uses

ECM is commonly employed in the transportation and aerospace industries. Turbine blades, vanes, disks, and transmission parts are made by ECM. Electrochemical machining is also used in the manufacture of precision guidance parts for missiles.

BIBLIOGRAPHY

"Electrolytic Machining" in *ECT* 2nd ed., Vol. 7, pp. 866–873, by Joseph Crawford, Ingersoll Milling Machine Company.

1. A. E. DeBarr and D. A. Oliver, *Electrochemical Machining*, Elsevier Scientific Publishing Co., Inc., New York, 1968.
2. J. F. Wilson, *Practice and Theory of Electrochemical Machining*, John Wiley & Sons, Inc., New York, 1971.
3. J. P. Hoare and M. A. LaBoda, "Electrochemical Machining" in E. Yeager and A. Salkind, eds., *Techniques of Electrochemistry*, Wiley-Interscience, a division of John Wiley & Sons, Inc., New York, 1978, Chapt. 2, p. 48.
4. J. P. Hoare and M. A. LaBoda, *Sci. Am.* **230**, 30 (Jan. 1974).
5. J. A. McGeough, *Principles of Electrochemical Machining*, Chapman and Hall Ltd., London, 1974; A. D. Davydov, *Elektrokhimiya* **11**, 809 (1975).
6. J. P. Hoare, *The Electrochemistry of Oxygen*, Wiley-Interscience, a division of John Wiley & Sons, Inc., New York, 1968, p. 357.
7. *Am. Mach.* **111**(22), 149 (1967).
8. R. R. Cole, *Int. J. Prod. Res.* **4**, 75 (1965).
9. K. Kawafune and co-workers, *Ann. CIRP* **15**, 443 (1967).
10. J. A. McGeough and H. Rasmussen, *Trans. ASME* **92B**, 400 (1970).
11. N. D. G. Mountford, *Trans. Inst. Met. Finish.* **40**, 171 (1960).
12. W. B. Kleiner, *Trans. ASAE* **72**, 123 (1964).
13. L. A. Williams, *Aircr. Prod.* **22**, 389 (1960); *Tool Engineer* **43**(12), 43 (1959); *Chem. Eng. News* **44**(34), 32 (1966).
14. C. L. Faust, *Trans. Inst. Met. Finish.* **40**, 171 (1963).
15. S. Ito and N. Shikata, *J. Mach. Lab. Japan* **12**, 50 (1966).
16. H. Kubeth and H. Heitmann, *Ind. Anz.* **46**, 975 (1963).
17. W. G. Clark and J. A. McGeough, *J. Appl. Electrochem.* **7**, 277 (1977).
18. R. De la Rue and C. W. Tobias, *J. Electrochem. Soc.* **106**, 827 (1959).
19. A. I. W. Moore, *PERA Res. Rept. 145*, Leicestershire, 1965.
20. C. W. Tobias, *J. Electrochem. Soc.* **106**, 833 (1959).
21. J. Hopenfeld and R. R. Cole, *Trans. ASME* **88B**, 455 (1966); **91B**, 755 (1969).
22. S. Ito, K. Chikamori, and F. Sakurai, *J. Mech. Lab. Japan* **12**, 37 (1966).
23. S. G. Bankoff, *Trans. ASME* **82C**, 265 (1960).
24. D. Landolt and co-workers, *J. Electrochem. Soc.* **116**, 1384 (1969); **117**, 839 (1970).
25. S. P. Lautrel and N. H. Cook, *Trans. ASME* **95B**, 992 (1973).
26. C. L. Faust and C. A. Snavely, *Iron Age* **186**(18), 77 (1960).
27. J. A. Gurklis, *DDC Rept. AD 613261*, Jan. 7, 1965; *Cobalt Engl. Ed.* **39**, 81 (1968).
28. J. Baunard, *J. Appl. Electrochem.* **4**, 117 (1974).
29. G. Rowden, *Metallurgia* **77**, 188 (1968).
30. W. A. Haggerty, *SAE Paper 680C*, April, 1963.
31. J. Bayer, M. A. Cummings, and A. U. Jollis, *DDC Rept. AD 450199*, Sept., 1964.
32. W. Konig, D. Pahl, and H. Degenhardt, *SME Paper MR 70-250*, 1970.
33. J. M. Fitz-Gerald, J. A. McGeough, and L. McL. Marsh, *J. Inst. Math. Its Appli.* **5**, 387, 409 (1969); **6**, 102 (1970).
34. H. Tipton, *I.E.E. Conference Electrical Methods Machining and Forming*, London **38**, 48 (1967).
35. J. A. McGeough, *Principles of Electrochemical Machining*, Chapman and Hall Ltd., London, 1974, p. 221.
36. U.S. Pats. 3,389,067 (June 18, 1968), 3,401,104 (Sept. 10, 1968), 3,421,987 (Jan. 14, 1969), and 3,429,791 (Feb. 25, 1969), M. A. LaBoda (to General Motors Corp.).
37. M. A. LaBoda and M. L. McMillan, *Electrochem. Technol.* **5**, 340, 346 (1967); U.S. Pat. 3,669,858 (June 13, 1972), M. A. LaBoda (to General Motors Corp.).
38. *Safe Handling and Use of Sodium Chlorate, Chemical Data Sheet, SD-42*, Manufacturing Chemists Association, revised 1952.
39. J. P. Hoare, *Nature (London)* **219**, 1034 (1968); *J. Electrochem. Soc.* **117**, 142 (1970).
40. J. P. Hoare and co-workers, *J. Electrochem. Soc.* **116**, 199 (1969).
41. M. A. LaBoda, J. P. Hoare, and S. E. Beacom, *Collect. Czech. Chem. Commun.* **36**, 680 (1971).
42. J. P. Hoare, A. J. Chartrand, and M. A. LaBoda, *J. Electrochem. Soc.* **120**, 1071 (1973).

43. K.-W. Mao, M. A. LaBoda, and J. P. Hoare, *J. Electrochem. Soc.* **119**, 419 (1972).
44. J. P. Hoare and K.-W. Mao, *J. Electrochem. Soc.* **120**, 1452 (1973).
45. J. P. Hoare and C. R. Wiese, *Corros. Sci.* **15**, 435 (1975).
46. K.-W. Mao, *J. Electrochem. Soc.* **120**, 1056 (1973).
47. K.-W. Mao and D. T. Chin, *J. Electrochem. Soc.* **121**, 191 (1974).
48. M. Datta and D. Landolt, *J. Electrochem. Soc.* **124**, 483 (1977).
49. J. P. Hoare and M. A. LaBoda in R. Baboian, ed., *Electrochemical Techniques for Corrosion*, NACE Publ., Houston, 1977, p. 106.
50. J. P. Hoare, K.-W. Mao, and A. J. Wallace, *Corrosion* **27**, 211 (1971).
51. M. A. LaBoda and co-workers, *J. Electrochem. Soc.* **120**, 643 (1973).
52. M. A. LaBoda and J. P. Hoare, *J. Electrochem. Soc.* **122**, 1489 (1975).

JAMES P. HOARE
MITCHELL A. LABODA
General Motors Research Laboratories

ELECTROMIGRATION

Electromigration or electrotransport refers to the net motion of atoms owing to the passage of an electrical current through a metallic conductor in either the solid or liquid state. Simple metals may be considered to be constituted of atoms stripped of their valence electrons, existing as ions surrounded by a sea of electrons. In the presence of an electric field, the ion cores are subjected to a force directly resulting from the field, as in an electrolytic solution. However, in electromigration there is, in addition to this direct field force, a force resulting from the friction between the numerous and rapidly moving electrons and the ion cores. This force, known as the electron wind force, is in the opposite direction to the field force in electron conductors, and for many metals is considerably greater than the field force. In metals with complex electronic structures, which are not electron conductors but are hole conductors, the concept of electron wind force is extended to refer to the friction resulting from the motion of the charge carriers regardless of their sign. Whereas in electrolytic solutions the current results entirely from the motion of ions, in usual electromigration phenomena the fraction of the current contributed by ion motion is quite negligible in comparison to that fraction of the current caused by the motion of the charge carriers, either electrons or holes.

The earliest known reference to electromigration is by Gerardin, who experimented with liquid metal alloys simultaneously with molten salts (1). Gerardin did not distinguish clearly between the two different types of phenomena, electrolysis and electromigration, yet his experiments were quite ingenious and deserve more than passing mention. He studied molten potassium uranate even though uranium had been isolated only twenty years previously. In liquid amalgams, where the main constituent is mercury, he found that the soluble elements, gold and potassium, moved towards the negative electrode. Working with molten plumbers' solder, presumably a Pb–Sn alloy, he observed that the metals collected at the two electrodes had different mechanical properties, implying different compositions. Particularly elegant was Gerardin's discovery that the passage of current through a liquid eutectic mixture of potassium and sodium caused solidification as a result of the separation of the two elements. About fifty years elapsed before the fundamental concept of friction, *reibung*, between atoms and charge carriers was introduced by Skaupy (2) in the early 1900s.

Current literature may be said to date from the middle of this century. A seminal set of experiments was performed on a series of Hume-Rothery phases belonging mostly to the copper-aluminum system (3). By showing that the direction of transport varied from phase to phase, they demonstrated the important role played by the sign of the charge carriers. Theoretical formulations of the electromigration force used a classical mechanics calculation of the momentum transfer from the charge carriers to atoms (4–5). Quantum mechanical calculations were used to arrive at essentially identical relations for the wind force (6). Recently, electromigration forces have been calculated for a large number of atoms through the use of pseudopotential approximations (7). Electromigration continues to be a subject of active interest for experimentalists, as well as theoreticians. In several books published during the last ten years the problems of electromigration are analyzed in detail and the current literature is thoroughly reviewed (8–11). A monograph on electromigration offers the double advantage of being quite up-to-date and of presenting various aspects of electromigration (eg, theory, applications to liquids and thin films, purification) in a concise tutorial manner (12).

In this article, after an analysis of some fundamental characteristics of electromigration phenomena, attention is focused on two areas where electromigration plays a technologically significant role. On the one hand, in the refining of some metals to a very high state of purity, electromigration has been found to be quite useful in cases where other techniques, such as zone refining (qv), have been found wanting. Electromigration in this case plays an important and very positive role. On the other hand, in the present electronic technology, electromigration is known for its deleterious effects causing failures in the thin film conducting lines which are an integral part of most electronic microcircuits (see Semiconductors).

Fundamental Aspects

Phenomenologically, electromigration results from a slight bias in the random motion of atoms and can be considered as a special case of diffusion (see Diffusion separation methods). Treatments of electromigration can be found in basic books on diffusion, eg, refs. 13 and 14. The velocity of the moving atoms v is given by the product of the mobility and the force, according to the Nernst-Einstein relation. This can be written as:

$$v_i = \frac{D_i}{kT} \cdot F_i \tag{1}$$

where D is the diffusion coefficient, F is the force on the moving atoms, k is the Boltzman constant, T is the temperature, and the subscript i refers to the atoms of the ith species. In diffusion problems the driving force is of the form $d\mu_i/dx$ which in many cases can be expressed more simply as $kT \cdot dc_i/dx$, in terms of the gradients in chemical potential or concentration. In considering electromigration phenomena it has become customary to write the driving force as:

$$F_i = Z_i^* \cdot Ee \tag{2}$$

which is to say that the force on the migrating atoms is given by the product of the electrical field E and an effective charge on the atoms $Z_i^* e$ (e is the charge of the electron). For the experimentalist, the problem is to obtain values of Z_i^* and D_i simultaneously, ie, to separate their individual values from their product, which is more directly accessible from the measurements than either Z^* or D. Theoreticians must try to calculate values of Z^* from first principles.

The theory of electromigration has been the subject of considerable activity in the last six years. Most of the published articles, however, do not contain results that would be directly accessible to experimentation; hence their technological importance remains rather limited. Only Sorbello has derived quantitative values for the electromigration forces (7). He has also written the only review article of the current theoretical literature (15). A quite thorough bibliography, which can be found in an article on electromigration in thin films (16), does not include the most recent publications (17–20). The calculations are quite complicated and beyond the scope of this article; only an outline of the problem is given here.

The Forces in Electromigration. The force on an atom in the presence of a current is usually considered to consist of two parts, one owing to the field F_e and one owing to the friction of the charge carriers F_{wd}. Of course this division is somewhat arbitrary and ideally one would hope to arrive at a single formulation. However, the consideration of those two types of force is intuitively suggestive. The force in equation 1 is of a statistical kind with a unique value averaged over a large number of atomic jumps. On a microscopic scale in a crystalline solid one should consider the changing configuration over a single jump path, calculate the electromigration forces for different positions along this path and find the accurate means of averaging these from one rest position to the next. For matrix atoms and for substitutional impurities, the problem is complicated by the presence of vacancies, which in most metals are required for any diffusive atomic motion. This is schematically represented in Figure 1. For an atom in position a to move to position c, it must move through a saddle point, position b, where it is surrounded by two half-vacancies as shown in Figure 1b. In pure gold it is estimated that although the electron wind force, expressed in arbitrary units, varies from 1 to 3.75 for positions a and b, respectively, it has an average value of 3 for the jump from a to c (7). For an interstitial impurity moving from position d to position e (Fig. 1a), the problem appears simpler and it is assumed that the force is constant. Unfortunately, from a practical point of view, the interstitial atoms that are of greatest technological importance are the elements hydrogen, oxygen, carbon and nitrogen

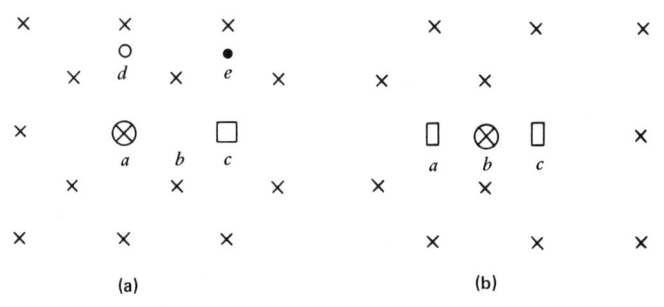

Figure 1. Schematic configuration of atoms moving from one position to the next in a single atomic jump.

in solution in transition metals. In those metals, the band structure of the electronic states is complex, the free-electron approximation does not hold, and theoretical calculations of the electromigration force are unavailable.

In general the electron wind force is expressed as:

$$Z^*_{wd} = - Z \cdot \gamma \cdot \frac{N \cdot \rho_d}{N_d \cdot \rho} \qquad (3)$$

where Z is the valency of the matrix atoms, γ is an averaging term, ρ_d is the specific resistivity of the mobile defects, N_d/N is the concentration of moving defects, and ρ is the resistivity of the metal. Although one is accustomed to think of the flow of electricity without the creation of mechanical forces between current and conductor, equation 3 upon reflection conforms to common physical sense. It indicates that the force on a defect is proportional to the contribution of this defect to the overall resistivity, as well as to the density of charge carriers; the $1/\rho$ term results from the competition between phonons and defects in scattering these charge carriers. Because of this matrix resistivity term, in cases where the electron wind force predominates, the absolute magnitude of Z^* should decrease with increasing temperature. This is in agreement with most experimental observations. In transition metals, where charge transport occurs simultaneously through electron and hole motion, the relative contributions of the two types of charge carriers to electromigration can be anticipated to change with temperature, so that Z^* may not vary as $1/T$. For substitutional alloy additions it is anticipated that ρ_d should vary as $(Z - z)^2$ where Z is the valency of the matrix atoms and z the valency of the solute atoms. It has been pointed out (6) that differences in the effective charge between lattice rest positions and saddle point positions cause the average value of z^*_{wd} to vary as $z^2 + (Z + z)^2$. The validity of this expression was demonstrated in a nice series of experiments conducted on a variety of alloying elements in Ag (21). Although it is usual to give values of Z^* as multiples of the unit electronic charge, F_{wd} can be described per unit of current density (7), which correctly reflects the proportionality between the flux of atoms and that of electrons or holes. For the overall effective charge one obtains an equation of the following form:

$$Z^* = Z \left(1 - \gamma \cdot \frac{N \cdot \rho_d}{N_d \cdot \rho} \right) \qquad (4)$$

There has been some experimental work done on electromigration in semiconductors but the literature on the subject is limited (see ref. 8). Experimental effects are small due to the relatively small values of D and of the wind force which is restricted by the small number of charge carriers. This remains true although their momentum is usually greater than that of their counterparts in metals as a result of the high mobility of the charge carriers in semiconductors. An example from present semiconductor technology lends some credence to the distinction between field and wind forces in electromigration. During the fabrication of Li drifted solid state detectors, constituted of the rectifying junction between p and n type Si, Li ions migrate under the application of a reverse potential (22). There the acting force is obviously equal to the direct potential (22). There the acting force is obviously the direct field force F_e, since the current is approximately zero. If drifting took place with the potential applied in the forward direction, the acting force would be the sum of F_e and F_{wd}.

In gold Z^*, the effective charge for Au atoms, is -7 (23) which indicates that the magnitude of F_{wd} is about eight times greater than F_e. In general, for simple metals, such as Cu, Al, or Au, the values of Z^* calculated from reasonable values of ρ_d are commensurate with those obtained from experiments. However, many of the interesting cases of electromigration involve metals with complex electron band structures where the theory is found wanting. Experimental results at times appear to defy explanation: for example, Ag and Au impurities in Pb are reported to migrate in opposite directions (24); in the electromigration of hydrogen and deuterium the effective charge on deuterium has always been found to be larger than that on hydrogen (25).

Common Experimental Techniques. For most solid materials the diffusion coefficients and the effective charges are sufficiently small so that experimental measurements of electromigration transport can be made only under circumstances that maximize the effect of both terms. In practice this means that electromigration experiments on bulk samples, as distinguished from thin films, are conducted at the highest current densities that are compatible with the requirement that the samples should not melt as a result of Joule heating. Generally, the current densities used are of the order of 10 kA/cm². Although one finds a variety of experimental techniques, it may be said that most of these conform to one of three different types which are described briefly.

The Drift Technique. In the drift technique, two cylinders of B are welded to a plane cross-section, and the cross-section is plated to a layer of A, in order to study the electromigration of elements A and B. The current is then made to flow across the length of A. Ordinary diffusion causes A to spread in opposite directions through the two cylinders of B; however, because of electromigration the center of mass of A will move from its original position as represented in Figure 2. The displacement d of the concentration profile of A with respect to the original weld joint (marked, eg, with an inert oxide) is a direct measure of electromigration velocity, and the spread in this concentration profile (Gaussian) gives the value of the diffusion coefficient. Provided that corrections can be made for the drift of the host element B, the drift technique yields simultaneous measurements of D and Z^*, which allows for considerable precision in the determination of Z^*. It has been used extensively in the study of the electromigration of substitutional alloy additions in silver (21). If the proper isotopes are available, A can be chosen to be an isotope of B, in which case the drift technique can

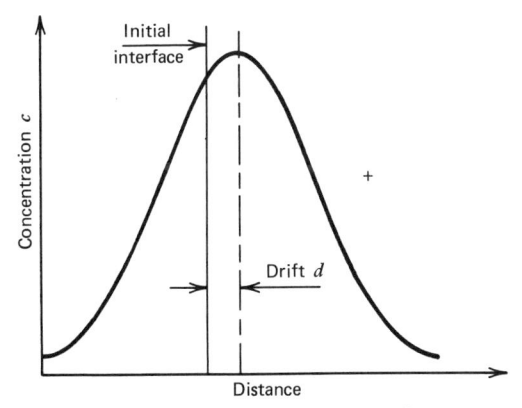

Figure 2. The diffusion spread of concentration and the associated electromigration displacement of the center of the concentration profile in the drift technique of investigation.

be used for the determination of self-electromigration, as has been done for Au (26).

The Marker Motion Technique. The drift technique requires that in the sample used for an electromigration experiment the central zone be maintained at a uniform temperature, which presents many experimental difficulties. In the marker motion technique, on the contrary, temperature gradients are used systematically to determine self-electromigration parameters. If a wire, held at its extremities by two cold electrodes, carries a high current, its temperature distribution will assume a parabolic curve as shown in Figure 3a. The electrotransport flux varies from zone to zone according to the temperature variations of the diffusion coefficient, $D = D_0 \exp(-Q/kT)$ where Q is the activation energy. As a result of the atomic flux changes, surface markers initially spaced equally along the wire are displaced with respect to one another. The effective charge also changes according to temperature; but both changes are small in comparison with variations in the diffusion coefficient. Variations in marker displacements along the length of the wire are graphically represented in Figure 3b. Measurement of these displacements yields the magnitude of Z^* and its variations as a function of temperature. Care must be exercised to make proper corrections for changes in the cross-section of the samples (including possible void formation) resulting

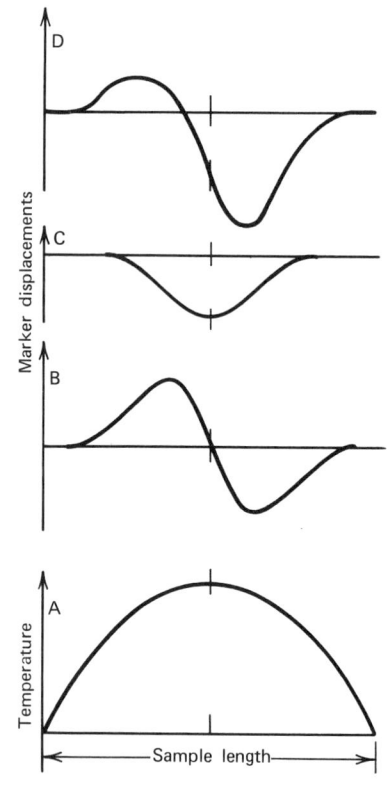

Figure 3. Typical conditions for the marker motion technique for investigating electromigration phenomena. A, the temperature distribution; B, the marker displacements owing to electromigration alone; C, the marker displacements caused by thermotransport alone; D, the marker displacements observed when electromigration and thermotransport occur simultaneously.

from electromigration. It is often observed, eg, with Pt (27), that the marker displacements are not symmetrical with respect to the center of the sample. This is generally owing to the Soret effect or thermotransport (28), the diffusion of matter in temperature gradients, which obviously occurs in opposite directions at both ends of the sample, as shown in Figure 3c. The specific contribution to the atomic flux resulting from thermotransport is given by the relation:

$$J = c \cdot \frac{D}{kT} \cdot \frac{Q^*}{T} \cdot \frac{dT}{dx} \tag{5}$$

where Q^* is the heat of transport. It is relatively easy to measure this effect by passing an a-c current of equal magnitude through a sample similar to the one used for electromigration tests. With ac the electromigration effect is zero, and the thermotransport remains the same as for dc. The difference between the two sets of measurements gives the effect of electrotransport alone. The accuracy of the electromigration determinations made by this technique depends on the precision of the measurements of the temperatures along the length of the sample. These measurements have to be made separately by means of thermocouples or an optical pyrometer (see Analytical methods; Temperature measurement). A number of investigators have used the marker motion technique to study electromigration in many materials, including Au (5) and Al (29). In the latter case, it was found that self-electromigration occurs more rapidly in commercial aluminum than in high purity material. The discovery of this enhancement effect in lattice electromigration throws light upon the contrary grain boundary effect that is discussed below (see Thin Films).

The Steady State Technique. The drift technique has been used to determine both impurities and self-electromigration, the marker motion technique is used exclusively for self-electromigration, whereas the steady-state technique has been used almost entirely for the study of interstitial solute electromigration. Although the steady state technique is needed mainly for the consideration of electromigration as a means of metal purification, it is analyzed here in some detail. Consider a sample of material A containing solute atoms B, which move toward the positive electrode under the influence of an electrical current. The mobility of B is much larger than that of A, so that the lattice of A can be assumed to provide a stable frame of reference for the motion of B. Initially the concentration c of B is uniform and equal to c_o. However, given that the B atoms cannot escape either into the positive electrode or to the atmosphere, electromigration causes the concentration of B to increase at the positive electrode. The resulting concentration gradient creates an atomic flux opposed to the electromigration flux, so that eventually a steady state occurs when the two opposing fluxes exactly counterbalance one another (see eqs. 1 and 2). At equilibrium the equality of the two fluxes can be written as:

$$c \cdot \frac{D}{kT} \cdot z^*Ee = c \cdot \frac{D}{kT} \cdot \frac{d\mu_b}{dx} \tag{6}$$

or

$$c \cdot \frac{D}{kT} \cdot z^*Ee = c \cdot \frac{D}{kT} \cdot \frac{dc}{dx} kT \tag{7}$$

Assuming that E and z^*, or their product, are constants independent of concentration, the integration of this latter expression over the length of the sample L gives:

$$\ln c = -\frac{z^*eE}{kT} x + \ln K \tag{8}$$

where K is given by the relation:

$$K = c_o L \frac{z^* E e}{kT} \left[1 - \exp\left(-\frac{z^* E e}{kT} L\right)\right] \tag{9}$$

For usual values of z^*, E, L, and T the quantity $[1 - \exp(-z^*EeL/kT)]$ is approximately equal to 1. In the literature concerning purification the terminology is slightly different: one refers to the electrical mobility, U, rather than to z^*:

$$z^* = \frac{UkT}{De} \tag{10}$$

Equations 8 and 9 are then written as:

$$\ln c = -\frac{UE}{D}x + \ln c_o L \frac{UE}{D} \tag{11}$$

The steady-state conditions are displayed in Figure 4. The maximum concentration is given by the preexponential part in equation 9. For increasing values of the driving force, as defined by the product of z^* and E, the slope of the $\ln c$ vs x curve increases. It can be seen that the slope in Figure 4 gives directly the value of z^* without reference to the corresponding value of D. However, use of the steady-state technique does not eliminate all need to consider kinetics: the time required to approach the steady-state conditions increases for long specimens and for small diffusion coefficients. In practice the steady-state technique has been used almost exclusively with fast diffusing solutes, mostly interstitial elements (eg, hydrogen (17), and a few substitutional ones), with unusually high diffusion coefficients.

Correlation and Vacancy Wind Effects, Alloy Additions. In usual diffusion experiments with pure metals, where the observed motion is that of an isotope (possibly radioactive) of the matrix atoms, the measurements do not yield values of D directly, but of the product Df, where f, the correlation factor, is smaller than 1. This is due to the fact that with a vacancy mechanism the diffusion of isotope atoms is not purely a random walk process since any one jump from a to b is likely to be followed by the opposite jump from b to a. In electromigration experiments with pure materials,

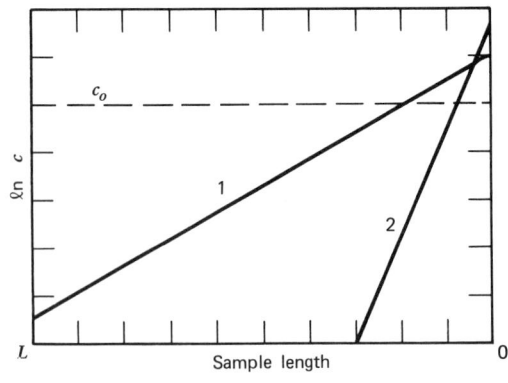

Figure 4. The concentration profile that obtains under steady state conditions. On the horizontal (length) axis the origin has been placed to the right. For the same diffusion coefficient the electromigration driving force, Z^*Ee, is four times higher for curve 2 than for curve 1.

however, the atomic flux is proportional to the diffusion coefficient itself, uncorrelated. Thus in the drift experiment the Gaussian spread is correlated, although the electromigration drift is not. The movements measured in pure samples by the marker displacement technique are not correlated. For a close-packed structure, such as that of Al, Cu or Au, the value of f is 0.78.

In considering alloy effects, one may start with the simple case of interstitial impurities; for these the atomic flux is uncorrelated. However, with substitutional impurities the problem is complicated. From a purely formal approach, with several mobile species, the atomic flux is given not only by the phenomenological coefficients L_{ii}, measuring the motion of species i in a gradient of this species, but also by the cross terms L_{ij}, measuring the motion of species i in a gradient of species j (30). Such cross terms give rise to the so-called vacancy flow effect. A physical, atomistic account of diffusion and electromigration in an alloy must take into consideration the association of vacancies with impurities and the ensuing effects of such associations on the frequency of elementary atomic jumps with various configurations (with respect to an impurity-vacancy pair). With usual diffusion in a dilute solution of B (with a concentration c) in A one finds that D, the coefficient of autodiffusion of A, is related to $D_{(0)}$, the coefficient in pure A, by the relation:

$$D = D_{(0)} (1 + bc) \qquad (12)$$

with b usually assuming positive values. It has been pointed out (31) that under electromigration conditions the enhancement factor b_{em} should generally be slightly different from b. Such a difference can be used to investigate details of diffusive atomic motion.

In many electromigration investigations, eg, the marker motion method, one cannot measure individual atomic fluxes, rather one obtains the overall atomic transport. In a close-packed binary alloy, that overall transport, given by the sum of the atomic fluxes J_a and J_b, is exactly equal (but of opposite sign) to the flux of vacancies J_v. The comparison between the electrotransport J_v in a dilute solution and that in a pure sample $J_{v(0)}$ has been treated in detail for a series of solvents: Al, Cu, Ag, and Au (32). Adopting a formulation similar to the one used for diffusion one may write:

$$J_v = J_{v(0)} (1 + \beta_{em} c) \qquad (13)$$

Like the diffusion enhancement factor b, the electromigration enhancement factor β_{em} is generally positive. For Cu, Zn, and Mg impurities in Al the respective values of β_{em} are calculated to be 100, 18, and 21. From this it can be anticipated that in commercial Al, with perhaps a total impurity content of 1 at %, the total electromigration effect could be twice as high as in pure Al (29).

Purification by Electromigration

General Considerations. The extensive and successful use of electrolysis methods for the purification of metals such as Pb and Cu stimulated attempts to use electromigration in purification of other metals. Many early reports on electromigration used the term solid state electrolysis although the two processes are basically different. One of the first studies (1940) of metal purification reported that oxygen in Zr moved in the same direction as the electrons so that one end of a sample was enriched in oxygen and the other was somewhat purified (33). A few of the later studies included: the application to Y, in which both nitrogen and oxygen and several substitutional solutes were found to move (34); purification of Ce with respect to metallic and nonmetallic impurities (35): electromigration results in Ce, U, and Pu (36); measurement of electromigration velocities and studies of purification in a number of transition, actinide, and lanthanide metals (37–51); and further work on several lanthanide metals (52–55) (see Actinides; Rare earth elements).

Purification by electromigration differs from classical chemical separation processes in that the redistribution of an impurity occurs within one phase rather than by partition between two different phases. For example, zone melting purification methods utilize the distribution between a liquid and a solid phase to segregate the impurities, whereas electromigration tilts the concentration profiles so that it is higher at one end of a single, either solid or liquid, phase (see Zone refining; Crystallization). The energy required to achieve separation, in an ideal solution, is the reverse of the free energy of mixing. The impurities do not need to be removed from the metal phase, only displaced within the phase, without modification of the total number of bonds and the ensuing enthalpy changes. Hence, metals with a very high affinity for impurities can be treated by electromigration methods. In a dilute solution the free energy change ΔG required to produce a change in concentration corresponds to the entropy of mixing only. For a change in impurity concentration by a factor of 100 at 27°C, the amount of free energy required would be:

$$\Delta G = 2.303 \text{ RT} \log 100 = 2.303 \times \frac{8.314 \text{ J}}{\text{mol} \cdot \text{K}} \times 300 \text{ K} \times \log 100$$

$$= 11.5 \text{ kJ/mol or } 0.12 \text{ eV/atom} \quad (14)$$

This is a relatively small amount of free energy and reflects the fact that only a redistribution of the impurity concentration is involved.

Electromigration purification has so far been a relatively slow process, restricted to small portions of a metal and requiring large amounts of current. It has been used mostly with very reactive metals that have such a high affinity for carbon, nitrogen, and oxygen that common chemical separation methods have failed. An example of the results that have been achieved is furnished by the work on Th (see Thorium). Carbon and nitrogen are held very tightly by Th and, indeed, electron-beam melting under high vacuum can preferentially vaporize Th and leave the liquid metal enriched in these impurities. By electrotransport purification the nitrogen concentration has been lowered from 60 ppm to <1 ppm, the carbon concentration from 40 ppm to <2 ppm and the oxygen concentration from 120 to <10 ppm. The same processing increased the resistance ratio, a measure of the purity of a metal, from 32 to >2800. The diffusion rates for interstitial elements are high enough so that metals with interstitial impurities can be treated in the solid state in a reasonable length of time. Fast-diffusing metallic solutes such as Fe in Sc have also been removed but, for solutes that diffuse at rates close to the rate of self-diffusion of the solvent atoms, prohibitively long times are required in solids. Such solutes do, of course, show much faster diffusion in the liquid state and hence could be purified by electromigration while molten, if all liquid-crucible interactions can be avoided. A helpful survey of the electromigration of interstitial atoms has been published recently (56). The interstitial impurities, carbon, oxygen, and nitrogen, have generally been found to migrate in the same direction with few exceptions, eg, in Ti (48), carbon migrates towards the positive electrode, nitrogen and oxygen in the opposite direction; and in Nb (57), the direction of motion of oxygen and carbon has been reported to vary with temperature (towards the negative electrode at low temperatures and towards the positive electrode at high temperatures). Yet as a whole in transition metals of the fourth, fifth and sixth periods the direction of interstitial motion varies regularly with the position of the host metal in the periodic table. Transport occurs towards the positive electrode in metals belonging to the IVb and Vb columns, and towards the negative electrode in metals belonging to the VIb, VIIb, and VIIIb columns.

With regard to purification it is quite fortunate that all interstitial impurities migrate in the same direction in a given host metal as this allows the removal of all these impurities in a single operation. However, the time required is dictated by the diffusion rate of the slowest moving impurity. Because the metal to be purified must be maintained at a relatively high temperature for many hours, the application of this purification method to even moderately volatile metals causes loss of metal by vaporization and melting at the reduced cross-sectional area. Most of the metals that have been purified were used in research investigations. For this purpose the small amount of metal processed per run and the high processing cost are not overwhelming considerations. Typically the rods to be purified of interstitial impurities have diameters of 2.5–8 mm and lengths of 60–160 mm; the times to purify such rods are 72–240 h. In several cases, rods of an impure metal were cut from a single crystal and then purified by electromigration. In other cases, the long time at high temperature resulted in extensive grain growth and the development of single crystals up to 2 cm long during electromigration.

Formal Basis of Purification. *The Steady State Concentration Profile.* The usual form of a specimen to be purified by electromigration is a rod through which a direct electric current is passed longitudinally. This simple geometric shape is both the easiest to model with mathematical relationships and also seems to be so far unsurpassed in achieving the highest purity. The migration of a solute in this specimen, if the solute does not escape, must result in an increase in concentration in one end and a depletion at the other. For a rod of uniform cross-section at constant temperature and with an initial uniform concentration, the concentration profile will change with time (58). Ultimately, a steady-state condition will be reached such that the flux owing to back diffusion will be equal to the electromigration flux. This leads to a concentration profile varying exponentially with distance as shown in Figure 4 and equations 8, 9, and 11.

The potential of electromigration as the basis of a purification method has been evaluated for a number of metal-impurity systems. From available electromigration and diffusion data the average purity of the purest half of a rod was calculated by use of realistic values of the length of the specimen and the electric field (59). In many systems the ratio of the electric mobility to the diffusion coefficient is large enough that very substantial purification can be achieved if the steady state can be reached. Many of the factors that affect the attainable purification and the time to reach the steady state are discussed in ref. 60. The steady-state concentration profile is shown in Figure 4, as a plot of ln c vs distance x. The term ln $c_o \, LUE/D$ in equation 11 is the intercept at $x = 0$ and represents the maximum concentration that can be achieved for a given quantity of impurity $c_o L$ and a given degree of segregation UE/D. The concentration then decreases from this maximum value according to a slope UE/D as the distance x from the impure end is increased.

Kinetics. An infinite time is required to reach exactly the concentration distribution characteristic of the steady state. However, a reasonable degree of purification can be obtained in a much shorter time (60). If the ratio UE/D is greater than 2, the electromigration velocity, being fast compared to diffusion, will control the time required to achieve purification. With a bar of length L the time t required for the averaged concentration in the purest three quarters of the bar to be within one relative percent of the steady state value is given by:

$$t = 1.25 \, L/UE \qquad (15)$$

If UE/D is appreciably less than unity, the diffusion behavior dictates the time to achieve the steady state. As a first approximation, the impurity concentration at the pure end will remain more than 5 relative percent above the steady state value when:

$$\frac{Dt}{L^2} \leq 0.4 \qquad (16)$$

Practical Aspects. In order to obtain the highest possible purity, the experimental conditions must be chosen after due quantitative consideration of a specific combination of an impurity and a metal. Only a qualitative treatment of the most important factors are summarized here and reference should be made to the more complete treatments that are cited. It is apparent from equation 11 and Figure 4 that the impurity concentration at any specific point becomes smaller as the distance of that point from the pile-up end increases. However, the distance x, from the pile-up end, can not be greater than the length L of the bar. Increasing the length of the bar increases the impurity concentration at the pile-up in direct proportion. Simultaneously, because of the exponential decrease in concentration with distance, an increased length of bar leads to a decreased impurity content in most of the bar. The increased length also, of course, means that a large quantity of material will be processed per run. However, the advantages are not without cost: inspection of equations 15 and 16 shows that the time required to reach the steady state is increased either directly proportionate to L if UE/D is large, or to L^2 under the less favorable purification condition when UE/D is small. The length of time to attain the steady state is not a trivial factor as it affects the capacity of an apparatus to produce purified metal. Unfortunately, the probability of a failure owing to accidents increases as the time increases.

The effect of the length of the electromigration zone on the concentration difference between the two ends plays an important role in the application to liquid metals and materials. With liquids it is obviously essential to maintain an appreciable length without any mixing by convection or otherwise. If even a slow fluid flow moves the liquid metal longitudinally, the separation produced by electromigration is negated. Except in very small diameter tubes, mixing in the liquid metal prevents the use of electromigration to augment the distribution obtained in direct zone melting purification (61–62). If it were possible to overcome the problem of mixing with liquid metals, the high diffusion coefficients and large electromigration mobilities (63) would make many separations quite practical; experimentation in space with a null gravitational field appears to offer some possibilities.

The electric field that produces the atomic movement of electromigration also produces a high current density in the metallic conductor. The current flow heats the rod to a temperature where diffusion and electromigration can occur. However, this heating also imposes a limit on the electric field that can be used with a given size and shape of conductor, since the Joule heat must be removed from the rod to prevent overheating and melting of the solid metal. A major portion of this heat is removed by radiation and convection at the surface and so higher electric fields and current densities can be achieved with smaller diameter cylindrical rods or with thin sheet shapes. Most of the work on purification reported to date has used electric fields in the range up to 0.2 V/cm and current densities of ca 0.2–2 kA/cm^2. Significant increases above these values would be possible, without reducing the diameter of the rods, only by increasing the rate of heat removal with heat pipes or liquid metal cooling (see Heat exchange technology). The electric fields that have been used are usually sufficient to produce a high degree of impurity pile-up and adequate purity. A principal advantage of increased electric fields would be in the shorter times required to reach the steady state and the decreased danger of contamination.

The transfer of impurities from the vacuum or inert gas surrounding the electromigration rod results in a higher concentration of impurities than would be predicted by equation 11. The final concentration results from a balance between the rate of impurity introduction from the environment and the rate of removal by electromigration. The velocity of migration is equal to UE; hence increasing E diminishes the importance of contamination. Similar results should be obtained by increasing the temperature, since U increases with temperature. However, in many systems the critical ratio U/D decreases with increasing temperature, in other systems it does not seem to change much, yet in some systems U/D changes sign with temperature (57,64–66). In any case, the rapid exponential increase of U with temperature is extremely important, since it reduces the time required for purification, and thereby reduces the effect of contamination. The major disadvantage of high temperatures is the increased chance of melting or excessive vaporization.

Usually the temperature in a real rod cannot be kept entirely constant along the length and typically the ends are somewhat cooler than the center. The thermal gradient at the purified end has several consequences. Impurities are removed from the cooler end more slowly that in the remainder of the rod; hence they continue to be introduced from this cold end reservoir into the more highly purified sections and raise the impurity concentration. This effect can be reduced by making double purification runs. During the first run the rod is substantially cleared of impurities except for the ends. These impure ends are then cropped and a second run completed with almost no impurities in the cooler ends. Thermomigration in the temperature gradients at the ends of a rod may play a role in the transfer of impurities from the cooler ends. This has apparently not been studied quantitatively. If the temperature at the junction from the electromigration rod to the current leads at each end is high enough, diffusion and electromigration may introduce impurities from this source, hence the end temperature usually is kept sufficiently low to prevent transport from the current leads.

At the impure end, the increase in impurity concentration may bring concentrations above the solubility limit and result in the precipitation of oxides, nitrides or carbides. This precipitation reduces the concentration of impurities in solution, which is useful in that it reduces the gradient that causes back-diffusion. Occasionally, because the amount of compound precipitated is so great, the electrical resistance and the temperature are raised, causing the material to melt; alternatively the precipitated compounds, being brittle, may allow the rod to break. Usually these problems are avoided with a purer starting material. In some instances, the high concentration of impurities may cause these solutes to diffuse into the current-lead material; this effect can be helpful if it does not result in deterioration of the current-lead. Electromigration that induces the pile-up of impurities at one end of a sample, and eventually the precipitation of these impurities, can cause the dissolution of precipitates at the other end. The two mechanisms, dissolution and precipitation, may occur simultaneously in a process which causes the apparent motion of precipitates. This latter phenomenon has been the object of a mathematical analysis (67), as well as the motion of precipitates in a temperature gradient (68). The dissolution of precipitates of Al_2Cu is thought to play a significant beneficial role in the behavior of Al–Cu thin film conductors (69–70).

Most of the study of purification by electromigration has been on interstitial solutes in solid metals. For many reactive metals it has been established as a simple and effective purification method for small quantities of materials. In solid metals fast-diffusing substitutional impurities can be removed but the time required for solutes with a normal coefficient of diffusion would be excessive. Liquid metals could be purified quite effectively if mixing in the liquid could be suppressed.

Thin Films

For the past century electromigration has stimulated the curiosity of theoreticians and the interest of many experimental scientists. On the practical side it is being used for the purification of a number of metallic elements. Yet it is fair to say that the most prevalent manifestations of electromigration have been the deleterious effects discovered in the current microcircuit technology. It was found that thin film conductors of aluminum used in planar silicon circuits can become discontinuous after prolonged passage of a direct current (71). This discovery set the stage for a period of intense activity in the investigation of electromigration phenomena that are almost entirely specific for thin films. The problems considered below are related exclusively to the metallic conductors that lead the electric current to, or away from transistors; they do not concern the behavior of the semiconducting devices themselves (see Electrical connectors).

Specific Characteristics. The difficulties encountered in electronic circuits arise from the needs to decrease the geometrical size of these circuits in order to achieve a low unit cost of fabrication and to satisfy the desire for high operational speeds, although signal levels cannot be reduced in similar proportions. Thus current densities of the order of 10–100 kA/cm^2 are commonly encountered in thin film conductors that may have a thickness of at least 0.1 μm and a width of 1–10 μm. Surprisingly, in thin films the increase in temperature can be quite negligible because of the rapid dissipation of heat by the substrates on which these thin film conductors are usually deposited, a condition that is particularly true for materials with a good thermal conductivity such as Si. With Al thin film conductors on oxidized Si substrates it is relatively easy to obtain current densities exceeding 1000 kA/cm^2 with temperature increments lower than 20°C. Whereas bulk electromigration phenomena occur at relatively low current densities and very high temperatures, with thin films one is concerned with extremely high current densities and relatively low temperatures.

Experimentalists investigating fundamental aspects of electromigration, or metallurgists purifying some samples, want to obtain results in perhaps a few days, but circuit designers are concerned with effects that may hinder the proper function of electronic devices after a few years (ie, with thin films the scale of time is stretched by a factor of more than 100).

The preceding sections considered electromigration in samples at high temperatures where the mechanism of transport is lattice diffusion. For all common metals such as Al, Au, and Cu which have a cubic symmetry this is an isotropic phenomenon; electromigration in anisotropic metals, eg, Mg (72), cannot be discussed within the scope of the present article. At the low temperatures that are relevant to electromigration in thin films, the dominating transport mechanism is grain-boundary diffusion, a phenomenon that varies from boundary to boundary according to the relative orientation of the adjacent grains even for cubic materials. Considering only the average diffusion in a polycrystalline material, one may grasp the difference between low and high temperature diffusion by reference to the Arrhenius plot in Figure 5, where for a constant driving force the logarithm of the total atomic flux is plotted against the reciprocal of the absolute temperature. At high temperatures the flux in a polycrystalline material is approximately equal to the flux in a single crystal. At low temperatures the atomic flux is much higher in a polycrystalline sample than in a single crystal; this effect is maximized for the smallest grain size. In general the transition occurs at temperatures about $T_m/2$, where T_m is the melting temperature expressed in kelvins. For Al this transition is ca 473 K so that, in circuits operating at tempera-

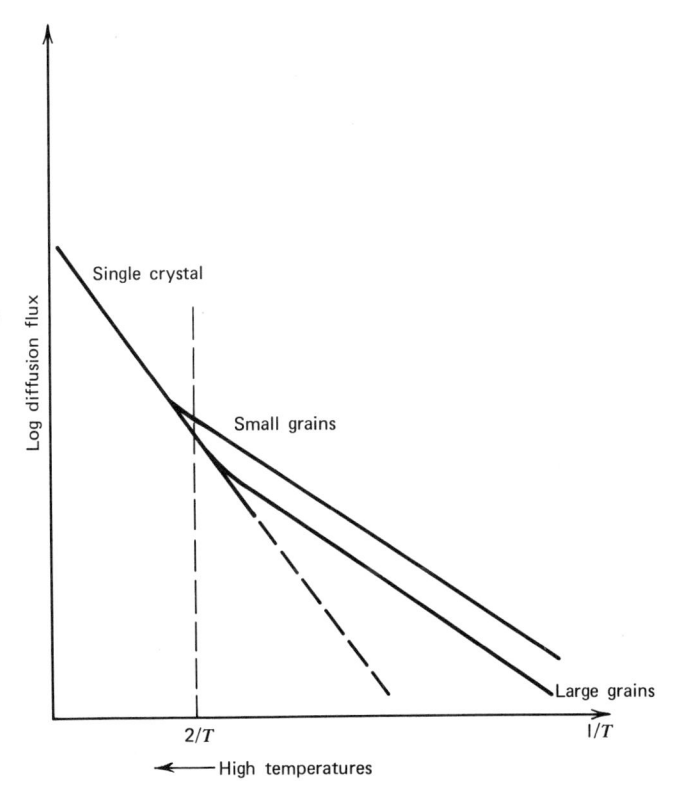

Figure 5. Arrhenius plot of the diffusive atomic flux in a single crystal and in polycrystalline samples.

tures <373 K, the transport mechanism is predominantly grain-boundary diffusion. Moreover, in thin films the effect of grain-boundaries is increased on account of the reduced grain size, often commensurate with the thickness of the films, and thereby smaller by a factor of ca 100 than in common bulk samples. Details of electromigration in thin films have been reviewed recently (16). The most prominent features are discussed below.

Failure-Formation and Failure Mechanisms. In Figure 6 one can see a discontinuity, a crack across the width of an Al thin film conductor, resulting from the passage of a current with a density of ca 2000 kA/cm^2 at a conductor temperature of 160°C for ca 1 wk. This crack is somewhat unusually straight, yet may be considered representative of similar cracks observed in thin films under different test conditions. However, some cracks may be so narrow as to remain undetectable, even with the use of a scanning electron microscope, or else may be hidden within a complex of mass accumulations, called hillocks, some of which are visible in Figures 6 and 7. Cracks causing electrical discontinuity are certainly the most dramatic failures caused by electromigration yet hillocks or whiskers, as shown in Figure 8 for an alloyed Al thin film but also detected in Pb–Sn solder points (73), may result in short circiuts between conducting lines lying side by side, or lines superimposed with a separating insulation layer (eg, sputtered silicon oxide). Both mass depletions (cracks and holes), and mass accumulations (hillocks and whiskers) result from some specific discontinuity or divergence in the atomic flux along the length of a conductor. With a homogeneous isothermal conductor, variations in cross section, and the ensuing variations in current density, would not of themselves cause either holes or hillocks.

In order to examine the problem of atomic flux divergence in some detail, the atomic current I_a at any one position along a polycrystalline conductor can be written as equation 17 for matrix atoms and equation 18 for alloy additions:

$$I_a = (ml)[\delta D'_o \exp(-Q'/kT)](1/kT)[Z^{*'}e(\rho J_e)] \qquad (17)$$

$$I_{ai} = (c_i\beta_i)(ml)[\delta D'_{oi} \exp(-Q_i'/kT)](1/kT)[z_i^{*'}e(\rho J_e)] \qquad (18)$$

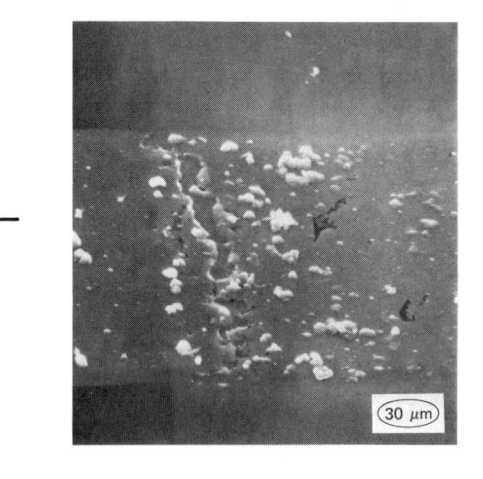

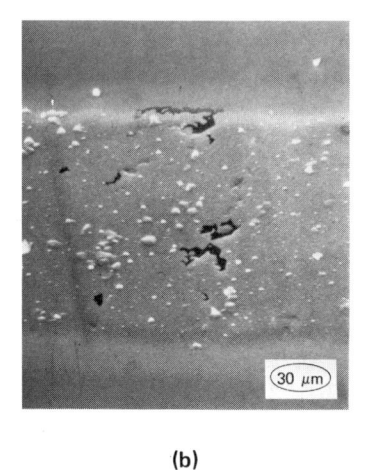

(a)   (b)

Figure 6. (a) A scanning electron microscope image of an electromigration crack across a thin film Al conductor. (b) Random damage formed elsewhere along the same conductor. The conductor had a width of 80 μm and a thickness of 0.3 μm. The crack caused failure after 160 h (ca 1 wk) at 170°C and 2×10^3 kA/cm^2.

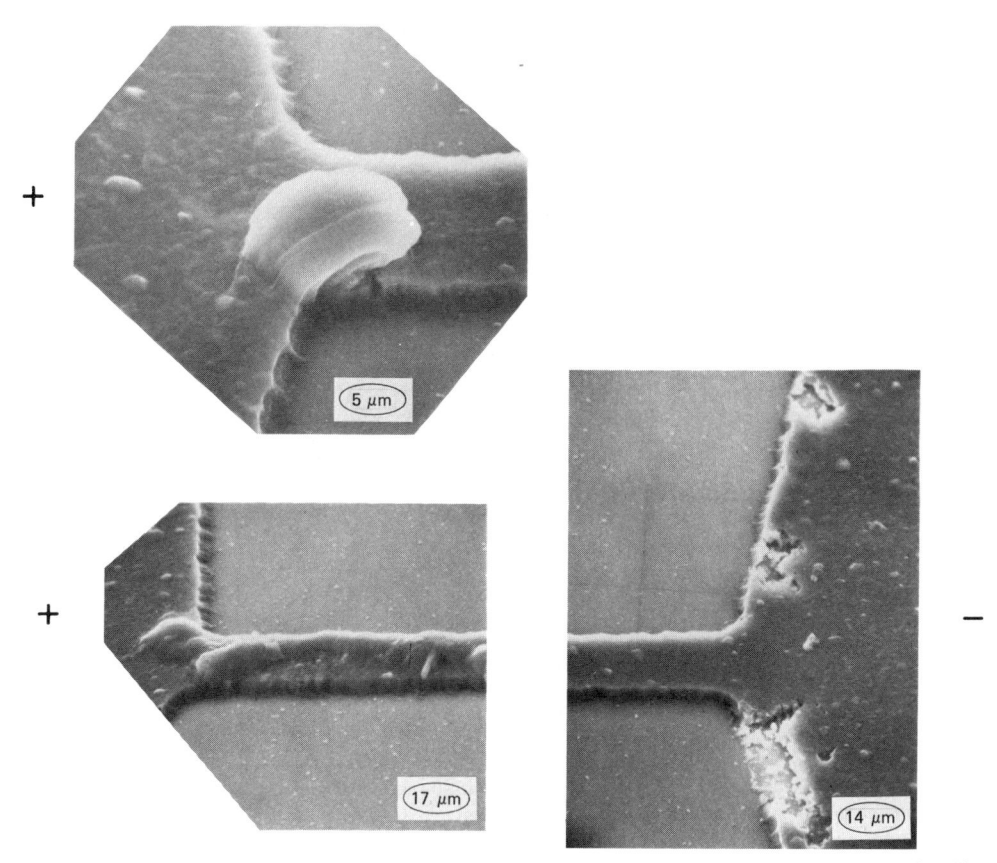

Figure 7. Scanning electron microscope pictures of damage formed at the end of Al alloy thin film conductors. Holes formed at the negative terminal, hillocks at the positive terminal.

In these relations m is the number of grain-boundaries in a given cross section, l is their average length and δ is their width, often assumed to be 1 nm, so that the product $ml\delta$ at one point along a conductor is the total grain-boundary section through which atoms are diffusing. For most thin films, with columnar grains normal to the substrate, m would be the number of grains across the width of a conductor, and l would be approximately equal to the film thickness. The primed quantities, eg, D'_o, relate to grain-boundary properties, not to those of the lattice. For alloy additions, c_i is the concentration in solution in the lattice at the temperature of the experiment and β_i is the grain boundary enrichment factor, a measure of the adsorption of impurity atoms on the grain boundaries, and J_e is the current density. The physical meaning of equations of the same type as 17 or 18 but for an homogeneous material, either a single crystal or a polycrystalline sample at high temperatures where grain-boundary transport is negligible, is directly related to the prevailing concepts of the effective charge Z^* or z^*. However, when considering grain-boundary transport, there arises a fundamental ambiguity in that the J_e term is by necessity the average current density, not the current density at, in, or along the grain-boundaries. Hence, the specific meaning to be attached to experimentally derived values of $Z^{*\prime}$ or $z^{*\prime}$ for grain boundaries is essentially flawed. In writing equations 17 or 18 one should add terms

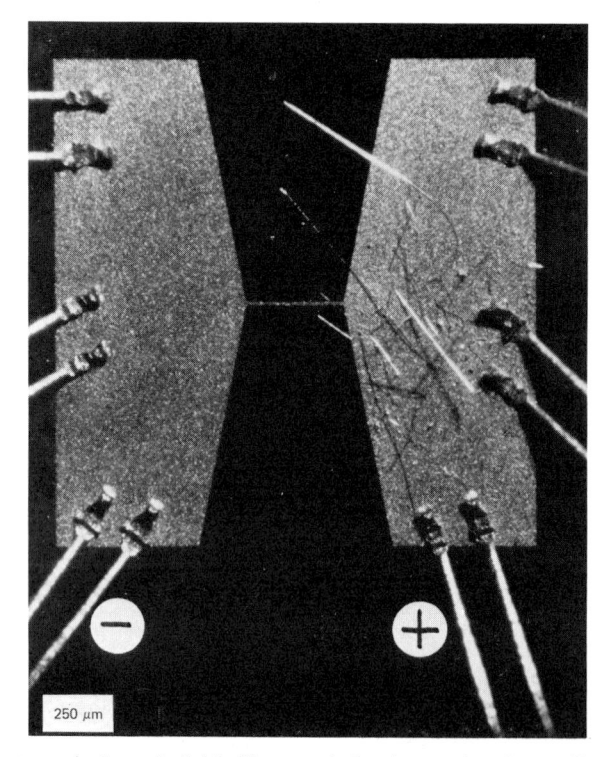

Figure 8. Photograph of a typical thin film sample for electromigration studies. The test conductor itself is the thin line (7 μm wide) with a length of 250 μm in the center of the image. Visible electrical connections are made to the large terminal areas. The material here is Al with 1 wt% Cr. Electromigration has caused the formation of many whiskers visible on the positive terminal after 8300 h of testing at 2×10^3 kA/cm² and 175°C.

owing to lattice transport; these have been ignored in the present case since even at temperatures as high as 175°C in thin films of Al the lattice transport is about one million times smaller than the transport along the grain-boundaries. Indeed, it has been experimentally shown that the electromigration effects that are of concern in the usual polycrystalline thin films are practically nonexistent in single crystal thin films of similar geometry (74). A complete analysis of all the diffusion terms contributing to electromigration in polycrystalline materials can be found in refs. 75–77.

Because of the rapid increase in the value of diffusion coefficients with temperature, temperature gradients constitute one of the main causes of atomic transport divergence along the length of thin film conductors. In experimental configurations such as in Figure 8, provided that the cross section of the conductors, their resistivity, and the current density are sufficiently small, it is relatively easy to obtain conditions where effects caused by temperature gradients are negligible. Failures in samples that are similar and tested under the same conditions are found to be randomly distributed along the length of the conductors. However, one must understand that this is a relative statement. If precautions are taken to eliminate, or at least reduce sufficiently other sources of transport divergence, the inevitable temperature gradients at the ends of conductors become manifest. Undoubtedly the defects shown in Figure 7, for samples where crack formation was delayed by alloying, are connected with such temperature gradients. In actual microcircuits, where heat is generated not only by the Joule effect in the conductors, but also in transistors and resistors, the distribution of temperature gradients and their effects can be complex.

Other sources of transport divergence along the length of conductors are related either to the geometry of the conductors, or more importantly to their microstructure. Obviously, significant changes in transport can be anticipated at the ends of conductors where there are changes in materials, eg, from Al to Si at transistor contacts, or from Al thin films to bonded wires or solder points at the other ends. The variations in the number of grain-boundaries m in equations 17 and 18 can be a cause of transport discontinuity. Even with an absolutely uniform grain size, and a perfectly regular conductor, the number m is bound to vary by unity from one position to another. Practically, films do not have a very uniform grain size and changes in grain sizes are often found where failures have occurred. These grain size considerations are quite important because the average grain diameter is often commensurate with the width of the conductors, a condition that becomes more acute and frequent with the continuous demands for ever-decreasing dimensions. The product $\delta D'_o$ is a measure of the structure of the boundaries, and of the density of mobile defects therein. This product, as well as the activation energy Q', varies widely from boundary to boundary, as a function of the angle between grains, even for grains with a common crystallographic axis, such as are encountered in films with a strong preferred-orientation (fiber texture). For alloying elements the matter is made worse since the enrichment factor β is also expected to be a strong function of the boundary structure. There have been no systematic studies of the value of the electromigration force in relation to grain boundary structures, yet it should be expected that this force, which is represented by the last term (in brackets) in equations 17 and 18, will vary from boundary to boundary, even if the obvious geometrical correction resulting from the angle between any random boundary and the direction of electrical current flow is accounted for. A steady-state flow requires that all of the factors considered above should balance each other equally along the length of a conductor. This is nearly impossible, hence electromigration failures in polycrystalline thin films are almost inevitable; their occurrence is only a matter of time. A particular configuration of importance in failure nucleation is the junction of three grains. The atomic flux along three connecting grain boundaries will be continuous only in the most exceptional set of circumstances (78). Normally such configurations, referred to as triple points, are a common cause of divergence, and some of the holes seen at random in Figure 6 are likely to be positioned at triple points. Whereas such holes are usually located at random, failure will occur along the length of a conductor where hole formation extends first across the whole width.

There have been few determinations of the values of the effective charge in grain-boundaries. Pending confirmation of the results that have been obtained, one may tentatively assume that the averaged grain-boundary values are not too different, perhaps within a factor of two, from the lattice values. Electromigration in Cu, Ag, and Au films has been reported to occur towards the negative electrode, in a direction opposite to that found in the lattice. These surprising results were the object of intense controversy since several investigators failed to obtain confirmation. It is most likely that impurity contamination from the substrates is at the origin of this unexpected reversal of electrotransport in thin films (79) (see also Corrosion).

Lifetime Tests, Failure Models. Technological necessities require that at a specified time the rate of failure in electronic devices should remain lower than some maximum, which can be chosen to be as small as desired. The selection of acceptable rate and time values varies according to the type of application considered. Practically, on account of time limitations this means that one must conduct accelerated failure tests at high temperatures and at high current densities as well as with a limited set of samples. The results thus obtained are extrapolated to normal use conditions (low current densities, low temperatures, long times, and very large device populations). Such extrapolations require an understanding of the laws relating failure times to current density and temperature and of the statistical distribution of these failure times. A detailed review of this problem has been presented in ref. 16; only a general outline is given here.

The samples to be tested are made from the materials generally used in thin film conductors: Al, various Al alloys, and Au almost always associated with another metal (not necessarily alloyed, but in superposed layers); and for special applications, nichrome, or Permalloy. Since failures occur randomly, not only with respect to geometry but also with respect to time, samples have to be tested in groups, for each test condition (for Al, current densities of the order of 10^3 kA/cm^2, and temperatures in a range near 200°C). It is usually found that the individual failure times obey a lognormal distribution, as seen in Figure 9 where the logarithms of the failure times are plotted

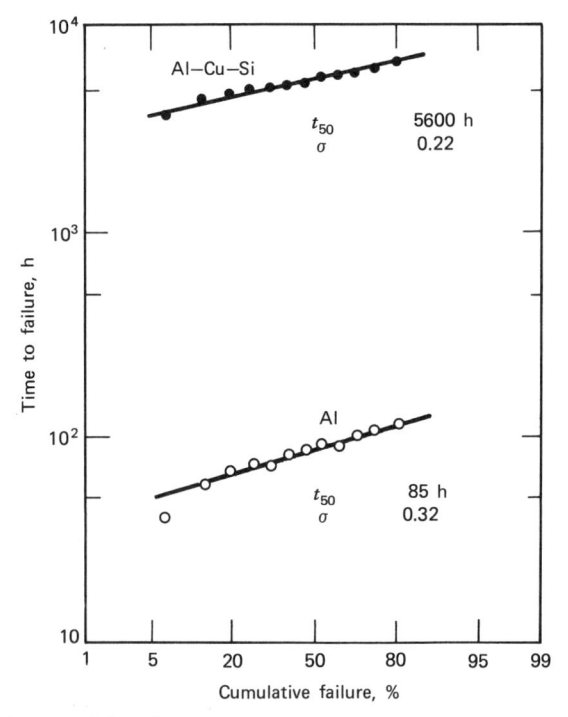

Figure 9. The lognormal distribution of failure times. The logarithm of the failure time is plotted (vertical axis) vs the order of the failure, or cumulative percentage, entered on a probability scale (horizontal axis). Comparison of the lower curve, for pure Al, with the upper curve, for Al, 4 wt% Cu, and 1.7 wt% Si (Al$_{56}$CuSi), demonstrate the beneficial effect of Cu alloy additions. Test conditions: 220°C, 800 kA/cm^2 (80).

against the rank of the individual failure on a probability scale. The individual points fall on a straight line most of the time which defines the distribution in terms of its slope. A measure of the distribution width is the standard deviation σ, and the intercept at 50% failure is the median failure time, t_{50}. A long median failure time and a small value of σ are desirable.

The reasons for the lognormal distribution of failure times are not clear. An analysis using numerical calculation techniques yielded the desired lognormal distribution without establishing in an explicit analytical way the relation between the initial elements of the model and the result (81). Long conductors lead to short t_{50}, but also small σ, since the probability of a configuration conducive to early failures increases with the length, and infinitely long conductors would have identical distributions of structural configurations (hence equal failure times). The relation between failure times, standard deviations and conductor length is schematically represented in Figure 10. Increasing the width of conductors will increase t_{50}, and probably decrease σ, yet little attention has been paid to the relationships involved, since the width of thin film conductors is usually dictated by considerations other than electromigration lifetimes.

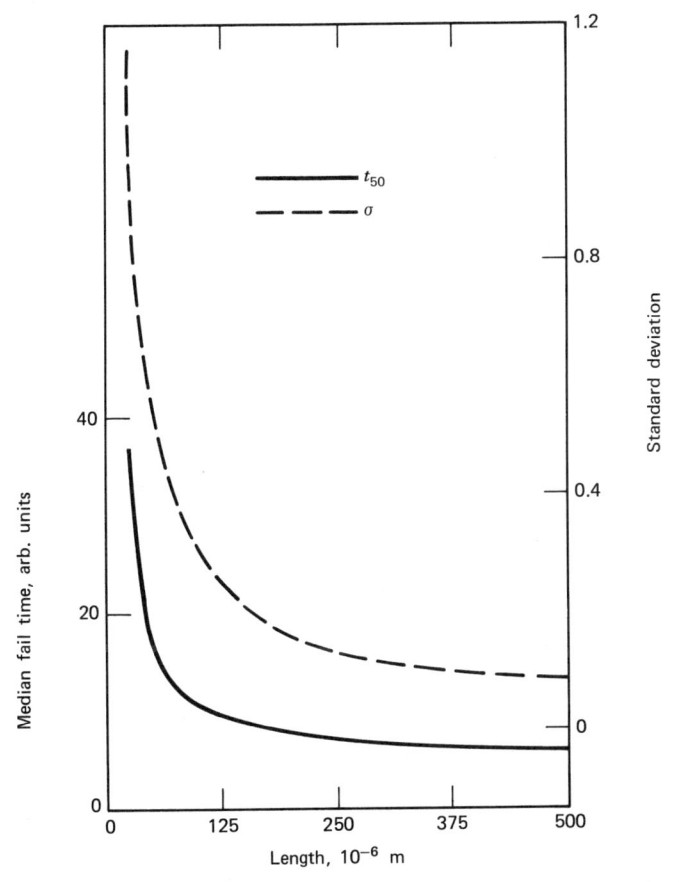

Figure 10. The relations between the length of thin film conductors (horizontal axis), the median failure time (left vertical axis) and the standard deviation (right vertical axis) (81).

As a first approximation the median lifetime of a set of conductors is inversely proportional to the intensity of the electrotransport. The relationship between lifetime, current density, and temperature is usually written:

$$t_{50} = A \exp(\Delta H/kT) J_e^{-n} \qquad (19)$$

or

$$\frac{t_{50}}{T} = B \exp(\Delta H/kT) J_e^{-n} \qquad (20)$$

where the proportionality terms A and B are functions of the structural and geometrical terms (eg, grain size, length, and width) discussed previously, ΔH is the activation energy for failure, and n is the current density exponent. The choice of expression 19 or 20 depends on the product of the temperature sensitive terms in the transport equations, $(1/T)Z^{*\prime}\rho$. Since the electron wind term in $Z^{*\prime}$ is inversely proportional to ρ (eq. 4), the temperature dependence of electrotransport depends on the preponderance of the field, or of the wind force, in $Z^{*\prime}$. The exponent n is unity only for small current densities or for extremely high heat dissipation configurations, conditions that minimize the increase in temperature resulting from the Joule heat caused by failure propagation. At best the temperature of conductors can be defined only at the start of a test, yet once a crack is nucleated, the resistance of the conductor will increase, and so will its temperature, which then becomes an unknown quantity. Consider two ideally identical conductors tested at two different current densities, J_e and $2\,J_e$; the conductors can be used as their own resistance thermometers and the two ambient temperatures can be adjusted perfectly to equal T of the two conductors under test at time zero. After a given lapse of time Δt_1, failure having progressed to a certain configuration, the resistance of the first conductor will have increased by ΔR, and its temperature by a corresponding amount ΔT_1. The second conductor, tested at a higher current density ($2\,J_e$) than the first, will reach the same configuration, since the two conductors are assumed to be perfectly identical, after a period Δt_2 smaller than Δt_1. At this point of time the resistance of this second conductor will have increased by ΔR also. However, since the increment in temperature can be anticipated to vary as the increment in Joule heat, namely as the square of the intensity, the temperature of the second conductor will have increased by a quantity ΔT_2 about equal to $4\,\Delta T_1$. Therefore, for identical failure configurations, failure will propagate in the conductor tested at $2\,J_e$ at a rate which is more than twice the rate that obtains in the other. This results in values of n that are greater than one, as seen in Figure 11, and creates obvious difficulties in extrapolating test data obtained at high current densities to use conditions at low current densities. For the same reasons the value of ΔH in equations 19 and 20 is equal to the activation energy for grain boundary diffusion Q' only for tests conducted under such conditions that Joule heat is negligible. In practical tests, these limitations are rarely obeyed and values of ΔH should be found to be smaller than Q' by increasing deviations with increasing current densities. At least in part, such considerations account for the dispersal of ΔH values reported for any one material. Further complications in the understanding of failure formation arise from current crowding in the immediate surroundings of a growing defect and from the ensuing increase in temperature. However, the relations between these effects and failure times have not been treated thoroughly in the open literature (see Materials reliability).

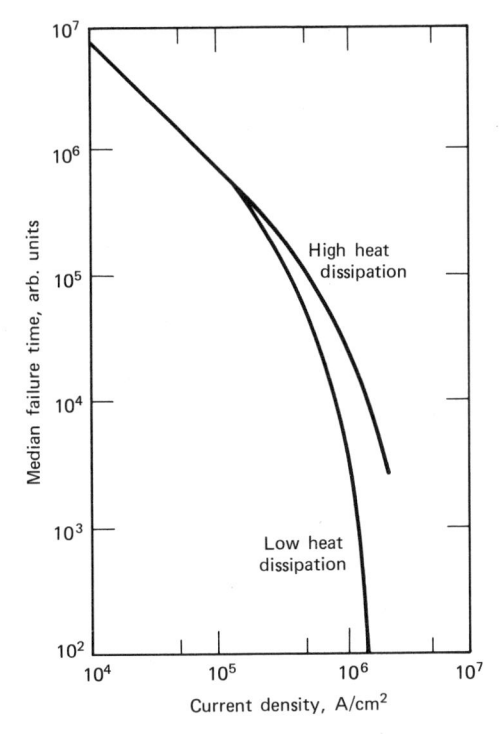

Figure 11. The effect of the current density on the median failure time. The current exponent n in the relation of t_{50} and J^{-n}, is equal to 1 only at low current densities (82).

Methods Used to Increase Electromigration Lifetimes. The most obvious way of increasing lifetimes is selection of a material with a low diffusion coefficient, yet this solution cannot be used frequently because of other technological requirements. The resistivity must be low and the metal must: (*1*) make ohmic contact to Si, (*2*) adhere well to silicon oxide, (*3*) not diffuse into Si and poison transistor junctions, and (*4*) preferably be easy to deposit and fabricate, usually by vacuum evaporation and photoetching techniques. Aluminum is almost ideally suited for this purpose. Gold, which is also used extensively, must be used in conjunction with other metals because of its poor adhesion to silicon dioxide. Materials selection cannot be justified entirely on the basis of the resistance to electromigration damage (see also Film deposition techniques).

Grain Size. Transport in thin film conductors can be reduced by increasing the grain size of the conductors (83), thus decreasing the number of paths for low temperature diffusion. The films are deposited on substrates at high temperatures. Problems with high values of σ, the width of the failure distribution, may be encountered when the grain size approaches the width of the conductors.

Dielectric Overlayers. Significant increases in lifetimes can be achieved by overcoating the conductors with a layer of a dielectric material, eg, fused glass (83), sputtered silicon oxide, or anodic aluminum oxide (84). Part of the effect may be owing to pressure hindering the formation of hillocks, and thereby reducing the formation of holes. Considering that damage is owing to the motion of vacancies from hillocks to holes, in a physically enclosed system pressure will develop at hillocks, and damage growth will cease when the force due to the pressure gradient is equal to the electromigration force. By modification of equation 7 this can be expressed as:

$$Z^{*\prime} E e = \frac{d\Omega P}{dx} \quad (21)$$

where P is the pressure and Ω is the molal volume of vacancies. Integration of this relation yields the value of the pressure required to prevent crack propagation:

$$P = \frac{Z^{*\prime} E e \lambda}{\Omega} \quad (22)$$

If λ, the distance between the source and sink of vacancies, is equal to the length of the conductors, the value of P is too high to be sustained by dielectric overlayers (85). However, much smaller values of P are obtained if λ is made as small as the grain size. Whereas failure occurs rapidly by transport from one side of a grain to another, an overlayer delays the process by requiring that transport should occur over greater distances, with a subsequent increase in the time for failure formation. The overlayer effect should be greatest for conductors with a small grain size such as may be encountered in thin resistive films of nichrome or permalloy.

Alloying. Quite significant increases in the lifetimes of Al thin film conductors are obtained with alloy additions of Cu (86). An increase in t_{50} by a factor close to 100 is seen in Figure 9 where failure times are displayed for both pure Al conductors and Al containing 4 wt % Cu and 1.7 wt % Si (80) (Si is added for other purposes and plays only a minor role in electromigration). This alloying effect is strictly a grain boundary effect since in lattice electromigration experiments impurities were found to increase the rate of Al transport (29). The diffusion rate of Al atoms along grain boundaries is reduced by the presence of Cu atoms adsorbed on these boundaries. The rate of electromigration of the Cu atoms themselves is relatively high, which requires additions of large amounts of Cu, up to ca 4 wt %, much above the solubility limit at the temperature of the tests, since the initially active Cu atoms are quickly removed from the grain boundaries. With excess Cu, precipitates of Al_2Cu are dissolved during electromigration testing, and are a continuous supply of Cu atoms. Ultimately failures usually occur in areas that have been depleted of Cu (69,80).

The effects observed with Cu, and also Mg additions to Al, have been reported for aluminum thin film conductors tested in a hydrogen atmosphere (87) (hydrogen is known as a grain-boundary-active element). Similar effects have also been achieved with Ta additions to Au. The existence of a suitable Au radioisotope allowed the observation that whereas, in the lattice, Ta additions increase the rate of Au autodiffusion, in grain-boundaries at low temperatures Ta decreases the rate of Au diffusion (88). Recently, a layer of a transition metal deposited within an Al film was allowed to react with the film to form a layer of an intermetallic compound (89). By suitable material selection and proper control of the annealing conditions, a continuous intermetallic layer, relatively inert to electromigration damage, may be produced. This layer can act as a bridge across cracks formed within the Al film, thus extending its useful life beyond the original value.

Thin Film Electromigration Experiments. Attention has been focused on problems related to failures because of their technological and economic importance. However, thin films have also been the subject of investigations aimed at the determination of fundamental electromigration parameters. In samples such as those shown in Figure

7 the volume of material transported can be estimated by careful examination of the scanning electron microscope images of holes and hillocks (these can be corroborated with electron microprobe data); the electromigration of matrix atoms can be evaluated therefrom. In alloyed films the electromigration of the alloying elements can similarly be obtained from microprobe analysis. Values of the activation energy for grain-boundary electromigration have been derived from kinetic resistance measurements. Among a number of significant investigations, two types of electromigration experiments on thin films are described below.

The Cross-Stripe Experiment. An otherwise uniform conductor is modified locally in order to observe a change of electromigration behavior specifically at the modified position. Figure 12, an example of such an experiment, shows an Al conductor locally implanted with oxygen ions. The area of implantation is visible as a perpendicular darkened zone, resulting from the polymerization of oil vapors in the implant accelerator. The formation of a crack in an area adjacent to the implant and on the side of the positive terminal provides evidence that diffusion is slower in the implant region than in the rest of the conductor. The cross-stripe experiment, originally conceived to illustrate the effect of Cu additions to Al (90), has been extended to the analysis of the electromigration of Cu in Al thin films (91). A dot of Cu is alloyed at some point along the length of an Al conductor. After a test of electromigration the concentration of Cu on each side of the dot is analyzed with an electron microprobe. The experiment is similar to the drift experiment (see Fig. 2) used with the lattice electromigration of impurity elements. However, the results are considerably different since in the lattice electromigration causes the drift of a symmetrical distribution of impurities, whereas with films where grain-boundary transport predominates, the

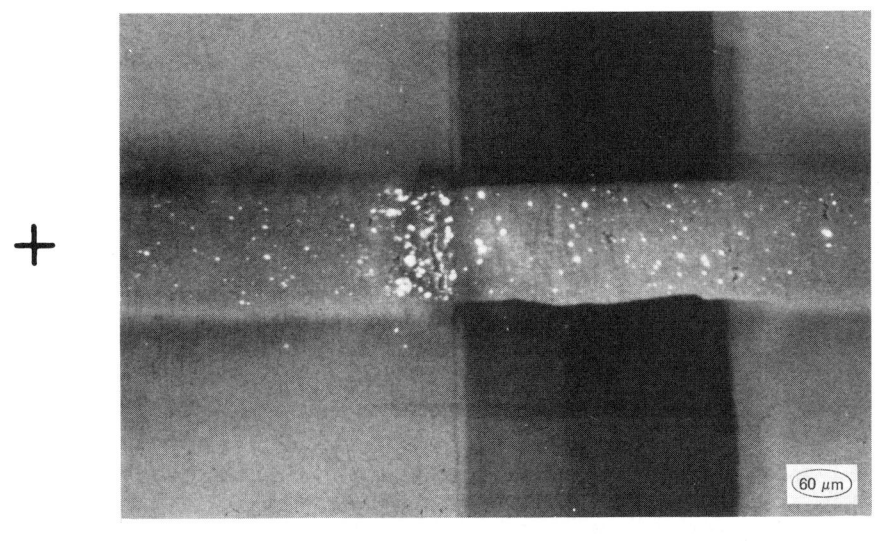

Figure 12. Photograph illustrating the cross-stripe experiment. Prior to electromigration testing part of this Al thin film conductor was implanted with oxygen ions. The area of implant is made visible by black hydrocarbon residues polymerized by the ion beam on to the underlying SiO_2. During testing a crack formed on the positive side in the region immediately adjacent to the implant area. The position of the crack shows that electromigration transport was reduced in the implant area relatively to the rest of the conductor. (In Al the atomic transport direction is that of electrons.)

distribution of Cu is found to be assymetrical with respect to the positive and negative electrodes (see in Fig. 13). Comparison of the two different slopes yields a value of ca −17 for $z^{*\prime}$ for Cu in Al grain-boundaries.

The Drift Velocity Experiment. The behavior of pure materials is investigated. The conductor A with a low resistivity is deposited over a thin layer B of a material with a higher resistivity as schematically represented in Figure 14. In areas where the good conductor is present, the current flows through it and causes electromigration therein. With electron conductors material is carried away from the negative edge, causing this edge to move and be transported to the positive edge. The displacement of the negative edge is a measure of the electromigration driving force. This technique has been successfully used with Au on Mo (92). With Al, the displacement is a function of the length of the conductors (93). This effect is probably caused by aluminum oxide and is probably related to the overlayer effect. Analysis of the results using a modified form of equation 22 (in terms of compressive stresses, rather than simply pressure) yields a value of $Z^{*\prime}$ for Al which seems to be too small, both with respect to values

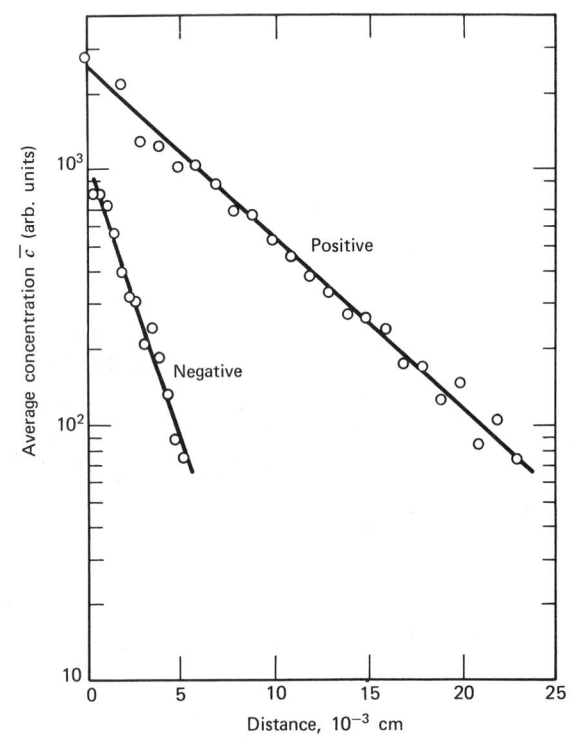

Figure 13. Cu electromigration in an Al thin film, cross-stripe experiment. After electromigration testing the concentration of Cu (averaged over the lattice and the grain boundaries, $\bar{c}$) is plotted as a function of distance on each side of the alloyed area. The contribution of grain boundary transport causes the distribution to become asymmetrical in contradistinction with the results of the comparable lattice experiment (see Fig. 2). Calculations of the electromigration parameters for the grain boundaries are based on a comparison of the slopes of the two sets of data, without requiring a determination of the absolute values of the concentration. For the case illustrated here the maximum concentration of Cu, at the intercepts with the vertical axis, is about 1 at %. For clarity the profile for the positive side has been shifted upward relatively to the other one (91).

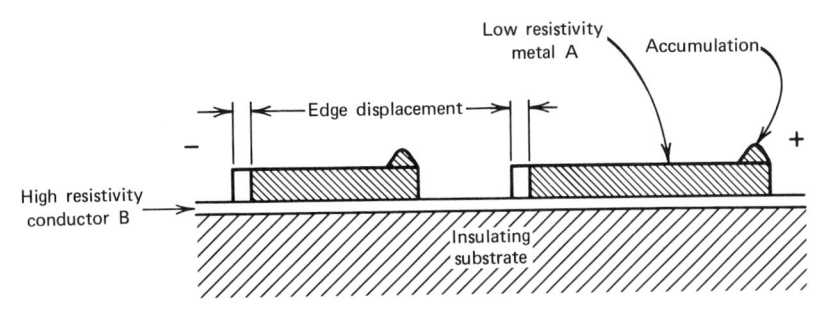

Figure 14. The drift velocity experiment. Differences in resistance cause the current to flow from the continuous high resistivity conductor B to the discrete low resistivity conductor A being investigated. The rate of motion of the edge at the point of current entrance allows the determination of the electromigration velocity in A.

obtained in bulk Al and to account for other experiments with Al thin films. This discussion emphasizes the relatively recent interest in electromigration phenomena in thin films, and the remaining paucity of confirmed measurements of transport parameters.

Pulse Effects, Crystallization. Hole nucleation may play a role in the behavior of thin film conductors subjected to very short current pulses, eg, in magnetic bubble devices (see Magnetic materials). The problem is complicated because of the related problems of heat flow and thermal fatigue. Yet it appears that the lifetimes observed under short pulse conditions are longer than would be anticipated on the basis of continuous current results, either because of a failure to nucleate holes, or because of the rapid resorption of holes having subcritical sizes (94–95).

The use of electromigration in crystal growth is not related to thin film phenomena in a physical sense, nevertheless it has potential applications in the electronic industry for the growth of epitaxial layers of GaAs (82) and other compounds. A detailed analysis of the process is extremely complex. In order to understand the phenomena resulting from the passage of current through a solid-liquid system one must consider in addition to electromigration a number of different effects: the dissipation of Joule heat, and the modifications of the thermal equilibrium owing to the Peltier effect (the heat, either positive or negative, generated by the flow of current through the thermocouple constituted by the solid–liquid interface) (see Thermoelectric energy conversion). The effects due to electromigration may be clouded by mixing by convection (owing to temperature and composition gradients), and by eddies (resulting from the nonparallelism of the current lines if the interface is not perfectly planar). Considering these difficulties, it is not surprising that experimental results are the object of some controversy (96). Yet conceptually it should be possible to use electromigration to cause a disproportionation of the liquid composition at the interface and thereby control the growth (or dissolution) conditions of the solid. Although the anticipation of practical results in the growth of compound crystals may be somewhat premature, at present confirmations of the existence of a real effect seem to be forthcoming not only with respect to GaAs (97) but also to the growth of $Hg_{1-x}Cd_xTe$ (98).

Activity Status

Although there has been recent interest in electromigration on the part of some theoreticians (see ref. 15, 17–20, and the bibliography in ref. 16), the level of activity in this field has undergone a decline in the last five years. Although a considerable amount of work used to be conducted in the USSR (4,11), in France (6,14,31), and in Scandinavia (10), the rate of publications on electromigration emanating from these parts of Europe has by now fallen almost to zero (with few exceptions, eg, ref. 99). The two schools that provide intellectual leadership remain those of Professor H. Huntington (5,12,25,27,29,72) at the Rensselaer Polytechnic Institute and of Professor H. Wever (3,9,56) at the Technical University of Berlin, FRG.

For the last fifteen years investigators simultaneously connected with the Department of Materials Science and Engineering, Iowa State University, and the Ames Laboratory of the U.S. Department of Energy have been in the forefront of work utilizing electromigration for the purification of metals (37,40,47,51,100). Recently a steady output of publications related to the purification of the rare earth elements (qv) has its source at the Centre of Materials Science, University of Birmingham (52–55). The great activity in the field of electromigration in thin films in the early 1970s was motivated by practical necessity and was largely confined to the laboratories of the electronic industry, almost exclusively in the United States (with a few exceptions, eg, Japan, ref. 84; Italy, ref. 101): Fairchild (71,80), Motorola (83), IBM (74,78,85,88) and Bell Telephone Laboratories (102). Now that different remedies for the failure of thin film conductors have been adopted, the flow of new publications has been greatly reduced although limited interest continues (16,89,94–95,103–105). Academic involvement in the study of electromigration in thin films has been centered in two locations: the Stevens Institute of Technology in Hoboken, N.J. (75–76) and the University of Florida in Gainesville (12,79).

Nomenclature

b = diffusion enhancement factor
b'_{em} = autodiffusion enhancement factor for electromigration
c = concentration
D = diffusion coefficient
e = electron charge
E = electrical field
f = correlation factor
F = force on moving atoms
F_e = force of electric field (direct field force)
F_{wd} = force of friction of the charge carriers (electron wind force)
h = Boltzmann constant
ΔH = activation energy for failure
I = current
J = atomic flux
J_v = flux of vacancies
l = length of grain-boundary
L = length of sample
m = number of grain-boundaries
n = current density exponent
P = pressure
Q = activation energy
Q^* = heat of transport

R = resistance
t = time
t_{50} = median failure time
T = temperature
T_m = melting temperature
U = electrical mobility
v = velocity
z = valency of solute atoms
Z = valency of matrix atoms
Z^* = effective charge of matrix atoms
β = grain boundary enrichment factor
β_{em} = electromigration enhancement factor
γ = averaging term
δ = width of grain-boundary
λ = distance between source and sink of vacancies
μ = chemical potential
ρ = resistivity of metal
ρ_d = specific resistivity of the mobile defects
σ = standard deviation
Ω = molal volume of vacancies
$'$ = use of prime, eg, $Z^{*\prime}$, Q' refers to grain boundary parameters

BIBLIOGRAPHY

"Electromigration" in ECT 2nd ed., Suppl. Vol., pp. 278–294, by H. B. Huntington, Rensselaer Polytechnic Institute.

1. M. Gerardin, *C. R. Acad. Sci.* **53**, 727 (1861).
2. F. Skaupy, *Verh. Dtsch. Phys. Ges.* **16**, 156 (1914).
3. W. Seith and H. Wever, *Z. Elektrochem.* **59**, 942 (1953).
4. V. B. Fiks, *Sov. Phys. Solid State* **1**, 14 (1959).
5. H. B. Huntington and A. R. Grone, *J. Phys. Chem. Solids* **20**, 76 (1961).
6. C. Bosvieux and J. Friedel, *J. Phys. Chem. Solids* **23**, 123 (1962).
7. R. S. Sorbello, *J. Phys. Chem. Solids* **34**, 937 (1973).
8. J. N. Pratt and R. G. Sellors, *Electrotransport in Metals and Alloys*, Trans. Tech SA, Riehen, Switz., 1973.
9. H. Wever, *Elektro-und Thermotransport in Metallen*, Johann Ambrosius Barth, Leipzig, GDR, 1973.
10. A. Lodding and T. Lagerwall, eds., *Atomic Transport in Solids and Liquids*, Verlag der Zeitschrift fur Naturforschung, Tubingen, FRG, 1971.
11. V. B. Fiks, *Ionic Conduction in Metals and Semiconductors, (Electrotransport)*, Nauka, Physico-Mathematical Literature, Moscow, U.S.S.R., 1969.
12. R. E. Hummel and H. B. Huntington, eds., *Electro- and Thermo-Transport in Metals and Alloys*, AIME, New York, 1977.
13. W. Jost, *Diffusion in Solids, Liquids, Gases*, Academic Press, Inc., New York, 1952, p. 324.
14. Y. Adda and J. Phillibert, *La Diffusion dans les Solides*, Presses Universitaires de France, Paris, Fr., 1966, p. 893.
15. R. S. Sorbello, "Basic Concepts of Electro- and Thermomigration; Driving Forces," in ref. 12, p. 2.
16. F. d'Heurle and P. Ho, "Electromigration" in J. Mayer, J. Poate, and K. N. Tu, eds., *Thin Films: Interdiffusion and Reactions,* John Wiley & Sons, Inc., New York, 1978, p. 243.
17. V. Erckmann and H. Wipf, *Phys. Rev. Lett.* **37**, 341 (1976).
18. R. Landauer, "Electrical Conductivity in Inhomogeneous Media" in J. C. Garland and D. B. Tanner, eds., *Electrical Transport and Optical Properties in Inhomogeneous Media*, AIP Conf. Proc. No. 40, American Institute of Physics, New York, 1978, p. 2.
19. R. S. Sorbello, "Microscopic Fields and Currents in d.c. Electrical Conductivity" in ref. 18, p. 355.
20. R. Landauer, *Phys. Rev. B* **16**, 4698 (1977).

21. N. Van Doan, *J. Phys. Chem. Solids* **31**, 2079 (1970).
22. E. M. Pell, *J. Appl. Phys.* **31**, 291 (1960).
23. Ch. Herzig and D. Cardis, *Appl. Phys.* **5**, 317 (1975).
24. Ch. Herzig and E. Stracke, *Phys. Status Solidi* **a27**, 25 (1975).
25. R. E. Einziger and H. B. Huntington, *J. Phys. Chem. Solids* **35**, 1563 (1974).
26. H. M. Gilder and D. Lazarus, *Phys. Rev.* **145**, 507 (1966).
27. H. B. Huntington and S. C. Ho, *J. Phys. Soc. Jpn.* **18**(Suppl. II), 202 (1963).
28. Th. Hehenkamp, "Thermotransport" in ref. 12, p. 68.
29. R. V. Penney, *J. Phys. Chem. Solids* **25**, 335 (1964).
30. P. G. Shewmon, *Diffusion in Solids*, McGraw-Hill Book Co., New York, 1963, p. 122; ref. 14, p. 52.
31. N. Van Doan, *J. Phys. Chem. Solids* **33**, 2161 (1972).
32. P. S. Ho, *Phys. Rev. B* **8**, 4534 (1973).
33. J. H. DeBoer and J. D. Fast, *Rec. Trav. Chim.* **59**, 161 (1940).
34. J. M. Williams and C. L. Huffine, *Nucl. Sci. Eng.* **9**, 500 (1961).
35. J. D. Marchant, E. S. Shedd, and T. A. Henrie, *Solid-State Electromigration of Impurities in Cerium Metal, BuMines RI-6894*, Washington, D.C., 1967.
36. R. H. Moore, F. M. Smith, and J. R. Morrey, *Trans. AIME* **233**, 1259 (1965).
37. O. N. Carlson, F. A. Schmidt, and D. T. Peterson, *J. Less-Common Met.* **10**, 1 (1966).
38. D. T. Peterson, F. A. Schmidt, and J. D. Verhoeven, *Trans. AIME* **236**, 1311 (1966).
39. F. A. Schmidt and J. C. Warner, *J. Less-Common Metals* **13**, 493 (1967).
40. D. T. Peterson and F. A. Schmidt, *J. Less-Common Met.* **18**, 111 (1969).
41. F. A. Schmidt, O. N. Carlson, and C. E. Swanson, *Met. Trans.* **1**, 1371 (1970).
42. D. T. Peterson and F. A. Schmidt, *J. Less-Common Met.* **24**, 223 (1971).
43. F. A. Schmidt and O. N. Carlson, *J. Less-Common Met.* **26**, 247 (1972).
44. D. T. Peterson and F. A. Schmidt, *J. Less-Common Met.* **29**, 321 (1972).
45. O. N. Carlson, F. A. Schmidt, and D. G. Alexander, *Met. Trans.* **3**, 1249 (1972).
46. O. N. Carlson, F. A. Schmidt, and J. C. Sever, *Met. Trans.* **4**, 2407 (1973).
47. O. N. Carlson, F. A. Schmidt, and D. T. Peterson, *J. Less-Common Met.* **39**, 277 (1975).
48. O. N. Carlson, F. A. Schmidt, and R. R. Lichtenberg, *Met. Trans.* **6A**, 725 (1975).
49. F. A. Schmidt and O. N. Carlson, *Met. Trans.* **7A**, 127 (1976).
50. F. A. Schmidt and O. N. Carlson, *J. Less-Common Met.* **50**, 237 (1976).
51. O. N. Carlson and F. A. Schmidt, *J. Less-Common Met.* **53**, 73 (1977).
52. R. G. Jordon and D. W. Jones, *J. Less-Common Met.* **31**, 125 (1973).
53. R. G. Jordon, D. W. Jones, and V. J. Hems, *J. Less-Common Met.* **42**, 101 (1975).
54. C. M. Muirhead and D. W. Jones, *J. Less-Common Met.* **50**, 73 (1976).
55. P. G. Mattocks and co-workers, *J. Less-Common Met.* **53**, 253 (1977).
56. H. Wever, "Electromigration of Interstitials" in ref. 12, p. 37.
57. E. Fromm, R. Kirchheim, and J. Mathuni, *J. Less-Common Met.* **43**, 211 (1975).
58. S. R. DeGroot, *Physica* **9**, 699 (1942).
59. J. D. Verhoeven, *J. Met.* **18**, 26 (1966).
60. D. T. Peterson, "Experimental Factors in the Purification of Metals by Electrotransport" in ref. 10, p. 104; D. T. Peterson, "Electromigration as a Purification Process" in ref. 12, p. 54.
61. J. D. Verhoeven, *Trans. AIME* **233**, 1156 (1965).
62. J. D. Verhoeven, *Trans. AIME* **239**, 694 (1967).
63. D. A. Rigney, "Electromigration in Liquid Metal Alloys" in ref. 12, p. 140.
64. Th. Hehenkamp, *Acta Met.* **14**, 887 (1966).
65. R. Kirchheim and E. Fromm, *Acta Met.* **22**, 1397 (1974).
66. J. Mathuni and co-workers. *Met. Trans.* **7A**, 977 (1976).
67. J. P. Stark, *Acta Met.* **26**, 1139 (1978).
68. *Ibid.*, 1133 (1978).
69. F. M. d'Heurle and co-workers, *J. Vac. Sci. Technol.* **9**, 289 (1972).
70. A. Gangulee and F. d'Heurle, *Thin Solid Films* **16**, 227 (1973).
71. A. Blech and H. Sello, *Physics of Failure in Electronics*, Vol. 5, Rome Air Development Center, Rome, New York, 1966, p. 496.
72. J. Wohgemuth, *J. Phys. Chem. Solids* **36**, 1025 (1975).
73. R. W. Berry and co-workers. *Appl. Phys. Lett.* **9**, 263 (1966).
74. F. d'Heurle and I. Ames, *Appl. Phys. Lett.* **16**, 80 (1970).
75. K. L. Tai and M. Ohring, *J. Appl. Phys.* **48**, 28 (1977).

76. *Ibid.*, 36 (1977).
77. P. S. Ho, *J. Appl. Phys.* **49,** 2735 (1978).
78. L. Berenbaum and R. Rosenberg, *Thin Solid Films* **4,** 187 (1969).
79. R. E. Hummel, B. K. Krumeich, and R. T. Detloff, *Appl. Phys. Lett.* **33,** 960 (1978); R. E. Hummel, "Electromigration in Thin Films" in ref. 12, p. 93.
80. A. J. Learn, *J. Electron. Mater.* **3,** 531 (1974).
81. M. Attardo, R. Rutledge, and R. Jack, *J. Appl. Phys.* **42,** 4343 (1971).
82. L. Jastrzebski and H. C. Gatos, *J. Crystal Growth* **42,** 309 (1977).
83. J. R. Black, *IEEE Trans.* **ED-16,** 338 (1969).
84. T. Satake and co-workers, *Jpn. J. Appl. Phys.* **12,** 518 (1973).
85. N. G. Ainslie, F. M. d'Heurle, and O. C. Wells, *Appl. Phys. Lett.* **20,** 172 (1972).
86. I. Ames, F. d'Heurle, and R. Horstmann, *IBM J. Res. Develop.* **14,** 461 (1970).
87. D.-Y. Shih and P. J. Ficalora, *Proceedings of 16th Reliability Physics Symposium*, IEEE, New York, 1978, p. 268.
88. D. Gupta and R. Rosenberg, *Thin Solid Films* **25,** 171 (1975).
89. J. K. Howard, J. F. White, and P. S. Ho, *J. Appl. Phys.* **49,** 4083 (1978).
90. J. K. Howard and R. F. Ross, *Appl. Phys. Lett.* **18,** 344 (1971).
91. P. S. Ho and J. K. Howard, *J. Appl. Phys.* **45,** 3229 (1974).
92. I. A. Blech and E. Kinsbron, *Thin Solid Films* **25,** 324 (1975).
93. I. A. Blech, *J. Appl. Phys.* **47,** 1203 (1976).
94. R. J. Miller, *Proceedings of 16th Reliability Physics Symposium*, IEEE, New York, 1978, p. 241.
95. E. Kinsbron and co-workers. *Proceedings of 16th Reliability Physics Symposium*, IEEE, New York, 1978, p. 248.
96. J. Daniele, *J. Electrochem. Soc.* **124,** 1143 (1977).
97. L. Jastrzebski, Y. Imamura, and H. C. Gatos, *J. Electrochem. Soc.* **125,** 1140 (1978).
98. P. E. Vanier, F. H. Pollak, and P. M. Raccah, *J. Electron. Mat.*, in press.
99. V. N. Grinyuk and G. F. Tikhinskii, *Fiz. Met. Metalloved.* **39** (3), 514 (1975).
100. F. A. Schmidt, R. J. Conzemius, and O. N. Carlson, *J. Less Common Met.* **59,** 53 (1978).
101. A. Bobbio, A. Ferro, and O. Saracio, *IEEE Trans.* **R-23,** 194 (1974).
102. H. Suhl and P. A. Turner, *J. Appl. Phys.* **44,** 4891 (1973).
103. J. R. Black, *Proceedings of 16th Reliability Physics Symposium*, IEEE, New York, 1978, p. 233.
104. A. Gangulee, F. d'Heurle, V. Ranieri and R. Fiorio, *J. Vac. Sci. Techn.* **16,** 156 (1979).
105. K. Reddy, F. Beniere, and D. Kostopulos, *J. Appl. Phys.* **50,** 2782 (1979).

FRANÇOIS M. D'HEURLE
IBM

DAVID T. PETERSON
Iowa State University

ELECTRONIC-GRADE CHEMICALS. See Fine chemicals.

ELECTRON SPIN RESONANCE. See Analytical methods.

ELECTROPHORETIC DEPOSITION. See Coating processes; Electromigration; and Powder coatings.

ELECTROPHOTOGRAPHY

It has been estimated that by the year 1980 almost 36% of all reprographic prints will be made by one or another electrophotographic process. The other major contender for this one trillion reproduction copy market is offset duplication (45%), the remainder being serviced by stencil (9%), spirit (8%), and an array of other reproduction technologies (2%). (see Reprography; Printing processes.)

Electrophotography is still a relatively young science. Although the first efforts to use photoconductive effects in producing an image date back to the 1920s, the field became active only after 1950 when a growing need for inexpensive photocopies stimulated the development of new technology based on electrostatic imaging (1). This application did not require portability, a seemingly unattainable goal, for electrophotographic cameras generally require external electrical connections as well as special illumination. Instead of competing with silver photography in compactness, photosensitivity, and pictorial rendition, electrophotography offers complementary advantages: instant delivery of inexpensive copies of text and graphics without wet chemical processing, high contrast prints of archival quality on plain (or nearly plain) paper, and a liberation from the need for dark storage of photosensitive material prior to use.

The principles and applications of electrophotography have been described in great detail in several monographs (2–4), numerous journal articles (5–7), and current publications. This article provides an overview of the main electrophotographic imaging processes known, with emphasis on those that are of practical importance and on materials-related topics within those processes that may be of special interest to chemical technology.

Electrophotography can be defined broadly as a process in which photons are captured to create an electrical image analogue of the original. This electrical analogue is, in turn, manipulated through a variety of steps that result in a physical image. As with other photographic schemes, in order for a process to be practical the amplification (image intensification) must form many visible elements in the product for each original photon. The primary photoreceptor must have sufficient sensitivity to the image radiation to record a signal. Light is lost through the optical system, and short exposure times are desirable. The total number of photons available in the image plane is far less than that needed to produce one molecule of final photoproduct from each photon. As in classical photography (qv) amplification is provided by a development process. By chemical or electronic means the latent electrical image is converted to a visible image. Just how much amplification is required can be calculated readily. With practical light sources, projection optics can furnish, at most, about 10^{12} photons/cm^2 to the photoreceptor in exposure times of one second or less. About 10^{17} light-absorbing molecules/cm^2 are required for an image of pleasing contrast (8). Since primary photochemical or photophysical events can yield, at most, one molecule of photoproduct for each absorbed photon (quantum yield $\Phi = 1$), photography requires at least a 10^5-fold amplification.

Imaging processes that lack amplification are plentiful and have found widespread applications in reprography, printing-plate making, photofabrication, contact duplication of transparencies, etc. These processes are unsuited for short-exposure projection systems (see Photoreactive polymers). Comprehensive reviews of these technologies can also be found in the literature (9–10).

In silver halide photography primary quantum yields are low ($\Phi \simeq 0.01$), and amplification in the development process ranges from 10^3 to 10^9. High amplification is achieved at the expense of resolution. In electrophotography the primary quantum yield is generally much higher. Under favorable conditions an absorbed photon has a 50–70% chance of forming a mobile electronic charge carrier—the primary unit of image information in all electrophotographic processes. Displacement of such carriers in an electric field produces a latent image, usually a charge pattern analogue of the original, that must be developed to become visible.

The necessary amplification in the case of projection imaging, focusing an illuminated original through a lens onto the photosensitive element, is obtained by some form of physical development, eg, by the attraction of relatively large, visible particles (8–10 μm dia) in response to only a few surface charges; by a change in electrostatic adhesion; by the electrostatically induced wrinkling of a smooth thermoplastic layer (see Deformation recording media); or by the electroplating of many metal ions on each image site. The free energy required to initiate image gain is supplied in the form of an electrostatic potential, typically applied across a dielectric layer before exposure or to a bias electrode (a closely spaced metal plate to which a potential is applied) during development. For example, on typical electrophotographic photoreceptors a charge density of about 10^{12} electrons/cm^2 suffices to attract a dense black layer of 10-μm toner particles, producing an optical absorbance of about 1.5. At a primary quantum yield, $\Phi = 0.5$, full exposure requires about 2×10^{12} photons/cm^2—an effective gain of about 10^5 over molecular systems such as diazo, photochromic (see Chromogenic materials), dye bleach, and photopolymers. Independent estimates of electrophotographic gain and discussions of the implications of variations in sensitivity, granularity, and sources of image noise are well documented (11–12).

Transfer Xerography

The most widely used form of electrophotography today is called transfer xerography. C. Carlson demonstrated in 1938 the basic principles of a new imaging process to produce inexpensive copies on plain paper (1,13). Since Carlson's first successful experiments, a mature technology called xerography (from the Greek: dry writing) has evolved—a technology that is the foundation of a multibillion dollar industry. Many new materials have replaced Carlson's original sulfur and anthracene photoreceptors and pigmented lycopodium powder toners. Fully automated copiers, duplicators, and printers have been developed, yet the basic process steps of transfer xerography remain essentially unchanged (2–3).

Overview of Process Steps. Figure 1 illustrates one cycle of transfer xerography starting at the nine o'clock position. Relative motion between the outer processing devices and the photoreceptor is maintained during most of the cycle. The moving element is usually the photoreceptor. Proceeding clockwise, the process steps are as follows.

Charging Step. A corona discharge device deposits gas ions on the photoreceptor surface, forming a blocking contact. These ions provide a uniform electric field across the photoreceptor and a uniform charge layer on the surface.

Exposure Step. The illuminated original is projected through a lens system and focused on the photoreceptor. Where light strikes the charged photoreceptor surface, surface charges are neutralized by increased conductivity across the photoreceptor

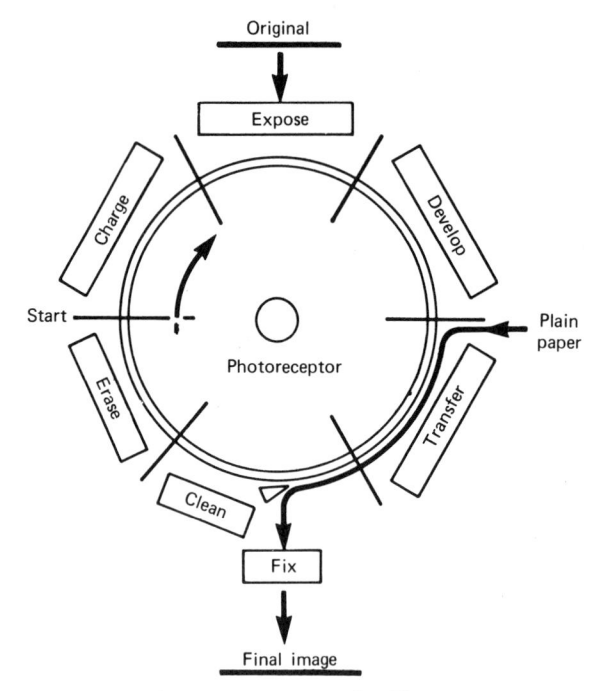

Figure 1. Schematic of one cycle in transfer xerography. The process steps take place on the surface of the photoreceptor, which is coated on the central imaging drum.

(photoconductivity). Unilluminated regions retain their charges. The resulting pattern of surface charge is the latent electrostatic image.

Development Step. A thermoplastic pigmented powder (toner) bearing a charge opposite to the surface charges on the photoreceptor is brought close to the photoreceptor so that toner particles can be attracted to the charged regions on the photoreceptor. The result is a physical image consisting of electrostatically held toner piles.

Transfer Step. A sheet of plain paper is brought into physical contact with the toned photoreceptor. A charge applied to the back side of the paper induces attraction of the toner image to the paper. This transfer of image preserves the right reading orientation of the original. The paper is stripped from the photoreceptor with the toner image clinging to it electrostatically.

Fixing Step. The toner pattern is permanently fused to the paper, eg, by means of a hot pressure roll or a radiant heater.

Although this concludes the image processing steps, two more steps are required in preparation for another cycle.

Cleaning Step. Because of incomplete transfer to paper, a toner image remains on the photoreceptor. This is wiped off with a brush, cloth or blade. A corona charge of reverse polarity aids in removing toner.

Erase Step. A uniform light source floods the photoreceptor to neutralize any residual charges from the previous image cycle, erasing the electrostatic image completely and conditioning the photoreceptor for another cycle.

A schematic view of a modern high speed xerographic duplicator capable of producing two images per second on plain paper is shown in Figure 2.

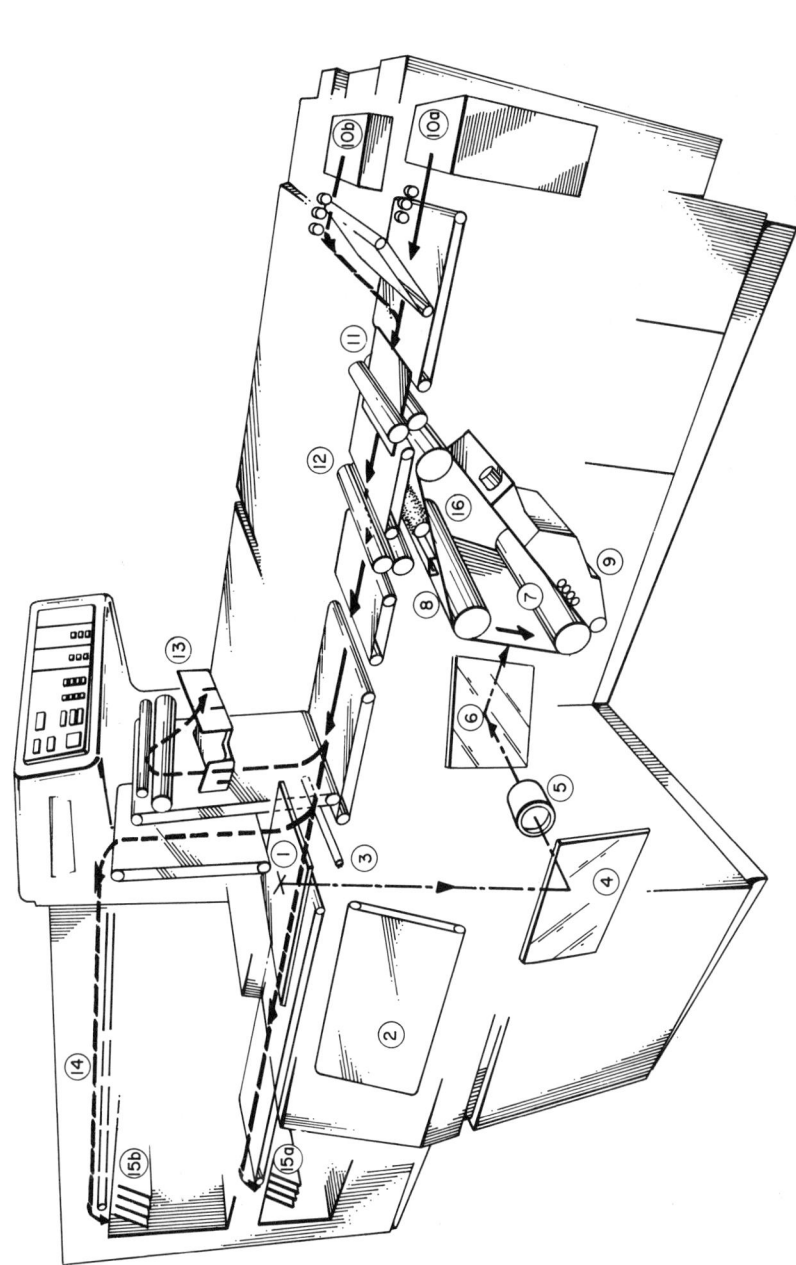

Figure 2. Transfer xerography, illustrated by a high-speed duplicator (the Xerox 9200) utilizing a photoreceptor belt. Documents are stored in tray [2] and fed automatically to platen [1], where each original is illuminated by flashlamps [3] and projected through optics [4,5,6] onto the photoconductive coating of the continuously moving belt [7]. A corona device [8] provides a uniform charge to the surface. A rack of magnetic brushes in the developer housing [9] applies the toner to the belt surface (see Fig. 9). Paper is fed from either stack [10a or 10b]. Roll [11] applies an electric charge to the back of the sheet to assist transfer of the toner image from the photoreceptor to the paper. The image is pressure fused to the paper by heated rolls [12]. The finished copy is delivered to tray [13] or sorted at [15a] or [15b]. A rotating brush [16], assisted by vacuum, removes residual toner from the belt before a new cycle starts. Duplicators of this type can turn out 2 copies per second. Courtesy of Xerox Corporation.

The Photoreceptor as the Central Device. In transfer xerography almost all process steps are carried out on the photoreceptor surface. The photoreceptor is reused, typically, between 10^4 and 10^6 times. At the beginning of each cycle it must be completely restored to its original state. If full-frame exposure is to be used, the photoreceptor must be a flexible belt to present a flat surface to the optical system. It must be seamless to allow random processing around its circumference. The surface of the photoreceptor must be resistant to abrasion by developer material, paper, and the cleaning device. Surface free energy should be low enough to facilitate cleaning of residual toner. Belt bases are made of either metal or metal-coated plastic. Rigid photoreceptors are cylindrical and usually fabricated or mounted on a metal substrate (see Photoreceptors for Transfer Xerography below).

Formation of the Latent Electrostatic Image. *Charging Process.* The first step necessary for creating a latent electrostatic image is the application of a uniform charge pattern on the surface of the photoreceptor; at the same time a uniform field from front to back is provided. Charges must be held on the surface, ie, they must be blocked and not passed immediately into the bulk of the surface (injected).

Corona discharge devices have proven the most reliable means of applying a stable, uniform charge layer to large area photoreceptors (14). Typically, these consist of one or more thin wires connected to a positive or negative high voltage supply and backed by a grounded shield that serves to stabilize the discharge (a corotron, Fig. 3a). A grid

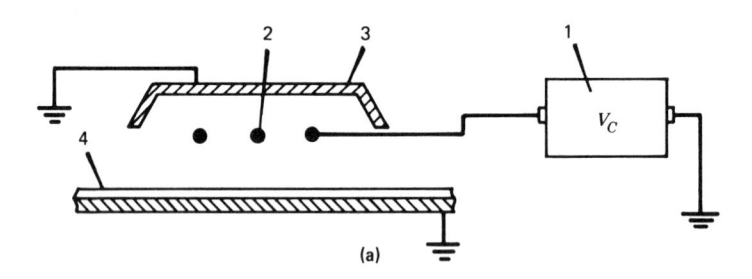

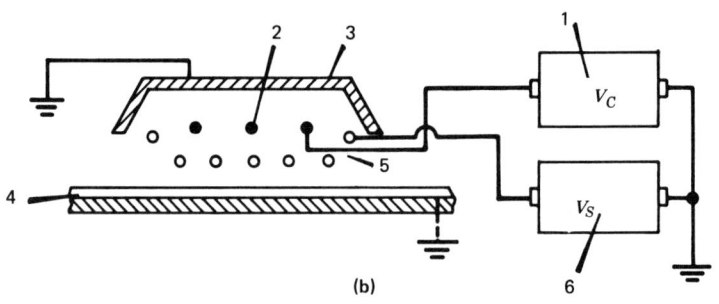

Figure 3. Two common corona-charging devices. (a) A three-wire corotron, connected to a high-voltage power supply of V_c [1], typically 6–7 kV, consists of three 75 µm diameter tungsten or platinum wires [2] and a grounded shield [3]. The photoreceptor surface [4] passes under the corotron. (b) A scorotron device, using three emitting wires [2] at V_c, and a 7-wire control grid [5] connected to a lower-voltage power supply [6] set to the screen potential (V_s) to be applied to the photoreceptor surface [4]. Courtesy of Focal Press, London (15).

of control wires may be added in order to limit the photoconductor potential; the resultant device is called a scorotron (Fig. 3b). Typically, the corona wires are operated at V_c = 6–8 kV, and the screen wires at V_s = 400–800 V.

Corona devices operate by ionizing the air surrounding the emitting wires and depositing gaseous ions, such as hydrated protons or CO_3^{2-} on the photoconductor surface (16–17). Once there, the ions must remain stably adsorbed without injecting their charge into the photoconductor and thus discharging the layer before it is illuminated. The detailed composition of the surface charge has been defined only for one type of photoconductor: in the case of zinc oxide–binder layers, it has been shown to comprise absorbed O_2^- ions (18).

Reliable contact electrification of large areas has proven difficult, partly because of roughness and chemical irregularities in the surface layers. Because the process involves localized air breakdown, it is sensitive to fluctuations in ambient humidity. A somewhat more successful alternative is induction charging. This involves field-induced injection of carriers from the conductive base through the photoconductive layer. These charges are finally trapped at the surface (Fig. 4) (19). Charging takes place during illumination to provide carriers (see below). Thus a latent image can be created in a single step.

If dielectric overcoating is used to protect the photoconductor surface, it can act as a separate charge-storage layer. Figure 5 shows one of several closely related photodielectric processes whereby selective photodischarge in the photoconductive layer allows the formation of a developable latent image across an overcoating of this type. The sequence of charging and illumination steps is readily altered to produce either positive or negative images from the same original. Transfer, fixing, and cleaning steps are similar to those for other forms of xerography (20–24).

Exposure Process. As in other photographic systems the image must be brought to the photosensitive surface with efficiency. Light sources are selected to match the spectral response of the photoreceptor and to provide adequate illumination of the original to be imaged. Incandescent lights, fluorescent bulbs with phosphors emitting various wavelengths, and xenon flash tubes are in common use. With stationary photoreceptor surfaces, full-frame exposure can be achieved by using shutters or intermittent illumination. Short-duration flash exposure allows the use of a moving photoreceptor. Automatic copying machines usually require continuous motion of

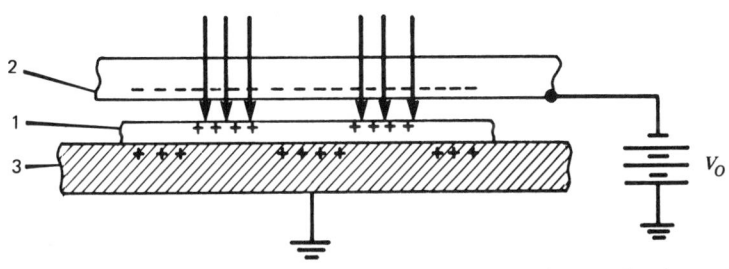

Figure 4. Induction charging. Image illumination (arrows) of a photoconductive surface [1] in an external field, applied by a closely spaced transparent electrode [2], is used to form a latent image in a single step. Mobile charges are injected from the base electrode [3] on the back of the photoreceptor. Rear illumination may be used if the photoreceptor backing is sufficiently transparent. Courtesy of Focal Press, London (19).

800 ELECTROPHOTOGRAPHY

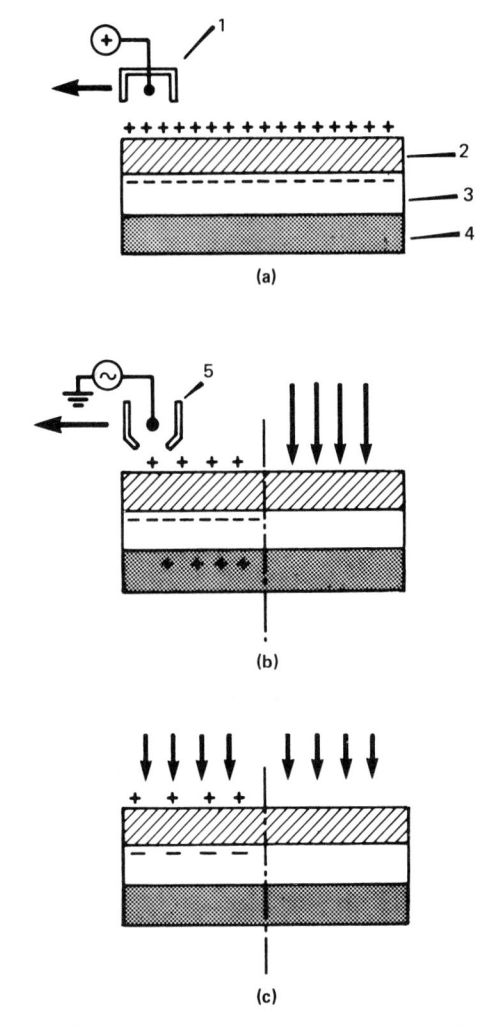

Figure 5. A typical photodielectric process. (**a**) Charge is deposited by a conventional corotron [1] on dielectric [2] while countercharge is injected into the photoconductive layer [3] from base [4]. (**b**) Image illumination (arrows) is applied through a transparent, grounded a-c corotron [5] to produce an equipotential surface whose charge density varies according to the image. (**c**) Uniform illumination produces a stable, developable, variable-potential image only across the dielectric layer (20).

the photoreceptor in order for all other process steps to go on at the same time. Here either the original moves synchronously with the photoreceptor (fixed lens optics) or the lens system moves to expose successive image segments on the moving photoreceptor (strip lens or scan optics).

For contact exposure the original must be fairly transparent and have image information on one side only. Reflex exposure, a special case of contact exposure, requires illumination through the back of the photoreceptor. It relies on differences in reflectivity on the document surface and requires inconveniently precise exposure control to produce prints of good contrast. Although these exposure methods eliminate the cost and complexity of projection optics, they have severe disadvantages and are, therefore, rarely used.

Xerographic Development. The electrostatic latent image can be made visible by a variety of techniques. Toner particles can be attracted only where electrostatic field lines extend above the surface of the receptor. In the absence of a development electrode, external field lines appear only along the periphery of charged areas (Fig. 6a), permitting toner attraction only to the outline of an area (25–27). Field lines can be extended outward across the entire charged surface by bringing a development electrode (Fig. 6b), connected to the back electrode of the photoreceptor, close to its surface. Charged toner particles introduced into this narrow space follow the external field lines and deposit more uniformly to provide solid area development.

For the development process to occur, toner particles must first be charged reproducibly. Conditions in the development zone must be carefully controlled to avoid dusting and contamination elsewhere. The field configuration is crucial to uniform development. Reviews of development processes can be found in more detail elsewhere (28–35). A description of the major development systems follows.

Powder Cloud Development. Figure 7a illustrates a powder-cloud box of a type used to develop very small potential differences along image edges, making it particularly suited to showing up minute contrast differences in soft-tissue xeroradiography (30–31). Toner is fed through charged nozzles and drifts as a cloud through a charged control grid onto the photoreceptor, following the existing field lines. Figure 7b depicts the transformation of the original through an electrostatic pattern to a final image. Note the enhancement of contrast along image edges.

Cascade Development. The charging of toner particles is usually accomplished by contact charging. Powder-cloud boxes use a charged metal nozzle through which the toner is forced. Contact charging can also be accomplished with the aid of a carrier.

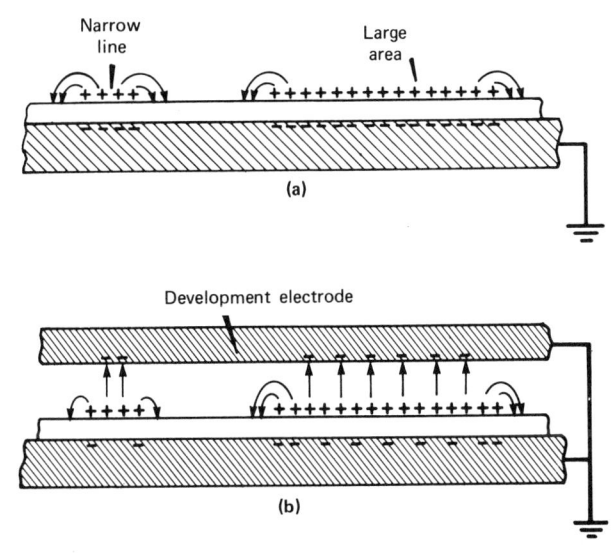

Figure 6. Field-line configurations of a latent electrostatic image (schematic). (a) In the absence of a development electrode, field lines extend only along the edges of latent charge patterns, producing edge enhancement in the developed image. (b) Addition of a closely spaced development electrode forces field lines to extend into the space above the photoreceptor. The field strength becomes everywhere proportional to the local photoreceptor potential, and the entire image area can be developed. Courtesy of Focal Press, London (25).

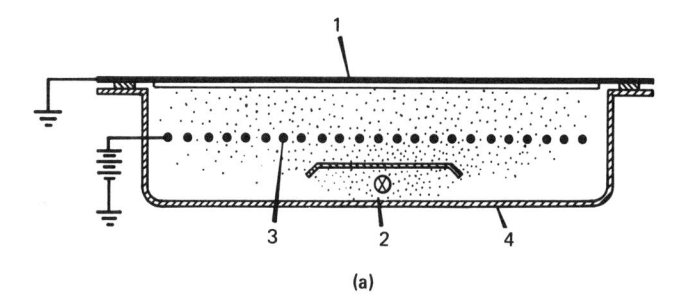

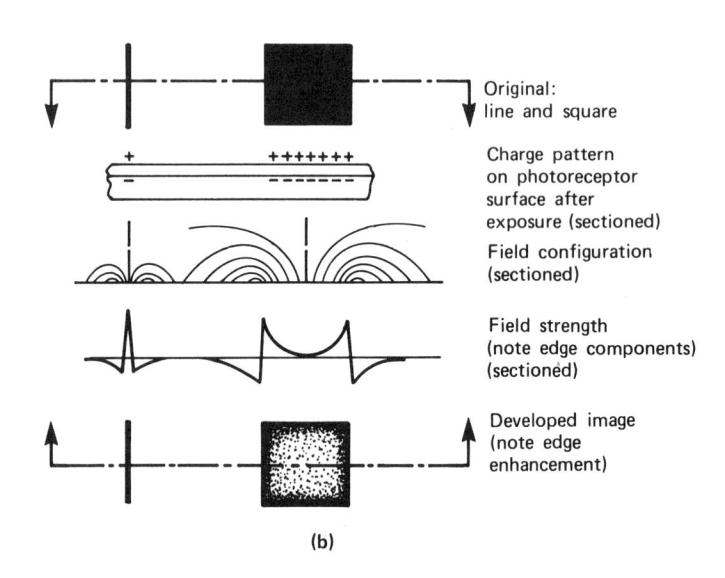

Figure 7. Powder cloud development. (a) Schematic of powder cloud development with xeroradiographic plate in place. Selenium plate [1] is face down; powder cloud generator [2] is a charged nozzle through which the toner enters the chamber. Open-wire control grid [3] is set at 1500 volts. Container [4] is typically 10 cm deep (33). (b) Image in various forms illustrating edge development effect (29).

Dielectric beads, about ten to fifty times the size of the toner particles, are mixed with the toner. Agitation of the mixture and the relative charge characteristics of bead and toner surfaces cause charge exchange. This process is called triboelectrification. The mixture of about 1–2 wt % toner and 98–99 wt % carrier is called developer. When this mixture is poured on the photoreceptor surface (Fig. 8), the beads bounce over the surface, giving up toner particles. Here again, edge enhancement is produced unless a closely spaced development electrode is provided.

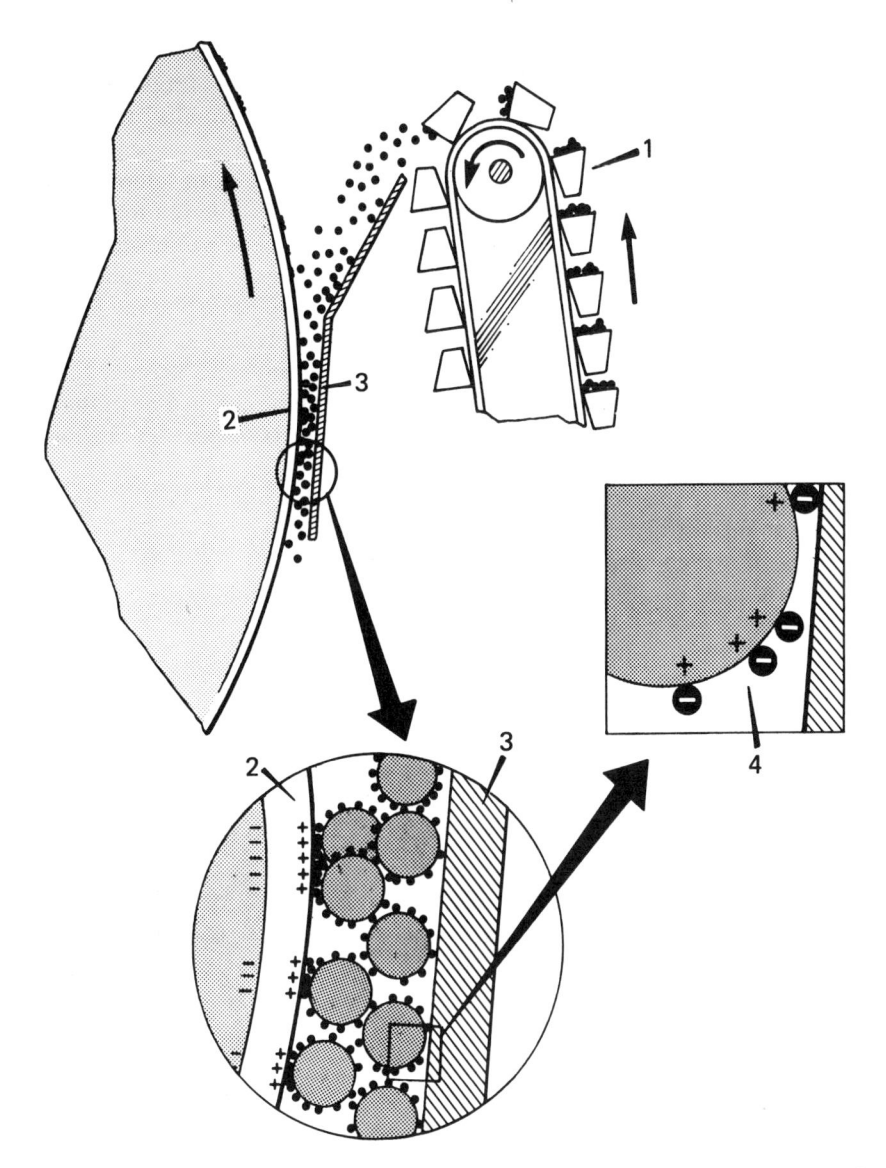

Figure 8. Cascade development. Buckets [1] on a carrier or belt deliver developer mix to photoreceptor drum [2]. Negatively charged toner particles are selectively attracted to positive sites on photoreceptor surface. A baffle or development electrode [3] may be added to minimize dusting. Detail [4] shows the relative size of 10-μm toner particles and a 100-μm carrier bead. Toner is charged by carrier via contact electrification in the agitated developer mass.

Magnetic Brush Development. The magnetic brush is an elegant method of providing the equivalent of a development electrode (32). Here, carrier beads are made of a magnetic material that also allows raising the effective dielectric constant of the developer mixture. In its simplest form the magnetic brush is a bar magnet that is dipped into a jar of magnetic developer and brushed across the photoreceptor. Grounding or biasing this device is an effective way to bring an electrode close to the photoreceptor surface during development. A multibrush structure used in a fast duplicator is shown schematically in Figure 9. Precise control of roll speeds, spacing, and bias voltage allows for adjustments of the final image quality (33–34). Linear development speeds of 50 cm/s can be readily achieved.

Touchdown or Impression Development. In this form of development, toner is precisely metered onto a roll that is covered with velvet or a microporous sponge (Fig. 10). The toner is precharged to keep it in place and to provide a uniform and opposite charge for the impression, or touchdown, step. The image is developed as the toned roll gently touches the photoreceptor surface in the development zone much like an inking roller in a printing process (28,36–37). Close control of roll spacing and bias determines image contrast and enables solid area development. The toner composition is chosen to be sufficiently conductive to allow it to become charged by induction from a metallic donor roll. It can also be made magnetic for transport by magnets inside the roll.

Image Reversal Development. In normal xerographic photocopying, the attraction of toner to an oppositely charged latent image takes place to develop a positive (or direct) image in which dark areas of the original are reproduced as dark areas on the copy (Fig. 11a). Negative (or reversal) images can be produced by reversing the charge

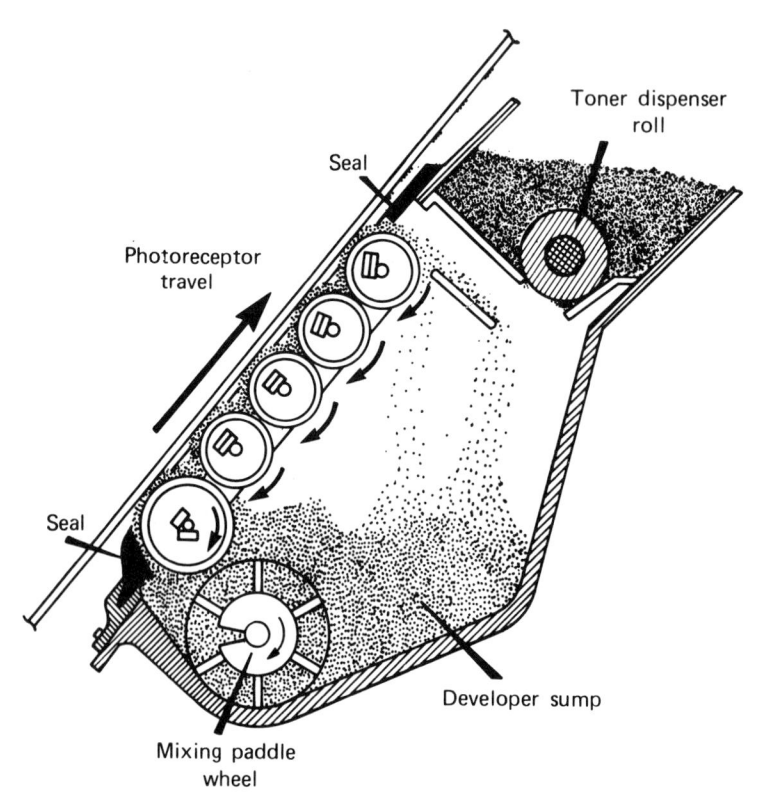

Figure 9. High speed magnetic brush development. All five rolls turn in the same direction. Their frictional surface drags the magnetic developer across the photoreceptor like a high-pile blanket. Stationary magnets inside the rolls provide the forces to align the magnetized carrier beads into bristles. Toner is periodically replenished to maintain optimal concentration in the mixture.

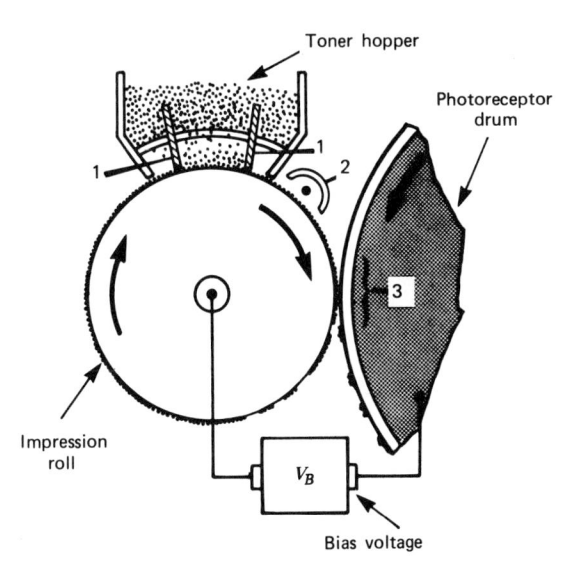

Figure 10. Touchdown or impression development. Toner is metered from the hopper onto the impression roll by blades [1] and is uniformly charged by corona [2]. Upon gentle contact in zone [3] the toner develops an image on the photoreceptor. Bias voltage V_B controls contrast and image quality. Courtesy of the Society of Photographic Scientists and Engineers (36).

on either toner or photoreceptor surface, so that the toner particles are repelled from latent image areas of the photoreceptor toward the development electrode. The latter is normally biased to approximately the potential of image areas; this cancels the net field in areas corresponding to dark portions of the image and causes maximum development in illuminated (discharged) areas (Fig. 11b). Bias can be adjusted to alter image contrast.

Transfer of Developed Image. To be useful, the developed toner image must be transferred to a receiving sheet, usually paper. With the aid of an externally applied field the toner particles are induced to jump to the paper as it separates from the photoreceptor (Fig. 12). As the paper separates, a slight lateral displacement of the toner particles occurs because of fringing fields. This limits the resolution in the final image. Adhesive-coated paper, originally used by Carlson, can capture the inherently higher image resolution on the photoreceptor.

A corotron usually charges the back of the paper during transfer (38). Alternatively, a semiconductive rubber roll can be brought in contact with the paper and a potential applied between its metal core and the photoreceptor backing (39). This biased transfer roll is particularly useful when multiple images must be transferred in registry onto a single sheet, as in a three-color copying device. Auxiliary charging devices are often used to increase the amount of toner transferred, to suppress the transfer of toner from unwanted areas on the photoreceptor such as image background, or to ease separation of the paper by an a-c corona discharge. Transfer to dielectric materials other than paper is possible. Plastic film foils yield transparencies, and offset masters can be made directly.

Fixing of Transferred Image. To become permanent the transferred toner image must be fixed to paper. A thermoplastic resin base in the toner powder permits fixing by fusing onto the paper. In noncontact fusing a radiant heat source provides energy

806 ELECTROPHOTOGRAPHY

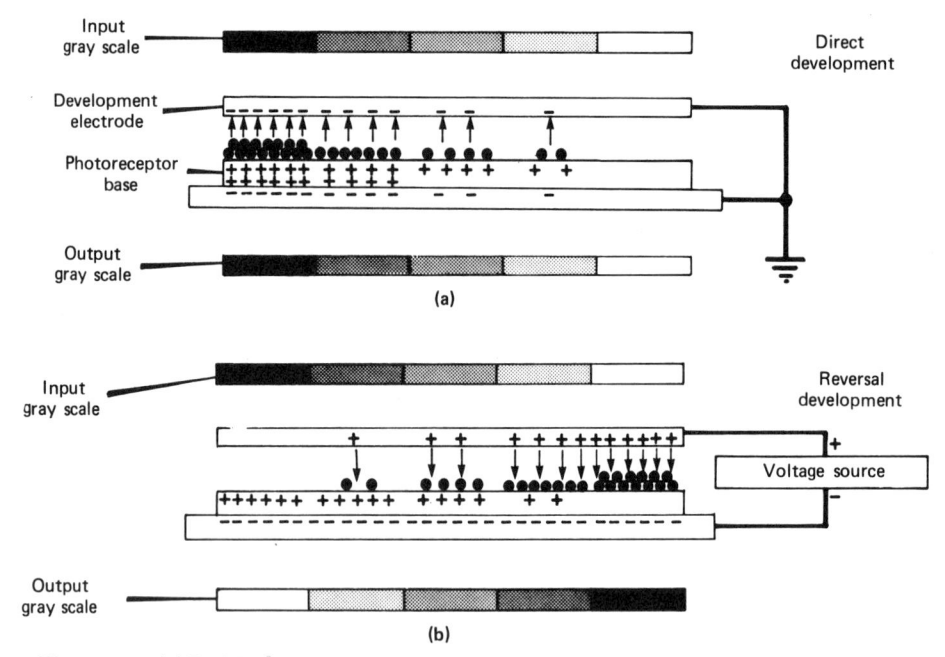

Figure 11. (a) Positive-to-positive reproduction of a gray scale by normal development. Negatively charged toner deposits selectively on positively charged areas of the photoreceptor, corresponding to dark areas of original. (b) Positive-to-negative reproduction of a gray scale by reversal development. Positively charged toner deposits selectively on discharged areas of the photoreceptor, corresponding to light areas of original. A bias field prevents deposition in charged areas.

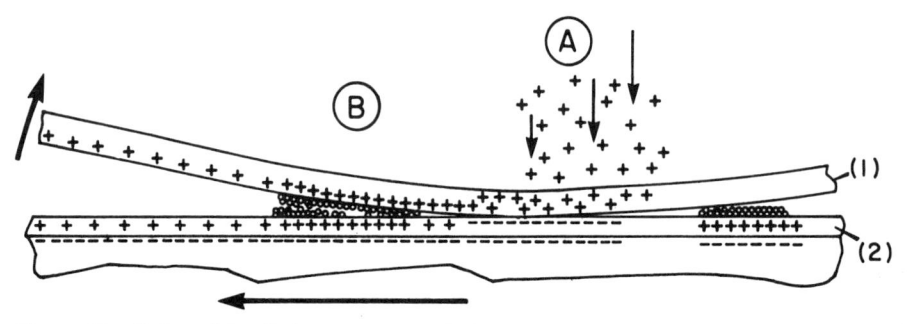

Figure 12. Enlarged detail of corona-actuated electrostatic transfer process. The back of paper [1] is charged positively by ions impinging at A; at point B the paper is separated from photoreceptor surface [2], pulling negatively charged toner off the surface. Courtesy of Focal Press, London (38).

to the toner image as the paper passes below it (40). Xenon flash lamps, rich in infrared radiation, permit selective absorption by the toner powder (41). In contact fusing, heated rollers with good release properties (eg, silicone or fluorocarbon elastomers), backed by a pressure roller, melt the toner and force it into the paper (42).

Pressure alone can accomplish fixing, but requires softer toners (39) or microencapsulated toners containing inks (43). Nonfusing toners can be fixed to a coated paper whose surface will soften upon heating or exposure to solvent vapors (44). Spray application of a fixative and adhesive coatings have also been used. Liquid-developed images (see below) are fixed by adsorption onto the paper which is usually coated. Because of the inertness of the materials used in transfer xerography, eg, stable polymers and carbon black, archival quality documents can be obtained.

Post-Imaging Process Steps. In cyclic transfer xerography the photoreceptor must be restored to its original condition after image transfer. Between 10 and 30% of the toned image and residual charge patterns remain on the receptor. Flooding the photoreceptor surface with light and applying neutralizing charges (45) aid in removing toner. Other aids are lubricants that are added to the toner (46) or applied directly to the photoreceptor surface (47). Both help form low friction surfaces, increasing transfer efficiency and cleanability. Fatty acid metal salts and fluorinated organics have been used for this purpose. Toner is removed by rotating brushes made of fur or of natural or synthetic fibers, and is collected by vacuum suction.

The photoreceptor can also be cleaned by a conforming wiper blade. In this case, lubricants are particularly helpful to prevent toner streaking. The removed toner falls into a reservoir, without the need for a vacuum. Disposable webs in contact with the toner residue on the photoreceptor surface also accomplish cleaning, but must be advanced periodically and eventually replaced (48).

Liquid-developed images in transfer xerography pose a special problem for cleaning because they often contain self-fixing agents. Photoreceptors in such processes can be cleaned by a spongelike roller soaked in the carrier liquid. The toner removed by this device is filtered and the liquid recycled (49).

Liquid Development. The most widely used liquid development scheme relies on electrophoretic particle migration (50). Pigment particles of about 1 μm in diameter are dispersed in a dielectric carrier liquid, usually a hydrocarbon. Polymeric and ionic additives control charges on the pigment, and fixatives give image permanence. Solid area development is possible with the aid of electrodes, bias voltages, or conductive additives. This development system offers very high resolution, approaching that of the latent image itself; 800 lines and spaces (or line pairs) per millimeter have been achieved. Application devices are simple, and heat-fixing can be avoided. One disadvantage of the system is carry-out of dielectric liquid on both photoreceptor and copy paper, which is both unavoidable and objectionable. In addition, transfer from and cleaning of the photoreceptor are difficult to achieve. Although liquid development has found its most prominent use in direct electrophotography, a successful liquid transfer system has been devised and marketed (51).

Another scheme, similar to impression development, avoids the carry-out of liquid altogether (52). A conductive, more viscous, true ink is metered into recesses of a finely patterned conductive roller or applicator. The ink is drawn out of the recesses in the applicator onto the photoreceptor in response to the external field of the latent image pattern. Since ink, rather than toner, forms the image, it can be transferred to paper much like in offset printing and no heat fixing is necessary. Resolution is limited by the applicator pattern.

Selective wetting of the photoreceptor surface (53) by an aqueous developer has also been accomplished. Here, the electrostatic pattern enhances the effective surface energy of the photoreceptor. As a result, the uniformly applied developer wets only the charged areas from which the image can be transferred to paper.

Materials for Transfer Xerography

Photoreceptors. The photoreceptor is an electronic device that forms the electrostatic image charge pattern in response to light. It typically comprises three layers (Fig. 13): a conductive base layer which can be omitted if the support of the photoreceptor is metallic; a thin dielectric barrier layer which is required only to prevent charge injection into the photoconductive layer; and the photoconductor layer on top. The conductive layer forms the back electrode. If a separate base layer is required, as on polymeric film supports, it consists of a sputtered or evaporated metal or conductive metal oxide layer well under 100 nm thick. Deliberately applied barriers may consist of inorganic oxides, sulfides, or thin polymeric films (54). In some cases the barrier layer can be omitted because dielectric oxides that form naturally on the base electrode constitute effective barriers.

The photoconductor itself must be capable of generating mobile carriers upon illumination and transporting them through the bulk of the layer. These mobile carriers must be absent initially or must be swept out during the charging step. Dark conductivity must be low so that little charge is lost between initial charging and development, usually 0.3–10 s, depending on the application. The corona charge applied to the surface provides the top electrode, believed to be in the form of adsorbed gas ions. Charge stability and uniformity have been achieved by applying a surface layer of electron donors (for positive charges) or acceptors (for negative charges) (54). The fact that corona charging provides a blocking electrode, preventing premature dark discharge, was recognized by Carlson (14). Figure 14 illustrates the time dependence of the surface potential of typical photoreceptors. The photoconductor layer generally acts as a capacitive device. The slope of the potential curve after the corona is turned off is a measure of dark discharge. The shape of the *light on* curve is the photoinduced discharge characteristic of the device. The residual potential, after illumination ceases, is an indication of trapped charges or fatigue. For cyclic operation of a photoreceptor it is important that the shapes of these curves remain essentially constant over time.

The optimal thickness of the photoconductor layer is determined by the potential required for development, typically 300–800 V. The exposure required to produce this potential measures the photosensitivity of the device. Since each absorbed photon can, at most, generate one conduction electron, the voltage contrast potential produced by illumination is, in principle, proportional to the dielectric thickness. Transport of carriers, however, becomes inefficient when thickness exceeds the mean trapping range for these carriers. Practical photoconductors are between 15 and 25 μm thick if they are made up of organic or binder-type layers of low dielectric constant ($K = 3$–5); selenium and its alloy layers ($K = 6$–9.5) are typically between 30 and 100 μm thick (59).

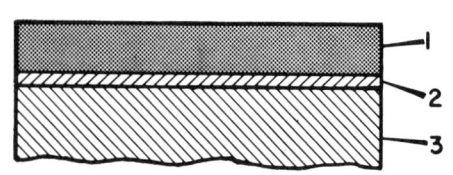

Figure 13. Cross section through typical xerographic photoreceptor. [1] Photoconductive layer, [2] injection barrier, [3] base electrode (not drawn to scale).

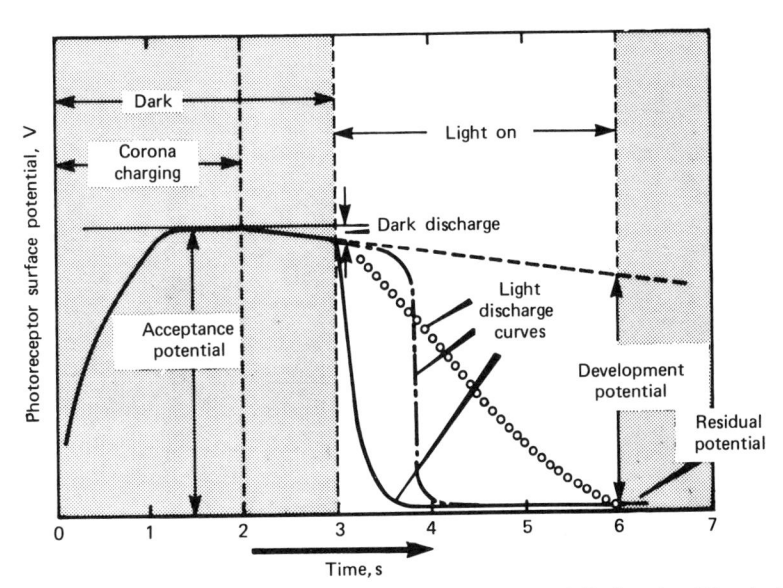

Figure 14. Surface potential characteristics during charging and discharging. The photo-induced discharge characteristics under steady illumination can vary substantially, depending on the photoconductor layer: —— bulk selenium (55) or PVK:TNF charge transfer complex (56); ○○○ negatively charged ZnO–binder Electrofax layer (57); · · · · · phthalocyanine–binder layer exhibiting an induction period prior to rapid discharge (58). (Curves are normalized for comparison.)

High photosensitivity requires effective light absorption in the desired spectral region (extinction coefficient 10^4–10^5 cm^{-1}); efficient generation of carriers, ie, separation of the photoexcited electron/hole pair in the applied field; and subsequent complete transport of the mobile charge through the photoconductive layer. Most photoconductor materials exhibit efficient transport for either electrons or holes. For front illumination, electron-transport photoconductors should be charged negatively, and hole-transport layers positively. True two-carrier photoconductors are rare.

Vitreous selenium, an amorphous mixture of octagonal rings and linear Se-polymers produced by vacuum evaporation, has long been the basis of the most commonly used photoreceptors (Fig. 15) (55–56,58,60–63). Tellurium may be added to extend red response. Arsenic alloys have been found to enhance both photosensitivity and stability against crystallization (60,64). Addition of halogen minimizes buildup of residual potentials (65). Dispersions of photoconductive pigments such as cadmium sulfide, zinc oxide, and phthalocyanine in dielectric binders may be used, provided the volume loading of pigment is sufficient to assure charge transport across the thin dielectric barrier between adjacent particles. The amount of pigment necessary to achieve optimal performance depends on the particular pigment and binder combination used. Crystal morphology of the pigment and dispersion or milling techniques both influence the behavior of the final layer. Binders are usually organic polymers in a solvent that is removed after coating the photosensitive layer on a conducting substrate (58,63,66). The particle–binder interface may also be doped with sensitizing dyes, electron acceptors, or so-called supersensitizers to enhance photoresponse, charge acceptance, and other desirable characteristics (67–68).

Organic photoconductors are of special interest to chemists because their behavior

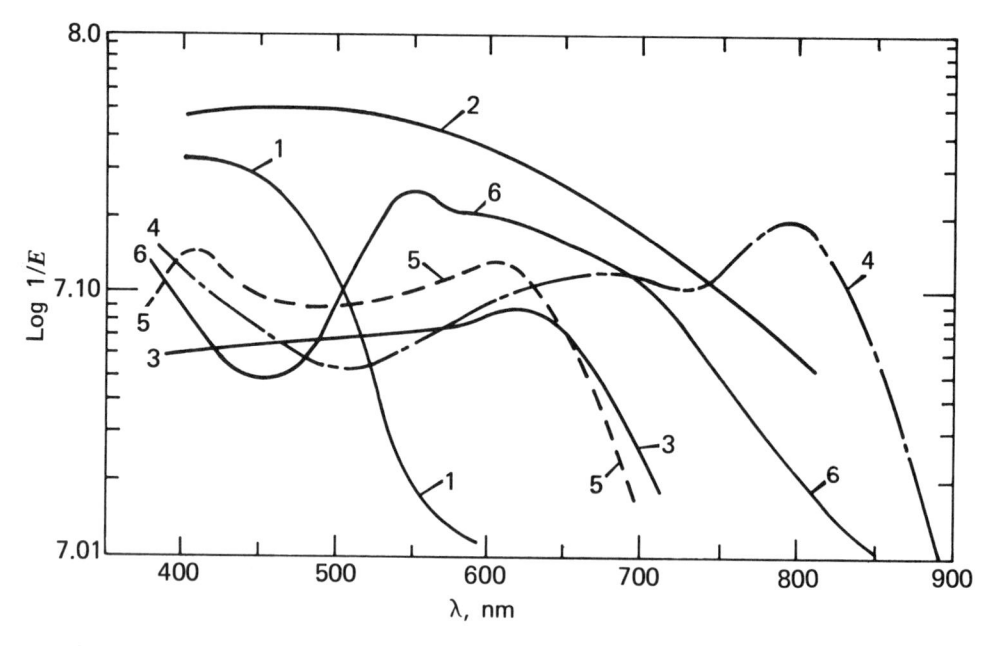

Figure 15. Relative spectral response of typical xerographic photoreceptors. Photosensitivity ($1/E$, in cm^2/J) is plotted as the reciprocal of the incident energy required for approximately 50% discharge of the initial surface voltage on the photoconductor. [1] selenium (60); [2] arsenic triselenide (61); [3] PVK:TNF (56), 1:1 molar complex of poly(vinyl carbazole) and trinitrofluorenone; [4] phthalocyanine-binder photoreceptor (58); [5] ZnO-binder coating dye-sensitized with bromphenol blue (62); [6] CdS-binder coating (65). (Note: J × 10^7 = erg.)

bridges the gap between photochemistry and solid state physics. Carlson used anthracene for some of his earliest experiments. Since then, many types of polynuclear aromatic and heterocyclic compounds have been shown to be useful, both in binder dispersion and as homogeneous polymers like poly(vinyl carbazole) (69–70). Although the inherent sensitivity of most of these materials lies in the ultraviolet, their response range can be extended into the visible by the addition of sensitizing dyes (71) and pigments (72), or by the formation of donor–acceptor complexes having photoactive charge transfer absorption bands in the visible (56,70,73). Carrier generation in these materials is strongly field dependent: typically, the quantum yield is proportional to the square of the field intensity (74).

The mechanism of dye sensitization, for years a subject in silver halide photography (75), has also been investigated in the context of electrophotography (see Dyes, sensitizing). The key question is whether the photoexcited dye sensitizes the uv-sensitive photoconductor by transferring neutral excitation energy or charge carriers. Experiments with double-layer photoreceptors (Fig. 16) have shown that the photogeneration of charge carriers takes place in the sensitizing layer and that these charges are injected into and transported across the underlying or overlaying photoconductor. Useful sensitizers (photogenerators) include selenium, and many organic dyes and pigments. Charge transport has been shown to be a purely molecular process—one that requires neither long-range order nor cooperative effects such as are invoked for band-model photoconductors (76). Electron donors, such as poly(vinyl carbazole),

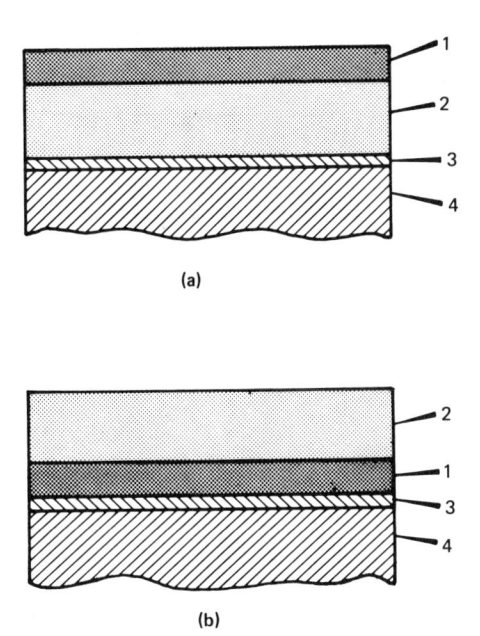

Figure 16. Active matrix photoreceptor with (**a**) generator layer on top, (**b**) at interface. [1] Generator (sensitizing) layer; [2] active transport layer; [3] blocking interface; [4] base electrode. (Layers not drawn to scale.)

are effective transport materials for positive charges; strong electron acceptors, such as trinitrofluorenone transport negative charges only (77). Spectroscopic measurements of the energy levels of the materials involved can be used to predict their ability to interact in this manner (77–78). This research has led to a new family of double-layer photoreceptors in which a relatively thin, visible-light-sensitive, carrier generator film is adjacent to a much thicker, transparent, polymeric, active transport layer. The former supplies photosensitivity; the latter, dielectric strength and voltage contrast. This allows greater freedom in designing photoreceptors that have desirable mechanical properties as well as high photosensitivity (79).

More recently a purely inorganic photoconductor has been developed, consisting of uniform arrays of hexagonal microcrystallites of cadmium sulfide (80), each about 70 nm in diameter and 350 nm deep, produced by a sputtering process (Fig. 17). The extremely thin layer sustains a uniform charge of about 30 V. Each crystallite is said to act as an independent photodiode. Coated on a substrate, such a layer serves as a panchromatic film, suitable for high-resolution liquid development.

Finally, the photoreceptor structure (shown in Fig. 5) in which an inert, protective dielectric overlies a photoconductive layer should be considered. In some applications (20) the electrostatic latent image is formed exclusively across the dielectric layer so that the photoconductor itself is not subjected to prolonged high field stress. This allows the use of some highly sensitive photoconductor materials, such as sintered cadmium sulfide, whose use in conventional single-layer photoreceptors would be precluded by an excessive dark-carrier concentration. Here, the imaging step requires simultaneous exposure and recharging which makes this process most suitable with slit scanning optics. As mentioned previously, these photoreceptors have the advantage

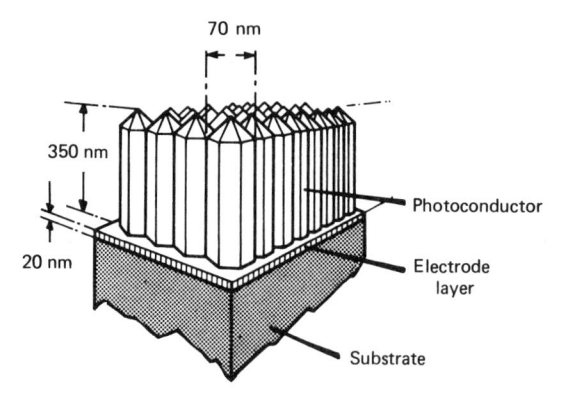

Figure 17. "Photodiode" arrays of sputtered cadmium sulfide crystallites. Courtesy of Coulter Systems Corporation.

that the photoactive material is fully protected against corona ions, abrasion, and other deleterious interactions with developer materials, paper, and cleaning devices, all necessary in the xerographic process.

Developer Materials. The toner particles used for developing xerographic images generally consist of 8–15 μm particles of a thermoplastic powder colored by a dispersion of 5–10% carbon black particles of less than 1 μm. Cyan, magenta, or yellow colorings may be substituted for use in color xerography (81–82) (see Color). Pigment concentration and dispersion must be adjusted to impart a conductivity to the toner mass that is appropriate to the development process. For efficient induction development (37) a conductivity greater than 10^{-4} S/cm is found to be desirable (83); most other development processes require the toner to retain charge applied by contact electrification for extended time periods. The toner thermoplastic is generally selected on the basis of its fusing characteristics; it must melt sharply at the lowest temperature consistent with stability to storage and to vigorous agitation that occurs in xerographic development chambers (84).

The developer mixture used for magnetic brush or cascade development, consisting typically of a 1–2 wt % concentration of toner, is dispersed in a carrier, a freely flowing mass of 100–500 μm-diameter steel, ferrite, or glass beads (33–35) (see Ferrites). The carrier beads are generally coated with an impact-resistant polymer of appropriate triboelectric characteristics designed to transfer a controlled amount of positive or negative charge to the toner by contact electrification. The physical–chemical effects responsible for charge transfer between insulators are as yet poorly understood (85). The literature suggests that the process involves, in part, electron exchange determined by the relative electrochemical potentials of the materials. However, charge exchange in dielectrics is a slow process, not likely to reach equilibrium (86). In a churning developer mass the situation is greatly complicated by abrasion and the impact transfer of ions, polymer and adsorbates (87), by the oxidation of polymer, and by localized electrical breakdown across the gaps between moving toner and carrier particles (88). Humidity has profound effects on triboelectric behavior (89). In addition, the dielectric relaxation time of the developer mass must be kept below the development time if effective solid area coverage is to be achieved; therefore, careful control of conductivity and dielectric constant are essential.

In a cycling process additional long-term interactions of the developer with other system components must be taken into account. The developer mass must not abrade, poison, or smudge the photoreceptor surface. It must be readily removable by a simple cleaning system; it should not be excessively degraded by impaction, heat, or corona oxidation. The toner particles must be readily transferable and fusible to the receiving sheet, but without being so soft as to coat the carrier beads. Finally, some blade-cleaning systems require the addition of lubricants to minimize van der Waals forces between the photoconductor surface and toner residues (46–47). Because so many conflicting requirements must be met, practical developer materials have been designed mostly by empirical methods, and the patent literature abounds in specific recipes (90). Specialty toners which contain magnetic pigments can be used for single-component development processes (83). Toners tailored to different fixing processes are described elsewhere (see above under Fixing of Transferred Image).

Direct Electrophotography

Electrofax Process. Coated-paper xerography was invented because of a chance observation: a faint image appeared on a magazine paper that had been accidentally charged, exposed, and dusted by toner powder without first having been contacted to a selenium plate. It was soon found that the surface sizing of the paper was photosensitive and contained zinc oxide. Thus coated-paper electrophotography (Electrofax) was born (91). Although this process avoids the cycling and cleaning requirements for a photoreceptor, a photosensitive coating, which becomes the final image, is needed on each sheet of paper. The toner is fused to the coating, avoiding the need for transfer and reducing the process steps from the six in transfer xerography to only four.

Commercial Electrofax paper base is generally treated with polymeric antistatic agents (92) to render it reliably conductive. The zinc oxide powder is dye-sensitized (71) by adsorption to extend its photosensitivity across the visible region. A uniform mixture of blue-, green-, and red-sensitive dyes may be used to keep the coating color a neutral light gray (93) while conferring panchromatic sensitivity. The key innovation in Electrofax was to produce images directly on the final sheet in fewer steps and make the photoreceptor an inexpensive consumable item.

Although Electrofax has found its widest use for low-volume black-and-white copying, at least one color copier has been marketed. Here, repeated exposures through filters alternate with dry or liquid development using subtractive color toners to produce full color prints. Paper-coating compositions based entirely on organics have appeared briefly, but ZnO-containing coatings still dominate the direct electrophotography field. Great improvements have been made in the feel of the paper and in reducing the coating weight.

Chargeless Electrophotography. Eliminating one more process step, charging, can still produce images. A form of chargeless electrophotography (94) is based on the creation of a small potential difference upon exposure. Photoelectrons selectively diffuse inward into the zinc oxide layer while holes remain trapped at the generation site, leaving a localized photovoltaic or Dember potential of a few tenths of a volt (95). A fine-particle liquid developer can develop these small potentials, offering extraordinarily high resolution. Unfortunately, development times of up to one minute are required, and the effect is strongly temperature-dependent. The process, therefore, has not been adaptable to automatic photocopiers.

Transparent Film Processes. Direct electrophotography on transparent organic photoconductor films has been used to produce transparencies. Since the transfer step is omitted, images resolving 200 line pairs/mm and better can be readily produced, particularly with electrophoretic liquid developers. This has led to analogue data recording and micrographic applications in which film or fiche can be generated by direct electrophotography and images can be added on the same film at a later time by repeating the process steps (96). Certain transparent organic photoconductor films can also be developed without toners; the photosensitive film itself or a surface coating is deformed according to an image charge pattern (97) (see Deformation recording media).

The preceding sections described the most widely practiced electrophotographic technologies. Discussion of many electrophotographic imaging systems that are less commonly used follows.

Transfer of Electrostatic Images (TESI)

In conventional xerography, the developed (toner) image is transferred to the receiving sheet, after which the photoreceptor must be erased and cleaned prior to reuse. An alternative that obviates the toner transfer and cleaning steps is the transfer of the electrostatic latent image to a suitable receiving sheet prior to development. Transfer of an electrostatic charge pattern can be accomplished either during photoreceptor exposure (ie, while the latent image is being formed) (Fig. 18) or after a stable charge pattern is formed on the photoreceptor by conventional charging and exposure. The names applied to these two process variants are simultaneous TESI and sequential TESI, respectively (99). In either case a dielectric-coated paper or film is pressed against the photoreceptor, generally in a suitable bias field. The image is made visible by toning the charged film or paper after it is pulled away from the

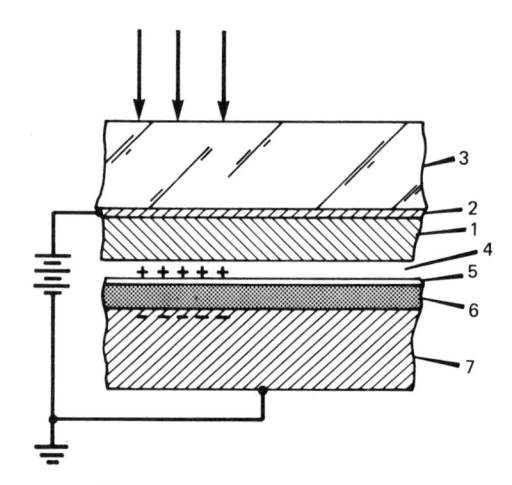

Figure 18. Simultaneous TESI. Image radiation (arrows) enters the photoconductor [1] through a transparent conductive-coated [2] glass backing [3]. Positive charge is transferred across the air gap [4] to a dielectric receiving sheet [5] on a resilient conductive backing pad [6] and metal support [7]. The threshold field across the air gap is not exceeded in dark areas, and no charge is transferred there. Courtesy of the Society of Photographic Scientists and Engineers (98).

photoreceptor. Toner is usually applied to the surface facing the photoreceptor, but it has been shown that a charge pattern can also be developed on the back of plastic film (100).

Although these processes were invented more than two decades ago, their commercial use has been delayed, perhaps because of the need for specially coated paper and the difficulty in coping with contact geometry and ambient humidity during charge transfer. Simultaneous TESI has recently been applied to a cathode ray tube (CRT) display terminal printer, and a compact copier using the sequential TESI process has recently been introduced (101).

Persistent Internal Polarization and Photoelectrets

An electrophotographic process that differs from xerography in the method used for establishing the electrostatic latent image, called persistent internal polarization (PIP), was invented in the United States (102–103) and independently under the name of photoelectret imaging in the Union of Soviet Socialist Republics (104). For xerography (Figs. 3 and 4), the external charge is applied to the photoconductor in the dark. In PIP, a polarizing field is applied to the photoconductor while it is illuminated to generate mobile charge carriers that are consequently separated and trapped within the layer.

Photosensitivity of PIP devices is very much lower than that of conventional xerography and there are no commercial applications of PIP to date.

Persistent Photoconductivity Imaging

A number of attempts have been made to reverse the usual xerographic sequence, by exposing the *neutral* photoconductor first and then charging it to form a latent image. This requires a photoconductor material in which photoexcited charge carriers retain mobility long enough to form a developable latent image after the charging step. Materials suggested for this purpose include doped zinc oxide, acid-treated organic charge-transfer complexes, and photolyzable organic compounds (105). None of these systems appears to have found practical use because most photoconductive dielectrics are incapable of separating the newly formed carriers at the generation site in the absence of an external field, a prerequisite for adequate sensitivity.

A process known as Magne-Dynamic is a more sophisticated approach to chargeless electrophotography (Fig. 19). An uncharged TiO_2–binder layer, coated on a metal-backed dielectric base, is exposed to an image pattern, then quickly developed with a positively biased brush of conductive magnetic toner loosely attached to a magnetized roller (83,106). Illumination not only renders the photoconductor temporarily conductive but also converts the contact between its surface and the subsequently applied toner mass from a blocking to an injecting state. In illuminated areas, positive charge is injected, and toner is repelled; in dark areas, toner is deposited by electrostatic attraction to the base. The toner image is thus developed on the positive charges and can be transferred to paper as in xerography. The latent image may be redeveloped several times without additional exposure. The photoreceptor can be erased by exposure to ultraviolet light, cleaned, and rested for several hours to be reused at a later time.

An ingenious extension of this process has been used for Color-in-Color imaging.

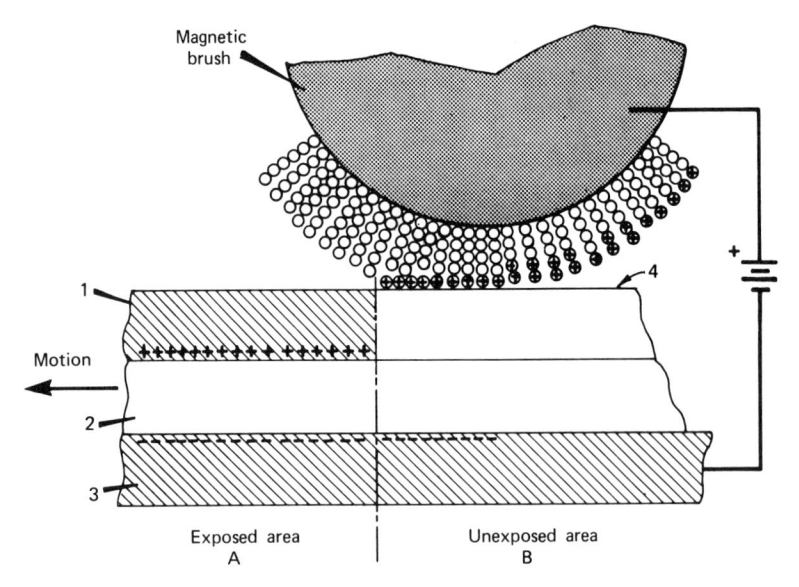

Figure 19. Magne-Dynamic process. The TiO$_2$–binder layer [1] becomes conductive upon exposure to light. Conductive toner particles brushed across region A inject their charges and are repelled. In region B toner is deposited because of the blocking nature of surface [4]. Dielectric layer [2] is backed by metal layer [3] (83).

A special photoreceptor web having a back coating of heat-sublimable dyes is used here. After exposure through an appropriately matched color filter, a toner image is developed directly on the TiO$_2$ front surface. The back of the photoreceptor is then pressed against a receiving sheet, while the front is uniformly irradiated with infrared. Toned areas absorb this radiation selectively, and transfer of heat to the back of the sheet deposits a corresponding dye pattern on the receiving paper by sublimation. The photoreceptor web is advanced to an adjacent section whose back side has been coated with a sublimable dye of another color and the process is repeated. Three photoreceptor sections are required for each color print (107).

Photoconductography

Electrolytic imaging is the only form of electrophotography in which chemical change plays a significant role. In photoconductography a conductivity pattern formed in a photoconductive insulator by light exposure is developed by electrochemical deposition. Photographic gain is achieved by drawing multiple charges through the photoconductor–electrolyte series circuit, or by secondary chemical amplification, or both.

Early attempts to use image illumination of various photoconductors to control the electrolytic coloration of paper and gelatin layers foundered for lack of photoconductor materials capable of adequate gain (108). A practical system (109) became possible when a way was found to dope zinc oxide in binder systems, making possible the injection of multiple electrons for each absorbed photon across the photoconductor layer. Electrolytic development provides a coulombic deposit of silver in exposed areas. Exposures of about 8×10^{16} photons/cm^2 are required, implying a system gain of only

2.0. The need for such intense exposures limits this process to projection printing from transparencies; it is not useful for ordinary photography (Fig. 20).

Generally, only a single image is formed on such a photoconductor, and the fatigue introduced by the long trapping life of the minority carriers has no adverse consequences. However, for color printing prompt reexposure is needed. In the Electro-color version of this process, electrolytically reducible ionic dye precursors are used instead of a silver salt (110).

Substantial chemical gain can be achieved if photoelectrolysis is used to deposit a pattern of development nuclei that then serve as the latent image for subsequent amplification, either by physical development (111) or by locally catalyzing photopolymerization (112). ASA speeds of 100–150, comparable to those of fast silver halide emulsions, have been claimed in the literature. These and other similar systems have been abandoned, probably because they require wet processing with unstable reagents.

Photoactive Pigment Electrophotography

Photoconductive Toner. A number of processes have been developed on the principles of photoconductivity of a pigment and the ability of such pigment particles to move under the influence of a field. The first such process relies on the photoconductive properties of a toner (113) which is laid down uniformly on a conductive plate and charged. Upon exposure to an image the discharged portion of the toner layer can be readily removed and the remaining toner pattern fused onto the plate. This imaging method has become common in marking ship's plates in Japan but has failed in the photocopying field.

A more successful approach to the formation of images by moving photoactive particles requires their containment in a liquid or plastic. In these vehicles a pigment

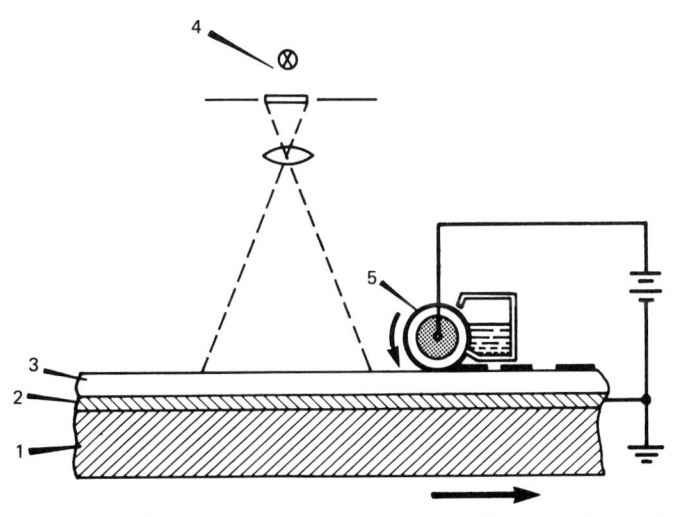

Figure 20. Photoelectrolytic imaging process. The imaging sheet, consisting of a paper base [1] laminated to aluminum foil [2] and coated with ZnO–binder mixture [3], is exposed from a microfilm projector [4], then developed by a biased sponge roll [5] wetted with silver salt solution (109).

818   ELECTROPHOTOGRAPHY

layer can be laid down more controllably and uniformly than by dusting dry toner powder.

Particle Migration Imaging. A totally new form of electrophotography uses the migration of microscopic selenium particles embedded in the top surface of a 10 μm plastic layer. The selenium moves through the layer under the influence of a field when the plastic is softened with a solvent. In fact, the plastic layer can be washed off completely, leaving the migrated selenium particles adhering to the support film (Fig. 21). Since only charged particles migrate, they form a high-contrast deep-red negative image (114). The process is strongly materials- and configuration-dependent. Dispersions of other photoconductive pigments, such as ZnO or phthalocyanine produce a positive image, ie, particles migrate preferentially in dark areas.

Photoelectrophoretic Imaging Processes. It has been known for many years (115) that electrically photosensitive particles dispersed in a dielectric fluid can move in response to illumination in an externally applied electric field. Photoconductive pigments such as zinc oxide and phthalocyanine can thus be caused to print out directly onto paper interposed between a backing electrode and a pigment suspension covered by a transparent imaging electrode. The normal image sense is negative. Illuminated

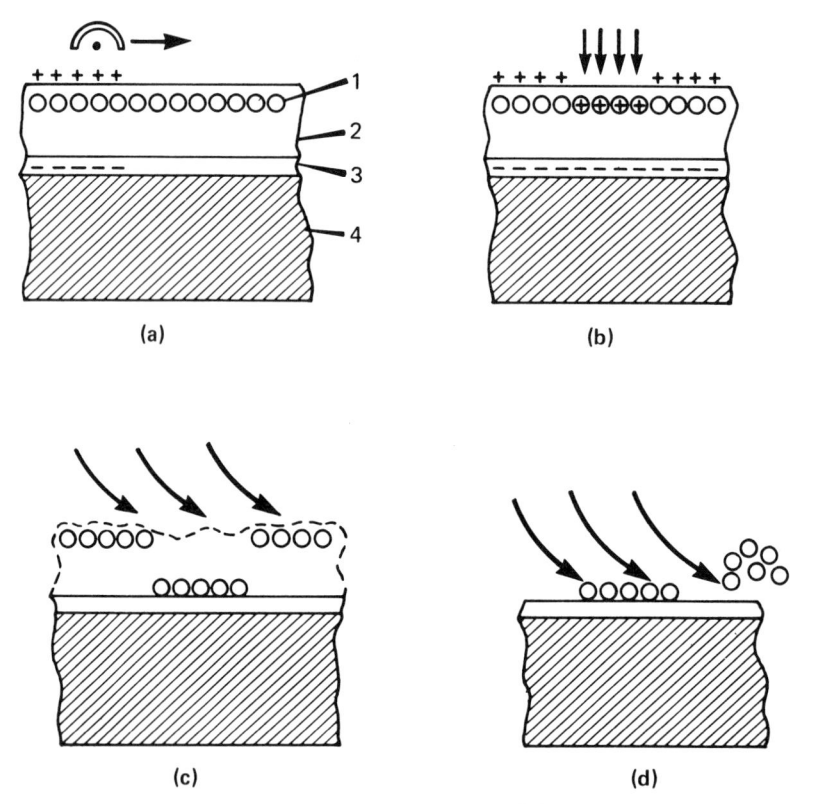

Figure 21. Particle migration imaging. Selenium particles [1] are imbedded just below the surface of a 10 μm plastic layer [2] on a transparent metallized [3] film base [4]. The composite film is: (**a**) charged; (**b**) exposed; (**c**) washed with a solvent to leave (**d**) a dark, firmly adhered image of selenium particles in illuminated areas of the film, forming a negative of the original. Courtesy of the Society of Photographic Scientists and Engineers (114).

particles exchange charge with the transparent electrode and are repelled towards the receiving sheet; dark particles remain in the suspension. This process, which is called photoelectrophoresis or PEP, has been applied to an experimental printer capable of converting light input to very clean black print on paper. Resolution in excess of 100 line pairs per millimeter has been demonstrated; depending on the pigment used, the photosensitivity is comparable to that of xerography (116).

Color Photoelectrophoresis. Perhaps the most challenging use of photoactive pigments and their electrophoretic motion is a single-step, full-color electrophotographic process (117) that uses the spectral response of individual photosensitive particles in a mixture of cyan, magenta, and yellow pigments. Under the influence of an electric field the particles migrate selectively according to illumination to produce instant full-color images. An oil suspension of roughly equal parts of electrically photosensitive cyan, magenta, and yellow pigments constitutes the ink. Phthalocyanines, quinacridones or azo-compounds, and naphthoquinones have been shown to work. During operation the ink is contained in a narrow gap between a transparent injecting electrode and a blocking back-electrode in the form of a roller. Exposure to a color positive original takes place while an electric field is applied. This produces a full-color positive copy on the transparent electrode. A complementary negative image remains on the roller (Fig. 22). This process has achieved considerable success in the laboratory in spite of many problems such as charge exchange between illuminated and dark particles. No commercial use has been announced to date.

Applications for Electrophotography

By far the most widespread application of electrophotography today is plain paper copying. In 1970 about one half of the 75×10^9 electrophotographic copies made worldwide were made by transfer xerography in 15% of the copying equipment in use. Coated paper machines (85%) accounted for the balance. It is estimated that by 1980 about 340×10^9 plain paper copies will be produced by 40% of the equipment, whereas

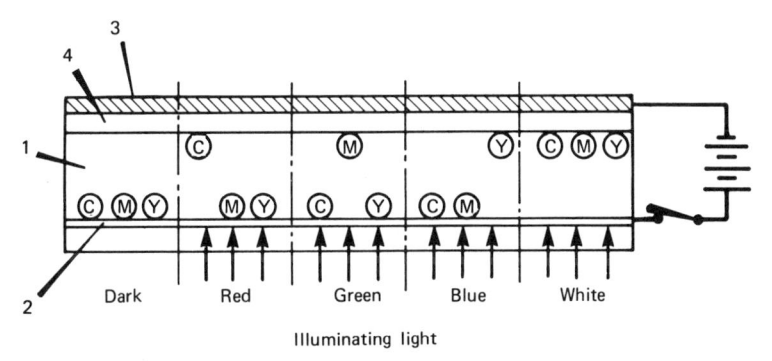

Figure 22. Color photoelectrophoresis. A dispersion of cyan C, magenta M, and yellow Y pigment particles in dielectric oil [1], the ink, is contained between a transparent injecting electrode [2] and a counter electrode [3] coated with a blocking layer [4]. Exposure to light absorbed by each type of particle causes it to be detached and move to the marking electrode, leaving a subtractive color behind on electrode [2]. Exposure to white light removes all particles; in dark areas a black mixture remains. The positive color image formed on the lower electrode can be transferred and fixed to paper, whereupon the electrode is reinked for the next imaging cycle. Courtesy of Focal Press, London (117).

820   ELECTROPHOTOGRAPHY

60% of the equipment will produce 60×10^9 coated paper copies. Fast xerographic duplicators can now produce two copies per second, sort and collate them into finished booklets. These machines have become competitive with short-run offset presses. Others can produce lithographic masters electrophotographically and transfer the masters to the impression cylinder of an integral printing press (see Printing processes).

Available computer printers can print on fanfold paper or single sheets at high speed (Fig. 23). Image information is projected directly onto the photoreceptor through a character mask or by means of a deflected and modulated laser beam. Forms may be overlaid by optical or electronic composition. These machines can be operated on-line from a computer or off-line from magnetic tape and disk drives. Such machines are bulky and expensive, but can handle very large volumes of individualized printing (500,000 or more copies per month).

Electrophotography has been adapted to making normal book or paper copies (hard copy) from display terminals. In this case the image information comes from

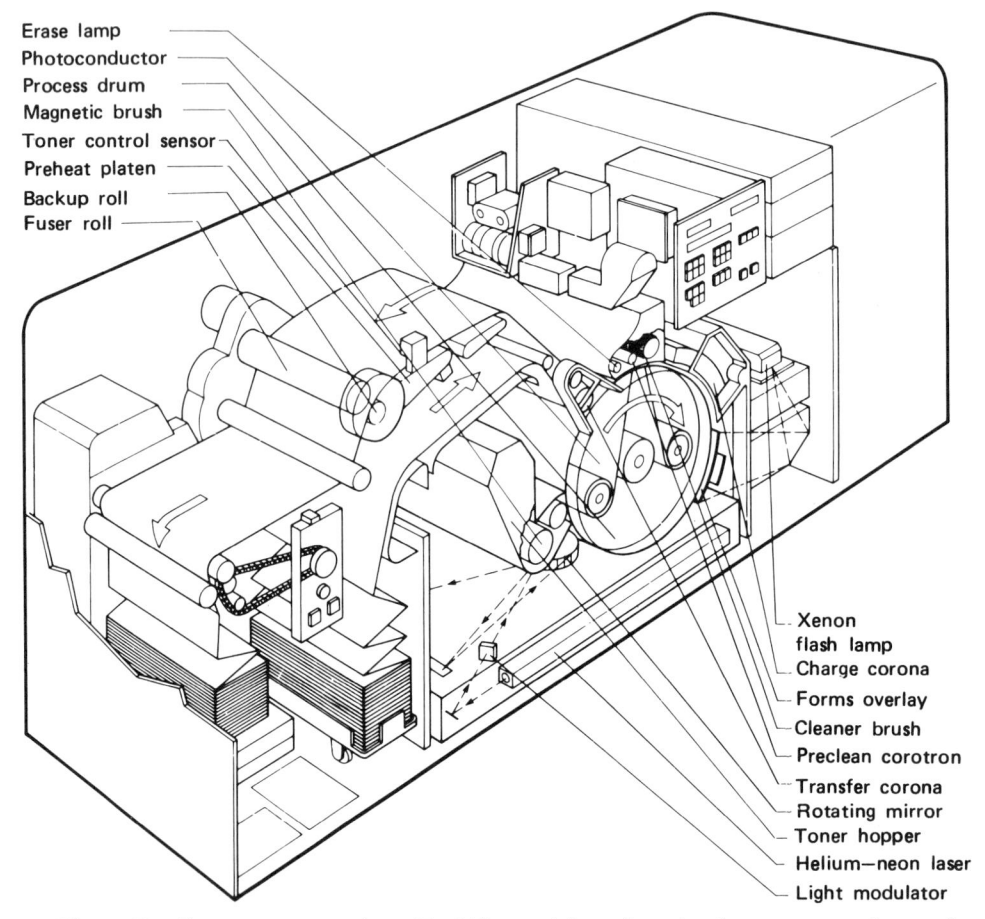

Figure 23. Computer output printer. Fanfold paper is brought to the photoreceptor at the transfer corona and fed out through the fuser station (arrows). This device prints up to 80 cm/s of paper fed through it. A modulated laser beam creates the image as a matrix of 6×7 dots/mm^2. Courtesy of the Society of Photographic Scientists and Engineers (118).

a cathode ray tube and the photoreceptor must be sensitive to the phosphor emission of the tube.

A uv-sensitive organic electrophotographic film has been specially designed for data recording (119). Microimaging on electrophotographic film can also be accomplished. A 35-mm slide processor, announced in Japan in 1971, uses a transparent organic photoconductive film. High resolution (125 line pairs per millimeter) images in microfiche format can be made at the rate of 7 images per minute by another device which also utilizes a dye-sensitized organic photoconductor (96). Both processes use liquid electrophoretic development. Similar electrophotographic materials render microimages visible by surface deformation (99) (see Deformation recording media). Several available reader–printers, based on electrophotography reproduce full-size hard copy from microfilm.

Two forms of color electrophotography have been described above. Another commercial machine uses standard xerography (Fig. 24). Three developer housings with appropriate color toners operate in sequence on the photoreceptor drum. Each image is transferred in registration to an electrically-biased roll before final transfer to paper. A slide projector attachment allows reproduction of color transparencies. Image screening is necessary for adequate tone reproduction; without screening only solid color tints reproduce well. A description of the slide printer equipment and a sample of its capability has been published (120).

A facsimile transmitter and receiver based on transfer xerography (121) can operate unattended over ordinary phone lines. A single laser beam is used both to scan the document for transmission and to provide a pulse-modulated signal on the photoreceptor when receiving.

X-ray sensitive selenium plates form the basis of a medical diagnostic system called xeroradiography (122). A fully automatic device precharges flat selenium plates contained in portable cassettes. After exposure the plates are processed automatically in a separate device. Either a positive or negative image can be developed in the powder cloud developer. The final image is transferred to a transparency or to paper and emerges from the processor. The high edge contrast of powder cloud development emphasizes small differences in soft tissue density that elude conventional radiography (see X-ray technology).

Advantages and Limitations of Electrophotography

The outstanding advantages of electrophotography are operator convenience, the ability to transfer images onto a variety of surfaces, the insensitivity of its recording media prior to charging, and the ability to manipulate the image electrically to suit particular applications. Reuse of photoreceptors and image add-on capabilities provide flexibility. As in other imaging systems tradeoffs must be made among sensitivity, image noise and resolution. The image gain in presently known electrophotographic systems is limited to 10^4 to 10^5 providing photographic speeds comparable to that of silver halide films with ASA ratings of 0.1–3. Dry toner xerography is most useful for high contrast photocopying. The inherently soft tone-reproduction curves and fine grain of other electrophotographic techniques suit them to pictorial rendition.

At this stage of its development, electrophotography is far from mature; it should be capable of much more in the future.

822 ELECTROPHOTOGRAPHY

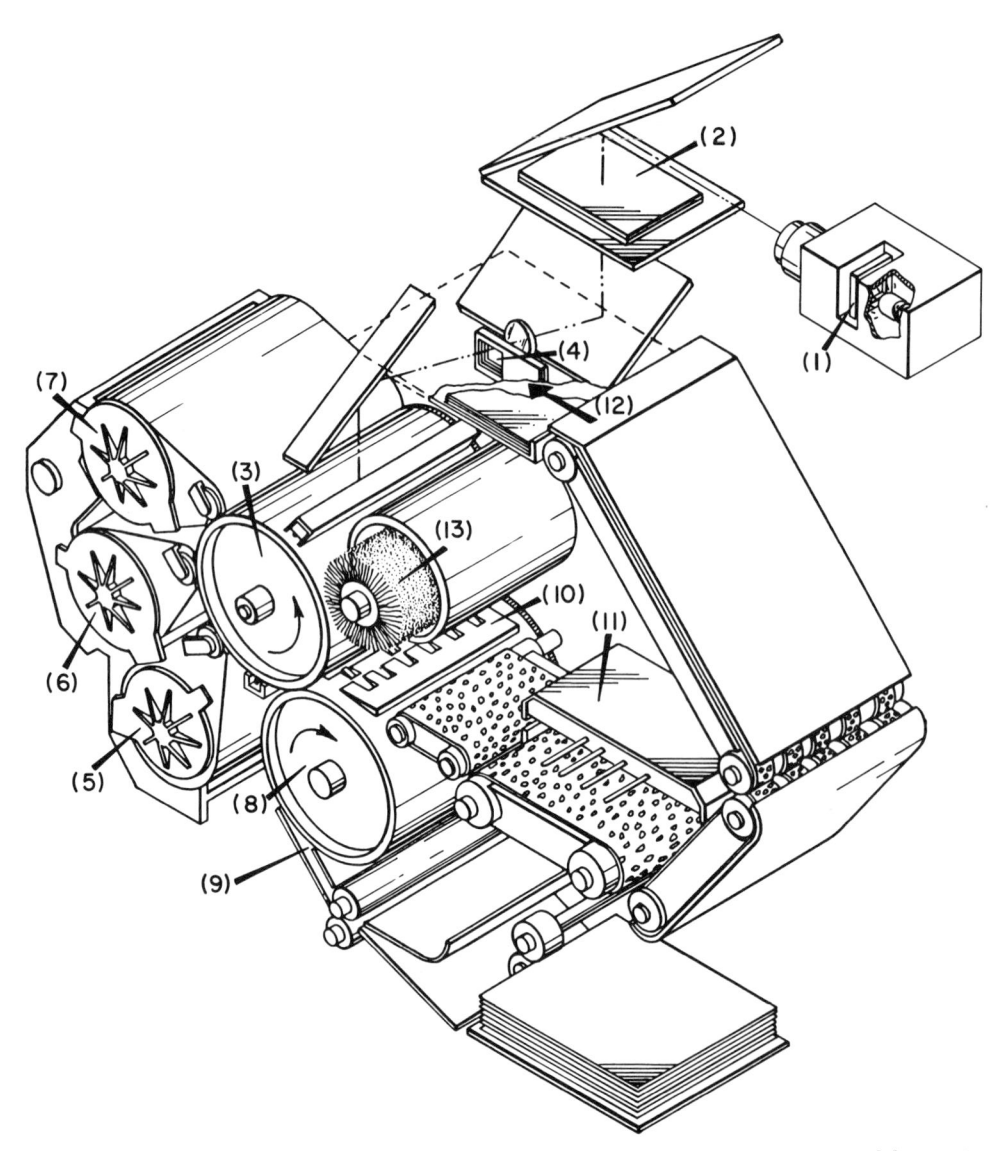

Figure 24. Color xerography. A color transparency [1] is projected through a screen [2] onto the panchromatic photoreceptor drum [3] three times in sequence through appropriate filters [4]. Each image is developed by the appropriate developer, cyan [5], magenta [6], and yellow [7], and transferred in registry onto the biased roll [8]. When all three toner images are completed, paper [9] is fed to the transfer roll. The transferred paper image emerges at [10] and is fused at [11]. The final print emerges at [12]. The brush [13] cleans the photoreceptor between images. Courtesy of Xerox Corporation.

BIBLIOGRAPHY

"Reprography" in *ECT* 2nd ed., Vol. 17, pp. 328–378, by Jack J. Bulloff, State University of New York at Albany.

1. C. F. Carlson, "Xerography" in P. A. Spencer, ed., *Progress in Photography: 1955–1958,* Focal Press, London, Eng., 1958.
2. J. H. Dessauer and H. E. Clark, *Xerography and Related Processes,* Focal Press, London, Eng., 1965.

3. R. M. Schaffert, *Electrophotography*, 2nd ed., Focal Press, London, Eng., 1975.
4. V. M. Fridkin, *Physical Principles of Electrophotography*, Focal Press, London, Eng., 1973.
5. J. W. Weigl, *Angew. Chem. Int. Ed.* **16**, 374 (1977).
6. J. H. Dessauer, G. R. Mott, and H. Bogdonoff, *Photogr. Eng.* **6**, 250 (1955).
7. R. B. Comizzoli, G. S. Moser, and D. A. Ross, *Proc. IEEE* **60**, 348 (1972).
8. J. W. Weigl, *J. Chem. Phys.* **24**, 370 (1956).
9. J. Kosar, *Light Sensitive Systems: Chemistry and Application of Non-Silver Halide Photographic Processes*, John Wiley & Sons, Inc., New York, 1965.
10. P. P. Hanson, *Photogr. Sci. Eng.* **14**, 438 (1970); R. J. Cox, *Non-Silver Photographic Processes*, Academic Press, Inc., New York, 1975.
11. R. Shaw, *J. Appl. Photogr. Eng.* **1**, 1 (1975); R. Goren, *J. Appl. Photogr. Eng.* **2**, 17 (1976).
12. F. Schmidlin, *IEEE Trans. Electron Dev.* **19**, 418 (1972).
13. C. F. Carlson, "History of Electrostatic Recording" in ref. 2, Chapt. I; J. H. Dessauer, *My Years with Xerox: The Billions Nobody Wanted*, Doubleday, Garden City, N.Y., 1971.
14. U.S. Pat. 2,588,699 (Mar. 11, 1952), C. F. Carlson; R. G. Vyverberg, "Charging Photoconductive Surfaces" in ref. 2, Chapt. VII; M. Levy and L. E. Walkup, "Introduction to the Xerographic Process" in ref. 2, Chapt. II.
15. Ref. 3, pp. 31, 234, 397, 441 ff.
16. M. M. Shahin, *J. Chem. Phys.* **45**, 26 (1966); *Photogr. Sci. Eng.* **15**, 22 (1971).
17. M. M. Shahin, in E. M. Pell and K. J. Teegarden, eds., *Electrophotography, Appl. Opt. Suppl.* **3**, 106 (1969).
18. W. Maenhout-Van der Vorst and F. Craenest, *Phys. Status Solidi* **5**, 357 (1964); K. Hauffe and R. Stechemesser, *Photogr. Sci. Eng.* **11**, 145 (1967).
19. R. G. Vyverberg in ref. 2, Chapt. VII, p. 201; U.S. Pat. 3,254,998 (June 7, 1966), F. A. Schwertz.
20. M. J. Mitsui, *IEEE Trans. Electron Dev.* **19**(4), 396 (1972).
21. U.S. Pat. 2,693,416 (Nov. 2, 1954), L. Butterfield (to Western Electric Co.).
22. U.S. Pats. 3,084,061 (Apr. 2, 1963); 3,234,019 (Feb. 8, 1966), R. H. Hall (to Xerox Corp.).
23. U.S. Pat. 3,124,456 (Mar. 10, 1964), T. H. Moore (to RCA).
24. P. Mark, *Photogr. Sci. Eng.* **18**, 254 (1974).
25. Ref. 3, p. 35.
26. R. M. Schaffert, *Photogr. Sci. Eng.* **6**, 197 (1962).
27. H. E. J. Neugebauer, *J. Appl. Opt.* **3**, 385 (1964).
28. R. W. Gundlach, "Development of Electrostatic Images" in ref. 2, Chapt. IX.
29. T. L. Thourson, *IEEE Trans. Electron Devices* **19**, 495 (1972).
30. J. T. Bickmore, "Aerosol Development" in ref. 2, Chapt. XI.
31. R. B. Lewis and H. M. Stark in W. F. Berg and K. Hauffe, eds., *Current Problems in Electrophotography*, de Gruyter, Berlin, 1972, p. 322.
32. U.S. Pats. 2,786,439, 2,786,441 (Mar. 26, 1957), C. J. Young (to RCA); 2,786,440 (Mar. 26, 1957), E. C. Giaimo (to RCA).
33. L. B. Schein in D. R. White and J. W. Weigl, eds., *Electrophotography, Second International Conference*, Society of Photographic Scientists and Engineers, Washington, D.C., 1974, p. 65.
34. L. B. Schein, *Photogr. Sci. Eng.* **19**, 255 (1975).
35. P. M. Cassiers and J. van Engeland, *Photogr. Sci. Eng.* **9**, 273 (1965).
36. S. L. Chang and C. V. Wilbur, "Impression Development" in ref. 35, p. 74.
37. U.S. Pats. 2,846,333 (Aug. 5, 1958), J. C. Wilson (to Xerox Corp.); 3,816,840 (June 11, 1974), A. R. Kotz (to Minnesota Mining & Manufacturing Co.); 3,455,276 (July 15, 1969), G. R. Anderson (to Minnesota Mining & Manufacturing Co.).
38. P. G. Andrus and F. W. Hudson, "Principles of Image Transfer and Fixation" in ref. 2, Chapt. XIV.
39. W. D. Bolton and W. E. Goetz, *Photogr. Eng.* **7**, 137 (1956).
40. U.S. Pat. 2,586,484 (Feb. 19, 1952), E. R. Sabel and E. Galbrecht (to Xerox Corp.).
41. U.S. Pats. 2,807,703 (Sept. 24, 1957), D. D. Roshon, Jr. (to IBM); 3,474,223 (Oct. 21, 1969), A. G. Leiga and co-workers (to Xerox Corp.).
42. U.S. Pats. 3,256,002 (June 14, 1966), F. W. Hudson (to Xerox Corp.); 3,268,351 (Aug. 23, 1966), W. G. VanDorn (to Xerox Corp.); 3,498,596 (Mar. 3, 1970), R. Moser (to Xerox Corp.); 3,539,161 (Nov. 10, 1970), J. F. Byrne (to Xerox Corp.).
43. U.S. Pat. 2,953,470 (Sept. 20, 1960), B. K. Green and L. Schleicher (to National Cash Register Co.).

824 ELECTROPHOTOGRAPHY

44. U.S. Pats. 2,684,301 (July 20, 1954), C. R. Mayo (to Xerox Corp.); 2,776,907 (Jan. 8, 1957), C. F. Carlson (to Batelle Development Corp.); 2,995,464 (Aug. 8, 1961), R. W. Gundlach (to Xerox Corp.).
45. U.S. Pat. 2,752,271 (June 26, 1956), L. E. Walkup and H. E. Carlton, Jr. (to Xerox Corp.).
46. U.S. Pat. 3,609,082 (Sept. 28, 1971), J. H. Moriconi and co-workers (to Xerox Corp.).
47. U.S. Pats. 3,646,866 (Mar. 7, 1972), E. Baltazzi and co-workers (to Addressograph-Multigraph Corp.); 3,649,262 (Mar. 14, 1972), S. Volkers (to Xerox Corp.).
48. U.S. Pats. 3,149,356 (Sept. 22, 1964), T. Murray and co-workers (to Xerox Corp.); 3,186,838 (June 1, 1965), W. P. Graff, Jr., and R. W. Gundlach (to Xerox Corp. and Bell & Howell Co.).
49. U.S. Pat. 3,598,487 (Aug. 10, 1971), M. Mizuguchi and co-workers (to Tokyo-Shibaura Electric Co.).
50. C. J. Claus and E. F. Mayer, "Liquid Development" in ref. 2, Chapt. XII; J. A. Dahlquist and I. Brodie, *J. Appl. Phys.* **40,** 3020 (1969); H. M. Stark and R. Menchel, *J. Appl. Phys.* **41,** 2905 (1970); J. Van Engeland, W. Verlinden, and J. Mariën, "Continuous Tone Electrophotography using Electrophoretic Development" in ref. 33, p. 80.
51. H. Tanaka and co-workers in ref. 33, p. 86.
52. U.S. Pat. 3,084,043 (Apr. 2, 1963), R. W. Gundlach (to Xerox Corp.).
53. U.S. Pats. 3,285,741 (Nov. 15, 1966), W. Gesierich, E. Weyde, and H. Haydn (to Agfa-Gevaert A.G.); 3,486,922 (Dec. 30, 1969), P. M. Cassiers (to Gevaert-Agfa N.V.); 3,276,424 (Oct. 4, 1966), G. Marx and D. Winkelmann (to Azoplate Corp.).
54. U.S. Pat. 2,901,348 (Aug. 25, 1959), J. H. Dessauer and H. E. Clark (to Xerox Corp.).
55. I. Chen and J. Mort, *J. Appl. Phys.* **43,** 1164 (1974).
56. R. M. Schaffert, *IBM J. Res. Dev.* **15,** 75 (1971).
57. R. Comizzoli and H. Kiess in ref. 31, p. 102.
58. J. W. Weigl and co-workers in ref. 31, p. 287; K. Hauffe and co-workers in ref. 31, p. 271.
59. W. E. Spear in ref. 17, p. 8; P. J. Warter in ref. 17, p. 65; H. Seki, *IEEE Trans. Electron. Devices* **19,** 421 (1972); D. M. Pai and S. W. Ing, *Phys. Rev.* **173,** 729 (1968); J. G. Edmund, *J. Non-Cryst. Solids* **1,** 39 (1968).
60. W. D. Hope and M. Levy, "Xerographic Photoreceptors Employing Selenium" in ref. 2, Chapt. IV; J. S. Berkes in ref. 33, p. 137.
61. E. Montrimas, S. Tauraitiene, and A. Tauraitis in ref. 31, p. 139.
62. W. Eckenback in ref. 31, p. 133.
63. C. Wood, "Xerographic Properties of Photoconductor–Binder Layers" in ref. 2, Chapt. V; ref. 3, pp. 318 ff.
64. U.S. Pat. 2,753,278 (July 3, 1956), W. Bixby and O. A. Ullrich (to Xerox Corp.); R. A. Fotland, *J. Appl. Phys.* **31,** 1558 (1960); U.S. Pat. 2,970,906 (Feb. 7, 1961), W. Bixby (to Xerox Corp.).
65. U.S. Pat. 3,639,120 (Feb. 1, 1972), C. Snelling (to Xerox Corp.).
66. U.S. Pats. 3,121,006, 3,121,007 (Feb. 11, 1964), A. E. Middleton and D. Reynolds (to Xerox Corp.).
67. C. J. Young and H. G. Greig, *RCA Rev.* **15,** 471 (1954); K. Hauffe, *Photogr. Sci. Eng.* **20,** 124 (1976).
68. E. Inoue and T. Yamaguchi, *Bull. Chem. Soc. Jpn.* **36,** 1573 (1963); U.S. Pat. 3,271,144 (Sept. 6, 1966), R. L. Clausen and D. K. Meyer (to 3M Corp.).
69. M. Smith and J. W. Weigl, "Organic Photoconductors" in ref. 2, Chapt. IV; J. H. Sharp and M. Smith in P. J. Holmes, ed., *Electrochemistry of Semiconductors,* Chapt. 8, Academic Press, Inc., New York, 1970; D. L. Stockman, *"The Present Status of Organic Photoconductors in Electrophotography"* in ref. 31, p. 194.
70. H. Hoegl, *J. Phys. Chem.* **69,** 755 (1965); U.S. Pats. 3,037,861 (July 5, 1962), H. Hoegl (to Kalle A.G.); 3,232,755 (Feb. 1, 1966), H. Hoegl and co-workers (to Azoplate Corp.); 3,287,120 (Nov. 22, 1966), H. Hoegl (to Azoplate Corp.).
71. J. W. Weigl in W. F. Berg, ed., *Photographic Science,* Focal Press, London, Eng., 1963, Sect. IV; H. Meier, W. Albrecht, and V. Tschirwitz in ref. 31, p. 163; K. Morimoto and Y. Murakami in ref. 17, p. 50; M. Ikeda and co-workers, *Photogr. Sci. Eng.* **19,** 60 (1975).
72. C. F. Hackett, *J. Chem. Phys.* **55,** 3178 (1971).
73. M. Hayashi, M. Kuroda, and A. Inami, *Bull. Chem. Soc. Jpn.* **39,** 1660 (1966); M. Lardon, E. Lell-Döller, and J. W. Weigl, *Mol. Cryst.* **2,** 241 (1967).
74. S. J. Fox in ref. 33, p. 170; P. J. Regensburger, *Photochem. Photobiol.* **8,** 429 (1968); D. M. Pai, *J. Chem. Phys.* **52,** 2285 (1970).
75. H. Meier, *Spectral Sensitization,* Focal Press, London, Eng., 1968.
76. W. Mehl and N. E. Wolff, *J. Phys. Chem. Solids* **25,** 1221 (1964).

77. J. W. Weigl, *Photochem. Photobiol.* **16,** 291 (1972); W. D. Gill, *J. Appl. Phys.* **43,** 5033 (1972); J. Mort and D. Pai, *Photoconductivity and Related Phenomena,* Elsevier Scientific Publishing Co., New York, 1976.
78. P. Nielsen, *Photogr. Sci. Eng.* **18,** 186 (1974); S. Kikuchi, Y. Takahashi, and T. Sakata in ref. 17, p. 42; P. J. Regensburger and N. L. Petruzzella, *J. Non-Cryst. Solids* **6,** 13 (1971).
79. U.S. Pats. 3,573,906 (Apr. 6, 1971), W. L. Goffe (to Xerox Corp.); 3,725,058 (Apr. 3, 1973), Y. Hayashi and M. Hasegawa (to Matsushita Electrical Industrial Co.); 3,837,851 (Sept. 24, 1974), M. D. Shattuck and W. J. Weiche (to IBM Corp.); 3,850,630 (Nov. 26, 1974), P. J. Regensburger and J. J. Jakubowski (to Xerox Corp.); 3,839,034 (Oct. 1, 1974), W. Wiedemann (to Kalle Aktiengesellschaft).
80. M. Kuehnle, *J. Appl. Photogr. Eng.* **4,** 155 (1978).
81. Ref. 3, p. 69.
82. J. S. Rydz and S. W. Johnson, *RCA Rev.* **19,** 465 (1958).
83. U.S. Pats. 3,563,734 (Feb. 16, 1971), B. L. Shely (to 3M Corp); 3,764,313 (Nov. 9, 1973), B. L. Shely (to Minnesota Mining & Manufacturing Co.); 3,639,245 (Feb. 1, 1972), R. B. Nelson (to Minnesota Mining & Manufacturing Co.).
84. J. M. O'Reilly and P. F. Erhardt, "Physical Properties of Toner Polymers" in ref. 33, p. 95.
85. S. Kittaka, *J. Phys. Soc. Jpn.* **14,** 523 (1959); H. W. Gibson, *J. Am. Chem. Soc.* **97,** 3832 (1975); M. W. Williams, *J. Macromol. Sci. Rev. Macromol. Chem. C* **14**(2), 251 (1976).
86. W. R. Harper, *Proc. Roy. Soc. London A* **218,** 111 (1953).
87. D. A. Hays and P. K. Watson, "Contact Charging of Polymers" in ref. 33, p. 108.
88. M. W. Williams, *IEEE Trans. Indust. Appl.* **IA-12,** 213 (1976).
89. E. H. Lehmann and G. R. Mott, "Characteristics of Cascade Development" in ref. 2, Chapt. X; C. R. Raschke in ref. 33, p. 104.
90. Ref. 3, pp. 69–73, 764 ff.
91. C. J. Young and H. G. Greig, *RCA Rev.* **15,** 469 (1954); J. A. Amick, *RCA Rev.* **20,** 753 (1959).
92. F. M. K. Werduschegg and H. E. Carr, *TAPPI* **50,** 26 (1967).
93. U.S. Pat. 3,403,023 (Sept. 24, 1968), H. R. Carrington and F. Gaesser (to GAF Corp.).
94. Austral. Pat. 243,184 (Feb. 4, 1963); Can. Pat. 692,101 (Aug. 4, 1964), K. A. Metcalfe and R. J. Wright (both to Commonwealth of Australia).
95. Ref. 3, p. 142.
96. D. P. Habib and J. D. Plumadore, *J. Microgr.* **7,** 249 (1974).
97. U.S. Pat. 3,547,628 (Dec. 15, 1970), N. E. Wolff (to RCA Corp.); N. E. Wolff, *RCS Rev.* **25,** 200 (1964); R. W. Gundlach and C. J. Claus, *Photogr. Sci. Eng.* **7,** 14 (1963); J. Gaynor and S. Aftergut, *Photogr. Sci. Eng.* **7,** 210 (1963).
98. R. L. Jepson and G. F. Day in ref. 35, p. 28.
99. U.S. Pats. 2,825,814 (Mar. 4, 1958), L. E. Walkup (to Xerox Corp.); 2,982,647 (May 2, 1961), C. F. Carlson and H. Bogdonoff (to Xerox Corp.); 2,833,648 (May 6, 1958), L. E. Walkup (to Xerox Corp.); 2,937,943 (May 24, 1960), L. E. Walkup (to Xerox Corp.).
100. W. Simm, *J. Appl. Photogr. Eng.* **4,** 123 (1978).
101. U.S. Pat. 3,827,801 (Aug. 6, 1974), S. Tanaka, Y. Enoguchi, and T. Fujiwara.
102. J. R. Freeman, H. P. Kallmann, and M. Silver, *Rev. Mod. Phys.* **33,** 553 (1961).
103. H. P. Kallmann and B. Rosenberg, *Phys. Rev.* **97,** 1596 (1955).
104. V. M. Fridkin and I. S. Zheludev, *Photoelectrets and the Electrophotographic Process,* Consultants Bureau, New York, 1961; V. M. Fridkin, *J. Opt. Soc. Am.* **50,** 545 (1960).
105. U.S. Pats. 2,845,348 (July 29, 1952), H. P. Kallmann (to U.S.A. Secretary of the Army); 2,976,144 (Mar. 21, 1961), A. Rose (to RCA); 3,010,883, 3,010,884 (Nov. 28, 1961), E. G. Johnson and B. W. Neher (to Minnesota Mining & Manufacturing Co.); 3,257,304 (June 21, 1966), E. Johnson (to 3M Corp.); 3,285,837 (Nov. 15, 1966), B. Neher (to 3M Corp.); 3,081,165 (Mar. 12, 1963), J. P. Ebert (to Xerox Corp.); J. A. Amick, *RCA Rev.* **20,** 765 (1959); P. M. Cassiers, *J. Photogr. Sci.* **10,** 57 (1962); A. H. Sporer, *Photogr. Sci. Eng.* **12,** 213 (1968).
106. "Magne-Dynamic Process Theory," *VHS-235 Duplicator,* instruction bulletin, 3M Corp.
107. D. H. Dybvig and co-workers, "Color-in-Color" in R. A. Eynard, ed., *Color: Theory and Imaging Systems,* Society of Photographic Scientists and Engineers, Washington, D.C., 1973, p. 403; U.S. Pat. 3,601,484 (Aug. 24, 1971), D. H. Dybvig, J. W. Ulseth, and J. A. Wiese (to 3M Corp.).
108. K. Wilcke, *Photogr. Korresp.* **57,** 173 (1920); J. Berchtold, *Sci. Ind. Photogr.* **26,** 465 (1955); J. J. Robillard, *Photogr. Sci. Eng.* **8,** 18 (1964).
109. U.S. Pats. 3,072,541 (Jan. 8, 1963), B. L. Shely and B. L. Clark (to 3M Corp.); 3,011,963 (Dec. 5, 1961), E. G. Johnson and B. W. Neher (to Minnesota Mining & Manufacturing Co.).

110. U.S. Pats, 3,172,827 (Mar. 9, 1964), V. Tulagin, R. F. Coles, and R. A. Miller (to 3M Corp.); 3,130,655 (Apr. 28, 1964), D. K. Meyer (to Minnesota Mining & Manufacturing Co.).
111. S. Tokumoto and co-workers, *Photogr. Sci. Eng.* **7,** 218 (1963).
112. M. C. Zerner, J. F. Sobieske, and H. A. Hodes, *Photogr. Sci. Eng.* **13,** 184 (1969).
113. Jpn. Pat. 42-2242 (March 26, 1966), R. W. Gundlach (to Fuji Xerox).
114. W. L. Goffe, *Photogr. Sci. Eng.* **15,** 304 (1971); S. Tutihasi, *Photogr. Sci. Eng.* **18,** 394 (1974); A. L. Pundsack, *Photogr. Sci. Eng.* **18,** 642 (1974).
115. U.S. Pats. 2,940,847 (June 14, 1960), E. K. Kaprelian; 3,100,426 (Aug. 13, 1963), E. K. Kaprelian; P. Cressman and G. C. Hartmann, *J. Chem. Phys.* **61,** 2740 (1974).
116. U.S. Pats. 3,741,760 (June 26, 1973), C. Snelling (to Xerox Corp.); 3,703,459 (Nov. 21, 1972), W. L. Little and R. H. Townsend (to Xerox Corp.).
117. V. Tulagin, *J. Opt. Soc. Am.* **59,** 328 (1969); U.S. Pats. 3,477,934 (Nov. 11, 1969), L. Carreira and V. Tulagin (to Xerox Corp.); 3,535,221 (Oct. 20, 1970), V. Tulagin (to Xerox Corp.); ref. 3, p. 187.
118. R. F. Wolter, *J. Appl. Photogr. Eng.* **4,** 151 (1978).
119. *Kodak Photoconductive Recording Film SO-101,* bulletin G-81, Eastman Kodak Corp., Rochester, N.Y., 1977.
120. L. D. Mailloux and J. E. Bollman, *J. Appl. Photogr. Eng.* **3,** 230 (1977).
121. *Xerox Facsimile Telecopier TC 200,* bulletin, Xerox Corp., Rochester, N.Y., 1977.
122. *Xeroradiography, Xerox 125 System,* bulletin, Xerox Corp., Rochester, N.Y., 1977.

NIKOLAUS E. WOLFF
Consultant, Hanover, N.H.

JOHN W. WEIGL
Xerox Corporation

ELECTROPLATING

Electroplating is the electrodeposition of an adherent metallic coating on an electrode in order to form a surface with properties or dimensions different from those of the basis metal. This ASTM definition (1) is broadened for purposes of the present article in two respects: the term adherent excludes electroforming, which is usually considered a branch of electroplating technology; and the term electrodeposition must be stretched to include electroless plating (qv) and immersion processes that do not employ electric current (see also Film deposition techniques; Metallic coatings).

As is usual, the practical and technological phases of electroplating matured much earlier than the science; electroplating with copper, silver, and gold was mentioned in the literature, and even patented (2), almost as soon as Faraday enunciated the laws of electrolysis. Until about World War I, electroplating remained essentially an art: solution compositions were closely guarded family secrets, and most electroplaters made no pretense of knowing even the fundamentals of electricity or chemistry. Most early uses of electroplating were predominantly decorative, and the physical and chemical properties of the product were not so important as its appearance. The demands of World Wars I and especially II for finishes with close tolerances and rigidly specified properties converted electroplating from a craft to a technology grounded in basic science and engineering. This development has been forwarded by advances in metallurgy and physical and electrochemistry, by improvements in sources of direct current and in measuring instruments, and by the rise, between the wars, of the chemical industry which made available reliable sources of pure chemicals (3). The electronics industry requires many electroplated finishes that must pass rigid specifications (4–5). Recent progress (1950–1975) has been reviewed (6).

Electroplating is a surface treatment. The material (work) being treated is made the cathode in an electroplating solution, or bath. Such baths are almost always aqueous solutions, so that only those metals that can be reduced from aqueous solutions of their salts can be electrodeposited. The only major exception at present is aluminum, which is plated on a semicommercial scale from organic electrolytes (7). Some of the refractory metals (eg, niobium, tantalum) can also be deposited from fused salts as coherent plates, but little commercial use has resulted from this development (8–9).

The thickness of deposit applied by electroplating varies with the application: from as little as 0.025 μm for decorative gold deposits through 25–50 μm for standard nickel–chromium plate on exterior automotive hardware, to 1 mm or more for electroforms.

The properties conferred by electroplating include improved corrosion resistance, appearance, frictional characteristics, wear resistance and hardness, solderability, specific electrical properties, and many others. Electroforming (including electrotyping) is used to manufacture articles that cannot be made as economically in any other way.

Table 1 lists U.S. consumption of various metals for electroplating in 1976. In that year total U.S. cadmium consumption was about 5625 metric tons, somewhat less than half of which was used in electroplating (11,13). An additional 4000 t of chromium over the amount listed in Table 1 was used in other forms of metal treating such as conversion coatings, anodizing, and plating on plastics. Data for other common plating metals such as copper and zinc are too fragmentary for quotation.

The Substrate in Electroplating

There is much more to electroplating than the final step of laying down a coating of the plating metal, and much more has to be considered than the properties of the plated metal. The final article will consist not of the deposit alone, but of deposit plus substrate. It is the properties of this combination that often determine the right metal to plate and the right solution to plate it from. A few examples may be illustrative.

Many metals, corrosion-resistant themselves, actually decrease the corrosion

Table 1. U.S. Consumption of Metals in Electroplating, 1976

Metal	Weight, t	Reference
tin	19,000	10
nickel	17,400	11
silver	270	11
gold	50	11
chromium	9,525	12

resistance of a composite when they are plated over a less noble metal. The galvanic couple set up by the contact of two dissimilar metals promotes corrosion. This factor is particularly important in plating upon aluminum, which is electrochemically more active than any metal with which it can be plated. For this reason applications for plating on aluminum must be carefully chosen, and plating procedures must be carried out with special care.

The plating process itself may have deleterious effects on the properties of the substrate. The best example of this is the embrittling effect of hydrogen on the high strength steels, particularly in zinc and cadmium plating (14–16), although almost any plating or preplating process can embrittle these materials. Tests for hydrogen embrittlement may be found in refs. 15 and 17.

The possibility that the deposit may in time, or at elevated temperatures, diffuse into the substrate and form an alloy with the substrate must be kept in mind (18). Such alloys may be brittle or have other undesirable properties. Where such effects must be avoided, a barrier layer of another metal, most often nickel, is interposed between the substrate and the final plate. In tin plating on brass for solderability, ASTM specifications (19) require a barrier coating of at least 2.5 μm of copper, bronze, or nickel to prevent migration of zinc to the surface.

Where the application involves exposure to high temperatures, relative coefficients of expansion of the substrate and the plate may have to be taken into account. Even the disposal of rejects may be a factor: some platers of zinc-base die castings will not consider tin plating, because common practice is merely to remelt rejected items, and there is a widespread suspicion of the effect of any tin at all in this alloy.

PREPARATION OF THE SUBSTRATE

Before a useful electroplate can be deposited on a surface, the surface must be in condition to receive it. Useful means, among other things, adherent; in electroforming the aim is just the opposite (see below). In either case the surface must be so prepared as to ensure the desired bond or lack of it. In the discussion that follows it is assumed that the goal is firm adherence to the substrate. The special treatments designed to ensure easy removal of the deposit must be left for treatises on electroforming (20–23).

The preplating treatments necessary to prepare the surface to accept an adherent deposit are generally aspects of cleaning, which is described in detail below. The ideal surface would be one consisting entirely of atoms of the metal to be plated upon, and having no foreign material at all. This is virtually impossible to attain, even in the laboratory. A practical definition, then, of a satisfactorily clean surface would be a surface containing no foreign material that interferes with the formation of an adherent deposit. In general, this connotes the removal of gross dirt and soil, heavy oxide or tarnish films, and in some cases surface skins of damaged metal produced by prior mechanical operations (see Metal surface treatments).

Cleaning. Choice of the proper cleaning or preparative cycle depends primarily upon the nature of the substrate to be prepared and the nature and amount of the soils to be removed (24). Ferrous and nonferrous metals generally require different types of cleaners. The more active metals such as aluminum require special techniques to prepare them for plating. And obviously, the more contaminated the surface the more cleaning it will require.

A typical cleaning cycle includes the following steps: (1) pickling to remove gross scale; (2) any mechanical preparation such as polishing or buffing; (3) cleaning to remove oils, greases, shop dirt, and polishing and buffing compounds; (4) rinsing; (5) acid dipping to remove oxide films; and (6) rinsing.

Because the nature of the soils to be removed in the cleaning step may vary widely, the cleaning step may correspondingly be anything from a simple dip in a mild detergent to a rather severe combination of several treatments.

Organic solvents are used to dissolve most oils and greases including those used to bind buffing and polishing compounds. One problem in using organic cleaners is that removal of oils and greases from a highly contaminated surface may merely leave the nonoily dirt dry and more firmly attached than ever. Although organic solvents may be used merely as dips, it is more common and effective to employ vapor degreasing, in which vapors of the solvent condense on the parts to be cleaned and run back into a pool of liquid solvent below them. Common solvents for this purpose are perchloroethylene, trichloroethylene, trichlorotrifluoroethane, methylene chloride, and 1,1,1-trichloroethane (methyl chloroform). In choosing a vapor degreasing solvent, EPA regulations and relative toxicity must be balanced against cost; perchlor and trichlor are more toxic than the others, but also less expensive. In any case, necessary precautions dictate that no one must enter a vapor degreaser for repair or cleaning without life-support and a partner outside for emergencies (see Chlorocarbons).

Alkaline cleaning generally follows organic cleaning. The cleaners, aqueous solutions of sodium compounds such as carbonate, silicate, phosphate, or hydroxide, usually also contain a surfactant (qv) and may contain chelators (see Chelating agents). The function of alkaline cleaners is to dislodge surface soil, principally by dispersing it in such form that it does not settle back on the work, and to some slight extent by saponifying the saponifiable oils and greases. Alkaline cleaning may be accomplished by one of three techniques, and sometimes by a combination of them: soak, cathodic or direct, and anodic or reverse. The last two are included under the heading of electrolytic cleaning. Alkaline cleaners are invariably applied at elevated temperatures.

In soak cleaning the part is immersed in the cleaner, possibly with mild agitation. Better results are usually obtained by electrolytic cleaning, which adds to the detergent action of the solution the scrubbing action of gas evolution on the surface of the work. In direct cleaning when the work is the cathode, twice as much gas, in this case hydrogen, is evolved as when the work is the anode (reverse cleaning), in which the gas evolved is oxygen. However, in cathodic cleaning impurities in the cleaner may be deposited on the surface, or hydrogen may embrittle the basis metal. Anodic cleaning, on the other hand, may form oxides on the surface of susceptible materials. Often the two are combined: A longer cathodic period is followed by a shorter period with the work as anode.

Formulations for cleaning solutions have been published (see Table 2) (25–26). A wide variety of proprietary formulations are also available. These can be tailored to the specific conditions and have the further advantage of available expert advice for solving particular cleaning problems.

The best cleaner will not do a good job if not properly maintained. Since the function of a cleaner is to remove dirt, the cleaner itself gets dirty; thus, at intervals the cleaner must be replaced with fresh solution.

Although emphasis here has been on alkaline cleaners, other types also have their areas of usefulness. Emulsion and di-phase cleaners combine, to some extent, the

Table 2. Cleaner Formulations[a]

Ingredient	Basis metal: Method of application:	Steel Electro	Copper Electro	Zinc Electro	Brass Electro	Aluminum Soak
sodium hydroxide, NaOH, %		50	25	20	10	
sodium metasilicate, Na_2SiO_3, %		40	40	40	40	40
sodium tripolyphosphate, $Na_5P_3O_{10}$, %[b]		5	10	10	10	40
sodium carbonate, Na_2CO_3, %		4	23	28	38	10
sodium bicarbonate, $NaHCO_3$, %						5
surfactant: 40% sodium linear alkylate sulfonate, %		1	2	2	2	5
cleaner concentration, g/L		60–120	30–60	30–45	25–45	30–60
current density, A/m^2		500–1000	200–500	200–500	150–500	
temperature, °C		80–boil	70–82	65–75	60–70	70–82

[a] Courtesy of McGraw-Hill Book Co. (26).
[b] Or chelating agent.

functions of organic solvents and alkaline cleaners. Alkaline cleaning itself is often divided into two steps: so-called precleaning, utilizing heavy-duty cleaners to remove the bulk of the soil, followed by milder alkalies to finish the job. No single prescription can serve for all cases. Cleaners may be applied as jets or sprays impinging on the work, as well as by the more usual method of immersion.

The water-break test is almost universally used as a test of adequate cleaning. Water-break is the gathering of a film of water into droplets or streaks, and its occurrence is a sign that the surface is not clean. Water will run off a clean surface in an unbroken sheet or film. Because the alkaline solution of a cleaner may mask this test, it should be applied after acid dipping or in later parts of the plating cycle. More sophisticated and delicate tests for surface contamination have been developed (27) but the water-break test remains in most general use.

The final test of good cleaning is the plating itself. Many problems in plating can be traced to improper cleaning, so that in troubleshooting in the plating shop, the cleaning cycle should be suspect until exonerated.

Rinsing. Adequate rinsing between all steps in cleaning and plating is of the utmost importance. Rinse waters should be clean and should not be allowed to become contaminated by drag-in of preceding solutions. Countercurrent rinsing is often employed as a means of conserving water while ensuring that the last rinse is relatively pure. Hot water is more efficient than cold for removing contaminants; on the other hand, it entails the risk that the work may dry before entering the next operation. Quality of the available water supply is often of importance; softened, deionized, or distilled water may be required for final rinses in many instances.

Adequate and efficient rinsing assumes additional importance in relation to waste disposal. To the extent that it cannot be fed back to the plating cycle, the last rinse constitutes effluent from the plant. Almost universally it must be treated to reduce harmful contaminants forbidden by EPA regulations, and the less of these it contains the cheaper waste treatment will be. Thus countercurrent rinsing, drain stations between tanks, and other means of reducing the concentration of contaminants in the rinses going to the sewer or watercourse are required (28–30).

Acid Dipping. Acid dipping, which generally follows cleaning, serves two purposes: it removes slight tarnish or oxide films formed in the cleaning step, and it neutralizes the alkaline film which even good rinsing cannot completely remove from the surface. It is thus particularly important when the plating solution is acid. The acid dip is usually a 10–30 vol % solution of hydrochloric or sulfuric acid, the latter solution being more common.

If plating is to be done in an alkaline solution, it is good practice to use an alkaline dip following the rinse after the acid dip. In general, when work is to proceed from an acid to an alkaline solution or vice versa, it is wise to insert a neutralizing dip between the two.

Drag-Out and Drag-In. Every solution in a plating cycle is contaminated, to a greater or less extent, by the solution that precedes it in the cycle. How serious this situation is depends on many factors, some controllable and some not. The shape of the parts and how they are positioned in the tank is of great importance, drain times vary, some solutions are more free-rinsing than others. Contamination caused by this drag-in may be serious or inconsequential: thus the introduction of a little alkali from a neutralizing dip into a cyanide plating bath is of little concern, but the introduction of chromate (from racks that have not been properly rinsed from a preceding cycle) into a nickel bath may cause havoc.

Drag-out also may be of great or little importance. Economic effects of the loss of a little copper sulfate solution are slight, but in gold plating every effort must be made to reclaim the metal from rinse waters. Drag-out can be positively beneficial, inasmuch as it helps to prevent the buildup of undesirable impurities in plating baths. Drag-out must also be considered in connection with the problem of waste disposal (see below).

Special Preparation Cycles. For plating on the common substrates such as ferrous metals and copper and its alloys, the preparation cycles briefly outlined above usually suffice. Other substrates require more specialized treatment. Among those requiring special preparation are aluminum and magnesium; zinc-base die castings; refractory metals like titanium, zirconium, tantalum, niobium, molybdenum, and tungsten; and nonmetallics like synthetic plastics, leather, wood, and plaster.

Aluminum and Magnesium. Aluminum and magnesium are so active both chemically and electrochemically that ordinary preparative cycles would leave them in an entirely unsatisfactory condition for electroplating. The usual methods of preparing these metals involve replacement of the naturally occurring oxide films with a thin film of zinc by the zincate process (31). Recent work, including independent investigations and cooperative studies by ASTM Committee B.08, has shown that other treatments offer improved adhesion, which has always been the principal problem in the performance of plated aluminum. In particular, a proprietary process (Alstan, M&T Chemicals, Inc.) involving a stannate rather than zincate dip has shown improved performance and is being adopted by industry. Several other alternatives also show promise and tests continue (32–35).

Plating on magnesium presents equally difficult problems; an ASTM Recommended Practice is available (36).

Zinc-Base Die Castings. Many of the problems in plating on zinc-base die castings have arisen from poor or porous castings, and as metallurgical techniques have improved these problems have diminished or disappeared (37). After cleaning, zinc die castings are almost invariably plated with copper, first striking in a bath such as shown in Table 3. This may be followed by any desired plate.

Table 3. Copper Strike Baths for Aluminum and Zinc Die Castings

	Aluminum[a]	Zinc-base[b]
copper cyanide, g/L	42	20–45
total NaCN, g/L	50	
free NaCN, g/L	5.7 max	10–20
Na_2CO_3, g/L	30	15–25
NaOH, g/L		3.8–7.5[c]
Rochelle salt, g/L	60	
pH (electrometric)	10.2–10.5	
temperature, °C	38–54	50–57
current density, A/m^2	260	270–650
time, min	2–5	2

[a] Ref. 31.
[b] Ref. 37.
[c] Optional.

Refractory Metals. The interest in plating on such metals as titanium, molybdenum, and niobium has arisen primarily from the needs of aerospace technology. Many of these metals have desirable high temperature properties but require protective coatings. Techniques have been worked out for electroplating on most of these metals, although some of the results leave something to be desired. Each of the metals poses its own particular problems of surface preparation; for further details see refs. 38–41.

Other Metals. ASTM recommended practices are available for preparation of several metals not considered above: low carbon steel (42); high carbon steel (43); iron castings (44); copper and its alloys (45); stainless steel (46); nickel and its alloys (47); and lead and its alloys (48).

Nonconductors. Electroplating on nonconducting materials—wood, leather, natural and synthetic plastics, etc—requires a conducting surface with at least minimal adhesion to the substrate. It is a very old art, almost as old as electroplating itself. Until about 1960 the principal uses of the process were sentimental, as in the bronzing of baby shoes, or artistic, such as providing reproductions of art objects. Processes employed included graphitizing the surface, applying bronze or other metallic paints, or chemical silvering by mixing a silver solution and a reducing solution from a two-nozzle gun so that the solutions mix just as they hit the surface to be silvered. Such methods provided poor adhesion: the article had to be completely encapsulated if the electroplate were to adhere satisfactorily.

Saubestre (49) dates the modern era of plating on synthetic plastics from about 1960; at that time platable plastics capable of receiving a deposit with adhesion sufficient for practical uses were developed. The first plastic to be plated on a commercial scale was ABS (acrylonitrile–butadiene–styrene) of special grade; somewhat later methods for plating on polypropylene were developed, and analogous processes are now available for polycarbonates, polysulfones, and others, although ABS and polypropylene still dominate the field. Although the adhesion obtained is not as strong as in plating on metals, it is satisfactory for many practical applications. The process competes directly with plating on zinc die castings, and the emphasis on weight saving in motor cars promotes the use of lower-density plastics.

The cycle for preparation of these plastics includes: cleaning, solvent treatment, conditioning, rendering the surface catalytic, and applying an electroless or autocatalytic deposit, usually nickel or copper (see Electroless plating). The only cleaning generally necessary is for removal of fingerprints or minor soils. Solvent treatment is required only if the plastic is hydrophobic; it is omitted with ABS and some other substrates. Conditioning, usually in a solution containing chromic and sulfuric acids, roughens the surface just enough to yield interlocking of the subsequent deposit, and changes the surface chemically to provide sites for chemical bonding. The catalytic surface is generally provided by adsorption of a reducing agent, normally stannous chloride, followed by immersion in a dilute solution of palladium chloride which is reduced to metallic palladium by the tin(II) on the surface. Although this can be accomplished in two steps, a one-step process containing both the stannous chloride and the palladium chloride in one solution has superseded the two-step method in most shops. When the surface has been thus rendered catalytic, electroless nickel or copper can be used to provide a conducting surface, after which electroplating takes place as usual.

Methods for avoiding the use of precious metals such as palladium have been proposed (50).

Plating on plastics is a vital step in the manufacture of printed-circuit boards (4–5,51) and is increasing in importance (52–53). A critical review of the patent literature to 1970 is available (54) as are monographs on the subject (55–56).

The Electroplating Process

The operations of electroplating, including the cleaning, rinsing, plating, and postplating treatments, can be carried on manually or with almost any desired degree of automation. Parts may be hung in the plating tank on wires or on racks; when many small parts are to be plated, they may be contained in wire baskets or, more commonly, in barrels that rotate in the plating tank. Movement from one operation to another may be by hand or by machine.

The necessary d-c power is derived from motor–generators or, increasingly, by rectifiers. Transistor rectifiers may be of three types: selenium, germanium, or most important, silicon (see Semiconductors). Power is conveyed to the plating tanks by bus bars; the anodes are hung into the tank from the positive bus bar, usually along the two sides (see Fig. 1) and the work to be plated from the negative or cathode bar, usually down the center. Tank voltage is read from a voltmeter, and current from an ammeter: these two instruments should be available for each plating tank. A third instrument, an ampere-hour meter, is often helpful as a means of regulating the thickness of deposit and for general control of the operations.

Some agitation of plating solutions or of the work is usually helpful. The oldest and simplest method consists of an operator merely swishing the work around at intervals, but automatic cathode-rod agitation is preferable. In some automatic plating machines the work is moved through the solution while it is plated. Provided clean air is used and the solution is not sensitive to oxygen or carbon dioxide, air agitation is convenient and efficient. Agitation may be provided by stirring or pumping the solution; the latter is often necessary when heat exchangers are used for temperature control or where continuous filtration is required. In barrel plating, agitation is supplied by the movement of the barrel.

Temperature control is almost always desirable in plating operations because

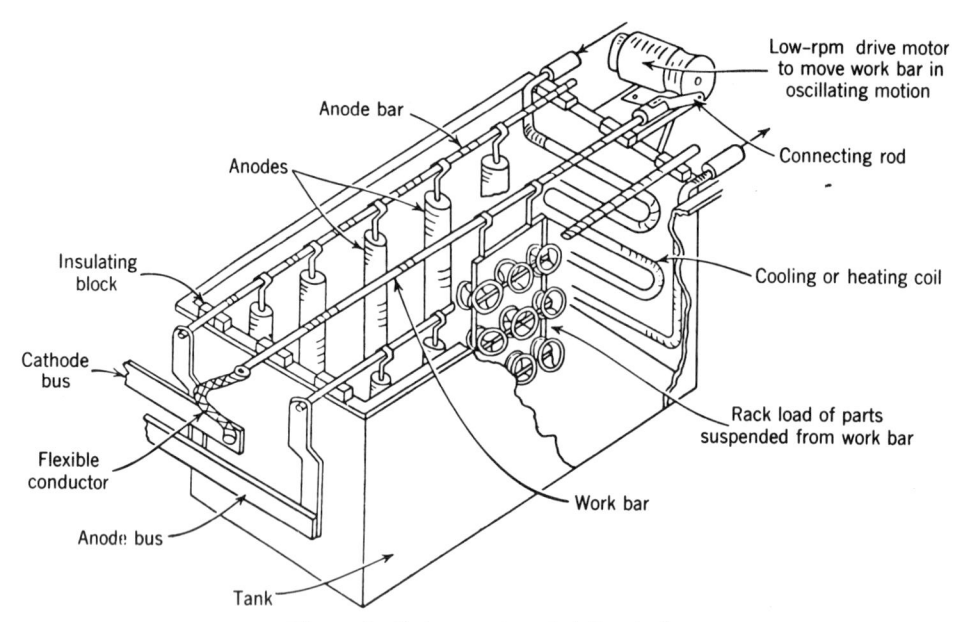

Figure 1. Cut-away view of plating tank.

the characteristics of plating solutions, of the deposit, or of both usually depend to a large extent on the temperature of operation. Heating or cooling coils in the tank itself may be used, or the solution may be circulated through a heat exchanger. Electric immersion heaters may be used. Occasionally plating tanks are heated by open gas flames beneath the tank.

Figure 1 shows a typical plating tank but there are many variations on this basic design. Instead of two anode and one cathode bars, there may be three anodes and two cathodes, etc. If tanks are of bare metal, they should be well-insulated from the floor, and other precautions should be taken against stray currents, which may be very troublesome and extremely difficult to trace.

In addition to the basic equipment—power source, plating, cleaning, and rinsing tanks, and bus bars—most plating installations require one or more of the following: filters, for either continuous or intermittent purification of solutions; drying facilities, which may range from a simple jet of compressed air to large ovens; racks of design appropriate to the part being plated, and a racking station where the work can be conveniently hung on these racks and unracked after plating; one or more stripping tanks for stripping faulty deposits so that parts can be reworked and for stripping the plating racks themselves; reclaim tanks if the drag-out is valuable and worth reclaiming; portable pumps for transferring solutions; and at least one empty tank so that a plating tank can be emptied and worked on.

In addition to the equipment required for electroplating and allied processes, the modern plating shop must include apparatus for waste treatment, metal recovery, or both (see Waste Disposal and Metal Recovery).

Continuous Plating. Electrolytic processes are well-suited to plating continuous coils of strip or wire. The substrate can be uncoiled, pickled, cleaned, plated, given postplating treatments, and recoiled in one continuous operation. The manufacture of electrolytic tinplate by the steel mills is the largest and best developed of these processes, but continuous plating of strip and wire with zinc, copper, and other metals is also extensively practiced. Equipment is highly specialized; solutions may be conventional or, like the Halogen process for tinplate, especially adapted to the application (57–58). It is not necessary that continuous plating be done on the enormous scale of the tinning lines of the steel mills; wire and narrow strip, as well as conduit and electrical terminals are economically plated for special applications on a very modest scale (59) (see Electrical connectors).

Materials of Construction. Plating tanks and auxiliary equipment are constructed of materials resistant to the particular solution involved. This usually means plain steel for alkaline solutions and rubber- or synthetic-lined steel for acid solutions. In using steel tanks the possibility of bipolar effects must be kept in mind, ie, since the tank is a conductor, it may become interposed in the electrical circuit. If the path anode–tank–cathode is electrically shorter than the path anode–cathode, this may have various unwanted side effects. In choosing linings for use with acid solutions, several other factors must also be considered: the heat resistance of the lining must be adequate for the temperature of operation of the solution; and some plating solutions, notably the bright nickels, are sensitive to the materials that may be leached out of the lining. Chromium plating solutions, because of their highly oxidizing nature, require special selection of lining materials; various lead alloys and brick are also commonly used.

Some of the refractory metals are being made use of to some extent in designing plating equipment: tantalum heating coils and titanium racks and anode baskets are examples. Advances in plastics are also having their effect, in improved lining materials with better heat resistance, plating barrels with higher mechanical strength, etc.

Economics. The economics of plating operations is a controversial and poorly understood subject. Except when the deposited metal is very expensive, eg, gold, the actual cost of the deposited metal is generally considered to be a minor factor in the total cost of plating, and power cost is even less important. It is, of course, cheaper to plate zinc than cadmium but the difference is not the ratio of their prices. Overhead and labor are the major factors; thus it is more important, economically, to deposit the required plate quickly and satisfactorily—to produce the most work in the least space with the fewest rejects—than to economize on supplies or power. It must be admitted, however, that cost-accounting procedures for the plating trade are not well developed, at least in the literature.

Safety. Hazards in plating operations arise from the nature of the materials routinely handled, many of them highly toxic. Thus certain minimum precautions are absolutely necessary. Many of the cleaning, plating, and pickling steps evolve fumes and spray, which must be vented without being permitted to come in contact with workers. Chromic acid spray is highly irritating to nasal mucosa, as is caustic spray; fumes from vapor degreasing are toxic; and many solutions cause dermatitic reactions. Fluoborates and fluorides must not come in contact with skin; cyanide solutions, normally alkaline, should never come in contact with acids. Most of the normal hazards of a plating room can be adequately handled by a combination of proper ventilation and appropriate protective clothing. Most state labor departments have definite codes specifying the maximum limits of various contaminants that are permitted in the air of the work place. Because of the low voltages employed, electrical hazards in plating are not usually of concern. Anodizing of aluminum and the high-voltage side of rectifiers constitute exceptions.

Waste Disposal and Metal Recovery. Present concern for pollution control has rendered waste control and metal recovery of prime importance in the minds of plating shop owners and operators. EPA regulations, when issued in final form, will have the force of law, and meeting these regulations will be expensive for many industries, including metal finishing. In the late 1970s more papers and reports on various methods of reducing or eliminating contaminants in the effluents have appeared in the journals devoted to the electroplating industry than on any other subject. In addition, symposia have been held at frequent intervals and continue to be scheduled (60–65).

It is not possible at present to make definitive statements concerning either EPA regulations or the best means of meeting them. The regulations are still tentative and subject to industry and public comment. The ways they can best be met are under continuing study from the standpoints of both effectiveness and economics.

Proposed EPA regulations are presented in Tables 4 and 5 (61–63). The optional limits (Table 4) for large plants apply if they use no strong chelating agents that would prevent the determination of some metals. The limits refer principally to discharges

Table 4. Proposed (1978) EPA Limits for Discharge into POTW Systems, in mg/L

Pollutant	Companies with less than 37,850 L (10,000 gal)/d discharge		Companies with more than 37,850 L/d discharge	
	1-d max	30-d average	1-d max	30-d average
CN[a]	2	0.8	0.2	0.08
CN, total			0.64	0.24
Cr^{6+}	0.25	0.09	0.25	0.09
Cu			4.6	2.0
Ni			3.6	1.8
Cr, total			4.2	1.6
Zn			3.4	1.5
Ag			1.0	0.34
Pb	0.8	0.4	0.8	0.4
Cd	1.0	0.5	1.0	0.5
Total metals[b]			7.5	3.9
pH	adjusted to 7.5–10		adjusted to 7.5–10	

[a] CN amenable to chlorination.
[b] Total of Cu, Ni, Zn, and total Cr; does not include Cd and Pb.

Table 5. Proposed EPA Optional Limits for Discharge into POTW Systems by Companies with Discharge Greater than 37,850 L (10,000 gal)/d, in mg/L

Pollutant	1-d max	30-d average
CN[a]	0.2	0.08
CN, total	0.64	0.24
Cr^{6+}	0.25	0.09
Pb	0.8	0.4
Cd	1.0	0.5
TSS[b]	15	10
pH	adjusted to 7.5–10	

[a] CN amenable to chlorination.
[b] Total suspended solids.

into municipal sewer systems (POTW, publicly owned treatment works). Little has appeared regarding discharge directly into watercourses.

There are two basic approaches to waste treatment: destruction and recovery. Original emphasis was on the former, with the objectionable metals being precipitated as sludges and disposed of to landfills, or similar means. As landfills become both less available and more restrictive in their acceptance of such sludges, emphasis has shifted to recovery of the metals for reuse in the plant or for sale to a refiner.

Although some attempts have been made to recover cyanides, in general cyanide is still treated by destructive methods. The usual method is chlorination by either chlorine gas or hypochlorites. The following reactions take place:

$$CN^- + Cl_2 \rightarrow CNCl + Cl^- \tag{1}$$

$$CNCl + 2\,OH^- \rightarrow CNO^- + Cl^- + H_2O \tag{2}$$

$$2\,CNO^- + 4\,OH^- + 3\,Cl_2 \rightarrow 6\,Cl^- + 2\,CO_2 + 2\,H_2O + N_2 \tag{3}$$

The final products are carbon dioxide and nitrogen, both innocuous. Reaction 1 is fast and takes place at all levels of pH; reaction 2 requires high pH; reaction 3, on the other hand, goes fastest at relatively neutral pH. It is, therefore, necessary to adjust the pH on a time schedule as the reactions proceed. In some cases it may not be necessary to carry out reaction 3 since cyanate, CNO^-, is far less toxic than cyanide and is a permitted waste in some jurisdictions.

Other methods of cyanide destruction have been proposed: electrolysis (66–68), which may also serve to recover some metals at the cathode while anodic oxidation is going on; use of peroxygen compounds (69), solvent extraction (70), oxidation on granulated carbon (71) and catalytic methods (72).

Hexavalent chromium, generated from chromium plating and other treatments such as conversion coatings and anodizing, is also highly toxic and strictly regulated. No satisfactory method has been found for precipitating Cr(VI); the usual method of disposal is to reduce it to Cr(III) using sulfur dioxide or sulfites. Occasionally, the ready availability and cheapness of spent pickle liquors containing ferrous ion make them attractive for this purpose. After reduction the Cr(III) can be precipitated as the hydrous oxide, similarly to other metals.

If the pollutant metals are to be precipitated as hydrous oxides and disposed of as sludges, pH control is required since different metals precipitate at different acidities, and some such as zinc are amphoteric. These hydrous oxides settle poorly and the sludges must be dewatered to some extent; flocculating agents are used in some plants. Any cheap alkali such as lime or caustic soda can be used for raising the pH, and sulfuric acid for lowering it. If both alkaline and acidic wastes are available, waste plus waste is proposed (73). Current shop practices and their success in waste control have been summarized (74–75). Precipitation as sulfides has been suggested (76–77). Usually cyanide destruction must precede precipitation of the contaminant metals.

The disadvantages of precipitation methods will be evident from the foregoing discussion, and attention has turned more and more to recovery of the valuable metals from effluents before their discharge. Among the methods proposed for recovery systems are reverse osmosis (qv) (RO) (78–83); evaporative recovery (84–86); ion exchange (qv) (87–88), and various combinations of two or more techniques (79,89). Most of these methods are technically feasible and the principal snag is their relatively high cost: large capital expense for equipment, energy for evaporation, membrane life for RO, etc.

838 ELECTROPLATING

Waste disposal and metal recovery systems are in general not offered by the conventional plating supply houses, but by firms that manufacture the osmotic membranes, evaporators, and ion-exchange equipment. Design is often left to consultants who specialize in the field. The problem is basically one of chemical engineering rather than electroplating; nevertheless, it is of paramount importance to the plating trade (see Recycling).

The requirement for maintaining low levels of cyanide in wastes has had a side effect on plating practice in that an active search for plating solutions that do not contain cyanide has been going on for several years, with some success. This is considered under the individual metals concerned.

Plating Solutions

Plating solutions are usually aqueous. Aluminum is plated on a semicommercial scale from organic electrolytes and the refractory metals, tungsten, molybdenum, tantalum, and niobium, have been deposited from fused electrolytes. Except for strictly laboratory procedures, these are the only present exceptions to the use of aqueous systems.

Every plating solution contains ingredients to perform at least the first, and usually several, of the following functions: (1) provide a source of ions of the metal(s) to be deposited; (2) form complexes with ions of the depositing metal; (3) provide conductivity; (4) stabilize the solution against hydrolysis or other forms of decomposition; (5) buffer the pH of the solution; (6) regulate the physical form of the deposit; (7) aid in anode corrosion (see below under Anodes); and (8) modify other properties peculiar to the solution involved.

Many compounds perform more than one of the stated functions. For example, a typical nickel-plating bath contains nickel sulfate and nickel chloride, both of which provide nickel ions and the necessary conductivity; the chloride aids in anode corrosion. Boric acid is added to buffer the solution; and small amounts of organic addition agents regulate the physical form and properties of the deposit.

The sections that follow describe briefly the processes for plating the commercial metals. All of the baths are characterized by certain properties about which the plater will want information for proper control of his operations.

Current Density Range. Current density is the average current in amperes divided by the area through which that current passes; the area is usually nominal area, since the true area for any but extremely smooth electrodes is seldom known. Units used in this regard are amperes per square meter A/m^2 (92.9 mA/ft^2).

Current densities at the anode and cathode are both important; they may differ considerably although not by so large a factor as to make necessary very great differences between the anode area and the cathode area. Most solutions exhibit a range of current densities within which deposits are satisfactory and outside of which they are not; for bright plating solutions this is called the bright range. Too high a current density at the anode may render it passive and too low a current density may cause the metal to dissolve in an unwanted valence state. Other factors being equal, the higher the current density obtainable, the faster the plating rate and thus, in general, the more economical the operation, but this generalization is subject to many individual modifications.

Throwing Power. Except in the special case of concentric anode and cathode, the current density over the electrodes varies from point to point. In general the areas on the cathode nearest to the anode receive a higher current density than those more remote. Thus more metal is plated on the projections than in the recesses. Many plating solutions, however, have the ability to moderate this condition to some degree. Throwing power may be defined as the improvement in metal distribution over primary current distribution on a cathode. (Primary current distribution is the distribution of current that depends only on the geometry of the cell.) This concept is qualitative rather than quantitative because all methods of measuring throwing power are open to various theoretical objections. In addition, throwing power is a function not only of the solution but also of many other operating variables. Nevertheless the property is real enough, and plating solutions may be categorized as having excellent, good, poor, or negative throwing power.

Acidity. Plating baths are either acid, neutral, or alkaline, and for most of them close control of the pH is essential to successful operation. Acid solutions include acid copper and tin and the various fluoborate baths; since most of these contain substantial amounts of free acid, pH control is not usually required. Neutral solutions (pH from about 5 to 8) include nickel and some of the pyrophosphate baths; and alkaline baths include most of the cyanide formulations as well as stannate tin. The last-named bath is controlled not by determining the pH but by titration for free alkali.

Anodes. Anodes in a plating bath perform two functions: they act as the positive electrode, introducing current into the solution; and, in most cases, they replenish the metal deposited at the cathode, thus maintaining the balance of the bath. Anodic replenishment is often the factor placing an upper limit on the usable current density. Use of inert anodes, with chemical replenishment, can avoid this difficulty. It is standard in chromium plating (but for other reasons) and has been used to some extent for other metals; see especially Tin Plating.

Anodes may be a source of difficulty if they produce slimes and sludges as they corrode. To keep the particles thus formed from reaching the cathode, anodes are often bagged in fabric such as cotton duck or nylon. Diaphragm cells are also in use.

Temperature. Control of temperature is important in almost all plating processes; each solution is characterized by a range of temperatures within which it gives best results. Temperature affects almost all the variables of the solution: conductivity, current efficiencies, nature of the deposit, and stability.

Current Efficiency. Faraday's laws predict the amount of metal that is deposited at the cathode or dissolved at the anode, but these amounts are not always realized; the deficiency is due to evolution of hydrogen at the cathode or oxygen at the anode, or to other side reactions. In practical plating operations, current efficiency is not usually of direct concern so long as it is known, but it is often important to equalize the efficiencies at cathode and anode so that the bath remains in balance.

Purity. Plating processes differ in sensitivity to the presence of impurities in the bath. Stannate tin is extremely tolerant of most impurities; bright nickel and most zinc-plating solutions are quite sensitive. Specific means are available for purifying most plating baths by chemical treatment; filtration through activated carbon is a generally useful method for removing organic contaminants. Another common technique for purifying solutions is the process known as dummying, which consists of plating for a period of time and usually at low current densities upon cathodes that are not intended for use, such as pieces of scrap sheet steel. This low-current-density electrolysis plates out metallic impurities and decomposes some organic contaminants.

A universal ingredient of all alkaline baths is carbonate, formed from decomposition of cyanides or pickup of CO_2 from the air. Carbonate may have deleterious effects in some baths if present at high levels. Several methods are available for removing excess carbonate; from baths formulated with sodium salts, advantage can be taken of the lowered solubility of sodium carbonate decahydrate at low temperatures, and it may be frozen out by cooling at about $0°C$. For potassium baths, methods based upon the addition of lime, calcium sulfate, or barium cyanide have been used. With alkaline stannate baths the carbonate is of little significance and can be allowed to build up until drag-out removes it as fast as it is formed.

Filtration can be used to remove solid particles that may interfere with satisfactory plating by being trapped in the deposit. Depending on circumstances, baths are filtered continuously during operation, or intermittently either as need arises or on a definite schedule. Filtration is often combined with treatment with activated carbon. Some baths, particularly if used only intermittently, may be clarified by settling; care is taken not to disturb the bottom few cm in the plating tank, and occasionally the tank is emptied and the sludge removed.

Bright Plating. Most deposits from simple plating solutions are mat unless the substrate is bright and the plate very thin. If the use is purely functional this is no drawback, but for decorative purposes a bright appearance is usually desired. This brightness was formerly obtained by mechanical treatment, buffing after plating, but today most metals can be bright as they come from the bath, and require little or no further treatment, at least as regards appearance. This bright appearance is produced by the addition to the plating bath of small amounts of one or more addition agents or brighteners, usually organic compounds. Most such agents are covered by patents and the processes are proprietary. The patent literature is voluminous, and many processes are marketed. Copper, silver, gold, zinc, cadmium, tin, brass, bronze, and lead–tin (solder) also can be plated bright; tin is the most recent metal to be added to the list. Bright acid tin baths are gradually replacing older formulations in general plating, though not for electrolytic tinplate. Since tin is seldom used decoratively, the advantages of a bright tin plate seem dubious, but there is no doubt that when confronted with a choice between a mat and a bright plate, the customer will choose the latter. The mechanism by which minute amounts of specific organic compounds so markedly change the character of the deposit has not been thoroughly elucidated, in spite of a great deal of effort and a copious literature.

Maintenance of Plating Baths. All plating baths require more or less chemical control. They must be analyzed, often or seldom, depending on individual circumstances, and adjustments made. For some baths a determination of specific gravity suffices for routine control; other baths require frequent and complete analysis.

A very useful tool in control is the Hull cell test. This patented cell consists of a trapezoidal box of nonconducting material with electrodes so arranged that by performing one plating experiment the operator can observe the characteristics of the deposit over a range of current densities and deduce the condition of the bath, whether adjustments are needed, and if so, of what kind. Any needed chemical additions should usually be made in liquid form, gradually, and with thorough stirring. Some automatic machines can be set up to feed additions continuously or intermittently on a preset schedule.

Individual Plating Baths

In addition to the bath compositions given below, most metals can be plated from proprietary solutions. In many cases, exemplified by the bright nickel processes, the principal ingredients are well known and standard but the additives are proprietary. The formulas presented should be regarded as typical, and many modifications are in satisfactory use; for the effect of changes in composition, impurities, and other details, monographs on electroplating should be consulted (25,90–92).

Cadmium. Cadmium affords good corrosion protection to steel and other substrates in marine atmospheres; it is readily solderable, and has a relatively pleasing appearance. It may be plated bright, semibright or mat, depending on the addition agents used. Corrosion resistance can be enhanced by chromate conversion coatings.

Cadmium is usually plated from cyanide solutions such as that shown in Table 6. Additional carbonate is formed during operation by decomposition of cyanide and absorption of CO_2 from the air. In amounts up to about 75 g/L this is not harmful, but excessive amounts should be eliminated.

In the plating of high strength steels, cadmium from cyanide baths embrittles the steel, and sometimes this cannot be relieved by the usual baking treatments. Although fluoborate solutions have been developed to avoid the cyanide problem (93), the popularity of plating with cadmium, one of the most harmful metals in the environment, is declining.

Chromium. Chromium is the final finish on the great majority of plated consumer items. As hard chromium it also has a host of engineering uses. The chromium-plating process is unique in several respects, and it vies with nickel in capturing the attention of research workers. Although chromium can be deposited from solutions of chromium(III) salts, such baths have not been practical for commercial plating purposes and solutions of chromic acid (chromium trioxide, CrO_3) are used almost exclusively (94). In order to permit the deposition of chromium, a small amount of another anion, typically sulfate, is essential. The ratio CrO_3–H_2SO_4 is usually about 100:1 (by weight), but for special purposes it can be varied from 50:1 to 150:1 or more.

There have been many relatively recent developments in chromium plating involving improvements in the plating process itself and greater knowledge of the role of chromium in the nickel–chromium composite, leading to dramatically improved performance.

Table 6. Cadmium-Plating Bath

	Still plating	Barrel plating
CdO, g/L	25–35	17–23
Cd metal, g/L	20–30	15–20
NaCN, g/L	90–120	70–90
NaOH, as formed by the reaction CdO + 4 NaCN + H_2O → $Na_2Cd(CN)_4$ + 2 NaOH		
addition agents	qs	qs
cathode current density, A/m²	50–500	
anode current density, A/m²	not >200	
temperature, °C	20–35	
throwing power	good	

(1) In addition to the conventional chromic acid–sulfate bath, there are now available several proprietary processes with such advantages as better throwing power, higher efficiency, and easier control. These advantages result at least in part from the use of catalyst anions other than sulfate, such as fluoride and fluosilicate.

(2) The final, very thin chromium deposit was long thought to be merely a tarnish preventive, having little effect on the corrosion resistance of the composite. Ample experience has demonstrated that the thickness and type of both the nickel and the chromium in the nickel–chromium composite have great effects on their corrosion behavior. The development of duplex nickel (see Nickel Plating) and the use of so-called microdiscontinuous chromium both improved the performance markedly. Conventional chromium plate cracks in a random pattern. The deliberate provision of controlled crack patterns, or discontinuities of controlled size and number, can delay corrosion significantly. The ASTM specification for nickel–chromium desposits for severe service (95) permits thinner nickel under microdiscontinuous chromium than when ordinary chromium is used. Microcracked chromium has at least 300 cracks per linear centimeter, with a minimum thickness of 0.8 μm; it is produced by modifications in the chromium-plating process. Microporous chromium must have a minimum thickness of 0.25 μm and contain a minimum of 10,000 pores/cm^2. It is produced by modifications in the nickel undercoat such that nonconducting particles such as alumina or barium sulfate are included in a thin nickel coating over the main deposit. Chromium does not plate over these inclusions, producing the microporosity. Microporous and microcracked chromium are subsumed under the name microdiscontinuous; they are about equivalent in performance.

For the reasons stated, the typical bath compositions given should be considered only as guides; for maximum performance the current literature and the vendors of proprietary processes should be consulted.

	Dilute bath 1	*Concentrated bath 2*
chromic acid, g/L	250	400
sulfuric acid, g/L	2.5	4

The current density and the temperature for bright plating are interrelated. A convenient chart is shown in Figure 2. In this figure the dot-and-dash line *A* circumscribes the bright plate area for dilute solutions and the dashed line *B* circumscribes the bright plate area for concentrated solutions.

In addition to its decorative uses, chromium plating is widely used as an engineering process, for its great hardness and excellent bearing properties. This so-called hard chromium plating is not different in principle from decorative plating, but because of the greater thicknesses employed certain problems are magnified: the long plating times entail the danger of etching the basis metal in low-current-density areas, and the close control of the physical and mechanical properties of the deposit requires more care in the entire cycle (96–97).

The benefit of the hardness of chromium deposits is not obtained unless the basis metal is sufficiently hard and the coating thickness satisfactory. Even a heavy deposit of chromium may be crushed or indented if applied over a soft basis metal such as copper. Table 7 gives basis metal hardnesses and thicknesses of chromium plate suggested for different applications.

Unlike most plating, chromium plating does not employ anodes of the depositing

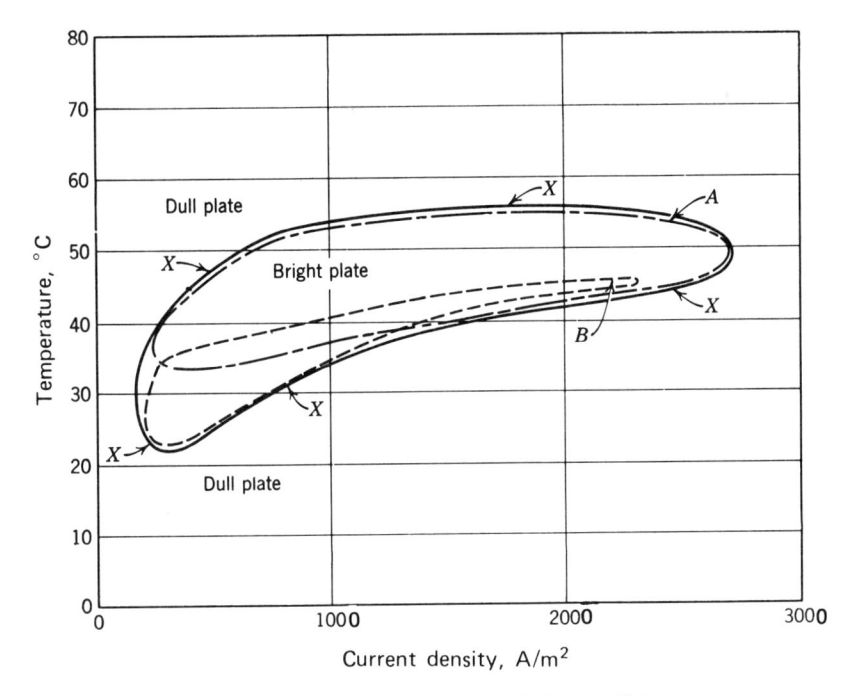

Figure 2. Chart of bright chromium plating conditions.

Table 7. Basis Metal Hardness and Thickness of Hard Chromium

Application	Hardness, Rockwell C	Chromium thickness, μm
drills	62–64	1.25–12.5
reamers	62–64	2.5–13
burnishing bars	60–62	12.5–75
drawing plugs or mandrels	60–62	38–200
drawing dies	60 inside, 45 outside	13–200
plastic molds	55–60	5–50
gages	48–58	2.5–38
pump shafts	55–62	12.5–75
rolls and drums		6–300
hydraulic rams		13–100
printing plates (engraved steel)		5–13

metal. Anodes are usually of lead–antimony or lead–tin alloy. Anode area is important principally because it determines the proportion of trivalent chromium in the bath.

Current efficiency of a chromium-plating solution is low: 8–12% for conventional baths; as high as 20% for some of the newer proprietary solutions. Efficiency tends to increase with current density, and this leads to the notoriously poor throwing power of these baths as compared with most other metals. The low efficiency also entails the copious evolution of hydrogen, causing spray which must be vented for safety of operators. Spray suppressants now available greatly alleviate this problem. Low efficiency and unusually small electrochemical equivalent (52/6 = 90 μg/C or 8.7 g/farad) combine to require abnormally long plating times for deposits of the thicknesses used in hard chromium plating. Table 8 indicates average plating conditions.

Table 8. Typical Chromium-Plating Conditions[a]

	Decorative plates	Hard chromium	
CrO_3, g/L	250–400	250	400
H_2SO_4, g/L	2.5–4	2.5	4
cathode current density, A/m²	1250–1750	3100	2200
temperature, °C	38–43	55	50
deposition rate, μm/h	8–13	25	13

[a] For conventional bath; for proprietaries consult vendor.

In spite of all the disadvantages inherent in chromium plating, it enjoys as widespread use as any plating process, proof enough of its unique properties.

Trivalent Baths. Although many early attempts to develop a chromium plating bath using Cr(III) were unsuccessful commercially (98), efforts have continued (99–100) and in 1978 it was claimed that a so-called trivalent bath had been in satisfactory commercial operation for about a year in several installations (101–102). No details of the proprietary solutions were disclosed, but operating conditions were discussed. Putative advantages of a trivalent bath would be higher electrochemical equivalent, perhaps better throwing and covering power, and the avoidance of the reduction step from Cr(VI) to Cr(III) in waste disposal. At present it cannot be predicted whether this development will continue to enjoy commercial acceptance; current literature must be consulted.

Black Chromium. Proprietary processes for deposition of so-called black chromium have been announced from time to time; the deposit is not chromium metal but an oxide mixture. The deposit has been proposed as a possible coating for solar heat collectors (103).

Cobalt. Cobalt plating is little used: for general purposes it seems to have no advantage over nickel and is more expensive. The magnetic properties of cobalt alloys have made them of interest in computer technology; alloys with nickel, molybdenum, phosphorus, iron, and tungsten in particular have been investigated. Typical plating baths are shown in Table 9 (104).

Copper. Copper, although sometimes used as a final finish, is more often applied as an undercoat for other metals. As such it has a wide variety of uses and is very extensively employed in the industry. Many different types of plating bath are available,

Table 9. Cobalt-Plating Solutions

	Sulfate	Chloride	Ammonium sulfate
$CoSO_4 \cdot 7H_2O$, g/L	330–565		
$CoCl_2 \cdot 6H_2O$, g/L	45 (optional)	300–430	
$Co(NH_4)_2(SO_4)_2 \cdot 6H_2O$, g/L			175–200
H_3BO_3, g/L	30–45	30–45	25–30
NaCl or KCl, g/L	17–25 (optional)		
H_2SO_4 or HCl to pH	3.0–5.0	2.3–4.0	5.0–5.2
temperature, °C	35–38	52–71	25
cathode current density, A/m²	220–500	500–650	100–300

both open and proprietary. Copper may be deposited mat, or with appropriate brightening agents, either semi- or fully bright. It is easily buffed to a high luster. It is widely used as a coating on steel that is to be subsequently plated with other deposits; by this means scratch marks and other defects in the steel are more readily covered than by working on the steel itself. Copper is universally used in plating on zinc-base die castings as the basis for further coatings.

The latest ASTM specification (95) permits somewhat less nickel if a bright, leveling acid copper solution is used. These proprietary processes, producing a bright and leveling deposit of copper, are relatively recent developments; the deposit is also of great value in the process of plating on plastics. A typical formulation is shown in Table 10.

The applications of the various copper-plating solutions overlap somewhat, but each has fairly well-defined areas of usefulness. The cyanide bath is used for copper-striking steel preparatory to further plating with acid copper and for the initial plate on zinc die castings. Its plating rate is rather slow but it is fairly easily controlled in operation. The Rochelle bath is similar to the cyanide but additions of Rochelle salts permit higher operating current densities. Still higher speeds of operation are possible in high-speed baths, in which some or all of sodium salts are replaced by potassium. Proprietary materials that claim advantages over Rochelle salt, and brighteners may also be used.

Current manipulation techniques, such as periodic reversal of the current, current interruption, and pulsed current (105–106), may be useful in improving the plating characteristics of some baths. Table 11 gives typical formulas.

Pyrophosphate solutions find applications in the plating of printed circuits. Their freedom from cyanide has made them of interest where waste disposal is a problem (see Table 12).

Although it cannot be used directly on steel, the copper sulfate bath is widely employed because it is inexpensive, plates rapidly, and is easily controlled. For building up heavy deposits the work is first "struck" in a cyanide bath and then transferred to the sulfate. Brighteners are available for this bath also. Fluoborate has advantages in speed of deposition, in spite of its higher initial cost.

Acid copper baths are extensively used in electroforming, including electrotyping, and in wire plating and many other engineering applications. Typical formulations are shown in Table 13. Physical and mechanical properties of electroformed copper have been reported (107–108).

Table 10. Typical Bright Acid Copper Bath

$CuSO_4 \cdot 5H_2O$, g/L	190–240
Cu metal	50–60
H_2SO_4, g/L	45–60
chloride, Cl^-	0.02–0.08
addition agent[a]	qs
temperature, °C	24–32
anode current density, A/m^2	15–300
cathode current density, A/m^2	300–600

[a] Proprietary.

846   ELECTROPLATING

Table 11. Representative Cyanide Copper Baths

	Strike	Rochelle	High efficiency[a]
CuCN, g/L	15	25	75
NaCN, g/L	23	35	95
KCN, g/L			115
Na_2CO_3, g/L	15	30	
NaOH, g/L			30
KOH, g/L			42
Rochelle salt, g/L		45	optional
by analysis			
Cu, g/L	11	19	53
free CN^-, g/L	6	6	10
pH		12.6	
temperature, °C	40–60	55–70	60–80
cathode current density, A/m^2	100–320	160–650	100–1100
anode current density, A/m^2	50–100	80–330	150–400
efficiency range, %:			
cathode	10–60	30–70	99+
anode	95–100	50–70	99+
limiting thickness, μm	2.5	13	no limit

[a] Na or K salts may be interchanged in equivalent molar amounts.

Table 12. Pyrophosphate Copper Bath

$Cu_2P_2O_7$, g/L	75
$K_4P_2O_7$, g/L	260
KNO_3, g/L	15
NH_3, g/L	2
pH	8.2–8.8
ratio $P_2O_7^{4-}/Cu^{2+}$	7–8
temperature, °C	50–60
cathode current density, A/m^2	100–800
agitation (required)	air
current efficiency, %	ca 100

Gold. Gold plating has become extremely important in the electronics and computer industries because of the low contact resistance, corrosion resistance, and solderability of the metal (5,109). It is still used for decorative purposes as well. Supplying materials and processes for gold plating is a very competitive business, and the number of special purpose baths is large. One supplier offers no less than 27 proprietary formulations, yielding deposits of various colors, hardnesses, and other properties, and containing small to fairly significant amounts of such alloying elements as antimony, tin, nickel, copper, silver, indium, and cobalt. Although it is possible for the plater to formulate solutions from published sources, few do; most rely on these proprietaries.

Gold has traditionally been plated from cyanide baths; since rinse reclamation is important for recovery of this precious metal, the cyanide disposal problem is of little consequence.

The cyanide baths can be roughly divided into three types: alkaline, neutral, and

Table 13. Acid Copper Plating

	Sulfate bath	Fluoborate bath
$CuSO_4 \cdot 5H_2O$, g/L	188	
H_2SO_4, g/L	74	
$Cu(BF_4)_2$, g/L		220–440
HBF_4, g/L		15–30
H_3BO_3, g/L		15–30
temperature, °C	30–45	18–50
cathode current density, A/m²	100–1000	100–4000[a]

[a] Very much higher current densities are obtainable under conditions of violent agitation.

acid. It was some time before it was realized that the gold cyanide complex, unlike most others used in plating, is stable at pH as low as 3; in fact the compound $HAu(CN)_2$ can be isolated. The acid baths, the latest addition to the gold plater's repertory, are probably the most versatile. A few noncyanide baths have been proposed, and one based on a sulfite complex has become important. Typical formulations of the three types of bath are shown in Table 14; many variations are possible.

The price of gold has stimulated research on possible substitutes, but few if any combine all the advantageous properties of gold. Among the substitutes proposed are palladium (also a precious metal but cheaper and less dense than gold, so that a given mass covers about twice the area at about half the cost) and the nickel–boron electroless deposit (see below). Platers have also resorted to spot-plating in which instead of plating an entire terminal, gold is put only where it is functional.

An ASTM specification (110) covers requirements for purity and thickness of gold plate for its engineering applications. A monograph on gold plating is available (111) (see Electrical connectors).

Table 14. Representative Gold-Plating Baths[a]

	Alkaline	Neutral	Acid	Strike
$KAu(CN)_2$, g/L	3–18	6–24	3–24	0.75–3
Au as metal	2–12	4–16	2–16	0.5–2
KCN, g/L	15–45			15–90
K_2CO_3, g/L	0–45			
K_2HPO_4, g/L	0–45	0–90	0–100	15–45
KOH, g/L	1–30			
brighteners[b]	qs	qs	qs	
chelators, g/L[c]		15–90	10–100	
temperature, °C	50–70	25–70	40–70	40–60
current density, A/m²	10–50	20–100	10–50	
anodes	Pt, SS[d]	Pt, SS[d]	Pt	SS[d]
pH	11–13	6–8	3–6	8–13

[a] Proprietaries generally preferred, see text.
[b] Omitted if mat finish desired.
[c] Proprietary.
[d] Stainless steel.

848   ELECTROPLATING

Indium. Indium is soft and tarnish resistant, and has uses in bearings, usually in combination with silver or lead. Indium can be plated from cyanide, sulfate, fluoborate, or sulfamate solutions (see Table 15). There appears to be no consensus as to which of the available baths is best. The metal is expensive and is used only when necessary.

Iron. Iron is not used for decorative or protective purposes, but does have minor electroforming and other engineering uses. It has been used for plating on aluminum pistons in at least one brand of automobile engine (112). Many baths have been proposed for iron plating; all contain iron in the iron(II) oxidation state. Typical formulations are shown in Table 16 but many variations are possible. The ferrous sulfate and ferrous ammonium sulfate baths can be operated at room temperature; the hot chloride bath requires elevated temperatures but can yield more ductile deposits. Physical and mechanical properties of the deposit depend strongly on the operating conditions.

Lead. Lead is plated from fluoborate solutions (see Table 17). It is used in battery parts and chemical construction, particularly for its resistance to sulfuric acid. It is more corrosion resistant than would be expected from its position in the emf series, because in most atmospheres pores tend to fill with insoluble corrosion products which prevent further attack. ASTM specifications are available for lead, and for the more widely used lead–tin (low tin) (113) and tin–lead (high tin or solder) alloy deposits (114). They are useful alloys in bearings, in the 90:10 lead–tin, and for solderability in the 40:60 range. Plating baths are similar to those for lead plating, with the addition of stannous fluoborate.

Nickel. Nickel shares with chromium the distinction of having been the subject of the most intensive and extensive research efforts, on all levels from the practical to the highly theoretical. It has almost innumerable uses, it is plated in all thicknesses

Table 15. Indium-Plating Baths

	High-pH cyanide	Sulfate	Fluoborate	Sulfamate
In as metal, g/L	15–30	20	72	20 min
InCl$_3$, g/L	30–55			
In$_2$(SO$_4$)$_3$, g/L		88		
In(SO$_3$NH$_2$)$_3$, g/L				70 min
In(BF$_4$)$_3$, g/L			235	
dextrose, g/L	30–40			8
KCN, g/L	140–160			
KOH, g/L	30–40			
NaCl, g/L				45
Na$_2$SO$_4$, g/L		10		
H$_3$BO$_3$, g/L			20–30	
NH$_4$BF$_4$, g/L			40–50	
HSO$_3$NH$_2$, g/L				25
pH		2–2.5	1	0.0–0.2
temperature, °C	20–30	20–30	20–30	20–30
current density, A/m^2	150–300	200	100–1600	200–1000
cathode efficiency, %	ca 50	30–70	50	90
anodes	steel	In + Pt[a]	In + Pt[a]	In

[a] In anode efficiency is higher than cathode efficiency, so some inert anode area is required.

Table 16. Iron-Plating Baths

	Sulfate	Double sulfate	Chloride	Sulfate-chloride	Fluoborate	Sulfamate
$FeSO_4 \cdot 7H_2O$, g/L	240			250		
$FeSO_4(NH_4)_2SO_4 \cdot 6H_2O$, g/L		250–360				
$FeCl_2 \cdot 4H_2O$, g/L			300–450	35		
$Fe(BF_4)_2$, g/L					225[a]	
$Fe(SO_3NH_2)_2$, g/L						305[a]
$CaCl_2$, g/L			300			
NaCl, g/L					10	
NH_4Cl, g/L				20		
$NH_4SO_3NH_2$, g/L						30–40
pH	2.8–3.5	2.8–3.5 or 4.0–5.5	1.2–1.8	3.5–5.5	3–3.4	2.7–3
temperature, °C	32–65	24–65	85 min	27–70	55–65	50–60
current density, A/m²	400–1000	200–1000	200–800	200–1000	200–900	500

[a] Added as commercial concentrate.

Table 17. Lead and Lead–Tin Alloy Plating

	Lead	7% Tin[a]	60% Tin (solder)
$Pb(BF_4)_2$, g/L[b]	220	160	53
(Pb)	120	88	30
$Sn(BF_4)_2$, g/L[b]		15	130
(Sn)		6	52
free HBF_4, g/L	30	100–200	100–200
free H_3BO_3, g/L	135		30
animal glue or gelatin, g/L	0.2		
peptone, g/L	0.2–1	0.5	5.0
temperature, °C	25–40	15–38	15–38
cathode current density, A/m²	50–500, avg 200[c]	300	300
anode current density, A/m²	100–300	150	

[a] For bearings.

[b] 100% basis; added as commercial concentrate.

[c] By doubling the concentration of this bath (except the glue or peptone), current density may be raised to 700 A/m².

from a mere flash to many millimeters, and there are available many different plating baths, each with its own advantages and limitations.

Most nickel-plating baths of today are based on the bath originally formulated by O. P. Watts and accordingly called the Watts bath (see Table 18). Hydrogen peroxide may be added to the Watts bath to prevent hydrogen pitting; this additive cannot be used in the bright and semibright modifications of the Watts bath, for these some type of surfactant such as sodium lauryl sulfate is substituted for the peroxide. For most decorative plating, the various bright, semibright, and leveling solutions are formed by addition of organic compounds in relatively small amounts to Watts solutions. These produce deposits that may be sufficiently bright as plated to require no buffing before final chromium plating. Other additives are capable of yielding deposits that can level minor scratches and other imperfections in the basis metal.

Table 18. Nickel Baths for Heavy Plating

Type and ingredients[a]	Concentration, g/L	pH (electrometric)	Temperature, °C	Cathode current density, A/m²[b]	Hardness[c] (Vickers)	Tensile strength, MPa[d]	Elongation, %	Residual stress, MPa[d]
Watts bath								
nickel sulfate, $NiSO_4 \cdot 6H_2O$	330	1.5–4.5	45–65	250–1000	140–160	380	30	125
nickel chloride, $NiCl_2 \cdot 6H_2O$	45							
boric acid, H_3BO_3	38							
hard								
nickel sulfate, $NiSO_4 \cdot 6H_2O$	180	5.6–5.9	43–60	200–1000	350–500	1030	5–8	300
ammonium chloride, NH_4Cl	25							
boric acid, H_3BO_3	30							
chloride								
nickel chloride, $NiCl_2 \cdot 6H_2O$	300	2.0	50–70	250–1000	230–260	690	20	275–345
boric acid, H_3BO_3	38							
chloride sulfate								
nickel sulfate, $NiSO_4 \cdot 6H_2O$	200	1.5–2.0	45	250–1000				
nickel chloride, $NiCl_2 \cdot 6H_2O$	175							
boric acid, H_3BO_3	40							
chloride acetate								
nickel chloride, $NiCl_2 \cdot 6H_2O$	135	4.5–4.9	30–50	200–1000	350	1380	10	
nickel acetate, $Ni(C_2H_3O_2)_2$	105							

Bath / Composition	g/L	pH	Temp	Current density	col5	col6	col7
nickel–cobalt		4.7	40	500	450–500		100–120
nickel sulfate, $NiSO_4 \cdot 6H_2O$	240						
nickel chloride, $NiCl_2 \cdot 6H_2O$	22.5						
boric acid, H_3BO_3	30						
ammonium sulfate, $(NH_4)_2SO_4$	1.5						
nickel formate, $Ni(CHO_2)_2 \cdot 2H_2O$	15						
cobalt sulfate, $CoSO_4 \cdot 7H_2O$	2.6						
fluoborate		2.0–3.5 (colorimetric)	40–80	400–1000	183	515	15–30	110
nickel (as $Ni(BF_4)_2$)	75						
free fluoboric acid HBF_4	4–37						
free boric acid, H_3BO_3	30						
sulfamate		3.0–5.0	40–60	200–3000	250–350	620	20–30	3.5
nickel sulfamate, $Ni(NH_2SO_3)_2$	450						
boric acid, H_3BO_3	30						
sulfamate chloride		3.5–4.2	40	200–250	190	745	15–20	10
nickel sulfamate, $Ni(NH_2SO_3)_2$	300						
nickel chloride, $NiCl_2 \cdot 6H_2O$	6						
boric acid, H_3BO_3	30						

[a] Antipitting agents are usually used, and other additives may be used as brighteners, levelers, stress reducers, etc.
[b] Higher current densities can be used with increasing agitation.
[c] For greater detail see ref. 108.
[d] To convert MPa to psi, multiply by 145.

The number of organic compounds that are capable of modifying the deposit in this way is legion, and the patent literature is extensive. Most compounds, however, are impractical because they have concomitant disadvantages, such as instability or deleterious effects on the physical or mechanical properties of the deposit, so that the number of commercially used brighteners is small relative to the number of patents. This field is almost entirely proprietary.

Many tests have shown conclusively that so-called duplex nickel—a layer of semibright, sulfur-free nickel followed by a thinner layer of fully bright nickel containing sulfur—has much better corrosion resistance than bright nickel alone; in fact for severe applications (such as exterior automotive brightwork) the ASTM specification (95) for nickel–chromium coatings requires duplex nickel as an undercoat for the final chromium (see Chromium). As stated, the performance of the composite plate is also improved by the use of microdiscontinuous chromium.

Nickel has a host of engineering uses, including especially electroforming (107), in which its physical and mechanical properties are of greater importance than its appearance. These properties can be varied over a fairly wide range by manipulation of the bath composition and operating conditions. In particular the sulfamate and fluoborate solutions are becoming more important, and are beginning to invade the field of decorative plating as well.

Monographs on nickel plating are available (115) and the International Nickel Co. (INCO) has published valuable pamphlets on the subject.

Platinum Group Metals. Of the platinum group metals, the most commonly used in plating is rhodium. In addition to its decorative applications, rhodium has properties that make it useful for engineering applications. These include corrosion resistance, stable electrical contact resistance, wear resistance, reflectivity, and heat resistance. Suggested thicknesses of rhodium for its engineering uses are detailed in an ASTM specification (116). Typical formulas are shown in Table 19. The metal content of these baths is derived from prepared solutions. Replenishment is also chemical, the anodes of the depositing metal being inert. In addition to the baths shown, baths based on mixed sulfate–phosphate systems can be used.

Platinum is plated from several types of solution, based on platinum "P" salt, which is diamminedinitroplatinum(II), or on other complexes. Platinum has found uses for production of platinized titanium or tantalum anodes. Such anodes are especially useful in chloride solutions, which would corrode most metals, and are of course cheaper than pure platinum (117–118).

Table 19. Rhodium-Plating Baths

	Sulfate[a]	Sulfate[b]	Phosphate[a]
Rh, as metal, g/L[c]	2	5.25	2
H_3PO_4, 85%, g/L			55–115
H_2SO_4, g/L	50–160	50–100	
temperature, °C	38–50	44–50	38–50
current density, A/m²	200–1000	1000–3000	200–1500
anodes	Pt	Pt	Pt

[a] Thin decorative.
[b] Heavy industrial.
[c] From commercial concentrate in appropriate acid.

Interest in palladium plating has been stimulated by the search for a cheaper substitute for gold. Palladium can be plated from various complex solutions (25), and its substitution for gold in some electronic applications has been reported (119–120).

Ruthenium and iridium can be plated but have found little use (121). Only one report of osmium plating on a laboratory scale has appeared (122), and there are no known uses for it in electroplating.

All the platinum metals except perhaps osmium can be plated from molten salts but the processes are in an exploratory stage (123).

Silver. Silver is deposited from cyanide baths exclusively; modern formulations differ only in detail from the earliest published plating bath, that of Elkington (2). In addition to its widespread use on flatware, hollow-ware, and in jewelry, silver has many applications that depend on its electrical and mechanical characteristics, being widely used in the electronics and related industries, and for bearings. Because of its nobility, silver will deposit by simple immersion on most substrates. Such immersion deposits are nonadherent and constitute a poor base for subsequent plating. Consequently, a two-step process is usually employed: a strike in a bath low in silver and high in cyanide to prevent this immersion deposit, followed by a regular silver-plating bath to build up the deposit to desired thickness. Two such strikes, the first containing copper, are recommended for plating on steel. Bath formulations are shown in Table 20.

A phenomenon known as silver migration has limited its used in miniature circuit boards: under a positive d-c potential within a damp resin, silver can migrate across insulation. On drying, silver will be found in the insulation, creating a leakage path. Silver plating is, therefore, forbidden in many military specifications for circuit boards.

Tin. For general plating, both alkaline stannate and acid sulfate or fluoborate baths are used for tin plating (Table 21) (see also Continuous Plating and Electrolytic Tinplate). The alkaline stannate bath can be formulated with either sodium or potassium salts, the latter being favored for their generally superior operating characteristics. The key to satisfactory operation of the stannate bath lies in proper control of the anode current density to ensure that the tin dissolves in the quadrivalent form as stannate. Under improper conditions the anode may dissolve as tin(II), which has immediate deleterious effects on the deposit.

Table 20. Representative Silver-Plating Baths

	Decorative	Industrial	Strike[a]	Strike[b]
AgCN, g/L	30–35	45–50	1.5–2.5	1.5–5
KCN, g/L	50–78	65–72	75–90	75–90
free KCN, g/L	35–50	45–50		
CuCN, g/L			1–15	
K_2CO_3, g/L		40–80		
KNO_3, g/L		40–60		
KOH, g/L		1–14		
brighteners	qs	qs		
temperature, °C	20–28	42–45	22–30	22–30
current density, A/m²	50–150	500–1000	150–300	150–300

[a] First strike for steel.
[b] Second strike for steel and strike for nonferrous metals.

Table 21. Tin-Plating Baths

	Stannate	Sulfate	Fluoborate
potassium stannate,[a] $K_2Sn(OH)_6$, g/L	100		
potassium hydroxide,[a] KOH, g/L	15		
stannous sulfate, $SnSO_4$, g/L		100	
sulfuric acid, H_2SO_4, g/L		100	
glue or gelatin, g/L		2	
2-naphthol, g/L		1	0.5
tin (as $Sn(F_4)_2$)			40
free HBF_4			40
gelatin			2
temperature, °C	65–90	20–30	20–40
current density[b], A/m²	300–600[c]	100–1000	200–1000[b]

[a] Sodium salts may be substituted, at some sacrifice in cathode efficiency and current density.

[b] Under conditions of violent agitation, as in high-speed strip and wire plating, much higher current densities are used.

[c] By increasing the concentration of the solution, much higher current densities can be obtained.

The use of tin anodes containing about 1% aluminum significantly increases the anode current density range; these are marketed as high-speed anodes; their mode of action has been investigated (124).

Potential anode difficulties can be avoided by the use of insoluble steel anodes, with chemical replenishment by a proprietary tin sol which contains only about $\frac{1}{4}$ as much potassium as stannate (125). These sols can be modified by the addition of a bismuth compound. The resulting deposit is 0.1–0.3% Bi, which inhibits or prevents the formation of gray tin, the low temperature nonmetallic allotrope of tin commonly called tin pest.

Although tin-plating baths long resisted efforts to produce a bright plate, this problem has been solved by several proprietary processes, using basically the sulfate baths with the addition of undisclosed additives. Mat tin plate can be brightened by melting the deposit and quenching it quickly, a process known as flow-brightening or reflowing. This process is used in general plating as well as for electrolytic tinplate.

Interest in tin plating has increased in the last few years by the demands of the electronics industry (19). Many investigations have been reported on its solderability under various conditions (126).

A phenomenon that has troubled the electronics industry is tin whiskers, thin filaments of metal that emanate from the deposit and may grow across miniature components to cause short circuits. Most metals can form these filamentary growths, but tin has been most widely investigated because of its use in electronics components (127–129). They are of no concern in ordinary applications, nor in high-voltage equipment, since the potential soon burns them; but when low voltages are involved they can cause failures. Various expedients, such as heat treating, have been used to overcome the problem; but since whisker growth is more or less random, large numbers of samples must be tested under field conditions and results are difficult to evaluate.

Zinc. Zinc offers so-called sacrificial protection of ferrous metals because it is anodic to the substrate; the latter is protected so long as some zinc remains in the immediate area; the presence of minor pinholes or discontinuities in the deposit is of little significance. Zinc is plated on continuous sheet and wire, on conduit, and on all types of hardware such as tools, nuts, bolts, and screws. The acid chloride, sulfate, and fluoborate baths are capable of higher plating speeds than cyanide and are used in strip and wire plating, but until recently the cyanide bath was almost exclusively used for general plating.

Because of the wide use and the fairly high cyanide concentration of the standard zinc plating bath, EPA regulations on cyanide in effluent stimulated efforts to reduce or eliminate the cyanide. Several answers have been proposed: use of more dilute baths, low cyanide baths, alkaline bath containing no cyanide, neutral chloride baths containing ammonium ions and chelating agents, and finally acid chloride baths (pH 4.7–5.5) that substitute potassium ion for the ammonium used in the neutral bath. These neutral and acid chloride baths are proprietary, requiring several additives, the nature of which is not disclosed. The acid baths have largely replaced neutral ones in practice. Typical formulas are listed in Tables 22 and 23 but many variations are possible (130–134).

Table 22. Alkaline Zinc Baths

	Standard	Dilute	Low cyanide[a]	Noncyanide[a]
zinc, g/L	45–60	7–20	5–15	5–15
total NaCN, g/L	90–150	7–50	5–15	
total NaOH, g/L	90–140	75–100	70–100	70–100
$NaCO_2$, g/L	30–75	20–100		
ratio NaCN–Zn	2.0–2.5	1.0–2.5		
brighteners[a]			qs	qs
cathode current density, A/m²	100–900			
anode current density, A/m²	30–450			
temperature, °C	20–50			

[a] Proprietary.

Table 23. Representative Acid and Neutral Chloride Zinc Baths

	I	II	III	Neutral[a]
$ZnSO_4 \cdot 7H_2O$, g/L	240	360	410	
zinc, g/L	55	82	93	24–50
NH_4Cl, g/L	15	30		
Na_2SO_4, g/L			75	
$NaC_2H_3O_2 \cdot 3H_2O$, g/L		15		
$Al_2(SO_4)_3 \cdot 18H_2O$, g/L	30			
$AlCl_3 \cdot 6H_2O$, g/L		20		
licorice, g/L	1			
glucose, g/L		120		
chloride, as Cl⁻, g/L				100–165
chelating agent[a]				45–90
pH				6.8–7.5

[a] Proprietaries.

ASTM specifications for zinc deposits on ferrous metals call for thicknesses of 5–25 μm of zinc, depending on the severity of the expected service (135).

Better appearance and performance can be obtained by chromate post-treatments of the zinc deposits (136); the mechanism of the formation of these conversion coatings has been investigated by many workers (137).

Zinc is deposited from aqueous solution only by virtue of a high hydrogen overvoltage since hydrogen would be preferentially deposited under equilibrium conditions. Some substrates with low hydrogen overvoltage, such as cast iron (because of the graphite inclusions), are electroplated with zinc only with great difficulty. These metals are often struck with tin or cadmium before zinc plating.

Alloy Plating. Alloy plating (138–139) is the simultaneous deposition of two or more metals. In general, electrodeposited alloys have the properties one would expect from a knowledge of the thermal equilibrium diagrams, but there are some notable exceptions. Although alloy plating has been a very attractive field for research (140), relatively few processes have been commercialized. As demand increases for deposits with properties unattainable in single metals, more processes are expected to emerge from the laboratory into production.

To the practicing plater the principal problems in alloy plating arise from the need for additional control. Since changes in the operating variables such as temperature, current density, and bath composition probably affect the two depositing metals in different degrees, it is usually necessary to maintain these variables within somewhat narrower ranges than in plating single metals. Except in the cases, usually confined to copper alloys, where an idea of the composition of the deposit can be gained from its color, it is also generally advisable to analyze the deposit to ensure that specified results are being obtained.

Complications in practical alloy plating increase almost exponentially as the number of codepositing metals is increased; thus very little has been accomplished with ternary or higher-order alloys. All of the practical processes, with minor proprietary exceptions, involve only two metals. Some alloys can be formed by depositing the two metals separately and subsequently interdiffusing them by the application of heat, but such processes are not alloy plating in the true sense and are not considered here. The following list of commercial alloy-plating processes covers those being carried out on a large scale.

Brass is no doubt the earliest commercially plated alloy. It is plated for decorative applications; its principal engineering use is to ensure good adhesion of rubber to steel wire in tire manufacture. All commercial baths are of the cyanide type. Table 24 shows representative formulas, but wide variations in composition are encountered. White brass has been used extensively as a nickel substitute during nickel shortages, and some use has survived in the manufacture of inexpensive toys and tubular furniture; it is also being used by one major automotive manufacturer.

Bronze, as the term is loosely used in the industry, may be an alloy of copper with zinc, cadmium, or tin, so long as the color matches that of wrought bronze. True bronze, however, is a copper–tin alloy, and bronzes in the 90:10 copper–tin range are plated fairly extensively for a variety of purposes. A typical formulation is a cyanide–stannate bath shown in Table 25. Anodes may be of copper, the tin content being maintained by additions of potassium stannate. When the tin content rises above about 18% the alloys are white. Proprietary processes are available.

Gold alloy plating has been mentioned under Gold.

Lead–tin alloy plating; see section on Lead.

Table 24. Brass Plating

	70/30	White
CuCN, g/L	53	16–20
Zn(CN)$_2$, g/L	30	35–40
total NaCN, g/L	90	52–60
free NaCN, g/L	7.5	4.5–6.5
Na$_2$CO$_3$, g/L	30	<37.5
NaOH, g/L		30–37.5
NH$_4$OH, g/L	5–13	
Na$_2$S, g/L		<0.23
Rochelle salt, g/L		1.5–2.2
pH	10.3–10.7	
temperature, °C	43–56	21–29
cathode current density, A/m^2	50–350	450
Cu in anodes, %	70	35

Table 25. Copper–Tin Alloy Plating

	Red bronze (10% Sn)
potassium stannate, g/L	60
potassium hydroxide, g/L	7.5
copper cyanide, g/L	40
potassium cyanide, g/L	90
free KCN, g/L	34
Rochelle salt[a]	45
brightener	if required, 2 mL/L
temperature, °C	60–70
cathode current density, A/m^2	200–1000
anodes	copper

[a] Proprietary addition agent at about 5 vol % gives improved efficiencies.

Nickel-base alloys are of interest for their magnetic properties in computer technology. Nickel–iron alloys are easily plated, and proprietary processes for plating them, containing up to about 35% iron, are marketed. Their only real advantage over nickel is cost. The alloy can be plated bright, and appears to be an acceptable substitute for bright nickel under chromium in applications where appearance is the principal requirement. For severe outdoor service the substitution has not been proved valid (141–144).

Tin–nickel is unusual; it has the composition 67:33 tin/nickel, corresponding to the compound NiSn, which is confirmed by x-ray data. As no such intermetallic compound appears on the thermal equilibrium diagram, it represents an exception to the statement that electrodeposited alloys correspond to thermally prepared ones. The deposit is semibright, fairly hard, but solderable, and has remarkable resistance to a long list of chemical reagents as well as to most atmospheric environments, particularly industrial. Its principal use has been in printed circuits and other communications equipment such as contacts. It can be used as a partial substitute for gold, under a thin coating of the latter, thus offering a significant cost saving (145–146). Tin–nickel is used to coat some watch parts and indoor appliances. The composition of the bath is shown in Table 26; other formulations have been published. An ASTM specification is available (147).

858 ELECTROPLATING

Table 26. Tin–Nickel Alloy Plating

stannous chloride, $SnCl_2$[a], g/L	50
nickel chloride, $NiCl_2.6H_2O$, g/L	300
ammonium bifluoride, NH_4HF_2, g/L	56
ammonium hydroxide, NH_4OH	to pH 2.0–2.5
temperature, °C	65–70
cathode current density, A/m²	100–300
anodes	nickel, bagged with nylon
anode current density, A/m²	to 500

[a] The dihydrate, tin crystals $SnCl_2.2H_2O$, may be used; 20% more is required.

Tin–cobalt alloys, often with a small amount of a third constituent, have been promoted (148–150) as substitutes for chromium plate in decorative applications. Advantages claimed are improved throwing power and current efficiency, and an appearance much like that of chromium. Corrosion performance has not been established.

Tin–zinc alloy plating has been suggested as a substitute for cadmium, since it exhibits to a considerable degree the corrosion resistance of zinc combined with the solderability of tin. Composition is usually 80:20 tin–zinc, deposited from a stannate–cyanide bath as shown in Table 27.

Tungsten has not been successfully plated from aqueous solutions, but its alloys with cobalt are easily plated and have interesting properties, which have not, however, led to significant commercial use.

Many rather exotic alloys have been reported, their sources being mainly the aerospace industries. Examples are too numerous to be cited in detail. They include, among many others, gold–molybdenum, chromium–molybdenum, silver–rhenium. Principal applications appear to be as solid film lubricants for high temperature service or for wear resistance under similar conditions.

Nonelectrolytic Plating Processes

There are many other ways than electroplating to deposit a coating of a metal on a substrate (see Film deposition techniques). Hot dipping, vacuum evaporation, chemical vapor deposition, and various aqueous processes not requiring current are some of the best developed. These last processes are sufficiently related to electroplating to be included in a discussion of the technology. Nonelectrolytic aqueous de-

Table 27. Tin–Zinc Alloy Plating

	Still	Barrel
potassium stannate, $K_2Sn(OH)_6$, g/L	120	95
zinc cyanide, $Zn(CN)_2$, g/L	11.3	15
potassium cyanide, KCN, g/L	30	34
free potassium hydroxide, KOH, g/L	7.5	11.3
temperature, °C	65 ± 2	
cathode current density, A/m²	100–800	
anodes	80:20 Sn–Zn; must be filmed	
anode current density, A/m²	150–250	

position includes immersion plating and chemical, autocatalytic, or what has come to be known as electroless plating (qv).

Immersion Plating. When the substrate metal is less noble than the plating metal, or can be made so by appropriate complexing agents in the solution, an immersion deposit may be formed, the prototype of which is the familiar $Fe + Cu^{2+} \rightarrow Cu + Fe^{2+}$. Many such immersion deposits are of no value because they are powdery or nonadherent, and in fact are to be avoided. On the other hand, some immersion processes have commercial use. Tin can be deposited on copper and its alloys from solutions containing a tin salt and a complexing agent for copper such as cyanide or thiourea. Two formulas are given in Table 28, and proprietary methods are available.

Tin is deposited on aluminum alloy pistons by immersion in alkaline stannate solutions. This process is practiced on a large scale by all the major automotive manufacturers. Copper–tin alloys are applied to steel wire (liquor finish) as a drawing lubricant and for color in such items as paper clips and hair pins.

Gold and some of the other precious metals are also frequently applied by immersion techniques.

In general, immersion deposits cannot be built up to thicknesses comparable to those obtainable by electrolytic methods, for as soon as the substrate is completely covered the reaction ceases. Nevertheless, some such deposits have appreciable thickness, entirely sufficient for the intended application. Immersion processes have the advantages of practically unlimited throwing power (limited only by access and renewal of the solution to the surface) and low capital cost, since they require no source of d-c power.

Autocatalytic Plating. Electroless plating (qv), as autocatalytic plating is better known, is defined (1) as deposition of a metallic coating by a controlled chemical reduction that is catalyzed by the metal or alloy being deposited. The term electroless plating has been carelessly applied to all processes that do not require an outside source of current, but the distinction between truly chemical reduction methods, ie, electroless plating, and electrochemical replacement such as immersion plating is a valuable one and should be maintained. The most widely used of the truly electroless processes is electroless nickel, in which nickel ions in the solution are reduced to the metal by a reductant (151). The deposition process is catalyzed by certain metallic surfaces, including the deposited metal itself; thus once initiated the deposition is autocatalytic. Fortunately, under proper control the reaction takes place only at the catalytic surface and not on the walls of the containing vessel or in the bulk of the solution.

The process has several advantages over electroplating: virtually unlimited throwing power, little or no excess deposit on high points, deposits of excellent chemical and physical properties, and the ability to coat surfaces such as those on the inside of tank cars which would be difficult or impossible to do with conventional electrolytic

Table 28. Immersion Tin on Copper

Alkaline		Acid	
potassium stannate, g/L	60	$SnCl_2$, g/L	4
KCN, g/L	120	thiourea, g/L	50
KOH, g/L	7.5	H_2SO_4, g/L	20
temperature, °C	30–65	temperature, °C	25
time, min	2–20	time, min	5–30

techniques. The principal disadvantage is high cost. The reducing agent, sodium hypophosphite, which is expensive, is consumed in substantial quantities, and the setup is often complicated, requiring exacting control. If ordinary electrolytic techniques will do the job, they are preferred, but electroless methods enlarge the range of possibilities.

Typical formulations contain nickel chloride, sodium hypophosphite, and one or more hydroxy acids such as lactic or glycolic. Operating temperatures range from 65 to 100°C. Several proprietary processes are available.

When sodium hypophosphite is used as the reductant, electroless nickel is not pure nickel but a nickel–phosphorus alloy containing 5–15% phosphorus.

Although sodium hypophosphite is the most widely used chemical reductant, solutions using sodium borohydride or amine boranes (especially dimethylaminoborane (DMAB)) are also employed, yielding deposits containing ca 0.3–10% boron, according to the bath used and the operating conditions.

Electroless nickel alloys have been tested against hard chromium for wear resistance on crank-shafts (152) and found to be somewhat superior. Hardness and wear resistance were reported (153–154) as well as the effect on corrosion resistance of the complexing agents used in the bath (155). Ternary and quaternary nickel alloys produced autocatalytically have been reported (156–157).

Next in importance to electroless nickel is electroless copper. This has been particularly useful in plating on nonmetallics and in printed circuitry. Typical baths contain copper sulfate, Rochelle salts, sodium hydroxide, formaldehyde as reducing agent, and other additives to increase plating rate and minimize spontaneous decomposition. As usual, many proprietaries are available. Electroless copper, widely used in printed circuitry (158) competes with electroless nickel in plating in plastics.

Electroless processes have been reported for gold, rhodium, palladium, cobalt, and silver (159).

Postplating Treatments

Postplating treatments (see Metal surface treatment) include chromate conversion coatings for zinc and cadmium, phosphate treatments for iron, and various passivating and brightening dips. In most general use are the chromate conversion coatings, which are designed to convert the surface from the naturally occurring oxide to one containing hexavalent chromium. Such treatments enhance the corrosion-protective value of the deposit and inhibit the formation of bulky and unsightly corrosion products; they also reduce the tendency of the surface to finger-stain on handling. Chromate conversion coatings impart various colors to the deposits, depending on details of the treating solutions; some of them can be dyed. Although formulas have been published, most chromate treatment is by proprietary processes.

Phosphate treatments are of three general types: iron, zinc, and manganese; the first two as bases for subsequent painting, the third for lubricity.

Specifications and Tests

As the capabilities of electroplating have expanded, there has been a great increase in the use of specifications. Almost universally, the electroplater is instructed to apply deposits of a certain minimum thickness. Other properties of the deposit often specified as well include hardness, tensile strength, corrosion resistance, adhesion, and others. Specifications call for tests to determine compliance. Many of these tests are highly specific to particular metals or are slight modifications of standard metallurgical procedures, and are not discussed here. Probably the three most widely useful tests are those for thickness of deposit, for corrosion resistance, and for adhesion.

ASTM tests relating to the testing of metallic coatings are listed in Table 29. The subject is considered in some detail in refs. 160–161.

Thickness. The thickness of a deposit on parts of any but the very simplest shape is not uniform. Most specifications call for a minimum thickness on significant surfaces, ie, either by direct reference in the specification or by the general rule that any functional or visible surface is significant.

As Table 29 indicates, there are many standard methods of thickness testing, and there are several others that have not been standardized but are useful in individual cases. Some of these are destructive of the part itself or of the coating; some are nondestructive (162). Perhaps the most widely used tests are the magnetic, coulometric, and beta backscatter. Microscopic examination of a cross section was formerly considered a referee method, but has somewhat gone out of favor since tests have shown that although a given laboratory can usually reproduce its own results, agreement among different laboratories is often poor. In addition, the method is time-consuming, completely destructive, and requires much more experience and expertise than most of the others. Many of the tests listed are embodied in commercially available instruments (see Nondestructive testing).

Table 29. ASTM Tests for Deposit Properties

ASTM designation	Abbreviated title	Type[a]
	Thickness	
B 555-75	Dropping Test	SD
B 556-71	Spot Test for Thin Chromium	SD
B 567-78	Beta Backscatter	ND
B 568-72 (1978)	X-Ray Spectrometry	ND
B 529-70	Eddy Current	ND[b]
B 530-75	Magnetic: Nickel Coatings	ND
B 499-75	Magnetic: Nonmagnetic Coatings, Magnetic Basis Metals	ND
B 487-75	Microscopic Cross Section	D
B 244-68 (1972)	Anodic Coatings on Aluminum, Eddy Current	ND
B 504-70 (1976)	Coulometric	SD
B 588-75	Double-Beam Interference Microscope	SD
	Ductility	
B 489-68 (1973)	Bend Test	D
B 490-68 (1975)	Micrometer Bend Test	D
	Adhesion	
B 571-72	Adhesion of Metallic Coatings	D
	Corrosion (accelerated)	
B 368-68 (1978)	Copper-Accelerated Acetic Acid Salt Spray (CASS)	D
B 380-65 (1978)	Corrodkote	D

[a] D = destructive; SD = semidestructive (destroys coating but not the part); ND = nondestructive.
[b] In process of revision.

Corrosion Resistance. The final criterion of corrosion resistance is actual service, but accelerated tests are needed both for predicting service life and for acceptance or rejection in production. Many such tests have been devised, but the correlation of some tests with service is questionable. No accelerated corrosion test can exactly duplicate the corrosive process that takes place during the life of the part while merely speeding it up. Correlation with service experience must be proved if the accelerated test is to have meaning.

Although the well-known neutral salt spray (fog) test (ASTM B 117) is widely used and specified, it is not a good measure of the corrosion resistance of plated coatings. It may, however, be useful as a quality control measure. The CASS and Corrodkote procedures are presently preferred, but their results must also be interpreted cautiously in comparing different coating systems. They have correlated fairly well with experience for nickel–chromium or copper–nickel–chromium deposits, but there are few data for other electrodeposits.

Adhesion. A plated coating must adhere to the substrate if it is to perform its function. Once a process has been set up and found to be satisfactory, tests for adhesion become a quality-control function, monitoring the proper control of the process. Quantitative measurement of the adhesive force is possible but is more a research tool than a routine test procedure (163); most shop tests are of a go-no-go type, and are crude but apparently adequate. They involve heating, bending, sawing, filing, hammering, or otherwise abusing the part and observing the nature of the break when failure occurs. Crude and unscientific as these procedures appear, they are useful in judging the quality of plated parts (164).

Physical and mechanical properties often specified, especially in engineering applications, include hardness, stress, and ductility. Tests for solderability (165), surface contour, and other properties have been devised.

Applications

Applications for plating are classified according to the principal function of the plate: thus the plating may be applied mainly for (1) appearance; (2) protection; (3) special surface properties; or (4) engineering or mechanical properties. These distinctions are not clear-cut: a purely decorative deposit must be, to some extent at least, protective as well, and other overlaps among the four named functions are common. Nevertheless the classification is convenient.

Decorative Plating. The most familiar plating applied for appearance is the so-called chromium plate (commonly but improperly referred to as chrome), which almost always consists of a composite coating having a very thin deposit of chromium as the final or topcoat. The undercoats, usually nickel or copper followed by nickel, have been relied upon for protective value, but it has been shown that the type and thickness of chromium play a very important role. Other metals having decorative applications include silver, gold, brass, bronze, nickel, copper, and rhodium; and occasionally, for special effects, lead, tin, and other metals may be used.

The list of articles plated for appearance is almost endless: automotive and aircraft parts, refrigerator hardware, electrical appliances, plumbing fixtures, office furniture, photographic equipment, golf clubs, firearms, handbag frames, pens and pencils, costume jewelry, office equipment—the list could be extended. Usually, protection of the basis metal is also involved.

Plating for Protection. Steel (qv), the most common structural metal, must be protected against rusting in almost all its applications. Paints and organic coatings are widely used for this purpose, as are zinc and cadmium electroplates. Both of these deposits protect the basis steel galvanically. That is, they corrode sacrificially, thus preventing electrochemical attack on the steel even when pores or scratches permit penetration of the corrodent (see Corrosion). Although both zinc and cadmium can be plated bright or can be brightened after plating by simple chemical treatment, this bright appearance does not last long in use, and neither metal would usually be chosen for applications where retention of pleasing appearance is important. So-called clear dip-conversion coatings make the appearance of zinc and cadmium electroplates sufficiently attractive for cheap hardware and related items.

Zinc can be applied to steel by methods other than electroplating, including hot galvanizing, Sherardizing, and Peen plating (see Metallic coatings). Judged by tonnage of product hot galvanizing is far more important than electroplating (often called electrogalvanizing). The protective value of zinc deposits depends almost entirely upon their thickness rather than on the method of application, but electroplating has some advantages in not forming a brittle alloy layer and in the absence of dross formation.

Zinc is the most economical plate that can be applied to steel for rust prevention. Thickness for thickness it is superior to cadmium in typical industrial atmospheres. Although cadmium is better in marine conditions by the thickness criterion, it is so much more expensive (about ten times recently) that the economics strongly favor zinc. Thus cadmium is used primarily in applications where two disadvantages of zinc are of importance: zinc forms bulky corrosion products that may interfere with the proper functioning of moving parts, and it is not easily soldered by usual techniques. Cadmium is superior to zinc in appearance, and its plating process is somewhat more easily controlled.

Because cadmium is extremely toxic to animal and human life, its concentration in effluents is rigidly controlled by EPA regulations; furthermore, the only commercially practical bath appears to be cyanide-based (although noncyanide formulations have been proposed, see under Cadmium Plating), and the tendency in the trade is to avoid cyanide systems for the same reason, even though methods for cyanide destruction are well developed. Several alternatives to cadmium plating have been proposed, including a tin–zinc alloy, a zinc–nickel alloy, and others (13,166). Zinc is used where possible in spite of its inferior performance in some applications. There seems little doubt that the use of cadmium in electroplating will decrease (167).

The use of tin to protect steel forms the basis of the very important tinplate and tin can industries. Tin does not protect steel galvanically under normal outdoor conditions as zinc and cadmium do, but under the special conditions inside the hermetically sealed can, in contact with the organic acids of foods, tin becomes sacrificially protective. An important property of tin in this use is its nontoxic nature. The manufacture of electrolytic tinplate represents the largest single application of electroplating. In 1977 United States production of electrolytic tinplate was about 4,228,000 t (10); the older hot-dip process is all but obsolete, since electrolytic methods permit the production of much lighter coatings of tin: a requirement if tinplate is to continue to compete economically with other container materials. Modern techniques, including the improvement of interior lacquers (enamels in the tinplate industry) have enabled these lighter coatings to perform as well as the former heavier ones. Other metals, such as copper, nickel, and chromium, are applied partly for protection and partly for appearance.

Special Surface Effects. In this category fall such uses of electroplating as rhodium plating for reflectance, silver and rhodium plating for electrical properties, and gold plating for good electrical contact (see Electrical connectors). One of the most important surface effects is increasing solderability by plating with tin. Gold, cadmium, or some tin alloys such as tin–lead are also used for solderability. Modern production methods in the electronic industry demand that components be practically instantaneously wetted by soft solders using only noncorrosive fluxes, and that the soldered joints be sound; further, they require that parts remain solderable for a reasonable period after fabrication and plating in order to simplify inventory problems. Copper and brass, though easily solderable when clean, quickly tarnish and acquire films not readily wetted by solder; plating with tin or other metals mentioned largely overcomes this problem (see Solders and brazing alloys).

As industry becomes more sophisticated in its use of materials, it can be expected that there will be fewer all-purpose metals and more application of specific materials for specific properties. This suggests that alloy plating will increase in usefulness, as special alloys are developed for special applications. This trend is already evident in the burgeoning development of magnetic alloys, which are applied not only by electroplating but also by vacuum evaporation and other techniques, for computer memories and similar devices.

Engineering Applications. Electroplates may be applied for their mechanical, physical, or chemical properties rather than primarily to protect or beautify the substrate. Such applications are generally termed engineering, and usually involve the application of much heavier deposits than other functions of plating. Metals of most interest in this connection are nickel and chromium; iron, copper, lead, silver, gold, and tin also have engineering uses.

The mechanical properties of nickel, ie, hardness, tensile strength, and stress, can be varied by controlling the conditions of deposition. These properties, combined with the generally good resistance of nickel to many corrosive chemicals, make it useful in a host of industrial applications.

The hardness and favorable frictional characteristics of chromium account for its use on cutting tools, files, dies, rolls for paper-making machinery, etc. The term hard chromium is used in the trade for such applications, although this chromium is not necessarily any harder than that applied for decorative purposes; heavy or thick chromium would be more descriptive. Chromium can also be plated with a special porous surface condition in which form it holds oil films well and is useful as a cylinder liner for internal combustion engines.

Worn or mismachined parts can be built up to proper dimensions by nickel or chromium plating; much expensive equipment is thus salvaged at a small fraction of the cost of replacing it.

Bearing properties can often be greatly improved by electroplating; porous chromium serves this purpose. Aluminum alloy pistons for internal combustion engines are tin-plated to prevent scoring of the cylinder walls by the abrasive aluminum oxide during the running-in period. Lead and its alloys, silver, and indium also find use in this general area (see Bearing materials).

Electroplates are applied for temporary use in metal treatments: steel parts are copper-plated in selected areas to prevent carburization at specific locations; tin–copper alloys are similarly used as a stop-off in the nitriding of steel.

Electroforming. Electroforming is a special engineering use of electroplating. In electroforming, the electrodeposited metal is usually much thicker than for other applications, up to 0.6 cm or more. In this manufacturing technique metal is deposited upon a form or mandrel, which is later separated from the deposit. Although this would seem to be a very expensive method for producing articles, it has been adapted to a few high-production items. Its principal applications arise from its unique capabilities: complicated shapes can be duplicated exactly with little or no scrap from machining; physical and mechanical properties of the electroformed metals can be closely controlled; and surface contour of the master is exactly reproduced down to the finest detail. This is illustrated by its use in the production of masters, positives, and stampers for phonograph records.

Mandrels for electroforming are of two types: temporary, those used to produce one electroform and then removed by melting or dissolving away from the electroform; and permanent, those that are removed from the electroform intact. Choice between them depends on many factors, including the shape of the electroform and the number of pieces to be produced. Temporary mandrels may be of low melting alloys or waxes that can be melted out, metals like aluminum or zinc that can be dissolved away chemically, or plaster and similar materials that can be easily broken away. Permanent mandrels are usually of stainless steel, chromium-plated steel, nickel alloys, or aluminum.

Success in electroforming depends primarily on two factors: extreme care in the plating operation, and skill and foresight in the design of mandrels to ensure their successful removal from the form (20–22,168).

BIBLIOGRAPHY

"Electroplating" in *ECT* 1st ed., Vol. 5, pp. 611–646, by C. L. Faust and W. H. Safranek, Battelle Memorial Institute; "Electroplating" in *ECT* 2nd ed., Vol. 8, pp. 36–74, by F. A. Lowenheim, M & T Chemicals, Inc.

1. *ASTM B 374-75: Definitions of Terms Relating to Electroplating.*
2. Brit. Pat. 8,447 (1840), G. Elkington and H. Elkington.
3. G. B. Hogaboom, *Met. Finish.* **51**(1), 72 (1953); F. A. Lowenheim, *Met. Finish.* **51**(1), 78 (1953).
4. C. F. Coombs, ed., *Printed Circuits Handbook,* McGraw-Hill Book Co., New York, 1967; 2nd ed. 1979.
5. *AES Plating in the Electronics Industry Symposia:* (a) First: Newark, N.J., 1966; (b) Second: Boston, Mass., 1969; (c) Third: Palo Alto, Calif., 1971; (d) Fourth: Indianapolis, Ind., 1973; (e) Fifth: New York, 1975; (f) Sixth: Chicago, Ill., 1977.
6. B. L. McKinney and C. L. Faust, *J. Electrochem. Soc.* **124**, 380C (1977).
7. W. B. Harding in F. A. Lowenheim, ed., *Modern Electroplating,* 3rd ed., John Wiley & Sons, Inc., New York, 1974, p. 63.
8. G. W. Mellors and S. Senderoff, *Plating* **51,** 972 (1964).
9. S. Senderoff in ref. 7, p. 473.
10. Private communication, Tin Research Institute, Columbus, Ohio.
11. *Minerals Yearbook,* Vol. 1, U.S. Bureau of Mines, Washington, D.C., 1975.
12. *National Academy of Sciences: Contingency Plans for Chromium Utilization, NMAB-335,* Printing and Publishing Office, National Academy of Sciences, 2101 Constitution Ave., Washington, D.C. 20418.
13. *Plating Surf. Finish.* **64**(11), 8 (1977).
14. H. J. Read, ed., *Hydrogen Embrittlement in Metal Finishing,* Reinhold Publishing Co., New York, 1961.
15. W. Beck, E. J. Jankowsky, and P. Fischer, *Hydrogen Stress Cracking of High Strength Steels, NADC-MA-7140,* Naval Air Development Center, Warminster, Pa., 1971, pp. 143–151.

16. N. C. Parthasaradhy, *Plating* **61,** 57 (1974).
17. *ASTM Spec. Tech. Pub. 543, Hydrogen Embrittlement Testing,* 1974.
18. J. C. Turn and E. L. Owen, *Plating* **61,** 1015 (1974).
19. *ASTM B 545-72, Standard Spec. for Electrodeposited Coatings of Tin.*
20. *AES/ASTM Electroforming Symposium, Atlanta, Ga., 1974.*
21. *ASTM B 450-67 (1972): Recommended Practice for Engineering Design of Electroformed Articles.*
22. *ASTM B 431-69: Recommended Practice for Processing of Mandrels for Electroforming.*
23. P. Spiro, *Electroforming,* 2nd ed., International Publications Services, New York, 1971.
24. *ASTM B 322-68 (1973), Recommended Practice for Cleaning Metals Prior to Electroplating.*
25. F. A. Lowenheim, *Electroplating: Fundamentals of Surface Finishing,* McGraw-Hill Book Co., New York, 1978.
26. *Ibid.,* p. 72.
27. H. B. Linford and E. B. Saubestre, *Plating* **38,** 713, 847 (1951); **40,** 489, 633 (1953).
28 J. B. Kushner, *Plating Surf. Finish.* **64**(8), 24 (1977).
29. J. B. Kushner in N. Hall, ed., *Metal Finishing Guidebook Directory,* Metal and Plastics Publications, Inc., Hackensack, N.J., annual.
30. Ref. 25, pp. 99–112.
31. *ASTM B 253-79, Recommended Practice for Preparation and Electroplating on Aluminum Alloys by the Zincate Process.*
32. *Plating Surf. Finish.* **64**(12), 16 (1977).
33. G. A. DiBari, *Plating Surf. Finish.* **64**(5), 58 (1977).
34. J. C. Jongkind, *Plating Surf. Finish.* **62,** 1135 (1975).
35. D. S. Lashmore, *Plating Surf. Finish.* **65**(4), 44 (1978).
36. *ASTM B 480-68 (1975), Recommended Practice for Preparation of Magnesium and Magnesium Alloys for Electroplating.*
37. *ANSI/ASTM B 252-69 (1977), Recommended Practice for Preparation of Zinc Alloy Die Castings for Electroplating.*
38. *ASTM B 629-77, Recommended Practice for Preparation of Molybdenum and Molybdenum Alloys for Electroplating.*
39. *ASTM B 481-68 (1973), Recommended Practice for Preparation of Titanium and Titanium Alloys for Electroplating;* E. W. Turns, J. W. Browning, and R. L. Jones, *Plating Surf. Finish.* **62,** 443 (1975).
40. *ASTM B 482-68 (1973), Recommended Practice for Preparation of Tungsten and Tungsten Alloys for Electroplating.*
41. J. G. Beach and C. L. Faust in ref. 7, pp. 618–635.
42. *ASTM B 183-79, Recommended Practice for Preparation of Low-Carbon Steel for Electroplating.*
43. *ASTM B 242-54 (1971), Recommended Practice for Preparation of High-carbon Steel for Electroplating.*
44. *ASTM B 320-60 (1971), Recommended Practice for Preparation of Iron Castings for Electroplating.*
45. *ASTM B 281-58 (1972), Recommended Practice for Preparation of Copper and Copper base Alloys for Electroplating.*
46. *ASTM B 254-79, Recommended Practice for Preparation and Electroplating on Stainless Steel.*
47. *ASTM B 558-79, Recommended Practice for Preparation of Nickel Alloys for Electroplating.*
48. *ASTM B 319-79, Recommended Practice for Preparation of Lead and Lead Alloys for Electroplating.*
49. E. B. Saubestre in ref. 7, p. 636.
50. U.S. Pat. 3,993,799 (Nov. 23, 1976), N. Feldstein (to Surface Technology, Inc.).
51. N. Feldstein, *Plating* **60,** 611 (1973).
52. G. A. DiBari, *Plating* **60,** 1252 (1973).
53. R. L. Cohen and R. L. Meek, *Plating Surf. Finish.* **63**(6), 47 (1976); R. L. Cohen, R. L. Meek, and K. W. West, *Plating Surf. Finish.* **63**(5), 52 (1976).
54. F. A. Lowenheim, *Metal Coating of Plastics,* Noyes Data Corp., Park Ridge, N.J., 1970.
55. W. Goldie, *Metallic Coating of Plastics,* Vols. 1 and 2, Electrochemical Publications Ltd., Ayr, Scotland, 1968–1969.

56. G. D. R. Jarrett and co-workers, *Plating on Plastics,* 2nd ed., International Publications Services, New York, 1971.
57. W. E. Hoare, E. S. Hedges, and B. T. K. Barry, *The Technology of Tinplate,* St. Martin's Press, New York, 1965.
58. J. B. Long, *Plating* **61,** 918 (1974).
59. *AES Continuous Plating Seminar,* Pittsburgh, Pa., 1974.
60. *First Annual Conference on Advanced Pollution Control for the Metal Finishing Industry, Lake Buena Vista, Fla., 1978, EPA 60018-78-010,* National Technical Information Service, Springfield, Va.
61. *Plating Surf. Finish.* **65**(2), 8 (1978); **65**(3), 25 (1978).
62. F. A. Stewart, *Met. Finish.* **76**(3), 67 (1978).
63. *Prod. Finish. (Cincinnati)* **42,** 70 (Mar. 1978).
64. *Prod. Finish. (Cincinnati)* **41,** 78 (Aug. 1977).
65. D. G. Foulke, *Plating Surf. Finish.* **62,** 980 (1975).
66. M. R. Hills, *Trans. Inst. Met. Finish.* **53,** 65 (1975).
67. D-T. Chin and B. Eckert, *Plating Surf. Finish.* **63**(10), 38 (1976).
68. I. Kennedy and S. Das Gupta in ref. 60, p. 49; *Finishers' Mgmt.* **23**(7), 7 (1978).
69. B. C. Lawes, L. B. Fournier, and O. B. Mathre, *Plating* **60,** 902 (1973).
70. F. L. Moore and W. S. Groenier, *Plating Surf. Finish.* **63**(8), 26 (1976).
71. D. C. Hoffmann, *Plating* **60,** 157 (1973).
72. M. Jola, *Plating Surf. Finish.* **63**(9), 42 (1976).
73. A. A. Cochran and L. C. George, *Plating Surf. Finish.* **63**(7), 38 (1976).
74. K. Y. Yost and D. R. Masarik, *Plating Surf. Finish.* **64**(1), 35 (1977).
75. P. S. Minor and P. A. Militello, *Plating Surf. Finish.* **64**(12), 20 (1977).
76. A. K. Robinson in ref. 60, p. 59.
77. M. C. Scott, *Prod. Finish. (Cincinnati)* **42**(11), 64 (Aug. 1978).
78. A. Golomb, *Plating* **60,** 482 (1973).
79. R. G. Donnelly and co-workers, *Plating* **61,** 432 (1974).
80. A. Golomb, *Plating* **61,** 931 (1974).
81. *Plating Surf. Finish.* **63**(4), 42 (1976).
82. R. J. Peterson and K. E. Cobian, *Plating Surf. Finish.* **63**(6), 51 (1976).
83. K. McNulty, P. R. Hoover, and R. L. Goldsmith in ref. 60, p. 66.
84. T. J. Kolesar, *Plating* **61,** 571 (1974).
85. H. S. Hartley in ref. 60, p. 86.
86. *Prod. Finish. (Cincinnati)* **42**(11), 42 (Aug. 1978).
87. R. Raman and E. L. Karlson, *Plating Surf. Finish.* **64**(8), 40 (1977).
88. A. R. Yeats, *Plating Surf. Finish.* **64**(4), 32, (1978).
89. H. S. Skrovonek and M. K. Stinson, *Plating Surf. Finish.* **64**(10), 32; (11), 24 (1977).
90. F. A. Lowenheim, ed., *Modern Electroplating,* 3rd ed., John Wiley & Sons, Inc., New York, 1974.
91. A. K. Graham, ed., *Electroplating Engineering Handbook,* 3rd ed., Van Nostrand Reinhold Co., New York, 1971.
92. N. Hall, ed., *Metal Finishing Guidebook Directory,* Metal and Plastics Publications, Inc., Hackensack, N.J., annual.
93. F. Mansfield and L. P. Street, *Plating* **61,** 850 (1974).
94. G. Dubpernell, *Electrodeposition of Chromium from Chromic Acid Baths,* Pergamon Press, New York, 1977.
95. *ASTM B 456-79, Specification for Electrodeposited Coatings of Nickel–Chromium.*
96. *ASTM B 177-68 (1978), Recommended Practice for Chromium Plating on Steel for Engineering Use.*
97. J. D. Greenwood, *Hard Chromium Plating,* 2nd ed., International Publications Services, New York, 1971.
98. G. Dubpernell in ref. 7, pp. 87–151.
99. V. E. Carter and I. R. A. Christie, *Trans. Inst. Met. Finish.* **51,** 41 (1973).
100. C. Barnes, J. J. B. Ward, and J. R. House, *Trans. Inst. Met. Finish.* **55,** 73 (1977).
101. L. Gianelos, *Plating Surf. Finish.* **65**(5), 17 (1978) (abstract only).
102. G. T. Robinson, *Prod. Finish. (Cincinnati)* **42**(11), 54 (Aug. 1978).
103. R. R. Sowell and D. M. Mattox, *Plating Surf. Finish.* **65**(1), 50 (1978).
104. R. Walker and B. Cruise, *Met. Fin.* **76**(5), 25; (6), 45 (1978).

868 ELECTROPLATING

105. C. C. Wan, H. Y. Cheh, and H. B. Linford, *Plating* **61**, 559 (1974).
106. H. Y. Cheh, H. B. Linford, and C. C. Wan, *Plating Surf. Finish.* **64**(5), 66 (1977).
107. *ASTM B 503-69: Recommended Practice for Use of Copper and Nickel Electroplating Solutions for Electroforming.*
108. W. H. Safranek, *The Properties of Electrodeposited Metals and Alloys,* American Elsevier Publishing Co., New York, 1974.
109. M. Antler and M. H. Drozdowicz, *Plating Surf. Finish.* **63**(9), 19 (1976).
110. *ASTM B 488-71, Specification for Electrodeposited Coatings of Gold for Engineering Use.*
111. F. H. Reid and W. Goldie, eds., *Gold Plating Technology,* Electrochemical Publications, Inc., Ayr, Scotland, 1974.
112. O. J. Klingenmaier, *Plating* **61**, 741 (1974).
113. *ASTM B 200-76, Specification for Electrodeposited Coatings of Lead and Lead–Tin Alloys on Steel and Ferrous Alloys.*
114. *ASTM B 579-73, Specification for Electrodeposited Coatings of Tin–Lead Alloy (Solder Plate).*
115. J. K. Dennis and T. E. Such, *Nickel and Chromium Plating,* Halsted Press, a division of John Wiley & Sons, Inc., New York, 1972; R. Brugger, *Nickel Plating,* International Publications Services, New York, 1970.
116. *ASTM B 634-78, Specification for Electrodeposited Coatings of Rhodium for Engineering Use.*
117. R. M. Skomoroski and co-workers, *Plating* **60**, 1115 (1973).
118. M. Kruger and D. R. Gabe, *Trans. Inst. Metal Finish.* **54**, 127 (1976).
119. J. McCaskie in ref. 5, (e), p. 5.
120. D. T. Napp in ref. 5 (e), p. 28.
121. E. A. Parker in ref. 92, pp. 261, 312 (1978 ed.).
122. J. N. Crosby, *Trans. Inst. Met. Finish.* **54**, 75 (1976).
123. W. B. Harding, *Plating Surf. Finish.* **64**(9), 48 (1977).
124. D. R. Gabe and P. Sripatr, *Trans. Inst. Met. Finish.* **51**, 141 (1973).
125. Ref. 7, p. 398.
126. W. G. Bader and R. G. Baker, *Plating* **60**, 242 (1973); D. Bernier, *Plating* **61**, 842 (1974).
127. A. Jafri, *Plating* **60**, 358 (1973).
128. S. C. Britton, *Trans. Inst. Metal Finish.* **52**, 95 (1974).
129. L. Zakraysek, *Plating Surf. Finish.* **64**(3), 38 (1977).
130. H. G. Todt, *Trans. Inst. Metal Finish.* **51**, 91 (1973).
131. M. J. Reidt and C. A. Boose, *Electrodeposition Surf. Treat.* **1**, 269 (1973).
132. H. G. Creutz and S. Martin, *Plating Surf. Finish.* **62**, 681 (1975); H. Geduld, *Plating Surf. Finish.* **62**, 687 (1975).
133. H. S. Schneider, *Plating Surf. Finish.* **64**(6), 52 (1977).
134. *AES Zinc Symposium, Syracuse, N.Y., October 1977, Prod. Finish. (Cincinnati)* **42**, 42 (Dec. 1977).
135. *ASTM B 633-78, Specification for Electrodeposited Coatings of Zinc on Iron and Steel.*
136. *ASTM B 201-68, Recommended Practice for Testing Chromate Coatings on Zinc and Cadmium Surfaces.*
137. L. F. G. Williams, *Surface Technol.* **4**, 355 (1976); **7**, 113 (1978).
138. A. Brenner, *Electrodeposition of Alloys,* Academic Press, Inc., New York, 1963.
139. A. Krohn and C. W. Bohn, *Electrodeposition Surf. Treat.* **1**, 199 (1973); updates 138.
140. E. Raub, *Plating Surf. Finish.* **63**(2), 29; (3), 30 (1976).
141. R. J. Clauss and R. A. Tremmel, *Plating* **60**, 811 (1973).
142. H. Chessin, E. J. Seyb, and P. D. Walker, *Plating Surf. Finish.* **63**(12), 32 (1976).
143. H. W. Reiner, *Plating Surf. Finish.* **63**(8), 13 (1976).
144. R. J. Clauss, R. A. Tremmel, and R. W. Klein, *Trans. Inst. Met. Finish.* **53**, 22 (1975).
145. M. Antler and co-workers, *Plating Surf. Finish.* **63**(7), 30 (1976).
146. M. Antler and M. H. Drozdowicz, *Plating Surf. Finish.* **65**(4), 39 (1978).
147. *ASTM B 605-75, Specification for Electrodeposited Coatings of Tin–Nickel Alloy.*
148. J. Hyner, *Plating Surf. Finish.* **64**(2), 32 (1977).
149. U.S. Pat. 4,035,249 (July 12, 1977), W. J. Wieczerniak.
150. U.S. Pat. 4,029,556 (June 14, 1977), E. Monaco and P. F. Monaco.
151. F. Pearlstein in ref. 7, pp. 710–747.
152. N. A. Tope, E. A. Baker, and B. C. Jackson, *Plating Surf. Finish.* **63**(10), 30 (1976).
153. K. Parker, *Plating* **61**, 834 (1974).

154. E. Johnson and F. Ogburn, *Surface Technol.* **4,** 161 (1976).
155. G. O. Mallory, *Plating* **61,** 1005 (1974).
156. G. O. Mallory, *Trans. Inst. Met. Finish.* **52,** 156 (1974).
157. G. O. Mallory, *Plating Surf. Finish.* **63**(6), 34 (1976).
158. C. J. Barrett, R. D. Rust, and R. J. Rhoda, *Plating Surf. Finish.* **65**(7), 36 (1978).
159. F. Pearlstein and R. F. Weightman, *Plating* **61,** 154 (1974).
160. R. Sard, H. Leidheiser Jr., and F. Ogburn, eds., *Properties of Electrodeposits, Their Measurement and Significance,* Electrochemical Society, Princeton, N.J., 1975.
161. A. Kutzelnigg, *Testing Metallic Coatings,* Robert Draper Ltd., Teddington, Eng., 1963.
162. *First AES Symposium on Thickness Testing of Surface Finishing,* AES, New York, 1978.
163. J. W. Dini and H. R. Johnson, *Met. Finish.* **75**(3), 42; (4), 48 (1977).
164. K. L. Mittal, ed., *ASTM STP 640, Adhesion Measurement of Thin Films, Thick Films, and Bulk Coatings,* 1978.
165. J. B. Long in ref. 160, p. 102; ref. 19, Appendix.
166. *Prod. Finish. (Cincinnati)* **42,** 66 (Dec. 1977).
167. P. Baeyens, *Galvanotechnik* **68,** 590 (1977).
168. Ref. 25, pp. 426–441.

General References

References 25, 90, 91, and 92 are general references.
ASTM specifications are published by that body in Part 9 of the *Annual Book of Standards,* available about July of each year; the latest revision should always be consulted; headquarters in Philadelphia, Pa.
The journal *Plating and Surface Finishing* is the new name of *Plating,* published by the American Electroplaters' Society (AES), Winter Park, Fla., which also publishes various symposia volumes.

FREDERICK A. LOWENHEIM
Consultant

ELECTROSTATIC PRECIPITATION. See Air pollution control methods.

ELECTROSTATIC SEALING

Conventionally, glass is sealed at a temperature that lowers its viscosity to <1 kPa·s (10^4 P) and permits the glass to flow with relative ease. When it contacts a suitably oxidized metal, the glass wets the surface and reacts with the metal oxide. For most glasses of interest, the appropriate sealing temperature is 800–1300°C (1). However, in many applications a substantially lower sealing temperature is desirable either to minimize distortion or because of the temperature limitations of the assembly. The former is of prime importance when the glass has an optical function or when it serves as a structural element that must maintain precision tolerances. The latter limitation frequently occurs in the assembly or hermetic packaging of temperature-sensitive electronic components.

In response to these requirements a sealing technique, often called field-assisted bonding or anodic bonding, has been developed in which parts are soldered together by means of a low temperature glass frit. Glass particles, suspended in a binder, are placed between the parts that are to be sealed. During an extended heating cycle the binder evaporates, and the particles fuse at a relatively low temperature so as to form an intermediate glass layer which, as in the conventional process, wets and reacts with the sealing surfaces. Although this technology represents a breakthrough in low temperature glass sealing, it has some drawbacks: it requires a long, carefully controlled heating cycle; for some geometries the complete removal of the binder presents difficulties; in a few applications the precision of dimensional tolerances is compromised during the fusing of the frit; and the thermal expansion characteristic of the fritted glass often fails to match those of the parts that are to be bonded (see also Glass; Laminated materials, glass).

In contrast to the above sealing methods, the technique described in this article is unusual in that the glass typically remains at a viscosity >1 TPa·s (10^{13} P) throughout the sealing process, and in that no foreign materials such as binders, fluxes, or adhesives are introduced. As a result, this technique combines the low temperature advantages of the frit method with the cleanliness of conventional glass sealing. The name electrostatic sealing (also field-assisted or anodic bonding) is derived from the central feature of the technique: the application of a voltage across the bonding surfaces to establish a high electric field at the interface (2).

Theory

Electrostatic sealing may be viewed as a two-step process: (1) electrostatic forces at the glass-metal interface bring the parts into intimate physical contact; and (2) bond formation between the glass and the metal. Although (1) is readily explained in terms of basic electrostatic theory, the mechanisms in (2) are not well understood at this time.

The first step is essential because even well-polished parts, when placed on top of each other, contact only at a few points. Over most of the interface they are separated by a gap of the order of 1 μm which gives rise to characteristic interference fringes. A voltage applied between the glass and the metal generates a strong electrostatic field in the gap. The corresponding electrostatic force tends to pull the parts into intimate contact. It should be noted that at low temperatures extremely high voltages would be required since most of the applied voltage would drop across the insulating glass plate. However, near the annealing point, the glasses under discussion are essentially solid electrolytes with sodium ions acting as the mobile charge carriers. As a result, the electrical resistance of the glass plate is small compared to that of the air gap and almost all of the applied voltage appears across the gap. If the gap is sufficiently small, a high electric field can be obtained at relatively low voltages. Usually the gap width at the interface is nonuniform. When the voltage is applied, a corresponding nonuniformity occurs in the electrostatic force; this force is greatest at smallest gap widths. Therefore, the glass and the metal usually do not contact simultaneously over the whole interface. Rather, contacting starts in one or two small regions and gradually spreads over the interface. This action is believed to be an advantage in that it minimizes the chance of trapping air bubbles at the glass–metal interface.

After the parts come into contact, the positive ions drift under the electric field from the glass–metal interface to the negative electrode where they are neutralized. As a result, a thin surface region in the glass adjacent to the positive metal face becomes depleted of sodium ions and tends to acquire a negative charge. Depending on the properties of the specific metal and glass, and perhaps on the temperature, a number of different processes can now take place:

(1) The applied voltage may pull into the glass positive metal ions which will neutralize the negative charge in the depleted glass layer. If the metal ions have a mobility comparable with that of sodium ions, the resistivity of the depleted layer does not differ significantly from the resistivity of the bulk of the glass. Hence, the high electric field at the glass–metal interface is reduced by orders of magnitude. The absence of a strong electrostatic attraction then permits the parts to separate. This situation, exemplified by the silver–Pyrex system, is not conducive to bonding (3).

(2) As in case (1), positive metal ions drift into the glass and tend to neutralize the negative charge in the depleted glass layer. However, the ions have a much lower mobility than sodium ions so that the resistivity in the depleted layer is correspondingly higher than in the bulk of the glass. Under these conditions, the electric field at the glass–metal interface remains high, and intimate contact is maintained between the parts. This situation has been observed by a combination of electrical and electron microprobe measurements for aluminum–Pyrex (3). Substitution of iron for sodium has also been detected by electron microprobe examination for the iron–sodium disilicate system (4) (see Analytical methods).

(3) The high electric field causes negatively charged oxygen in the depleted glass layer to drift towards the metal. Under favorable conditions this process may result in the oxidation of the metal (3).

(4) As an alternative explanation for the formation of oxide films during electrostatic sealing, adsorbed water vapor dissociates in the presence of the field and the resulting hydrogen ions drift into the depleted glass layer while the hydroxyl ions react with the metal (5).

It should be noted that mechanisms (2)–(4) stipulate a high field region in the glass in which current flow does not obey Ohm's law but is proportional to the exponential of the field. Under constant voltage conditions, it is evident that, with continued transport of sodium ions, the depleted region widens and the field, inversely proportional to the width, is gradually reduced. Hence, one should expect a steady decrease in current with time. For the aluminum–Pyrex and silicon–Pyrex systems, the predicted relation between current and field has been confirmed. The predicted decay of current as a function of time is invariably observed during electrostatic sealing. Mechanism (1) represents the only case in which the electric field in the sodium-depleted region is relatively small. Under these conditions, as represented by the silver–Pyrex system, no decay in current with time is observed.

Although it is undeniable that ion migration takes place during field-assisted sealing, it has not yet been unequivocally established whether the migration is instrumental or merely incidental to bond formation. Irrespective of the precise function of the migration, it appears in many glass–metal systems to promote an interface structure that resembles the preferred structure for strong bonding in conventional sealing, ie, metal–metal oxide–glass boundary layer which contains a small percentage of the metal–bulk glass (6).

The Technique

The electrostatic process has been applied to the sealing of glass to metals and semiconductors (qv). Since the methodology is identical, semiconductors for simplicity are included under the metals. Basically the method consists of the following steps. The parts that are to be sealed are machined so that their faces are smooth and conform to each other as closely as possible. As shown in Figure 1, the clean faces are then placed in contact and heated to a temperature at which the viscosity of the glass approaches the annealing range (1–32 TPa·s or (1–32) × 10^{13} P). If a d-c voltage of 400–1000 V is applied for 1–5 min across the glass–metal interface with the metal positive with respect to the glass, many glass–metal systems form strong, hermetic seals. The experimental ranges cited above are typical. From the discussion below of the various process parameters, it can be seen that under suitable conditions a substantial extension of the ranges is feasible.

Any glass–metal system that can be sealed at all by the electrostatic method has invariably been found to be sealable at temperatures in the annealing range (ie, between the strain point and the annealing point). However, sealing below and above the annealing range is usually possible. For instance, for Pyrex (Corning #7740) the annealing range is 515–565°C, yet Pyrex has been successfully sealed to silicon at 300°C and also around 700°C. By contrast, conventional sealing requires temperatures >1200°C.

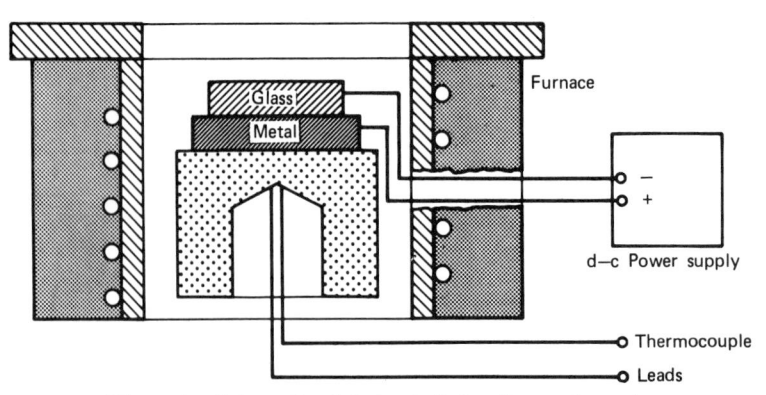

Figure 1. Schematic of electrostatic bonding equipment.

The lower the desired sealing temperature, the more important is proper surface preparation. As stated above, it is essential that the faces to be sealed are smooth and conform to each other. The required finish can readily be produced by lapping and polishing to a surface finish of typically <30 nm. To facilitate these operations, it is usually advisable to select a plane shape. The parts to be assembled must be cleaned to ensure that the faces are free of dust particles or other obstacles that could obstruct intimate contact between the glass and the metal during the sealing operation.

With proper procedures the surfaces should, prior to sealing, be in sufficiently close contact to give rise to interference fringes with a spacing of ca 5 mm. The reason for exceptional surface preparation is that electrostatic seals are made at temperatures at which the viscosity of the glass is high and plastic flow is minimal. By contrast, the viscosity of glass sealed by conventional methods is ten billion times lower, allowing the glass to flow readily around relatively large nonuniformities.

An important requirement in conventional glass sealing is the preoxidation of the metal surface. When surfaces are highly polished as required for electrostatic sealing, there is no such requirement, although a very thin oxide film tends to form on certain metals and alloys incidental to sealing. However, in the case of Kovar it was found that a mild preoxidation permitted sealing to surfaces with surface roughness of up to 150 nm (7).

Electrostatic sealing is preferably performed under constant voltage conditions with the current limited to a few mA. With 200–1000 V and the glass near the annealing temperature, the current rapidly decays from the mA to the μA range. As discussed below, sometimes it is advantageous to apply the voltage at a relatively low temperature and maintain the voltage while the parts heat up to the sealing temperature. The maximum useful voltage is determined by the requirement that dielectric breakdown in the glass as well as gaseous breakdown at the edges of a seal should be avoided as far as possible since the localized discharges tend to damage the polished surfaces.

The voltage may be applied to the glass in a variety of ways. For areas smaller than a 1 cm^2, a point contact with a wire is found to be sufficient. For larger areas, a conductive film or a metal plate in contact with the glass surface is adequate (since such a plate is at negative potential with respect to the glass, it will not seal). For cylindrical surfaces, a wire may be wrapped around at a distance of ca 1 cm from the seal. Although these methods result in excellent electrical contact, they all tend to leave the glass marked to varying degrees. Some of the marks are small pits resulting from sparking between the electrode and the glass. In other cases, marks apparently result from electrochemical interaction. If it is important that the glass face remain undamaged, an additional plate of polished glass may be inserted between the metal electrode and the glass that is to be sealed. The plate serves as a glass electrode that suppresses sparking and other damage.

Once the parts reach the sealing temperature, seals usually may be completed in less than five minutes. Sealing is rapid if faces are properly prepared, the temperature is near the annealing range, and the sealing voltage is greater than a few hundred volts.

As in conventional glass–metal sealing, it is normally necessary that the thermal expansion characteristics of the glass closely match those of the metal between room temperature and the sealing temperature. This requirement is unnecessary, however, if the metal is a film or a thin, annealed foil. Some degree of thermal mismatch may also be tolerated for very small surfaces. Seals have been made in a variety of atmospheres including room air, forming gas (90% N, 10% H), nitrogen, oxygen, and at reduced pressure of ca 1.3 kPa (10 mm Hg).

Table 1 lists several glass–metal systems that have been successfully sealed. Some of these metals, notably silicon and Kovar, seal readily over a very wide range of parameters. Other metals require greater care, usually because they form poorly adhering oxides during the sealing process. Although the oxides bond well to the glass, the seals are nevertheless mechanically weak, with failure occurring at the metal-oxide interface. One possible remedy is development of a suitable preoxidation procedure. Three simple approaches that are successful in minimizing the growth of the oxide are: (*1*) Sealing in an oxygen-deficient or a reducing atmosphere. (*2*) Initial application of the bonding voltage at a very low temperature in room air while the parts are heated to the bonding temperature (7). One of the effects of the applied voltage is formation of an electrostatic attractive force that pulls the parts together, presumably impedes the access of oxygen to the metal surface, and produces a thinner but apparently more cohesive oxide film. (*3*) Covering the polished metal surface prior to sealing with a vacuum-deposited film of a material that is easily bonded to glass irrespective of ambient conditions. Evaporated films of aluminum, silicon monoxide, and nichrome, and sputtered films of silicon are particularly suitable in this approach. Essentially the same technique can be utilized for the sealing of glass to glass, ie, a film of one of the above materials is deposited on the polished glass face before it is sealed to a second glass part, thus forming the structure of a glass–metal film on glass. A complementary procedure has been described for silicon-to-silicon seals by means of an intermediary layer of sputtered Pyrex (8).

Table 1. Selected Materials That Have Been Sealed by the Electrostatic Process

Material	Bonded to glass[a]
Kovar (iron alloy)	7052
	7056
molybdenum	7052
titanium	0080
tantalum	7052
Carpenter 49 (iron alloy)	Mosaic F/O[b]
	Corning F/O[b]
Niromet 44 (iron alloy)	Mosaic F/O[b]
	Corning F/O[b]
NiSpan (iron alloy)	7740
silicon	7740
	7070
germanium	7052
Al film on 7740/7052/0120	7740/7052/0120
Al film on CerVIT (ceramic)	CerVIT
SiO film on 7740/7056	7740/7056
Si film on quartz	quartz
NiCr film on 7740	7740
SiO$_2$ film on silicon	SiO$_2$ film on silicon

[a] Numbers refer to Corning glasses.
[b] F/O = fiber optic (see Fiber optics).

Seal Properties

Seals are evaluated with regard to strength, hermeticity, and resistance to thermal shock. The tensile strength of electrostatic glass seals was investigated in detail for the system Kovar 7052 glass (7). The surface finish of the bonding faces was ca 10 nm; the Kovar was not preoxidized, and sealing was accomplished with 1000 V applied at ca 100°C in air and maintained while the parts were further heated and kept at 500°C for 10 min. Under these conditions, 80% of the seals failed at 10–20 MPa (1500–3000 psi), average strength amounted to 16 MPa (2300 psi), and 70% of the failures were due to fracture of the glass. When the surface roughness of the Kovar parts was increased from 10 to 150 nm, the strength of the seals dropped by a factor of two, with failure almost invariably occurring in the oxide film. Glass seals to nonpreoxidized Kovar parts were compared with similar seals to parts preoxidized in air at 350°C for 15 min. Unlike the seals to nonpreoxidized parts, seals to preoxidized parts exhibited no reduction in seal strength with increasing surface roughness. Furthermore, the preoxidized seals failed by glass fracture irrespective of surface roughness.

Experiments were made to test the stability of the seals and, if possible, to increase their strength. Although none of the tests produced significant changes in seal strength, a brief summary may be of interest.

For some time it was thought that, as in conventional sealing, an extended annealing treatment should increase the seal strength. Yet despite repeated efforts no significant change in strength could be established. Nor were any changes in strength observed when Kovar #7052 seals fabricated at 400°C were heated to 700°C in various ambient gases, ie, to 200°C above the annealing point. The irreversibility of the completed seals was tested as follows. Parts sealed by the electrostatic process, before being allowed to cool, were subjected to a reversal in the polarity of the sealing voltage. As in other tests, there was no indication of any change in seal strength.

The hermeticity of the seals was evaluated by a Helium Leak Detector with a sensitivity of 5×10^{-10} cm^3 He/s. Provided the surface finish was <20 nm and the faces were properly cleaned, the electrostatic process generally produced hermetic seals. With seals to soft metals such as aluminum and Kovar, leakage problems were occasionally encountered from scratches that were inadvertently produced during cleaning. Surprisingly, seals to Kovar surfaces with a roughness of as much as 120 nm proved to be consistently hermetic provided the surfaces were preoxidized and free of scratches.

The effects of thermal cycling and thermal shock were investigated as a part of a program that compared electrostatic sealing with frit sealing (9). The test vehicle was fiber-optic (F/O) face plates (see Fiber optics) of 2.54 cm dia that were sealed to flanges of Niromet 44. Five thermal cycles between −25°C and 125°C had no effect on the seals. In destructive thermal shock tests, seals were rapidly transferred from hot to cold water baths. Fractures first appeared at a temperature difference of 50°C, but 67% of the seals survived at 70°C difference. The distribution of failures was similar for both sealing methods (see Materials reliability).

Uses

Although there are broad applications for the electrostatic sealing method, only a few examples that demonstrate its benefits are given here. Since sealing temperatures are low and macro-distortion of the glass is minimal, the method is useful in the mounting of optical components. An application of electrostatic sealing has been described in which elliptical quartz windows are attached at the precisely defined Brewster angle to quartz laser systems (10) (see Lasers). Quartz-to-quartz sealing was accomplished by means of an intermediary silicon film at 850°C. This method of mounting resulted in excellent optical performance of the laser. Similarly, distortion-free optical performance was obtained from the previously discussed fiber optics that were sealed to matching flanges at 600°C with 600 V. In a structural application, electrostatic sealing was used for the mounting of accurately spaced metal electrodes. The latter were Kovar rings that were spaced by precision-machined 7052-glass tubes. Since the spacers were bonded to the electrodes by hermetic butt seals, they could serve as integral sections of vacuum tubes (see Vacuum technology). An interesting feature of this procedure was the ability to make a number of seals simultaneously. An application to solid state electronics involved sealing cover glass to silicon solar cells by means of the electrostatic process (11) (see Photovoltaic cells).

The application that has been investigated most widely is concerned with the mounting of silicon pressure transducers. Conventionally, this has been accomplished by organic adhesives (qv) that formed strong bonds at low temperatures. However, mechanical creep and hysteresis effects in the adhesives seriously degrade the performance of the transducers. An additional drawback of adhesives is their limited temperature capabilities. Very significant improvements in hysteresis, sensitivity, and temperature compensation can be obtained by the substitution of electrostatic glass seals for epoxy bonds (12).

Although the technological merits of electrostatic sealing can be defined with some precision, its economic advantages and disadvantages are still in the process of being evaluated. The general simplicity of the method, and in particular the low sealing temperature, the short heating cycle, and the absence of a liquid phase are obviously attractive features in that they provide equipment designers with great flexibility. At least in the early stages of production, equipment requirements are probably quite modest. On the other hand, it appears that the required surface preparation can be labor-intensive and therefore rather costly.

A detailed analysis of the inherent costs of polishing is not attempted here, but the following points appear to be pertinent: (*1*) when parts require polished surfaces for reasons other than sealing, the electrostatic process becomes feasible without additional surface treatment; (*2*) lapping and polishing are batch processes whose costs are proportional to area; thus it would seem that electrostatic sealing is most economical for small parts; and (*3*) in view of the work described earlier with relatively rough, preoxidized Kovar parts, the possibility should not be excluded that methods will eventually be discovered that will permit a significant relaxation of the current stringent surface conditions.

The electrostatic sealing process is commercially utilized by a number of companies under licensing agreements with P. R. Mallory & Co. Inc. which holds the applicable patents (13). As far as is known, one of the most popular applications is in the construction of pressure transducers (see Semiconductors). As discussed earlier, the sealing method offers unique technical advantages in this area, and the cost of surface preparation is modest because of the small size of the areas that have to be bonded.

BIBLIOGRAPHY

1. *Properties of Selected Commercial Glasses,* Corning Glass Works, Corning, N.Y., 1963.
2. G. Wallis and D. I. Pomerantz, *J. Appl. Phys.* **40,** 3946 (1969).
3. G. Wallis, *J. Am. Ceram. Soc.* **53,** 563 (1970).
4. M. P. Borom, *J. Am. Ceram. Soc.* **56,** 254 (1973).
5. P. B. DeNee, *J. Appl. Phys.* **40,** 5396 (1969).
6. J. A. Park and R. M. Fulrath, *J. Am. Ceram. Soc.* **45,** 592 (1962).
7. G. Wallis, J. Dorsey, and J. Beckett, *Am. Ceram. Soc. Bull.* **50,** 948 (1971).
8. A. D. Brooks, R. P. Donovan, and C. A. Hardesty, *J. Electrochem. Soc.* **119,** 545 (1971).
9. G. Wallis, *J. Automat. Eng.* **79,** 15 (1971).
10. B. Smith, *IEEE J. Quantum Electron.* **9,** 546 (1973).
11. A. R. Kirkpatrick, *Contract #F 33615-74-C-2001,* Air Force Aero Propulsion Laboratory, Wright-Patterson Air Force Base, Ohio, 1975.
12. E. E. Orth and P. Cannon, *Seventh Transducer Workshop,* Secretariat Range Commanders Council, White Sands Missile Range, N.M., Apr. 4–6, 1972, p. 119.
13. U.S. Pat. 3,397,278 (Aug. 13, 1968), D. I. Pomerantz (to P. R. Mallory & Co. Inc.).

GEORGE WALLIS
P. R. Mallory & Co. Inc.

EMBEDDING

The embedment of objects, the complete encasement of objects in a medium, practiced for centuries, has today become a science involving large numbers of scientists and engineers in all parts of the industrial world. The objectives of embedding may be either functional or decorative. Both are covered in this article with the greater emphasis on functional embedding. The most important functional area is the embedding of electrical and electronic circuitry. Embedding in electronics is a highly technical and specialized field, using very high performance materials and processes that are elaborate, precise, and tightly regulated, often by electronic controls.

Embedding technology, in its present form, began in the 1940s. Early embedding practices used waxes and bitumens, materials that are largely inadequate for modern needs because of their performance limitations in thermal stability, and in electrical and mechanical parameters. The rapid development in synthetic resins has produced today's high performance embedding materials. The early synthetic resins used for embedding had serious shortcomings. The first of these was a phenolic casting material developed by Baekeland in 1906. The use of phenolics has not become widespread because the acidic catalyst used in their manufacture is detrimental to electronic circuitry and components. The inherent brittleness in the phenolic resins leads also to an undesirable tendency to crack during temperature cycling.

Since functional embedding, mostly electrical and electronics, is by far the largest application of embedding, the presentation of embedding here provides information useful for that area. However, much of the information is equally important for nonfunctional or nonelectrical embedding. A later section of this article reviews some of the unique factors related to nonfunctional or nonelectrical applications.

Terminology for Embedding

The terminology covering the area of embedding is extensive, and unfortunately, there is little uniformity in the use of some of these terms. Nonspecialists may, for example, incorrectly use the terms "potting," "molding," or "encapsulating" to apply to the entire field. There are accurate and correct terms, however, each referring to the attainment of embedding by a different process. Thus embedding, which is the complete encasement of an object in a medium, can be achieved by several processes. The correct process-related terms are as follows:

Casting: this embedding process consists of pouring a catalyzed or hardenable liquid into a mold. The hardened cast part takes the shape of the mold and the mold is removed for reuse.

Potting: this embedding process is similar to casting except that the catalyzed or hardenable liquid is poured into a shell or housing which remains as an integral part of the unit.

Impregnating: this embedding process consists of completely immersing a part in a liquid so that the interstices are thoroughly soaked and wetted; the process is usually accelerated by vacuum or pressure, or both.

Encapsulating: this embedding process consists of coating (usually by dipping) a part with a curable or hardenable coating; coatings are relatively thick compared with varnish coatings.

Transfer Molding: this embedding process involves the transfer of a catalyzed or hardenable material, under pressure, from a pot or container into the mold which contains the part to be embedded.

Embedding Process Considerations

Often, a given embedded product can be made by two or more of the various embedding processes. Thus, an analysis of the important comparative advantages of the methods is required. This section presents some of the most important of these considerations. The comparison is summarized in Table 1 (see also Plastics technology).

Casting. For casting applications, the design of the mold and the design of the assembly should provide for minimum internal stresses during curing and for proper final dimensions after allowing for shrinkage.

In casting processes, the mold is cleaned thoroughly and a suitable release agent is applied, for many of the resins used normally adhere to the walls of the mold (see Abherents). The mold is then positioned around the part, and any points where leakage of the liquid resin from the mold might occur are sealed with a material such as cellulose acetate butyrate if they cannot be conveniently gasketed against leakage. Next, the resin and catalyst are mixed and are poured slowly into the mold so as to avoid air entrapment during the pouring.

The entire assembly is allowed to cure as required by the resin–catalyst system either at room temperature under its own exothermic heat or in an oven at some higher temperature. Finally, the part is separated from the mold. A typical cycle for producing embedded assemblies by the cast resin technique is shown in Figure 1.

Table 1. Comparison Summary For Embedding Processes[a]

Process	Advantages	Limitations	Material requirements	Applications
casting	requires a minimum of equipment and facilities; is ideal for short runs	for large-volume runs, molds, mold handling and maintenance; can be expensive; assemblies must be positioned so they do not touch the mold during casting; patching or correcting surface defects can be difficult	viscosity must be controlled so that the embedding material flows completely around all parts in the assembly at the processing temperature and pressure	most mechanical or electro-mechanical assemblies within certain size limitations can be cast
potting	excellent for large-volume runs; tooling is minimal; presence of a shell or housing assures no exposed components as can occur in casting	some materials do not adhere to shell or housing; electrical shorting to the housing can occur if the housing is metal	same material requirements as for casting except that materials that bond to the shells or housings are required	most mechanical or electro-mechanical assemblies, subject to certain size limitations and housing complexity limitations
impregnating	the most positive method for obtaining total embedding in deep or dense assembly sections such as transformer coils	requires vacuum or pressure equipment which can be costly; in curing, the impregnating material tends to run out of the assembly creating internal voids unless an encapsulating coating has first been applied to the outside of the assembly	low viscosity materials are required for the most efficient and most thorough impregnation	dense assemblies which must be thoroughly soaked; electrical coils are primary examples
encapsulating	requires a minimum of equipment and facilities	obtaining a uniform, drip-free coating is difficult; specialized equipment for applying encapsulating coatings by spray techniques overcomes this problem, however	must be both high viscosity and thixotropic so that they do not run off during the cure	parts requiring a thick outer coating, such as transformers

Table 1. (continued)

Process	Advantages	Limitations	Material requirements	Applications
transfer molding	economical for large-volume operations	initial facility and mold costs are high; requires care so that parts of assemblies are not exposed; some pressure is required and processing temperatures are often higher than for other embedding operations	should be moldable at the lowest possible pressure and temperature and should cure in the shortest possible time for lowest processing cost	for embedding small electronic assemblies in large-volume operations

[a] Refs. 1–2.

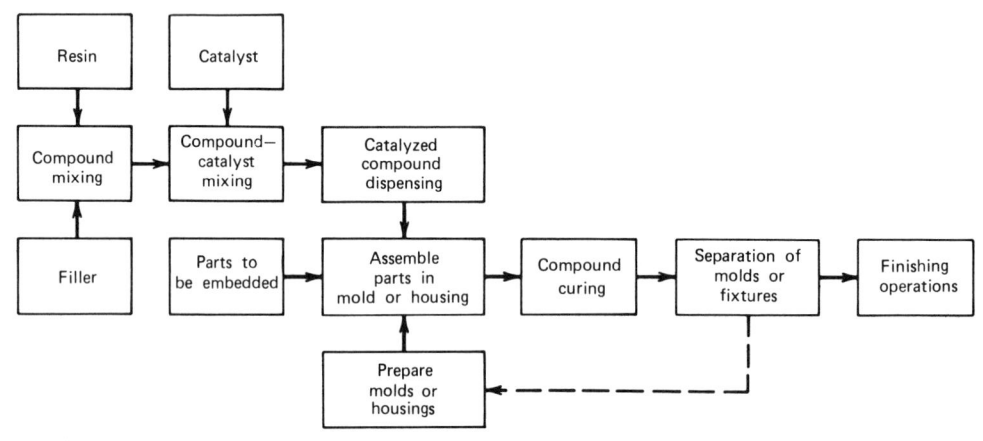

Figure 1. Typical process sequence for producing embedded electronic assemblies by casting (1–2).

The two types of molds in common use are metal and plastic, (Table 2). The relatively high thermal conductivity of metal molds is an advantage if oven curing of the resin is used. The heat is rapidly and uniformly transferred into the resin. The insulating properties of plastics may require that parts processed in plastic molds be preheated before the liquid resin is poured.

Potting. The potting process is similar to casting, except that the assembly to be potted is positioned in a can, shell, or other container. Since the shell or can will not be separated from the finished part, no mold-release treatment is necessary. However, if the container is metal and if the assembly is electrical, it is usually necessary to place a sheet of insulating material between the assembly and the can. The higher the applied voltage, the more stringent is this requirement.

Housings for potted assemblies are made of metal or plastic (Table 3). The same

Table 2. Considerations in Selection of Molds To Be Used in Casting[a]

Mold material and fabrications	Advantages	Disadvantages
machined steel	good dimensional control; can be made for complex shapes and insert patterns; good heat transfer; surfaces can be polished	assembly sometimes difficult; can corrode; usually requires mold release
machined aluminum	same as machined steel except more easily machined	same as machined steel, except for corrosion; easily damaged, because of softness of metal
cast aluminum	none over machined aluminum	same as machined aluminum; surface finish and tolerances usually not as good as machined aluminum; complex molds not as accurate as machined metal
sprayed metal[b]	none over machined metal; good surface possible	use usually limited to simple forms; not always easy to control mold quality; number of quality parts per mold limited; requires mold release
dip molded[b] (slush casting)	same as sprayed-metal molds	same as sprayed metal molds
cast epoxy	good dimensional control; surface can be polished; can be made for inserts and multiple-part molds; long life and low maintenance	dimensional control not quite as good as in machined-metal molds; requires mold release and cleaning; low thermal conductivity compared with metals
cast plastisols	parts easily removed from molds; molds are easy to make	short useful life; poor dimensional control
cast RTV silicone rubber	same as for plastisols; better life than plastisols	poor dimensional control, though better than plastisols
machined TFE fluorocarbon	no mold release required; convenient to make for short runs and simple shapes; withstands high-temperature cures	poor dimensional control
machined polyethylene and polypropylene	same as listed for TFE fluorocarbon except temperature capability and lower cost	poor dimensional control
molded polyethylene and polypropylene	same as listed for TFE fluorocarbon except temperature capability and lower cost	poor dimensional control

[a] Refs. 1–2.
[b] Although sprayed-metal molds and dip-molded molds are similar, differences in method of making these two types may give one an advantage over the other in specific cases.

882 EMBEDDING

Table 3. Considerations in Selection of Housing To Be Used in Potting[a]

Housing or container	Advantages	Disadvantages
steel	many standard sizes available; easily plated for solderability; good thermal conductivity; easily cleaned by vapor degreasing; good adhesive bond formed with most resins; easily painted; flame resistant	can corrode in salt spray and humidity; fitting of lids sometimes a problem; cut off of resin-filled can is sometimes difficult; possibility of electrical shorting
aluminum	same as for steel except plating ease; lightweight; corrosion resistant	same as for steel except aluminum is more corrosion-resistant; not easily soldered
molded thermosets (epoxy, alkyd, phenolic, diallyl phthalate, etc)	many standard sizes available; good insulator; corrosion resistant; color or identification can be molded in; terminals can sometimes be molded in; cut off of resin-filled shell easier than for metal cans; same type of material can be used for shell and filling resin, resulting in good compatibility	does not always adhere to resin, especially if silicone mold releases used to make shell; sealing of leakage joints can be difficult; physically weaker than steel, especially in thin sections; molding flash can cause fitting problems; cleaning of resin spillage can break shells
molded thermoplastics (nylon, polyethylene, polystyrene, etc)	same as listed for thermosets except last two items; often less prone to cracking than thermosetting shells, although this depends on resiliency of material	same as first three items listed in thermosets; adhesion can be poor due to excellent release characteristics of most thermoplastics; shell can distort from heat; cut off can be a problem, due to melting or softening of thermoplastics under mechanically generated heat

[a] Refs. 1–2.

comparisons of these materials apply to housings as to molds with respect to curing and thermal conductivity. Since the housing remains an integral part of the embedded unit, however, some additional considerations are in order. For instance, corrosion resistance in humidity and salt spray can be important, particularly if metal housings are used. Also, there may be standard metal containers available for certain types of products, such as transformers, which can reduce tooling costs. Thus, the choice between metals and plastics can be more important for potted units than for cast units, since the latter requires only a temporary mold. Comparisons of casting and potting techniques are shown in Table 4.

Impregnating. In the impregnating process, the liquid resin is forced into all the interstices of the component or assembly, after which the resin is cured or hardened. This can be an independent operation, or it can be used in conjunction with encapsulating, casting, or potting. Impregnation differs from encapsulation in that encapsulation produces only a coating, with little or no resin penetration into the assembly. Penetration is most important for certain electrical parts such as transformers.

Impregnation is sometimes accomplished by centrifugal means. The part is po-

Table 4. Comparison Considerations For Casting and Potting in Relation to Important Process and Product Parameters[a]

Parameter	Casting process	Potting process
skin thickness	difficult to control; components can become exposed in high-component-density packages	controlled minimum wall or skin thickness, due to thickness of shell or housing
surface appearance	cavities and surface blemishes often require reworking	established by surface appearance of shell or housing, though problems can arise if resin spillage not controlled
repairability	resin exposed for easy access	shell or housing must be removed and replaced
handling	handling and transfer of unhoused assembly can reduce yield	most handling of unembedded unit can be in housing
assembly	if molds are not well maintained, or it unit fits tightly into mold, handling can cause breakage of components	assembly is simplified since new shells or housings are always used, and wall thickness is controlled
manufacturing-cycle efficiency	production rate usually limited by quantity of molds	output not limited by tools
tool preparation and maintenance	relatively expensive	costs are minimal

[a] Refs. 1–2.

sitioned in a mold, the mold filled the resin, and the entire assembly spun at a high velocity.

If both encapsulation and impregnation are required, as in some transformer applications, the encapsulation dip coating is usually applied first. A hole is left in the coating so that the low-viscosity impregnating resin can be forced in after the shell created by encapsulation has hardened. This procedure provides a container, thus eliminating drain off of the impregnating material during its hardening or curing cycle.

Transfer Molding. Although most resin-embedded electronic packages were once produced by casting or potting techniques, the use of transfer molding is becoming widespread. Transfer molding offers advantages in economy and increased production rates for those assemblies that adapt to this technique and that are produced in large quantities. The advantages include a large reduction in the number of processing steps (over casting or potting), as shown in Figure 2, and a shorter curing time of the embedding compound. Transfer-molding materials cure in minutes; liquid casting and potting resins require hours.

Major limitations of the transfer-molding process are: (1) The assembly must be able to withstand pressures of 345–1724 kPa (50 to 250 psi). (2) The assembly must be able to take the curing temperature of 120–175°C. (3) Production volume must be large enough to justify equipment expenditures.

In transfer molding, a dry, solid molding compound, usually in powder or pellet form, is heated in a molding press to the point of becoming flowable or liquid, at which time it flows (is transferred) under pressure into a mold cavity containing the assembly to be embedded, as shown in Figure 3. The plastic remains in the heated mold for a short time until curing is completed.

Potting method

Assemble module to cover

Assemble box to cover and bond

Insert shims to control pin length

Dip to apply strippable coating

Prepare potting compound

Pour potting compound

Cure (5 h)

Remove strippable coating

Cut off excess gate

Transfer-molding method

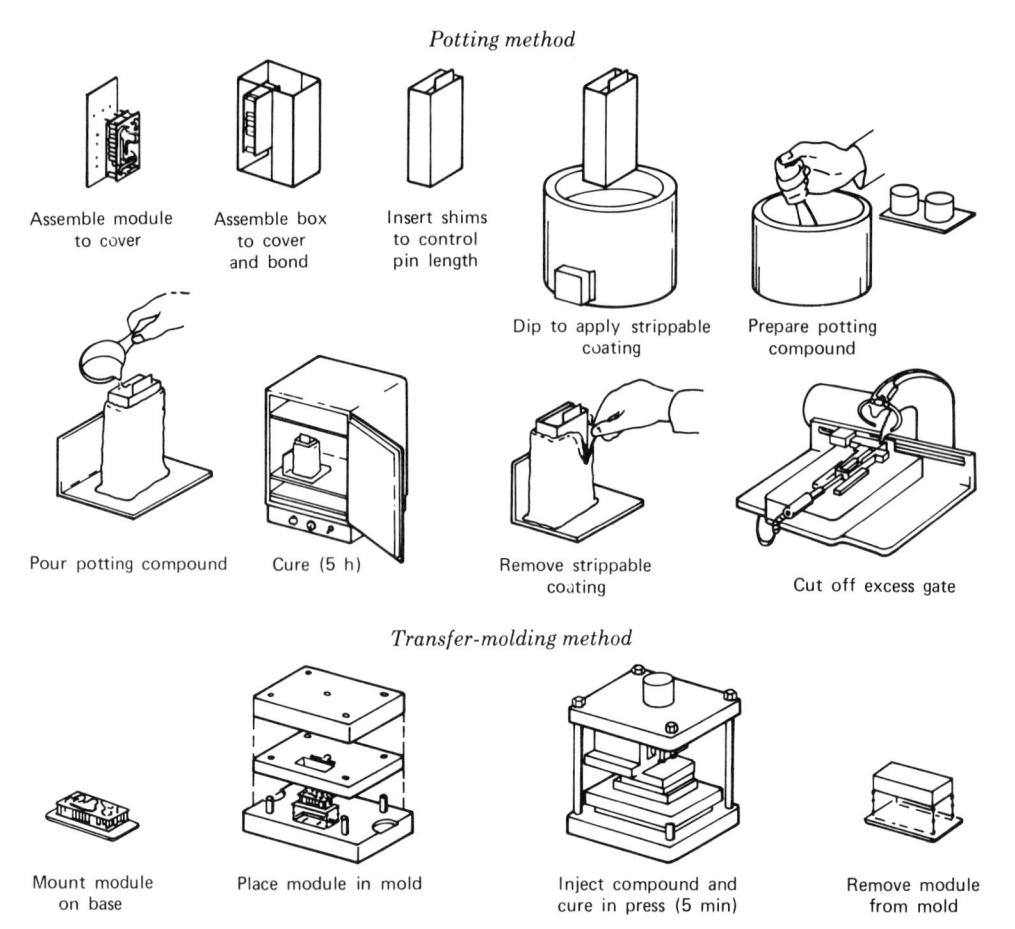

Mount module on base

Place module in mold

Inject compound and cure in press (5 min)

Remove module from mold

Figure 2. Comparison of processing steps for embedding electronic modules by potting and by transfer molding (1–2).

The transfer mold shown in Figure 3 embeds two similar parts, but a larger number of parts can be molded simultaneously. The fact that multicavity molds are common in transfer molding is one of the economies for large-volume runs. Many cavities can be filled as rapidly as a single cavity, thereby reducing the cost per part. Although mold cost increases as the number of cavities increases, it does not increase proportionately. Overall mold cost per part produced can be further reduced by incorporating cavities of different shapes into the same mold in proportion to the production volumes required, or by using mold inserts to vary the cavity configuration as required by changing production needs.

The small gate scar that remains on the finished part at the point where the molding compound goes out of the runners from the transfer cylinder into the cavity is usually unobjectionable.

Primary Embedding Materials. Although waxes and bitumens are still used occasionally, most materials used for embedding are plastics. Liquid or easily liquefied plastics are most commonly used. These plastic resins are most readily fabricated into complex structures, and they provide the proper physical characteristics required for embedding. As a class plastics are insulators. They therefore provide the necessary

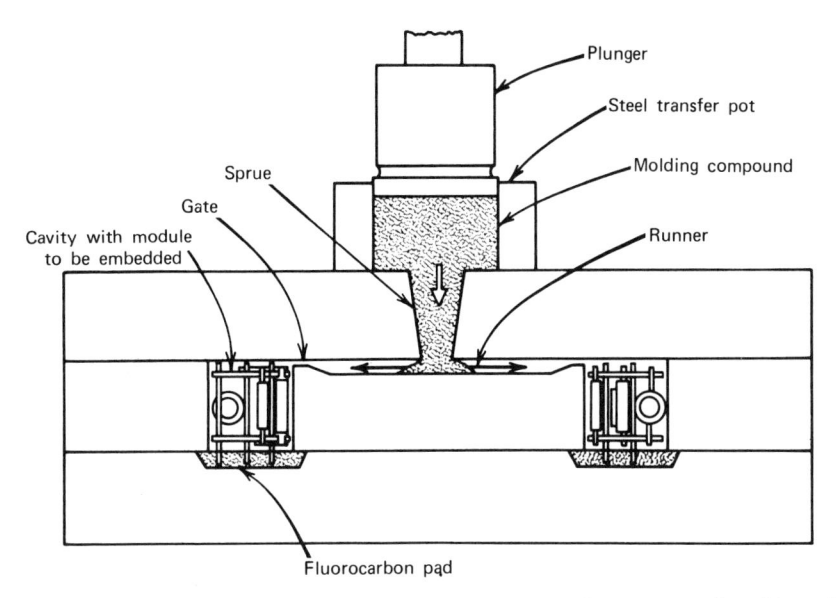

Figure 3. Typical transfer mold assembly showing flow of molding compound used for embedding (1–2).

insulation required in electronic and electrical applications. Useful plastics fall into two major classes, thermoplastics and thermosets.

Thermoplastics are characterized by their ability to flow upon the application of heat and substantial pressure. They become rigid again in a reversible manner upon subsequent cooling. Thermoplastics are polymers made up essentially of difunctional units. Simple chains formed of difunctional units can slip past each other to a limited extent upon heating or stressing; upon cooling or the relief of stress, the cohesive forces between molecules becomes predominant, and the plastic becomes rigid again. Typical examples of thermoplastics are polyethylene (see Olefin polymers) and polystyrene (see Styrene plastics).

Thermosets contain sizable portions of multifunctional units between which the cohesive forces are insufficient to prevent flow. These materials are liquids or solids at room temperature. Upon the addition of hardeners, ie, curing agents, and the application of heat cure, they harden into solids. Enough cross-links are formed so that flow is no longer possible. These changes are not reversible. Typical examples of thermosets are the phenolics (see Phenolic resins) and epoxies. Rubbery polymers are composed of chains that, because of the way the difunctional units are held together, tend to coil into helices that act as elastic springs, hence the name elastomers (see Elastomers, synthetic). In this case, the cross-linking reaction is commonly known as vulcanization. All degrees of rigidity and flexibility are available, since most of the thermosetting resins can be modified or obtained in flexibilities almost equal to those of elastomeric materials, and even softer than the low-melt thermoplastic materials.

As noted above, plastic resins most commonly used for embedding are the liquid resins and compounds, or those that can easily be liquefied by moderate heat or pressure, or both. There are many chemical types, each having different end properties.

Most liquid resins cure by heat or curing-agent influences, or both, give off heat during the curing process, and are thermosetting. Resin viscosity or fluidity, and the time-temperature curve (exotherm) for the exothermic reaction, which vary with each resin system, are key properties describing the nature of the individual resin through the curing cycle.

Viscosity. Flow properties of resins are important because of the need for flow and penetration at atmospheric or low pressures. If the viscosity is high, the formulation is difficult to pour and does not flow properly around inserts or components, thus allowing internal cavities to form. A high viscosity resin is usually too thick to allow evacuation of entrapped air, which also promotes cavity formations. High viscosity also makes mixing difficult. On the other hand, a resin whose viscosity is too low may leak through openings in the mold or container.

For most embedding applications, there is an optimum range of viscosity. For impregnating operations, extremely low viscosities, 100 mPa·s (= cP) or less, are desirable because complete impregnation of the parts under vacuum is required. In practice, however, impregnation is often achieved with viscosities considerably higher (up to 1000 mPa·s or higher). However, the higher the viscosity, the longer the cycling time or the higher the vacuum required for complete impregnation.

For embedment operations such as casting or potting, there is no lower limit on the viscosity, provided the mold or container is tight enough to prevent leakage. Usually, however, if impregnation is not required and if the components are not packed tightly, viscosities in the range of 1000–5000 mPa·s are satisfactory for casting and potting operations.

An encapsulation coating requires a thixotropic (nonflowing) material with an extremely high viscosity, because the part is dipped into the compound and the coated part cured without the use of a mold or container. The coating must not flow off during the curing operation.

Viscosity usually can be lowered by heating the resin, as shown in Figure 4, or by adding diluents, and it can be raised by adding fillers (qv). Not all resin viscosities show so strong a temperature dependence as that in Figure 4; silicones, for example, have a relatively flat viscosity curve (see Rheological measurements).

Exothermic Properties. Most polymeric resins used for embedment form by an exothermic reaction. It is essential that the exothermic properties of a particular system be known and that they be controlled. Too much heat may cause resin cracking during cure, or may adversely affect heat-sensitive components.

Three characteristics are commonly used for control measurements of these exothermic properties: gel time, peak temperature, and time-to-peak temperature. These quantities are measured from a single graphic plot of temperature versus time for a given resin-catalyst-curing agent system. A typical exothermic curve for a polyester resin is shown in Figure 5. Although the shapes of these curves vary widely from system to system, the curve for a given system should be closely reproducible. Exothermic curves also vary with mass of resin, as shown in Figure 6.

Gel time is the interval from the time the exothermic reaction reaches 65°C to the time when it is 5.5°C above the bath temperature (Fig. 5). The reason for starting the timing at 65°C rather than from the time at which the catalyst and resin are initially mixed is that it is not always practical to have the temperature of the ingredients precisely the same when the reaction starts, during and immediately after the initial mixing. The common base point is used to assure better reproducibility. Gelation

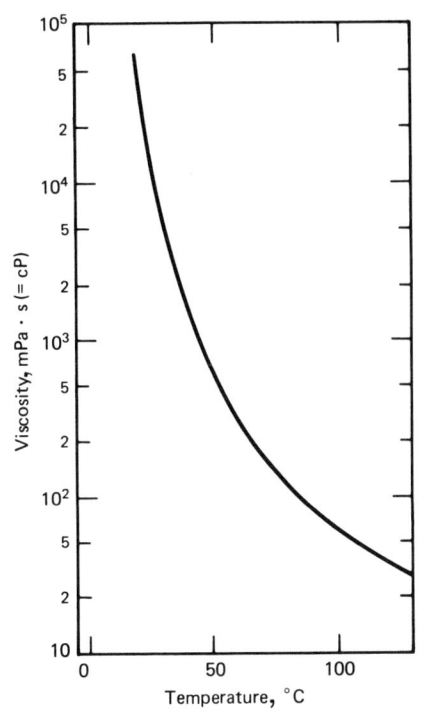

Figure 4. Viscosity–temperature relationship for a bisphenol epoxy resin (1–2).

usually occurs by the time the temperature has slightly exceeded the bath temperature, for this type of resin system. A practical indication of gel time is the time interval from first mixing to semisolid consistency.

The temperature rise is much greater after gelation has occurred than before gelation. This tendency is common for exothermic polymerizations. All changes in the resin, curing agent, or curing cycle vary this curve, which was nearly flat or very steep.

Transfer-Molding Resins. Although many plastic materials can be transfer-molded, most materials require transfer pressures too high to allow embedment of delicate electronic assemblies without damaging the components or distorting the position of the assembly in the mold. Development of plastic materials that can be transferred at low pressures has made possible the embedment of fragile assemblies by transfer molding. The most widely used materials are epoxies, although low-pressure molding materials have been developed from silicones, diallyl phthalates, phenolics, and alkyds (see Alkyd resins), the first two enjoying wide usage. Even though these materials are handled quite differently, their post-cure properties are similar to those of cured liquid resins of the same chemical type.

THERMOSETTING EMBEDDING RESINS AND FILLED COMPOUNDS

Many chemically distinct embedding resins are available, and there are many variations in each group. The most important of these are discussed in the following sections. Typical mechanical, physical, thermal, and electrical properties of several of these classes are shown in Tables 5 and 6.

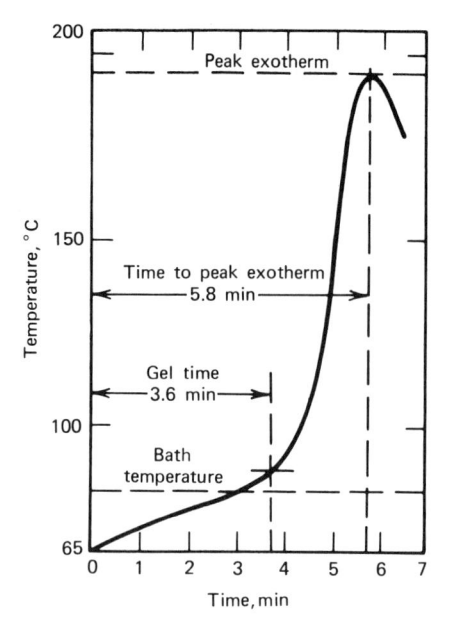

Figure 5. Exothermic reaction curve for a typical polyester resin, using benzoyl peroxide catalyst, and cured at 80°C (1–2).

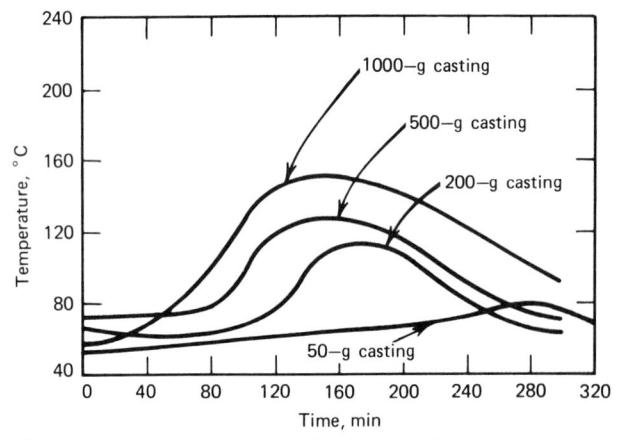

Figure 6. Exothermic curves, as a function of resin mass, for bisphenol epoxy and 5% piperidine curing agent (catalytic type) cured at 60°C (1–2).

Epoxies. The most used of the embedding resins are the epoxies, in many types and modifications (see Epoxy resins). All classes of epoxies have certain outstanding characteristics important in electronic assemblies. Chief among these properties are low shrinkage, excellent adhesion, excellent resistance to most environmental extremes, and ease of application for casting, potting, or encapsulation.

The original class of epoxies, the bisphenols, are the work-horses of the electronics industry. They are available as solids and as liquids over a wide viscosity range. The resins are syrupy; they are available in viscosities from 4000 mPa·s to semisolids at

Table 5. Typical Mechanical and Physical Properties of Several Common Embedding Resins[a]

Material	Tensile strength, kPa[b]	Elongation, %	Compr. strength,[b] kPa	Impact strength, Izod (J/cm of notch)[c]	Hardness	Linear shrinkage during cure, %	Water absorption, wt %
Epoxy							
rigid, unfilled	62,060	3	137,900	1.7220	Rockwell M 100	0.3	0.12
rigid, filled	68,950	2	172,400	1.3776	Rockwell M 110	0.1	0.07
flexible, unfilled	34,480	50	55,160	10.332	Shore D 50	0.9	0.38
flexible, filled	27,580	40	68,950	6.888	Shore D 65	0.6	0.32
Polyester							
rigid, unfilled	68,950	3	172,400	10.332	Rockwell M 100	2.2	0.35
flexible, unfilled	10,340	100		24.108	Shore A 90	3.0	1.5
silicone flexible, unfilled	3,447	175		no break	Shore A 40	0.4	0.12
urethane flexible, unfilled	3,447	300	137,900	no break	Shore A 70	2.0	0.65

[a] Refs. 1–2.
[b] To convert kPa to psi, multiply by 0.145.
[c] To convert J/cm to ft lb/in., divide by 3.44.

room temperature. Their viscosity can be reduced, of course, by heating. Other important classes of epoxies are cycloaliphatic diepoxides, novolac epoxies, and hydantoin epoxies.

Cured properties of bisphenol epoxies, as well as of other epoxies, are controlled by the type of curing agent used with the resin. Major types of curing agents are aliphatic amines, aromatic amines, catalytic curing agents, and acid anhydrides. These are discussed in Table 7 (see also Insulation, electric).

Although epoxy resins constitute the largest volume usage for embedding electronic packages, other resins are also important. These include silicones, urethanes, polyesters, thermosetting hydrocarbons, thermosetting acrylics, and polysulfides. Also, foams and low-density resins are widely used for electronic packaging for weight-reduction purposes.

Silicones. The silicone resins are convenient to use, they are available over a wide range of viscosities, and most of them can be cured either at room temperature or at low temperatures (see Silicon compounds, silicones). Silicones maintain their properties over a wide temperature range, generally from approximately −65 to 200°C and, in some cases, up to 300°C. Their excellent electrical properties, particularly low loss factors, and the fact that they generate little or no exothermic heat are additional advantages.

Three classes of silicones are used for embedding applications: RTV silicones and flexible resins; silicone gels; and rigid, solventless resins.

Table 6. Typical Thermal and Electrical Properties of Several Common Embedding Resins[a]

Material	Heat-distortion temp, °C	Thermal shock per MIL-I-16923	Coefficient of thermal expansion, ppm/°C	Thermal conductivity, W/(m·K)	Dissipation factor[b]	Dielectric constant[b]	Volume[c] resistivity, Ω·cm	Dielectric[d] strength, V/m	Arc resistance, s
epoxy									
rigid, unfilled	140	fails	55	16.7×10^{-2}	0.006	4.2	10^{15}	17.7×10^6	85
rigid, filled	140	marginal	30	62.8×10^{-2}	0.02	4.7	10^{15}	17.7×10^6	150
flexible, unfilled	<RT	passes	100	16.7×10^{-2}	0.03	3.9	10^{15}	13.8×10^6	120
flexible, filled	<RT	passes	70	50.2×10^{-2}	0.05	4.1	3×10^{15}	14.2×10^6	130
polyester									
rigid, unfilled	120	fails	75	16.7×10^{-2}	0.017	3.7	10^{14}	17.3×10^6	125
flexible, unfilled	<RT	passes	130	16.7×10^{-2}	0.10	6.0	5×10^{12}	12.8×10^6	135
silicone									
flexible, unfilled	<RT	passes	400	20.9×10^{-2}	0.001	4.0	2×10^{15}	21.7×10^6	120
urethane									
flexible, unfilled	<RT	passes	150	20.9×10^{-2}	0.016	5.2	2×10^{12}	15.8×10^6	15.8×10^6

[a] Refs. 1–2.
[b] Dissipation factor and dielectric constant are at 60 Hz and room temperature.
[c] Volume resistivity is at 500 V d-c.
[d] Dielectric strength is short time; to convert V/m to V/mil, multiply by 2.54×10^{-5}.

Table 7. Curing Agents For Epoxy Resins[a]

Curing agent type	Characteristics	Typical materials
aliphatic amines	Aliphatic amines allow curing epoxy resins at room temperature and thus are widely used. Resins cured with aliphatic amines, however, usually develop the highest temperatures during the curing reaction and therefore the mass of material that can be cured is limited. Epoxy resins cured with aliphatic amines have the greatest tendency toward degradation of electrical and physical properties at elevated temperatures.	diethylenetriamine (DETA), triethylenetetramine (TETA)
aromatic amines	Aromatic amine-cured epoxies usually have a longer working life than do aliphatic amine-cured epoxies. Aromatic amines usually require an elevated-temperature cure. Many of these curing agents are solid and must be melted into the epoxy, making them relatively difficult to use. The cured resin systems, however, can be used at temperatures considerably above those safe for aliphatic amine-cured resin systems.	*meta*-phenylenediamine (MPDA), methylenedianiline (MDA), diaminodiphenyl sulfone (DDS or DADS)
catalytic curing agents	Catalytic curing agents have a longer working life than aliphatic amine curing agents, and, like the aromatic amines, normally require curing at 95°C or above. In some cases, the exothermic reaction is critically affected by mass of resin mixture.	piperidine, boron trifluoride-ethylamine complex, benzyldimethylamine (BDMA)
acid anhydrides	The recent development of liquid acid anhydrides provides curing agents that are easy to work with, have lower toxicity than amines, and offer optimum high-temperature properties of the cured resin. These curing agents are becoming more and more widely used.	nadic methyl anhydride (NMA), dodecenylsuccinic anhydride (DDSA), hexahydrophthalic anhydride (HHPA), alkendic anhydride

[a] Refs. 1–2.

RTV Silicones and Flexible Resins. RTV silicones and flexible resins are by far the most widely used silicones. These flexible materials have excellent thermal-shock resistance and low internal curing stresses. Some can be cured at room temperatures and others at relatively low baking temperatures. Most are pigmented or colored, but several clear, flexible resins are available. These clear silicone materials are increasingly used because they have most of the good properties of the pigmented materials as well as optical clarity. The combination of flexibility and clarity facilitates cutting and repair when needed.

The cure of some clear, flexible resins is inhibited by contact with certain materials. Notable inhibiting materials are sulfur-vulcanized rubber and certain RTV silicone rubbers. This problem can usually be overcome with a coating of noninhibiting material on the component.

Silicone Gels. Silicone gels, as the name implies, exist in a gel state after being cured. Although these materials are very tough, they are usually used in a can or case. Silicone gels have the interesting advantage of allowing test probes to be inserted through the gel for electrical checking of circuits and components. After the test probes are withdrawn, the memory of the gel is sufficient to heal the portion that has been broken by the probes.

Rigid Solventless Silicones. Rigid solventless silicones are not used as widely as the other silicones because their resistance to thermal shock and cracking is less than that of the flexible materials, and because the rigid solventless resins are not as convenient to work with as the room-temperature curing materials. However, where the general properties of silicones are desired and rigidity is preferred to flexibility, the rigid solventless resins should be considered.

Polyesters. Polyester resins are widely used as embedding materials because of the wide range of flexibility and viscosity they offer (see Polyesters). These materials are popular because of their low cost and overall good electrical characteristics, especially low electrical losses. However, polyesters are inferior to epoxies with regard to adhesion, shrinkage, shock resistance, and humidity resistance.

Polyester encapsulating materials are formulated from: (*1*) a base resin, unsaturated polyester, (*2*) a monomer, and (*3*) a curing agent. A number of basic polyester resins are available. Depending on how they are prepared, these are generally referred to as general purpose, flexible, and fire-resistant resins, respectively. Some polyester resins incorporate bisphenol A or isophthalic anhydride in their structure for improved properties. Monomers are generally incorporated in polyester formulations to achieve low viscosity, low cost, and varied end properties. Properties of polyesters are affected not only by the type of monomer but also by its concentration in the formulation. Higher percentages of monomer generally cause greater shrinkage and more prevalent cracking. Curing systems do not greatly affect the end properties of the polyester, but they do affect storage life and processing conditions. Depending on the choice of catalyst and the reactivity of the polyester, cure cycles can vary from very rapid at room temperature to extended high temperature cures. The effect of fillers on polyesters is similar to that on epoxies, but more discretion must be used in their choice. Certain fillers are capable of prematurely gelling the polyester or of hindering its cure. Calcium carbonate is the most widely used filler for polyesters.

Polyurethanes. Cured polyurethanes are generally very tough and have outstanding abrasion resistance (see Urethane polymers). Formulations can be varied to produce soft elastomers or tough solids. Tear strength is high compared to that of other flexible materials. Polyurethanes offer excellent shock absorption because of their viscoelastic nature, and they produce very low internal stress. For these reasons they are often used on delicate electronic devices. Urethanes retain their good mechanical properties over a wide temperature range (cryogenic temperatures to 155°C for a short term, and 130°C continuous exposure). But they have limited life in high temperature, high humidity environments. Urethanes are also resistant to a wide range of solvents, offer excellent resistance to oxygen aging and have good electrical properties. Adhesion to most substrates is better than that of other encapsulation materials.

Cured polyurethanes are reaction products of an isocyanate, a polyol, generally polyether or polyester, and a curing agent. Polyester-based urethanes generally have better flexibility; but the use of polyethers offers better chemical resistance and hydrolytic stability. Curing agent selection influences curing characteristics as well as end properties. Diamines are the most common curing agents for polyurethanes.

Polysulfides. Liquid polysulfide resins are widely used in potting electrical connectors (qv). The cured polysulfide rubber is flexible and has excellent resistance to solvents, oxidation, ozone, and weathering (see Polymers containing sulfur, polysulfides). Gas permeability is low, and electrical insulation properties are good at temperatures between −53 and +150°C. At 25°C, cured polysulfide rubber has a volume resistivity of 10^9 Ω-cm and a 1-MHz dielectric constant of 7.5. Polysulfide rubber resins are the same chemical class as the polysulfide rubber resins used in modifying epoxy resins.

Polybutadienes. Polybutadienes are a recently developed group of thermosetting materials that have excellent electrical properties, high thermal stability, and outstanding resistance to water and aqueous solutions. Molding compounds are formulated from essentially all-hydrocarbon butadiene–styrene copolymers, polybutadienes, and blends of these two. They are high viscosity resins that are cured by peroxides, and exhibit high exotherms when formed. Disadvantages of these resins are brittleness and high shrinkage in curing. Currently, few manufacturers produce polybutadienes (see Elastomers, synthetic).

Low-Density Foams. Since most liquid resins can be made into low density foams by addition of selective foaming or blowing agents, most of the resins discussed previously are available in formulations that can be foamed in place. Epoxy and silicone foams are used in many embedding applications, but the polyurethanes are by far the most widely used foams. Urethane foams do not require blowing agents since gas is liberated during polymerization. These foams cure at room temperature or a low baking temperature and are relatively easy to work with, particularly the prepolymer foams.

Generally, foams have lower electrical losses, lower dielectric strength, lower thermal conductivity, and less mechanical strength than high density resins. Changes in these properties are usually almost directly proportional to foam density (see Foamed plastics).

Allylic Resins. Diallyl phthalate and diallyl isophthalate are the most widely used allylic resins. Although casting formulations are possible, transfer molding compounds comprise the largest embedding applications. These resins are noted for their excellent electrical properties and retention of those properties under environmental extremes. High insulation values are maintained in high humidity and up to 175°C for diallyl phthalate and 205°C for diallyl isophthalate (see Allyl monomers and polymers).

The allylic resins are generally filled to enhance their dimensional stability and mechanical properties. Diallyl phthalates offer very low after-shrinkage and good chemical resistance. Freedom from ionic and other corrosive impurities make them compatible with sensitive semiconductor elements. The electrical loss characteristics and dielectric strength do not vary as greatly with temperature as do those of epoxy and phenolic compounds. Allylic resins have high resistivity compared to other plastics.

Fillers. Fillers play a most important role in the use of resins for embedding. Fillers (qv) are additives, usually inert, capable of modifying nearly any basic resin properties in the direction desired. Fillers overcome many of the limitations of the basic resins. The proper use of fillers can produce major changes in properties such as thermal conductivity, coefficient of thermal expansion, shrinkage, thermal-shock resistance, density, reaction exotherms, viscosity, and cost. Although fillers can be used with all resins, they are used mainly with thermosetting resins, owing to the basic brittleness of unfilled thermosets, and the ease with which they can be mixed into liquid thermosets.

Costs and effects on resin properties of the more commonly used fillers are given in Table 8. Because of the large number of materials and suppliers available, this listing is not comprehensive.

Table 8. Costs and Effects on Resin Properties of Commonly Used Fillers[a]

		Property increase						Property decrease					
Type of filler[b]	Approx cost, ¢/kg	Thermal conductivity	Thermal-shock resistance	Impact resistance	Compressive strength	Arc resistance	Machinability	Electrical conductivity	Cost	Cracking	Exotherm	Coefficient of expansion	Density	Shrinkage
Bulk														
sand	2.204	X			X				c		X	X		X
silica	2.20–4.41	X			X				c		X	X		X
talc	2.20–8.82						c		X		X			X
clay	2.20–6.61		X		X		c		X	X	X	X		X
calcium carbonate	1.10–11.02						c		X		X			X
calcium sulfate (anhydrous)	4.41–8.82	X				X	c		X		X	X		X
Reinforcing														
mica	6.61–19.84		X	c					X	X				
asbestos	4.41–11.02		X	c						X				
wollastonite	4.41–6.61	X	X	c					X	X				
chopped glass	99–80		X	c						X				
wood flour	20.5		X	X			X	c	X				c	
sawdust			X	X			X	c	X				c	
Specialty														
quartz	4.41–11.02	X		X	X					X		c		X
aluminum	11.02–33.10	X	X		X						X	c		X
hydrate alumina	6.61–13.22					c								
Li-Al silicate		X			X					X	X	c		X
beryl		X			X				X	X		c	X	
graphite	13.22–66.12						X	c						
powder metals		X	X	X	X		X	X	X	X	X	X		X
low-density spheres	165.30–330.60											c		

[a] Refs. 1–2.
[b] Particle size of fillers are 74 μm (200-mesh) or finer, except for sand, hollow spheres, and reinforcing fillers that depend on particle configuration for the desired effect.
[c] Denotes most significant property of each filler listed.

Product Design With Thermosetting Embedding Materials. Table 9 matches some of the most important design objectives for embedded products with the best available embedding materials.

WAXES AND THERMOPLASTIC EMBEDDING MATERIALS

Owing to the low temperature stability of waxes and bitumens, these materials have only limited embedding application. The use of most thermoplastic resins is also restricted by the high molding pressures and temperatures required in their processing. The materials having some use are described briefly below.

Waxes. Wax is probably the original encapsulating and embedding material. It has excellent electrical properties and its use is extremely simple (see Waxes). Most waxes are fluid at 95°C for dipping or casting. Their low melting temperature limits their utility to approximately 52°C. Asphalts and tars have similar properties and are even less expensive. Cellulose esters (see Cellulose derivatives) are used in the same way as the waxes. Mechanically, they are stronger but their moisture absorption is high, and they are inferior to wax electrically.

Hydrocarbons. The hydrocarbon polymers, polyethylene, polystyrene, and polybutadiene, all have outstanding electrical properties, particularly in high-frequency applications where their low dielectric constants, 2.3–2.5, and outstanding low loss tangents, 0.0003–0.0005, make them the chosen materials. Polystyrene resins are not widely used for embedding applications because of certain practical limitations. They are generally useful only below 125°C; they have a long cure cycle and high shrinkage upon cure, causing a tendency to crack during the curing operation. Their normal expansion is high and curing is air-inhibited, causing the exposed surface to remain tacky.

Vinyls. Poly(vinyl chloride) and its copolymers are used in encapsulation techniques (see Vinyl polymers). Conformal coatings are possible using plastisols or organosols. Plastisol applications make use of a suspension of finely divided particles of polymeric liquid plasticizer. The component is dipped into the plastisol and the adhering layer is baked to form a solid impervious layer.

An organosol is a true solution of the plastic in solvent. It is usually applied by brushing or spraying. The solvent is subsequently removed in the baking operation. Organosol coatings must be much thinner than plastisol coatings owing to the danger of film rupture by solvent vaporization.

Decorative Embedding

The vast majority of embedding applications serve electrical or mechanical functional objectives, but decorative embedding also enjoys widespread use. Examples are primarily decorative items embedded in an unfilled resin having some degree of transparency, although nontransparent art objects are also cast using materials similar to those used in functional embedding. Also, scientific specimens are often embedded for cross-sectioning investigations.

Decorative embedding presents some unique limitations on selection of material and process. For use in decorative embedding other than nontransparent art objects, the transparency requirement for most decorative embedding poses several major limitations. Most embedding materials are not adequately transparent, especially in

Table 9. Selection of Embedding Materials For Important Product Design Objectives[a]

Design objective	Material candidates
adhesion of resin to assembly	Resins differ in their adhesive tendencies. Excellent bonding is usually obtained with the epoxies and urethanes. If adhesion of the base resin is a problem, primers can often be used to advantage. Cleanliness of parts is also important.
low electrical loss and/or low dielectric constant	Silicones and thermosetting hydrocarbons are outstanding among embedding resins in providing low electrical losses and low dielectric constants. Silicones are noted for their retention of good electrical properties even at elevated temperatures and high frequencies. Many other resins with low electrical values at room temperature improve as temperature is increased, particularly above 100°C.
thermal stability	Some of the most thermally stable resin systems are silicones, novolac epoxies, anhydride-cured epoxies, aromatic-amine-cured epoxies, and thermosetting acrylic resins. Most of the higher temperature resin formulations exhibit low weight loss in prolonged heat aging at temperatures to 200°C. There are many differences among these high-temperature resins, however, in retention of physical and electrical properties. Thus, it must be decided which of these high-temperature properties are most important. Thermal stability, both with respect to weight loss and retention of mechanical properties at elevated temperatures, can often be increased by addition of reinforcing fillers such as glass fibers.
cost	Polyesters are perhaps the best candidates where lowest material cost is required. Other possibilities however, are the low-density foams, which have a low cost per unit volume despite their higher basic resin cost, and epoxy resins heavily loaded with low-cost fillers.
room-temperature cure	Room-temperature curing formulations are available in silicone-rubbers and resins, urethanes, polyesters, epoxies, and polysulfides. Although room-temperature curing is mandatory in many applications, optimum resin properties are usually obtained by heat curing. In some instances, however, extended curing time at room temperature is equivalent to a heat cure.
low-temperature flexibility	Silicone rubbers and flexible silicone resins are the best materials for this requirement.
rigidity	Epoxy and polyester resins are prime candidates where rigidity is required. But some rigid formulations are either brittle or have cracking tendencies. Hence some compromise between hardness, toughness, brittleness, and crack resistance must be made.
flexibility	The most flexible materials are RTV silicone rubbers, urethane resins, and polysulfide rubbers. Flexible resins are produced from normally rigid resins by addition of various flexibilizers or modification of the base material, and resins such as epoxies and polyesters are available in flexible formulations. Resins are available in nearly any degree of flexibility or hardness desired.
clarity	Water-clear resins are available in the epoxy, polyester, and silicone materials. Clear epoxies are normally rigid; clear silicones are soft or flexible. Most other resins are not water clear but are amber or light colored so that parts can be seen through the cured, unfilled resins.
repairability	Repairability is easiest with the flexible materials because they can be cut easily. Especially repairable are the silicone gels. Rigid resins can be repaired by softening or dissolving in solvents or by heat softening.
low weight	Low-density foams, especially urethane foams, are most common, both in rigid and flexible formulations; many other resins, especially

Table 9. (*continued*)

Design objective	Material candidates
	epoxies and silicones, are also available in low-density foam formulations. Foams are normally available in densities from (0.9–9 kg/m^3). In addition to low-density foams, low-density resin systems (containing low-density fillers) are available in all embedding resins. Density of these resin systems is, of course, considerably higher than that of foams. However, physical and environmental properties are usually much better.
high thermal conductivity	Differences in thermal conductivities of base resins are slight. Hence high-thermal conductivity embedding materials are those incorporating large amounts of filler, especially large-particle fillers such as coarse sand, or aluminum oxide, magnesium oxide, or beryllium oxide. There is definitely a limit to the thermal conductivity obtainable in base resins, and thermal conductivity does not increase in proportion to the thermal conductivity of fillers used.

[a] Refs. 1–2.

thick sections. Most exhibit some light brown color. Second, even those that are transparent do not have a high transparency, so that cloudiness or a tinted appearance occurs in other than rather thin sections. Third, some that do have high transparency are used with monomers or solvents that can dissolve, attack, or discolor the object being embedded, especially biological or organic specimens. In this class are clear polyesters, which contain styrene or other monomers, and acrylics, which are used in liquid solvent carriers. A fourth basic limitation in decorative embedding is that, without close controls, the curing, baking, or polymerization process often involves heat which may darken either the embedding material or the object being embedded. It is usually desirable to use either a room temperature or low baking temperature embedding material, and an embedding material with lowest possible temperature rise during cure.

There are other problems in decorative embedding which demands use of embedding materials used at the lowest possible temperatures. These problems are (1) thermal expansion of the embedding material and embedded specimen, and (2) shrinkage of embedding material during hardening or cure. Both of these factors, especially in combination, readily result in cracking of the embedding material. The cracking problem is aggravated when high differentials of thermal expansion exist between embedding material and embedded specimen, and when the embedded specimen has sharp edges and corners, which result in stress point. Fillers are used in functional embedding to reduce this problem, but the transparency requirement prohibits use of fillers for decorative embedding.

Although critical limitations exist for decorative embedding, compatible combinations of embedding materials and embedded objects do exist. Through proper investigation and analysis, most requirements can be met. The problems, however, are very real, and must be both fully considered during investigation and fully controlled during manufacturing if quality products are to result.

Standards and Controls for Embedding

Embedding, like all industrial processes, requires a well-planned set of manufacturing standards. Frequently, the importance of the embedment process in the overall manufacturing procedure is overlooked because the material cost of the embedding resin may be insignificant relative to the cost of the electronic assembly. The importance of proper controls cannot be stressed too strongly from the standpoint of higher yield, lower cost, and increased efficiency, and from the standpoint of safety. Frequently, the potential hazard of handling embedding resins is not nearly so well recognized in an electronics plant as it would be in a chemical plant.

Quality control should cover both manufacturing and production control, including process specifications covering manufacturing instructions, in-process tests, safety procedures, well-organized record keeping, and any specialized manufacturing precautions and final product control, including tests designed to ensure that the product meets customer requirements and good monitoring and record keeping of final product properties to eliminate many processing and product problems before they arise. Several of the specifications and standard test methods most useful in assuring material and product control are presented in Table 10.

Finally, the key to reliable embedding techniques lies in material selection and maintaining the material quality by extensive testing. Applicable publications include MIL, FED, and ASTM specifications, as well as SPI (Society of the Plastics Industry), NEMA (National Electrical Manufacturers' Association), and U.S. Underwriters' Laboratories Standards. In most cases, material suppliers will provide a more exact definition of the quality control procedures for these materials and their applicable government and industrial specifications.

Table 10. Some Specifications and Standards for Embedding

U.S. Military Specifications
    MIL-I-27,27A
    MIL-I-6923D
    MIL-I-17023C
    MIL-T-5422, 5422B
    MIL-STD-202, 202A
    MIL-R-10509C
    MIL-E-5272A
    MIL-S-8516
U.S. Federal Specification
    FED-L-P-406b
American Society For Testing and Materials Standard Methods
    ASTM D 149, D 149-44, D 149-59
    ASTM D 150-59T
    ASTM D 257-57T
    ASTM D 648
Institute of Electrical and Electronic Engineers Standards
    IEEE No. 1
    IEEE No. 50
National Association of Electrical Manufacturers' Classification on temperature
    classifications for electrical insulation

BIBLIOGRAPHY

"Embedding" in *ECT* 2nd ed., Vol. 8, pp. 102–116, by F. J. Modic and D. A. Barsness, General Electric Company.

1. C. A. Harper, *Handbook of Plastics and Elastomers,* McGraw-Hill Book Co., Inc., New York, 1975.
2. C. A. Harper, *Handbook of Materials and Processes for Electronics,* McGraw-Hill Book Co., Inc., New York, 1973.

General References

M. C. Volk, J. W. Lefforge, and R. Stetson, *Electrical Encapsulation,* Reinhold Publishing Corp., New York, 1962; see especially Chapt. 5, p. 54.
C. A. Harper, *Electronic Packaging with Resins,* McGraw-Hill Book Co., Inc., New York, 1961.
H. Lee and K. Neville, *Epoxy Resins,* McGraw-Hill Book Co., Inc., New York, 1957.
J. R. Lawrence, *Polyester Resins,* Reinhold Publishing Corp., New York, 1960.
M. B. Horn, *Acrylic Resins,* Reinhold Publishing Corp., New York, 1960.
L. E. Neilsen, *Mechanical Properties of Polymers,* Reinhold Publishing Corp., New York, 1962.
The Encyclopedia of Plastics Equipment, Reinhold Publishing Corp., New York, 1960.
Modern Plastics Encyclopedia, Modern Plastics, New York, annual publication.
J. Delmonte, *Plastics in Engineering,* 3rd ed., Penton Publishing Co., Cleveland, Ohio, 1949.
H. R. Simonds, *Source Book of the New Plastics,* Reinhold Publishing Corp., New York, annual publication.
C. A. Harper, *Plastics for Electronics,* Kiver Publications, Inc., Chicago, 1964.
C. G. Clark, "Potting, Embedment, and Encapsulation," *Space Aeronautics* (Dec. 1961).
J. Dexter, "Using Silicones to Meet Performance Demands in Electronic Equipment," *Mach. Des.* (May 24, 1962).
J. W. Hawkins, "Silicones—Coatings, Encapsulants, Potting, Embedding," *Electronic Design News* (July 1962).
F. L. Koved, "Encapsulating to Military Specifications," *Electron. Ind.* (July, 1963).
C. V. Lundberg, "A Guide to Potting and Encapsulating Materials," *Mater. Des. Eng.* (May, June 1960).
"Properties of Encapsulating Compounds," *Electron. Prod.* (April, 1963).
D. C. R. Miller, "High-Temperature Flexible Potting Resins Offer Unique Properties," *Electronics and Communications* (Oct. 1962).
J. M. Segarra, "A New Embedding Procedure for the Preservation of Pathological Specimens, Using Clear Silicone Potting Compounds," *Am. J. Clin. Pathol.* **40,** 655 (Dec. 1963).
H. L. Uglione, "Evaluation of Polyurethans, Polysulfides, and Epoxies, for Connector Potting and Molding Applications," *Insulation* (Apr. 1963).
"Potting and Encapsulation Technology Update," *Circuits Manufacturing* (Aug. 1977).

Journals

Electrical Design News (monthly); *Electronic Design* (weekly); *Electronics* (weekly); *Insulation* (monthly); *Materials in Design Engineering* (moithly); *Modern Packaging* (monthly); *Modern Plastics* (monthly); *Plastics World* (monthly); *Product Engineering* (weekly); *SPE Journal* (monthly); *Electronic Packaging and Production* (monthly); and *Circuits Manufacturing* (monthly).

CHARLES A. HARPER
Westinghouse Electric Corporation

EMULSIONS

An emulsion is a mixture of two or more immiscible liquids, one being present in the other in the form of droplets. Industrial emulsions frequently contain a solid as a third ingredient. Strictly speaking, a mixture of a solid dispersed in a liquid is a suspension. A variation in which molten waxes are emulsified is called a wax emulsion, although it is in fact a suspension or dispersion at room temperature. In the classic emulsion, the oil may either be dispersed in the water (oil-in-water or o/w emulsion) or the water dispersed in the oil (water-in-oil, w/o, or inverse emulsion).

This terminology is important because the emulsion characteristically assumes the properties of the external phase, a key factor in emulsion formulation and design. For example, an oil-in-water emulsion can be diluted with water or dried by evaporation leaving the other ingredients as a film.

Emulsions are found in nature, two of the principal examples being milk and rubber latex. These emulsions are stabilized by natural emulsifying agents. In a like fashion, commercial emulsions require emulsifying agents.

Emulsions are used in a variety of fields such as textiles, leather, and metal treatment; foods, cosmetics, pharmaceuticals, and paints; in agricultural chemicals, polymerizations, cleaning and polishing, and ore and petroleum recovery.

Emulsions are inherently unstable systems and the risk of deteriorating during storage is greater than with a nonemulsified product. Emulsion technology, though seemingly based on simple interface principles, is highly complex, especially when dynamic and static conditions are considered (see Cosmetics; Food processing; Textiles; Leather; Pharmaceuticals; Insect control technology; Paint; Polymers; Polishes; Petroleum).

Properties

Type of Emulsion. The emulsion may be o/w or w/o, designating the continuous and discontinuous or internal phases. In general, o/w emulsions conduct electricity, are dilutable with water, feel more like water, dry (lose water) rapidly, can be washed away (off the skin, etc), are more corrosive, and exhibit the aqueous properties of the continuous phase. On the other hand, w/o emulsions conduct electricity poorly if at all, may be diluted with oil or solvents, feel more like oil, resist drying or loss of water, although they lose a volatile solvent readily, are difficult to wash away, are less corrosive, and, in general, depending upon the oil phase, exhibit the properties of the continuous oil phase.

Dual emulsions, in which the internal phase is an inverse emulsion, are occasionally prepared accidentally. Because the optimum conditions for the primary and secondary emulsions are in conflict, dual emulsions generally lack the stability required for commercial preparations, and hence are rarely used.

The choice of emulsifier, which is of prime importance, is determined by the ionic nature, chemical type, HLB (hydrophile–lipophile balance), and the amount used.

Volume Ratio of External to Internal Phase. A low-internal-phase (LIP) emulsion assumes the overall characteristics of the external phase. Very dilute (LIP) emulsions are often difficult to handle because of the dilution effect on the emulsifier. On the other hand, high-internal-phase (HIP) emulsions exhibit higher apparent viscosities as the internal phase volume ratio increases and finally a thick, nonflowing paste is formed. Increased viscosity may improve stability somewhat.

The external-phase properties determine dispersibility (in water or oil) and electrical conductivity, and, in combination with other basic properties, viscosity, wetting characteristics, and feel.

The internal-phase properties are much less important with respect to the properties of the final emulsion. They are related to inversion characteristics and ease of preparation. The difference in specific gravity of external and internal phases is a key factor in creaming.

Particle Size. The particle size of a liquid emulsion is related to the method of preparation, the energy input, the viscosity difference between the phases, and the type and amount of surfactant used. With reference to small-particle-size formation, emulsions may be classified into low emulsifier formulas requiring appreciable mechanical input, and high emulsifier formulas that require only moderate mechanical effort. Energy input is an important variable (1). Particle size generally decreases with vigorous agitation, smaller viscosity difference between the two phases, and the use of a larger amount of the proper surfactant.

The coarsest and the smallest particle-size emulsions may be prepared in similar fashion, in which the mechanical work input is less than in all other methods, ie, by simple stirring. In some instances, a fine particle size may be achieved by the inversion technique described below.

At present, emulsion particle size is best determined (2–4) by photomicroscopic techniques, particle size counting (5), and a particle size distribution curve or profile. The particle size of o/w emulsions may be studied by means of a Coulter counter, a device that measures the change in conductivity of the continuous phase as each particle passes through a minute orifice. All present measuring systems require extensive dilution, thus the measurement is not being made on the original emulsion. Dilution with 0.15% Tween 80 in water stabilizes the emulsion during particle size evaluation (see Size measurement of particles).

Particle size determination is not a commonly employed tool in industrial studies because it is complex and time-consuming and needs to be repeated several times especially in the evaluation of many closely related formulas. Particle size is often estimated by the appearance of a thin film (a form of light scattering). Table 1 describes the relationship of particle size to the appearance of an emulsion film as it drains from the side of a jar.

In an emulsion, the larger the particle, the greater is the tendency to coalescence and further increase the particle size. Thus fine particles promote stable emulsions. Coalescence may be retarded by an emulsifier or gum which provides a protective colloid action. Increasing the viscosity of the continuous phase also mechanically retards coalescence.

Table 1. Effect of Particle Size of Dispersed Phase on Emulsion Appearance

Particle size	Appearance
macroglobules	two phases may be distinguished
greater than 1 μm	milky-white emulsion
1–0.1 μm	blue–white emulsion, especially a thin layer
0.1–0.05 μm	gray semitransparent, dries bright
0.05 μm and smaller	transparent, dries bright

The particle size of an emulsion may be reduced by (*1*) increasing the amount of emulsifier, (*2*) improving its HLB, (*3*) preparing the emulsion by phase inversion to provide an extended internal phase at the time of inversion to the final emulsion type, and (*4*) improved agitation.

Average particle size and particle size distribution may be used as a product specification.

Characteristics

Emulsion properties and characteristics are not necessarily related to the properties of the major ingredients and can frequently be tailor-made to suit application requirements. They are built into an emulsion during formulation. Table 2 summarizes emulsion characteristics and relates them to their basic origins.

Characteristics of emulsions must be considered for at least three and possibly more stages in the life of the product: (*1*) preparation, (*2*) storage and shipping, and (*3*) utilization. Other factors that may be included are: preparation at site, packaging, conditions after application (utilization), and others.

Table 2. Characteristics of Emulsions

Characteristics	Reasons
appearance	
clarity	
clear	small particle size; matched refractive indexes
translucent	medium particle size
opaque	large particle size; unmatched refractive indexes
color	
white	large particle size; unmatched refractive indexes
gray	medium-small particle size; unmatched refractive indexes
colors	colors in continuous phase
viscosity	
thick (high)	HIP emulsion; small particle size, thickeners in outside phase
thin	LIP emulsion with no thickener
dispersibility	
in water	o/w
in oil	w/o
ease of preparation	
high, easy	emulsifier, solution level; low viscosity concentrate
low, difficulties	low emulsifier level
re-emulsification	emulsifier selection; emulsifier level
stability	
high, good	emulsifier selection; emulsifier moderately high
low, poor	low emulsifier levels; emulsifier selected for other property
stable to electrolytes	emulsifier selection
on evaporation (o/w)	emulsifier selection; emulsifier level
spoilage	preservative selection; sterile packaging
wetting-spreading	
high	emulsion type; emulsifier selection
low	emulsifier selection
particle size	
small	emulsifier selection; emulsifier level
large	emulsifier level

Appearance. The word emulsion suggests a milky-white fluid, although emulsions may actually range from a white milk through gray translucence to sparkling clarity, at times even showing a range of colors. Appearance is governed by (1) particle size, (2) difference in refractive index of the two phases, (3) colors, and (4) presence of solids.

Clarity or transparency may be gained either by having both phases of the same refractive index or by dispersing the internal phase in such small particles that refraction does not occur because the particle size of the emulsion is much smaller than the wavelength of light (see Table 1).

Even in milky-white emulsions variations can be observed, some with a greyish, others with a bluish cast. This effect is noted in emulsions that are in a thin film; eg, when draining from the wall of a jar. As seen from Table 1, the appearance is related to particle size and thus appearance can be used in some instances as a particle size test.

Viscosity. Emulsions can be thin or thick fluids, pastes, or gels and may exhibit thixotropy or dilatency. Viscosity is influenced by (1) the characteristics of the external phase, including additives, (2) the volume ratio of the two phases, and (3) the particle size. Note that the type of emulsion is not regarded as a major influence on viscosity despite the common belief that o/w emulsions are thinner than w/o. This is true only so far as the oils frequently used are more viscous than water. The viscosity of an emulsion is essentially the viscosity of the external phase as long as it represents more than half of the total volume (see Rheological measurements).

The viscosity of the continuous phase may be increased by adding thickeners or gelling agents that are compatible with the emulsifier. Many thickeners, such as CMC (sodium carboxymethyl cellulose), methyl cellulose (see Cellulose derivatives), and natural gums (qv) or clays (qv) may often be added with little or no change in the basic emulsifier. If the thickener or gelling agent is a surfactant in its own right, the overall balance of the emulsifier probably requires readjustment. Emulsion viscosity can often be reduced by increasing the proportion of the continuous phase, usually water. The adding of polar solvents, such as alcohol or acetone that may reduce viscosity usually causes a marked reduction in emulsion stability. Presumably the emulsifier, being more soluble in the polar solvent, is extracted from the interface which is then weakened. Thickening or thinning of the dispersed phase usually has little or no effect upon the overall viscosity. In normally fluid o/w polymer emulsions, viscosity differences can be obtained by varying the nature of the adsorbed water structure around each particle by means of a change in surfactant or electrolyte concentration (6) (see Soap; Surfactants).

Thickening agents, particularly of the surfactant type, may either have a simple thickening or a thixotropic effect. A thixotropic product may appear to be essentially gel-like in character until it is agitated mildly, when it becomes fluid. Upon standing, it becomes gel-like once again. Emulsions may also be formulated that have gel-like characteristics which, upon stirring, become liquid. This is not thixotropy but merely an initial thickening action.

Changing the ratio of the external and internal phases has a marked effect. As the proportion of internal phase is increased beyond 50 vol %, the emulsion viscosity increases to a point where the emulsion is no longer fluid. When the volume of the internal phase exceeds that of the external phase, the emulsion particles become crowded. Under these conditions, particle size, particle charges, and similar relationships assume greater importance in determining the apparent viscosity. Theoretically, the maximum volume that can be occupied by uniform spherical particles for the dispersed phase of an emulsion is 74% of the total volume. Emulsions may be prepared that have an internal phase as high as 99%. In these cases, considerable distortion from the usual spherical particle shape of the dispersed phase occurs. The actual crowding and distortion that occur at high concentrations of the internal phase may be seen for o/w emulsions in Figure 1(**e**–**h**) and for w/o emulsions in Figure 3(**a**–**h**). Corresponding viscosity changes may be seen in Figure 1(**a**–**d**) and Figures 2 and 3(**a**–**d**). These figures further illustrate that, contrary to general opinion, viscous or nonfluid emulsions are not all of the w/o type. Equivalent apparent viscosities may be built up with either type.

Structural viscosity is also characteristic of pigment suspensions and is often substantially reduced by the addition of emulsifiers. The viscosity of emulsions and suspensions frequently increases with aging.

Emulsion viscosity may be lowered by increasing the proportion of the continuous phase or decreasing the viscosity of the continuous phase. To lower the viscosity of suspensions, the addition of various types of surface-active agents is frequently effective, probably because of flocculation or deflocculation action.

Emulsion viscosity is increased by (*1*) adding thickeners such as soap gels, lipophilic fatty acid esters, gums, or alumina gel to the continuous phase; (*2*) raising the proportion of the internal phase; (*3*) reducing the particle size or clumping of existing particles; and (*4*) incorporating fine air particles as a third phase.

The preparation of cosmetic lotions is a good example of viscosity control. The object is to prepare a lotion that appears heavy, it must have a high apparent viscosity, although it must remain fluid on long standing. These requirements are met by a formula containing approximately 90% continuous phase (water). The desired results are obtained by a proper balance of emulsifiers and viscosity builders (soaps or lipophilic fatty acid esters). A major difficulty encountered in these formulations is that the gel structure changes under varying storage conditions, and the product frequently sets to a semisolid that does not pour (see Cosmetics).

Viscosity provides a useful control for emulsions, though care must be taken to assure that the samples have identical histories.

Stability. Considered here is the physical stability, not chemical changes resulting from oxidation, etc. Although the aim is usually good stability, specific instability may be designed into a formula if desired. Instability of emulsions occurs by creaming (settling) and coalescence. The two actions must be considered separately even though they frequently occur side by side.

Creaming (or sedimentation (qv)) is exhibited if the two phases differ in specific gravity and if the emulsion particles are so large initially, or grow to such a size, that they are not responsive to Brownian movement. If no coalescence or agglomeration occurs, the emulsion may be redispersed—a shake-well-before-using system—and provide a satisfactory product.

When an emulsion creams, globules of the internal phase become more and more crowded in the cream layer and show an increased tendency to coalesce, leading to complete phase separation. Creaming may be sufficiently complete to effect a separation of most of the continuous phase without coalescence. Total phase separation is a result of agglomeration and coalescence of the dispersed phase.

Creaming may be reduced by (*1*) adjustment of the specific gravity of one or both

Figure 1. O/w emulsions: (**a**–**d**) the increase in viscosity of an o/w emulsion with an increase in the proportion of oil (dispersed phase); (**e**–**h**) photomicrographs (×250) of emulsions **a**–**d**, respectively.

Figure 2. Amounts of water (left) and oil (right) used in preparing w/o emulsion shown in Figure 3.

phases, (2) lowering of particle size, (3) changing the particle charge, and (4) increasing the viscosity of the continuous phase.

Coalescence of emulsion particles is often irreversible and is a case of true instability. It may be reversible if a relatively large proportion of emulsifier is employed and if the emulsifier has been chosen so as to afford the greatest ease of emulsification. Most emulsions that are prepared with considerable mechanical effort are not reversible. Coalescence occurs when the interfacial film ruptures at the point of juncture of two particles of the discontinuous or internal phase. The choice of emulsifier (assuming that it is adjusted to the proper HLB) is of utmost importance to provide stability against coalescence. Coalescence may be reduced by (1) reduction of creaming, (2) increasing the charge on the particles, and (3) increasing the viscosity of the continuous phase.

Stability to a freeze–thaw cycle is required for many products; ie, destruction of the interface by ice crystals formed during freezing is avoided. Most likely this is a direct physical action. Careful choice of emulsifier and use of an adequate amount can markedly reduce this form of emulsion destruction.

Stability to elevated temperature is an important requirement. If viscosity is reduced by elevated temperature, a reduction of stability can certainly be expected. Stability at elevated temperatures may also be affected by changes in solubility of the emulsifiers in one or the other phase, thus altering their distribution and affecting the interface. Temperature stability is also related to the melting point of the dispersed phase. If the temperature is raised above the melting point of the dispersed phase, conditions will be different from those for which the emulsion was designed. Incorporation of air in an emulsion may result in greatly reduced stability.

Under certain circumstances, stability is actually not desirable; eg, in insecticide emulsions. The insecticide is often applied from an emulsion and should break as soon as possible and deposit the active ingredients. Some salad dressings and hair formulations are designed with an oil phase that separates on a few hours' standing, with reemulsification on shaking. This allows the customer to see the oil content of the product.

Emulsion stability for practical purposes usually requires, in addition to resistance

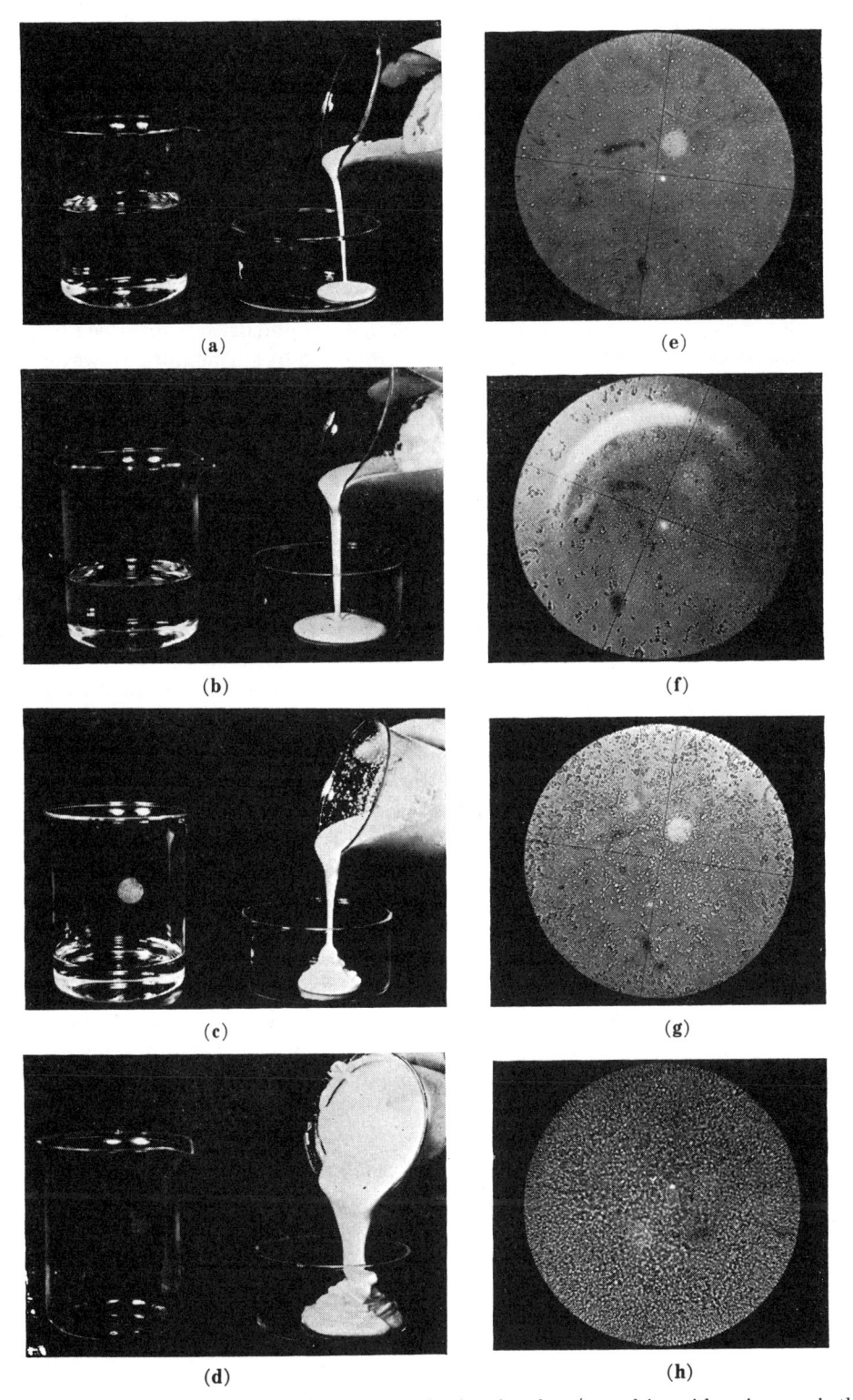

Figure 3. W/o emulsions: (**a–d**) the increase in viscosity of a w/o emulsion with an increase in the proportion of water (dispersed phase) as shown by the amounts of water left in the beaker in each instance, ie, 75, 50, 25, and 0%; (**e–h**) photomicrographs (×250) of emulsions **a–d**, respectively.

to creaming and coalescence, the maintenance of a given viscosity. Viscosity may be affected by bacterial action which can also destroy many surfactants. Cleanliness and addition of preservatives may solve these problems.

In some instances, desired demulsification may be effected by destroying the emulsifier. This is possible with the aid of acidic soaps or polyvalent salts, or a combination of cationic and anionic emulsifiers.

Wetting and Spreading. Many emulsions must be formulated to wet and spread on a variety of surfaces, eg, polish on floors (or painted surfaces), insecticides on leaves, cosmetics on skin, and paint on plaster, wallboard, wood, and dried paints (see Polishes; Paint; Insect control technology). Wetting is promoted by surfactants. However, the optimum surfactant for emulsification is seldom, if ever, the best wetting agent. Slight modifications and additions usually suffice to provide a combination of emulsification and good wetting. Choice of the additives depends upon the system, but low-HLB laurates, nonionic silicone copolymers, and fluorocarbon surfactants usually improve wetting most efficiently.

Wetting and penetration should be considered separately since the wetting (low surface and interfacial tension) beyond the necessary minimum reduces capillary attraction, and hence gives poor penetration.

Point of Inversion. Studies of emulsions of two immiscible liquids (7) without an emulsifier showed that at a certain ratio of the two liquids inversion takes place, and the external phase becomes the internal phase. The phase volume ratio at the point of inversion was found to be equal to the square root of the ratio of the viscosities of the two components.

However, an emulsifier exerts a controlling effect on the inversion (8–9). The inversion phase ratio seems to depend on the concentration of the emulsifier, its chemical nature, and its HLB. At a certain HLB inversion occurs most easily. On changing the HLB to either higher or lower values, the stability of o/w or w/o emulsions increases, reaches a maximum, and then decreases. Inversion and especially phase inversion temperatures are related to the choice of emulsifier (10).

Emulsion Concentrates. With a given formula, the ease of formation of an emulsion is controlled by the choice of emulsifier which is based on both HLB values and chemical type. Emulsifiability may be improved by matching the viscosity of the oil with the aqueous phase, increasing the emulsifier content and, if salts of acids are part of the formula, adding them with part of the aqueous phase toward the end of the process after a primary emulsion is formed.

In formulating an emulsion concentrate, its viscosity may be sufficiently high that dilution with water or a solvent presents problems, chiefly because of the wide difference in viscosity. In this case, high temperatures are employed or the concentrate is prediluted sufficiently to avoid viscosity disparity.

Dispersibility and Electrical Conductivity. The dispersibility of an emulsion is determined by the continuous phase; thus if the continuous phase is water soluble, the emulsion may be diluted with water. If, on the other hand, the continuous phase is oil soluble, the emulsion may be diluted with oil. The ease with which an emulsion may be diluted is increased by decreasing the viscosity of the continuous phase.

The electrical conductivity of an emulsion depends upon the conductivity of the continuous phase.

Feel. The initial feel of an emulsion is usually related to the feel of the external phase; thus an o/w emulsion feels like water, whereas a w/o emulsion feels oily. This feel is, of course, modified by the viscosity of the emulsion, because the greater the viscosity the greater the tendency to feel emollient or even oily. After an emulsion is applied, eg, in cosmetics, the feel changes with the evaporation of water, and depending upon the smoothness with which the emulsion inverts, it becomes more or less oily. The smoothness of inversion of a cosmetic emulsion does not influence the eventual feel, but it does influence the feel at the moment of inversion during application. If the inversion is not accomplished smoothly, the emulsion may be said to weep, sweat, or more accurately to break. As would be expected, the feel of an emulsion after inversion is largely influenced by the nonvolatile ingredients.

Drying Rate. The drying rate or loss of water depends on the type of emulsion. An o/w emulsion loses water until it is dry or in equilibrium with atmospheric rh. A w/o emulsion loses water very slowly.

Formulation

After the need or desirability for an emulsion form of product is established, the formulation is designed according to the considerations outlined below.

Major Ingredients. The major ingredient is usually determined by the intended use of the emulsion. For example, a polish based on a blend of waxes or oils is related to the surface to be protected. A weed killer (see Herbicides) or insecticide emulsion is related to the pesticide to be used. A metal-cutting oil is based on a selected mineral or vegetable oil combined with an extreme pressure additive to obtain the desired lubrication (qv). Cosmetic creams are based on oil and wax blends chosen to give the desired feel and cleansing action.

In addition to oil-soluble ingredients, water-soluble or other hydrophile compounds may be needed for a variety of reasons. For example, a cosmetic antiperspirant cream contains the water-soluble antiperspirant as the primary ingredient, and oil and wax as secondary ingredients. Hydraulic fluids (qv) contain aqueous corrosion inhibitors. Many o/w emulsions contain hydrophilic thickeners.

Emulsion Type. The choice of o/w or w/o is a pivotal decision, controlling emulsifier selection and formulation. The decision is based on use needs, such as dilution, feel, washability, and the like.

Type of emulsifiers generally refers to anionic, cationic, amphoteric, and nonionic characteristics. Anionic emulsifiers are generally less expensive than nonionic, cationic, or amphoteric emulsifiers. Nonionic emulsifiers are the least irritating followed by amphoteric and anionics, with cationic emulsifiers being the most irritating. Foaming is strongest with some anionics, less with amphoterics, and least with nonionics (see Foams). Foaming action, if desired, is usually aided by specific agents (see also Defoamers).

Some emulsifiers are incompatible. Thus anionic and cationic emulsifiers are generally not used together because one precipitates the other. Either cationic or anionics may be used with nonionics and amphoterics.

Particle Size. Particle size is a function of the efficacy of the emulsifier, the amount used, and the manufacturing techniques (dilution, energy input, equipment, temperature, cooling, etc). To achieve economic emulsion stability, it is desirable to attain the minimal practical particle size with the least amount of emulsifier.

Hydrophile–Lipophile Balance (HLB) of Surfactants. The hydrophile–lipophile balance is an expression of the relative simultaneous attraction of an emulsifier for water and for oil (or for the two phases of the emulsion system being considered). It is determined by the chemical composition and extent of ionization of a given emulsifier. For example, propylene glycol monostearate (pure) has a low HLB (strongly lipophilic); a polyoxyethylene monostearate ($H(OC_2H_4)_n OOC(CH_2)_{16}CH_3$) having a long polyoxyethylene chain has a high HLB (hydrophilic); and sodium stearate ($CH_3(CH_2)_{16}COONa$) has a very high HLB (strongly hydrophilic), since it ionizes and thus provides an even stronger hydrophilic tendency.

The HLB of an emulsifier determines the type of an emulsion that tends to be formed. However, it is an indication of the behavior characteristics and not of emulsifier efficiency. Thus emulsifiers with low HLB values tend to make w/o emulsions. For any specific problem, both the best HLB and the best chemical class of emulsifiers must be found.

The HLB number, for most nonionic emulsifiers, is merely an indication of the percentage weight of the hydrophilic portion of the molecule. If a nonionic emulsifier were 100% hydrophilic (which, of course, does not exist), it would be assigned an HLB value of 20, the factor $1/5$ having been adopted because of the convenience of handling smaller numbers.

The HLB values for most nonionic emulsifiers can be calculated from either theoretical composition or analytical data. The theoretical composition method may lead to considerable error, since many surfactants are unfortunately known by names that do not properly reflect their actual composition. Thus data obtained by actually analyzing the emulsifier are usually a better basis for determining HLB values (11) than a possibly erroneous commercial name. This is especially true among the nonionics.

Determination. For many nonionic emulsifiers, HLB values may be calculated by one of the following methods:

(*1*) HLB = $E/5$, in which E is the wt % of hydrophilic content of the molecule (or wt % oxyethylene for ethylene oxide condensation products) (12).

For example, in a polyoxyethylene stearate, in which the oxyethylene content was determined to be 85%, HLB = 85/5 = 17. Likewise, in a polyoxyethylene oleyl alcohol condensation product in which the oxyethylene content was determined to be 25%, HLB = 25/5 = 5.

(*2*) HLB = $20(1 - S/A)$, where S is the saponification number (13) of the emulsifier (ester type). The saponification value or number is a measure of the amount of alkali required to saponify a definite weight of fat, and is commonly expressed as the number of milligrams of potassium hydroxide required to saponify one gram of fat. In this equation, A is the acid number (14) of the fatty acid moiety of the surfactant. The fatty acid is separated from the emulsifier by saponification with excess alkali, made acid with inorganic acid, extracted from the aqueous phase with hexane, and recovered by evaporation of the solvent. The acid number is determined on the recovered acid and is the neutralization value or equivalent expressed as the number of milligrams of potassium hydroxide required to neutralize one gram of fat.

For example, in a typical commercial-grade glycerol monostearate in which the saponification number was determined to be 175 and the extracted acids had an acid number of 200, the HLB = $20(1 - 175/200) = 2.5$. In the same way, a sorbitan monolaurate having a saponification number 164 and prepared from acids having an acid number of 290, has an HLB of 8.7 determined by the equation HLB = 20(1 − 164/290) = 8.7. This example may be complicated by the addition of ethylene oxide, whereupon the saponification number becomes 48.5 whereas the acid number of the fatty acid moiety remains 290. Under these conditions, the HLB value calculation is HLB = 20(1 − 48.5/290) = 16.7. Even with blended nonionics, eg, a blend of mono- and diglycerides and a polyoxyethylene sorbitan monooleate, analysis reveals a saponification number of 150 and an acid number (of the extracted combined acids) of 203. In this instance, the calculation is HLB = 20(1 − 150/203) = 5.2.

Although the formulas given above are satisfactory for many nonionic emulsifiers, certain other nonionic types exhibit behavior which is apparently unrelated to their composition; for example, those based on propylene oxide or butylene oxide or those containing nitrogen or sulfur. Furthermore, the environment in which an ionic surfactant exists influences its apparent HLB, ie, pH, salt content, and type, etc. Anionic surfactants can range from low HLB (polyvalent metal soaps) to high HLB (typical monovalent soaps, sodium lauryl sulfate, etc). Likewise, cationic and amphoteric surfactants may cover a wide range of apparent HLB.

Thus the HLB values of these special nonionics, and of all ionics, must be estimated by experimental methods so that their HLB values are aligned with those of the nonionic emulsifiers. An experimentally determined HLB value for such an emulsifier does not necessarily indicate the percentage weight of its hydrophilic portion; for example, it is found experimentally that the HLB of pure sodium lauryl sulfate is about 40, which does not mean that it is 200% hydrophilic but merely that it shows an apparent HLB of 40 when used in combination with other emulsifiers.

The experimental method of HLB determination (15), although not precise, briefly consists of blending the unknown emulsifier in varying ratios with an emulsifier of known HLB, and using the blend to emulsify an oil of known required HLB. The best performing blend is assumed to have an HLB value approximately equal to the required HLB of the oil so that the HLB value of the unknown can be calculated. In practice, a large number of experimental emulsions must be made from which an average HLB value for the unknown is finally calculated.

Needless to say, such a procedure can be difficult and time-consuming. However, the lack of knowledge of an exact HLB number for an emulsifier is not necessarily a serious disadvantage.

A rough estimate of HLB can be made from the water solubility of the emulsifier, and in many instances, this is adequate for screening work. Although this method is not infallible, it can be used to approximate the HLB of many emulsifiers according to their solubility or dispersibility characteristics as shown in Table 3. It must be pointed out that the HLB of an emulsifier is not absolutely related to solubility, in

Table 3. HLB by Dispersibility

Dispersion characteristics	HLB range
no dispersibility in water	1–4
poor dispersion	3–6
milky dispersion after vigorous agitation	6–8
stable milky dispersion	8–10
translucent to clear dispersion	10–13
clear solution	13+

water or in oils; thus two emulsifiers of similar HLB may exhibit differences in solubilities.

HLB Values. The HLB values of many surfactants have been published (16–17) (see Table 4). Some of the published data are based on calculations that yield HLB values in excess of 20 units for nonionic surfactants and these appear to be in error.

Experimental determination of HLB and correlation with other physical properties has been attempted by a variety of methods, none of which are applicable to all surfactants.

The HLB value is related to emulsion performance. With increasing HLB, the performance factor increases to a maximum and then decreases. The HLB value at which the maximum occurs is called the required HLB for the oil in that system (o/w or w/o). The maximum can be related to emulsification, stability, ease of emulsification, wetting, dispersing, foaming, or any desired property. The optimum for different functions does not necessarily occur at the same HLB but can vary depending upon the chemical nature of the emulsifier and the ingredients. Test emulsions are prepared to observe the optimum HLB for each of the desired functions.

A controllable range of HLB is usually obtained by blending low and high HLB emulsifiers. A list of applications and some suggested emulsifier blends and their HLB values has been published (18). The blending operation can achieve not only the optimum HLB but it can also allow the inclusion of emulsifiers that aid other desired functions.

Table 5 presents the required HLB values of a number of oils and waxes. This table can be used to estimate a required HLB.

The required HLB can be determined experimentally on any matched pair of emulsifiers one lipophilic and one hydrophilic, of known HLB values (eg, Atlas Span 60 HLB = 4.71, and Tween 60, HLB = 14.9). A trial run is first made so the selection of emulsifiers at this point need not be perfect for the particular formula.

The first series can consist of seven test emulsions, using a different mixture of the above emulsifier samples in each. An excess of emulsifier (ie, 10–20 wt% of the oil phase) is used and dissolved or intimately dispersed into the oil phase, melting ingredients together if necessary.

Although simple mixing of ingredients and emulsifiers might be sufficient at this point, it is best that for each of the seven emulsions, preparation methods be used that are as nearly identical as possible to intended plant methods.

Using appropriate methods for comparison and evaluation, based on the product requirements, including emulsion type (o/w or w/o), one or more of the seven emulsifier combinations will fairly quickly give a better emulsion than the others, even though not necessarily a very good one. If all the emulsions seem fairly good, with not much noticeable difference, then repeat the seven tests using less emulsifier. Conversely, if all the emulsions are poor and show no great variation, repeat the tests with a higher emulsifier content.

More often than not, the emulsions can be compared for stability (separation of ingredients), perhaps in a matter of minutes, perhaps overnight, or after heating or freeze–thaw cycles. However, it is entirely possible that the criterion for a good emulsion might be clarity or viscosity, or ease of preparation or application. Whatever the index for judgment might be, these preliminary tests enable selection of an approximate workable HLB range for the emulsifier system. This preliminary test series provides an estimate of the required HLB for the oil–wax phase of the type of emulsion desired.

Table 4. HLB Values[a] and Selected Trade Names for Surfactants

Chemical designation and CAS Registry No.	Type	HLB value
oleic acid [112-80-1]	N	1.0
lanolin alcohols [61788-49-6]	N	1.0
acetylated sucrose diester	N	1.0
ethylene glycol distearate [627-83-8]	N	1.3
acetylated monoglycerides	N	1.5
sorbitan trioleate [26266-58-6]	N	1.8
glycerol dioleate [25637-84-7]	N	1.8
sorbitan tristearate [26658-19-5]	N	2.1
ethylene glycol monostearate [111-60-4]	N	2.9
sucrose distearate [27195-16-0]	N	3.0
decaglycerol decaoleate [11094-60-3]	N	3.0
propylene glycol monostearate [1323-39-3]	N	3.4
glycerol monooleate [25496-72-4]	N	3.4
diglycerine sesquioleate	N	3.5
sorbitan sesquioleate [8007-43-0]	N	3.7
glycerol monostearate [31566-31-1]	N	3.8
acetylated monoglycerides (stearate)	N	3.8
decaglycerol octaoleate [66734-10-9]	N	4.0
diethylene glycol monostearate [106-11-6]	N	4.3
sorbitan monooleate [1333-68-2]	N	4.3
propylene glycol monolaurate [10108-22-2]	N	4.5
high molecular weight fatty amine blend	C	4.5
POE[b] (1.5) nonyl phenol (ether) [9016-45-9]	N	4.6
sorbitan monostearate [1338-41-6]	N	4.7
POE (2) oleyl alcohol (ether) [25190-05-0]	N	4.9
POE (2) stearyl alcohol (ether) [9005-00-9]	N	4.9
POE sorbitol beeswax derivative	N	5.0
PEG[c] 200 distearate [9005-08-7]	N	5.0
calcium stearoxyl-2-lactylate [5793-94-2]	A	5.1
glycerol monolaurate [27215-38-9]	N	5.2
POE (2) octyl alcohol (ether) [27252-75-1]	N	5.3
sodium-O-stearyllactate [18200-72-1]	A	5.7
decaglycerol tetraoleate	N	6.0
PEG 300 dilaurate [9005-02-1]	N	6.3
sorbitan monopalmitate [26266-57-9]	N	6.7
N,N-dimethylstearamide [3886-90-6]	N	7.0
PEG 400 distearate [9005-08-7]	N	7.2
high molecular weight amine blend	C	7.5
POE (5) lanolin alcohol (ether) [61790-91-8]	N	7.7
polyethylene glycol ether of linear alcohol	N	7.7
POE octylphenol (ether) [9002-93-1]	N	7.8
soya lecithin [8020-84-6]	N	8.0
diacetylated tartaric acid esters of monoglycerides	N	8.0
POE (4) stearic acid (monoester) [9004-99-3]	N	8.0
sodium stearoyllactylate [18200-72-1]	A	8.3
sorbitan monolaurate [1338-43-8]	N	8.6
POE (4) nonylphenol (ether) [9016-45-9]	N	8.9
calcium dodecylbenzene sulfonate [26264-06-2]	A	9
isopropyl ester of lanolin fatty acids	N	9.0
POE (4) tridecyl alcohol (ether) [24938-91-8]	N	9.3
POE (4) lauryl alcohol (ether) [9002-92-0]	N	9.5
POP[d]/POE condensate	N	9.5
POE (5) sorbitan monooleate [9005-65-6]	N	10.0
POE (40) sorbitol hexaoleate [9011-29-4]	N	10.2
PEG 400 dilaurate [9005-02-1]	N	10.4
POE (5) nonylphenol (ether) [9016-45-9]	N	10.5

Table 4. (*continued*)

Chemical designation and CAS Registry No.	Type	HLB value
POE (20) sorbitan tristearate [9005-71-4]	N	10.5
POP/POE condensate [9003-11-6]	N	10.6
POE (6) nonylphenol (ether) [9016-45-9]	N	10.9
glycerol monostearate-self emulfsifying [31566-31-1]	A	11.0
POE (20) lanolin (ether and ester)	N	11.0
POE (20) sorbitan trioleate [9005-70-3]	N	11.0
POE (8) stearic acid (monoester) [9004-99-3]	N	11.1
POE (50) sorbitol hexaoleate [9011-29-4]	N	11.4
POE (6) tridecyl alcohol (ether) [24938-91-8]	N	11.4
PEG 400 monostearate [9004-99-3]	N	11.7
alkyl aryl sulfonate	A	11.7
triethanolamine oleate soap [2717-15-9]	A	12
POE (8) nonylphenol (ether) [9016-45-9]	N	12.3
POE (10) stearyl alcohol (ether) [9005-00-9]	N	12.4
POE (8) tridecyl alcohol (ether) [24938-91-8]	N	12.7
POP/POE condensate	N	12.7
POE (8) lauric acid (monoester) [9004-81-3]	N	12.8
POE (10) cetyl alcohol (ether) [9004-95-9]	N	12.9
acetylated POE (10) lanolin	N	13.0
POE (20) glycerol monostearate [53195-79-2]	N	13.1
PEG 400 monolaurate [9004-81-3]	N	13.1
POE (16) lanolin alcohol (ether) [61790-81-6]	N	13.2
POE (4) sorbitan monolaurate [9005-64-5]	N	13.3
POE (10) nonylphenol (ether) [9016-45-9]	N	13.3
POE (15) tall oil fatty acids (ester)	N	13.4
POE (10) octylphenol (ether) [9002-93-1]	N	13.6
PEG 600 monostearate [9004-99-3]	N	13.6
POP/POE condensate	N	13.8
tertiary amines: POE fatty amines	C	13.9
POE (24) cholesterol [27321-96-6]	N	14.0
POE (14) nonylphenol (ether) [9016-45-9]	N	14.4
POE (12) lauryl alcohol (ether) [9002-92-0]	N	14.5
POE (20) sorbitan monostearate [9005-67-8]	N	14.9
sucrose monolaurate [25339-99-5]	N	15.0
POE (20) sorbitan monooleate [9005-65-6]	N	15.0
POE (16) lanolin alcohols (ether) [8051-96-5]	N	15.0
acetylated POE (9) lanolin [68784-35-0]	N	15.0
POE (20) stearyl alcohol (ether) [9005-00-9]	N	15.3
POE (20) oleyl alcohol (ether) [25190-05-0]	N	15.3
PEG 1000 monooleate [9004-96-0]	N	15.4
POE (20) tallow amine [61790-82-7]	C	15.5
POE (20) sorbitan monopalmitate [9005-66-7]	N	15.6
POE (20) cetyl alcohol (ether) [9004-95-9]	N	15.7
POE (25) propylene glycol monostearate [37231-60-0]	N	16.0
POE (20) nonylphenol (ether) [9016-45-9]	N	16.0
PEG (1000) monolaurate [9004-81-3]	N	16.5
POP/POE condensate	N	16.8
POE (20) sorbitan monolaurate [9005-64-5]	N	16.9
POE (23) lauryl alcohol (ether) [9002-92-0]	N	16.9
POE (40) stearic acid (monoester) [9004-99-3]	N	16.9
POE (50) lanolin (ether and ester) [61790-81-6]	N	17.0
POE (25) soyasterol [68648-64-6]	N	17.0
POE (30) nonylphenol (ether) [9016-45-9]	N	17.1
PEG 4000 distearate [9005-08-7]	N	17.3
POE (50) stearic acid (monoester) [9004-99-3]	N	17.9
sodium oleate [143-91-1]	A	18.0

Table 4. (*continued*)

Chemical designation and CAS Registry No.	Type	HLB value
POE (70) dinonylphenol (ether) [9014-93-1]	N	18.0
POE (20) castor oil (ether, ester) [61791-12-6]	N	18.1
POP/POE condensate	N	18.7
potassium oleate [143-18-0]	A	20
N-cetyl-N-ethyl morpholinium ethyl sulfate (35%) [78-21-7]	C	30
ammonium lauryl sulfate [2235-54-3]	A	31
triethanolamine lauryl sulfate [139-96-8]	A	34
sodium alkyl sulfate	A	40

Chemical designation	Trade names	Manufacturers
alkyl/alkoxy sulfates	Auirol	Henkel, Inc.
	Duponol	Dupont, Inc.
	Stepanol	Stepen Chemical Co.
alkyl aryl sulfonates	Alkanol	Dupont, Inc.
	Conco	Continental Chemical Co.
	Atlas G-3300	ICI Americas, Inc.
glycerol fatty acid esters	Aldo	Glyco Chemicals, Inc.
	Atmos	ICI Americas Inc.
	Kessco	Armak Div, Akzona Inc.
	Myverol	Eastman Chemical Products, Inc.
	Witconol	Witco Chemical Corp.
lanolin-based derivatives	Amerchol	Amerchol Div, CPC Intnatl., Inc.
	Atlas G-1441, G-1790	ICI Americas, Inc.
	Cleavlan	Emery Industries, Inc.
POE alkyl phenols	Igepal	GAF Corp.
	Sterox	Monsanto Co.
	Triton	Rohm & Haas Co.
POE amines	Ethomeen	Armak Div, Akzona Inc.
	Atcor MC; Atlas G-3780A	ICI Americas, Inc.
POE fatty acids (esters), also PEG fatty acid esters	Emulsogen	American Hoechst Corp.
	Emulphor	GAF Corp.
	Myrj; Renex	ICI Americas, Inc.
	Witconol	Witco Chemical Corp.
POE fatty alcohols (ethers)	Brij	ICI Americas, Inc.
	Emery	Emery Industries, Inc.
	Ethosperse	Armak Div, Akzona Inc.
	Promulgin	Robinson-Wagner Co., Inc.
POE fatty esters and oils	Aldesperse	Glyco Chemicals, Inc.
	Emerest	Emery Industries, Inc.
	Synlube	Milliken Chemical Co.
	Tween	ICI Americas, Inc.
POP/POE condensates, block polymers	Pluronic	BASF Wyandotte Corp.
	Synperonics	ICI Americas, Inc.
sorbitan fatty acid esters	Accosperse	Armstrong Chemical Co., Inc.
	Emsorb	Emery Industries, Inc.
	Lonzest	Lonza, Inc.
	Span	ICI Americas, Inc.
succinate, sulfo derivatives	Aerosol	American Cyanamid Co.
	Tergitol	Union Carbide Corp.

[a] Refs. 16 and 17.
[b] POE = polyoxyethylene [25609-81-8].
[c] PEG = poly(ethylene glycol) [25322-68-3].
[d] POP = polyoxypropylene [34465-52-6].

Table 5. Required HLB Values for Emulsification of Oils and Waxes

Compound	CAS Registry No.	HLB	Compound	CAS Registry No.	HLB
\multicolumn{6}{c}{O/w emulsion}					
acetophenone	[98-86-2]	14	dimethyl silicone	[9016-00-6]	9
dimer acid[a]	[61788-89-4]	14	ethylaniline	[103-69-5]	13
isostearic acid	[2724-58-5]	15–16	ethyl benzoate	[93-89-0]	13
lauric acid	[143-07-7]	16	fenchone	[1196-79-5]	12
linoleic acid	[60-33-3]	16	isopropyl myristate	[110-27-0]	12
oleic acid	[112-80-11]	17	isopropyl palmitate	[142-91-6]	12
ricinoleic acid	[141-22-0]	16	kerosene	[8008-20-6]	12
cetyl alcohol	[36653-82-4]	16	lanolin, anhydrous	[8006-54-1]	12
decyl alcohol	[25339-17-7]	15	lard	[61789-99-9]	5
hexadecyl alcohol	[36653-82-4]	11–12	laurylamine	[124-22-1]	12
isodecyl alcohol	[25339-17-7]	14	menhaden oil	[8002-50-4]	12
lauryl alcohol	[112-53-8]	14	methyl phenyl silicone	[42557-10-8]	7
oleyl alcohol	[143-28-2]	14	methyl silicone	[9076-37-3]	11
stearyl alcohol	[112-92-5]	15–16	mineral oil, aromatic	[8012-95-1]	12
tridecyl alcohol	[112-70-9]	14	mineral oil, paraffinic	[8012-95-1]	10
Arlamol E	[25231-24-4]	7	mineral spirits	[8030-30-6]	14
beeswax	[8012-89-3]	9	mink oil	[8023-74-3]	9
benzene	[71-43-2]	15	nitrobenzene	[98-53-3]	13
benzonitrile	[100-47-0]	14	nonylphenol	[25154-52-3]	14
bromobenzene	[108-86-1]	13	ortho-dichlorobenzene	[95-50-1]	13
butyl stearate	[123-95-5]	11	palm oil		7
carbon tetrachloride	[56-23-5]	16	paraffin wax	[8002-74-2]	10
carnauba wax	[8015-86-9]	15	petrolatum	[8009-03-8]	7–8
castor oil	[8001-79-4]	14	petroleum naphtha	[8030-30-6]	14
ceresine wax		8	pine oil	[8002-09-3]	16
chlorinated paraffin	[8029-39-8]	12–14	polyethylene wax	[9002-88-4]	15
chlorobenzene	[108-90-7]	13	propylene tetramer	[9003-07-0]	14
cocoa butter		6	rapeseed oil	[8002-13-9]	7
corn oil	[8001-30-7]	8	safflower oil		7
cottonseed oil	[8001-29-4]	6	soybean oil		6
cyclohexane	[110-82-7]	15	styrene	[100-42-5]	15
decahydronaphthalene	[91-17-8]	15	tallow	[61789-97-7]	6
decyl acetate	[112-30-1]	11	toluene	[108-88-3]	15
diethylaniline	[91-66-7]	14	trichlorotrifluoroethane	[76-13-1]	14
diisooctyl phthalate	[27554-26-3]	13	tricresyl phosphate	[1330-78-1]	17
diisopropylbenzene	[25321-09-9]	15	xylene	[1330-20-7]	14
\multicolumn{6}{c}{W/o emulsion}					
gasoline		7	mineral oil		6
kerosene	[8008-20-6]	6	stearyl alcohol	[112-92-5]	7

[a] See dimer acids.

Emulsifier. Selection of the emulsifier type begins with the choice between ionic and nonionic. Based on required HLB, this selection is then broadened to include members of the vast array of surfactants available. Following the experimental procedure outlined above, suppose it was found that a blend of Span 60 and Tween 60 (both of which are stearates), at an HLB of 12, gave a better emulsion than any other

HLB of these two emulsifiers in the above test series. That particular HLB of about 12 will be best for any emulsifier type. Now it must be determined whether some other blend at HLB 12 (eg, laurates, palmitates, or oleates) might not be better or more efficient than the stearates. (In any case, it will have an HLB of about 12.) By blending two emulsifiers, the exact HLB needed can be attained rather than trying a single emulsifier having an HLB that does not quite match. Moreover, the emulsifier chemical type can be selected to suit the oil or other active ingredients, instead of having to limit or adjust the active ingredients to suit the emulsifier.

The most stable emulsion systems usually consist of blends of two or more emulsifiers, one portion having lipophilic tendencies, the other hydrophilic. (For example, glycerol monostearate, self-emulsifying grade, is actually a blend of lipophilic nonself-emulsifying glycerol monostearate with a hydrophilic soap or other substance to make it more water soluble.) Only in relatively rare instances is a single emulsifier suitable, even though it might have the exact HLB needed.

The chemical class of emulsifier is at times dictated by application requirements. For example, a paper chemical emulsion must be precipitated with alum and, therefore, a soap is indicated as the emulsifier; if acid stability is required, a nonionic emulsifier is suitable; food emulsions require FDA-approved emulsifiers, etc.

After making the preliminary screening based on these premises, a variety of low-HLB emulsifiers of different chemical types is chosen for which corresponding high-HLB emulsifiers are available. A blend is then made in a weight ratio so that its HLB is 12, the required HLB for the emulsion, and test emulsions are prepared. An evaluation of these usually shows wide differences for different chemical types and a selection is easily made based on the criteria for the emulsion system.

Emulsifier selection is completed by determining the optimum blend of one or more of the best chemical types. The low and high HLB emulsifiers of the optimum blend are reblended to mixtures having incremental HLB values close to the indicated required HLB. This last fine tuning is necessary because HLB values and required HLB values are not precise since they are based on a combination of approximate calculations and on empirical tests.

As just demonstrated, the HLB system reduces but does not eliminate the trial and error in emulsifier selection. An attempt to further reduce or eliminate this trial and error aspect makes use of the cohesive energy rates system (19). By mathematical matching of cohesive energies of the ingredients, a match of chemical type is indicated. An appreciable amount of information is required, including solubility parameter or surface tension, molecular weight, and density, and the correlations are not easy. Mixtures, present in all commercial applications, add further complications. The complexity relative to the original HLB concept has resulted in minimal utilization.

Another system of HLB calculations is based on phase-inversion temperature (PIT) (20). These data correlate well with HLB values and show a possible direction for further study. Furthermore, emulsification and solubilization data have been correlated (21–22) and some interesting ideas, including volume-relative concepts, have been advanced. However, HLB value calculations and correlations have been published (23–24) that appear to be in conflict with the original calculations.

Solubilization can be increased by high ratios of emulsifier to oil (ranging from 1:1 to 5:1 or more). The oil and emulsifier should be soluble one in the other and dilution is generally made at room temperature by adding oil and emulsifier to water (for o/w systems). The selection of the optimum HLB within 0.1 unit sometimes increases efficiency dramatically, allowing reduction of the ratio of emulsifier to oil. Co-solvents, such as alcohol, are often beneficial, and wax or solid solubilization is possible but chilling must be rapid to avoid particle growth as the emulsion is cooled through the melting point.

Emulsifiable Concentrates. Moderate amounts of emulsifier are needed ranging from 5 to 20% emulsifier by weight of the oil–wax mixture). The emulsifier blend must be soluble in the oil, or capable of being integrated into the oil (eg, with alcohols, glycol polyethers, fatty acids, small amounts of water, etc). The emulsifier must be chosen to balance ease of emulsification and emulsion stability and, if necessary, ease of reemulsification). In general, emulsifiable concentrates are formulated to be poured into water.

Emulsions with minimum emulsifier concentration are prepared with vigorous mechanical action (with a homogenizer, colloid mill, ultrasonics (qv), etc) in stages. Small amounts of oil are added to water and emulsifier, milled, recycled with more oil, etc, until the final oil level is attained. The HLB required at this point is higher than usual and must be checked. Particle size is largely determined by the type of emulsification equipment employed and the number of passes through the equipment.

Substantive Emulsions. Particle charge is required so that deposition occurs on a chosen substrate (anionic or cationic emulsifier is the usual route).

Stability Requirements. Emulsion stability usually involves, in addition to resistance to creaming and coalescence, the maintenance of a given viscosity or fluidity. Viscosity without the incorporation of thickeners depends upon the viscosity of the continuous phase (for LIP emulsions) and the amount of internal phase (for HIP emulsions). For economic as well as technical reasons, most commercial emulsions are o/w and have low oil (internal) phase levels. Thus, in order to increase viscosity, the external phase, ie, water, must be thickened. Suitable formulations include thickeners such as soap gels, gums, starches, proteins, polymers, and the like. Generally they require higher than usual levels of mechanical agitation.

Bacterial action may affect thickeners and destroy surfactants. This can be avoided by clean conditions and the use of preservatives.

Stable Wax Emulsions. Agglomeration of wax particles on agitation, which may destroy the emulsion, can be avoided by buffering with an organic ester (25) (see Waxes).

Controllable Stability. Breakable emulsions are based on surfactant selection; these can be destroyed (1) by acids or polyvalent salts, (2) conflicting anionic–cationic surfactants, (3) special surfactants.

Bacterial Stability. Since emulsions are a combination of fatty materials and water, sometimes containing carbohydrates, they are good substrates for bacterial growth. Rarely are emulsions packaged under sterile conditions and preservatives are usually required (see Industrial antimicrobial agents). An emulsifiable concentrate, such as an agricultural pesticide, where the emulsion is prepared and immediately used, may need no preservatives. Storage conditions such as provided for ice cream where preservation is achieved by pasteurization and freezing, may suffice. Cosmetic creams, where the consumer introduces a wide range of bacteria with each use, require preservation. Also in metal-cutting fluids a stable preservative must be incorporated.

Contamination of emulsions probably occurs most frequently through water supplies, and less so from oils or waxes and emulsifiers, which are surprisingly clean as manufactured. Some of the more common preservatives employed in emulsions include acetic, propionic, and benzoic acids and their salts and esters, and aldehydes and biguanides.

Preparation

A good manufacturing process is based on a properly developed laboratory procedure. Scale-up introduces problems of agitation, incorporation of air (particularly via vortices in large tanks), surface–volume ratios, addition rates of ingredients (especially at inversion points), cooling rates, and raw material control.

Laboratory. Small-scale laboratory and pilot-plant models of planetary mixers, motor-driven propellers, turbines, colloid mills, and homogenizers are available. In addition, electric egg beaters, high-speed blender food mixers, and shaking machines are used. Agitation is usually much more vigorous and efficient than that in plant-scale equipment which must be considered when applying results from one to the other.

In preparing an emulsion of moderate viscosity in a typical laboratory set up, the actual work input may be surprisingly high. In equipment scale-up, surface-to-volume ratios, peripheral speeds of agitators, and tendencies to maelstrom and foam production differ, as well as heating and cooling rates.

Laboratory preparations should duplicate plant conditions as far as possible and err, if at all, on the side of too little energy input. A batch kettle, either heated or unheated, is similar to a beaker–motor propeller (slow motor speed is preferred and a simple baffle is usually an improvement). A plant homogenizer resembles a hand homogenizer or a gear pump with a pressure relief valve. Emulsification of a concentrate is most uniformly handled by mechanical timer-controlled shaking (26).

A Waring blender, when used to prepare emulsions, imparts enormous amounts of energy and incorporates large quantities of air. It often gives results that are not achieved in subsequent plant scale-up.

Another major problem in the laboratory occurs when emulsions are prepared hot and then cooled. A laboratory beaker of emulsion will cool from 60 to 30°C in a few minutes. A 3900-L (1000-gal) tank requires much longer, even with forced cooling. For emulsions that contain waxy components, the rate of cooling through the melting range can be all important. Hence the best cooling schedule should be determined, then the permissible deviation, and finally the plant conditions.

Emulsions formulated to produce smaller particle size via inversion techniques require particular attention at the inversion point. A smooth and complete inversion produces the smallest particle size. Major controlling factors include temperatures and addition rate of the inverting phase (eventually the external phase). Since slow addition of this phase generally is desired for finest particle size, little difficulty is encountered; cycle time in the batch kettle has to be considered.

The selection of raw materials of high quality and consistency is guided by the products used during laboratory formulation. However, the plant management may wish to substitute similar raw stocks. In some instances, particularly with natural products (eg, beeswax), a laboratory preparation of the emulsion must be made with similar raw materials and tested for critical properties. Major ingredients, such as mineral oils, are usually purchased according to specifications. The quality of the water is most critical. It may differ between laboratory and plant, with seasons, and it cannot be monitored. Bacteria and metal ions may be present. Thus variations in the water can produce disturbing effects that are frequently blamed on other ingredients.

The same emulsion from the same supplier should be used, even if two suppliers publish identical specifications. For oils and waxes, normal raw material quality control is adequate. For the water supply, standards should be established. For emulsifiers, normal raw material quality control *plus* identification with an approved supplier is essential.

Plant. Emulsions are produced either by batch or by continuous processes. Based on the laboratory tests, a specific procedure should be established for preparing an emulsion on plant scale. Apparently small deviations may result in a totally different product.

Addition of Ingredients. Order and rate of addition may affect the acceptability of the product. An apt illustration is the preparation of an o/w emulsion by the inversion technique. Oil phase and emulsifier are blended in a tank. Water is then slowly added to the oil phase with stirring. The initially hazy oil and emulsifier blend usually clears at first and then becomes cloudy again. As more water is added, the emulsion assumes a milky cast while the viscosity increases. At some point, called the inversion point, the viscosity suddenly decreases. At this point the emulsion has changed from w/o to o/w. Further additions of water may than be made rapidly. If the oil is initially added to the water, a poor emulsion results unless enough of the right emulsifier is employed. The inversion technique is used to prepare o/w emulsions containing less emulsifier.

Specific procedures must be worked out for each formulation. Generally, all the oils and oil-soluble ingredients are best combined as the oil phase. Polyols are added with the water but salts are best added with the last half to quarter of the water after a good primary emulsion has been established.

Temperature. In most instances of liquid–liquid emulsification, ambient temperature is preferred. With some equipment, heat will be generated during the emulsification step and must be removed.

Heating of ingredients may be necessary to effect emulsification; it may also be necessary to pasteurize or sterilize the product. For wax emulsions, heating is necessary. The wax or the oil phase should be heated to at least 5°C over the melting point of the highest-melting wax and the aqueous phase should be heated to at least 2–3°C over the temperature of the oil phase. Shocking the wax by cooling during mixing is thus avoided.

The rate of cooling a wax emulsion, especially at the melting point of the wax, is critical. Each emulsion must be studied to determine the best cooling rate. A cooling board or a heat-exchange unit may accelerate the cooling.

When an emulsion system is heated, emulsifier solubility is altered and this may change its emulsifying properties. In addition, care must be taken to avoid unwanted chemical reactions.

Equipment. The purpose of the emulsification equipment, whether simple or complex, is to break up or disperse the internal phase in the external phase, so that the particle size of the resulting emulsion is sufficiently small to retard coalescence and resulting breakdown of the emulsion for the required time of stability. The major concerns in the choice of emulsifying equipment are the apparent viscosity in all stages of manufacture, the amount of mechanical energy input required, and heat exchange demands. Preparation of emulsions is greatly affected by the type of agitation (see also Mixing).

Stirring. Manual stirring is the simplest form of agitation. Depending upon the selection of emulsifiers and the ingredients to be emulsified, either stable or semistable, large or small particle-size emulsions may be prepared. For manual stirring, a formulation of maximum dispersibility is required. Ease of dispersibility is not related to stability.

For many viscous emulsions containing high solids, soap gels, resinous materials, etc, a mechanically rotated paddle or anchor type agitator is best.

The planetary stirrer has been developed for emulsification in the food industry. It has a rotating paddle and at the same time the axis about which it rotates follows a circular orbit. In this way a large batch of heavy batter may be intimately mixed. These same mixers are used at higher speeds with a wire whisk for aeration and whipping of low viscosity emulsions because of excessive aeration.

Aeration. Stirring by means of bubbling air or gas through a liquid is not much more efficient than manual stirring unless extremely large volumes of gas are used. A modification of this system, consisting of the injection of live steam into a tank, is usually much more efficient because of the condensation of the steam and the resulting cavitation or steam-hammer effect. The use of air or steam is most practical in low viscosity systems.

Propeller Agitation. In this popular type of equipment, one or more propellers are mounted on a common shaft in a mixing tank. Modifications include variation in the location of the propellers in the tank and the use of two or more propeller shafts, and complex propellers. Propeller agitation is most satisfactory for low and medium viscosity emulsions. When properly used with adequate emulsifying agents, propeller agitation results in finer particle size than homogenization of milling with smaller quantities of emulsifiers.

Turbine Agitation. The inclusion of fixed baffles either on the tank wall or adjacent to the propellers, as in a turbine rotor and stator increases the efficiency of agitation considerably. However, turbine agitators are preferred since baffle plates in a tank frequently result in areas of little or no agitation. Turbine-type agitators are available in various sizes, speeds, and rotor–stator clearances, and in many modified designs. Turbine-type systems may be designed to give a very high degree of shearing action and may be used with fluids of higher viscosity than propellers. At high viscosities, the gross agitation of the batch may be insufficient.

Colloid Mill. The colloid mill (Fig. 4) may actually be considered as a modification of a turbine, although in this case the clearance between the rotor and stator is of the order of a few micrometers. With such small clearances, an extremely high shearing action occurs. The product from a colloid mill usually has a uniform particle size, no doubt because of the fixed clearance between the rotor and stator. Owing to the tremendous shearing forces applied to the emulsion, a significant temperature rise occurs during emulsification and in most cases external cooling must be employed. Milling may be done on fluids and pastes (with positive feed for the latter). Rate of throughput varies inversely with the viscosity.

Homogenization. In a homogenizer, emulsification is effected by forcing the two phases past one or two spring-seated valves (Fig. 5) at pressures of 3.45–34.5 MPa (500–5000 psi). Emulsification occurs not only while the components pass under the valve seat but also when the emulsion impinges against the retaining wall that surrounds the valve. Homogenizers are also built with more than one stage of emulsification, ie, with successive relief valves. This is of value when the high pressure ho-

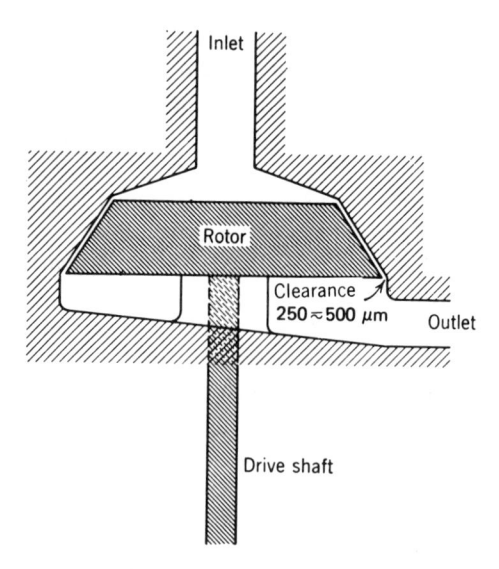

Figure 4. A colloid mill.

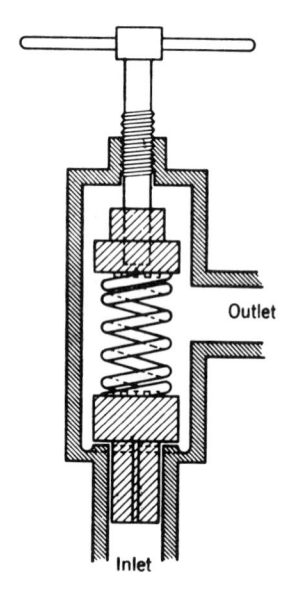

Figure 5. A homogenizer.

mogenization promotes clumping of the fine particles of emulsion. The second stage of homogenization, at a lower pressure, breaks up the clumps and produces a lower viscosity product. Using comparable ingredients, homogenizers usually give an emulsion of finer average particle size than colloid mills, although the particle size is not as uniform. The temperature rise during homogenization is not very large, between 5 and 20°C. However, the actual temperature rise throughout the homogenizer and pump may be only 5–20°C or it may be as much as 25–50°C, depending upon the type of pump employed. A piston pump gives a lower temperature rise than a gear pump. Owing to clearances in the gear pump, a certain quantity of liquid continually by-passes the pump and is partially homogenized before reaching the homogenizer head. Homogenizers handle liquids or pastes and the throughput rate is little affected by viscosity.

Ultrasonics (qv). The high-frequency or ultrasonic emulsifier (Fig. 6) is best suited for low viscosity liquids, though it has been successful with systems having viscosities as high as salad dressings and dye–paste emulsions. A laboratory size is available.

Ultrasonic energy is developed either mechanically (Fig. 6) or electrically. In the former case, a pump forces the combined phases past a tuned vane which vibrates and produces energy via cavitation. Pressures range from 1 to 3.5 MPa (150–500 psi) in the chamber surrounding the tuned vane.

Electrically, an oscillator generator causes ultrasonic vibration in either a magnetostriction device or a piezoelectric crystal. Major difficulties in transfer of this energy to the liquid have retarded extensive commercial use of this form of ultrasonic energy in emulsification.

Miscellaneous. The units discussed above are frequently used in combinations and special modifications, eg, the Oakes mixer to make cake batter for the food industry. Table 6 gives the characteristics of several types of agitators. The power requirement is lowest for the propeller mixer, followed by the turbine and homogenizer, and is highest for the colloid mill. Pebble and ball mills and other slow-speed grinding equipment are used for some emulsions and extensively in pigment suspension.

Applications

The combination properties and characteristics of a given emulsion should be directly related to its use. The reasons favoring emulsions are mostly economic, safety, and esthetic. Appearance (desired opacity or translucence) and physical form (lotion, ointment, or liquid) are additional reasons. In general, water as a diluent or vehicle is less expensive than a solvent and is less hazardous.

However, emulsions are more difficult to manufacture, ship (avoiding freezing and excessive temperature), and package (to avoid corrosion) than solutions or powders. The inherent lack of stability is a drawback for many commercial uses. Some utilize it, others require indefinite stability. Control of stability, in concentrates, dilutions, and emulsifications is of prime importance.

The wide variations of applications are based on a surprisingly small group of functions related to the basic properties. A prime reason for using an emulsion is the

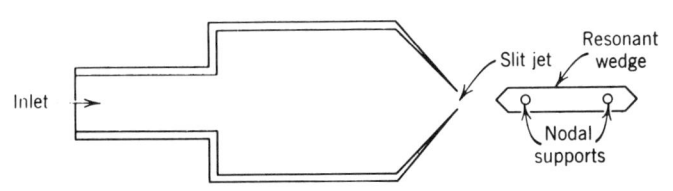

Figure 6. Schematic diagram of a Pohlman whistle. The nodal supports of the blade are separated by a distance equal to half the wavelength of the characteristic vibration of the system.

924 EMULSIONS

Table 6. Emulsion Equipment

Type	Agitator speed, rpm	Mechanical energy input	Usable viscosity range	Heat-exchange demand range[a]
anchor agitator	slow	low	best for high viscosity	fair
wall-scraping anchor agitator[b]	slow	low	best for medium and high viscosity	good
propeller mixer[c]	medium	low–medium	best for low–medium viscosity	fair–poor
votator	medium	low–medium	best for high viscosity	excellent
rotating cage mills[d]	high	high	best for low–medium viscosity	fair
rotating disk mills[e]	high	high	best for low–medium viscosity	fair
homogenizer	slow	high	low–moderately high	fair–poor
colloid mill	high	high	low–medium	fair–poor
mechanical ultrasonic[f]	high	low–medium	low–medium	fair–good
electric ultrasonic	not critical	low–medium	low–medium	fair

[a] Without auxiliary equipment.
[b] Usually in pairs; counter-rotating.
[c] Lightning mixer.
[d] Premier dispersator.
[e] Cowles dissolver or Hockmeyer dispersator.
[f] Rapasonic.

desire to provide a system or method for delivering a material under controlled conditions, for practical or esthetic reasons. This may be film deposition, dilution, spreading, or some similar action, but the prime action is one of distribution or delivery of a given ingredient.

Emulsion design as well as emulsion requirements follow industry patterns, but the technology is based on generally applicable properties and formulations. For example, formulations for the cosmetic industry differ from those for metal industries and the food industry; however, the basic technology is the same.

Table 7 shows the ten major industries using surfactants and the functions most often required by these applications (except for soaps and household detergents).

Most commercial emulsions are o/w, and in their final form most are LIP emulsions. A majority of emulsions contain a petroleum product as one of the principal ingredients, the second most important class of nonaqueous ingredients being vegetable oils and waxes. The largest volume emulsifiers are anionic (soaps), although nonionics are involved in a wider variety of formulations.

Packaging, Shipping, and Storage

Emulsions have to be packaged tightly to avoid loss of water and protect against rust. Emulsions often contain ingredients that attack can and drum linings, especially at an oil–water interface. Either durable can liners or plastic film inserts should be used. With suitable plastic liners, metal or glass containers can be replaced by cardboard or fiber packs.

Water loss may effect emulsion instability and result in phase separation, lumping,

Table 7. Functions of Emulsions

Principal applications of surfactants	Detergent action	Emulsification	Wetting–spreading[a]	Foaming	Film deposition	Particle size control	Solids suspension
textiles and leather processing	scouring agents	finishes					
ore flotation			frothers				
oil drilling operations			secondary recovery sprays				drilling muds
agricultural sprays		sprays					
cosmetics	shampoos	creams, lotions		shampoos			
elastomers		emulsified polymers				emulsified polymers	
foods							
lubricants	detergent oils	marine lubricants	corrosion resistant oils		corrosion resistant oils		
paint					latex paints		paints
pharmaceuticals		emulsified medicines					

[a] Includes phase displacement.

or clumping. The latter can also be caused by improper cooking; hence, weight loss or water analyses must be used to detect the actual cause.

In shipping and storage, high temperatures have to be avoided, since they affect the apparent viscosity, the solubility of the surfactant, and, therefore, the emulsification efficiency. Special formulations are needed for resistance to high temperatures as well as freezing and thawing. In fact, most commercial emulsions are tested and rated as being stable to freeze–thaw cycles.

During shipment, vibration can result in agglomeration of particles, particularly in wax emulsion. This can be minimized by adding weak electrolyte to the aqueous phase.

Evaluation and Analysis

Quality control starts with repetitive ease of preparation, either in the plant for a product sold as an emulsion, or of the emulsifiable concentrate for a product sold for emulsification by the consumer. The purpose of quality control is to check the development of ease of emulsification. Preparation ease rarely improves but usually deteriorates.

For performance evaluation one or more of the following items should be checked: emulsion type; stability under ambient conditions, at high temperature, in freeze–thaw cycles, under agitation (eg, in shipping containers), in shearing action (eg, pumping), with water loss (eg, w/o cosmetic cream), on concentration (ie, behavior of a cosmetic or paint on evaporation of water), on dilution (to a use or test concentration), on addition of extraneous materials; viscosity or rheology and change on dilution (to use concentrations); appearance, color, clarity–opacity, etc; particle size and its change on dilution, heating, etc; pH (if o/w); odor and taste; bacterial content (behavior toward preservatives); particle charge; chemical sensitivity (to salts, acids, etc); and one or more performance tests.

Analysis is usually based on water content. Further chemical analyses are only carried out under special circumstances.

General specifications for emulsions cannot be established because of the wide variety of ingredients and uses involved. A suitable specification would probably include water content, emulsion type, pH (if o/w), one or two key performance tests, and analysis of the emulsion for any legally regulated ingredient, as required for good manufacturing practice. A representative batch sample should be kept for the anticipated life of the product.

Accelerated stability tests at 40–50°C and above are used extensively but are not reliable. The stability of a good emulsion at room temperature cannot be checked by quality control since several years time may evolve before instability is observed. A relationship between temperature and stability can be established but must be viewed with caution. If the emulsion is heated high enough to effect a phase change (ie, to melt a wax phase), the surfactant relative to a liquid–liquid system may differ markedly from a liquid/solid system. Furthermore, at higher temperatures, the solubility of the emulsifier in both phases changes and this affects the emulsifier distribution, the amount at the interface, and the interface properties. Thus the emulsion tested behaves not in the same manner as originally at room temperature.

A true measure of stability would be a change in particle size distribution with time (or other selected parameters), a time-consuming and expensive test. Alternatively, particle size change is estimated by the change in ratio of spectral absorption at two wavelengths (26).

Measurement of creaming or sedimentation is a frequently used test procedure. Although a number of methods can detect the oil content without destroying the cream (radio frequency modification, nmr (27), etc), the visual observation is satisfactory for most formulations.

Analyses of products for the purpose of duplication is more sophisticated. The two phases have to be separated, extracted, and analyzed.

Safety, Environmental Aspects, and Government Regulations

Safety. Most emulsifiers are of low toxicity. The major variable is the oil or wax being emulsified and any addition to the aqueous phase. Emulsions, especially of fine particle size, can promote contact with and in some cases alter absorption of oils or waxes and thus minor toxic reactions can be promoted (eg, skin lotions or aerosols may cause dermatitis).

Toxicity is mainly of concern in the food and medicinal fields (salad dressings, batters, vitamin concentrates, etc). All ingredients added to foods and pharmaceuticals are subject to FDA regulations. Thus formulations within these regulations assure a wide margin of safety.

Environmental Aspects. Emulsions can aid in abating pollution. The emulsifier in formulations that enter the agro–eco-system must be biodegradable. O/w paint and ink emulsions reduce atmospheric solvent pollution by reduced solvent use. Tank cleaners can be formulated with breakable emulsions so that the oil washed from the tank can be recycled. Emulsion concentrates have been formulated to aid in the dispersion of oil spills. However, governmental regulations limit the use of these oil spill dispersion formulas to specific situations.

Government Regulations. Table 8 gives the government regulations for several major emulsion applications. Most closely regulated are emulsions utilized in or related to foods. Only a few foods are specified by the food standards (eg, chocolate, ice cream, and mayonnaise) which define the ingredients, and in most instances, provide limits on their amounts. All nonstandardized foods must list their ingredients or order of concentration.

Food ingredients, such as emulsifiers, must pass stringent toxicological tests. Furthermore, the need of the ingredient, to accomplish a defined purpose in the food must be proved. Assuming all ingredients satisfy the toxicological tests, a short term test of the final product will usually demonstrate no untoward synergism of the ingredients and the data may be submitted to FDA for their approval. All ingredients must be either GRAS or added via a granted petition.

Pharmaceuticals, require particular attention as to efficacy of medicinally active ingredients. Otherwise, oral pharmaceuticals are tested similarly to food emulsions. Topical pharmaceuticals, except for the medicinally active ingredients and cosmetic applications are closely related. Possible problems are skin and eye irritation and allergies. Similar criteria are important for many industrial applications, such as cutting oils, where extensive skin contact with the machine operator or emulsion user may occur.

Pesticides represent a special problem because the active ingredients are intended toxicants. Their formulation often represents a balance between risk and benefit. The

Table 8. Major Emulsifier Applications and Governmental Regulations

Application	Human contact route	U.S. Government regulations[a]	Requirements[a]
food, general	ingestion	Food Standard Food Additive Petition, or GRAS (list of ingredients[b])[c]	must demonstrate efficacy as well as safety
pharmaceuticals general, oral	ingestion	NDA[b] and IND[c]	must demonstrate efficacy as well as safety
general, topical	skin surface	NDA[b] and IND[c]	must demonstrate efficacy as well as safety
pesticides (including biocides)	ingestion of residues; skin surface	FIFRA[c]	must demonstrate efficacy as well as human and environmental safety
miscellaneous household uses (polishes, cleaners, etc)	skin surface	TSCA	ingredients must be in inventory or reported via PMN
miscellaneous industrial uses (coolant, cutting oil, cleaning, etc)	skin surface	TSCA	ingredients must be in inventory or reported via PMN
textiles	skin surface	TSCA	ingredients must be in inventory or reported via PMN
leather	skin surface	TSCA	ingredients must be in inventory or reported via PMN
elastomers	skin surface[d]	TSCA	ingredients must be in inventory or reported via PMN
paints	skin surface[e]	TSCA	ingredients must be in inventory or reported via PMN
inks	skin surface	TSCA	ingredients must be in inventory or reported via PMN
lubricants (emulsifier)	skin surface	TSCA	ingredients must be in inventory or reported via PMN
cosmetics	skin surface	CTFA	voluntary demonstration of human safety

[a] GRAS, Generally Recognized As Safe; NDA, New Drug Application; IND, Investigative New Drug; FIFRA, *Federal Insecticide, Fungicide, and Rodenticide Act*; TSCA, *Toxic Substance Control Act*; CTFA, Cosmetics, Toiletries and Fragrance Act; PMN, Pre Manufacture Notice of TSCA.
[b] Except for standardized foods.
[c] To assure efficacy as well as safety.
[d] Films may be used for food wraps, FDA.
[e] Paints may be accidentally ingested.

surfactants employed are rarely, if ever, comparable in toxicity or skin irritation to the active ingredients.

Foods, pharmaceuticals, cosmetics, and pesticides have been discussed separately since practically all other products are subject to the *Toxic Substance Control Act* (TSCA), the above items being specifically eliminated from the regulation. For those formulas that do come under TSCA, the control of the mixture proportions is important. All ingredients of a formula should be registered with EPA and part of their accepted inventory. If any ingredient is considered toxic, special treatment is necessary. Generally, most household and industrial emulsions are used in a way so that none is ingested and skin contact is negligible.

BIBLIOGRAPHY

"Emulsions" in *ECT* 1st ed., Vol. 5, pp. 692–718, by W. C. Griffin, Atlas Powder Company; "Emulsions" in *ECT* 2nd ed., Vol. 8, pp. 117–154, by W. C. Griffin, Atlas Chemical Industries, Inc.

1. P. Becher, *J. Colloid Interface Sci.* **24**(1), (1967).
2. S. J. Berkman, *Phys. Chem.* **39**, 527 (1935).
3. A. King and L. N. Mukherjee, *J. Soc. Chem. Ind.* **58**, 243 (1939).
4. J. O. Sibree, *Trans. Faraday Soc.* **27**, 161 (1931).
5. P. Becher, "Particle Size," *Am. Chem. Soc. Symp.* Pittsburgh, Pa.
6. C. F. Fryling, *J. Colloid Sci.* **18**, 713 (1963).
7. F. H. Haynie, Jr., R. A. Moses, and G. C. Yeh, *Am. Inst. Chem. Eng. J.* **19**, 260 (1964).
8. E. G. Cockbain and J. H. Schulman, *Trans. Faraday Soc.* **36**, 661 (1940).
9. P. Becher, *J. Soc. Cosmet. Chem.* **9**, 141 (1958).
10. H. Arai and K. Shinoda, *J. Colloid Interface Sci.* **25**(3), (1967).
11. W. C. Griffin, *J. Soc. Cosmet. Chem.* **5**, 249 (1954).
12. P. W. Morgan, *Ind. Eng. Chem. Anal. Ed.* **18**, 500 (1946).
13. *AOCS Cd 3-25*, American Oil Chemists' Society, New York.
14. *AOCS Cd 6-38, AOCS L3a-57*, American Oil Chemists' Society, New York.
15. W. C. Griffin, *Proc. Sci. Sec. Toilet Goods Assoc.* **6**(6), 43 (1946).
16. *McCutcheon's Detergents & Emulsions*, North American ed., annual; J. M. Sumé and J. A. Casterà *Galenica Acta* **24**, 3 (1971).
17. J. T. Davies, "A Quantitative Kinetic Theory of Emulsion, Type I, Physical Chemistry of the Emulsifying Agent," *Proc. 2nd Int. Congr. Surface Activ.* **1**, 426 (1957); A. Leo, C. Hansch, and D. Elkins, "Partition Coefficients and their Uses," *Chem. Rev.* **71**, 525 (1971).
18. *The HLB System*, ICI United States, Inc., Wilmington, Del., 1976.
19. A. Beerbower and M. W. Hill, *Deterg. Emulsif.*, 223 (1971).
20. K. Shinoda, "The Comparison Between the PIT System and the HLB-Value System to Emulsifier Selection," *5th International Congress on Surface-Active Substances, Barcelona, Spain, 1968*.
21. T. J. Lin, *J. Soc. Cosmet. Chem.* **28**, 457 (1977).
22. L. Marzall *Cosmet. Toiletr.* **91**, 21 (1976).
23. G. Rimlinger, *Parfums Cosmet. Savons Fr.* **14**, 479 (1971).
24. I. R. Schmolka, *Nonionic Surfactants* **1**, 300 (1967).
25. R. W. Behrens and W. C. Griffin, *Anal. Chem.* **24**, 1076 (1952); R. W. Behrens *J. Agric. Food Chem.* **6**(1), 20 (1958).
26. K. Horie and co-workers, *Cosmet. Toiletr.* **93**, 53 (1978).
27. J. Trumbetas, J. A. Fioriti, and R. J. Sims, *J. Am. Oil Chem. Soc.* **55**, 248 (1978).

General References

P. Becher, *Emulsions: Theory and Practice*, Reinhold Publishing Corp., New York, 1965.
P. Becher, ed., *Encyclopedia of Emulsion Technology*, Marcel Dekker, Inc., New York, 1979.
S. Friberg, *Food Emulsions*, Marcel Dekker, Inc., New York.

K. J. Lissant, *Emulsions and Emulsion Technology,* Parts I and II, Marcel Dekker, Inc., New York, 1974.
L. M. Prince, *Microemulsions: Theory and Practice,* Academic Press, Inc., New York, 1977.
M. J. Schick, *Nonionic Surfactants,* Vol. 1, Marcel Dekker, Inc., New York, 1967.
P. Sherman, *Emulsion Science,* Academic Press, London, Eng., and New York, 1968.
A. L. Smith, *Theory and Practice of Emulsion Technology,* Academic Press, London, Eng., and New York, 1976.
J. C. Johnson, "Emulsifiers and Emulsifying Techniques 1979," *Chem. Technol. Rev.* (125), 16 (1979).

WILLIAM C. GRIFFIN
ICI United States Inc.